9.25, 9.40. **Ch 10:** 10.14, 10.15, 10.16, 10.17, 10.18, 10.19, 10.23, 10.24, 10.31, 10.56, 10.57, 10.58, 10.59. **Ch 11:** 11.40, 11.41, 11.42, 11.43, 11.44, 11.45, 11.46, 11.47, 11.48, 11.49, 11.50. **Ch 12:** 12.8, 12.10, 12.12, 12.14, 12.25, 12.34, 12.35, 12.36, 12.37, 12.51, 12.52, 12.53. **Ch 13:** 13.6, 13.7, 13.25, 13.26, 13.27, 13.36, 13.37, 13.38, 13.39, 13.40, 13.41, 13.42, 13.43, 13.44, 13.45. **Ch 14:** 14.17. **Ch 15:** 15.10, 15.11, 15.13, 15.22, 15.34, 15.35, 15.36, 15.38, 15.42, 15.43, 15.47, 15.48. **Ch 16:** 16.5, 16.11, 16.12, 16.17, 16.20, 16.39, 16.41, 16.64, 16.74.

### Business and Consumer Behavior

**Ch 1:** 1.39, 1.66, 1.67, 1.69, 1.156. **Ch 2:** 2.3, 2.4, 2.19, 2.29, 2.37, 2.39, 2.62, 2.66, 2.84, 2.90, 2.124, 2.129, 2.134, 2.138, 2.161, 2.162, 2.163, 2.164. **Ch 3:** 3.3, 3.11, 3.15, 3.16, 3.25, 3.28, 3.30, 3.32, 3.35, 3.36, 3.41, 3.48, 3.54, 3.57, 3.60, 3.61, 3.70, 3.78, 3.83, 3.89, 3.107, 3.114, 3.117, 3.120, 3.125, 3.130. **Ch 4:** 4.5, 4.14, 4.15, 4.32, 4.53, 4.54, 4.55, 4.75, 4.77, 4.87, 4.89, 4.90, 4.106, 4.107, 4.108, 4.109, 4.111, 4.135, 4.144. **Ch 5:** 5.20, 5.55, 5.56, 5.60, 5.69, 5.71, 5.77, 5.78. **Ch 6:** 6.7, 6.8, 6.20, 6.21, 6.31, 6.34, 6.35, 6.42, 6.53, 6.54, 6.55, 6.72, 6.100, 6.112, 6.115, 6.117, 6.123, 6.124, 6.125, 6.132. **Ch 7:** 7.8, 7.9, 7.14, 7.34, 7.38, 7.42, 7.45, 7.63, 7.65, 7.66, 7.67. 7.68, 7.74, 7.75, 7.80, 7.83, 7.87, 7.93, 7.101, 7.103, 7.104, 7.110, 7.125, 7.141, 7.142. **Ch 8:** 8.22, 8.30, 8.31, 8.34, 8.35, 8.36, 8.37, 8.38, 8.51, 8.64, 8.65, 8.66, 8.67, 8.70, 8.71, 8.80. **Ch 9:** 9.30, 9.31, 9.36. **Ch 10:** 10.37, 10.38, 10.52. **Ch 11:** 11.17. **Ch 12:** 12.24, 12.34, 12.36, 12.40, 12.41, 12.63, 12.69. **Ch 13:** 13.14, 13.15, 13.17, 13.18, 13.19, 13.20, 13.33, 13.34. **Ch 14:** 14.3, 14.4, 14.5, 14.6, 14.7, 14.8, 14.9, 14.11, 14.12, 14.13, 14.14, 14.15, 14.16, 14.18, 14.19, 14.20, 14.21, 14.22, 14.23, 14.24, 14.31, 14.33, 14.34, 14.36, 14.37, 14.38. **Ch 15:** 15.17, 15.28, 15.40, 15.41. **Ch 16:** 16.1, 16.15, 16.18, 16.19, 16.22, 16.26, 16.28, 16.31, 16.40, 16.42, 16.43, 16.54, 16.55, 16.58, 16.67, 16.73, 16.79, 16.84, 16.85. **Ch 17:** 17.5, 17.14.

### College Life

**Ch 1:** 1.1, 1.2, 1.3, 1.5, 1.6, 1.7, 1.8, 1.9, 1.10, 1.11, 1.12, 1.13, 1.14, 1.27, 1.37, 1.38, 1.41, 1.47, 1.48, 1.49, 1.50, 1.51, 1.52, 1.53, 1.55, 1.61, 1.72, 1.116, 1.146, 1.154, 1.157, 1.160. **Ch 2:** 2.1, 2.6, 2.7, 2.8, 2.9, 2.30, 2.31, 2.31, 2.33, 2.49, 2.50, 2.58, 2.59, 2.60, 2.61, 2.78, 2.79, 2.105, 2.106, 2.107, 2.108, 2.109, 2.110, 2.120, 2.154. **Ch 3:** 3.1, 3.2, 3.20, 3.50, 3.51, 3.54, 3.68, 3.69, 3.79, 3.83, 3.123. **Ch 4:** 4.10, 4.46, 4.60, 4.65, 4.97, 4.99, 4.101, 4.102, 4.103, 4.104, 4.105, 4.112, 4.113, 4.114, 4.115. **Ch 5:** 5.2, 5.18, 5.25, 5.28, 5.73, 5.77. **Ch 6:** 6.1, 6.2, 6.3, 6.5, 6.6, 6.14, 6.15, 6.16, 6.19, 6.23, 6.24, 6.36, 6.52, 6.63, 6.69, 6.72. **Ch 7:** 7.1, 7.2, 7.3, 7.6, 7.38. **Ch 8:** 8.3, 8.5, 8.6, 8.11, 8.12, 8.13, 8.16, 8.19, 8.20, 8.23, 8.32, 8.33, 8.43, 8.46, 8.47, 8.52, 8.74, 8.76, 8.83. **Ch 9:** 9.4, 9.7, 9.8, 9.17, 9.21, 9.23, 9.24, 9.34, 9.35, 9.41. **Ch 10:** 10.10, 10.11, 10.42, 10.43. **Ch 12:** 12.9, 12.10, 12.11, 12.12, 12.14, 12.23, 12.26. **Ch 13:** 13.22, 13.23, 13.24. **Ch 16:** 16.83. **Ch 17:** 17.1, 17.2, 17.3, 17.7, 17.8, 17.11.

### Demographics and Characteristics of People

**Ch 1:** 1.4, 1.15, 1.16, 1.17, 1.20, 1.21, 1.22, 1.62, 1.65, 1.68, 1.82, 1.83, 1.84, 1.117, 1.159, 1.161, 1.169, 1.172. **Ch 2:** 2.10, 2.14, 2.92, 2.111, 2.112, 2.113, 2.114, 2.115, 2.116, 2.120, 2.126. **Ch 3:** 3.76, 3.112, 3.113. **Ch 4:** 4.11, 4.21, 4.22, 4.28, 4.29, 4.30, 4.34, 4.37, 4.42, 4.43, 4.44, 4.45, 4.52, 4.59, 4.84, 4.89, 4.90, 4.106, 4.107, 4.108, 4.116, 4.117, 4.118, 4.119, 4.120, 4.121, 4.124, 4.141, 4.143. **Ch 5:** 5.4, 5.21, 5.31, 5.68. **Ch 6:** 6.116. **Ch 7:** 7.130, 7.131. **Ch 8:** 8.26. **Ch 9:** 9.7, 9.8. **Ch 16:** 16.26.

### Economics and Finance

**Ch 1:** 1.170, 1.172. **Ch 2:** 2.12, 2.27, 2.28, 2.36, 2.38, 2.48, 2.70, 2.71, 2.92, 2.103, 2.142, 2.151, 2.152. **Ch 3:** 3.40, 3.49, 3.87. **Ch 4:** 4.39, 4.91, 4.92, 4.93, 4.110, 4.136. **Ch 5:** 5.57, 5.62. **Ch** [illegible] **10** [illegible] 10. [illegible] 11. [illegible] 16.4 [illegible]

### Education and Child Development

**Ch 1:** 1.25, 1.43, 1.44, 1.45, 1.74, 1.99, 1.100, 1.101, 1.102, 1.103, 1.104, 1.105, 1.106, 1.126, 1.127, 1.128, 1.129, 1.130, 1.131, 1.132, 1.133, 1.134, 1.135, 1.146, 1.153, 1.157, 1.167, 1.173. **Ch 2:** 2.11, 2.13, 2.35, 2.51, 2.65, 2.74, 2.91, 2.122, 2.127, 2.128, 2.131, 2.135, 2.155, 2.156, 2.159, 2.160, 2.161. **Ch 3:** 3.9, 3.12, 3.13, 3.14, 3.17, 3.19, 3.93, 3.94, 3.127. **Ch 4:** 4.46, 4.73, 4.97, 4.99, 4.103, 4.104, 4.105, 4.122, 4.138. **Ch 5:** 5.28, 5.32, 5.48, 5.50, 5.66. **Ch 6:** 6.5, 6.6, 6.54, 6.64, 6.65, 6.69, 6.95, 6.96, 6.113. **Ch 7:** 7.111, 7.127, 7.128, 7.138, 7.139. **Ch 9:** 9.29, 9.35. **Ch 10:** 10.10, 10.11, 10.30, 10.32, 10.34, 10.53, 10.54, 10.55, 10.61. **Ch 11:** 11.1, 11.3, 11.5, 11.6, 11.27, 11.28, 11.29, 11.30. **Ch 12:** 12.26, 12.42, 12.43, 12.44, 12.61, 12.62, 12.64. **Ch 13:** 13.5, 13.7, 13.12, 13.22, 13.23, 13.24, 13.46, 13.47, 13.48, 13.49. **Ch 14:** 14.39, 14.40, 14.41, 14.42. **Ch 15:** 15.7, 15.8, 15.9, 15.12, 15.14, 15.26. **Ch 16:** 16.4, 16.9, 16.14, 16.32, 16.51, 16.53, 16.56, 16.59. **Ch 17:** 17.75.

### Ethics

**Ch 3:** 3.96, 3.97, 3.98, 3.99, 3.100, 3.101, 3.102, 3.103, 3.104, 3.105, 3.106, 3.107, 3.108, 3.109, 3.110, 3.111, 3.112, 3.129, 3.130, 3.131, 3.132, 3.133.

### Health and Nutrition

**Ch 1:** 1.23, 1.24, 1.35, 1.57, 1.58, 1.59, 1.70, 1.92, 1.112, 1.114, 1.136, 1.137, 1.138, 1.139, 1.140. **Ch 2:** 2.9, 2.20, 2.21, 2.53, 2.54, 2.55, 2.68, 2.73, 2.82, 2.83, 2.87, 2.89, 2.93, 2.121, 2.129, 2.130, 2.132, 2.133, 2.136, 2.137, 2.145, 2.146, 2.166. **Ch 3:** 3.7, 3.24, 3.26, 3.27, 3.29, 3.31, 3.37, 3.38, 3.39, 3.45, 3.88, 3.100, 3.101, 3.104, 3.106, 3.108, 3.109, 3.116, 3.124, 3.126, 3.131. **Ch 4:** 4.6, 4.21, 4.22, 4.29, 4.34, 4.42, 4.43, 4.44, 4.45,

4.52, 4.84, 4.101, 4.102, 4.125, 4.126, 4.127, 4.128, 4.129, 4.143. **Ch 5:** 5.47, 5.51, 5.53, 5.65. **Ch 6:** 6.17, 6.18, 6.22, 6.25, 6.29, 6.30, 6.37, 6.60, 6.63, 6.70, 6.73, 6.92, 6.121. **Ch 7:** 7.26, 7.27, 7.28, 7.30, 7.31, 7.32, 7.33, 7.37, 7.39, 7.40, 7.46, 7.47, 7.51, 7.61, 7.62, 7.63, 7.64, 7.67, 7.68, 7.76, 7.78, 7.79, 7.85, 7.86, 7.88, 7.94, 7.99, 7.100, 7.102, 7.107, 7.117, 7.118, 7.119, 7.120, 7.121, 7.126, 7.136, 7.137. **Ch 8:** 8.23, 8.24, 8.27, 8.28, 8.46, 8.47, 8.54. **Ch 9:** 9.4, 9.11, 9.21, 9.39. **Ch 10:** 10.13, 10.25, 10.26, 10.27, 10.28, 10.29, 10.35, 10.36, 10.50, 10.60, 10.62, 10.63. **Ch 11:** 11.13, 11.14, 11.15, 11.19, 11.20, 11.21, 11.34, 11.35, 11.36, 11.37, 11.38, 11.39. **Ch 12:** 12.9, 12.11, 12.13, 12.21, 12.22, 12.29, 12.30, 12.31, 12.32, 12.33, 12.39, 12.45, 12.46, 12.47, 12.48, 12.49, 12.50, 12.58, 12.59, 12.60, 12.68. **Ch 13:** 13.7, 13.8, 13.21, 13.28, 13.29, 13.30, 13.36. **Ch 14:** 14.25, 14.27, 14.29, 14.35. **Ch 15:** 15.15, 15.16, 15.23, 15.24, 15.27, 15.29, 15.30, 15.31, 15.32, 15.33, 15.37, 15.39, 15.44, 15.45, 15.46. **Ch 16:** 16.7, 16.10, 16.64, 16.65, 16.66, 16.68, 16.69, 16.70, 16.71, 16.72, 16.81, 16.82, 16.83. **Ch 17:** 17.50.

### Humanities and Social Sciences

**Ch 1:** 1.26, 1.60, 1.85, 1.124, 1.125. **Ch 2:** 2.17, 2.41, 2.47, 2.50, 2.67, 2.75, 2.123, 2.125, 2.134, 2.157, 2.158, 2.162, 2.163. **Ch 3:** 3.8, 3.55, 3.61, 3.63, 3.64, 3.66, 3.71, 3.72, 3.73, 3.74, 3.75, 3.80, 3.85, 3.86, 3.98, 3.100, 3.101, 3.105, 3.110, 3.111, 3.112, 3.128, 3.132, 3.133. **Ch 4:** 4.24, 4.66, 4.148. **Ch 5:** 5.18, 5.22, 5.24, 5.25, 5.26, 5.27, 5.29, 5.72, 5.76. **Ch 6:** 6.53, 6.62, 6.68, 6.93. **Ch 7:** 7.29, 7.41, 7.43, 7.48, 7.129. **Ch 8:** 8.17, 8.18, 8.25, 8.44, 8.45, 8.53, 8.55, 8.56, 8.69, 8.72, 8.73, 8.81. **Ch 9:** 9.6, 9.9, 9.10, 9.15, 9.16, 9.18, 9.19, 9.22, 9.27, 9.28, 9.33, 9.38. **Ch 10:** 10.46. **Ch 11:** 11.31, 11.32, 11.33. **Ch 12:** 12.27, 12.28, 12.38, 12.66, 12.67. **Ch 13:** 13.9, 13.10, 13.12. **Ch 14:** 14.26, 14.28, 14.30. **Ch 15:** 15.25. **Ch 16:** 16.23, 16.50, 16.77, 16.78, 16.86, 16.87.

### International

**Ch 2:** 2.14, 2.15, 2.23, 2.25, 2.29, 2.37, 2.87, 2.88, 2.89, 2.126, 2.146, 2.151, 2.152, 2.156. **Ch 3:** 3.28, 3.48, 3.78, 3.86, 3.88, 3.104, 3.116. **Ch 4:** 4.22, 4.24, 4.59. **Ch 6:** 6.60, 6.62, 6.70. **Ch 9:** 9.37. **Ch 10:** 10.46, 10.47, 10.48, 10.49, 10.56, 10.57, 10.58. **Ch 11:** 11.31, 11.32, 11.33. **Ch 12:** 12.27, 12.28, 12.35, 12.38, 12.49, 12.50, 12.66, 12.67. **Ch 13:** 13.12. **Ch 15:** 15.13, 15.30, 15.38, 15.43, 15.47. **Ch 16:** 16.64, 16.65, 16.66. **Ch 17:** 17.22.

### Manufacturing, Products, and Processes

**Ch 1:** 1.145. **Ch 2:** 2.119, 2.144, 2.147. **Ch 3:** 3.33. **Ch 4:** 4.81. **Ch 5:** 5.40, 5.43, 5.45, 5.54, 5.58, 5.59, 5.63, 5.74, 5.75. **Ch 6:** 6.123. **Ch 7:** 7.11, 7.50, 7.76, 7.78, 7.79, 7.94, 7.107. **Ch 8:** 8.68. **Ch 9:** 9.5, 9.32. **Ch 10:** 10.21, 10.22. **Ch 11:** 11.51, 11.52, 11.53, 11.54, 11.55, 11.56, 11.57, 11.58, 11.59, 11.60. **Ch 12:** 12.29, 12.30, 12.31, 12.32, 12.33, 12.54, 12.55, 12.56, 12.57, 12.58, 12.59, 12.60. **Ch 13:** 13.31, 13.32. **Ch 14:** 14.32. **Ch 17:** 17.9, 17.10, 17.12, 17.13, 17.15, 17.16, 17.17, 17.18, 17.19, 17.20, 17.26, 17.31, 17.32, 17.34, 17.35, 17.36, 17.37, 17.38, 17.39, 17.40, 17.41, 17.42, 17.43, 17.44, 17.45, 17.46, 17.53, 17.54, 17.55, 17.56, 17.59, 17.60, 17.61, 17.65, 17.66, 17.67, 17.68, 17.69, 17.70, 17.71, 17.72, 17.73, 17.74, 17.76, 17.77, 17.79, 17.83, 17.84, 17.85, 17.86, 17.87.

### Motor Vehicles and Fuel

**Ch 1:** 1.146. **Ch 2:** 2.16, 2.22, 2.44, 2.45, 2.52, 2.86, 2.97, 2.99. **Ch 3:** 3.4. **Ch 4:** 4.12, 4.13, 4.109, 4.111. **Ch 6:** 6.26, 6.27, 6.71. **Ch 7:** 7.24, 7.35, 7.49. **Ch 8:** 8.14, 8.15, 8.58, 8.59. **Ch 10:** 10.2, 10.3, 10.4, 10.12. **Ch 16:** 16.24, 16.60.

### Physical Sciences

**Ch 1:** 1.40, 1.73, 1.94, 1.147. **Ch 2:** 2.24, 2.40, 2.69, 2.72, 2.77, 2.94. **Ch 6:** 6.32, 6.61. **Ch 10:** 10.39, 10.40, 10.41. **Ch 16:** 16.44, 16.48.

### Sports and Leisure

**Ch 1:** 1.46, 1.81, 1.143, 1.165, 1.166. **Ch 2:** 2.19, 2.23, 2.25, 2.39, 2.62, 2.95, 2.98, 2.100, 2.139, 2.140, 2.148, 2.150, 2.165. **Ch 3:** 3.43, 3.47, 3.121. **Ch 4:** 4.2, 4.3, 4.4, 4.7, 4.9, 4.18, 4.27, 4.31, 4.32, 4.38, 4.50, 4.56, 4.58, 4.72, 4.80, 4.83, 4.96, 4.98, 4.100, 4.133, 4.134. 4.137, 4.139, 4.140, 4.142. **Ch 5:** 5.23, 5.30, 5.42, 5.44, 5.52. **Ch 6:** 6.13, 6.14, 6.16, 6.128. **Ch 8:** 8.11, 8.12, 8.43, 8.54, 8.61, 8.82. **Ch 9:** 9.23, 9.24. **Ch 10:** 10.62, 10.63. **Ch 11:** 11.18. **Ch 12:** 12.23, 12.47, 12.48. **Ch 13:** 13.8. **Ch 14:** 14.1, 14.35. **Ch 15:** 15.1, 15.2, 15.3, 15.4, 15.5, 15.6, 15.18, 15.19, 15.20, 15.21. **Ch 16:** 16.6, 16.9, 16.13, 16.16, 16.37, 16.38, 16.47, 16.62, 16.76. **Ch 17:** 17.4, 17.49.

### Technology and the Internet

**Ch 1:** 1.19, 1.158, 1.162. **Ch 3:** 3.5, 3.6, 3.7, 3.10, 3.12, 3.14, 3.34, 3.42, 3.50, 3.51. **Ch 4:** 4.19, 4.20, 4.26, 4.33, 4.35, 4.36, 4.51, 4.74, 4.78, 4.123, 4.145. **Ch 5:** 5.1, 5.13, 5.14, 5.15, 5.16, 5.17, 5.42, 5.44. **Ch 6:** 6.9, 6.23, 6.24. **Ch 7:** 7.54, 7.55, 7.63, 7.74, 7.75, 7.87, 7.101. **Ch 8:** 8.1, 8.2, 8.4, 8.13, 8.41, 8.49, 8.50, 8.58, 8.59, 8.60, 8.62, 8.65, 8.66, 8.67. **Ch 9:** 9.26. **Ch 10:** 10.44. **Ch 11:** 11.17, 11.22, 11.23, 11.34, 11.25, 11.26. **Ch 12:** 12.10, 12.12, 12.13, 12.14, 12.65. **Ch 13:** 13.50. **Ch 14:** 14.21, 14.22, 14.23, 14.24, 14.33, 14.34. **Ch 15:** 15.12. **Ch 16:** 16.6, 16.9, 16.13, 16.16, 16.38.

# Introduction to the
# Practice of Statistics

## Authors' note about the cover

*Introduction to the Practice of Statistics* emphasizes the use of graphical and numerical summaries to understand data. The front cover shows a painting entitled *0 to 9*, by the American artist Jasper Johns in 1961. In this work, the structure of the painting is determined by number sequence, just as our graphical summaries are determined by the numerical calculations that we perform when we analyze data. Can you find all of the digits in the painting?

# Introduction to the Practice of Statistics

SIXTH EDITION

*EXTENDED VERSION*

**DAVID S. MOORE**

**GEORGE P. McCABE**

**BRUCE A. CRAIG**

*Purdue University*

W. H. Freeman and Company
New York

| | |
|---|---|
| *Senior Publisher:* | CRAIG BLEYER |
| *Publisher:* | RUTH BARUTH |
| *Development Editors:* | SHONA BURKE, ANNE SCANLAN-ROHRER |
| *Senior Media Editor:* | ROLAND CHEYNEY |
| *Assistant Editor:* | BRIAN TEDESCO |
| *Editorial Assistant:* | KATRINA WILHELM |
| *Marketing Coordinator:* | DAVE QUINN |
| *Photo Editor:* | CECILIA VARAS |
| *Photo Researcher:* | ELYSE RIEDER |
| *Cover and Text Designer:* | VICKI TOMASELLI |
| *Senior Project Editor:* | MARY LOUISE BYRD |
| *Illustrations:* | INTEGRE TECHNICAL PUBLISHING CO. |
| *Production Manager:* | JULIA DE ROSA |
| *Composition:* | INTEGRE TECHNICAL PUBLISHING CO. |
| *Printing and Binding:* | RR DONNELLEY |



Library of Congress Control Number: 2007938575

ISBN-13: 978-1-4292-1623-4
ISBN-10: 1-4292-1623-9 (Extended Version, Casebound)

ISBN-13: 978-1-4292-1622-7
ISBN-10: 1-4292-1622-0 (Casebound)

ISBN-13: 978-1-4292-1621-0
ISBN-10: 1-4292-1621-2 (Paperback)



Printed in the United States of America

Second printing

W. H. Freeman and Company
41 Madison Avenue
New York, NY 10010
Houndmills, Basingstoke RG21 6XS, England
www.whfreeman.com

# Brief Contents

# Contents



Sections marked with an asterisk are optional.

# PART III Topics in Inference

# TO TEACHERS: *About This Book*

Statistics is the science of data. *Introduction to the Practice of Statistics* (*IPS*) is an introductory text based on this principle. We present the most-used methods of basic statistics in a way that emphasizes working with data and mastering statistical reasoning. *IPS* is elementary in mathematical level but conceptually rich in statistical ideas and serious in its aim to help students think about data and use statistical methods with understanding.

Some schematic history will help place *IPS* in the universe of texts for a first course in statistics for students from a variety of disciplines. Traditional texts were almost entirely devoted to methods of inference, with quick coverage of means, medians, and histograms as a preliminary. No doubt this reflected the fact that inference is the only part of statistics that has a mathematical theory behind it. Several innovative books aimed at nontraditional audiences pioneered a broader approach that paid more attention to design of samples and experiments, the messiness of real data, and discussion of real-world statistical studies and controversies. All were written by widely known statisticians whose main business was not writing textbooks. *The Nature of Statistics* (Wallis and Roberts) has passed away, but *Statistics* (Freedman and collaborators) and *Statistics: Concepts and Controversies* (Moore) remain alive and well. None of these books tried to meet the needs of a typical first course because their audiences did not need full coverage of standard statistical methods.

*IPS* was the first book to successfully combine attention to broader content and reasoning with comprehensive presentation of the most-used statistical methods. It reflects the consensus among statisticians—even stronger now than when the first edition appeared—concerning the content of an introduction to our discipline. This consensus is expressed in a report from the joint curriculum committee of the American Statistical Association and the Mathematical Association of America[1] and in discussions in leading journals.[2] *IPS* has been successful for several reasons:

1. *IPS* examines the nature of modern statistical practice at a level suitable for beginners. Attention to data analysis and data production as well as to probability and inference is "new" only in the world of textbooks. Users of statistical methods have always paid attention to all of these. Contemporary research in statistics, driven by advances in computing, puts more emphasis on sophisticated "looking at data" and on data-analytic ways of thinking. Formal inference remains important and receives careful treatment, but it appears as part of a larger picture.

2. *IPS* has a logical overall progression, so data analysis and data production strengthen the presentation of inference rather than stand apart from it. We stress that data analysis is an essential preliminary to inference because inference requires clean data. The most useful "goodness of fit" procedure, for example, is the normal quantile plot presented in Chapter 1 and used frequently in the inference chapters. We emphasize that when you do formal statistical inference, you are acting as if your data come from properly randomized data production. We use random samples and experimental randomization to motivate the need for probability as a language for inference.

3. *IPS* presents data analysis as more than a collection of techniques for exploring data. We integrate techniques with discussion of systematic ways of thinking about data. We also work hard to make data-analytic thinking accessible to beginners by presenting a series of simple principles: always plot your data; look for overall patterns and deviations from them; when looking at the overall pattern of a distribution for one variable, consider shape, center, and spread; for relations between two variables, consider form, direction, and strength; always ask whether a relationship between variables is influenced by other variables lurking in the background. Inference is similarly treated as more than a collection of methods. We warn students about pitfalls in clear cautionary discussions—about regression and correlation, experiments, sample surveys, confidence intervals, and significance tests. Our goal throughout *IPS* is to present principles and techniques together in a way that is accessible to beginners and lays a foundation for students who will go on to more advanced study.
4. *IPS* integrates discussion of techniques, reasoning, and practice using real examples to drive the exposition. Students learn the technique of least-squares regression and how to interpret the regression slope. But they also learn the conceptual ties between regression and correlation, the importance of looking for influential observations (always plot your data), and to beware of averaged data and the restricted-range effect.
5. *IPS* is aware of current developments both in statistical science and in teaching statistics. For example, the first edition already favored the version of the two-sample $t$ procedures that does not assume equal population variances and discussed the great difference in robustness between standard tests for means and for variances. In the fourth edition, we introduced the modified ("plus four") confidence intervals for proportions that are shown by both computational studies[3] and theory[4] to be superior to the standard intervals for all but very large samples. Brief optional "Beyond the Basics" sections give quick overviews of topics such as density estimation, scatterplot smoothers, nonlinear regression, and data mining. Chapter 16 on resampling methods offers an extended introduction to one of the most important recent advances in statistical methodology.

The title of the book expresses our intent to introduce readers to statistics as it is used in practice. Statistics in practice is concerned with gaining understanding from data; it focuses on problem solving rather than on methods that may be useful in specific settings. A text cannot fully imitate practice because it must teach specific methods in a logical order and must use data that are not the reader's own. Nonetheless, our interest and experience in applying statistics have influenced the nature of *IPS* in several ways.

**Statistical Thinking** Statistics is interesting and useful because it provides strategies and tools for using data to gain insight into real problems. As the continuing revolution in computing automates the details of doing calculations and making graphs, an emphasis on statistical concepts and on insight from data becomes both more practical for students and teachers and more important for users who must supply what is not automated. No student should complete a first statistics course, for example, without a firm grasp of the distinction between observational studies and experiments and of why randomized comparative experiments are the gold standard for evidence of causation.

We have seen many statistical mistakes, but few have involved simply getting a calculation wrong. We therefore ask students to learn to explore data, always starting with plots, to think about the context of the data and the design of the study that produced the data, the possible influence of wild observations on conclusions, and the reasoning that lies behind standard methods of inference. Users of statistics who form these habits from the beginning are well prepared to learn and use more advanced methods.

**Data** Data are numbers with a context, as we say in "To Students: What Is Statistics?" A newborn who weighs 10.3 pounds is a big baby, and the birth weight could not plausibly be 10.3 ounces or 10.3 kilograms. Because context makes numbers meaningful, our examples and exercises use real data with real contexts that we briefly describe. Calculating the mean of five numbers is arithmetic, not statistics. We hope that the presence of background information, even in exercises intended for routine drill, will encourage students to always consider the meaning of their calculations as well as the calculations themselves. Note in this connection that a calculation or a graph or "reject $H_0$" is rarely a full answer to a statistical problem. We strongly encourage requiring students always to state a brief conclusion in the context of the problem. This helps build data sense as well as the communication skills that employers value.

**Mathematics** Although statistics is a mathematical science, it is not a field of mathematics and should not be taught as if it were. A fruitful mathematical theory (based on probability, which *is* a field of mathematics) underlies some parts of basic statistics, but by no means all. The distinction between observation and experiment, for example, is a core statistical idea that is ignored by the theory.[5] Mathematically trained teachers, rightly resisting a formula-based approach, sometimes identify conceptual understanding with mathematical understanding. When teaching statistics, we must emphasize statistical ideas and recognize that mathematics is not the only vehicle for conceptual understanding. *IPS* requires only the ability to read and use equations without having each step parsed. We require no algebraic derivations, let alone calculus. Because this is a *statistics* text, it is richer in ideas and requires more thought than the low mathematical level suggests.

**Calculators and Computers** Statistical calculations and graphics are in practice automated by software. We encourage instructors to use software of their choice or a graphing calculator that includes functions for both data analysis and basic inference. *IPS* includes some topics that reflect the dominance of software in practice, such as normal quantile plots and the version of the two-sample $t$ procedures that does not require equal variances. Several times we display the output of multiple software systems for the same problem. The point is that a student who knows the basics can interpret almost any output. Students like this reassurance, and it helps focus their attention on understanding rather than reading output.

**Judgment** Statistics in practice requires judgment. It is easy to list the mathematical assumptions that justify use of a particular procedure, but not so easy to decide when the procedure can be safely used in practice. Because judgment develops through experience, an introductory course should present clear guidelines and not make unreasonable demands on the judgment of students. We have given guidelines—for example, on using the $t$ procedures for comparing two means but avoiding the $F$ procedures for comparing two variances—that we follow ourselves. Similarly, many exercises require students to use some judgment and (equally important) to explain their choices in words. Many students would prefer to stick to calculating, and many statistics texts allow them to. Requiring more will do them much good in the long run.

**Teaching Experiences** We have successfully used *IPS* in courses taught to quite diverse student audiences. For general undergraduates from mixed disciplines, we cover Chapters 1 to 8 and Chapter 9, 10, or 12, omitting all optional material. For a quantitatively strong audience—sophomores planning to major in actuarial science or statistics—we move more quickly. We add Chapters 10 and 11 to the core material in Chapters 1 to 8 and include most optional content. We de-emphasize Chapter 4 (probability) because these students will take a probability course later in their program, though we make intensive use of software for simulating probabilities as well as for statistical analysis. The third group we teach contains beginning graduate students in such fields as education, family studies, and retailing. These mature but sometimes quantitatively unprepared students read the entire text (Chapters 11 and 13 lightly), again with reduced emphasis on Chapter 4 and some parts of Chapter 5. In all cases, beginning with data analysis and data production (Part I) helps students overcome their fear of statistics and builds a sound base for studying inference. We find that *IPS* can be flexibly adapted to quite varied audiences by paying attention to our clear designation of some material as optional and by varying the chapters assigned.

## The Sixth Edition: *What's New?*

- **Co-author** We are delighted to welcome Professor Bruce Craig to the *Introduction to the Practice of Statistics* author team. Bruce is currently Director of the Statistical Consulting Service at Purdue University and is an outstanding teacher. His vast experience consulting and collaborating with individuals who use statistical methods in their work provides him with perspective on the field of statistics that resonates with the approach of this text.

- **Ethics** Chapter 3 now contains a new section (3.4) on ethics. We believe that this topic is a very important part of the undergraduate curriculum and that a course in statistics is an ideal forum to stimulate thought and discussion about ethical issues.

- **Text Organization** Logistic Regression, previously treated in Chapter 16, now appears in Chapter 14. Similarly, Bootstrap Methods and Permutation Tests has moved to Chapter 16. This change is in line with the increasing importance of logistic regression in statistical practice. In response to suggestions from current *IPS* users, we have moved the material on data analysis for two-way tables from Chapter 9 back to Chapter 2 (Section 2.5). In addition, the large sample confidence procedures are now the featured methods for one and two proportions in Chapter 9, and the plus-four have been moved to Beyond the Basics sections, a more appropriate location. The table of contents follows what we consider to be the best ordering of the topics from a pedagogical point of view. However, the text chapters are generally written to enable instructors to teach the material in the order they prefer.

- **Design** A new design incorporates colorful, revised figures throughout to aid students' understanding of text material. Photographs related to chapter examples and exercises make connections to real-life applications and provide a visual context for topics.
- **Exercises and Examples** Exercises and examples are labeled to help instructors and students easily identify key topics and application areas. The number of total exercises has increased by 15%. Approximately half the total exercises are new or revised to reflect current data and a variety of topics. *IPS* examples and exercises cover a wide range of application areas. An application index is provided for instructors to easily select and assign content related to specific fields.
- **Use Your Knowledge Exercises** Short exercises designed to reinforce key concepts now appear throughout each chapter. These exercises are listed, with page numbers, at the end of each section for easy reference.

LOOK BACK

- **Look Back** At key points in the text Look Back margin notes direct the reader to the first explanation of a topic, providing page numbers for easy reference.

In addition to the new Sixth Edition enhancements, *IPS* has retained the successful pedagogical features from previous editions:

- **Caution** Warnings in the text, signaled by a caution icon, help students avoid common errors and misconceptions.

- **Challenge Exercises** More challenging exercises are signaled with an icon. Challenge exercises are varied: some are mathematical, some require open-ended investigation, and so on.

- **Applets** Applet icons are used throughout the text to signal where related, interactive statistical applets can be found on the text Web site (www.whfreeman.com/ips6e) and CD-ROM.
- **Statistics in Practice** Formerly found at the opening of each chapter, these accounts by professionals who use statistics on the job are now located on the *IPS* Web site and CD-ROM.
- **CrunchIt! Statistical Software** Developed by Webster West of Texas A&M University, CrunchIt! is an easy-to-use program for students and offers capabilities well beyond those needed for a first course. CrunchIt! output, along with other statistical software output, is integrated throughout the text. Access to CrunchIt! is available online through an access-code–protected Web site. Access codes are available in every new copy of *IPS* 6e or can be purchased online.

## Acknowledgments

We are pleased that the first five editions of *Introduction to the Practice of Statistics* have helped move the teaching of introductory statistics in a direction supported by most statisticians. We are grateful to the many colleagues and students who have provided helpful comments, and we hope that they will find this new edition another step forward. In particular, we would like to thank the following colleagues who offered specific comments on the new edition:

Mary A. Bergs
*Mercy College of Northwest Ohio*
John F. Brewster
*University of Manitoba*
Karen Buro
*Grant MacEwan College*
Smiley W. Cheng
*University of Manitoba*
Gerarda Darlington
*University of Guelph*
Linda Dawson
*University of Washington, Tacoma*
Michael Evans
*University of Toronto*
Mary Gray
*American University*
Rick Gumina
*Colorado State University*
Patricia Humphrey
*Georgia Southern University*
Mohammad Kazemi
*The University of North Carolina at Charlotte*
Jeff Kollath
*Oregon State University*
William J. Kubik
*Hanover College*
Charles Liberty
*Keene State College*
Brian Macpherson
*University of Manitoba*
Henry Mesa
*Portland Community College*
Helen Noble
*San Diego State University*
Richard Numrich
*College of Southern Nevada*
Becky Parker
*Montana State University*
Maria Consuelo C. Pickle
*St. Petersburg College*
German J. Pliego
*University of St. Thomas*
Philip Protter
*Cornell University*
John G. Reid
*Mount Saint Vincent University*
Diann Reischman
*Grand Valley State University*
Shane Rollans
*Thompson Rivers University*
Nancy Roper
*Portland Community College*
Teri Rysz
*University of Cincinnati*
Engin Sungur
*University of Minnesota, Morris*
Todd Swanson
*Hope College*
Anthony L. Truog
*University of Wisconsin, Whitewater*
Augustin Vukov
*University of Toronto*
Erwin Walker
*Clemson University*
Nathan Wetzel
*University of Wisconsin, Stevens Point*

The professionals at W. H. Freeman and Company, in particular Mary Louise Byrd, Ruth Baruth, and Shona Burke, have contributed greatly to the success of *IPS*. Additionally, we would like to thank Anne Scanlan-Rohrer, Pam Bruton, Jackie Miller, and Darryl Nester for their valuable contributions to the Sixth Edition. Most of all, we are grateful to the many people in varied disciplines and occupations with whom we have worked to gain understanding from data. They have provided both material for this book and the experience that enabled us to write it. What the eminent statistician John Tukey called "the real problems experience and the real data experience" have shaped our view of statistics, convincing us of the need for beginning instruction to focus on data and concepts, building intellectual skills that transfer to more elaborate settings and remain essential when all details are automated. We hope that users and potential users of statistical techniques will find this emphasis helpful.

## Media and Supplements

### For Students

NEW! STATSPORTAL

**portals.bfwpub.com/ips6e** (Access code required. Available packaged with *Introduction to the Practice of Statistics*, Sixth Edition, or for purchase online.) StatsPortal is the digital gateway to *IPS* 6e, designed to enrich the course and enhance students' study skills through a collection of Web-based tools. StatsPortal integrates a rich suite of diagnostic, assessment, tutorial, and enrichment features, enabling students to master statistics at their own pace. It is organized around three main teaching and learning components:

- **Interactive eBook** offers a complete and customizable online version of the text, fully integrated with all the media resources available with *IPS* 6e. The eBook allows students to quickly search the text, highlight key areas, and add notes about what they're reading. Similarly, instructors can customize the eBook to add, hide, and reorder content, add their own material, and highlight key text for students.

- **Resources** organizes all the resources for *IPS* 6e into one location for students' ease of use. These resources include the following:

  - **StatTutor Tutorials** offer over 150 audio-multimedia tutorials tied directly to the textbook, including videos, applets, and animations.

  - **Statistical Applets** are 16 interactive applets to help students master key statistical concepts.

  - **CrunchIt! Statistical Software** allows users to analyze data from any Internet location. Designed with the novice user in mind, the software is not only easily accessible but also easy to use. CrunchIt! offers all the basic statistical routines covered in the introductory statistics courses. **CrunchIt!** statistical software is available via an access-code protected Web site. Access codes are available in every new copy of *IPS* 6e or can be purchased online.

  - **Stats@Work Simulations** put students in the role of statistical consultants, helping them better understand statistics interactively within the context of real-life scenarios. Students are asked to interpret and analyze data presented to them in report form, as well as to interpret current event news stories. All tutorials are graded and offer helpful hints and feedback.

  - **EESEE Case Studies** developed by The Ohio State University Statistics Department provide students with a wide variety of timely, real examples with real data. Each case study is built around several thought-provoking questions that make students think carefully about the statistical issues raised by the stories. **EESEE** case studies are available via an access-code-protected Web site. Access codes are available in every new copy of *IPS* 6e or can be purchased online.

- **Podcast Chapter Summary** provides students with an audio version of chapter summaries to download and review on an mp3 player.
- **Data Sets** are available in ASCII, Excel, JMP, Minitab, TI, SPSS, and S-Plus formats.
- **Online Tutoring with SMARTHINKING** is available for homework help from specially trained, professional educators.
- **Student Study Guide with Selected Solutions** includes explanations of crucial concepts and detailed solutions to key text problems with step-by-step models of important statistical techniques.
- **Statistical Software Manuals** for TI-83/84, Minitab, Excel, JMP, and SPSS provide instruction, examples, and exercises using specific statistical software packages.
- **Interactive Table Reader** allows students to use statistical tables interactively to seek the information they need.

**Resources (instructors only)**

- **Instructor's Guide with Full Solutions** includes worked-out solutions to all exercises, teaching suggestions, and chapter comments.
- **Test Bank** contains complete solutions for textbook exercises.
- **Lecture PowerPoint slides** offer a detailed lecture presentation of statistical concepts covered in each chapter of *IPS*.

- **Assignments** organizes assignments and guides instructors through an easy-to-create assignment process providing access to questions from the Test Bank, Web Quizzes, and Exercises from *IPS* 6e. The Assignment Center enables instructors to create their own assignments from a variety of question types for self-graded assignments. This powerful assignment manager allows instructors to select their preferred policies in regard to scheduling, maximum attempts, time limitations, feedback, and more!

## Online Study Center: www.whfreeman.com/osc/ips6e

(Access code required. Available for purchase online.) In addition to all the offerings available on the Companion Web site, the OSC offers:

- **StatTutor Tutorials**
- **Stats@Work Simulations**
- **Study Guide with Selected Solutions**
- **Statistical Software Manuals**

## Companion Web site: www.whfreeman.com/ips6e

Seamlessly integrates topics from the text. On this open-access Web site, students can find the following:

- **Interactive Statistical Applets** that allow students to manipulate data and see the corresponding results graphically.

- **Data Sets** in ASCII, Excel, JMP, Minitab, TI, SPSS, and S-Plus formats.
- **Interactive Exercises and Self-Quizzes** to help students prepare for tests.
- **Optional Companion Chapters 14, 15, 16, and 17,** covering logistic regression, nonparametric tests, bootstrap methods and permutation tests, and statistics for quality control and capability.
- **Supplementary Exercises** for every chapter.

**Interactive Student CD-ROM** Included with every new copy of *IPS*, the CD contains access to the companion chapters, applets, and data sets also found on the Companion Web site.

**Special Software Packages** Student versions of JMP, Minitab, S-PLUS, and SPSS are available on a CD-ROM packaged with the textbook. This software is not sold separately and must be packaged with a text or a manual. Contact your W. H. Freeman representative for information or visit www.whfreeman.com.

**NEW! SMARTHINKING Online Tutoring** (access code required) W. H. Freeman and Company is partnering with SMARTHINKING to provide students with free online tutoring and homework help from specially trained, professional educators. Twelve-month subscriptions are available for packaging with *IPS*.

**Printed Study Guide** prepared by Michael A. Fligner of The Ohio State University offers students explanations of crucial concepts in each section of *IPS*, plus detailed solutions to key text problems and stepped-through models of important statistical techniques. ISBN 1-4292-1473-2

## For Instructors

The **Instructor's Web site** www.whfreeman.com/ips6e requires user registration as an instructor and features all the student Web materials plus:

- **Instructor version of EESEE** (Electronic Encyclopedia of Statistical Examples and Exercises), with solutions to the exercises in the student version and **CrunchIt!** statistical software.
- **Instructor's Guide,** including full solutions to all exercises in .pdf format.
- **PowerPoint slides** containing all textbook figures and tables.
- **Lecture PowerPoint slides** offering a detailed lecture presentation of statistical concepts covered in each chapter of *IPS*.
- **Full answers to the Supplementary Exercises** on the student Web site.

**Instructor's Guide with Solutions** by Darryl Nester, Bluffton University. This printed guide includes full solutions to all exercises and provides video and Internet resources and sample examinations. It also contains brief discussions of the *IPS* approach for each chapter. ISBN 1-4292-1472-4

**Test Bank** by Brian Macpherson, University of Manitoba. The test bank contains hundreds of multiple-choice questions to generate quizzes and tests. Available in print as well as electronically on CD-ROM (for Windows and Mac), where questions can be downloaded, edited, and resequenced to suit the instructor's needs.
Printed Version, ISBN 1-4292-1471-6
Computerized (CD) Version, ISBN 1-4292-1859-2

**Enhanced Instructor's Resource CD-ROM** Allows instructors to search and export (by key term or chapter) all the material from the student CD, plus:

- All text images and tables
- Statistical applets and data sets
- Instructor's Guide with full solutions
- PowerPoint files and lecture slides
- Test bank files

ISBN 1-4292-1503-8

**Course Management Systems** W. H. Freeman and Company provides courses for Blackboard, WebCT (Campus Edition and Vista), and Angel course management systems. They are completely integrated courses that you can easily customize and adapt to meet your teaching goals and course objectives. On request, Freeman also provides courses for users of Desire2Learn and Moodle. Visit www.bfwpub.com/lmc for more information.

i-clicker

**i-clicker** is a new two-way radio-frequency classroom response solution developed by educators for educators. University of Illinois physicists Tim Stelzer, Gary Gladding, Mats Selen, and Benny Brown created the i-clicker system after using competing classroom response solutions and discovering they were neither classroom-appropriate nor student-friendly. Each step of i-clicker's development has been informed by teaching and learning. i-clicker is superior to other systems from both pedagogical and technical standpoints. To learn more about packaging i-clicker with this textbook, please contact your local sales rep or visit www.iclicker.com.

# TO STUDENTS: *What Is Statistics?*

Statistics is the science of collecting, organizing, and interpreting numerical facts, which we call *data*. We are bombarded by data in our everyday lives. The news mentions imported car sales, the latest poll of the president's popularity, and the average high temperature for today's date. Advertisements claim that data show the superiority of the advertiser's product. All sides in public debates about economics, education, and social policy argue from data. A knowledge of statistics helps separate sense from nonsense in the flood of data.

The study and collection of data are also important in the work of many professions, so training in the science of statistics is valuable preparation for a variety of careers. Each month, for example, government statistical offices release the latest numerical information on unemployment and inflation. Economists and financial advisors, as well as policymakers in government and business, study these data in order to make informed decisions. Doctors must understand the origin and trustworthiness of the data that appear in medical journals. Politicians rely on data from polls of public opinion. Business decisions are based on market research data that reveal consumer tastes. Engineers gather data on the quality and reliability of manufactured products. Most areas of academic study make use of numbers, and therefore also make use of the methods of statistics.

## Understanding from Data

*The goal of statistics is to gain understanding from data*. To gain understanding, we often operate on a set of numbers—we average or graph them, for example. But we must do more, because data are not just numbers; they are numbers that have some context that helps us understand them.

You read that low birth weight is a major reason why infant mortality in the United States is higher than in most other advanced nations. The report goes on to say that 7.8% of children born in the United States have low birth weight, and that 13.4% of black infants have low birth weight.[1] To make sense of these numbers you must know what counts as low birth weight (less than 2500 grams, or 5.5 pounds) and have some feeling for the weights of babies. You probably recognize that 5.5 pounds is small, that 7.5 pounds (3400 grams) is about average, and that 10 pounds (4500 grams) is a big baby.

Another part of the context is the source of the data. How do we know that 7.8% of American babies have low birth weight or that the average weight of newborns is about 3400 grams? The data come from the National Center for Health Statistics, a government office to which the states report information from all birth certificates issued each month. These are the most complete data available about births in the United States.

When you do statistical problems—even straightforward textbook problems—don't just graph or calculate. Think about the context and state your conclusions in the specific setting of the problem. As you are learning how to do statistical calculations and graphs, remember that the goal of statistics is not calculation for its own sake but gaining understanding from numbers. The calculations and graphs can be automated by a calculator or software, but you must supply the understanding. This book presents only the most common specific procedures for statistical analysis. A thorough grasp of the principles of statistics will enable you to quickly learn more advanced methods as needed. Always keep in mind, however, that a fancy computer analysis carried out without attention to basic principles will often produce elaborate nonsense. As you read, seek to understand the principles, as well as the necessary details of methods and recipes.

## The Rise of Statistics

Historically, the ideas and methods of statistics developed gradually as society grew interested in collecting and using data for a variety of applications. The earliest origins of statistics lie in the desire of rulers to count the number of inhabitants or measure the value of taxable land in their domains. As the physical sciences developed in the seventeenth and eighteenth centuries, the importance of careful measurements of weights, distances, and other physical quantities grew. Astronomers and surveyors striving for exactness had to deal with variation in their measurements. Many measurements should be better than a single measurement, even though they vary among themselves. How can we best combine many varying observations? Statistical methods that are still important were invented to analyze scientific measurements.

By the nineteenth century, the agricultural, life, and behavioral sciences also began to rely on data to answer fundamental questions. How are the heights of parents and children related? Does a new variety of wheat produce higher yields than the old, and under what conditions of rainfall and fertilizer? Can a person's mental ability and behavior be measured just as we measure height and reaction time? Effective methods for dealing with such questions developed slowly and with much debate.[2]

As methods for producing and understanding data grew in number and sophistication, the new discipline of statistics took shape in the twentieth century. Ideas and techniques that originated in the collection of government data, in the study of astronomical or biological measurements, and in the attempt to understand heredity or intelligence came together to form a unified "science of data." That science of data—statistics—is the topic of this text.

## The Organization of This Book

Part I of this book, called "Looking at Data," concerns data analysis and data production. The first two chapters deal with statistical methods for organizing and describing data. These chapters progress from simpler to more complex data. Chapter 1 examines data on a single variable; Chapter 2 is devoted to relationships among two or more variables. You will learn both how to examine data produced by others and how to organize and summarize your own data. These summaries will be first graphical, then numerical, and then, when appropriate, in the form of a mathematical model that gives a compact description of the overall pattern of the data. Chapter 3 outlines arrangements (called "designs") for producing data that answer specific questions. The principles presented in this chapter will help you to design proper samples and experiments and to evaluate such investigations in your field of study.

Part II, consisting of Chapters 4 to 8, introduces statistical inference—formal methods for drawing conclusions from properly produced data. Statistical inference uses the language of probability to describe how reliable its conclusions are, so some basic facts about probability are needed to understand inference. Probability is the subject of Chapters 4 and 5. Chapter 6, perhaps the most important chapter in the text, introduces the reasoning of statistical inference. Effective inference is based on good procedures for producing data (Chapter 3), careful examination of the data (Chapters 1 and 2), and an understanding of the nature of statistical inference as discussed in Chapter 6. Chapters 7 and 8 describe some of the most common specific methods of inference for drawing conclusions about means and proportions from one and two samples.

The five shorter chapters in Part III introduce somewhat more advanced methods of inference, dealing with relations in categorical data, regression and correlation, and analysis of variance. Supplement chapters, available on the book-companion CD and Web site, present additional statistical topics.

## What Lies Ahead

*Introduction to the Practice of Statistics* is full of data from many different areas of life and study. Many exercises ask you to express briefly some understanding gained from the data. In practice, you would know much more about the background of the data you work with and about the questions you hope the data will answer. No textbook can be fully realistic. But it is important to form the habit of asking, "What do the data tell me?" rather than just concentrating on making graphs and doing calculations.

You should have some help in automating many of the graphs and calculations. You should certainly have a calculator with basic statistical functions. Look for key words such as "two-variable statistics" or "regression" when you shop for a calculator. More advanced (and more expensive) calculators will do much more, including some statistical graphs. You may be asked to use software as well. There are many kinds of statistical software, from spreadsheets to large programs for advanced users of statistics. The kind of computing available to learners varies a great deal from place to place—but the big ideas of statistics don't depend on any particular level of access to computing.

Because graphing and calculating are automated in statistical practice, the most important assets you can gain from the study of statistics are an understanding of the big ideas and the beginnings of good judgment in working with data. Ideas and judgment can't (at least yet) be automated. They guide you in telling the computer what to do and in interpreting its output. This book tries to explain the most important ideas of statistics, not just teach methods. Some examples of big ideas that you will meet are "always plot your data," "randomized comparative experiments," and "statistical significance."

You learn statistics by doing statistical problems. Practice, practice, practice. Be prepared to work problems. The basic principle of learning is persistence. Being organized and persistent is more helpful in reading this book than knowing lots of math. The main ideas of statistics, like the main ideas of any important subject, took a long time to discover and take some time to master. The gain will be worth the pain.

# ABOUT THE AUTHORS

**David S. Moore** is Shanti S. Gupta Distinguished Professor of Statistics, Emeritus, at Purdue University and was 1998 president of the American Statistical Association. He received his A.B. from Princeton and his Ph.D. from Cornell, both in mathematics. He has written many research papers in statistical theory and served on the editorial boards of several major journals. Professor Moore is an elected fellow of the American Statistical Association and of the Institute of Mathematical Statistics and an elected member of the International Statistical Institute. He has served as program director for statistics and probability at the National Science Foundation.

In recent years, Professor Moore has devoted his attention to the teaching of statistics. He was the content developer for the Annenberg/Corporation for Public Broadcasting college-level telecourse *Against All Odds: Inside Statistics* and for the series of video modules *Statistics: Decisions through Data*, intended to aid the teaching of statistics in schools. He is the author of influential articles on statistics education and of several leading texts. Professor Moore has served as president of the International Association for Statistical Education and has received the Mathematical Association of America's national award for distinguished college or university teaching of mathematics.

**George P. McCabe** is the Associate Dean for Academic Affairs and a Professor of Statistics at Purdue University. In 1966 he received a B.S. degree in mathematics from Providence College and in 1970 a Ph.D. in mathematical statistics from Columbia University. His entire professional career has been spent at Purdue with sabbaticals at Princeton, the Commonwealth Scientific and Industrial Research Organization (CSIRO) in Melbourne, Australia, the University of Berne (Switzerland), the National Institute of Standards and Technology (NIST) in Boulder, Colorado, and the National University of Ireland in Galway. Professor McCabe is an elected fellow of the American Statistical Association and was 1998 chair of its section on Statistical Consulting. He has served on the editorial boards of several statistics journals. He has consulted with many major corporations and has testified as an expert witness on the use of statistics in several cases.

Professor McCabe's research interests have focused on applications of statistics. Much of his recent work has been focused on problems in nutrition, including nutrient requirements, calcium metabolism, and bone health. He is author or coauthor of over 150 publications in many different journals.

**Bruce A. Craig** is Professor of Statistics and Director of the Statistical Consulting Service at Purdue University. He received his B.S. in mathematics and economics from Washington University in St. Louis and his Ph.D. in statistics from the University of Wisconsin–Madison. He is an active member of the American Statistical Association and will be chair of its section on Statistical Consulting in 2009. He also is an active member of the Eastern North American Region of the International Biometrics Society and was elected by the voting membership to the Regional Committee between 2003 and 2006. Professor Craig serves on the editorial board of several statistical journals and serves on many data and safety monitoring boards, including Purdue's IRB.

Professor Craig's research interests focus on the development of novel statistical methodology to address research questions in the life sciences. Areas of current interest are protein structure determination, diagnostic testing, and animal abundance estimation. In 2005, he was named Purdue University Faculty Scholar.

# DATA TABLE INDEX

# BEYOND THE BASICS INDEX

CHAPTER

# 1

# Looking at Data—Distributions

Students planning a referendum on college fees. See Example 1.1.



## Introduction

*Statistics is the science of learning from data.* Data are numerical facts. Here is an example of a situation where students used the results of a referendum to convince their university Board of Trustees to make a decision.

EXAMPLE

**1.1 Students vote for service learning scholarships.** According to the National Service-Learning Clearinghouse: "Service-learning is a teaching and learning strategy that integrates meaningful community service with instruction and reflection to enrich the learning experience, teach civic responsibility, and strengthen communities."[1] University of Illinois at Urbana–Champaign students decided that they wanted to become involved in this national movement. They proposed a $15.00 per semester Legacy of Service and Learning Scholarship fee. Each year, $10.00 would be invested in an endowment and $5.00 would be used to fund current-use scholarships. In a referendum, students voted 3785 to 2977 in favor of the proposal. On April 11, 2006, the university Board of Trustees approved the proposal. Approximately $370,000 in current-use scholarship funds will be generated each year, and with the endowment, it is expected that in 20 years there will be more than a million dollars per year for these scholarships.

To learn from data, we need more than just the numbers. The numbers in a medical study, for example, mean little without some knowledge of the goals of the study and of what blood pressure, heart rate, and other measurements contribute to those goals. That is, *data are numbers with a context,* and we need to understand the context if we are to make sense of the numbers. On the other hand, measurements from the study's several hundred subjects are of little value to even the most knowledgeable medical expert until the tools of statistics organize, display, and summarize them. We begin our study of statistics by mastering the art of examining data.

## Variables

Any set of data contains information about some group of *individuals.* The information is organized in *variables.*

### INDIVIDUALS AND VARIABLES

**Individuals** are the objects described in a set of data. Individuals are sometimes people. When the objects that we want to study are not people, we often call them **cases.**

A **variable** is any characteristic of an individual. A variable can take different values for different individuals.

EXAMPLE

**1.2 Data for students in a statistics class.** Figure 1.1 shows part of a data set for students enrolled in an introductory statistics class. Each row gives the data on one student. The values for the different variables are in the columns. This data set has eight variables. ID is an identifier for each student. Exam1, Exam2, Homework, Final, and Project give the points earned, out of a total of 100 possible, for each of these course requirements. Final grades are based on a possible 200 points for each exam and the final, 300 points for Homework, and 100 points for Project. TotalPoints is the variable that gives the composite score. It is computed by adding 2 times Exam1, Exam2, and Final, 3 times Homework plus 1 times Project. Grade is the grade earned in the course. This instructor used cut-offs of 900, 800, 700, etc. for the letter grades.

Microsoft Excel

| | A | B | C | D | E | F | G | H |
|---|---|---|---|---|---|---|---|---|
| 1 | ID | Exam1 | Exam2 | Homework | Final | Project | TotalPoints | Grade |
| 2 | 101 | 89 | 94 | 88 | 87 | 95 | 899 | B |
| 3 | 102 | 78 | 84 | 90 | 89 | 94 | 866 | B |
| 4 | 103 | 71 | 80 | 75 | 79 | 95 | 780 | C |
| 5 | 104 | 95 | 98 | 97 | 96 | 93 | 962 | A |
| 6 | 105 | 79 | 88 | 85 | 88 | 96 | 861 | B |

**FIGURE 1.1** Spreadsheet for Example 1.2.

spreadsheet

The display in Figure 1.1 is from an Excel **spreadsheet.** Most statistical software packages use similar spreadsheets and many are able to import Excel spreadsheets.

## USE YOUR KNOWLEDGE

**1.1 Read the spreadsheet.** Refer to Figure 1.1. Give the values of the variables Exam1, Exam2, and Final for the student with ID equal to 104.

**1.2 Calculate the grade.** A student whose data do not appear on the spreadsheet scored 88 on Exam1, 85 on Exam2, 77 for Homework, 90 on the Final, and 80 on the Project. Find TotalPoints for this student and give the grade earned.

Spreadsheets are very useful for doing the kind of simple computations that you did in Exercise 1.2. You can type in a formula and have the same computation performed for each row.

Note that the names we have chosen for the variables in our spreadsheet do not have spaces. For example, we could have used the name "Exam 1" for the first exam score rather than Exam1. In many statistical software packages, however, spaces are not allowed in variable names. For this reason, when creating spreadsheets for eventual use with statistical software, it is best to avoid spaces in variable names. Another convention is to use an underscore (_) where you would normally use a space. For our data set, we could use Exam_1, Exam_2, and Final_Exam.

In practice, any set of data is accompanied by background information that helps us understand the data. When you plan a statistical study or explore data from someone else's work, ask yourself the following questions:

1. **Why? What purpose** do the data have? Do we hope to answer some specific questions? Do we want to draw conclusions about individuals other than those for whom we actually have data?
2. **Who?** What **individuals** do the data describe? **How many** individuals appear in the data?
3. **What?** How many **variables** do the data contain? What are the **exact definitions** of these variables? Some variables have units. Weights, for example, might be recorded in pounds, in thousands of pounds, or in kilograms. For these kinds of variables, you need to know the **unit of measurement.**

EXAMPLE

**1.3 Individuals and variables.** The data set in Figure 1.1 was constructed to keep track of the grades for students in an introductory statistics course. The individuals are the students in the class. There are 8 variables in this data set. These include an identifier for each student and scores for the various course requirements. There are no units for ID and grade. The other variables all have "points" as the unit.

Some variables, like gender and college major, simply place individuals into categories. Others, like height and grade point average, take numerical values for which we can do arithmetic. It makes sense to give an average salary for a company's employees, but it does not make sense to give an "average" gender. We can, however, count the numbers of female and male employees and do arithmetic with these counts.

### CATEGORICAL AND QUANTITATIVE VARIABLES

A **categorical variable** places an individual into one of two or more groups or categories.

A **quantitative variable** takes numerical values for which arithmetic operations such as adding and averaging make sense.

The **distribution** of a variable tells us what values it takes and how often it takes these values.

EXAMPLE

**1.4 Variables for students in a statistics course.** Suppose the data for the students in the introductory statistics class were also to be used to study relationships between student characteristics and success in the course. For this purpose, we might want to use a data set like the spreadsheet in Figure 1.2. Here, we have decided to focus on the TotalPoints and Grade as the outcomes of interest. Other variables of interest have been included: Gender, PrevStat (whether or not the student has taken a statistics course previously), and Year (student classification as first, second, third, or fourth year). ID is a categorical variable, total points is a quantitative variable, and the remaining variables are all categorical.

Microsoft Excel

| | A | B | C | D | E | F |
|---|---|---|---|---|---|---|
| 1 | ID | TotalPoints | Grade | Gender | PrevStat | Year |
| 2 | 101 | 899 | A | F | Yes | 4 |
| 3 | 102 | 866 | B | M | Yes | 3 |
| 4 | 103 | 780 | C | M | No | 3 |
| 5 | 104 | 962 | A | M | No | 1 |
| 6 | 105 | 861 | B | F | No | 4 |

**FIGURE 1.2** Spreadsheet for Example 1.4.

In our example, the possible values for the grade variable are A, B, C, D, and F. When computing grade point averages, many colleges and universities translate these letter grades into numbers using $A = 4$, $B = 3$, $C = 2$, $D = 1$, and $F = 0$. The transformed variable with numeric values is considered to be quantitative because we can average the numerical values across different courses to obtain a grade point average.

Sometimes, experts argue about numerical scales such as this. They ask whether or not the difference between an A and a B is the same as the difference between a D and an F. Similarly, many questionnaires ask people to respond on a 1 to 5 scale with 1 representing strongly agree, 2 representing agree, etc. Again we could ask about whether or not the five possible values for this scale are equally spaced in some sense. From a practical point of view, the averages that can be computed when we convert categorical scales such as these to numerical values frequently provide a very useful way to summarize data.

## USE YOUR KNOWLEDGE

**1.3** **Apartment rentals.** A data set lists apartments available for students to rent. Information provided includes the monthly rent, whether or not cable is included free of charge, whether or not pets are allowed, the number of bedrooms, and the distance to the campus. Describe the individuals or cases in the data set, give the number of variables, and specify whether each variable is categorical or quantitative.

## Measurement: know your variables

The context of data includes an understanding of the variables that are recorded. Often the variables in a statistical study are easy to understand: height in centimeters, study time in minutes, and so on. But each area of work also has its own special variables. A psychologist uses the Minnesota Multiphasic Personality Inventory (MMPI), and a physical fitness expert measures "VO2 max," the volume of oxygen consumed per minute while exercising at your maximum capacity. Both of these variables are measured with special **instruments.** VO2 max is measured by exercising while breathing into a mouthpiece connected to an apparatus that measures oxygen consumed. Scores on the MMPI are based on a long questionnaire, which is also an instrument. Part of mastering your field of work is learning what variables are important and how they are best measured. Because details of particular measurements usually require knowledge of the particular field of study, we will say little about them.

instrument

CAUTION

*Be sure that each variable really does measure what you want it to. A poor choice of variables can lead to misleading conclusions.* Often, for example, the **rate** at which something occurs is a more meaningful measure than a simple count of occurrences.

rate

EXAMPLE

**1.5 Accidents for passenger cars and motorcycles.** The government's Fatal Accident Reporting System says that 27,102 passenger cars were involved in fatal accidents in 2002. Only 3339 motorcycles had fatal accidents that year.[2] Does this mean that motorcycles are safer than cars? Not at all—there are many more cars than motorcycles, so we expect cars to have a higher *count* of fatal accidents.

A better measure of the dangers of driving is a *rate,* the number of fatal accidents divided by the number of vehicles on the road. In 2002, passenger cars had about 21 fatal accidents for each 100,000 vehicles registered. There were about 67 fatal accidents for each 100,000 motorcycles registered. The rate for motorcycles is more than three times the rate for cars. Motorcycles are, as we might guess, much more dangerous than cars.

## 1.1 Displaying Distributions with Graphs

exploratory data analysis

Statistical tools and ideas help us examine data in order to describe their main features. This examination is called **exploratory data analysis.** Like an explorer crossing unknown lands, we want first to simply describe what we see. Here are two basic strategies that help us organize our exploration of a set of data:

- Begin by examining each variable by itself. Then move on to study the relationships among the variables.
- Begin with a graph or graphs. Then add numerical summaries of specific aspects of the data.

We will follow these principles in organizing our learning. This chapter presents methods for describing a single variable. We will study relationships among several variables in Chapter 2. Within each chapter, we will begin with graphical displays, then add numerical summaries for more complete description.

### Graphs for categorical variables

The values of a categorical variable are labels for the categories, such as "female" and "male." The **distribution** of a categorical variable lists the categories and gives either the **count** or the **percent** of individuals who fall in each category. For example, how well educated are 30-something young adults? Here is the distribution of the highest level of education for people aged 25 to 34 years:[3]

| Education | Count (millions) | Percent |
|---|---|---|
| Less than high school | 4.6 | 12.1 |
| High school graduate | 11.6 | 30.5 |
| Some college | 7.4 | 19.5 |
| Associate degree | 3.3 | 8.7 |
| Bachelor's degree | 8.6 | 22.6 |
| Advanced degree | 2.5 | 6.6 |

Are you surprised that only 29.2% of young adults have at least a bachelor's degree?

bar graph

pie chart

The graphs in Figure 1.3 display these data. The **bar graph** in Figure 1.3(a) quickly compares the sizes of the six education groups. The heights of the bars show the percents in the six categories. The **pie chart** in Figure 1.3(b) helps us see what part of the whole each group forms. For example, the "Bachelor's" slice makes up 22.6% of the pie because 22.6% of young adults have a bachelor's degree but no higher degree. We have moved that slice out to call attention to it. Because pie charts lack a scale, we have added the percents to the labels for the slices. *Pie charts require that you include all the categories that make up a whole. Use them only when you want to emphasize each category's relation to the whole.* Bar graphs are easier to read and are also more flexible. For example, you can use a bar graph to compare the numbers of students at your college majoring in biology, business, and political science. A pie chart cannot make this comparison because not all students fall into one of these three majors.

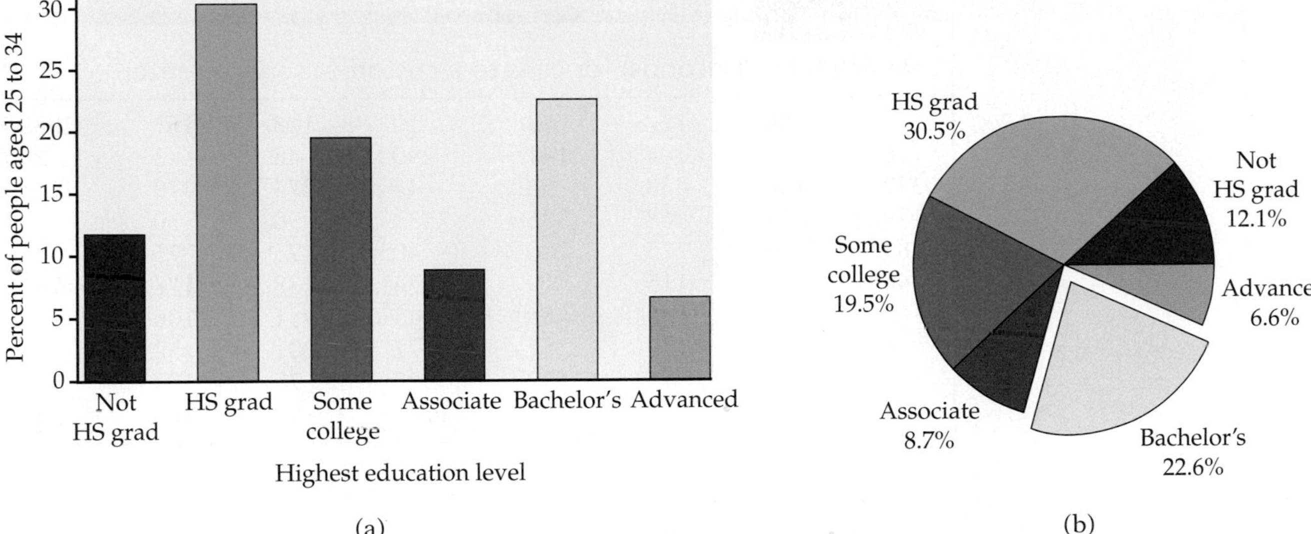


**FIGURE 1.3** (a) Bar graph of the educational attainment of people aged 25 to 34 years. (b) Pie chart of the education data, with bachelor's degree holders emphasized.

## USE YOUR KNOWLEDGE

**1.4 Read the pie chart.** Refer to Figure 1.3(b). What percent of young adults have either an associate degree or a bachelor's degree?

Bar graphs and pie charts help an audience grasp a distribution quickly. They are, however, of limited use for data analysis because it is easy to understand data on a single categorical variable, such as highest level of education, without a graph. We will move on to quantitative variables, where graphs are essential tools.

## Data analysis in action: don't hang up on me

Many businesses operate call centers to serve customers who want to place an order or make an inquiry. Customers want their requests handled thoroughly. Businesses want to treat customers well, but they also want to avoid wasted time on the phone. They therefore monitor the length of calls and encourage their representatives to keep calls short. Here is an example of the difficulties this policy can cause.

EXAMPLE

**1.6 Individuals and variables for the customer service center.** We have data on the length of all 31,492 calls made to the customer service center of a small bank in a month. Table 1.1 displays the lengths of the first 80 calls. The file for the complete data set is *eg01-004,* which you can find on the text CD and Web site.[4]

Take a look at the data in Table 1.1. The numbers are meaningless without some background information. The *individuals* are calls made to the bank's call center. The *variable* recorded is the length of each call. The *units* are

**TABLE 1.1**

**Service times (seconds) for calls to a customer service center**

| | | | | | | | |
|---|---|---|---|---|---|---|---|
| 77 | 289 | 128 | 59 | 19 | 148 | 157 | 203 |
| 126 | 118 | 104 | 141 | 290 | 48 | 3 | 2 |
| 372 | 140 | 438 | 56 | 44 | 274 | 479 | 211 |
| 179 | 1 | 68 | 386 | 2631 | 90 | 30 | 57 |
| 89 | 116 | 225 | 700 | 40 | 73 | 75 | 51 |
| 148 | 9 | 115 | 19 | 76 | 138 | 178 | 76 |
| 67 | 102 | 35 | 80 | 143 | 951 | 106 | 55 |
| 4 | 54 | 137 | 367 | 277 | 201 | 52 | 9 |
| 700 | 182 | 73 | 199 | 325 | 75 | 103 | 64 |
| 121 | 11 | 9 | 88 | 1148 | 2 | 465 | 25 |

seconds. We see that the call lengths vary a great deal. The longest call lasted 2631 seconds, almost 44 minutes. More striking is that 8 of these 80 calls lasted less than 10 seconds. What's going on?

Figure 1.4 is a histogram of the lengths of all 31,492 calls. We did not plot the few lengths greater than 1200 seconds (20 minutes). As expected, the graph shows that most calls last between about a minute and 5 minutes, with some lasting much longer when customers have complicated problems. More striking is the fact that 7.6% of all calls are no more than 10 seconds long. It turned out that the bank penalized representatives whose average call length was too long—so some representatives just hung up on customers in order to bring their average length down. Neither the customers nor the bank were happy about this. The bank changed its policy, and later data showed that calls under 10 seconds had almost disappeared.

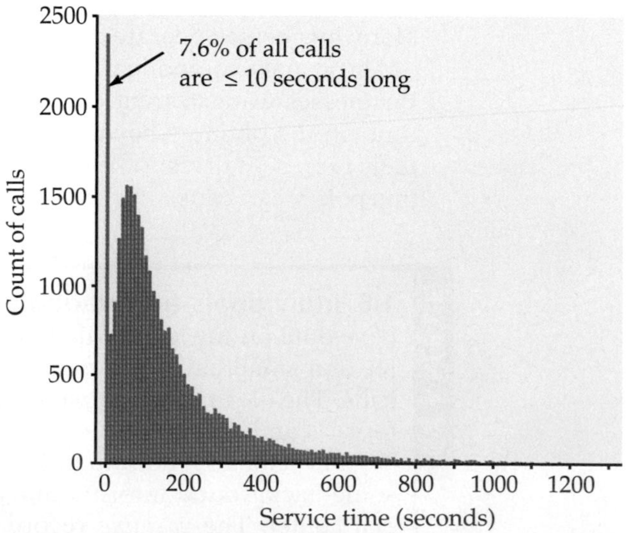


**FIGURE 1.4** The distribution of call lengths for 31,492 calls to a bank's customer service center, for Example 1.6. The data show a surprising number of very short calls. These are mostly due to representatives deliberately hanging up in order to bring down their average call length.

tails

The extreme values of a distribution are in the **tails** of the distribution. The high values are in the upper, or right, tail and the low values are in the lower, or left, tail. The overall pattern in Figure 1.4 is made up of the many moderate call lengths and the long right tail of more lengthy calls. The striking departure from the overall pattern is the surprising number of very short calls in the left tail.

Our examination of the call center data illustrates some important principles:

- After you understand the background of your data (individuals, variables, units of measurement), the first thing to do is almost always **plot your data.**
- When you look at a plot, look for an **overall pattern** and also for any **striking departures** from the pattern.

We now turn to the kinds of graphs that are used to describe the distribution of a quantitative variable. We will explain how to make the graphs by hand, because knowing this helps you understand what the graphs show. However, making graphs by hand is so tedious that software is almost essential for effective data analysis unless you have just a few observations.

## Stemplots

A *stemplot* (also called a stem-and-leaf plot) gives a quick picture of the shape of a distribution while including the actual numerical values in the graph. Stemplots work best for small numbers of observations that are all greater than 0.

### STEMPLOT

To make a **stemplot:**

**1.** Separate each observation into a **stem** consisting of all but the final (rightmost) digit and a **leaf,** the final digit. Stems may have as many digits as needed, but each leaf contains only a single digit.

**2.** Write the stems in a vertical column with the smallest at the top, and draw a vertical line at the right of this column.

**3.** Write each leaf in the row to the right of its stem, in increasing order out from the stem.

EXAMPLE

**1.7 Literacy of men and women.** The Islamic world is attracting increased attention in Europe and North America. Table 1.2 shows the percent of men and women at least 15 years old who were literate in 2002 in the major Islamic nations. We omitted countries with populations less than 3 million. Data for a few nations, such as Afghanistan and Iraq, are not available.[5]

To make a stemplot of the percents of females who are literate, use the first digits as stems and the second digits as leaves. Algeria's 60% literacy rate, for example, appears as the leaf 0 on the stem 6. Figure 1.5 shows the steps in making the plot.

**TABLE 1.2**

**Literacy rates (percent) in Islamic nations**

| Country | Female percent | Male percent | Country | Female percent | Male percent |
|---|---|---|---|---|---|
| Algeria | 60 | 78 | Morocco | 38 | 68 |
| Bangladesh | 31 | 50 | Saudi Arabia | 70 | 84 |
| Egypt | 46 | 68 | Syria | 63 | 89 |
| Iran | 71 | 85 | Tajikistan | 99 | 100 |
| Jordan | 86 | 96 | Tunisia | 63 | 83 |
| Kazakhstan | 99 | 100 | Turkey | 78 | 94 |
| Lebanon | 82 | 95 | Uzbekistan | 99 | 100 |
| Libya | 71 | 92 | Yemen | 29 | 70 |
| Malaysia | 85 | 92 | | | |

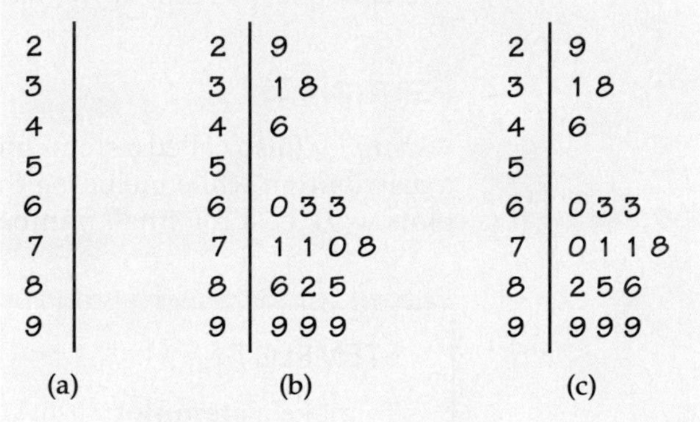

```
2 |          2 | 9          2 | 9
3 |          3 | 1 8        3 | 1 8
4 |          4 | 6          4 | 6
5 |          5 |            5 |
6 |          6 | 0 3 3      6 | 0 3 3
7 |          7 | 1 1 0 8    7 | 0 1 1 8
8 |          8 | 6 2 5      8 | 2 5 6
9 |          9 | 9 9 9      9 | 9 9 9
  (a)            (b)            (c)
```

**FIGURE 1.5** Making a stemplot of the data in Example 1.7. (a) Write the stems. (b) Go through the data and write each leaf on the proper stem. For example, the values on the 8 stem are 86, 82, and 85 in the order of the table. (c) Arrange the leaves on each stem in order out from the stem. The 8 stem now has leaves 2 5 6.

cluster

The overall pattern of the stemplot is irregular, as is often the case when there are only a few observations. There do appear to be two **clusters** of countries. The plot suggests that we might ask what explains the variation in literacy. For example, why do the three central Asian countries (Kazakhstan, Tajikistan, and Uzbekistan) have very high literacy rates?

## USE YOUR KNOWLEDGE

**1.5** **Make a stemplot.** Here are the scores on the first exam in an introductory statistics course for 30 students in one section of the course:

| | | | | | | | | | | | | | | |
|---|---|---|---|---|---|---|---|---|---|---|---|---|---|---|
| 80 | 73 | 92 | 85 | 75 | 98 | 93 | 55 | 80 | 90 | 92 | 80 | 87 | 90 | 72 |
| 65 | 70 | 85 | 83 | 60 | 70 | 90 | 75 | 75 | 58 | 68 | 85 | 78 | 80 | 93 |

Use these data to make a stemplot. Then use the stemplot to describe the distribution of the first-exam scores for this course.

back-to-back stemplot

When you wish to compare two related distributions, a **back-to-back stemplot** with common stems is useful. The leaves on each side are ordered out from the common stem. Here is a back-to-back stemplot comparing the distributions of female and male literacy rates in the countries of Table 1.2.

| Female | | Male |
|---:|:---:|:---|
| 9 | 2 | |
| 81 | 3 | |
| 6 | 4 | |
| | 5 | 0 |
| 330 | 6 | 88 |
| 8110 | 7 | 08 |
| 652 | 8 | 3459 |
| 999 | 9 | 22456 |
| | 10 | 000 |

The values on the left are the female percents, as in Figure 1.5, but ordered out from the stem from right to left. The values on the right are the male percents. It is clear that literacy is generally higher among males than among females in these countries.

splitting stems
trimming

*Stemplots do not work well for large data sets, where each stem must hold a large number of leaves.* Fortunately, there are two modifications of the basic stemplot that are helpful when plotting the distribution of a moderate number of observations. You can double the number of stems in a plot by **splitting each stem** into two: one with leaves 0 to 4 and the other with leaves 5 through 9. When the observed values have many digits, it is often best to **trim** the numbers by removing the last digit or digits before making a stemplot. You must use your judgment in deciding whether to split stems and whether to trim, though statistical software will often make these choices for you. Remember that the purpose of a stemplot is to display the shape of a distribution. If a stemplot has fewer than about five stems, you should usually split the stems unless there are few observations. If there are many stems with no leaves or only one leaf, trimming will reduce the number of stems. Here is an example that makes use of both of these modifications.

EXAMPLE

**1.8 Stemplot for length of service calls.** Return to the 80 customer service call lengths in Table 1.1. To make a stemplot of this distribution, we first trim the call lengths to tens of seconds by dropping the last digit. For example, 56 seconds trims to 5 and 143 seconds trims to 14. (We might also round to the nearest 10 seconds, but trimming is faster than rounding if you must do it by hand.)

We can then use tens of seconds as our leaves, with the digits to the left forming stems. This gives us the single-digit leaves that a stemplot requires. For example, 56 trimmed to 5 becomes leaf 5 on the 0 stem; 143 trimmed to 14 becomes leaf 4 on the 1 stem.

Because we have 80 observations, we split the stems. Thus, 56 trimmed to 5 becomes leaf 5 on the second 0 stem, along with all leaves 5 to 9. Leaves

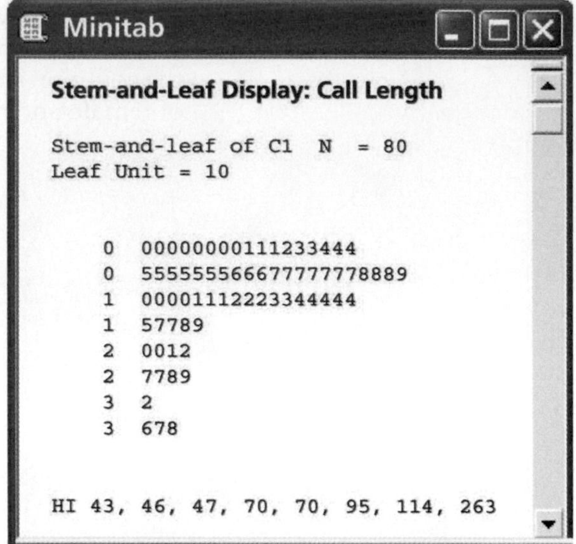


**FIGURE 1.6** Stemplot from Minitab of the 80 call lengths in Table 1.1, for Example 1.8. The software has trimmed the data by removing the last digit. It has also split stems and listed the highest observations apart from the plot.

0 to 4 go on the first 0 stem. Figure 1.6 is a stemplot of these data made by software. The software automatically did what we suggest: trimmed to tens of seconds and split stems. To save space, the software also listed the largest values as "HI" rather than create stems all the way up to 26. The stemplot shows the overall pattern of the distribution, with many short to moderate lengths and some very long calls.

## Histograms

Stemplots display the actual values of the observations. This feature makes stemplots awkward for large data sets. Moreover, the picture presented by a stemplot divides the observations into groups (stems) determined by the number system rather than by judgment. Histograms do not have these limitations. A **histogram** breaks the range of values of a variable into classes and displays only the count or percent of the observations that fall into each class. You can choose any convenient number of classes, but you should always choose classes of equal width. Histograms are slower to construct by hand than stemplots and do not display the actual values observed. For these reasons we prefer stemplots for small data sets. The construction of a histogram is best shown by example. Any statistical software package will of course make a histogram for you.

histogram

EXAMPLE

**1.9 Distribution of IQ scores.** You have probably heard that the distribution of scores on IQ tests is supposed to be roughly "bell-shaped." Let's look at some actual IQ scores. Table 1.3 displays the IQ scores of 60 fifth-grade students chosen at random from one school.[6]

1. Divide the range of the data into classes of equal width. The scores in Table 1.3 range from 81 to 145, so we choose as our classes

**TABLE 1.3**

IQ test scores for 60 randomly chosen fifth-grade students

| | | | | | | | | | |
|---|---|---|---|---|---|---|---|---|---|
| 145 | 139 | 126 | 122 | 125 | 130 | 96 | 110 | 118 | 118 |
| 101 | 142 | 134 | 124 | 112 | 109 | 134 | 113 | 81 | 113 |
| 123 | 94 | 100 | 136 | 109 | 131 | 117 | 110 | 127 | 124 |
| 106 | 124 | 115 | 133 | 116 | 102 | 127 | 117 | 109 | 137 |
| 117 | 90 | 103 | 114 | 139 | 101 | 122 | 105 | 97 | 89 |
| 102 | 108 | 110 | 128 | 114 | 112 | 114 | 102 | 82 | 101 |

$$75 \leq \text{IQ score} < 85$$

$$85 \leq \text{IQ score} < 95$$

$$\vdots$$

$$145 \leq \text{IQ score} < 155$$

Be sure to specify the classes precisely so that each individual falls into exactly one class. A student with IQ 84 would fall into the first class, but IQ 85 falls into the second.

frequency
frequency table

2. Count the number of individuals in each class. These counts are called **frequencies,** and a table of frequencies for all classes is a **frequency table.**

| Class | Count | Class | Count |
|---|---|---|---|
| 75 to 84 | 2 | 115 to 124 | 13 |
| 85 to 94 | 3 | 125 to 134 | 10 |
| 95 to 104 | 10 | 135 to 144 | 5 |
| 105 to 114 | 16 | 145 to 154 | 1 |

3. Draw the histogram. First, on the horizontal axis mark the scale for the variable whose distribution you are displaying. That's IQ score. The scale runs from 75 to 155 because that is the span of the classes we chose. The vertical axis contains the scale of counts. Each bar represents a class. The base of the bar covers the class, and the bar height is the class count. There is no horizontal space between the bars unless a class is empty, so that its bar has height zero. Figure 1.7 is our histogram. It does look roughly "bell-shaped."

Large sets of data are often reported in the form of frequency tables when it is not practical to publish the individual observations. In addition to the frequency (count) for each class, we may be interested in the fraction or percent of the observations that fall in each class. A histogram of percents looks just like a frequency histogram such as Figure 1.7. Simply relabel the vertical scale to read in percents. Use histograms of percents for comparing several distributions that have different numbers of observations.

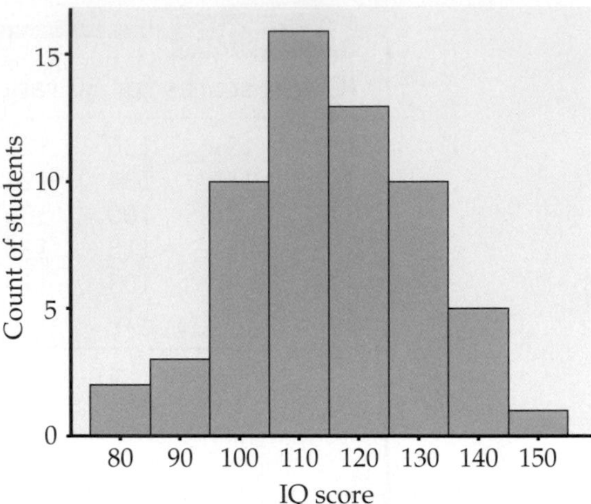


**FIGURE 1.7** Histogram of the IQ scores of 60 fifth-grade students, for Example 1.9.

## USE YOUR KNOWLEDGE

**1.6 Make a histogram.** Refer to the first-exam scores from Exercise 1.5. Use these data to make a histogram using classes 50–59, 60–69, etc. Compare the histogram with the stemplot as a way of describing this distribution. Which do you prefer for these data?

Our eyes respond to the *area* of the bars in a histogram. Because the classes are all the same width, area is determined by height and all classes are fairly represented. There is no one right choice of the classes in a histogram. Too few classes will give a "skyscraper" graph, with all values in a few classes with tall bars. Too many will produce a "pancake" graph, with most classes having one or no observations. Neither choice will give a good picture of the shape of the distribution. You must use your judgment in choosing classes to display the shape. Statistical software will choose the classes for you. The software's choice is often a good one, but you can change it if you want.

*You should be aware that the appearance of a histogram can change when you change the classes.* Figure 1.8 is a histogram of the customer service call lengths

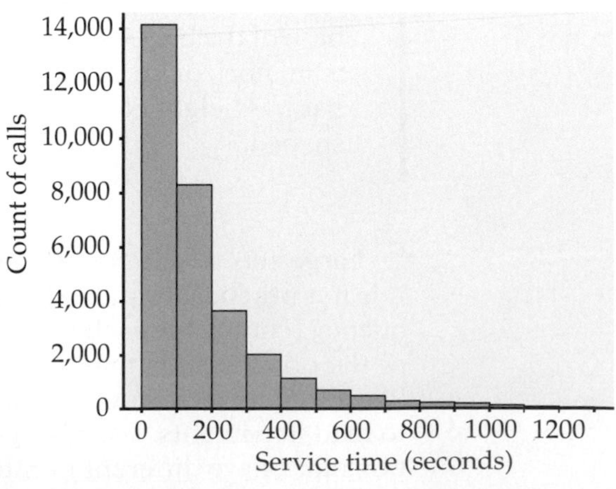


**FIGURE 1.8** The "default" histogram produced by software for the call lengths in Example 1.6. This choice of classes hides the large number of very short calls that is revealed by the histogram of the same data in Figure 1.4.

that are also displayed in Figure 1.4. It was produced by software with no special instructions from the user. The software's "default" histogram shows the overall shape of the distribution, but it hides the spike of very short calls by lumping all calls of less than 100 seconds into the first class. We produced Figure 1.4 by asking for smaller classes after Table 1.1 suggested that very short calls might be a problem. Software automates making graphs, but it can't replace thinking about your data. The histogram function in the *One-Variable Statistical Calculator* applet on the text CD and Web site allows you to change the number of classes by dragging with the mouse, so that it is easy to see how the choice of classes affects the histogram.

## USE YOUR KNOWLEDGE

**1.7** **Change the classes in the histogram.** Refer to the first-exam scores from Exercise 1.5 and the histogram you produced in Exercise 1.6. Now make a histogram for these data using classes 40–59, 60–79, and 80–100. Compare this histogram with the one that you produced in Exercise 1.6.

**1.8** **Use smaller classes.** Repeat the previous exercise using classes 55–59, 60–64, 65–69, etc.

Although histograms resemble bar graphs, their details and uses are distinct. A histogram shows the distribution of counts or percents among the values of a single variable. A bar graph compares the size of different items. The horizontal axis of a bar graph need not have any measurement scale but simply identifies the items being compared. Draw bar graphs with blank space between the bars to separate the items being compared. Draw histograms with no space, to indicate that all values of the variable are covered. *Some spreadsheet programs, which are not primarily intended for statistics, will draw histograms as if they were bar graphs, with space between the bars. Often, you can tell the software to eliminate the space to produce a proper histogram.*

## Examining distributions

Making a statistical graph is not an end in itself. The purpose of the graph is to help us understand the data. After you make a graph, always ask, "What do I see?" Once you have displayed a distribution, you can see its important features as follows.

### EXAMINING A DISTRIBUTION

In any graph of data, look for the **overall pattern** and for striking **deviations** from that pattern.

You can describe the overall pattern of a distribution by its **shape, center,** and **spread.**

An important kind of deviation is an **outlier,** an individual value that falls outside the overall pattern.

In Section 1.2, we will learn how to describe center and spread numerically. For now, we can describe the center of a distribution by its *midpoint,* the value with roughly half the observations taking smaller values and half taking larger values. We can describe the spread of a distribution by giving the *smallest and largest values.* Stemplots and histograms display the shape of a distribution in the same way. Just imagine a stemplot turned on its side so that the larger values lie to the right. Some things to look for in describing shape are:

modes
unimodal

- Does the distribution have one or several major peaks, called **modes**? A distribution with one major peak is called **unimodal.**

symmetric
skewed

- Is it approximately symmetric or is it skewed in one direction? A distribution is **symmetric** if the values smaller and larger than its midpoint are mirror images of each other. It is **skewed to the right** if the right tail (larger values) is much longer than the left tail (smaller values).

Some variables commonly have distributions with predictable shapes. Many biological measurements on specimens from the same species and sex—lengths of bird bills, heights of young women—have symmetric distributions. Money amounts, on the other hand, usually have right-skewed distributions. There are many moderately priced houses, for example, but the few very expensive mansions give the distribution of house prices a strong right-skew.

EXAMPLE

**1.10 Examine the histogram.** What does the histogram of IQ scores (Figure 1.7) tell us? **Shape:** The distribution is *roughly symmetric* with a *single peak* in the center. We don't expect real data to be perfectly symmetric, so we are satisfied if the two sides of the histogram are roughly similar in shape and extent. **Center:** You can see from the histogram that the midpoint is not far from 110. Looking at the actual data shows that the midpoint is 114. **Spread:** The spread is from 81 to 145. There are no outliers or other strong deviations from the symmetric, unimodal pattern.

The distribution of call lengths in Figure 1.8, on the other hand, is strongly *skewed to the right.* The midpoint, the length of a typical call, is about 115 seconds, or just under 2 minutes. The spread is very large, from 1 second to 28,739 seconds.

The longest few calls are *outliers.* They stand apart from the long right tail of the distribution, though we can't see this from Figure 1.8, which omits the largest observations. The longest call lasted almost 8 hours—that may well be due to equipment failure rather than an actual customer call.

## USE YOUR KNOWLEDGE

**1.9 Describe the first-exam scores.** Refer to the first-exam scores from Exercise 1.5. Use your favorite graphical display to describe the shape, the center, and the spread of these data. Are there any outliers?

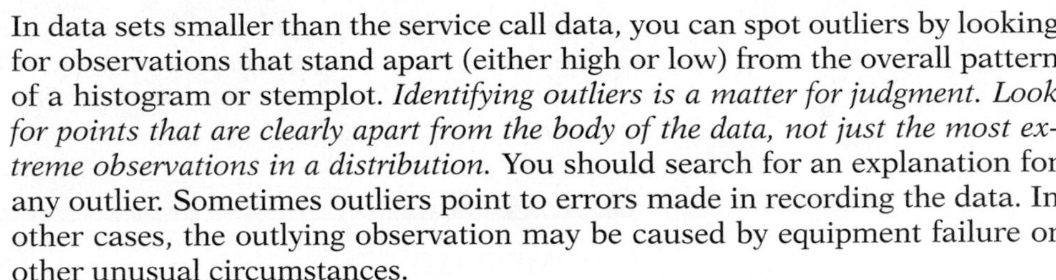

## Dealing with outliers

In data sets smaller than the service call data, you can spot outliers by looking for observations that stand apart (either high or low) from the overall pattern of a histogram or stemplot. *Identifying outliers is a matter for judgment. Look for points that are clearly apart from the body of the data, not just the most extreme observations in a distribution.* You should search for an explanation for any outlier. Sometimes outliers point to errors made in recording the data. In other cases, the outlying observation may be caused by equipment failure or other unusual circumstances.

EXAMPLE

**1.11 Semiconductor wires.** Manufacturing an electronic component requires attaching very fine wires to a semiconductor wafer. If the strength of the bond is weak, the component may fail. Here are measurements on the breaking strength (in pounds) of 23 connections:[7]

| | | | | | | | |
|---|---|---|---|---|---|---|---|
| 0 | 0 | 550 | 750 | 950 | 950 | 1150 | 1150 |
| 1150 | 1150 | 1150 | 1250 | 1250 | 1350 | 1450 | 1450 |
| 1450 | 1550 | 1550 | 1550 | 1850 | 2050 | 3150 | |

Figure 1.9 is a histogram of these data. We expect the breaking strengths of supposedly identical connections to have a roughly symmetric overall pattern, showing chance variation among the connections. Figure 1.9 does show a symmetric pattern centered at about 1250 pounds—but it also shows three *outliers* that stand apart from this pattern, two low and one high.

The engineers were able to explain all three outliers. The two low outliers had strength 0 because the bonds between the wire and the wafer were not made. The high outlier at 3150 pounds was a measurement error. Further study of the data can simply omit the three outliers. One immediate finding is that the variation in breaking strength is too large—550 pounds to 2050 pounds when we ignore the outliers. The process of bonding wire to wafer must be improved to give more consistent results.

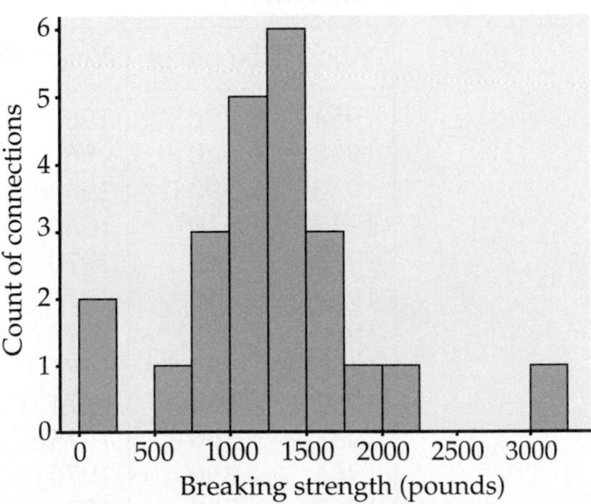


**FIGURE 1.9** Histogram of a distribution with both low and high outliers, for Example 1.11.

## Time plots

Whenever data are collected over time, it is a good idea to plot the observations in time order. *Displays of the distribution of a variable that ignore time order, such as stemplots and histograms, can be misleading when there is systematic change over time.*

### TIME PLOT

A **time plot** of a variable plots each observation against the time at which it was measured. Always put time on the horizontal scale of your plot and the variable you are measuring on the vertical scale. Connecting the data points by lines helps emphasize any change over time.

EXAMPLE

**1.12 Water from the Mississippi River.** Table 1.4 lists the volume of water discharged by the Mississippi River into the Gulf of Mexico for each year from 1954 to 2001.[8] The units are cubic kilometers of water—the Mississippi is a big river. Both graphs in Figure 1.10 describe these data. The histogram in Figure 1.10(a) shows the distribution of the volume discharged. The histogram is symmetric and unimodal, with center near 550 cubic kilometers. We might think that the data show just chance year-to-year fluctuation in river level about its long-term average.

Figure 1.10(b) is a time plot of the same data. For example, the first point lies above 1954 on the "Year" scale at height 290, the volume of water discharged by the Mississippi in 1954. The time plot tells a more interesting story

**TABLE 1.4**

Yearly discharge of the Mississippi River (in cubic kilometers of water)

| Year | Discharge | Year | Discharge | Year | Discharge | Year | Discharge |
|---|---|---|---|---|---|---|---|
| 1954 | 290 | 1966 | 410 | 1978 | 560 | 1990 | 680 |
| 1955 | 420 | 1967 | 460 | 1979 | 800 | 1991 | 700 |
| 1956 | 390 | 1968 | 510 | 1980 | 500 | 1992 | 510 |
| 1957 | 610 | 1969 | 560 | 1981 | 420 | 1993 | 900 |
| 1958 | 550 | 1970 | 540 | 1982 | 640 | 1994 | 640 |
| 1959 | 440 | 1971 | 480 | 1983 | 770 | 1995 | 590 |
| 1960 | 470 | 1972 | 600 | 1984 | 710 | 1996 | 670 |
| 1961 | 600 | 1973 | 880 | 1985 | 680 | 1997 | 680 |
| 1962 | 550 | 1974 | 710 | 1986 | 600 | 1998 | 690 |
| 1963 | 360 | 1975 | 670 | 1987 | 450 | 1999 | 580 |
| 1964 | 390 | 1976 | 420 | 1988 | 420 | 2000 | 390 |
| 1965 | 500 | 1977 | 430 | 1989 | 630 | 2001 | 580 |

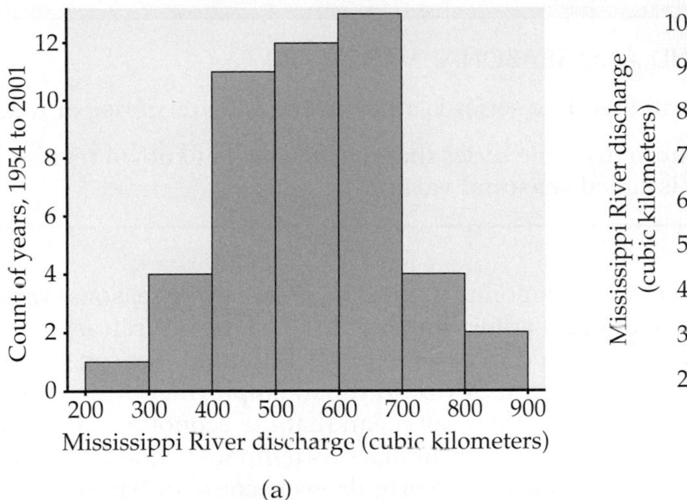


(a)

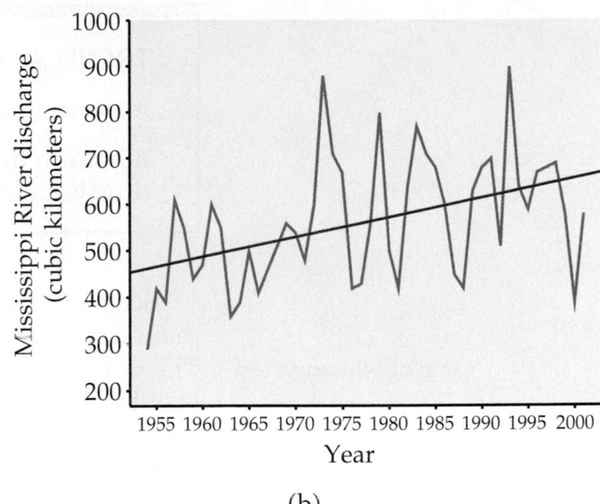


(b)

**FIGURE 1.10** (a) Histogram of the volume of water discharged by the Mississippi River over the 48 years from 1954 to 2001, for Example 1.12. Data are from Table 1.4. (b) Time plot of the volume of water discharged by the Mississippi River for the years 1954 to 2001. The line shows the trend toward increasing river flow, a trend that cannot be seen in the histogram in Figure 1.10(a).

trend

than the histogram. There is a great deal of year-to-year variation, but there is also a clear increasing **trend** over time. That is, there is a long-term rise in the volume of water discharged. The line on the graph is a "trend line" calculated from the data to describe this trend. The trend reflects climate change: rainfall and river flows have been increasing over most of North America.

time series

Many interesting data sets are **time series,** measurements of a variable taken at regular intervals over time. Government, economic, and social data are often published as time series. Some examples are the monthly unemployment rate and the quarterly gross domestic product. Weather records, the demand for electricity, and measurements on the items produced by a manufacturing process are other examples of time series. Time plots can reveal the main features of a time series.

## BEYOND THE BASICS

### Decomposing Time Series*

When you examine a time plot, again look first for overall patterns and then for striking deviations from those patterns. Here are two important types of overall patterns to look for in a time series.

*"Beyond the Basics" sections briefly discuss supplementary topics. Your software may make some of these topics available to you. For example, the results plotted in Figures 1.11 to 1.13 come from the Minitab statistical software.

### TREND AND SEASONAL VARIATION

A **trend** in a time series is a persistent, long-term rise or fall.

A pattern in a time series that repeats itself at known regular intervals of time is called **seasonal variation.**

seasonally adjusted

Because many economic time series show strong seasonal variation, government agencies often adjust for this variation before releasing economic data. The data are then said to be **seasonally adjusted.** Seasonal adjustment helps avoid misinterpretation. A rise in the unemployment rate from December to January, for example, does not mean that the economy is slipping. Unemployment almost always rises in January as temporary holiday help is laid off and outdoor employment in the North drops because of bad weather. The seasonally adjusted unemployment rate reports an increase only if unemployment rises more than normal from December to January.

EXAMPLE

**1.13 Gasoline prices.** Figure 1.11 is a time plot of the average retail price of regular gasoline each month for the years 1990 to 2003.[9] The prices are *not* seasonally adjusted. You can see the upward spike in prices due to the 1990 Iraqi invasion of Kuwait, the drop in 1998 when an economic crisis in Asia reduced demand for fuel, and rapid price increases in 2000 and 2003 due to instability in the Middle East and OPEC production limits. These deviations are so large that overall patterns are hard to see.

There is nonetheless a clear *trend* of increasing price. Much of this trend just reflects inflation, the rise in the overall price level during these years. In addition, a close look at the plot shows *seasonal variation,* a regular rise and fall that recurs each year. Americans drive more in the summer vacation season, so the price of gasoline rises each spring, then drops in the fall as demand goes down.

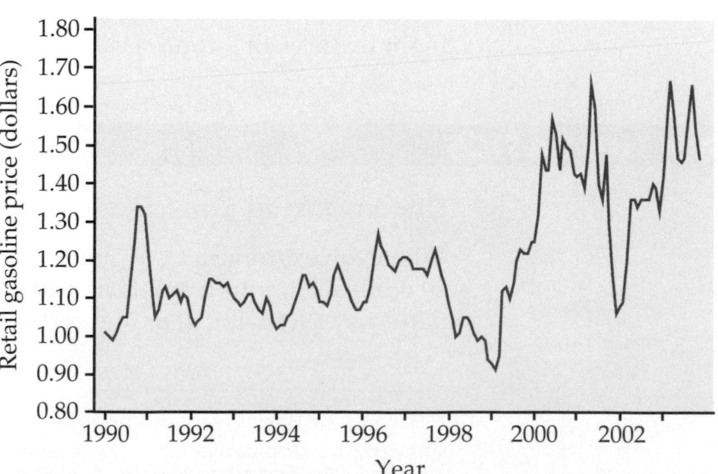


**FIGURE 1.11** Time plot of the average monthly price of regular gasoline from 1990 to 2003, for Example 1.13.

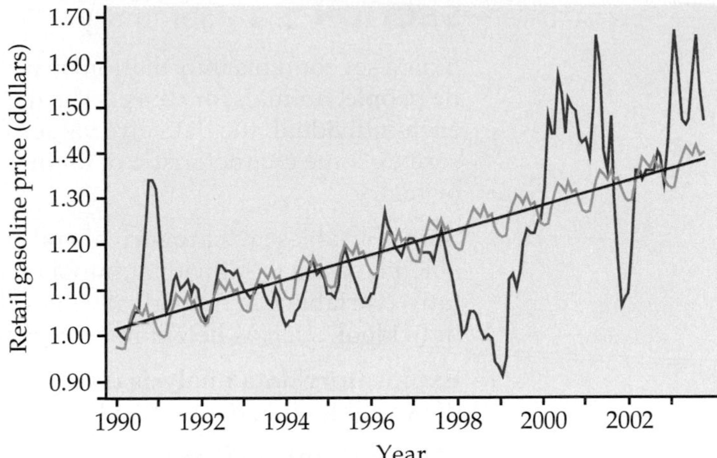


**FIGURE 1.12** Time plot of gasoline prices with a trend line and seasonal variation added. These are overall patterns extracted from the data by software.

Statistical software can help us examine a time series by "decomposing" the data into systematic patterns, such as trends and seasonal variation, and the *residuals* that remain after we remove these patterns. Figure 1.12 superimposes the trend and seasonal variation on the time plot of gasoline prices. The red line shows the increasing trend. The seasonal variation appears as the colored line that regularly rises and falls each year. This is an average of the seasonal pattern for all the years in the original data, automatically extracted by software.

The trend and seasonal variation in Figure 1.12 are overall patterns in the data. Figure 1.13 is a plot of what remains when we subtract both the trend and the seasonal variation from the original data. That is, Figure 1.13 emphasizes the deviations from the pattern. In the case of gasoline prices, the deviations are large (as much as 30 cents both up and down). It is clear that we can't use trend and seasonal variation to predict gasoline prices at all accurately.

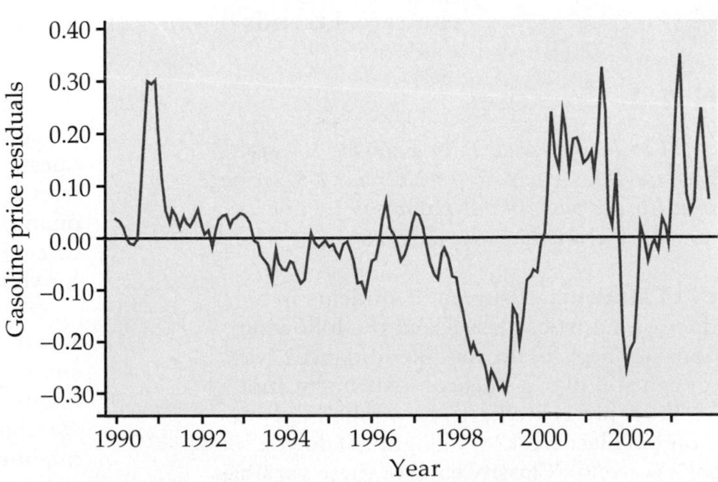


**FIGURE 1.13** The residuals that remain when we subtract both trend and seasonal variation from monthly gasoline prices.

## SECTION 1.1 Summary

A data set contains information on a collection of **individuals.** Individuals may be people, animals, or things. The data for one individual make up a **case.** For each individual, the data give values for one or more **variables.** A variable describes some characteristic of an individual, such as a person's height, gender, or salary.

Some variables are **categorical** and others are **quantitative.** A categorical variable places each individual into a category, such as male or female. A quantitative variable has numerical values that measure some characteristic of each individual, such as height in centimeters or annual salary in dollars.

**Exploratory data analysis** uses graphs and numerical summaries to describe the variables in a data set and the relations among them.

The **distribution** of a variable tells us what values it takes and how often it takes these values.

**Bar graphs** and **pie charts** display the distributions of categorical variables. These graphs use the counts or percents of the categories.

**Stemplots** and **histograms** display the distributions of quantitative variables. Stemplots separate each observation into a **stem** and a one-digit **leaf.** Histograms plot the **frequencies** (counts) or the percents of equal-width classes of values.

When examining a distribution, look for **shape, center,** and **spread** and for clear **deviations** from the overall shape.

Some distributions have simple shapes, such as **symmetric** or **skewed.** The number of **modes** (major peaks) is another aspect of overall shape. Not all distributions have a simple overall shape, especially when there are few observations.

**Outliers** are observations that lie outside the overall pattern of a distribution. Always look for outliers and try to explain them.

When observations on a variable are taken over time, make a **time plot** that graphs time horizontally and the values of the variable vertically. A time plot can reveal **trends** or other changes over time.

## SECTION 1.1 Exercises

*For Exercises 1.1 to 1.2, see page 3; for Exercise 1.3, see page 5; for Exercise 1.4, see page 7; for Exercise 1.5, see page 10; for Exercise 1.6, see page 14; for Exercises 1.7 and 1.8, see page 15; and for Exercise 1.9, see page 16.*

**1.10 Survey of students.** A survey of students in an introductory statistics class asked the following questions: (a) age; (b) do you like to dance? (yes, no); (c) can you play a musical instrument (not at all, a little, pretty well); (d) how much did you spend on food last week? (e) height; (f) do you like broccoli? (yes, no). Classify each of these variables as categorical or quantitative and give reasons for your answers.

**1.11 What questions would you ask?** Refer to the previous exercise. Make up your own survey questions with at least six questions. Include at least two categorical variables and at least two quantitative variables. Tell which variables are categorical and which are quantitative. Give reasons for your answers.

**1.12 Study habits of students.** You are planning a survey to collect information about the study habits of college students. Describe two categorical variables and two quantitative variables that you might measure for each student. Give the units of measurement for the quantitative variables.

**1.13 Physical fitness of students.** You want to measure the "physical fitness" of college students. Describe several variables you might use to measure fitness. What instrument or instruments does each measurement require?

**1.14 Choosing a college or university.** Popular magazines rank colleges and universities on their "academic quality" in serving undergraduate students. Describe five variables that you would like to see measured for each college if you were choosing where to study. Give reasons for each of your choices.

**1.15 Favorite colors.** What is your favorite color? One survey produced the following summary of responses to that question: blue, 42%; green, 14%; purple, 14%; red, 8%; black, 7%; orange, 5%; yellow, 3%; brown, 3%; gray, 2%; and white, 2%.[10] Make a bar graph of the percents and write a short summary of the major features of your graph.

**1.16 Least-favorite colors.** Refer to the previous exercise. The same study also asked people about their least-favorite color. Here are the results: orange, 30%; brown, 23%; purple, 13%; yellow, 13%; gray, 12%; green, 4%; white, 4%; red, 1%; black, 0%; and blue, 0%. Make a bar graph of these percents and write a summary of the results.

**1.17 Ages of survey respondents.** The survey about color preferences reported the age distribution of the people who responded. Here are the results:

| Age group (years) | 1–18 | 19–24 | 25–35 | 36–50 | 51–69 | 70 and over |
|---|---|---|---|---|---|---|
| Count | 10 | 97 | 70 | 36 | 14 | 5 |

(a) Add the counts and compute the percents for each age group.

(b) Make a bar graph of the percents.

(c) Describe the distribution.

(d) Explain why your bar graph is not a histogram.

**1.18 Garbage.** The formal name for garbage is "municipal solid waste." The table at the top of the next column gives a breakdown of the materials that made up American municipal solid waste.[11]

(a) Add the weights for the nine materials given, including "Other." Each entry, including the total, is separately rounded to the nearest tenth. So the sum and the total may differ slightly because of **roundoff error.**

| Material | Weight (million tons) | Percent of total |
|---|---|---|
| Food scraps | 25.9 | 11.2 |
| Glass | 12.8 | 5.5 |
| Metals | 18.0 | 7.8 |
| Paper, paperboard | 86.7 | 37.4 |
| Plastics | 24.7 | 10.7 |
| Rubber, leather, textiles | 15.8 | 6.8 |
| Wood | 12.7 | 5.5 |
| Yard trimmings | 27.7 | 11.9 |
| Other | 7.5 | 3.2 |
| Total | 231.9 | 100.0 |

(b) Make a bar graph of the percents. The graph gives a clearer picture of the main contributors to garbage if you order the bars from tallest to shortest.

(c) If you use software, also make a pie chart of the percents. Comparing the two graphs, notice that it is easier to see the small differences among "Food scraps," "Plastics," and "Yard trimmings" in the bar graph.

**1.19 Spam.** Email spam is the curse of the Internet. Here is a compilation of the most common types of spam:[12]

| Type of spam | Percent |
|---|---|
| Adult | 14.5 |
| Financial | 16.2 |
| Health | 7.3 |
| Leisure | 7.8 |
| Products | 21.0 |
| Scams | 14.2 |

Make two bar graphs of these percents, one with bars ordered as in the table (alphabetical) and the other with bars in order from tallest to shortest. Comparisons are easier if you order the bars by height. A bar graph ordered from tallest to shortest bar is sometimes called a **Pareto chart,** after the Italian economist who recommended this procedure.

**1.20 Women seeking graduate and professional degrees.** The table on the next page gives the percents of women among students seeking various graduate and professional degrees:[13]

(a) Explain clearly why we cannot use a pie chart to display these data.

(b) Make a bar graph of the data. (Comparisons are easier if you order the bars by height.)

| Degree | Percent female |
|---|---|
| Master's in business administration | 39.8 |
| Master's in education | 76.2 |
| Other master of arts | 59.6 |
| Other master of science | 53.0 |
| Doctorate in education | 70.8 |
| Other PhD degree | 54.2 |
| Medicine (MD) | 44.0 |
| Law | 50.2 |
| Theology | 20.2 |

**1.21 An aging population.** The population of the United States is aging, though less rapidly than in other developed countries. Here is a stemplot of the percents of residents aged 65 and over in the 50 states, according to the 2000 census. The stems are whole percents and the leaves are tenths of a percent.

| | |
|---|---|
| 5 | 7 |
| 6 | |
| 7 | |
| 8 | 5 |
| 9 | 679 |
| 10 | 6 |
| 11 | 02233677 |
| 12 | 0011113445789 |
| 13 | 00012233345568 |
| 14 | 034579 |
| 15 | 36 |
| 16 | |
| 17 | 6 |

(a) There are two outliers: Alaska has the lowest percent of older residents, and Florida has the highest. What are the percents for these two states?

(b) Ignoring Alaska and Florida, describe the shape, center, and spread of this distribution.

**1.22 Split the stems.** Make another stemplot of the percent of residents aged 65 and over in the states other than Alaska and Florida by splitting stems 8 to 15 in the plot from the previous exercise. Which plot do you prefer? Why?

**1.23 Diabetes and glucose.** People with diabetes must monitor and control their blood glucose level. The goal is to maintain "fasting plasma glucose" between about 90 and 130 milligrams per deciliter (mg/dl). Here are the fasting plasma glucose levels for 18 diabetics enrolled in a diabetes control class, five months after the end of the class:[14]

| | | | | | | | | |
|---|---|---|---|---|---|---|---|---|
| 141 | 158 | 112 | 153 | 134 | 95 | 96 | 78 | 148 |
| 172 | 200 | 271 | 103 | 172 | 359 | 145 | 147 | 255 |

Make a stemplot of these data and describe the main features of the distribution. (You will want to trim and also split stems.) Are there outliers? How well is the group as a whole achieving the goal for controlling glucose levels?

**1.24 Compare glucose of instruction and control groups.** The study described in the previous exercise also measured the fasting plasma glucose of 16 diabetics who were given individual instruction on diabetes control. Here are the data:

| | | | | | | | |
|---|---|---|---|---|---|---|---|
| 128 | 195 | 188 | 158 | 227 | 198 | 163 | 164 |
| 159 | 128 | 283 | 226 | 223 | 221 | 220 | 160 |

Make a back-to-back stemplot to compare the class and individual instruction groups. How do the distribution shapes and success in achieving the glucose control goal compare?

**1.25 Vocabulary scores of seventh-grade students.** Figure 1.14 displays the scores of all 947 seventh-grade students in the public schools of Gary, Indiana, on the vocabulary part of the Iowa Test of Basic Skills.[15] Give a brief description of the overall pattern (shape, center, spread) of this distribution.

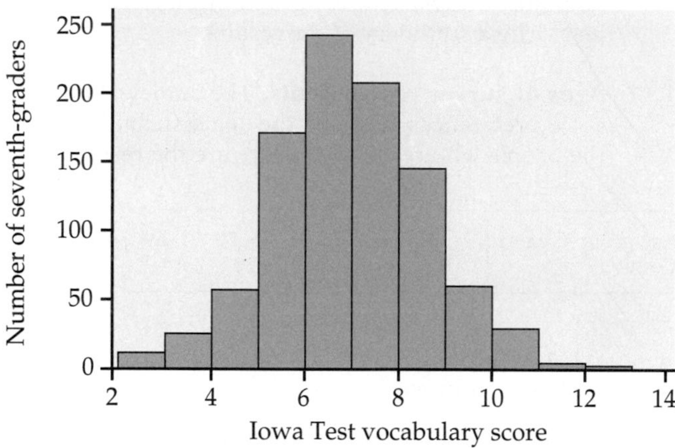


**FIGURE 1.14** Histogram of the Iowa Test of Basic Skills vocabulary scores of seventh-grade students in Gary, Indiana, for Exercise 1.25.

**1.26 Shakespeare's plays.** Figure 1.15 is a histogram of the lengths of words used in Shakespeare's plays. Because there are so many words in the plays, we use a histogram of percents. What is the overall shape of this distribution? What does this shape say about word lengths in Shakespeare? Do you expect other authors to have word length distributions of the same general shape? Why?

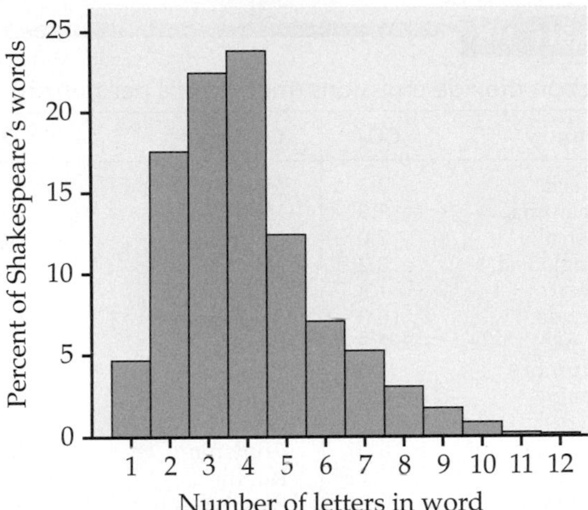


**FIGURE 1.15** Histogram of lengths of words used in Shakespeare's plays, for Exercise 1.26.

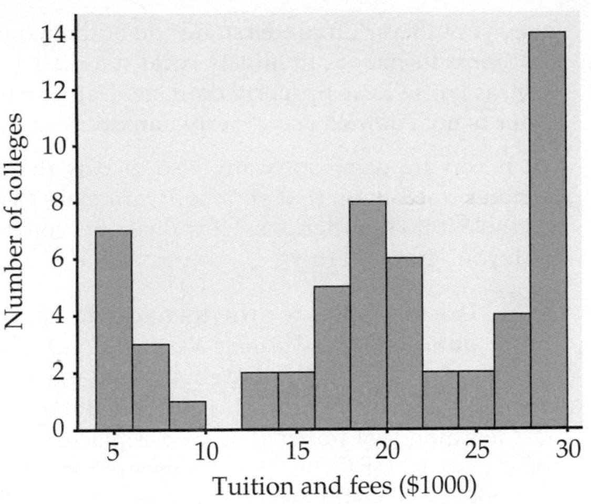


**FIGURE 1.16** Histogram of the tuition and fees charged by four-year colleges in Massachusetts, for Exercise 1.27.

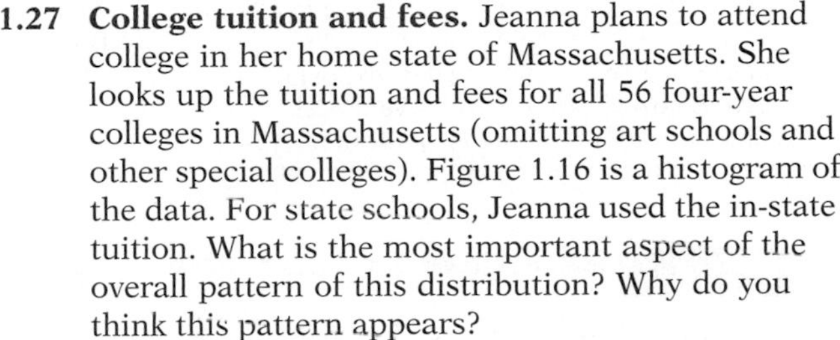

**1.27 College tuition and fees.** Jeanna plans to attend college in her home state of Massachusetts. She looks up the tuition and fees for all 56 four-year colleges in Massachusetts (omitting art schools and other special colleges). Figure 1.16 is a histogram of the data. For state schools, Jeanna used the in-state tuition. What is the most important aspect of the overall pattern of this distribution? Why do you think this pattern appears?

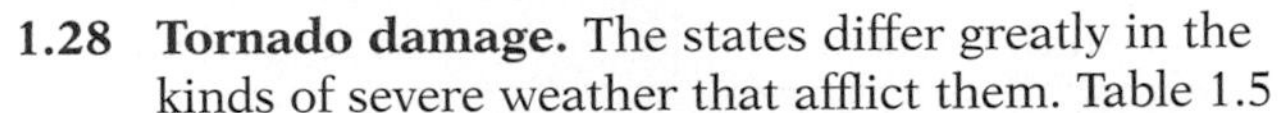

**1.28 Tornado damage.** The states differ greatly in the kinds of severe weather that afflict them. Table 1.5 shows the average property damage caused by tornadoes per year over the period from 1950 to 1999 in each of the 50 states and Puerto Rico.[16] (To adjust for the changing buying power of the dollar over time, all damages were restated in 1999 dollars.)

(a) What are the top five states for tornado damage? The bottom five?

(b) Make a histogram of the data, by hand or using software, with classes "0 ≤ damage < 10," "10 ≤ damage < 20," and so on. Describe the shape, center, and spread of the distribution. Which states

**TABLE 1.5**

Average property damage per year due to tornadoes

| State | Damage ($millions) | State | Damage ($millions) | State | Damage ($millions) |
|---|---|---|---|---|---|
| Alabama | 51.88 | Louisiana | 27.75 | Ohio | 44.36 |
| Alaska | 0.00 | Maine | 0.53 | Oklahoma | 81.94 |
| Arizona | 3.47 | Maryland | 2.33 | Oregon | 5.52 |
| Arkansas | 40.96 | Massachusetts | 4.42 | Pennsylvania | 17.11 |
| California | 3.68 | Michigan | 29.88 | Puerto Rico | 0.05 |
| Colorado | 4.62 | Minnesota | 84.84 | Rhode Island | 0.09 |
| Connecticut | 2.26 | Mississippi | 43.62 | South Carolina | 17.19 |
| Delaware | 0.27 | Missouri | 68.93 | South Dakota | 10.64 |
| Florida | 37.32 | Montana | 2.27 | Tennessee | 23.47 |
| Georgia | 51.68 | Nebraska | 30.26 | Texas | 88.60 |
| Hawaii | 0.34 | Nevada | 0.10 | Utah | 3.57 |
| Idaho | 0.26 | New Hampshire | 0.66 | Vermont | 0.24 |
| Illinois | 62.94 | New Jersey | 2.94 | Virginia | 7.42 |
| Indiana | 53.13 | New Mexico | 1.49 | Washington | 2.37 |
| Iowa | 49.51 | New York | 15.73 | West Virginia | 2.14 |
| Kansas | 49.28 | North Carolina | 14.90 | Wisconsin | 31.33 |
| Kentucky | 24.84 | North Dakota | 14.69 | Wyoming | 1.78 |

may be outliers? (To understand the outliers, note that most tornadoes in largely rural states such as Kansas cause little property damage. Damage to crops is not counted as property damage.)

(c) If you are using software, also display the "default" histogram that your software makes when you give it no instructions. How does this compare with your graph in (b)?

**1.29** **Use an applet for the tornado damage data.** The *One-Variable Statistical Calculator* applet on the text CD and Web site will make stemplots and histograms. It is intended mainly as a learning tool rather than as a replacement for statistical software. The histogram function is particularly useful because you can change the number of classes by dragging with the mouse. The tornado damage data from Table 1.5 are available in the applet. Choose this data set and go to the "Histogram" tab.

(a) Sketch the default histogram that the applet first presents. If the default graph does not have nine classes, drag it to make a histogram with nine classes and sketch the result. This should agree with your histogram in part (b) of the previous exercise.

(b) Make a histogram with one class and also a histogram with the greatest number of classes that the applet allows. Sketch the results.

(c) Drag the graph until you find the histogram that you think best pictures the data. How many classes did you choose? Sketch your final histogram.

**1.30** **Carbon dioxide from burning fuels.** Burning fuels in power plants or motor vehicles emits carbon dioxide ($CO_2$), which contributes to global warming. Table 1.6 displays $CO_2$ emissions per person from countries with population at least 20 million.[17]

(a) Why do you think we choose to measure emissions per person rather than total $CO_2$ emissions for each country?

(b) Display the data of Table 1.6 in a graph. Describe the shape, center, and spread of the distribution. Which countries are outliers?

**1.31** **California temperatures.** Table 1.7 contains data on the mean annual temperatures (degrees Fahrenheit) for the years 1951 to 2000 at two locations in California: Pasadena and Redding.[18] Make time plots of both time series and compare their main features. You can see why discussions of climate change often bring disagreement.

**1.32** **What do you miss in the histogram?** Make a histogram of the mean annual temperatures

**TABLE 1.6**

Carbon dioxide emissions (metric tons per person)

| Country | $CO_2$ | Country | $CO_2$ |
|---|---|---|---|
| Algeria | 2.3 | Mexico | 3.7 |
| Argentina | 3.9 | Morocco | 1.0 |
| Australia | 17.0 | Myanmar | 0.2 |
| Bangladesh | 0.2 | Nepal | 0.1 |
| Brazil | 1.8 | Nigeria | 0.3 |
| Canada | 16.0 | Pakistan | 0.7 |
| China | 2.5 | Peru | 0.8 |
| Columbia | 1.4 | Tanzania | 0.1 |
| Congo | 0.0 | Philippines | 0.9 |
| Egypt | 1.7 | Poland | 8.0 |
| Ethiopia | 0.0 | Romania | 3.9 |
| France | 6.1 | Russia | 10.2 |
| Germany | 10.0 | Saudi Arabia | 11.0 |
| Ghana | 0.2 | South Africa | 8.1 |
| India | 0.9 | Spain | 6.8 |
| Indonesia | 1.2 | Sudan | 0.2 |
| Iran | 3.8 | Thailand | 2.5 |
| Iraq | 3.6 | Turkey | 2.8 |
| Italy | 7.3 | Ukraine | 7.6 |
| Japan | 9.1 | United Kingdom | 9.0 |
| Kenya | 0.3 | United States | 19.9 |
| Korea, North | 9.7 | Uzbekistan | 4.8 |
| Korea, South | 8.8 | Venezuela | 5.1 |
| Malaysia | 4.6 | Vietnam | 0.5 |

**TABLE 1.7**

Mean annual temperatures (°F) in two California cities

| | Mean Temperature | | | Mean Temperature | |
|---|---|---|---|---|---|
| Year | Pasadena | Redding | Year | Pasadena | Redding |
| 1951 | 62.27 | 62.02 | 1976 | 64.23 | 63.51 |
| 1952 | 61.59 | 62.27 | 1977 | 64.47 | 63.89 |
| 1953 | 62.64 | 62.06 | 1978 | 64.21 | 64.05 |
| 1954 | 62.88 | 61.65 | 1979 | 63.76 | 60.38 |
| 1955 | 61.75 | 62.48 | 1980 | 65.02 | 60.04 |
| 1956 | 62.93 | 63.17 | 1981 | 65.80 | 61.95 |
| 1957 | 63.72 | 62.42 | 1982 | 63.50 | 59.14 |
| 1958 | 65.02 | 64.42 | 1983 | 64.19 | 60.66 |
| 1959 | 65.69 | 65.04 | 1984 | 66.06 | 61.72 |
| 1960 | 64.48 | 63.07 | 1985 | 64.44 | 60.50 |
| 1961 | 64.12 | 63.50 | 1986 | 65.31 | 61.76 |
| 1962 | 62.82 | 63.97 | 1987 | 64.58 | 62.94 |
| 1963 | 63.71 | 62.42 | 1988 | 65.22 | 63.70 |
| 1964 | 62.76 | 63.29 | 1989 | 64.53 | 61.50 |
| 1965 | 63.03 | 63.32 | 1990 | 64.96 | 62.22 |
| 1966 | 64.25 | 64.51 | 1991 | 65.60 | 62.73 |
| 1967 | 64.36 | 64.21 | 1992 | 66.07 | 63.59 |
| 1968 | 64.15 | 63.40 | 1993 | 65.16 | 61.55 |
| 1969 | 63.51 | 63.77 | 1994 | 64.63 | 61.63 |
| 1970 | 64.08 | 64.30 | 1995 | 65.43 | 62.62 |
| 1971 | 63.59 | 62.23 | 1996 | 65.76 | 62.93 |
| 1972 | 64.53 | 63.06 | 1997 | 66.72 | 62.48 |
| 1973 | 63.46 | 63.75 | 1998 | 64.12 | 60.23 |
| 1974 | 63.93 | 63.80 | 1999 | 64.85 | 61.88 |
| 1975 | 62.36 | 62.66 | 2000 | 66.25 | 61.58 |

at Pasadena for the years 1951 to 2000. (Data appear in Table 1.7.) Describe the distribution of temperatures. Then explain why this histogram misses very important facts about temperatures in Pasadena.

1.33 CAUTION **Change the scale of the axis.** The impression that a time plot gives depends on the scales you use on the two axes. If you stretch the vertical axis and compress the time axis, change appears to be more rapid. Compressing the vertical axis and stretching the time axis make change appear slower. Make two more time plots of the data for Pasadena in Table 1.7, one that makes mean temperature appear to increase very rapidly and one that shows only a slow increase. The moral of this exercise is: *pay close attention to the scales when you look at a time plot.*

1.34 **Fish in the Bering Sea.** "Recruitment," the addition of new members to a fish population, is an important measure of the health of ocean ecosystems. Here are data on the recruitment of rock sole in the Bering Sea between 1973 and 2000:[19]

| Year | Recruitment (millions) | Year | Recruitment (millions) |
|---|---|---|---|
| 1973 | 173 | 1987 | 4700 |
| 1974 | 234 | 1988 | 1702 |
| 1975 | 616 | 1989 | 1119 |
| 1976 | 344 | 1990 | 2407 |
| 1977 | 515 | 1991 | 1049 |
| 1978 | 576 | 1992 | 505 |
| 1979 | 727 | 1993 | 998 |
| 1980 | 1411 | 1994 | 505 |
| 1981 | 1431 | 1995 | 304 |
| 1982 | 1250 | 1996 | 425 |
| 1983 | 2246 | 1997 | 214 |
| 1984 | 1793 | 1998 | 385 |
| 1985 | 1793 | 1999 | 445 |
| 1986 | 2809 | 2000 | 676 |

(a) Make a graph to display the distribution of rock sole recruitment, then describe the pattern and any striking deviations that you see.

(b) Make a time plot of recruitment and describe its pattern. As is often the case with time series data, a time plot is needed to understand what is happening.

1.35 **Thinness in Asia.** Asian culture does not emphasize thinness, but young Asians are often influenced by Western culture. In a study of concerns about weight among young Korean women, researchers administered the Drive for Thinness scale (a questionnaire) to 264 female college students in Seoul, South Korea.[20] Drive for Thinness measures excessive concern with weight and dieting and fear of weight gain. Roughly speaking, a score of 15 is typical of Western women with eating disorders but is unusually high (90th percentile) for other Western women. Graph the data and describe the shape, center, and spread of the distribution of Drive for Thinness scores for these Korean students. Are there any outliers?

1.36 CHALLENGE **Acidity of rainwater.** Changing the choice of classes can change the appearance of a histogram. Here is an example in which a small shift in the classes, with no change in the number of classes, has an important effect on the histogram. The data are the acidity levels (measured by pH) in 105 samples of rainwater. Distilled water has pH 7.00. As the water becomes more acidic, the pH goes down. The pH of rainwater is important to environmentalists because of the problem of acid rain.[21]

| | | | | | | | |
|---|---|---|---|---|---|---|---|
| 4.33 | 4.38 | 4.48 | 4.48 | 4.50 | 4.55 | 4.59 | 4.59 |
| 4.61 | 4.61 | 4.75 | 4.76 | 4.78 | 4.82 | 4.82 | 4.83 |
| 4.86 | 4.93 | 4.94 | 4.94 | 4.94 | 4.96 | 4.97 | 5.00 |
| 5.01 | 5.02 | 5.05 | 5.06 | 5.08 | 5.09 | 5.10 | 5.12 |
| 5.13 | 5.15 | 5.15 | 5.15 | 5.16 | 5.16 | 5.16 | 5.18 |
| 5.19 | 5.23 | 5.24 | 5.29 | 5.32 | 5.33 | 5.35 | 5.37 |
| 5.37 | 5.39 | 5.41 | 5.43 | 5.44 | 5.46 | 5.46 | 5.47 |
| 5.50 | 5.51 | 5.53 | 5.55 | 5.55 | 5.56 | 5.61 | 5.62 |
| 5.64 | 5.65 | 5.65 | 5.66 | 5.67 | 5.67 | 5.68 | 5.69 |
| 5.70 | 5.75 | 5.75 | 5.75 | 5.76 | 5.76 | 5.79 | 5.80 |
| 5.81 | 5.81 | 5.81 | 5.81 | 5.85 | 5.85 | 5.90 | 5.90 |
| 6.00 | 6.03 | 6.03 | 6.04 | 6.04 | 6.05 | 6.06 | 6.07 |
| 6.09 | 6.13 | 6.21 | 6.34 | 6.43 | 6.61 | 6.62 | 6.65 |
| 6.81 | | | | | | | |

(a) Make a histogram of pH with 14 classes, using class boundaries 4.2, 4.4, . . . , 7.0. How many modes does your histogram show? More than one mode suggests that the data contain groups that have different distributions.

(b) Make a second histogram, also with 14 classes, using class boundaries 4.14, 4.34, . . . , 6.94. The classes are those from (a) moved 0.06 to the left. How many modes does the new histogram show?

(c) Use your software's histogram function to make a histogram without specifying the number of classes or their boundaries. How does the software's default histogram compare with those in (a) and (b)?

1.37 CHALLENGE **Identify the histograms.** A survey of a large college class asked the following questions:

1. Are you female or male? (In the data, male = 0, female = 1.)

2. Are you right-handed or left-handed? (In the data, right = 0, left = 1.)

3. What is your height in inches?

4. How many minutes do you study on a typical weeknight?

Figure 1.17 shows histograms of the student responses, in scrambled order and without scale markings. Which histogram goes with each variable? Explain your reasoning.

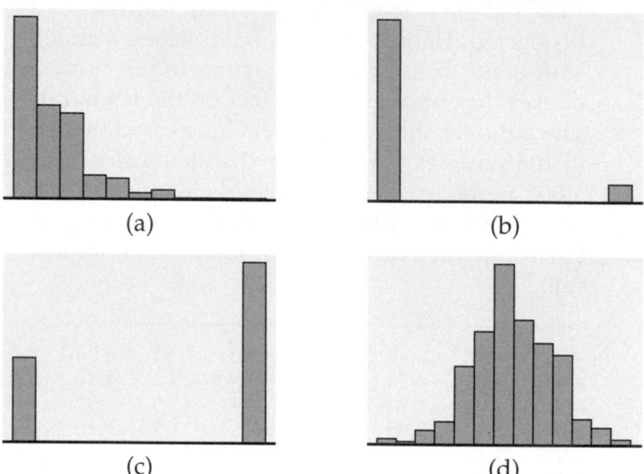

**FIGURE 1.17** Match each histogram with its variable, for Exercise 1.37.

**1.38 Sketch a skewed distribution.** Sketch a histogram for a distribution that is skewed to the left. Suppose that you and your friends emptied your pockets of coins and recorded the year marked on each coin. The distribution of dates would be skewed to the left. Explain why.

**1.39 Oil wells.** How much oil the wells in a given field will ultimately produce is key information in deciding whether to drill more wells. Here are the estimated total amounts of oil recovered from 64 wells in the Devonian Richmond Dolomite area of the Michigan basin, in thousands of barrels:[22]

| | | | | | | | |
|---|---|---|---|---|---|---|---|
| 21.7 | 53.2 | 46.4 | 42.7 | 50.4 | 97.7 | 103.1 | 51.9 |
| 43.4 | 69.5 | 156.5 | 34.6 | 37.9 | 12.9 | 2.5 | 31.4 |
| 79.5 | 26.9 | 18.5 | 14.7 | 32.9 | 196.0 | 24.9 | 118.2 |
| 82.2 | 35.1 | 47.6 | 54.2 | 63.1 | 69.8 | 57.4 | 65.6 |
| 56.4 | 49.4 | 44.9 | 34.6 | 92.2 | 37.0 | 58.8 | 21.3 |
| 36.6 | 64.9 | 14.8 | 17.6 | 29.1 | 61.4 | 38.6 | 32.5 |
| 12.0 | 28.3 | 204.9 | 44.5 | 10.3 | 37.7 | 33.7 | 81.1 |
| 12.1 | 20.1 | 30.5 | 7.1 | 10.1 | 18.0 | 3.0 | 2.0 |

Graph the distribution and describe its main features.

**1.40 The density of the earth.** In 1798 the English scientist Henry Cavendish measured the density of the earth by careful work with a torsion balance. The variable recorded was the density of the earth as a multiple of the density of water. Here are Cavendish's 29 measurements:[23]

| | | | | | | | |
|---|---|---|---|---|---|---|---|
| 5.50 | 5.61 | 4.88 | 5.07 | 5.26 | 5.55 | 5.36 | 5.29 |
| 5.58 | 5.65 | 5.57 | 5.53 | 5.62 | 5.29 | 5.44 | 5.34 |
| 5.79 | 5.10 | 5.27 | 5.39 | 5.42 | 5.47 | 5.63 | 5.34 |
| 5.46 | 5.30 | 5.75 | 5.68 | 5.85 | | | |

Present these measurements graphically by either a stemplot or a histogram and explain the reason for your choice. Then briefly discuss the main features of the distribution. In particular, what is your estimate of the density of the earth based on these measurements?

**1.41 Time spent studying.** Do women study more than men? We asked the students in a large first-year college class how many minutes they studied on a typical weeknight. Here are the responses of random samples of 30 women and 30 men from the class:

| Women | | | | | Men | | | | |
|---|---|---|---|---|---|---|---|---|---|
| 180 | 120 | 180 | 360 | 240 | 90 | 120 | 30 | 90 | 200 |
| 120 | 180 | 120 | 240 | 170 | 90 | 45 | 30 | 120 | 75 |
| 150 | 120 | 180 | 180 | 150 | 150 | 120 | 60 | 240 | 300 |
| 200 | 150 | 180 | 150 | 180 | 240 | 60 | 120 | 60 | 30 |
| 120 | 60 | 120 | 180 | 180 | 30 | 230 | 120 | 95 | 150 |
| 90 | 240 | 180 | 115 | 120 | 0 | 200 | 120 | 120 | 180 |

(a) Examine the data. Why are you not surprised that most responses are multiples of 10 minutes? We eliminated one student who claimed to study 30,000 minutes per night. Are there any other responses you consider suspicious?

(b) Make a back-to-back stemplot of these data. Report the approximate midpoints of both groups. Does it appear that women study more than men (or at least claim that they do)?

**1.42 Guinea pigs.** Table 1.8 gives the survival times in days of 72 guinea pigs after they were injected with tubercle bacilli in a medical experiment.[24] Make a suitable graph and describe the shape, center, and spread of the distribution of survival times. Are there any outliers?

**1.43 Grades and self-concept.** Table 1.9 presents data on 78 seventh-grade students in a rural midwestern school.[25] The researcher was interested in the relationship between the students' "self-concept"

**TABLE 1.8**

Survival times (days) of guinea pigs in a medical experiment

| | | | | | | | | | |
|---|---|---|---|---|---|---|---|---|---|
| 43 | 45 | 53 | 56 | 56 | 57 | 58 | 66 | 67 | 73 |
| 74 | 79 | 80 | 80 | 81 | 81 | 81 | 82 | 83 | 83 |
| 84 | 88 | 89 | 91 | 91 | 92 | 92 | 97 | 99 | 99 |
| 100 | 100 | 101 | 102 | 102 | 102 | 103 | 104 | 107 | 108 |
| 109 | 113 | 114 | 118 | 121 | 123 | 126 | 128 | 137 | 138 |
| 139 | 144 | 145 | 147 | 156 | 162 | 174 | 178 | 179 | 184 |
| 191 | 198 | 211 | 214 | 243 | 249 | 329 | 380 | 403 | 511 |
| 522 | 598 | | | | | | | | |

and their academic performance. The data we give here include each student's grade point average (GPA), score on a standard IQ test, and gender, taken from school records. Gender is coded as F for female and M for male. The students are identified only by an observation number (OBS). The missing OBS numbers show that some students dropped out of the study. The final variable is each student's score on the Piers-Harris Children's Self-Concept Scale, a psychological test administered by the researcher.

(a) How many variables does this data set contain? Which are categorical variables and which are quantitative variables?

(b) Make a stemplot of the distribution of GPA, after rounding to the nearest tenth of a point.

**TABLE 1.9**

Educational data for 78 seventh-grade students

| OBS | GPA | IQ | Gender | Self-concept | OBS | GPA | IQ | Gender | Self-concept |
|---|---|---|---|---|---|---|---|---|---|
| 001 | 7.940 | 111 | M | 67 | 043 | 10.760 | 123 | M | 64 |
| 002 | 8.292 | 107 | M | 43 | 044 | 9.763 | 124 | M | 58 |
| 003 | 4.643 | 100 | M | 52 | 045 | 9.410 | 126 | M | 70 |
| 004 | 7.470 | 107 | M | 66 | 046 | 9.167 | 116 | M | 72 |
| 005 | 8.882 | 114 | F | 58 | 047 | 9.348 | 127 | M | 70 |
| 006 | 7.585 | 115 | M | 51 | 048 | 8.167 | 119 | M | 47 |
| 007 | 7.650 | 111 | M | 71 | 050 | 3.647 | 97 | M | 52 |
| 008 | 2.412 | 97 | M | 51 | 051 | 3.408 | 86 | F | 46 |
| 009 | 6.000 | 100 | F | 49 | 052 | 3.936 | 102 | M | 66 |
| 010 | 8.833 | 112 | M | 51 | 053 | 7.167 | 110 | M | 67 |
| 011 | 7.470 | 104 | F | 35 | 054 | 7.647 | 120 | M | 63 |
| 012 | 5.528 | 89 | F | 54 | 055 | 0.530 | 103 | M | 53 |
| 013 | 7.167 | 104 | M | 54 | 056 | 6.173 | 115 | M | 67 |
| 014 | 7.571 | 102 | F | 64 | 057 | 7.295 | 93 | M | 61 |
| 015 | 4.700 | 91 | F | 56 | 058 | 7.295 | 72 | F | 54 |
| 016 | 8.167 | 114 | F | 69 | 059 | 8.938 | 111 | F | 60 |
| 017 | 7.822 | 114 | F | 55 | 060 | 7.882 | 103 | F | 60 |
| 018 | 7.598 | 103 | F | 65 | 061 | 8.353 | 123 | M | 63 |
| 019 | 4.000 | 106 | M | 40 | 062 | 5.062 | 79 | M | 30 |
| 020 | 6.231 | 105 | F | 66 | 063 | 8.175 | 119 | M | 54 |
| 021 | 7.643 | 113 | M | 55 | 064 | 8.235 | 110 | M | 66 |
| 022 | 1.760 | 109 | M | 20 | 065 | 7.588 | 110 | M | 44 |
| 024 | 6.419 | 108 | F | 56 | 068 | 7.647 | 107 | M | 49 |
| 026 | 9.648 | 113 | M | 68 | 069 | 5.237 | 74 | F | 44 |
| 027 | 10.700 | 130 | F | 69 | 071 | 7.825 | 105 | M | 67 |
| 028 | 10.580 | 128 | M | 70 | 072 | 7.333 | 112 | F | 64 |
| 029 | 9.429 | 128 | M | 80 | 074 | 9.167 | 105 | M | 73 |
| 030 | 8.000 | 118 | M | 53 | 076 | 7.996 | 110 | M | 59 |
| 031 | 9.585 | 113 | M | 65 | 077 | 8.714 | 107 | F | 37 |
| 032 | 9.571 | 120 | F | 67 | 078 | 7.833 | 103 | F | 63 |
| 033 | 8.998 | 132 | F | 62 | 079 | 4.885 | 77 | M | 36 |
| 034 | 8.333 | 111 | F | 39 | 080 | 7.998 | 98 | F | 64 |
| 035 | 8.175 | 124 | M | 71 | 083 | 3.820 | 90 | M | 42 |
| 036 | 8.000 | 127 | M | 59 | 084 | 5.936 | 96 | F | 28 |
| 037 | 9.333 | 128 | F | 60 | 085 | 9.000 | 112 | F | 60 |
| 038 | 9.500 | 136 | M | 64 | 086 | 9.500 | 112 | F | 70 |
| 039 | 9.167 | 106 | M | 71 | 087 | 6.057 | 114 | M | 51 |
| 040 | 10.140 | 118 | F | 72 | 088 | 6.057 | 93 | F | 21 |
| 041 | 9.999 | 119 | F | 54 | 089 | 6.938 | 106 | M | 56 |

(c) Describe the shape, center, and spread of the GPA distribution. Identify any suspected outliers from the overall pattern.

(d) Make a back-to-back stemplot of the rounded GPAs for female and male students. Write a brief comparison of the two distributions.

**1.44** **Describe the IQ scores.** Make a graph of the distribution of IQ scores for the seventh-grade students in Table 1.9. Describe the shape, center, and spread of the distribution, as well as any outliers. IQ scores are usually said to be centered at 100. Is the midpoint for these students close to 100, clearly above, or clearly below?

**1.45** **Describe the self-concept scores.** Based on a suitable graph, briefly describe the distribution of self-concept scores for the students in Table 1.9. Be sure to identify any suspected outliers.

**1.46** **The Boston Marathon.** Women were allowed to enter the Boston Marathon in 1972. The following table gives the times (in minutes, rounded to the nearest minute) for the winning women from 1972 to 2006.

| Year | Time | Year | Time | Year | Time | Year | Time |
|---|---|---|---|---|---|---|---|
| 1972 | 190 | 1981 | 147 | 1990 | 145 | 1999 | 143 |
| 1973 | 186 | 1982 | 150 | 1991 | 144 | 2000 | 146 |
| 1974 | 167 | 1983 | 143 | 1992 | 144 | 2001 | 144 |
| 1975 | 162 | 1984 | 149 | 1993 | 145 | 2002 | 141 |
| 1976 | 167 | 1985 | 154 | 1994 | 142 | 2003 | 145 |
| 1977 | 168 | 1986 | 145 | 1995 | 145 | 2004 | 144 |
| 1978 | 165 | 1987 | 146 | 1996 | 147 | 2005 | 145 |
| 1979 | 155 | 1988 | 145 | 1997 | 146 | 2006 | 143 |
| 1980 | 154 | 1989 | 144 | 1998 | 143 | | |

Make a graph that shows change over time. What overall pattern do you see? Have times stopped improving in recent years? If so, when did improvement end?

## 1.2 Describing Distributions with Numbers

Interested in a sporty car? Worried that it may use too much gas? The Environmental Protection Agency lists most such vehicles in its "two-seater" or "minicompact" categories. Table 1.10 gives the city and highway gas mileage for cars in these groups.[26] (The mileages are for the basic engine and transmission combination for each car.) We want to compare two-seaters with minicompacts and city mileage with highway mileage. We can begin with graphs, but numerical summaries make the comparisons more specific.

A brief description of a distribution should include its *shape* and numbers describing its *center* and *spread*. We describe the shape of a distribution based on inspection of a histogram or a stemplot. Now we will learn specific ways to use numbers to measure the center and spread of a distribution. We can calculate these numerical measures for any quantitative variable. But to interpret measures of center and spread, and to choose among the several measures we will learn, you must think about the shape of the distribution and the meaning of the data. The numbers, like graphs, are aids to understanding, not "the answer" in themselves.

### Measuring center: the mean

Numerical description of a distribution begins with a measure of its center or average. The two common measures of center are the *mean* and the *median*. The mean is the "average value" and the median is the "middle value." These are two different ideas for "center," and the two measures behave differently. We need precise recipes for the mean and the median.

**TABLE 1.10**

Fuel economy (miles per gallon) for 2004 model vehicles

| Two-Seater Cars | | | Minicompact Cars | | |
|---|---|---|---|---|---|
| Model | City | Highway | Model | City | Highway |
| Acura NSX | 17 | 24 | Aston Martin Vanquish | 12 | 19 |
| Audi TT Roadster | 20 | 28 | Audi TT Coupe | 21 | 29 |
| BMW Z4 Roadster | 20 | 28 | BMW 325CI | 19 | 27 |
| Cadillac XLR | 17 | 25 | BMW 330CI | 19 | 28 |
| Chevrolet Corvette | 18 | 25 | BMW M3 | 16 | 23 |
| Dodge Viper | 12 | 20 | Jaguar XK8 | 18 | 26 |
| Ferrari 360 Modena | 11 | 16 | Jaguar XKR | 16 | 23 |
| Ferrari Maranello | 10 | 16 | Lexus SC 430 | 18 | 23 |
| Ford Thunderbird | 17 | 23 | Mini Cooper | 25 | 32 |
| Honda Insight | 60 | 66 | Mitsubishi Eclipse | 23 | 31 |
| Lamborghini Gallardo | 9 | 15 | Mitsubishi Spyder | 20 | 29 |
| Lamborghini Murcielago | 9 | 13 | Porsche Cabriolet | 18 | 26 |
| Lotus Esprit | 15 | 22 | Porsche Turbo 911 | 14 | 22 |
| Maserati Spyder | 12 | 17 | | | |
| Mazda Miata | 22 | 28 | | | |
| Mercedes-Benz SL500 | 16 | 23 | | | |
| Mercedes-Benz SL600 | 13 | 19 | | | |
| Nissan 350Z | 20 | 26 | | | |
| Porsche Boxster | 20 | 29 | | | |
| Porsche Carrera 911 | 15 | 23 | | | |
| Toyota MR2 | 26 | 32 | | | |

### THE MEAN $\bar{x}$

To find the **mean** $\bar{x}$ of a set of observations, add their values and divide by the number of observations. If the $n$ observations are $x_1, x_2, \ldots, x_n$, their mean is

$$\bar{x} = \frac{x_1 + x_2 + \cdots + x_n}{n}$$

or, in more compact notation,

$$\bar{x} = \frac{1}{n} \sum x_i$$

The $\sum$ (capital Greek sigma) in the formula for the mean is short for "add them all up." The bar over the $x$ indicates the mean of all the $x$-values. Pronounce the mean $\bar{x}$ as "x-bar." This notation is so common that writers who are discussing data use $\bar{x}$, $\bar{y}$, etc. without additional explanation. The subscripts on the observations $x_i$ are just a way of keeping the $n$ observations separate. They do not necessarily indicate order or any other special facts about the data.

EXAMPLE

**1.14 Highway mileage for two-seaters.** The mean highway mileage for the 21 two-seaters in Table 1.10 is

$$\bar{x} = \frac{x_1 + x_2 + \cdots + x_n}{n}$$
$$= \frac{24 + 28 + 28 + \cdots + 32}{21}$$
$$= \frac{518}{21} = 24.7 \text{ miles per gallon}$$

In practice, you can key the data into your calculator and hit the $\bar{x}$ key.

## USE YOUR KNOWLEDGE

**1.47 Find the mean.** Here are the scores on the first exam in an introductory statistics course for 10 students:

80 73 92 85 75 98 93 55 80 90

Find the mean first-exam score for these students.

The data for Example 1.14 contain an outlier: the Honda Insight is a hybrid gas-electric car that doesn't belong in the same category as the 20 gasoline-powered two-seater cars. If we exclude the Insight, the mean highway mileage drops to 22.6 mpg. The single outlier adds more than 2 mpg to the mean highway mileage. This illustrates an important weakness of the mean as a measure of center: *the mean is sensitive to the influence of a few extreme observations.* These may be outliers, but a skewed distribution that has no outliers will also pull the mean toward its long tail. Because the mean cannot resist the influence of extreme observations, we say that it is not a **resistant measure** of center. A measure that is resistant does more than limit the influence of outliers. Its value does not respond strongly to changes in a few observations, no matter how large those changes may be. The mean fails this requirement because we can make the mean as large as we wish by making a large enough increase in just one observation.

resistant measure

## Measuring center: the median

We used the midpoint of a distribution as an informal measure of center in the previous section. The *median* is the formal version of the midpoint, with a specific rule for calculation.

### THE MEDIAN *M*

The **median *M*** is the midpoint of a distribution. Half the observations are smaller than the median and the other half are larger than the median. Here is a rule for finding the median:

1. Arrange all observations in order of size, from smallest to largest.

2. If the number of observations $n$ is odd, the median $M$ is the center observation in the ordered list. Find the location of the median by counting $(n + 1)/2$ observations up from the bottom of the list.

3. If the number of observations $n$ is even, the median $M$ is the mean of the two center observations in the ordered list. The location of the median is again $(n + 1)/2$ from the bottom of the list.

Note that the formula $(n + 1)/2$ does *not* give the median, just the location of the median in the ordered list. Medians require little arithmetic, so they are easy to find by hand for small sets of data. Arranging even a moderate number of observations in order is tedious, however, so that finding the median by hand for larger sets of data is unpleasant. Even simple calculators have an $\bar{x}$ button, but you will need computer software or a graphing calculator to automate finding the median.

EXAMPLE

**1.15 Find the median.** To find the median highway mileage for 2004 model two-seater cars, arrange the data in increasing order:

13 15 16 16 17 19 20 22 23 23 **23** 24 25 25 26 28 28 28 29 32 66

Be sure to list *all* observations, even if they repeat the same value. The median is the bold 23, the 11th observation in the ordered list. You can find the median by eye—there are 10 observations to the left and 10 to the right. Or you can use the recipe $(n + 1)/2 = 22/2 = 11$ to locate the median in the list.

What happens if we drop the Honda Insight? The remaining 20 cars have highway mileages

13 15 16 16 17 19 20 22 23 **23 23** 24 25 25 26 28 28 28 29 32

Because the number of observations $n = 20$ is even, there is no center observation. There is a center pair—the bold pair of 23s have 9 observations to their left and 9 to their right. The median $M$ is the mean of the center pair, which is 23. The recipe $(n + 1)/2 = 21/2 = 10.5$ for the position of the median in the list says that the median is at location "ten and one-half," that is, halfway between the 10th and 11th observations.

You see that the median is more resistant than the mean. Removing the Honda Insight did not change the median at all. Even if we mistakenly enter the Insight's mileage as 660 rather than 66, the median remains 23. The very high value is simply one observation to the right of center. The *Mean and Median* applet on the text CD and Web site is an excellent way to compare the resistance of $M$ and $\bar{x}$. See Exercises 1.75 to 1.77 for use of this applet.

## USE YOUR KNOWLEDGE

**1.48 Find the median.** Here are the scores on the first exam in an introductory statistics course for 10 students:

80 73 92 85 75 98 93 55 80 90

Find the median first-exam score for these students.

## Mean versus median

The median and mean are the most common measures of the center of a distribution. The mean and median of a symmetric distribution are close together. If the distribution is exactly symmetric, the mean and median are exactly the same. In a skewed distribution, the mean is farther out in the long tail than is the median. The endowment for a college or university is money set aside and invested. The income from the endowment is usually used to support various programs. The distribution of the sizes of the endowments of colleges and universities is strongly skewed to the right. Most institutions have modest endowments, but a few are very wealthy. The median endowment of colleges and universities in a recent year was \$70 million—but the mean endowment was over \$320 million. The few wealthy institutions pulled the mean up but did not affect the median. *Don't confuse the "average" value of a variable (the mean) with its "typical" value, which we might describe by the median.*

CAUTION

We can now give a better answer to the question of how to deal with outliers in data. First, look at the data to identify outliers and investigate their causes. You can then correct outliers if they are wrongly recorded, delete them for good reason, or otherwise give them individual attention. The three outliers in Figure 1.9 (page 17) can all be dropped from the data once we discover why they appear. If you have no clear reason to drop outliers, you may want to use resistant methods, so that outliers have little influence over your conclusions. The choice is often a matter for judgment. The government's fuel economy guide lists the Honda Insight with the other two-seaters in Table 1.10. We might choose to report median rather than mean gas mileage for all two-seaters to avoid giving too much influence to one car model. In fact, we think that the Insight doesn't belong, so we will omit it from further analysis of these data.

## Measuring spread: the quartiles

A measure of center alone can be misleading. Two nations with the same median family income are very different if one has extremes of wealth and poverty and the other has little variation among families. A drug with the correct mean concentration of active ingredient is dangerous if some batches are much too high and others much too low. We are interested in the *spread* or *variability* of incomes and drug potencies as well as their centers. **The simplest useful numerical description of a distribution consists of both a measure of center and a measure of spread.**

We can describe the spread or variability of a distribution by giving several percentiles. The median divides the data in two; half of the observations are above the median and half are below the median. We could call the median the 50th percentile. The upper **quartile** is the median of the upper half of the data. Similarly, the lower quartile is the median of the lower half of the data. With the median, the quartiles divide the data into four equal parts; 25% of the data are in each part.

quartile

percentile

We can do a similar calculation for any percent. The ***p*th percentile** of a distribution is the value that has $p$ percent of the observations fall at or below it. To calculate a percentile, arrange the observations in increasing order and count up the required percent from the bottom of the list. Our definition of percentiles is a bit inexact because there is not always a value with exactly $p$

percent of the data at or below it. We will be content to take the nearest observation for most percentiles, but the quartiles are important enough to require an exact rule.

### THE QUARTILES $Q_1$ AND $Q_3$

To calculate the quartiles:

**1.** Arrange the observations in increasing order and locate the median $M$ in the ordered list of observations.

**2.** The **first quartile $Q_1$** is the median of the observations whose position in the ordered list is to the left of the location of the overall median.

**3.** The **third quartile $Q_3$** is the median of the observations whose position in the ordered list is to the right of the location of the overall median.

EXAMPLE

**1.16 Find the median and the quartiles.** The highway mileages of the 20 gasoline-powered two-seater cars, arranged in increasing order, are

13 15 16 16 17 19 20 22 23 23 | 23 24 25 25 26 28 28 28 29 32

The median is midway between the center pair of observations. We have marked its position in the list by |. The first quartile is the median of the 10 observations to the left of the position of the median. Check that its value is $Q_1 = 18$. Similarly, the third quartile is the median of the 10 observations to the right of the |. Check that $Q_3 = 27$.

When there is an odd number of observations, the median is the unique center observation, and the rule for finding the quartiles excludes this center value. The highway mileages of the 13 minicompact cars in Table 1.10 are (in order)

19 22 23 23 23 26 **26** 27 28 29 29 31 32

The median is the bold 26. The first quartile is the median of the 6 observations falling to the left of this point in the list, $Q_1 = 23$. Similarly, $Q_3 = 29$.

We find other percentiles more informally if we are working without software. For example, we take the 90th percentile of the 13 minicompact mileages to be the 12th in the ordered list, because $0.90 \times 13 = 11.7$, which we round to 12. The 90th percentile is therefore 31 mpg.

### USE YOUR KNOWLEDGE

**1.49 Find the quartiles.** Here are the scores on the first-exam in an introductory statistics course for 10 students:

80 73 92 85 75 98 93 55 80 90

Find the quartiles for these first-exam scores.

EXAMPLE

**1.17 Results from software.** Statistical software often provides several numerical measures in response to a single command. Figure 1.18 displays such output from the CrunchIt! and Minitab software for the highway mileages of two-seater cars (without the Honda Insight).

Both tell us that there are 20 observations and give the mean, median, quartiles, and smallest and largest data values. Both also give other measures, some of which we will meet soon. CrunchIt! is basic online software that offers no choice of output. Minitab allows you to choose the descriptive measures you want from a long list.

The quartiles from CrunchIt! agree with our values from Example 1.16. But Minitab's quartiles are a bit different. For example, our rule for hand calculation gives first quartile $Q_1 = 18$. Minitab's value is $Q_1 = 17.5$. *There are several rules for calculating quartiles, which often give slightly different values. The differences are always small. For describing data, just report the values that your software gives.*

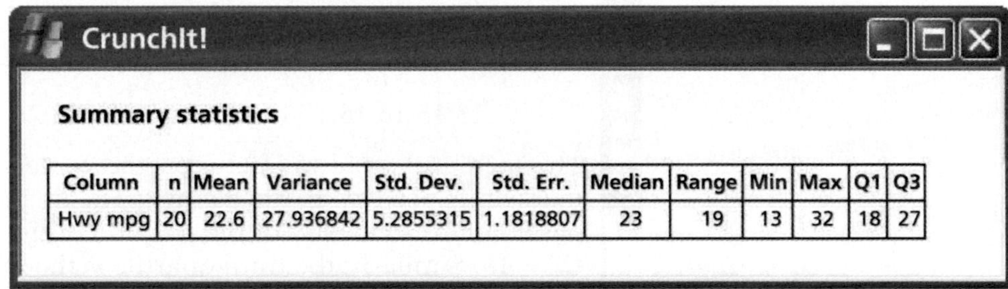
CrunchIt!

**Summary statistics**

| Column | n | Mean | Variance | Std. Dev. | Std. Err. | Median | Range | Min | Max | Q1 | Q3 |
|---|---|---|---|---|---|---|---|---|---|---|---|
| Hwy mpg | 20 | 22.6 | 27.936842 | 5.2855315 | 1.1818807 | 23 | 19 | 13 | 32 | 18 | 27 |

(a)

Minitab

**Descriptive Statistics: Hwy mpg**

```
          Total
Variable  Count   Mean StDev Variance Minimum     Q1 Median     Q3 Maximum
Hwy mpg      20  22.60  5.29    27.94   13.00  17.50  23.00  27.50   32.00


Variable Range IQR
Hwy mpg  19.00 10.00
```

(b)

**FIGURE 1.18** Numerical descriptions of the highway gas mileage of two-seater cars from software, for Example 1.17. (a) CrunchIt! (b) Minitab.

## The five-number summary and boxplots

In Section 1.1, we used the smallest and largest observations to indicate the spread of a distribution. These single observations tell us little about the distribution as a whole, but they give information about the tails of the distribution that is missing if we know only $Q_1$, $M$, and $Q_3$. To get a quick summary of both center and spread, combine all five numbers.

### THE FIVE-NUMBER SUMMARY

The **five-number summary** of a set of observations consists of the smallest observation, the first quartile, the median, the third quartile, and the largest observation, written in order from smallest to largest. In symbols, the five-number summary is

$$\text{Minimum} \quad Q_1 \quad M \quad Q_3 \quad \text{Maximum}$$

These five numbers offer a reasonably complete description of center and spread. The five-number summaries for highway gas mileages are

13 18 23 27 32

for two-seaters and

19 23 26 29 32

for minicompacts. The median describes the center of the distribution; the quartiles show the spread of the center half of the data; the minimum and maximum show the full spread of the data.

## USE YOUR KNOWLEDGE

**1.50 Find the five-number summary.** Here are the scores on the first exam in an introductory statistics course for 10 students:

80 73 92 85 75 98 93 55 80 90

Find the five-number summary for these first-exam scores.

The five-number summary leads to another visual representation of a distribution, the *boxplot*. Figure 1.19 shows boxplots for both city and highway gas mileages for our two groups of cars.

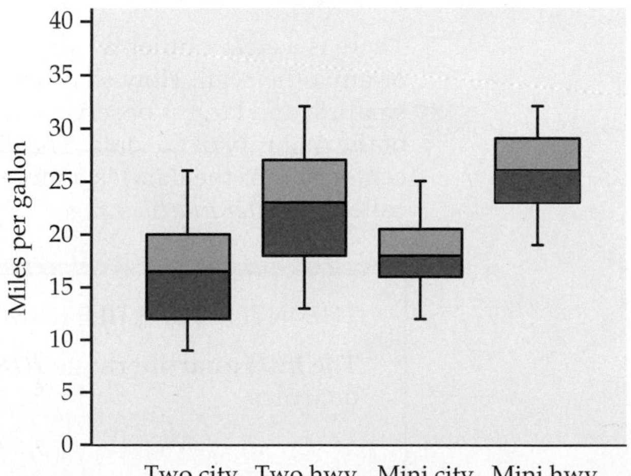


**FIGURE 1.19** Boxplots of the highway and city gas mileages for cars classified as two-seaters and as minicompacts by the Environmental Protection Agency.

BOXPLOT

A **boxplot** is a graph of the five-number summary.

- A central box spans the quartiles $Q_1$ and $Q_3$.
- A line in the box marks the median $M$.
- Lines extend from the box out to the smallest and largest observations.

When you look at a boxplot, first locate the median, which marks the center of the distribution. Then look at the spread. The quartiles show the spread of the middle half of the data, and the extremes (the smallest and largest observations) show the spread of the entire data set.

USE YOUR KNOWLEDGE

**1.51 Make a boxplot.** Here are the scores on the first exam in an introductory statistics course for 10 students:

80 73 92 85 75 98 93 55 80 90

Make a boxplot for these first-exam scores.

Boxplots are particularly effective for comparing distributions as we did in Figure 1.19. We see at once that city mileages are lower than highway mileages. The minicompact cars have slightly higher median gas mileages than the two-seaters, and their mileages are markedly less variable. In particular, the low gas mileages of the Ferraris and Lamborghinis in the two-seater group pull the group minimum down.

## The 1.5 × *IQR* rule for suspected outliers

Look again at the 80 service center call lengths in Table 1.1 (page 8). Figure 1.6 (page 12) is a stemplot of their distribution. You can check that the five-number summary is

1 54.5 103.5 200 2631

There is a clear outlier, a call lasting 2631 seconds, more than twice the length of any other call. How shall we describe the spread of this distribution? The smallest and largest observations are extremes that do not describe the spread of the majority of the data. The distance between the quartiles (the range of the center half of the data) is a more resistant measure of spread. This distance is called the *interquartile range.*

THE INTERQUARTILE RANGE *IQR*

The **interquartile range *IQR*** is the distance between the first and third quartiles,

$$IQR = Q_3 - Q_1$$

For our data on service call lengths, $IQR = 200 - 54.5 = 145.5$. The quartiles and the $IQR$ are not affected by changes in either tail of the distribution. They are therefore resistant, because changes in a few data points have no further effect once these points move outside the quartiles. However, *no single numerical measure of spread, such as IQR, is very useful for describing skewed distributions.* The two sides of a skewed distribution have different spreads, so one number can't summarize them. We can often detect skewness from the five-number summary by comparing how far the first quartile and the minimum are from the median (left tail) with how far the third quartile and the maximum are from the median (right tail). The interquartile range is mainly used as the basis for a rule of thumb for identifying suspected outliers.

### THE 1.5 × *IQR* RULE FOR OUTLIERS

Call an observation a suspected outlier if it falls more than $1.5 \times IQR$ above the third quartile or below the first quartile.

EXAMPLE

**1.18 Outliers for call length data.** For the call length data in Table 1.1,

$$1.5 \times IQR = 1.5 \times 145.5 = 218.25$$

Any values below $54.5 - 218.25 = -163.75$ or above $200 + 218.25 = 418.25$ are flagged as possible outliers. There are no low outliers, but the 8 longest calls are flagged as possible high outliers. Their lengths are

438 465 479 700 700 951 1148 2631

Statistical software often uses the $1.5 \times IQR$ rule. For example, the stemplot in Figure 1.6 lists these 8 observations separately. Boxplots drawn by software are often **modified boxplots** that plot suspected outliers individually. Figure 1.20 is a modified boxplot of the call length data. The lines extend out from the central box only to the smallest and largest observations that are not flagged by the $1.5 \times IQR$ rule. The 8 largest call lengths are plotted as individual points, though 2 of them are identical and so do not appear separately.

modified boxplot

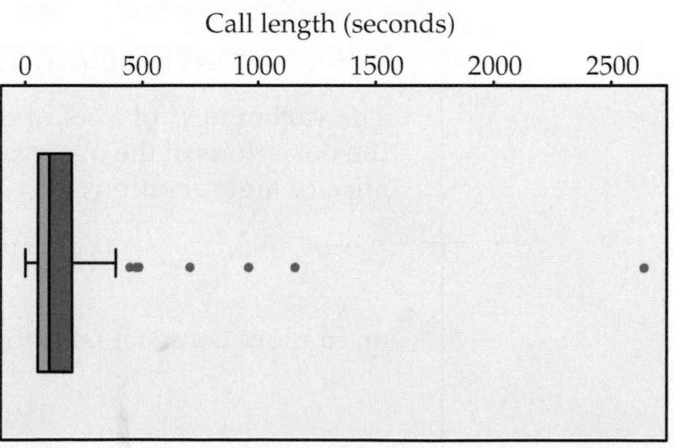


**FIGURE 1.20** Modified boxplot of the call lengths in Table 1.1, for Example 1.18.

The distribution of call lengths is very strongly skewed. We may well decide that only the longest call is truly an outlier in the sense of deviating from the overall pattern of the distribution. The other 7 calls are just part of the long right tail. The $1.5 \times IQR$ rule does not remove the need to look at the distribution and use judgment. It is useful mainly to call our attention to unusual observations.

## USE YOUR KNOWLEDGE

**1.52 Find the *IQR*.** Here are the scores on the first exam in an introductory statistics course for 10 students:

80 73 92 85 75 98 93 55 80 90

Find the interquartile range and use the $1.5 \times IQR$ rule to check for outliers. How low would the lowest score need to be for it to be an outlier according to this rule?

The stemplot in Figure 1.6 and the modified boxplot in Figure 1.20 tell us much more about the distribution of call lengths than the five-number summary or other numerical measures. The routine methods of statistics compute numerical measures and draw conclusions based on their values. These methods are very useful, and we will study them carefully in later chapters. But they cannot be applied blindly, by feeding data to a computer program, because *statistical measures and methods based on them are generally meaningful only for distributions of sufficiently regular shape.* This principle will become clearer as we progress, but it is good to be aware at the beginning that quickly resorting to fancy calculations is the mark of a statistical amateur. Look, think, and choose your calculations selectively.

## Measuring spread: the standard deviation

The five-number summary is not the most common numerical description of a distribution. That distinction belongs to the combination of the mean to measure center and the *standard deviation* to measure spread. The standard deviation measures spread by looking at how far the observations are from their mean.

### THE STANDARD DEVIATION $s$

The **variance $s^2$** of a set of observations is the average of the squares of the deviations of the observations from their mean. In symbols, the variance of $n$ observations $x_1, x_2, \ldots, x_n$ is

$$s^2 = \frac{(x_1 - \overline{x})^2 + (x_2 - \overline{x})^2 + \cdots + (x_n - \overline{x})^2}{n - 1}$$

or, in more compact notation,

$$s^2 = \frac{1}{n-1} \sum (x_i - \overline{x})^2$$

The **standard deviation** $s$ is the square root of the variance $s^2$:

$$s = \sqrt{\frac{1}{n-1}\sum(x_i - \overline{x})^2}$$

The idea behind the variance and the standard deviation as measures of spread is as follows: The deviations $x_i - \overline{x}$ display the spread of the values $x_i$ about their mean $\overline{x}$. Some of these deviations will be positive and some negative because some of the observations fall on each side of the mean. In fact, *the sum of the deviations of the observations from their mean will always be zero.* Squaring the deviations makes them all positive, so that observations far from the mean in either direction have large positive squared deviations. The variance is the average squared deviation. Therefore, $s^2$ and $s$ will be large if the observations are widely spread about their mean, and small if the observations are all close to the mean.

EXAMPLE

**1.19 Metabolic rate.** A person's metabolic rate is the rate at which the body consumes energy. Metabolic rate is important in studies of weight gain, dieting, and exercise. Here are the metabolic rates of 7 men who took part in a study of dieting. (The units are calories per 24 hours. These are the same calories used to describe the energy content of foods.)

1792 1666 1362 1614 1460 1867 1439

Enter these data into your calculator or software and verify that

$$\overline{x} = 1600 \text{ calories} \qquad s = 189.24 \text{ calories}$$

Figure 1.21 plots these data as dots on the calorie scale, with their mean marked by an asterisk (∗). The arrows mark two of the deviations from the mean. If you were calculating $s$ by hand, you would find the first deviation as

$$x_1 - \overline{x} = 1792 - 1600 = 192$$

Exercise 1.70 asks you to calculate the seven deviations, square them, and find $s^2$ and $s$ directly from the deviations. Working one or two short examples by hand helps you understand how the standard deviation is obtained. In practice you will use either software or a calculator that will find $s$ from keyed-in data. The two software outputs in Figure 1.18 both give the variance and standard deviation for the highway mileage data.

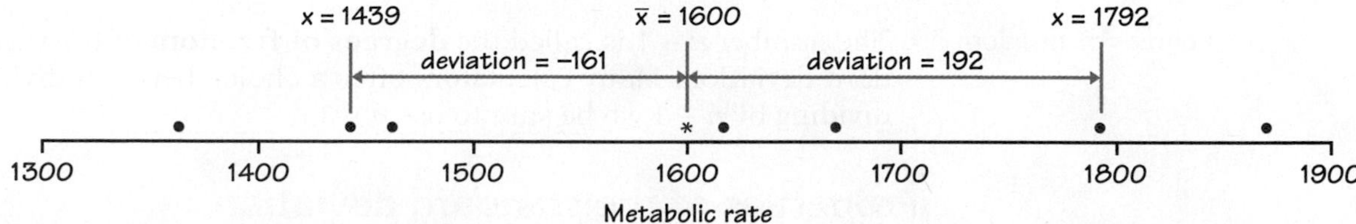


**FIGURE 1.21** Metabolic rates for seven men, with the mean (∗) and the deviations of two observations from the mean, for Example 1.19.

## USE YOUR KNOWLEDGE

**1.53 Find the variance and the standard deviation.** Here are the scores on the first exam in an introductory statistics course for 10 students:

80 73 92 85 75 98 93 55 80 90

Find the variance and the standard deviation for these first-exam scores.

The idea of the variance is straightforward: it is the average of the squares of the deviations of the observations from their mean. The details we have just presented, however, raise some questions.

*Why do we square the deviations?*

- First, the sum of the squared deviations of any set of observations from their mean is the smallest that the sum of squared deviations from any number can possibly be. This is not true of the unsquared distances. So squared deviations point to the mean as center in a way that distances do not.
- Second, the standard deviation turns out to be the natural measure of spread for a particularly important class of symmetric unimodal distributions, the *Normal distributions.* We will meet the Normal distributions in the next section. We commented earlier that the usefulness of many statistical procedures is tied to distributions of particular shapes. This is distinctly true of the standard deviation.

*Why do we emphasize the standard deviation rather than the variance?*

- One reason why is that $s$, not $s^2$, is the natural measure of spread for Normal distributions.
- There is also a more general reason to prefer $s$ to $s^2$. Because the variance involves squaring the deviations, it does not have the same unit of measurement as the original observations. The variance of the metabolic rates, for example, is measured in squared calories. Taking the square root remedies this. The standard deviation $s$ measures spread about the mean in the original scale.

*Why do we average by dividing by $n - 1$ rather than $n$ in calculating the variance?*

- Because the sum of the deviations is always zero, the last deviation can be found once we know the other $n - 1$. So we are not averaging $n$ unrelated numbers. Only $n - 1$ of the squared deviations can vary freely, and we average by dividing the total by $n - 1$.

degrees of freedom

- The number $n - 1$ is called the **degrees of freedom** of the variance or standard deviation. Many calculators offer a choice between dividing by $n$ and dividing by $n - 1$, so be sure to use $n - 1$.

## Properties of the standard deviation

Here are the basic properties of the standard deviation $s$ as a measure of spread.

PROPERTIES OF THE STANDARD DEVIATION

- $s$ measures spread about the mean and should be used only when the mean is chosen as the measure of center.
- $s = 0$ only when there is *no spread.* This happens only when all observations have the same value. Otherwise, $s > 0$. As the observations become more spread out about their mean, $s$ gets larger.
- $s$, like the mean $\bar{x}$, is not resistant. A few outliers can make $s$ very large.

USE YOUR KNOWLEDGE

**1.54 A standard deviation of zero.** Construct a data set with 5 cases that has a variable with $s = 0$.

*The use of squared deviations renders s even more sensitive than $\bar{x}$ to a few extreme observations.* For example, dropping the Honda Insight from our list of two-seater cars reduces the mean highway mileage from 24.7 mpg to 22.6 mpg. It cuts the standard deviation more than half, from 10.8 mpg with the Insight to 5.3 mpg without it. Distributions with outliers and strongly skewed distributions have large standard deviations. The number $s$ does not give much helpful information about such distributions.

## Choosing measures of center and spread

How do we choose between the five-number summary and $\bar{x}$ and $s$ to describe the center and spread of a distribution? Because the two sides of a strongly skewed distribution have different spreads, no single number such as $s$ describes the spread well. The five-number summary, with its two quartiles and two extremes, does a better job.

CHOOSING A SUMMARY

The five-number summary is usually better than the mean and standard deviation for describing a skewed distribution or a distribution with strong outliers. Use $\bar{x}$ and $s$ only for reasonably symmetric distributions that are free of outliers.

EXAMPLE

**1.20 Standard deviation as a measure of risk.** A central principle in the study of investments is that taking bigger risks is rewarded by higher returns, at least on the average over long periods of time. It is usual in finance to measure risk by the standard deviation of returns, on the grounds that investments whose returns vary a lot from year to year are less predictable and therefore more risky than those whose returns don't vary much. Compare, for example, the approximate mean and standard deviation of the annual percent returns on American common stocks and U.S. Treasury bills over the period from 1950 to 2003:

| Investment | Mean return | Standard deviation |
|---|---|---|
| Common stocks | 13.2% | 17.6% |
| Treasury bills | 5.0% | 2.9% |

Stocks are risky. They went up more than 13% per year on the average during this period, but they dropped almost 28% in the worst year. The large standard deviation reflects the fact that stocks have produced both large gains and large losses. When you buy a Treasury bill, on the other hand, you are lending money to the government for one year. You know that the government will pay you back with interest. That is much less risky than buying stocks, so (on the average) you get a smaller return.

Are $\bar{x}$ and $s$ good summaries for distributions of investment returns? Figure 1.22 displays stemplots of the annual returns for both investments. (Because stock returns are so much more spread out, a back-to-back stemplot does not work well. The stems in the stock stemplot are tens of percents; the stems for bills are percents. The lowest returns are −28% for stocks and 0.9% for bills.) You see that returns on Treasury bills have a right-skewed distribution. Convention in the financial world calls for $\bar{x}$ and $s$ because some parts of investment theory use them. For describing this right-skewed distribution, however, the five-number summary would be more informative.

```
-2 | 8 1
-1 | 9 1 1 1 1 0
-0 | 9 6 4 3
 0 | 0 0 0 1 2 3 8 9 9
 1 | 1 3 3 4 4 6 6 6 7 8
 2 | 0 1 1 2 3 4 4 4 5 7 7 9 9
 3 | 0 1 1 2 3 4 6 7
 4 | 5
 5 | 0
```

(a)

```
 0 | 9
 1 | 0 2 5 5 6 6 6 8
 2 | 1 5 7 7 9
 3 | 0 1 1 3 5 5 8 9 9
 4 | 2 4 7 7 8
 5 | 1 1 2 2 2 5 6 6 7 8
 6 | 2 4 5 6 9
 7 | 2 7 8
 8 | 0 4 8
 9 | 8
10 | 4 5
11 | 3
12 |
13 |
14 | 7
```

(b)

**FIGURE 1.22** Stemplots of annual returns for stocks and Treasury bills, 1950 to 2003, for Example 1.20. (a) Stock returns, in whole percents. (b) Treasury bill returns, in percents and tenths of a percent.

*Remember that a graph gives the best overall picture of a distribution. Numerical measures of center and spread report specific facts about a distribution, but they do not describe its entire shape.* Numerical summaries do not disclose the presence of multiple modes or gaps, for example. **Always plot your data.**

## Changing the unit of measurement

The same variable can be recorded in different units of measurement. Americans commonly record distances in miles and temperatures in degrees Fahrenheit, while the rest of the world measures distances in kilometers and temperatures in degrees Celsius. Fortunately, it is easy to convert numerical descriptions of a distribution from one unit of measurement to another. This is true because a change in the measurement unit is a *linear transformation* of the measurements.

### LINEAR TRANSFORMATIONS

A **linear transformation** changes the original variable $x$ into the new variable $x_{\text{new}}$ given by an equation of the form

$$x_{\text{new}} = a + bx$$

Adding the constant $a$ shifts all values of $x$ upward or downward by the same amount. In particular, such a shift changes the origin (zero point) of the variable. Multiplying by the positive constant $b$ changes the size of the unit of measurement.

EXAMPLE

### 1.21 Change the units.

(a) If a distance $x$ is measured in kilometers, the same distance in miles is

$$x_{\text{new}} = 0.62x$$

For example, a 10-kilometer race covers 6.2 miles. This transformation changes the units without changing the origin—a distance of 0 kilometers is the same as a distance of 0 miles.

(b) A temperature $x$ measured in degrees Fahrenheit must be reexpressed in degrees Celsius to be easily understood by the rest of the world. The transformation is

$$x_{\text{new}} = \frac{5}{9}(x - 32) = -\frac{160}{9} + \frac{5}{9}x$$

Thus, the high of 95°F on a hot American summer day translates into 35°C. In this case

$$a = -\frac{160}{9} \quad \text{and} \quad b = \frac{5}{9}$$

This linear transformation changes both the unit size and the origin of the measurements. The origin in the Celsius scale (0°C, the temperature at which water freezes) is 32° in the Fahrenheit scale.

*Linear transformations do not change the shape of a distribution.* If measurements on a variable $x$ have a right-skewed distribution, any new variable $x_{\text{new}}$

obtained by a linear transformation $x_{new} = a + bx$ (for $b > 0$) will also have a right-skewed distribution. If the distribution of $x$ is symmetric and unimodal, the distribution of $x_{new}$ remains symmetric and unimodal.

Although a linear transformation preserves the basic shape of a distribution, the center and spread will change. Because linear changes of measurement scale are common, we must be aware of their effect on numerical descriptive measures of center and spread. Fortunately, the changes follow a simple pattern.

EXAMPLE

**1.22 Use scores to find the points.** In an introductory statistics course, homework counts for 300 points out of a total of 1000 possible points for all course requirements. During the semester there were 12 homework assignments and each was given a grade on a scale of 0 to 100. The maximum total score for the 12 homework assignments is therefore 1200. To convert the homework scores to final grade points, we need to convert the scale of 0 to 1200 to a scale of 0 to 300. We do this by multiplying the homework scores by 300/1200. In other words, we divide the homework scores by 4. Here are the homework scores and the corresponding final grade points for 5 students:

| Student | 1 | 2 | 3 | 4 | 5 |
|---|---|---|---|---|---|
| Score | 1056 | 1080 | 900 | 1164 | 1020 |
| Points | 264 | 270 | 225 | 291 | 255 |

These two sets of numbers measure the same performance on homework for the course. Since we obtained the points by dividing the scores by 4, the mean of the points will be the mean of the scores divided by 4. Similarly, the standard deviation of points will be the standard deviation of the scores divided by 4.

## USE YOUR KNOWLEDGE

**1.55 Calculate the points for a student.** Use the setting of Example 1.22 to find the points for a student whose score is 950.

Here is a summary of the rules for linear transformations:

### EFFECT OF A LINEAR TRANSFORMATION

To see the effect of a linear transformation on measures of center and spread, apply these rules:

- Multiplying each observation by a positive number $b$ multiplies both measures of center (mean and median) and measures of spread (interquartile range and standard deviation) by $b$.

- Adding the same number $a$ (either positive or negative) to each observation adds $a$ to measures of center and to quartiles and other percentiles but does not change measures of spread.

In Example 1.22, when we converted from score to points, we described the transformation as dividing by 4. The multiplication part of the summary of the effect of a linear transformation applies to this case because division by 4 is the same as multiplication by 0.25. Similarly, the second part of the summary applies to subtraction as well as addition because subtraction is simply the addition of a negative number.

The measures of spread $IQR$ and $s$ do not change when we add the same number $a$ to all of the observations because adding a constant changes the location of the distribution but leaves the spread unaltered. You can find the effect of a linear transformation $x_{\text{new}} = a + bx$ by combining these rules. For example, if $x$ has mean $\bar{x}$, the transformed variable $x_{\text{new}}$ has mean $a + b\bar{x}$.

## SECTION 1.2 Summary

A numerical summary of a distribution should report its **center** and its **spread** or **variability.**

The **mean** $\bar{x}$ and the **median** $M$ describe the center of a distribution in different ways. The mean is the arithmetic average of the observations, and the median is their midpoint.

When you use the median to describe the center of the distribution, describe its spread by giving the **quartiles.** The **first quartile** $Q_1$ has one-fourth of the observations below it, and the **third quartile** $Q_3$ has three-fourths of the observations below it.

The **interquartile range** is the difference between the quartiles. It is the spread of the center half of the data. The **1.5 x** $IQR$ **rule** flags observations more than $1.5 \times IQR$ beyond the quartiles as possible outliers.

The **five-number summary** consisting of the median, the quartiles, and the smallest and largest individual observations provides a quick overall description of a distribution. The median describes the center, and the quartiles and extremes show the spread.

**Boxplots** based on the five-number summary are useful for comparing several distributions. The box spans the quartiles and shows the spread of the central half of the distribution. The median is marked within the box. Lines extend from the box to the extremes and show the full spread of the data. In a **modified boxplot,** points identified by the $1.5 \times IQR$ rule are plotted individually.

The **variance** $s^2$ and especially its square root, the **standard deviation** $s$, are common measures of spread about the mean as center. The standard deviation $s$ is zero when there is no spread and gets larger as the spread increases.

A **resistant measure** of any aspect of a distribution is relatively unaffected by changes in the numerical value of a small proportion of the total number of observations, no matter how large these changes are. The median and quartiles are resistant, but the mean and the standard deviation are not.

The mean and standard deviation are good descriptions for symmetric distributions without outliers. They are most useful for the Normal distributions introduced in the next section. The five-number summary is a better exploratory summary for skewed distributions.

**Linear transformations** have the form $x_{new} = a + bx$. A linear transformation changes the origin if $a \neq 0$ and changes the size of the unit of measurement if $b > 0$. Linear transformations do not change the overall shape of a distribution. A linear transformation multiplies a measure of spread by $b$ and changes a percentile or measure of center $m$ into $a + bm$.

Numerical measures of particular aspects of a distribution, such as center and spread, do not report the entire shape of most distributions. In some cases, particularly distributions with multiple peaks and gaps, these measures may not be very informative.

## SECTION 1.2 Exercises

*For Exercise 1.47, see page 32; for Exercise 1.48, see page 33; for Exercise 1.49, see page 35; for Exercises 1.50, see page 37; for Exercise 1.51, see page 38; for Exercise 1.52, see page 40; for Exercise 1.53, see page 42; for Exercise 1.54, see page 43; and for Exercise 1.55, see page 46.*

**1.56 Longleaf pine trees.** The Wade Tract in Thomas County, Georgia, is an old-growth forest of longleaf pine trees (*Pinus palustris*) that has survived in a relatively undisturbed state since before the settlement of the area by Europeans. A study collected data about 584 of these trees.[27] One of the variables measured was the diameter at breast height (DBH). This is the diameter of the tree at 4.5 feet and the units are centimeters (cm). Only trees with DBH greater than 1.5 cm were sampled. Here are the diameters of a random sample of 40 of these trees:

| | | | | | | | | | |
|---|---|---|---|---|---|---|---|---|---|
| 10.5 | 13.3 | 26.0 | 18.3 | 52.2 | 9.2 | 26.1 | 17.6 | 40.5 | 31.8 |
| 47.2 | 11.4 | 2.7 | 69.3 | 44.4 | 16.9 | 35.7 | 5.4 | 44.2 | 2.2 |
| 4.3 | 7.8 | 38.1 | 2.2 | 11.4 | 51.5 | 4.9 | 39.7 | 32.6 | 51.8 |
| 43.6 | 2.3 | 44.6 | 31.5 | 40.3 | 22.3 | 43.3 | 37.5 | 29.1 | 27.9 |

(a) Find the five-number summary for these data.

(b) Make a boxplot.

(c) Make a histogram.

(d) Write a short summary of the major features of this distribution. Do you prefer the boxplot or the histogram for these data?

**1.57 Blood proteins in children from Papua New Guinea.** C-reactive protein (CRP) is a substance that can be measured in the blood. Values increase substantially within 6 hours of an infection and reach a peak within 24 to 48 hours after. In adults, chronically high values have been linked to an increased risk of cardiovascular disease. In a study of apparently healthy children aged 6 to 60 months in Papua New Guinea, CRP was measured in 90 children.[28] The units are milligrams per liter (mg/l). Here are the data from a random sample of 40 of these children:

| | | | | | | | | | |
|---|---|---|---|---|---|---|---|---|---|
| 0.00 | 3.90 | 5.64 | 8.22 | 0.00 | 5.62 | 3.92 | 6.81 | 30.61 | 0.00 |
| 73.20 | 0.00 | 46.70 | 0.00 | 0.00 | 26.41 | 22.82 | 0.00 | 0.00 | 3.49 |
| 0.00 | 0.00 | 4.81 | 9.57 | 5.36 | 0.00 | 5.66 | 0.00 | 59.76 | 12.38 |
| 15.74 | 0.00 | 0.00 | 0.00 | 0.00 | 9.37 | 20.78 | 7.10 | 7.89 | 5.53 |

(a) Find the five-number summary for these data.

(b) Make a boxplot.

(c) Make a histogram.

(d) Write a short summary of the major features of this distribution. Do you prefer the boxplot or the histogram for these data?

**1.58** CHALLENGE **Transform the blood proteins values.** Refer to the previous exercise. With strongly skewed distributions such as this, we frequently reduce the skewness by taking a log transformation. We have a bit of a problem here, however, because some of the data are recorded as 0.00 and the logarithm of zero is not defined. For this variable, the value 0.00 is recorded whenever the amount of CRP in the blood is below the level that the measuring instrument is capable of detecting. The usual procedure in this circumstance is to add a small number to each observation before taking the logs. Transform these data by adding 1 to each observation and then taking the logarithm. Use the questions in the previous exercise as a guide to your analysis and prepare a summary contrasting this analysis with the one that you performed in the previous exercise.

**1.59** CHALLENGE **Vitamin A deficiency in children from Papua New Guinea.** In the Papua New Guinea study that provided the data for the previous two exercises, the researchers also measured serum retinol. A low value of this variable can be an indicator of vitamin A deficiency. Below are the data on the same sample of 40 children from this study. The units are micromoles per liter ($\mu$mol/l).

| | | | | | | | | | |
|---|---|---|---|---|---|---|---|---|---|
| 1.15 | 1.36 | 0.38 | 0.34 | 0.35 | 0.37 | 1.17 | 0.97 | 0.97 | 0.67 |
| 0.31 | 0.99 | 0.52 | 0.70 | 0.88 | 0.36 | 0.24 | 1.00 | 1.13 | 0.31 |
| 1.44 | 0.35 | 0.34 | 1.90 | 1.19 | 0.94 | 0.34 | 0.35 | 0.33 | 0.69 |
| 0.69 | 1.04 | 0.83 | 1.11 | 1.02 | 0.56 | 0.82 | 1.20 | 0.87 | 0.41 |

Analyze these data. Use the questions in the previous two exercises as a guide.

**1.60 Luck and puzzle solving.** Children in a psychology study were asked to solve some puzzles and were then given feedback on their performance. Then they were asked to rate how luck played a role in determining their scores.[29] This variable was recorded on a 1 to 10 scale with 1 corresponding to very lucky and 10 corresponding to very unlucky. Here are the scores for 60 children:

| | | | | | | | | | | | | | | |
|---|---|---|---|---|---|---|---|---|---|---|---|---|---|---|
| 1 | 10 | 1 | 10 | 1 | 1 | 10 | 5 | 1 | 1 | 8 | 1 | 10 | 2 | 1 |
| 9 | 5 | 2 | 1 | 8 | 10 | 5 | 9 | 10 | 10 | 9 | 6 | 10 | 1 | 5 |
| 1 | 9 | 2 | 1 | 7 | 10 | 9 | 5 | 10 | 10 | 10 | 1 | 8 | 1 | 6 |
| 10 | 1 | 6 | 10 | 10 | 8 | 10 | 3 | 10 | 8 | 1 | 8 | 10 | 4 | 2 |

Use numerical and graphical methods to describe these data. Write a short report summarizing your work.

**1.61 College tuition and fees.** Figure 1.16 (page 25) is a histogram of the tuition and fees charged by the 56 four-year colleges in the state of Massachusetts. Here are those charges (in dollars), arranged in increasing order:

| | | | | | | | |
|---|---|---|---|---|---|---|---|
| 4,123 | 4,186 | 4,324 | 4,342 | 4,557 | 4,884 | 5,397 | 6,129 |
| 6,963 | 6,972 | 8,232 | 13,584 | 13,612 | 15,500 | 15,934 | 16,230 |
| 16,696 | 16,700 | 17,044 | 17,500 | 18,550 | 18,750 | 19,145 | 19,300 |
| 19,410 | 19,700 | 19,700 | 19,910 | 20,234 | 20,400 | 20,640 | 20,875 |
| 21,165 | 21,302 | 22,663 | 23,550 | 24,324 | 25,840 | 26,965 | 27,522 |
| 27,544 | 27,904 | 28,011 | 28,090 | 28,420 | 28,420 | 28,900 | 28,906 |
| 28,950 | 29,060 | 29,338 | 29,392 | 29,600 | 29,624 | 29,630 | 29,875 |

Find the five-number summary and make a boxplot. What distinctive feature of the histogram do these summaries miss? Remember that numerical summaries are not a substitute for looking at the data.

**1.62 Outliers in percent of older residents.** The stemplot in Exercise 1.21 (page 24) displays the distribution of the percents of residents aged 65 and over in the 50 states. Stemplots help you find the five-number summary because they arrange the observations in increasing order.

(a) Give the five-number summary of this distribution.

(b) Does the $1.5 \times IQR$ rule identify Alaska and Florida as suspected outliers? Does it also flag any other states?

**1.63 Tornados and property damage.** Table 1.5 (page 25) shows the average property damage caused by tornadoes over a 50-year period in each of the states. The distribution is strongly skewed to the right.

(a) Give the five-number summary. Explain why you can see from these five numbers that the distribution is right-skewed.

(b) A histogram or stemplot suggests that a few states are outliers. Show that there are *no* suspected outliers according to the $1.5 \times IQR$ rule. You see once again that a rule is not a substitute for plotting your data.

(c) Find the mean property damage. Explain why the mean and median differ so greatly for this distribution.

**1.64 Carbon dioxide emissions.** Table 1.6 (page 26) gives carbon dioxide ($CO_2$) emissions per person for countries with population at least 20 million. The distribution is strongly skewed to the right. The United States and several other countries appear to be high outliers.

(a) Give the five-number summary. Explain why this summary suggests that the distribution is right-skewed.

(b) Which countries are outliers according to the $1.5 \times IQR$ rule? Make a stemplot or histogram of the data. Do you agree with the rule's suggestions about which countries are and are not outliers?

**1.65 Median versus mean for net worth.** A report on the assets of American households says that the median net worth of households headed by someone aged less than 35 years is \$11,600. The mean net worth of these same young households is \$90,700.[30] What explains the difference between these two measures of center?

**1.66 Mean versus median for oil wells.** Exercise 1.39 (page 28) gives data on the total oil recovered from 64 wells. Your graph in that exercise shows that the distribution is clearly right-skewed.

(a) Find the mean and median of the amounts recovered. Explain how the relationship between the mean and the median reflects the shape of the distribution.

(b) Give the five-number summary and explain briefly how it reflects the shape of the distribution.

**1.67 Mean versus median.** A small accounting firm pays each of its five clerks \$35,000, two junior accountants \$80,000 each, and the firm's owner \$320,000. What is the mean salary paid at this firm? How many of the employees earn less than the mean? What is the median salary?

**1.68 Be careful about how you treat the zeros.** In computing the median income of any group, some federal agencies omit all members of the group who had no income. Give an example to show that the reported median income of a group can go down even though the group becomes economically better off. Is this also true of the mean income?

**1.69 How does the median change?** The firm in Exercise 1.67 gives no raises to the clerks and junior accountants, while the owner's take increases to \$455,000. How does this change affect the mean? How does it affect the median?

**1.70 Metabolic rates.** Calculate the mean and standard deviation of the metabolic rates in Example 1.19 (page 41), showing each step in detail. First find the mean $\bar{x}$ by summing the 7 observations and dividing by 7. Then find each of the deviations $x_i - \bar{x}$ and their squares. Check that the deviations have sum 0. Calculate the variance as an average of the squared deviations (remember to divide by $n - 1$). Finally, obtain $s$ as the square root of the variance.

**1.71** CHALLENGE **Hurricanes and losses.** A discussion of extreme weather says: "In most states, hurricanes occur infrequently. Yet, when a hurricane hits, the losses can be catastrophic. Average annual losses are not a meaningful measure of damage from rare but potentially catastrophic events."[31] Why is this true?

**1.72 Distributions for time spent studying.** Exercise 1.41 (page 28) presented data on the nightly study time claimed by first-year college men and women. The most common methods for formal comparison of two groups use $\bar{x}$ and $s$ to summarize the data. We wonder if this is appropriate here. Look at your back-to-back stemplot from Exercise 1.41, or make one now if you have not done so.

(a) What kinds of distributions are best summarized by $\bar{x}$ and $s$? It isn't easy to decide whether small data sets with irregular distributions fit the criteria. We will learn a better tool for making this decision in the next section.

(b) Each set of study times appears to contain a high outlier. Are these points flagged as suspicious by the $1.5 \times IQR$ rule? How much does removing the outlier change $\bar{x}$ and $s$ for each group? The presence of outliers makes us reluctant to use the mean and standard deviation for these data unless we remove the outliers on the grounds that these students were exaggerating.

**1.73 The density of the earth.** Many standard statistical methods that you will study in Part II of this book are intended for use with distributions that are symmetric and have no outliers. These methods start with the mean and standard deviation, $\bar{x}$ and $s$. Two examples of scientific data for which standard methods should work well are the pH measurements in Exercise 1.36 (page 27) and Cavendish's measurements of the density of the earth in Exercise 1.40 (page 28).

(a) Summarize each of these data sets by giving $\bar{x}$ and $s$.

(b) Find the median for each data set. Is the median quite close to the mean, as we expect it to be in these examples?

**1.74 IQ scores.** Many standard statistical methods that you will study in Part II of this book are intended for use with distributions that are symmetric and have no outliers. These methods start with the mean and standard deviation, $\bar{x}$ and $s$. For example, standard methods would typically be used for the IQ and GPA data in Table 1.9 (page 29).

(a) Find $\bar{x}$ and $s$ for the IQ data. In large populations, IQ scores are standardized to have mean 100 and standard deviation 15. In what way does the distribution of IQ among these students differ from the overall population?

(b) Find the median IQ score. It is, as we expect, close to the mean.

(c) Find the mean and median for the GPA data. The two measures of center differ a bit. What feature of the data (see your stemplot in Exercise 1.43 or make a new stemplot) explains the difference?

**1.75** APPLET **Mean and median for two observations.** The *Mean and Median* applet allows you to place observations on a line and see their mean and median visually. Place two observations on the line, by clicking below it. Why does only one arrow appear?

**1.76** APPLET **Mean and median for three observations.** In the *Mean and Median* applet, place three observations on the line by clicking below it, two close together near the center of the line and one somewhat to the right of these two.

(a) Pull the single rightmost observation out to the right. (Place the cursor on the point, hold down a mouse button, and drag the point.) How does the mean behave? How does the median behave? Explain briefly why each measure acts as it does.

(b) Now drag the rightmost point to the left as far as you can. What happens to the mean? What happens to the median as you drag this point past the other two (watch carefully)?

**1.77** APPLET **Mean and median for five observations.** Place five observations on the line in the *Mean and Median* applet by clicking below it.

(a) Add one additional observation *without changing the median.* Where is your new point?

(b) Use the applet to convince yourself that when you add yet another observation (there are now seven in all), the median does not change no matter where you put the seventh point. Explain why this must be true.

**1.78 Hummingbirds and flowers.** Different varieties of the tropical flower *Heliconia* are fertilized by different species of hummingbirds. Over time, the lengths of the flowers and the form of the hummingbirds' beaks have evolved to match each other. Here are data on the lengths in millimeters of three varieties of these flowers on the island of Dominica:[32]

| *H. bihai* | | | | | | | |
|---|---|---|---|---|---|---|---|
| 47.12 | 46.75 | 46.81 | 47.12 | 46.67 | 47.43 | 46.44 | 46.64 |
| 48.07 | 48.34 | 48.15 | 50.26 | 50.12 | 46.34 | 46.94 | 48.36 |

| *H. caribaea* red | | | | | | | |
|---|---|---|---|---|---|---|---|
| 41.90 | 42.01 | 41.93 | 43.09 | 41.47 | 41.69 | 39.78 | 40.57 |
| 39.63 | 42.18 | 40.66 | 37.87 | 39.16 | 37.40 | 38.20 | 38.07 |
| 38.10 | 37.97 | 38.79 | 38.23 | 38.87 | 37.78 | 38.01 | |

| *H. caribaea* yellow | | | | | | | |
|---|---|---|---|---|---|---|---|
| 36.78 | 37.02 | 36.52 | 36.11 | 36.03 | 35.45 | 38.13 | 37.1 |
| 35.17 | 36.82 | 36.66 | 35.68 | 36.03 | 34.57 | 34.63 | |

Make boxplots to compare the three distributions. Report the five-number summaries along with your graph. What are the most important differences among the three varieties of flower?

**1.79 Compare the three varieties of flowers.** The biologists who collected the flower length data in the previous exercise compared the three *Heliconia* varieties using statistical methods based on $\bar{x}$ and $s$.

(a) Find $\bar{x}$ and $s$ for each variety.

(b) Make a stemplot of each set of flower lengths. Do the distributions appear suitable for use of $\bar{x}$ and $s$ as summaries?

**1.80** CHALLENGE **Effects of logging in Borneo.** "Conservationists have despaired over destruction of tropical rainforest by logging, clearing, and burning." These words begin a report on a statistical study of the effects of logging in Borneo. Researchers compared forest plots that had never been logged (Group 1) with similar plots nearby that had been logged 1 year earlier (Group 2) and 8 years earlier (Group 3). All plots were 0.1 hectare in area. Here are the counts of trees for plots in each group:[33]

| | | | | | | | | | | | | |
|---|---|---|---|---|---|---|---|---|---|---|---|---|
| Group 1: | 27 | 22 | 29 | 21 | 19 | 33 | 16 | 20 | 24 | 27 | 28 | 19 |
| Group 2: | 12 | 12 | 15 | 9 | 20 | 18 | 17 | 14 | 14 | 2 | 17 | 19 |
| Group 3: | 18 | 4 | 22 | 15 | 18 | 19 | 22 | 12 | 12 | | | |

Give a complete comparison of the three distributions, using both graphs and numerical summaries. To what extent has logging affected the count of trees? The researchers used an analysis based on $\bar{x}$ and $s$. Explain why this is reasonably well justified.

**1.81** CHALLENGE **Running and heart rate.** How does regular running affect heart rate? The RUNNERS data set, described in detail in the Data Appendix, contains heart rates for four groups of people:

Sedentary females

Sedentary males

Female runners (at least 15 miles per week)

Male runners (at least 15 miles per week)

The heart rates were measured after 6 minutes of exercise on a treadmill. There are 200 subjects in each group. Give a complete comparison of the four distributions, using both graphs and numerical summaries. How would you describe the effect of running on heart rate? Is the effect different for men and women?

*The WORKERS data set, described in the Data Appendix, contains the sex, level of education, and income of 71,076 people between the ages of 25 and 64 who were employed full-time in 2001.*

*The boxplots in Figure 1.23 compare the distributions of income for people with five levels of education. This figure is a variation on the boxplot idea: because large data sets often contain very extreme observations, the lines extend from the central box only to the 5th and 95th percentiles. Exercises 1.82 to 1.84 concern these data.*

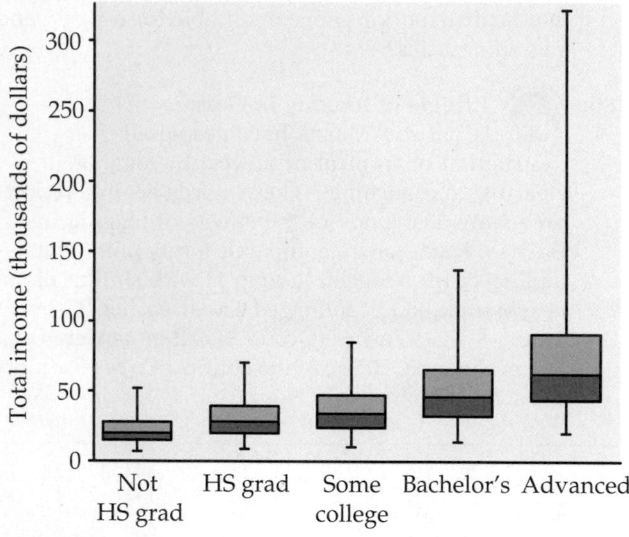


**FIGURE 1.23** Boxplots comparing the distributions of income for employed people aged 25 to 64 years with five different levels of education. The lines extend from the quartiles to the 5th and 95th percentiles.

**1.82 Income for people with bachelor's degrees.** The data include 14,959 people whose highest level of education is a bachelor's degree.

(a) What is the position of the median in the ordered list of incomes (1 to 14,959)? From the boxplot, about what is the median income of people with a bachelor's degree?

(b) What is the position of the first and third quartiles in the ordered list of incomes for these people? About what are the numerical values of $Q_1$ and $Q_3$?

(c) You answered (a) and (b) from a boxplot that omits the lowest 5% and the highest 5% of incomes. Explain why leaving out these values has only a very small effect on the median and quartiles.

**1.83 Find the 5th and 95th percentiles.** About what are the positions of the 5th and 95th percentiles in the ordered list of incomes of the 14,959 people with a bachelor's degree? Incomes outside this range do not appear in the boxplot. About what are the numerical values of the 5th and 95th percentiles of income? (For comparison, the largest income among all 14,959 people was $481,720. That one person made this much tells us less about the group than does the 95th percentile.)

**1.84 How does income change with education?** Write a brief description of how the distribution of income changes with the highest level of education reached. Be sure to discuss center, spread, and skewness. Give some specifics read from the graph to back up your statements.

**1.85** CHALLENGE **Shakespeare's plays.** Look at the histogram of lengths of words in Shakespeare's plays, Figure 1.15 (page 25). The heights of the bars tell us what percent of words have each length. What is the median length of words used by Shakespeare? Similarly, what are the quartiles? Give the five-number summary for Shakespeare's word lengths.

**1.86** CHALLENGE **Create a data set.** Create a set of 5 positive numbers (repeats allowed) that have median 10 and mean 7. What thought process did you use to create your numbers?

**1.87 Create another data set.** Give an example of a small set of data for which the mean is larger than the third quartile.

**1.88** CHALLENGE **Deviations from the mean sum to zero.** Use the definition of the mean $\bar{x}$ to show that the sum of the deviations $x_i - \bar{x}$ of the observations from their mean is always zero. This is one reason why the variance and standard deviation use squared deviations.

**1.89** CHALLENGE **A standard deviation contest.** This is a standard deviation contest. You must choose four numbers from the whole numbers 0 to 20, with repeats allowed.

(a) Choose four numbers that have the smallest possible standard deviation.

(b) Choose four numbers that have the largest possible standard deviation.

(c) Is more than one choice possible in either (a) or (b)? Explain.

**1.90 Does your software give incorrect answers?** This exercise requires a calculator with a standard deviation button or statistical software on a computer. The observations

20,001 20,002 20,003

have mean $\bar{x} = 20{,}002$ and standard deviation $s = 1$. Adding a 0 in the center of each number, the next set becomes

200,001 200,002 200,003

The standard deviation remains $s = 1$ as more 0s are added. Use your calculator or computer to calculate the standard deviation of these numbers, adding extra 0s until you get an incorrect answer. How soon did you go wrong? This demonstrates that calculators and computers cannot handle an arbitrary number of digits correctly.

**1.91 Guinea pigs.** Table 1.8 (page 29) gives the survival times of 72 guinea pigs in a medical study. Survival times—whether of cancer patients after treatment or of car batteries in everyday use—are almost always right-skewed. Make a graph to verify that this is true of these survival times. Then give a numerical summary that is appropriate for such data. Explain in simple language, to someone who knows no statistics, what your summary tells us about the guinea pigs.

**1.92 Weight gain.** A study of diet and weight gain deliberately overfed 16 volunteers for eight weeks. The mean increase in fat was $\bar{x} = 2.39$ kilograms and the standard deviation was $s = 1.14$ kilograms. What are $\bar{x}$ and $s$ in pounds? (A kilogram is 2.2 pounds.)

**1.93 Compare three varieties of flowers.** Exercise 1.78 reports data on the lengths in millimeters of flowers of three varieties of *Heliconia.* In Exercise 1.79 you found the mean and standard deviation for each variety. Starting from the $\bar{x}$- and $s$-values in millimeters, find the means and standard deviations in inches. (A millimeter is 1/1000 of a meter. A meter is 39.37 inches.)

**1.94** CHALLENGE **The density of the earth.** Henry Cavendish (see Exercise 1.40, page 28) used $\bar{x}$ to summarize his 29 measurements of the density of the earth.

(a) Find $\bar{x}$ and $s$ for his data.

(b) Cavendish recorded the density of the earth as a multiple of the density of water. The density of water is almost exactly 1 gram per cubic centimeter, so his measurements have these units. In American units, the density of water is 62.43 pounds per cubic foot. This is the weight of a cube of water measuring 1 foot (that is, 30.48 cm) on each side. Express Cavendish's first result for the earth (5.50 g/cm$^3$) in pounds per cubic foot. Then find $\bar{x}$ and $s$ in pounds per cubic foot.

**1.95 Guinea pigs.** Find the **quintiles** (the 20th, 40th, 60th, and 80th percentiles) of the guinea pig survival times in Table 1.8 (page 29). For quite large sets of data, the quintiles or the **deciles** (10th, 20th, 30th, etc. percentiles) give a more detailed summary than the quartiles.

**1.96** CHALLENGE **Changing units from inches to centimeters.** Changing the unit of length from inches to centimeters multiplies each length by 2.54 because there are 2.54 centimeters in an inch. This change of units multiplies our usual measures of spread by 2.54. This is true of *IQR* and the standard deviation. What happens to the variance when we change units in this way?

**1.97 A different type of mean.** The **trimmed mean** is a measure of center that is more resistant than the mean but uses more of the available information than the median. To compute the 10% trimmed mean, discard the highest 10% and the lowest 10% of the observations and compute the mean of the remaining 80%. Trimming eliminates the effect of a small number of outliers. Compute the 10% trimmed mean of the guinea pig survival time data in Table 1.8 (page 29). Then compute the 20% trimmed mean. Compare the values of these measures with the median and the ordinary untrimmed mean.

**1.98** CHALLENGE **Changing units from centimeters to inches.** Refer to Exercise 1.56. Change the measurements from centimeters to inches by multiplying each value by 0.39. Answer the questions from the previous exercise and explain the effect of the transformation on these data.

# 1.3 Density Curves and Normal Distributions

We now have a kit of graphical and numerical tools for describing distributions. What is more, we have a clear strategy for exploring data on a single quantitative variable:

1. Always plot your data: make a graph, usually a stemplot or a histogram.
2. Look for the overall pattern and for striking deviations such as outliers.
3. Calculate an appropriate numerical summary to briefly describe center and spread.

Technology has expanded the set of graphs that we can choose for Step 1. It is possible, though painful, to make histograms by hand. Using software, clever algorithms can describe a distribution in a way that is not feasible by hand, by fitting a smooth curve to the data in addition to or instead of a histogram. The curves used are called **density curves.** Before we examine density curves in detail, here is an example of what software can do.

EXAMPLE

**1.23 Density curves of pH and survival times.** Figure 1.24 illustrates the use of density curves along with histograms to describe distributions.[34] Figure 1.24(a) shows the distribution of the acidity (pH) of rainwater, from Exercise 1.36 (page 27). That exercise illustrates how the choice of classes can change the shape of a histogram. The density curve and the software's default histogram agree that the distribution has a single peak and is approximately symmetric.

Figure 1.24(b) shows a strongly skewed distribution, the survival times of guinea pigs from Table 1.8 (page 29). The histogram and density curve agree on the overall shape and on the "bumps" in the long right tail. The density curve shows a higher peak near the single mode of the distribution. The histogram divides the observations near the mode into two classes, thus reducing the peak.

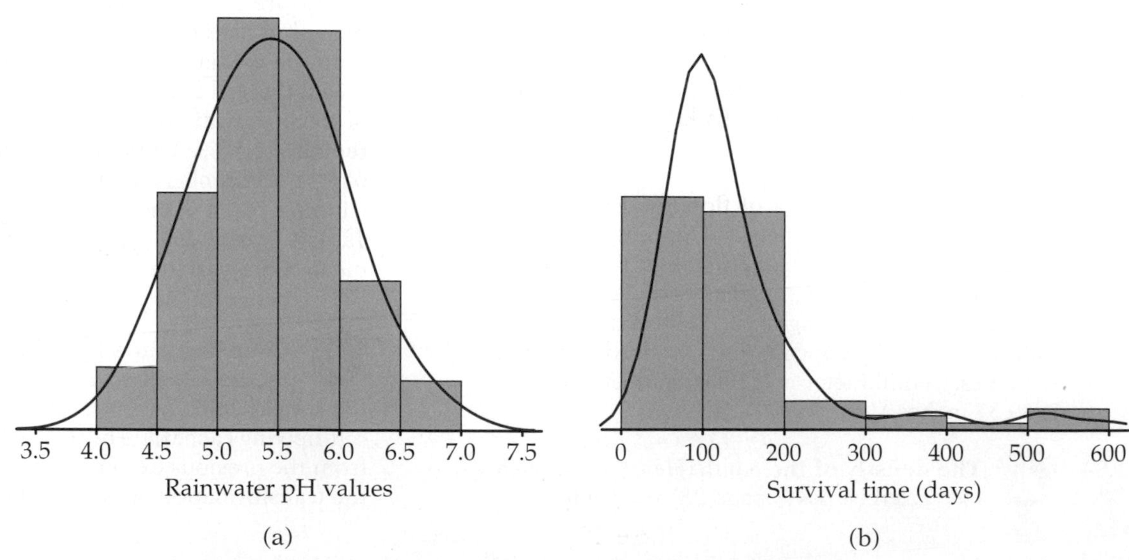


**FIGURE 1.24** (a) The distribution of pH values measuring the acidity of 105 samples of rainwater, for Example 1.23. The roughly symmetric distribution is pictured by both a histogram and a density curve. (b) The distribution of the survival times of 72 guinea pigs in a medical experiment, for Example 1.23. The right-skewed distribution is pictured by both a histogram and a density curve.

In general, software that draws density curves describes the data in a way that is less arbitrary than choosing classes for a histogram. A smooth density curve is, however, an idealization that pictures the overall pattern of the data but ignores minor irregularities as well as any outliers. We will concentrate, not on general density curves, but on a special class, the bell-shaped Normal curves.

## Density curves

One way to think of a density curve is as a smooth approximation to the irregular bars of a histogram. Figure 1.25 shows a histogram of the scores of all 947 seventh-grade students in Gary, Indiana, on the vocabulary part of the Iowa Test of Basic Skills. Scores of many students on this national test have a very regular distribution. The histogram is symmetric, and both tails fall off quite smoothly from a single center peak. There are no large gaps or obvious outliers. The curve drawn through the tops of the histogram bars in Figure 1.25 is a good description of the overall pattern of the data.

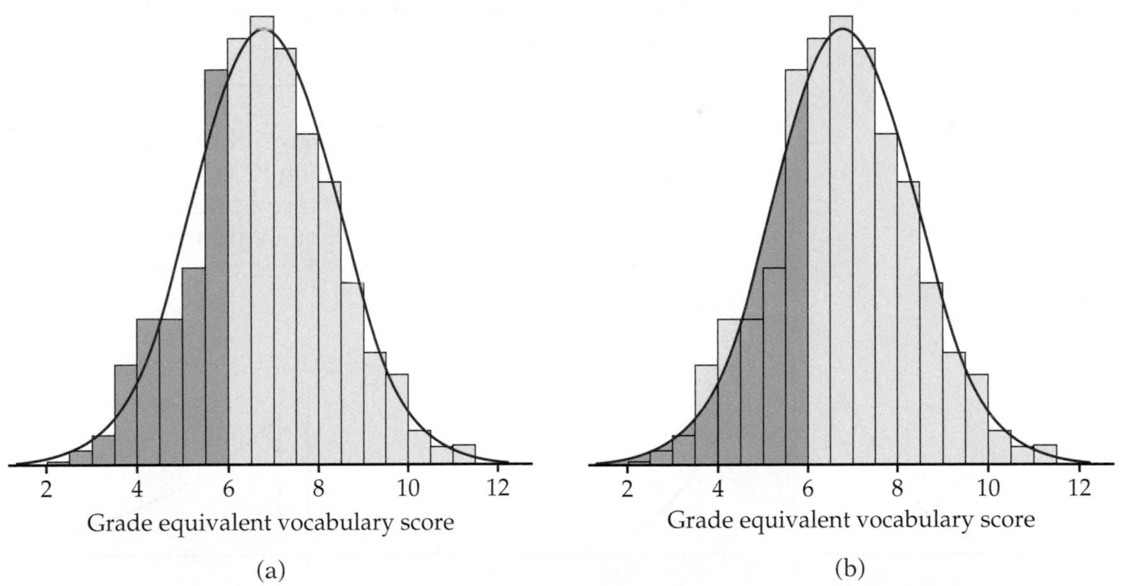


**FIGURE 1.25** (a) The distribution of Iowa Test vocabulary scores for Gary, Indiana, seventh-graders. The shaded bars in the histogram represent scores less than or equal to 6.0. The proportion of such scores in the data is 0.303. (b) The shaded area under the Normal density curve also represents scores less than or equal to 6.0. This area is 0.293, close to the true 0.303 for the actual data.

EXAMPLE

**1.24 Vocabulary scores.** In a histogram, the *areas* of the bars represent either counts or proportions of the observations. In Figure 1.25(a) we have shaded the bars that represent students with vocabulary scores 6.0 or lower. There are 287 such students, who make up the proportion $287/947 = 0.303$ of all Gary seventh-graders. The shaded bars in Figure 1.25(a) make up proportion 0.303 of the total area under all the bars. If we adjust the scale so that the total area of the bars is 1, the area of the shaded bars will be 0.303.

In Figure 1.25(b), we have shaded the *area under the curve* to the left of 6.0. Adjust the scale so that the total area under the curve is exactly 1.

Areas under the curve then represent proportions of the observations. That is, *area = proportion.* The curve is then a density curve. The shaded area under the density curve in Figure 1.25(b) represents the proportion of students with score 6.0 or lower. This area is 0.293, only 0.010 away from the histogram result. You can see that areas under the density curve give quite good approximations of areas given by the histogram.

DENSITY CURVE

A **density curve** is a curve that

- is always on or above the horizontal axis and
- has area exactly 1 underneath it.

A density curve describes the overall pattern of a distribution. The area under the curve and above any range of values is the proportion of all observations that fall in that range.

The density curve in Figure 1.25 is a *Normal curve.* Density curves, like distributions, come in many shapes. Figure 1.26 shows two density curves, a symmetric Normal density curve and a right-skewed curve. A density curve of an appropriate shape is often an adequate description of the overall pattern of a distribution. Outliers, which are deviations from the overall pattern, are not described by the curve.

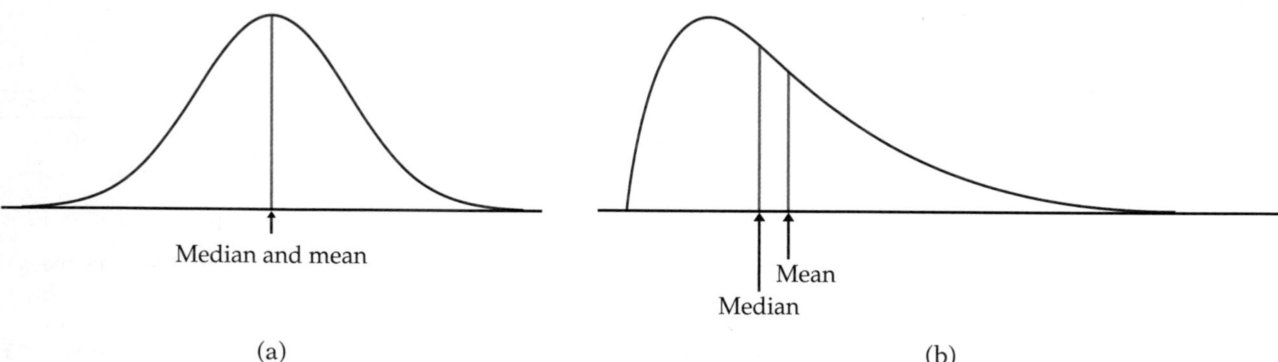


**FIGURE 1.26** (a) A symmetric density curve with its mean and median marked. (b) A right-skewed density curve with its mean and median marked.

## Measuring center and spread for density curves

Our measures of center and spread apply to density curves as well as to actual sets of observations, but only some of these measures are easily seen from the curve. A **mode** of a distribution described by a density curve is a peak point of the curve, the location where the curve is highest. Because areas under a density curve represent proportions of the observations, the **median** is the point with half the total area on each side. You can roughly locate the **quartiles** by dividing the area under the curve into quarters as accurately as possible by eye. The *IQR* is then the distance between the first and third quartiles. There are mathematical ways of calculating areas under curves. These allow us to locate the median and quartiles exactly on any density curve.

What about the mean and standard deviation? The mean of a set of observations is their arithmetic average. If we think of the observations as weights strung out along a thin rod, the mean is the point at which the rod would balance. This fact is also true of density curves. The mean is the point at which the curve would balance if it were made out of solid material. Figure 1.27 illustrates this interpretation of the mean. We have marked the mean and median on the density curves in Figure 1.26. A symmetric curve, such as the Normal curve in Figure 1.26(a), balances at its center of symmetry. Half the area under a symmetric curve lies on either side of its center, so this is also the median. For a right-skewed curve, such as that shown in Figure 1.26(b), the small area in the long right tail tips the curve more than the same area near the center. The mean (the balance point) therefore lies to the right of the median. It is hard to locate the balance point by eye on a skewed curve. There are mathematical ways of calculating the mean for any density curve, so we are able to mark the mean as well as the median in Figure 1.26(b). The standard deviation can also be calculated mathematically, but it can't be located by eye on most density curves.

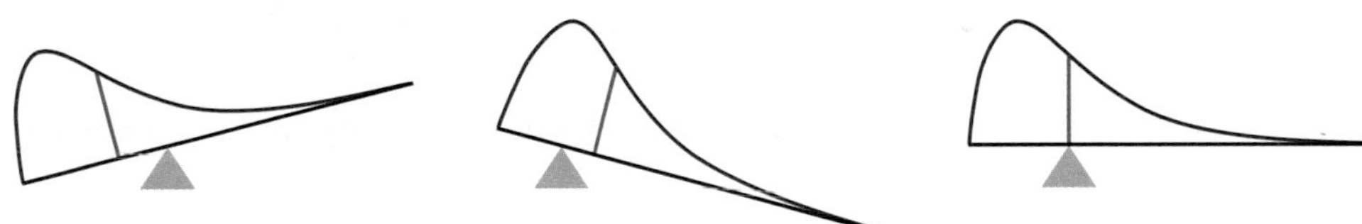

**FIGURE 1.27** The mean of a density curve is the point at which it would balance.

## MEDIAN AND MEAN OF A DENSITY CURVE

The **median** of a density curve is the equal-areas point, the point that divides the area under the curve in half.

The **mean** of a density curve is the balance point, at which the curve would balance if made of solid material.

The median and mean are the same for a symmetric density curve. They both lie at the center of the curve. The mean of a skewed curve is pulled away from the median in the direction of the long tail.

A density curve is an idealized description of a distribution of data. For example, the symmetric density curve in Figure 1.25 is exactly symmetric, but the histogram of vocabulary scores is only approximately symmetric. We therefore need to distinguish between the mean and standard deviation of the density curve and the numbers $\bar{x}$ and $s$ computed from the actual observations. The usual notation for the mean of an idealized distribution is $\mu$ (the Greek letter mu). We write the standard deviation of a density curve as $\sigma$ (the Greek letter sigma).

mean $\mu$

standard deviation $\sigma$

## Normal distributions

Normal curves

One particularly important class of density curves has already appeared in Figures 1.25 and 1.26(a). These density curves are symmetric, unimodal, and bell-shaped. They are called **Normal curves,** and they describe *Normal distributions.* All Normal distributions have the same overall shape. The exact density curve for a particular Normal distribution is specified by giving its mean $\mu$ and its standard deviation $\sigma$. The mean is located at the center of the symmetric curve and is the same as the median. Changing $\mu$ without changing $\sigma$ moves the Normal curve along the horizontal axis without changing its spread. The standard deviation $\sigma$ controls the spread of a Normal curve. Figure 1.28 shows two Normal curves with different values of $\sigma$. The curve with the larger standard deviation is more spread out.

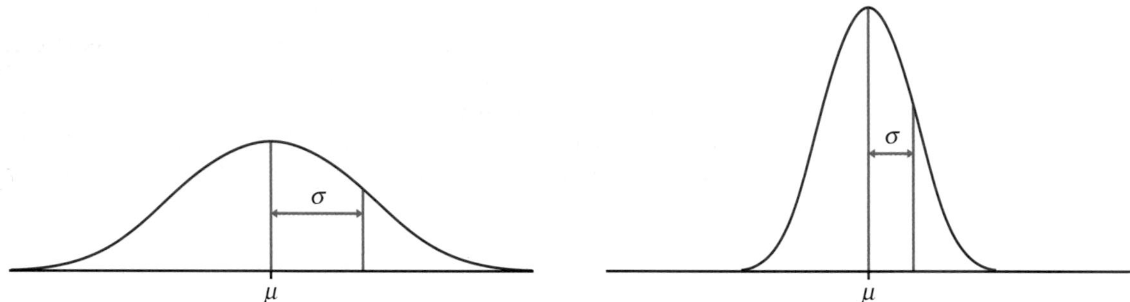


**FIGURE 1.28** Two Normal curves, showing the mean $\mu$ and standard deviation $\sigma$.

The standard deviation $\sigma$ is the natural measure of spread for Normal distributions. Not only do $\mu$ and $\sigma$ completely determine the shape of a Normal curve, but we can locate $\sigma$ by eye on the curve. Here's how. As we move out in either direction from the center $\mu$, the curve changes from falling ever more steeply

to falling ever less steeply

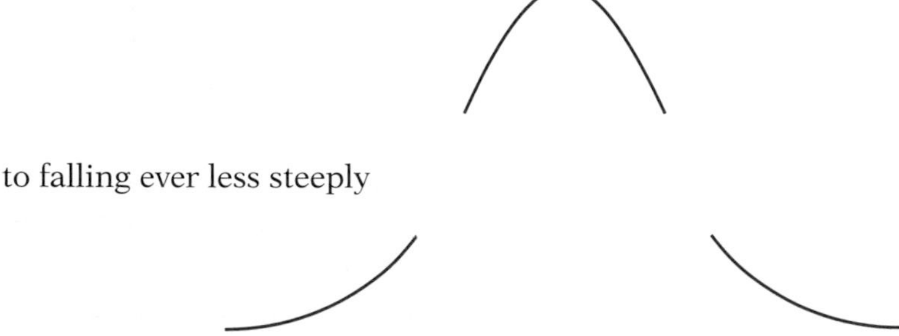

*The points at which this change of curvature takes place are located at distance $\sigma$ on either side of the mean $\mu$.* You can feel the change as you run your finger along a Normal curve, and so find the standard deviation. Remember that $\mu$ and $\sigma$ alone do not specify the shape of most distributions, and that the shape of density curves in general does not reveal $\sigma$. These are special properties of Normal distributions.

There are other symmetric bell-shaped density curves that are not Normal. The Normal density curves are specified by a particular equation. The height of the density curve at any point $x$ is given by

$$\frac{1}{\sigma\sqrt{2\pi}}e^{-\frac{1}{2}\left(\frac{x-\mu}{\sigma}\right)^2}$$

We will not make direct use of this fact, although it is the basis of mathematical work with Normal distributions. Notice that the equation of the curve is completely determined by the mean $\mu$ and the standard deviation $\sigma$.

Why are the Normal distributions important in statistics? Here are three reasons. First, Normal distributions are good descriptions for some distributions of *real data*. Distributions that are often close to Normal include scores on tests taken by many people (such as the Iowa Test of Figure 1.25), repeated careful measurements of the same quantity, and characteristics of biological populations (such as lengths of baby pythons and yields of corn). Second, Normal distributions are good approximations to the results of many kinds of *chance outcomes,* such as tossing a coin many times. Third, and most important, we will see that many *statistical inference* procedures based on Normal distributions work well for other roughly symmetric distributions. HOWEVER . . . *even though many sets of data follow a Normal distribution, many do not.* Most income distributions, for example, are skewed to the right and so are not Normal. Non-Normal data, like non-Normal people, not only are common but are sometimes more interesting than their Normal counterparts.

## The 68–95–99.7 rule

Although there are many Normal curves, they all have common properties. Here is one of the most important.

### THE 68–95–99.7 RULE

In the Normal distribution with mean $\mu$ and standard deviation $\sigma$:

- Approximately **68%** of the observations fall within $\sigma$ of the mean $\mu$.
- Approximately **95%** of the observations fall within $2\sigma$ of $\mu$.
- Approximately **99.7%** of the observations fall within $3\sigma$ of $\mu$.

Figure 1.29 illustrates the 68–95–99.7 rule. By remembering these three numbers, you can think about Normal distributions without constantly making detailed calculations.

EXAMPLE

**1.25 Heights of young women.** The distribution of heights of young women aged 18 to 24 is approximately Normal with mean $\mu = 64.5$ inches and standard deviation $\sigma = 2.5$ inches. Figure 1.30 shows what the 68–95–99.7 rule says about this distribution.

Two standard deviations is 5 inches for this distribution. The 95 part of the 68–95–99.7 rule says that the middle 95% of young women are between $64.5 - 5$ and $64.5 + 5$ inches tall, that is, between 59.5 inches and 69.5 inches. This fact is exactly true for an exactly Normal distribution. It is approximately

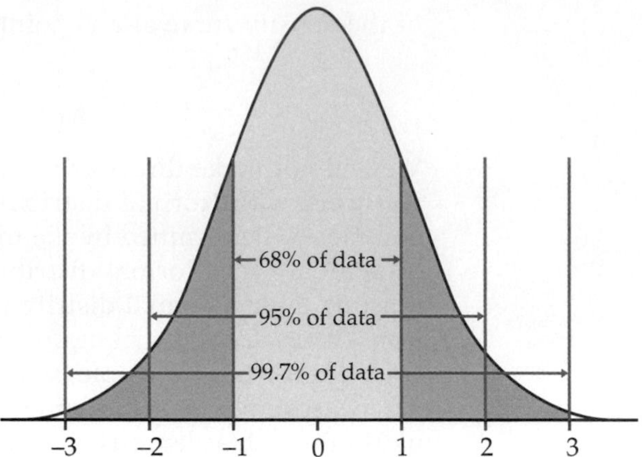


**FIGURE 1.29** The 68–95–99.7 rule for Normal distributions.

true for the heights of young women because the distribution of heights is approximately Normal.

The other 5% of young women have heights outside the range from 59.5 to 69.5 inches. Because the Normal distributions are symmetric, half of these women are on the tall side. So the tallest 2.5% of young women are taller than 69.5 inches.

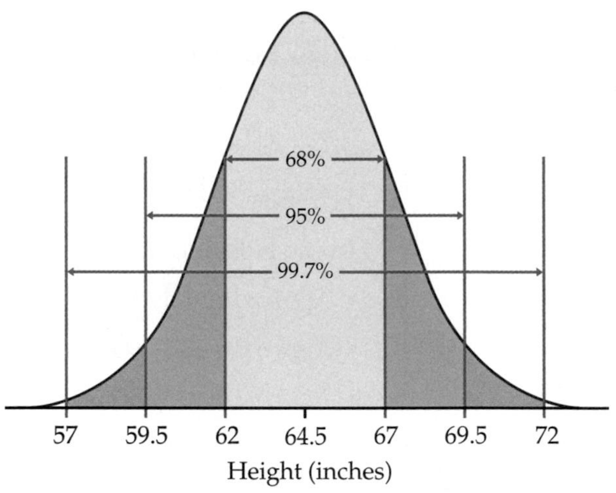


**FIGURE 1.30** The 68–95–99.7 rule applied to the heights of young women, for Example 1.25.

$N(\mu, \sigma)$

Because we will mention Normal distributions often, a short notation is helpful. We abbreviate the Normal distribution with mean $\mu$ and standard deviation $\sigma$ as $\boldsymbol{N(\mu, \sigma)}$. For example, the distribution of young women's heights is $N(64.5, 2.5)$.

## USE YOUR KNOWLEDGE

**1.99** **Test scores.** Many states have programs for assessing the skills of students in various grades. The Indiana Statewide Testing for Educational Progress (ISTEP) is one such program.[35] In a recent year 76,531 tenth-grade Indiana students took the English/language arts exam. The mean score was 572 and the standard deviation was 51.

Assuming that these scores are approximately Normally distributed, $N(572, 51)$, use the 68–95–99.7 rule to give a range of scores that includes 95% of these students.

**1.100** **Use the 68–95–99.7 rule.** Refer to the previous exercise. Use the 68–95–99.7 rule to give a range of scores that includes 99.7% of these students.

## Standardizing observations

As the 68–95–99.7 rule suggests, all Normal distributions share many properties. In fact, all Normal distributions are the same if we measure in units of size $\sigma$ about the mean $\mu$ as center. Changing to these units is called *standardizing*. To standardize a value, subtract the mean of the distribution and then divide by the standard deviation.

### STANDARDIZING AND $z$-SCORES

If $x$ is an observation from a distribution that has mean $\mu$ and standard deviation $\sigma$, the **standardized value** of $x$ is

$$z = \frac{x - \mu}{\sigma}$$

A standardized value is often called a **$z$-score**.

A $z$-score tells us how many standard deviations the original observation falls away from the mean, and in which direction. Observations larger than the mean are positive when standardized, and observations smaller than the mean are negative.

EXAMPLE

**1.26 Find some $z$-scores.** The heights of young women are approximately Normal with $\mu = 64.5$ inches and $\sigma = 2.5$ inches. The $z$-score for height is

$$z = \frac{\text{height} - 64.5}{2.5}$$

A woman's standardized height is the number of standard deviations by which her height differs from the mean height of all young women. A woman 68 inches tall, for example, has $z$-score

$$z = \frac{68 - 64.5}{2.5} = 1.4$$

or 1.4 standard deviations above the mean. Similarly, a woman 5 feet (60 inches) tall has $z$-score

$$z = \frac{60 - 64.5}{2.5} = -1.8$$

or 1.8 standard deviations less than the mean height.

## USE YOUR KNOWLEDGE

**1.101 Find the $z$-score.** Consider the ISTEP scores (see Exercise 1.99), which we can assume are approximately Normal, $N(572, 51)$. Give the $z$-score for a student who received a score of 600.

**1.102 Find another $z$-score.** Consider the ISTEP scores, which we can assume are approximately Normal, $N(572, 51)$. Give the $z$-score for a student who received a score of 500. Explain why your answer is negative even though all of the test scores are positive.

We need a way to write variables, such as "height" in Example 1.25, that follow a theoretical distribution such as a Normal distribution. We use capital letters near the end of the alphabet for such variables. If $X$ is the height of a young woman, we can then shorten "the height of a young woman is less than 68 inches" to "$X < 68$." We will use lowercase $x$ to stand for any specific value of the variable $X$.

We often standardize observations from symmetric distributions to express them in a common scale. We might, for example, compare the heights of two children of different ages by calculating their $z$-scores. The standardized heights tell us where each child stands in the distribution for his or her age group.

Standardizing is a linear transformation that transforms the data into the standard scale of $z$-scores. We know that a linear transformation does not change the shape of a distribution, and that the mean and standard deviation change in a simple manner. In particular, *the standardized values for any distribution always have mean 0 and standard deviation 1.*

If the variable we standardize has a Normal distribution, standardizing does more than give a common scale. It makes all Normal distributions into a single distribution, and this distribution is still Normal. Standardizing a variable that has any Normal distribution produces a new variable that has the *standard Normal distribution.*

### THE STANDARD NORMAL DISTRIBUTION

The **standard Normal distribution** is the Normal distribution $N(0, 1)$ with mean 0 and standard deviation 1.

If a variable $X$ has any Normal distribution $N(\mu, \sigma)$ with mean $\mu$ and standard deviation $\sigma$, then the standardized variable

$$Z = \frac{X - \mu}{\sigma}$$

has the standard Normal distribution.

## Normal distribution calculations

Areas under a Normal curve represent proportions of observations from that Normal distribution. There is no formula for areas under a Normal curve. Calculations use either software that calculates areas or a table of areas. The table cumulative proportion

and most software calculate one kind of area: **cumulative proportions.** A cumulative proportion is the proportion of observations in a distribution that lie at or below a given value. When the distribution is given by a density curve, the cumulative proportion is the area under the curve to the left of a given value. Figure 1.31 shows the idea more clearly than words do.

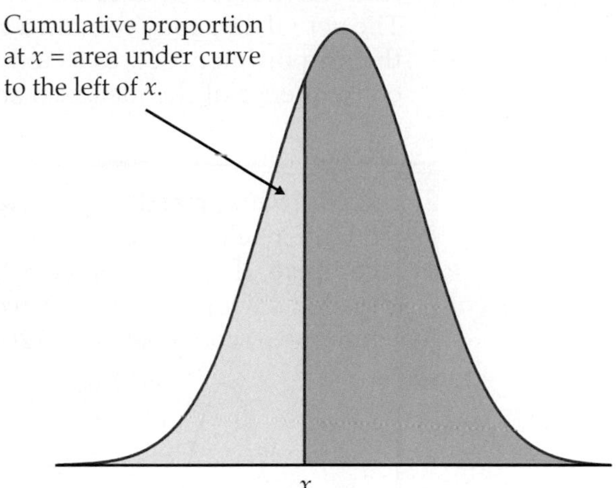


**FIGURE 1.31** The *cumulative proportion* for a value $x$ is the proportion of all observations from the distribution that are less than or equal to $x$. This is the area to the left of $x$ under the Normal curve.

The key to calculating Normal proportions is to match the area you want with areas that represent cumulative proportions. Then get areas for cumulative proportions either from software or (with an extra step) from a table. The following examples show the method in pictures.

EXAMPLE

**1.27 The NCAA standard for SAT scores.** The National Collegiate Athletic Association (NCAA) requires Division I athletes to get a combined score of at least 820 on the SAT Mathematics and Verbal tests to compete in their first college year. (Higher scores are required for students with poor high school grades.) The scores of the 1.4 million students in the class of 2003 who took the SATs were approximately Normal with mean 1026 and standard deviation 209. What proportion of all students had SAT scores of at least 820?

Here is the calculation in pictures: the proportion of scores above 820 is the area under the curve to the right of 820. That's the total area under the curve (which is always 1) minus the cumulative proportion up to 820.

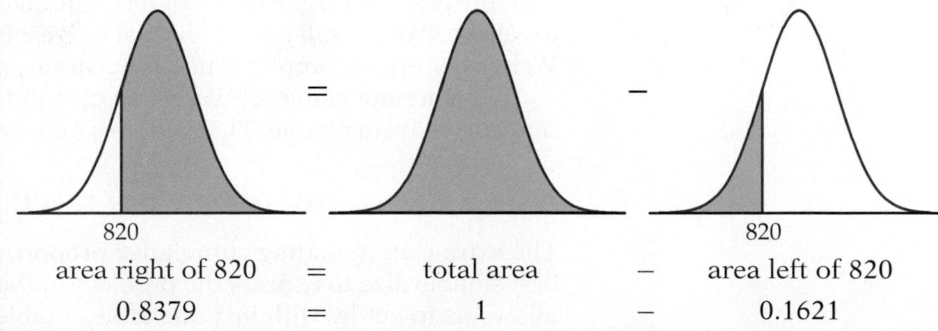

That is, the proportion of all SAT takers who would be NCAA qualifiers is 0.8379, or about 84%.

There is *no* area under a smooth curve and exactly over the point 820. Consequently, the area to the right of 820 (the proportion of scores > 820) is the same as the area at or to the right of this point (the proportion of scores ≥ 820). The actual data may contain a student who scored exactly 820 on the SAT. That the proportion of scores exactly equal to 820 is 0 for a Normal distribution is a consequence of the idealized smoothing of Normal distributions for data.

EXAMPLE

**1.28 NCAA partial qualifiers.** The NCAA considers a student a "partial qualifier" eligible to practice and receive an athletic scholarship, but not to compete, if the combined SAT score is at least 720. What proportion of all students who take the SAT would be partial qualifiers? That is, what proportion have scores between 720 and 820? Here are the pictures:

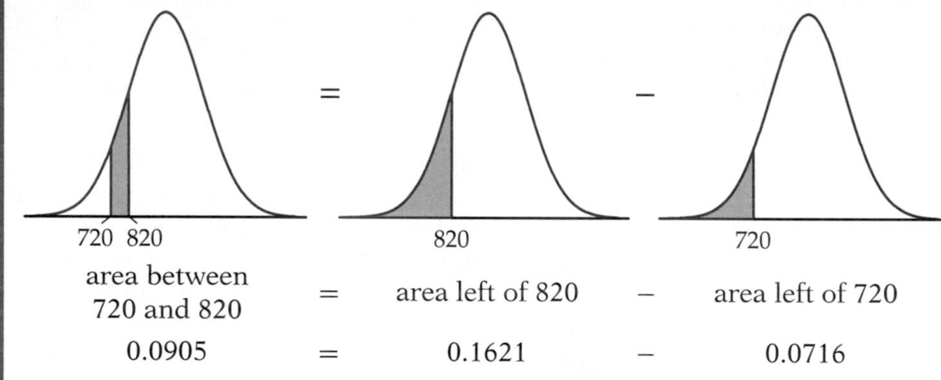


About 9% of all students who take the SAT have scores between 720 and 820.

How do we find the numerical values of the areas in Examples 1.27 and 1.28? If you use software, just plug in mean 1026 and standard deviation 209. Then ask for the cumulative proportions for 820 and for 720. (Your software will probably refer to these as "cumulative probabilities." We will learn in Chapter 4 why the language of probability fits.) If you make a sketch of the area you want, you will never go wrong.

You can use the *Normal Curve* applet on the text CD and Web site to find Normal proportions. The applet is more flexible than most software—it will find any Normal proportion, not just cumulative proportions. The applet is an excellent way to understand Normal curves. But, because of the limitations of Web browsers, the applet is not as accurate as statistical software.

If you are not using software, you can find cumulative proportions for Normal curves from a table. That requires an extra step, as we now explain.

## Using the standard Normal table

The extra step in finding cumulative proportions from a table is that we must first standardize to express the problem in the standard scale of $z$-scores. This allows us to get by with just one table, a table of *standard Normal cumulative proportions.* Table A in the back of the book gives cumulative proportions for the standard Normal distribution. Table A also appears on the inside front cover. The pictures at the top of the table remind us that the entries are cumulative proportions, areas under the curve to the left of a value $z$.

EXAMPLE

**1.29 Find the proportion from z.** What proportion of observations on a standard Normal variable $Z$ take values less than 1.47?

*Solution:* To find the area to the left of 1.47, locate 1.4 in the left-hand column of Table A, then locate the remaining digit 7 as .07 in the top row. The entry opposite 1.4 and under .07 is 0.9292. This is the cumulative proportion we seek. Figure 1.32 illustrates this area.

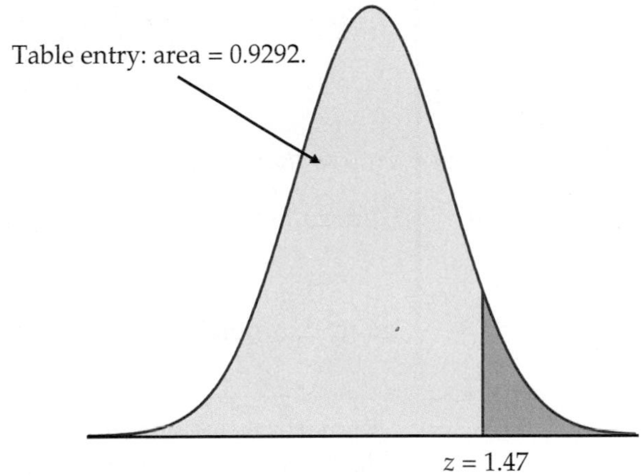


**FIGURE 1.32** The area under a standard Normal curve to the left of the point $z = 1.47$ is 0.9292, for Example 1.29. Table A gives areas under the standard Normal curve.

Now that you see how Table A works, let's redo the NCAA Examples 1.27 and 1.28 using the table.

EXAMPLE

**1.30 Find the proportion from x.** What proportion of all students who take the SAT have scores of at least 820? The picture that leads to the answer is exactly the same as in Example 1.27. The extra step is that we first standardize in order to read cumulative proportions from Table A. If $X$ is SAT score, we want the proportion of students for which $X \geq 820$.

1. *Standardize.* Subtract the mean, then divide by the standard deviation, to transform the problem about $X$ into a problem about a standard Normal $Z$:

$$X \geq 820$$

$$\frac{X - 1026}{209} \geq \frac{820 - 1026}{209}$$

$$Z \geq -0.99$$

2. *Use the table.* Look at the pictures in Example 1.27. From Table A, we see that the proportion of observations less than $-0.99$ is 0.1611. The area to the right of $-0.99$ is therefore $1 - 0.1611 = 0.8389$. This is about 84%.

The area from the table in Example 1.30 (0.8389) is slightly less accurate than the area from software in Example 1.27 (0.8379) because we must round $z$ to two places when we use Table A. The difference is rarely important in practice.

EXAMPLE

**1.31 Proportion of partial qualifiers.** What proportion of all students who take the SAT would be partial qualifiers in the eyes of the NCAA? That is, what proportion of students have SAT scores between 720 and 820? First, sketch the areas, exactly as in Example 1.28. We again use $X$ as shorthand for an SAT score.

1. *Standardize.*

$$\begin{aligned} 720 \leq \quad & X < 820 \\ \frac{720 - 1026}{209} \leq \quad & \frac{X - 1026}{209} < \frac{820 - 1026}{209} \\ -1.46 \leq \quad & Z < -0.99 \end{aligned}$$

2. *Use the table.*

$$\begin{aligned} \text{area between } -1.46 \text{ and } -0.99 &= (\text{area left of } -0.99) - (\text{area left of } -1.46) \\ &= 0.1611 - 0.0721 = 0.0890 \end{aligned}$$

As in Example 1.28, about 9% of students would be partial qualifiers.

Sometimes we encounter a value of $z$ more extreme than those appearing in Table A. For example, the area to the left of $z = -4$ is not given directly in the table. The $z$-values in Table A leave only area 0.0002 in each tail unaccounted for. For practical purposes, we can act as if there is zero area outside the range of Table A.

## USE YOUR KNOWLEDGE

**1.103 Find the proportion.** Consider the ISTEP scores, which are approximately Normal, $N(572, 51)$. Find the proportion of students who have scores less than 600. Find the proportion of students who have scores greater than or equal to 600. Sketch the relationship between these two calculations using pictures of Normal curves similar to the ones given in Example 1.27.

**1.104 Find another proportion.** Consider the ISTEP scores, which are approximately Normal, $N(572, 51)$. Find the proportion of students who have scores between 600 and 650. Use pictures of Normal curves similar to the ones given in Example 1.28 to illustrate your calculations.

## Inverse Normal calculations

Examples 1.25 to 1.29 illustrate the use of Normal distributions to find the proportion of observations in a given event, such as "SAT score between 720 and

820." We may instead want to find the observed value corresponding to a given proportion.

Statistical software will do this directly. Without software, use Table A backward, finding the desired proportion in the body of the table and then reading the corresponding $z$ from the left column and top row.

EXAMPLE

**1.32 How high for the top 10%?** Scores on the SAT Verbal test in recent years follow approximately the $N(505, 110)$ distribution. How high must a student score in order to place in the top 10% of all students taking the SAT?

Again, the key to the problem is to draw a picture. Figure 1.33 shows that we want the score $x$ with area above it 0.10. That's the same as area below $x$ equal to 0.90.

Statistical software has a function that will give you the $x$ for any cumulative proportion you specify. The function often has a name such as "inverse cumulative probability." Plug in mean 505, standard deviation 110, and cumulative proportion 0.9. The software tells you that $x = 645.97$. We see that a student must score at least 646 to place in the highest 10%.

Without software, first find the standard score $z$ with cumulative proportion 0.9, then "unstandardize" to find $x$. Here is the two-step process:

1. *Use the table.* Look in the body of Table A for the entry closest to 0.9. It is 0.8997. This is the entry corresponding to $z = 1.28$. So $z = 1.28$ is the standardized value with area 0.9 to its left.

2. *Unstandardize* to transform the solution from $z$ back to the original $x$ scale. We know that the standardized value of the unknown $x$ is $z = 1.28$. So $x$ itself satisfies

$$\frac{x - 505}{110} = 1.28$$

Solving this equation for $x$ gives

$$x = 505 + (1.28)(110) = 645.8$$

This equation should make sense: it finds the $x$ that lies 1.28 standard deviations above the mean on this particular Normal curve. That is the

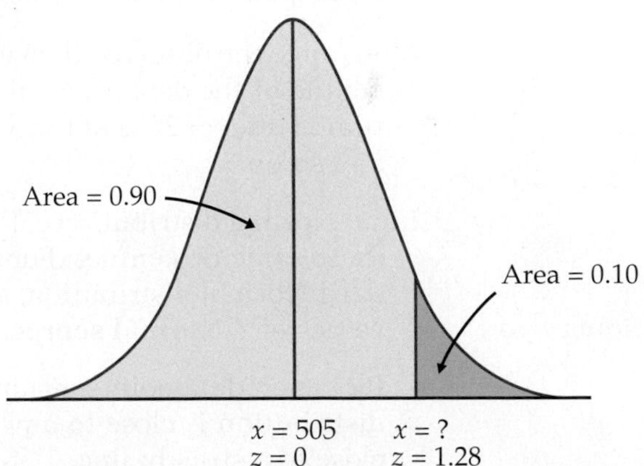


**FIGURE 1.33** Locating the point on a Normal curve with area 0.10 to its right, for Example 1.32. The result is $x = 646$, or $z = 1.28$ in the standard scale.

"unstandardized" meaning of $z = 1.28$. The general rule for unstandardizing a $z$-score is

$$x = \mu + z\sigma$$

## USE YOUR KNOWLEDGE

**1.105 What score is needed to be in the top 5%?** Consider the ISTEP scores, which are approximately Normal, $N(572, 51)$. How high a score is needed to be in the top 5% of students who take this exam?

**1.106 Find the score that 60% of students will exceed.** Consider the ISTEP scores, which are approximately Normal, $N(572, 51)$. Sixty percent of the students will score above $x$ on this exam. Find $x$.

## Normal quantile plots

The Normal distributions provide good descriptions of some distributions of real data, such as the Gary vocabulary scores. The distributions of some other common variables are usually skewed and therefore distinctly non-Normal. Examples include economic variables such as personal income and gross sales of business firms, the survival times of cancer patients after treatment, and the service lifetime of mechanical or electronic components. While experience can suggest whether or not a Normal distribution is plausible in a particular case, it is risky to assume that a distribution is Normal without actually inspecting the data.

A histogram or stemplot can reveal distinctly non-Normal features of a distribution, such as outliers (the breaking strengths in Figure 1.9, page 17), pronounced skewness (the survival times in Figure 1.24(b), page 54), or gaps and clusters (the Massachusetts college tuitions in Figure 1.16, page 25). If the stemplot or histogram appears roughly symmetric and unimodal, however, we need a more sensitive way to judge the adequacy of a Normal model. The most useful tool for assessing Normality is another graph, the **Normal quantile plot.**

Normal quantile plot

Here is the basic idea of a Normal quantile plot. The graphs produced by software use more sophisticated versions of this idea. It is not practical to make Normal quantile plots by hand.

1. Arrange the observed data values from smallest to largest. Record what percentile of the data each value occupies. For example, the smallest observation in a set of 20 is at the 5% point, the second smallest is at the 10% point, and so on.

2. Do Normal distribution calculations to find the values of $z$ corresponding to these same percentiles. For example, $z = -1.645$ is the 5% point of the standard Normal distribution, and $z = -1.282$ is the 10% point. We call these values of $Z$ **Normal scores.**

Normal scores

3. Plot each data point $x$ against the corresponding Normal score. If the data distribution is close to any Normal distribution, the plotted points will lie close to a straight line.

Any Normal distribution produces a straight line on the plot because standardizing turns any Normal distribution into a standard Normal distribution. Standardizing is a linear transformation that can change the slope and intercept of the line in our plot but cannot turn a line into a curved pattern.

### USE OF NORMAL QUANTILE PLOTS

If the points on a Normal quantile plot lie close to a straight line, the plot indicates that the data are Normal. Systematic deviations from a straight line indicate a non-Normal distribution. Outliers appear as points that are far away from the overall pattern of the plot.

Figures 1.34 to 1.36 are Normal quantile plots for data we have met earlier. The data $x$ are plotted vertically against the corresponding standard Normal $z$-score plotted horizontally. The $z$-score scale extends from $-3$ to 3 because almost all of a standard Normal curve lies between these values. These figures show how Normal quantile plots behave.

EXAMPLE

**1.33 Breaking strengths are Normal.** Figure 1.34 is a Normal quantile plot of the breaking strengths in Example 1.11 (page 17). Lay a transparent straightedge over the center of the plot to see that most of the points lie close to a straight line. A Normal distribution describes these points quite well. The only substantial deviations are short horizontal runs of points. Each run represents repeated observations having the same value—there are five measurements at 1150, for example. This phenomenon is called **granularity.** It is caused by the limited precision of the measurements and does not represent an important deviation from Normality.

granularity

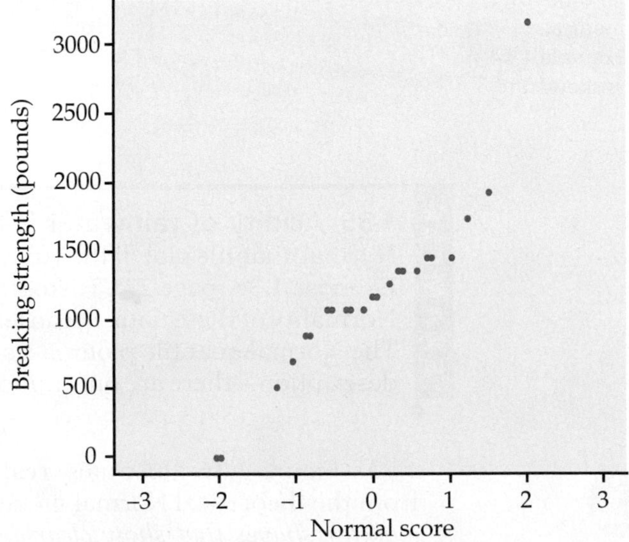


**FIGURE 1.34** Normal quantile plot of the breaking strengths of wires bonded to a semiconductor wafer, for Example 1.33. This distribution has a Normal shape except for outliers in both tails.

The high outlier at 3150 pounds lies above the line formed by the center of the data—it is farther out in the high direction than we expect Normal data to be.

The two low outliers at 0 lie below the line—they are suspiciously far out in the low direction. Compare Figure 1.34 with the histogram of these data in Figure 1.9 (page 17).

EXAMPLE

**1.34 Survival times are not Normal.** Figure 1.35 is a Normal quantile plot of the guinea pig survival times from Table 1.8 (page 29). Figure 1.24(b) (page 54) shows that this distribution is strongly skewed to the right.

To see the right-skewness in the Normal quantile plot, draw a line through the leftmost points, which correspond to the smaller observations. The larger observations fall systematically above this line. That is, the right-of-center observations have larger values than in a Normal distribution. *In a right-skewed distribution, the largest observations fall distinctly above a line drawn through the main body of points.* Similarly, left-skewness is evident when the smallest observations fall below the line. Unlike Figure 1.34, there are no individual outliers.

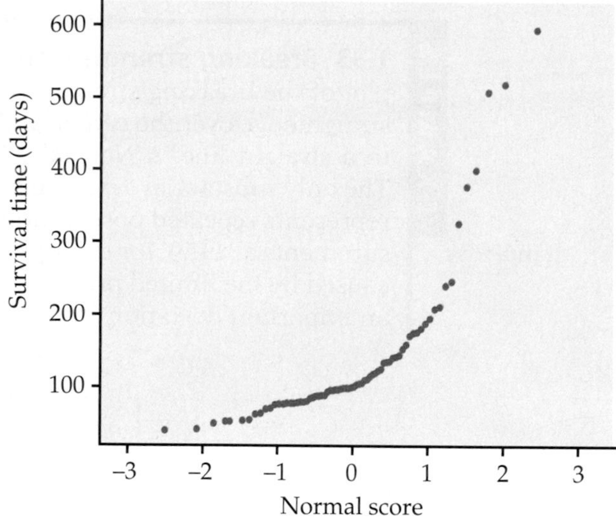


**FIGURE 1.35** Normal quantile plot of the survival times of guinea pigs in a medical experiment, for Example 1.34. This distribution is skewed to the right.

EXAMPLE

**1.35 Acidity of rainwater is approximately Normal.** Figure 1.36 is a Normal quantile plot of the 105 acidity (pH) measurements of rainwater from Exercise 1.36 (page 27). Histograms don't settle the question of approximate Normality of these data, because their shape depends on the choice of classes. The Normal quantile plot makes it clear that a Normal distribution is a good description—there are only minor wiggles in a generally straight-line pattern.

As Figure 1.36 illustrates, real data almost always show some departure from the theoretical Normal model. *When you examine a Normal quantile plot, look for shapes that show clear departures from Normality. Don't overreact to*

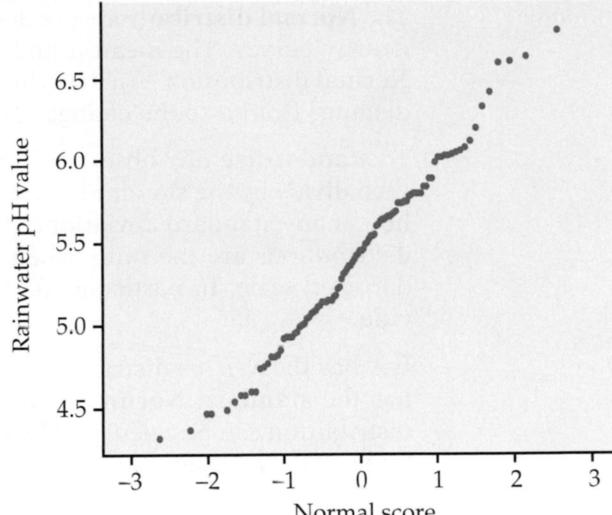


**FIGURE 1.36** Normal quantile plot of the acidity (pH) values of 105 samples of rainwater, for Example 1.35. This distribution is approximately Normal.

*minor wiggles in the plot.* When we discuss statistical methods that are based on the Normal model, we will pay attention to the sensitivity of each method to departures from Normality. Many common methods work well as long as the data are approximately Normal and outliers are not present.

## BEYOND THE BASICS

### Density Estimation

A density curve gives a compact summary of the overall shape of a distribution. Many distributions do not have the Normal shape. There are other families of density curves that are used as mathematical models for various distribution shapes. Modern software offers a more flexible option, illustrated by the two graphs in Figure 1.24 (page 54). A **density estimator** does not start with any specific shape, such as the Normal shape. It looks at the data and draws a density curve that describes the overall shape of the data. Density estimators join stemplots and histograms as useful graphical tools for exploratory data analysis.

density estimator

## SECTION 1.3 Summary

The overall pattern of a distribution can often be described compactly by a **density curve.** A density curve has total area 1 underneath it. Areas under a density curve give proportions of observations for the distribution.

The **mean** $\mu$ (balance point), the **median** (equal-areas point), and the **quartiles** can be approximately located by eye on a density curve. The **standard deviation** $\sigma$ cannot be located by eye on most density curves. The mean and median are equal for symmetric density curves, but the mean of a skewed curve is located farther toward the long tail than is the median.

The **Normal distributions** are described by bell-shaped, symmetric, unimodal density curves. The mean $\mu$ and standard deviation $\sigma$ completely specify the Normal distribution $N(\mu, \sigma)$. The mean is the center of symmetry, and $\sigma$ is the distance from $\mu$ to the change-of-curvature points on either side.

To **standardize** any observation $x$, subtract the mean of the distribution and then divide by the standard deviation. The resulting **$z$-score** $z = (x - \mu)/\sigma$ says how many standard deviations $x$ lies from the distribution mean. All Normal distributions are the same when measurements are transformed to the standardized scale. In particular, all Normal distributions satisfy the **68–95–99.7 rule.**

If $X$ has the $N(\mu, \sigma)$ distribution, then the standardized variable $Z = (X - \mu)/\sigma$ has the **standard Normal distribution** $N(0, 1)$. Proportions for any Normal distribution can be calculated by software or from the **standard Normal table** (Table A), which gives the **cumulative proportions** of $Z < z$ for many values of $z$.

The adequacy of a Normal model for describing a distribution of data is best assessed by a **Normal quantile plot,** which is available in most statistical software packages. A pattern on such a plot that deviates substantially from a straight line indicates that the data are not Normal.

## SECTION 1.3 Exercises

*For Exercises 1.99 and 1.100, see pages 60 and 61; for Exercises 1.101 and 1.102, see page 62; for Exercises 1.103 and 1.104, see page 66; and for Exercises 1.105 and 1.106, see page 68.*

**1.107 Sketch some density curves.** Sketch density curves that might describe distributions with the following shapes:

(a) Symmetric, but with two peaks (that is, two strong clusters of observations).

(b) Single peak and skewed to the right.

**1.108 A uniform distribution.** If you ask a computer to generate "random numbers" between 0 and 1, you will get observations from a **uniform distribution.** Figure 1.37 graphs the density curve for a uniform distribution. Use areas under this density curve to answer the following questions.

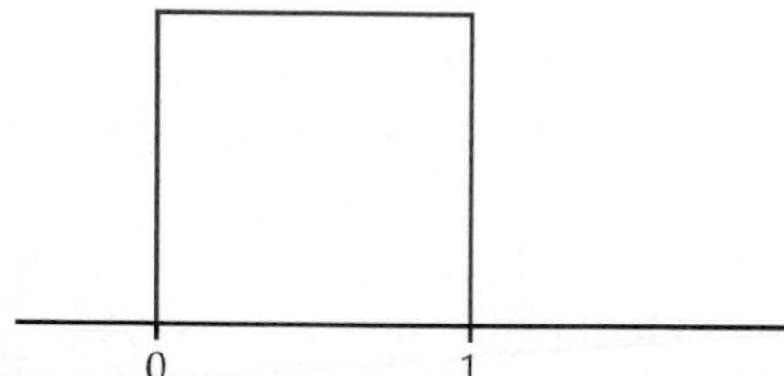


**FIGURE 1.37** The density curve of a uniform distribution, for Exercise 1.108.

(a) Why is the total area under this curve equal to 1?

(b) What proportion of the observations lie below 0.35?

(c) What proportion of the observations lie between 0.35 and 0.65?

**1.109 Use a different range for the uniform distribution.** Many random number generators allow users to specify the range of the random numbers to be produced. Suppose that you specify that the outcomes are to be distributed uniformly between 0 and 4. Then the density curve of the outcomes has constant height between 0 and 4, and height 0 elsewhere.

(a) What is the height of the density curve between 0 and 4? Draw a graph of the density curve.

(b) Use your graph from (a) and the fact that areas under the curve are proportions of outcomes to find the proportion of outcomes that are less than 1.

(c) Find the proportion of outcomes that lie between 0.5 and 2.5.

**1.110 Find the mean, the median, and the quartiles.** What are the mean and the median of the uniform distribution in Figure 1.37? What are the quartiles?

**1.111 Three density curves.** Figure 1.38 displays three density curves, each with three points marked on

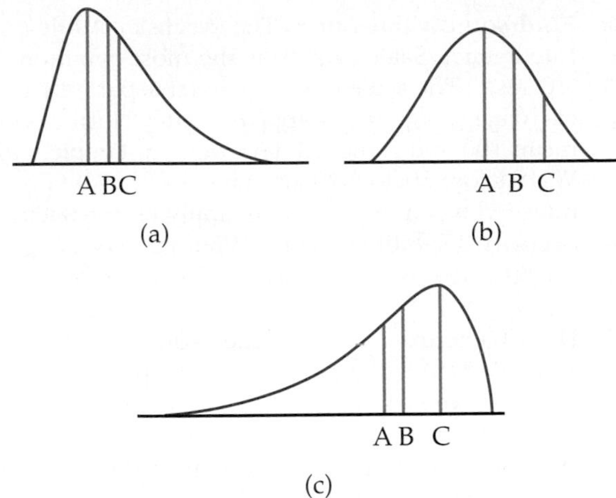


**FIGURE 1.38** Three density curves, for Exercise 1.111.

it. At which of these points on each curve do the mean and the median fall?

**1.112 Length of pregnancies.** The length of human pregnancies from conception to birth varies according to a distribution that is approximately Normal with mean 266 days and standard deviation 16 days. Draw a density curve for this distribution on which the mean and standard deviation are correctly located.

**1.113** APPLET **Use the Normal Curve applet.** The 68–95–99.7 rule for Normal distributions is a useful approximation. You can use the *Normal Curve* applet on the text CD and Web site to see how accurate the rule is. Drag one flag across the other so that the applet shows the area under the curve between the two flags.

(a) Place the flags one standard deviation on either side of the mean. What is the area between these two values? What does the 68–95–99.7 rule say this area is?

(b) Repeat for locations two and three standard deviations on either side of the mean. Again compare the 68–95–99.7 rule with the area given by the applet.

**1.114 Pregnancies and the 68–95–99.7 rule.** The length of human pregnancies from conception to birth varies according to a distribution that is approximately Normal with mean 266 days and standard deviation 16 days. Use the 68–95–99.7 rule to answer the following questions.

(a) Between what values do the lengths of the middle 95% of all pregnancies fall?

(b) How short are the shortest 2.5% of all pregnancies? How long do the longest 2.5% last?

**1.115 Horse pregnancies are longer.** Bigger animals tend to carry their young longer before birth. The length of horse pregnancies from conception to birth varies according to a roughly Normal distribution with mean 336 days and standard deviation 3 days. Use the 68–95–99.7 rule to answer the following questions.

(a) Almost all (99.7%) horse pregnancies fall in what range of lengths?

(b) What percent of horse pregnancies are longer than 339 days?

**1.116 Binge drinking survey.** One reason that Normal distributions are important is that they describe how the results of an opinion poll would vary if the poll were repeated many times. About 20% of college students say they are frequent binge drinkers. Think about taking many randomly chosen samples of 1600 students. The proportions of college students in these samples who say they are frequent binge drinkers will follow the Normal distribution with mean 0.20 and standard deviation 0.01. Use this fact and the 68–95–99.7 rule to answer these questions.

(a) In many samples, what percent of samples give results above 0.2? Above 0.22?

(b) In a large number of samples, what range contains the central 95% of proportions of students who say they are frequent binge drinkers?

**1.117 Heights of women.** The heights of women aged 20 to 29 are approximately Normal with mean 64 inches and standard deviation 2.7 inches. Men the same age have mean height 69.3 inches with standard deviation 2.8 inches. What are the $z$-scores for a woman 6 feet tall and a man 6 feet tall? What information do the $z$-scores give that the actual heights do not?

**1.118** APPLET **Use the Normal Curve applet.** Use the *Normal Curve* applet for the standard Normal distribution to say how many standard deviations above and below the mean the quartiles of any Normal distribution lie.

**1.119 Acidity of rainwater.** The Normal quantile plot in Figure 1.36 (page 71) shows that the acidity (pH) measurements for rainwater samples in Exercise 1.36 are approximately Normal. How well do these scores satisfy the 68–95–99.7 rule?

To find out, calculate the mean $\bar{x}$ and standard deviation $s$ of the observations. Then calculate the percent of the 105 measurements that fall between $\bar{x} - s$ and $\bar{x} + s$ and compare your result with 68%. Do the same for the intervals covering two and three standard deviations on either side of the mean. (The 68–95–99.7 rule is exact for any theoretical Normal distribution. It will hold only approximately for actual data.)

**1.120 Find some proportions.** Using either Table A or your calculator or software, find the proportion of observations from a standard Normal distribution that satisfies each of the following statements. In each case, sketch a standard Normal curve and shade the area under the curve that is the answer to the question.

(a) $Z < 1.65$

(b) $Z > 1.65$

(c) $Z > -0.76$

(d) $-0.76 < Z < 1.65$

**1.121 Find more proportions.** Using either Table A or your calculator or software, find the proportion of observations from a standard Normal distribution for each of the following events. In each case, sketch a standard Normal curve and shade the area representing the proportion.

(a) $Z \leq -1.9$

(b) $Z \geq -1.9$

(c) $Z > 1.55$

(d) $-1.9 < Z < 1.55$

**1.122 Find some values of $z$.** Find the value $z$ of a standard Normal variable $Z$ that satisfies each of the following conditions. (If you use Table A, report the value of $z$ that comes closest to satisfying the condition.) In each case, sketch a standard Normal curve with your value of $z$ marked on the axis.

(a) 25% of the observations fall below $z$.

(b) 35% of the observations fall above $z$.

**1.123 Find more values of $z$.** The variable $Z$ has a standard Normal distribution.

(a) Find the number $z$ that has cumulative proportion 0.85.

(b) Find the number $z$ such that the event $Z > z$ has proportion 0.40.

**1.124 Find some values of $z$.** The Wechsler Adult Intelligence Scale (WAIS) is the most common "IQ test." The scale of scores is set separately for each age group and is approximately Normal with mean 100 and standard deviation 15. People with WAIS scores below 70 are considered mentally retarded when, for example, applying for Social Security disability benefits. What percent of adults are retarded by this criterion?

**1.125 High IQ scores.** The Wechsler Adult Intelligence Scale (WAIS) is the most common "IQ test." The scale of scores is set separately for each age group and is approximately Normal with mean 100 and standard deviation 15. The organization MENSA, which calls itself "the high IQ society," requires a WAIS score of 130 or higher for membership. What percent of adults would qualify for membership?

*There are two major tests of readiness for college, the ACT and the SAT. ACT scores are reported on a scale from 1 to 3. The distribution of ACT scores for more than 1 million students in a recent high school graduating class was roughly Normal with mean $\mu = 20.8$ and standard deviation $\sigma = 4.8$. SAT scores are reported on a scale from 400 to 1600. The SAT scores for 1.4 million students in the same graduating class were roughly Normal with mean $\mu = 1026$ and standard deviation $\sigma = 209$. Exercises 1.126 to 1.135 are based on this information.*

**1.126 Compare an SAT score with an ACT score.** Tonya scores 1320 on the SAT. Jermaine scores 28 on the ACT. Assuming that both tests measure the same thing, who has the higher score? Report the $z$-scores for both students.

**1.127 Make another comparison.** Jacob scores 17 on the ACT. Emily scores 680 on the SAT. Assuming that both tests measure the same thing, who has the higher score? Report the $z$-scores for both students.

**1.128 Find the ACT equivalent.** Jose scores 1380 on the SAT. Assuming that both tests measure the same thing, what score on the ACT is equivalent to Jose's SAT score?

**1.129 Find the SAT equivalent.** Maria scores 29 on the ACT. Assuming that both tests measure the same thing, what score on the SAT is equivalent to Maria's ACT score?

**1.130 Find the SAT percentile.** Reports on a student's ACT or SAT usually give the percentile as well as the actual score. The percentile is just the cumulative proportion stated as a percent: the percent of all scores that were lower than this one. Tonya scores 1320 on the SAT. What is her percentile?

**1.131 Find the ACT percentile.** Reports on a student's ACT or SAT usually give the percentile as well as the actual score. The percentile is just the cumulative proportion stated as a percent: the percent of all scores that were lower than this one. Jacob scores 17 on the ACT. What is his percentile?

**1.132 How high is the top 10%?** What SAT scores make up the top 10% of all scores?

**1.133 How low is the bottom 20%?** What SAT scores make up the bottom 20% of all scores?

**1.134 Find the ACT quartiles.** The quartiles of any distribution are the values with cumulative proportions 0.25 and 0.75. What are the quartiles of the distribution of ACT scores?

**1.135 Find the SAT quintiles.** The quintiles of any distribution are the values with cumulative proportions 0.20, 0.40, 0.60, and 0.80. What are the quintiles of the distribution of SAT scores?

**1.136 Proportion of women with high cholesterol.** Too much cholesterol in the blood increases the risk of heart disease. Young women are generally less afflicted with high cholesterol than other groups. The cholesterol levels for women aged 20 to 34 follow an approximately Normal distribution with mean 185 milligrams per deciliter (mg/dl) and standard deviation 39 mg/dl.[36]

(a) Cholesterol levels above 240 mg/dl demand medical attention. What percent of young women have levels above 240 mg/dl?

(b) Levels above 200 mg/dl are considered borderline high. What percent of young women have blood cholesterol between 200 and 240 mg/dl?

**1.137 Proportion of men with high cholesterol.** Middle-aged men are more susceptible to high cholesterol than the young women of the previous exercise. The blood cholesterol levels of men aged 55 to 64 are approximately Normal with mean 222 mg/dl and standard deviation 37 mg/dl. What percent of these men have high cholesterol (levels above 240 mg/dl)? What percent have borderline high cholesterol (between 200 and 240 mg/dl)?

**1.138 Diagnosing osteoporosis.** Osteoporosis is a condition in which the bones become brittle due to loss of minerals. To diagnose osteoporosis, an elaborate apparatus measures bone mineral density (BMD). BMD is usually reported in standardized form. The standardization is based on a population of healthy young adults. The World Health Organization (WHO) criterion for osteoporosis is a BMD 2.5 standard deviations below the mean for young adults. BMD measurements in a population of people similar in age and sex roughly follow a Normal distribution.

(a) What percent of healthy young adults have osteoporosis by the WHO criterion?

(b) Women aged 70 to 79 are of course not young adults. The mean BMD in this age is about −2 on the standard scale for young adults. Suppose that the standard deviation is the same as for young adults. What percent of this older population has osteoporosis?

**1.139 Length of pregnancies.** The length of human pregnancies from conception to birth varies according to a distribution that is approximately Normal with mean 266 days and standard deviation 16 days.

(a) What percent of pregnancies last less than 240 days (that's about 8 months)?

(b) What percent of pregnancies last between 240 and 270 days (roughly between 8 months and 9 months)?

(c) How long do the longest 20% of pregnancies last?

**1.140** CHALLENGE **Quartiles for Normal distributions.** The quartiles of any distribution are the values with cumulative proportions 0.25 and 0.75.

(a) What are the quartiles of the standard Normal distribution?

(b) Using your numerical values from (a), write an equation that gives the quartiles of the $N(\mu, \sigma)$ distribution in terms of $\mu$ and $\sigma$.

(c) The length of human pregnancies from conception to birth varies according to a distribution that is approximately Normal with mean 266 days and standard deviation 16 days. Apply your result from (b): what are the quartiles of the distribution of lengths of human pregnancies?

**1.141** CHALLENGE ***IQR* for Normal distributions.** Continue your work from the previous exercise. The interquartile range *IQR* is the distance between the first and third quartiles of a distribution.

(a) What is the value of the *IQR* for the standard Normal distribution?

(b) There is a constant $c$ such that $IQR = c\sigma$ for any Normal distribution $N(\mu, \sigma)$. What is the value of $c$?

**1.142** CHALLENGE **Outliers for Normal distributions.** Continue your work from the previous two exercises. The percent of the observations that are suspected outliers according to the $1.5 \times IQR$ rule is the same for any Normal distribution. What is this percent?

**1.143** **Heart rates of runners.** Figure 1.39 is a Normal quantile plot of the heart rates of the 200 male runners in the study described in Exercise 1.81 (page 51). The distribution is close to Normal. How can you see this? Describe the nature of the small deviations from Normality that are visible in the plot.

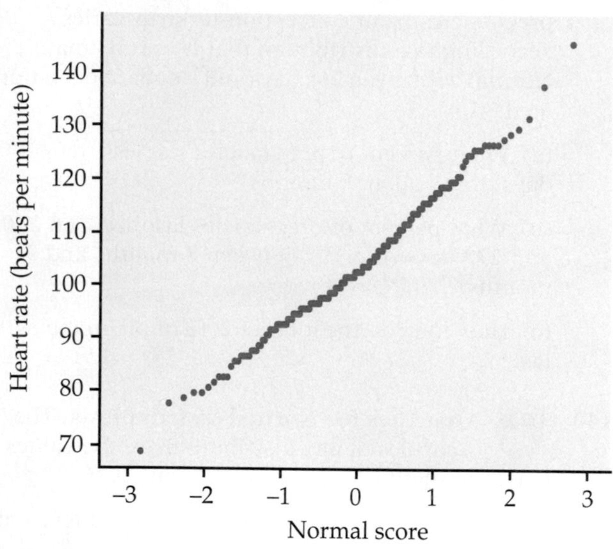


**FIGURE 1.39** Normal quantile plot of the heart rates of 200 male runners, for Exercise 1.143.

**1.144** **Carbon dioxide emissions.** Figure 1.40 is a Normal quantile plot of the emissions of carbon dioxide ($CO_2$) per person in 48 countries, from Table 1.6 (page 26). In what way is this distribution non-Normal? Comparing the plot with Table 1.6, which countries would you call outliers?

**1.145** **Electrical meters.** The distance between two mounting holes is important to the performance of an electrical meter. The manufacturer measures this distance regularly for quality control purposes, recording the data as thousandths of an inch more than 0.600 inches. For example, 0.644 is recorded as 44. Figure 1.41 is a Normal quantile plot of the distances for the last 90 electrical meters measured.[37] Is the overall shape of the distribution approximately Normal? Why does the plot have a "stair-step" appearance?

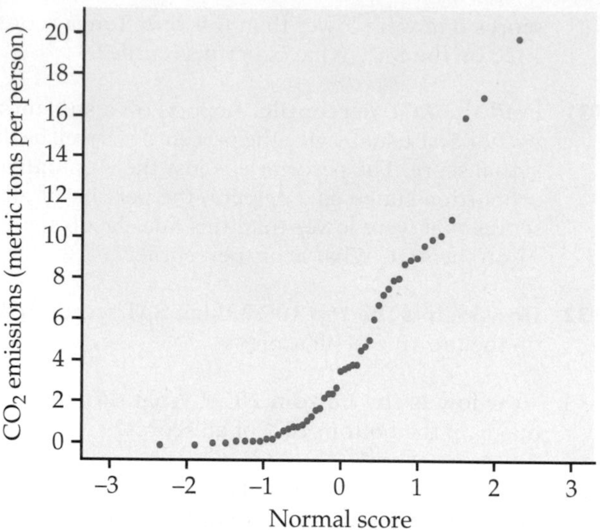


**FIGURE 1.40** Normal quantile plot of $CO_2$ emissions in 48 countries, for Exercise 1.144.

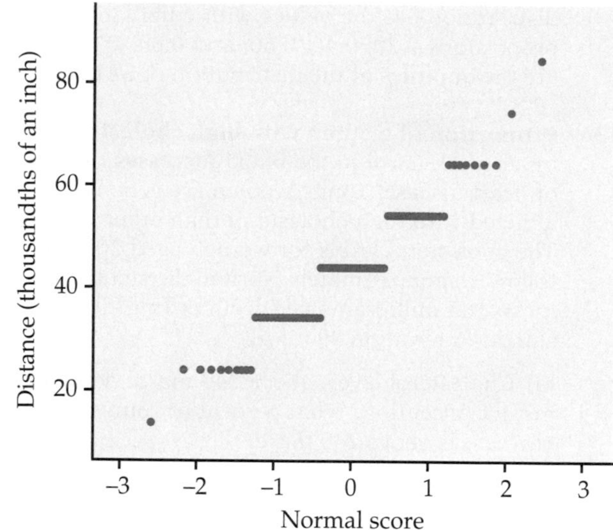


**FIGURE 1.41** Normal quantile plot of distance between mounting holes, for Exercise 1.145.

**1.146** CHALLENGE **Four Normal quantile plots.** Figure 1.42 shows four Normal quantile plots for data that you have seen before, without scales for the variables plotted. In scrambled order, they are:

1. The IQ scores in the histogram of Figure 1.7 (page 14).

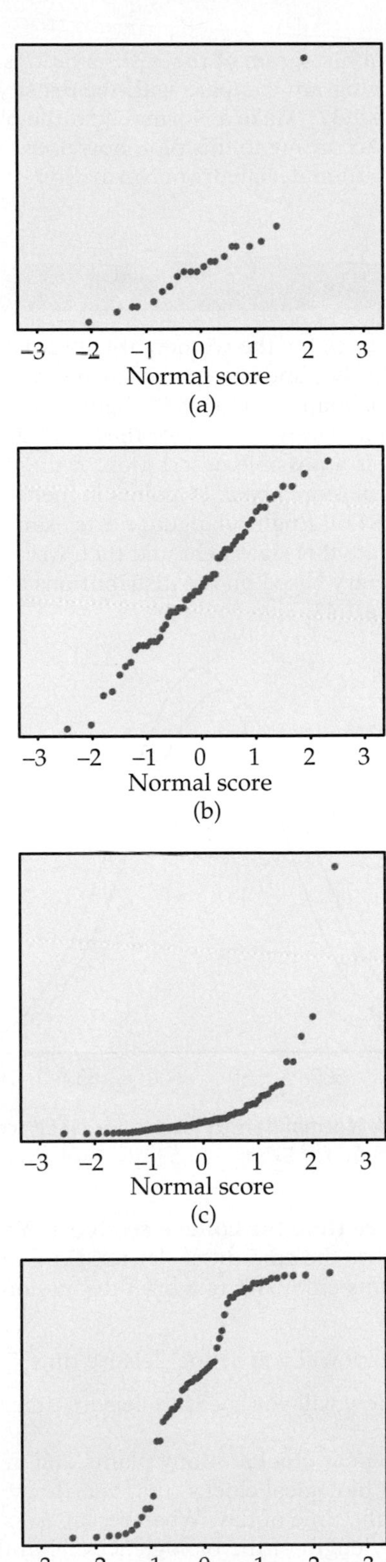


**FIGURE 1.42** Four Normal quantile plots, for Exercise 1.146.

2. The tuition and fee charges of Massachusetts colleges in the histogram of Figure 1.16 (page 25).

3. The highway gas mileages of two-seater cars, including the Honda Insight, from Table 1.10 (page 31).

4. The 80 customer service call lengths from Table 1.1, displayed in the stemplot of Figure 1.6 (page 12).

Which Normal quantile plot goes with each data set? Explain the reasons for your choices.

*The remaining exercises for this section require the use of software that will make Normal quantile plots.*

**1.147 Density of the earth.** We expect repeated careful measurements of the same quantity to be approximately Normal. Make a Normal quantile plot for Cavendish's measurements in Exercise 1.40 (page 28). Are the data approximately Normal? If not, describe any clear deviations from Normality.

**1.148 Three varieties of flowers.** The study of tropical flowers and their hummingbird pollinators (Exercise 1.78, page 51) measured lengths for three varieties of *Heliconia* flowers. We expect that such biological measurements will have roughly Normal distributions.

(a) Make Normal quantile plots for each of the three flower varieties. Which distribution is closest to Normal?

(b) The other two distributions show the same kind of mild deviation from Normality. In what way are these distributions non-Normal?

**1.149 Logging in Borneo.** The study of the effects of logging on tree counts in the Borneo rain forest (Exercise 1.80, page 51) used statistical methods that are based on Normal distributions. Make Normal quantile plots for each of the three groups of forest plots. Are the three distributions roughly Normal? What are the most prominent deviations from Normality that you see?

**1.150 Use software to generate some data.** Use software to generate 100 observations from the standard Normal distribution. Make a histogram of these observations. How does the shape of the histogram compare with a Normal density curve? Make a Normal quantile plot of the data. Does the plot suggest any important deviations from Normality? (Repeating this exercise several times is a good way to become familiar with how histograms and Normal quantile plots look when data actually are close to Normal.)

**1.151 Use software to generate more data.** Use software to generate 100 observations from the uniform distribution described in Exercise 1.108. Make a histogram of these observations. How does the histogram compare with the density curve in Figure 1.37? Make a Normal quantile plot of your data. According to this plot, how does the uniform distribution deviate from Normality?

## CHAPTER 1 Exercises

**1.152 Park space and population.** Below are data on park and open space in several U.S. cities with high population density.[38] In this table, population is reported in thousands of people, and park and open space is called open space, with units of acres.

| City | Population | Open space |
|---|---|---|
| Baltimore | 651 | 5,091 |
| Boston | 589 | 4,865 |
| Chicago | 2,896 | 11,645 |
| Long Beach | 462 | 2,887 |
| Los Angeles | 3,695 | 29,801 |
| Miami | 362 | 1,329 |
| Minneapolis | 383 | 5,694 |
| New York | 8,008 | 49,854 |
| Oakland | 399 | 3,712 |
| Philadelphia | 1,518 | 10,685 |
| San Francisco | 777 | 5,916 |
| Washington, D.C. | 572 | 7,504 |

(a) Make a bar graph for population. Describe what you see in the graph.

(b) Do the same for open space.

(c) For each city, divide the open space by population. This gives rates: acres of open space per thousand residents.

(d) Make a bar graph of the rates.

(e) Redo the bar graph that you made in part (d) by ordering the cities by their open space to population rate.

(f) Which of the two bar graphs in (d) and (e) do you prefer? Give reasons for your answer.

**1.153 Compare two Normal curves.** In Exercise 1.99, we worked with the distribution of ISTEP scores on the English/language arts portion of the exam for tenth-graders. We used the fact that the distribution of scores for the 76,531 students who took the exam was approximately $N(572, 51)$. These students were classified in a variety of ways, and summary statistics were reported for these different subgroups. When classified by gender, the scores for the women are approximately $N(579, 49)$, and the scores for the men are approximately $N(565, 55)$. Figure 1.43 gives the Normal density curves for these two distributions. Here is a possible description of these data: women score about 14 points higher than men on the ISTEP English/language arts exam. Critically evaluate this statement and then write your own summary based on the distributions displayed in Figure 1.43.

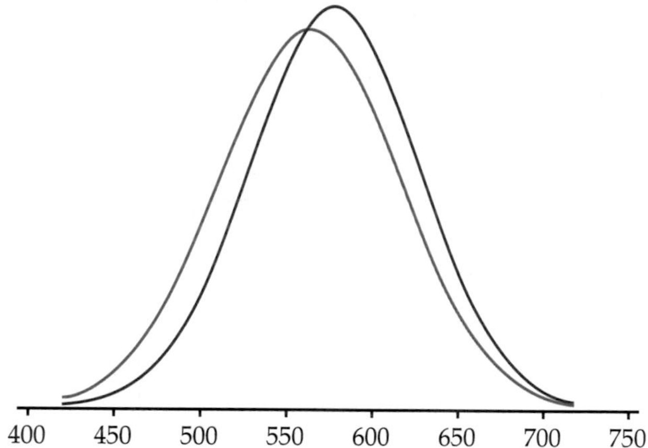


**FIGURE 1.43** Normal density curves for ISTEP scores of women and men, for Example 1.53.

**1.154 Leisure time for college students.** You want to measure the amount of "leisure time" that college students enjoy. Write a brief discussion of two issues:

(a) How will you define "leisure time"?

(b) How will you measure leisure time?

**1.155 Biological clocks.** Many plants and animals have "biological clocks" that coordinate activities with the time of day. When researchers looked at the length of the biological cycle in the plant *Arabidopsis* by measuring leaf movements, they found that the length of the cycle is not always 24 hours. Further study discovered that cycle length changes systematically with north-south location.

**TABLE 1.11**

**Biological clock cycle lengths for a plant species in different locations**

| | | | | | | | | | |
|---|---|---|---|---|---|---|---|---|---|
| 23.89 | 23.72 | 23.74 | 24.35 | 25.05 | 24.56 | 23.69 | 22.33 | 23.79 | 22.12 |
| 25.39 | 23.08 | 25.64 | 23.98 | 25.84 | 25.46 | 24.37 | 24.13 | 24.40 | 24.74 |
| 24.44 | 24.82 | 23.56 | 24.96 | 24.21 | 23.85 | 24.57 | 23.44 | 23.64 | 24.23 |
| 24.01 | 24.58 | 25.57 | 23.73 | 24.11 | 23.21 | 25.08 | 24.03 | 24.62 | 23.51 |
| 23.21 | 23.41 | 23.69 | 22.97 | 24.65 | 24.65 | 24.29 | 23.89 | 25.08 | 23.89 |
| 24.95 | 23.09 | 23.21 | 24.66 | 23.88 | 25.33 | 24.38 | 24.68 | 25.34 | 25.22 |
| 23.45 | 23.39 | 25.43 | 23.16 | 23.95 | 23.25 | 24.72 | 24.89 | 24.88 | 24.71 |
| 23.58 | 25.98 | 24.28 | 24.25 | 23.16 | 24.19 | 27.22 | 23.77 | 26.21 | 24.33 |
| 24.34 | 24.89 | 24.32 | 24.14 | 24.00 | 23.48 | 25.81 | 24.99 | 24.18 | 22.73 |
| 24.18 | 23.95 | 24.48 | 23.89 | 24.24 | 24.96 | 24.58 | 24.29 | 24.31 | 23.64 |
| 23.87 | 23.68 | 24.87 | 23.00 | 23.48 | 24.26 | 23.34 | 25.11 | 24.69 | 24.97 |
| 24.64 | 24.49 | 23.61 | 24.07 | 26.60 | 24.91 | 24.76 | 25.09 | 26.56 | 25.13 |
| 24.81 | 25.63 | 25.63 | 24.69 | 24.41 | 23.79 | 22.88 | 22.00 | 23.33 | 25.12 |
| 24.00 | 24.31 | 23.03 | 24.51 | 28.55 | 22.96 | 23.61 | 24.72 | 24.04 | 25.18 |
| 24.30 | 24.22 | 24.39 | 24.73 | 24.68 | 24.14 | 24.57 | 24.42 | 25.62 | |

Table 1.11 contains cycle lengths for 149 locations around the world.[39] Describe the distribution of cycle lengths with a histogram and numerical summaries. In particular, how much variation is there among locations?

**1.156 Product preference.** Product preference depends in part on the age, income, and gender of the consumer. A market researcher selects a large sample of potential car buyers. For each consumer, she records gender, age, household income, and automobile preference. Which of these variables are categorical and which are quantitative?

**1.157 Distance-learning courses.** The 222 students enrolled in distance-learning courses offered by a college ranged from 18 to 64 years of age. The mode of their ages was 19. The median age was 31.[40] Explain how this can happen.

**1.158 Internet service.** Late in 2003, there were 77.4 million residential subscribers to Internet service in the United States. The numbers of subscribers claimed by the top 10 providers of service were as follows.[41] (There is some doubt about the accuracy of these claims.)

| Service provider | Subscribers (millions) | Service provider | Subscribers (millions) |
|---|---|---|---|
| America Online | 24.7 | SBC | 3.1 |
| MSN | 8.7 | Verizon | 2.1 |
| United Online | 5.2 | Cox | 1.8 |
| EarthLink | 5.0 | Charter | 1.5 |
| Comcast | 4.9 | BellSouth | 1.3 |

Display these data in a graph. How many subscribers do the many smaller providers have? Add an "Other" entry in your graph. Business people looking at this graph see an industry that offers opportunities for larger companies to take over.

**1.159 Weights are not Normal.** The heights of people of the same sex and similar ages follow Normal distributions reasonably closely. Weights, on the other hand, are not Normally distributed. The weights of women aged 20 to 29 have mean 141.7 pounds and median 133.2 pounds. The first and third quartiles are 118.3 pounds and 157.3 pounds. What can you say about the shape of the weight distribution? Explain your reasoning.

**1.160 What graph would you use?** What type of graph or graphs would you plan to make in a study of each of the following issues?

(a) What makes of cars do students drive? How old are their cars?

(b) How many hours per week do students study? How does the number of study hours change during a semester?

(c) Which radio stations are most popular with students?

(d) When many students measure the concentration of the same solution for a chemistry course laboratory assignment, do their measurements follow a Normal distribution?

**1.161 Household size and household income.** Rich and poor households differ in ways that go beyond income. Figure 1.44 displays histograms that compare the distributions of household size (number of people) for low-income and high-income households in 2002.[42] Low-income households had incomes less than \$15,000, and

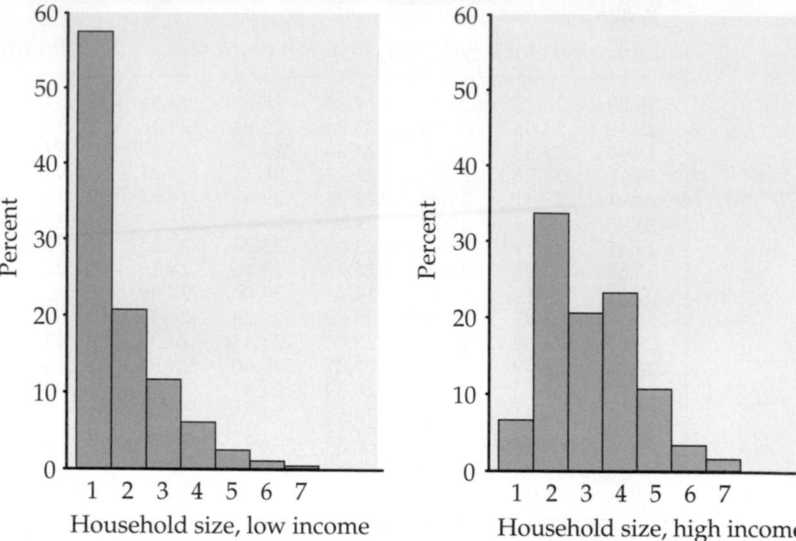


**FIGURE 1.44** The distributions of household size for households with incomes less than $15,000 (*left*) and households with incomes of at least $100,000 (*right*), for Exercise 1.161.

high-income households had incomes of at least $100,000.

(a) About what percent of each group of households consisted of two people?

(b) What are the important differences between these two distributions? What do you think explains these differences?

**1.162 Spam filters.** A university department installed a spam filter on its computer system. During a 21-day period, 6693 messages were tagged as spam. How much spam you get depends on what your online habits are. Here are the counts for some students and faculty in this department (with log-in IDs changed, of course):

| ID | Count | ID | Count | ID | Count | ID | Count |
|---|---|---|---|---|---|---|---|
| AA | 1818 | BB | 1358 | CC | 442 | DD | 416 |
| EE | 399 | FF | 389 | GG | 304 | HH | 251 |
| II | 251 | JJ | 178 | KK | 158 | LL | 103 |

All other department members received fewer than 100 spam messages. How many did the others receive in total? Make a graph and comment on what you learn from these data.

**1.163 Two distributions.** If two distributions have exactly the same mean and standard deviation, must their histograms have the same shape? If they have the same five-number summary, must their histograms have the same shape? Explain.

**1.164 By-products from DDT.** By-products from the pesticide DDT were major threats to the survival of birds of prey until use of DDT was banned at the end of 1972. Can time plots show the effect of the ban? Here are two sets of data for bald eagles nesting in the forests of northwestern Ontario.[43] The data set below gives the mean number of young per breeding area.

| Year | 1966 | 1967 | 1968 | 1969 | 1970 | 1971 | 1972 | 1973 |
|---|---|---|---|---|---|---|---|---|
| Young | 1.26 | 0.73 | 0.89 | 0.84 | 0.54 | 0.60 | 0.54 | 0.78 |
| Year | 1974 | 1975 | 1976 | 1977 | 1978 | 1979 | 1980 | 1981 |
| Young | 0.46 | 0.77 | 0.86 | 0.96 | 0.82 | 0.98 | 0.93 | 1.12 |

The following data are measurements of the chemical DDE (the by-product of DDT that most threatens birds of prey) from bald eagle eggs in the same area of Canada. These are in parts per million (ppm). There are often several measurements per year.

| Year | 1967 | 1967 | 1968 | 1971 | 1971 | 1972 | 1976 |
|---|---|---|---|---|---|---|---|
| DDE | 44 | 95 | 121 | 125 | 95 | 87 | 13.3 |
| Year | 1976 | 1976 | 1976 | 1976 | 1977 | 1977 | 1980 |
| DDE | 16.4 | 50.4 | 59.8 | 56.4 | 0.6 | 23.8 | 16.6 |
| Year | 1980 | 1980 | 1981 | 1981 | 1981 | | |
| DDE | 14.5 | 24.0 | 7.8 | 48.2 | 53.4 | | |

Make time plots of eagle young and of mean DDE concentration in eggs. How does the effect of banning DDT appear in your plots?

**1.165 Babe Ruth and Mark McGwire.** Babe Ruth hit 60 home runs in 1927, a record that stood until Mark McGwire hit 70 in 1998. A proper comparison of Ruth and McGwire should include their historical context. Here are the number of home runs by the major league leader for each year in baseball history, 1876 to 2003, in order from left to right. Make a time plot. (Be sure to add the scale of years.)

| | | | | | | | | | | | | | | | | |
|---|---|---|---|---|---|---|---|---|---|---|---|---|---|---|---|---|
| 5 | 3 | 4 | 9 | 6 | 7 | 7 | 10 | 27 | 11 | 11 | 17 | 14 | 20 | 14 | 16 | 13 |
| 19 | 18 | 17 | 13 | 11 | 15 | 25 | 12 | 16 | 16 | 13 | 10 | 9 | 12 | 10 | 12 | 9 |
| 10 | 21 | 14 | 19 | 19 | 24 | 12 | 12 | **11** | **29** | **54** | **59** | 42 | **41** | **46** | 39 | **47** |
| **60** | **54** | **46** | 56 | **46** | 58 | 48 | 49 | 36 | 49 | 46 | 58 | 35 | 43 | 37 | 36 | 34 |
| 33 | 28 | 44 | 51 | 40 | 54 | 47 | 42 | 37 | 47 | 49 | 51 | 52 | 44 | 47 | 46 | 41 |
| 61 | 49 | 45 | 49 | 52 | 49 | 44 | 44 | 49 | 45 | 48 | 40 | 44 | 36 | 38 | 38 | 52 |
| 46 | 48 | 48 | 31 | 39 | 40 | 43 | 40 | 40 | **49** | 42 | 47 | 51 | 44 | 43 | 46 | 43 |
| 50 | **52** | **58** | **70** | **65** | 50 | 73 | 57 | 47 | | | | | | | | |

(a) Describe the effect of World War II (1942 to 1945 seasons).

(b) Ruth led in the 11 years in boldface between 1918 and 1931. McGwire led in the 5 boldface years between 1987 and 1999. Briefly compare the achievements of Ruth and McGwire in the context of their times.

**1.166 Barry Bonds.** The single-season home run record was broken by Barry Bonds of the San Francisco Giants in 2001, when he hit 73 home runs. Here are Bonds's home run totals from 1986 (his first year) to 2003:

| | | | | | | | | |
|---|---|---|---|---|---|---|---|---|
| 16 | 25 | 24 | 19 | 33 | 25 | 34 | 46 | 37 |
| 33 | 42 | 40 | 37 | 34 | 49 | 73 | 46 | 45 |

Make a stemplot of these data. Bonds's record year is a high outlier. How do his career mean and median number of home runs change when we drop the record 73? What general fact about the mean and median does your result illustrate?

**1.167** CHALLENGE **Norms for reading scores.** Raw scores on behavioral tests are often transformed for easier comparison. A test of reading ability has mean 75 and standard deviation 10 when given to third-graders. Sixth-graders have mean score 82 and standard deviation 11 on the same test. To provide separate "norms" for each grade, we want scores in each grade to have mean 100 and standard deviation 20.

(a) What linear transformation will change third-grade scores $x$ into new scores $x_{new} = a + bx$ that have the desired mean and standard deviation? (Use $b > 0$ to preserve the order of the scores.)

(b) Do the same for the sixth-grade scores.

(c) David is a third-grade student who scores 78 on the test. Find David's transformed score. Nancy is a sixth-grade student who scores 78. What is her transformed score? Who scores higher within his or her grade?

(d) Suppose that the distribution of scores in each grade is Normal. Then both sets of transformed scores have the $N(100, 20)$ distribution. What percent of third-graders have scores less than 78? What percent of sixth-graders have scores less than 78?

**1.168 Damage caused by tornados.** The average damage caused by tornadoes in the states (Table 1.5, page 25) and the estimated amount of oil recovered from different oil wells (Exercise 1.39, page 28) both have right-skewed distributions. Choose one of these data sets. Make a Normal quantile plot. How is the skewness of the distribution visible in the plot? Based on the plot, which observations (if any) would you call outliers?

**1.169 Proportions older than 65.** We know that the distribution of the percents of state residents over 65 years of age has a low outlier (Alaska) and a high outlier (Florida). The stemplot in Exercise 1.21 (page 24) looks unimodal and roughly symmetric.

(a) Sketch what a Normal quantile plot would look like for a distribution that is Normal except for two outliers, one in each direction.

(b) If your software includes Normal quantile plots, make a plot of the percent-over-65 data and discuss what you see.

**1.170 Returns on stocks.** Returns on common stocks are "heavy tailed." That is, they have more values far from the center in both the low and the high tails than a Normal distribution would have. However, average returns for many individual stocks over longer periods of time become more Normal.

(a) Sketch the appearance of a Normal quantile plot for a distribution having roughly Normal center and heavy tails. Explain the reasoning behind your sketch.

(b) The data include the annual returns for the years 1950 to 2003, pictured in the stemplot in Figure 1.22(a). If your software allows, make a Normal quantile plot of these returns. Is the distribution clearly heavy tailed? Are there other clear deviations from Normality?

**1.171 Use software to generate some data.** Most statistical software packages have routines for generating values of variables having specified distributions. Use your statistical software to generate 25 observations from the $N(20, 5)$ distribution. Compute the mean and standard deviation $\bar{x}$ and $s$ of the 25 values you obtain. How close are $\bar{x}$ and $s$ to the $\mu$ and $\sigma$ of the distribution from which the observations were drawn? Repeat 19 more times the process of generating 25 observations from the $N(20, 5)$ distribution and recording $\bar{x}$ and $s$. Make a stemplot of the 20 values of $\bar{x}$ and another stemplot of the 20 values of $s$. Make Normal quantile plots of both sets of data. Briefly describe each of these distributions. Are they symmetric or skewed? Are they roughly Normal? Where are their centers? (The distributions of measures like $\bar{x}$ and $s$ when repeated sets of observations are made from the same theoretical distribution will be very important in later chapters.)

**1.172 Distribution of income.** Each March, the Bureau of Labor Statistics collects detailed information about more than 50,000 randomly selected households. The WORKERS data set contains data on 71,076 people from the March 2002 survey. All of these people were between 25 and 64 years of age and worked throughout the year. The Data Appendix describes this data set in detail. Describe the distribution of incomes for these people. Use graphs and numbers, and briefly state your main findings. Because this is a very large randomly selected sample, your results give a good description of incomes for all working Americans aged 25 to 64.

**1.173 SAT mathematics scores and grade point averages.** The CSDATA data set described in the Data Appendix contains information on 234 computer science students. We are interested in comparing the SAT Mathematics scores and grade point averages of female students with those of male students. Make two sets of side-by-side boxplots to carry out these comparisons. Write a brief discussion of the male-female comparisons. Then make Normal quantile plots of grade point averages and SAT Math scores separately for men and women. Which students are clear outliers? Which of the four distributions are approximately Normal if we ignore outliers?

CHAPTER

2

# Looking at Data—Relationships

Do large breeds of dogs have shorter lives? See Example 2.1.



## Introduction

In Chapter 1 we learned to use graphical and numerical methods to describe the distribution of a single variable. Many of the interesting examples of the use of statistics involve relationships between pairs of variables. Learning ways to describe relationships with graphical and numerical methods is the focus of this chapter.

EXAMPLE

**2.1 Large breeds of dogs have shorter lives.** Purebred dogs from breeds that are large tend to have shorter life spans than purebred dogs from breeds that are small. For example, one study found that miniature poodles lived an average of 9.3 years while Great Danes lived an average of only 4.6 years.[1] Irish wolfhounds have sometimes been referred to by the nickname "the heartbreak breed" because of their short life span relative to other breeds.[2]

We are particularly interested in situations where two variables are related in some way. To study relationships, we measure both variables on the same individuals or cases.

USE YOUR KNOWLEDGE

2.1 **Relationship between first test and final exam.** You want to study the relationship between the score on the first test and the score on the final exam for the 35 students enrolled in an elementary statistics class. Who are the individuals for your study?

We use the term *associated* to describe the relationship between two variables, such as breed and life span in Example 2.1. Here is another example where two variables are associated.

EXAMPLE

**2.2 Size and price of a coffee beverage.** You visit a local Starbucks to buy a Mocha Frappuccino©. The barista explains that this blended coffee beverage comes in three sizes and asks if you want a Tall, a Grande, or a Venti. The prices are \$3.15, \$3.65, and \$4.15, respectively. There is a clear association between the size of the Mocha Frappuccino and its price.

ASSOCIATION BETWEEN VARIABLES

Two variables measured on the same cases are **associated** if knowing the value of one of the variables tells you something about the values of the other variable that you would not know without this information.

In the Mocha Frappuccino example, knowing the size tells you the exact price, so the association here is very strong. Many statistical associations, however, are simply overall tendencies that allow exceptions. Although smokers on the average die earlier than nonsmokers, some people live to 90 while smoking three packs a day. Knowing that a person smokes tells us that the person is in a group of people who are more likely to die at a younger age than people in the group of nonsmokers. The association here is much weaker than the one in the Mocha Frappuccino example.

## Examining relationships

When you examine the relationship between two or more variables, first ask the preliminary questions that are familiar from Chapter 1:

- What *individuals or cases* do the data describe?
- What *variables* are present? How are they measured?
- Which variables are *quantitative* and which are *categorical?*

EXAMPLE

**2.3 Cases and variable types.** In Example 2.1 the cases are dog breeds. The type of dog breed is a categorical variable, and the average life span is a quantitative variable. In Example 2.2 the cases are the containers of coffee. Size is a categorical variable with values Tall, Grande, and Venti. Price is a quantitative variable.

## USE YOUR KNOWLEDGE

**2.2** **Suppose we used breed size?** Suppose that for the dog breed example we were able to obtain some measure of average size for each of the breeds. If we replaced type of dog breed with the average breed size, how would this change the explanation in Example 2.3?

**2.3** **Replace names by ounces.** In the Mocha Frappuccino example, the variable size is categorical, with Tall, Grande, and Venti as the possible values. Suppose you converted these values to the number of ounces: Tall is 12 ounces, Grande is 16 ounces, and Venti is 24 ounces. For studying the relationship between ounces and price, describe the cases and the variables, and state whether each is quantitative or categorical.

When you examine the relationship between two variables, a new question becomes important:

- Is your purpose simply to explore the nature of the relationship, or do you hope to show that one of the variables can explain variation in the other? That is, are some of the variables *response variables* and others *explanatory variables?*

### RESPONSE VARIABLE, EXPLANATORY VARIABLE

A **response variable** measures an outcome of a study. An **explanatory variable** explains or causes changes in the response variables.

It is easiest to identify explanatory and response variables when we actually set values of one variable in order to see how it affects another variable.

EXAMPLE

**2.4 Beer drinking and blood alcohol levels.** How does drinking beer affect the level of alcohol in our blood? The legal limit for driving in most states is 0.08%. Student volunteers at Ohio State University drank different numbers of cans of beer. Thirty minutes later, a police officer measured their blood alcohol content. Number of beers consumed is the explanatory variable, and percent of alcohol in the blood is the response variable.

When you don't set the values of either variable but just observe both variables, there may or may not be explanatory and response variables. Whether there are depends on how you plan to use the data.

EXAMPLE

**2.5 Student loans.** A college student aid officer looks at the findings of the National Student Loan Survey. She notes data on the amount of debt of recent graduates, their current income, and how stressful they feel about college debt. She isn't interested in predictions but is simply trying to understand the situation of recent college graduates.

A sociologist looks at the same data with an eye to using amount of debt and income, along with other variables, to explain the stress caused by college debt. Now amount of debt and income are explanatory variables, and stress level is the response variable.

In many studies, the goal is to show that changes in one or more explanatory variables actually *cause* changes in a response variable. But many explanatory-response relationships do not involve direct causation. The SAT scores of high school students help predict the students' future college grades, but high SAT scores certainly don't cause high college grades.

Some of the statistical techniques in this chapter require us to distinguish explanatory from response variables; others make no use of this distinction. You will often see explanatory variables called **independent variables** and response variables called **dependent variables.** The idea behind this language is that response variables depend on explanatory variables. Because the words "independent" and "dependent" have other meanings in statistics that are unrelated to the explanatory-response distinction, we prefer to avoid those words.

independent variable
dependent variable

Most statistical studies examine data on more than one variable. Fortunately, statistical analysis of several-variable data builds on the tools used for examining individual variables. The principles that guide our work also remain the same:

- Start with a graphical display of the data.
- Look for overall patterns and deviations from those patterns.
- Based on what you see, use numerical summaries to describe specific aspects of the data.

## 2.1 Scatterplots

The most useful graph for displaying the relationship between two quantitative variables is a *scatterplot.*

### SCATTERPLOT

A **scatterplot** shows the relationship between two quantitative variables measured on the same individuals. The values of one variable appear on the horizontal axis, and the values of the other variable appear on the vertical axis. Each individual in the data appears as the point in the plot fixed by the values of both variables for that individual.

Always plot the explanatory variable, if there is one, on the horizontal axis (the $x$ axis) of a scatterplot. As a reminder, we usually call the explanatory variable $x$ and the response variable $y$. If there is no explanatory-response distinction, either variable can go on the horizontal axis.

EXAMPLE

**2.6 SAT scores.** More than a million high school seniors take the SAT college entrance examination each year. We sometimes see the states "rated" by the average SAT scores of their seniors. For example, Illinois students average 1179 on the SAT, which looks better than the 1038 average of Massachusetts students. Rating states by SAT scores makes little sense, however, because average SAT score is largely explained by what percent of a state's students take the SAT. The scatterplot in Figure 2.1 allows us to see how the mean SAT score in each state is related to the percent of that state's high school seniors who take the SAT.[3]

Each point on the plot represents a single individual—that is, a single state. Because we think that the percent taking the exam influences mean score, percent taking is the explanatory variable and we plot it horizontally. For example, 20% of West Virginia high school seniors take the SAT, and their mean score is 1032. West Virginia appears as the point (20, 1032) in the scatterplot, above 20 on the $x$ axis and to the right of 1032 on the $y$ axis.

We see at once that state average SAT score is closely related to the percent of students who take the SAT. Illinois has a high mean score, but only 11% of Illinois seniors take the SAT. In Massachusetts, on the other hand, 82% of seniors take the exam.

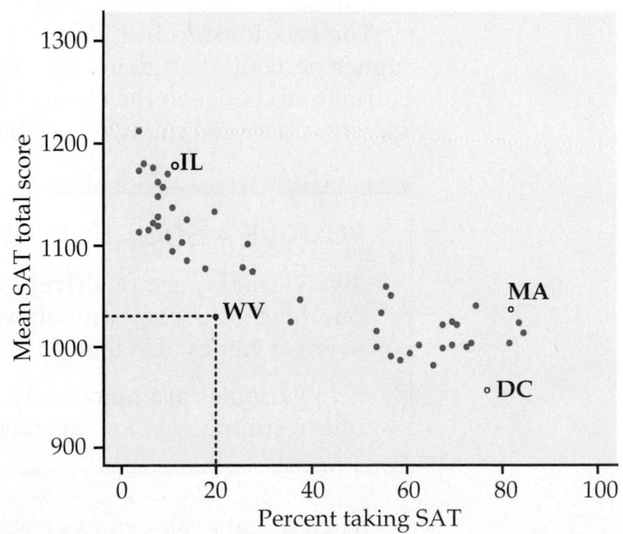


**FIGURE 2.1** State mean SAT scores plotted against the percent of high school seniors in each state who take the SAT exams, for Example 2.6. The point for West Virginia (20% take the SAT, mean score 1032) is highlighted.

## Interpreting scatterplots

To look more closely at a scatterplot such as Figure 2.1, apply the strategies of exploratory analysis learned in Chapter 1.

> **EXAMINING A SCATTERPLOT**
>
> In any graph of data, look for the **overall pattern** and for striking **deviations** from that pattern.
>
> You can describe the overall pattern of a scatterplot by the **form, direction,** and **strength** of the relationship.
>
> An important kind of deviation is an **outlier,** an individual value that falls outside the overall pattern of the relationship.

clusters

Figure 2.1 shows an interesting *form:* there are two distinct **clusters** of states. In one cluster, more than half of high school seniors take the SAT, and the mean scores are low. Fewer than 40% of seniors in states in the other cluster take the SAT—fewer than 20% in most of these states—and these states have higher mean scores.

Clusters in a graph suggest that the data describe several distinct kinds of individuals. The two clusters in Figure 2.1 do in fact describe two distinct sets of states. There are two common college entrance examinations, the SAT and the ACT. Each state tends to prefer one or the other. In ACT states (the left cluster in Figure 2.1), most students who take the SAT are applying to selective out-of-state colleges. This select group performs well. In SAT states (the right cluster), many seniors take the SAT, and this broader group has a lower mean score.

There are no clear *outliers* in Figure 2.1, but each cluster does include a state whose mean SAT score is lower than we would expect from the percent of its students who take the SAT. These points are West Virginia in the cluster of ACT states and the District of Columbia (a city rather than a state) in the cluster of SAT states.

The relationship in Figure 2.1 also has a clear *direction:* states in which a higher percent of students take the SAT tend to have lower mean scores. This is true both between the clusters and within each cluster. This is a *negative association* between the two variables.

> **POSITIVE ASSOCIATION, NEGATIVE ASSOCIATION**
>
> Two variables are **positively associated** when above-average values of one tend to accompany above-average values of the other and below-average values also tend to occur together.
>
> Two variables are **negatively associated** when above-average values of one accompany below-average values of the other, and vice versa.

When a scatterplot shows distinct clusters, it is often useful to describe the overall pattern separately within each cluster. The *form* of the relationship in

linear relationship

the ACT states is roughly **linear.** That is, the points roughly follow a straight line. The *strength* of a relationship in a scatterplot is determined by how closely the points follow a clear form. The linear pattern among the ACT states is moderately strong because the points show only modest scatter about the straight-line pattern. In summary, the ACT states in Figure 2.1 show a moderately strong negative linear relationship. The cluster of SAT states shows a much weaker relationship between percent taking the SAT and mean SAT score.

## USE YOUR KNOWLEDGE

**2.4** **Make a scatterplot.** In our Mocha Frappuccino example, the 12-ounce drink costs \$3.15, the 16-ounce drink costs \$3.65, and the 24-ounce drink costs \$4.15. Explain which variable should be used as the explanatory variable and make a scatterplot. Describe the scatterplot and the association between these two variables.

## Adding categorical variables to scatterplots

The Census Bureau groups the states into four broad regions, named Midwest, Northeast, South, and West. We might ask about regional patterns in SAT exam scores. Figure 2.2 repeats part of Figure 2.1, with an important difference. We have plotted only the Northeast and Midwest groups of states, using the plot symbol "e" for the northeastern states and the symbol "m" for the midwestern states.

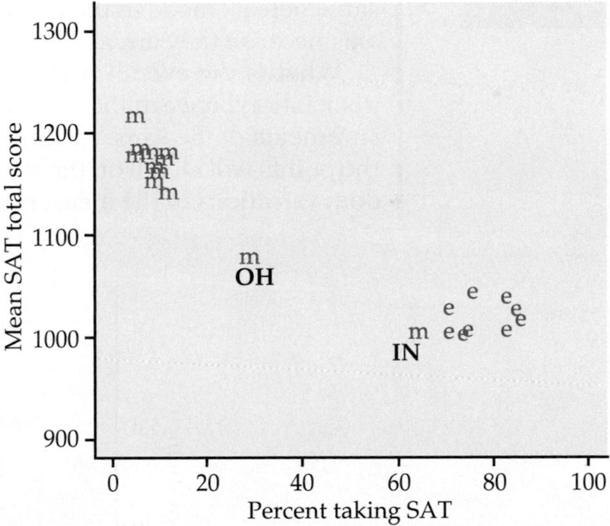


**FIGURE 2.2** State mean SAT scores and percent taking the SAT for the northeastern states (plot symbol "e") and the midwestern states (plot symbol "m").

The regional comparison is striking. The 9 northeastern states are all SAT states—in fact, at least 70% of high school graduates in each of these states take the SAT. The 12 midwestern states are mostly ACT states. In 10 of these states, the percent taking the SAT is between 4% and 11%. One midwestern state is clearly an outlier within the region. Indiana is an SAT state (63% take the SAT) that falls close to the northeastern cluster. Ohio, where 28% take the SAT, also lies outside the midwestern cluster.

In dividing the states into regions, we introduced a third variable into the scatterplot. "Region" is a categorical variable that has four values, although we plotted data from only two of the four regions. The two regions are displayed by the two different plotting symbols.[4]

**CATEGORICAL VARIABLES IN SCATTERPLOTS**

To add a categorical variable to a scatterplot, use a different plot color or symbol for each category.

## More examples of scatterplots

Experience in examining scatterplots is the foundation for more detailed study of relationships among quantitative variables. Here is an example with a pattern different from that in Figure 2.1.

EXAMPLE

**2.7 The Trans-Alaska Oil Pipeline.** The Trans-Alaska Oil Pipeline is a tube formed from 1/2-inch-thick steel that carries oil across 800 miles of sensitive arctic and subarctic terrain. The pipe and the welds that join pipe segments were carefully examined before installation. How accurate are field measurements of the depth of small defects? Figure 2.3 compares the results of measurements on 100 defects made in the field with measurements of the same defects made in the laboratory.[5] We plot the laboratory results on the $x$ axis because they are a standard against which we compare the field results.

What is the overall pattern of this scatterplot? There is a positive linear association between the two variables. This is what we expect from two measurements of the same quantity. If field and laboratory measurements agree, the points will all fall on the $y = x$ line drawn on the plot, except for small random variations in the measurements. In fact, we see that the points for larger

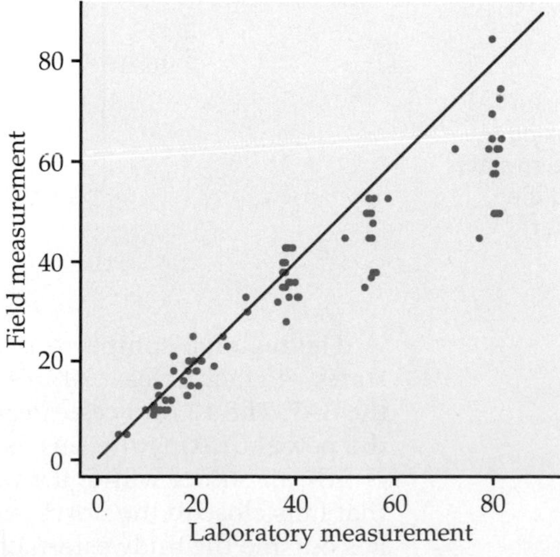


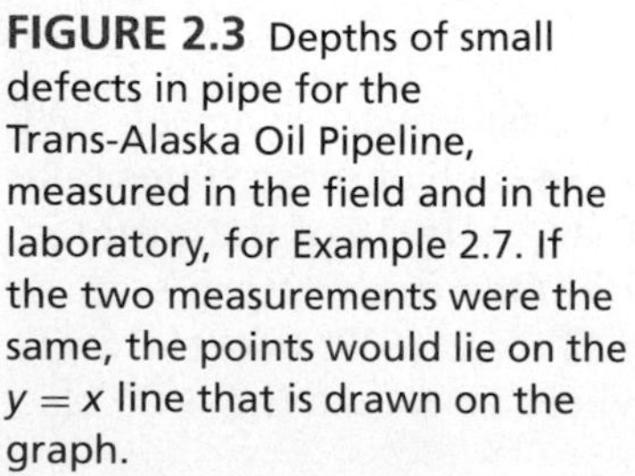

**FIGURE 2.3** Depths of small defects in pipe for the Trans-Alaska Oil Pipeline, measured in the field and in the laboratory, for Example 2.7. If the two measurements were the same, the points would lie on the $y = x$ line that is drawn on the graph.

defects fall systematically below this line. That is, the field measurements are too small compared with the laboratory results as a standard. This is an important finding that can be used to adjust future field measurements.

Another part of the overall pattern is that the strength of the linear relationship decreases as the size of the defects increases. Field data show more variation (vertical spread in the plot) for large defect sizes than for small sizes. An increase in the spread in a response variable as the size of the response increases is a common pattern. It implies that predictions of the response based on the overall pattern will be less accurate for large responses.

Did you notice a fine point of graphing technique? Because both $x$ and $y$ measure the same thing, the graph is square and the same scales appear on both axes.

Some scatterplots appear quite different from the cloud of points in Figure 2.1 and the linear pattern in Figure 2.3. This is true, for example, in experiments in which measurements of a response variable are taken at a few selected levels of the explanatory variable. The following example illustrates the use of scatterplots in this setting.

EXAMPLE

**2.8 Predators and prey.** Here is one way in which nature regulates the size of animal populations: high population density attracts predators, who remove a higher proportion of the population than when the density of the prey is low. One study looked at kelp perch and their common predator, the kelp bass. The researcher set up four large circular pens on sandy ocean bottom in southern California. He chose young perch at random from a large group and placed 10, 20, 40, and 60 perch in the four pens. Then he dropped the nets protecting the pens, allowing bass to swarm in, and counted the perch left after 2 hours. Here are data on the proportions of perch eaten in four repetitions of this setup:[6]

| Perch | Proportion killed | | | |
|---|---|---|---|---|
| 10 | 0.0 | 0.1 | 0.3 | 0.3 |
| 20 | 0.2 | 0.3 | 0.3 | 0.6 |
| 40 | 0.075 | 0.3 | 0.6 | 0.725 |
| 60 | 0.517 | 0.55 | 0.7 | 0.817 |

The scatterplot in Figure 2.4 displays the results of this experiment. Because number of perch in a pen is the explanatory variable, we plot it horizontally as the $x$ variable. The proportion of perch eaten by bass is the response variable $y$. Notice that there are two identical responses in the 10-perch group and also in the 20-perch group. These pairs of observations occupy the same points on the plot, so we use a different symbol for points that represent two observations. *Most software does not alert you to repeated values in your data when making scatterplots.* This can affect the impression the plot creates, especially when there are just a few points.

CAUTION

The vertical spread of points above each pen size shows the variation in proportions of perch eaten by bass. To see the overall pattern behind this

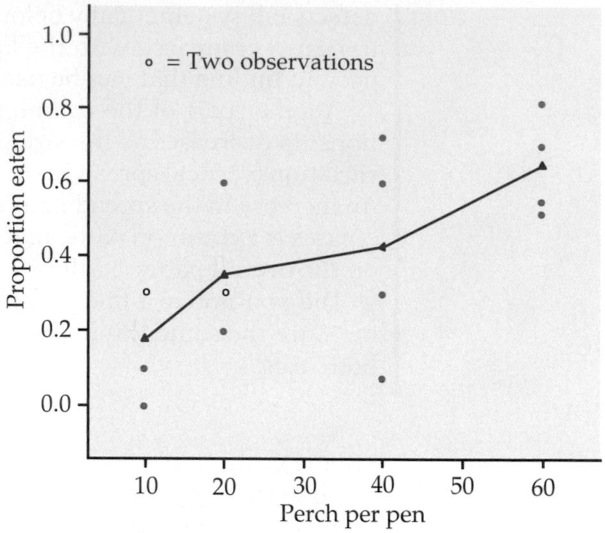

**FIGURE 2.4** Data from an experiment in ecology; proportion of perch eaten by bass plotted against the number of perch present, for Example 2.8. The lines connect the mean responses (triangles) for each group.

variation, plot the mean response for each pen size. In Figure 2.4, these means are marked by triangles and joined by line segments. There is a clear positive association between number of prey present and proportion eaten by predators. Moreover, the relationship is not far from linear.

## BEYOND THE BASICS

### Scatterplot Smoothers

A scatterplot provides a complete picture of the relationship between two quantitative variables. A complete picture is often too detailed for easy interpretation, so we try to describe the plot in terms of an overall pattern and deviations from that pattern. Though we can often do this by eye, more systematic methods of extracting the overall pattern are helpful. This is called **smoothing** a scatterplot. Example 2.9 suggests how to proceed when we are plotting a response variable $y$ against an explanatory variable $x$. We smoothed Figure 2.4 by averaging the $y$-values separately for each $x$-value. Though not all scatterplots have many $y$-values at the same value of $x$, as did Figure 2.4, modern software provides scatterplot smoothers that form an overall pattern by looking at the $y$-values for points in the neighborhood of each $x$-value. Smoothers use *resistant* calculations, so they are not affected by outliers in the plot.

smoothing

EXAMPLE

**2.9 Dummies in motorcycle crashes.** Crash a motorcycle into a wall. The rider, fortunately, is a dummy with an instrument to measure acceleration (change of velocity) mounted in its head. Figure 2.5 plots the acceleration of the dummy's head against time in milliseconds.[7] Acceleration is measured in g's, or multiples of the acceleration due to gravity at the earth's surface. The motorcycle approaches the wall at a constant speed (acceleration near 0). As it hits, the dummy's head snaps forward and decelerates violently (negative

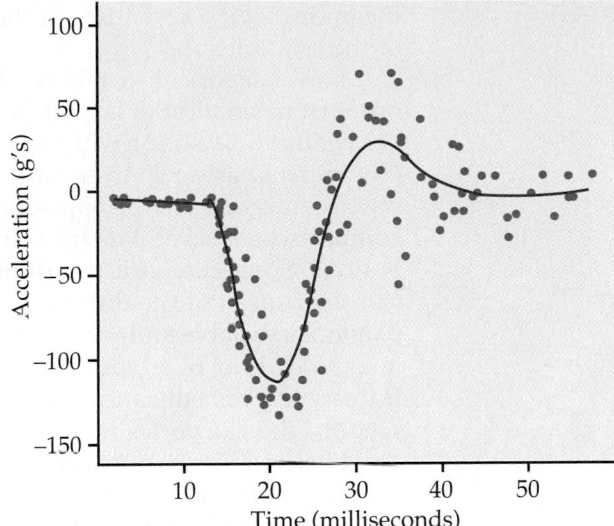


**FIGURE 2.5** Smoothing a scatterplot, for Example 2.9. Time plot of the acceleration of the head of a crash dummy as a motorcycle hits a wall, with the overall pattern calculated by a scatterplot smoother.

acceleration reaching more than 100 g's), then snaps back again (up to 75 g's) and wobbles a bit before coming to rest.

The scatterplot has a clear overall pattern, but it does not obey a simple form such as linear. Moreover, the strength of the pattern varies, from quite strong at the left of the plot to weaker (much more scatter) at the right. A scatterplot smoother deals with this complexity quite effectively and draws a line on the plot to represent the overall pattern.

## Categorical explanatory variables

Scatterplots display the association between two quantitative variables. To display a relationship between a categorical explanatory variable and a quantitative response variable, make a side-by-side comparison of the distributions of the response for each category. We have already met some tools for such comparisons:

- A back-to-back stemplot compares two distributions. See the comparison of literacy rates (the quantitative response) for females and males (two categories) on page 11.
- Side-by-side boxplots compare any number of distributions. See the comparison of gas mileage (the quantitative response) for minicompact and two-seater cars on the highway and in the city (four categories) in Figure 1.19 (page 37).

You can also use a type of scatterplot to display the association between a categorical explanatory variable and a quantitative response. Suppose, for example, that the prey-predator study of Example 2.8 had compared four species of prey rather than four densities of prey. The plot in Figure 2.4 remains helpful if we mark the prey species as A, B, C, and D at equal intervals on the horizontal axis in place of the count of perch per pen. A graph of the mean or median responses at the four locations still shows the overall nature of the relationship.

Many categorical variables, like prey species or type of car, have no natural order from smallest to largest. In such situations we cannot speak of a positive or negative association with the response variable. If the mean responses in our plot increase as we go from left to right, we could make them decrease by writing the categories in the opposite order. The plot simply presents a side-by-side comparison of several distributions. The categorical variable labels the distributions. Some categorical variables do have a least-to-most order, however. We can then speak of the direction of the association between the categorical explanatory variable and the quantitative response. Look again at the boxplots of income by level of education in Figure 1.23, on page 52. Although the Census Bureau records education in categories, such as "did not graduate from high school," the categories have an order from less education to more education. The boxes in Figure 1.23 are arranged in order of increasing education. They show a positive association between education and income: people with more education tend to have higher incomes.

## SECTION 2.1 Summary

To study relationships between variables, we must measure the variables on the same group of individuals or cases.

If we think that a variable $x$ may explain or even cause changes in another variable $y$, we call $x$ an **explanatory variable** and $y$ a **response variable.**

A **scatterplot** displays the relationship between two quantitative variables. Mark values of one variable on the horizontal axis ($x$ axis) and values of the other variable on the vertical axis ($y$ axis). Plot each individual's data as a point on the graph.

Always plot the explanatory variable, if there is one, on the $x$ axis of a scatterplot. Plot the response variable on the $y$ axis.

Plot points with different colors or symbols to see the effect of a categorical variable in a scatterplot.

In examining a scatterplot, look for an overall pattern showing the **form, direction,** and **strength** of the relationship, and then for **outliers** or other deviations from this pattern.

**Form: Linear relationships,** where the points show a straight-line pattern, are an important form of relationship between two variables. Curved relationships and **clusters** are other forms to watch for.

**Direction:** If the relationship has a clear direction, we speak of either **positive association** (high values of the two variables tend to occur together) or **negative association** (high values of one variable tend to occur with low values of the other variable).

**Strength:** The **strength** of a relationship is determined by how close the points in the scatterplot lie to a simple form such as a line.

To display the relationship between a categorical explanatory variable and a quantitative response variable, make a graph that compares the distributions of the response for each category of the explanatory variable.

## SECTION 2.1 Exercises

*For Exercise 2.1, see page 84; for Exercises 2.2 and 2.3, see page 85; and for Exercise 2.4, see page 89.*

**2.5 Average temperatures.** Here are the average temperatures in degrees for Lafayette, Indiana, during the months of February through May:

| Month | February | March | April | May |
|---|---|---|---|---|
| Temperature (degrees F) | 30 | 41 | 51 | 62 |

(a) Explain why month should be the explanatory variable for examining this relationship.

(b) Make a scatterplot and describe the relationship.

**2.6 Relationship between first test and final exam.** How strong is the relationship between the score on the first test and the score on the the final exam in an elementary statistics course? Here are data for eight students from such a course:

| First-test score | 153 | 144 | 162 | 149 | 127 | 118 | 158 | 153 |
|---|---|---|---|---|---|---|---|---|
| Final-exam score | 145 | 140 | 145 | 170 | 145 | 175 | 170 | 160 |

(a) Which variable should play the role of the explanatory variable in describing this relationship?

(b) Make a scatterplot and describe the relationship.

(c) Give some possible reasons why this relationship is so weak.

**2.7 Relationship between second test and final exam.** Refer to the previous exercise. Here are the data for the second test and the final exam for the same students:

| Second-test score | 158 | 162 | 144 | 162 | 136 | 158 | 175 | 153 |
|---|---|---|---|---|---|---|---|---|
| Final-exam score | 145 | 140 | 145 | 170 | 145 | 175 | 170 | 160 |

(a) Explain why you should use the second-test score as the explanatory variable.

(b) Make a scatterplot and describe the relationship.

(c) Why do you think the relationship between the second-test score and the final-exam score is stronger than the relationship between the first-test score and the final-exam score?

**2.8 Add an outlier to the plot.** Refer to the previous exercise. Add a ninth student whose scores on the second test and final exam would lead you to classify the additional data point as an outlier. Highlight the outlier on your scatterplot and describe the performance of the student on the second exam and final exam and why that leads to the conclusion that the result is an outlier. Give a possible reason for the performance of this student.

**2.9 Explanatory and response variables.** In each of the following situations, is it more reasonable to simply explore the relationship between the two variables or to view one of the variables as an explanatory variable and the other as a response variable? In the latter case, which is the explanatory variable and which is the response variable?

(a) The weight of a child and the age of the child from birth to 10 years.

(b) High school English grades and high school math grades.

(c) The rental price of apartments and the number of bedrooms in the apartment.

(d) The amount of sugar added to a cup of coffee and how sweet the coffee tastes.

(e) The student evaluation scores for an instructor and the student evaluation scores for the course.

**2.10 Parents' income and student loans.** How well does the income of a college student's parents predict how much the student will borrow to pay for college? We have data on parents' income and college debt for a sample of 1200 recent college graduates. What are the explanatory and response variables? Are these variables categorical or quantitative? Do you expect a positive or negative association between these variables? Why?

**2.11 Reading ability and IQ.** A study of reading ability in schoolchildren chose 60 fifth-grade children at random from a school. The researchers had the children's scores on an IQ test and on a test of reading ability.[8] Figure 2.6 (on page 96) plots reading test score (response) against IQ score (explanatory).

(a) Explain why we should expect a positive association between IQ and reading score for children in the same grade. Does the scatterplot show a positive association?

(b) A group of four points appear to be outliers. In what way do these children's IQ and reading scores deviate from the overall pattern?

(c) Ignoring the outliers, is the association between IQ and reading scores roughly linear? Is it very strong? Explain your answers.

**2.12 Treasury bills and common stocks.** What is the relationship between returns from buying Treasury bills and returns from buying common stocks? The stemplots in Figure 1.22 (page 44) show the two

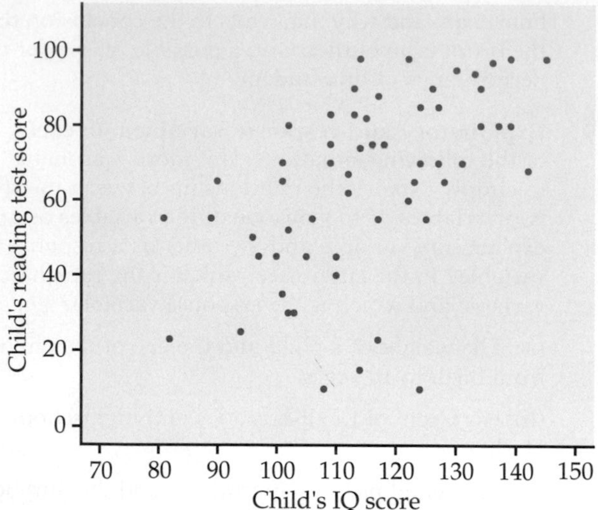


**FIGURE 2.6** IQ and reading test scores for 60 fifth-grade children, for Exercise 2.11.

individual distributions of percent returns. To see the relationship, we need a scatterplot. Figure 2.7 plots the annual returns on stocks for the years 1950 to 2003 against the returns on Treasury bills for the same years.

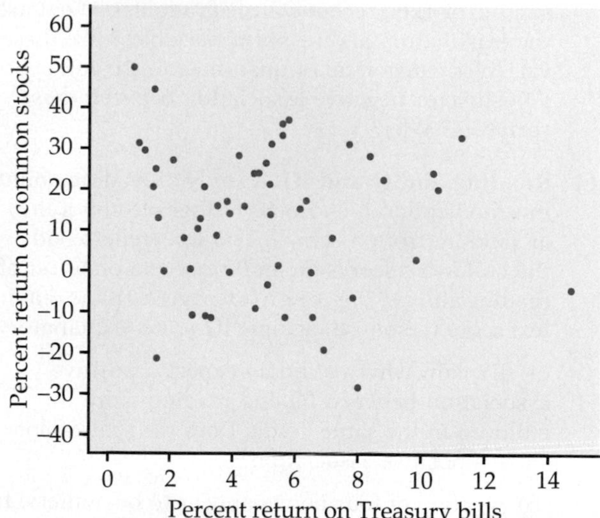


**FIGURE 2.7** Percent return on Treasury bills and common stocks for the years 1950 to 2003, for Exercise 2.12.

(a) The best year for stocks during this period was 1954. The worst year was 1974. About what were the returns on stocks in those two years?

(b) Treasury bills are a measure of the general level of interest rates. The years around 1980 saw very high interest rates. Treasury bill returns peaked in 1981. About what was the percent return that year?

(c) Some people say that high Treasury bill returns tend to go with low returns on stocks. Does such a pattern appear clearly in Figure 2.7? Does the plot have any clear pattern?

**2.13 Can children estimate their reading ability?** The main purpose of the study cited in Exercise 2.11 was to ask whether schoolchildren can estimate their own reading ability. The researchers had the children's scores on a test of reading ability. They asked each child to estimate his or her reading level, on a scale from 1 (low) to 5 (high). Figure 2.8 is a scatterplot of the children's estimates (response) against their reading scores (explanatory).

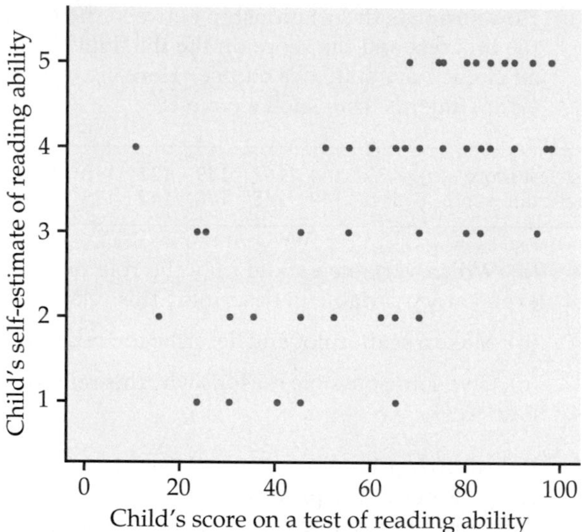


**FIGURE 2.8** Reading test scores for 60 fifth-grade children and the children's estimates of their own reading levels, for Exercise 2.13.

(a) What explains the "stair-step" pattern in the plot?

(b) Is there an overall positive association between reading score and self-estimate?

(c) There is one clear outlier. What is this child's self-estimated reading level? Does this appear to over- or underestimate the level as measured by the test?

**2.14 Literacy of men and women.** Table 1.2 (page 10) shows the percent of men and women at least 15 years old who were literate in 2002 in the major Islamic nations for which data were available. Make a scatterplot of these data, taking male literacy as the explanatory variable. Describe the direction, form, and strength of the relationship. Are there any identical observations that plot as the same point? Are there any clear outliers?

**2.15 Small falcons in Sweden.** Often the percent of an animal species in the wild that survive to breed again is lower following a successful breeding season. This is part of nature's self-regulation, tending to keep population size stable. A study of merlins (small falcons) in northern Sweden observed the number of breeding pairs in an isolated area and the percent of males (banded for identification) who returned the next breeding season. Here are data for nine years:[9]

| Pairs | 28 | 29 | 29 | 29 | 30 | 32 | 33 | 38 | 38 |
|---|---|---|---|---|---|---|---|---|---|
| Percent | 82 | 83 | 70 | 61 | 69 | 58 | 43 | 50 | 47 |

(a) Why is the response variable the *percent* of males that return rather than the *number* of males that return?

(b) Make a scatterplot. To emphasize the pattern, also plot the mean response for years with 29 and 38 breeding pairs and draw lines connecting the mean responses for the six values of the explanatory variable.

(c) Describe the pattern. Do the data support the theory that a smaller percent of birds survive following a successful breeding season?

**2.16 City and highway gas mileage.** Table 1.10 (page 31) gives the city and highway gas mileages for minicompact and two-seater cars. We expect a positive association between the city and highway mileages of a group of vehicles. We have already seen that the Honda Insight is a different type of car, so omit it as you work with these data.

(a) Make a scatterplot that shows the relationship between city and highway mileage, using city mileage as the explanatory variable. Use different plotting symbols for the two types of cars.

(b) Interpret the plot. Is there a positive association? Is the form of the plot roughly linear? Is the form of the relationship similar for the two types of car? What is the most important difference between the two types?

**2.17 Social rejection and pain.** We often describe our emotional reaction to social rejection as "pain." A clever study asked whether social rejection causes activity in areas of the brain that are known to be activated by physical pain. If it does, we really do experience social and physical pain in similar ways. Subjects were first included and then deliberately excluded from a social activity while increases in blood flow in their brains were measured. After each activity, the subjects filled out questionnaires that assessed how excluded they felt.

Below are data for 13 subjects.[10] The explanatory variable is "social distress" measured by each subject's questionnaire score after exclusion relative to the score after inclusion. (So values greater than 1 show the degree of distress caused by exclusion.) The response variable is activity in the anterior cingulate cortex, a region of the brain that is activated by physical pain.

| Subject | Social distress | Brain activity | Subject | Social distress | Brain activity |
|---|---|---|---|---|---|
| 1 | 1.26 | −0.055 | 8 | 2.18 | 0.025 |
| 2 | 1.85 | −0.040 | 9 | 2.58 | 0.027 |
| 3 | 1.10 | −0.026 | 10 | 2.75 | 0.033 |
| 4 | 2.50 | −0.017 | 11 | 2.75 | 0.064 |
| 5 | 2.17 | −0.017 | 12 | 3.33 | 0.077 |
| 6 | 2.67 | 0.017 | 13 | 3.65 | 0.124 |
| 7 | 2.01 | 0.021 | | | |

Plot brain activity against social distress. Describe the direction, form, and strength of the relationship, as well as any outliers. Do the data suggest that brain activity in the "pain" region is directly related to the distress from social exclusion?

**2.18 Biological clocks.** Many plants and animals have "biological clocks" that coordinate activities with the time of day. When researchers looked at the length of the biological cycles in the plant *Arabidopsis* by measuring leaf movements, they found that the length of the cycle is not always 24 hours. The researchers suspected that the plants adapt their clocks to their north-south position. Plants don't know geography, but they do respond to light, so the researchers looked at the relationship between the plants' cycle lengths and the length of the day on June 21 at their locations. The data file has data on cycle length and day length, both in hours, for 146 plants.[11] Plot cycle length as the response variable against day length as the explanatory variable. Does there appear to be a positive association? Is it a strong association? Explain your answers.

**2.19 Business revenue and team value in the NBA.** Management theory says that the value of a business should depend on its operating income, the income produced by the business after taxes. (Operating income excludes income from sales of assets and investments, which don't reflect the actual business.) Total revenue, which ignores costs, should be less

**TABLE 2.1**

NBA teams as businesses

| Team | Value ($millions) | Revenue ($millions) | Income ($millions) |
|---|---|---|---|
| Los Angeles Lakers | 447 | 149 | 22.8 |
| New York Knicks | 401 | 160 | 13.5 |
| Chicago Bulls | 356 | 119 | 49.0 |
| Dallas Mavericks | 338 | 117 | −17.7 |
| Philadelphia 76ers | 328 | 109 | 2.0 |
| Boston Celtics | 290 | 97 | 25.6 |
| Detroit Pistons | 284 | 102 | 23.5 |
| San Antonio Spurs | 283 | 105 | 18.5 |
| Phoenix Suns | 282 | 109 | 21.5 |
| Indiana Pacers | 280 | 94 | 10.1 |
| Houston Rockets | 278 | 82 | 15.2 |
| Sacramento Kings | 275 | 102 | −16.8 |
| Washington Wizards | 274 | 98 | 28.5 |
| Portland Trail Blazers | 272 | 97 | −85.1 |
| Cleveland Cavaliers | 258 | 72 | 3.8 |
| Toronto Raptors | 249 | 96 | 10.6 |
| New Jersey Nets | 244 | 94 | −1.6 |
| Utah Jazz | 239 | 85 | 13.8 |
| Miami Heat | 236 | 91 | 7.9 |
| Minnesota Timberwolves | 230 | 85 | 6.9 |
| Memphis Grizzlies | 227 | 63 | −19.7 |
| Denver Nuggets | 218 | 75 | 7.9 |
| New Orleans Hornets | 216 | 80 | 21.9 |
| Los Angeles Clippers | 208 | 72 | 15.9 |
| Atlanta Hawks | 202 | 78 | −8.4 |
| Orlando Magic | 199 | 80 | 13.1 |
| Seattle Supersonics | 196 | 70 | 2.4 |
| Golden State Warriors | 188 | 70 | 7.8 |
| Milwaukee Bucks | 174 | 70 | −15.1 |

important. Table 2.1 shows the values, operating incomes, and revenues of an unusual group of businesses: the teams in the National Basketball Association (NBA).[12] Professional sports teams are generally privately owned, often by very wealthy individuals who may treat their team as a source of prestige rather than as a business.

(a) Plot team value against revenue. There are several outliers. Which teams are these, and in what way are they outliers? Is there a positive association between value and revenue? Is the pattern roughly linear?

(b) Now plot value against operating income. Are the same teams outliers? Does revenue or operating income better predict the value of an NBA team?

**2.20 Two problems with feet.** Metatarsus adductus (call it MA) is a turning in of the front part of the foot that is common in adolescents and usually corrects itself. Hallux abducto valgus (call it HAV) is a deformation of the big toe that is not common in youth and often requires surgery. Perhaps the severity of MA can help predict the severity of HAV. Table 2.2 gives data on 38 consecutive patients who came to a medical center for HAV surgery.[13] Using X-rays, doctors measured the angle of deformity for both MA and HAV. They speculated that there is a positive association—more serious MA is associated with more serious HAV.

**TABLE 2.2**

Two measurements of foot deformities

| HAV angle | MA angle | HAV angle | MA angle | HAV angle | MA angle |
|---|---|---|---|---|---|
| 28 | 18 | 21 | 15 | 16 | 10 |
| 32 | 16 | 17 | 16 | 30 | 12 |
| 25 | 22 | 16 | 10 | 30 | 10 |
| 34 | 17 | 21 | 7 | 20 | 10 |
| 38 | 33 | 23 | 11 | 50 | 12 |
| 26 | 10 | 14 | 15 | 25 | 25 |
| 25 | 18 | 32 | 12 | 26 | 30 |
| 18 | 13 | 25 | 16 | 28 | 22 |
| 30 | 19 | 21 | 16 | 31 | 24 |
| 26 | 10 | 22 | 18 | 38 | 20 |
| 28 | 17 | 20 | 10 | 32 | 37 |
| 13 | 14 | 18 | 15 | 21 | 23 |
| 20 | 20 | 26 | 16 | | |

(a) Make a scatterplot of the data in Table 2.2. (Which is the explanatory variable?)

(b) Describe the form, direction, and strength of the relationship between MA angle and HAV angle. Are there any clear outliers in your graph?

(c) Do you think the data confirm the doctors' speculation?

**2.21 Body mass and metabolic rate.** Metabolic rate, the rate at which the body consumes energy, is important in studies of weight gain, dieting, and exercise. The table below gives data on the lean body mass and resting metabolic rate for 12 women and 7 men who are subjects in a study of dieting. Lean body mass, given in kilograms, is a person's weight leaving out all fat. Metabolic rate is measured in calories burned per 24 hours, the same calories

| Subject | Sex | Mass | Rate | Subject | Sex | Mass | Rate |
|---|---|---|---|---|---|---|---|
| 1 | M | 62.0 | 1792 | 11 | F | 40.3 | 1189 |
| 2 | M | 62.9 | 1666 | 12 | F | 33.1 | 913 |
| 3 | F | 36.1 | 995 | 13 | M | 51.9 | 1460 |
| 4 | F | 54.6 | 1425 | 14 | F | 42.4 | 1124 |
| 5 | F | 48.5 | 1396 | 15 | F | 34.5 | 1052 |
| 6 | F | 42.0 | 1418 | 16 | F | 51.1 | 1347 |
| 7 | M | 47.4 | 1362 | 17 | F | 41.2 | 1204 |
| 8 | F | 50.6 | 1502 | 18 | M | 51.9 | 1867 |
| 9 | F | 42.0 | 1256 | 19 | M | 46.9 | 1439 |
| 10 | M | 48.7 | 1614 | | | | |

**TABLE 2.3**

World record times for the 10,000-meter run

| Men | | | | Women | |
|---|---|---|---|---|---|
| Record year | Time (seconds) | Record year | Time (seconds) | Record year | Time (seconds) |
| 1912 | 1880.8 | 1962 | 1698.2 | 1967 | 2286.4 |
| 1921 | 1840.2 | 1963 | 1695.6 | 1970 | 2130.5 |
| 1924 | 1835.4 | 1965 | 1659.3 | 1975 | 2100.4 |
| 1924 | 1823.2 | 1972 | 1658.4 | 1975 | 2041.4 |
| 1924 | 1806.2 | 1973 | 1650.8 | 1977 | 1995.1 |
| 1937 | 1805.6 | 1977 | 1650.5 | 1979 | 1972.5 |
| 1938 | 1802.0 | 1978 | 1642.4 | 1981 | 1950.8 |
| 1939 | 1792.6 | 1984 | 1633.8 | 1981 | 1937.2 |
| 1944 | 1775.4 | 1989 | 1628.2 | 1982 | 1895.3 |
| 1949 | 1768.2 | 1993 | 1627.9 | 1983 | 1895.0 |
| 1949 | 1767.2 | 1993 | 1618.4 | 1983 | 1887.6 |
| 1949 | 1761.2 | 1994 | 1612.2 | 1984 | 1873.8 |
| 1950 | 1742.6 | 1995 | 1603.5 | 1985 | 1859.4 |
| 1953 | 1741.6 | 1996 | 1598.1 | 1986 | 1813.7 |
| 1954 | 1734.2 | 1997 | 1591.3 | 1993 | 1771.8 |
| 1956 | 1722.8 | 1997 | 1587.8 | | |
| 1956 | 1710.4 | 1998 | 1582.7 | | |
| 1960 | 1698.8 | 2004 | 1580.3 | | |

used to describe the energy content of foods. The researchers believe that lean body mass is an important influence on metabolic rate.

(a) Make a scatterplot of the data, using different symbols or colors for men and women.

(b) Is the association between these variables positive or negative? What is the form of the relationship? How strong is the relationship? Does the pattern of the relationship differ for women and men? How do the male subjects as a group differ from the female subjects as a group?

**2.22 Fuel consumption and speed.** How does the fuel consumption of a car change as its speed increases? Below are data for a British Ford Escort. Speed is measured in kilometers per hour, and fuel consumption is measured in liters of gasoline used per 100 kilometers traveled.[14]

| Speed (km/h) | Fuel used (liters/100 km) | Speed (km/h) | Fuel used (liter/100 km) |
|---|---|---|---|
| 10 | 21.00 | 90 | 7.57 |
| 20 | 13.00 | 100 | 8.27 |
| 30 | 10.00 | 110 | 9.03 |
| 40 | 8.00 | 120 | 9.87 |
| 50 | 7.00 | 130 | 10.79 |
| 60 | 5.90 | 140 | 11.77 |
| 70 | 6.30 | 150 | 12.83 |
| 80 | 6.95 | | |

(a) Make a scatterplot. (Which variable should go on the $x$ axis?)

(b) Describe the form of the relationship. In what way is it not linear? Explain why the form of the relationship makes sense.

(c) It does not make sense to describe the variables as either positively associated or negatively associated. Why not?

(d) Is the relationship reasonably strong or quite weak? Explain your answer.

**2.23 World records for the 10K.** Table 2.3 shows the progress of world record times (in seconds) for the 10,000-meter run up to mid-2004.[15] Concentrate on the women's world record times. Make a scatterplot with year as the explanatory variable. Describe the pattern of improvement over time that your plot displays.

**2.24** CHALLENGE **How do icicles grow?** How fast do icicles grow? Japanese researchers measured the growth of icicles in a cold chamber under various conditions of temperature, wind, and water flow.[16] Table 2.4 contains data produced under two sets of conditions. In both cases, there was no wind and the temperature was set at −11°C. Water flowed over the icicle at a higher rate (29.6 milligrams per second) in Run 8905 and at a slower rate (11.9 mg/s) in Run 8903.

**TABLE 2.4**

Growth of icicles over time

| Run 8903 | | | | Run 8905 | | | |
|---|---|---|---|---|---|---|---|
| Time (min) | Length (cm) | Time (min) | Length (cm) | Time (min) | Length (cm) | Time (min) | Length (cm) |
| 10 | 0.6 | 130 | 18.1 | 10 | 0.3 | 130 | 10.4 |
| 20 | 1.8 | 140 | 19.9 | 20 | 0.6 | 140 | 11.0 |
| 30 | 2.9 | 150 | 21.0 | 30 | 1.0 | 150 | 11.9 |
| 40 | 4.0 | 160 | 23.4 | 40 | 1.3 | 160 | 12.7 |
| 50 | 5.0 | 170 | 24.7 | 50 | 3.2 | 170 | 13.9 |
| 60 | 6.1 | 180 | 27.8 | 60 | 4.0 | 180 | 14.6 |
| 70 | 7.9 | | | 70 | 5.3 | 190 | 15.8 |
| 80 | 10.1 | | | 80 | 6.0 | 200 | 16.2 |
| 90 | 10.9 | | | 90 | 6.9 | 210 | 17.9 |
| 100 | 12.7 | | | 100 | 7.8 | 220 | 18.8 |
| 110 | 14.4 | | | 110 | 8.3 | 230 | 19.9 |
| 120 | 16.6 | | | 120 | 9.6 | 240 | 21.1 |

(a) Make a scatterplot of the length of the icicle in centimeters versus time in minutes, using separate symbols for the two runs.

(b) Write a careful explanation of what your plot shows about the growth of icicles.

**2.25 Records for men and women in the 10K.** Table 2.3 shows the progress of world record times (in seconds) for the 10,000-meter run for both men and women.

(a) Make a scatterplot of world record time against year, using separate symbols for men and women. Describe the pattern for each sex. Then compare the progress of men and women.

(b) Women began running this long distance later than men, so we might expect their improvement to be more rapid. Moreover, it is often said that men have little advantage over women in distance running as opposed to sprints, where muscular strength plays a greater role. Do the data appear to support these claims?

**2.26 Worms and plant growth.** To demonstrate the effect of nematodes (microscopic worms) on plant growth, a botanist introduces different numbers of nematodes into 16 planting pots. He then transplants a tomato seedling into each pot. Here are data on the increase in height of the seedlings (in centimeters) 14 days after planting:[17]

| Nematodes | Seedling growth | | | |
|---|---|---|---|---|
| 0 | 10.8 | 9.1 | 13.5 | 9.2 |
| 1,000 | 11.1 | 11.1 | 8.2 | 11.3 |
| 5,000 | 5.4 | 4.6 | 7.4 | 5.0 |
| 10,000 | 5.8 | 5.3 | 3.2 | 7.5 |

(a) Make a scatterplot of the response variable (growth) against the explanatory variable (nematode count). Then compute the mean growth for each group of seedlings, plot the means against the nematode counts, and connect these four points with line segments.

(b) Briefly describe the conclusions about the effects of nematodes on plant growth that these data suggest.

**2.27 Mutual funds.** Fidelity Investments, like other large mutual funds companies, offers many "sector funds" that concentrate their investments in narrow segments of the stock market. These funds often rise or fall by much more than the market as a whole. We can group them by broader market sector to compare returns. Here are percent total returns for 23 Fidelity "Select Portfolios" funds for the year 2003, a year in which stocks rose sharply:[18]

| Market sector | Fund returns (percent) | | | | | | |
|---|---|---|---|---|---|---|---|
| Consumer | 23.9 | 14.1 | 41.8 | 43.9 | 31.1 | | |
| Financial services | 32.3 | 36.5 | 30.6 | 36.9 | 27.5 | | |
| Technology | 26.1 | 62.7 | 68.1 | 71.9 | 57.0 | 35.0 | 59.4 |
| Natural resources | 22.9 | 7.6 | 32.1 | 28.7 | 29.5 | 19.1 | |

(a) Make a plot of total return against market sector (space the four market sectors equally on the horizontal axis). Compute the mean return for each sector, add the means to your plot, and connect the means with line segments.

(b) Based on the data, which of these market sectors were the best places to invest in 2003? Hindsight is wonderful.

(c) Does it make sense to speak of a positive or negative association between market sector and total return?

**2.28** CHALLENGE **Mutual funds in another year.** The data for 2003 in the previous exercise make sector funds look attractive. Stocks rose sharply in 2003, after falling sharply in 2002 (and also in 2001 and 2000). Let's look at the percent returns for both 2003 and 2002 for these same 23 funds. Here they are:

| 2002 return | 2003 return | 2002 return | 2003 return | 2002 return | 2003 return |
|---|---|---|---|---|---|
| −17.1 | 23.9 | −0.7 | 36.9 | −37.8 | 59.4 |
| −6.7 | 14.1 | −5.6 | 27.5 | −11.5 | 22.9 |
| −21.1 | 41.8 | −26.9 | 26.1 | −0.7 | 36.9 |
| −12.8 | 43.9 | −42.0 | 62.7 | 64.3 | 32.1 |
| −18.9 | 31.1 | −47.8 | 68.1 | −9.6 | 28.7 |
| −7.7 | 32.3 | −50.5 | 71.9 | −11.7 | 29.5 |
| −17.2 | 36.5 | −49.5 | 57.0 | −2.3 | 19.1 |
| −11.4 | 30.6 | −23.4 | 35.0 | | |

Do a careful graphical analysis of these data: side-by-side comparison of the distributions of returns in 2002 and 2003 and also a description of the relationship between the returns of the same funds in these two years. What are your most important findings? (The outlier is Fidelity Gold Fund.)

## 2.2 Correlation

A scatterplot displays the form, direction, and strength of the relationship between two quantitative variables. Linear (straight-line) relations are particularly important because a straight line is a simple pattern that is quite common. We say a linear relationship is strong if the points lie close to a straight line, and weak if they are widely scattered about a line. Our eyes are not good judges of how strong a relationship is. The two scatterplots in Figure 2.9 depict exactly the same data, but the plot on the right is drawn smaller in a large field. The plot on the left seems to show a stronger relationship. Our eyes can be fooled by changing the plotting scales or the amount of white space around the cloud of points in a scatterplot.[19] We need to follow our strategy for data analysis by using a numerical measure to supplement the graph. *Correlation* is the measure we use.

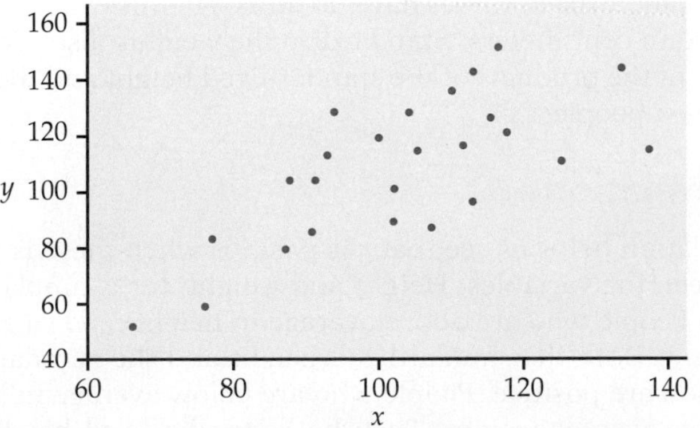


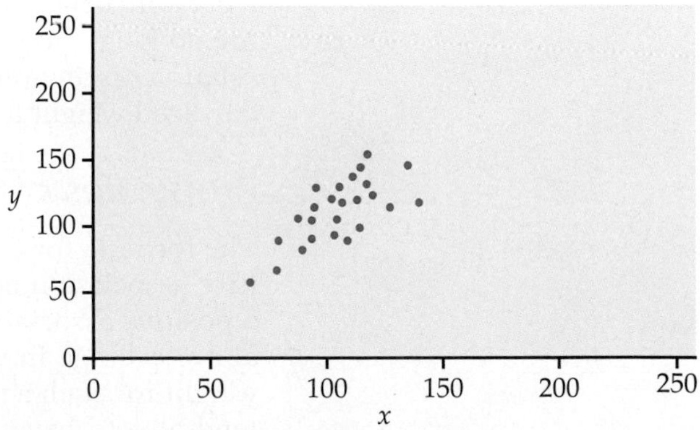


**FIGURE 2.9** Two scatterplots of the same data. The linear pattern in the plot on the right appears stronger because of the surrounding space.

## The correlation $r$

We have data on variables $x$ and $y$ for $n$ individuals. Think, for example, of measuring height and weight for $n$ people. Then $x_1$ and $y_1$ are your height and your weight, $x_2$ and $y_2$ are my height and my weight, and so on. For the $i$th individual, height $x_i$ goes with weight $y_i$. Here is the definition of correlation.

### CORRELATION

The **correlation** measures the direction and strength of the linear relationship between two quantitative variables. Correlation is usually written as $r$.

Suppose that we have data on variables $x$ and $y$ for $n$ individuals. The means and standard deviations of the two variables are $\overline{x}$ and $s_x$ for the $x$-values, and $\overline{y}$ and $s_y$ for the $y$-values. The correlation $r$ between $x$ and $y$ is

$$r = \frac{1}{n-1} \sum \left( \frac{x_i - \overline{x}}{s_x} \right) \left( \frac{y_i - \overline{y}}{s_y} \right)$$

As always, the summation sign $\sum$ means "add these terms for all the individuals." The formula for the correlation $r$ is a bit complex. It helps us see what correlation is but is not convenient for actually calculating $r$. In practice you should use software or a calculator that finds $r$ from keyed-in values of two variables $x$ and $y$. Exercise 2.29 asks you to calculate a correlation step-by-step from the definition to solidify its meaning.

The formula for $r$ begins by standardizing the observations. Suppose, for example, that $x$ is height in centimeters and $y$ is weight in kilograms and that we have height and weight measurements for $n$ people. Then $\overline{x}$ and $s_x$ are the mean and standard deviation of the $n$ heights, both in centimeters. The value

$$\frac{x_i - \overline{x}}{s_x}$$

is the standardized height of the $i$th person, familiar from Chapter 1. The standardized height says how many standard deviations above or below the mean a person's height lies. Standardized values have no units—in this example, they are no longer measured in centimeters. Standardize the weights also. The correlation $r$ is an average of the products of the standardized height and the standardized weight for the $n$ people.

## Properties of correlation

The formula for correlation helps us see that $r$ is positive when there is a positive association between the variables. Height and weight, for example, have a positive association. People who are above average in height tend to also be above average in weight. Both the standardized height and the standardized weight for such a person are positive. People who are below average in height tend also to have below-average weight. Then both standardized height and standardized weight are negative. In both cases, the products in the formula for $r$ are mostly positive and so $r$ is positive. In the same way, we can see that

$r$ is negative when the association between $x$ and $y$ is negative. More detailed study of the formula gives more detailed properties of $r$. Here is what you need to know in order to interpret correlation:

- Correlation makes no use of the distinction between explanatory and response variables. It makes no difference which variable you call $x$ and which you call $y$ in calculating the correlation.

- *Correlation requires that both variables be quantitative, so that it makes sense to do the arithmetic indicated by the formula for r.* We cannot calculate a correlation between the incomes of a group of people and what city they live in, because city is a categorical variable.
- Because $r$ uses the standardized values of the observations, $r$ does not change when we change the units of measurement of $x$, $y$, or both. Measuring height in inches rather than centimeters and weight in pounds rather than kilograms does not change the correlation between height and weight. The correlation $r$ itself has no unit of measurement; it is just a number.
- Positive $r$ indicates positive association between the variables, and negative $r$ indicates negative association.
- The correlation $r$ is always a number between $-1$ and 1. Values of $r$ near 0 indicate a very weak linear relationship. The strength of the relationship increases as $r$ moves away from 0 toward either $-1$ or 1. Values of $r$ close to $-1$ or 1 indicate that the points lie close to a straight line. The extreme values $r = -1$ and $r = 1$ occur only when the points in a scatterplot lie exactly along a straight line.

- Correlation measures the strength of only the linear relationship between two variables. *Correlation does not describe curved relationships between variables, no matter how strong they are.*

- *Like the mean and standard deviation, the correlation is not resistant: r is strongly affected by a few outlying observations.* Use $r$ with caution when outliers appear in the scatterplot.

The scatterplots in Figure 2.10 illustrate how values of $r$ closer to 1 or $-1$ correspond to stronger linear relationships. To make the essential meaning of $r$ clear, the standard deviations of both variables in these plots are equal and the horizontal and vertical scales are the same. In general, it is not so easy to guess the value of $r$ from the appearance of a scatterplot. Remember that changing the plotting scales in a scatterplot may mislead our eyes, but it does not change the standardized values of the variables and therefore cannot change the correlation. To explore how extreme observations can influence $r$, use the *Correlation and Regression* applet available on the text CD and Web site.

EXAMPLE

**2.10 Scatterplots and correlations.** The real data we have examined also illustrate the behavior of correlation.

Figure 2.1 (page 87), despite the clusters, shows a quite strong negative linear association between the percent of a state's high school seniors who take the SAT exam and their mean SAT score. The correlation is $r = -0.877$.

Figure 2.3 (page 90) shows a strong positive linear association between the two measurements of defect depth. The correlation is $r = 0.944$. That the

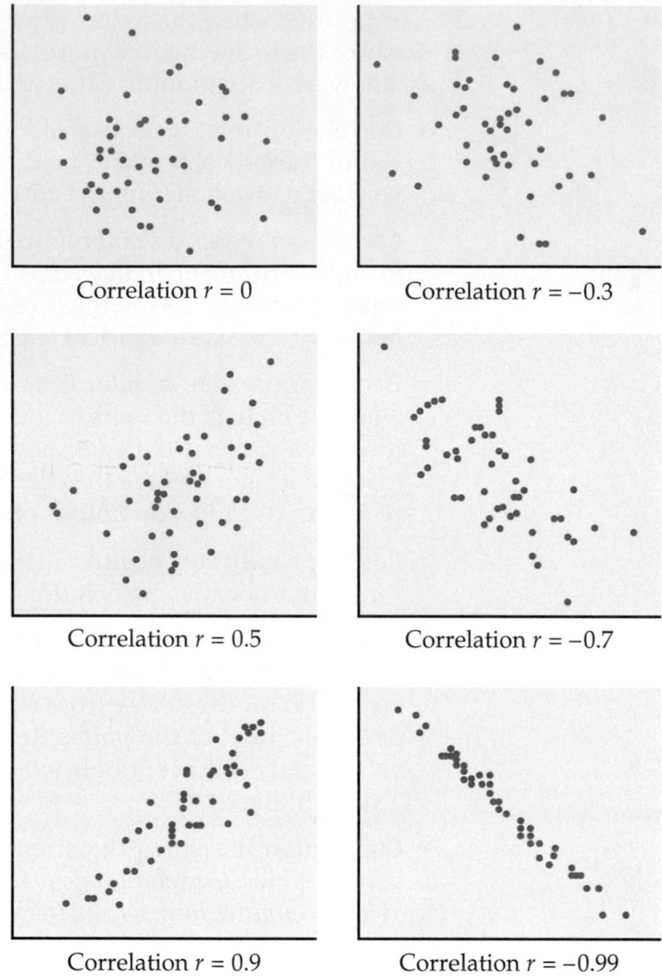


**FIGURE 2.10** How the correlation $r$ measures the direction and strength of a linear association.

pattern doesn't follow the $y = x$ line drawn on the graph doesn't matter—correlation measures closeness to whatever line describes the data, not to a line that we specify in advance.

Figure 2.7 (page 96) shows a very weak relationship between returns on Treasury bills and on common stocks. We expect a small negative $r$, and calculation gives $r = -0.113$.

The correlation between time and acceleration for the motorcycle crash data graphed in Figure 2.5 (page 93) is $r = 0.296$. Because the relationship is not at all linear, $r$ provides no useful information. *Always plot your data before calculating common statistical measures such as correlation.*

Finally, remember that **correlation is not a complete description of two-variable data,** even when the relationship between the variables is linear. You should give the means and standard deviations of both $x$ and $y$ along with the correlation. (Because the formula for correlation uses the means and standard deviations, these measures are the proper choices to accompany a correlation.) Conclusions based on correlations alone may require rethinking in the light of a more complete description of the data.

EXAMPLE

**2.11 Scoring of figure skating in the Olympics.** Until a scandal at the 2002 Olympics brought change, figure skating was scored by judges on a scale from 0.0 to 6.0. The scores were often controversial. We have the scores awarded by two judges, Pierre and Elena, to many skaters. How well do they agree? We calculate that the correlation between their scores is $r = 0.9$. But the mean of Pierre's scores is 0.8 point lower than Elena's mean.

These facts do not contradict each other. They are simply different kinds of information. The mean scores show that Pierre awards lower scores than Elena. But because Pierre gives *every* skater a score about 0.8 point lower than Elena, the correlation remains high. Adding the same number to all values of either $x$ or $y$ does not change the correlation. If both judges score the same skaters, the competition is scored consistently because Pierre and Elena agree on which performances are better than others. The high $r$ shows their agreement. But if Pierre scores some skaters and Elena others, we must add 0.8 points to Pierre's scores to arrive at a fair comparison.

## SECTION 2.2 Summary

The **correlation $r$** measures the direction and strength of the linear (straight line) association between two quantitative variables $x$ and $y$. Although you can calculate a correlation for any scatterplot, $r$ measures only linear relationships.

Correlation indicates the direction of a linear relationship by its sign: $r > 0$ for a positive association and $r < 0$ for a negative association.

Correlation always satisfies $-1 \leq r \leq 1$ and indicates the strength of a relationship by how close it is to $-1$ or 1. Perfect correlation, $r = \pm 1$, occurs only when the points lie exactly on a straight line.

Correlation ignores the distinction between explanatory and response variables. The value of $r$ is not affected by changes in the unit of measurement of either variable. Correlation is not resistant, so outliers can greatly change the value of $r$.

## SECTION 2.2 Exercises

**2.29 Coffee prices and deforestation.** Coffee is a leading export from several developing countries. When coffee prices are high, farmers often clear forest to plant more coffee trees. Here are data for five years on prices paid to coffee growers in Indonesia and the rate of deforestation in a national park that lies in a coffee-producing region:[20]

| Price (cents per pound) | Deforestation (percent) |
|---|---|
| 29 | 0.49 |
| 40 | 1.59 |
| 54 | 1.69 |
| 55 | 1.82 |
| 72 | 3.10 |

(a) Make a scatterplot. Which is the explanatory variable? What kind of pattern does your plot show?

(b) Find the correlation $r$ step-by-step. That is, find the mean and standard deviation of the two variables. Then find the five standardized values for each variable and use the formula for $r$. Explain how your value for $r$ matches your graph in (a).

(c) Now enter these data into your calculator or software and use the correlation function to find $r$. Check that you get the same result as in (b).

**2.30 First test and final exam.** In Exercise 2.6 you looked at the relationship between the score on the first test and the score on the final exam in an elementary statistics course. The data for eight students from such a course are presented in the following table.

| First-test score | 153 | 144 | 162 | 149 | 127 | 118 | 158 | 153 |
|---|---|---|---|---|---|---|---|---|
| Final-exam score | 145 | 140 | 145 | 170 | 145 | 175 | 170 | 160 |

(a) Find the correlation between these two variables.

(b) In Exercise 2.6 we noted that the relationship between these two variables is weak. Does your calculation of the correlation support this statement? Explain your answer.

**2.31 Second test and final exam.** Refer to the previous exercise. Here are the data for the second test and the final exam for the same students:

| Second-test score | 158 | 162 | 144 | 162 | 136 | 158 | 175 | 153 |
|---|---|---|---|---|---|---|---|---|
| Final-exam score | 145 | 140 | 145 | 170 | 145 | 175 | 170 | 160 |

(a) Find the correlation between these two variables.

(b) In Exercise 2.7 we noted that the relationship between these two variables is stronger than the relationship between the two variables in the previous exercise. How do the values of the correlations that you calculated support this statement? Explain your answer.

**2.32 The effect of an outlier.** Refer to the previous exercise. Add a ninth student whose scores on the second test and final exam would lead you to classify the additional data point as an outlier. Recalculate the correlation with this additional case and summarize the effect it has on the value of the correlation.

**2.33 The effect of a different point.** Examine the data in Exercise 2.31 and add a ninth student who has low scores on the second test and the final exam, and fits the overall pattern of the other scores in the data set. Calculate the correlation and compare it with the correlation that you calculated in Exercise 2.31. Write a short summary of your findings.

**2.34 Perch and bass.** Figure 2.4 (page 92) displays the positive association between number of prey (perch) present in an area and the proportion eaten by predators (bass).

(a) Do you think the correlation between these variables is closest to $r = 0.1$, $r = 0.6$, or $r = 0.9$? Explain the reason for your guess.

(b) Calculate the correlation. Was your guess correct?

**2.35 IQ and reading scores.** Figure 2.6 (page 96) displays the positive association between the IQ scores of fifth-grade students and their reading scores. Do you think the correlation between these variables is closest to $r = 0.1$, $r = 0.6$, or $r = 0.9$? Explain the reason for your guess.

**2.36** CHALLENGE **Mutual funds.** Mutual fund reports often give correlations to describe how the prices of different investments are related. You look at the correlations between three Fidelity funds and the Standard & Poor's 500 stock index, which describes stocks of large U.S. companies. The three funds are Dividend Growth (stocks of large U.S. companies), Small Cap Stock (stocks of small U.S. companies), and Emerging Markets (stocks in developing countries). For 2003, the three correlations are $r = 0.35$, $r = 0.81$, and $r = 0.98$.[21]

(a) Which correlation goes with each fund? Explain your answer.

(b) The correlations of the three funds with the index are all positive. Does this tell you that stocks went up in 2003? Explain your answer.

**2.37 Coffee prices in dollars or euros.** Coffee is currently priced in dollars. If it were priced in euros, and the dollar prices in Exercise 2.29 were translated into the equivalent prices in euros, would the correlation between coffee price and percent deforestation change? Explain your answer.

**2.38 Mutual funds.** Exercise 2.28 (page 101) gives data on the returns from 23 Fidelity "sector funds" in 2002 (a down-year for stocks) and 2003 (an up-year).

(a) Make a scatterplot if you did not do so in Exercise 2.28. Fidelity Gold Fund, the only fund with a positive return in both years, is an extreme outlier.

(b) To demonstrate that correlation is not resistant, find $r$ for all 23 funds and then find $r$ for the 22 funds other than Gold. Explain from Gold's position in your plot why omitting this point makes $r$ more negative.

**2.39 NBA teams.** Table 2.1 (page 98) gives the values of the 29 teams in the National Basketball Association, along with their total revenues and operating incomes. You made scatterplots of value against both explanatory variables in Exercise 2.19.

(a) Find the correlations of team value with revenue and with operating income. Do you think that the two values of $r$ provide a good first comparison of what the plots show about predicting value?

(b) Portland is an outlier in the plot of value against income. How does $r$ change when you remove Portland? Explain from the position of this point why the change has the direction it does.

**2.40 Correlations measure strong and weak linear associations.** Your scatterplots for Exercises 2.18 (page 97) and 2.24 (Table 2.4, page 100) illustrate a quite weak linear association and a very strong linear association. Find the correlations that go with these plots. It isn't surprising that a laboratory experiment on physical behavior (the icicles) gives a much stronger correlation than field data on living things (the biological clock). How strong a correlation must be to interest scientists depends on the field of study.

**2.41 Heights of people who date each other.** A student wonders if tall women tend to date taller men than do short women. She measures herself, her dormitory roommate, and the women in the adjoining rooms; then she measures the next man each woman dates. Here are the data (heights in inches):

| | | | | | | |
|---|---|---|---|---|---|---|
| Women ($x$) | 66 | 64 | 66 | 65 | 70 | 65 |
| Men ($y$) | 72 | 68 | 70 | 68 | 71 | 65 |

(a) Make a scatterplot of these data. Based on the scatterplot, do you expect the correlation to be positive or negative? Near $\pm 1$ or not?

(b) Find the correlation $r$ between the heights of the men and women.

(c) How would $r$ change if all the men were 6 inches shorter than the heights given in the table? Does the correlation tell us whether women tend to date men taller than themselves?

(d) If heights were measured in centimeters rather than inches, how would the correlation change? (There are 2.54 centimeters in an inch.)

(e) If every woman dated a man exactly 3 inches taller than herself, what would be the correlation between male and female heights?

**2.42 An interesting set of data.** Make a scatterplot of the following data.

| | | | | | | |
|---|---|---|---|---|---|---|
| $x$ | 1 | 2 | 3 | 4 | 10 | 10 |
| $y$ | 1 | 3 | 3 | 5 | 1 | 11 |

Use your calculator to show that the correlation is about 0.5. What feature of the data is responsible for reducing the correlation to this value despite a strong straight-line association between $x$ and $y$ in most of the observations?

**2.43** 

**Use the applet.** You are going to use the *Correlation and Regression* applet to make different scatterplots with 10 points that have correlation close to 0.8. *Many patterns can have the same correlation. Always plot your data before you trust a correlation.*

(a) Stop after adding the first 2 points. What is the value of the correlation? Why does it have this value no matter where the 2 points are located?

(b) Make a lower-left to upper-right pattern of 10 points with correlation about $r = 0.8$. (You can drag points up or down to adjust $r$ after you have 10 points.) Make a rough sketch of your scatterplot.

(c) Make another scatterplot, this time with 9 points in a vertical stack at the left of the plot. Add one point far to the right and move it until the correlation is close to 0.8. Make a rough sketch of your scatterplot.

(d) Make yet another scatterplot, this time with 10 points in a curved pattern that starts at the lower left, rises to the right, then falls again at the far right. Adjust the points up or down until you have a quite smooth curve with correlation close to 0.8. Make a rough sketch of this scatterplot also.

**2.44 Gas mileage and speed.** Exercise 2.22 (page 99) gives data on gas mileage against speed for a small car. Make a scatterplot if you have not already done so, then find the correlation $r$. Explain why $r$ is close to zero despite a strong relationship between speed and gas used.

**2.45 City and highway gas mileage.** Table 1.10 (page 31) gives the city and highway gas mileages for 21 two-seater cars, including the Honda Insight gas-electric hybrid car.

(a) Make a scatterplot of highway mileage $y$ against city mileage $x$ for all 21 cars. There is a strong positive linear association. The Insight lies far from the other points. Does the Insight extend the linear pattern of the other cars, or is it far from the line they form?

(b) Find the correlation between city and highway mileages both without and with the Insight. Based on your answer to (a), explain why $r$ changes in this direction when you add the Insight.

**2.46** **Use the applet.** Go to the *Correlation and Regression* applet. Click on the scatterplot to create a group of 10 points in the lower-left corner of the scatterplot with a strong straight-line negative pattern (correlation about $-0.9$).

(a) Add one point at the upper left that is in line with the first 10. How does the correlation change?

(b) Drag this last point down until it is opposite the group of 10 points. How small can you make the correlation? Can you make the correlation positive? *A single outlier can greatly strengthen or weaken a correlation. Always plot your data to check for outlying points.*

**2.47 What is the correlation?** Suppose that women always married men 2 years older than themselves. Draw a scatterplot of the ages of 5 married couples, with the wife's age as the explanatory variable. What is the correlation $r$ for your data? Why?

**2.48** CHALLENGE **High correlation does not mean that the values are the same.** Investment reports often include correlations. Following a table of correlations among mutual funds, a report adds, "Two funds can have perfect correlation, yet different levels of risk. For example, Fund A and Fund B may be perfectly correlated, yet Fund A moves 20% whenever Fund B moves 10%." Write a brief explanation, for someone who knows no statistics, of how this can happen. Include a sketch to illustrate your explanation.

**2.49 Student ratings of teachers.** A college newspaper interviews a psychologist about student ratings of the teaching of faculty members. The psychologist says, "The evidence indicates that the correlation between the research productivity and teaching rating of faculty members is close to zero." The paper reports this as "Professor McDaniel said that good researchers tend to be poor teachers, and vice versa." Explain why the paper's report is wrong. Write a statement in plain language (don't use the word "correlation") to explain the psychologist's meaning.

**2.50 What's wrong?** Each of the following statements contains a blunder. Explain in each case what is wrong.

(a) "There is a high correlation between the gender of American workers and their income."

(b) "We found a high correlation ($r = 1.09$) between students' ratings of faculty teaching and ratings made by other faculty members."

(c) "The correlation between planting rate and yield of corn was found to be $r = 0.23$ bushel."

**2.51** CHALLENGE **IQ and GPA.** Table 1.9 (page 29) reports data on 78 seventh-grade students. We expect a positive association between IQ and GPA. Moreover, some people think that self-concept is related to school performance. Examine in detail the relationships between GPA and the two explanatory variables IQ and self-concept. Are the relationships roughly linear? How strong are they? Are there unusual points? What is the effect of removing these points?

**2.52** CHALLENGE **Effect of a change in units.** Consider again the correlation $r$ between the speed of a car and its gas consumption, from the data in Exercise 2.22 (page 99).

(a) Transform the data so that speed is measured in miles per hour and fuel consumption in gallons per mile. (There are 1.609 kilometers in a mile and 3.785 liters in a gallon.) Make a scatterplot and find the correlation for both the original and the transformed data. How did the change of units affect your results?

(b) Now express fuel consumption in miles per gallon. (So each value is $1/x$ if $x$ is gallons per mile.) Again make a scatterplot and find the correlation. How did this change of units affect your results?

(*Lesson:* The effects of a linear transformation of the form $x_{\text{new}} = a + bx$ are simple. The effects of a nonlinear transformation are more complex.)

## 2.3 Least-Squares Regression

Correlation measures the direction and strength of the linear (straight-line) relationship between two quantitative variables. If a scatterplot shows a linear relationship, we would like to summarize this overall pattern by drawing a line on the scatterplot. A *regression line* summarizes the relationship between two variables, but only in a specific setting: when one of the variables helps explain or predict the other. That is, regression describes a relationship between an explanatory variable and a response variable.

**REGRESSION LINE**

A **regression line** is a straight line that describes how a response variable $y$ changes as an explanatory variable $x$ changes. We often use a regression line to **predict** the value of $y$ for a given value of $x$. Regression, unlike correlation, requires that we have an explanatory variable and a response variable.

EXAMPLE

**2.12 Fidgeting and fat gain.** Does fidgeting keep you slim? Some people don't gain weight even when they overeat. Perhaps fidgeting and other "nonexercise activity" (NEA) explains why—the body might spontaneously increase nonexercise activity when fed more. Researchers deliberately overfed 16 healthy young adults for 8 weeks. They measured fat gain (in kilograms) and, as an explanatory variable, increase in energy use (in calories) from activity other than deliberate exercise—fidgeting, daily living, and the like. Here are the data:[22]

| | | | | | | | | |
|---|---|---|---|---|---|---|---|---|
| NEA increase (cal) | −94 | −57 | −29 | 135 | 143 | 151 | 245 | 355 |
| Fat gain (kg) | 4.2 | 3.0 | 3.7 | 2.7 | 3.2 | 3.6 | 2.4 | 1.3 |
| NEA increase (cal) | 392 | 473 | 486 | 535 | 571 | 580 | 620 | 690 |
| Fat gain (kg) | 3.8 | 1.7 | 1.6 | 2.2 | 1.0 | 0.4 | 2.3 | 1.1 |

Figure 2.11 is a scatterplot of these data. The plot shows a moderately strong negative linear association with no outliers. The correlation is $r = -0.7786$. People with larger increases in nonexercise activity do indeed gain less fat. A line drawn through the points will describe the overall pattern well.

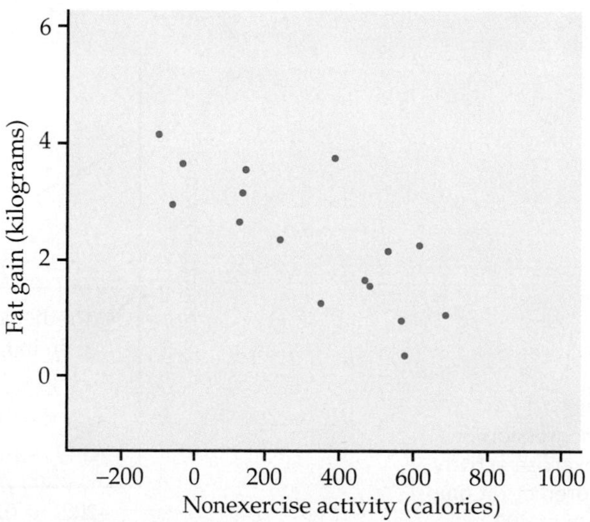


**FIGURE 2.11** Fat gain after 8 weeks of overeating, plotted against the increase in nonexercise activity over the same period, for Example 2.12.

## Fitting a line to data

fitting a line

When a scatterplot displays a linear pattern, we can describe the overall pattern by drawing a straight line through the points. Of course, no straight line passes exactly through all of the points. **Fitting a line** to data means drawing a line that comes as close as possible to the points. The equation of a line fitted to the data gives a compact description of the dependence of the response variable $y$ on the explanatory variable $x$.

### STRAIGHT LINES

Suppose that $y$ is a response variable (plotted on the vertical axis) and $x$ is an explanatory variable (plotted on the horizontal axis). A straight line relating $y$ to $x$ has an equation of the form

$$y = b_0 + b_1 x$$

In this equation, $b_1$ is the **slope,** the amount by which $y$ changes when $x$ increases by one unit. The number $b_0$ is the **intercept,** the value of $y$ when $x = 0$.

EXAMPLE

**2.13 Regression line for fat gain.** Any straight line describing the nonexercise activity data has the form

$$\text{fat gain} = b_0 + (b_1 \times \text{NEA increase})$$

In Figure 2.12 we have drawn the regression line with the equation

$$\text{fat gain} = 3.505 - (0.00344 \times \text{NEA increase})$$

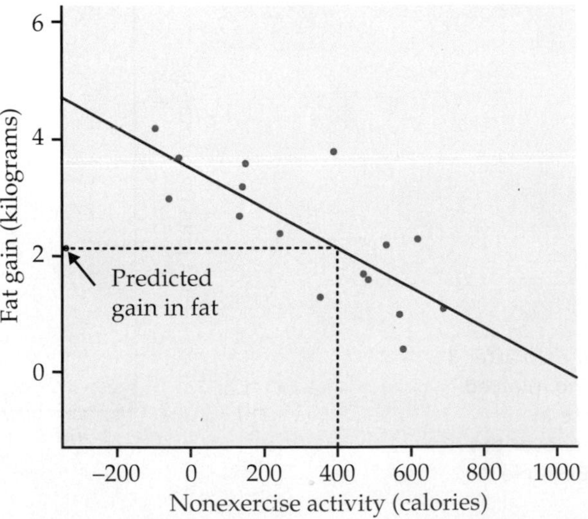


**FIGURE 2.12** A regression line fitted to the nonexercise activity data and used to predict fat gain for an NEA increase of 400 calories.

The figure shows that this line fits the data well. The slope $b_1 = -0.00344$ tells us that fat gained goes down by 0.00344 kilogram for each added calorie of NEA.

The slope $b_1$ of a line $y = b_0 + b_1 x$ is the *rate of change* in the response $y$ as the explanatory variable $x$ changes. The slope of a regression line is an important numerical description of the relationship between the two variables. For Example 2.13, the intercept, $b_0 = 3.505$ kilograms, is the estimated fat gain if NEA does not change when a person overeats.

**USE YOUR KNOWLEDGE**

**2.53 Plot the data with the line.** Make a sketch of the data in Example 2.12 and plot the line

$$\text{fat gain} = 4.505 - (0.00344 \times \text{NEA increase})$$

on your sketch. Explain why this line does not give a good fit to the data.

## Prediction

prediction

We can use a regression line to **predict** the response $y$ for a specific value of the explanatory variable $x$.

EXAMPLE

**2.14 Prediction for fat gain.** Based on the linear pattern, we want to predict the fat gain for an individual whose NEA increases by 400 calories when she overeats. To use the fitted line to predict fat gain, go "up and over" on the graph in Figure 2.12. From 400 calories on the $x$ axis, go up to the fitted line and over to the $y$ axis. The graph shows that the predicted gain in fat is a bit more than 2 kilograms.

If we have the equation of the line, it is faster and more accurate to substitute $x = 400$ in the equation. The predicted fat gain is

$$\text{fat gain} = 3.505 - (0.00344 \times 400) = 2.13 \text{ kilograms}$$

The accuracy of predictions from a regression line depends on how much scatter about the line the data show. In Figure 2.12, fat gains for similar increases in NEA show a spread of 1 or 2 kilograms. The regression line summarizes the pattern but gives only roughly accurate predictions.

**USE YOUR KNOWLEDGE**

**2.54 Predict the fat gain.** Use the regression equation in Example 2.13 to predict the fat gain for a person whose NEA increases by 600 calories.

EXAMPLE

**2.15 Is this prediction reasonable?** Can we predict the fat gain for someone whose nonexercise activity increases by 1500 calories when she overeats? We can certainly substitute 1500 calories into the equation of the line. The prediction is

$$\text{fat gain} = 3.505 - (0.00344 \times 1500) = -1.66 \text{ kilograms}$$

That is, we predict that this individual loses fat when she overeats. This prediction is not trustworthy. Look again at Figure 2.12. An NEA increase of 1500 calories is far outside the range of our data. We can't say whether increases this large ever occur, or whether the relationship remains linear at such extreme values. Predicting fat gain when NEA increases by 1500 calories *extrapolates* the relationship beyond what the data show.

EXTRAPOLATION

**Extrapolation** is the use of a regression line for prediction far outside the range of values of the explanatory variable $x$ used to obtain the line. Such predictions are often not accurate.

USE YOUR KNOWLEDGE

**2.55 Would you use the regression equation to predict?** Consider the following values for NEA increase: −400, 200, 500, 1000. For each, decide whether you would use the regression equation in Example 2.13 to predict fat gain or whether you would be concerned that the prediction would not be trustworthy because of extrapolation. Give reasons for your answers.

## Least-squares regression

Different people might draw different lines by eye on a scatterplot. This is especially true when the points are widely scattered. We need a way to draw a regression line that doesn't depend on our guess as to where the line should go. No line will pass exactly through all the points, but we want one that is as close as possible. We will use the line to predict $y$ from $x$, so we want a line that is as close as possible to the points in the *vertical* direction. That's because the prediction errors we make are errors in $y$, which is the vertical direction in the scatterplot.

The line in Figure 2.12 predicts 2.13 kilograms of fat gain for an increase in nonexercise activity of 400 calories. If the actual fat gain turns out to be 2.3 kilograms, the error is

$$\begin{aligned}\text{error} &= \text{observed gain} - \text{predicted gain}\\ &= 2.3 - 2.13 = 0.17 \text{ kilograms}\end{aligned}$$

Errors are positive if the observed response lies above the line, and negative if the response lies below the line. We want a regression line that makes these prediction errors as small as possible. Figure 2.13 illustrates the idea. For clarity, the plot shows only three of the points from Figure 2.12, along with the line, on an expanded scale. The line passes below two of the points and above one of them. The vertical distances of the data points from the line appear as vertical line segments. A "good" regression line makes these distances as small as possible. There are many ways to make "as small as possible" precise. The most common is the *least-squares* idea. The line in Figures 2.12 and 2.13 is in fact the least-squares regression line.

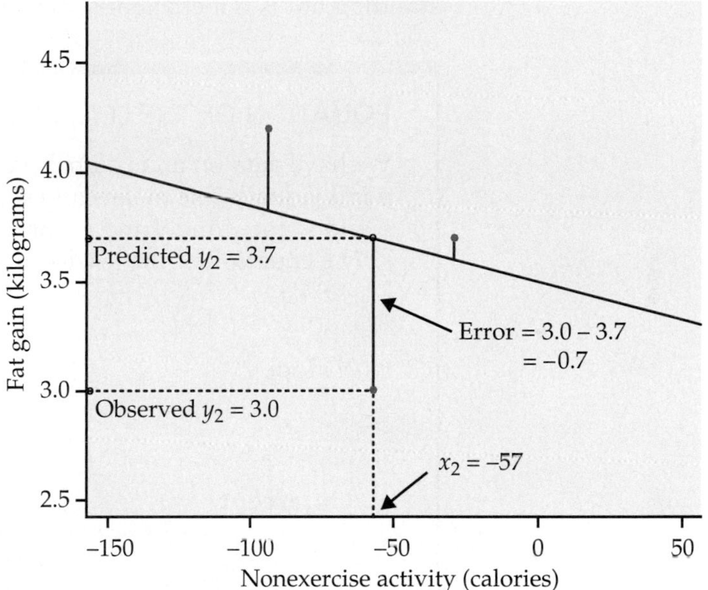


**FIGURE 2.13** The least-squares idea: make the errors in predicting $y$ as small as possible by minimizing the sum of their squares.

### LEAST-SQUARES REGRESSION LINE

The **least-squares regression line of $y$ on $x$** is the line that makes the sum of the squares of the vertical distances of the data points from the line as small as possible.

Here is the least-squares idea expressed as a mathematical problem. We represent $n$ observations on two variables $x$ and $y$ as

$$(x_1, y_1),\ (x_2, y_2),\ \ldots,\ (x_n, y_n)$$

If we draw a line $y = b_0 + b_1 x$ through the scatterplot of these observations, the line predicts the value of $y$ corresponding to $x_i$ as $\hat{y}_i = b_0 + b_1 x_i$. We write $\hat{y}$ (read "y-hat") in the equation of a regression line to emphasize that the line gives a *predicted* response $\hat{y}$ for any $x$. The predicted response will usually not be exactly the same as the actually *observed* response $y$. The method of least squares chooses the line that makes the sum of the squares of these errors as small as possible. To find this line, we must find the values of the intercept $b_0$

and the slope $b_1$ that minimize

$$\sum(\text{error})^2 = \sum(y_i - b_0 - b_1x_i)^2$$

for the given observations $x_i$ and $y_i$. For the NEA data, for example, we must find the $b_0$ and $b_1$ that minimize

$$(-94 - b_0 - 4.2b_1)^2 + (-57 - b_0 - 3.0b_1)^2 + \cdots + (690 - b_0 - 1.1b_1)^2$$

These values are the intercept and slope of the least-squares line.

You will use software or a calculator with a regression function to find the equation of the least-squares regression line from data on $x$ and $y$. We will therefore give the equation of the least-squares line in a form that helps our understanding but is not efficient for calculation.

### EQUATION OF THE LEAST-SQUARES REGRESSION LINE

We have data on an explanatory variable $x$ and a response variable $y$ for $n$ individuals. The means and standard deviations of the sample data are $\bar{x}$ and $s_x$ for $x$ and $\bar{y}$ and $s_y$ for $y$, and the correlation between $x$ and $y$ is $r$. The equation of the least-squares regression line of $y$ on $x$ is

$$\hat{y} = b_0 + b_1x$$

with **slope**

$$b_1 = r\frac{s_y}{s_x}$$

and **intercept**

$$b_0 = \bar{y} - b_1\bar{x}$$

EXAMPLE

**2.16 Check the calculations.** Verify from the data in Example 2.12 that the mean and standard deviation of the 16 increases in NEA are

$$\bar{x} = 324.8 \text{ calories} \quad \text{and} \quad s_x = 257.66 \text{ calories}$$

The mean and standard deviation of the 16 fat gains are

$$\bar{y} = 2.388 \text{ kg} \quad \text{and} \quad s_y = 1.1389 \text{ kg}$$

The correlation between fat gain and NEA increase is $r = -0.7786$. The least-squares regression line of fat gain $y$ on NEA increase $x$ therefore has slope

$$b_1 = r\frac{s_y}{s_x} = -0.7786\frac{1.1389}{257.66}$$
$$= -0.00344 \text{ kg per calorie}$$

and intercept

$$b_0 = \bar{y} - b_1\bar{x} = 2.388 - (-0.00344)(324.8)$$
$$= 3.505 \text{ kg}$$

The equation of the least-squares line is

$$\hat{y} = 3.505 - 0.00344x$$

*When doing calculations like this by hand, you may need to carry extra decimal places in the preliminary calculations to get accurate values of the slope and intercept.* Using software or a calculator with a regression function eliminates this worry.

## Interpreting the regression line

The slope $b_1 = -0.00344$ kilograms per calorie in Example 2.16 is the change in fat gain as NEA increases. The units "kilograms of fat gained per calorie of NEA" come from the units of $y$ (kilograms) and $x$ (calories). Although the correlation does not change when we change the units of measurement, the equation of the least-squares line does change. The slope in grams per calorie would be 1000 times as large as the slope in kilograms per calorie, because there are 1000 grams in a kilogram. The small value of the slope, $b_1 = -0.00344$, does not mean that the effect of increased NEA on fat gain is small—it just reflects the choice of kilograms as the unit for fat gain. *The slope and intercept of the least-squares line depend on the units of measurement—you can't conclude anything from their size.*

The expression $b_1 = rs_y/s_x$ for the slope says that, along the regression line, **a change of one standard deviation in $x$ corresponds to a change of $r$ standard deviations in $y$.** When the variables are perfectly correlated ($r = 1$ or $r = -1$), the change in the predicted response $\hat{y}$ is the same (in standard deviation units) as the change in $x$. Otherwise, when $-1 < r < 1$, the change in $\hat{y}$ is less than the change in $x$. As the correlation grows less strong, the prediction $\hat{y}$ moves less in response to changes in $x$.

**The least-squares regression line always passes through the point $(\bar{x}, \bar{y})$** on the graph of $y$ against $x$. Check that when you substitute $\bar{x} = 324.8$ into the equation of the regression line in Example 2.16, the result is $\hat{y} = 2.388$, equal to the mean of $y$. So the least-squares regression line of $y$ on $x$ is the line with slope $rs_y/s_x$ that passes through the point $(\bar{x}, \bar{y})$. We can describe regression entirely in terms of the basic descriptive measures $\bar{x}$, $s_x$, $\bar{y}$, $s_y$, and $r$. If both $x$ and $y$ are standardized variables, so that their means are 0 and their standard deviations are 1, then the regression line has slope $r$ and passes through the origin.

Figure 2.14 displays the basic regression output for the nonexercise activity data from two statistical software packages. Other software produces very similar output. You can find the slope and intercept of the least-squares line, calculated to more decimal places than we need, in both outputs. The software also provides information that we do not yet need, including some that we trimmed from Figure 2.14. Part of the art of using software is to ignore the extra information that is almost always present. Look for the results that you need. Once you understand a statistical method, you can read output from almost any software.

## Correlation and regression

Least-squares regression looks at the distances of the data points from the line only in the $y$ direction. So the two variables $x$ and $y$ play different roles in regression.

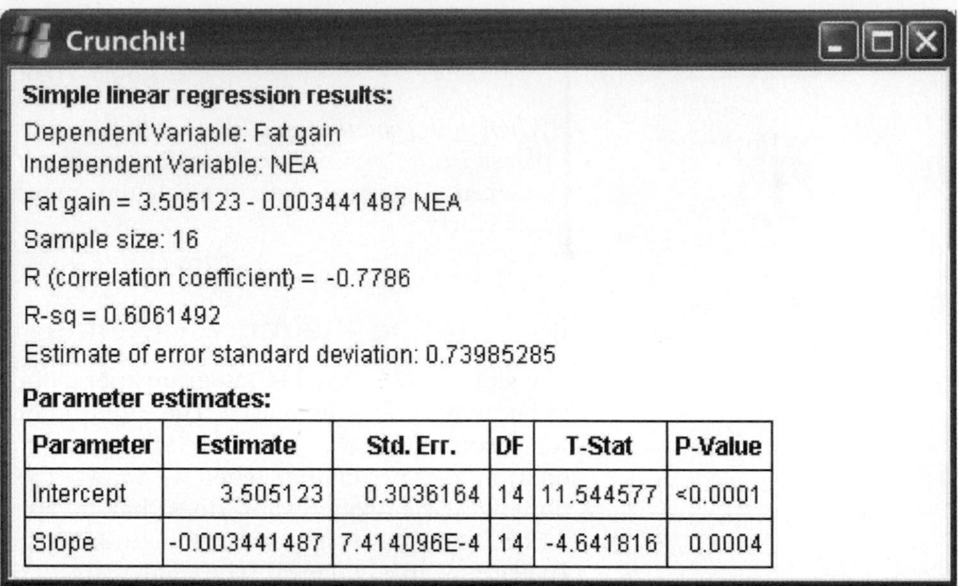

CrunchIt!

**Simple linear regression results:**

Dependent Variable: Fat gain
Independent Variable: NEA
Fat gain = 3.505123 - 0.003441487 NEA
Sample size: 16
R (correlation coefficient) = -0.7786
R-sq = 0.6061492
Estimate of error standard deviation: 0.73985285

**Parameter estimates:**

| Parameter | Estimate | Std. Err. | DF | T-Stat | P-Value |
|---|---|---|---|---|---|
| Intercept | 3.505123 | 0.3036164 | 14 | 11.544577 | <0.0001 |
| Slope | -0.003441487 | 7.414096E-4 | 14 | -4.641816 | 0.0004 |

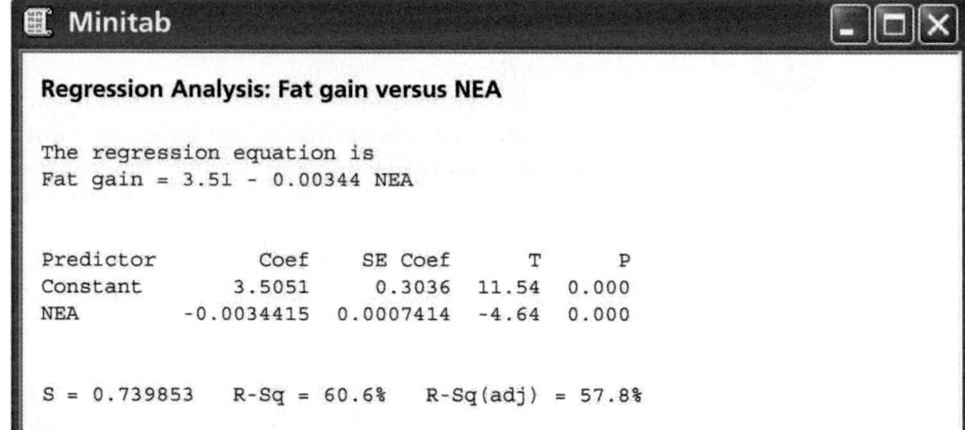

Minitab

**Regression Analysis: Fat gain versus NEA**

```
The regression equation is
Fat gain = 3.51 - 0.00344 NEA


Predictor        Coef    SE Coef      T      P
Constant       3.5051     0.3036  11.54  0.000
NEA        -0.0034415  0.0007414  -4.64  0.000


S = 0.739853   R-Sq = 60.6%   R-Sq(adj) = 57.8%
```

**FIGURE 2.14** Regression results for the nonexercise activity data from two statistical software packages. Other software produces similar output.

EXAMPLE

**2.17 The universe is expanding.** Figure 2.15 is a scatterplot of data that played a central role in the discovery that the universe is expanding. They are the distances from the earth of 24 spiral galaxies and the speed at which these galaxies are moving away from us, reported by the astronomer Edwin Hubble in 1929.[23] There is a positive linear relationship, $r = 0.7842$. More distant galaxies are moving away more rapidly. Astronomers believe that there is in fact a perfect linear relationship, and that the scatter is caused by imperfect measurements.

The two lines on the plot are the two least-squares regression lines. The regression line of velocity on distance is solid. The regression line of distance on velocity is dashed. *Although there is only one correlation between velocity and distance, regression of velocity on distance and regression of distance on velocity give different lines. In doing regression, you must choose which variable is explanatory.*

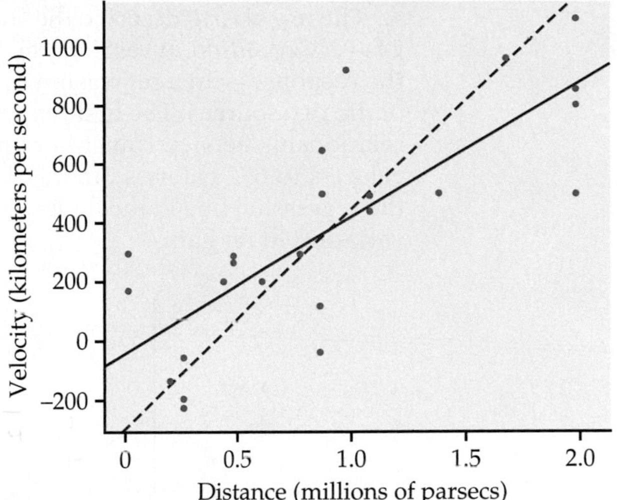


**FIGURE 2.15** Hubble's data on the velocity and distance of 24 galaxies, for Example 2.17. The lines are the least-squares regression lines of velocity on distance (solid) and of distance on velocity (dashed).

Even though the correlation $r$ ignores the distinction between explanatory and response variables, there is a close connection between correlation and regression. We saw that the slope of the least-squares line involves $r$. Another connection between correlation and regression is even more important. In fact, the numerical value of $r$ as a measure of the strength of a linear relationship is best interpreted by thinking about regression. Here is the fact we need.

### $r^2$ IN REGRESSION

The **square of the correlation, $r^2$,** is the fraction of the variation in the values of $y$ that is explained by the least-squares regression of $y$ on $x$.

The correlation between NEA increase and fat gain for the 16 subjects in Example 2.12 is $r = -0.7786$. Because $r^2 = 0.606$, the straight-line relationship between NEA and fat gain explains about 61% of the vertical scatter in fat gains in Figure 2.12. When you report a regression, give $r^2$ as a measure of how successfully the regression explains the response. Both software outputs in Figure 2.14 include $r^2$, either in decimal form or as a percent. When you see a correlation, square it to get a better feel for the strength of the association. Perfect correlation ($r = -1$ or $r = 1$) means the points lie exactly on a line. Then $r^2 = 1$ and all of the variation in one variable is accounted for by the linear relationship with the other variable. If $r = -0.7$ or $r = 0.7$, $r^2 = 0.49$ and about half the variation is accounted for by the linear relationship. In the $r^2$ scale, correlation $\pm 0.7$ is about halfway between 0 and $\pm 1$.

## USE YOUR KNOWLEDGE

**2.56 What fraction of the variation is explained?** Consider the following correlations: −0.9, −0.5, −0.3, 0, 0.3, 0.5, and 0.9. For each, give the fraction of the variation in $y$ that is explained by the least-squares regression of $y$ on $x$. Summarize what you have found from performing these calculations.

The use of $r^2$ to describe the success of regression in explaining the response $y$ is very common. It rests on the fact that there are two sources of variation in the responses $y$ in a regression setting. Figure 2.16 gives a rough visual picture of the two sources. The first reason for the variation in fat gains is that there is a relationship between fat gain $y$ and increase in NEA $x$. As $x$ increases from $-94$ calories to 690 calories among the 16 subjects, it pulls fat gain $y$ with it along the regression line in the figure. The linear relationship explains this part of the variation in fat gains.

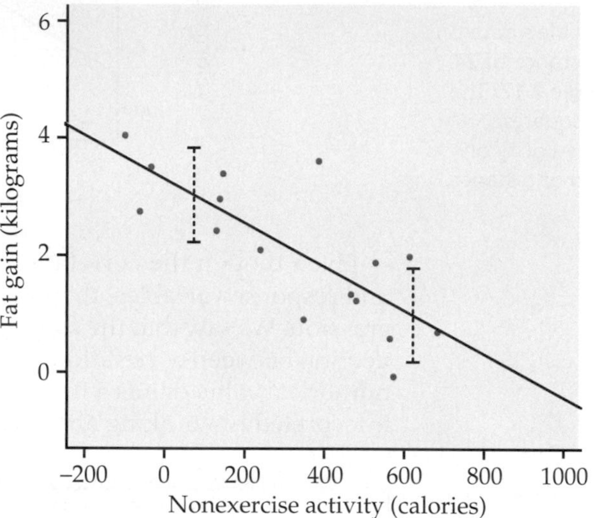


**FIGURE 2.16** Explained and unexplained variation in regression. As $x$ increases, it pulls $y$ with it along the line. That is the variation explained by the regression. The scatter of the data points above and below the line, suggested by the dashed segments, is not explained by the regression.

The fat gains do not lie exactly on the line, however, but are scattered above and below it. This is the second source of variation in $y$, and the regression line tells us nothing about how large it is. The vertical dashed lines in Figure 2.16 show a rough average for the spread in $y$ when we fix a value of $x$. We use $r^2$ to measure variation along the line as a fraction of the total variation in the fat gains. In Figure 2.16, about 61% of the variation in fat gains among the 16 subjects is due to the straight-line tie between $y$ and $x$. The remaining 39% is vertical scatter in the observed responses remaining after the line has fixed the predicted responses.

## *Understanding $r^2$

Here is a more specific interpretation of $r^2$. The fat gains $y$ in Figure 2.16 range from 0.4 kilograms to 4.2 kilograms. The variance of these responses, a measure of how variable they are, is

$$\text{variance of observed values } y = 1.297$$

Much of this variability is due to the fact that as $x$ increases from $-94$ calories to 690 calories it pulls height $y$ along with it. If the only variability in the observed responses were due to the straight-line dependence of fat gain on NEA, the observed gains would lie exactly on the regression line. That is, they would be the same as the predicted gains $\hat{y}$. We can compute the predicted gains by substitut-

*This explanation is optional reading.

ing the NEA values for each subject into the equation of the least-squares line. Their variance describes the variability in the predicted responses. The result is

$$\text{variance of predicted values } \hat{y} = 0.786$$

This is what the variance would be if the responses fell exactly on the line, that is, if the linear relationship explained 100% of the observed variation in $y$. Because the responses don't fall exactly on the line, the variance of the predicted values is smaller than the variance of the observed values. Here is the fact we need:

$$r^2 = \frac{\text{variance of predicted values } \hat{y}}{\text{variance of observed values } y}$$

$$= \frac{0.786}{1.297} = 0.606$$

This fact is always true. The squared correlation gives the variance the responses would have if there were no scatter about the least-squares line as a fraction of the variance of the actual responses. This is the exact meaning of "fraction of variation explained" as an interpretation of $r^2$.

These connections with correlation are special properties of least-squares regression. They are not true for other methods of fitting a line to data. One reason that least squares is the most common method for fitting a regression line to data is that it has many convenient special properties.

## BEYOND THE BASICS

### Transforming Relationships

How is the weight of an animal's brain related to the weight of its body? Figure 2.17 is a scatterplot of brain weight against body weight for 96 species of mam-

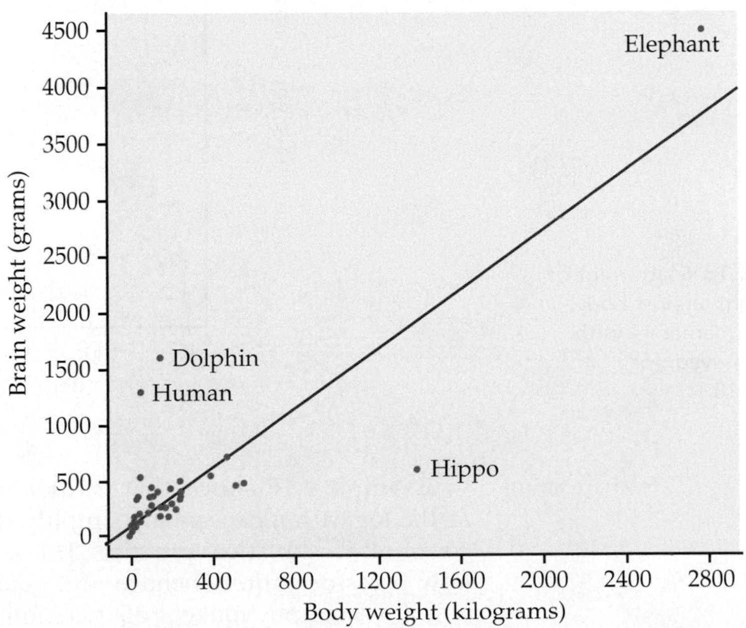


**FIGURE 2.17** Scatterplot of brain weight against body weight for 96 species of mammals.

mals.[24] The line is the least-squares regression line for predicting brain weight from body weight. The outliers are interesting. We might say that dolphins and humans are smart, hippos are dumb, and African elephants are just big. That's because dolphins and humans have larger brains than their body weights suggest, hippos have smaller brains, and the elephant is much heavier than any other mammal in both body and brain.

EXAMPLE

**2.18 Brain weight and body weight.** The plot in Figure 2.17 is not very satisfactory. Most mammals are so small relative to elephants and hippos that their points overlap to form a blob in the lower-left corner of the plot. The correlation between brain weight and body weight is $r = 0.86$, but this is misleading. If we remove the elephant, the correlation for the other 95 species is $r = 0.50$. Figure 2.18 is a scatterplot of the data with the four outliers removed to allow a closer look at the other 92 observations. We can now see that the relationship is not linear. It bends to the right as body weight increases.

Biologists know that data on sizes often behave better if we take *logarithms* before doing more analysis. Figure 2.19 plots the logarithm of brain weight against the logarithm of body weight for all 96 species. The effect is almost magical. There are no longer any extreme outliers or very influential observations. The pattern is very linear, with correlation $r = 0.96$. The vertical spread about the least-squares line is similar everywhere, so that predictions of brain weight from body weight will be about equally precise for any body weight (in the log scale).

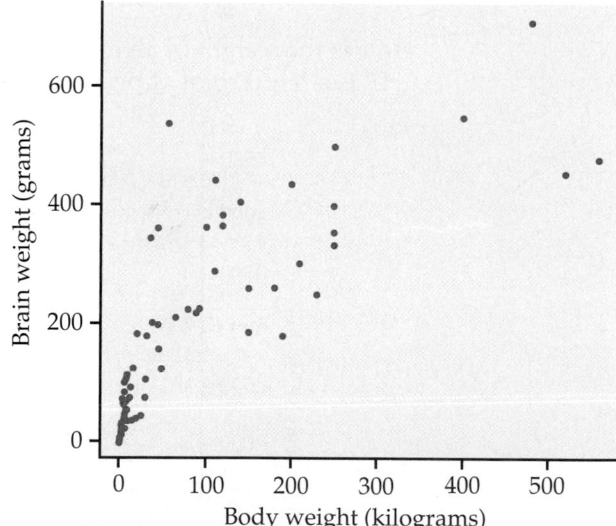


**FIGURE 2.18** Scatterplot of brain weight against body weight for mammals, with outliers removed, for Example 2.18.

transforming

Example 2.18 shows that **transforming** data by applying a function such as the logarithm can greatly simplify statistical analysis. Transforming data is common in statistical practice. There are systematic principles that describe how transformations behave and guide the search for transformations that will, for example, make a distribution more Normal or a curved relationship

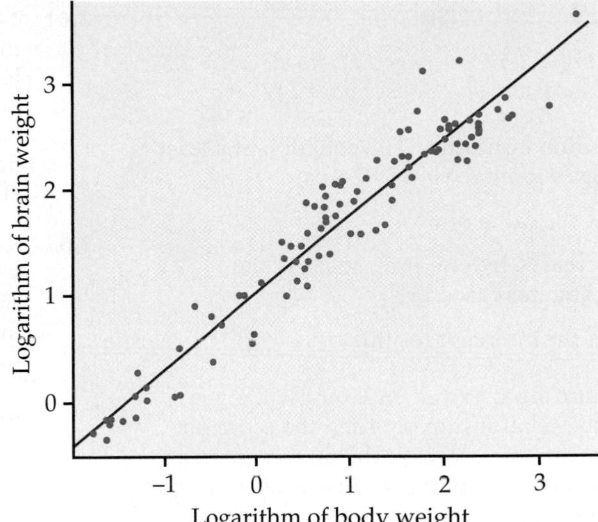


**FIGURE 2.19** Scatterplot of the logarithm of brain weight against the logarithm of body weight for 96 species of mammals, for Example 2.18.

more linear. You can read about these principles in the extra material entitled *Transforming Relationships* available on the text CD and Web site.

## SECTION 2.3 Summary

A **regression line** is a straight line that describes how a response variable $y$ changes as an explanatory variable $x$ changes.

The most common method of fitting a line to a scatterplot is least squares. The **least-squares regression line** is the straight line $\hat{y} = b_0 + b_1 x$ that minimizes the sum of the squares of the vertical distances of the observed $y$-values from the line.

You can use a regression line to **predict** the value of $y$ for any value of $x$ by substituting this $x$ into the equation of the line. **Extrapolation** beyond the range of $x$-values spanned by the data is risky.

The **slope** $b_1$ of a regression line $\hat{y} = b_0 + b_1 x$ is the rate at which the predicted response $\hat{y}$ changes along the line as the explanatory variable $x$ changes. Specifically, $b_1$ is the change in $\hat{y}$ when $x$ increases by 1. The numerical value of the slope depends on the units used to measure $x$ and $y$.

The **intercept** $b_0$ of a regression line $\hat{y} = b_0 + b_1 x$ is the predicted response $\hat{y}$ when the explanatory variable $x = 0$. This prediction is of no statistical use unless $x$ can actually take values near 0.

The least-squares regression line of $y$ on $x$ is the line with slope $b_1 = rs_y/s_x$ and intercept $b_0 = \bar{y} - b_1\bar{x}$. This line always passes through the point $(\bar{x}, \bar{y})$.

**Correlation and regression** are closely connected. The correlation $r$ is the slope of the least-squares regression line when we measure both $x$ and $y$ in standardized units. The square of the correlation $r^2$ is the fraction of the variance of one variable that is explained by least-squares regression on the other variable.

## SECTION 2.3 Exercises

*For Exercises 2.53 and 2.54, see page 111; for Exercise 2.55, see page 112; and for Exercise 2.56, see page 117.*

**2.57 The regression equation.** The equation of a least-squares regression line is $y = 10 + 5x$.

(a) What is the value of $y$ for $x = 5$?

(b) If $x$ increases by one unit, what is the corresponding increase in $y$?

(c) What is the intercept for this equation?

**2.58 First test and final exam.** In Exercise 2.6 you looked at the relationship between the score on the first test and the score on the final exam in an elementary statistics course. Here are data for eight students from such a course:

| First-test score | 153 | 144 | 162 | 149 | 127 | 118 | 158 | 153 |
|---|---|---|---|---|---|---|---|---|
| Final-exam score | 145 | 140 | 145 | 170 | 145 | 175 | 170 | 160 |

(a) Plot the data with the first-test scores on the $x$ axis and the final-exam scores on the $y$ axis.

(b) Find the least-squares regression line for predicting the final-exam score using the first-test score.

(c) Graph the least-squares regression line on your plot.

**2.59 Second test and final exam.** Refer to the previous exercise. Here are the data for the second test and the final exam for the same students:

| Second-test score | 158 | 162 | 144 | 162 | 136 | 158 | 175 | 153 |
|---|---|---|---|---|---|---|---|---|
| Final-exam score | 145 | 140 | 145 | 170 | 145 | 175 | 170 | 160 |

(a) Plot the data with the second-test scores on the $x$ axis and the final-exam scores on the $y$ axis.

(b) Find the least-squares regression line for predicting the final-exam score using the second-test score.

(c) Graph the least-squares regression line on your plot.

**2.60 The effect of an outlier.** Refer to the previous exercise. Add a ninth student whose scores on the second test and final exam would lead you to classify the additional data point as an outlier. Recalculate the least-squares regression line with this additional case and summarize the effect it has on the least-squares regression line.

**2.61 The effect of a different point.** Examine the data in Exercise 2.31 and add a ninth student who has low scores on the second test and the final exam, and fits the overall pattern of the other scores in the data set. Recalculate the least-squares regression line with this additional case and summarize the effect it has on the least-squares regression line.

**2.62 Revenue and value of NBA teams.** Table 2.1 (page 98) gives the values of the 29 teams in the National Basketball Association, along with their operating incomes and revenues. Plots and correlations show that revenue predicts team value much better than does operating income. The least-squares regression line for predicting value from revenue is

$$\text{value} = 21.4 + (2.59 \times \text{revenue})$$

(a) What is the slope of this line? Express in simple language what the slope says about the relationship of value to revenue.

(b) The Los Angeles Lakers are the NBA's most valuable team, valued at $447 million, with $149 million in revenue. Use the line to predict the value of the Lakers from their revenue. What is the error in this prediction?

(c) The correlation between revenue and team value is $r = 0.9265$. What does the correlation say about the success of the regression line in predicting the values of the 29 teams?

**2.63 Water discharged by the Mississippi River.** Figure 1.10(b) (page 19) is a time plot of the volume of water discharged by the Mississippi River for the years 1954 to 2001. Water volume is recorded in cubic kilometers. The trend line on the plot is the least-squares regression line. The equation of this line is

$$\text{water discharged} = -7792 + (4.2255 \times \text{year})$$

(a) How much (on the average) does the volume of water increase with each passing year?

(b) What does the equation say about the volume of water flowing out of the Mississippi in the year 1780? Why is this extrapolation clearly nonsense?

(c) What is the predicted volume discharged in 1990 (round to the nearest cubic kilometer)? What is the prediction error for 1990?

(d) Can you see evidence of the great floods of 1973 and 1993, even on the plot of annual water discharged? Explain.

**2.64** CAUTION **Perch and bass.** Example 2.8 (page 91) gives data from an experiment in ecology. Figure 2.4 is the scatterplot of proportion of perch eaten by bass against the number of perch in a pen before the bass were let in. There is a roughly linear pattern. The least-squares line for predicting proportion eaten from initial count of perch is

$$\text{proportion eaten} = 0.120 + (0.0086 \times \text{count})$$

(a) When 10 more perch are added to a pen, what happens to the proportion that are eaten (according to the line)? Explain your answer.

(b) If there are no perch in a pen, what proportion does the line predict will be eaten? Explain why this prediction is nonsense. What is wrong with using the regression line to predict $y$ when $x = 0$? *You see that the intercept, though it is needed to draw the line, may have no statistical interpretation if $x = 0$ is outside the range of the data.*

**2.65 Progress in math scores.** Every few years, the National Assessment of Educational Progress asks a national sample of eighth-graders to perform the same math tasks. The goal is to get an honest picture of progress in math. Here are the last few national mean scores, on a scale of 0 to 500:[25]

| Year | 1990 | 1992 | 1996 | 2000 | 2003 | 2005 |
|---|---|---|---|---|---|---|
| Score | 263 | 268 | 272 | 273 | 278 | 279 |

(a) Make a time plot of the mean scores, by hand. This is just a scatterplot of score against year. There is a slow linear increasing trend.

(b) Find the regression line of mean score on time step-by-step. First calculate the mean and standard deviation of each variable and their correlation (use a calculator with these functions). Then find the equation of the least-squares line from these. Draw the line on your scatterplot. What percent of the year-to-year variation in scores is explained by the linear trend?

(c) Now use software or the regression function on your calculator to verify your regression line.

**2.66 The Trans-Alaska Oil Pipeline.** Figure 2.3 (page 90) plots field measurements on the depth of 100 small defects in the Trans-Alaska Oil Pipeline against laboratory measurements of the same defects. Drawing the $y = x$ line on the graph shows that field measurements tend to be too low for larger defect depths.

(a) Find the equation of the least-squares regression line for predicting field measurement from laboratory measurement. Make a scatterplot with this line drawn on it. How does the least-squares line differ from the $y = x$ line?

(b) What is the slope of the $y = x$ line? What is the slope of the regression line? Say in simple language what these slopes mean.

**2.67 Social exclusion and pain.** Exercise 2.17 (page 97) gives data from a study that shows that social exclusion causes "real pain." That is, activity in the area of the brain that responds to physical pain goes up as distress from social exclusion goes up. Your scatterplot in Exercise 2.17 shows a moderately strong linear relationship.

(a) What is the equation of the least-squares regression line for predicting brain activity from social distress score? Make a scatterplot with this line drawn on it.

(b) On your plot, show the "up and over" lines that predict brain activity for social distress score 2.0. Use the equation of the regression line to get the predicted brain activity level. Verify that it agrees with your plot.

(c) What percent of the variation in brain activity among these subjects is explained by the straight-line relationship with social distress score?

**2.68 Problems with feet.** Your scatterplot in Exercise 2.20 (page 98) suggests that the severity of the mild foot deformity called MA can help predict the severity of the more serious deformity called HAV. Table 2.2 (page 98) gives data for 38 young patients.

(a) Find the equation of the least-squares regression line for predicting HAV angle from MA angle. Add this line to the scatterplot you made in Exercise 2.20.

(b) A new patient has MA angle 25 degrees. What do you predict this patient's HAV angle to be?

(c) Does knowing MA angle allow doctors to predict HAV angle accurately? Explain your answer from the scatterplot, then calculate a numerical measure to support your finding.

**2.69 Growth of icicles.** Table 2.4 (page 100) gives data on the growth of icicles at two rates of water flow. You examined these data in Exercise 2.24. Use least-squares regression to estimate the rate (centimeters per minute) at which icicles grow at these two flow rates. How does flow rate affect growth?

**2.70 Mutual funds.** Exercise 2.28 (page 101) gives the returns of 23 Fidelity "sector funds" for the years

2002 and 2003. These mutual funds invest in narrow segments of the stock market. They often rise faster than the overall market in up-years, such as 2003, and fall faster than the market in down-years, such as 2002. A scatterplot shows that Fidelity Gold Fund—the only fund that went up in 2002—is an outlier. In Exercise 2.38, you showed that this outlier has a strong effect on the correlation. The least-squares line, like the correlation, is not resistant.

(a) Find the equations of two least-squares lines for predicting 2003 return from 2002 return, one for all 23 funds and one omitting Fidelity Gold Fund. Make a scatterplot with both lines drawn on it. The two lines are very different.

(b) Starting with the least-squares idea, explain why adding Fidelity Gold Fund to the other 22 funds moves the line in the direction that your graph shows.

**2.71 Stocks and Treasury bills.** The scatterplot in Figure 2.7 (page 96) suggests that returns on common stocks may be somewhat lower in years with high interest rates. Here is part of the output from software for the regression of stock returns on the Treasury bill returns for the same years:

```
Stock = 16.639318 - 0.67974913 Tbill
Sample size: 54
R (correlation coefficient) = -0.113
R-sq = 0.01275773
Estimate of error standard deviation: 17.680649
```

If you knew the return on Treasury bills for next year, do you think you could predict the return on stocks quite accurately? Use both the scatterplot in Figure 2.7 and a number from the regression output to justify your answer.

**2.72 Icicle growth.** Find the mean and standard deviation of the times and icicle lengths for the data on Run 8903 in Table 2.4 (page 100). Find the correlation between the two variables. Use these five numbers to find the equation of the regression line for predicting length from time. Verify that your result agrees with that in Exercise 2.69. Use the same five numbers to find the equation of the regression line for predicting the time an icicle has been growing from its length. What units does the slope of each of these lines have?

**2.73 Metabolic rate and lean body mass.** Compute the mean and the standard deviation of the metabolic rates and lean body masses in Exercise 2.21 (page 98) and the correlation between these two variables. Use these values to find the slope of the regression line of metabolic rate on lean body mass. Also find the slope of the regression line of lean body mass on metabolic rate. What are the units for each of the two slopes?

**2.74 IQ and self-concept.** Table 1.9 (page 29) reports data on 78 seventh-grade students. We want to know how well each of IQ score and self-concept score predicts GPA using least-squares regression. We also want to know which of these explanatory variables predicts GPA better. Give numerical measures that answer these questions, and explain your answers.

**2.75 Heights of husbands and wives.** The mean height of American women in their early twenties is about 64.5 inches and the standard deviation is about 2.5 inches. The mean height of men the same age is about 68.5 inches, with standard deviation about 2.7 inches. If the correlation between the heights of husbands and wives is about $r = 0.5$, what is the equation of the regression line of the husband's height on the wife's height in young couples? Draw a graph of this regression line. Predict the height of the husband of a woman who is 67 inches tall.

**2.76** CHALLENGE **A property of the least-squares regression line.** Use the equation for the least- squares regression line to show that this line always passes through the point $(\bar{x}, \bar{y})$.

**2.77 Icicle growth.** The data for Run 8903 in Table 2.4 (page 100) describe how the length $y$ in centimeters of an icicle increases over time $x$. Time is measured in minutes.

(a) What are the numerical values and units of measurement for each of $\bar{x}$, $s_x$, $\bar{y}$, $s_y$, and the correlation $r$ between $x$ and $y$?

(b) There are 2.54 centimeters in an inch. If we measure length $y$ in inches rather than in centimeters, what are the new values of $\bar{y}$, $s_y$, and the correlation $r$?

(c) If we measure length $y$ in inches rather than in centimeters, what is the new value of the slope $b_1$ of the least-squares line for predicting length from time?

**2.78 Predict final-exam scores.** In Professor Friedman's economics course the correlation between the students' total scores before the final examination and their final-examination scores is $r = 0.55$. The pre-exam totals for all students in the course have mean 270 and standard deviation 30. The final-exam scores have mean 70 and standard deviation 9. Professor Friedman has lost Julie's final exam but knows that her total before the exam was 310. He decides to predict her final-exam score from her pre-exam total.

(a) What is the slope of the least-squares regression line of final-exam scores on pre-exam total scores in this course? What is the intercept?

(b) Use the regression line to predict Julie's final-exam score.

(c) Julie doesn't think this method accurately predicts how well she did on the final exam. Calculate $r^2$ and use the value you get to argue that her actual score could have been much higher or much lower than the predicted value.

**2.79** CHALLENGE **Class attendance and grades.** A study of class attendance and grades among first-year students at a state university showed that in general students who attended a higher percent of their classes earned higher grades. Class attendance explained 16% of the variation in grade index among the students. What is the numerical value of the correlation between percent of classes attended and grade index?

**2.80** CHALLENGE **Pesticide decay.** Fenthion is a pesticide used to control the olive fruit fly. There are government limits on the amount of pesticide residue that can be present in olive products. Because the pesticide decays over time, producers of olive oil might simply store the oil until the fenthion has decayed. The simple exponential decay model says that the concentration $C$ of pesticide remaining after time $t$ is

$$C = C_0 e^{-kt}$$

where $C_0$ is the initial concentration and $k$ is a constant that determines the rate of decay. This model is a straight line if we take the logarithm of the concentration:

$$\log C = \log C_0 - kt$$

(The logarithm here is the natural logarithm, not the common logarithm with base 10.) Here are data on the concentration (milligrams of fenthion per kilogram of oil) in specimens of Greek olive oil:[26]

| Days stored | Concentration | | | | |
|---|---|---|---|---|---|
| 28 | 0.99 | 0.99 | 0.96 | 0.95 | 0.93 |
| 84 | 0.96 | 0.94 | 0.91 | 0.91 | 0.90 |
| 183 | 0.89 | 0.87 | 0.86 | 0.85 | 0.85 |
| 273 | 0.87 | 0.86 | 0.84 | 0.83 | 0.83 |
| 365 | 0.83 | 0.82 | 0.80 | 0.80 | 0.79 |

(a) Plot the natural logarithm of concentration against days stored. Notice that there are several pairs of identical data points. Does the pattern suggest that the model of simple exponential decay describes the data reasonably well, at least over this interval of time? Explain your answer.

(b) Regress the logarithm of concentration on time. Use your result to estimate the value of the constant $k$.

**2.81** CHALLENGE **The decay product is toxic.** Unfortunately, the main product of the decay of the pesticide fenthion is fenthion sulfoxide, which is also toxic. Here are data on the total concentration of fenthion and fenthion sulfoxide in the same specimens of olive oil described in the previous exercise:

| Days stored | Concentration | | | | |
|---|---|---|---|---|---|
| 28 | 1.03 | 1.03 | 1.01 | 0.99 | 0.99 |
| 84 | 1.05 | 1.04 | 1.00 | 0.99 | 0.99 |
| 183 | 1.03 | 1.02 | 1.01 | 0.98 | 0.98 |
| 273 | 1.07 | 1.06 | 1.03 | 1.03 | 1.02 |
| 365 | 1.06 | 1.02 | 1.01 | 1.01 | 0.99 |

(a) Plot concentration against days stored. Your software may fill the available space in the plot, which in this case hides the pattern. Try a plot with vertical scale from 0.8 to 1.2. Be sure your plot takes note of the pairs of identical data points.

(b) What is the slope of the least-squares line for predicting concentration of fenthion and fenthion sulfoxide from days stored? Explain why this value agrees with the graph.

(c) What do the data say about the idea of reducing fenthion in olive oil by storing the oil before selling it?

## 2.4 Cautions about Correlation and Regression

Correlation and regression are among the most common statistical tools. They are used in more elaborate form to study relationships among many variables, a situation in which we cannot see the essentials by studying a single scatterplot.

We need a firm grasp of the use and limitations of these tools, both now and as a foundation for more advanced statistics.

## Residuals

A regression line describes the overall pattern of a linear relationship between an explanatory variable and a response variable. Deviations from the overall pattern are also important. In the regression setting, we see deviations by looking at the scatter of the data points about the regression line. The vertical distances from the points to the least-squares regression line are as small as possible in the sense that they have the smallest possible sum of squares. Because they represent "left-over" variation in the response after fitting the regression line, these distances are called *residuals*.

### RESIDUALS

A **residual** is the difference between an observed value of the response variable and the value predicted by the regression line. That is,

$$\begin{aligned} \text{residual} &= \text{observed } y - \text{predicted } y \\ &= y - \hat{y} \end{aligned}$$

EXAMPLE

**2.19 Residuals for fat gain.** Example 2.12 (page 109) describes measurements on 16 young people who volunteered to overeat for 8 weeks. Those whose nonexercise activity (NEA) spontaneously rose substantially gained less fat than others. Figure 2.20(a) is a scatterplot of these data. The pattern is linear. The least-squares line is

$$\text{fat gain} = 3.505 - (0.00344 \times \text{NEA increase})$$

One subject's NEA rose by 135 calories. That subject gained 2.7 kilograms of fat. The predicted gain for 135 calories is

$$\hat{y} = 3.505 - (0.00344 \times 135) = 3.04 \text{ kg}$$

The residual for this subject is therefore

$$\begin{aligned} \text{residual} &= \text{observed } y - \text{predicted } y \\ &= y - \hat{y} \\ &= 2.7 - 3.04 = -0.34 \text{ kg} \end{aligned}$$

Most regression software will calculate and store residuals for you.

## USE YOUR KNOWLEDGE

**2.82 Find the predicted value and the residual.** Another individual in the NEA data set has NEA increase equal to 143 calories and fat gain

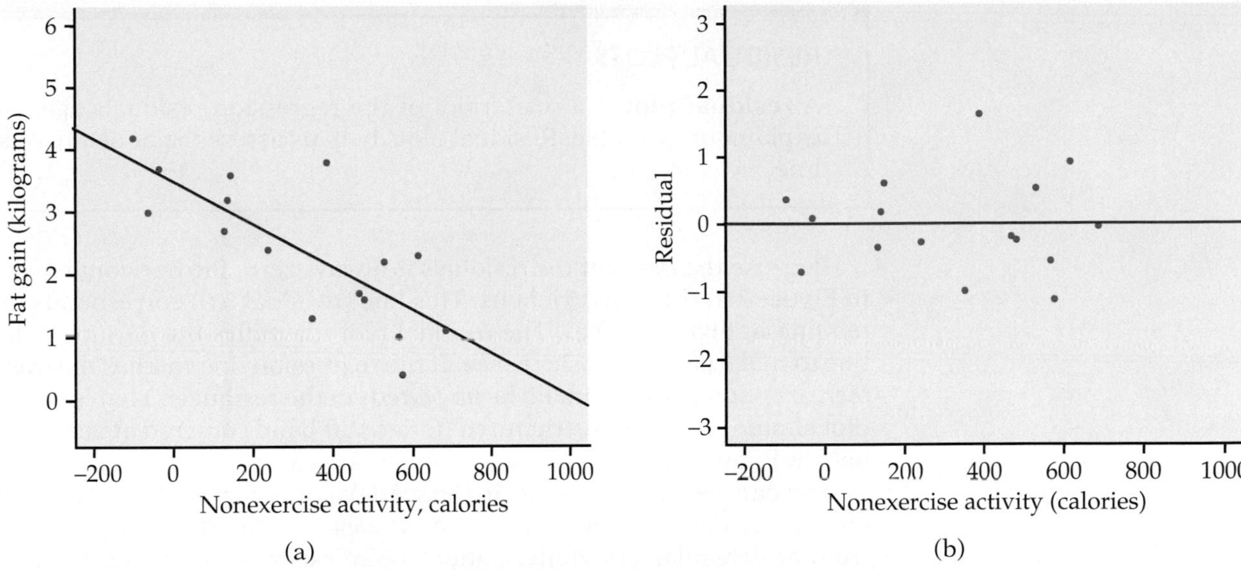


**FIGURE 2.20** (a) Scatterplot of fat gain versus increase in nonexercise activity, with the least-squares line, for Example 2.19. (b) Residual plot for the regression displayed in Figure 2.20(a). The line at $y = 0$ marks the mean of the residuals.

equal to 3.2 kg. Find the predicted value of fat gain for this individual and then calculate the residual. Explain why this residual is positive.

Because the residuals show how far the data fall from our regression line, examining the residuals helps assess how well the line describes the data. Although residuals can be calculated from any model fitted to the data, the residuals from the least-squares line have a special property: **the mean of the least-squares residuals is always zero.**

## USE YOUR KNOWLEDGE

**2.83 Find the sum of the residuals.** Here are the 16 residuals for the NEA data rounded to two decimal places:

| | | | | | | | |
|---|---|---|---|---|---|---|---|
| 0.37 | −0.70 | 0.10 | −0.34 | 0.19 | 0.61 | −0.26 | −0.98 |
| 1.64 | −0.18 | −0.23 | 0.54 | −0.54 | −1.11 | 0.93 | −0.03 |

Find the sum of these residuals. Note that the sum is not exactly zero because of roundoff error.

You can see the residuals in the scatterplot of Figure 2.20(a) by looking at the vertical deviations of the points from the line. The *residual plot* in Figure 2.20(b) makes it easier to study the residuals by plotting them against the explanatory variable, increase in NEA.

## RESIDUAL PLOTS

A **residual plot** is a scatterplot of the regression residuals against the explanatory variable. Residual plots help us assess the fit of a regression line.

Because the mean of the residuals is always zero, the horizontal line at zero in Figure 2.20(b) helps orient us. This line (residual = 0) corresponds to the fitted line in Figure 2.20(a). The residual plot magnifies the deviations from the line to make patterns easier to see. If the regression line catches the overall pattern of the data, there should be *no pattern* in the residuals. That is, the residual plot should show an unstructured horizontal band centered at zero. The residuals in Figure 2.20(b) do have this irregular scatter.

You can see the same thing in the scatterplot of Figure 2.20(a) and the residual plot of Figure 2.20(b). It's just a bit easier in the residual plot. Deviations from an irregular horizontal pattern point out ways in which the regression line fails to catch the overall pattern. For example, if the overall pattern in the scatterplot is curved rather than straight, the residuals will magnify the curved pattern, moving up and down rather than straight across. Exercise 2.86 is an example of this. Here is a different kind of example.

EXAMPLE

**2.20 Patterns in the Trans-Alaska Oil Pipeline residuals.** Figure 2.3 (page 90) plots field measurements on the depth of 100 small defects in the Trans-Alaska Oil Pipeline against laboratory measurements of the same defects. The $y = x$ line on the graph shows that field measurements tend to be too low for larger defect depths. The least-squares regression line for predicting field result from lab result, unlike the $y = x$ line, goes through the center of the points. Figure 2.21 is the residual plot for these data.

Although the horizontal line at zero does go through the middle of the points, the residuals are more spread out both above and below the line as we

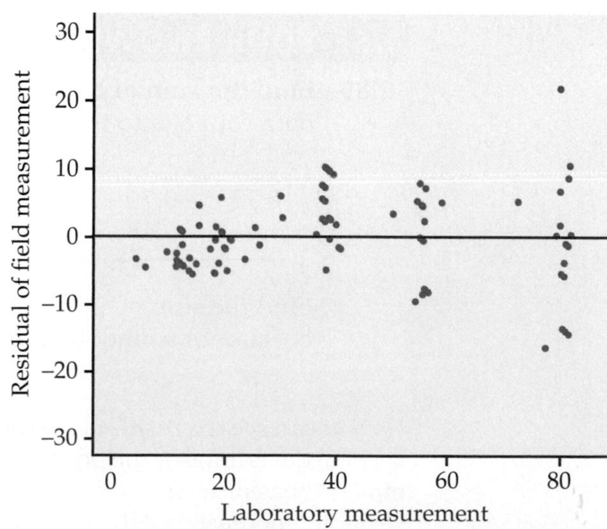


**FIGURE 2.21** Residual plot for the regression of field measurements of Alaska pipeline defects on laboratory measurements of the same defects, for Example 2.20.

move to the right. The field measurements are more variable as the true defect depth measured in the lab increases. There is indeed a straight-line pattern, but the regression line doesn't catch the important fact that the variability of field measurements increases with defect depth. The scatterplot makes this clear, and the residual plot magnifies the picture.

## Outliers and influential observations

When you look at scatterplots and residual plots, look for striking individual points as well as for an overall pattern. Here is an example of data that contain some unusual cases.

EXAMPLE

**2.21 Diabetes and blood sugar.** People with diabetes must manage their blood sugar levels carefully. They measure their fasting plasma glucose (FPG) several times a day with a glucose meter. Another measurement, made at regular medical checkups, is called HbA. This is roughly the percent of red blood cells that have a glucose molecule attached. It measures average exposure to glucose over a period of several months. Table 2.5 gives data on both HbA and FPG for 18 diabetics five months after they had completed a diabetes education class.[27]

Because both FPG and HbA measure blood glucose, we expect a positive association. The scatterplot in Figure 2.22 shows a surprisingly weak relationship, with correlation $r = 0.4819$. The line on the plot is the least-squares regression line for predicting FPG from HbA. Its equation is

$$\hat{y} = 66.4 + 10.41x$$

It appears that one-time measurements of FPG can vary quite a bit among people with similar long-term levels, as measured by HbA.

Two unusual cases are marked in Figure 2.22. Subjects 15 and 18 are unusual in different ways. Subject 15 has dangerously high FPG and lies far from the regression line in the $y$ direction. Subject 18 is close to the line but far out

**TABLE 2.5**

Two measures of glucose level in diabetics

| Subject | HbA (%) | FPG (mg/ml) | Subject | HbA (%) | FPG (mg/ml) | Subject | HbA (%) | FPG (mg/ml) |
|---|---|---|---|---|---|---|---|---|
| 1 | 6.1 | 141 | 7 | 7.5 | 96 | 13 | 10.6 | 103 |
| 2 | 6.3 | 158 | 8 | 7.7 | 78 | 14 | 10.7 | 172 |
| 3 | 6.4 | 112 | 9 | 7.9 | 148 | 15 | 10.7 | 359 |
| 4 | 6.8 | 153 | 10 | 8.7 | 172 | 16 | 11.2 | 145 |
| 5 | 7.0 | 134 | 11 | 9.4 | 200 | 17 | 13.7 | 147 |
| 6 | 7.1 | 95 | 12 | 10.4 | 271 | 18 | 19.3 | 255 |

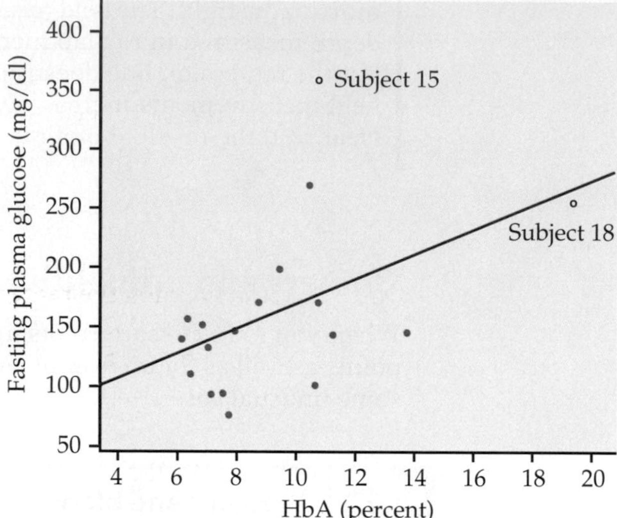


**FIGURE 2.22** Scatterplot of fasting plasma glucose against HbA (which measures long-term blood glucose), with the least-squares line, for Example 2.21.

in the $x$ direction. The residual plot in Figure 2.23 confirms that Subject 15 has a large residual and that Subject 18 does not.

Points that are outliers in the $x$ direction, like Subject 18, can have a strong influence on the position of the regression line. Least-squares lines make the sum of squares of the vertical distances to the points as small as possible. A point that is extreme in the $x$ direction with no other points near it pulls the line toward itself.

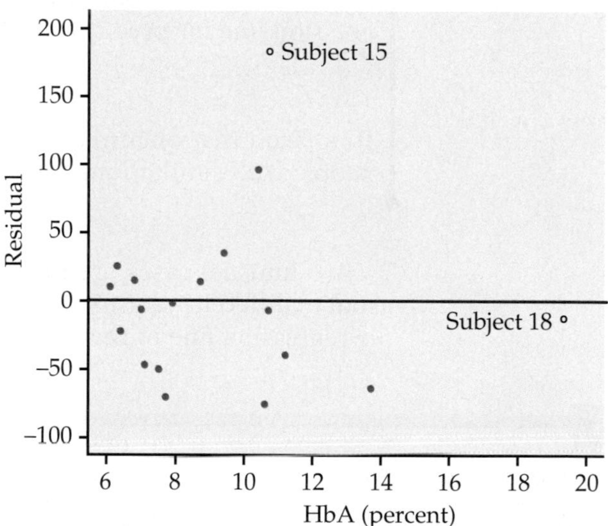


**FIGURE 2.23** Residual plot for the regression of FPG on HbA. Subject 15 is an outlier in $y$. Subject 18 is an outlier in $x$ that may be influential but does not have a large residual.

## OUTLIERS AND INFLUENTIAL OBSERVATIONS IN REGRESSION

An **outlier** is an observation that lies outside the overall pattern of the other observations. Points that are outliers in the $y$ direction of a scatterplot have large regression residuals, but other outliers need not have large residuals.

An observation is **influential** for a statistical calculation if removing it would markedly change the result of the calculation. Points that are outliers in the $x$ direction of a scatterplot are often influential for the least-squares regression line.

Influence is a matter of degree—how much does a calculation change when we remove an observation? It is difficult to assess influence on a regression line without actually doing the regression both with and without the suspicious observation. A point that is an outlier in $x$ is often influential. But if the point happens to lie close to the regression line calculated from the other observations, then its presence will move the line only a little and the point will not be influential. The influence of a point that is an outlier in $y$ depends on whether there are many other points with similar values of $x$ that hold the line in place. Figures 2.22 and 2.23 identify two unusual observations. How influential are they?

EXAMPLE

**2.22 Influential observations.** Subjects 15 and 18 both influence the correlation between FPG and HbA, in opposite directions. Subject 15 weakens the linear pattern; if we drop this point, the correlation increases from $r = 0.4819$ to $r = 0.5684$. Subject 18 extends the linear pattern; if we omit this subject, the correlation drops from $r = 0.4819$ to $r = 0.3837$.

To assess influence on the least-squares line, we recalculate the line leaving out a suspicious point. Figure 2.24 shows three least-squares lines. The solid line is the regression line of FPG on HbA based on all 18 subjects. This is the same line that appears in Figure 2.22. The dotted line is calculated from all subjects except Subject 18. You see that point 18 does pull the line down toward itself. But the influence of Subject 18 is not very large—the dotted and solid lines are close together for HbA values between 6 and 14, the range of all except Subject 18.

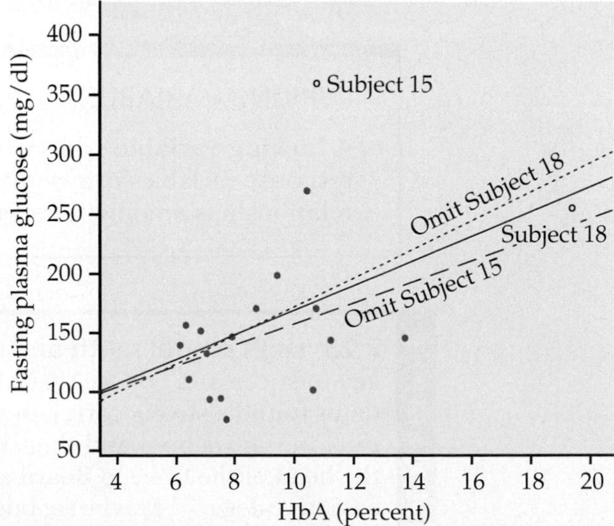


**FIGURE 2.24** Three regression lines for predicting FPG from HbA, for Example 2.22. The solid line uses all 18 subjects. The dotted line leaves out Subject 18. The dashed line leaves out Subject 15. "Leaving one out" calculations are the surest way to assess influence.

The dashed line omits Subject 15, the outlier in $y$. Comparing the solid and dashed lines, we see that Subject 15 pulls the regression line up. The influence is again not large, but it exceeds the influence of Subject 18.

The best way to see how points that are extreme in $x$ can influence the regression line is to use the *Correlation and Regression* applet on the text CD and Web site. As Exercise 2.102 demonstrates, moving one point can pull the line to almost any position on the graph.

We did not need the distinction between outliers and influential observations in Chapter 1. A single large salary that pulls up the mean salary $\bar{x}$ for a group of workers is an outlier because it lies far above the other salaries. It is also influential because the mean changes when it is removed. In the regression setting, however, not all outliers are influential. Because influential observations draw the regression line toward themselves, we may not be able to spot them by looking for large residuals.

## Beware the lurking variable

Correlation and regression are powerful tools for measuring the association between two variables and for expressing the dependence of one variable on the other. These tools must be used with an awareness of their limitations. We have seen that:

- Correlation measures *only linear association,* and fitting a straight line makes sense only when the overall pattern of the relationship is linear. Always plot your data before calculating.
- *Extrapolation* (using a fitted model far outside the range of the data that we used to fit it) often produces unreliable predictions.
- Correlation and least-squares regression are *not resistant.* Always plot your data and look for potentially influential points.

Another caution is even more important: the relationship between two variables can often be understood only by taking other variables into account. *Lurking variables* can make a correlation or regression misleading.

### LURKING VARIABLE

A **lurking variable** is a variable that is not among the explanatory or response variables in a study and yet may influence the interpretation of relationships among those variables.

EXAMPLE

**2.23 High school math and success in college.** Is high school math the key to success in college? A College Board study of 15,941 high school graduates found a strong correlation between how much math minority students took in high school and their later success in college. News articles quoted the head of the College Board as saying that "math is the gatekeeper for success in college."[28] Maybe so, but we should also think about lurking variables.

Minority students from middle-class homes with educated parents no doubt take more high school math courses. They also are more likely to have a stable family, parents who emphasize education and can pay for college, and so on. These students would succeed in college even if they took fewer math courses. The family background of the students is a lurking variable that probably explains much of the relationship between math courses and college success.

EXAMPLE

**2.24 Imports and spending for health care.** Figure 2.25 displays a strong positive linear association. The correlation between these variables is $r = 0.9749$. Because $r^2 = 0.9504$, regression of $y$ on $x$ will explain 95% of the variation in the values of $y$.

The explanatory variable in Figure 2.25 is the dollar value of goods imported into the United States in the years between 1990 and 2001. The response variable is private spending on health in the same years. There is no economic relationship between these variables. The strong association is due entirely to the fact that both imports and health spending grew rapidly in these years. The common year for each point is a lurking variable. Any two variables that both increase over time will show a strong association. This does not mean that one variable explains or influences the other. In this example, the scatterplot and correlation are correct as exercises in following recipes, but they shed no light on any real situation.

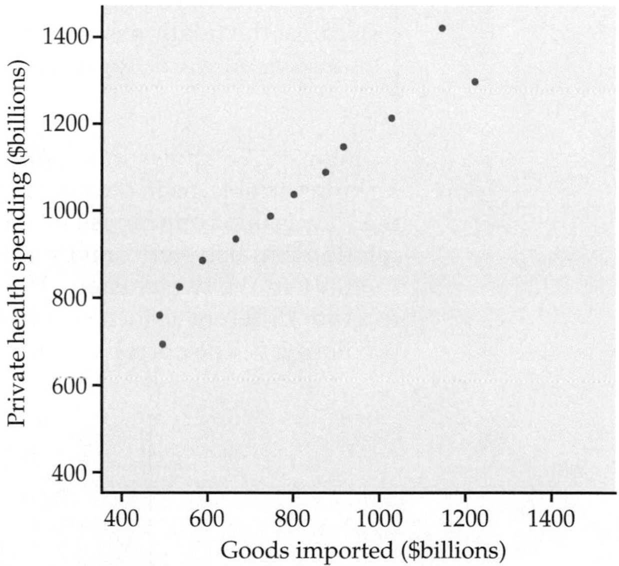


**FIGURE 2.25** The relationship between private spending on health and the value of goods imported in the same year, for Example 2.24.

Correlations such as that in Example 2.24 are sometimes called "nonsense correlations." The correlation is real. What is nonsense is the suggestion that the variables are directly related so that changing one of the variables *causes* changes in the other. The question of causation is important enough to merit separate treatment in Section 2.6. For now, just remember that an association between two variables $x$ and $y$ can reflect many types of relationship among $x$, $y$, and one or more lurking variables.

**ASSOCIATION DOES NOT IMPLY CAUSATION**

An association between an explanatory variable $x$ and a response variable $y$, even if it is very strong, is not by itself good evidence that changes in $x$ actually cause changes in $y$.

Lurking variables sometimes create a correlation between $x$ and $y$, as in Examples 2.23 and 2.24. They can also hide a true relationship between $x$ and $y$, as the following example illustrates.

EXAMPLE

**2.25 Overcrowding and indoor toilets.** A study of housing conditions and health in the city of Hull, England, measured a large number of variables for each of the wards into which the city is divided. Two of the variables were an index $x$ of overcrowding and an index $y$ of the lack of indoor toilets. Because $x$ and $y$ are both measures of inadequate housing, we expect a high correlation. Yet the correlation was only $r = 0.08$. How can this be? Investigation disclosed that some poor wards were dominated by public housing. These wards had high values of $x$ but low values of $y$ because public housing always includes indoor toilets. Other poor wards lacked public housing, and in these wards high values of $x$ were accompanied by high values of $y$. Because the relationship between $x$ and $y$ differed in the two types of wards, analyzing all wards together obscured the nature of the relationship.[29]

Figure 2.26 shows in simplified form how groups formed by a categorical lurking variable, as in the housing example, can make the correlation $r$ misleading. The groups appear as clusters of points in the scatterplot. There is a strong relationship between $x$ and $y$ within each of the clusters. In fact, $r = 0.85$ and $r = 0.91$ in the two clusters. However, because similar values of $x$ correspond to quite different values of $y$ in the two clusters, $x$ alone is of little value for predicting $y$. The correlation for all points displayed is therefore low: $r = 0.14$.

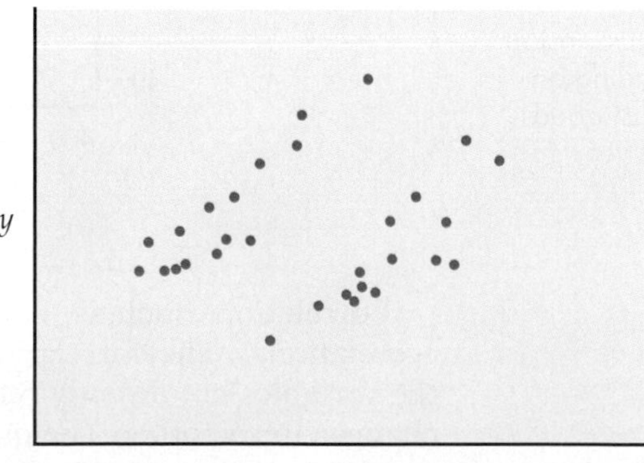


**FIGURE 2.26** This scatterplot has a low $r$ even though there is a strong correlation within each of the two clusters.

This example is another reminder to plot the data rather than simply calculate numerical measures such as the correlation.

## Beware correlations based on averaged data

Regression or correlation studies sometimes work with averages or other measures that combine information from many individuals. For example, if we plot the average height of young children against their age in months, we will see a very strong positive association with correlation near 1. But individual children of the same age vary a great deal in height. A plot of height against age for individual children will show much more scatter and lower correlation than the plot of average height against age.

CAUTION

*A correlation based on averages over many individuals is usually higher than the correlation between the same variables based on data for individuals.* This fact reminds us again of the importance of noting exactly what variables a statistical study involves.

## The restricted-range problem

A regression line is often used to predict the response $y$ to a given value $x$ of the explanatory variable. Successful prediction does not require a cause-and-effect relationship. If both $x$ and $y$ respond to the same underlying unmeasured variables, $x$ may help us predict $y$ even though $x$ has no direct influence on $y$. For example, the scores of SAT exams taken in high school help predict college grades. There is no cause-and-effect tie between SAT scores and college grades. Rather, both reflect a student's ability and knowledge.

How well do SAT scores, perhaps with the help of high school grades, predict college GPA? We can use the correlation $r$ and its square to get a rough answer. There is, however, a subtle difficulty.

EXAMPLE

**2.26 SAT scores and GPA.** Combining several studies for students graduating from college since 1980, the College Board reports these correlations between explanatory variables and the overall GPA of college students:

| SAT Math and Verbal | High school grades | SAT plus grades |
|---|---|---|
| $r = 0.36$ | $r = 0.42$ | $r = 0.52$ |

Because $0.52^2 = 0.27$, we see that SAT scores plus students' high school records explain about 27% of the variation in GPA among college students.

The subtle problem? Colleges differ greatly in the range of students they attract. Almost all students at Princeton have high SAT scores and did well in high school. At Generic State College, most students are in the middle range of SAT scores and high school performance. Both sets of students receive the full spread of grades. We suspect that if Princeton admitted weaker students they would get lower grades, and that the typical Princeton student would get very high grades at Generic State. This is the *restricted-range problem:* the data do not contain information on the full range of both explanatory and

response variables. *When data suffer from restricted range, r and $r^2$ are lower than they would be if the full range could be observed.*

Thus, $r = 0.52$ understates the actual ability of SAT scores and high school grades to predict college GPA. One investigator found 21 colleges that enrolled the full range of high school graduates. Sure enough, for these colleges, $r = 0.65$.[30]

Did you notice that the correlations in Example 2.26 involve more than one explanatory variable? It is common to use several explanatory variables together to predict a response. This is called *multiple regression.* Each $r$ in the example is a *multiple correlation coefficient,* whose square is the proportion of variation in the response explained by the multiple regression. Chapter 11 introduces multiple regression.

## BEYOND THE BASICS

### Data Mining

Chapters 1 and 2 of this book are devoted to the important aspect of statistics called *exploratory data analysis* (EDA). We use graphs and numerical summaries to examine data, searching for patterns and paying attention to striking deviations from the patterns we find. In discussing regression, we advanced to using the pattern we find (in this case, a linear pattern) for prediction.

Suppose now that we have a truly enormous data base, such as all purchases recorded by the cash register scanners of a national retail chain during the past week. Surely this treasure chest of data contains patterns that might guide business decisions. If we could see clearly the types of activewear preferred in large California cities and compare the preferences of small Midwest cities—right now, not at the end of the season—we might improve profits in both parts of the country by matching stock with demand. This sounds much like EDA, and indeed it is. Exploring really large data bases in the hope of finding useful patterns is called **data mining.** Here are some distinctive features of data mining:

data mining

- When you have 100 gigabytes of data, even straightforward calculations and graphics become impossibly time-consuming. So efficient algorithms are very important.
- The structure of the data base and the process of storing the data (the fashionable term is *data warehousing*), perhaps by unifying data scattered across many departments of a large corporation, require careful consideration.
- Data mining requires automated tools that work based on only vague queries by the user. The process is too complex to do step-by-step as we have done in EDA.

All of these features point to the need for sophisticated computer science as a basis for data mining. Indeed, data mining is often thought of as a part of computer science. Yet many statistical ideas and tools—mostly tools for dealing with multidimensional data, not the sort of thing that appears in a first statistics course—are very helpful. Like many modern developments, data mining crosses the boundaries of traditional fields of study.

Do remember that the perils we associate with blind use of correlation and regression are yet more perilous in data mining, where the fog of an immense data base prevents clear vision. Extrapolation, ignoring lurking variables, and confusing association with causation are traps for the unwary data miner.

## SECTION 2.4 Summary

You can examine the fit of a regression line by plotting the **residuals,** which are the differences between the observed and predicted values of $y$. Be on the lookout for points with unusually large residuals and also for nonlinear patterns and uneven variation about the line.

Also look for **influential observations,** individual points that substantially change the regression line. Influential observations are often outliers in the $x$ direction, but they need not have large residuals.

Correlation and regression must be **interpreted with caution.** Plot the data to be sure that the relationship is roughly linear and to detect outliers and influential observations.

**Lurking variables** may explain the relationship between the explanatory and response variables. Correlation and regression can be misleading if you ignore important lurking variables.

We cannot conclude that there is a cause-and-effect relationship between two variables just because they are strongly associated. **High correlation does not imply causation.**

**A correlation based on averages** is usually higher than if we used data for individuals. A correlation based on data with a **restricted range** is often lower than would be the case if we could observe the full range of the variables.

## SECTION 2.4 Exercises

*For Exercise 2.82, see page 126; for Exercise 2.83, see page 127.*

**2.84** **Price and ounces.** In Example 2.2 (page 84) and Exercise 2.3 (page 85) we examined the relationship between the price and the size of a Mocha Frappuccino©. The 12-ounce Tall drink costs \$3.15, the 16-ounce Grande is \$3.65, and the 24-ounce Venti is \$4.15.

(a) Plot the data and describe the relationship. (Explain why you should plot size in ounces on the $x$ axis.)

(b) Find the least-squares regression line for predicting the price using size. Add the line to your plot.

(c) Draw a vertical line from the least-squares line to each data point. This gives a graphical picture of the residuals.

(d) Find the residuals and verify that they sum to zero.

(e) Plot the residuals versus size. Interpret this plot.

**2.85** **Average monthly temperatures.** Here are the average monthly temperatures for Chicago, Illinois:

| Month | 1 | 2 | 3 | 4 | 5 | 6 |
|---|---|---|---|---|---|---|
| Temperature (°F) | 21.0 | 25.4 | 37.2 | 48.6 | 58.9 | 68.6 |
| Month | 7 | 8 | 9 | 10 | 11 | 12 |
| Temperature (°F) | 73.2 | 71.7 | 64.4 | 52.8 | 40.0 | 26.6 |

In this table, months are coded as integers, with January corresponding to 1 and December corresponding to 12.

(a) Plot the data with month on the $x$ axis and temperature on the $y$ axis. Describe the relationship.

(b) Find the least-squares regression line and add it to the plot. Does the line give a good fit to the data? Explain your answer.

(c) Calculate the residuals and plot them versus month. Describe the pattern and explain what the residual plot tells you about the relationship between temperature and month in Chicago.

(d) Do you think you would find a similar pattern if you plotted the same kind of data for another city?

(e) Would your answer to part (d) change if the other city was Melbourne, Australia? Explain why or why not.

**2.86 Fuel consumption and speed.** Exercise 2.22 (page 99) gives data on the fuel consumption $y$ of a car at various speeds $x$. The relationship is strongly curved: fuel used decreases with increasing speed at low speeds, then increases again as higher speeds are reached. The equation of the least-squares regression line for these data is

$$\hat{y} = 11.058 - 0.01466x$$

The residuals, in the same order as the observations, are

| | | | | | | | |
|---|---|---|---|---|---|---|---|
| 10.09 | 2.24 | −0.62 | −2.47 | −3.33 | −4.28 | −3.73 | −2.94 |
| −2.17 | −1.32 | −0.42 | 0.57 | 1.64 | 2.76 | 3.97 | |

(a) Make a scatterplot of the observations and draw the regression line on your plot. The line is a poor description of the curved relationship.

(b) Check that the residuals have sum zero (up to roundoff error).

(c) Make a plot of the residuals against the values of $x$. Draw a horizontal line at height zero on your plot. The residuals show the same pattern about this line as the data points show about the regression line in the scatterplot in (a).

**2.87 Growth of infants in Egypt.** A study of nutrition in developing countries collected data from the Egyptian village of Nahya. Here are the mean weights (in kilograms) for 170 infants in Nahya who were weighed each month during their first year of life:[31]

| Age (months) | 1 | 2 | 3 | 4 | 5 | 6 | 7 | 8 | 9 | 10 | 11 | 12 |
|---|---|---|---|---|---|---|---|---|---|---|---|---|
| Weight (kg) | 4.3 | 5.1 | 5.7 | 6.3 | 6.8 | 7.1 | 7.2 | 7.2 | 7.2 | 7.2 | 7.5 | 7.8 |

(a) Plot weight against time.

(b) A hasty user of statistics enters the data into software and computes the least-squares line without plotting the data. The result is

```
The regression equation is
weight = 4.88 + 0.267 age
```

Plot this line on your graph. Is it an acceptable summary of the overall pattern of growth? Remember that you can calculate the least-squares line for *any* set of two-variable data. It's up to you to decide if it makes sense to fit a line.

(c) Fortunately, the software also prints out the residuals from the least-squares line. In order of age along the rows, they are

| | | | | | |
|---|---|---|---|---|---|
| −0.85 | −0.31 | 0.02 | 0.35 | 0.58 | 0.62 |
| 0.45 | 0.18 | −0.08 | −0.35 | −0.32 | −0.28 |

Verify that the residuals have sum zero (except for roundoff error). Plot the residuals against age and add a horizontal line at zero. Describe carefully the pattern that you see.

**2.88 Pesticide in olive oil.** Exercise 2.80 gives data on the concentration of the pesticide fenthion in Greek olive oil that has been stored for various lengths of time. The exponential decay model used to describe how concentration decreases over time proposes a curved relationship between storage time and concentration. Do the residuals from fitting a regression line show a curved pattern? The least-squares line for predicting concentration is

$$\hat{y} = 0.965 - 0.00045x$$

(a) The first batch of olive oil was stored for 28 days and had fenthion concentration 0.99 mg/kg. What is the predicted concentration for this batch? What is the residual?

(b) The residuals, arranged as in the data table in Exercise 2.88, are:

| Days stored | Residual | | | | |
|---|---|---|---|---|---|
| 28 | 0.0378 | 0.0378 | 0.0078 | −0.0022 | −0.0222 |
| 84 | 0.0329 | 0.0129 | −0.0171 | −0.0171 | −0.0271 |
| 183 | 0.0072 | −0.0128 | −0.0228 | −0.0328 | −0.0328 |
| 273 | 0.0275 | 0.0175 | −0.0025 | −0.0125 | −0.0125 |
| 365 | 0.0286 | 0.0186 | −0.0014 | −0.0014 | −0.0114 |

Check that your residual from (a) agrees (up to roundoff error) with the value 0.0378 given here. Verify that the residuals sum to zero (again up to roundoff error).

(c) Make a residual plot. Is a curved pattern visible? Is the curve very strong? (Software often makes the pattern hard to see because it fills the entire plot area. Try a plot with vertical scale from −0.1 to 0.1.)

**2.89 Effect of using means.** Your plot for Exercise 2.87 shows that the increase of the mean weight of children in Nahya is very linear during the first 5 months of life. The correlation of age and weight is $r = 0.9964$ for the first 5 months. Weight in these data is the mean for 170 children. Explain why the correlation between age and weight for the 170 individual children would surely be much smaller.

**2.90 A test for job applicants.** Your company gives a test of cognitive ability to job applicants before deciding whom to hire. Your boss has asked you to use company records to see if this test really helps predict the performance ratings of employees. Explain carefully to your boss why the restricted-range problem may make it difficult to see a strong relationship between test scores and performance ratings.

**2.91** CHALLENGE CAUTION **A lurking variable.** The effect of a lurking variable can be surprising when individuals are divided into groups. In recent years, the mean SAT score of all high school seniors has increased. But the mean SAT score has decreased for students at each level of high school grades (A, B, C, and so on). Explain how grade inflation in high school (the lurking variable) can account for this pattern. *A relationship that holds for each group within a population need not hold for the population as a whole. In fact, the relationship can even change direction.*

**2.92** CHALLENGE **Another example.** Here is another example of the group effect cautioned about in the previous exercise. Explain how as a nation's population grows older mean income can go down for workers in each age group, yet still go up for all workers.

**2.93 Basal metabolic rate.** Careful statistical studies often include examination of potential lurking variables. This was true of the study of the effect of nonexercise activity (NEA) on fat gain (Example 2.12, page 109), our lead example in Section 2.3. Overeating may lead our bodies to spontaneously increase NEA (fidgeting and the like). Our bodies might also spontaneously increase their basal metabolic rate (BMR), which measures energy use while resting. If both energy uses increased, regressing fat gain on NEA alone would be misleading. Here are data on BMR and fat gain for the same 16 subjects whose NEA we examined earlier:

| BMR increase (cal) | 117 | 352 | 244 | −42 | −3 | 134 | 136 | −32 |
|---|---|---|---|---|---|---|---|---|
| Fat gain (kg) | 4.2 | 3.0 | 3.7 | 2.7 | 3.2 | 3.6 | 2.4 | 1.3 |
| BMR increase (cal) | −99 | 9 | −15 | −70 | 165 | 172 | 100 | 35 |
| Fat gain (kg) | 3.8 | 1.7 | 1.6 | 2.2 | 1.0 | 0.4 | 2.3 | 1.1 |

The correlation between NEA and fat gain is $r = -0.7786$. The slope of the regression line for predicting fat gain from NEA is $b_1 = -0.00344$ kilogram per calorie. What are the correlation and slope for BMR and fat gain? Explain why these values show that BMR has much less effect on fat gain than does NEA.

**2.94 Gas chromatography.** Gas chromatography is a technique used to detect very small amounts of a substance, for example, a contaminant in drinking water. Laboratories use regression to calibrate such techniques. The data below show the results of five measurements for each of four amounts of the substance being investigated.[32] The explanatory variable $x$ is the amount of substance in the specimen, measured in nanograms (ng), units of $10^{-9}$ gram. The response variable $y$ is the reading from the gas chromatograph.

| Amount (ng) | Response | | | | |
|---|---|---|---|---|---|
| 0.25 | 6.55 | 7.98 | 6.54 | 6.37 | 7.96 |
| 1.00 | 29.7 | 30.0 | 30.1 | 29.5 | 29.1 |
| 5.00 | 211 | 204 | 212 | 213 | 205 |
| 20.00 | 929 | 905 | 922 | 928 | 919 |

(a) Make a scatterplot of these data. The relationship appears to be approximately linear, but the wide variation in the response values makes it hard to see detail in this graph.

(b) Compute the least-squares regression line of $y$ on $x$, and plot this line on your graph.

(c) Now compute the residuals and make a plot of the residuals against $x$. It is much easier to see deviations from linearity in the residual plot. Describe carefully the pattern displayed by the residuals.

**2.95 Golf scores.** The following table presents the golf scores of 11 members of a women's golf team in two rounds of tournament play.

| Player | 1 | 2 | 3 | 4 | 5 | 6 | 7 | 8 | 9 | 10 | 11 |
|---|---|---|---|---|---|---|---|---|---|---|---|
| Round 1 | 89 | 90 | 87 | 95 | 86 | 81 | 105 | 83 | 88 | 91 | 79 |
| Round 2 | 94 | 85 | 89 | 89 | 81 | 76 | 89 | 87 | 91 | 88 | 80 |

(a) Plot the data with the Round 1 scores on the $x$ axis and the Round 2 scores on the $y$ axis. There is a generally linear pattern except for one potentially influential observation. Circle this observation on your graph.

(b) Here are the equations of two least-squares lines. One of them is calculated from all 11 data points and the other omits the influential observation.

$$\hat{y} = 20.49 + 0.754x$$

$$\hat{y} = 50.01 + 0.410x$$

Draw both lines on your scatterplot. Which line omits the influential observation? How do you know this?

**2.96** **Climate change.** Drilling down beneath a lake in Alaska yields chemical evidence of past changes in climate. Biological silicon, left by the skeletons of single-celled creatures called diatoms, measures the abundance of life in the lake. A rather complex variable based on the ratio of certain isotopes relative to ocean water gives an indirect measure of moisture, mostly from snow. As we drill down, we look farther into the past. Here are data from 2300 to 12,000 years ago:[33]

| Isotope (%) | Silicon (mg/g) | Isotope (%) | Silicon (mg/g) | Isotope (%) | Silicon (mg/g) |
|---|---|---|---|---|---|
| −19.90 | 97 | −20.71 | 154 | −21.63 | 224 |
| −19.84 | 106 | −20.80 | 265 | −21.63 | 237 |
| −19.46 | 118 | −20.86 | 267 | −21.19 | 188 |
| −20.20 | 141 | −21.28 | 296 | −19.37 | 337 |

(a) Make a scatterplot of silicon (response) against isotope (explanatory). Ignoring the outlier, describe the direction, form, and strength of the relationship. The researchers say that this and relationships among other variables they measured are evidence for cyclic changes in climate that are linked to changes in the sun's activity.

(b) The researchers single out one point: "The open circle in the plot is an outlier that was excluded in the correlation analysis." Circle this outlier on your graph. What is the correlation with and without this point? The point strongly influences the correlation.

(c) Is the outlier also strongly influential for the regression line? Calculate and draw on your graph two regression lines, and discuss what you see.

**2.97** **City and highway gas mileage.** Table 1.10 (page 31) gives the city and highway gas mileages for 21 two-seater cars, including the Honda Insight gas-electric hybrid car. In Exercise 2.45 you investigated the influence of the Insight on the correlation between city and highway mileage.

(a) Make a scatterplot of highway mileage (response) against city mileage (explanatory) for all 21 cars.

(b) Use software or a graphing calculator to find the regression line for predicting highway mileage from city mileage and also the 21 residuals for this regression. Make a residual plot with a horizontal line at zero. (The "stacks" in the plot are due to the fact that mileage is measured only to the nearest mile per gallon.)

(c) Which car has the largest positive residual? The largest negative residual?

(d) The Honda Insight, an extreme outlier, does not have the largest residual in either direction. Why is this not surprising?

**2.98** **Stride rate of runners.** Runners are concerned about their form when racing. One measure of form is the stride rate, the number of steps taken per second. As running speed increases, the stride rate should also increase. In a study of 21 of the best American female runners, researchers measured the stride rate for different speeds. The following table gives the speeds (in feet per second) and the mean stride rates for these runners:[34]

| Speed | 15.86 | 16.88 | 17.50 | 18.62 | 19.97 | 21.06 | 22.11 |
|---|---|---|---|---|---|---|---|
| Stride rate | 3.05 | 3.12 | 3.17 | 3.25 | 3.36 | 3.46 | 3.55 |

(a) Plot the data with speed on the $x$ axis and stride rate on the $y$ axis. Does a straight line adequately describe these data?

(b) Find the equation of the regression line of stride rate on speed. Draw this line on your plot.

(c) For each of the speeds given, obtain the predicted value of the stride rate and the residual. Verify that the residuals add to zero.

(d) Plot the residuals against speed. Describe the pattern. What does the plot indicate about the adequacy of the linear fit? Are there any potentially influential observations?

**2.99** **City and highway gas mileage.** Continue your work in Exercise 2.97. Find the regression line for predicting highway mileage from city mileage for the 20 two-seater cars other than the Honda Insight. Draw both regression lines on your scatterplot. Is the Insight very influential for the least-squares line? (Look at the position of the lines for city mileages between 10 and 30 MPG, values that cover most cars.) What explains your result?

**2.100** **Stride rate and running speed.** Exercise 2.98 gives data on the mean stride rate of a group of 21 elite female runners at various running speeds. Find the correlation between speed and stride rate. Would you expect this correlation to increase or decrease if we had data on the individual stride rates of all 21 runners at each speed? Why?

**2.101** APPLET CAUTION **Use the applet.** It isn't easy to guess the position of the least-squares line by eye. Use the *Correlation and Regression* applet to compare a line you draw with the least-squares line. Click on the scatterplot to create a group of 15 to 20 points from lower left to upper right with a clear positive straight-line pattern (correlation around 0.7). Click the "Draw line" button and use the mouse to draw a line through the middle of the cloud of points from lower left to upper right. Note the "thermometer" that appears above the plot. The blue portion is the sum of the squared vertical distances from the points in the plot to the least-squares line. The green portion is the "extra" sum of squares for your line—it shows by how much your line misses the smallest possible sum of squares.

(a) You drew a line by eye through the middle of the pattern. Yet the right-hand part of the bar is probably almost entirely green. What does that tell you?

(b) Now click the "Show least-squares line" box. Is the slope of the least-squares line smaller (the new line is less steep) or larger (line is steeper) than that of your line? If you repeat this exercise several times, you will consistently get the same result. *The least-squares line minimizes the vertical distances of the points from the line. It is not the line through the "middle" of the cloud of points.* This is one reason why it is hard to draw a good regression line by eye.

**2.102** APPLET **Use the applet.** Go to the *Correlation and Regression* applet. Click on the scatterplot to create a group of 10 points in the lower-left corner of the scatterplot with a strong straight-line pattern (correlation about −0.9). In Exercise 2.46 you started here to see that correlation $r$ is not resistant. Now click the "Show least-squares line" box to display the regression line.

(a) Add one point at the upper left that is far from the other 10 points but exactly on the regression line. Why does this outlier have no effect on the line even though it changes the correlation?

(b) Now drag this last point down until it is opposite the group of 10 points. You see that one end of the least-squares line chases this single point, while the other end remains near the middle of the original group of 10. What makes the last point so influential?

**2.103** **Education and income.** There is a strong positive correlation between years of education and income for economists employed by business firms. (In particular, economists with doctorates earn more than economists with only a bachelor's degree.) There is also a strong positive correlation between years of education and income for economists employed by colleges and universities. But when all economists are considered, there is a *negative* correlation between education and income. The explanation for this is that business pays high salaries and employs mostly economists with bachelor's degrees, while colleges pay lower salaries and employ mostly economists with doctorates. Sketch a scatterplot with two groups of cases (business and academic) that illustrates how a strong positive correlation within each group and a negative overall correlation can occur together. (*Hint:* Begin by studying Figure 2.26.)

**2.104** **Dangers of not looking at a plot.** Table 2.6 presents four sets of data prepared by the statistician Frank Anscombe to illustrate the dangers of calculating without first plotting the data.[35]

(a) Without making scatterplots, find the correlation and the least-squares regression line for all four data sets. What do you notice? Use the regression line to predict $y$ for $x = 10$.

(b) Make a scatterplot for each of the data sets and add the regression line to each plot.

(c) In which of the four cases would you be willing to use the regression line to describe the dependence of $y$ on $x$? Explain your answer in each case.

**TABLE 2.6**

**Four data sets for exploring correlation and regression**

| | | | | | | | | | | | |
|---|---|---|---|---|---|---|---|---|---|---|---|
| **Data Set A** | | | | | | | | | | | |
| $x$ | 10 | 8 | 13 | 9 | 11 | 14 | 6 | 4 | 12 | 7 | 5 |
| $y$ | 8.04 | 6.95 | 7.58 | 8.81 | 8.33 | 9.96 | 7.24 | 4.26 | 10.84 | 4.82 | 5.68 |
| **Data Set B** | | | | | | | | | | | |
| $x$ | 10 | 8 | 13 | 9 | 11 | 14 | 6 | 4 | 12 | 7 | 5 |
| $y$ | 9.14 | 8.14 | 8.74 | 8.77 | 9.26 | 8.10 | 6.13 | 3.10 | 9.13 | 7.26 | 4.74 |
| **Data Set C** | | | | | | | | | | | |
| $x$ | 10 | 8 | 13 | 9 | 11 | 14 | 6 | 4 | 12 | 7 | 5 |
| $y$ | 7.46 | 6.77 | 12.74 | 7.11 | 7.81 | 8.84 | 6.08 | 5.39 | 8.15 | 6.42 | 5.73 |
| **Data Set D** | | | | | | | | | | | |
| $x$ | 8 | 8 | 8 | 8 | 8 | 8 | 8 | 8 | 8 | 8 | 19 |
| $y$ | 6.58 | 5.76 | 7.71 | 8.84 | 8.47 | 7.04 | 5.25 | 5.56 | 7.91 | 6.89 | 12.50 |

## 2.5 Data Analysis for Two-Way Tables

LOOK BACK
**quantitative and categorical variables, page 4**

When we study relationships between two variables, one of the first questions we ask is whether each variable is quantitative or categorical. For two quantitative variables, we use a scatterplot to examine the relationship, and we fit a line to the data if the relationship is approximately linear. If one of the variables is quantitative and the other is categorical, we can use the methods in Chapter 1 to describe the distribution of the quantitative variable for each value of the categorical variable. This leaves us with the situation where both variables are categorical. In this section we discuss methods for studying these relationships.

Some variables—such as gender, race, and occupation—are inherently categorical. Other categorical variables are created by grouping values of a quantitative variable into classes. Published data are often reported in grouped form to save space. To describe categorical data, we use the *counts* (frequencies) or *percents* (relative frequencies) of individuals that fall into various categories.

### The two-way table

two-way table

A key idea in studying relationships between two variables is that both variables must be measured on the same individuals or cases. When both variables are categorical, the raw data are summarized in a **two-way table** that gives counts of observations for each combination of values of the two categorical variables. Here is an example.

EXAMPLE

**2.27 Binge drinking by college students.** Alcohol abuse has been described by college presidents as the number one problem on campus, and it is an important cause of death in young adults. How common is it? A survey of 17,096 students in U.S. four-year colleges collected information on drinking behavior and alcohol-related problems.[36] The researchers defined "frequent binge drinking" as having five or more drinks in a row three or more times in the past two weeks. Here is the two-way table classifying students by gender and whether or not they are frequent binge drinkers:

**Two-way table for frequent binge drinking and gender**

| | Gender | |
|---|---|---|
| **Frequent binge drinker** | **Men** | **Women** |
| Yes | 1630 | 1684 |
| No | 5550 | 8232 |

We see that there are 1630 male students who are frequent binge drinkers and 5550 male students who are not.

## USE YOUR KNOWLEDGE

**2.105 Read the table.** How many female students are binge drinkers? How many are not?

For the binge-drinking example, we could view gender as an explanatory variable and frequent binge drinking as a response variable. This is why we put gender in the columns (like the $x$ axis in a regression) and frequent binge drinking in the rows (like the $y$ axis in a regression). We call binge drinking the **row variable** because each horizontal row in the table describes the drinking behavior. Gender is the **column variable** because each vertical column describes one gender group. Each combination of values for these two variables is called a **cell.** For example, the cell corresponding to women who are not frequent binge drinkers contains the number 8232. This table is called a 2 × 2 table because there are 2 rows and 2 columns.

row and column variables

cell

To describe relationships between two categorical variables, we compute different types of percents. Our job is easier if we expand the basic two-way table by adding various totals. We illustrate the idea with our binge-drinking example.

EXAMPLE

**2.28 Add the margins to the table.** We expand the table in Example 2.27 by adding the totals for each row, for each column, and the total number of all of the observations. Here is the result:

**Two-way table for frequent binge drinking and gender**

| | Gender | | |
|---|---|---|---|
| **Frequent binge drinker** | **Men** | **Women** | **Total** |
| Yes | 1,630 | 1,684 | 3,314 |
| No | 5,550 | 8,232 | 13,782 |
| Total | 7,180 | 9,916 | 17,096 |

In this study there are 7180 male students. The total number of binge drinkers is 3314 and the total number of individuals in the study is 17,096.

## USE YOUR KNOWLEDGE

**2.106 Read the margins of the table.** How many women are subjects in the binge-drinking study? What is the total number of students who are not binge drinkers?

In this example, be sure that you understand how the table is obtained from the raw data. Think about a data file with one line per subject. There would be 17,096 lines or records in this data set. In the two-way table, each individual is counted once and only once. As a result, the sum of the counts in the table is the total number of individuals in the data set. *Most errors in the use of categorical-data methods come from a misunderstanding of how these tables are constructed.*

## Joint distribution

joint distribution

We are now ready to compute some proportions that help us understand the data in a two-way table. Suppose that we are interested in the men who are binge drinkers. The proportion of these is simply 1630 divided by 17,096, or 0.095. We would estimate that 9.5% of college students are male frequent binge drinkers. For each cell, we can compute a proportion by dividing the cell entry by the total sample size. The collection of these proportions is the **joint distribution** of the two categorical variables.

EXAMPLE

**2.29 The joint distribution.** For the binge-drinking example, the joint distribution of binge drinking and gender is

**Joint distribution of frequent binge drinking and gender**

| | Gender | |
|---|---|---|
| Frequent binge drinker | Men | Women |
| Yes | 0.095 | 0.099 |
| No | 0.325 | 0.482 |

Because this is a distribution, the sum of the proportions should be 1. For this example the sum is 1.001. The difference is due to roundoff error.

## USE YOUR KNOWLEDGE

**2.107 Explain the computation.** Explain how the entry for the women who are not binge drinkers in Example 2.29 is computed from the table in Example 2.28.

From the joint distribution we see that the proportions of men and women frequent binge drinkers are similar in the population of college students. For the men we have 9.5%; the women are slightly higher at 9.9%. Note, however, that the proportion of women who are not frequent binge drinkers is also higher than the proportion of men. One reason for this is that there are more women in the sample than men. To understand this set of data we will need to do some additional calculations. Let's look at the distribution of gender.

## Marginal distributions

marginal distribution

When we examine the distribution of a single variable in a two-way table, we are looking at a **marginal distribution.** There are two marginal distributions, one for each categorical variable in the two-way table. They are very easy to compute.

EXAMPLE

**2.30 The marginal distribution of gender.** Look at the table in Example 2.28. The total numbers of men and women are given in the bottom row, labeled "Total." Our sample has 7180 men and 9916 women. To find the marginal distribution of gender we simply divide these numbers by the total sample size, 17,096. The marginal distribution of gender is

| Marginal distribution of gender | | |
|---|---|---|
| | Men | Women |
| Proportion | 0.420 | 0.580 |

Note that the proportions sum to 1; there is no roundoff error.

Often we prefer to use percents rather than proportions. Here is the marginal distribution of gender described with percents:

| Marginal distribution of gender | | |
|---|---|---|
| | Men | Women |
| Percent | 42.0% | 58.0% |

Which form do you prefer?

The other marginal distribution for this example is the distribution of binge drinking.

EXAMPLE

**2.31 The marginal distribution in percents.** Here is the marginal distribution of the frequent-binge-drinking variable (in percents):

| Marginal distribution of frequent binge drinking | | |
|---|---|---|
| | Yes | No |
| Percent | 19.4% | 80.6% |

**USE YOUR KNOWLEDGE**

**2.108 Explain the marginal distribution.** Explain how the marginal distribution of frequent binge drinking given in Example 2.31 is computed from the entries in the table given in Example 2.28.

LOOK BACK
**bar graphs and pie charts, page 6**

Each marginal distribution from a two-way table is a distribution for a single categorical variable. We can use a bar graph or a pie chart to display such a distribution. For our two-way table, we will be content with numerical summaries: for example, 58% of these college students are women, and 19.4% of the students are frequent binge drinkers. When we have more rows or columns, the graphical displays are particularly useful.

## Describing relations in two-way tables

The table in Example 2.29 contains much more information than the two marginal distributions of gender alone and frequent binge drinking alone. We need to do a little more work to examine the relationship. *Relationships among categorical variables are described by calculating appropriate percents from the counts given.* What percents do you think we should use to describe the relationship between gender and frequent binge drinking?

EXAMPLE

**2.32 Women who are frequent binge drinkers.** What percent of the women in our sample are frequent binge drinkers? This is the count of the women who are frequent binge drinkers as a percent of the number of women in the sample:

$$\frac{1684}{9916} = 0.170 = 17.0\%$$

**USE YOUR KNOWLEDGE**

**2.109 Find the percent.** Show that the percent of men who are frequent binge drinkers is 22.7%.

Recall that when we looked at the joint distribution of gender and binge drinking, we found that among all college students in the sample, 9.5% were male frequent binge drinkers and 9.9% were female frequent binge drinkers. The percents are fairly similar because the counts for these two groups, 1630 and 1684, are close. The calculations that we just performed, however, give us a different view. When we look separately at women and men, we see that the proportions of frequent binge drinkers are somewhat different, 17.0% for women versus 22.7% for men.

## Conditional distributions

In Example 2.32 we looked at the women alone and examined the distribution of the other categorical variable, frequent binge drinking. Another way to say this is that we conditioned on the value of gender being female. Similarly, we can condition on the value of gender being male. When we condition on the value of one variable and calculate the distribution of the other variable, we obtain a **conditional distribution.** Note that in Example 2.32 we calculated only the percent for frequent binge drinking. The complete conditional distribution gives the proportions or percents for all possible values of the conditioning variable.

conditional distribution

EXAMPLE

**2.33 Conditional distribution of binge drinking for women.** For women, the conditional distribution of the binge-drinking variable in terms of percents is

| Conditional distribution of binge drinking for women | | |
|---|---|---|
| | **Yes** | **No** |
| Percent | 17.0% | 83.0% |

Note that we have included the percents for both of the possible values, Yes and No, of the binge-drinking variable. These percents sum to 100%.

## USE YOUR KNOWLEDGE

**2.110 A conditional distribution.** Perform the calculations to show that the conditional distribution of binge drinking for men is

| Conditional distribution of binge drinking for men | | |
|---|---|---|
| | **Yes** | **No** |
| Percent | 22.7% | 77.3% |

Comparing the conditional distributions (Example 2.33 and Exercise 2.110) reveals the nature of the association between gender and frequent binge drinking. In this set of data the men are more likely to be frequent binge drinkers than the women.

Bar graphs can help us to see relationships between two categorical variables. No single graph (such as a scatterplot) portrays the form of the relationship between categorical variables, and no single numerical measure (such as the correlation) summarizes the strength of an association. Bar graphs are flexible enough to be helpful, but you must think about what comparisons you want to display. For numerical measures, we must rely on well-chosen percents or on more advanced statistical methods.[37]

*A two-way table contains a great deal of information in compact form. Making that information clear almost always requires finding percents. You must decide which percents you need.* Of course, we prefer to use software to compute the joint, marginal, and conditional distributions.

EXAMPLE

**2.34 Software output.** Figure 2.27 gives computer output for the data in Example 2.27 using SPSS, Minitab, and SAS. There are minor variations among software packages, but these are typical of what is usually produced. Each cell in the 2 × 2 table has four entries. These are the count (the number of observations in the cell), the conditional distributions for rows and columns, and the joint distribution. Note that all of these are expressed as percents rather than proportions. Marginal totals and distributions are given in the rightmost column and the bottom row.

Most software packages order the row and column labels numerically or alphabetically. In general, it is better to use words rather than numbers for the column labels. This sometimes involves some additional work but it avoids the kind of confusion that can result when you forget the real values associated with each numerical value. You should verify that the entries in Figure 2.27 correspond to the calculations that we performed in Examples 2.29 to 2.33. In addition, verify the calculations for the conditional distributions of gender for each value of the frequent-binge-drinking variable.

## Simpson's paradox

As is the case with quantitative variables, the effects of lurking variables can strongly influence relationships between two categorical variables. Here is an example that demonstrates the surprises that can await the unsuspecting consumer of data.

SPSS

BINGE * GENDER Crosstabulation

| | | | GENDER | | Total |
|---|---|---|---|---|---|
| | | | Male | Female | |
| BINGE | Yes | Count | 1630 | 1684 | 3314 |
| | | % within BINGE | 49.2% | 50.8% | 100.0% |
| | | % within GENDER | 22.7% | 17.0% | 19.4% |
| | | % of Total | 9.5% | 9.9% | 19.4% |
| | No | Count | 5550 | 8232 | 13782 |
| | | % within BINGE | 40.3% | 59.7% | 100.0% |
| | | % within GENDER | 77.3% | 83.0% | 80.6% |
| | | % of Total | 32.5% | 48.2% | 80.6% |
| Total | | Count | 7180 | 9916 | 17096 |
| | | % within BINGE | 42.0% | 58.0% | 100.0% |
| | | % within GENDER | 100.0% | 100.0% | 100.0% |
| | | % of Total | 42.0% | 58.0% | 100.0% |

**FIGURE 2.27** Computer output for the binge-drinking study in Example 2.34. *(continued)*

Minitab

| | no | yes | All |
|---|---|---|---|
| men | 5550 | 1630 | 7180 |
| | 77.30 | 22.70 | 100.00 |
| | 40.27 | 49.19 | 42.00 |
| | 32.46 | 9.53 | 42.00 |
| women | 8232 | 1684 | 9916 |
| | 83.02 | 16.98 | 100.00 |
| | 59.73 | 50.81 | 58.00 |
| | 48.15 | 9.85 | 58.00 |
| All | 13782 | 3314 | 17096 |
| | 80.62 | 19.38 | 100.00 |
| | 100.00 | 100.00 | 100.00 |
| | 80.62 | 19.38 | 100.00 |

Cell Contents --
Count
% of Row
% of Col
% of Tbl

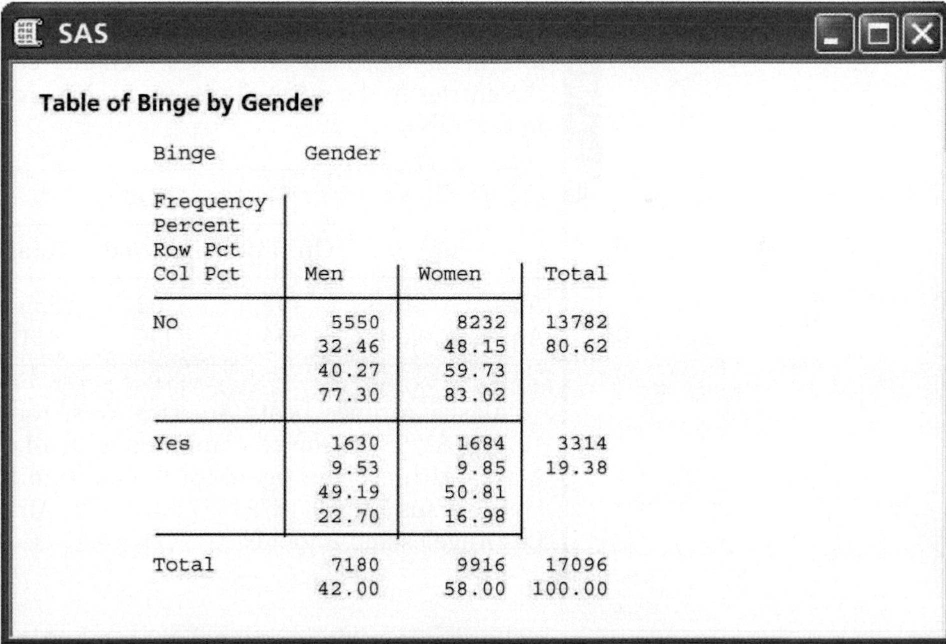

SAS

**Table of Binge by Gender**

Binge　　Gender

| Frequency<br>Percent<br>Row Pct<br>Col Pct | Men | Women | Total |
|---|---|---|---|
| No | 5550 | 8232 | 13782 |
| | 32.46 | 48.15 | 80.62 |
| | 40.27 | 59.73 | |
| | 77.30 | 83.02 | |
| Yes | 1630 | 1684 | 3314 |
| | 9.53 | 9.85 | 19.38 |
| | 49.19 | 50.81 | |
| | 22.70 | 16.98 | |
| Total | 7180 | 9916 | 17096 |
| | 42.00 | 58.00 | 100.00 |

**FIGURE 2.27** *(Continued)* Computer output for the binge-drinking study in Example 2.34.

EXAMPLE

**2.35 Flight delays.** Air travelers would like their flights to arrive on time. Airlines collect data on on-time arrivals and report them to the government. Following are one month's data for flights from two western cities for two airlines:

| | Alaska Airlines | America West |
|---|---|---|
| On time | 718 | 5534 |
| Delayed | 74 | 532 |
| Total | 792 | 6066 |

Alaska Airlines is delayed 9.3% (74/792) of the time, and America West is delayed only 8.8% (532/6066) of the time. It seems that you should choose America West to minimize delays.

Some cities are more likely to have delays than others, however. If we consider which city the flights left from in our analysis, we obtain a more complete picture of the two airlines' on-time flights.

EXAMPLE

**2.36 Is there a difference between Los Angeles and Phoenix?** Here are the data broken down by which city each flight left from.[38] Check that the entries in the original two-way table are just the sums of the city entries in this table.

| | Los Angeles | | | Phoenix | | |
|---|---|---|---|---|---|---|
| | On time | Delayed | Total | On time | Delayed | Total |
| Alaska Airlines | 497 | 62 | 559 | 221 | 12 | 233 |
| America West | 694 | 117 | 811 | 4840 | 415 | 5255 |

Alaska Airlines beats America West for flights from Los Angeles: only 11.1% (62/559) delayed compared with 14.4% (117/811) for America West. Alaska Airlines wins again for flights from Phoenix, with 5.2% (12/233) delayed versus 7.9% (415/5255). So Alaska Airlines is the better choice for both Los Angeles and Phoenix.

The city of origin for each flight is a lurking variable when we compare the late-flight percents of the two airlines. When we ignore the lurking variable, America West seems better, even though Alaska Airlines does better for each city. How can Alaska Airlines do better for both cities, yet do worse overall? Let's look at the data. Seventy-one percent (559/792) of Alaska Airlines flights are from Los Angeles, where there are more delays for both airlines. On the other hand, 87% of the America West flights are from Phoenix, where there are few delays. *The original two-way table, which did not take account of the city of origin for each flight, was misleading.* This example illustrates *Simpson's paradox.*

> **SIMPSON'S PARADOX**
>
> An association or comparison that holds for all of several groups can reverse direction when the data are combined to form a single group. This reversal is called **Simpson's paradox.**

The lurking variables in Simpson's paradox are categorical. That is, they break the observations into groups, such as the city of origin for the airline flights. *Simpson's paradox is an extreme form of the fact that observed associations can be misleading when there are lurking variables.*

## The perils of aggregation

three-way table

The flight data in Example 2.36 are given in a **three-way table** that reports the frequencies of each combination of levels of three categorical variables: city, airline, and delayed or not. We present a three-way table as two or more two-way tables, one for each level of the third variable. In Example 2.36, there is a separate table of airline versus delayed or not for each city. We can obtain the original two-way table by adding the corresponding entries in these tables. This is called **aggregating** the data. Aggregation has the effect of ignoring the city variable, which then becomes a lurking variable. *Conclusions that seem obvious when we look only at aggregated data can become debatable when we examine lurking variables.*

aggregation

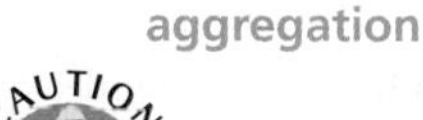

## SECTION 2.5 Summary

A **two-way table** of counts organizes data about two categorical variables. Values of the **row variable** label the rows that run across the table, and values of the **column variable** label the columns that run down the table. Two-way tables are often used to summarize large amounts of data by grouping outcomes into categories.

The **joint distribution** of the row and column variables is found by dividing the count in each cell by the total number of observations.

The **row totals** and **column totals** in a two-way table give the **marginal distributions** of the two variables separately. It is clearer to present these distributions as percents of the table total. Marginal distributions do not give any information about the relationship between the variables.

To find the **conditional distribution** of the row variable for one specific value of the column variable, look only at that one column in the table. Find each entry in the column as a percent of the column total.

There is a conditional distribution of the row variable for each column in the table. Comparing these conditional distributions is one way to describe the association between the row and the column variables. It is particularly useful when the column variable is the explanatory variable. When the row variable is explanatory, find the conditional distribution of the column variable for each row and compare these distributions.

**Bar graphs** are a flexible means of presenting categorical data. There is no single best way to describe an association between two categorical variables.

We present data on three categorical variables in a **three-way table,** printed as separate two-way tables for each level of the third variable. A comparison between two variables that holds for each level of a third variable can be changed or even reversed when the data are **aggregated** by summing over all levels of the third variable. **Simpson's paradox** refers to the reversal of a comparison by aggregation. It is an example of the potential effect of lurking variables on an observed association.

## SECTION 2.5 Exercises

*For Exercise 2.105, see page 143; for Exercises 2.106 and 2.107, see page 144; for Exercises 2.108 and 2.109, see page 146; and for Exercise 2.110, see page 147.*

**2.111 Full-time and part-time college students.** The Census Bureau provides estimates of numbers of people in the United States classified in various ways.[39] Let's look at college students. The following table gives us data to examine the relation between age and full-time or part-time status. The numbers in the table are expressed as thousands of U.S. college students.

**U.S. college students by age and status**

| | Status | |
|---|---|---|
| **Age** | **Full-time** | **Part-time** |
| 15–19 | 3388 | 389 |
| 20–24 | 5238 | 1164 |
| 25–34 | 1703 | 1699 |
| 35 and over | 762 | 2045 |

(a) What is the U.S. Census Bureau estimate of the number of full-time college students aged 15 to 19?

(b) Give the joint distribution of age and status for this table.

(c) What is the marginal distribution of age? Display the results graphically.

(d) What is the marginal distribution of status? Display the results graphically.

**2.112 Condition on age.** Refer to the previous exercise. Find the conditional distribution of status for each of the four age categories. Display the distributions graphically and summarize their differences and similarities.

**2.113 Condition on status.** Refer to the previous two exercises. Compute the conditional distribution of age for each of the two status categories. Display the distributions graphically and write a short paragraph describing the distributions and how they differ.

**2.114 Enrollment of recent high school graduates.** The table below gives some census data concerning the enrollment status of recent high school graduates aged 16 to 24 years.[40] The table entries are in thousands of students.

**Enrollment and gender**

| Status | Men | Women |
|---|---|---|
| Two-year college, full-time | 890 | 969 |
| Two-year college, part-time | 340 | 403 |
| Four-year college, full-time | 2897 | 3321 |
| Four-year college, part-time | 249 | 383 |
| Graduate school | 306 | 366 |
| Vocational school | 160 | 137 |

(a) How many male recent high school graduates aged 16 to 24 years were enrolled full-time in two-year colleges?

(b) Give the marginal distribution of gender for these students. Display the results graphically.

(c) What is the marginal distribution of status for these students? Display the results graphically.

**2.115 Condition on status.** Refer to the previous exercise. Find the conditional distribution of gender for each status. Describe the distributions graphically and write a short summary comparing the major features of these distributions.

**2.116 Condition on gender.** Refer to the previous two exercises. Find the conditional distribution of status for each gender. Describe the distributions graphically and write a short summary comparing the major features of these distributions.

**2.117 Complete the table.** Here are the row and column totals for a two-way table with two rows and two columns:

| | | |
|---|---|---|
| $a$ | $b$ | 200 |
| $c$ | $d$ | 200 |
| 200 | 200 | 400 |

Find *two different* sets of counts $a$, $b$, $c$, and $d$ for the body of the table that give these same totals. This shows that the relationship between two variables cannot be obtained from the two individual distributions of the variables.

**2.118 Construct a table with no association.** Construct a $3 \times 3$ table of counts where there is no apparent association between the row and column variables.

**2.119 Survey response rates.** A market research firm conducted a survey of companies in its state. They mailed a questionnaire to 300 small companies, 300 medium-sized companies, and 300 large companies. The rate of nonresponse is important in deciding how reliable survey results are. Here are the data on response to this survey:

| | Response | | |
|---|---|---|---|
| **Size of company** | **Yes** | **No** | **Total** |
| Small | 175 | 125 | 300 |
| Medium | 145 | 155 | 300 |
| Large | 120 | 180 | 300 |

(a) What was the overall percent of nonresponse?

(b) Describe how nonresponse is related to the size of the business. (Use percents to make your statements precise.)

(c) Draw a bar graph to compare the nonresponse percents for the three size categories.

(d) Using the total number of responses as a base, compute the percent of responses that come from each of small, medium, and large businesses.

(e) The sampling plan was designed to obtain equal numbers of responses from small, medium, and large companies. In preparing an analysis of the survey results, do you think it would be reasonable to proceed as if the responses represented companies of each size equally?

**2.120 Career plans of young women and men.** A study of the career plans of young women and men sent questionnaires to all 722 members of the senior class in the College of Business Administration at the University of Illinois. One question asked which major within the business program the student had chosen.[41] Here are the data from the students who responded:

| | Gender | |
|---|---|---|
| **Major** | **Female** | **Male** |
| Accounting | 68 | 56 |
| Administration | 91 | 40 |
| Economics | 5 | 6 |
| Finance | 61 | 59 |

(a) Describe the differences between the distributions of majors for women and men with percents, with a graph, and in words.

(b) What percent of the students did not respond to the questionnaire? The nonresponse weakens conclusions drawn from these data.

**2.121 Treatment for cocaine addiction.** Cocaine addiction can be difficult to overcome. Since addicts derive pleasure from the drug, one proposed aid is to provide an antidepressant drug. A 3-year study with 72 chronic cocaine users compared an antidepressant drug called desipramine with lithium and a placebo. (Lithium is a standard drug to treat cocaine addiction. A placebo is a tablet with no effects that tastes and looks like the antidepressant drug. It is used so that the effect of being in the study but not taking the antidepressant drug can be seen.) One-third of the subjects, chosen at random, received each treatment.[42] Here are the results:

| | Cocaine relapse? | |
|---|---|---|
| **Treatment** | **Yes** | **No** |
| Desipramine | 10 | 14 |
| Lithium | 18 | 6 |
| Placebo | 20 | 4 |

Compare the effectiveness of the three treatments in preventing relapse. Use percents and draw a bar graph. Write a brief summary of your conclusions.

## 2.6 The Question of Causation

In many studies of the relationship between two variables, the goal is to establish that changes in the explanatory variable *cause* changes in the response variable. Even when a strong association is present, the conclusion that this association is due to a causal link between the variables is often hard to find. What ties between two variables (and others lurking in the background) can explain an observed association? What constitutes good evidence for causation? We begin our consideration of these questions with a set of examples. In each case, there is a clear association between an explanatory variable $x$ and a response variable $y$. Moreover, the association is positive whenever the direction makes sense.

EXAMPLE

**2.37 Observed associations.** Here are some examples of observed association between $x$ and $y$:

1. $x$ = mother's body mass index
   $y$ = daughter's body mass index
2. $x$ = amount of the artificial sweetener saccharin in a rat's diet
   $y$ = count of tumors in the rat's bladder
3. $x$ = a student's SAT score as a high school senior
   $y$ = a student's first-year college grade point average
4. $x$ = monthly flow of money into stock mutual funds
   $y$ = monthly rate of return for the stock market
5. $x$ = whether a person regularly attends religious services
   $y$ = how long the person lives
6. $x$ = the number of years of education a worker has
   $y$ = the worker's income

### Explaining association: causation

Figure 2.28 shows in outline form how a variety of underlying links between variables can explain association. The dashed double-arrow line represents an observed association between the variables $x$ and $y$. Some associations

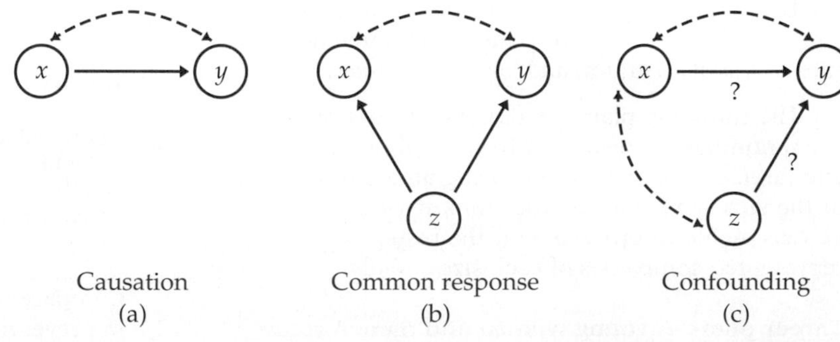


**FIGURE 2.28** Some possible explanations for an observed association. The dashed double-arrow lines show an association. The solid arrows show a cause-and-effect link. The variable $x$ is explanatory, $y$ is a response variable, and $z$ is a lurking variable.

are explained by a direct cause-and-effect link between these variables. The first diagram in Figure 2.28 shows "$x$ causes $y$" by a solid arrow running from $x$ to $y$.

Items 1 and 2 in Example 2.37 are examples of direct causation. Thinking about these examples, however, shows that "causation" is not a simple idea.

EXAMPLE

**2.38 Body mass index of mothers and daughters.** A study of Mexican American girls aged 9 to 12 years recorded body mass index (BMI), a measure of weight relative to height, for both the girls and their mothers. People with high BMI are overweight or obese. The study also measured hours of television, minutes of physical activity, and intake of several kinds of food. The strongest correlation ($r = 0.506$) was between the BMI of daughters and the BMI of their mothers.[43]

Body type is in part determined by heredity. Daughters inherit half their genes from their mothers. There is therefore a direct causal link between the BMI of mothers and daughters. Yet the mothers' BMIs explain only 25.6% (that's $r^2$ again) of the variation among the daughters' BMIs. Other factors, such as diet and exercise, also influence BMI. *Even when direct causation is present, it is rarely a complete explanation of an association between two variables.*

The best evidence for causation comes from experiments that actually change $x$ while holding all other factors fixed. If $y$ changes, we have good reason to think that $x$ caused the change in $y$.

EXAMPLE

**2.39 Saccharin and bladder tumors in rats.** Experiments show that large amounts of saccharin—about 5% of the entire diet—cause bladder tumors in rats. Should we avoid saccharin as a replacement for sugar in food? Rats are not people. Although we can't experiment with people, studies of people who consume different amounts of saccharin fail to find an association between saccharin and bladder tumors.[44] *Even well-established causal relations may not generalize to other settings.*

## Explaining association: common response

"Beware the lurking variable" is good advice when thinking about an association between two variables. The second diagram in Figure 2.28 illustrates **common response.** The observed association between the variables $x$ and $y$ is explained by a lurking variable $z$. Both $x$ and $y$ change in response to changes in $z$. This common response creates an association even though there may be no direct causal link between $x$ and $y$.

common response

The third and fourth items in Example 2.37 illustrate how common response can create an association.

EXAMPLE

**2.40 SAT scores and college grades.** Students who are smart and who have learned a lot tend to have both high SAT scores and high college grades. The positive correlation is explained by this common response to students' ability and knowledge.

EXAMPLE

**2.41 Stock market performance and investments in mutual funds.** There is a strong positive correlation between how much money individuals add to mutual funds each month and how well the stock market does the same month. Is the new money driving the market up? The correlation may be explained in part by common response to underlying investor sentiment: when optimism reigns, individuals send money to funds and large institutions also invest more. The institutions would drive up prices even if individuals did nothing. In addition, what causation there is may operate in the other direction: when the market is doing well, individuals rush to add money to their mutual funds.[45]

## Explaining association: confounding

We noted in Example 2.38 that inheritance no doubt explains part of the association between the body mass indexes (BMIs) of daughters and their mothers. Can we use $r$ or $r^2$ to say how much inheritance contributes to the daughters' BMIs? No. It may well be that mothers who are overweight also set an example of little exercise, poor eating habits, and lots of television. Their daughters pick up these habits to some extent, so the influence of heredity is mixed up with influences from the girls' environment. We call this mixing of influences *confounding*.

### CONFOUNDING

Two variables are **confounded** when their effects on a response variable cannot be distinguished from each other. The confounded variables may be either explanatory variables or lurking variables.

When many variables interact with each other, confounding of several variables often prevents us from drawing conclusions about causation. The third diagram in Figure 2.28 illustrates confounding. Both the explanatory variable $x$ and the lurking variable $z$ may influence the response variable $y$. Because $x$ is confounded with $z$, we cannot distinguish the influence of $x$ from the influence of $z$. We cannot say how strong the direct effect of $x$ on $y$ is. In fact, it can be hard to say if $x$ influences $y$ at all.

The last two associations in Example 2.37 (Items 5 and 6) are explained in part by confounding.

EXAMPLE

**2.42 Religion and a long life.** Many studies have found that people who are active in their religion live longer than nonreligious people. But people who attend church or mosque or synagogue also take better care of themselves than nonattenders. They are less likely to smoke, more likely to exercise, and less likely to be overweight. The effects of these good habits are confounded with the direct effects of attending religious services.

EXAMPLE

**2.43 Education and income.** It is likely that more education is a cause of higher income—many highly paid professions require advanced education. However, confounding is also present. People who have high ability and come from prosperous homes are more likely to get many years of education than people who are less able or poorer. Of course, people who start out able and rich are more likely to have high earnings even without much education. We can't say how much of the higher income of well-educated people is actually caused by their education.

Many observed associations are at least partly explained by lurking variables. Both common response and confounding involve the influence of a lurking variable (or variables) $z$ on the response variable $y$. The distinction between these two types of relationship is less important than the common element, the influence of lurking variables. The most important lesson of these examples is one we have already emphasized: **even a very strong association between two variables is not by itself good evidence that there is a cause-and-effect link between the variables.**

## Establishing causation

How can a direct causal link between $x$ and $y$ be established? The best method—indeed, the only fully compelling method—of establishing causation is to conduct a carefully designed experiment in which the effects of possible lurking variables are controlled. Chapter 3 explains how to design convincing experiments.

Many of the sharpest disputes in which statistics plays a role involve questions of causation that cannot be settled by experiment. Does gun control reduce violent crime? Does living near power lines cause cancer? Has "outsourcing" work to overseas locations reduced overall employment in the United States? All of these questions have become public issues. All concern associations among variables. And all have this in common: they try to pinpoint cause and effect in a setting involving complex relations among many interacting variables. Common response and confounding, along with the number of potential lurking variables, make observed associations misleading. Experiments are not possible for ethical or practical reasons. We can't assign some people to live near power lines or compare the same nation with and without strong gun controls.

EXAMPLE

**2.44 Power lines and leukemia.** Electric currents generate magnetic fields. So living with electricity exposes people to magnetic fields. Living near power lines increases exposure to these fields. Really strong fields can disturb living cells in laboratory studies. Some people claim that the weaker fields we experience if we live near power lines cause leukemia in children.

It isn't ethical to do experiments that expose children to magnetic fields. It's hard to compare cancer rates among children who happen to live in more and less exposed locations because leukemia is rare and locations vary in many ways other than magnetic fields. We must rely on studies that compare children who have leukemia with children who don't.

A careful study of the effect of magnetic fields on children took five years and cost $5 million. The researchers compared 638 children who had leukemia and 620 who did not. They went into the homes and actually measured the magnetic fields in the children's bedrooms, in other rooms, and at the front door. They recorded facts about nearby power lines for the family home and also for the mother's residence when she was pregnant. Result: no evidence of more than a chance connection between magnetic fields and childhood leukemia.[46]

"No evidence" that magnetic fields are connected with childhood leukemia doesn't prove that there is no risk. It says only that a careful study could not find any risk that stands out from the play of chance that distributes leukemia cases across the landscape. Critics continue to argue that the study failed to measure some lurking variables, or that the children studied don't fairly represent all children. Nonetheless, a carefully designed study comparing children with and without leukemia is a great advance over haphazard and sometimes emotional counting of cancer cases.

EXAMPLE

**2.45 Smoking and lung cancer.** Despite the difficulties, it is sometimes possible to build a strong case for causation in the absence of experiments. The evidence that smoking causes lung cancer is about as strong as nonexperimental evidence can be.

Doctors had long observed that most lung cancer patients were smokers. Comparison of smokers and similar nonsmokers showed a very strong association between smoking and death from lung cancer. Could the association be due to common response? Might there be, for example, a genetic factor that predisposes people both to nicotine addiction and to lung cancer? Smoking and lung cancer would then be positively associated even if smoking had no direct effect on the lungs. Or perhaps confounding is to blame. It might be that smokers live unhealthy lives in other ways (diet, alcohol, lack of exercise) and that some other habit confounded with smoking is a cause of lung cancer. How were these objections overcome?

Let's answer this question in general terms: What are the criteria for establishing causation when we cannot do an experiment?

- *The association is strong.* The association between smoking and lung cancer is very strong.
- *The association is consistent.* Many studies of different kinds of people in many countries link smoking to lung cancer. That reduces the chance that a lurking variable specific to one group or one study explains the association.
- *Higher doses are associated with stronger responses.* People who smoke more cigarettes per day or who smoke over a longer period get lung cancer more often. People who stop smoking reduce their risk.
- *The alleged cause precedes the effect in time.* Lung cancer develops after years of smoking. The number of men dying of lung cancer rose as smoking became more common, with a lag of about 30 years. Lung cancer kills more men than any other form of cancer. Lung cancer was rare among women until women began to smoke. Lung cancer in women rose along with smoking, again with a lag of about 30 years, and has now passed breast cancer as the leading cause of cancer death among women.
- *The alleged cause is plausible.* Experiments with animals show that tars from cigarette smoke do cause cancer.

Medical authorities do not hesitate to say that smoking causes lung cancer. The U.S. surgeon general states that cigarette smoking is "the largest avoidable cause of death and disability in the United States."[47] The evidence for causation is overwhelming—but it is not as strong as the evidence provided by well-designed experiments.

## SECTION 2.6 Summary

Some observed associations between two variables are due to a **cause-and-effect** relationship between these variables, but others are explained by **lurking variables.**

The effect of lurking variables can operate through **common response** if changes in both the explanatory and response variables are caused by changes in lurking variables. **Confounding** of two variables (either explanatory or lurking variables) means that we cannot distinguish their effects on the response variable.

That an association is due to causation is best established by an **experiment** that changes the explanatory variable while controlling other influences on the response.

In the absence of experimental evidence, be cautious in accepting claims of causation. Good evidence of causation requires a strong association that appears consistently in many studies, a clear explanation for the alleged causal link, and careful examination of possible lurking variables.

## SECTION 2.6 Exercises

**2.122 Online courses.** Many colleges offer online versions of some courses that are also taught in the classroom. It often happens that the students who enroll in the online version do better than the classroom students on the course exams.This does not show that online instruction is more effective than classroom teaching, because the kind of people who sign up for online courses are often quite different from the classroom students.

Suggest some student characteristics that you think could be confounded with online versus classroom. Use a diagram like Figure 2.28(c) (page 154) to illustrate your ideas.

**2.123 Marriage and income.** Data show that men who are married, and also divorced or widowed men, earn quite a bit more than men who have never been married. This does not mean that a man can raise his income by getting married. Suggest several lurking variables that you think are confounded with marital status and that help explain the association between marital status and income. Use a diagram like Figure 2.28(c) to illustrate your ideas.

**2.124 CEO compensation and layoffs.** "Based on an examination of twenty-two companies that announced large layoffs during 1994, Downs found a strong (.31) correlation between the size of the layoffs and the compensation of the CEOs."[48] This correlation is probably explained by common response to a lurking variable, the size of the company as measured by its number of employees. Explain how common response could create the observed correlation. Use a diagram like Figure 2.28(b) to illustrate your explanation.

**2.125 Exercise and self-confidence.** A college fitness center offers an exercise program for staff members who choose to participate. The program assesses each participant's fitness, using a treadmill test, and also administers a personality questionnaire. There is a moderately strong positive correlation between fitness score and score for self-confidence. Is this good evidence that improving fitness increases self-confidence? Explain why or why not.

**2.126** CHALLENGE **Health and income.** An article entitled "The Health and Wealth of Nations" says: "The positive correlation between health and income per capita is one of the best-known relations in international development. This correlation is commonly thought to reflect a causal link running from income to health.... Recently, however, another intriguing possibility has emerged: that the health-income correlation is partly explained by a causal link running the other way—from health to income."[49]

Explain how higher income in a nation can cause better health. Then explain how better health can cause higher national income. There is no simple way to determine the direction of the link.

**2.127 Music and academic performance.** The Kalamazoo (Michigan) Symphony once advertised a "Mozart for Minors" program with this statement: "Question: Which students scored 51 points higher in verbal skills and 39 points higher in math? Answer: Students who had experience in music."[50] In fact, good academic performance and early exposure to classical music are in part common responses to lurking variables. What background information about students could explain the association? Use a diagram like Figure 2.28(b) to show the situation.

**2.128 Coaching for the SAT.** A study finds that high school students who take the SAT, enroll in an SAT coaching course, and then take the SAT a second time raise their SAT Mathematics scores from a mean of 521 to a mean of 561.[51] What factors other than "taking the course causes higher scores" might explain this improvement?

**2.129 Computer chip manufacturing and miscarriages.** A study showed that women who work in the production of computer chips have abnormally high numbers of miscarriages. The union claimed that exposure to chemicals used in production caused the miscarriages. Another possible explanation is that these workers spend most of their work time standing up. Illustrate these relationships in a diagram like those in Figure 2.28.

**2.130 Hospital size and length of stay.** A study shows that there is a positive correlation between the size of a hospital (measured by its number of beds $x$) and the median number of days $y$ that patients remain in the hospital. Does this mean that you can shorten a hospital stay by choosing a small hospital? Use a diagram like those in Figure 2.28 to explain the association.

**2.131 Watching TV and low grades.** Children who watch many hours of television get lower grades in school on the average than those who watch less TV. Explain clearly why this fact does not show that watching TV *causes* poor grades. In particular, suggest some other variables that may be confounded with heavy TV viewing and may contribute to poor grades.

**2.132 Artificial sweeteners.** People who use artificial sweeteners in place of sugar tend to be heavier than people who use sugar. Does this mean that artificial sweeteners cause weight gain? Give a more plausible explanation for this association.

**2.133 Exercise and mortality.** A sign in a fitness center says, "Mortality is halved for men over 65 who walk at least 2 miles a day."

(a) Mortality is eventually 100% for everyone. What do you think "mortality is halved" means?

(b) Assuming that the claim is true, explain why this fact does not show that exercise *causes* lower mortality.

**2.134 Self-esteem and work performance.** People who do well tend to feel good about themselves. Perhaps helping people feel good about themselves will help them do better in their jobs and in life. Raising self-esteem became for a time a goal in many schools and companies. Can you think of explanations for the association between high self-esteem and good performance other than "self-esteem causes better work"?

## CHAPTER 2 Exercises

**2.135 Graduation rates.** One of the factors used to evaluate undergraduate programs is the proportion of incoming students who graduate. This quantity, called the graduation rate, can be predicted by other variables such as the SAT or ACT scores and the high school record of the incoming students. One of the components that *U.S. News & World Report* uses when evaluating colleges is the difference between the actual graduation rate and the rate predicted by a regression equation.[52] In this chapter, we call this quantity the residual. Explain why the residual is a better measure to evaluate college graduation rates than the raw graduation rate.

**2.136** CHALLENGE **Eating fruits and vegetables and smoking.** The Centers for Disease Prevention and Control (CDC) Behavioral Risk Factor Surveillance System (BRFSS) collects data related to health conditions and risk behaviors.[53] Aggregated data by state are in the BRFSS data set described in the Data Appendix. Figure 2.29 is really a plot of two of the BRFSS variables. Fruits & Vegetables is the percent of adults in the state who report eating at least five servings of fruits and vegetables per day; Smoking is the percent who smoke every day. Table 2.7 (on page 162) gives the data for this exercise.

(a) Describe the relationship between Fruits & Vegetables and Smoking. Explain why you might expect this type of association.

(b) Find the correlation between the two variables.

(c) For Utah, 22.1% eat at least five servings of fruits and vegetables per day and 8.5% smoke every day. Find Utah on the plot and describe its position relative to the other states.

(d) For California, the percents are 28.9% for Fruits & Vegetables and 9.8% for Smoking. Find California on the plot and describe its position relative to the other states.

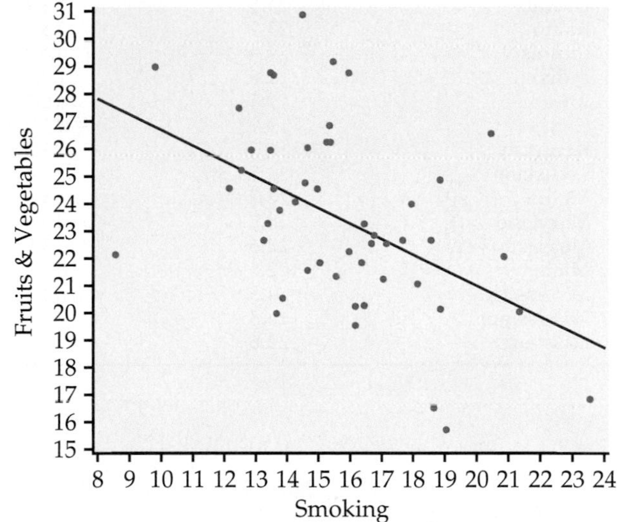


**FIGURE 2.29** Fruits & Vegetables versus Smoking with least-squares regression line, for Exercise 2.136.

(e) Pick your favorite state and write a short summary of its position relative to states that you would consider to be similar. Then use Table 2.7 to determine if your idea is supported by the data. Summarize your results.

**2.137** CHALLENGE **Eating fruits and vegetables and education.** Refer to the previous exercise. The BRFSS data set contains a variable called EdCollege, the proportion of adults who have completed college.

(a) Plot the data with FruitVeg5 on the $x$ axis and EdCollege on the $y$ axis. Describe the overall pattern of the data.

(b) Add the least-squares regression line to your plot. Does the line give a summary of the overall pattern? Explain your answer.

**TABLE 2.7**

**Fruit and vegetable consumption and smoking**

| State | Fruits & Vegetables (percent) | Smoking (percent) | State | Fruits & Vegetables (percent) | Smoking (percent) |
|---|---|---|---|---|---|
| Alabama | 20.1 | 18.8 | Montana | 24.7 | 14.5 |
| Alaska | 24.8 | 18.8 | Nebraska | 20.2 | 16.1 |
| Arizona | 23.7 | 13.7 | Nevada | 22.5 | 16.6 |
| Arkansas | 21.0 | 18.1 | New Hampshire | 29.1 | 15.4 |
| California | 28.9 | 9.8 | New Jersey | 25.9 | 12.8 |
| Colorado | 24.5 | 13.5 | New Mexico | 21.5 | 14.6 |
| Connecticut | 27.4 | 12.4 | New York | 26.0 | 14.6 |
| Delaware | 21.3 | 15.5 | North Carolina | 22.5 | 17.1 |
| Florida | 26.2 | 15.2 | North Dakota | 21.8 | 15.0 |
| Georgia | 23.2 | 16.4 | Ohio | 22.6 | 17.6 |
| Hawaii | 24.5 | 12.1 | Oklahoma | 15.7 | 19.0 |
| Idaho | 23.2 | 13.3 | Oregon | 25.9 | 13.4 |
| Illinois | 24.0 | 14.2 | Pennsylvania | 23.9 | 17.9 |
| Indiana | 22.0 | 20.8 | Rhode Island | 26.8 | 15.3 |
| Iowa | 19.5 | 16.1 | South Carolina | 21.2 | 17.0 |
| Kansas | 19.9 | 13.6 | South Dakota | 20.5 | 13.8 |
| Kentucky | 16.8 | 23.5 | Tennessee | 26.5 | 20.4 |
| Louisiana | 20.2 | 16.4 | Texas | 22.6 | 13.2 |
| Maine | 28.7 | 15.9 | Utah | 22.1 | 8.5 |
| Maryland | 28.7 | 13.4 | Vermont | 30.8 | 14.4 |
| Massachusetts | 28.6 | 13.5 | Virginia | 26.2 | 15.3 |
| Michigan | 22.8 | 16.7 | Washington | 25.2 | 12.5 |
| Minnesota | 24.5 | 14.9 | West Virginia | 20.0 | 21.3 |
| Mississippi | 16.5 | 18.6 | Wisconsin | 22.2 | 15.9 |
| Missouri | 22.6 | 18.5 | Wyoming | 21.8 | 16.3 |

(c) Pick out a few states and use their position in the graph to write a short summary of how they compare with other states.

(d) Can you conclude that earning a college degree will cause you to eat five servings of fruits and vegetables per day? Explain your answer.

**2.138 Predicting text pages.** The editor of a statistics text would like to plan for the next edition. A key variable is the number of pages that will be in the final version. Text files are prepared by the authors using a word processor called LaTeX, and separate files contain figures and tables. For the previous edition of the text, the number of pages in the LaTeX files can easily be determined, as well as the number of pages in the final version of the text. The table presents the data.

(a) Plot the data and describe the overall pattern.

(b) Find the equation of the least-squares regression line and add the line to your plot.

(c) Find the predicted number of pages for the next edition if the number of LaTeX pages is 62.

| Chapter | 1 | 2 | 3 | 4 | 5 | 6 | 7 | 8 | 9 | 10 | 11 | 12 | 13 |
|---|---|---|---|---|---|---|---|---|---|---|---|---|---|
| LaTeX pages | 77 | 73 | 59 | 80 | 45 | 66 | 81 | 45 | 47 | 43 | 31 | 46 | 26 |
| Text pages | 99 | 89 | 61 | 82 | 47 | 68 | 87 | 45 | 53 | 50 | 36 | 52 | 19 |

(d) Write a short report for the editor explaining to her how you constructed the regression equation and how she could use it to estimate the number of pages in the next edition of the text.

**2.139** CHALLENGE **Points scored in women's basketball games.** Use the Internet to find the scores for the past season's women's basketball team at a college of your choice. Is there a relationship between the points scored by your chosen team and the points scored by the opponent? Summarize the data and write a report on your findings.

**2.140** CHALLENGE **Look at the data for men.** Refer to the previous exercise. Analyze the data for the men's team from the same college and compare your results with those for the women.

**2.141 Endangered animals and habitat.** Endangered animal species often live in isolated patches of habitat. If the population size in a patch varies a lot (due to weather, for example), the species is more likely to disappear from that patch in a bad year. Here is a general question: Is there less variation in population size when a patch of habitat has more diverse vegetation? If so, maintaining habitat diversity can help protect endangered species.

A researcher measured the variation over time in the population of a cricket species in 45 habitat patches. He also measured the diversity of each patch.[54] He reported his results by giving the least-squares equation

$$\text{population variation} = 84.4 - 0.13 \times \text{diversity}$$

along with the fact that $r^2 = 0.34$. Do these results support the idea that more diversity goes with less variation in population size? Is the relationship very strong or only moderately strong?

**2.142 Stock prices and earnings.** In the long run, the price of a company's stock ought to parallel changes in the company's earnings. Table 2.8 gives data on the annual growth rates in earnings and in stock prices (both in percent) for major industry groups as set by Standard & Poor's.[55]

(a) Make a graph showing how earnings growth explains growth in stock price. Does it appear to be true that (on the average in the long run) stock price growth parallels earnings growth?

(b) What percent of the variation in stock price growth among industry groups can be explained by the linear relationship with earnings growth?

(c) If stock prices exactly followed earnings, the slope of the least-squares line for predicting price growth from earnings growth would be 1. Explain why. What is the slope of the least-squares line for these data?

(d) What is the correlation between earnings growth and price growth? If we had data on all of the individual companies in these 20 industries, would the correlation be higher or lower? Why?

**2.143** CHALLENGE **Monkey calls.** The usual way to study the brain's response to sounds is to have subjects listen to "pure tones." The response to recognizable sounds may differ. To compare responses, researchers anesthetized macaque monkeys. They fed pure tones and also monkey calls directly to their brains by inserting electrodes. Response to the stimulus was measured by the firing rate (electrical spikes per second) of neurons in various areas of the brain. Table 2.9 (on page 164) contains the responses for 37 neurons.[56]

(a) One notable finding is that responses to monkey calls are generally stronger than responses to pure tones. Give a numerical measure that supports this finding.

(b) Make a scatterplot of monkey call response against pure tone response (explanatory variable). Find the least-squares line and add it to your plot. Mark on your plot the point with the largest residual (positive or negative) and also a point that is an outlier in the $x$ direction.

(c) How influential are each of these points for the correlation $r$?

(d) How influential are each of these points for the regression line?

**2.144** CHALLENGE **Plywood strength.** How strong is a building material such as plywood? To be specific, support a 24-inch by 2-inch strip

**TABLE 2.8**

Percent growth in stock price and earnings for industry groups

| Industry | Earnings growth (%) | Price growth (%) | Industry | Earnings growth (%) | Price growth (%) |
|---|---|---|---|---|---|
| Auto | 3.3 | 2.9 | Oil: international | 7.7 | 7.7 |
| Banks | 8.6 | 6.5 | Oil equipment/services | 10.1 | 10.8 |
| Chemicals | 6.6 | 3.1 | Railroad | 6.6 | 6.6 |
| Computers | 10.2 | 5.3 | Retail: food | 6.9 | 6.9 |
| Drugs | 11.3 | 10.0 | Retail: department stores | 10.1 | 9.5 |
| Electrical equipment | 8.5 | 8.2 | Soft drinks | 12.7 | 12.0 |
| Food | 7.6 | 6.5 | Steel | −1.0 | −1.6 |
| Household products | 9.7 | 10.1 | Tobacco | 12.3 | 11.7 |
| Machinery | 5.1 | 4.7 | Utilities: electric | 2.8 | 1.4 |
| Oil: domestic | 7.4 | 7.3 | Utilities: gas | 5.2 | 6.2 |

TABLE 2.9

Neuron response to tones and monkey calls

| Tone | Call | Tone | Call | Tone | Call | Tone | Call |
|---|---|---|---|---|---|---|---|
| 474 | 500 | 145 | 42 | 71 | 134 | 35 | 103 |
| 256 | 138 | 141 | 241 | 68 | 65 | 31 | 70 |
| 241 | 485 | 129 | 194 | 59 | 182 | 28 | 192 |
| 226 | 338 | 113 | 123 | 59 | 97 | 26 | 203 |
| 185 | 194 | 112 | 182 | 57 | 318 | 26 | 135 |
| 174 | 159 | 102 | 141 | 56 | 201 | 21 | 129 |
| 176 | 341 | 100 | 118 | 47 | 279 | 20 | 193 |
| 168 | 85 | 74 | 62 | 46 | 62 | 20 | 54 |
| 161 | 303 | 72 | 112 | 41 | 84 | 19 | 66 |
| 150 | 208 | | | | | | |

of plywood at both ends and apply force in the middle until the strip breaks. The modulus of rupture (MOR) is the force needed to break the strip. We would like to be able to predict MOR without actually breaking the wood. The modulus of elasticity (MOE) is found by bending the wood without breaking it. Both MOE and MOR are measured in pounds per square inch. Here are data for 32 specimens of the same type of plywood:[57]

| MOE | MOR | MOE | MOR | MOE | MOR |
|---|---|---|---|---|---|
| 2,005,400 | 11,591 | 1,720,930 | 10,232 | 1,558,770 | 11,565 |
| 1,166,360 | 8,542 | 1,355,960 | 8,395 | 2,212,310 | 15,317 |
| 1,842,180 | 12,750 | 1,411,210 | 10,654 | 1,747,010 | 11,794 |
| 2,088,370 | 14,512 | 1,842,630 | 10,223 | 1,791,150 | 11,413 |
| 1,615,070 | 9,244 | 1,984,690 | 13,499 | 2,535,170 | 13,920 |
| 1,938,440 | 11,904 | 2,181,910 | 12,702 | 1,355,720 | 9,286 |
| 2,047,700 | 11,208 | 1,559,700 | 11,209 | 1,646,010 | 8,814 |
| 2,037,520 | 12,004 | 2,372,660 | 12,799 | 1,472,310 | 6,326 |
| 1,774,850 | 10,541 | 1,580,930 | 12,062 | 1,488,440 | 9,214 |
| 1,457,020 | 10,314 | 1,879,900 | 11,357 | 2,349,090 | 13,645 |
| 1,959,590 | 11,983 | 1,594,750 | 8,889 | | |

Can we use MOE to predict MOR accurately? Use the data to write a discussion of this question.

**2.145 Distribution of the residuals.** Some statistical methods require that the residuals from a regression line have a Normal distribution. The residuals for the nonexercise activity example are given in Exercise 2.83 (page 127). Is their distribution close to Normal? Make a Normal quantile plot to find out.

**2.146 Asian culture and thinness.** Asian culture does not emphasize thinness, but young Asians are often influenced by Western culture. In a study of concerns about weight among young Korean women, researchers administered the Drive for Thinness scale (a questionnaire) to 264 female college students in Seoul, South Korea. This scale measures excessive concern with weight and dieting and fear of weight gain. In Exercise 1.35 (page 27), you examined the distribution of Drive for Thinness scores among these college women. The study looked at several explanatory variables. One was "Body Dissatisfaction," also measured by a questionnaire. Use the data set for this example for your work.

(a) Make a scatterplot of Drive for Thinness (response) against Body Dissatisfaction. The appearance of the plot is a result of the fact that both variables take only whole-number values. Such variables are common in the social and behavioral sciences.

(b) Add the least-squares line to your plot. The line shows a linear relationship. How strong is this relationship? Body Dissatisfaction was more strongly correlated with Drive for Thinness than any of the other explanatory variables examined. Rather weak relationships are common in social and behavioral sciences, because individuals vary a great deal. Using several explanatory variables together improves prediction of the response. This is *multiple regression,* discussed in Chapter 11.

**2.147 Solar heating panels and gas consumption.** To study the energy savings due to adding solar heating panels to a house, researchers measured the natural-gas consumption of the house for more than a year, then installed solar panels and observed the natural-gas consumption for almost two years. The explanatory variable $x$ is degree-days per day during the several weeks covered by each observation, and the response variable $y$ is gas consumption (in hundreds of cubic feet) per day during the same period. Figure 2.30 plots $y$

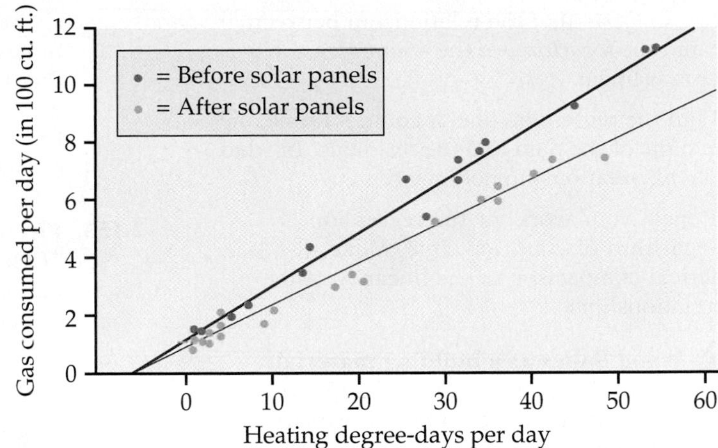


**FIGURE 2.30** The regression of residential natural-gas consumption on heating degree-days before and after installation of solar heating panels, for Exercise 2.147.

against $x$, with separate symbols for observations taken before and after the installation of the solar panels.[58] The least-squares regression lines were computed separately for the before and after data and are drawn on the plot. The regression lines are

$$\text{Before:} \quad \hat{y} = 1.089 + 0.189x$$

$$\text{After:} \quad \hat{y} = 0.853 + 0.157x$$

(a) Does the scatterplot suggest that a straight line is an appropriate description of the relationship between degree-days and natural-gas consumption? Do any individual observations appear to have large residuals or to be highly influential?

(b) About how much additional natural gas was consumed per day for each additional degree-day before the panels were added? After the panels were added?

(c) The daily average temperature during January in this location is about 30°, which corresponds to 35 degree-days per day. Use the regression lines to predict daily gas usage for a day with 35 degree-days before and after installation of the panels.

(d) The Energy Information Agency says that natural gas cost consumers about $1.20 per 100 cubic feet in the fall of 2003. At this rate, how much money do the solar panels save in the 31 days of January?

**2.148 Running speed and stride rate.** The following table gives data on the relationship between running speed (feet per second) and stride rate (steps taken per second) for elite female runners.[59]

| Speed | 15.86 | 16.88 | 17.50 | 18.62 | 19.97 | 21.06 | 22.11 |
|---|---|---|---|---|---|---|---|
| Stride rate | 3.05 | 3.12 | 3.17 | 3.25 | 3.36 | 3.46 | 3.55 |

Here are the corresponding data from the same source for male runners:

| Speed | 15.86 | 16.88 | 17.50 | 18.62 | 19.97 | 21.06 | 22.11 |
|---|---|---|---|---|---|---|---|
| Stride rate | 2.92 | 2.98 | 3.03 | 3.11 | 3.22 | 3.31 | 3.41 |

(a) Plot the data for both groups on one graph using different symbols to distinguish between the points for females and those for males.

(b) Suppose now that the data came to you without identification as to gender. Compute the least-squares line from all of the data and plot it on your graph.

(c) Compute the residuals from this line for each observation. Make a plot of the residuals against speed. How does the fact that the data come from two distinct groups show up in the residual plot?

**2.149** CHALLENGE **Marine bacteria and X-rays.** Expose marine bacteria to X-rays for time periods from 1 to 15 minutes. Here are the number of surviving bacteria (in hundreds) on a culture plate after each exposure time:[60]

| Time | Count | Time | Count | Time | Count |
|---|---|---|---|---|---|
| 1 | 355 | 6 | 106 | 11 | 36 |
| 2 | 211 | 7 | 104 | 12 | 32 |
| 3 | 197 | 8 | 60 | 13 | 21 |
| 4 | 166 | 9 | 56 | 14 | 19 |
| 5 | 142 | 10 | 38 | 15 | 15 |

Theory suggests that the relationship between time and the *logarithm* of the count of surviving bacteria is linear.

(a) Find the regression line of count on time and make plots of the data and the residuals. In what way is this relationship nonlinear?

(b) Repeat your work for the regression of the logarithm of count on time. Make a numerical comparison of the linearity of the two relationships.

**2.150** CHALLENGE **Wood flakes as a building material.** Wood scientists are interested in replacing solid-wood building material by less expensive products made from wood flakes. They collected the following data to examine the relationship between the length (in inches) and the strength (in pounds per square inch) of beams made from wood flakes:[61]

| Length | 5 | 6 | 7 | 8 | 9 | 10 | 11 | 12 | 13 | 14 |
|---|---|---|---|---|---|---|---|---|---|---|
| Strength | 446 | 371 | 334 | 296 | 249 | 254 | 244 | 246 | 239 | 234 |

(a) Make a scatterplot that shows how the length of a beam affects its strength.

(b) Describe the overall pattern of the plot. Are there any outliers?

(c) Fit a least-squares line to the entire set of data. Graph the line on your scatterplot. Does a straight line adequately describe these data?

(d) The scatterplot suggests that the relation between length and strength can be described by *two* straight lines, one for lengths less than 9 inches and another for lengths 9 inches or greater. Fit least-squares lines to these two subsets of the data, and draw the lines on your plot. Do they describe the data adequately? What question would you now ask the wood experts?

**2.151 Global investing.** One reason to invest abroad is that markets in different countries don't move in step. When American stocks go down, foreign stocks may go up. So an investor who holds both bears less risk. That's the theory. Now we read: "The correlation between changes in American and European share prices has risen from 0.4 in the mid-1990s to 0.8 in 2000."[62] Explain to an investor who knows no statistics why this fact reduces the protection provided by buying European stocks.

**2.152 Stock prices in Europe and the U.S.** The same article that claims that the correlation between changes in stock prices in Europe and the United States was 0.8 in 2000 goes on to say: "Crudely, that means that movements on Wall Street can explain 80% of price movements in Europe." Is this true? What is the correct percent explained if $r = 0.8$?

**2.153 Firefighters and fire damage.** Someone says, "There is a strong positive correlation between the number of firefighters at a fire and the amount of damage the fire does. So sending lots of firefighters just causes more damage." Explain why this reasoning is wrong.

**2.154** CHALLENGE **Midterm-exam scores and final-exam scores.** We expect that students who do well on the midterm exam in a course will usually also do well on the final exam. Gary Smith of Pomona College looked at the exam scores of all 346 students who took his statistics class over a 10-year period.[63] The least-squares line for predicting final-exam score from midterm-exam score was $\hat{y} = 46.6 + 0.41x$.

Octavio scores 10 points above the class mean on the midterm. How many points above the class mean do you predict that he will score on the final? (*Hint:* What is the predicted final-exam score for the class mean midterm score $\bar{x}$?) This is an example of **regression to the mean,** the phenomenon that gave "regression" its name: students who do well on the midterm will on the average do less well on the final, but still above the class mean.)

**2.155 SAT scores and grade point averages.** Can we predict college grade point average from SAT scores and high school grades? The CSDATA data set described in the Data Appendix contains information on this issue for a large group of computer science students. We will look only at SAT Mathematics scores as a predictor of later college GPA, using the variables SATM and GPA from CSDATA. Make a scatterplot, obtain $r$ and $r^2$, and draw on your plot the least-squares regression line for predicting GPA from SATM. Then write a brief discussion of the ability of SATM alone to predict GPA. (In Chapter 11 we will see how combining several explanatory variables improves our ability to predict.)

**2.156 University degrees in Asia.** Asia has become a major competitor of the United States and Western Europe in education as well as economics. Following are counts of first university degrees in science and engineering in the three regions:[64]

| Field | Region: United States | Western Europe | Asia |
|---|---|---|---|
| Engineering | 61,941 | 158,931 | 280,772 |
| Natural science | 111,158 | 140,126 | 242,879 |
| Social science | 182,166 | 116,353 | 236,018 |

Direct comparison of counts of degrees would require us to take into account Asia's much larger population. We can, however, compare the distribution of degrees by field of study in the three regions. Do this using calculations and graphs, and write a brief summary of your findings.

**2.157** CHALLENGE **Motivation to participate in volunteer service.** A study examined patterns and characteristics of volunteer service for young people from high school through early adulthood.[65] Here are some data that can be used to compare males and females on participation in unpaid volunteer service or community service and motivation for participation:

| | Participants | | | |
|---|---|---|---|---|
| | Motivation | | | |
| Gender | Strictly voluntary | Court-ordered | Other | Nonparticipants |
| Men | 31.9% | 2.1% | 6.3% | 59.7% |
| Women | 43.7% | 1.1% | 6.5% | 48.7% |

Note that the percents in each row sum to 100%. Graphically compare the volunteer service profiles for men and women. Describe any differences that are striking.

**2.158** CHALLENGE **Look at volunteers only.** Refer to the previous exercise. Recompute the table for volunteers only. To do this, take the entries for each motivation and divide by the percent of volunteers. Do this separately for each gender. Verify that the percents sum to 100% for each gender. Give a graphical summary to compare the motivation of men and women who are volunteers. Compare this with your summary in the previous exercise, and write a short paragraph describing similarities and differences in these two views of the data.

**2.159 An example of Simpson's paradox.** Mountain View University has professional schools in business and law. Here is a three-way table of applicants to these professional schools, categorized by gender, school, and admission decision:[66]

| Business | | | Law | | |
|---|---|---|---|---|---|
| | Admit | | | Admit | |
| Gender | Yes | No | Gender | Yes | No |
| Male | 400 | 200 | Male | 90 | 110 |
| Female | 200 | 100 | Female | 200 | 200 |

(a) Make a two-way table of gender by admission decision for the combined professional schools by summing entries in the three-way table.

(b) From your two-way table, compute separately the percents of male and female applicants admitted. Male applicants are admitted to Mountain View's professional schools at a higher rate than female applicants.

(c) Now compute separately the percents of male and female applicants admitted by the business school and by the law school.

(d) Explain carefully, as if speaking to a skeptical reporter, how it can happen that Mountain View appears to favor males when this is not true within each of the professional schools.

**2.160 Construct an example with four schools.** Refer to the previous exercise. Make up a similar table for a hypothetical university having four different schools that illustrates the same point. Carefully summarize your table with the appropriate percents.

**2.161** CHALLENGE **Class size and class level.** A university classifies its classes as either "small" (fewer than 40 students) or "large." A dean sees that 62% of Department A's classes are small, while Department B has only 40% small classes. She wonders if she should cut Department A's budget and insist on larger classes. Department A responds to the dean by pointing out that classes for third- and fourth-year students tend to be smaller than classes for first- and second-year students. The three-way table below gives the counts of classes by department, size, and student audience. Write a short report for the dean that summarizes these data. Start by computing the percents of small classes in the two departments, and include other numerical and graphical comparisons as needed. The following table presents the numbers of classes to be analyzed.

| | Department A | | | Department B | | |
|---|---|---|---|---|---|---|
| Year | Large | Small | Total | Large | Small | Total |
| First | 2 | 0 | 2 | 18 | 2 | 20 |
| Second | 9 | 1 | 10 | 40 | 10 | 50 |
| Third | 5 | 15 | 20 | 4 | 16 | 20 |
| Fourth | 4 | 16 | 20 | 2 | 14 | 16 |

**2.162 Sexual imagery in magazine ads.** In what ways do advertisers in magazines use sexual imagery to appeal to youth? One study classified each of 1509 full-page or larger ads as "not sexual" or "sexual," according to the amount and style of the clothing of the male or female model in the ad. The ads were also classified according to the target readership of the magazine.[67] Here is the two-way table of counts:

| | Magazine readership | | | |
|---|---|---|---|---|
| Model clothing | Women | Men | General interest | Total |
| Not sexual | 351 | 514 | 248 | 1113 |
| Sexual | 225 | 105 | 66 | 396 |
| Total | 576 | 619 | 314 | 1509 |

(a) Summarize the data numerically and graphically.

(b) All of the ads were taken from the March, July, and November issues of six magazines in one year. Discuss how this fact influences your interpretation of the results.

**2.163 Age of the intended readership.** The ads in the study described in the previous exercise were also classified according to the age group of the intended readership. Here is a summary of the data:

| | Magazine readership age group | |
|---|---|---|
| Model clothing | Young adult | Mature adult |
| Not sexual (percent) | 72.3% | 76.1% |
| Sexual (percent) | 27.7% | 23.9% |
| Number of ads | 1006 | 503 |

Using parts (a) and (b) of the previous exercise as a guide, analyze these data and write a report summarizing your work.

**2.164 Identity theft.** A study of identity theft looked at how well consumers protect themselves from this increasingly prevalent crime. The behaviors of 61 college students were compared with the behaviors of 59 nonstudents.[68] One of the questions was "When asked to create a password, I have used either my mother's maiden name, or my pet's name, or my birth date, or the last four digits of my social security number, or a series of consecutive numbers." For the students, 22 agreed with this statement while 30 of the nonstudents agreed.

(a) Display the data in a two-way table and analyze the data. Write a short summary of your results.

(b) The students in this study were junior and senior college students from two sections of a course in Internet marketing at a large northeastern university. The nonstudents were a group of individuals who were recruited to attend commercial focus groups on the West Coast conducted by a lifestyle marketing organization. Discuss how the method of selecting the subjects in this study relates to the conclusions that can be drawn from it.

**2.165** CHALLENGE **Athletes and gambling.** A survey of student athletes that asked questions about gambling behavior classified students according to the National Collegiate Athletic Association (NCAA) division.[69] For male student athletes, the percents who reported wagering on collegiate sports are given here along with the numbers of respondents in each division:

| Division | I | II | III |
|---|---|---|---|
| Percent | 17.2% | 21.0% | 24.4% |
| Number | 5619 | 2957 | 4089 |

(a) Analyze the data. Give details and a short summary of your conclusion.

(b) The percents in the table above are given in the NCAA report, but the numbers of male student athletes in each division who responded to the survey question are estimated based on other information in the report. To what extent do you think this has an effect on the results?

(c) Some student athletes may be reluctant to provide this kind of information, even in a survey where there is no possibility that they can be identified. Discuss how this fact may affect your conclusions.

2.166 CHALLENGE **Health conditions and risk behaviors.** The data set BRFSS described in the Data Appendix gives several variables related to health conditions and risk behaviors as well as demographic information for the 50 states and the District of Columbia. Pick at least three pairs of variables to analyze. Write a short report on your findings.

CHAPTER 3

# Producing Data

A magazine article says that men need Pilates exercise more than women. Read the Introduction to learn more.



## Introduction

In Chapters 1 and 2 we learned some basic tools of *data analysis.* We used graphs and numbers to describe data. When we do **exploratory data analysis,** we rely heavily on plotting the data. We look for patterns that suggest interesting conclusions or questions for further study. However, *exploratory analysis alone can rarely provide convincing evidence for its conclusions, because striking patterns we find in data can arise from many sources.*

exploratory data analysis

### Anecdotal data

It is tempting to simply draw conclusions from our own experience, making no use of more broadly representative data. A magazine article about Pilates says that men need this form of exercise even more than women. The article describes the benefits that two men received from taking Pilates classes. A newspaper ad states that a particular brand of windows are "considered to be the best" and says that "now is the best time to replace your windows and doors." These types of stories, or *anecdotes,* sometimes provide quantitative data. However, this type of data does not give us a sound basis for drawing conclusions.

ANECDOTAL EVIDENCE

**Anecdotal evidence** is based on haphazardly selected individual cases, which often come to our attention because they are striking in some way. These cases need not be representative of any larger group of cases.

USE YOUR KNOWLEDGE

**3.1 Final Fu.** Your friends are big fans of "Final Fu," MTV2's martial arts competition. To what extent do you think you can generalize your preferences for this show to all students at your college?

**3.2 Describe an example.** Find an example from some recent experience where anecdotal evidence is used to draw a conclusion that is not justified. Describe the example and explain why it cannot be used in this way.

**3.3 Preference for Jolt Cola.** Jamie is a hard-core computer programmer. He and all his friends prefer Jolt Cola (caffeine equivalent to two cups of coffee) to either Coke or Pepsi (caffeine equivalent to less than one cup of coffee).[1] Explain why Jamie's experience is not good evidence that most young people prefer Jolt to Coke or Pepsi.

**3.4 Automobile seat belts.** When the discussion turns to the pros and cons of wearing automobile seat belts, Herman always brings up the case of a friend who survived an accident because he was not wearing a seat belt. The friend was thrown out of the car and landed on a grassy bank, suffering only minor injuries, while the car burst into flames and was destroyed. Explain briefly why this anecdote does not provide good evidence that it is safer not to wear seat belts.

## Available data

Occasionally, data are collected for a particular purpose but can also serve as the basis for drawing sound conclusions about other research questions. We use the term **available data** for this type of data.

available data

AVAILABLE DATA

**Available data** are data that were produced in the past for some other purpose but that may help answer a present question.

The library and the Internet can be good sources of available data. Because producing new data is expensive, we all use available data whenever possible. However, the clearest answers to present questions often require that data be produced to answer those specific questions. Here are two examples:

EXAMPLE

**3.1 Causes of death.** If you visit the National Center for Health Statistics Web site, www.cdc.gov/nchs, you will learn that accidents are the most common cause of death among people aged 20 to 24, accounting for over 40% of all deaths. Homicide is next, followed by suicide. AIDS ranks seventh, behind heart disease and cancer, at 1% of all deaths. The data also show that it is dangerous to be a young man: the overall death rate for men aged 20 to 24 is three times that for women, and the death rate from homicide is more than five times higher among men.

EXAMPLE

**3.2 Math skills of children.** At the Web site of the National Center for Education Statistics, nces.ed.gov/nationsreportcard/mathematics, you will find full details about the math skills of schoolchildren in the latest National Assessment of Educational Progress (Figure 3.1). Mathematics scores have slowly but steadily increased since 1990. All racial/ethnic groups, both men and women, and students in most states are getting better in math.

Many nations have a single national statistical office, such as Statistics Canada (www.statcan.ca) or Mexico's INEGI (www.inegi.gob.mx). More than 70 different U.S. agencies collect data. You can reach most of them through the government's FedStats site (www.fedstats.gov).

USE YOUR KNOWLEDGE

**3.5 Find some available data.** Visit the Internet and find an example of available data that is interesting to you. Explain how the data were collected and what questions the study was designed to answer.

A survey of college athletes is designed to estimate the percent who gamble. Do restaurant patrons give higher tips when their server repeats their order carefully? The validity of our conclusions from the analysis of data collected to address these issues rests on a foundation of carefully collected data. In this chapter, we will develop the skills needed to produce trustworthy data and to judge the quality of data produced by others. The techniques for producing data we will study require no formulas, but they are among the most important ideas in statistics. Statistical designs for producing data rely on either *sampling* or *experiments*.

## Sample surveys and experiments

How have the attitudes of Americans, on issues ranging from abortion to work, changed over time? **Sample surveys** are the usual tool for answering questions like these.

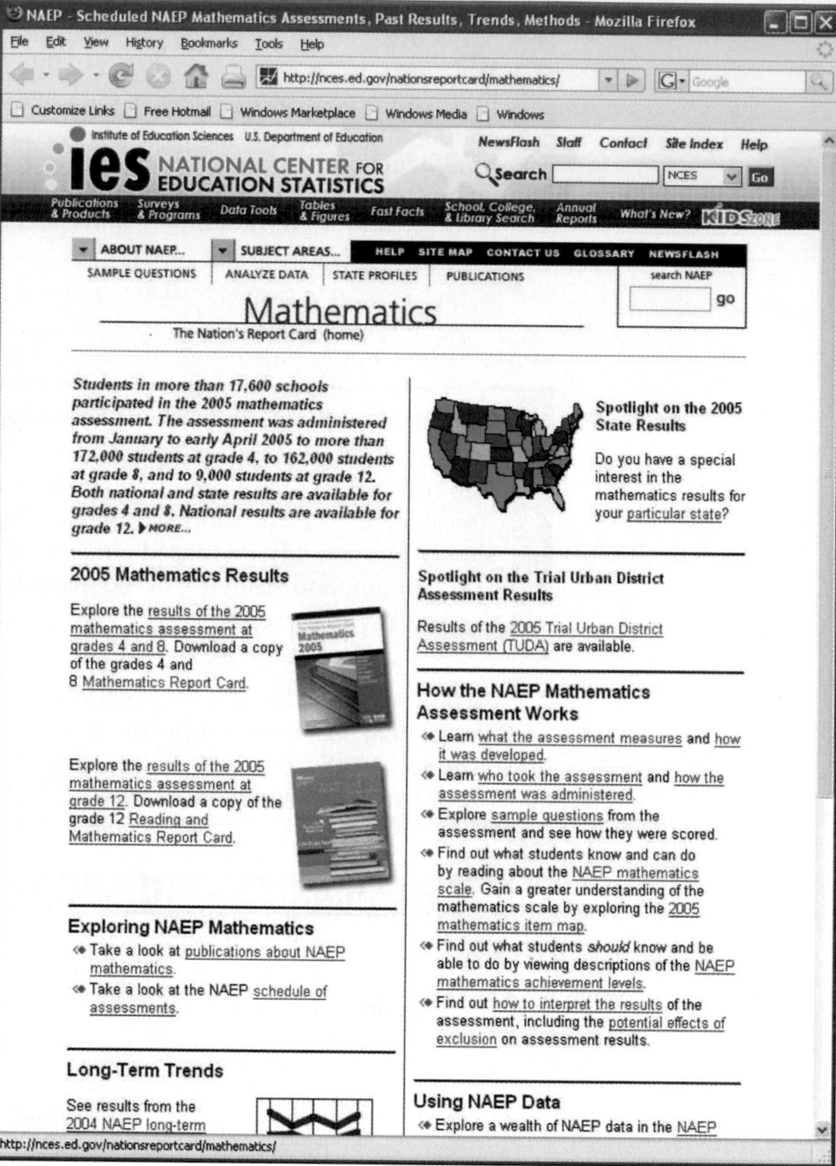


**FIGURE 3.1** The Web sites of government statistical offices are prime sources of data. Here is the home page of the National Assessment of Educational Progress.

EXAMPLE

**3.3 The General Social Survey.** One of the most important sample surveys is the General Social Survey (GSS) conducted by the NORC, a national organization for research and computing affiliated with the University of Chicago.[2] The GSS interviews about 3000 adult residents of the United States every second year.

sample
population

The GSS selects a **sample** of adults to represent the larger **population** of all English-speaking adults living in the United States. The idea of *sampling* is to study a part in order to gain information about the whole. Data are often produced by sampling a population of people or things. Opinion polls, for example, report the views of the entire country based on interviews with a sample of about 1000 people. Government reports on employment and unemployment are produced from a monthly sample of about 60,000 households. The quality of manufactured items is monitored by inspecting small samples each hour or each shift.

## USE YOUR KNOWLEDGE

**3.6 Find a sample survey.** Use the Internet or some printed material to find an example of a sample survey that interests you. Describe the population, how the sample was collected, and some of the conclusions.

In all of our examples, the expense of examining every item in the population makes sampling a practical necessity. Timeliness is another reason for preferring a sample to a **census,** which is an attempt to contact every individual in the entire population. We want information on current unemployment and public opinion next week, not next year. Moreover, a carefully conducted sample is often more accurate than a census. Accountants, for example, sample a firm's inventory to verify the accuracy of the records. Attempting to count every last item in the warehouse would be not only expensive but inaccurate. Bored people do not count carefully.

census

If conclusions based on a sample are to be valid for the entire population, a sound design for selecting the sample is required. Sampling designs are the topic of Section 3.2.

A sample survey collects information about a population by selecting and measuring a sample from the population. The goal is a picture of the population, disturbed as little as possible by the act of gathering information. Sample surveys are one kind of *observational study.*

### OBSERVATION VERSUS EXPERIMENT

In an **observational study** we observe individuals and measure variables of interest but do not attempt to influence the responses.

In an **experiment** we deliberately impose some treatment on individuals and we observe their responses.

## USE YOUR KNOWLEDGE

**3.7 Cell phones and brain cancer.** One study of cell phones and the risk of brain cancer looked at a group of 469 people who have brain cancer. The investigators matched each cancer patient with a person of the same sex, age, and race who did not have brain cancer, then asked about use of cell phones.[3] Result: "Our data suggest that use of handheld cellular telephones is not associated with risk of brain cancer." Is this an observational study or an experiment? Why? What are the explanatory and response variables?

**3.8 Violent acts on prime-time TV.** A typical hour of prime-time television shows three to five violent acts. Linking family interviews and police records shows a clear association between time spent watching TV as a child and later aggressive behavior.[4]

(a) Explain why this is an observational study rather than an experiment. What are the explanatory and response variables?

(b) Suggest several lurking variables describing a child's home life that may be related to how much TV he or she watches. Explain why this makes it difficult to conclude that more TV *causes* more aggressive behavior.

intervention

An observational study, even one based on a statistical sample, is a poor way to determine what will happen if we change something. The best way to see the effects of a change is to do an **intervention**—where we actually impose the change. When our goal is to understand cause and effect, experiments are the only source of fully convincing data.

EXAMPLE

**3.4 Child care and behavior.** A study of child care enrolled 1364 infants in 1991 and planned to follow them through their sixth year in school. Twelve years later, the researchers published an article finding that "the more time children spent in child care from birth to age four-and-a-half, the more adults tended to rate them, both at age four-and-a-half and at kindergarten, as less likely to get along with others, as more assertive, as disobedient, and as aggressive."[5]

What can we conclude from this study? If parents choose to use child care, are they more likely to see these undesirable behaviors in their children?

EXAMPLE

**3.5 Is there a cause and effect relationship?** Example 3.4 describes an observational study. Parents made all child care decisions and the study did not attempt to influence them. A summary of the study stated, "The study authors noted that their study was not designed to prove a cause and effect relationship. That is, the study cannot prove whether spending more time in child care causes children to have more problem behaviors."[6] Perhaps employed parents who use child care are under stress and the children react to their parents' stress. Perhaps single parents are more likely to use child care. Perhaps parents are more likely to place in child care children who already have behavior problems.

We can imagine an experiment that would remove these difficulties. From a large group of young children, choose some to be placed in child care and others to remain at home. This is an experiment because the treatment (child care or not) is imposed on the children. Of course, this particular experiment is neither practical nor ethical.

In Examples 3.4 and 3.5, we say that the effect of child care on behavior is **confounded** with (mixed up with) other characteristics of families who use child care. Observational studies that examine the effect of a single variable on an outcome can be misleading when the effects of the explanatory variable are confounded with those of other variables. Because experiments allow us to isolate the effects of specific variables, we generally prefer them. Here is an example.

confounded

EXAMPLE

**3.6 A dietary behavior experiment.** An experiment was designed to examine the effect of a 30-minute instructional session in a Food Stamp office on the dietary behavior of low-income women.[7] A group of women were randomly assigned to either the instructional session or no instruction. Two months later, data were collected on several measures of their behavior.

Experiments usually require some sort of randomization, as in this example. We begin the discussion of statistical designs for data collection in Section 3.1 with the principles underlying the design of experiments.

## USE YOUR KNOWLEDGE

**3.9 Software for teaching biology.** An educational software company wants to compare the effectiveness of its computer animation for teaching cell biology with that of a textbook presentation. The company tests the biological knowledge of each of a group of first-year college students, then randomly divides them into two groups. One group uses the animation, and the other studies the text. The company retests all the students and compares the increase in understanding of cell biology in the two groups. Is this an experiment? Why or why not? What are the explanatory and response variables?

**3.10 Find an experiment.** Use the Internet or some printed material to find an example of an experiment that interests you. Describe how the experiment was conducted and some of the conclusions.

Statistical techniques for producing data are the foundation for formal **statistical inference,** which answers specific questions with a known degree of confidence. In Section 3.3, we discuss some basic ideas related to inference.

statistical inference

Should an experiment or sample survey that could possibly provide interesting and important information always be performed? How can we safeguard the privacy of subjects in a sample survey? What constitutes the mistreatment of people or animals who are studied in an experiment? These are questions of **ethics.** In Section 3.4, we address ethical issues related to the design of studies and the analysis of data.

ethics

## 3.1 Design of Experiments

A study is an experiment when we actually do something to people, animals, or objects in order to observe the response. Here is the basic vocabulary of experiments.

> **EXPERIMENTAL UNITS, SUBJECTS, TREATMENT**
>
> The individuals on which the experiment is done are the **experimental units.** When the units are human beings, they are called **subjects.** A specific experimental condition applied to the units is called a **treatment.**

factors

level of a factor

Because the purpose of an experiment is to reveal the response of one variable to changes in other variables, the distinction between explanatory and response variables is important. The explanatory variables in an experiment are often called **factors.** Many experiments study the joint effects of several factors. In such an experiment, each treatment is formed by combining a specific value (often called a **level**) of each of the factors.

EXAMPLE

**3.7 Are smaller class sizes better?** Do smaller classes in elementary school really benefit students in areas such as scores on standard tests, staying in school, and going on to college? We might do an observational study that compares students who happened to be in smaller and larger classes in their early school years. Small classes are expensive, so they are more common in schools that serve richer communities. Students in small classes tend to also have other advantages: their schools have more resources, their parents are better educated, and so on. Confounding makes it impossible to isolate the effects of small classes.

The Tennessee STAR program was an experiment on the effects of class size. It has been called "one of the most important educational investigations ever carried out." The *subjects* were 6385 students who were beginning kindergarten. Each student was assigned to one of three *treatments:* regular class (22 to 25 students) with one teacher, regular class with a teacher and a full-time teacher's aide, and small class (13 to 17 students). These treatments are levels of a single *factor,* the type of class. The students stayed in the same type of class for four years, then all returned to regular classes. In later years, students from the small classes had higher scores on standard tests, were less likely to fail a grade, had better high school grades, and so on. The benefits of small classes were greatest for minority students.[8]

Example 3.7 illustrates the big advantage of experiments over observational studies. **In principle, experiments can give good evidence for causation.** In an experiment, we study the specific factors we are interested in, while controlling the effects of lurking variables. All the students in the Tennessee STAR program followed the usual curriculum at their schools. Because students were assigned to different class types within their schools, school resources and family backgrounds were not confounded with class type. The only systematic difference was the type of class. When students from the small classes did better than those in the other two types, we can be confident that class size made the difference.

EXAMPLE

**3.8 Repeated exposure to advertising.** What are the effects of repeated exposure to an advertising message? The answer may depend both on the length of the ad and on how often it is repeated. An experiment investigated this question using undergraduate students as *subjects*. All subjects viewed a 40-minute television program that included ads for a digital camera. Some subjects saw a 30-second commercial; others, a 90-second version. The same commercial was shown either 1, 3, or 5 times during the program.

This experiment has two *factors:* length of the commercial, with 2 levels, and repetitions, with 3 levels. The 6 combinations of one level of each factor form 6 *treatments*. Figure 3.2 shows the layout of the treatments. After viewing, all of the subjects answered questions about their recall of the ad, their attitude toward the camera, and their intention to purchase it. These are the *response variables*.[9]

| | | Factor B Repetitions | | |
|---|---|---|---|---|
| | | 1 time | 3 times | 5 times |
| Factor A Length | 30 seconds | 1 | 2 | 3 |
| | 90 seconds | 4 | 5 | 6 |

**FIGURE 3.2** The treatments in the study of advertising, for Example 3.8. Combining the levels of the two factors forms six treatments.

Example 3.8 shows how experiments allow us to study the combined effects of several factors. The interaction of several factors can produce effects that could not be predicted from looking at the effects of each factor alone. Perhaps longer commercials increase interest in a product, and more commercials also increase interest, but if we both make a commercial longer and show it more often, viewers get annoyed and their interest in the product drops. The two-factor experiment in Example 3.8 will help us find out.

## USE YOUR KNOWLEDGE

**3.11 Food for a trip to the moon.** Storing food for long periods of time is a major challenge for those planning for human space travel beyond the moon. One problem is that exposure to radiation decreases the length of time that food can be stored. One experiment examined the effects of nine different levels of radiation on a particular type of fat, or lipid.[10] The amount of oxidation of the lipid is the measure of the extent of the damage due to the radiation. Three samples are exposed to each radiation level. Give the experimental units, the treatments, and the response variable. Describe the factor and its levels. There are many different types of lipids. To what extent do you think the results of this experiment can be generalized to other lipids?

**3.12 Learning how to draw.** A course in computer graphics technology requires students to learn multiview drawing concepts. This topic is traditionally taught using supplementary material printed on paper. The instructor of the course believes that a Web-based interactive drawing program will be more effective in increasing the drawing skills of the students.[11] The 50 students who are enrolled in the course will be randomly assigned to either the paper-based instruction or the Web-based instruction. A standardized drawing test will be given before and after the instruction. Explain why this study is an experiment and give the experimental units, the treatments, and the response variable. Describe the factor and its levels. To what extent do you think the results of this experiment can be generalized to other settings?

## Comparative experiments

Laboratory experiments in science and engineering often have a simple design with only a single treatment, which is applied to all of the experimental units. The design of such an experiment can be outlined as

**Treatment ⟶ Observe response**

For example, we may subject a beam to a load (treatment) and measure its deflection (observation). We rely on the controlled environment of the laboratory to protect us from lurking variables. When experiments are conducted in the field or with living subjects, such simple designs often yield invalid data. That is, we cannot tell whether the response was due to the treatment or to lurking variables. A medical example will show what can go wrong.

EXAMPLE

**3.9 Gastric freezing.** "Gastric freezing" is a clever treatment for ulcers in the upper intestine. The patient swallows a deflated balloon with tubes attached, then a refrigerated liquid is pumped through the balloon for an hour. The idea is that cooling the stomach will reduce its production of acid and so relieve ulcers. An experiment reported in the *Journal of the American Medical Association* showed that gastric freezing did reduce acid production and relieve ulcer pain. The treatment was safe and easy and was widely used for several years. The design of the experiment was

**Gastric freezing ⟶ Observe pain relief**

placebo effect

The gastric freezing experiment was poorly designed. The patients' response may have been due to the **placebo effect.** A placebo is a dummy treatment. Many patients respond favorably to any treatment, even a placebo. This may be due to trust in the doctor and expectations of a cure or simply to the fact that medical conditions often improve without treatment. The response to a dummy treatment is the placebo effect.

A later experiment divided ulcer patients into two groups. One group was treated by gastric freezing as before. The other group received a placebo treatment in which the liquid in the balloon was at body temperature rather than freezing. The results: 34% of the 82 patients in the treatment group improved, but so did 38% of the 78 patients in the placebo group. This and other properly designed experiments showed that gastric freezing was no better than a placebo, and its use was abandoned.[12]

The first gastric freezing experiment gave misleading results because the effects of the explanatory variable were confounded with the placebo effect. We can defeat confounding by *comparing* two groups of patients, as in the second gastric freezing experiment. The placebo effect and other lurking variables now operate on both groups. The only difference between the groups is the actual effect of gastric freezing. The group of patients who received a sham treatment is called a **control group,** because it enables us to control the effects of outside variables on the outcome. Control is the first basic principle of statistical design of experiments. Comparison of several treatments in the same environment is the simplest form of control.

control group

*Uncontrolled experiments in medicine and the behavioral sciences can be dominated by such influences as the details of the experimental arrangement, the selection of subjects, and the placebo effect.* The result is often *bias.*

BIAS

The design of a study is **biased** if it systematically favors certain outcomes.

An uncontrolled study of a new medical therapy, for example, is biased in favor of finding the treatment effective because of the placebo effect. It should not surprise you to learn that uncontrolled studies in medicine give new therapies a much higher success rate than proper comparative experiments. Well-designed experiments usually compare several treatments.

USE YOUR KNOWLEDGE

**3.13 Does using statistical software improve exam scores?** An instructor in an elementary statistics course wants to know if using a new statistical software package will improve students' final-exam scores. He asks for volunteers and about half of the class agrees to work with the new software. He compares the final-exam scores of the students who used the new software with the scores of those who did not. Discuss possible sources of bias in this study.

## Randomization

experiment design

The **design of an experiment** first describes the response variable or variables, the factors (explanatory variables), and the layout of the treatments, with comparison as the leading principle. Figure 3.2 illustrates this aspect of the design of a study of response to advertising. The second aspect of design is the rule used to assign the experimental units to the treatments. Comparison of the effects of several treatments is valid only when all treatments are applied to similar groups of experimental units. If one corn variety is planted on more fertile ground, or if one cancer drug is given to more seriously ill patients, comparisons among treatments are meaningless. Systematic differences among the groups of experimental units in a comparative experiment cause bias. How can we assign experimental units to treatments in a way that is fair to all of the treatments?

Experimenters often attempt to match groups by elaborate balancing acts. Medical researchers, for example, try to match the patients in a "new drug" experimental group and a "standard drug" control group by age, sex, physical condition, smoker or not, and so on. Matching is helpful but not adequate—there are too many lurking variables that might affect the outcome. The experimenter is unable to measure some of these variables and will not think of others until after the experiment. Some important variables, such as how advanced a cancer patient's disease is, are so subjective that an experimenter might bias the study by, for example, assigning more advanced cancer cases to a promising new treatment in the unconscious hope that it will help them.

*The statistician's remedy is to rely on chance to make an assignment that does not depend on any characteristic of the experimental units and that does not rely on the judgment of the experimenter in any way.* The use of chance can be combined with matching, but the simplest design creates groups by chance alone. Here is an example.

EXAMPLE

**3.10 Cell phones and driving.** Does talking on a hands-free cell phone distract drivers? Undergraduate students "drove" in a high-fidelity driving simulator equipped with a hands-free cell phone. The car ahead brakes: how quickly does the subject respond? Twenty students (the control group) simply drove. Another 20 (the experimental group) talked on the cell phone while driving.

This experiment has a single factor (cell phone use) with two levels. The researchers must divide the 40 student subjects into two groups of 20. To do this in a completely unbiased fashion, put the names of the 40 students in a hat, mix them up, and draw 20. These students form the experimental group and the remaining 20 make up the control group. Figure 3.3 outlines the design of this experiment.[13]

randomization

The use of chance to divide experimental units into groups is called **randomization.** The design in Figure 3.3 combines comparison and ran-

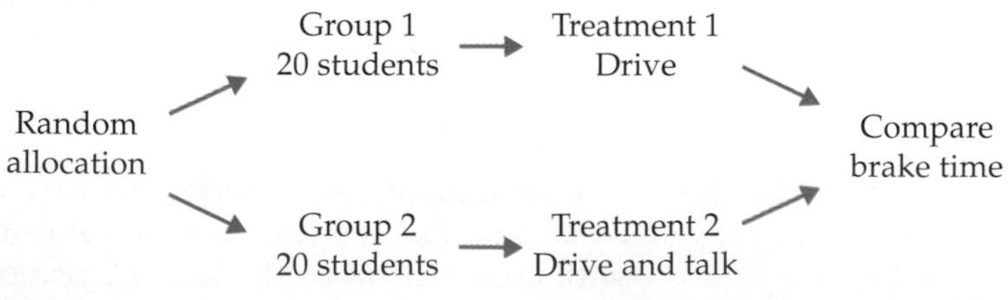


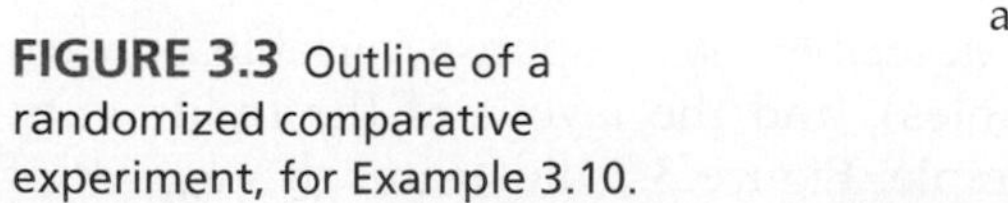
**FIGURE 3.3** Outline of a randomized comparative experiment, for Example 3.10.

domization to arrive at the simplest randomized comparative design. This "flowchart" outline presents all the essentials: randomization, the sizes of the groups and which treatment they receive, and the response variable. There are, as we will see later, statistical reasons for generally using treatment groups about equal in size.

## USE YOUR KNOWLEDGE

**3.14 Diagram the drawing experiment.** Refer to Exercise 3.12 (page 180). Draw a diagram similar to Figure 3.3 that describes the computer graphics drawing experiment.

**3.15 Diagram the food for Mars experiment.** Refer to Exercise 3.11 (page 179). Draw a diagram similar to Figure 3.3 that describes the food for space travel experiment.

## Randomized comparative experiments

The logic behind the randomized comparative design in Figure 3.3 is as follows:

- Randomization produces two groups of subjects that we expect to be similar in all respects before the treatments are applied.
- Comparative design helps ensure that influences other than the cell phone operate equally on both groups.
- Therefore, differences in average brake reaction time must be due either to talking on the cell phone or to the play of chance in the random assignment of subjects to the two groups.

That "either-or" deserves more comment. We cannot say that *any* difference in the average reaction times of the experimental and control groups is caused by talking on the cell phone. There would be some difference even if both groups were treated the same, because the natural variability among people means that some react faster than others. Chance can assign the faster-reacting students to one group or the other, so that there is a chance difference between the groups. We would not trust an experiment with just one subject in each group, for example. The results would depend too much on which group got lucky and received the subject with quicker reactions. If we assign many students to each group, however, the effects of chance will average out. There will be little difference in the average reaction times in the two groups unless talking on the cell phone causes a difference. "Use enough subjects to reduce chance variation" is the third big idea of statistical design of experiments.

### PRINCIPLES OF EXPERIMENTAL DESIGN

The basic principles of statistical design of experiments are

**1. Compare** two or more treatments. This will control the effects of lurking variables on the response.

**2. Randomize**—use impersonal chance to assign experimental units to treatments.

3. **Repeat** each treatment on many units to reduce chance variation in the results.

We hope to see a difference in the responses so large that it is unlikely to happen just because of chance variation. We can use the laws of probability, which give a mathematical description of chance behavior, to learn if the treatment effects are larger than we would expect to see if only chance were operating. If they are, we call them *statistically significant.*

STATISTICAL SIGNIFICANCE

An observed effect so large that it would rarely occur by chance is called **statistically significant.**

You will often see the phrase "statistically significant" in reports of investigations in many fields of study. It tells you that the investigators found good evidence for the effect they were seeking. The cell phone study, for example, reported statistically significant evidence that talking on a cell phone increases the mean reaction time of drivers when the car in front of them brakes.

## How to randomize

The idea of randomization is to assign subjects to treatments by drawing names from a hat. In practice, experimenters use software to carry out randomization. Most statistical software will choose 20 out of a list of 40 at random, for example. The list might contain the names of 40 human subjects. The 20 chosen form one group, and the 20 that remain form the second group. The *Simple Random Sample* applet on the text CD and Web site makes it particularly easy to choose treatment groups at random.

You can randomize without software by using a *table of random digits.* Thinking about random digits helps you to understand randomization even if you will use software in practice. Table B at the back of the book is a table of random digits.

RANDOM DIGITS

A **table of random digits** is a list of the digits 0, 1, 2, 3, 4, 5, 6, 7, 8, 9 that has the following properties:

1. The digit in any position in the list has the same chance of being any one of 0, 1, 2, 3, 4, 5, 6, 7, 8, 9.

2. The digits in different positions are independent in the sense that the value of one has no influence on the value of any other.

You can think of Table B as the result of asking an assistant (or a computer) to mix the digits 0 to 9 in a hat, draw one, then replace the digit drawn, mix again, draw a second digit, and so on. The assistant's mixing and drawing saves us the work of mixing and drawing when we need to randomize. Table B begins with the digits 19223950340575628713. To make the table easier to read, the digits appear in groups of five and in numbered rows. The groups and rows have no meaning—the table is just a long list of digits having the properties 1 and 2 described above.

Our goal is to use random digits for experimental randomization. We need the following facts about random digits, which are consequences of the basic properties 1 and 2:

- Any *pair* of random digits has the same chance of being any of the 100 possible pairs: 00, 01, 02, ..., 98, 99.
- Any *triple* of random digits has the same chance of being any of the 1000 possible triples: 000, 001, 002, ..., 998, 999.
- ...and so on for groups of four or more random digits.

EXAMPLE

**3.11 Randomize the students.** In the cell phone experiment of Example 3.10, we must divide 40 students at random into two groups of 20 students each.

*Step 1: Label.* Give each student a numerical label, using as few digits as possible. Two digits are needed to label 40 students, so we use labels

01, 02, 03, ..., 39, 40

It is also correct to use labels 00 to 39 or some other choice of 40 two-digit labels.

*Step 2: Table.* Start anywhere in Table B and read two-digit groups. Suppose we begin at line 130, which is

69051 64817 87174 09517 84534 06489 87201 97245

The first 10 two-digit groups in this line are

69 05 16 48 17 87 17 40 95 17

Each of these two-digit groups is a label. The labels 00 and 41 to 99 are not used in this example, so we ignore them. The first 20 labels between 01 and 40 that we encounter in the table choose students for the experimental group. Of the first 10 labels in line 130, we ignore four because they are too high (over 40). The others are 05, 16, 17, 17, 40, and 17. The students labeled 05, 16, 17, and 40 go into the experimental group. Ignore the second and third 17s because that student is already in the group. Run your finger across line 130 (and continue to the following lines) until you have chosen 20 students. They are the students labeled

05, 16, 17, 40, 20, 19, 32, 04, 25, 29,
37, 39, 31, 18, 07, 13, 33, 02, 36, 23

You should check at least the first few of these. These students form the experimental group. The remaining 20 are the control group.

As Example 3.11 illustrates, randomization requires two steps: assign labels to the experimental units and then use Table B to select labels at random. Be sure that all labels are the same length so that all have the same chance to be chosen. Use the shortest possible labels—one digit for 9 or fewer individuals, two digits for 10 to 100 individuals, and so on. Don't try to scramble the labels as you assign them. Table B will do the required randomizing, so assign labels in any convenient manner, such as in alphabetical order for human subjects. You can read digits from Table B in any order—along a row, down a column, and so on—because the table has no order. As an easy standard practice, we recommend reading along rows.

It is easy to use statistical software or Excel to randomize. Here are the steps:

*Step 1: Label.* The first step, assigning labels to the experimental units, is similar to the procedure we described above. One difference, however, is that we are not restricted to using numerical labels. Any system where each experimental unit has a unique label identifier will work.

Step 2: *Use the computer.* Once we have the labels, we then create a data file with the labels and generate a random number for each label. In Excel, this can be done with the RAND() function. Finally, we sort the entire data set based on the random numbers. Groups are formed by selecting units in order from the sorted list.

This process is essentially the same as writing the labels on a deck of cards, shuffling the cards, and dealing them out one at a time.

EXAMPLE

**3.12 Using software for randomization.** Let's do a randomization similar to the one we did in Example 3.11, but this time using Excel. Here we will use 10 experimental units. We will assign 5 to the treatment group and 5 to the control group. We first create a data set with the numbers 1 to 10 in the first column. See Figure 3.4(a). Then we use RAND() to generate 10 random numbers in the second column. See Figure 3.4(b). Finally, we sort the data set based on the numbers in the second column. See Figure 3.4(c). The first 5 labels (8, 5, 9, 4, and 6) are assigned to the experimental group. The remaining 5 labels (3, 10, 7, 2, and 1) correspond to the control group.

Excel (a)

| | A | B |
|---|---|---|
| 1 | 1 | |
| 2 | 2 | |
| 3 | 3 | |
| 4 | 4 | |
| 5 | 5 | |
| 6 | 6 | |
| 7 | 7 | |
| 8 | 8 | |
| 9 | 9 | |
| 10 | 10 | |
| 11 | | |

Excel (b)

| | A | B |
|---|---|---|
| 1 | 1 | 0.925672 |
| 2 | 2 | 0.893959 |
| 3 | 3 | 0.548247 |
| 4 | 4 | 0.349591 |
| 5 | 5 | 0.118440 |
| 6 | 6 | 0.390180 |
| 7 | 7 | 0.760262 |
| 8 | 8 | 0.077044 |
| 9 | 9 | 0.348467 |
| 10 | 10 | 0.601167 |

Excel (c)

| | A | B |
|---|---|---|
| 1 | 8 | 0.077044 |
| 2 | 5 | 0.118440 |
| 3 | 9 | 0.348467 |
| 4 | 4 | 0.349591 |
| 5 | 6 | 0.390180 |
| 6 | 3 | 0.548247 |
| 7 | 10 | 0.601167 |
| 8 | 7 | 0.760262 |
| 9 | 2 | 0.893959 |
| 10 | 1 | 0.925672 |
| 11 | | |

**FIGURE 3.4** Randomization of 10 experimental units using a computer, for Example 3.12. (a) Labels. (b) Random numbers. (c) Sorted list of labels.

completely randomized design

When all experimental units are allocated at random among all treatments, as in Example 3.11, the experimental design is **completely randomized.** Completely randomized designs can compare any number of treatments. The treatments can be formed by levels of a single factor or by more than one factor.

EXAMPLE

**3.13 Randomization of the TV commercial experiment.** Figure 3.2 (page 179) displays six treatments formed by the two factors in an experiment on response to a TV commercial. Suppose that we have 150 students who are willing to serve as subjects. We must assign 25 students at random to each group. Figure 3.5 outlines the completely randomized design.

To carry out the random assignment, label the 150 students 001 to 150. (Three digits are needed to label 150 subjects.) Enter Table B and read three-digit groups until you have selected 25 students to receive Treatment 1 (a 30-second ad shown once). If you start at line 140, the first few labels for Treatment 1 subjects are 129, 048, and 003.

Continue in Table B to select 25 more students to receive Treatment 2 (a 30-second ad shown 3 times). Then select another 25 for Treatment 3 and so on until you have assigned 125 of the 150 students to Treatments 1 through 5. The 25 students who remain get Treatment 6. The randomization is straightforward, but very tedious to do by hand. We recommend the *Simple Random Sample* applet. Exercise 3.35 shows how to use the applet to do the randomization for this example.

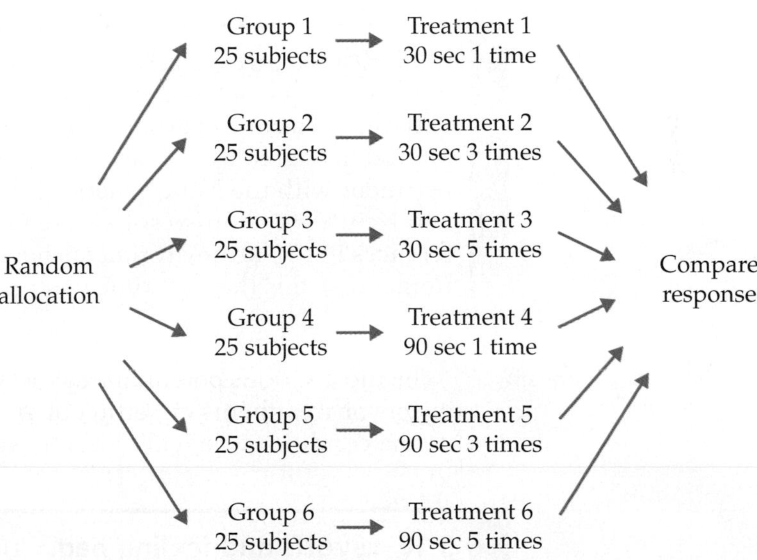


**FIGURE 3.5** Outline of a completely randomized design comparing six treatments, for Example 3.13.

## USE YOUR KNOWLEDGE

**3.16 Do the randomization.** Use computer software to carry out the randomization in Example 3.13.

## Cautions about experimentation

The logic of a randomized comparative experiment depends on our ability to treat all the experimental units identically in every way except for the actual treatments being compared. Good experiments therefore require careful attention to details. For example, the subjects in the second gastric freezing experiment (Example 3.9) all got the same medical attention during the study.

double-blind

Moreover, the study was **double-blind**—neither the subjects themselves nor the medical personnel who worked with them knew which treatment any subject had received. The double-blind method avoids unconscious bias by, for example, a doctor who doesn't think that "just a placebo" can benefit a patient.

CAUTION

*Many—perhaps most—experiments have some weaknesses in detail. The environment of an experiment can influence the outcomes in unexpected ways.* Although experiments are the gold standard for evidence of cause and effect, really convincing evidence usually requires that a number of studies in different places with different details produce similar results. Here are some brief examples of what can go wrong.

EXAMPLE

**3.14 Placebo for a marijuana experiment.** A study of the effects of marijuana recruited young men who used marijuana. Some were randomly assigned to smoke marijuana cigarettes, while others were given placebo cigarettes. This failed: the control group recognized that their cigarettes were phony and complained loudly. It may be quite common for blindness to fail because the subjects can tell which treatment they are receiving.[14]

EXAMPLE

**3.15 Knock out genes.** To study genetic influence on behavior, experimenters "knock out" a gene in one group of mice and compare their behavior with that of a control group of normal mice. The results of these experiments often don't agree as well as hoped, so investigators did exactly the same experiment with the same genetic strain of mice in Oregon, Alberta (Canada), and New York. Many results were very different.[15] It appears that small differences in the lab environments have big effects on the behavior of the mice. Remember this the next time you read that our genes control our behavior.

lack of realism

The most serious potential weakness of experiments is **lack of realism.** The subjects or treatments or setting of an experiment may not realistically duplicate the conditions we really want to study. Here is an example.

EXAMPLE

**3.16 Layoffs and feeling bad.** How do layoffs at a workplace affect the workers who remain on the job? Psychologists asked student subjects to proofread text for extra course credit, then "let go" some of the workers (who were actually accomplices of the experimenters). Some subjects were told that those let go had performed poorly (Treatment 1). Others were told that not all could be kept and that it was just luck that they were kept and others let go (Treatment 2). We can't be sure that the reactions of the students are the same as those of workers who survive a layoff in which other workers lose their jobs. Many behavioral science experiments use student subjects in a campus setting. Do the conclusions apply to the real world?

Lack of realism can limit our ability to apply the conclusions of an experiment to the settings of greatest interest. Most experimenters want to generalize their conclusions to some setting wider than that of the actual experiment. *Statistical analysis of an experiment cannot tell us how far the results will generalize to other settings.* Nonetheless, the randomized comparative experiment, because of its ability to give convincing evidence for causation, is one of the most important ideas in statistics.

## Matched pairs designs

Completely randomized designs are the simplest statistical designs for experiments. They illustrate clearly the principles of control, randomization, and repetition. However, completely randomized designs are often inferior to more elaborate statistical designs. In particular, matching the subjects in various ways can produce more precise results than simple randomization.

matched pairs design

The simplest use of matching is a **matched pairs design,** which compares just two treatments. The subjects are matched in pairs. For example, an experiment to compare two advertisements for the same product might use pairs of subjects with the same age, sex, and income. The idea is that matched subjects are more similar than unmatched subjects, so that comparing responses within a number of pairs is more efficient than comparing the responses of groups of randomly assigned subjects. Randomization remains important: which one of a matched pair sees the first ad is decided at random. One common variation of the matched pairs design imposes both treatments on the same subjects, so that each subject serves as his or her own control. Here is an example.

EXAMPLE

**3.17 Matched pairs for the cell phone experiment.** Example 3.10 describes an experiment on the effects of talking on a cell phone while driving. The experiment compared two treatments, driving in a simulator and driving in the simulator while talking on a hands-free cell phone. The response variable is the time the driver takes to apply the brake when the car in front brakes suddenly. In Example 3.10, 40 student subjects were assigned at random, 20 students to each treatment. This is a completely randomized design, outlined in Figure 3.3. Subjects differ in driving skill and reaction times. The completely randomized design relies on chance to create two similar groups of subjects.

In fact, the experimenters used a matched pairs design in which all subjects drove both with and without using the cell phone. They compared each individual's reaction times with and without the phone. If all subjects drove first with the phone and then without it, the effect of talking on the cell phone would be confounded with the fact that this is the first run in the simulator. The proper procedure requires that all subjects first be trained in using the simulator, that the *order* in which a subject drives with and without the phone be random, and that the two drives be on separate days to reduce the chance that the results of the second treatment will be influenced by the first treatment.

The completely randomized design uses chance to decide which 20 subjects will drive with the cell phone. The other 20 drive without it. The matched pairs design uses chance to decide which 20 subjects will drive first with and then without the cell phone. The other 20 drive first without and then with the phone.

## Block designs

The matched pairs design of Example 3.17 uses the principles of comparison of treatments, randomization, and repetition on several experimental units. However, the randomization is not complete (all subjects randomly assigned to treatment groups) but restricted to assigning the order of the treatments for each subject. *Block designs* extend the use of "similar subjects" from pairs to larger groups.

### BLOCK DESIGN

A **block** is a group of experimental units or subjects that are known before the experiment to be similar in some way that is expected to affect the response to the treatments. In a **block design,** the random assignment of units to treatments is carried out separately within each block.

Block designs can have blocks of any size. A block design combines the idea of creating equivalent treatment groups by matching with the principle of forming treatment groups at random. Blocks are another form of *control.* They control the effects of some outside variables by bringing those variables into the experiment to form the blocks. Here are some typical examples of block designs.

EXAMPLE

**3.18 Blocking in a cancer experiment.** The progress of a type of cancer differs in women and men. A clinical experiment to compare three therapies for this cancer therefore treats sex as a blocking variable. Two separate randomizations are done, one assigning the female subjects to the treatments and the other assigning the male subjects. Figure 3.6 outlines the design of this experiment. Note that there is no randomization involved in making up the blocks. They are groups of subjects who differ in some way (sex in this case) that is apparent before the experiment begins.

EXAMPLE

**3.19 Blocking in an agriculture experiment.** The soil type and fertility of farmland differ by location. Because of this, a test of the effect of tillage type (two types) and pesticide application (three application schedules) on soybean yields uses small fields as blocks. Each block is divided into six plots, and the six treatments are randomly assigned to plots separately within each block.

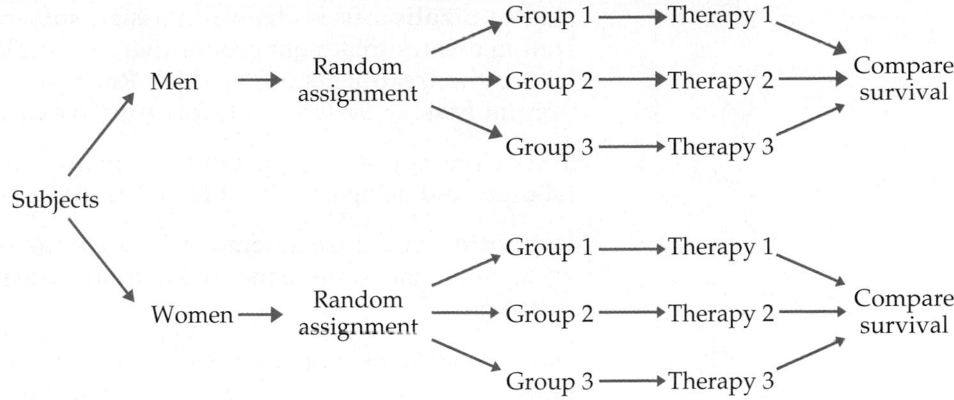


**FIGURE 3.6** Outline of a block design, for Example 3.18. The blocks consist of male and female subjects. The treatments are three therapies for cancer.

EXAMPLE

**3.20 Blocking in an education experiment.** The Tennessee STAR class size experiment (Example 3.7) used a block design. It was important to compare different class types in the same school because the children in a school come from the same neighborhood, follow the same curriculum, and have the same school environment outside class. In all, 79 schools across Tennessee participated in the program. That is, there were 79 blocks. New kindergarten students were randomly placed in the three types of class separately within each school.

Blocks allow us to draw separate conclusions about each block, for example, about men and women in the cancer study in Example 3.18. Blocking also allows more precise overall conclusions because the systematic differences between men and women can be removed when we study the overall effects of the three therapies. The idea of blocking is an important additional principle of statistical design of experiments. A wise experimenter will form blocks based on the most important unavoidable sources of variability among the experimental units. Randomization will then average out the effects of the remaining variation and allow an unbiased comparison of the treatments.

## SECTION 3.1 Summary

In an experiment, one or more **treatments** are imposed on the **experimental units** or **subjects.** Each treatment is a combination of **levels** of the explanatory variables, which we call **factors.**

The **design** of an experiment refers to the choice of treatments and the manner in which the experimental units or subjects are assigned to the treatments.

The basic principles of statistical design of experiments are **control, randomization,** and **repetition.**

The simplest form of control is **comparison.** Experiments should compare two or more treatments in order to prevent **confounding** the effect of a treatment with other influences, such as lurking variables.

**Randomization** uses chance to assign subjects to the treatments. Randomization creates treatment groups that are similar (except for chance variation) before the treatments are applied. Randomization and comparison together prevent **bias,** or systematic favoritism, in experiments.

You can carry out randomization by giving numerical labels to the experimental units and using a **table of random digits** to choose treatment groups.

**Repetition** of the treatments on many units reduces the role of chance variation and makes the experiment more sensitive to differences among the treatments.

Good experiments require attention to detail as well as good statistical design. Many behavioral and medical experiments are **double-blind. Lack of realism** in an experiment can prevent us from generalizing its results.

In addition to comparison, a second form of control is to restrict randomization by forming **blocks** of experimental units that are similar in some way that is important to the response. Randomization is then carried out separately within each block.

**Matched pairs** are a common form of blocking for comparing just two treatments. In some matched pairs designs, each subject receives both treatments in a random order. In others, the subjects are matched in pairs as closely as possible, and one subject in each pair receives each treatment.

## SECTION 3.1 Exercises

*For Exercises 3.1 to 3.4, see page 172; for Exercise 3.5, see page 173; for Exercises 3.6 and 3.7, see page 175; for Exercise 3.8, see page 176; for Exercises 3.9 and 3.10, see page 177; for Exercises 3.11 and 3.12, see pages 179 and 180; for Exercise 3.13, see page 181; for Exercises 3.14 and 3.15, see page 183; and for Exercise 3.16, see page 187.*

**3.17 What is needed?** Explain what is deficient in each of the following proposed experiments and explain how you would improve the experiment.

(a) Two forms of a lab exercise are to be compared. There are 10 rows in the classroom. Students who sit in the first 5 rows of the class are given the first form, and students who sit in the last 5 rows are given the second form.

(b) The effectiveness of a leadership program for high school students is evaluated by examining the change in scores on a standardized test of leadership skills.

(c) An innovative method for teaching introductory biology courses is examined by using the traditional method in the fall zoology course and the new method in the spring botany course.

**3.18 What is wrong?** Explain what is wrong with each of the following randomization procedures and describe how you would do the randomization correctly.

(a) A list of 50 subjects is entered into a computer file and then sorted by last name. The subjects are assigned to five treatments by taking the first 10 subjects for Treatment 1, the next 10 subjects for Treatment 2, and so forth.

(b) Eight subjects are to be assigned to two treatments, four to each. For each subject, a coin is tossed. If the coin comes up heads, the subject is assigned to the first treatment; if the coin comes up tails, the subject is assigned to the second treatment.

(c) An experiment will assign 80 rats to four different treatment conditions. The rats arrive from the supplier in batches of 20 and the treatment lasts two weeks. The first batch of 20 rats is randomly assigned to one of the four treatments, and data for these rats are collected. After a one-week break, another batch of 20 rats arrives and is assigned to one of the three remaining treatments. The process continues until the last batch of rats is given the treatment that has not been assigned to the three previous batches.

**3.19 Evaluate a new teaching method.** A teaching innovation is to be evaluated by randomly assigning students to either the traditional approach or the new approach. The change in a standardized test score is the response variable. Explain how this experiment should be done in a double-blind fashion.

**3.20 Can you change attitudes toward binge drinking?** A experiment designed to change attitudes about binge drinking is to be performed using college students as subjects. Discuss some variables that you might use if you were to use a block design for this experiment.

**3.21 Compost tea.** Compost tea is rich in micro-organisms that help plants grow. It is made by soaking compost in water.[16] Design a comparative experiment that will provide evidence about whether or not compost tea works for a particular type of plant that interests you. Be sure to provide all details regarding your experiment, including the response variable or variables that you will measure.

**3.22** CHALLENGE **Measuring water quality in streams and lakes.** Water quality of streams and lakes is an issue of concern to the public. Although trained professionals typically are used to take reliable measurements, many volunteer groups are gathering and distributing information based on data that they collect.[17] You are part of a team to train volunteers to collect accurate water quality data. Design an experiment to evaluate the effectiveness of the training. Write a summary of your proposed design to present to your team. Be sure to include all of the details that they will need to evaluate your proposal.

*For each of the experimental situations described in Exercises 3.23 to 3.25, identify the experimental units or subjects, the factors, the treatments, and the response variables.*

**3.23 How well do pine trees grow in shade?** Ability to grow in shade may help pines in the dry forests of Arizona resist drought. How well do these pines grow in shade? Investigators planted pine seedlings in a greenhouse in either full light or light reduced to 5% of normal by shade cloth. At the end of the study, they dried the young trees and weighed them.

**3.24 Will the students do more exercise and eat better?** Most American adolescents don't eat well and don't exercise enough. Can middle schools increase physical activity among their students? Can they persuade students to eat better? Investigators designed a "physical activity intervention" to increase activity in physical education classes and during leisure periods throughout the school day. They also designed a "nutrition intervention" that improved school lunches and offered ideas for healthy home-packed lunches. Each participating school was randomly assigned to one of the interventions, both interventions, or no intervention. The investigators observed physical activity and lunchtime consumption of fat.

**3.25 Refusals in telephone surveys.** How can we reduce the rate of refusals in telephone surveys? Most people who answer at all listen to the interviewer's introductory remarks and then decide whether to continue. One study made telephone calls to randomly selected households to ask opinions about the next election. In some calls, the interviewer gave her name, in others she identified the university she was representing, and in still others she identified both herself and the university. For each type of call, the interviewer either did or did not offer to send a copy of the final survey results to the person interviewed. Do these differences in the introduction affect whether the interview is completed?

**3.26 Does aspirin prevent strokes and heart attacks?** The Bayer Aspirin Web site claims that "Nearly five decades of research now link aspirin to the prevention of stroke and heart attacks." The most important evidence for this claim comes from the Physicians' Health Study, a large medical experiment involving 22,000 male physicians. One group of about 11,000 physicians took an aspirin every second day, while the rest took a placebo. After several years the study found that subjects in the aspirin group had significantly fewer heart attacks than subjects in the placebo group.

(a) Identify the experimental subjects, the factor and its levels, and the response variable in the Physicians' Health Study.

(b) Use a diagram to outline a completely randomized design for the Physicians' Health Study.

(c) What does it mean to say that the aspirin group had "significantly fewer heart attacks"?

**3.27 Chronic tension headaches.** Doctors identify "chronic tension-type headaches" as headaches that occur almost daily for at least six months. Can antidepressant medications or stress management training reduce the number and severity of these headaches? Are both together more effective than either alone? Investigators compared four treatments: antidepressant alone, placebo alone, antidepressant plus stress management, and placebo plus stress management. Outline the design of the experiment. The headache sufferers named below have agreed to participate in the study. Use software or Table B at line 151 to randomly assign the subjects to the treatments.

| | | | | |
|---|---|---|---|---|
| Anderson | Archberger | Bezawada | Cetin | Cheng |
| Chronopoulou | Codrington | Daggy | Daye | Engelbrecht |
| Guha | Hatfield | Hua | Kim | Kumar |
| Leaf | Li | Lipka | Lu | Martin |
| Mehta | Mi | Nolan | Olbricht | Park |
| Paul | Rau | Saygin | Shu | Tang |
| Towers | Tyner | Vassilev | Wang | Watkins |
| Xu | | | | |

**3.28 Smoking marijuana and willingness to work.** How does smoking marijuana affect willingness to work? Canadian researchers persuaded people who used marijuana to live for 98 days in a "planned environment." The subjects earned money by weaving belts. They used their earnings to pay for meals and other consumption and could keep any money left over. One group smoked two potent marijuana cigarettes every evening. The other group smoked two weak marijuana cigarettes. All subjects could buy more cigarettes but were given strong or weak cigarettes, depending on their group. Did the weak and strong groups differ in work output and earnings?[18]

(a) Outline the design of this experiment.

(b) Here are the names of the 20 subjects. Use software or Table B at line 101 to carry out the randomization your design requires.

| | | | | |
|---|---|---|---|---|
| Becker | Brifcani | Chen | Crabill | Cunningham |
| Dicklin | Fein | Gorman | Knapp | Lucas |
| McCarty | Merkulyeva | Mitchell | Ponder | Roe |
| Saeed | Seele | Truong | Wayman | Woodley |

**3.29 Eye cataracts.** Eye cataracts are responsible for over 40% of blindness worldwide. Can drinking tea regularly slow the growth of cataracts? We can't experiment on people, so we use rats as subjects. Researchers injected 21 young rats with a substance that causes cataracts. One group of the rats also received black tea extract; a second group received green tea extract; and a third got a placebo, a substance with no effect on the body. The response variable was the growth of cataracts over the next six weeks. Yes, both tea extracts did slow cataract growth.[19]

(a) Outline the design of this experiment.

(b) Use software or Table B, starting at line 120, to assign rats to treatments.

**3.30 Guilt among workers who survive a layoff.** Workers who survive a layoff of other employees at their location may suffer from "survivor guilt." A study of survivor guilt and its effects used as subjects 90 students who were offered an opportunity to earn extra course credit by doing proofreading. Each subject worked in the same cubicle as another student, who was an accomplice of the experimenters. At a break midway through the work, one of three things happened:

Treatment 1: The accomplice was told to leave; it was explained that this was because she performed poorly.

Treatment 2: It was explained that unforeseen circumstances meant there was only enough work for one person. By "chance," the accomplice was chosen to be laid off.

Treatment 3: Both students continued to work after the break.

The subjects' work performance after the break was compared with performance before the break.[20]

(a) Outline the design of this completely randomized experiment.

(b) If you are using software, randomly assign the 90 students to the treatments. If not, use Table B at line 153 to choose the first four subjects for Treatment 1.

**3.31 Diagram the exercise and eating experiment.** Twenty-four public middle schools agree to participate in the experiment described in Exercise 3.24. Use a diagram to outline a completely randomized design for this experiment. Then do the randomization required to assign schools to treatments. If you use Table B, start at line 160.

**3.32 Price cuts on athletic shoes.** Stores advertise price reductions to attract customers. What type of price cut is most attractive? Market researchers prepared ads for athletic shoes announcing different levels of discounts (20%, 40%, 60%, or 80%). The student subjects who read the ads were also given "inside information" about the fraction of shoes on sale (25%, 50%, 75%, or 100%). Each subject then rated the attractiveness of the sale on a scale of 1 to 7.[21]

(a) There are two factors. Make a sketch like Figure 3.2 (page 179) that displays the treatments formed by all combinations of levels of the factors.

(b) Outline a completely randomized design using 96 student subjects. Use software or Table B at line 111 to choose the subjects for the first treatment.

**3.33 Treatment of clothing fabrics.** A maker of fabric for clothing is setting up a new line to "finish" the raw fabric. The line will use either metal rollers or natural-bristle rollers to raise the surface of the fabric; a dyeing cycle time of either 30 minutes or 40 minutes; and a temperature of either 150° or 175° Celsius. An experiment will compare all combinations of these choices. Four specimens of fabric will be subjected to each treatment and scored for quality.

(a) What are the factors and the treatments? How many individuals (fabric specimens) does the experiment require?

(b) Outline a completely randomized design for this experiment. (You need not actually do the randomization.)

**3.34** APPLET **Use the simple random sample applet.** You can use the *Simple Random Sample* applet to choose a treatment group at random once you have labeled the subjects. Example 3.11 (page 185) uses Table B to choose 20 students from a group of 40 for the treatment group in a study of the effect of cell phones on driving. Use the applet to choose the 20 students for the experimental group. Which students did you choose? The remaining 20 students make up the control group.

**3.35** APPLET **Use the simple random sample applet.** The *Simple Random Sample* applet allows you to randomly assign experimental units to more than two groups without difficulty. Example 3.13 (page 187) describes a randomized comparative experiment in which 150 students are randomly assigned to six groups of 25.

(a) Use the applet to randomly choose 25 out of 150 students to form the first group. Which students are in this group?

(b) The population hopper now contains the 125 students that were not chosen, in scrambled order. Click "Sample" again to choose 25 of these remaining students to make up the second group. Which students were chosen?

(c) Click "Sample" three more times to choose the third, fourth, and fifth groups. Don't take the time to write down these groups. Check that there are only 25 students remaining in the population hopper. These subjects get Treatment 6. Which students are they?

**3.36** CHALLENGE **Effectiveness of price discounts.** Experiments with more than one factor allow insight into interactions between the factors. A study of the attractiveness of advertised price discounts had two factors: percent of all goods on sale (25%, 50%, 75%, or 100%) and whether the discount was stated precisely as 60% off or as a range, 50% to 70% off. Subjects rated the attractiveness of the sale on a scale of 1 to 7. Figure 3.7 shows the mean ratings for the eight treatments formed from the two factors.[22] Based on these results, write a careful description of how percent on sale and precise discount versus range of discounts influence the attractiveness of a sale.

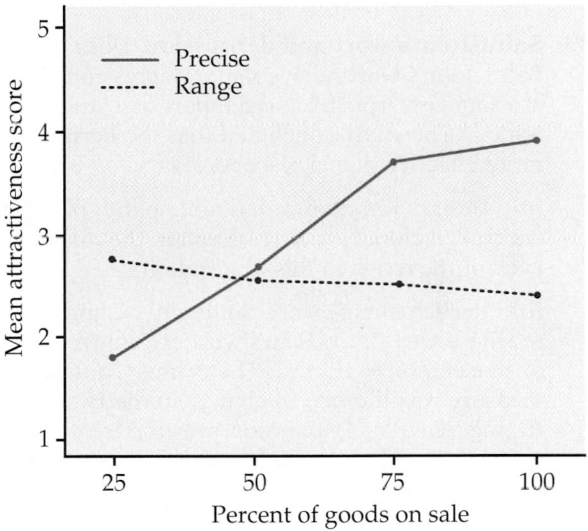


**FIGURE 3.7** Mean responses to eight treatments in an experiment with two factors, showing interaction between the factors, for Exercise 3.36.

**3.37** CHALLENGE **Health benefits of bee pollen.** "Bee pollen is effective for combating fatigue, depression, cancer, and colon disorders." So says a Web site that offers the pollen for sale. We wonder if bee pollen really does prevent colon disorders. Here are two ways to study this question. Explain why the first design will produce more trustworthy data.

1. Find 400 women who do not have colon disorders. Assign 200 to take bee pollen capsules and the other 200 to take placebo capsules that are identical in appearance. Follow both groups for 5 years.

2. Find 200 women who take bee pollen regularly. Match each with a woman of the same age, race, and occupation who does not take bee pollen. Follow both groups for 5 years.

**3.38 Treatment of pain for cancer patients.** Health care providers are giving more attention to relieving the pain of cancer patients. An article in the journal

*Cancer* surveyed a number of studies and concluded that controlled-release morphine tablets, which release the painkiller gradually over time, are more effective than giving standard morphine when the patient needs it.[23] The "methods" section of the article begins: "Only those published studies that were controlled (i.e., randomized, double blind, and comparative), repeated-dose studies with CR morphine tablets in cancer pain patients were considered for this review." Explain the terms in parentheses to someone who knows nothing about medical trials.

**3.39 Saint-John's-wort and depression.** Does the herb Saint-John's-wort relieve major depression? Here are some excerpts from the report of a study of this issue.[24] The study concluded that the herb is no more effective than a placebo.

(a) "Design: Randomized, double-blind, placebo-controlled clinical trial...." Explain the meaning of each of the terms in this description.

(b) "Participants ...were randomly assigned to receive either Saint-John's-wort extract ($n = 98$) or placebo ($n = 102$).... The primary outcome measure was the rate of change in the Hamilton Rating Scale for Depression over the treatment period." Based on this information, use a diagram to outline the design of this clinical trial.

**3.40 The Monday effect on stock prices.** Puzzling but true: stocks tend to go down on Mondays. There is no convincing explanation for this fact. A recent study looked at this "Monday effect" in more detail, using data on the daily returns of stocks on several U.S. exchanges over a 30-year period. Here are some of the findings:

*To summarize, our results indicate that the well-known Monday effect is caused largely by the Mondays of the last two weeks of the month. The mean Monday return of the first three weeks of the month is, in general, not significantly different from zero and is generally significantly higher than the mean Monday return of the last two weeks. Our finding seems to make it more difficult to explain the Monday effect.*[25]

A friend thinks that "significantly" in this article has its plain English meaning, roughly "I think this is important." Explain in simple language what "significantly higher" and "not significantly different from zero" actually tell us here.

**3.41 Five-digit zip codes and delivery time of mail.** Does adding the five-digit postal zip code to an address really speed up delivery of letters? Does adding the four more digits that make up "zip + 4" speed delivery yet more? What about mailing a letter on Monday, Thursday, or Saturday? Describe the design of an experiment on the speed of first-class mail delivery. For simplicity, suppose that all letters go from you to a friend, so that the sending and receiving locations are fixed.

**3.42** APPLET **Use the simple random sample applet.** The *Simple Random Sample* applet can demonstrate how randomization works to create similar groups for comparative experiments. Suppose that (unknown to the experimenters) the 20 even-numbered students among the 40 subjects for the cell phone study in Example 3.11 (page 185) have fast reactions, and that the odd-numbered students have slow reactions. We would like the experimental and control groups to contain similar numbers of the fast reactors. Use the applet to choose 10 samples of size 20 from the 40 students. (Be sure to click "Reset" after each sample.) Record the counts of even-numbered students in each of your 10 samples. You see that there is considerable chance variation but no systematic bias in favor of one or the other group in assigning the fast-reacting students. Larger samples from larger populations will on the average do a better job of making the two groups equivalent.

**3.43 Does oxygen help football players?** We often see players on the sidelines of a football game inhaling oxygen. Their coaches think this will speed their recovery. We might measure recovery from intense exercise as follows: Have a football player run 100 yards three times in quick succession. Then allow three minutes to rest before running 100 yards again. Time the final run. Because players vary greatly in speed, you plan a matched pairs experiment using 20 football players as subjects. Describe the design of such an experiment to investigate the effect of inhaling oxygen during the rest period. Why should each player's two trials be on different days? Use Table B at line 140 to decide which players will get oxygen on their first trial.

**3.44 Carbon dioxide in the atmosphere.** The concentration of carbon dioxide ($CO_2$) in the atmosphere is increasing rapidly due to our use of fossil fuels. Because plants use $CO_2$ to fuel photosynthesis, more $CO_2$ may cause trees and other plants to grow faster. An elaborate apparatus allows researchers to pipe extra $CO_2$ to a 30-meter circle of forest. We want to compare the growth in base area of trees in treated and untreated areas to see if extra $CO_2$ does in fact increase growth. We can afford to treat 3 circular areas.[26]

(a) Describe the design of a completely randomized experiment using 6 well-separated 30-meter circular areas in a pine forest. Sketch the forest area with the 6 circles and carry out the randomization your design calls for.

(b) Regions within the forest may differ in soil fertility. Describe a matched pairs design using three pairs of circles that will reduce the extra variation due to different fertility. Sketch the forest area with the new arrangement of circles and carry out the randomization your design calls for.

3.45 CHALLENGE **Calcium and the bones of young girls.** Calcium is important to the bone development of young girls. To study how the bodies of young girls process calcium, investigators used the setting of a summer camp. Calcium was given in Hawaiian Punch at either a high or a low level. The camp diet was otherwise the same for all girls. Suppose that there are 50 campers.

(a) Outline a completely randomized design for this experiment.

(b) Describe a matched pairs design in which each girl receives both levels of calcium (with a "washout period" between). What is the advantage of the matched pairs design over the completely randomized design?

(c) The same randomization can be used in different ways for both designs. Label the subjects 01 to 50. You must choose 25 of the 50. Use Table B at line 110 to choose just the first 5 of the 25. How are the 25 subjects chosen treated in the completely randomized design? How are they treated in the matched pairs design?

3.46 CHALLENGE **Random digits.** Table B is a table of random digits. Which of the following statements are true of a table of random digits, and which are false? Explain your answers.

(a) There are exactly four 0s in each row of 40 digits.

(b) Each pair of digits has chance 1/100 of being 00.

(c) The digits 0000 can never appear as a group, because this pattern is not random.

3.47 **Vitamin C for ultramarathon runners.** An ultramarathon, as you might guess, is a footrace longer than the 26.2 miles of a marathon. Runners commonly develop respiratory infections after an ultramarathon. Will taking 600 milligrams of vitamin C daily reduce these infections? Researchers randomly assigned ultramarathon runners to receive either vitamin C or a placebo. Separately, they also randomly assigned these treatments to a group of nonrunners the same age as the runners. All subjects were watched for 14 days after the big race to see if infections developed.[27]

(a) What is the name for this experimental design?

(b) Use a diagram to outline the design.

(c) The report of the study said:

*Sixty-eight percent of the runners in the placebo group reported the development of symptoms of upper respiratory tract infection after the race; this was significantly more than that reported by the vitamin C–supplemented group (33%).*

Explain to someone who knows no statistics why "significantly more" means there is good reason to think that vitamin C works.

## 3.2 Sampling Design

A political scientist wants to know what percent of college-age adults consider themselves conservatives. An automaker hires a market research firm to learn what percent of adults aged 18 to 35 recall seeing television advertisements for a new sport utility vehicle. Government economists inquire about average household income. In all these cases, we want to gather information about a large group of individuals. We will not, as in an experiment, impose a treatment in order to observe the response. Also, time, cost, and inconvenience forbid contacting every individual. In such cases, we gather information about only part of the group—a *sample*—in order to draw conclusions about the whole. **Sample surveys** are an important kind of observational study.

sample survey

**POPULATION AND SAMPLE**

The entire group of individuals that we want information about is called the **population.**

A **sample** is a part of the population that we actually examine in order to gather information.

Notice that "population" is defined in terms of our desire for knowledge. If we wish to draw conclusions about all U.S. college students, that group is our population even if only local students are available for questioning. The sample is the part from which we draw conclusions about the whole. The **design** of a sample survey refers to the method used to choose the sample from the population.

sample design

EXAMPLE

**3.21 The Reading Recovery program.** The Reading Recovery (RR) program has specially trained teachers work one-on-one with at-risk first-grade students to help them learn to read. A study was designed to examine the relationship between the RR teachers' beliefs about their ability to motivate students and the progress of the students whom they teach.[28] The National Data Evaluation Center (NDEC) Web site (`www.ndec.us`) says that there are 13,823 RR teachers. The researchers send a questionnaire to a random sample of 200 of these. The population consists of all 13,823 RR teachers, and the sample is the 200 that were randomly selected.

Unfortunately, our idealized framework of population and sample does not exactly correspond to the situations that we face in many cases. In Example 3.21, the list of teachers was prepared at a particular time in the past. It is very likely that some of the teachers on the list are no longer working as RR teachers today. New teachers have been trained in RR methods and are not on the list. In spite of these difficulties, we still view the list as the population. Also, we do not expect to get a response from every teacher in our random sample. We may have out-of-date addresses for some who are still working as RR teachers, and some teachers may choose not to respond to our survey questions.

In reporting the results of a sample survey it is important to include all details regarding the procedures used. Follow-up mailings or phone calls to those who do not initially respond can help increase the response rate. The proportion of the original sample who actually provide usable data is called the **response rate** and should be reported for all surveys. If only 150 of the teachers who were sent questionnaires provided usable data, the response rate would be 150/200, or 75%.

response rate

**USE YOUR KNOWLEDGE**

**3.48 Job satisfaction in Mongolian universities.** A educational research team wanted to examine the relationship between faculty participation in decision making and job satisfaction in Mongolian public universities. They are planning to randomly select 300 faculty members from a list of 2500 faculty members in these universities. The Job Descriptive Index (JDI) will be used to measure job satisfaction, and the Conway Adaptation of the Alutto-Belasco Decisional Participation Scale will be used to measure decision participation. Describe the population and the sample for this study. Can you determine the response rate?

**3.49 Taxes and forestland usage.** A study was designed to assess the impact of taxes on forestland usage in part of the Upper Wabash River Watershed in Indiana.[29] A survey was sent to 772 forest owners from this region and 348 were returned. Consider the population, the sample, and the response rate for this study. Describe these based on the information given and indicate any additional information that you would need to give a complete answer.

Poor sample designs can produce misleading conclusions. Here is an example.

EXAMPLE

**3.22 Sampling pieces of steel.** A mill produces large coils of thin steel for use in manufacturing home appliances. The quality engineer wants to submit a sample of 5-centimeter squares to detailed laboratory examination. She asks a technician to cut a sample of 10 such squares. Wanting to provide "good" pieces of steel, the technician carefully avoids the visible defects in the coil material when cutting the sample. The laboratory results are wonderful but the customers complain about the material they are receiving.

Online opinion polls are particularly vulnerable to bias because the sample who respond are not representative of the population at large. Here is an example that also illustrates how the results of such polls can be manipulated.

EXAMPLE

**3.23 The American Family Association.** The American Family Association (AFA) is a conservative group that claims to stand for "traditional family values." It regularly posts online poll questions on its Web site—just click on a response to take part. Because the respondents are people who visit this site, the poll results always support AFA's positions. Well, almost always. In 2004, AFA's online poll asked about the heated issue of allowing same-sex marriage. Soon, email lists and social-network sites favored mostly by young liberals pointed to the AFA poll. Almost 850,000 people responded, and 60% of them favored legalization of same-sex marriage. AFA claimed that homosexual rights groups had skewed its poll.

As the AFA poll illustrates, you can't always trust poll results. People who take the trouble to respond to an open invitation are not representative of the entire adult population. That's true of regular visitors to AFA's site, of the activists who made a special effort to vote in the marriage poll, and of the people who bother to respond to write-in, call-in, or online polls in general.

In both Examples 3.22 and 3.23, the sample was selected in a manner that guaranteed that it would not be representative of the entire population. These sampling schemes display *bias,* or systematic error, in favoring some parts of the population over others. Online polls use *voluntary response samples,* a particularly common form of biased sample.

### VOLUNTARY RESPONSE SAMPLE

A **voluntary response sample** consists of people who choose themselves by responding to a general appeal. Voluntary response samples are biased because people with strong opinions, especially negative opinions, are most likely to respond.

The remedy for bias in choosing a sample is to allow impersonal chance to do the choosing, so that there is neither favoritism by the sampler (as in Example 3.22) nor voluntary response (as in Example 3.23). Random selection of a sample eliminates bias by giving all individuals an equal chance to be chosen, just as randomization eliminates bias in assigning experimental subjects.

## Simple random samples

The simplest sampling design amounts to placing names in a hat (the population) and drawing out a handful (the sample). This is *simple random sampling.*

### SIMPLE RANDOM SAMPLE

A **simple random sample (SRS)** of size $n$ consists of $n$ individuals from the population chosen in such a way that every set of $n$ individuals has an equal chance to be the sample actually selected.

Each treatment group in a completely randomized experimental design is an SRS drawn from the available experimental units. We select an SRS by labeling all the individuals in the population and using software or a table of random digits to select a sample of the desired size, just as in experimental randomization. Notice that an SRS not only gives each individual an equal chance to be chosen (thus avoiding bias in the choice) but gives every possible sample an equal chance to be chosen. There are other random sampling designs that give each individual, but not each sample, an equal chance. One such design, systematic random sampling, is described in Exercise 3.64.

EXAMPLE

**3.24 Spring break destinations.** A campus newspaper plans a major article on spring break destinations. The authors intend to call a few randomly chosen resorts at each destination to ask about their attitudes toward groups of students as guests. Here are the resorts listed in one city. The first step is to label the members of this population as shown.

| | | | | | | | |
|---|---|---|---|---|---|---|---|
| 01 | Aloha Kai | 08 | Captiva | 15 | Palm Tree | 22 | Sea Shell |
| 02 | Anchor Down | 09 | Casa del Mar | 16 | Radisson | 23 | Silver Beach |
| 03 | Banana Bay | 10 | Coconuts | 17 | Ramada | 24 | Sunset Beach |
| 04 | Banyan Tree | 11 | Diplomat | 18 | Sandpiper | 25 | Tradewinds |
| 05 | Beach Castle | 12 | Holiday Inn | 19 | Sea Castle | 26 | Tropical Breeze |
| 06 | Best Western | 13 | Lime Tree | 20 | Sea Club | 27 | Tropical Shores |
| 07 | Cabana | 14 | Outrigger | 21 | Sea Grape | 28 | Veranda |

Now enter Table B, and read two-digit groups until you have chosen three resorts. If you enter at line 185, Banana Bay (03), Palm Tree (15), and Cabana (07) will be called.

Most statistical software will select an SRS for you, eliminating the need for Table B. The *Simple Random Sample* applet on the text CD and Web site is a convenient way to automate this task.

Excel can do the job in a way similar to what we used when we randomized experimental units to treatments in designed experiments. There are four steps:

1. Create a data set with all of the elements of the population in the first column.

2. Assign a random number to each element of the population; put these in the second column.

3. Sort the data set by the random number column.

4. The simple random sample is obtained by taking elements in the sorted list until the desired sample size is reached.

We illustrate the procedure with a simplified version of Example 3.24.

EXAMPLE

**3.25 Select a random sample.** Suppose that the population from Example 3.24 is only the first two rows of the display given there:

| | | | |
|---|---|---|---|
| Aloha Kai | Captiva | Palm Tree | Sea Shell |
| Anchor Down | Casa del Mar | Radisson | Silver Beach |

Note that we do not need the numerical labels to identify the individuals in the population. Suppose that we want to select a simple random sample of three resorts from this population. Figure 3.8(a) gives the spreadsheet with the population names. The random numbers generated by the RAND() function are given in the second column in Figure 3.8(b). The sorted data set is given in Figure 3.8(c). We have added a third column to the speadsheet to indicate which resorts were selected for our random sample. They are Captiva, Radisson, and Silver Beach.

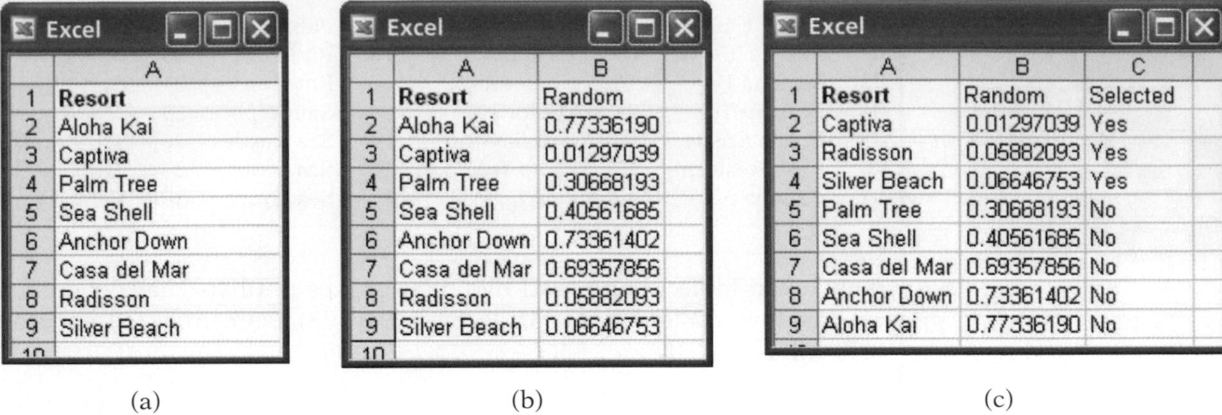

| | A |
|---|---|
| 1 | **Resort** |
| 2 | Aloha Kai |
| 3 | Captiva |
| 4 | Palm Tree |
| 5 | Sea Shell |
| 6 | Anchor Down |
| 7 | Casa del Mar |
| 8 | Radisson |
| 9 | Silver Beach |
| 10 | |

(a)

| | A | B |
|---|---|---|
| 1 | **Resort** | Random |
| 2 | Aloha Kai | 0.77336190 |
| 3 | Captiva | 0.01297039 |
| 4 | Palm Tree | 0.30668193 |
| 5 | Sea Shell | 0.40561685 |
| 6 | Anchor Down | 0.73361402 |
| 7 | Casa del Mar | 0.69357856 |
| 8 | Radisson | 0.05882093 |
| 9 | Silver Beach | 0.06646753 |
| 10 | | |

(b)

| | A | B | C |
|---|---|---|---|
| 1 | **Resort** | Random | Selected |
| 2 | Captiva | 0.01297039 | Yes |
| 3 | Radisson | 0.05882093 | Yes |
| 4 | Silver Beach | 0.06646753 | Yes |
| 5 | Palm Tree | 0.30668193 | No |
| 6 | Sea Shell | 0.40561685 | No |
| 7 | Casa del Mar | 0.69357856 | No |
| 8 | Anchor Down | 0.73361402 | No |
| 9 | Aloha Kai | 0.77336190 | No |

(c)

**FIGURE 3.8** Selection of a simple random sample of resorts, for Example 3.25.

## USE YOUR KNOWLEDGE

**3.50 Ringtones for cell phones.** You decide to change the ringtones for your cell phone by choosing 2 from a list of the 10 most popular ringtones.[30] Here is the list:

| | | | |
|---|---|---|---|
| Super Mario Brothers Theme | Sexy Love | Ms. New Booty | Ridin' Rims |
| I Write Sins Not Tragedies | Gasolina | My Humps | The Pink Panther |
| Down | Agarrala | | |

Select your two ringtones using a simple random sample.

**3.51 Listen to three songs.** The walk to your statistics class takes about 10 minutes, about the amount of time needed to listen to three songs on your iPod. You decide to take a simple random sample of songs from a Billboard list of Rock Songs.[31] Here is the list:

| | | | |
|---|---|---|---|
| Miss Murder | Animal I Have Become | Steady, As She Goes | Dani California |
| The Kill (Bury Me) | Original Fire | When You Were Young | MakeD—Sure |
| Vicarious | The Diary of Jane | | |

Select the three songs for your iPod using a simple random sample.

## Stratified samples

The general framework for designs that use chance to choose a sample is a *probability sample.*

### PROBABILITY SAMPLE

A **probability sample** is a sample chosen by chance. We must know what samples are possible and what chance, or probability, each possible sample has.

Some probability sampling designs (such as an SRS) give each member of the population an *equal* chance to be selected. This may not be true in more elaborate sampling designs. In every case, however, the use of chance to select the sample is the essential principle of statistical sampling.

Designs for sampling from large populations spread out over a wide area are usually more complex than an SRS. For example, it is common to sample important groups within the population separately, then combine these samples. This is the idea of a *stratified sample.*

### STRATIFIED RANDOM SAMPLE

To select a **stratified random sample,** first divide the population into groups of similar individuals, called **strata.** Then choose a separate SRS in each stratum and combine these SRSs to form the full sample.

Choose the strata based on facts known before the sample is taken. For example, a population of election districts might be divided into urban, suburban, and rural strata. A stratified design can produce more exact information than an SRS of the same size by taking advantage of the fact that individuals in the same stratum are similar to one another. Think of the extreme case in which all individuals in each stratum are identical: just one individual from each stratum is then enough to completely describe the population. Strata for sampling are similar to blocks in experiments. We have two names because the idea of grouping similar units before randomizing arose separately in sampling and in experiments.

EXAMPLE

**3.26 A stratified sample of dental claims.** A dentist is suspected of defrauding insurance companies by describing some dental procedures incorrectly on claim forms and overcharging for them. An investigation begins by examining a sample of his bills for the past three years. Because there are five suspicious types of procedures, the investigators take a stratified sample. That is, they randomly select bills for each of the five types of procedures separately.

## Multistage samples

Another common means of restricting random selection is to choose the sample in stages. This is common practice for national samples of households or people. For example, data on employment and unemployment are gathered by the government's Current Population Survey, which conducts interviews in about 60,000 households each month. The cost of sending interviewers to the widely scattered households in an SRS would be too high. Moreover, the government wants data broken down by states and large cities. The Current Population Survey therefore uses a **multistage sampling design.** The final sample consists of clusters of nearby households that an interviewer can easily visit.

multistage sample

Most opinion polls and other national samples are also multistage, though interviewing in most national samples today is done by telephone rather than in person, eliminating the economic need for clustering. The Current Population Survey sampling design is roughly as follows:[32]

Stage 1. Divide the United States into 2007 geographical areas called Primary Sampling Units, or PSUs. PSUs do not cross state lines. Select a sample of 754 PSUs. This sample includes the 428 PSUs with the largest population and a stratified sample of 326 of the others.

Stage 2. Divide each PSU selected into smaller areas called "blocks." Stratify the blocks using ethnic and other information and take a stratified sample of the blocks in each PSU.

Stage 3. Sort the housing units in each block into clusters of four nearby units. Interview the households in a probability sample of these clusters.

Analysis of data from sampling designs more complex than an SRS takes us beyond basic statistics. But the SRS is the building block of more elaborate designs, and analysis of other designs differs more in complexity of detail than in fundamental concepts.

## Cautions about sample surveys

Random selection eliminates bias in the choice of a sample from a list of the population. Sample surveys of large human populations, however, require much more than a good sampling design.[33] To begin, we need an accurate and complete list of the population. Because such a list is rarely available, most samples suffer from some degree of *undercoverage.* A sample survey of households, for example, will miss not only homeless people but prison inmates and students in dormitories. An opinion poll conducted by telephone will miss the 6% of American households without residential phones. The results of national sample surveys therefore have some bias if the people not covered—who most often are poor people—differ from the rest of the population.

A more serious source of bias in most sample surveys is *nonresponse,* which occurs when a selected individual cannot be contacted or refuses to cooperate. Nonresponse to sample surveys often reaches 50% or more, even with careful planning and several callbacks. Because nonresponse is higher in urban areas, most sample surveys substitute other people in the same area to avoid favoring rural areas in the final sample. If the people contacted differ from those who are rarely at home or who refuse to answer questions, some bias remains.

### UNDERCOVERAGE AND NONRESPONSE

**Undercoverage** occurs when some groups in the population are left out of the process of choosing the sample.

**Nonresponse** occurs when an individual chosen for the sample can't be contacted or does not cooperate.

EXAMPLE

**3.27 Nonresponse in the Current Population Survey.** How bad is nonresponse? The Current Population Survey (CPS) has the lowest nonresponse rate of any poll we know: only about 4% of the households in the CPS sample refuse to take part and another 3% or 4% can't be contacted. People are more likely to respond to a government survey such as the CPS, and the CPS contacts its sample in person before doing later interviews by phone.

The General Social Survey (Figure 3.9) is the nation's most important social science research survey. The GSS also contacts its sample in person, and it is run by a university. Despite these advantages, its most recent survey had a 30% rate of nonresponse.

What about polls done by the media and by market research and opinion-polling firms? We don't know their rates of nonresponse, because they won't say. That itself is a bad sign. The Pew Research Center for People and the Press designed a careful telephone survey and published the results: out of 2879 households called, 1658 were never at home, refused, or would not finish the interview. That's a nonresponse rate of 58%.[34]

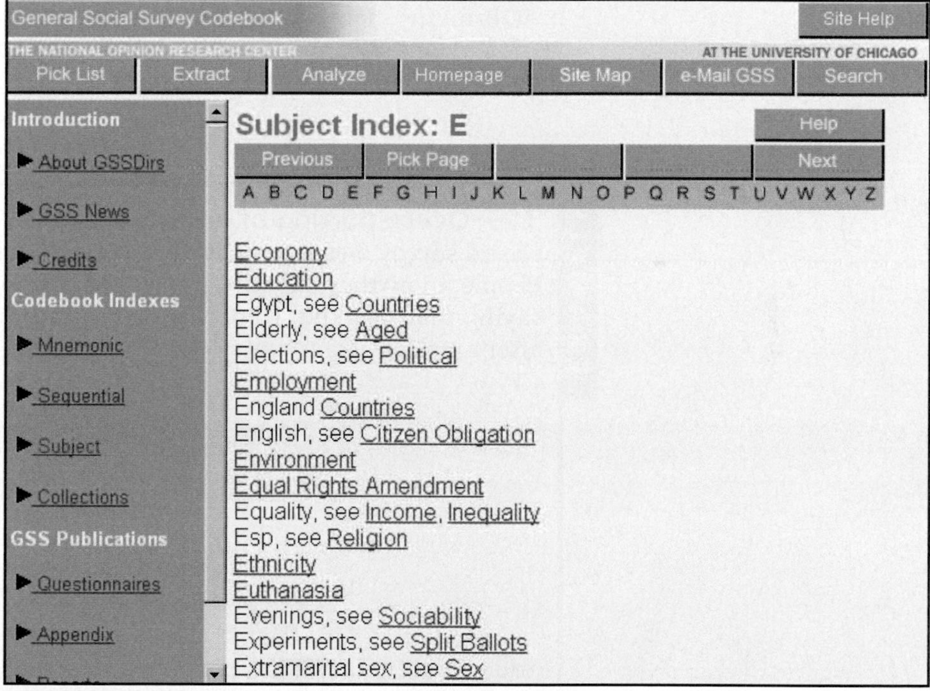


**FIGURE 3.9** Part of the subject index for the General Social Survey (GSS). The GSS has assessed attitudes on a wide variety of topics since 1972. Its continuity over time makes the GSS a valuable source for studies of changing attitudes.

Most sample surveys, and almost all opinion polls, are now carried out by telephone. This and other details of the interview method can affect the results.

EXAMPLE

**3.28 How should the data be collected?** A Pew Research Center Poll has asked about belief in God for many years. In response to the statement "I never doubt the existence of God," subjects are asked to choose from the responses

completely agree   mostly agree   mostly disagree   completely disagree

In 1990, subjects were interviewed in person and were handed a card with the four responses on it. In 1991, the poll switched to telephone interviews. In 1990, 60% said "completely agree," in line with earlier years. In 1991, 71% completely agreed. The increase is probably explained by the effect of hearing "completely agree" read first by the interviewer.[35]

response bias

The behavior of the respondent or of the interviewer can cause **response bias** in sample results. Respondents may lie, especially if asked about illegal or unpopular behavior. The race or sex of the interviewer can influence responses to questions about race relations or attitudes toward feminism. Answers to questions that ask respondents to recall past events are often inaccurate because of faulty memory. For example, many people "telescope" events in the past, bringing them forward in memory to more recent time periods. "Have you visited a dentist in the last 6 months?" will often elicit a "Yes" from someone who last visited a dentist 8 months ago.[36]

EXAMPLE

**3.29 Overreporting of voter behavior.** "One of the most frequently observed survey measurement errors is the overreporting of voting behavior."[37] People know they should vote, so those who didn't vote tend to save face by saying that they did. Here are the data from a typical sample of 663 people after an election:

| | | What they said: | |
|---|---|---|---|
| | | I voted | I didn't |
| What they did: | Voted | 358 | 13 |
| | Didn't vote | 120 | 172 |

You can see that 478 people (72%) said that they voted, but only 371 people (56%) actually did vote.

wording of questions

The **wording of questions** is the most important influence on the answers given to a sample survey. Confusing or leading questions can introduce strong bias, and even minor changes in wording can change a survey's outcome. Here are some examples.

EXAMPLE

**3.30 The form of the question is important.** In response to the question "Are you heterosexual, homosexual, or bisexual?" in a social science research survey, one woman answered, "It's just me and my husband, so bisexual." The issue is serious, even if the example seems silly: reporting about sexual behavior is difficult because people understand and misunderstand sexual terms in many ways.

How do Americans feel about government help for the poor? Only 13% think we are spending too much on "assistance to the poor," but 44% think we are spending too much on "welfare." How do the Scots feel about the movement to become independent from England? Well, 51% would vote for "independence for Scotland," but only 34% support "an independent Scotland separate from the United Kingdom." It seems that "assistance to the poor" and "independence" are nice, hopeful words. "Welfare" and "separate" are negative words.[38]

CAUTION

The statistical design of sample surveys is a science, but this science is only part of the art of sampling. Because of nonresponse, response bias, and the difficulty of posing clear and neutral questions, you should hesitate to fully trust reports about complicated issues based on surveys of large human populations. *Insist on knowing the exact questions asked, the rate of nonresponse, and the date and method of the survey before you trust a poll result.*

## SECTION 3.2 Summary

A sample survey selects a **sample** from the **population** of all individuals about which we desire information. We base conclusions about the population on data about the sample.

The **design** of a sample refers to the method used to select the sample from the population. **Probability sampling designs** use impersonal chance to select a sample.

The basic probability sample is a **simple random sample (SRS).** An SRS gives every possible sample of a given size the same chance to be chosen.

Choose an SRS by labeling the members of the population and using a **table of random digits** to select the sample. Software can automate this process.

To choose a **stratified random sample,** divide the population into **strata,** groups of individuals that are similar in some way that is important to the response. Then choose a separate SRS from each stratum and combine them to form the full sample.

**Multistage samples** select successively smaller groups within the population in stages, resulting in a sample consisting of clusters of individuals. Each stage may employ an SRS, a stratified sample, or another type of sample.

Failure to use probability sampling often results in **bias,** or systematic errors in the way the sample represents the population. **Voluntary response** samples, in which the respondents choose themselves, are particularly prone to large bias.

In human populations, even probability samples can suffer from bias due to **undercoverage** or **nonresponse,** from **response bias** due to the behavior of the interviewer or the respondent, or from misleading results due to **poorly worded questions.**

## SECTION 3.2 Exercises

*For Exercises 3.48 and 3.49, see pages 198 and 199; and for Exercises 3.50 and 3.51, see page 202.*

**3.52 What's wrong?** Explain what is wrong in each of the following scenarios.

(a) The population consists of all individuals selected in a simple random sample.

(b) In a poll of an SRS of residents in a local community, respondents are asked to indicate the level of their concern about the dangers of dihydrogen monoxide, a substance that is a major component of acid rain and in its gaseous state can cause severe burns. (*Hint:* Ask a friend who is majoring in chemistry about this substance or search the Internet for information about it.)

(c) Students in a class are asked to raise their hands if they have cheated on an exam one or more times within the past year.

**3.53 What's wrong?** Explain what is wrong with each of the following random selection procedures and explain how you would do the randomization correctly.

(a) To determine the reading level of an introductory statistics text, you evaluate all of the written material in the third chapter.

(b) You want to sample student opinions about a proposed change in procedures for changing majors. You hand out questionnaires to 100 students as they arrive for class at 7:30 A.M.

(c) A population of subjects is put in alphabetical order and a simple random sample of size 10 is taken by selecting the first 10 subjects in the list.

**3.54 Importance of students as customers.** A committee on community relations in a college town plans to survey local businesses about the importance of students as customers. From telephone book listings, the committee chooses 150 businesses at random. Of these, 73 return the questionnaire mailed by the committee. What is the population for this sample survey? What is the sample? What is the rate (percent) of nonresponse?

**3.55 Popularity of news personalities.** A Gallup Poll conducted telephone interviews with 1001 U.S. adults aged 18 and over on July 24–27, 2006. One of the questions asked whether the respondents had a favorable or an unfavorable opinion of 17 news personalities. Diane Sawyer received the highest rating, with 80% of the respondents giving her a favorable rating.[39]

(a) What is the population for this sample survey? What was the sample size?

(b) The report on the survey states that 8% of the respondents either never heard of Sawyer or had no opinion about her. When they included only those who provided an opinion, Sawyer's approval percent rose to 88% and she was still at the top of the list. Charles Gibson, on the other hand, was ranked eighth on the original list, with a 55% favorable rating. When only those providing an opinion were counted, his rank rose to second, with 87% approving. Discuss the advantages and disadvantages of the two different ways of reporting the approval percent. State which one you prefer and why.

**3.56 Identify the populations.** For each of the following sampling situations, identify the population as exactly as possible. That is, say what kind of individuals the population consists of and say exactly which individuals fall in the population. If the information given is not complete, complete the description of the population in a reasonable way.

(a) A college has changed its core curriculum and wants to obtain detailed feedback information from the students during each of the first 12 weeks of the coming semester. Each week, a random sample of 5 students will be selected to be interviewed.

(b) The American Community Survey (ACS) will replace the census "long form" starting with the 2010 census. The main part of the ACS contacts 250,000 addresses by mail each month, with follow-up by phone and in person if there is no response. Each household answers questions about their housing, economic, and social status.

(c) An opinion poll contacts 1161 adults and asks them, "Which political party do you think has better ideas for leading the country in the twenty-first century?"

**3.57 Interview residents of apartment complexes.** You are planning a report on apartment living in a college town. You decide to select 5 apartment complexes at random for in-depth interviews with residents. Select a simple random sample of 5 of the following apartment complexes. If you use Table B, start at line 137.

| | | |
|---|---|---|
| Ashley Oaks | Country View | Mayfair Village |
| Bay Pointe | Country Villa | Nobb Hill |
| Beau Jardin | Crestview | Pemberly Courts |
| Bluffs | Del-Lynn | Peppermill |
| Brandon Place | Fairington | Pheasant Run |
| Briarwood | Fairway Knolls | Richfield |
| Brownstone | Fowler | Sagamore Ridge |
| Burberry | Franklin Park | Salem Courthouse |
| Cambridge | Georgetown | Village Manor |
| Chauncey Village | Greenacres | Waterford Court |
| Country Squire | Lahr House | Williamsburg |

**3.58 Using GIS to identify mint field conditions.** A Geographic Information System (GIS) is to be used to distinguish different conditions in mint fields. Ground observations will be used to classifiy regions of each field as either healthy mint, diseased mint, or weed-infested mint. The GIS divides mint-growing areas into regions called pixels. An experimental area contains 200 pixels. For a random sample of 25 pixels, ground measurements will be made to determine the status of the mint, and these observations will be compared with information obtained by the GIS. Select the random sample. If you use Table B, start at line 112 and choose only the first 5 pixels in the sample.

**3.59** APPLET **Use the simple random sample applet.** After you have labeled the individuals in a population, the *Simple Random Sample* applet automates the task of choosing an SRS. Use the applet to choose the sample in the previous exercise.

**3.60** APPLET **Use the simple random sample applet.** There are approximately 371 active telephone area codes covering Canada, the United States, and some Caribbean areas. (More are created regularly.) You want to choose an SRS of 25 of these area codes for a study of available telephone numbers. Label the codes 001 to 371 and use the *Simple Random Sample* applet to choose your sample. (If you use Table B, start at line 120 and choose only the first 5 codes in the sample.)

**3.61 Census tracts.** The Census Bureau divides the entire country into "census tracts" that contain about 4000 people. Each tract is in turn divided into small "blocks," which in urban areas are bounded by local streets. An SRS of blocks from a census tract is often the next-to-last stage in a multistage sample. Figure 3.10 shows part of census tract 8051.12, in Cook County, Illinois, west of Chicago. The 44 blocks in this tract are divided into three "block groups." Group 1 contains 6 blocks numbered 1000 to 1005;

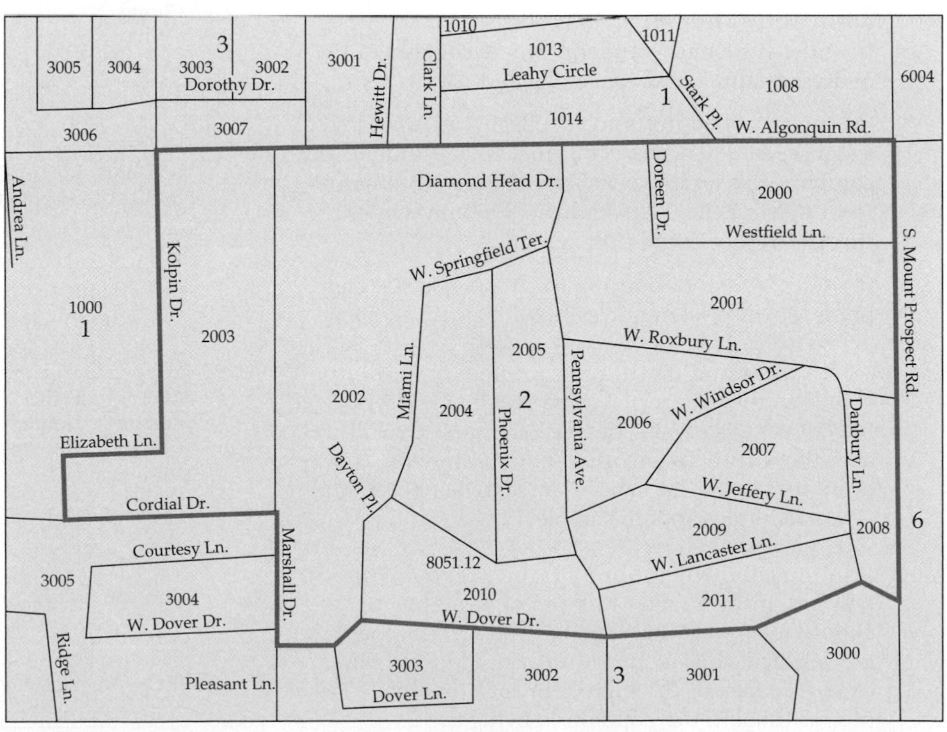


**FIGURE 3.10** Census blocks in Cook County, Illinois, for Exercises 3.61 and 3.63. The outlined area is a block group.

Group 2 (outlined in Figure 3.10) contains 12 blocks numbered 2000 to 2011; Group 3 contains 26 blocks numbered 3000 to 3025. Use Table B, beginning at line 135, to choose an SRS of 5 of the 44 blocks in this census tract. Explain carefully how you labeled the blocks.

**3.62 Repeated use of Table B.** In using Table B repeatedly to choose samples or do randomization for experiments, you should not always begin at the same place, such as line 101. Why not?

**3.63 A stratified sample.** Exercise 3.61 asks you to choose an SRS of blocks from the census tract pictured in Figure 3.10. You might instead choose a stratified sample of one block from the 6 blocks in Group 1, two from the 12 blocks in Group 2, and three from the 26 blocks in Group 3. Choose such a sample, explaining carefully how you labeled blocks and used Table B.

**3.64 Systematic random samples. Systematic random samples** are often used to choose a sample of apartments in a large building or dwelling units in a block at the last stage of a multistage sample. An example will illustrate the idea of a systematic sample. Suppose that we must choose 4 addresses out of 100. Because 100/4 = 25, we can think of the list as four lists of 25 addresses. Choose 1 of the first 25 at random, using Table B. The sample contains this address and the addresses 25, 50, and 75 places down the list from it. If 13 is chosen, for example, then the systematic random sample consists of the addresses numbered 13, 38, 63, and 88.

(a) A study of dating among college students wanted a sample of 200 of the 9000 single male students on campus. The sample consisted of every 45th name from a list of the 9000 students. Explain why the survey chooses every 45th name.

(b) Use Table B at line 125 to choose the starting point for this systematic sample.

**3.65** CHALLENGE **Systematic random samples versus simple random samples.** The previous exercise introduces systematic random samples. Explain carefully why a systematic random sample *does* give every individual the same chance to be chosen but is *not* a simple random sample.

**3.66 Random digit telephone dialing.** An opinion poll in California uses random digit dialing to choose telephone numbers at random. Numbers are selected separately within each California area code. The size of the sample in each area code is proportional to the population living there.

(a) What is the name for this kind of sampling design?

(b) California area codes, in rough order from north to south, are

| | | | | | | | | | | | | |
|---|---|---|---|---|---|---|---|---|---|---|---|---|
| 530 | 707 | 916 | 209 | 415 | 925 | 510 | 650 | 408 | 831 | 805 | 559 | 760 |
| 661 | 818 | 213 | 626 | 323 | 562 | 709 | 310 | 949 | 909 | 858 | 619 | |

Another California survey does not call numbers in all area codes but starts with an SRS of 10 area codes. Choose such an SRS. If you use Table B, start at line 122.

**3.67 Stratified samples of forest areas.** Stratified samples are widely used to study large areas of forest. Based on satellite images, a forest area in the Amazon basin is divided into 14 types. Foresters studied the four most commercially valuable types: alluvial climax forests of quality levels 1, 2, and 3, and mature secondary forest. They divided the area of each type into large parcels, chose parcels of each type at random, and counted tree species in a 20- by 25-meter rectangle randomly placed within each parcel selected. Here is some detail:

| Forest type | Total parcels | Sample size |
|---|---|---|
| Climax 1 | 36 | 4 |
| Climax 2 | 72 | 7 |
| Climax 3 | 31 | 3 |
| Secondary | 42 | 4 |

Choose the stratified sample of 18 parcels. Be sure to explain how you assigned labels to parcels. If you use Table B, start at line 140.

**3.68 Select club members to go to a convention.** A club has 30 student members and 10 faculty members. The students are

| | | | | |
|---|---|---|---|---|
| Abel | Fisher | Huber | Moran | Reinmann |
| Carson | Golomb | Jimenez | Moskowitz | Santos |
| Chen | Griswold | Jones | Neyman | Shaw |
| David | Hein | Kiefer | O'Brien | Thompson |
| Deming | Hernandez | Klotz | Pearl | Utts |
| Elashoff | Holland | Liu | Potter | Vlasic |

and the faculty members are

| | | | | |
|---|---|---|---|---|
| Andrews | Fernandez | Kim | Moore | Rabinowitz |
| Besicovitch | Gupta | Lightman | Phillips | Yang |

The club can send 5 students and 3 faculty members to a convention and decides to choose those who will go by random selection. Select a stratified random sample of 5 students and 3 faculty members.

**3.69** CHALLENGE **Stratified samples for alcohol attitudes.** At a party there are 30 students over age 21 and 20 students under age 21. You choose at random 3 of those over 21 and separately choose at random 2 of those under 21 to interview about attitudes toward alcohol. You have given every student at the party the same chance to be interviewed: what is that chance? Why is your sample not an SRS?

**3.70** **Stratified samples for accounting audits.** Accountants use stratified samples during audits to verify a company's records of such things as accounts receivable. The stratification is based on the dollar amount of the item and often includes 100% sampling of the largest items. One company reports 5000 accounts receivable. Of these, 100 are in amounts over $50,000; 500 are in amounts between $1000 and $50,000; and the remaining 4400 are in amounts under $1000. Using these groups as strata, you decide to verify all of the largest accounts and to sample 5% of the midsize accounts and 1% of the small accounts. How would you label the two strata from which you will sample? Use Table B, starting at line 115, to select the first 5 accounts from each of these strata.

**3.71** **Nonresponse in telephone surveys.** A common form of nonresponse in telephone surveys is "ring-no-answer." That is, a call is made to an active number but no one answers. The Italian National Statistical Institute looked at nonresponse to a government survey of households in Italy during the periods January 1 to Easter and July 1 to August 31. All calls were made between 7 and 10 P.M., but 21.4% gave "ring-no-answer" in one period versus 41.5% "ring-no-answer" in the other period.[40] Which period do you think had the higher rate of no answers? Why? Explain why a high rate of nonresponse makes sample results less reliable.

**3.72** **The sampling frame.** The list of individuals from which a sample is actually selected is called the **sampling frame.** Ideally, the frame should list every individual in the population, but in practice this is often difficult. A frame that leaves out part of the population is a common source of undercoverage.

(a) Suppose that a sample of households in a community is selected at random from the telephone directory. What households are omitted from this frame? What types of people do you think are likely to live in these households? These people will probably be underrepresented in the sample.

(b) It is usual in telephone surveys to use random digit dialing equipment that selects the last four digits of a telephone number at random after being given the area code and the exchange (the first three digits). Which of the households that you mentioned in your answer to (a) will be included in the sampling frame by random digit dialing?

**3.73** **The Excite Poll.** The Excite Poll can be found online at `poll.excite.com`. The question appears on the screen, and you simply click buttons to vote "Yes," "No," "Not sure," or "Don't care." On July 22, 2006, the question was "Do you agree or disagree with proposed legislation that would discontinue the U.S. penny coin?" In all, 631 said "Yes," another 564 said "No," and the remaining 65 indicated that they were not sure.

(a) What is the sample size for this poll?

(b) Compute the percent of responses in each of the possible response categories.

(c) Discuss the poll in terms of the population and sample framework that we have studied in this chapter.

**3.74** **Survey questions.** Comment on each of the following as a potential sample survey question. Is the question clear? Is it slanted toward a desired response?

(a) "Some cell phone users have developed brain cancer. Should all cell phones come with a warning label explaining the danger of using cell phones?"

(b) "Do you agree that a national system of health insurance should be favored because it would provide health insurance for everyone and would reduce administrative costs?"

(c) "In view of escalating environmental degradation and incipient resource depletion, would you favor economic incentives for recycling of resource-intensive consumer goods?"

**3.75** **Use of a budget surplus.** In 2000, when the federal budget showed a large surplus, the Pew Research Center asked two questions of random samples of adults. Both questions stated that Social Security would be "fixed." Here are the uses suggested for the remaining surplus:

*Should the money be used for a tax cut, or should it be used to fund new government programs?*

*Should the money be used for a tax cut, or should it be spent on programs for education, the environment, health care, crime-fighting and military defense?*

One of these questions drew 60% favoring a tax cut; the other, only 22%. Which wording pulls respondents toward a tax cut? Why?

3.76 CHALLENGE **How many children are in your family?** A teacher asks her class, "How many children are there in your family, including yourself?" The mean response is about 3 children. According to the 2000 census, families that have children average 1.86 children. Why is a sample like this biased toward higher outcomes?

3.77 CHALLENGE **Bad survey questions.** Write your own examples of bad sample survey questions.

(a) Write a biased question designed to get one answer rather than another.

(b) Write a question that is confusing, so that it is hard to answer.

3.78 CHALLENGE **Economic attitudes of Spaniards.** Spain's Centro de Investigaciones Sociológicos carried out a sample survey on the economic attitudes of Spaniards.[41] Of the 2496 adults interviewed, 72% agreed that "Employees with higher performance must get higher pay." On the other hand, 71% agreed that "Everything a society produces should be distributed among its members as equally as possible and there should be no major differences." Use these conflicting results as an example in a short explanation of why opinion polls often fail to reveal public attitudes clearly.

## 3.3 Toward Statistical Inference

A market research firm interviews a random sample of 2500 adults. Result: 66% find shopping for clothes frustrating and time-consuming. That's the truth about the 2500 people in the sample. What is the truth about the almost 220 million American adults who make up the population? Because the sample was chosen at random, it's reasonable to think that these 2500 people represent the entire population fairly well. So the market researchers turn the *fact* that 66% of the *sample* find shopping frustrating into an *estimate* that about 66% of *all adults* feel this way. That's a basic move in statistics: use a fact about a sample to estimate the truth about the whole population. We call this **statistical inference** because we infer conclusions about the wider population from data on selected individuals. To think about inference, we must keep straight whether a number describes a sample or a population. Here is the vocabulary we use.

statistical inference

### PARAMETERS AND STATISTICS

A **parameter** is a number that describes the **population.** A parameter is a fixed number, but in practice we do not know its value.

A **statistic** is a number that describes a **sample.** The value of a statistic is known when we have taken a sample, but it can change from sample to sample. We often use a statistic to estimate an unknown parameter.

EXAMPLE

**3.31 Attitudes toward shopping.** Are attitudes toward shopping changing? Sample surveys show that fewer people enjoy shopping than in the past. A survey by the market research firm Yankelovich Clancy Shulman asked a nationwide random sample of 2500 adults if they agreed or disagreed that "I like buying new clothes, but shopping is often frustrating and time-consuming." Of the respondents, 1650, or 66%, said they agreed.[42] The

proportion of the sample who agree is

$$\hat{p} = \frac{1650}{2500} = 0.66 = 66\%$$

The number $\hat{p} = 0.66$ is a *statistic*. The corresponding *parameter* is the proportion (call it $p$) of all adult U.S. residents who would have said "Agree" if asked the same question. We don't know the value of the parameter $p$, so we use the statistic $\hat{p}$ to estimate it.

**USE YOUR KNOWLEDGE**

**3.79 Sexual harassment of college students.** A recent survey of 2036 undergraduate college students aged 18 to 24 reports that 62% of college students say they have encountered some type of sexual harassment while at college.[43] Describe the sample and the population for this setting.

**3.80 Web polls.** If you connect to the Web site `worldnetdaily.com/polls/`, you will be given the opportunity to give your opinion about a different question of public interest each day. Can you apply the ideas about populations and samples that we have just discussed to this poll? Explain why or why not.

## Sampling variability

If Yankelovich took a second random sample of 2500 adults, the new sample would have different people in it. It is almost certain that there would not be exactly 1650 positive responses. That is, the value of the statistic $\hat{p}$ will vary from sample to sample. This basic fact is called **sampling variability:** the value of a statistic varies in repeated random sampling. Could it happen that one random sample finds that 66% of adults find clothes shopping frustrating and a second random sample finds that only 42% feel this way? Random samples eliminate *bias* from the act of choosing a sample, but they can still be wrong because of the *variability* that results when we choose at random. If the variation when we take repeat samples from the same population is too great, we can't trust the results of any one sample.

sampling variability

We are saved by the second great advantage of random samples. The first advantage is that choosing at random eliminates favoritism. That is, random sampling attacks bias. The second advantage is that if we take lots of random samples of the same size from the same population, the variation from sample to sample will follow a predictable pattern. **All of statistical inference is based on one idea: to see how trustworthy a procedure is, ask what would happen if we repeated it many times.**

To understand why sampling variability is not fatal, we ask, "What would happen if we took many samples?" Here's how to answer that question:

- Take a large number of samples from the same population.
- Calculate the sample proportion $\hat{p}$ for each sample.
- Make a histogram of the values of $\hat{p}$.

- Examine the distribution displayed in the histogram for shape, center, and spread, as well as outliers or other deviations.

In practice it is too expensive to take many samples from a large population such as all adult U.S. residents. But we can imitate many samples by using random digits. Using random digits from a table or computer software to imitate chance behavior is called **simulation.**

simulation

EXAMPLE

**3.32 Simulate a random sample.** We will simulate drawing simple random samples (SRSs) of size 100 from the population of all adult U.S. residents. Suppose that in fact 60% of the population find clothes shopping time-consuming and frustrating. Then the true value of the parameter we want to estimate is $p = 0.6$. (Of course, we would not sample in practice if we already knew that $p = 0.6$. We are sampling here to understand how sampling behaves.)

We can imitate the population by a table of random digits, with each entry standing for a person. Six of the ten digits (say 0 to 5) stand for people who find shopping frustrating. The remaining four digits, 6 to 9, stand for those who do not. Because all digits in a random number table are equally likely, this assignment produces a population proportion of frustrated shoppers equal to $p = 0.6$. We then imitate an SRS of 100 people from the population by taking 100 consecutive digits from Table B. The statistic $\hat{p}$ is the proportion of 0s to 5s in the sample.

Here are the first 100 entries in Table B with digits 0 to 5 highlighted:

| **1**9**223** | 9**5034** | **05**7**5**6 | **2**87**13** | 96**40**9 | **12531** | **42544** | 8**2**8**53** |
|---|---|---|---|---|---|---|---|
| 7**3**676 | **4**7**150** | 99**400** | **01**9**2**7 | **2**77**54** | **42**6**4**8 | 8**2425** | **3**6**2**9**0** |
| **454**67 | 7**1**7**0**9 | 77**55**8 | **000**9**5** | | | | |

There are 64 digits between 0 and 5, so $\hat{p} = 64/100 = 0.64$. A second SRS based on the second 100 entries in Table B gives a different result, $\hat{p} = 0.55$. The two sample results are different, and neither is equal to the true population value $p = 0.6$. That's sampling variability.

## Sampling distributions

Simulation is a powerful tool for studying chance. Now that we see how simulation works, it is faster to abandon Table B and to use a computer programmed to generate random numbers.

EXAMPLE

**3.33 Take many random samples.** Figure 3.11 illustrates the process of choosing many samples and finding the sample proportion $\hat{p}$ for each one. Follow the flow of the figure from the population at the left, to choosing an SRS and finding the $\hat{p}$ for this sample, to collecting together the $\hat{p}$'s from many samples. The histogram at the right of the figure shows the distribution of the values of $\hat{p}$ from 1000 separate SRSs of size 100 drawn from a population with $p = 0.6$.

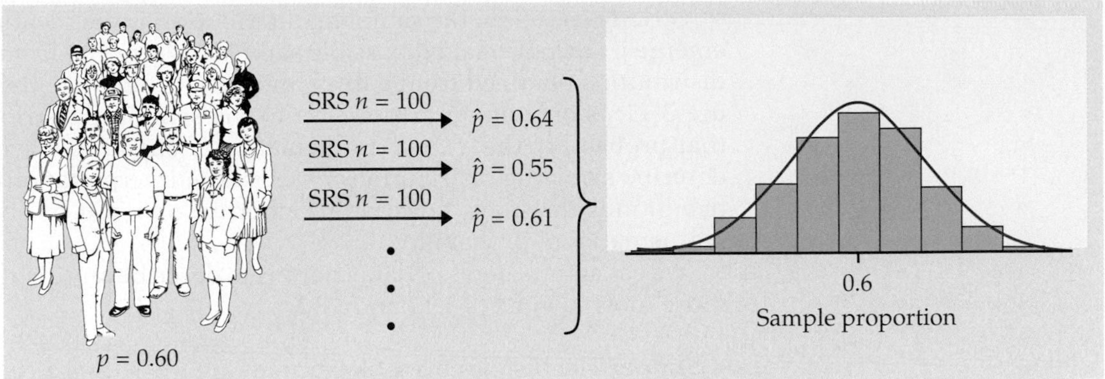


**FIGURE 3.11** The results of many SRSs have a regular pattern. Here, we draw 1000 SRSs of size 100 from the same population. The population proportion is $p = 0.60$. The histogram shows the distribution of the 1000 sample proportions.

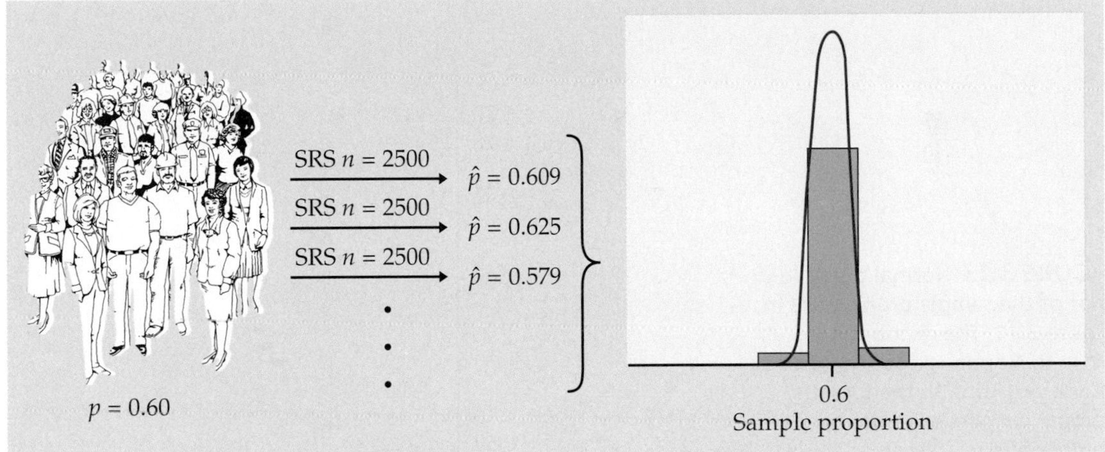


**FIGURE 3.12** The distribution of sample proportions for 1000 SRSs of size 2500 drawn from the same population as in Figure 3.11. The two histograms have the same scale. The statistic from the larger sample is less variable.

Of course, Yankelovich interviewed 2500 people, not just 100. Figure 3.12 is parallel to Figure 3.11. It shows the process of choosing 1000 SRSs, each of size 2500, from a population in which the true proportion is $p = 0.6$. The 1000 values of $\hat{p}$ from these samples form the histogram at the right of the figure. Figures 3.11 and 3.12 are drawn on the same scale. Comparing them shows what happens when we increase the size of our samples from 100 to 2500. These histograms display the *sampling distribution* of the statistic $\hat{p}$ for two sample sizes.

### SAMPLING DISTRIBUTION

The **sampling distribution** of a statistic is the distribution of values taken by the statistic in all possible samples of the same size from the same population.

Strictly speaking, the sampling distribution is the ideal pattern that would emerge if we looked at all possible samples of size 100 from our population. A distribution obtained from a fixed number of trials, like the 1000 trials in Figure 3.11, is only an approximation to the sampling distribution. We will see that probability theory, the mathematics of chance behavior, can sometimes describe sampling distributions exactly. The interpretation of a sampling distribution is the same, however, whether we obtain it by simulation or by the mathematics of probability.

We can use the tools of data analysis to describe any distribution. Let's apply those tools to Figures 3.11 and 3.12.

- **Shape:** The histograms look Normal. Figure 3.13 is a Normal quantile plot of the values of $\hat{p}$ for our samples of size 100. It confirms that the distribution in Figure 3.11 is close to Normal. The 1000 values for samples of size 2500 in Figure 3.12 are even closer to Normal. The Normal curves drawn through the histograms describe the overall shape quite well.

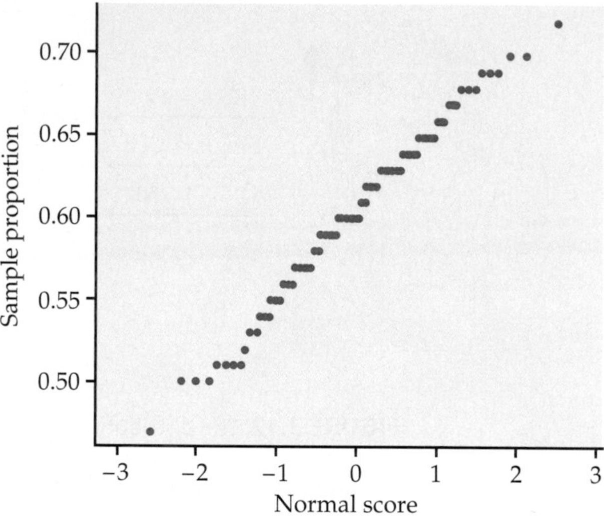


**FIGURE 3.13** Normal quantile plot of the sample proportions in Figure 3.11. The distribution is close to Normal except for some granularity due to the fact that sample proportions from a sample of size 100 can take only values that are multiples of 0.01. Because a plot of 1000 points is hard to read, this plot presents only every 10th value.

- **Center:** In both cases, the values of the sample proportion $\hat{p}$ vary from sample to sample, but the values are centered at 0.6. Recall that $p = 0.6$ is the true population parameter. Some samples have a $\hat{p}$ less than 0.6 and some greater, but there is no tendency to be always low or always high. That is, $\hat{p}$ has no **bias** as an estimator of $p$. This is true for both large and small samples. (Want the details? The mean of the 1000 values of $\hat{p}$ is 0.598 for samples of size 100 and 0.6002 for samples of size 2500. The median value of $\hat{p}$ is exactly 0.6 for samples of both sizes.)

- **Spread:** The values of $\hat{p}$ from samples of size 2500 are much less spread out than the values from samples of size 100. In fact, the standard deviations are 0.051 for Figure 3.11 and 0.0097, or about 0.01, for Figure 3.12.

Although these results describe just two sets of simulations, they reflect facts that are true whenever we use random sampling.

USE YOUR KNOWLEDGE

**3.81 Effect of sample size on the sampling distribution.** You are planning a study and are considering taking an SRS of either 200 or 400 observations. Explain how the sampling distribution would differ for these two scenarios.

## Bias and variability

Our simulations show that a sample of size 2500 will almost always give an estimate $\hat{p}$ that is close to the truth about the population. Figure 3.12 illustrates this fact for just one value of the population proportion, but it is true for any population. Samples of size 100, on the other hand, might give an estimate of 50% or 70% when the truth is 60%.

Thinking about Figures 3.11 and 3.12 helps us restate the idea of bias when we use a statistic like $\hat{p}$ to estimate a parameter like $p$. It also reminds us that variability matters as much as bias.

BIAS AND VARIABILITY

**Bias** concerns the center of the sampling distribution. A statistic used to estimate a parameter is **unbiased** if the mean of its sampling distribution is equal to the true value of the parameter being estimated.

The **variability of a statistic** is described by the spread of its sampling distribution. This spread is determined by the sampling design and the sample size $n$. Statistics from larger probability samples have smaller spreads.

We can think of the true value of the population parameter as the bull's-eye on a target, and of the sample statistic as an arrow fired at the bull's-eye. Bias and variability describe what happens when an archer fires many arrows at the target. *Bias* means that the aim is off, and the arrows land consistently off the bull's-eye in the same direction. The sample values do not center about the population value. Large *variability* means that repeated shots are widely scattered on the target. Repeated samples do not give similar results but differ widely among themselves. Figure 3.14 shows this target illustration of the two types of error.

Notice that small variability (repeated shots are close together) can accompany large bias (the arrows are consistently away from the bull's-eye in one direction). And small bias (the arrows center on the bull's-eye) can accompany large variability (repeated shots are widely scattered). A good sampling scheme, like a good archer, must have both small bias and small variability. Here's how we do this.

MANAGING BIAS AND VARIABILITY

**To reduce bias,** use random sampling. When we start with a list of the entire population, simple random sampling produces unbiased estimates—the values of a statistic computed from an SRS neither consistently overestimate nor consistently underestimate the value of the population parameter.

**To reduce the variability** of a statistic from an SRS, use a larger sample. You can make the variability as small as you want by taking a large enough sample.

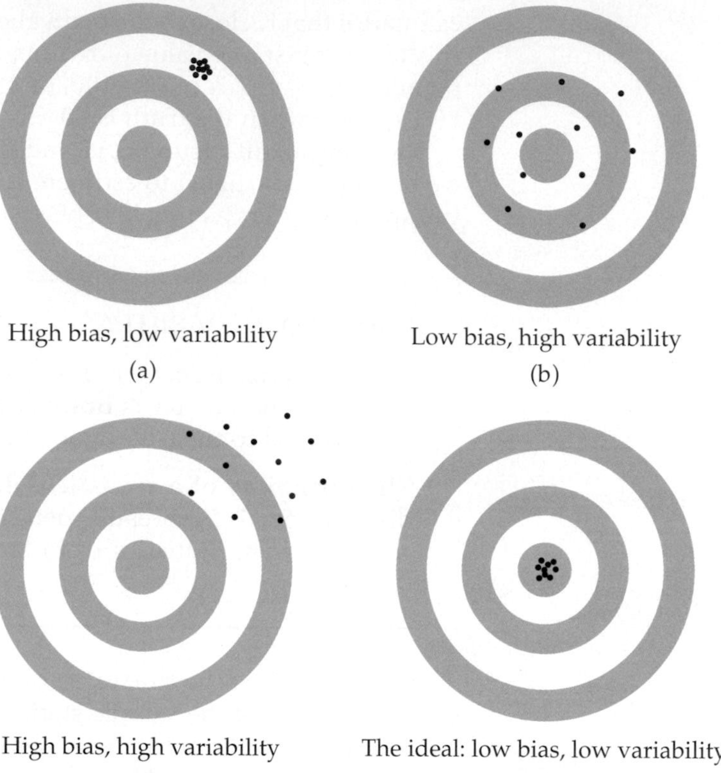


**FIGURE 3.14** Bias and variability in shooting arrows at a target. Bias means the archer systematically misses in the same direction. Variability means that the arrows are scattered.

In practice, Yankelovich takes only one sample. We don't know how close to the truth an estimate from this one sample is because we don't know what the truth about the population is. But *large random samples almost always give an estimate that is close to the truth.* Looking at the pattern of many samples shows that we can trust the result of one sample. The Current Population Survey's sample of 60,000 households estimates the national unemployment rate very accurately. Of course, only probability samples carry this guarantee. The American Family Association's voluntary response sample (Example 3.23, page 199) is worthless even though 850,000 people responded. Using a probability sampling design and taking care to deal with practical difficulties reduce bias in a sample. The size of the sample then determines how close to the population truth the sample result is likely to fall. Results from a sample survey usually come with a **margin of error** that sets bounds on the size of the likely error. The margin of error directly reflects the variability of the sample statistic, so it is smaller for larger samples. We will describe the details in later chapters.

margin of error

## Sampling from large populations

Yankelovich's sample of 2500 adults is only about 1 out of every 90,000 adults in the United States. Does it matter whether we sample 1-in-100 individuals in the population or 1-in-90,000?

> **POPULATION SIZE DOESN'T MATTER**
>
> The variability of a statistic from a random sample does not depend on the size of the population, as long as the population is at least 100 times larger than the sample.

Why does the size of the population have little influence on the behavior of statistics from random samples? To see why this is plausible, imagine sampling harvested corn by thrusting a scoop into a lot of corn kernels. The scoop doesn't know whether it is surrounded by a bag of corn or by an entire truckload. As long as the corn is well mixed (so that the scoop selects a random sample), the variability of the result depends only on the size of the scoop.

The fact that the variability of sample results is controlled by the size of the sample has important consequences for sampling design. An SRS of size 2500 from the 220 million adult residents of the United States gives results as precise as an SRS of size 2500 from the 665,000 adult inhabitants of San Francisco. This is good news for designers of national samples but bad news for those who want accurate information about the citizens of San Francisco. If both use an SRS, both must use the same size sample to obtain equally trustworthy results.

## Why randomize?

Why randomize? The act of randomizing guarantees that the results of analyzing our data are subject to the laws of probability. The behavior of statistics is described by a sampling distribution. The form of the distribution is known, and in many cases is approximately Normal. Often the center of the distribution lies at the true parameter value, so that the notion that randomization eliminates bias is made more precise. The spread of the distribution describes the variability of the statistic and can be made as small as we wish by choosing a large enough sample. In a randomized experiment, we can reduce variability by choosing larger groups of subjects for each treatment.

These facts are at the heart of formal statistical inference. Later chapters will have much to say in more technical language about sampling distributions and the way statistical conclusions are based on them. What any user of statistics must understand is that all the technical talk has its basis in a simple question: *What would happen if the sample or the experiment were repeated many times?* The reasoning applies not only to an SRS but also to the complex sampling designs actually used by opinion polls and other national sample surveys. The same conclusions hold as well for randomized experimental designs. The details vary with the design but the basic facts are true whenever randomization is used to produce data.

Remember that proper statistical design is not the only aspect of a good sample or experiment. *The sampling distribution shows only how a statistic*

*varies due to the operation of chance in randomization. It reveals nothing about possible bias due to undercoverage or nonresponse in a sample, or to lack of realism in an experiment.* The actual error in estimating a parameter by a statistic can be much larger than the sampling distribution suggests. What is worse, there is no way to say how large the added error is. The real world is less orderly than statistics textbooks imply.

## BEYOND THE BASICS

### Capture-Recapture Sampling

Sockeye salmon return to reproduce in the river where they were hatched four years earlier. How many salmon survived natural perils and heavy fishing to make it back this year? How many mountain sheep are there in Colorado? Are migratory songbird populations in North America decreasing or holding their own? These questions concern the size of animal populations. Biologists address them with a special kind of repeated sampling, called *capture-recapture sampling.*

EXAMPLE

**3.34 Estimate the number of least flycatchers.** You are interested in the number of least flycatchers migrating along a major route in the north-central United States. You set up "mist nets" that capture the birds but do not harm them. The birds caught in the net are fitted with a small aluminum leg band and released. Last year you banded and released 200 least flycatchers. This year you repeat the process. Your net catches 120 least flycatchers, 12 of which have tags from last year's catch.

The proportion of your second sample that have bands should estimate the proportion in the entire population that are banded. So if $N$ is the unknown number of least flycatchers, we should have approximately

$$\text{proportion banded in sample} = \text{proportion banded in population}$$

$$\frac{12}{120} = \frac{200}{N}$$

Solve for $N$ to estimate that the total number of flycatchers migrating while your net was up this year is approximately

$$N = 200 \times \frac{120}{12} = 2000$$

The capture-recapture idea extends the use of a sample proportion to estimate a population proportion. The idea works well if both samples are SRSs from the population and the population remains unchanged between samples. In practice, complications arise because, for example, some of the birds tagged last year died before this year's migration. Variations on capture-recapture samples are widely used in wildlife studies and are now finding other applications. One way to estimate the census undercount in a district is to consider the census as "capturing and marking" the households that respond. Census workers then visit the district, take an SRS of households, and see how many of those counted by the census show up in the sample. Capture-recapture estimates the total count of households in the district. As with estimating wildlife populations, there are many practical pitfalls. Our final word is as before: the real world is less orderly than statistics textbooks imply.

## SECTION 3.3 Summary

A number that describes a population is a **parameter.** A number that can be computed from the data is a **statistic.** The purpose of sampling or experimentation is usually **inference:** use sample statistics to make statements about unknown population parameters.

A statistic from a probability sample or randomized experiment has a **sampling distribution** that describes how the statistic varies in repeated data production. The sampling distribution answers the question "What would happen if we repeated the sample or experiment many times?" Formal statistical inference is based on the sampling distributions of statistics.

A statistic as an estimator of a parameter may suffer from **bias** or from high **variability.** Bias means that the center of the sampling distribution is not equal to the true value of the parameter. The variability of the statistic is described by the spread of its sampling distribution. Variability is usually reported by giving a **margin of error** for conclusions based on sample results.

Properly chosen statistics from randomized data production designs have no bias resulting from the way the sample is selected or the way the experimental units are assigned to treatments. We can reduce the variability of the statistic by increasing the size of the sample or the size of the experimental groups.

## SECTION 3.3 Exercises

*For Exercises 3.79 and 3.80, see page 213; and for Exercise 3.81, see page 217.*

**3.82 What's wrong?** State what is wrong in each of the following scenarios.

(a) A sampling distribution describes the distribution of some characteristic in a population.

(b) A statistic will have a large amount of bias whenever it has high variability.

(c) The variability of a statistic based on a small sample from a population will be the same as the variability of a large sample from the same population.

**3.83 Describe the population and the sample.** For each of the following situations, describe the population and the sample.

(a) A survey of 17,096 students in U.S. four-year colleges reported that 19.4% were binge drinkers.

(b) In a study of work stress, 100 restaurant workers were asked about the impact of work stress on their personal lives.

(c) A tract of forest has 584 longleaf pine trees. The diameters of 40 of these trees were measured.

**3.84 Bias and variability.** Figure 3.15 (on page 222) shows histograms of four sampling distributions of statistics intended to estimate the same parameter. Label each distribution relative to the others as high or low bias and as high or low variability.

**3.85 Opinions of Hispanics.** A New York Times News Service article on a poll concerned with the opinions of Hispanics includes this paragraph:

*The poll was conducted by telephone from July 13 to 27, with 3,092 adults nationwide, 1,074 of whom described themselves as Hispanic. It has a margin of sampling error of plus or minus three percentage points for the entire poll and plus or minus four*

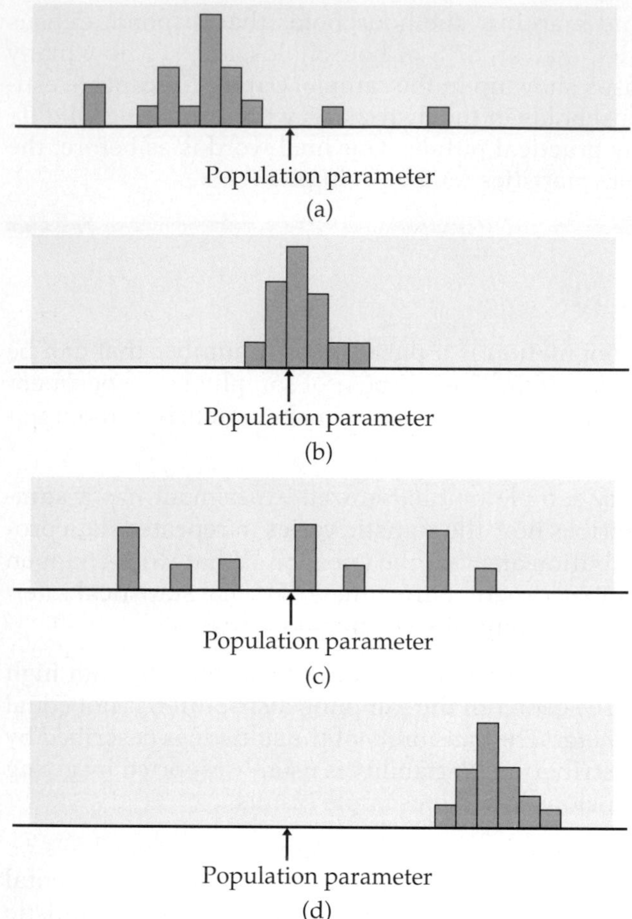


**FIGURE 3.15** Determine which of these sampling distributions displays high or low bias and high or low variability, for Exercise 3.84.

*percentage points for Hispanics. Sample sizes for most Hispanic nationalities, like Cubans or Dominicans, were too small to break out the results separately.*[44]

(a) Why is the "margin of sampling error" larger for Hispanics than for all 3092 respondents?

(b) Why would a very small sample size prevent a responsible news organization from breaking out results for Cubans?

**3.86 Gallup Canada polls.** Gallup Canada bases its polls of Canadian public opinion on telephone samples of about 1000 adults, the same sample size as Gallup uses in the United States. Canada's population is about one-ninth as large as that of the United States, so the percent of adults that Gallup interviews in Canada is nine times as large as in the United States. Does this mean that the margin of error for a Gallup Canada poll is smaller? Explain your answer.

**3.87 Real estate ownership.** An agency of the federal government plans to take an SRS of residents in each state to estimate the proportion of owners of real estate in each state's population. The populations of the states range from less than 500,000 people in Wyoming to about 35 million in California.

(a) Will the variability of the sample proportion vary from state to state if an SRS of size 2000 is taken in each state? Explain your answer.

(b) Will the variability of the sample proportion change from state to state if an SRS of 1/10 of 1% (0.001) of the state's population is taken in each state? Explain your answer.

**3.88 The health care system in Ontario.** The Ministry of Health in the Canadian province of Ontario wants to know whether the national health care system is achieving its goals in the province. The ministry conducted the Ontario Health Survey, which interviewed a probability sample of 61,239 adults who live in Ontario.[45]

(a) What is the population for this sample survey? What is the sample?

(b) The survey found that 76% of males and 86% of females in the sample had visited a general practitioner at least once in the past year. Do you think these estimates are close to the truth about the entire population? Why?

*The remaining exercises demonstrate the idea of a sampling distribution. Sampling distributions are the basis for statistical inference. We strongly recommend doing some of these exercises.*

**3.89** APPLET **Use the probability applet.** The *Probability* applet simulates tossing a coin, with the advantage that you can choose the true long-term proportion, or probability, of a head. Example 3.33 discusses sampling from a population in which proportion $p = 0.6$ (the parameter) find shopping frustrating. Tossing a coin with probability $p = 0.6$ of a head simulates this situation: each head is a person who finds shopping frustrating, and each tail is a person who does not. Set the "Probability of heads" in the applet to 0.6 and the number of tosses to 25. This simulates an SRS of size 25 from this population. By alternating between "Toss" and "Reset" you can take many samples quickly.

(a) Take 50 samples, recording the number of heads in each sample. Make a histogram of the 50 sample proportions (count of heads divided by 25). You are constructing the sampling distribution of this statistic.

(b) Another population contains only 20% who approve of legal gambling. Take 50 samples of size 25 from this population, record the number in each sample who approve, and make a histogram of the 50 sample proportions. How do the centers of your two histograms reflect the differing truths about the two populations?

**3.90** CHALLENGE **Use the statistical software for simulations.** Statistical software can speed simulations. We are interested in the sampling distribution of the proportion $\hat{p}$ of people who find shopping frustrating in an SRS from a population in which proportion $p$ find shopping frustrating. Here, $p$ is a parameter and $\hat{p}$ is a statistic used to estimate $p$. We will see in Chapter 5 that "binomial" is the key word to look for in the software menus. For example, in CrunchIt! go to "Simulate data" in the "Data" menu, and choose "Binomial."

(a) Set $n = 50$ and $p = 0.6$ and generate 100 binomial observations. These are the counts for 100 SRSs of size 50 when 60% of the population finds shopping frustrating. Save these counts and divide them by 50 to get values of $\hat{p}$ from 100 SRSs. Make a stemplot of the 100 values of $\hat{p}$.

(b) Repeat this process with $p = 0.3$, representing a population in which only 30% of people find shopping frustrating. Compare your two stemplots. How does changing the parameter $p$ affect the center and spread of the sampling distribution?

(c) Now generate 100 binomial observations with $n = 200$ and $p = 0.6$. This simulates 100 SRSs, each of size 200. Obtain the 100 sample proportions $\hat{p}$ and make a stemplot. Compare this with your stemplot from (a). How does changing the sample size $n$ affect the center and spread of the sampling distribution?

**3.91 Use Table B for a simulation.** We can construct a sampling distribution by hand in the case of a very small sample from a very small population. The population contains 10 students. Here are their scores on an exam:

| Student | 0 | 1 | 2 | 3 | 4 | 5 | 6 | 7 | 8 | 9 |
|---|---|---|---|---|---|---|---|---|---|---|
| Score | 82 | 62 | 80 | 58 | 72 | 73 | 65 | 66 | 74 | 62 |

The parameter of interest is the mean score, which is 69.4. The sample is an SRS of $n = 4$ students drawn from this population. The students are labeled 0 to 9 so that a single random digit from Table B chooses one student for the sample.

(a) Use Table B to draw an SRS of size 4 from this population. Write the four scores in your sample and calculate the mean $\bar{x}$ of the sample scores. This statistic is an estimate of the population parameter.

(b) Repeat this process 9 more times. Make a histogram of the 10 values of $\bar{x}$. You are constructing the sampling distribution of $\bar{x}$. Is the center of your histogram close to 69.4? (Ten repetitions give only a crude approximation to the sampling distribution. If possible, pool your work with that of other students—using different parts of Table B—to obtain several hundred repetitions and make a histogram of the values of $\bar{x}$. This histogram is a better approximation to the sampling distribution.)

**3.92** APPLET **Use the simple random sample applet.** The *Simple Random Sample* applet can illustrate the idea of a sampling distribution. Form a population labeled 1 to 100. We will choose an SRS of 10 of these numbers. That is, in this exercise, the numbers themselves are the population, not just labels for 100 individuals. The mean of the whole numbers 1 to 100 is 50.5. This is the parameter, the mean of the population.

(a) Use the applet to choose an SRS of size 10. Which 10 numbers were chosen? What is their mean? This is a statistic, the sample mean $\bar{x}$.

(b) Although the population and its mean 50.5 remain fixed, the sample mean changes as we take more samples. Take another SRS of size 10. (Use the "Reset" button to return to the original population before taking the second sample.) What are the 10 numbers in your sample? What is their mean? This is another value of $\bar{x}$.

(c) Take 8 more SRSs from this same population and record their means. You now have 10 values of the sample mean $\bar{x}$ from 10 SRSs of the same size from the same population. Make a histogram of the 10 values and mark the population mean 50.5 on the horizontal axis. Are your 10 sample values roughly centered at the population value? (If you kept going forever, your $\bar{x}$-values would form the sampling distribution of the sample mean; the population mean would indeed be the center of this distribution.)

**3.93 Analyze simple random samples.** The CSDATA data set contains the college grade point averages (GPAs) of all 224 students in a university entering class who planned to major in computer science. This is our population. Statistical software can take repeated samples to illustrate sampling variability.

(a) Using software, describe this population with a histogram and with numerical summaries. In particular, what is the mean GPA in the population? This is a parameter.

(b) Choose an SRS of 20 members from this population. Make a histogram of the GPAs in the sample and find their mean. The sample mean is a statistic. Briefly compare the distributions of GPA in the sample and in the population.

(c) Repeat the process of choosing an SRS of size 20 four more times (five in all). Record the five histograms of your sample GPAs. Does it seem reasonable to you from this small trial that an SRS will usually produce a sample that is generally representative of the population?

**3.94 Simulate the sampling distribution of the mean.** Continue the previous exercise, using software to illustrate the idea of a sampling distribution.

(a) Choose 20 more SRSs of size 20 in addition to the 5 you have already chosen. Don't make histograms of these latest samples—just record the mean GPA for each sample. Make a histogram of the 25 sample means. This histogram is a rough approximation to the sampling distribution of the mean.

(b) One sign of bias would be that the distribution of the sample means was systematically on one side of the true population mean. Mark the population mean GPA on your histogram of the 25 sample means. Is there a clear bias?

(c) Find the mean and standard deviation of your 25 sample means. We expect that the mean will be close to the true mean of the population. Is it? We also expect that the standard deviation of the sampling distribution will be smaller than the standard deviation of the population. Is it?

**3.95 Toss a coin.** Coin tossing can illustrate the idea of a sampling distribution. The population is all outcomes (heads or tails) we would get if we tossed a coin forever. The parameter $p$ is the proportion of heads in this population. We suspect that $p$ is close to 0.5. That is, we think the coin will show about one-half heads in the long run. The sample is the outcomes of 20 tosses, and the statistic $\hat{p}$ is the proportion of heads in these 20 tosses (count of heads divided by 20).

(a) Toss a coin 20 times and record the value of $\hat{p}$.

(b) Repeat this sampling process 9 more times. Make a stemplot of the 10 values of $\hat{p}$. You are constructing the sampling distribution of $\hat{p}$. Is the center of this distribution close to 0.5? (Ten repetitions give only a crude approximation to the sampling distribution. If possible, pool your work with that of other students to obtain several hundred repetitions and make a histogram of the values of $\hat{p}$.)

## 3.4 Ethics

The production and use of data, like all human endeavors, raise ethical questions. We won't discuss the telemarketer who begins a telephone sales pitch with "I'm conducting a survey." Such deception is clearly unethical. It enrages legitimate survey organizations, which find the public less willing to talk with them. Neither will we discuss those few researchers who, in the pursuit of professional advancement, publish fake data. There is no ethical question here—faking data to advance your career is just wrong. It will end your career when uncovered. But just how honest must researchers be about real, unfaked data? Here is an example that suggests the answer is "More honest than they often are."

EXAMPLE

**3.35 Provide all of the critical information.** Papers reporting scientific research are supposed to be short, with no extra baggage. Brevity can allow the researchers to avoid complete honesty about their data. Did they choose their subjects in a biased way? Did they report data on only some of their subjects? Did they try several statistical analyses and report only the ones that looked best? The statistician John Bailar screened more than 4000 medical papers in more than a decade as consultant to the *New England*

*Journal of Medicine.* He says, "When it came to the statistical review, it was often clear that critical information was lacking, and the gaps nearly always had the practical effect of making the authors' conclusions look stronger than they should have."[46] The situation is no doubt worse in fields that screen published work less carefully.

The most complex issues of data ethics arise when we collect data from people. The ethical difficulties are more severe for experiments that impose some treatment on people than for sample surveys that simply gather information. Trials of new medical treatments, for example, can do harm as well as good to their subjects. Here are some basic standards of data ethics that must be obeyed by any study that gathers data from human subjects, whether sample survey or experiment.

### BASIC DATA ETHICS

The organization that carries out the study must have an **institutional review board** that reviews all planned studies in advance in order to protect the subjects from possible harm.

All individuals who are subjects in a study must give their **informed consent** before data are collected.

All individual data must be kept **confidential.** Only statistical summaries for groups of subjects may be made public.

The law requires that studies funded by the federal government obey these principles. But neither the law nor the consensus of experts is completely clear about the details of their application.

## Institutional review boards

The purpose of an institutional review board is not to decide whether a proposed study will produce valuable information or whether it is statistically sound. The board's purpose is, in the words of one university's board, "to protect the rights and welfare of human subjects (including patients) recruited to participate in research activities." The board reviews the plan of the study and can require changes. It reviews the consent form to be sure that subjects are informed about the nature of the study and about any potential risks. Once research begins, the board monitors its progress at least once a year.

The most pressing issue concerning institutional review boards is whether their workload has become so large that their effectiveness in protecting subjects drops. When the government temporarily stopped human-subject research at Duke University Medical Center in 1999 due to inadequate protection of subjects, more than 2000 studies were going on. That's a lot of review work. There are shorter review procedures for projects that involve only minimal risks to subjects, such as most sample surveys. When a board is overloaded, there is a temptation to put more proposals in the minimal-risk category to speed the work.

## USE YOUR KNOWLEDGE

*The exercises in this section on Ethics are designed to help you think about the issues that we are discussing and to formulate some opinions. In general there are no wrong or right answers but you need to give reasons for your answers.*

**3.96 Do these proposals involve minimal risk?** You are a member of your college's institutional review board. You must decide whether several research proposals qualify for lighter review because they involve only minimal risk to subjects. Federal regulations say that "minimal risk" means the risks are no greater than "those ordinarily encountered in daily life or during the performance of routine physical or psychological examinations or tests." That's vague. Which of these do you think qualifies as "minimal risk"?

(a) Draw a drop of blood by pricking a finger in order to measure blood sugar.

(b) Draw blood from the arm for a full set of blood tests.

(c) Insert a tube that remains in the arm, so that blood can be drawn regularly.

**3.97 Who should be on an institutional review board?** Government regulations require that institutional review boards consist of at least five people, including at least one scientist, one nonscientist, and one person from outside the institution. Most boards are larger, but many contain just one outsider.

(a) Why should review boards contain people who are not scientists?

(b) Do you think that one outside member is enough? How would you choose that member? (For example, would you prefer a medical doctor? A member of the clergy? An activist for patients' rights?)

## Informed consent

Both words in the phrase "informed consent" are important, and both can be controversial. Subjects must be *informed* in advance about the nature of a study and any risk of harm it may bring. In the case of a sample survey, physical harm is not possible. The subjects should be told what kinds of questions the survey will ask and about how much of their time it will take. Experimenters must tell subjects the nature and purpose of the study and outline possible risks. Subjects must then *consent* in writing.

EXAMPLE

**3.36 Who can give informed consent?** Are there some subjects who can't give informed consent? It was once common, for example, to test new vaccines on prison inmates who gave their consent in return for good-behavior credit. Now we worry that prisoners are not really free to refuse, and the law forbids most medical experiments in prisons.

Very young children can't give fully informed consent, so the usual procedure is to ask their parents. A study of new ways to teach reading is about to start at a local elementary school, so the study team sends consent forms home to parents. Many parents don't return the forms. Can their children take part in the study because the parents did not say "No," or should we allow only children whose parents returned the form and said "Yes"?

What about research into new medical treatments for people with mental disorders? What about studies of new ways to help emergency room patients who may be unconscious or have suffered a stroke? In most cases, there is not time even to get the consent of the family. Does the principle of informed consent bar realistic trials of new treatments for unconscious patients?

These are questions without clear answers. Reasonable people differ strongly on all of them. There is nothing simple about informed consent.[47]

The difficulties of informed consent do not vanish even for capable subjects. Some researchers, especially in medical trials, regard consent as a barrier to getting patients to participate in research. They may not explain all possible risks; they may not point out that there are other therapies that might be better than those being studied; they may be too optimistic in talking with patients even when the consent form has all the right details. On the other hand, mentioning every possible risk leads to very long consent forms that really are barriers. "They are like rental car contracts," one lawyer said. Some subjects don't read forms that run five or six printed pages. Others are frightened by the large number of possible (but unlikely) disasters that might happen and so refuse to participate. Of course, unlikely disasters sometimes happen. When they do, lawsuits follow and the consent forms become yet longer and more detailed.

## Confidentiality

Ethical problems do not disappear once a study has been cleared by the review board, has obtained consent from its subjects, and has actually collected data about the subjects. It is important to protect the subjects' privacy by keeping all data about individuals confidential. The report of an opinion poll may say what percent of the 1500 respondents felt that legal immigration should be reduced. It may not report what *you* said about this or any other issue.

anonymity

Confidentiality is not the same as **anonymity.** Anonymity means that subjects are anonymous—their names are not known even to the director of the study. Anonymity is rare in statistical studies. Even where anonymity is possible (mainly in surveys conducted by mail), it prevents any follow-up to improve nonresponse or inform subjects of results.

Any breach of confidentiality is a serious violation of data ethics. The best practice is to separate the identity of the subjects from the rest of the data at once. Sample surveys, for example, use the identification only to check on who did or did not respond. In an era of advanced technology, however, it is no longer enough to be sure that each individual set of data protects people's privacy. The government, for example, maintains a vast amount of information about citizens in many separate data bases—census responses, tax returns, Social Security information, data from surveys such as the Current Population Survey, and so on. Many of these data bases can be searched by computers for statistical studies. A clever computer search of several data bases might be able, by combining information, to identify you and learn a great deal about you even if your name and other identification have been removed from the data available for search. A colleague from Germany once remarked that "female full professor of statistics with PhD from the United States" was enough to identify her among all the citizens of Germany. Privacy and confidentiality of data are hot issues among statisticians in the computer age.

EXAMPLE

**3.37 Data collected by the government.** Citizens are required to give information to the government. Think of tax returns and Social Security contributions. The government needs these data for administrative purposes—to see if we paid the right amount of tax and how large a Social Security benefit we are owed when we retire. Some people feel that individuals should be able to forbid any other use of their data, even with all identification removed. This would prevent using government records to study, say, the ages, incomes, and household sizes of Social Security recipients. Such a study could well be vital to debates on reforming Social Security.

USE YOUR KNOWLEDGE

**3.98 How can we obtain informed consent?** A researcher suspects that traditional religious beliefs tend to be associated with an authoritarian personality. She prepares a questionnaire that measures authoritarian tendencies and also asks many religious questions. Write a description of the purpose of this research to be read by subjects in order to obtain their informed consent. You must balance the conflicting goals of not deceiving the subjects as to what the questionnaire will tell about them and of not biasing the sample by scaring off religious people.

**3.99 Should we allow this personal information to be collected?** In which of the circumstances below would you allow collecting personal information without the subjects' consent?

(a) A government agency takes a random sample of income tax returns to obtain information on the average income of people in different occupations. Only the incomes and occupations are recorded from the returns, not the names.

(b) A social psychologist attends public meetings of a religious group to study the behavior patterns of members.

(c) A social psychologist pretends to be converted to membership in a religious group and attends private meetings to study the behavior patterns of members.

## Clinical trials

Clinical trials are experiments that study the effectiveness of medical treatments on actual patients. Medical treatments can harm as well as heal, so clinical trials spotlight the ethical problems of experiments with human subjects. Here are the starting points for a discussion:

- Randomized comparative experiments are the only way to see the true effects of new treatments. Without them, risky treatments that are no better than placebos will become common.
- Clinical trials produce great benefits, but most of these benefits go to future patients. The trials also pose risks, and these risks are borne by the subjects of the trial. So we must balance future benefits against present risks.
- Both medical ethics and international human rights standards say that "the interests of the subject must always prevail over the interests of science and society."

The quoted words are from the 1964 Helsinki Declaration of the World Medical Association, the most respected international standard. The most outrageous examples of unethical experiments are those that ignore the interests of the subjects.

EXAMPLE

**3.38 The Tuskegee study.** In the 1930s, syphilis was common among black men in the rural South, a group that had almost no access to medical care. The Public Health Service Tuskegee study recruited 399 poor black sharecroppers with syphilis and 201 others without the disease in order to observe how syphilis progressed when no treatment was given. Beginning in 1943, penicillin became available to treat syphilis. The study subjects were not treated. In fact, the Public Health Service prevented any treatment until word leaked out and forced an end to the study in the 1970s.

The Tuskegee study is an extreme example of investigators following their own interests and ignoring the well-being of their subjects. A 1996 review said, "It has come to symbolize racism in medicine, ethical misconduct in human research, paternalism by physicians, and government abuse of vulnerable people." In 1997, President Clinton formally apologized to the surviving participants in a White House ceremony.[48]

Because "the interests of the subject must always prevail," medical treatments can be tested in clinical trials only when there is reason to hope that they will help the patients who are subjects in the trials. Future benefits aren't enough to justify experiments with human subjects. Of course, if there is already strong evidence that a treatment works and is safe, it is unethical *not* to give it. Here are the words of Dr. Charles Hennekens of the Harvard Medical School, who directed the large clinical trial that showed that aspirin reduces the risk of heart attacks:

> *There's a delicate balance between when to do or not do a randomized trial. On the one hand, there must be sufficient belief in the agent's potential to justify exposing half the subjects to it. On the other hand, there must be sufficient doubt about its efficacy to justify withholding it from the other half of subjects who might be assigned to placebos.*[49]

Why is it ethical to give a control group of patients a placebo? Well, we know that placebos often work. What is more, placebos have no harmful side effects. So in the state of balanced doubt described by Dr. Hennekens, the placebo group may be getting a better treatment than the drug group. If we *knew* which treatment was better, we would give it to everyone. When we don't know, it is ethical to try both and compare them. Here are some harder questions about placebos, with arguments on both sides.

EXAMPLE

**3.39 Is it ethical to use a placebo?** You are testing a new drug. Is it ethical to give a placebo to a control group if an effective drug already exists?

**Yes:** The placebo gives a true baseline for the effectiveness of the new drug. There are three groups: new drug, best existing drug, and placebo. Every clinical trial is a bit different, and not even genuinely effective treatments work in every setting. The placebo control helps us see if the study is flawed so that even the best existing drug does not beat the placebo. Sometimes the placebo wins, so the doubt needed to justify its use is present. Placebo controls are ethical except for life-threatening conditions.

**No:** It isn't ethical to deliberately give patients an inferior treatment. We don't know whether the new drug is better than the existing drug, so it is ethical to give both in order to find out. If past trials showed that the existing drug is better than a placebo, it is no longer right to give patients a placebo. After all, the existing drug includes the placebo effect. A placebo group is ethical only if the existing drug is an older one that did not undergo proper clinical trials or doesn't work well or is dangerous.

## USE YOUR KNOWLEDGE

**3.100 Is this study ethical?** Researchers on aging proposed to investigate the effect of supplemental health services on the quality of life of older people. Eligible patients on the rolls of a large medical clinic were to be randomly assigned to treatment and control groups. The treatment group would be offered hearing aids, dentures, transportation, and other services not available without charge to the control group. The review board felt that providing these services to some but not other persons in the same institution raised ethical questions. Do you agree?

**3.101 Should the treatments be given to everyone?** Effective drugs for treating AIDS are very expensive, so most African nations cannot afford to give them to large numbers of people. Yet AIDS is more common in parts of Africa than anywhere else. Several clinical trials are looking at ways to prevent pregnant mothers infected with HIV from passing the infection to their unborn children, a major source of HIV infections in Africa. Some people say these trials are unethical because they do not give effective AIDS drugs to their subjects, as would be required in rich nations. Others reply that the trials are looking for treatments that can work in the real world in Africa and that they promise benefits at least to the children of their subjects. What do you think?

## Behavioral and social science experiments

When we move from medicine to the behavioral and social sciences, the direct risks to experimental subjects are less acute, but so are the possible benefits to the subjects. Consider, for example, the experiments conducted by psychologists in their study of human behavior.

EXAMPLE

**3.40 Personal space.** Psychologists observe that people have a "personal space" and get annoyed if others come too close to them. We don't like strangers to sit at our table in a coffee shop if other tables are available, and we see people move apart in elevators if there is room to do so. Americans tend to require more personal space than people in most other cultures. Can violations of personal space have physical, as well as emotional, effects?

Investigators set up shop in a men's public rest room. They blocked off urinals to force men walking in to use either a urinal next to an experimenter (treatment group) or a urinal separated from the experimenter (control group). Another experimenter, using a periscope from a toilet stall, measured how long the subject took to start urinating and how long he kept at it.[50]

This personal space experiment illustrates the difficulties facing those who plan and review behavioral studies.

- There is no risk of harm to the subjects, although they would certainly object to being watched through a periscope. What should we protect subjects from when physical harm is unlikely? Possible emotional harm? Undignified situations? Invasion of privacy?
- What about informed consent? The subjects in Example 3.40 did not even know they were participating in an experiment. Many behavioral experiments rely on hiding the true purpose of the study. The subjects would change their behavior if told in advance what the investigators were looking for. Subjects are asked to consent on the basis of vague information. They receive full information only after the experiment.

The "Ethical Principles" of the American Psychological Association require consent unless a study merely observes behavior in a public place. They allow deception only when it is necessary to the study, does not hide information that might influence a subject's willingness to participate, and is explained to subjects as soon as possible. The personal space study (from the 1970s) does not meet current ethical standards.

We see that the basic requirement for informed consent is understood differently in medicine and psychology. Here is an example of another setting with yet another interpretation of what is ethical. The subjects get no information and give no consent. They don't even know that an experiment may be sending them to jail for the night.

EXAMPLE

**3.41 Domestic violence.** How should police respond to domestic-violence calls? In the past, the usual practice was to remove the offender and order him to stay out of the household overnight. Police were reluctant to make arrests because the victims rarely pressed charges. Women's groups argued that arresting offenders would help prevent future violence even if no charges were filed. Is there evidence that arrest will reduce future offenses? That's a question that experiments have tried to answer.

A typical domestic-violence experiment compares two treatments: arrest the suspect and hold him overnight, or warn the suspect and release him. When police officers reach the scene of a domestic-violence call, they calm the participants and investigate. Weapons or death threats require an arrest. If the facts permit an arrest but do not require it, an officer radios headquarters for instructions. The person on duty opens the next envelope in a file prepared in advance by a statistician. The envelopes contain the treatments in random order. The police either arrest the suspect or warn and release him, depending on the contents of the envelope. The researchers then watch police records and visit the victim to see if the domestic violence reoccurs.

The first such experiment appeared to show that arresting domestic-violence suspects does reduce their future violent behavior. As a result of this evidence, arrest has become the common police response to domestic violence.

The domestic-violence experiments shed light on an important issue of public policy. Because there is no informed consent, the ethical rules that govern clinical trials and most social science studies would forbid these experiments. They were cleared by review boards because, in the words of one domestic-violence researcher, "These people became subjects by committing acts that allow the police to arrest them. You don't need consent to arrest someone."

## SECTION 3.4 Summary

Approval of an **institutional review board** is required for studies that involve human or animals as subjects.

Human subjects must give **informed consent** if they are to participate in experiments.

Data on human subjects must be kept **confidential.**

## SECTION 3.4 Exercises

*For Exercises 3.96 and 3.97, see page 226; for Exercises 3.98 and 3.99, see page 228; and for Exercises 3.100 and 3.101, see page 230.*

**3.102 What is wrong?** Explain what is wrong in each of the following scenarios.

(a) Clinical trials are always ethical as long as they randomly assign patients to the treatments.

(b) The job of an institutional review board is complete when they decide to allow a study to be conducted.

(c) A treatment that has no risk of physical harm to subjects is always ethical.

**3.103 Serving as an experimental subject for extra credit.** Students taking Psychology 001 are required to serve as experimental subjects. Students in Psychology 002 are not required to serve, but they are given extra credit if they do so. Students in Psychology 003 are required either to sign up as subjects or to write a term paper. Serving as an experimental subject may be educational, but current ethical standards frown on using "dependent subjects" such as prisoners or charity medical patients. Students are certainly somewhat dependent on their teachers. Do you object to any of these course policies? If so, which ones, and why?

**3.104 Informed consent to take blood samples.** Researchers from Yale, working with medical teams in Tanzania, wanted to know how common infection with the AIDS virus is among pregnant women in that country. To do this, they planned to test blood samples drawn from pregnant women.

Yale's institutional review board insisted that the researchers get the informed consent of each woman and tell her the results of the test. This is the usual procedure in developed nations. The Tanzanian government did not want to tell the women why blood was drawn or tell them the test results. The government feared panic if many people turned out to have an incurable disease for which the country's medical system could not provide care. The study was canceled. Do you think that Yale was right to apply its usual standards for protecting subjects?

**3.105 The General Social Survey.** One of the most important nongovernment surveys in the United States is the National Opinion Research Center's General Social Survey. The GSS regularly monitors public opinion on a wide variety of political and social issues. Interviews are conducted in person in the subject's home. Are a subject's responses to GSS questions anonymous, confidential, or both? Explain your answer.

**3.106 Anonymity and confidentiality in health screening.** Texas A&M, like many universities, offers free screening for HIV, the virus that causes AIDS. The announcement says, "Persons who sign up for the HIV Screening will be assigned a number so that they do not have to give their name." They can learn the results of the test by telephone, still without giving their name. Does this practice offer *anonymity* or just *confidentiality?*

**3.107 Anonymity and confidentiality in mail surveys.** Some common practices may appear to offer anonymity while actually delivering only confidentiality. Market researchers often use mail surveys that do not ask the respondent's identity but contain hidden codes on the questionnaire that identify the respondent. A false claim of anonymity is clearly unethical. If only confidentiality is promised, is it also unethical to say nothing about the identifying code, perhaps causing respondents to believe their replies are anonymous?

**3.108 Use of stored blood.** Long ago, doctors drew a blood specimen from you as part of treating minor anemia. Unknown to you, the sample was stored. Now researchers plan to use stored samples from you and many other people to look for genetic factors that may influence anemia. It is no longer possible to ask your consent. Modern technology can read your entire genetic makeup from the blood sample.

(a) Do you think it violates the principle of informed consent to use your blood sample if your name is on it but you were not told that it might be saved and studied later?

(b) Suppose that your identity is not attached. The blood sample is known only to come from (say) "a 20-year-old white female being treated for anemia." Is it now OK to use the sample for research?

(c) Perhaps we should use biological materials such as blood samples only from patients who have agreed to allow the material to be stored for later use in research. It isn't possible to say in advance what kind of research, so this falls short of the usual standard for informed consent. Is it nonetheless acceptable, given complete confidentiality and the fact that using the sample can't physically harm the patient?

**3.109 Testing vaccines.** One of the most important goals of AIDS research is to find a vaccine that will protect against HIV. Because AIDS is so common in parts of Africa, that is the easiest place to test a vaccine. It is likely, however, that a vaccine would be so expensive that it could not (at least at first) be widely used in Africa. Is it ethical to test in Africa if the benefits go mainly to rich countries? The treatment group of subjects would get the vaccine, and the placebo group would later be given the vaccine if it proved effective. So the actual subjects would benefit—it is the future benefits that would go elsewhere. What do you think?

**3.110 Political polls.** The presidential election campaign is in full swing, and the candidates have hired polling organizations to take regular polls to find out what the voters think about the issues. What information should the pollsters be required to give out?

(a) What does the standard of informed consent require the pollsters to tell potential respondents?

(b) The standards accepted by polling organizations also require giving respondents the name and address of the organization that carries out the poll. Why do you think this is required?

(c) The polling organization usually has a professional name such as "Samples Incorporated," so respondents don't know that the poll is being paid for by a political party or candidate. Would revealing the sponsor to respondents bias the poll? Should the sponsor always be announced whenever poll results are made public?

**3.111 Should poll results be made public?** Some people think that the law should require that all political poll results be made public. Otherwise, the possessors of poll results can use the information to their own advantage. They can act on the information, release only selected parts of it, or time the release for best effect. A candidate's organization replies that they are paying for the poll in order to gain information for their own use, not to amuse the public. Do you favor requiring complete disclosure of political poll results? What about other private surveys, such as market research surveys of consumer tastes?

**3.112 The 2000 census.** The 2000 census long form asked 53 detailed questions, for example:

*Do you have COMPLETE plumbing facilities in this house, apartment, or mobile home; that is, 1) hot and and cold piped water, 2) a flush toilet, and 3) a bathtub or shower?*

The form also asked your income in dollars, broken down by source, and whether any "physical, mental, or emotional condition" causes you difficulty in "learning, remembering, or concentrating." Some members of Congress objected to these questions, even though Congress had approved them.

Give brief arguments on both sides of the debate over the long form: the government has legitimate uses for such information, but the questions seem to invade people's privacy.

## CHAPTER 3 Exercises

**3.113 Select a random sample of workers.** The WORKERS data set contains information about 14,959 people aged 25 to 64 whose highest level of education is a bachelor's degree.

(a) In order to select an SRS of these people, how would you assign labels?

(b) Use Table B at line 185 to choose the first 3 members of the SRS.

**3.114 Cash bonuses for the unemployed.** Will cash bonuses speed the return to work of unemployed people? The Illinois Department of Employment Security designed an experiment to find out. The subjects were 10,065 people aged 20 to 54 who were filing claims for unemployment insurance. Some were offered \$500 if they found a job within 11 weeks and held it for at least 4 months. Others could tell potential employers that the state would pay the employer \$500 for hiring them. A control group got neither kind of bonus.[51]

(a) Suggest a few response variables of interest to the state and outline the design of the experiment.

(b) How will you label the subjects for random assignment? Use Table B at line 167 to choose the first 3 subjects for the first treatment.

**3.115 Name the designs.** What is the name for each of these study designs?

(a) A study to compare two methods of preserving wood started with boards of southern white pine. Each board was ripped from end to end to form two edge-matched specimens. One was assigned to Method A; the other to Method B.

(b) A survey on youth and smoking contacted by telephone 300 smokers and 300 nonsmokers, all 14 to 22 years of age.

(c) Does air pollution induce DNA mutations in mice? Starting with 40 male and 40 female mice, 20 of each sex were housed in a polluted industrial area downwind from a steel mill. The other 20 of each sex were housed at an unpolluted rural location 30 kilometers away.

**3.116 Prostate treatment study using Canada's national health records.** A large observational study used records from Canada's national health care system to compare the effectiveness of two ways to treat prostate disease. The two treatments are traditional surgery and a new method that does not require surgery. The records described many patients whose doctors had chosen one or the other method. The study found that patients treated by the new method were significantly more likely to die within 8 years.[52]

(a) Further study of the data showed that this conclusion was wrong. The extra deaths among patients who received the new treatment could be explained by lurking variables. What lurking variables might be confounded with a doctor's choice of surgical or nonsurgical treatment?

(b) You have 300 prostate patients who are willing to serve as subjects in an experiment to compare the two methods. Use a diagram to outline the design of a randomized comparative experiment.

**3.117 Price promotions and consumers' expectations.** A researcher studying the effect of price promotions on consumers' expectations makes up two different histories of the store price of a hypothetical brand of laundry detergent for the past year. Students in a marketing course view one or the other price history on a computer.

Some students see a steady price, while others see regular promotions that temporarily cut the price. Then the students are asked what price they would expect to pay for the detergent. Is this study an experiment? Why? What are the explanatory and response variables?

**3.118 What type of study?** What is the best way to answer each of the questions below: an experiment, a sample survey, or an observational study that is not a sample survey? Explain your choices.

(a) Are people generally satisfied with how things are going in the country right now?

(b) Do college students learn basic accounting better in a classroom or using an online course?

(c) How long do your teachers wait on the average after they ask their class a question?

**3.119 Choose the type of study.** Give an example of a question about college students, their behavior, or their opinions that would best be answered by

(a) a sample survey.

(b) an observational study that is not a sample survey.

(c) an experiment.

**3.120 Compare the burgers.** Do consumers prefer the taste of a cheeseburger from McDonald's or from Wendy's in a blind test in which neither burger is identified? Describe briefly the design of a matched pairs experiment to investigate this question. How will you use randomization?

**3.121 Bicycle gears.** How does the time it takes a bicycle rider to travel 100 meters depend on which gear is used and how steep the course is? It may be, for example, that higher gears are faster on the level but lower gears are faster on steep inclines. Discuss the design of a two-factor experiment to investigate this issue, using one bicycle with three gears and one rider. How will you use randomization?

**3.122** CHALLENGE **Design an experiment.** The previous two exercises illustrate the use of statistically designed experiments to answer questions that arise in everyday life. Select a question of interest to you that an experiment might answer and carefully discuss the design of an appropriate experiment.

**3.123** CHALLENGE **Design a survey.** You want to investigate the attitudes of students at your school about the faculty's commitment to teaching. The student government will pay the costs of contacting about 500 students.

(a) Specify the exact population for your study; for example, will you include part-time students?

(b) Describe your sample design. Will you use a stratified sample?

(c) Briefly discuss the practical difficulties that you anticipate; for example, how will you contact the students in your sample?

**3.124 Compare two doses of a drug.** A drug manufacturer is studying how a new drug behaves in patients. Investigators compare two doses: 5 milligrams (mg) and 10 mg. The drug can be administered by injection, by a skin patch, or by intravenous drip. Concentration in the blood after 30 minutes (the response variable) may depend both on the dose and on the method of administration.

(a) Make a sketch that describes the treatments formed by combining dosage and method. Then use a diagram to outline a completely randomized design for this two-factor experiment.

(b) "How many subjects?" is a tough issue. We will explain the basic ideas in Chapter 6. What can you say now about the advantage of using larger groups of subjects?

**3.125 Discolored french fries.** Few people want to eat discolored french fries. Potatoes are kept refrigerated before being cut for french fries to prevent spoiling and preserve flavor. But immediate processing of cold potatoes causes discoloring due to complex chemical reactions. The potatoes must therefore be brought to room temperature before processing. Design an experiment in which tasters will rate the color and flavor of french fries prepared from several groups of potatoes. The potatoes will be fresh picked or stored for a month at room temperature or stored for a month refrigerated. They will then be sliced and cooked either immediately or after an hour at room temperature.

(a) What are the factors and their levels, the treatments, and the response variables?

(b) Describe and outline the design of this experiment.

(c) It is efficient to have each taster rate fries from all treatments. How will you use randomization in presenting fries to the tasters?

**3.126 Would the results be different for men and women?** The drug that is the subject of the experiment in Exercise 3.124 may behave differently in men and women. How would you modify your experimental design to take this into account?

**3.127** CHALLENGE **Informed consent.** The requirement that human subjects give their informed consent to participate in an experiment can greatly reduce the number of available subjects. For example, a study of new teaching methods asks the consent of parents for their children to be randomly assigned to be taught by either a new method or the standard method. Many parents do not return the forms, so their children must continue to be taught by the standard method. Why is it not correct to consider these children as part of the control group along with children who are randomly assigned to the standard method?

**3.128** CHALLENGE **Two ways to ask sensitive questions.** Sample survey questions are usually read from a computer screen. In a Computer Aided Personal Interview (CAPI), the interviewer reads the questions and enters the responses. In a Computer Aided Self Interview (CASI), the interviewer stands aside and the respondent reads the questions and enters responses. One method almost always shows a higher percent of subjects admitting use of illegal drugs. Which method? Explain why.

**3.129 Your institutional review board.** Your college or university has an institutional review board that screens all studies that use human subjects. Get a copy of the document that describes this board (you can probably find it online).

(a) According to this document, what are the duties of the board?

(b) How are members of the board chosen? How many members are not scientists? How many members are not employees of the college? Do these members have some special expertise, or are they simply members of the "general public"?

**3.130 Use of data produced by the government.** Data produced by the government are often available free or at low cost to private users. For example, satellite weather data produced by the U.S. National Weather Service are available free to TV stations for their weather reports and to anyone on the Web. *Opinion 1:* Government data should be available to everyone at minimal cost. European governments, on the other hand, charge TV stations for weather data. *Opinion 2:* The satellites are expensive, and the TV stations are making a profit from their weather services, so they should share the cost. Which opinion do you support, and why?

**3.131 Should we ask for the consent of the parents?** The Centers for Disease Control and Prevention, in a survey of teenagers, asked the subjects if they were sexually active. Those who said "Yes" were then asked,

*How old were you when you had sexual intercourse for the first time?*

Should consent of parents be required to ask minors about sex, drugs, and other such issues, or is consent of the minors themselves enough? Give reasons for your opinion.

**3.132 A theft experiment.** Students sign up to be subjects in a psychology experiment. When they arrive, they are told that interviews are running late and are taken to a waiting room. The experimenters then stage a theft of a valuable object left in the waiting room. Some subjects are alone with the thief, and others are in pairs—these are the treatments being compared. Will the subject report the theft? The students had agreed to take part in an unspecified study, and the true nature of the experiment is explained to them afterward. Do you think this study is ethically OK?

**3.133 A cheating experiment.** A psychologist conducts the following experiment: she measures the attitude of subjects toward cheating, then has them play a game rigged so that winning without cheating is impossible. The computer that organizes the game also records—unknown to the subjects—whether or not they cheat. Then attitude toward cheating is retested. Subjects who cheat tend to change their attitudes to find cheating more acceptable. Those who resist the temptation to cheat tend to condemn cheating more strongly on the second test of attitude. These results confirm the psychologist's theory. This experiment tempts subjects to cheat. The subjects are led to believe that they can cheat secretly when in fact they are observed. Is this experiment ethically objectionable? Explain your position.

CHAPTER

4

# Probability: The Study of Randomness

Annie Duke, professional poker player, with a large stack of chips at the World Series of Poker. See Example 4.2 to learn more about probability and Texas hold 'em.



## Introduction

The reasoning of statistical inference rests on asking, "How often would this method give a correct answer if I used it very many times?" When we produce data by random sampling or randomized comparative experiments, the laws of probability answer the question "What would happen if we did this many times?" Games of chance like Texas hold 'em are exciting because the outcomes are determined by the rules of probability.

## 4.1 Randomness

Toss a coin, or choose an SRS. The result can't be predicted in advance, because the result will vary when you toss the coin or choose the sample repeatedly. But there is nonetheless a regular pattern in the results, a pattern that emerges clearly only after many repetitions. This remarkable fact is the basis for the idea of probability.

LOOK BACK
**sampling distributions, page 214**

EXAMPLE

**4.1 Toss a coin 5000 times.** When you toss a coin, there are only two possible outcomes, heads or tails. Figure 4.1 shows the results of tossing a coin 5000 times twice. For each number of tosses from 1 to 5000, we have plotted the proportion of those tosses that gave a head. Trial A (solid line) begins tail, head, tail, tail. You can see that the proportion of heads for Trial A starts at 0 on the first toss, rises to 0.5 when the second toss gives a head, then falls to 0.33 and 0.25 as we get two more tails. Trial B, on the other hand, starts with five straight heads, so the proportion of heads is 1 until the sixth toss.

The proportion of tosses that produce heads is quite variable at first. Trial A starts low and Trial B starts high. As we make more and more tosses, however, the proportions of heads for both trials get close to 0.5 and stay there. If we made yet a third trial at tossing the coin a great many times, the proportion of heads would again settle down to 0.5 in the long run. We say that 0.5 is the *probability* of a head. The probability 0.5 appears as a horizontal line on the graph.

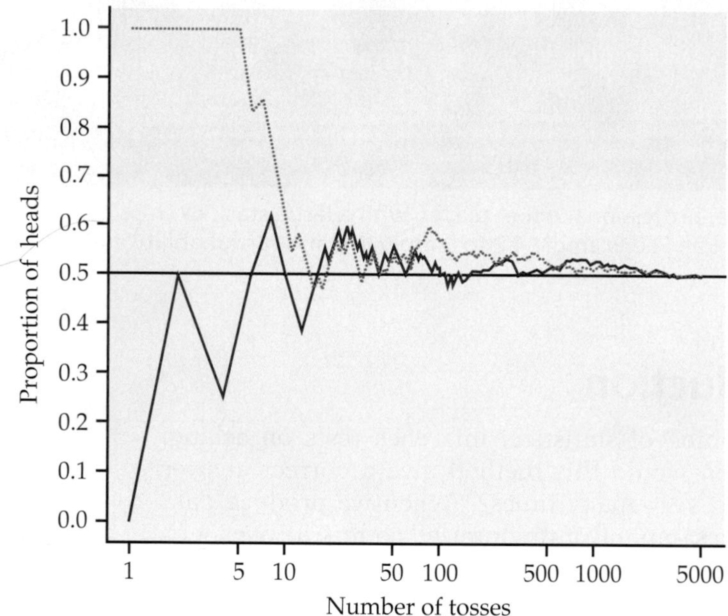


**FIGURE 4.1** The proportion of tosses of a coin that give a head varies as we make more tosses. Eventually, however, the proportion approaches 0.5, the probability of a head. This figure shows the results of two trials of 5000 tosses each.

The *Probability* applet on the text Web site animates Figure 4.1. It allows you to choose the probability of a head and simulate any number of tosses of a coin with that probability. Try it. You will see that the proportion of heads gradually settles down close to the chosen probability. Equally important, you will also see that the proportion in a small or moderate number of tosses can be far from the probability. *Probability describes only what happens in the long run. Most people expect chance outcomes to show more short-term regularity than is actually true.*

EXAMPLE

**4.2 Texas hold 'em.** In the card game Texas hold 'em, each player is dealt two cards. After a round of betting, three "community" cards, which can be used by any player, are dealt, followed by another round of betting. Then two additional community cards are dealt, with a round of betting after each. The best poker hand wins. The last community card turned is called the river. Suppose that you are dealt an ace and a king. The probability that you will get another ace or king by the river, that is, after the five community cards are dealt, is about 0.5. This means that about half of the time that you hold these cards, you will finish with a hand that has at least a pair of kings or a pair of aces.

## The language of probability

"Random" in statistics is not a synonym for "haphazard" but a description of a kind of order that emerges in the long run. We often encounter the unpredictable side of randomness in our everyday experience, but we rarely see enough repetitions of the same random phenomenon to observe the long-term regularity that probability describes. You can see that regularity emerging in Figure 4.1. In the very long run, the proportion of tosses that give a head is 0.5. This is the intuitive idea of probability. Probability 0.5 means "occurs half the time in a very large number of trials."

### RANDOMNESS AND PROBABILITY

We call a phenomenon **random** if individual outcomes are uncertain but there is nonetheless a regular distribution of outcomes in a large number of repetitions.

The **probability** of any outcome of a random phenomenon is the proportion of times the outcome would occur in a very long series of repetitions.

Real coins have bumps and imperfections that make the probability of heads a little different from 0.5. The probability might be 0.499999 or 0.500002. We call a coin **fair** if the probability of heads is exactly 0.5. For our study of probability in this chapter, we will assume that we know the actual values of probabilities. Thus, we assume things like fair coins, even though we know that real coins are not exactly fair. We do this to learn what kinds of outcomes we are likely to see when we make such assumptions. When we study statistical inference in later chapters, we look at the situation from the opposite point of view: given that we have observed certain outcomes, what can we say about the probabilities that generated these outcomes?

fair coin

### USE YOUR KNOWLEDGE

**4.1** **Use Table B.** We can use the random digits in Table B in the back of the text to simulate tossing a fair coin. Start at line 109 and read the numbers from left to right. If the number is 0, 1, 2, 3, or 4, you will say that the coin toss resulted in a head; if the number is a 5, 6, 7, 8, or 9, the outcome is tails. Use the first 20 random digits on line 109 to simulate 20 tosses of a fair coin. What is the actual proportion of heads in your simulated sample? Explain why you did not get exactly 10 heads.

Probability describes what happens in very many trials, and we must actually observe many trials to pin down a probability. In the case of tossing a coin, some diligent people have in fact made thousands of tosses.

EXAMPLE

**4.3 Many tosses of a coin.** The French naturalist Count Buffon (1707–1788) tossed a coin 4040 times. Result: 2048 heads, or proportion 2048/4040 = 0.5069 for heads.

Around 1900, the English statistician Karl Pearson heroically tossed a coin 24,000 times. Result: 12,012 heads, a proportion of 0.5005.

While imprisoned by the Germans during World War II, the South African statistician John Kerrich tossed a coin 10,000 times. Result: 5067 heads, proportion of heads 0.5067.

## Thinking about randomness

That some things are random is an observed fact about the world. The outcome of a coin toss, the time between emissions of particles by a radioactive source, and the sexes of the next litter of lab rats are all random. So is the outcome of a random sample or a randomized experiment. Probability theory is the branch of mathematics that describes random behavior. Of course, we can never observe a probability exactly. We could always continue tossing the coin, for example. Mathematical probability is an idealization based on imagining what would happen in an indefinitely long series of trials.

The best way to understand randomness is to observe random behavior—not only the long-run regularity but the unpredictable results of short runs. You can do this with physical devices such as coins and dice, but software simulations of random behavior allow faster exploration. As you explore randomness, remember:

independence

- You must have a long series of **independent** trials. That is, the outcome of one trial must not influence the outcome of any other. Imagine a crooked gambling house where the operator of a roulette wheel can stop it where she chooses—she can prevent the proportion of "red" from settling down to a fixed number. These trials are not independent.
- The idea of probability is empirical. Simulations start with given probabilities and imitate random behavior, but we can estimate a real-world probability only by actually observing many trials.
- Nonetheless, simulations are very useful because we need long runs of trials. In situations such as coin tossing, the proportion of an outcome often requires several hundred trials to settle down to the probability of that out-

come. The kinds of physical random devices suggested in the exercises are too slow for this. Short runs give only rough estimates of a probability.

## The uses of probability

Probability theory originated in the study of games of chance. Tossing dice, dealing shuffled cards, and spinning a roulette wheel are examples of deliberate randomization. In that respect, they are similar to random sampling. Although games of chance are ancient, they were not studied by mathematicians until the sixteenth and seventeenth centuries. It is only a mild simplification to say that probability as a branch of mathematics arose when seventeenth-century French gamblers asked the mathematicians Blaise Pascal and Pierre de Fermat for help. Gambling is still with us, in casinos and state lotteries. We will make use of games of chance as simple examples that illustrate the principles of probability.

Careful measurements in astronomy and surveying led to further advances in probability in the eighteenth and nineteenth centuries because the results of repeated measurements are random and can be described by distributions much like those arising from random sampling. Similar distributions appear in data on human life span (mortality tables) and in data on lengths or weights in a population of skulls, leaves, or cockroaches.[1] Now, we employ the mathematics of probability to describe the flow of traffic through a highway system, the Internet, or a computer processor; the genetic makeup of individuals or populations; the energy states of subatomic particles; the spread of epidemics or rumors; and the rate of return on risky investments. Although we are interested in probability because of its usefulness in statistics, the mathematics of chance is important in many fields of study.

## SECTION 4.1 Summary

A **random phenomenon** has outcomes that we cannot predict but that nonetheless have a regular distribution in very many repetitions.

The **probability** of an event is the proportion of times the event occurs in many repeated trials of a random phenomenon.

## SECTION 4.1 Exercises

*For Exercise 4.1, see page 239.*

**4.2** **Is music playing on the radio?** Turn on your favorite music radio station 10 times at least 10 minutes apart. Each time record whether or not music is playing. Calculate the number of times music is playing divided by 10. This number is an estimate of the probability that music is playing when you turn on this station. It is also an estimate of the proportion of time that music is playing on this station.

**4.3** **Wait 5 seconds between each observation.** Refer to the previous exercise. Explain why you would not want to wait only 5 seconds between each time you turn the radio station on.

**4.4** **Winning at craps.** The game of craps starts with a "come-out" roll where the shooter rolls a pair of dice. If the total is 7 or 11, the shooter wins immediately (there are ways that the shooter can win on later rolls if other numbers are rolled on the come-out roll). Roll a pair of dice 25 times and estimate the probability that the shooter wins immediately on the come-out roll. For a pair of perfectly made dice, the probability is 0.2222.

**4.5** **The color of candy.** It is reasonable to think that packages of M&M's Milk Chocolate Candies are filled at the factory with candies chosen at random from the very large number produced. So a package of M&M's contains a number of repetitions of a random phenomenon: choosing a candy at random and noting its color. What is the probability that an M&M's Milk Chocolate Candy is green? To find out, buy one or more packs. How many candies did you examine? How many were green? What is your estimate of the probability that a randomly chosen candy is green?

4.6 **Side effects of eyedrops.** You go to the doctor and she prescribes a medicine for an eye infection that you have. Suppose that the probability of a serious side effect from the medicine is 0.00001. Explain in simple terms what this number means.

4.7 APPLET **Simulate free throws.** The basketball player Shaquille O'Neal makes about half of his free throws over an entire season. Use Table B or the *Probability* applet to simulate 100 free throws shot independently by a player who has probability 0.5 of making each shot.

(a) What percent of the 100 shots did he hit?

(b) Examine the sequence of hits and misses. How long was the longest run of shots made? Of shots missed? (Sequences of random outcomes often show runs longer than our intuition thinks likely.)

4.8 APPLET **Use the *Probability* applet.** The idea of probability is that the *proportion* of heads in many tosses of a balanced coin eventually gets close to 0.5. But does the actual *count* of heads get close to one-half the number of tosses? Let's find out. Set the "Probability of heads" in the *Probability* applet to 0.5 and the number of tosses to 40. You can extend the number of tosses by clicking "Toss" again to get 40 more. Don't click "Reset" during this exercise.

(a) After 40 tosses, what is the proportion of heads? What is the count of heads? What is the difference between the count of heads and 20 (one-half the number of tosses)?

(b) Keep going to 120 tosses. Again record the proportion and count of heads and the difference between the count and 60 (half the number of tosses).

(c) Keep going. Stop at 240 tosses and again at 480 tosses to record the same facts. Although it may take a long time, the laws of probability say that the proportion of heads will always get close to 0.5 and also that the difference between the count of heads and half the number of tosses will always grow without limit.

4.9 APPLET **A question about dice.** Here is a question that a French gambler asked the mathematicians Fermat and Pascal at the very beginning of probability theory: what is the probability of getting at least one six in rolling four dice? The *Law of Large Numbers* applet allows you to roll several dice and watch the outcomes. (Ignore the title of the applet for now.) Because simulation—just like real random phenomena—often takes very many trials to estimate a probability accurately, let's simplify the question: is this probability clearly greater than 0.5, clearly less than 0.5, or quite close to 0.5? Use the applet to roll four dice until you can confidently answer this question. You will have to set "Rolls" to 1 so that you have time to look at the four up-faces. Keep clicking "Roll dice" to roll again and again. How many times did you roll four dice? What percent of your rolls produced at least one six?

## 4.2 Probability Models

The idea of probability as a proportion of outcomes in very many repeated trials guides our intuition but is hard to express in mathematical form. A description of a random phenomenon in the language of mathematics is called a **probability model.** To see how to proceed, think first about a very simple random phenomenon, tossing a coin once. When we toss a coin, we cannot know the outcome in advance. What do we know? We are willing to say that the outcome will be either heads or tails. Because the coin appears to be balanced, we believe that each of these outcomes has probability 1/2. This description of coin tossing has two parts:

probability model

- A list of possible outcomes
- A probability for each outcome

This two-part description is the starting point for a probability model. We will begin by describing the outcomes of a random phenomenon and then learn how to assign probabilities to the outcomes.

## Sample spaces

A probability model first tells us what outcomes are possible.

> SAMPLE SPACE
>
> The **sample space $S$** of a random phenomenon is the set of all possible outcomes.

The name "sample space" is natural in random sampling, where each possible outcome is a sample and the sample space contains all possible samples. To specify $S$, we must state what constitutes an individual outcome and then state which outcomes can occur. We often have some freedom in defining the sample space, so the choice of $S$ is a matter of convenience as well as correctness. The idea of a sample space, and the freedom we may have in specifying it, are best illustrated by examples.

EXAMPLE

**4.4 Sample space for tossing a coin.** Toss a coin. There are only two possible outcomes, and the sample space is

$$S = \{\text{heads, tails}\}$$

or, more briefly, $S = \{\text{H, T}\}$.

EXAMPLE

**4.5 Sample space for random digits.** Let your pencil point fall blindly into Table B of random digits. Record the value of the digit it lands on. The possible outcomes are

$$S = \{0, 1, 2, 3, 4, 5, 6, 7, 8, 9\}$$

EXAMPLE

**4.6 Sample space for tossing a coin four times.** Toss a coin four times and record the results. That's a bit vague. To be exact, record the results of each of the four tosses in order. A typical outcome is then HTTH. Counting shows that there are 16 possible outcomes. The sample space $S$ is the set of all 16 strings of four H's and T's.

Suppose that our only interest is the number of heads in four tosses. Now we can be exact in a simpler fashion. The random phenomenon is to toss a coin four times and count the number of heads. The sample space contains only five outcomes:

$$S = \{0, 1, 2, 3, 4\}$$

This example illustrates the importance of carefully specifying what constitutes an individual outcome.

Although these examples seem remote from the practice of statistics, the connection is surprisingly close. Suppose that in conducting an opinion poll you select four people at random from a large population and ask each if he or she favors reducing federal spending on low-interest student loans. The answers are "Yes" or "No." The possible outcomes—the sample space—are exactly as in Example 4.4 if we replace heads by "Yes" and tails by "No." Similarly, the possible outcomes of an SRS of 1500 people are the same in principle as the possible outcomes of tossing a coin 1500 times. One of the great advantages of mathematics is that the essential features of quite different phenomena can be described by the same mathematical model.

## USE YOUR KNOWLEDGE

**4.10 When do you study?** A student is asked on which day of the week he or she spends the most time studying. What is the sample space?

The sample spaces described above all correspond to categorical variables where we can list all of the possible values. Other sample spaces correspond to quantitative variables. Here is an example.

EXAMPLE

**4.7 Using software.** Most statistical software has a function that will generate a random number between 0 and 1. The sample space is

$$S = \{\text{all numbers between 0 and 1}\}$$

This $S$ is a mathematical idealization. Any specific random number generator produces numbers with some limited number of decimal places so that, strictly speaking, not all numbers between 0 and 1 are possible outcomes. For example, Minitab generates random numbers like 0.736891, with six decimal places. The entire interval from 0 to 1 is easier to think about. It also has the advantage of being a suitable sample space for different software systems that produce random numbers with different numbers of digits.

## USE YOUR KNOWLEDGE

**4.11 Sample space for heights.** You record the height in inches of a randomly selected student. What is the sample space?

A sample space $S$ lists the possible outcomes of a random phenomenon. To complete a mathematical description of the random phenomenon, we must also give the probabilities with which these outcomes occur.

The true long-term proportion of any outcome—say, "exactly 2 heads in four tosses of a coin"—can be found only empirically, and then only approximately. How then can we describe probability mathematically? Rather than immediately attempting to give "correct" probabilities, let's confront the easier task of laying down rules that any assignment of probabilities must satisfy. We need to assign probabilities not only to single outcomes but also to sets of outcomes.

EVENT

An **event** is an outcome or a set of outcomes of a random phenomenon. That is, an event is a subset of the sample space.

EXAMPLE

**4.8 Exactly 2 heads in four tosses.** Take the sample space $S$ for four tosses of a coin to be the 16 possible outcomes in the form HTHH. Then "exactly 2 heads" is an event. Call this event $A$. The event $A$ expressed as a set of outcomes is

$$A = \{\text{HHTT, HTHT, HTTH, THHT, THTH, TTHH}\}$$

In a probability model, events have probabilities. What properties must any assignment of probabilities to events have? Here are some basic facts about any probability model. These facts follow from the idea of probability as "the long-run proportion of repetitions on which an event occurs."

1. **Any probability is a number between 0 and 1.** Any proportion is a number between 0 and 1, so any probability is also a number between 0 and 1. An event with probability 0 never occurs, and an event with probability 1 occurs on every trial. An event with probability 0.5 occurs in half the trials in the long run.
2. **All possible outcomes together must have probability 1.** Because every trial will produce an outcome, the sum of the probabilities for all possible outcomes must be exactly 1.
3. **If two events have no outcomes in common, the probability that one or the other occurs is the sum of their individual probabilities.** If one event occurs in 40% of all trials, a different event occurs in 25% of all trials, and the two can never occur together, then one or the other occurs on 65% of all trials because 40% + 25% = 65%.
4. **The probability that an event does not occur is 1 minus the probability that the event does occur.** If an event occurs in (say) 70% of all trials, it fails to occur in the other 30%. The probability that an event occurs and the probability that it does not occur always add to 100%, or 1.

## Probability rules

Formal probability uses mathematical notation to state Facts 1 to 4 more concisely. We use capital letters near the beginning of the alphabet to denote events.

If $A$ is any event, we write its probability as $P(A)$. Here are our probability facts in formal language. As you apply these rules, remember that they are just another form of intuitively true facts about long-run proportions.

### PROBABILITY RULES

**Rule 1.** The probability $P(A)$ of any event $A$ satisfies $0 \leq P(A) \leq 1$.

**Rule 2.** If $S$ is the sample space in a probability model, then $P(S) = 1$.

**Rule 3.** Two events $A$ and $B$ are **disjoint** if they have no outcomes in common and so can never occur together. If $A$ and $B$ are disjoint,

$$P(A \text{ or } B) = P(A) + P(B)$$

This is the **addition rule for disjoint events.**

**Rule 4.** The **complement** of any event $A$ is the event that $A$ does not occur, written as $A^c$. The **complement rule** states that

$$P(A^c) = 1 - P(A)$$

You may find it helpful to draw a picture to remind yourself of the meaning of complements and disjoint events. A picture like Figure 4.2 that shows the sample space $S$ as a rectangular area and events as areas within $S$ is called a **Venn diagram.** The events $A$ and $B$ in Figure 4.2 are disjoint because they do not overlap. As Figure 4.3 shows, the complement $A^c$ contains exactly the outcomes that are not in $A$.

Venn diagram

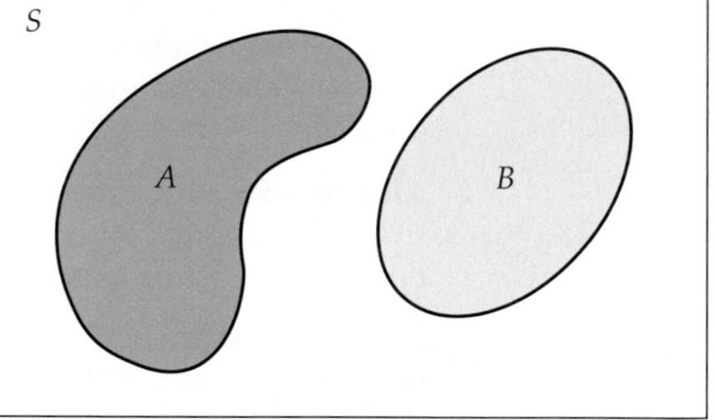


**FIGURE 4.2** Venn diagram showing disjoint events $A$ and $B$. Disjoint events have no common outcomes.

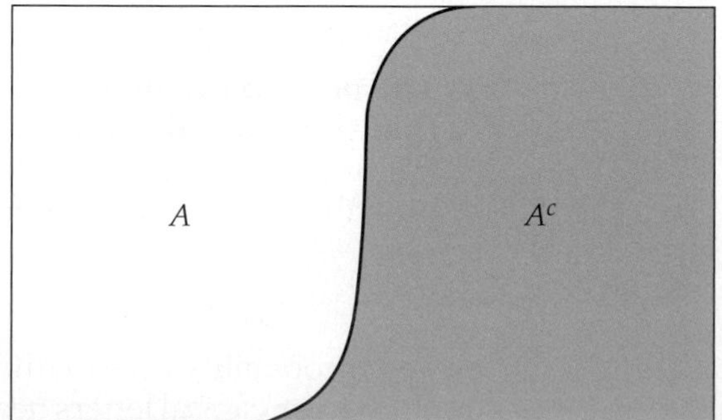


**FIGURE 4.3** Venn diagram showing the complement $A^c$ of an event $A$. The complement consists of all outcomes that are not in $A$.

EXAMPLE

**4.9 Cell phones and accidents.** Some states are considering laws that will ban the use of cell phones while driving because they believe that the ban will reduce phone-related car accidents. One study classified these types of accidents by the day of the week when they occurred.[2] For this example, we use the values from this study as our probability model. Here are the probabilities:

| Day | Sun. | Mon. | Tues. | Wed. | Thur. | Fri. | Sat. |
|---|---|---|---|---|---|---|---|
| Probability | 0.03 | 0.19 | 0.18 | 0.23 | 0.19 | 0.16 | 0.02 |

Each probability is between 0 and 1. The probabilities add to 1 because these outcomes together make up the sample space $S$. Our probability model corresponds to picking a phone-related accident at random and asking on what day of the week it occurred.

Let's use the probability rules 3 and 4 to find some probabilities of events for days when phone-related accidents occur.

EXAMPLE

**4.10 Accidents on weekends.** What is the probability that an accident occurs on a weekend, that is, Saturday or Sunday? Because an accident can occur on Saturday or Sunday but it cannot occur on both days of the week, these two events are disjoint. Using Rule 3, we find

$$\begin{aligned} P(\text{Saturday or Sunday}) &= P(\text{Saturday}) + P(\text{Sunday}) \\ &= 0.02 + 0.03 = 0.05 \end{aligned}$$

The chance that an accident occurs on a Saturday or Sunday is 5%. Suppose we want to find the probability that a phone-related accident occurs on a weekday.

EXAMPLE

**4.11 Use the complement rule.** To solve this problem, we could use Rule 3 and add the probabilities for Monday, Tuesday, Wednesday, Thursday, and Friday. However, it is easier to use the probability that we already calculated for weekends and Rule 4. The event that the accident occurs on a weekday is the complement of the event that the accident occurs on a weekend. Using our notation for events, we have

$$\begin{aligned} P(\text{weekday}) &= 1 - P(\text{weekend}) \\ &= 1 - 0.05 = 0.95 \end{aligned}$$

We see that 95% of phone-related accidents occur on weekdays.

USE YOUR KNOWLEDGE

**4.12 Phone-related accidents on Monday or Friday.** Find the probability that a phone-related accident occurred on a Monday or a Friday.

**4.13 Not on Wednesday.** Find the probability that a phone-related accident occurred on a day other than a Wednesday.

## Assigning probabilities: finite number of outcomes

The individual outcomes of a random phenomenon are always disjoint. So the addition rule provides a way to assign probabilities to events with more than one outcome: start with probabilities for individual outcomes and add to get probabilities for events. This idea works well when there are only a finite (fixed and limited) number of outcomes.

PROBABILITIES IN A FINITE SAMPLE SPACE

Assign a probability to each individual outcome. These probabilities must be numbers between 0 and 1 and must have sum 1.

The probability of any event is the sum of the probabilities of the outcomes making up the event.

EXAMPLE

**4.12 Benford's law.** Faked numbers in tax returns, payment records, invoices, expense account claims, and many other settings often display patterns that aren't present in legitimate records. Some patterns, like too many round numbers, are obvious and easily avoided by a clever crook. Others are more subtle. It is a striking fact that the first digits of numbers in legitimate records often follow a distribution known as **Benford's law.** Here it is (note that a first digit can't be 0):[3]

Benford's law

| First digit | 1 | 2 | 3 | 4 | 5 | 6 | 7 | 8 | 9 |
|---|---|---|---|---|---|---|---|---|---|
| Probability | 0.301 | 0.176 | 0.125 | 0.097 | 0.079 | 0.067 | 0.058 | 0.051 | 0.046 |

Benford's law usually applies to the first digits of the sizes of similar quantities, such as invoices, expense account claims, and county populations. Investigators can detect fraud by comparing the first digits in records such as invoices paid by a business with these probabilities.

EXAMPLE

**4.13 Find some probabilities for Benford's law.** Consider the events

$$A = \{\text{first digit is 1}\}$$

$$B = \{\text{first digit is 6 or greater}\}$$

From the table of probabilities,

$$P(A) = P(1) = 0.301$$

$$P(B) = P(6) + P(7) + P(8) + P(9)$$

$$= 0.067 + 0.058 + 0.051 + 0.046 = 0.222$$

Note that $P(B)$ is not the same as the probability that a first digit is strictly greater than 6. The probability $P(6)$ that a first digit is 6 is included in "6 or greater" but not in "greater than 6."

## USE YOUR KNOWLEDGE

**4.14 Benford's law.** Using the probabilities for Benford's law, find the probability that a first digit is anything other than 1.

**4.15 Use the addition rule.** Use the addition rule with the probabilities for the events $A$ and $B$ from Example 4.13 to find the probability that a first digit is either 1 or 6 or greater.

Be careful to apply the addition rule only to disjoint events.

EXAMPLE

**4.14 Apply the addition rule to Benford's law.** Check that the probability of the event $C$ that a first digit is odd is

$$P(C) = P(1) + P(3) + P(5) + P(7) + P(9) = 0.609$$

The probability

$$P(B \text{ or } C) = P(1) + P(3) + P(5) + P(6) + P(7) + P(8) + P(9) = 0.727$$

is *not* the sum of $P(B)$ and $P(C)$, because events $B$ and $C$ are not disjoint. Outcomes 7 and 9 are common to both events.

## Assigning probabilities: equally likely outcomes

Assigning correct probabilities to individual outcomes often requires long observation of the random phenomenon. In some circumstances, however, we are willing to assume that individual outcomes are equally likely because of some balance in the phenomenon. Ordinary coins have a physical balance that should make heads and tails equally likely, for example, and the table of random digits comes from a deliberate randomization.

EXAMPLE

**4.15 First digits that are equally likely.** You might think that first digits are distributed "at random" among the digits 1 to 9 in business records. The 9 possible outcomes would then be equally likely. The sample space for a single digit is

$$S = \{1, 2, 3, 4, 5, 6, 7, 8, 9\}$$

Because the total probability must be 1, the probability of each of the 9 outcomes must be 1/9. That is, the assignment of probabilities to outcomes is

| First digit | 1 | 2 | 3 | 4 | 5 | 6 | 7 | 8 | 9 |
|---|---|---|---|---|---|---|---|---|---|
| Probability | 1/9 | 1/9 | 1/9 | 1/9 | 1/9 | 1/9 | 1/9 | 1/9 | 1/9 |

The probability of the event $B$ that a randomly chosen first digit is 6 or greater is

$$\begin{aligned} P(B) &= P(6) + P(7) + P(8) + P(9) \\ &= \frac{1}{9} + \frac{1}{9} + \frac{1}{9} + \frac{1}{9} = \frac{4}{9} = 0.444 \end{aligned}$$

Compare this with the Benford's law probability in Example 4.13. A crook who fakes data by using "random" digits will end up with too many first digits 6 or greater and too few 1s and 2s.

In Example 4.15 all outcomes have the same probability. Because there are 9 equally likely outcomes, each must have probability 1/9. Because exactly 4 of the 9 equally likely outcomes are 6 or greater, the probability of this event is 4/9. In the special situation where all outcomes are equally likely, we have a simple rule for assigning probabilities to events.

## EQUALLY LIKELY OUTCOMES

If a random phenomenon has $k$ possible outcomes, all equally likely, then each individual outcome has probability $1/k$. The probability of any event $A$ is

$$\begin{aligned} P(A) &= \frac{\text{count of outcomes in } A}{\text{count of outcomes in } S} \\ &= \frac{\text{count of outcomes in } A}{k} \end{aligned}$$

Most random phenomena do not have equally likely outcomes, so the general rule for finite sample spaces is more important than the special rule for equally likely outcomes.

## USE YOUR KNOWLEDGE

**4.16 Possible outcomes for rolling a die.** A die has six sides with 1 to 6 "spots" on the sides. Give the probability distribution for the six possible outcomes that can result when a perfect die is rolled.

## Independence and the multiplication rule

Rule 3, the addition rule for disjoint events, describes the probability that *one or the other* of two events $A$ and $B$ will occur in the special situation when $A$ and $B$ cannot occur together because they are disjoint. Our final rule describes the probability that *both* events $A$ and $B$ occur, again only in a special situation. More general rules appear in Section 4.5, but in our study of statistics we will need only the rules that apply to special situations.

Suppose that you toss a balanced coin twice. You are counting heads, so two events of interest are

$$A = \{\text{first toss is a head}\}$$

$$B = \{\text{second toss is a head}\}$$

The events $A$ and $B$ are not disjoint. They occur together whenever both tosses give heads. We want to compute the probability of the event $\{A \text{ and } B\}$ that *both* tosses are heads. The Venn diagram in Figure 4.4 illustrates the event $\{A \text{ and } B\}$ as the overlapping area that is common to both $A$ and $B$.

The coin tossing of Buffon, Pearson, and Kerrich described in Example 4.3 makes us willing to assign probability 1/2 to a head when we toss a coin. So

$$P(A) = 0.5$$

$$P(B) = 0.5$$

What is $P(A \text{ and } B)$? Our common sense says that it is 1/4. The first coin will give a head half the time and then the second will give a head on half of those trials, so both coins will give heads on $1/2 \times 1/2 = 1/4$ of all trials in the long run. This reasoning assumes that the second coin still has probability 1/2 of a head after the first has given a head. This is true—we can verify it by tossing two coins many times and observing the proportion of heads on the second toss after the first toss has produced a head. We say that the events "head on the first toss" and "head on the second toss" are *independent*. Here is our final probability rule.

> THE MULTIPLICATION RULE FOR INDEPENDENT EVENTS
>
> **Rule 5.** Two events $A$ and $B$ are **independent** if knowing that one occurs does not change the probability that the other occurs. If $A$ and $B$ are independent,
>
> $$P(A \text{ and } B) = P(A)P(B)$$
>
> This is the **multiplication rule for independent events.**

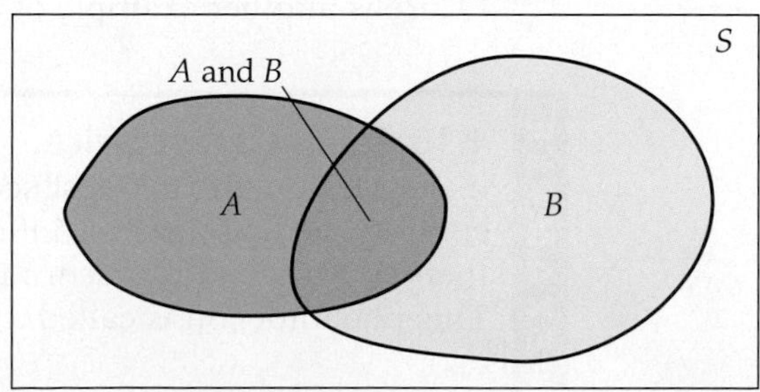


**FIGURE 4.4** Venn diagram showing the event $\{A \text{ and } B\}$. This event consists of outcomes common to $A$ and $B$.

Our definition of independence is rather informal. We will make this informal idea precise in Section 4.5. In practice, though, we rarely need a precise definition of independence, because independence is usually *assumed* as part of a probability model when we want to describe random phenomena that seem to be physically unrelated to each other. Here is an example of independence.

EXAMPLE

**4.16 Coins do not have memory.** Because a coin has no memory and most coin tossers cannot influence the fall of the coin, it is safe to assume that successive coin tosses are independent. For a balanced coin this means that after we see the outcome of the first toss, we still assign probability 1/2 to heads on the second toss.

## USE YOUR KNOWLEDGE

**4.17 Two tails in two tosses.** What is the probability of obtaining two tails on two tosses of a fair coin?

Here is an example of a situation where there are dependent events.

EXAMPLE

**4.17 Dependent events in cards.** The colors of successive cards dealt from the same deck are not independent. A standard 52-card deck contains 26 red and 26 black cards. For the first card dealt from a shuffled deck, the probability of a red card is $26/52 = 0.50$ because the 52 possible cards are equally likely. Once we see that the first card is red, we know that there are only 25 reds among the remaining 51 cards. The probability that the second card is red is therefore only $25/51 = 0.49$. Knowing the outcome of the first deal changes the probabilities for the second.

## USE YOUR KNOWLEDGE

**4.18 The probability of a second ace.** A deck of 52 cards contains 4 aces, so the probability that a card drawn from this deck is an ace is 4/52. If we know that the first card drawn is an ace, what is the probability that the second card drawn is also an ace? Using the idea of independence, explain why this probability is not 4/52.

Here is another example of a situation where events are dependent.

EXAMPLE

**4.18 Taking a test twice.** If you take an IQ test or other mental test twice in succession, the two test scores are not independent. The learning that occurs on the first attempt influences your second attempt. If you learn a lot, then your second test score might be a lot higher than your first test score. This phenomenon is called a carry-over effect.

When independence is part of a probability model, the multiplication rule applies. Here is an example.

EXAMPLE

**4.19 Mendel's peas.** Gregor Mendel used garden peas in some of the experiments that revealed that inheritance operates randomly. The seed color of Mendel's peas can be either green or yellow. Two parent plants are "crossed" (one pollinates the other) to produce seeds. Each parent plant carries two genes for seed color, and each of these genes has probability 1/2 of being passed to a seed. The two genes that the seed receives, one from each parent, determine its color. The parents contribute their genes independently of each other.

Suppose that both parents carry the $G$ and the $Y$ genes. The seed will be green if both parents contribute a $G$ gene; otherwise it will be yellow. If $M$ is the event that the male contributes a $G$ gene and $F$ is the event that the female contributes a $G$ gene, then the probability of a green seed is

$$\begin{aligned} P(M \text{ and } F) &= P(M)P(F) \\ &= (0.5)(0.5) = 0.25 \end{aligned}$$

In the long run, 1/4 of all seeds produced by crossing these plants will be green.

*The multiplication rule applies only to independent events; you cannot use it if events are not independent.* Here is a distressing example of misuse of the multiplication rule.

EXAMPLE

**4.20 Sudden infant death syndrome.** Sudden infant death syndrome (SIDS) causes babies to die suddenly (often in their cribs) with no explanation. Deaths from SIDS have been greatly reduced by placing babies on their backs, but as yet no cause is known.

When more than one SIDS death occurs in a family, the parents are sometimes accused. One "expert witness" popular with prosecutors in England told juries that there is only a 1 in 73 million chance that two children in the same family could have died naturally. Here's his calculation: the rate of SIDS in a nonsmoking middle-class family is 1 in 8500. So the probability of two deaths is

$$\frac{1}{8500} \times \frac{1}{8500} = \frac{1}{72{,}250{,}000}$$

Several women were convicted of murder on this basis, without any direct evidence that they had harmed their children.

As the Royal Statistical Society said, this reasoning is nonsense. It assumes that SIDS deaths in the same family are independent events. The cause of SIDS is unknown: "There may well be unknown genetic or environmental factors that predispose families to SIDS, so that a second case within the family becomes much more likely."[4] The British government decided to review the cases of 258 parents convicted of murdering their babies.

The multiplication rule $P(A \text{ and } B) = P(A)P(B)$ holds if $A$ and $B$ are independent but not otherwise. The addition rule $P(A \text{ or } B) = P(A) + P(B)$ holds if $A$ and $B$ are disjoint but not otherwise. Resist the temptation to use these simple formulas when the circumstances that justify them are not present. *You must also be certain not to confuse disjointness and independence. Disjoint events cannot be independent.* If $A$ and $B$ are disjoint, then the fact that $A$ occurs tells us that $B$ cannot occur—look again at Figure 4.2. Unlike disjointness or complements, independence cannot be pictured by a Venn diagram, because it involves the probabilities of the events rather than just the outcomes that make up the events.

## Applying the probability rules

If two events $A$ and $B$ are independent, then their complements $A^c$ and $B^c$ are also independent and $A^c$ is independent of $B$. Suppose, for example, that 75% of all registered voters in a suburban district are Republicans. If an opinion poll interviews two voters chosen independently, the probability that the first is a Republican and the second is not a Republican is $(0.75)(0.25) = 0.1875$. The multiplication rule also extends to collections of more than two events, provided that all are independent. Independence of events $A$, $B$, and $C$ means that no information about any one or any two can change the probability of the remaining events. The formal definition is a bit messy. Fortunately, independence is usually assumed in setting up a probability model. We can then use the multiplication rule freely, as in this example.

By combining the rules we have learned, we can compute probabilities for rather complex events. Here is an example.

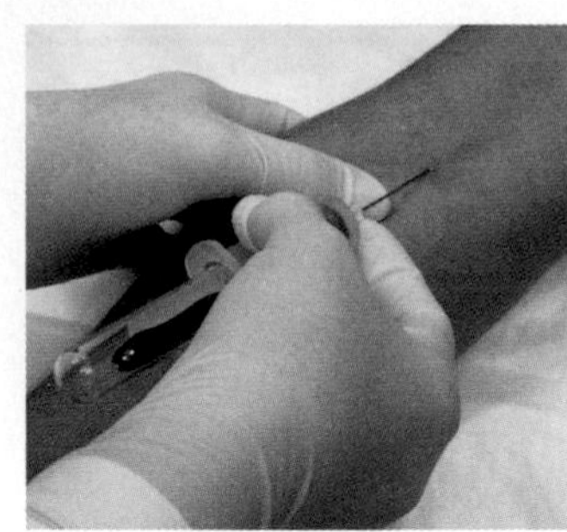

EXAMPLE

**4.21 HIV testing.** Many people who come to clinics to be tested for HIV, the virus that causes AIDS, don't come back to learn the test results. Clinics now use "rapid HIV tests" that give a result in a few minutes. Applied to people who have no HIV antibodies, one rapid test has probability about 0.004 of producing a false-positive (that is, of falsely indicating that antibodies are present).[5] If a clinic tests 200 people who are free of HIV antibodies, what is the probability that at least one false-positive will occur?

It is reasonable to assume as part of the probability model that the test results for different individuals are independent. The probability that the test is positive for a single person is 0.004, so the probability of a negative result is $1 - 0.004 = 0.996$ by the complement rule. The probability of at least one false-positive among the 200 people tested is therefore

$$\begin{aligned} P(\text{at least one positive}) &= 1 - P(\text{no positives}) \\ &= 1 - P(\text{200 negatives}) \\ &= 1 - 0.996^{200} \\ &= 1 - 0.4486 = 0.5514 \end{aligned}$$

The probability is greater than 1/2 that at least one of the 200 people will test positive for HIV, even though no one has the virus.

## SECTION 4.2 Summary

A **probability model** for a random phenomenon consists of a sample space $S$ and an assignment of probabilities $P$.

The **sample space S** is the set of all possible outcomes of the random phenomenon. Sets of outcomes are called **events.** $P$ assigns a number $P(A)$ to an event $A$ as its probability.

The **complement $A^c$** of an event $A$ consists of exactly the outcomes that are not in $A$. Events $A$ and $B$ are **disjoint** if they have no outcomes in common. Events $A$ and $B$ are **independent** if knowing that one event occurs does not change the probability we would assign to the other event.

Any assignment of probability must obey the rules that state the basic properties of probability:

**Rule 1.** $0 \leq P(A) \leq 1$ for any event $A$.

**Rule 2.** $P(S) = 1$.

**Rule 3. Addition rule:** If events $A$ and $B$ are **disjoint,** then $P(A \text{ or } B) = P(A) + P(B)$.

**Rule 4. Complement rule:** For any event $A$, $P(A^c) = 1 - P(A)$.

**Rule 5. Multiplication rule:** If events $A$ and $B$ are **independent,** then $P(A \text{ and } B) = P(A)P(B)$.

## SECTION 4.2 Exercises

*For Exercises 4.10 and 4.11, see page 244; for Exercises 4.12 and 4.13, see page 248; for Exercises 4.14 and 4.15, see page 249; for Exercise 4.16, see page 250; and for Exercises 4.17 and 4.18, see page 252.*

**4.19 Evaluating Web page designs.** You are a Web page designer and you set up a page with five different links. A user of the page can click on one of the links or he or she can leave that page. Describe the sample space for the outcome of a visitor to your Web page.

**4.20 Record the length of time spent on the page.** Refer to the previous exercise. You also decide to measure the length of time a visitor spends on your page. Give the sample space for this measure.

**4.21 Distribution of blood types.** All human blood can be "ABO-typed" as one of O, A, B, or AB, but the distribution of the types varies a bit among groups of people. Here is the distribution of blood types for a randomly chosen person in the United States:

| Blood type | A | B | AB | O |
|---|---|---|---|---|
| U.S. probability | 0.40 | 0.11 | 0.04 | ? |

(a) What is the probability of type O blood in the United States?

(b) Maria has type B blood. She can safely receive blood transfusions from people with blood types O and B. What is the probability that a randomly chosen American can donate blood to Maria?

**4.22 Blood types in China.** The distribution of blood types in China differs from the U.S. distribution given in the previous exercise:

| Blood type | A | B | AB | O |
|---|---|---|---|---|
| China probability | 0.27 | 0.26 | 0.12 | 0.35 |

Choose an American and a Chinese at random, independently of each other. What is the probability that both have type O blood? What is the probability that both have the same blood type?

**4.23 Are the probabilities legitimate?** In each of the following situations, state whether or not the given assignment of probabilities to individual outcomes is legitimate, that is, satisfies the rules of probability. If not, give specific reasons for your answer.

(a) Choose a college student at random and record gender and enrollment status: $P(\text{female full-time}) = 0.46$, $P(\text{female part-time}) = 0.54$, $P(\text{male full-time}) = 0.44$, $P(\text{male part-time}) = 0.56$.

(b) Deal a card from a shuffled deck: $P(\text{clubs}) = 12/52$, $P(\text{diamonds}) = 12/52$, $P(\text{hearts}) = 12/52$, $P(\text{spades}) = 16/52$.

(c) Roll a die and record the count of spots on the up-face: $P(1) = 1/3$, $P(2) = 1/6$, $P(3) = 0$, $P(4) = 1/3$, $P(5) = 1/6$, $P(6) = 0$.

**4.24 French and English in Canada.** Canada has two official languages, English and French. Choose a Canadian at random and ask, "What is your mother tongue?" Here is the distribution of responses, combining many separate languages from the broad Asian/Pacific region:[6]

| Language | English | French | Asian/Pacific | Other |
|---|---|---|---|---|
| Probability | ? | 0.23 | 0.07 | 0.11 |

(a) What probability should replace "?" in the distribution?

(b) What is the probability that a Canadian's mother tongue is not English?

**4.25 Education levels of young adults.** Choose a young adult (age 25 to 34 years) at random. The probability is 0.12 that the person chosen did not complete high school, 0.31 that the person has a high school diploma but no further education, and 0.29 that the person has at least a bachelor's degree.

(a) What must be the probability that a randomly chosen young adult has some education beyond high school but does not have a bachelor's degree?

(b) What is the probability that a randomly chosen young adult has at least a high school education?

**4.26 Spam topics.** A majority of email messages are now "spam." Choose a spam email message at random. Here is the distribution of topics:[7]

| Topic | Adult | Financial | Health | Leisure | Products | Scams |
|---|---|---|---|---|---|---|
| Probability | 0.145 | 0.162 | 0.073 | 0.078 | 0.210 | 0.142 |

(a) What is the probability that a spam email does not concern one of these topics?

(b) Corinne is particularly annoyed by spam offering "adult" content (that is, pornography) and scams. What is the probability that a randomly chosen spam email falls into one or the other of these categories?

**4.27** CHALLENGE **Loaded dice.** There are many ways to produce crooked dice. To *load* a die so that 6 comes up too often and 1 (which is opposite 6) comes up too seldom, add a bit of lead to the filling of the spot on the 1 face. Because the spot is solid plastic, this works even with transparent dice. If a die is loaded so that 6 comes up with probability 0.2 and the probabilities of the 2, 3, 4, and 5 faces are not affected, what is the assignment of probabilities to the six faces?

**4.28 Race in the census.** The 2000 census allowed each person to choose from a long list of races. That is, in the eyes of the Census Bureau, you belong to whatever race you say you belong to. "Hispanic/Latino" is a separate category; Hispanics may be of any race. If we choose a resident of the United States at random, the 2000 census gives these probabilities:

| | Hispanic | Not Hispanic |
|---|---|---|
| Asian | 0.000 | 0.036 |
| Black | 0.003 | 0.121 |
| White | 0.060 | 0.691 |
| Other | 0.062 | 0.027 |

Let $A$ be the event that a randomly chosen American is Hispanic, and let $B$ be the event that the person chosen is white.

(a) Verify that the table gives a legitimate assignment of probabilities.

(b) What is $P(A)$?

(c) Describe $B^c$ in words and find $P(B^c)$ by the complement rule.

(d) Express "the person chosen is a non-Hispanic white" in terms of events $A$ and $B$. What is the probability of this event?

**4.29 Rh blood types.** Human blood is typed as O, A, B, or AB and also as Rh-positive or Rh-negative. ABO type and Rh-factor type are independent because they are governed by different genes. In the American population, 84% of people are Rh-positive. Use the information about ABO type in Exercise 4.21 to give the probability distribution of blood type (ABO and Rh) for a randomly chosen person.

**4.30 Are the events independent?** Exercise 4.28 assigns probabilities for the ethnic background of a randomly chosen resident of the United States. Let $A$ be the event that the person chosen is Hispanic, and let $B$ be the event that he or she is white. Are events $A$ and $B$ independent? How do you know?

**4.31 Roulette.** A roulette wheel has 38 slots, numbered 0, 00, and 1 to 36. The slots 0 and 00 are colored green, 18 of the others are red, and 18 are black. The dealer spins the wheel and at the same time rolls a small ball along the wheel in the opposite direction. The wheel is carefully balanced so that the ball is equally likely to land in any slot when the wheel slows. Gamblers can bet on various combinations of numbers and colors.

(a) What is the probability that the ball will land in any one slot?

(b) If you bet on "red," you win if the ball lands in a red slot. What is the probability of winning?

(c) The slot numbers are laid out on a board on which gamblers place their bets. One column of numbers on the board contains all multiples of 3, that is, 3, 6, 9, ..., 36. You place a "column bet" that wins if any of these numbers comes up. What is your probability of winning?

**4.32 Winning the lottery.** A state lottery's Pick 3 game asks players to choose a three-digit number, 000 to 999. The state chooses the winning three-digit number at random, so that each number has probability 1/1000. You win if the winning number contains the digits in your number, in any order.

(a) Your number is 456. What is your probability of winning?

(b) Your number is 212. What is your probability of winning?

**4.33 PINs.** The personal identification numbers (PINs) for automatic teller machines usually consist of four digits. You notice that most of your PINs have at least one 0, and you wonder if the issuers use lots of 0s to make the numbers easy to remember. Suppose that PINs are assigned at random, so that all four-digit numbers are equally likely.

(a) How many possible PINs are there?

(b) What is the probability that a PIN assigned at random has at least one 0?

**4.34 Universal blood donors.** People with type O-negative blood are universal donors. That is, any patient can receive a transfusion of O-negative blood. Only 7% of the American population have O-negative blood. If 10 people appear at random to give blood, what is the probability that at least 1 of them is a universal donor?

**4.35 Disappearing Internet sites.** Internet sites often vanish or move, so that references to them can't be followed. In fact, 13% of Internet sites referenced in papers in major scientific journals are lost within two years after publication.[8] If a paper contains seven Internet references, what is the probability that all seven are still good two years later? What specific assumptions did you make in order to calculate this probability?

**4.36 Random digit dialing.** Most sample surveys use random digit dialing equipment to call residential telephone numbers at random. The telephone-polling firm Zogby International reports that the probability that a call reaches a live person is 0.2.[9] Calls are independent.

(a) A polling firm places 5 calls. What is the probability that none of them reaches a person?

(b) When calls are made to New York City, the probability of reaching a person is only 0.08. What is the probability that none of 5 calls made to New York City reaches a person?

**4.37 Is this calculation correct?** Government data show that 6% of the American population are at least 75 years of age and that about 51% are women. Explain why it is wrong to conclude that because $(0.06)(0.51) = 0.0306$ about 3% of the population are women aged 75 or over.

**4.38 Colored dice.** Here's more evidence that our intuition about chance behavior is not very accurate. A six-sided die has four green and two red faces, all equally probable. Psychologists asked students to say which of these color sequences is most likely to come up at the beginning of a long set of rolls of this die:

RGRRR
RGRRRG
GRRRRR

More than 60% chose the second sequence.[10] What is the correct probability of each sequence?

**4.39 Random walks and stock prices.** The "random walk" theory of securities prices holds that price movements in disjoint time periods are independent of each other. Suppose that we record only whether the price is up or down each year, and that the probability that our portfolio rises in price in any one year is 0.65. (This probability is approximately correct for a portfolio containing equal dollar amounts of all common stocks listed on the New York Stock Exchange.)

(a) What is the probability that our portfolio goes up for 3 consecutive years?

(b) If you know that the portfolio has risen in price 2 years in a row, what probability do you assign to the event that it will go down next year?

(c) What is the probability that the portfolio's value moves in the same direction in both of the next 2 years?

**4.40** CHALLENGE **Axioms of probability.** Show that any assignment of probabilities to events that obeys Rules 2 and 3 on page 246 automatically obeys the complement rule (Rule 4). This implies that a mathematical treatment of probability can start from just Rules 1, 2, and 3. These rules are sometimes called *axioms* of probability.

**4.41** CHALLENGE **Independence of complements.** Show that if events $A$ and $B$ obey the multiplication rule, $P(A \text{ and } B) = P(A)P(B)$, then $A$ and the complement $B^c$ of $B$ also obey the multiplication rule, $P(A \text{ and } B^c) = P(A)P(B^c)$. That is, if events $A$ and $B$ are independent, then $A$ and $B^c$ are also independent. (*Hint:* Start by drawing a Venn diagram and noticing that the events "$A$ and $B$" and "$A$ and $B^c$" are disjoint.)

***Mendelian inheritance.*** *Some traits of plants and animals depend on inheritance of a single gene. This is called Mendelian inheritance, after Gregor Mendel (1822–1884). Exercises 4.42 to 4.45 are based on the following information about Mendelian inheritance of blood type.*

*Each of us has an ABO blood type, which describes whether two characteristics called A and B are present. Every human being has two blood type alleles (gene forms), one inherited from our mother and one from our father. Each of these alleles can be A, B, or O. Which two we inherit determines our blood type. The following table shows what our blood type is for each combination of two alleles:*

| Alleles inherited | Blood type |
|---|---|
| A and A | A |
| A and B | AB |
| A and O | A |
| B and B | B |
| B and O | B |
| O and O | O |

*We inherit each of a parent's two alleles with probability 0.5. We inherit independently from our mother and father.*

**4.42 Blood types of children.** Hannah and Jacob both have alleles A and B.

(a) What blood types can their children have?

(b) What is the probability that their next child has each of these blood types?

**4.43 Parents with alleles B and O.** Nancy and David both have alleles B and O.

(a) What blood types can their children have?

(b) What is the probability that their next child has each of these blood types?

**4.44 Two children.** Jennifer has alleles A and O. José has alleles A and B. They have two children. What is the probability that both children have blood type A? What is the probability that both children have the same blood type?

**4.45 Three children.** Jasmine has alleles A and O. Joshua has alleles B and O.

(a) What is the probability that a child of these parents has blood type O?

(b) If Jasmine and Joshua have three children, what is the probability that all three have blood type O? What is the probability that the first child has blood type O and the next two do not?

## 4.3 Random Variables

Sample spaces need not consist of numbers. When we toss a coin four times, we can record the outcome as a string of heads and tails, such as HTTH. In statistics, however, we are most often interested in numerical outcomes such as the count of heads in the four tosses. It is convenient to use a shorthand notation: Let $X$ be the number of heads. If our outcome is HTTH, then $X = 2$. If the next outcome is TTTH, the value of $X$ changes to $X = 1$. The possible values of $X$ are 0, 1, 2, 3, and 4. Tossing a coin four times will give $X$ one of these possible values. Tossing four more times will give $X$ another and probably different value. We call $X$ a *random variable* because its values vary when the coin tossing is repeated.

RANDOM VARIABLE

A **random variable** is a variable whose value is a numerical outcome of a random phenomenon.

We usually denote random variables by capital letters near the end of the alphabet, such as $X$ or $Y$. Of course, the random variables of greatest interest to us are outcomes such as the mean $\overline{x}$ of a random sample, for which we will keep the familiar notation.[11] As we progress from general rules of probability toward statistical inference, we will concentrate on random variables. When a random variable $X$ describes a random phenomenon, the sample space $S$ just lists the possible values of the random variable. We usually do not mention $S$ separately. There remains the second part of any probability model, the assignment of probabilities to events. There are two main ways of assigning probabilities to the values of a random variable. The two types of probability models that result will dominate our application of probability to statistical inference.

## Discrete random variables

We have learned several rules of probability but only one method of assigning probabilities: state the probabilities of the individual outcomes and assign probabilities to events by summing over the outcomes. The outcome probabilities must be between 0 and 1 and have sum 1. When the outcomes are numerical, they are values of a random variable. We will now attach a name to random variables having probability assigned in this way.[12]

DISCRETE RANDOM VARIABLE

A **discrete random variable** $X$ has a finite number of possible values. The **probability distribution** of $X$ lists the values and their probabilities:

| Value of $X$ | $x_1$ | $x_2$ | $x_3$ | $\cdots$ | $x_k$ |
|---|---|---|---|---|---|
| Probability | $p_1$ | $p_2$ | $p_3$ | $\cdots$ | $p_k$ |

The probabilities $p_i$ must satisfy two requirements:

**1.** Every probability $p_i$ is a number between 0 and 1.

**2.** $p_1 + p_2 + \cdots + p_k = 1$.

Find the probability of any event by adding the probabilities $p_i$ of the particular values $x_i$ that make up the event.

EXAMPLE

**4.22 Grade distributions.** North Carolina State University posts the grade distributions for its courses online.[13] Students in one section of English 210 in the spring 2006 semester received 31% A's, 40% B's, 20% C's, 4% D's, and 5% F's. Choose an English 210 student at random. To "choose at random" means to give every student the same chance to be chosen. The student's grade on a four-point scale (with A = 4) is a random variable $X$.

The value of $X$ changes when we repeatedly choose students at random, but it is always one of 0, 1, 2, 3, or 4. Here is the distribution of $X$:

| Value of $X$ | 0 | 1 | 2 | 3 | 4 |
|---|---|---|---|---|---|
| Probability | 0.05 | 0.04 | 0.20 | 0.40 | 0.31 |

The probability that the student got a B or better is the sum of the probabilities of an A and a B. In the language of random variables,

$$\begin{aligned} P(X \geq 3) &= P(X = 3) + P(X = 4) \\ &= 0.40 + 0.31 = 0.71 \end{aligned}$$

## USE YOUR KNOWLEDGE

**4.46 Will the course satisfy the requirement?** Refer to Example 4.22. Suppose that a grade of D or F in English 210 will not count as satisfying a requirement for a major in linguistics. What is the probability that a randomly selected student will not satisfy this requirement?

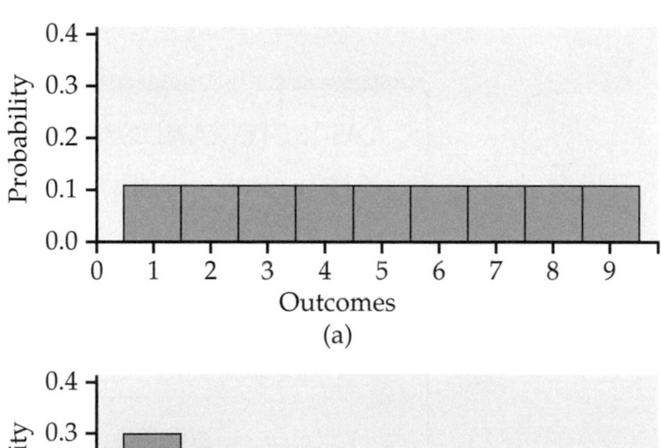


**FIGURE 4.5** Probability histograms for (a) random digits 1 to 9 and (b) Benford's law. The height of each bar shows the probability assigned to a single outcome.

probability histogram

We can use histograms to show probability distributions as well as distributions of data. Figure 4.5 displays **probability histograms** that compare the probability model for random digits for business records (Example 4.15) with the model given by Benford's law (Example 4.12). The height of each bar shows the probability of the outcome at its base. Because the heights are probabilities, they add to 1. As usual, all the bars in a histogram have the same width. So the areas also display the assignment of probability to outcomes. Think of these histograms as idealized pictures of the results of very many trials. The histograms make it easy to quickly compare the two distributions.

EXAMPLE

**4.23 Number of heads in four tosses of a coin.** What is the probability distribution of the discrete random variable $X$ that counts the number of heads in four tosses of a coin? We can derive this distribution if we make two reasonable assumptions:

- The coin is balanced, so it is fair and each toss is equally likely to give H or T.
- The coin has no memory, so tosses are independent.

The outcome of four tosses is a sequence of heads and tails such as HTTH. There are 16 possible outcomes in all. Figure 4.6 lists these outcomes along with the value of $X$ for each outcome. The multiplication rule for independent events tells us that, for example,

$$P(\text{HTTH}) = \frac{1}{2} \times \frac{1}{2} \times \frac{1}{2} \times \frac{1}{2} = \frac{1}{16}$$

Each of the 16 possible outcomes similarly has probability 1/16. That is, these outcomes are equally likely.

The number of heads $X$ has possible values 0, 1, 2, 3, and 4. These values are *not* equally likely. As Figure 4.6 shows, there is only one way that $X = 0$ can occur: namely, when the outcome is TTTT. So

$$P(X = 0) = \frac{1}{16} = 0.0625$$

The event $\{X = 2\}$ can occur in six different ways, so that

$$P(X = 2) = \frac{\text{count of ways } X = 2 \text{ can occur}}{16}$$

$$= \frac{6}{16} = 0.375$$

We can find the probability of each value of $X$ from Figure 4.6 in the same way. Here is the result:

| Value of $X$ | 0 | 1 | 2 | 3 | 4 |
|---|---|---|---|---|---|
| Probability | 0.0625 | 0.25 | 0.375 | 0.25 | 0.0625 |

**FIGURE 4.6** Possible outcomes in four tosses of a coin, for Example 4.23. The outcomes are arranged by the values of the random variable $X$, the number of heads.

| | | HTTH | | |
|---|---|---|---|---|
| | | HTHT | | |
| | HTTT | THTH | HHHT | |
| | THTT | HHTT | HHTH | |
| | TTHT | THHT | HTHH | |
| TTTT | TTTH | TTHH | THHH | HHHH |
| X = 0 | X = 1 | X = 2 | X = 3 | X = 4 |

Figure 4.7 is a probability histogram for the distribution in Example 4.23. The probability distribution is exactly symmetric. The probabilities (bar heights) are idealizations of the proportions after very many tosses of four coins. The actual distribution of proportions observed would be nearly symmetric but is unlikely to be exactly symmetric.

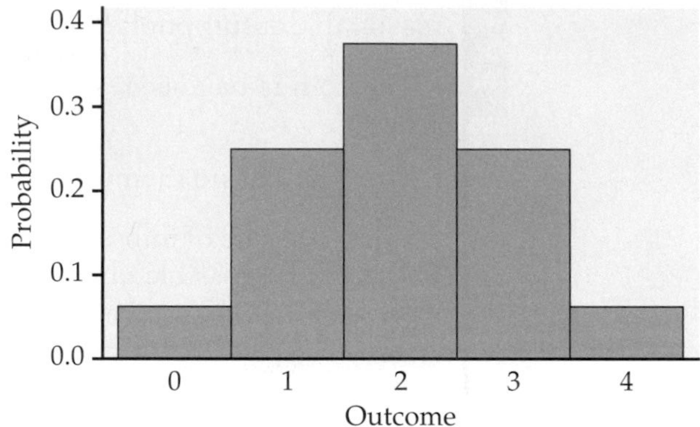


**FIGURE 4.7** Probability histogram for the number of heads in four tosses of a coin.

EXAMPLE

**4.24 Probability of at least two heads.** Any event involving the number of heads observed can be expressed in terms of $X$, and its probability can be found from the distribution of $X$. For example, the probability of tossing at least two heads is

$$P(X \geq 2) = 0.375 + 0.25 + 0.0625 = 0.6875$$

The probability of at least one head is most simply found by use of the complement rule:

$$\begin{aligned} P(X \geq 1) &= 1 - P(X = 0) \\ &= 1 - 0.0625 = 0.9375 \end{aligned}$$

Recall that tossing a coin $n$ times is similar to choosing an SRS of size $n$ from a large population and asking a yes-or-no question. We will extend the results of Example 4.23 when we return to sampling distributions in the next chapter.

## USE YOUR KNOWLEDGE

**4.47 Two tosses of a fair coin.** Find the probability distribution for the number of heads that appear in two tosses of a fair coin.

## Continuous random variables

When we use the table of random digits to select a digit between 0 and 9, the result is a discrete random variable. The probability model assigns probability 1/10 to each of the 10 possible outcomes. Suppose that we want to choose a number at random between 0 and 1, allowing *any* number between 0 and 1 as the outcome. Software random number generators will do this. You can visualize such a random number by thinking of a spinner (Figure 4.8) that turns freely on its axis and slowly comes to a stop. The pointer can come to rest anywhere on a circle that is marked from 0 to 1. The sample space is now an entire interval of numbers:

$$S = \{\text{all numbers } x \text{ such that } 0 \leq x \leq 1\}$$

How can we assign probabilities to events such as $\{0.3 \leq x \leq 0.7\}$? As in the case of selecting a random digit, we would like all possible outcomes to be equally likely. But we cannot assign probabilities to each individual value of $x$ and then sum, because there are infinitely many possible values. Instead, we use a new way of assigning probabilities directly to events—as *areas under a density curve.* Any density curve has area exactly 1 underneath it, corresponding to total probability 1.

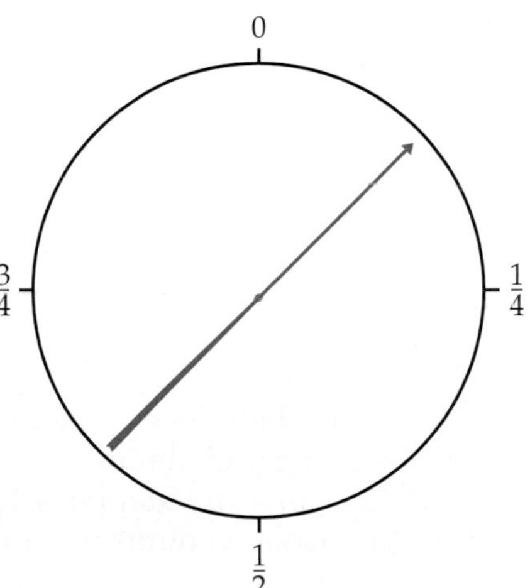


**FIGURE 4.8** A spinner that generates a random number between 0 and 1.

EXAMPLE

uniform distribution

**4.25 Uniform random numbers.** The random number generator will spread its output uniformly across the entire interval from 0 to 1 as we allow it to generate a long sequence of numbers. The results of many trials are represented by the density curve of a **uniform distribution.** This density curve appears in red in Figure 4.9. It has height 1 over the interval from 0 to 1, and height 0 everywhere else. The area under the density curve is 1: the area of a square with base 1 and height 1. The probability of any event is the area under the density curve and above the event in question.

As Figure 4.9(a) illustrates, the probability that the random number generator produces a number $X$ between 0.3 and 0.7 is

$$P(0.3 \leq X \leq 0.7) = 0.4$$

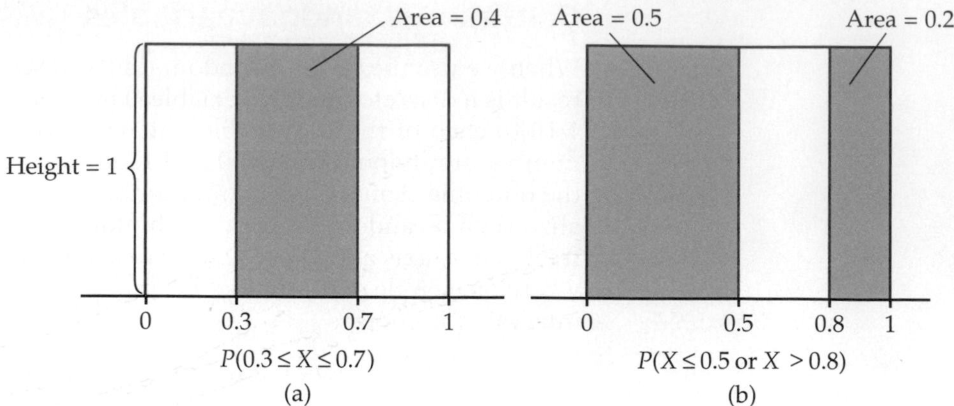


**FIGURE 4.9** Assigning probabilities for generating a random number between 0 and 1, for Example 4.25. The probability of any interval of numbers is the area above the interval and under the density curve.

because the area under the density curve and above the interval from 0.3 to 0.7 is 0.4. The height of the density curve is 1, and the area of a rectangle is the product of height and length, so the probability of any interval of outcomes is just the length of the interval.

Similarly,

$$P(X \leq 0.5) = 0.5$$

$$P(X > 0.8) = 0.2$$

$$P(X \leq 0.5 \text{ or } X > 0.8) = 0.7$$

Notice that the last event consists of two nonoverlapping intervals, so the total area above the event is found by adding two areas, as illustrated by Figure 4.9(b). This assignment of probabilities obeys all of our rules for probability.

Probability as area under a density curve is a second important way of assigning probabilities to events. Figure 4.10 illustrates this idea in general form. We call $X$ in Example 4.25 a *continuous random variable* because its values are not isolated numbers but an entire interval of numbers.

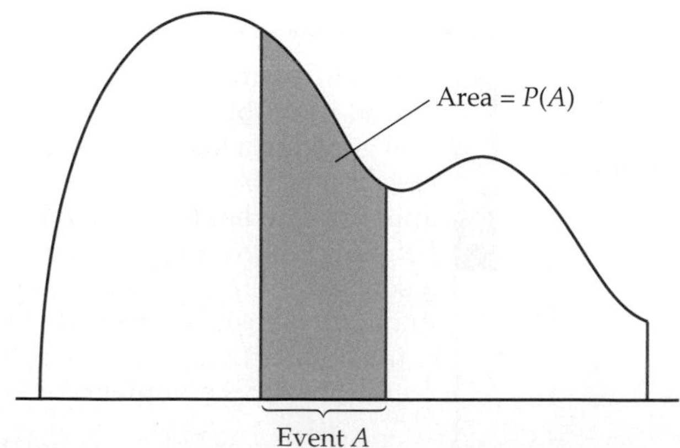


**FIGURE 4.10** The probability distribution of a continuous random variable assigns probabilities as areas under a density curve. The total area under any density curve is 1.

## USE YOUR KNOWLEDGE

**4.48 Find the probability.** For the uniform distribution described in Example 4.25, find the probability that $X$ is between 0.1 and 0.4.

### CONTINUOUS RANDOM VARIABLE

A **continuous random variable** $X$ takes all values in an interval of numbers. The **probability distribution** of $X$ is described by a density curve. The probability of any event is the area under the density curve and above the values of $X$ that make up the event.

The probability model for a continuous random variable assigns probabilities to intervals of outcomes rather than to individual outcomes. In fact, **all continuous probability distributions assign probability 0 to every individual outcome.** Only intervals of values have positive probability. To see that this is true, consider a specific outcome such as $P(X = 0.8)$ in the context of Example 4.25. The probability of any interval is the same as its length. The point 0.8 has no length, so its probability is 0.

Although this fact may seem odd, it makes intuitive, as well as mathematical, sense. The random number generator produces a number between 0.79 and 0.81 with probability 0.02. An outcome between 0.799 and 0.801 has probability 0.002. A result between 0.799999 and 0.800001 has probability 0.000002. You see that as we approach 0.8, the probability gets closer to 0. To be consistent, the probability of outcome *exactly* equal to 0.8 must be 0. Because there is no probability exactly at $X = 0.8$, the two events $\{X > 0.8\}$ and $\{X \geq 0.8\}$ have the same probability. We can ignore the distinction between $>$ and $\geq$ when finding probabilities for continuous (but not discrete) random variables.

## Normal distributions as probability distributions

The density curves that are most familiar to us are the Normal curves. Because any density curve describes an assignment of probabilities, *Normal distributions are probability distributions.* Recall that $N(\mu, \sigma)$ is our shorthand for the Normal distribution having mean $\mu$ and standard deviation $\sigma$. In the language of random variables, if $X$ has the $N(\mu, \sigma)$ distribution, then the standardized variable

$$Z = \frac{X - \mu}{\sigma}$$

is a standard Normal random variable having the distribution $N(0, 1)$.

EXAMPLE

**4.26 Cheating.** Students are reluctant to report cheating by other students. A sample survey puts this question to an SRS of 400 undergraduates: "You witness two students cheating on a quiz. Do you go to the professor?" Suppose that, if we could ask all undergraduates, 12% would answer "Yes."[14]

The proportion $p = 0.12$ is a *parameter* that describes the population of all undergraduates. The proportion $\hat{p}$ of the sample who answer "Yes" is a *statistic* used to estimate $p$. The statistic $\hat{p}$ is a random variable because repeating the SRS would give a different sample of 400 undergraduates and a different value of $\hat{p}$. We will see in the next chapter that $\hat{p}$ has approximately the $N(0.12, 0.016)$ distribution. The mean 0.12 of this distribution is the same as the population parameter because $\hat{p}$ is an unbiased estimate of $p$. The standard deviation is controlled mainly by the size of the sample.

What is the probability that the survey result differs from the truth about the population by more than 2 percentage points? Because $p = 0.12$, the survey misses by more than 2 percentage points if $\hat{p} < 0.10$ or $\hat{p} > 0.14$. Figure 4.11 shows this probability as an area under a Normal density curve. You can find it by software or by standardizing and using Table A. Let's start with the complement rule,

$$P(\hat{p} < 0.10 \text{ or } \hat{p} > 0.14) = 1 - P(0.10 \leq \hat{p} \leq 0.14)$$

From Table A,

$$\begin{aligned} P(0.10 \leq \hat{p} \leq 0.14) &= P\left(\frac{0.10 - 0.12}{0.016} \leq \frac{\hat{p} - 0.12}{0.016} \leq \frac{0.14 - 0.12}{0.016}\right) \\ &= P(-1.25 \leq Z \leq 1.25) \\ &= 0.8944 - 0.1056 = 0.7888 \end{aligned}$$

The probability we seek is therefore $1 - 0.7888 = 0.2112$. About 21% of sample results will be off by more than 2 percentage points. The arrangement of this calculation is familiar from our earlier work with Normal distributions. Only the language of probability is new.

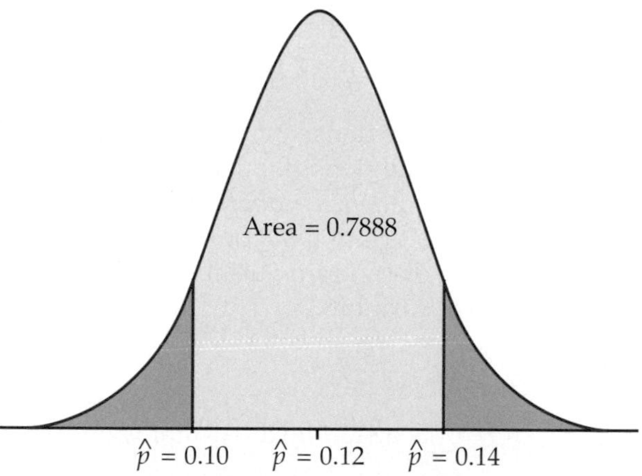


**FIGURE 4.11** Probability in Example 4.26 as area under a Normal density curve.

We began this chapter with a general discussion of the idea of probability and the properties of probability models. Two very useful specific types of probability models are distributions of discrete and continuous random variables. In our study of statistics we will employ only these two types of probability models.

## SECTION 4.3 Summary

A **random variable** is a variable taking numerical values determined by the outcome of a random phenomenon. The **probability distribution** of a random variable $X$ tells us what the possible values of $X$ are and how probabilities are assigned to those values.

A random variable $X$ and its distribution can be **discrete** or **continuous.**

A **discrete random variable** has finitely many possible values. The probability distribution assigns each of these values a probability between 0 and 1 such that the sum of all the probabilities is exactly 1. The probability of any event is the sum of the probabilities of all the values that make up the event.

A **continuous random variable** takes all values in some interval of numbers. A **density curve** describes the probability distribution of a continuous random variable. The probability of any event is the area under the curve and above the values that make up the event.

**Normal distributions** are one type of continuous probability distribution.

You can picture a probability distribution by drawing a **probability histogram** in the discrete case or by graphing the density curve in the continuous case.

## SECTION 4.3 Exercises

*For Exercise 4.46, see page 260; for Exercise 4.47, see page 262; and for Exercise 4.48, see page 265.*

**4.49 Discrete or continuous.** In each of the following situations decide if the random variable is discrete or continuous and give a reason for your answer.

(a) Your Web page has five different links and a user can click on one of the links or can leave the page. You record the length of time that a user spends on the Web page before clicking one of the links or leaving the page.

(b) The number of hits on your Web page.

(c) The yearly income of a visitor to your Web page.

**4.50 Texas hold 'em.** The game of Texas hold 'em starts with each player receiving two cards. Here is the probability distribution for the number of aces in two-card hands:

| Number of aces | 0 | 1 | 2 |
|---|---|---|---|
| Probability | 0.559 | 0.382 | 0.059 |

(a) Verify that this assignment of probabilities satisfies the requirement that the sum of the probabilities for a discrete distribution must be 1.

(b) Make a probability histogram for this distribution.

(c) What is the probability that a hand contains at least one ace? Show two different ways to calculate this probability.

**4.51 Spell-checking software.** Spell-checking software catches "nonword errors," which result in a string of letters that is not a word, as when "the" is typed as "teh." When undergraduates are asked to write a 250-word essay (without spell-checking), the number $X$ of nonword errors has the following distribution:

| Value of $X$ | 0 | 1 | 2 | 3 | 4 |
|---|---|---|---|---|---|
| Probability | 0.1 | 0.3 | 0.3 | 0.2 | 0.1 |

(a) Sketch the probability distribution for this random variable.

(b) Write the event "at least one nonword error" in terms of $X$. What is the probability of this event?

(c) Describe the event $X \leq 2$ in words. What is its probability? What is the probability that $X < 2$?

**4.52 Length of human pregnancies.** The length of human pregnancies from conception to birth varies according to a distribution that is approximately Normal with mean 266 days and standard deviation 16 days. Call the length of a randomly chosen pregnancy $Y$.

(a) Make a sketch of the density curve for this random variable.

(b) What is $P(Y > 300)$?

**4.53 Owner-occupied and rented housing units.** How do rented housing units differ from units occupied by their owners? Here are the distributions of the number of rooms for owner-occupied units and renter-occupied units in San Jose, California:[15]

| **Rooms** | 1 | 2 | 3 | 4 | 5 |
|---|---|---|---|---|---|
| Owned | 0.003 | 0.002 | 0.023 | 0.104 | 0.210 |
| Rented | 0.008 | 0.027 | 0.287 | 0.363 | 0.164 |
| **Rooms** | 6 | 7 | 8 | 9 | 10 |
| Owned | 0.224 | 0.197 | 0.149 | 0.053 | 0.035 |
| Rented | 0.093 | 0.039 | 0.013 | 0.003 | 0.003 |

Make probability histograms of these two distributions, using the same scales. What are the most important differences between the distributions for owner-occupied and rented housing units?

**4.54 Households and families in government data.** In government data, a household consists of all occupants of a dwelling unit, while a family consists of two or more persons who live together and are related by blood or marriage. So all families form households, but some households are not families. Here are the distributions of household size and of family size in the United States:

| Number of persons | 1 | 2 | 3 | 4 | 5 | 6 | 7 |
|---|---|---|---|---|---|---|---|
| Household probability | 0.27 | 0.33 | 0.16 | 0.14 | 0.06 | 0.03 | 0.01 |
| Family probability | 0 | 0.44 | 0.22 | 0.20 | 0.09 | 0.03 | 0.02 |

Make probability histograms for these two discrete distributions, using the same scales. What are the most important differences between the sizes of households and families?

**4.55 Find the probabilities.** Let the random variable $X$ be the number of rooms in a randomly chosen owner-occupied housing unit in San Jose, California. Exercise 4.53 gives the distribution of $X$.

(a) Express "the unit has 6 or more rooms" in terms of $X$. What is the probability of this event?

(b) Express the event $\{X > 6\}$ in words. What is its probability?

(c) What important fact about discrete random variables does comparing your answers to (a) and (b) illustrate?

**4.56 Tossing two dice.** Some games of chance rely on tossing two dice. Each die has six faces, marked with 1, 2, . . . , 6 spots called pips. The dice used in casinos are carefully balanced so that each face is equally likely to come up. When two dice are tossed, each of the 36 possible pairs of faces is equally likely to come up. The outcome of interest to a gambler is the sum of the pips on the two up-faces. Call this random variable $X$.

(a) Write down all 36 possible pairs of faces.

(b) If all pairs have the same probability, what must be the probability of each pair?

(c) Write the value of $X$ next to each pair of faces and use this information with the result of (b) to give the probability distribution of $X$. Draw a probability histogram to display the distribution.

(d) One bet available in craps wins if a 7 or an 11 comes up on the next roll of two dice. What is the probability of rolling a 7 or an 11 on the next roll?

(e) Several bets in craps lose if a 7 is rolled. If any outcome other than 7 occurs, these bets either win or continue to the next roll. What is the probability that anything other than a 7 is rolled?

**4.57** CHALLENGE **Nonstandard dice.** Nonstandard dice can produce interesting distributions of outcomes. You have two balanced, six-sided dice. One is a standard die, with faces having 1, 2, 3, 4, 5, and 6 spots. The other die has three faces with 0 spots and three faces with 6 spots. Find the probability distribution for the total number of spots $Y$ on the up-faces when you roll these two dice.

**4.58** CHALLENGE **Dungeons & Dragons.** Role-playing games like Dungeons & Dragons use many different types of dice, usually having either 4, 6, 8, 10, 12, or 20 sides. Roll a balanced 8-sided die and a balanced 6-sided die and add the spots on the up-faces. Call the sum $X$. What is the probability distribution of the random variable $X$?

**4.59 Foreign-born residents of California.** The Census Bureau reports that 27% of California residents are foreign-born. Suppose that you choose three Californians at random, so that each has probability 0.27 of being foreign-born and the three are independent of each other. Let the random variable $W$ be the number of foreign-born people you chose.

(a) What are the possible values of $W$?

(b) Look at your three people in order. There are eight possible arrangements of foreign (F) and domestic (D) birth. For example, FFD means the first two are foreign-born and the third is not. All eight arrangements are equally likely. What is the probability of each one?

(c) What is the value of $W$ for each arrangement in (b)? What is the probability of each possible value of $W$? (This is the distribution of a Yes/No response for an SRS of size 3. In principle, the same idea works for an SRS of any size.)

**4.60 Select the members of a student advisory board.** Weary of the low turnout in student elections, a college administration decides to choose an SRS of three students to form an advisory board that represents student opinion. Suppose that 40% of all students oppose the use of student fees to fund student interest groups, and that the opinions of the three students on the board are independent. Then the probability is 0.4 that each opposes the funding of interest groups.

(a) Call the three students A, B, and C. What is the probability that A and B support funding and C opposes it?

(b) List all possible combinations of opinions that can be held by students A, B, and C. (*Hint:* There are eight possibilities.) Then give the probability of each of these outcomes. Note that they are not equally likely.

(c) Let the random variable $X$ be the number of student representatives who oppose the funding of interest groups. Give the probability distribution of $X$.

(d) Express the event "a majority of the advisory board opposes funding" in terms of $X$ and find its probability.

**4.61 Uniform random numbers.** Let $X$ be a random number between 0 and 1 produced by the idealized uniform random number generator described in Example 4.25 and Figure 4.9. Find the following probabilities:

(a) $P(X < 0.4)$

(b) $P(X \leq 0.4)$

(c) What important fact about continuous random variables does comparing your answers to (a) and (b) illustrate?

**4.62 Find the probabilities.** Let the random variable $X$ be a random number with the uniform density curve in Figure 4.9. Find the following probabilities:

(a) $P(X \geq 0.35)$

(b) $P(X = 0.35)$

(c) $P(0.35 < X < 1.35)$

(d) $P(0.15 \leq X \leq 0.25 \text{ or } 0.8 \leq X \leq 0.9)$

(e) The probability that $X$ is not in the interval 0.3 to 0.7.

**4.63 Uniform numbers between 0 and 2.** Many random number generators allow users to specify the range of the random numbers to be produced. Suppose that you specify that the range is to be all numbers between 0 and 2. Call the random number generated $Y$. Then the density curve of the random variable $Y$ has constant height between 0 and 2, and height 0 elsewhere.

(a) What is the height of the density curve between 0 and 2? Draw a graph of the density curve.

(b) Use your graph from (a) and the fact that probability is area under the curve to find $P(Y \leq 1.5)$.

(c) Find $P(0.6 < Y < 1.7)$.

(d) Find $P(Y \geq 0.9)$.

**4.64 The sum of two uniform random numbers.** Generate *two* random numbers between 0 and 1 and take $Y$ to be their sum. Then $Y$ is a continuous random variable that can take any value between 0 and 2. The density curve of $Y$ is the triangle shown in Figure 4.12.

(a) Verify by geometry that the area under this curve is 1.

(b) What is the probability that $Y$ is less than 1? (Sketch the density curve, shade the area that represents the probability, then find that area. Do this for (c) also.)

(c) What is the probability that $Y$ is less than 1.5?

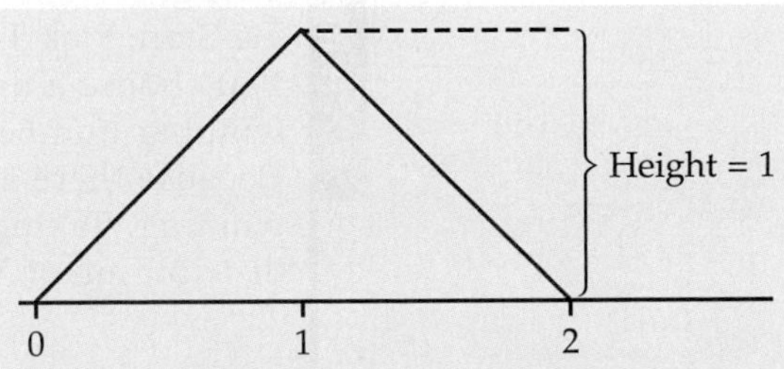


**FIGURE 4.12** The density curve for the sum $Y$ of two random numbers, for Exercise 4.64.

**4.65 How many close friends?** How many close friends do you have? Suppose that the number of close friends adults claim to have varies from person to person with mean $\mu = 9$ and standard deviation $\sigma = 2.5$. An opinion poll asks this question of an SRS of 1100 adults. We will see in the next chapter that in this situation the sample mean response $\bar{x}$ has approximately the Normal distribution with mean 9 and standard deviation 0.075. What is $P(8 \leq \bar{x} \leq 10)$, the probability that the statistic $\bar{x}$ estimates the parameter $\mu$ to within $\pm 1$?

**4.66 Normal approximation for a sample proportion.** A sample survey contacted an SRS of 663 registered voters in Oregon shortly after an election and asked respondents whether they had voted. Voter records show that 56% of registered voters had actually voted. We will see in the next chapter that in this situation the proportion $\hat{p}$ of the sample who voted has approximately the Normal distribution with mean $\mu = 0.56$ and standard deviation $\sigma = 0.019$.

(a) If the respondents answer truthfully, what is $P(0.52 \leq \hat{p} \leq 0.60)$? This is the probability that the statistic $\hat{p}$ estimates the parameter 0.56 within plus or minus 0.04.

(b) In fact, 72% of the respondents said they had voted ($\hat{p} = 0.72$). If respondents answer truthfully, what is $P(\hat{p} \geq 0.72)$? This probability is so small that it is good evidence that some people who did not vote claimed that they did vote.

## 4.4 Means and Variances of Random Variables

The probability histograms and density curves that picture the probability distributions of random variables resemble our earlier pictures of distributions of data. In describing data, we moved from graphs to numerical measures such as means and standard deviations. Now we will make the same move to expand our descriptions of the distributions of random variables. We can speak of the mean winnings in a game of chance or the standard deviation of the randomly varying number of calls a travel agency receives in an hour. In this section we will learn more about how to compute these descriptive measures and about the laws they obey.

### The mean of a random variable

The mean $\bar{x}$ of a set of observations is their ordinary average. The mean of a random variable $X$ is also an average of the possible values of $X$, but with an essential change to take into account the fact that not all outcomes need be equally likely. An example will show what we must do.

EXAMPLE

**4.27 The Tri-State Pick 3 lottery.** Most states and Canadian provinces have government-sponsored lotteries. Here is a simple lottery wager, from the Tri-State Pick 3 game that New Hampshire shares with Maine and Vermont. You choose a three-digit number, 000 to 999. The state chooses a three-digit winning number at random and pays you \$500 if your number is chosen. Because there are 1000 three-digit numbers, you have probability 1/1000 of winning. Taking $X$ to be the amount your ticket pays you, the probability distribution of $X$ is

| Payoff $X$ | \$0 | \$500 |
|---|---|---|
| Probability | 0.999 | 0.001 |

What is your average payoff from many tickets? The ordinary average of the two possible outcomes \$0 and \$500 is \$250, but that makes no sense as the average because \$500 is much less likely than \$0. In the long run you receive \$500 once in every 1000 tickets and \$0 on the remaining 999 of 1000 tickets. The long-run average payoff is

$$\$500\frac{1}{1000} + \$0\frac{999}{1000} = \$0.50$$

or fifty cents. That number is the mean of the random variable $X$. (Tickets cost \$1, so in the long run the state keeps half the money you wager.)

If you play Tri-State Pick 3 several times, we would as usual call the mean of the actual amounts you win $\bar{x}$. The mean in Example 4.27 is a different quantity—it is the long-run average winnings you expect if you play a very large number of times.

## USE YOUR KNOWLEDGE

**4.67 Find the mean of the probability distribution.** You toss a fair coin. If the outcome is heads, you win \$1.00; if the outcome is tails, you win nothing. Let $X$ be the amount that you win in a single toss of a coin. Find the probability distribution of this random variable and its mean.

Just as probabilities are an idealized description of long-run proportions, the mean of a probability distribution describes the long-run average outcome. We can't call this mean $\bar{x}$, so we need a different symbol. The common symbol for the **mean of a probability distribution** is $\mu$, the Greek letter mu. We used $\mu$ in Chapter 1 for the mean of a Normal distribution, so this is not a new notation. We will often be interested in several random variables, each having a different probability distribution with a different mean. To remind ourselves that we are talking about the mean of $X$, we often write $\mu_X$ rather than simply $\mu$. In Example 4.27, $\mu_X = \$0.50$. Notice that, as often happens, the mean is not a possible value of $X$. You will often find the mean of a random variable $X$ called the **expected value** of $X$. This term can be misleading, for we don't necessarily expect one observation on $X$ to be close to its expected value.

mean $\mu$

expected value

The mean of any discrete random variable is found just as in Example 4.27. It is an average of the possible outcomes, but a weighted average in which each outcome is weighted by its probability. Because the probabilities add to 1, we have total weight 1 to distribute among the outcomes. An outcome that occurs half the time has probability one-half and gets one-half the weight in calculating the mean. Here is the general definition.

### MEAN OF A DISCRETE RANDOM VARIABLE

Suppose that $X$ is a discrete random variable whose distribution is

| Value of $X$ | $x_1$ | $x_2$ | $x_3$ | $\cdots$ | $x_k$ |
|---|---|---|---|---|---|
| Probability | $p_1$ | $p_2$ | $p_3$ | $\cdots$ | $p_k$ |

To find the **mean** of $X$, multiply each possible value by its probability, then add all the products:

$$\mu_X = x_1p_1 + x_2p_2 + \cdots + x_kp_k$$
$$= \sum x_ip_i$$

EXAMPLE

**4.28 The mean of equally likely first digits.** If first digits in a set of data all have the same probability, the probability distribution of the first digit $X$ is then

| First digit $X$ | 1 | 2 | 3 | 4 | 5 | 6 | 7 | 8 | 9 |
|---|---|---|---|---|---|---|---|---|---|
| Probability | 1/9 | 1/9 | 1/9 | 1/9 | 1/9 | 1/9 | 1/9 | 1/9 | 1/9 |

The mean of this distribution is

$$\mu_X = 1 \times \frac{1}{9} + 2 \times \frac{1}{9} + 3 \times \frac{1}{9} + 4 \times \frac{1}{9} + 5 \times \frac{1}{9} + 6 \times \frac{1}{9}$$
$$+ 7 \times \frac{1}{9} + 8 \times \frac{1}{9} + 9 \times \frac{1}{9}$$
$$= 45 \times \frac{1}{9} = 5$$

Suppose the random digits in Example 4.28 had a different probability distribution. In Example 4.12 (page 248) we described Benford's law as a probability distribution that describes first digits of numbers in many real situations. Let's calculate the mean for Benford's law.

EXAMPLE

**4.29 The mean of first digits following Benford's law.** Here is the distribution of the first digit for data that follow Benford's law. We use the letter $V$ for this random variable to distinguish it from the one that we studied in Example 4.28. The distribution of $V$ is

| First digit $V$ | 1 | 2 | 3 | 4 | 5 | 6 | 7 | 8 | 9 |
|---|---|---|---|---|---|---|---|---|---|
| Probability | 0.301 | 0.176 | 0.125 | 0.097 | 0.079 | 0.067 | 0.058 | 0.051 | 0.046 |

The mean of $V$ is

$$\mu_V = (1)(0.301) + (2)(0.176) + (3)(0.125) + (4)(0.097) + (5)(0.079)$$
$$+ (6)(0.067) + (7)(0.058) + (8)(0.051) + (9)(0.046)$$
$$= 3.441$$

The mean reflects the greater probability of smaller first digits under Benford's law than when first digits 1 to 9 are equally likely.

Figure 4.13 locates the means of $X$ and $V$ on the two probability histograms. Because the discrete uniform distribution of Figure 4.13(a) is symmetric, the mean lies at the center of symmetry. We can't locate the mean of the right-skewed distribution of Figure 4.13(b) by eye—calculation is needed.

What about continuous random variables? The probability distribution of a continuous random variable $X$ is described by a density curve. Chapter 1 (page 57) showed how to find the mean of the distribution: it is the point at which the area under the density curve would balance if it were made out of solid material. The mean lies at the center of symmetric density curves such as the Normal curves. Exact calculation of the mean of a distribution with a skewed density curve requires advanced mathematics.[16] The idea that the mean is the balance point of the distribution applies to discrete random variables as well, but in the discrete case we have a formula that gives us this point.

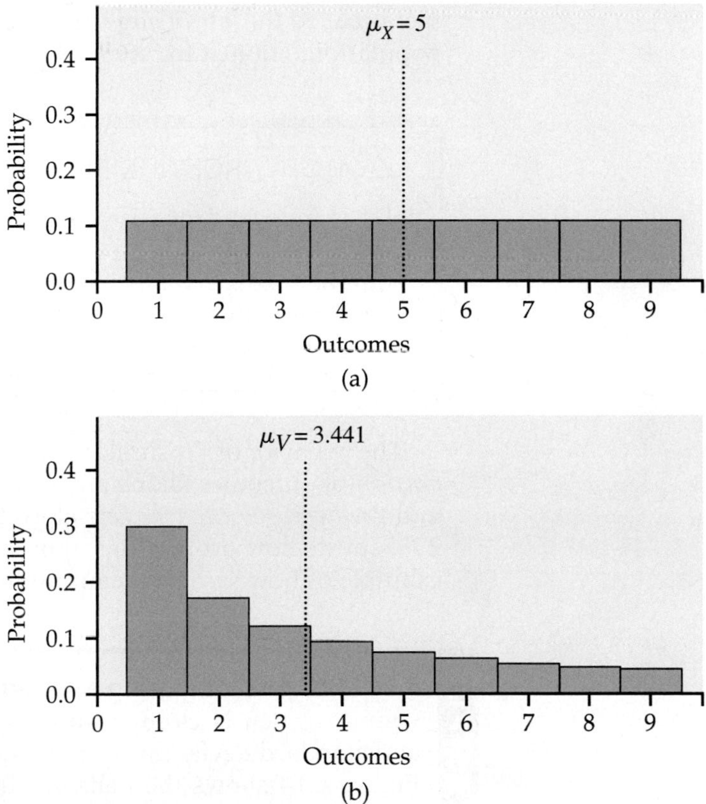


**FIGURE 4.13** Locating the mean of a discrete random variable on the probability histogram for (a) digits between 1 and 9 chosen at random; (b) digits between 1 and 9 chosen from records that obey Benford's law.

## Statistical estimation and the law of large numbers

We would like to estimate the mean height $\mu$ of the population of all American women between the ages of 18 and 24 years. This $\mu$ is the mean $\mu_X$ of the random variable $X$ obtained by choosing a young woman at random and measuring her height. To estimate $\mu$, we choose an SRS of young women and use

LOOK BACK
**sampling distributions, page 214**

the sample mean $\bar{x}$ to estimate the unknown population mean $\mu$. In the language of Section 3.3 (page 212), $\mu$ is a *parameter* and $\bar{x}$ is a *statistic*. Statistics obtained from probability samples are random variables because their values vary in repeated sampling. The sampling distributions of statistics are just the probability distributions of these random variables.

It seems reasonable to use $\bar{x}$ to estimate $\mu$. An SRS should fairly represent the population, so the mean $\bar{x}$ of the sample should be somewhere near the mean $\mu$ of the population. Of course, we don't expect $\bar{x}$ to be exactly equal to $\mu$, and we realize that if we choose another SRS, the luck of the draw will probably produce a different $\bar{x}$.

If $\bar{x}$ is rarely exactly right and varies from sample to sample, why is it nonetheless a reasonable estimate of the population mean $\mu$? We gave one answer in Section 3.4: $\bar{x}$ is unbiased and we can control its variability by choosing the sample size. Here is another answer: if we keep on adding observations to our random sample, the statistic $\bar{x}$ is *guaranteed* to get as close as we wish to the parameter $\mu$ and then stay that close. We have the comfort of knowing that if we can afford to keep on measuring more women, eventually we will estimate the mean height of all young women very accurately. This remarkable fact is called the *law of large numbers*. It is remarkable because it holds for *any* population, not just for some special class such as Normal distributions.

### LAW OF LARGE NUMBERS

Draw independent observations at random from any population with finite mean $\mu$. Decide how accurately you would like to estimate $\mu$. As the number of observations drawn increases, the mean $\bar{x}$ of the observed values eventually approaches the mean $\mu$ of the population as closely as you specified and then stays that close.

The behavior of $\bar{x}$ is similar to the idea of probability. In the long run, the *proportion* of outcomes taking any value gets close to the *probability* of that value, and the *average outcome* gets close to the distribution *mean*. Figure 4.1 (page 238) shows how proportions approach probability in one example. Here is an example of how sample means approach the distribution mean.

EXAMPLE

**4.30 Heights of young women.** The distribution of the heights of all young women is close to the Normal distribution with mean 64.5 inches and standard deviation 2.5 inches. Suppose that $\mu = 64.5$ were exactly true. Figure 4.14 shows the behavior of the mean height $\bar{x}$ of $n$ women chosen at random from a population whose heights follow the $N(64.5, 2.5)$ distribution. The graph plots the values of $\bar{x}$ as we add women to our sample. The first woman drawn had height 64.21 inches, so the line starts there. The second had height 64.35 inches, so for $n = 2$ the mean is

$$\bar{x} = \frac{64.21 + 64.35}{2} = 64.28$$

This is the second point on the line in the graph.

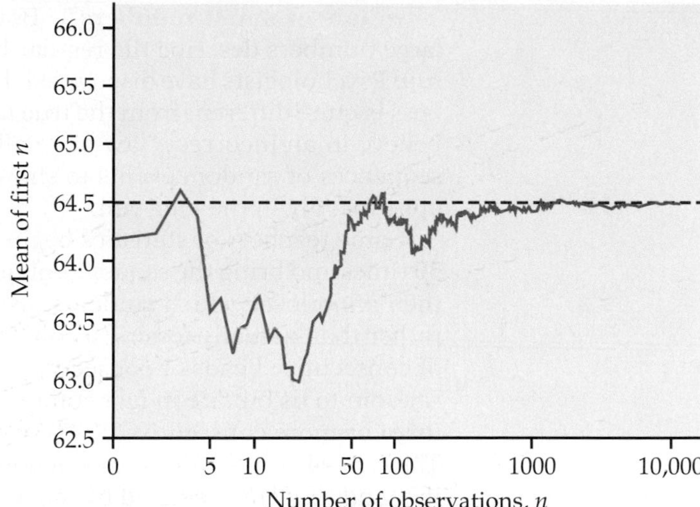


**FIGURE 4.14** The law of large numbers in action. As we take more observations, the sample mean always approaches the mean of the population.

At first, the graph shows that the mean of the sample changes as we take more observations. Eventually, however, the mean of the observations gets close to the population mean $\mu = 64.5$ and settles down at that value. The law of large numbers says that this *always* happens.

## USE YOUR KNOWLEDGE

**4.68 Use the *Law of Large Numbers* applet.** The *Law of Large Numbers* applet animates a graph like Figure 4.14. Use it to better understand the law of large numbers by making a similar graph.

The mean $\mu$ of a random variable is the average value of the variable in two senses. By its definition, $\mu$ is the average of the possible values, weighted by their probability of occurring. The law of large numbers says that $\mu$ is also the long-run average of many independent observations on the variable. The law of large numbers can be proved mathematically starting from the basic laws of probability.

## Thinking about the law of large numbers

The law of large numbers says broadly that the average results of many independent observations are stable and predictable. The gamblers in a casino may win or lose, but the casino will win in the long run because the law of large numbers says what the average outcome of many thousands of bets will be. An insurance company deciding how much to charge for life insurance and a fast-food restaurant deciding how many beef patties to prepare also rely on the fact that averaging over many individuals produces a stable result. It is worth the effort to think a bit more closely about so important a fact.

**The "law of small numbers"** Both the rules of probability and the law of large numbers describe the regular behavior of chance phenomena *in the long run.* Psychologists have discovered that our intuitive understanding of randomness is quite different from the true laws of chance.[17] For example, most people believe in an incorrect "law of small numbers." That is, we expect even short sequences of random events to show the kind of average behavior that in fact appears only in the long run.

Some teachers of statistics begin a course by asking students to toss a coin 50 times and bring the sequence of heads and tails to the next class. The teacher then announces which students just wrote down a random-looking sequence rather than actually tossing a coin. The faked tosses don't have enough "runs" of consecutive heads or consecutive tails. Runs of the same outcome don't look random to us but are in fact common. For example, the probability of a run of three or more consecutive heads or tails in just 10 tosses is greater than 0.8.[18] The runs of consecutive heads or consecutive tails that appear in real coin tossing (and that are predicted by the mathematics of probability) seem surprising to us. Because we don't expect to see long runs, we may conclude that the coin tosses are not independent or that some influence is disturbing the random behavior of the coin.

EXAMPLE

**4.31 The "hot hand" in basketball.** Belief in the law of small numbers influences behavior. If a basketball player makes several consecutive shots, both the fans and her teammates believe that she has a "hot hand" and is more likely to make the next shot. This is doubtful. Careful study suggests that runs of baskets made or missed are no more frequent in basketball than would be expected if each shot were independent of the player's previous shots. Baskets made or missed are just like heads and tails in tossing a coin. (Of course, some players make 30% of their shots in the long run and others make 50%, so a coin-toss model for basketball must allow coins with different probabilities of a head.) Our perception of hot or cold streaks simply shows that we don't perceive random behavior very well.[19]

*Our intuition doesn't do a good job of distinguishing random behavior from systematic influences. This is also true when we look at data. We need statistical inference to supplement exploratory analysis of data because probability calculations can help verify that what we see in the data is more than a random pattern.*

**How large is a large number?** The law of large numbers says that the actual mean outcome of many trials gets close to the distribution mean $\mu$ as more trials are made. It doesn't say how many trials are needed to guarantee a mean outcome close to $\mu$. That depends on the *variability* of the random outcomes. The more variable the outcomes, the more trials are needed to ensure that the mean outcome $\bar{x}$ is close to the distribution mean $\mu$. Casinos understand this: the outcomes of games of chance are variable enough to hold the interest of gamblers. Only the casino plays often enough to rely on the law of large numbers. Gamblers get entertainment; the casino has a business.

## BEYOND THE BASICS

### More Laws of Large Numbers

The law of large numbers is one of the central facts about probability. It helps us understand the mean $\mu$ of a random variable. It explains why gambling casinos and insurance companies make money. It assures us that statistical estimation will be accurate if we can afford enough observations. The basic law of large numbers applies to independent observations that all have the same distribution. Mathematicians have extended the law to many more general settings. Here are two of these.

**Is there a winning system for gambling?** Serious gamblers often follow a system of betting in which the amount bet on each play depends on the outcome of previous plays. You might, for example, double your bet on each spin of the roulette wheel until you win—or, of course, until your fortune is exhausted. Such a system tries to take advantage of the fact that you have a memory even though the roulette wheel does not. Can you beat the odds with a system based on the outcomes of past plays? No. Mathematicians have established a stronger version of the law of large numbers that says that, if you do not have an infinite fortune to gamble with, your long-run average winnings $\mu$ remain the same as long as successive trials of the game (such as spins of the roulette wheel) are independent.

**What if observations are not independent?** You are in charge of a process that manufactures video screens for computer monitors. Your equipment measures the tension on the metal mesh that lies behind each screen and is critical to its image quality. You want to estimate the mean tension $\mu$ for the process by the average $\bar{x}$ of the measurements. Alas, the tension measurements are not independent. If the tension on one screen is a bit too high, the tension on the next is more likely to also be high. Many real-world processes are like this—the process stays stable in the long run, but observations made close together are likely to be both above or both below the long-run mean. Again the mathematicians come to the rescue: as long as the dependence dies out fast enough as we take measurements farther and farther apart in time, the law of large numbers still holds.

## Rules for means

You are studying flaws in the painted finish of refrigerators made by your firm. Dimples and paint sags are two kinds of surface flaw. Not all refrigerators have the same number of dimples: many have none, some have one, some two, and so on. You ask for the average number of imperfections on a refrigerator. The inspectors report finding an average of 0.7 dimples and 1.4 sags per refrigerator. How many total imperfections of both kinds (on the average) are there on a refrigerator? That's easy: if the average number of dimples is 0.7 and the average number of sags is 1.4, then counting both gives an average of $0.7 + 1.4 = 2.1$ flaws.

In more formal language, the number of dimples on a refrigerator is a random variable $X$ that varies as we inspect one refrigerator after another. We know only that the mean number of dimples is $\mu_X = 0.7$. The number of paint sags is a second random variable $Y$ having mean $\mu_Y = 1.4$. (As usual, the subscripts keep straight which variable we are talking about.) The total number of both dimples and sags is another random variable, the sum $X + Y$. Its mean $\mu_{X+Y}$ is the average number of dimples and sags together. It is just the sum of the individual means $\mu_X$ and $\mu_Y$. That's an important rule for how means of random variables behave.

Here's another rule. The crickets living in a field have mean length 1.2 inches. What is the mean in centimeters? There are 2.54 centimeters in an inch, so the length of a cricket in centimeters is 2.54 times its length in inches. If we multiply every observation by 2.54, we also multiply their average by 2.54. The mean in centimeters must be $2.54 \times 1.2$, or about 3.05 centimeters. More formally, the length in inches of a cricket chosen at random from the field is a random variable $X$ with mean $\mu_X$. The length in centimeters is $2.54X$, and this new random variable has mean $2.54\mu_X$.

The point of these examples is that means behave like averages. Here are the rules we need.

### RULES FOR MEANS

**Rule 1.** If $X$ is a random variable and $a$ and $b$ are fixed numbers, then

$$\mu_{a+bX} = a + b\mu_X$$

**Rule 2.** If $X$ and $Y$ are random variables, then

$$\mu_{X+Y} = \mu_X + \mu_Y$$

EXAMPLE

**4.32 Sales of cars, trucks, and SUVs.** Linda is a sales associate at a large auto dealership. At her commission rate of 25% of gross profit on each vehicle she sells, Linda expects to earn \$350 for each car sold and \$400 for each truck or SUV sold. Linda motivates herself by using probability estimates of her sales. For a sunny Saturday in April, she estimates her car sales as follows:

| Cars sold | 0 | 1 | 2 | 3 |
|---|---|---|---|---|
| Probability | 0.3 | 0.4 | 0.2 | 0.1 |

Linda's estimate of her truck or SUV sales is

| Vehicles sold | 0 | 1 | 2 |
|---|---|---|---|
| Probability | 0.4 | 0.5 | 0.1 |

Take $X$ to be the number of cars Linda sells and $Y$ the number of trucks or SUVs. The means of these random variables are

$$\mu_X = (0)(0.3) + (1)(0.4) + (2)(0.2) + (3)(0.1)$$
$$= 1.1 \text{ cars}$$
$$\mu_Y = (0)(0.4) + (1)(0.5) + (2)(0.1)$$
$$= 0.7 \text{ trucks or SUVs}$$

Linda's earnings, at $350 per car and $400 per truck or SUV, are

$$Z = 350X + 400Y$$

Combining Rules 1 and 2, her mean earnings are

$$\mu_Z = 350\mu_X + 400\mu_Y$$
$$= (350)(1.1) + (400)(0.7) = \$665$$

This is Linda's best estimate of her earnings for the day. It's a bit unusual for individuals to use probability estimates, but they are a common tool for business planners.

personal probability

The probabilities in Example 4.32 are **personal probabilities** that describe Linda's informed opinion about her sales in the coming weekend. Although personal probabilities need not be based on observing many repetitions of a random phenomenon, they must obey the rules of probability if they are to make sense. Personal probability extends the usefulness of probability models to one-time events, but remember that they are subject to the follies of human opinion. Overoptimism is common: 40% of college students think that they will eventually reach the top 1% in income.

## USE YOUR KNOWLEDGE

**4.69** **Find $\mu_Y$.** The random variable $X$ has mean $\mu_X = 10$. If $Y = 15 + 8X$, what is $\mu_Y$?

**4.70** **Find $\mu_W$.** The random variable $U$ has mean $\mu_U = 20$ and the random variable $V$ has mean $\mu_V = 20$. If $W = 0.5U + 0.5V$, find $\mu_W$.

## The variance of a random variable

The mean is a measure of the center of a distribution. A basic numerical description requires in addition a measure of the spread or variability of the distribution. The variance and the standard deviation are the measures of spread that accompany the choice of the mean to measure center. Just as for the mean, we need a distinct symbol to distinguish the variance of a random variable from the variance $s^2$ of a data set. We write the variance of a random variable $X$ as $\sigma_X^2$. Once again the subscript reminds us which variable we have in mind. The definition of the variance $\sigma_X^2$ of a random variable is similar to the definition of the sample variance $s^2$ given in Chapter 1. That is, the variance is an average value of the squared deviation $(X - \mu_X)^2$ of the variable $X$ from its mean $\mu_X$. As for the mean, the average we use is a weighted average in which each outcome is weighted by its probability in order to take account of outcomes that are not equally likely. Calculating this weighted average is straightforward for discrete random variables but requires advanced mathematics in the continuous case. Here is the definition.

### VARIANCE OF A DISCRETE RANDOM VARIABLE

Suppose that $X$ is a discrete random variable whose distribution is

| Value of $X$ | $x_1$ | $x_2$ | $x_3$ | $\cdots$ | $x_k$ |
|---|---|---|---|---|---|
| Probability | $p_1$ | $p_2$ | $p_3$ | $\cdots$ | $p_k$ |

and that $\mu_X$ is the mean of $X$. The **variance** of $X$ is

$$\begin{aligned}\sigma_X^2 &= (x_1 - \mu_X)^2 p_1 + (x_2 - \mu_X)^2 p_2 + \cdots + (x_k - \mu_X)^2 p_k \\ &= \sum (x_i - \mu_X)^2 p_i\end{aligned}$$

The **standard deviation** $\sigma_X$ of $X$ is the square root of the variance.

EXAMPLE

**4.33 Find the mean and the variance.** In Example 4.32 we saw that the number $X$ of cars that Linda hopes to sell has distribution

| Cars sold | 0 | 1 | 2 | 3 |
|---|---|---|---|---|
| Probability | 0.3 | 0.4 | 0.2 | 0.1 |

We can find the mean and variance of $X$ by arranging the calculation in the form of a table. Both $\mu_X$ and $\sigma_X^2$ are sums of columns in this table.

| $x_i$ | $p_i$ | $x_i p_i$ | $(x_i - \mu_X)^2 p_i$ |
|---|---|---|---|
| 0 | 0.3 | 0.0 | $(0 - 1.1)^2(0.3) = 0.363$ |
| 1 | 0.4 | 0.4 | $(1 - 1.1)^2(0.4) = 0.004$ |
| 2 | 0.2 | 0.4 | $(2 - 1.1)^2(0.2) = 0.162$ |
| 3 | 0.1 | 0.3 | $(3 - 1.1)^2(0.1) = 0.361$ |
| | | $\mu_X = 1.1$ | $\sigma_X^2 = 0.890$ |

We see that $\sigma_X^2 = 0.89$. The standard deviation of $X$ is $\sigma_X = \sqrt{0.89} = 0.943$. The standard deviation is a measure of the variability of the number of cars Linda sells. As in the case of distributions for data, the standard deviation of a probability distribution is easiest to understand for Normal distributions.

## USE YOUR KNOWLEDGE

**4.71 Find the variance and the standard deviation.** The random variable $X$ has the following probability distribution:

| Value of $X$ | 0 | 2 |
|---|---|---|
| Probability | 0.5 | 0.5 |

Find the variance $\sigma_X^2$ and the standard deviation $\sigma_X$ for this random variable.

## Rules for variances and standard deviations

What are the facts for variances that parallel Rules 1 and 2 for means? *The mean of a sum of random variables is always the sum of their means, but this addition rule is true for variances only in special situations.* To understand why, take $X$ to be the percent of a family's after-tax income that is spent and $Y$ the percent that is saved. When $X$ increases, $Y$ decreases by the same amount. Though $X$ and $Y$ may vary widely from year to year, their sum $X + Y$ is always 100% and does not vary at all. It is the association between the variables $X$ and $Y$ that prevents their variances from adding. If random variables are independent, this kind of association between their values is ruled out and their variances do add. Two random variables $X$ and $Y$ are **independent** if knowing that any event involving $X$ alone did or did not occur tells us nothing about the occurrence of any event involving $Y$ alone. Probability models often assume independence when the random variables describe outcomes that appear unrelated to each other. You should ask in each instance whether the assumption of independence seems reasonable.

independence

When random variables are not independent, the variance of their sum depends on the **correlation** between them as well as on their individual variances. In Chapter 2, we met the correlation $r$ between two observed variables measured on the same individuals. We defined (page 102) the correlation $r$ as an average of the products of the standardized $x$ and $y$ observations. The correlation between two random variables is defined in the same way, once again using a weighted average with probabilities as weights. We won't give the details—it is enough to know that the correlation between two random variables has the same basic properties as the correlation $r$ calculated from data. We use $\rho$, the Greek letter rho, for the correlation between two random variables. The correlation $\rho$ is a number between $-1$ and 1 that measures the direction and strength of the linear relationship between two variables. **The correlation between two independent random variables is zero.**

correlation

Returning to family finances, if $X$ is the percent of a family's after-tax income that is spent and $Y$ the percent that is saved, then $Y = 100 - X$. This is a perfect linear relationship with a negative slope, so the correlation between $X$ and $Y$ is $\rho = -1$. With the correlation at hand, we can state the rules for manipulating variances.

## RULES FOR VARIANCES AND STANDARD DEVIATIONS

**Rule 1.** If $X$ is a random variable and $a$ and $b$ are fixed numbers, then

$$\sigma^2_{a+bX} = b^2\sigma^2_X$$

**Rule 2.** If $X$ and $Y$ are independent random variables, then

$$\sigma^2_{X+Y} = \sigma^2_X + \sigma^2_Y$$
$$\sigma^2_{X-Y} = \sigma^2_X + \sigma^2_Y$$

This is the **addition rule for variances of independent random variables.**

**Rule 3.** If $X$ and $Y$ have correlation $\rho$, then

$$\sigma^2_{X+Y} = \sigma^2_X + \sigma^2_Y + 2\rho\sigma_X\sigma_Y$$
$$\sigma^2_{X-Y} = \sigma^2_X + \sigma^2_Y - 2\rho\sigma_X\sigma_Y$$

This is the **general addition rule for variances of random variables.**

To find the standard deviation, take the square root of the variance.

*Because a variance is the average of squared deviations from the mean, multiplying $X$ by a constant $b$ multiplies $\sigma^2_X$ by the square of the constant.* Adding a constant $a$ to a random variable changes its mean but does not change its variability. The variance of $X + a$ is therefore the same as the variance of $X$. Because the square of $-1$ is 1, the addition rule says that the variance of a difference of independent random variables is the *sum* of the variances. For independent random variables, the difference $X - Y$ is more variable than either $X$ or $Y$ alone because variations in both $X$ and $Y$ contribute to variation in their difference.

As with data, we prefer the standard deviation to the variance as a measure of the variability of a random variable. *Rule 2 for variances implies that standard deviations of independent random variables do not add. To combine standard deviations, use the rules for variances.* For example, the standard deviations of $2X$ and $-2X$ are both equal to $2\sigma_X$ because this is the square root of the variance $4\sigma^2_X$.

EXAMPLE

**4.34 Payoff in the Tri-State Pick 3 lottery.** The payoff $X$ of a \$1 ticket in the Tri-State Pick 3 game is \$500 with probability 1/1000 and 0 the rest of the time. Here is the combined calculation of mean and variance:

| $x_i$ | $p_i$ | $x_ip_i$ | $(x_i - \mu_X)^2p_i$ | |
|---|---|---|---|---|
| 0 | 0.999 | 0 | $(0 - 0.5)^2(0.999) =$ | 0.24975 |
| 500 | 0.001 | 0.5 | $(500 - 0.5)^2(0.001) =$ | 249.50025 |
| | | $\mu_X = 0.5$ | $\sigma^2_X =$ | 249.75 |

The mean payoff is 50 cents. The standard deviation is

$$\sigma_X = \sqrt{249.75} = \$15.80.$$

It is usual for games of chance to have large standard deviations because large variability makes gambling exciting.

If you buy a Pick 3 ticket, your winnings are $W = X - 1$ because the dollar you paid for the ticket must be subtracted from the payoff. Let's find the mean and variance for this random variable.

EXAMPLE

**4.35 Winnings in the Tri-State Pick 3 lottery.** By the rules for means, the mean amount you win is

$$\mu_W = \mu_X - 1 = -\$0.50$$

That is, you lose an average of 50 cents on a ticket. The rules for variances remind us that the variance and standard deviation of the winnings $W = X - 1$ are the same as those of $X$. Subtracting a fixed number changes the mean but not the variance.

Suppose now that you buy a \$1 ticket on each of two different days. The payoffs $X$ and $Y$ on the two tickets are independent because separate drawings are held each day. Your total payoff is $X + Y$. Let's find the mean and standard deviation for this payoff.

EXAMPLE

**4.36 Two tickets.** The mean for the payoff for the two tickets is

$$\mu_{X+Y} = \mu_X + \mu_Y = \$0.50 + \$0.50 = \$1.00$$

Because $X$ and $Y$ are independent, the variance of $X + Y$ is

$$\sigma^2_{X+Y} = \sigma^2_X + \sigma^2_Y = 249.75 + 249.75 = 499.5$$

The standard deviation of the total payoff is

$$\sigma_{X+Y} = \sqrt{499.5} = \$22.35$$

This is not the same as the sum of the individual standard deviations, which is \$15.80 + \$15.80 = \$31.60. Variances of independent random variables add; standard deviations do not.

When we add random variables that are correlated, we need to use the correlation for the calculation of the variance, but not for the calculation of the mean. Here is an example.

EXAMPLE

**4.37 The SAT Math score and the SAT Verbal score are dependent.** Scores on the Mathematics part of the SAT college entrance exam in a recent year had mean 519 and standard deviation 115. Scores on the Verbal part of the SAT had mean 507 and standard deviation 111. What are the mean and standard deviation of total SAT score?[20]

Think of choosing one student's scores at random. Expressed in the language of random variables,

| | | |
|---|---|---|
| SAT Math score $X$ | $\mu_X = 519$ | $\sigma_X = 115$ |
| SAT Verbal score $Y$ | $\mu_Y = 507$ | $\sigma_Y = 111$ |

The total score is $X + Y$. The mean is easy:

$$\mu_{X+Y} = \mu_X + \mu_Y = 519 + 507 = 1026$$

The variance and standard deviation of the total *cannot be computed* from the information given. SAT Verbal and Math scores are not independent, because students who score high on one exam tend to score high on the other also. Therefore, Rule 2 does not apply. We need to know $\rho$, the correlation between $X$ and $Y$, to apply Rule 3.

The correlation between SAT Math and Verbal scores was $\rho = 0.71$. By Rule 3,

$$\begin{aligned}\sigma^2_{X+Y} &= \sigma^2_X + \sigma^2_Y + 2\rho\sigma_X\sigma_Y \\ &= (115)^2 + (111)^2 + (2)(0.71)(115)(111) \\ &= 43{,}672\end{aligned}$$

The variance of the sum $X + Y$ is greater than the sum of the variances $\sigma^2_X + \sigma^2_Y$ because of the positive correlation between SAT Math scores and SAT Verbal scores. That is, $X$ and $Y$ tend to move up together and down together, which increases the variability of their sum. Find the standard deviation from the variance,

$$\sigma_{X+Y} = \sqrt{43{,}672} = 209$$

Total SAT scores had mean 1026 and standard deviation 209.

There are situations where we need to combine several of our rules to find means and standard deviations. Here is an example.

EXAMPLE

**4.38 Investing in Treasury bills and an index fund.** Zadie has invested 20% of her funds in Treasury bills and 80% in an "index fund" that represents all U.S. common stocks. The rate of return of an investment over a time period is the percent change in the price during the time period, plus any income received. If $X$ is the annual return on T-bills and $Y$ the annual return on stocks, the portfolio rate of return is

$$R = 0.2X + 0.8Y$$

The returns $X$ and $Y$ are random variables because they vary from year to year. Based on annual returns between 1950 and 2003, we have

| | | |
|---|---|---|
| $X$ = annual return on T-bills | $\mu_X = 5.0\%$ | $\sigma_X = 2.9\%$ |
| $Y$ = annual return on stocks | $\mu_Y = 13.2\%$ | $\sigma_Y = 17.6\%$ |
| Correlation between $X$ and $Y$ | $\rho = -0.11$ | |

Stocks had higher returns than T-bills on the average, but the standard deviations show that returns on stocks varied much more from year to year. That is, the risk of investing in stocks is greater than the risk for T-bills because their returns are less predictable.

For the return $R$ on Zadie's portfolio of 20% T-bills and 80% stocks,

$$R = 0.2X + 0.8Y$$

$$\begin{aligned}\mu_R &= 0.2\mu_X + 0.8\mu_Y \\ &= (0.2 \times 5.0) + (0.8 \times 13.2) = 11.56\%\end{aligned}$$

To find the variance of the portfolio return, combine Rules 1 and 3. Use the fact that, for example, the variance of $0.2X$ is $(0.2)^2$ times the variance of $X$. Also use the fact that changing scales does not change the correlation, so that the correlation between $0.2X$ and $0.8Y$ is the same as the correlation between $X$ and $Y$.

$$\begin{aligned}\sigma_R^2 &= \sigma_{0.2X}^2 + \sigma_{0.8Y}^2 + 2\rho\sigma_{0.2X}\sigma_{0.8Y} \\ &= (0.2)^2\sigma_X^2 + (0.8)^2\sigma_Y^2 + 2\rho(0.2 \times \sigma_X)(0.8 \times \sigma_Y) \\ &= (0.2)^2(2.9)^2 + (0.8)^2(17.6)^2 + (2)(-0.11)(0.2 \times 2.9)(0.8 \times 17.6) \\ &= 196.786 \\ \sigma_R &= \sqrt{196.786} = 14.03\%\end{aligned}$$

The portfolio has a smaller mean return than an all-stock portfolio, but it is also less risky. That's why Zadie put some funds into Treasury bills.

## SECTION 4.4 Summary

The probability distribution of a random variable $X$, like a distribution of data, has a **mean** $\mu_X$ and a **standard deviation** $\sigma_X$.

The **law of large numbers** says that the average of the values of $X$ observed in many trials must approach $\mu$.

The **mean** $\mu$ is the balance point of the probability histogram or density curve. If $X$ is discrete with possible values $x_i$ having probabilities $p_i$, the mean is the average of the values of $X$, each weighted by its probability:

$$\mu_X = x_1p_1 + x_2p_2 + \cdots + x_kp_k$$

The **variance** $\sigma_X^2$ is the average squared deviation of the values of the variable from their mean. For a discrete random variable,

$$\sigma_X^2 = (x_1 - \mu)^2p_1 + (x_2 - \mu)^2p_2 + \cdots + (x_k - \mu)^2p_k$$

The **standard deviation** $\sigma_X$ is the square root of the variance. The standard deviation measures the variability of the distribution about the mean. It is easiest to interpret for Normal distributions.

The mean and variance of a continuous random variable can be computed from the density curve, but to do so requires more advanced mathematics.

The means and variances of random variables obey the following rules. If $a$ and $b$ are fixed numbers, then

$$\mu_{a+bX} = a + b\mu_X$$
$$\sigma^2_{a+bX} = b^2\sigma^2_X$$

If $X$ and $Y$ are any two random variables having correlation $\rho$, then

$$\mu_{X+Y} = \mu_X + \mu_Y$$
$$\sigma^2_{X+Y} = \sigma^2_X + \sigma^2_Y + 2\rho\sigma_X\sigma_Y$$
$$\sigma^2_{X-Y} = \sigma^2_X + \sigma^2_Y - 2\rho\sigma_X\sigma_Y$$

If $X$ and $Y$ are **independent**, then $\rho = 0$. In this case,

$$\sigma^2_{X+Y} = \sigma^2_X + \sigma^2_Y$$
$$\sigma^2_{X-Y} = \sigma^2_X + \sigma^2_Y$$

To find the standard deviation, take the square root of the variance.

## SECTION 4.4 Exercises

*For Exercise 4.67, see page 271; for Exercise 4.68, see page 275; for Exercises 4.69 and 4.70, see page 279; and for Exercise 4.71, see page 281.*

**4.72 Mean of the distribution for the number of aces.** In Exercise 4.50 you examined the probability distribution for the number of aces when you are dealt two cards in the game of Texas hold 'em. Let $X$ represent the number of aces in a randomly selected deal of two cards in this game. Here is the probability distribution for the random variable $X$:

| Value of $X$ | 0 | 1 | 2 |
|---|---|---|---|
| Probability | 0.559 | 0.382 | 0.059 |

Find $\mu_X$, the mean of the probability distribution of $X$.

**4.73 Mean of the grade distribution.** Example 4.22 gives the distribution of grades (A = 4, B = 3, and so on) in English 210 at North Carolina State University as

| Value of $X$ | 0 | 1 | 2 | 3 | 4 |
|---|---|---|---|---|---|
| Probability | 0.05 | 0.04 | 0.20 | 0.40 | 0.31 |

Find the average (that is, the mean) grade in this course.

**4.74 Mean of the distributions of errors.** Typographical and spelling errors can be either "nonword errors" or "word errors." A nonword error is not a real word, as when "the" is typed as "teh." A word error is a real word, but not the right word, as when "lose" is typed as "loose." When undergraduates are asked to write a 250-word essay (without spell-checking), the number of nonword errors has the following distribution:

| Errors | 0 | 1 | 2 | 3 | 4 |
|---|---|---|---|---|---|
| Probability | 0.1 | 0.3 | 0.3 | 0.2 | 0.1 |

The number of word errors has this distribution:

| Errors | 0 | 1 | 2 | 3 |
|---|---|---|---|---|
| Probability | 0.4 | 0.3 | 0.2 | 0.1 |

What are the mean numbers of nonword errors and word errors in an essay?

**4.75 Means of the numbers of rooms in housing units.** How do rented housing units differ from units occupied by their owners? Exercise 4.53 (page 268) gives the distributions of the number of rooms for owner-occupied units and renter-occupied units in San Jose, California. Find the mean number of rooms for both types of housing unit. How do the means reflect the differences between the distributions that you found in Exercise 4.53?

**4.76 Find the mean of the sum.** Figure 4.12 (page 269) displays the density curve of the sum $Y = X_1 + X_2$ of two independent random numbers, each uniformly distributed between 0 and 1.

(a) The mean of a continuous random variable is the balance point of its density curve. Use this fact to find the mean of $Y$ from Figure 4.12.

(b) Use the same fact to find the means of $X_1$ and $X_2$. (They have the density curve pictured in Figure 4.9, page 264.) Verify that the mean of $Y$ is the sum of the mean of $X_1$ and the mean of $X_2$.

**4.77 Standard deviations of numbers of rooms in housing units.** Which of the two distributions of room counts appears more spread out in the probability histograms you made in Exercise 4.53 (page 268)? Why? Find the standard deviation for both distributions. The standard deviation provides a numerical measure of spread.

**4.78 The effect of correlation.** Find the mean and standard deviation of the total number of errors (nonword errors plus word errors) in an essay if the error counts have the distributions given in Exercise 4.74 and

(a) the counts of nonword and word errors are independent.

(b) students who make many nonword errors also tend to make many word errors, so that the correlation between the two error counts is 0.4.

**4.79 Means and variances of sums.** The rules for means and variances allow you to find the mean and variance of a sum of random variables without first finding the distribution of the sum, which is usually much harder to do.

(a) A single toss of a balanced coin has either 0 or 1 head, each with probability 1/2. What are the mean and standard deviation of the number of heads?

(b) Toss a coin four times. Use the rules for means and variances to find the mean and standard deviation of the total number of heads.

(c) Example 4.23 (page 261) finds the distribution of the number of heads in four tosses. Find the mean and standard deviation from this distribution. Your results in (b) and (c) should agree.

**4.80** CHALLENGE **Toss a 4-sided die twice.** Role-playing games like Dungeons & Dragons use many different types of dice. Suppose that a four-sided die has faces marked 1, 2, 3, 4. The intelligence of a character is determined by rolling this die twice and adding 1 to the sum of the spots. The faces are equally likely and the two rolls are independent. What is the average (mean) intelligence for such characters? How spread out are their intelligences, as measured by the standard deviation of the distribution?

**4.81 A mechnanical assembly.** A mechanical assembly (Figure 4.15) consists of a rod with a bearing on each end. The three parts are manufactured independently, and all vary a bit from part to part. The length of the rod has mean 12 centimeters (cm) and standard deviation 0.004 millimeters (mm). The length of a bearing has mean 2 cm and standard deviation 0.001 mm. What are the mean and standard deviation of the total length of the assembly?

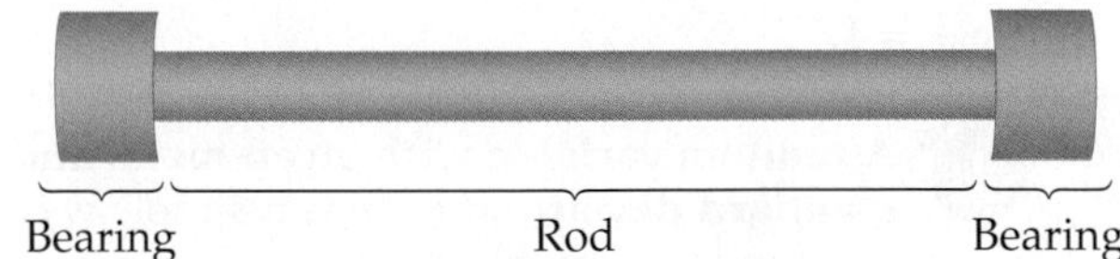


**FIGURE 4.15** Sketch of a mechanical assembly, for Exercise 4.81.

**4.82 Sums of Normal random variables.** Continue your work in the previous exercise. Dimensions of mechanical parts are often roughly Normal. According to the 68–95–99.7 rule, 95% of rods have lengths within $\pm d_1$ of 12 cm and 95% of bearings have lengths within $\pm d_2$ of 2 cm.

(a) What are the values of $d_1$ and $d_2$? These are often called the "natural tolerances" of the parts.

(b) Statistical theory says that any sum of independent Normal random variables has a

Normal distribution. So the total length of the assembly is roughly Normal. What is the natural tolerance for the total length? It is *not* $d_1 + 2d_2$, because standard deviations don't add.

**4.83 Will you assume independence?** In which of the following games of chance would you be willing to assume independence of $X$ and $Y$ in making a probability model? Explain your answer in each case.

(a) In blackjack, you are dealt two cards and examine the total points $X$ on the cards (face cards count 10 points). You can choose to be dealt another card and compete based on the total points $Y$ on all three cards.

(b) In craps, the betting is based on successive rolls of two dice. $X$ is the sum of the faces on the first roll, and $Y$ the sum of the faces on the next roll.

**4.84 Transform the distribution of heights from centimeters to inches.** A report of the National Center for Health Statistics says that the heights of 20-year-old men have mean 176.8 centimeters (cm) and standard deviation 7.2 cm. There are 2.54 centimeters in an inch. What are the mean and standard deviation in inches?

**4.85 What happens when the correlation is 1?** We know that variances add if the random variables involved are uncorrelated ($\rho = 0$), but not otherwise. The opposite extreme is perfect positive correlation ($\rho = 1$). Show by using the general addition rule for variances that in this case the standard deviations add. That is, $\sigma_{X+Y} = \sigma_X + \sigma_Y$ if $\rho_{XY} = 1$.

**4.86 A random variable with given mean and standard deviation.** Here is a simple way to create a random variable $X$ that has mean $\mu$ and standard deviation $\sigma$: $X$ takes only the two values $\mu - \sigma$ and $\mu + \sigma$, each with probability 0.5. Use the definition of the mean and variance for discrete random variables to show that $X$ does have mean $\mu$ and standard deviation $\sigma$.

***Insurance.*** *The business of selling insurance is based on probability and the law of large numbers. Consumers buy insurance because we all face risks that are unlikely but carry high cost. Think of a fire destroying your home. So we form a group to share the risk: we all pay a small amount, and the insurance policy pays a large amount to those few of us whose homes burn down. The insurance company sells many policies, so it can rely on the law of large numbers. Exercises 4.87 to 4.90 explore aspects of insurance.*

**4.87 Fire insurance.** An insurance company looks at the records for millions of homeowners and sees that the mean loss from fire in a year is $\mu = \$300$ per person. (Most of us have no loss, but a few lose their homes. The \$300 is the average loss.) The company plans to sell fire insurance for \$300 plus enough to cover its costs and profit. Explain clearly why it would be stupid to sell only 10 policies. Then explain why selling thousands of such policies is a safe business.

**4.88 Mean and standard deviation for 10 and for 12 policies.** In fact, the insurance company sees that in the entire population of homeowners, the mean loss from fire is $\mu = \$300$ and the standard deviation of the loss is $\sigma = \$400$. What are the mean and standard deviation of the average loss for 10 policies? (Losses on separate policies are independent.) What are the mean and standard deviation of the average loss for 12 policies?

**4.89 Life insurance.** According to the current Commissioners' Standard Ordinary mortality table, adopted by state insurance regulators in December 2002, a 25-year-old man has these probabilities of dying during the next five years:[21]

| Age at death | 25 | 26 | 27 | 28 | 29 |
|---|---|---|---|---|---|
| Probability | 0.00039 | 0.00044 | 0.00051 | 0.000057 | 0.00060 |

(a) What is the probability that the man does not die in the next five years?

(b) An online insurance site offers a term insurance policy that will pay \$100,000 if a 25-year-old man dies within the next 5 years. The cost is \$175 per year. So the insurance company will take in \$875 from this policy if the man does not die within five years. If he does die, the company must pay \$100,000. Its loss depends on how many premiums were paid, as follows:

| Age at death | 25 | 26 | 27 | 28 | 29 |
|---|---|---|---|---|---|
| Loss | \$99,825 | \$99,650 | \$99,475 | \$99,300 | \$99,125 |

What is the insurance company's mean cash intake from such polices?

**4.90 Risk for one versus thousands of life insurance policies.** It would be quite risky for you to insure the life of a 25-year-old friend under the terms of Exercise 4.89. There is a high probability that your friend would live and you would gain \$875 in premiums. But if he were to die, you would lose almost \$100,000. Explain carefully why selling insurance is not risky for an insurance company that insures many thousands of 25-year-old men.

***Portfolio analysis.*** *Here are the means, standard deviations, and correlations for the annual returns from three Fidelity mutual funds for the 10 years ending in February 2004.*[22] *Because there are three random variables, there are three correlations. We use subscripts to show which pair of random variables a correlation refers to.*

$W$ = annual return on 500 Index Fund

$X$ = annual return on Investment Grade Bond Fund

$Y$ = annual return on Diversified International Fund

$$\mu_W = 11.12\% \qquad \sigma_W = 17.46\%$$

$$\mu_X = 6.46\% \qquad \sigma_X = 4.18\%$$

$$\mu_Y = 11.10\% \qquad \sigma_Y = 15.62\%$$

Correlations

$$\rho_{WX} = -0.22 \qquad \rho_{WY} = 0.56 \qquad \rho_{XY} = -0.12$$

*Exercises 4.91 to 4.93 make use of these historical data.*

**4.91 Investing in a mix of U.S. stocks and foreign stocks.** Many advisers recommend using roughly 20% foreign stocks to diversify portfolios of U.S. stocks. You see that the 500 Index (U.S. stocks) and Diversified International (foreign stocks) Funds had almost the same mean returns. A portfolio of 80% 500 Index and 20% Diversified International will deliver this mean return with less risk. Verify this by finding the mean and standard deviation of returns on this portfolio. (Example 4.38, page 284, shows how to find the mean and standard deviation for the portfolio.)

**4.92 The effect of correlation.** Diversification works better when the investments in a portfolio have small correlations. To demonstrate this, suppose that returns on 500 Index Fund and Diversified International Fund had the means and standard deviations we have given but were uncorrelated ($\rho_{WY} = 0$). Show that the standard deviation of a portfolio that combines 80% 500 Index with 20% Diversified International is then smaller than your result from the previous exercise. What happens to the mean return if the correlation is 0?

**4.93 A portfolio with three investments.** Portfolios often contain more than two investments. The rules for means and variances continue to apply, though the arithmetic gets messier. A portfolio containing proportions $a$ of 500 Index Fund, $b$ of Investment Grade Bond Fund, and $c$ of Diversified International Fund has return $R = aW + bX + cY$. Because $a$, $b$, and $c$ are the proportions invested in the three funds, $a + b + c = 1$. The mean and variance of the portfolio return $R$ are

$$\mu_R = a\mu_W + b\mu_X + c\mu_Y$$

$$\sigma_R^2 = a^2\sigma_W^2 + b^2\sigma_X^2 + c^2\sigma_Y^2 + 2ab\rho_{WX}\sigma_W\sigma_X + 2ac\rho_{WY}\sigma_W\sigma_Y + 2bc\rho_{XY}\sigma_X\sigma_Y$$

A basic well-diversified portfolio has 60% in 500 Index, 20% in Investment Grade Bond, and 20% in Diversified International. What are the (historical) mean and standard deviation of the annual returns for this portfolio? What does an investor gain by choosing this diversified portfolio over 100% U.S. stocks? What does the investor lose (at least in this time period)?

## 4.5 General Probability Rules*

Our study of probability has concentrated on random variables and their distributions. Now we return to the laws that govern any assignment of probabilities. The purpose of learning more laws of probability is to be able to give probability models for more complex random phenomena. We have already met and used five rules.

*This section extends the rules of probability discussed in Section 4.2. This material is not needed for understanding the statistical methods in later chapters. It can therefore be omitted if desired.

RULES OF PROBABILITY

**Rule 1.** $0 \leq P(A) \leq 1$ for any event $A$

**Rule 2.** $P(S) = 1$

**Rule 3. Addition rule:** If $A$ and $B$ are **disjoint** events, then

$$P(A \text{ or } B) = P(A) + P(B)$$

**Rule 4. Complement rule:** For any event $A$,

$$P(A^c) = 1 - P(A)$$

**Rule 5. Multiplication rule:** If $A$ and $B$ are **independent** events, then

$$P(A \text{ and } B) = P(A)P(B)$$

## General addition rules

Probability has the property that if $A$ and $B$ are disjoint events, then $P(A \text{ or } B) = P(A) + P(B)$. What if there are more than two events, or if the events are not disjoint? These circumstances are covered by more general addition rules for probability.

UNION

The **union** of any collection of events is the event that at least one of the collection occurs.

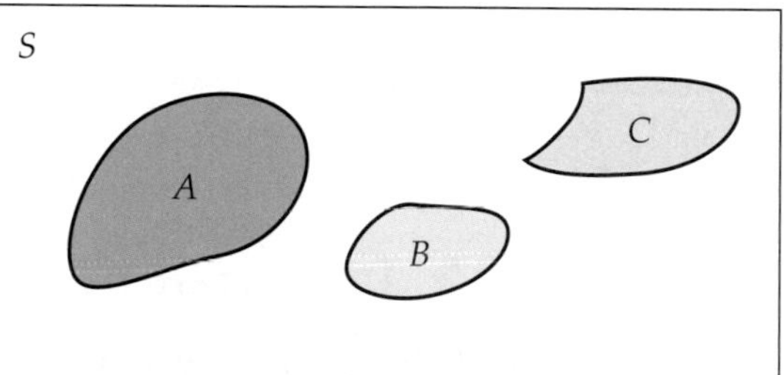


**FIGURE 4.16** The addition rule for disjoint events: $P(A \text{ or } B \text{ or } C) = P(A) + P(B) + P(C)$ when events $A$, $B$, and $C$ are disjoint.

For two events $A$ and $B$, the union is the event $\{A \text{ or } B\}$ that $A$ or $B$ or both occur. From the addition rule for two disjoint events we can obtain rules for more general unions. Suppose first that we have several events—say, $A$, $B$, and $C$—that are disjoint in pairs. That is, no two can occur simultaneously. The Venn diagram in Figure 4.16 illustrates three disjoint events. The addition rule for two disjoint events extends to the following law:

**ADDITION RULE FOR DISJOINT EVENTS**

If events $A$, $B$, and $C$ are disjoint in the sense that no two have any outcomes in common, then

$$P(\text{one or more of } A, B, C) = P(A) + P(B) + P(C)$$

This rule extends to any number of disjoint events.

EXAMPLE

**4.39 Probabilities as areas.** Generate a random number $X$ between 0 and 1. What is the probability that the first digit after the decimal point will be odd? The random number $X$ is a continuous random variable whose density curve has constant height 1 between 0 and 1 and is 0 elsewhere. The event that the first digit of $X$ is odd is the union of five disjoint events. These events are

$$0.10 \leq X < 0.20$$
$$0.30 \leq X < 0.40$$
$$0.50 \leq X < 0.60$$
$$0.70 \leq X < 0.80$$
$$0.90 \leq X < 1.00$$

Figure 4.17 illustrates the probabilities of these events as areas under the density curve. Each area is 0.1. The union of the five therefore has probability equal to the sum, or 0.5. As we should expect, a random number is equally likely to begin with an odd or an even digit.

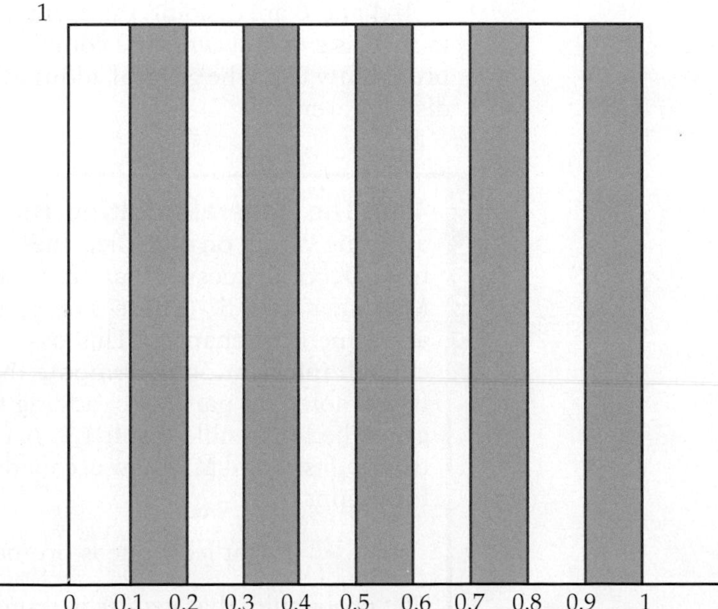


**FIGURE 4.17** The probability that the first digit after the decimal point of a random number is odd is the sum of the probabilities of the 5 disjoint events shown. See Example 4.39.

## USE YOUR KNOWLEDGE

**4.94 Probability that you roll a 3 or a 5.** If you roll a die, the probability of each of the six possible outcomes (1, 2, 3, 4, 5, 6) is 1/6. What is the probability that you roll a 3 or a 5?

If events $A$ and $B$ are not disjoint, they can occur simultaneously. The probability of their union is then *less* than the sum of their probabilities. As Figure 4.18 suggests, the outcomes common to both are counted twice when we add probabilities, so we must subtract this probability once. Here is the addition rule for the union of any two events, disjoint or not.

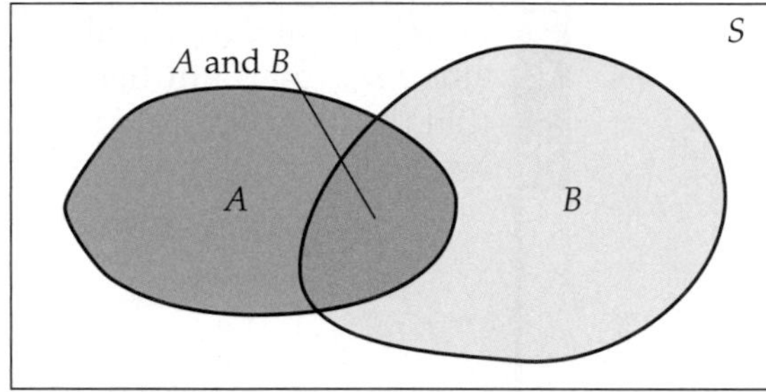


**FIGURE 4.18** The union of two events that are not disjoint. The general addition rule says that $P(A \text{ or } B) = P(A) + P(B) - P(A \text{ and } B)$.

### GENERAL ADDITION RULE FOR UNIONS OF TWO EVENTS

For any two events $A$ and $B$,

$$P(A \text{ or } B) = P(A) + P(B) - P(A \text{ and } B)$$

If $A$ and $B$ are disjoint, the event $\{A \text{ and } B\}$ that both occur has no outcomes in it. This *empty event* is the complement of the sample space $S$ and must have probability 0. So the general addition rule includes Rule 3, the addition rule for disjoint events.

EXAMPLE

**4.40 The general addition rule.** Deborah and Matthew are anxiously awaiting word on whether they have been made partners of their law firm. Deborah guesses that her probability of making partner is 0.7 and that Matthew's is 0.5. (These are personal probabilities reflecting Deborah's assessment of chance.) This assignment of probabilities does not give us enough information to compute the probability that at least one of the two is promoted. In particular, adding the individual probabilities of promotion gives the impossible result 1.2. If Deborah also guesses that the probability that *both* she and Matthew are made partners is 0.3, then by the addition rule for unions

$$P(\text{at least one is promoted}) = 0.7 + 0.5 - 0.3 = 0.9$$

The probability that *neither* is promoted is then 0.1 by the complement rule.

**USE YOUR KNOWLEDGE**

**4.95 Probability that your roll is even or greater than 4.** If you roll a die, the probability of each of the six possible outcomes (1, 2, 3, 4, 5, 6) is 1/6. What is the probability that your roll is even or greater than 4?

Venn diagrams are a great help in finding probabilities for unions because you can just think of adding and subtracting areas. Figure 4.19 shows some events and their probabilities for Example 4.40. What is the probability that Deborah is promoted and Matthew is not? The Venn diagram shows that this is the probability that Deborah is promoted minus the probability that both are promoted, $0.7 - 0.3 = 0.4$. Similarly, the probability that Matthew is promoted and Deborah is not is $0.5 - 0.3 = 0.2$. The four probabilities that appear in the figure add to 1 because they refer to four disjoint events whose union is the entire sample space.

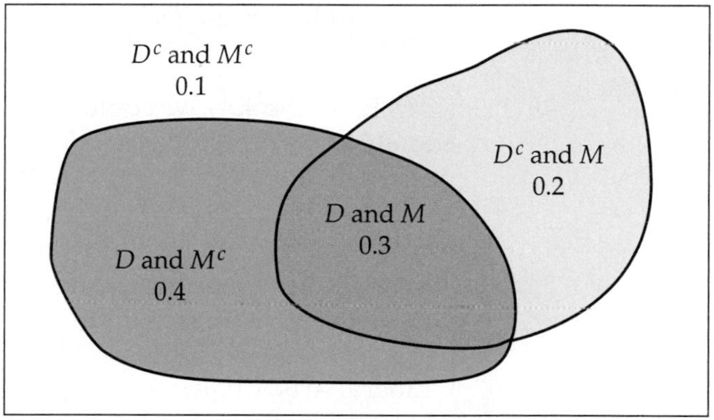


$D$ = Deborah is made partner
$M$ = Matthew is made partner

**FIGURE 4.19** Venn diagram and probabilities for Example 4.40.

## Conditional probability

The probability we assign to an event can change if we know that some other event has occurred. This idea is the key to many applications of probability.

EXAMPLE

**4.41 Probability of being dealt an ace.** Slim is a professional poker player. He stares at the dealer, who prepares to deal. What is the probability that the card dealt to Slim is an ace? There are 52 cards in the deck. Because the deck was carefully shuffled, the next card dealt is equally likely to be any of the cards that Slim has not seen. Four of the 52 cards are aces. So

$$P(\text{ace}) = \frac{4}{52} = \frac{1}{13}$$

This calculation assumes that Slim knows nothing about any cards already dealt. Suppose now that he is looking at 4 cards already in his hand, and that one of them is an ace. He knows nothing about the other 48 cards except that exactly 3 aces are among them. Slim's probability of being dealt an ace *given what he knows* is now

$$P(\text{ace} \mid 1 \text{ ace in 4 visible cards}) = \frac{3}{48} = \frac{1}{16}$$

Knowing that there is 1 ace among the 4 cards Slim can see changes the probability that the next card dealt is an ace.

conditional probability

The new notation $P(A \mid B)$ is a **conditional probability.** That is, it gives the probability of one event (the next card dealt is an ace) under the condition that we know another event (exactly 1 of the 4 visible cards is an ace). You can read the bar | as "given the information that."

## USE YOUR KNOWLEDGE

**4.96 The probability of another ace.** Suppose two of the four cards in Slim's hand are aces. What is the probability that the next card dealt to him is an ace?

In Example 4.41 we could find probabilities because we are willing to use an equally likely probability model for a shuffled deck of cards. Here is an example based on data.

EXAMPLE

**4.42 College course grades.** Students at the University of New Harmony received 10,000 course grades last semester. Table 4.1 breaks down these grades by which school of the university taught the course. The schools are Liberal Arts, Engineering and Physical Sciences, and Health and Human Services. (Table 4.1 is based closely on grade distributions at an actual university, simplified a bit for clarity.[23])

It is common knowledge that college grades are lower in engineering and the physical sciences (EPS) than in liberal arts and social sciences. Consider the two events

$$A = \text{the grade comes from an EPS course}$$

$$B = \text{the grade is below a B}$$

**TABLE 4.1**

Grades awarded at a university, by school

| | Grade level | | | |
|---|---|---|---|---|
| | A | B | Below B | Total |
| Liberal Arts | 2,142 | 1,890 | 2,268 | 6,300 |
| Engineering and Physical Sciences | 368 | 432 | 800 | 1,600 |
| Health and Human Services | 882 | 630 | 588 | 2,100 |
| Total | 3,392 | 2,952 | 3,656 | 10,000 |

There are 10,000 grades, of which 3656 are below B. Choosing at random gives each grade an equal chance, so the probability of choosing a grade below a B is

$$P(B) = \frac{3,656}{10,000} = 0.3656$$

To find the *conditional* probability that a grade is below a B, *given the information* that it comes from the EPS school, look only at the "Engineering and Physical Sciences" row. The EPS grades are all in this row, so the information given says that only this row is relevant. The conditional probability is

$$P(B \mid A) = \frac{800}{1600} = 0.5$$

The conditional probability that a grade is below a B, given that we know it comes from an EPS course, is much higher than the probability for a randomly chosen grade.

*It is easy to confuse probabilities and conditional probabilities involving the same events.* For example, Table 4.1 says that

$$P(A) = \frac{1600}{10,000} = 0.16$$

$$P(A \text{ and } B) = \frac{800}{10,000} = 0.08$$

$$P(B \mid A) = \frac{800}{1600} = 0.5$$

Be sure you understand how we found these three results. There is a relationship among these three probabilities. The probability that a grade is both from EPS *and* below a B is the product of the probabilities that it is from EPS and that it is below a B, *given* that it is from EPS. That is,

$$\begin{aligned} P(A \text{ and } B) &= P(A) \times P(B \mid A) \\ &= \frac{1600}{10,000} \times \frac{800}{1600} \\ &= \frac{800}{10,000} = 0.08 \quad \text{(as before)} \end{aligned}$$

Try to think your way through this in words: First, the grade is from EPS; then, given that it is from EPS, it is below B. We have just discovered the fundamental multiplication rule of probability.

**MULTIPLICATION RULE**

The probability that both of two events $A$ and $B$ happen together can be found by

$$P(A \text{ and } B) = P(A)P(B \mid A)$$

Here $P(B \mid A)$ is the conditional probability that $B$ occurs, given the information that $A$ occurs.

## USE YOUR KNOWLEDGE

**4.97 Select a grade from the population.** Refer to Table 4.1 and consider selecting a single grade from this population.

(a) What is the probability that the grade is from Health and Human Services? 0.21

(b) What is the probability that the grade is an A? 0.3392

(c) What is the probability that the grade is an A, given that it is from Health and Human Services? 0.42

(d) Explain why your answers to (b) and (c) are not the same.

EXAMPLE

**4.43 Downloading music from the Internet.** The multiplication rule is just common sense made formal. For example, 29% of Internet users download music files, and 67% of downloaders say they don't care if the music is copyrighted.[24] So the percent of Internet users who download music (event $A$) *and* don't care about copyright (event $B$) is 67% of the 29% who download, or

$$(0.67)(0.29) = 0.1943 = 19.43\%$$

The multiplication rule expresses this as

$$\begin{aligned} P(A \text{ and } B) &= P(A) \times P(B \mid A) \\ &= (0.29)(0.67) = 0.1943 \end{aligned}$$

EXAMPLE

**4.44 Probability of a favorable draw.** Slim is still at the poker table. At the moment, he wants very much to draw two diamonds in a row. As he sits at the table looking at his hand and at the upturned cards on the table, Slim sees 11 cards. Of these, 4 are diamonds. The full deck contains 13 diamonds among its 52 cards, so 9 of the 41 unseen cards are diamonds. To find Slim's probability of drawing two diamonds, first calculate

$$\begin{aligned} P(\text{first card diamond}) &= \frac{9}{41} \\ P(\text{second card diamond} \mid \text{first card diamond}) &= \frac{8}{40} \end{aligned}$$

Slim finds both probabilities by counting cards. The probability that the first card drawn is a diamond is 9/41 because 9 of the 41 unseen cards are diamonds. If the first card is a diamond, that leaves 8 diamonds among the 40 remaining cards. So the *conditional* probability of another diamond is 8/40.

The multiplication rule now says that

$$P(\text{both cards diamonds}) = \frac{9}{41} \times \frac{8}{40} = 0.044$$

Slim will need luck to draw his diamonds.

## USE YOUR KNOWLEDGE

**4.98 The probability that the next two cards are diamonds.** In the setting of Exercise 4.44, suppose Slim sees 25 cards and the only diamonds are the 3 in his hand. What is the probability that the next 2 cards dealt to Slim will be diamonds? This outcome would give him 5 cards from the same suit, a hand that is called a flush.

If $P(A)$ and $P(A \text{ and } B)$ are given, we can rearrange the multiplication rule to produce a *definition* of the conditional probability $P(B \mid A)$ in terms of unconditional probabilities.

### DEFINITION OF CONDITIONAL PROBABILITY

When $P(A) > 0$, the **conditional probability** of $B$ given $A$ is

$$P(B \mid A) = \frac{P(A \text{ and } B)}{P(A)}$$

*Be sure to keep in mind the distinct roles in $P(B \mid A)$ of the event $B$ whose probability we are computing and the event $A$ that represents the information we are given.* The conditional probability $P(B \mid A)$ makes no sense if the event $A$ can never occur, so we require that $P(A) > 0$ whenever we talk about $P(B \mid A)$.

EXAMPLE

**4.45 A conditional probability.** What is the conditional probability that a grade at the University of New Harmony is an A, given that it comes from a liberal arts course? We see from Table 4.1 that

$$P(\text{liberal arts course}) = \frac{6300}{10{,}000} = 0.63$$

$$P(\text{A grade } \textit{and} \text{ liberal arts course}) = \frac{2142}{10{,}000} = 0.2142$$

The definition of conditional probability therefore says that

$$P(\text{A grade} \mid \text{liberal arts course}) = \frac{P(\text{A grade } \textit{and} \text{ liberal arts course})}{P(\text{liberal arts course})} = \frac{0.2142}{0.63} = 0.34$$

Note that this agrees with the result obtained from the "Liberal Arts" row of Table 4.1:

$$P(\text{A grade} \mid \text{liberal arts course}) = \frac{2142}{6300} = 0.34$$

## USE YOUR KNOWLEDGE

**4.99 Find the conditional probability.** Refer to Table 4.1. What is the conditional probability that a grade is a B, given that it comes from Engineering and Physical Sciences? Find the answer by dividing two numbers from Table 4.1 and using the multiplication rule according to the method in Example 4.45.

## General multiplication rules

The definition of conditional probability reminds us that in principle all probabilities, including conditional probabilities, can be found from the assignment of probabilities to events that describe random phenomena. More often, however, conditional probabilities are part of the information given to us in a probability model, and the multiplication rule is used to compute $P(A \text{ and } B)$. This rule extends to more than two events.

The union of a collection of events is the event that *any* of them occur. Here is the corresponding term for the event that *all* of them occur.

### INTERSECTION

The **intersection** of any collection of events is the event that *all* of the events occur.

To extend the multiplication rule to the probability that all of several events occur, the key is to condition each event on the occurrence of *all* of the preceding events. For example, the intersection of three events $A$, $B$, and $C$ has probability

$$P(A \text{ and } B \text{ and } C) = P(A)P(B \mid A)P(C \mid A \text{ and } B)$$

EXAMPLE

**4.46 High school athletes and professional careers.** Only 5% of male high school basketball, baseball, and football players go on to play at the college level. Of these, only 1.7% enter major league professional sports. About 40% of the athletes who compete in college and then reach the pros have a career of more than 3 years.[25] Define these events:

$$A = \{\text{competes in college}\}$$

$$B = \{\text{competes professionally}\}$$

$$C = \{\text{pro career longer than 3 years}\}$$

What is the probability that a high school athlete competes in college and then goes on to have a pro career of more than 3 years? We know that

$$P(A) = 0.05$$

$$P(B \mid A) = 0.017$$

$$P(C \mid A \text{ and } B) = 0.4$$

The probability we want is therefore

$$\begin{aligned} P(A \text{ and } B \text{ and } C) &= P(A)P(B \mid A)P(C \mid A \text{ and } B) \\ &= 0.05 \times 0.017 \times 0.4 = 0.00034 \end{aligned}$$

Only about 3 of every 10,000 high school athletes can expect to compete in college and have a professional career of more than 3 years. High school students would be wise to concentrate on studies rather than on unrealistic hopes of fortune from pro sports.

## Tree diagrams

Probability problems often require us to combine several of the basic rules into a more elaborate calculation. Here is an example that illustrates how to solve problems that have several stages.

EXAMPLE

**4.47 Online chat rooms.** Online chat rooms are dominated by the young. Teens are the biggest users. If we look only at adult Internet users (aged 18 and over), 47% of the 18 to 29 age group chat, as do 21% of the 30 to 49 age group and just 7% of those 50 and over. To learn what percent of all Internet users participate in chat, we also need the age breakdown of users. Here it is: 29% of adult Internet users are 18 to 29 years old (event $A_1$), another 47% are 30 to 49 (event $A_2$), and the remaining 24% are 50 and over (event $A_3$).[26]

tree diagram

What is the probability that a randomly chosen user of the Internet participates in chat rooms (event $C$)? To find out, use the **tree diagram** in Figure 4.20 to organize your thinking. Each segment in the tree is one stage of the problem. Each complete branch shows a path through the two stages. The probability written on each segment is the conditional probability of an Internet user following that segment, given that he or she has reached the node from which it branches.

Starting at the left, an Internet user falls into one of the three age groups. The probabilities of these groups

$$P(A_1) = 0.29 \qquad P(A_2) = 0.47 \qquad P(A_3) = 0.24$$

mark the leftmost branches in the tree. Conditional on being 18 to 29 years old, the probability of participating in chat is $P(C \mid A_1) = 0.47$. So the conditional probability of *not* participating is

$$P(C^c \mid A_1) = 1 - 0.47 = 0.53$$

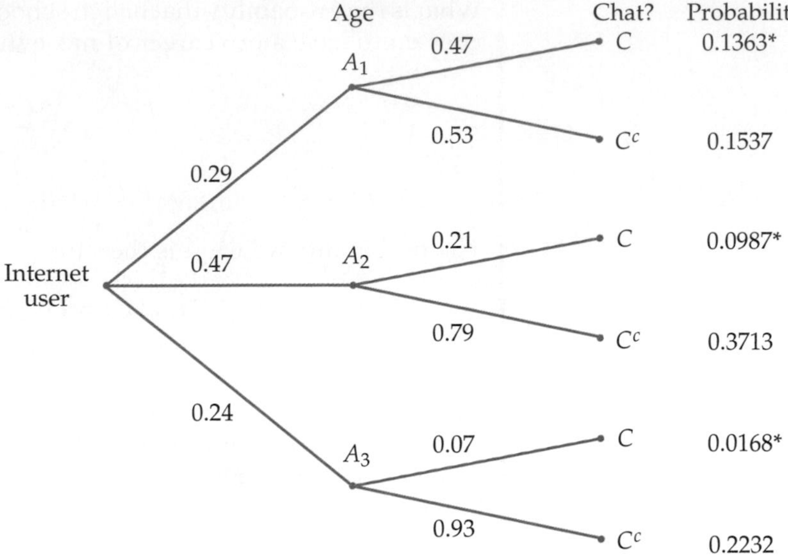


**FIGURE 4.20** Tree diagram for Example 4.47. The probability $P(C)$ is the sum of the probabilities of the three branches marked with asterisks ($*$).

These conditional probabilities mark the paths branching out from the $A_1$ node in Figure 4.20. The other two age group nodes similarly lead to two branches marked with the conditional probabilities of chatting or not. The probabilities on the branches from any node add to 1 because they cover all possibilities, given that this node was reached.

There are three disjoint paths to $C$, one for each age group. By the addition rule, $P(C)$ is the sum of their probabilities. The probability of reaching $C$ through the 18 to 29 age group is

$$\begin{aligned} P(C \text{ and } A_1) &= P(A_1)P(C \mid A_1) \\ &= 0.29 \times 0.47 = 0.1363 \end{aligned}$$

Follow the paths to $C$ through the other two age groups. The probabilities of these paths are

$$P(C \text{ and } A_2) = P(A_2)P(C \mid A_2) = (0.47)(0.21) = 0.0987$$

$$P(C \text{ and } A_3) = P(A_3)P(C \mid A_3) = (0.24)(0.07) = 0.0168$$

The final result is

$$P(C) = 0.1363 + 0.0987 + 0.0168 = 0.2518$$

About 25% of all adult Internet users take part in chat rooms.

It takes longer to explain a tree diagram than it does to use it. Once you have understood a problem well enough to draw the tree, the rest is easy. Tree diagrams combine the addition and multiplication rules. The multiplication rule says that the probability of reaching the end of any complete branch is the product of the probabilities written on its segments. The probability of any outcome, such as the event $C$ that an adult Internet user takes part in chat rooms, is then found by adding the probabilities of all branches that are part of that event.

## USE YOUR KNOWLEDGE

**4.100 Draw a tree diagram.** Refer to Slim's chances of a flush in Exercise 4.98. Draw a tree diagram to describe the outcomes for the two cards that he will be dealt. At the first stage, his draw can be a diamond or a non-diamond. At the second stage, he has the same possible outcomes but the probabilities are different.

## Bayes's rule

There is another kind of probability question that we might ask in the context of thinking about online chat. What percent of adult chat room participants are aged 18 to 29?

EXAMPLE

**4.48 Conditional versus unconditional probabilities.** In the notation of Example 4.47 this is the conditional probability $P(A_1 \mid C)$. Start from the definition of conditional probability and then apply the results of Example 4.47:

$$P(A_1 \mid C) = \frac{P(A_1 \text{ and } C)}{P(C)}$$

$$= \frac{0.1363}{0.2518} = 0.5413$$

Over half of adult chat room participants are between 18 and 29 years old. Compare this conditional probability with the original information (unconditional) that 29% of adult Internet users are between 18 and 29 years old. Knowing that a person chats increases the probability that he or she is young.

We know the probabilities $P(A_1)$, $P(A_2)$, and $P(A_3)$ that give the age distribution of adult Internet users. We also know the conditional probabilities $P(C \mid A_1)$, $P(C \mid A_2)$, and $P(C \mid A_3)$ that a person from each age group chats. Example 4.47 shows how to use this information to calculate $P(C)$. The method can be summarized in a single expression that adds the probabilities of the three paths to $C$ in the tree diagram:

$$P(C) = P(A_1)P(C \mid A_1) + P(A_2)P(C \mid A_2) + P(A_3)P(C \mid A_3)$$

In Example 4.48 we calculated the "reverse" conditional probability $P(A_1 \mid C)$. The denominator 0.2518 in that example came from the expression just above. Put in this general notation, we have another probability law.

### BAYES'S RULE

Suppose that $A_1, A_2, \ldots, A_k$ are disjoint events whose probabilities are not 0 and add to exactly 1. That is, any outcome is in exactly one of these events. Then if $C$ is any other event whose probability is not 0 or 1,

$$P(A_i \mid C) = \frac{P(C \mid A_i)P(A_i)}{P(C \mid A_1)P(A_1) + P(C \mid A_2)P(A_2) + \cdots + P(A_k)P(C \mid A_k)}$$

The numerator in Bayes's rule is always one of the terms in the sum that makes up the denominator. The rule is named after Thomas Bayes, who wrestled with arguing from outcomes like $C$ back to the $A_i$ in a book published in 1763. It is far better to think your way through problems like Examples 4.47 and 4.48 rather than memorize these formal expressions.

## Independence again

The conditional probability $P(B \mid A)$ is generally not equal to the unconditional probability $P(B)$. That is because the occurrence of event $A$ generally gives us some additional information about whether or not event $B$ occurs. If knowing that $A$ occurs gives no additional information about $B$, then $A$ and $B$ are independent events. The formal definition of independence is expressed in terms of conditional probability.

### INDEPENDENT EVENTS

Two events $A$ and $B$ that both have positive probability are **independent** if

$$P(B \mid A) = P(B)$$

This definition makes precise the informal description of independence given in Section 4.2. We now see that the multiplication rule for independent events, $P(A \text{ and } B) = P(A)P(B)$, is a special case of the general multiplication rule, $P(A \text{ and } B) = P(A)P(B \mid A)$, just as the addition rule for disjoint events is a special case of the general addition rule.

## SECTION 4.5 Summary

The **complement** $A^c$ of an event $A$ contains all outcomes that are not in $A$. The **union** $\{A \text{ or } B\}$ of events $A$ and $B$ contains all outcomes in $A$, in $B$, or in both $A$ and $B$. The **intersection** $\{A \text{ and } B\}$ contains all outcomes that are in both $A$ and $B$, but not outcomes in $A$ alone or $B$ alone.

The **conditional probability** $P(B \mid A)$ of an event $B$, given an event $A$, is defined by

$$P(B \mid A) = \frac{P(A \text{ and } B)}{P(A)}$$

when $P(A) > 0$. In practice, conditional probabilities are most often found from directly available information.

The essential general rules of elementary probability are

**Legitimate values:** $0 \leq P(A) \leq 1$ for any event $A$

**Total probability 1:** $P(S) = 1$

**Complement rule:** $P(A^c) = 1 - P(A)$

**Addition rule:** $P(A \text{ or } B) = P(A) + P(B) - P(A \text{ and } B)$

**Multiplication rule:** $P(A \text{ and } B) = P(A)P(B \mid A)$

If $A$ and $B$ are **disjoint,** then $P(A \text{ and } B) = 0$. The general addition rule for unions then becomes the special addition rule, $P(A \text{ or } B) = P(A) + P(B)$.

$A$ and $B$ are **independent** when $P(B \mid A) = P(B)$. The multiplication rule for intersections then becomes $P(A \text{ and } B) = P(A)P(B)$.

In problems with several stages, draw a **tree diagram** to organize use of the multiplication and addition rules.

## SECTION 4.5 Exercises

*For Exercise 4.94, see page 292; for Exercise 4.95, see page 293; for Exercise 4.96, see page 294; for Exercise 4.97, see page 296; for Exercise 4.98, see page 297; for Exercise 4.99, see page 298; and for Exercise 4.100, see page 301.*

**4.101 Binge drinking and gender.** In a college population, students are classified by gender and whether or not they are frequent binge drinkers. Here are the probabilities:

| | Men | Women |
|---|---|---|
| Binge drinker | 0.11 | 0.12 |
| Not binge drinker | 0.32 | 0.45 |

(a) Verify that the sum of the probabilities is 1.

(b) What is the probability that a randomly selected student is not a binge drinker?

(c) What is the probability that a randomly selected male student is not a binge drinker?

(d) Explain why your answers to (b) and (c) are different. Use language that would be understood by someone who has not studied the material in this chapter.

**4.102 Find some probabilities.** Refer to the previous exercise.

(a) Find the probability that a randomly selected student is a male binge drinker, and find the probability that a randomly selected student is a female binge drinker.

(b) Find the probability that a student is a binge drinker, given that the student is male, and find the probability that a student is a binge drinker, given that the student is female.

(c) Your answer for part (a) gives a higher probability for females, while your answer for part (b) gives a higher probability for males. Interpret your answers in terms of the question of whether there are gender differences in binge-drinking behavior. Decide which comparison you prefer and explain the reasons for your preference.

**4.103 Attendance at 2-year and 4-year colleges.** In a large national population of college students, 61% attend 4-year institutions and the rest attend 2-year institutions. Males make up 44% of the students in the 4-year institutions and 41% of the students in the 2-year institutions.

(a) Find the four probabilities for each combination of gender and type of institution in the following table. Be sure that your probabilities sum to 1.

| | Men | Women |
|---|---|---|
| 4-year institution | | |
| 2-year institution | | |

(b) Consider randomly selecting a female student from this population. What is the probability that she attends a 4-year institution?

**4.104 Draw a tree diagram.** Refer to the previous exercise. Draw a tree diagram to illustrate the probabilities in a situation where you first identify the type of institution attended and then identify the gender of the student.

**4.105 Draw a different tree diagram for the same setting.** Refer to the previous two exercises. Draw a tree diagram to illustrate the probabilities in a situation where you first identify the gender of the student and then identify the type of institution attended. Explain why the probabilities in this tree diagram are different from those that you used in the previous exercise.

**4.106 Education and income.** Call a household prosperous if its income exceeds \$100,000. Call the household educated if the householder completed college. Select an American household at random, and let $A$ be the event that the selected household is prosperous and $B$ the event that it is educated. According to the Current Population Survey, $P(A) = 0.138$, $P(B) = 0.261$, and the probability that a household is both prosperous and educated is $P(A \text{ and } B) = 0.082$. What is the probability $P(A \text{ or } B)$ that the household selected is either prosperous or educated?

**4.107 Find a conditional probability.** In the setting of the previous exercise, what is the conditional probability that a household is prosperous, given that it is educated? Explain why your result shows that events $A$ and $B$ are not independent.

**4.108 Draw a Venn diagram.** Draw a Venn diagram that shows the relation between the events $A$ and $B$ in Exercise 4.106. Indicate each of the following events on your diagram and use the information in Exercise 4.106 to calculate the probability of each event. Finally, describe in words what each event is.

(a) $\{A \text{ and } B\}$

(b) $\{A^c \text{ and } B\}$

(c) $\{A \text{ and } B^c\}$

(d) $\{A^c \text{ and } B^c\}$

**4.109 Sales of cars and light trucks.** Motor vehicles sold to individuals are classified as either cars or light trucks (including SUVs) and as either domestic or imported. In a recent year, 69% of vehicles sold were light trucks, 78% were domestic, and 55% were domestic light trucks. Let $A$ be the event that a vehicle is a car and $B$ the event that it is imported. Write each of the following events in set notation and give its probability.

(a) The vehicle is a light truck.

(b) The vehicle is an imported car.

**4.110 Income tax returns.** In 2004, the Internal Revenue Service received 312,226,042 individual tax returns. Of these, 12,757,005 reported an adjusted gross income of at least \$100,000, and 240,128 reported at least \$1 million.[27] If you know that a randomly chosen return shows an income of \$100,000 or more, what is the conditional probability that the income is at least \$1 million?

**4.111 Conditional probabilities and independence.** Using the information in Exercise 4.109, answer these questions.

(a) Given that a vehicle is imported, what is the conditional probability that it is a light truck?

(b) Are the events "vehicle is a light truck" and "vehicle is imported" independent? Justify your answer.

**4.112 Job offers.** Julie is graduating from college. She has studied biology, chemistry, and computing and hopes to work as a forensic scientist applying her science background to crime investigation. Late one night she thinks about some jobs she has applied for. Let $A$, $B$, and $C$ be the events that Julie is offered a job by

$A$ = the Connecticut Office of the Chief Medical Examiner

$B$ = the New Jersey Division of Criminal Justice

$C$ = the federal Disaster Mortuary Operations Response Team

Julie writes down her personal probabilities for being offered these jobs:

| | | |
|---|---|---|
| $P(A) = 0.7$ | $P(B) = 0.5$ | $P(C) = 0.3$ |
| $P(A \text{ and } B) = 0.3$ | $P(A \text{ and } C) = 0.1$ | |
| $P(B \text{ and } C) = 0.1$ | $P(A \text{ and } B \text{ and } C) = 0$ | |

Make a Venn diagram of the events $A$, $B$, and $C$. As in Figure 4.19 (page 293), mark the probabilities of every intersection involving these events and their complements. Use this diagram for Exercises 4.113 to 4.115.

**4.113 Find the probability of at least one offer.** What is the probability that Julie is offered at least one of the three jobs?

**4.114 Find the probability of another event.** What is the probability that Julie is offered both the Connecticut and New Jersey jobs, but not the federal job?

**4.115 Find a conditional probability.** If Julie is offered the federal job, what is the conditional probability that she is also offered the New Jersey job? If Julie is offered the New Jersey job, what is the conditional probability that she is also offered the federal job?

**4.116 Academic degrees and gender.** Here are the projected numbers (in thousands) of earned degrees in the United States in the 2010–2011 academic year, classified by level and by the sex of the degree recipient:[28]

| | Bachelor's | Master's | Professional | Doctorate |
|---|---|---|---|---|
| Female | 933 | 402 | 51 | 26 |
| Male | 661 | 260 | 44 | 26 |

(a) Convert this table to a table giving the probabilities for selecting a degree earned and classifying the recipient by gender and the degree by the levels given above.

(b) If you choose a degree recipient at random, what is the probability that the person you choose is a woman?

(c) What is the conditional probability that you choose a woman, given that the person chosen received a professional degree?

(d) Are the events "choose a woman" and "choose a professional degree recipient" independent? How do you know?

**4.117 Find some probabilities.** The previous exercise gives the projected number (in thousands) of earned degrees in the United States in the 2010–2011 academic year. Use these data to answer the following questions.

(a) What is the probability that a randomly chosen degree recipient is a man?

(b) What is the conditional probability that the person chosen received a bachelor's degree, given that he is a man?

(c) Use the multiplication rule to find the probability of choosing a male bachelor's degree recipient. Check your result by finding this probability directly from the table of counts.

***Working.*** *In the language of government statistics, you are "in the labor force" if you are available for work and either working or actively seeking work. The unemployment rate is the proportion of the labor force (not of the entire population) who are unemployed. Here are data from the Current Population Survey for the civilian population aged 25 years and over. The table entries are counts in thousands of people.[29] Exercises 4.118 to 4.121 concern these data.*

| Highest education | Total population | In labor force | Employed |
|---|---|---|---|
| Did not finish high school | 28,021 | 12,623 | 11,552 |
| High school but no college | 59,844 | 38,210 | 36,249 |
| Some college, but no bachelor's degree | 46,777 | 33,928 | 32,429 |
| College graduate | 51,568 | 40,414 | 39,250 |

**4.118 Find the unemployment rates.** Find the unemployment rate for people with each level of education. How does the unemployment rate change with education? Explain carefully why your results show that level of education and being employed are not independent.

**4.119 Conditional probabilities and independence.**

(a) What is the probability that a randomly chosen person 25 years of age or older is in the labor force?

(b) If you know that the person chosen is a college graduate, what is the conditional probability that he or she is in the labor force?

(c) Are the events "in the labor force" and "college graduate" independent? How do you know?

**4.120 Find some conditional probabilities.** You know that a person is employed. What is the conditional probability that he or she is a college graduate? You know that a second person is a college graduate. What is the conditional probability that he or she is employed?

**4.121** CHALLENGE **A lurking variable.** Beware the lurking variable. The low labor force participation rate of people who did not finish high school is explained by the confounding of education level with a variable that lurks behind the "aged 25 years and over" restriction for these data. Explain this confounding.

**4.122** CHALLENGE **Gender and majors.** The probability that a randomly chosen student at the University of New Harmony is a woman is 0.62. The probability that the student is studying education is 0.17. The conditional probability that the student is a woman, given that the student is studying education, is 0.8. What is the conditional probability that the student is studying education, given that she is a woman?

**4.123 Spelling errors.** As explained in Exercise 4.74 (page 286), spelling errors in a text can be either nonword errors or word errors. Nonword errors make up 25% of all errors. A human proofreader will catch 90% of nonword errors and 70% of word errors. What percent of all errors will the proofreader catch? (Draw a tree diagram to organize the information given.)

**4.124** CHALLENGE **Mathematics degrees and gender.** Of the 16,071 degrees in mathematics given by U.S. colleges and universities in a recent year, 73% were bachelor's degrees, 21% were master's degrees, and the rest were doctorates. Moreover, women earned 48% of the bachelor's degrees, 42% of the master's degrees, and 29% of the doctorates.[30] You choose a mathematics degree at random and find that it was awarded to a woman. What is the probability that it is a bachelor's degree?

***Genetic counseling.*** *Conditional probabilities and Bayes's rule are a basis for counseling people who may have genetic defects that can be passed to their children. Exercises 4.125 to 4.129 concern genetic counseling settings.*

**4.125 Albinism.** People with albinism have little pigment in their skin, hair, and eyes. The gene that governs albinism has two forms (called alleles), which we denote by $a$ and $A$. Each person has a pair of these genes, one inherited from each parent. A child inherits one of each parent's two alleles, independently with probability 0.5. Albinism is a recessive trait, so a person is albino only if the inherited pair is $aa$.

(a) Beth's parents are not albino but she has an albino brother. This implies that both of Beth's parents have type $Aa$. Why?

(b) Which of the types $aa$, $Aa$, $AA$ could a child of Beth's parents have? What is the probability of each type?

(c) Beth is not albino. What are the conditional probabilities for Beth's possible genetic types, given this fact? (Use the definition of conditional probability.)

**4.126 Find some conditional probabilities.** Beth knows the probabilities for her genetic types from part (c) of the previous exercise. She marries Bob, who is albino. Bob's genetic type must be $aa$.

(a) What is the conditional probability that a child of Beth and Bob is non-albino if Beth has type $Aa$? What is the conditional probability of a non-albino child if Beth has type $AA$?

(b) Beth and Bob's first child is non-albino. What is the conditional probability that Beth is a carrier, type $Aa$?

**4.127 Cystic fibrosis.** Cystic fibrosis is a lung disorder that often results in death. It is inherited but can be inherited only if both parents are carriers of an abnormal gene. In 1989, the CF gene that is abnormal in carriers of cystic fibrosis was identified. The probability that a randomly chosen person of European ancestry carries an abnormal CF gene is 1/25. (The probability is less in other ethnic groups.) The CF20m test detects most but not all harmful mutations of the CF gene. The test is positive for 90% of people who are carriers. It is (ignoring human error) never positive for people who are not carriers. Jason tests positive. What is the probability that he is a carrier?

**4.128 Use Bayes's rule.** Refer to the previous exercise. Jason knows that he is a carrier of cystic fibrosis. His wife, Julianne, has a brother with cystic fibrosis, which means the probability is 2/3 that she is a carrier. If Julianne is a carrier, each child she has with Jason has probability 1/4 of having cystic fibrosis. If she is not a carrier, her children cannot have the disease. Jason and Julianne have one child, who does not have cystic fibrosis. This information reduces the probability that Julianne is a carrier. Use Bayes's rule to find the conditional probability that Julianne is a carrier, given that she and Jason have one child who does not have cystic fibrosis.

**4.129 Muscular dystrophy.** Muscular dystrophy is an incurable muscle-wasting disease. The most common and serious type, called DMD, is caused by a sex-linked recessive mutation. Specifically: women can be carriers but do not get the disease; a son of a carrier has probability 0.5 of having DMD; a daughter has probability 0.5 of being a carrier. As many as 1/3 of DMD cases, however, are due to spontaneous mutations in sons of mothers who are not carriers. Toni has one son, who has DMD.

In the absence of other information, the probability is 1/3 that the son is the victim of a spontaneous mutation and 2/3 that Toni is a carrier. There is a screening test called the CK test that is positive with probability 0.7 if a woman is a carrier and with probability 0.1 if she is not. Toni's CK test is positive. What is the probability that she is a carrier?

## CHAPTER 4 Exercises

**4.130 Toss a pair of dice two times.** Consider tossing a pair of fair dice two times. For each of the following pairs of events, tell whether they are disjoint, independent, or neither.

(a) $A = 7$ on the first roll, $B = 6$ or less on the first roll.

(b) $A = 7$ on the first roll, $B = 6$ or less on the second roll.

(c) $A = 6$ or less on the second roll, $B = 5$ or less on the first roll.

(d) $A = 6$ or less on the second roll, $B = 5$ or less on the second roll.

**4.131 Find the probabilities.** Refer to the previous exercise. Find the probabilities for each event.

**4.132 Some probability distributions.** Here is a probability distribution for a random variable $X$:

| Value of $X$ | 1 | 2 | 3 |
|---|---|---|---|
| Probability | 0.2 | 0.6 | 0.2 |

(a) Find the mean and standard deviation for this distribution.

(b) Construct a different probability distribution with the same possible values, the same mean, and a larger standard deviation. Show your work and report the standard deviation of your new distribution.

(c) Construct a different probability distribution with the same possible values, the same mean, and a smaller standard deviation. Show your work and report the standard deviation of your new distribution.

**4.133 A fair bet at craps.** Almost all bets made at gambling casinos favor the house. In other words, the difference between the amount bet and the mean of the distribution of the payoff is a positive number. An exception is "taking the odds" at the game of craps, a bet that a player can make under certain circumstances. The bet becomes available when a shooter throws a 4, 5, 6, 8, 9, or 10 on the initial roll. This number is called the "point"; when a point is rolled, we say that a point has been established. If a 4 is the point, an odds bet can be made that wins if a 4 is rolled before a 7 is rolled. The probability of winning this bet is 0.5 and the payoff for a \$10 bet is \$20. The same probability of winning and payoff apply for an odds bet on a 10. For an initial roll of 5 or 9, the odds bet has a winning probability of 2/3 and the payoff for a \$10 bet is \$15. Similarly, when the initial roll is 6 or 8, the odds bet has a winning probability of 5/6 and the payoff for a \$10 bet is \$12. Find the mean of the payoff distribution for each of these bets. Then confirm that the bets are fair by showing that the difference between amount bet and the mean of the distribution of the payoff is zero.

**4.134 An ancient Korean drinking game.** An ancient Korean drinking game involves a 14-sided die. The players roll the die in turn and must submit to whatever humiliation is written on the up-face: something like "Keep still when tickled on face." Six of the 14 faces are squares. Let's call them A, B, C, D, E, and F for short. The other eight faces are triangles, which we will call 1, 2, 3, 4, 5, 6, 7, and 8. Each of the squares is equally likely. Each of the triangles is also equally likely, but the triangle probability differs from the square probability. The probability of getting a square is 0.72. Give the probability model for the 14 possible outcomes.

**4.135 Wine tasters.** Two wine tasters rate each wine they taste on a scale of 1 to 5. From data on their ratings of a large number of wines, we obtain the following probabilities for both tasters' ratings of a randomly chosen wine:

| | Taster 2 | | | | |
|---|---|---|---|---|---|
| Taster 1 | 1 | 2 | 3 | 4 | 5 |
| **1** | 0.03 | 0.02 | 0.01 | 0.00 | 0.00 |
| **2** | 0.02 | 0.07 | 0.06 | 0.02 | 0.01 |
| **3** | 0.01 | 0.05 | 0.25 | 0.05 | 0.01 |
| **4** | 0.00 | 0.02 | 0.05 | 0.20 | 0.02 |
| **5** | 0.00 | 0.01 | 0.01 | 0.02 | 0.06 |

(a) Why is this a legitimate assignment of probabilities to outcomes?

(b) What is the probability that the tasters agree when rating a wine?

(c) What is the probability that Taster 1 rates a wine higher than 3? What is the probability that Taster 2 rates a wine higher than 3?

**4.136 Profits from an investment.** Rotter Partners is planning a major investment. The amount of profit

$X$ is uncertain but a probabilistic estimate gives the following distribution (in millions of dollars):

| Profit | 1 | 1.5 | 2 | 4 | 10 |
|---|---|---|---|---|---|
| Probability | 0.4 | 0.2 | 0.2 | 0.1 | 0.1 |

(a) Find the mean profit $\mu_X$ and the standard deviation $\sigma_X$ of the profit.

(b) Rotter Partners owes its source of capital a fee of \$200,000 plus 10% of the profits $X$. So the firm actually retains

$$Y = 0.9X - 0.2$$

from the investment. Find the mean and standard deviation of $Y$.

**4.137 Prizes for grocery store customers.** A grocery store gives its customers cards that may win them a prize when matched with other cards. The back of the card announces the following probabilities of winning various amounts if a customer visits the store 10 times:

| Amount | \$1000 | \$250 | \$100 | \$10 |
|---|---|---|---|---|
| Probability | 1/10,000 | 1/1000 | 1/100 | 1/20 |

(a) What is the probability of winning nothing?

(b) What is the mean amount won?

(c) What is the standard deviation of the amount won?

**4.138** CHALLENGE **SAT scores.** The College Board finds that the distribution of students' SAT scores depends on the level of education their parents have. Children of parents who did not finish high school have SAT Math scores $X$ with mean 445 and standard deviation 106. Scores $Y$ of children of parents with graduate degrees have mean 566 and standard deviation 109. Perhaps we should standardize to a common scale for equity. Find positive numbers $a$, $b$, $c$, and $d$ such that $a + bX$ and $c + dY$ both have mean 500 and standard deviation 100.

**4.139** CHALLENGE **Lottery tickets.** Joe buys a ticket in the TriState Pick 3 lottery every day, always betting on 956. He will win something if the winning number contains 9, 5, and 6 in any order. Each day, Joe has probability 0.006 of winning, and he wins (or not) independently of other days because a new drawing is held each day. What is the probability that Joe's first winning ticket comes on the 20th day?

**4.140** CHALLENGE **Slot machines.** Slot machines are now video games, with winning determined by electronic random number generators. In the old days, slot machines were like this: you pull the lever to spin three wheels; each wheel has 20 symbols, all equally likely to show when the wheel stops spinning; the three wheels are independent of each other. Suppose that the middle wheel has 8 bells among its 20 symbols, and the left and right wheels have 1 bell each.

(a) You win the jackpot if all three wheels show bells. What is the probability of winning the jackpot?

(b) What is the probability that the wheels stop with exactly 2 bells showing?

*The following exercises require familiarity with the material presented in the optional Section 4.5.*

**4.141 Higher education at 2-year and 4-year institutions.** The following table gives the counts of U.S. institutions of higher education classified as public or private and as 2-year or 4-year:[31]

| | Public | Private |
|---|---|---|
| 2-year | 639 | 1894 |
| 4-year | 1061 | 622 |

Convert the counts to probabilities and summarize the relationship between these two variables using conditional probabilities.

**4.142 Odds bets at craps.** Refer to the odds bets at craps in Exercise 4.133. Suppose that whenever the shooter has an initial roll of 4, 5, 6, 8, 9, or 10, you take the odds. Here are the probabilities for these initial rolls:

| Point | 4 | 5 | 6 | 8 | 9 | 10 |
|---|---|---|---|---|---|---|
| Probability | 3/36 | 4/36 | 5/36 | 5/36 | 4/36 | 3/36 |

Draw a tree diagram with the first stage showing the point rolled and the second stage showing whether the point is again rolled before a 7 is rolled. Include a first-stage branch showing the outcome that a point is not established. In this case, the amount bet is zero and the distribution of the winnings is the special random variable that has $P(X = 0) = 1$. For the combined betting system where the player always makes a \$10 odds bet when it is available, show that the game is fair.

**4.143 Weights and heights of children adjusted for age.** The idea of conditional probabilities has many interesting applications, including the idea of a conditional distribution. For example, the National Center for Health Statistics produces distributions for weight and height for children while conditioning on other variables. Visit the Web site `cdc.gov/growthcharts/` and describe the different ways that weight and height distributions are conditioned on other variables.

**4.144 Wine tasting.** In the setting of Exercise 4.135, Taster 1's rating for a wine is 3. What is the conditional probability that Taster 2's rating is higher than 3?

**4.145 Internet usage patterns of students and other adults.** Students have different patterns of Internet use than other adults. Among adult Internet users, 4.1% are full-time students and another 2.9% are part-time students. Students are much more likely to access the Internet from someplace other than work or home: 58% of full-time students do so, as do 30% of part-time students, but only 21% of other users do so.[32] What percent of all adult users reach the Internet from someplace other than home or work?

**4.146 An interesting case of independence.** Independence of events is not always obvious. Toss two balanced coins independently. The four possible combinations of heads and tails in order each have probability 0.25. The events

$A$ = head on the first toss

$B$ = both tosses have the same outcome

may seem intuitively related. Show that $P(B \mid A) = P(B)$, so that $A$ and $B$ are in fact independent.

**4.147 Find some conditional probabilities.** Choose a point at random in the square with sides $0 \le x \le 1$ and $0 \le y \le 1$. This means that the probability that the point falls in any region within the square is the area of that region. Let $X$ be the $x$ coordinate and $Y$ the $y$ coordinate of the point chosen. Find the conditional probability $P(Y < 1/2 \mid Y > X)$. (*Hint:* Sketch the square and the events $Y < 1/2$ and $Y > X$.)

**4.148** CHALLENGE **Sample surveys for sensitive issues.** It is difficult to conduct sample surveys on sensitive issues because many people will not answer questions if the answers might embarrass them. **Randomized response** is an effective way to guarantee anonymity while collecting information on topics such as student cheating or sexual behavior. Here is the idea. To ask a sample of students whether they have plagiarized a term paper while in college, have each student toss a coin in private. If the coin lands heads *and* they have not plagiarized, they are to answer "No." Otherwise, they are to give "Yes" as their answer. Only the student knows whether the answer reflects the truth or just the coin toss, but the researchers can use a proper random sample with follow-up for nonresponse and other good sampling practices.

Suppose that in fact the probability is 0.3 that a randomly chosen student has plagiarized a paper. Draw a tree diagram in which the first stage is tossing the coin and the second is the truth about plagiarism. The outcome at the end of each branch is the answer given to the randomized-response question. What is the probability of a "No" answer in the randomized-response poll? If the probability of plagiarism were 0.2, what would be the probability of a "No" response on the poll? Now suppose that you get 39% "No" answers in a randomized-response poll of a large sample of students at your college. What do you estimate to be the percent of the population who have plagiarized a paper?

CHAPTER

5

# Sampling Distributions

The heights of young women are approximately Normal. See Example 5.1.



## Introduction

Statistical inference draws conclusions about a population or process on the basis of data. The data are summarized by *statistics* such as means, proportions, and the slopes of least-squares regression lines. When the data are produced by random sampling or randomized experimentation, a statistic is a random variable that obeys the laws of probability theory. *Sampling distributions* of statistics provide the link between probability and data. A sampling distribution shows how a statistic would vary in repeated data production. That is, a sampling distribution is a probability distribution that answers the question "What would happen if we did this many times?" The sampling distribution tells us about the results we are likely to see if, for example, we survey a sample of 2000 college students. In Section 3.3 we simulated a large number of random samples to illustrate the idea of a sampling distribution.


LOOK BACK
**sampling distribution, page 215**

**THE DISTRIBUTION OF A STATISTIC**

A statistic from a random sample or randomized experiment is a random variable. The probability distribution of the statistic is its **sampling distribution.**

Probability distributions also play a second role in statistical inference. Any quantity that can be measured for each member of a population is described by the distribution of its values for all members of the population. This is the context in which we first met distributions, as density curves that provide models for the overall pattern of data. Imagine choosing one individual at random from the population. The results of repeated choices have a probability distribution that is the distribution of the population.

LOOK BACK
**density curves, page 56**

EXAMPLE

**5.1 Heights of young women.** The distribution of heights of women between the ages of 18 and 24 is approximately Normal with mean 64.5 inches and standard deviation 2.5 inches. Select a woman at random and measure her height. The result is a random variable $X$. We don't know the height of a randomly chosen woman, but we do know that in repeated sampling $X$ will have the same $N(64.5, 2.5)$ distribution that describes the pattern of heights in the entire population. We call $N(64.5, 2.5)$ the *population distribution.*

**POPULATION DISTRIBUTION**

The **population distribution** of a variable is the distribution of its values for all members of the population. The population distribution is also the probability distribution of the variable when we choose one individual at random from the population.

The population of all women between the ages of 18 and 24 actually exists, so that we can in principle draw an SRS from it. Sometimes our population of interest does not actually exist. For example, suppose we are interested in studying final-exam scores in a statistics course. We have the scores for the 37 students in the course this semester. For the purposes of statistical inference, we might want to consider these 37 students as part of a hypothetical population of similar students who would take this course. In this sense, these students represent not only themselves but also a larger population of similar students. The key idea is to think of the observations that you have as coming from a population with a probability distribution.

LOOK BACK
**SRS, page 200**

To progress from discussing probability as a topic in itself to probability as a foundation for inference, we start by studying the sampling distributions of some common statistics. In each case, the sampling distribution depends on both the population distribution and the way we collect the data from the population.

# 5.1 Sampling Distributions for Counts and Proportions

We begin our study of sampling distributions with the simplest case of a random variable, where there are only two possible outcomes. Here is an example.

EXAMPLE

**5.2 Parents put too much pressure on their children.** A sample survey asks 2000 college students whether they think that parents put too much pressure on their children. We would like to view the responses of these students as representative of a larger population of students who hold similar beliefs. That is, we will view the responses of the sampled students as an SRS from a population.

When there are only two possible outcomes for a random variable, we can summarize the results by giving the count for one of the possible outcomes. We let $n$ represent the sample size and we use $X$ to represent the random variable that gives the count for the outcome of interest.

EXAMPLE

**5.3 The random variable of interest.** In our sample survey of college students, $n = 2000$ and $X$ is the number of students who think that parents put too much pressure on their children. Suppose $X = 840$. The random variable of interest is $X$ and its value is 840.

In our example, we chose the random variable $X$ to be the number of students who think that parents put too much pressure on their children. We could have chosen $X$ to be the number of students who do not think that parents put too much pressure on their children. The choice is yours. Often we make the choice based on how we would like to describe the results in a written summary. Which choice do you prefer in this example?

To interpret the meaning of the random variable $X$ in this setting, we need to know the sample size $n$. The conclusion we would draw about student opinions in our survey would be quite different if we had observed $X = 840$ from a sample of size $n = 1000$.

sample proportion

When a random variable has two possible outcomes, we can use the **sample proportion,** $\hat{p} = X/n$, as a summary.

EXAMPLE

**5.4 The sample proportion.** The sample proportion of students surveyed who think that parents put too much pressure on their children is

$$\hat{p} = \frac{840}{2000} = 0.42$$

USE YOUR KNOWLEDGE

**5.1** **Use of the Internet to find a place to live.** A poll of 1500 college students asked whether or not they had used the Internet to find a place to live sometime within the past year. There were 525 students who answered "Yes"; the other 975 answered "No."

(a) What is $n$?

(b) Choose one of the two possible outcomes to define the random variable, $X$. Give a reason for your choice.

(c) What is the value of $X$?

(d) Find the sample proportion, $\hat{p}$.

**5.2** **Seniors who have taken a statistics course.** In a random sample of 200 senior students from your college, 40% reported that they had taken a statistics course. Give $n$, $X$, and $\hat{p}$ for this setting.

Sample counts and sample proportions are common statistics. This section describes their sampling distributions.

## The binomial distributions for sample counts

The distribution of a count $X$ depends on how the data are produced. Here is a simple but common situation.

THE BINOMIAL SETTING

1. There are a fixed number $n$ of observations.
2. The $n$ observations are all independent.
3. Each observation falls into one of just two categories, which for convenience we call "success" and "failure."
4. The probability of a success, call it $p$, is the same for each observation.

Think of tossing a coin $n$ times as an example of the binomial setting. Each toss gives either heads or tails. The outcomes of successive tosses are independent. If we call heads a success, then $p$ is the probability of a head and remains the same as long as we toss the same coin. The number of heads we count is a random variable $X$. The distribution of $X$, and more generally of the count of successes in any binomial setting, is completely determined by the number of observations $n$ and the success probability $p$.

BINOMIAL DISTRIBUTIONS

The distribution of the count $X$ of successes in the binomial setting is called the **binomial distribution** with parameters $n$ and $p$. The parameter $n$ is the number of observations, and $p$ is the probability of a success on any one observation. The possible values of $X$ are the whole numbers from 0 to $n$. As an abbreviation, we say that $X$ is $B(n, p)$.

The binomial distributions are an important class of discrete probability distributions. Later in this section we will learn how to assign probabilities to outcomes and how to find the mean and standard deviation of binomial distributions. *The most important skill for using binomial distributions is the ability to recognize situations to which they do and don't apply.*

EXAMPLE

**5.5 Two binomial examples.**

(a) Genetics says that children receive genes from their parents independently. Each child of a particular pair of parents has probability 0.25 of having type O blood. If these parents have 3 children, the number who have type O blood is the count $X$ of successes in 3 independent trials with probability 0.25 of a success on each trial. So $X$ has the $B(3, 0.25)$ distribution.

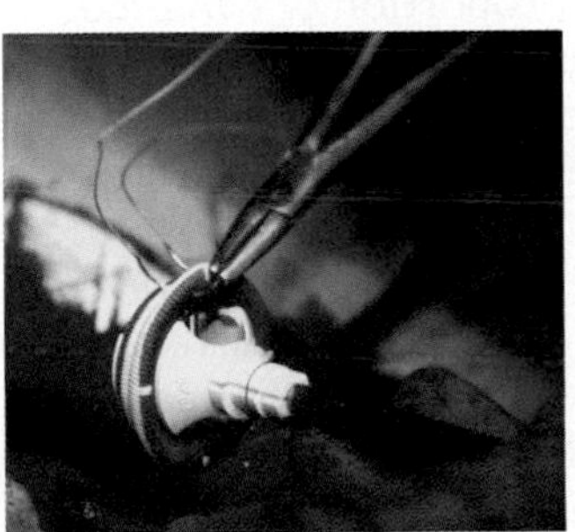

(b) Engineers define reliability as the probability that an item will perform its function under specific conditions for a specific period of time. Replacement heart valves made of animal tissue, for example, have probability 0.77 of performing well for 15 years.[1] The probability of failure is therefore 0.23. It is reasonable to assume that valves in different patients fail (or not) independently of each other. The number of patients in a group of 500 who will need another valve replacement within 15 years has the $B(500, 0.23)$ distribution.

## USE YOUR KNOWLEDGE

**5.3** **Toss a coin.** Toss a fair coin 20 times. Give the distribution of $X$, and the number of heads that you observe.

**5.4** **Genetics and blood types.** Genetics says that children receive genes from their parents independently. Suppose each child of a particular pair of parents has probability 0.25 of having type O blood. If these parents have 4 children, what is the distribution of the number who have type O blood? Explain your answer.

## Binomial distributions in statistical sampling

The binomial distributions are important in statistics when we wish to make inferences about the proportion $p$ of "successes" in a population. Here is a typical example.

EXAMPLE

**5.6 Audits of financial records.** The financial records of businesses may be audited by state tax authorities to test compliance with tax laws. It is too time-consuming to examine all sales and purchases made by a company during the period covered by the audit. Suppose the auditor examines an SRS of 150 sales records out of 10,000 available. One issue is whether each sale was correctly classified as subject to state sales tax or not. Suppose that 800 of the 10,000 sales are incorrectly classified. Is the count $X$ of misclassified records in the sample a binomial random variable?

*Choosing an SRS from a population is not quite a binomial setting.* Removing one record in Example 5.6 changes the proportion of bad records in the remaining population, so the state of the second record chosen is not independent of the first. Because the population is large, however, removing a few items has a very small effect on the composition of the remaining population. Successive inspection results are very nearly independent. The population proportion of misclassified records is

$$p = \frac{800}{10,000} = 0.08$$

If the first record chosen is bad, the proportion of bad records remaining is $799/9999 = 0.079908$. If the first record is good, the proportion of bad records left is $800/9999 = 0.080008$. These proportions are so close to 0.08 that for practical purposes we can act as if removing one record has no effect on the proportion of misclassified records remaining. We act as if the count $X$ of misclassified sales records in the audit sample has the binomial distribution $B(150, 0.08)$.

Populations like that described in Example 5.6 often contain a relatively small number of items with very large values. An SRS taken from such a population will likely include very few items of this type. Therefore, it is common to use a stratified sample in settings like this. Strata are defined based on dollar value, and within each stratum, an SRS is taken. The results are then combined to obtain an estimate for the entire population.

LOOK BACK
**stratified sample, page 203**

SAMPLING DISTRIBUTION OF A COUNT

A population contains proportion $p$ of successes. If the population is much larger than the sample, the count $X$ of successes in an SRS of size $n$ has approximately the binomial distribution $B(n, p)$.

The accuracy of this approximation improves as the size of the population increases relative to the size of the sample. As a rule of thumb, we will use the binomial sampling distribution for counts when the population is at least 20 times as large as the sample.

## Finding binomial probabilities: software and tables

We will later give a formula for the probability that a binomial random variable takes any of its values. In practice, you will rarely have to use this formula for calculations. Some calculators and most statistical software packages calculate binomial probabilities.

EXAMPLE

**5.7 The probability of exactly 10 misclassified sales records.** In the audit setting of Example 5.6, what is the probability that the audit finds exactly 10 misclassified sales records? What is the probability that the audit finds no more than 10 misclassified records? Figure 5.1 shows the output from one statistical software system. You see that if the count $X$ has the $B(150, 0.08)$ distribution,

$$P(X = 10) = 0.106959 \qquad P(X \leq 10) = 0.338427$$

It was easy to request these calculations in the software's menus. The output supplies more decimal places than we need and uses labels that may not be helpful (for example, "Probability Density Function" when the distribution is discrete, not continuous). But, as usual with software, we can ignore distractions and find the results we need.

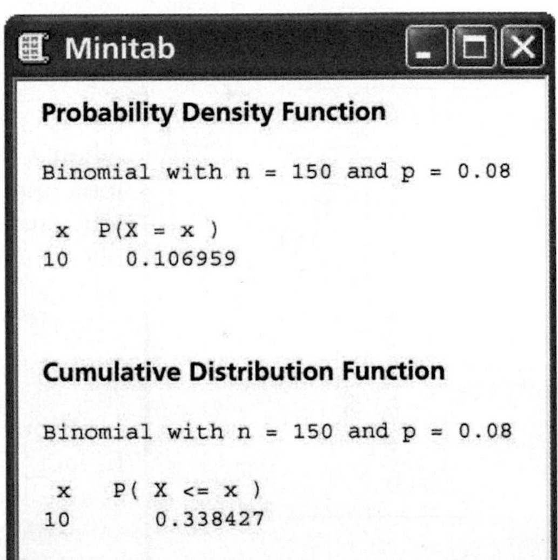


**FIGURE 5.1** Binomial probabilities: output from the Minitab statistical software, for Example 5.7.

If you do not have suitable computing facilities, you can still shorten the work of calculating binomial probabilities for some values of $n$ and $p$ by looking up probabilities in Table C in the back of this book. The entries in the table are the probabilities $P(X = k)$ of individual outcomes for a binomial random variable $X$.

EXAMPLE

**5.8 The probability histogram.** Suppose that the audit in Example 5.6 chose just 15 sales records. What is the probability that no more than 1 of the 15 is misclassified? The count $X$ of misclassified records in the sample has approximately the $B(15, 0.08)$ distribution. Figure 5.2 is a probability histogram for this distribution. The distribution is strongly skewed. Although $X$ can take any whole-number value from 0 to 15, the probabilities of values larger than 5 are so small that they do not appear in the histogram.

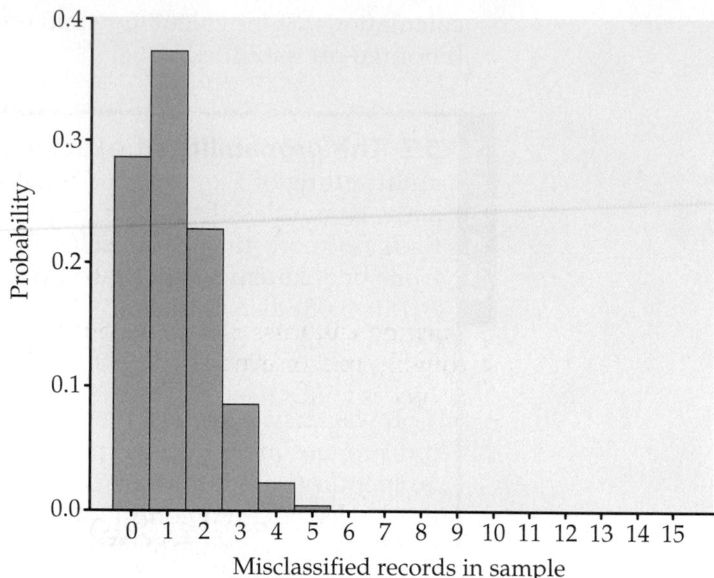


**FIGURE 5.2** Probability histogram for the binomial distribution with $n = 15$ and $p = 0.08$, for Example 5.8.

| | | $p$ |
|---|---|---|
| $n$ | $k$ | .08 |
| 15 | 0 | .2863 |
| | 1 | .3734 |
| | 2 | .2273 |
| | 3 | .0857 |
| | 4 | .0223 |
| | 5 | .0043 |
| | 6 | .0006 |
| | 7 | .0001 |
| | 8 | |
| | 9 | |

We want to calculate

$$P(X \leq 1) = P(X = 0) + P(X = 1)$$

when $X$ has the $B(15, 0.08)$ distribution. To use Table C for this calculation, look opposite $n = 15$ and under $p = 0.08$. This part of the table appears at the left. The entry opposite each $k$ is $P(X = k)$. Blank entries are 0 to four decimal places, so we have omitted them here. You see that

$$P(X \leq 1) = P(X = 0) + P(X = 1)$$
$$= 0.2863 + 0.3734 = 0.6597$$

About two-thirds of all samples will contain no more than 1 bad record. In fact, almost 29% of the samples will contain no bad records. The sample of size 15 cannot be trusted to provide adequate evidence about misclassified sales records. A larger number of observations is needed.

The values of $p$ that appear in Table C are all 0.5 or smaller. When the probability of a success is greater than 0.5, restate the problem in terms of the number of failures. The probability of a failure is less than 0.5 when the probability of a success exceeds 0.5. When using the table, always stop to ask whether you must count successes or failures.

EXAMPLE

**5.9 She makes 75% of her free throws.** Corinne is a basketball player who makes 75% of her free throws over the course of a season. In a key game, Corinne shoots 12 free throws and misses 5 of them. The fans think that she failed because she was nervous. Is it unusual for Corinne to perform this poorly?

To answer this question, assume that free throws are independent with probability 0.75 of a success on each shot. (Studies of long sequences of free throws have found no evidence that they are dependent, so this is a reasonable assumption.) Because the probability of making a free throw is greater than 0.5, we count misses in order to use Table C. The probability of a miss is $1 - 0.75$, or 0.25. The number $X$ of misses in 12 attempts has the $B(12, 0.25)$ distribution.

We want the probability of missing 5 or more. This is

$$P(X \geq 5) = P(X = 5) + P(X = 6) + \cdots + P(X = 12)$$
$$= 0.1032 + 0.0401 + \cdots + 0.0000 = 0.1576$$

Corinne will miss 5 or more out of 12 free throws about 16% of the time, or roughly one of every six games. While below her average level, this performance is well within the range of the usual chance variation in her shooting.

## USE YOUR KNOWLEDGE

**5.5 Find the probabilities.**

(a) Suppose $X$ has the $B(4, 0.3)$ distribution. Use software or Table C to find $P(X = 0)$ and $P(X \geq 3)$.

(b) Suppose $X$ has the $B(4, 0.7)$ distribution. Use software or Table C to find $P(X = 4)$ and $P(X \leq 1)$.

(c) Explain the relationship between your answers to parts (a) and (b) of this exercise.

## Binomial mean and standard deviation

If a count $X$ has the $B(n, p)$ distribution, what are the mean $\mu_X$ and the standard deviation $\sigma_X$? We can guess the mean. If Corinne makes 75% of her free throws, the mean number made in 12 tries should be 75% of 12, or 9. That's $\mu_X$ when $X$ is $B(12, 0.75)$. Intuition suggests more generally that the mean of the $B(n, p)$ distribution should be $np$. Can we show that this is correct and also obtain a short formula for the standard deviation? Because binomial distributions are discrete probability distributions, we could find the mean and variance by using the definitions in Section 4.4. Here is an easier way.

LOOK BACK
**means and variances of random variables, page 270**

A binomial random variable $X$ is the count of successes in $n$ independent observations that each have the same probability $p$ of success. Let the random variable $S_i$ indicate whether the $i$th observation is a success or failure by taking the values $S_i = 1$ if a success occurs and $S_i = 0$ if the outcome is a failure. The $S_i$ are independent because the observations are, and each $S_i$ has the same simple distribution:

| Outcome | 1 | 0 |
|---|---|---|
| Probability | $p$ | $1 - p$ |

LOOK BACK
**mean and variance of a discrete random variable, page 271**

From the definition of the mean of a discrete random variable, we know that the mean of each $S_i$ is

$$\mu_S = (1)(p) + (0)(1 - p) = p$$

Similarly, the definition of the variance shows that $\sigma_S^2 = p(1-p)$. Because each $S_i$ is 1 for a hit and 0 for a miss, to find the total number of hits $X$ we add the $S_i$'s:

$$X = S_1 + S_2 + \cdots + S_n$$

Apply the addition rules for means and variances to this sum. To find the mean of $X$ we add the means of the $S_i$'s:

$$\begin{aligned}\mu_X &= \mu_{S_1} + \mu_{S_2} + \cdots + \mu_{S_n} \\ &= n\mu_S = np\end{aligned}$$

Similarly, the variance is $n$ times the variance of a single $S$, so that

$$\sigma_X^2 = np(1-p)$$

The standard deviation $\sigma_X$ is the square root of the variance. Here is the result.

### BINOMIAL MEAN AND STANDARD DEVIATION

If a count $X$ has the binomial distribution $B(n, p)$, then

$$\mu_X = np$$
$$\sigma_X = \sqrt{np(1-p)}$$

EXAMPLE

**5.10 The Helsinki Heart Study.** The Helsinki Heart Study asked whether the anticholesterol drug gemfibrozil reduces heart attacks. In planning such an experiment, the researchers must be confident that the sample sizes are large enough to enable them to observe enough heart attacks. The Helsinki study planned to give gemfibrozil to about 2000 men aged 40 to 55 and a placebo to another 2000. The probability of a heart attack during the five-year period of the study for men this age is about 0.04. What are the mean and standard deviation of the number of heart attacks that will be observed in one group if the treatment does not change this probability?

There are 2000 independent observations, each having probability $p =$ 0.04 of a heart attack. The count $X$ of heart attacks has the $B(2000, 0.04)$ distribution, so that

$$\mu_X = np = (2000)(0.04) = 80$$
$$\sigma_X = \sqrt{np(1-p)} = \sqrt{(2000)(0.04)(0.96)} = 8.76$$

The expected number of heart attacks is large enough to permit conclusions about the effectiveness of the drug. In fact, there were 84 heart attacks among the 2035 men actually assigned to the placebo, quite close to the mean. The gemfibrozil group of 2046 men suffered only 56 heart attacks. This is evidence that the drug does reduce the chance of a heart attack.

## Sample proportions

What proportion of a company's sales records have an incorrect sales tax classification? What percent of adults favor stronger laws restricting firearms? In statistical sampling we often want to estimate the **proportion** $p$ of "successes" in a population. Our estimator is the sample proportion of successes:

$$\hat{p} = \frac{\text{count of successes in sample}}{\text{size of sample}}$$
$$= \frac{X}{n}$$

*Be sure to distinguish between the proportion $\hat{p}$ and the count $X$.* The count takes whole-number values between 0 and $n$, but a proportion is always a number between 0 and 1. In the binomial setting, the count $X$ has a binomial distribution. The proportion $\hat{p}$ does *not* have a binomial distribution. We can, however, do probability calculations about $\hat{p}$ by restating them in terms of the count $X$ and using binomial methods.

EXAMPLE

**5.11 Do you like buying clothes?** A sample survey asks a nationwide random sample of 2500 adults if they agree or disagree that "I like buying new clothes, but shopping is often frustrating and time-consuming." Suppose that 60% of all adults would agree if asked this question. What is the probability that the sample proportion who agree is at least 58%?

The count $X$ who agree has the binomial distribution $B(2500, 0.6)$. The sample proportion $\hat{p} = X/2500$ does *not* have a binomial distribution, because it is not a count. We can translate any question about a sample proportion $\hat{p}$ into a question about the count $X$. Because 58% of 2500 is 1450,

$$P(\hat{p} \geq 0.58) = P(X \geq 1450)$$
$$= P(X = 1450) + P(X = 1451) + \cdots + P(X = 2500)$$

This is a rather elaborate calculation. We must add more than 1000 binomial probabilities. Software tells us that $P(\hat{p} \geq 0.58) = 0.9802$. Because some software packages cannot handle an $n$ as large as 2500, we need another way to do this calculation.

LOOK BACK
rules for means, page 278
rules for variances, page 282

As a first step, find the mean and standard deviation of a sample proportion. We know the mean and standard deviation of a sample count, so apply the rules from Section 4.4 for the mean and variance of a constant times a random variable. Here is the result.

### MEAN AND STANDARD DEVIATION OF A SAMPLE PROPORTION

Let $\hat{p}$ be the sample proportion of successes in an SRS of size $n$ drawn from a large population having population proportion $p$ of successes.

The mean and standard deviation of $\hat{p}$ are

$$\mu_{\hat{p}} = p$$

$$\sigma_{\hat{p}} = \sqrt{\frac{p(1-p)}{n}}$$

The formula for $\sigma_{\hat{p}}$ is exactly correct in the binomial setting. It is approximately correct for an SRS from a large population. We will use it when the population is at least 20 times as large as the sample.

EXAMPLE

**5.12 The mean and the standard deviation.** The mean and standard deviation of the proportion of the survey respondents in Example 5.11 who find shopping frustrating are

$$\mu_{\hat{p}} = p = 0.6$$

$$\sigma_{\hat{p}} = \sqrt{\frac{p(1-p)}{n}} = \sqrt{\frac{(0.6)(0.4)}{2500}} = 0.0098$$

## USE YOUR KNOWLEDGE

**5.6 Find the mean and the standard deviation.** If we toss a fair coin 100 times, the number of heads is a random variable that is binomial.

(a) Find the mean and the standard deviation of the sample proportion.

(b) Is your answer to part (a) the same as the mean and the standard deviation of the sample count? Explain your answer.

LOOK BACK
**unbiased statistic, page 217**

The fact that the mean of $\hat{p}$ is $p$ states in statistical language that the sample proportion $\hat{p}$ in an SRS is an *unbiased estimator* of the population proportion $p$. When a sample is drawn from a new population having a different value of the population proportion $p$, the sampling distribution of the unbiased estimator $\hat{p}$ changes so that its mean moves to the new value of $p$. We observed this fact empirically in Section 3.4 and have now verified it from the laws of probability.

The variability of $\hat{p}$ about its mean, as described by the variance or standard deviation, gets smaller as the sample size increases. So a sample proportion from a large sample will usually lie quite close to the population proportion $p$. We observed this in the simulation experiment on page 214 in Section 3.3. Now we have discovered exactly how the variability decreases: the standard deviation is $\sqrt{p(1-p)/n}$. *The* $\sqrt{n}$ *in the denominator means that the sample size must be multiplied by 4 if we wish to divide the standard deviation in half.*

## Normal approximation for counts and proportions

Using simulation, we discovered in Section 3.4 that the sampling distribution of a sample proportion $\hat{p}$ is close to Normal. Now we know that the distribution of $\hat{p}$ is that of a binomial count divided by the sample size $n$. This seems at first to be a contradiction. To clear up the matter, look at Figure 5.3. This is a probability histogram of the exact distribution of the proportion of frustrated shoppers $\hat{p}$, based on the binomial distribution $B(2500, 0.6)$. There are hundreds of narrow bars, one for each of the 2501 possible values of $\hat{p}$. Most have probabilities too small to show in a graph. *The probability histogram looks very Normal!* In fact, both the count $X$ and the sample proportion $\hat{p}$ are approximately Normal in large samples.

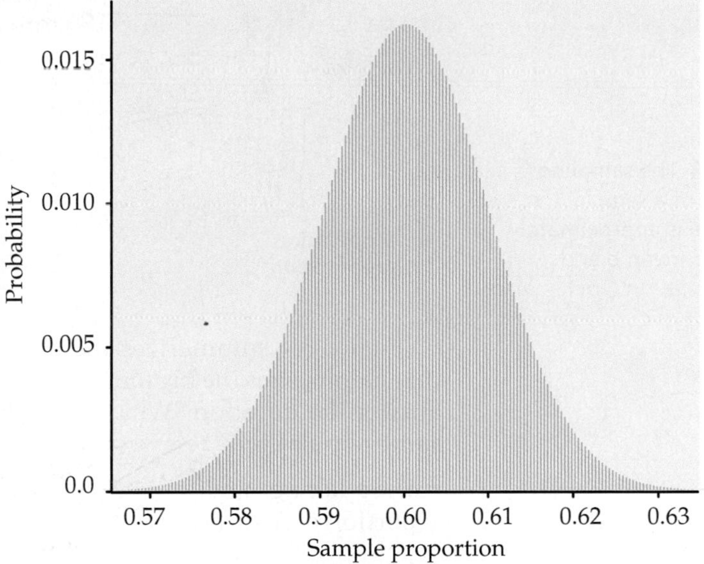


**FIGURE 5.3** Probability histogram of the sample proportion $\hat{p}$ based on a binomial count with $n = 2500$ and $p = 0.6$. The distribution is very close to Normal.

> **NORMAL APPROXIMATION FOR COUNTS AND PROPORTIONS**
>
> Draw an SRS of size $n$ from a large population having population proportion $p$ of successes. Let $X$ be the count of successes in the sample and $\hat{p} = X/n$ be the sample proportion of successes. When $n$ is large, the sampling distributions of these statistics are approximately Normal:
>
> $$X \text{ is approximately } N(np, \sqrt{np(1-p)})$$
>
> $$\hat{p} \text{ is approximately } N\left(p, \sqrt{\frac{p(1-p)}{n}}\right)$$
>
> As a rule of thumb, we will use this approximation for values of $n$ and $p$ that satisfy $np \geq 10$ and $n(1-p) \geq 10$.

These Normal approximations are easy to remember because they say that $\hat{p}$ and $X$ are Normal, with their usual means and standard deviations. Whether or not you use the Normal approximations should depend on how accurate your calculations need to be. For most statistical purposes great accuracy is not required. Our "rule of thumb" for use of the Normal approximations reflects this judgment.

The accuracy of the Normal approximations improves as the sample size $n$ increases. They are most accurate for any fixed $n$ when $p$ is close to 1/2, and least accurate when $p$ is near 0 or 1. You can compare binomial distributions with their Normal approximations by using the *Normal Approximation to Binomial* applet. This applet allows you to change $n$ or $p$ while watching the effect on the binomial probability histogram and the Normal curve that approximates it.

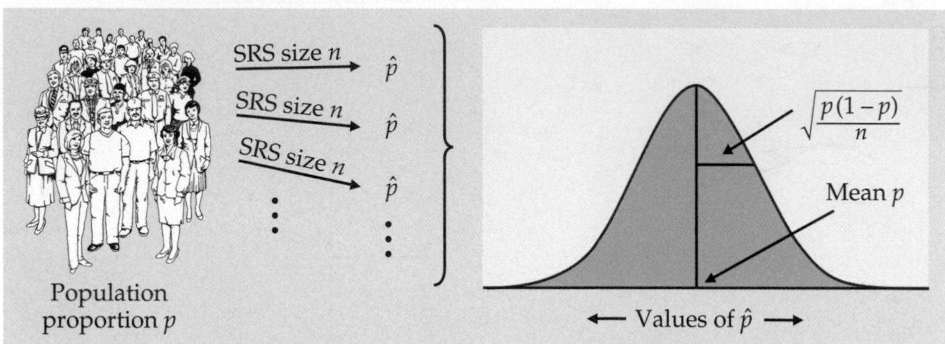


**FIGURE 5.4** The sampling distribution of a sample proportion $\hat{p}$ is approximately Normal with mean $p$ and standard deviation $\sqrt{p(1-p)/n}$.

Figure 5.4 summarizes the distribution of a sample proportion in a form that emphasizes the big idea of a sampling distribution. Sampling distributions answer the question "What would happen if we took many samples from the same population?"

- Keep taking random samples of size $n$ from a population that contains proportion $p$ of successes.
- Find the sample proportion $\hat{p}$ for each sample.
- Collect all the $\hat{p}$'s and display their distribution.

That's the sampling distribution of $\hat{p}$.

EXAMPLE

**5.13 Compare the Normal approximation with the exact calculation.** Let's compare the Normal approximation for the calculation of Example 5.11 with the exact calculation from software. We want to calculate $P(\hat{p} \geq 0.58)$ when the sample size is $n = 2500$ and the population proportion is $p = 0.6$. Example 5.12 shows that

$$\mu_{\hat{p}} = p = 0.6$$

$$\sigma_{\hat{p}} = \sqrt{\frac{p(1-p)}{n}} = 0.0098$$

Act as if $\hat{p}$ were Normal with mean 0.6 and standard deviation 0.0098. The approximate probability, as illustrated in Figure 5.5, is

$$P(\hat{p} \geq 0.58) = P\left(\frac{\hat{p} - 0.6}{0.0098} \geq \frac{0.58 - 0.6}{0.0098}\right)$$
$$\doteq P(Z \geq -2.04) = 0.9793$$

That is, about 98% of all samples have a sample proportion that is at least 0.58. Because the sample was large, this Normal approximation is quite accurate. It misses the software value 0.9802 by only 0.0009.

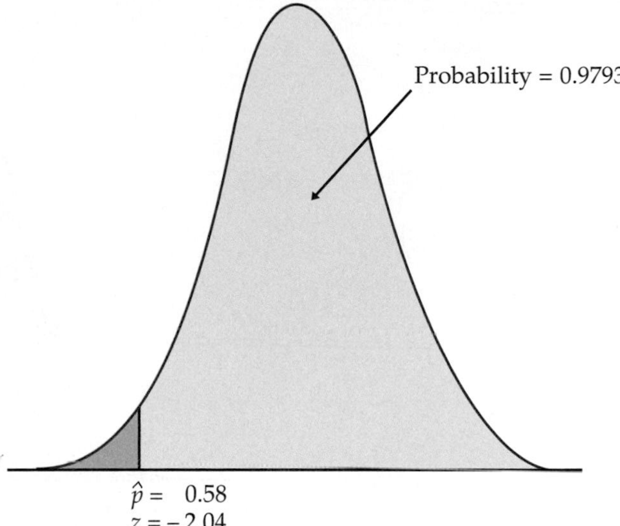


**FIGURE 5.5** The Normal probability calculation for Example 5.13.

EXAMPLE

**5.14 Using the Normal appoximation.** The audit described in Example 5.6 examined an SRS of 150 sales records for compliance with sales tax laws. In fact, 8% of all the company's sales records have an incorrect sales tax classification. The count $X$ of bad records in the sample has approximately the $B(150, 0.08)$ distribution.

According to the Normal approximation to the binomial distributions, the count $X$ is approximately Normal with mean and standard deviation

$$\mu_X = np = (150)(0.08) = 12$$

$$\sigma_X = \sqrt{np(1-p)} = \sqrt{(150)(0.08)(0.92)}$$
$$= 3.3226$$

The Normal approximation for the probability of no more than 10 misclassified records is the area to the left of $X = 10$ under the Normal curve. Using Table A,

$$P(X \leq 10) = P\left(\frac{X - 12}{3.3226} \leq \frac{10 - 12}{3.3226}\right)$$
$$\doteq P(Z \leq -0.60) = 0.2743$$

Software tells us that the actual binomial probability that no more than 10 of the records in the sample are misclassified is $P(X \leq 10) = 0.3384$. The Normal approximation is only roughly accurate. Because $np = 12$, this combination of $n$ and $p$ is close to the border of the values for which we are willing to use the approximation.

The distribution of the count of bad records in a sample of 15 is distinctly non-Normal, as Figure 5.2 showed. When we increase the sample size to 150, however, the shape of the binomial distribution becomes roughly Normal.

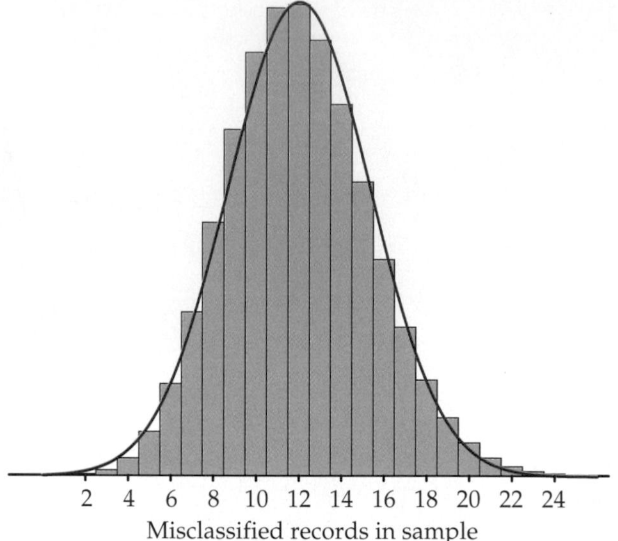


**FIGURE 5.6** Probability histogram and Normal approximation for the binomial distribution with $n = 150$ and $p = 0.08$, for Example 5.14.

Figure 5.6 displays the probability histogram of the binomial distribution with the density curve of the approximating Normal distribution superimposed. Both distributions have the same mean and standard deviation, and both the area under the histogram and the area under the curve are 1. The Normal curve fits the histogram reasonably well. Look closely: the histogram is slightly skewed to the right, a property that the symmetric Normal curve can't match.

**USE YOUR KNOWLEDGE**

**5.7** **Use the Normal approximation.** Suppose we toss a fair coin 100 times. Use the Normal approximation to find the probability that the sample proportion is

(a) between 0.4 and 0.6. (b) between 0.45 and 0.55.

## The continuity correction*

Figure 5.7 illustrates an idea that greatly improves the accuracy of the Normal approximation to binomial probabilities. The binomial probability $P(X \leq 10)$ is the area of the histogram bars for values 0 to 10. The bar for $X = 10$ actually extends from 9.5 to 10.5. Because the discrete binomial distribution puts probability only on whole numbers, the probabilities $P(X \leq 10)$ and $P(X \leq 10.5)$ are the same. The Normal distribution spreads probability continuously, so these two Normal probabilities are different. The Normal approximation is more accurate if we consider $X = 10$ to extend from 9.5 to 10.5, matching the bar in the probability histogram.

The event $\{X \leq 10\}$ includes the outcome $X = 10$. Figure 5.7 shades the area under the Normal curve that matches all the histogram bars for outcomes 0 to

*This material can be omitted if desired.

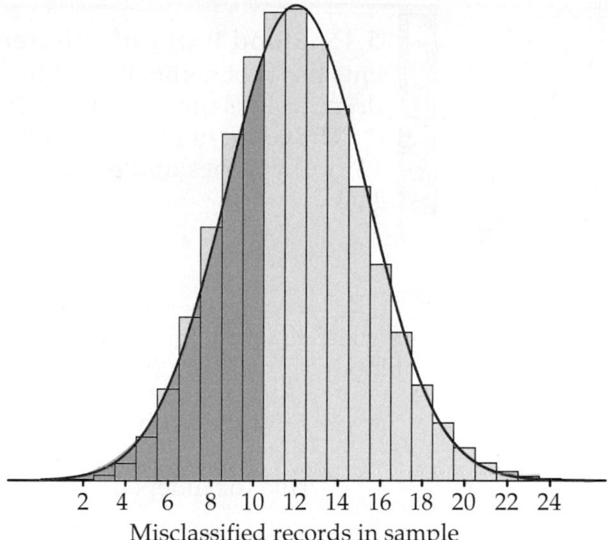


**FIGURE 5.7** Area under the Normal approximation curve for the probability in Example 5.14.

10, bounded on the right not by 10, but by 10.5. So $P(X \leq 10)$ is calculated as $P(X \leq 10.5)$. On the other hand, $P(X < 10)$ excludes the outcome $X = 10$, so we exclude the entire interval from 9.5 to 10.5 and calculate $P(X \leq 9.5)$ from the Normal table. Here is the result of the Normal calculation in Example 5.14 improved in this way:

$$\begin{aligned} P(X \leq 10) &= P(X \leq 10.5) \\ &= P\left(\frac{X - 12}{3.3226} \leq \frac{10.5 - 12}{3.3226}\right) \\ &\doteq P(Z \leq -0.45) = 0.3264 \end{aligned}$$

The improved approximation misses the binomial probability by only 0.012. Acting as though a whole number occupies the interval from 0.5 below to 0.5 above the number is called the **continuity correction** to the Normal approximation. If you need accurate values for binomial probabilities, try to use software to do exact calculations. If no software is available, use the continuity correction unless $n$ is very large. Because most statistical purposes do not require extremely accurate probability calculations, we do not emphasize use of the continuity correction.

continuity correction

## Binomial formula*

We can find a formula for the probability that a binomial random variable takes any value by adding probabilities for the different ways of getting exactly that many successes in $n$ observations. Here is the example we will use to show the idea.

---

*The formula for binomial probabilities is useful in many settings, but we will not need it in our study of statistical inference. This section can therefore be omitted if desired.

EXAMPLE

**5.15 Blood types of children.** Each child born to a particular set of parents has probability 0.25 of having blood type O. If these parents have 5 children, what is the probability that exactly 2 of them have type O blood?

The count of children with type O blood is a binomial random variable $X$ with $n = 5$ tries and probability $p = 0.25$ of a success on each try. We want $P(X = 2)$.

Because the method doesn't depend on the specific example, we will use "S" for success and "F" for failure. In Example 5.15, "S" would stand for type O blood. Do the work in two steps.

*Step 1:* Find the probability that a specific 2 of the 5 tries give successes, say the first and the third. This is the outcome SFSFF. The multiplication rule for independent events tells us that

$$\begin{aligned} P(\text{SFSFF}) &= P(\text{S})P(\text{F})P(\text{S})P(\text{F})P(\text{F}) \\ &= (0.25)(0.75)(0.25)(0.75)(0.75) \\ &= (0.25)^2(0.75)^3 \end{aligned}$$

*Step 2:* Observe that the probability of *any one* arrangement of 2 S's and 3 F's has this same probability. That's true because we multiply together 0.25 twice and 0.75 three times whenever we have 2 S's and 3 F's. The probability that $X = 2$ is the probability of getting 2 S's and 3 F's in any arrangement whatsoever. Here are all the possible arrangements:

| | | | | |
|---|---|---|---|---|
| SSFFF | SFSFF | SFFSF | SFFFS | FSSFF |
| FSFSF | FSFFS | FFSSF | FFSFS | FFFSS |

There are 10 of them, all with the same probability. The overall probability of 2 successes is therefore

$$P(X = 2) = 10(0.25)^2(0.75)^3 = 0.2637$$

The pattern of this calculation works for any binomial probability. To use it, we need to be able to count the number of arrangements of $k$ successes in $n$ observations without actually listing them. We use the following fact to do the counting.

BINOMIAL COEFFICIENT

The number of ways of arranging $k$ successes among $n$ observations is given by the **binomial coefficient**

$$\binom{n}{k} = \frac{n!}{k!\,(n-k)!}$$

for $k = 0, 1, 2, \ldots, n$.

factorial

The formula for binomial coefficients uses the **factorial** notation. The factorial $n!$ for any positive whole number $n$ is

$$n! = n \times (n-1) \times (n-2) \times \cdots \times 3 \times 2 \times 1$$

Also, $0! = 1$. Notice that the larger of the two factorials in the denominator of a binomial coefficient will cancel much of the $n!$ in the numerator. For example, the binomial coefficient we need for Example 5.15 is

$$\begin{aligned}\binom{5}{2} &= \frac{5!}{2!\,3!} \\ &= \frac{(5)(4)(3)(2)(1)}{(2)(1) \times (3)(2)(1)} \\ &= \frac{(5)(4)}{(2)(1)} = \frac{20}{2} = 10\end{aligned}$$

This agrees with our previous calculation.

*The notation* $\binom{n}{k}$ *is not related to the fraction* $\frac{n}{k}$. A helpful way to remember its meaning is to read it as "binomial coefficient $n$ choose $k$." Binomial coefficients have many uses in mathematics, but we are interested in them only as an aid to finding binomial probabilities. The binomial coefficient $\binom{n}{k}$ counts the number of ways in which $k$ successes can be distributed among $n$ observations. The binomial probability $P(X = k)$ is this count multiplied by the probability of any specific arrangement of the $k$ successes. Here is the formula we seek.

### BINOMIAL PROBABILITY

If $X$ has the binomial distribution $B(n, p)$ with $n$ observations and probability $p$ of success on each observation, the possible values of $X$ are $0, 1, 2, \ldots, n$. If $k$ is any one of these values, the **binomial probability** is

$$P(X = k) = \binom{n}{k} p^k (1-p)^{n-k}$$

Here is an example of the use of the binomial probability formula.

EXAMPLE

**5.16 Using the binomial probability formula.** The number $X$ of misclassified sales records in the auditor's sample in Example 5.8 has the $B(15, 0.08)$ distribution. The probability of finding no more than 1 misclassified record is

$$\begin{aligned}P(X \leq 1) &= P(X = 0) + P(X = 1) \\ &= \binom{15}{0}(0.08)^0(0.92)^{15} + \binom{15}{1}(0.08)^1(0.92)^{14} \\ &= \frac{15!}{0!\,15!}(1)(0.2863) + \frac{15!}{1!\,14!}(0.08)(0.3112)\end{aligned}$$

$$= (1)(1)(0.2863) + (15)(0.08)(0.3112)$$
$$= 0.2863 + 0.3734 = 0.6597$$

The calculation used the facts that $0! = 1$ and that $a^0 = 1$ for any number $a \neq 0$. The result agrees with that obtained from Table C in Example 5.8.

## USE YOUR KNOWLEDGE

**5.8** **A bent coin.** A coin is slightly bent, and as a result the probability of a head is 0.52. Suppose that you toss the coin four times.

(a) Use the binomial formula to find the probability of 3 or more heads.

(b) Compare your answer with the one that you would obtain if the coin were fair.

## SECTION 5.1 Summary

A **count** $X$ of successes has the **binomial distribution** $B(n, p)$ in the **binomial setting:** there are $n$ trials, all independent, each resulting in a success or a failure, and each having the same probability $p$ of a success.

**Binomial probabilities** are most easily found by software. There is an exact formula that is practical for calculations when $n$ is small. Table C contains binomial probabilities for some values of $n$ and $p$. For large $n$, you can use the Normal approximation.

The binomial distribution $B(n, p)$ is a good approximation to the **sampling distribution of the count of successes** in an SRS of size $n$ from a large population containing proportion $p$ of successes. We will use this approximation when the population is at least 20 times larger than the sample.

The mean and standard deviation of a **binomial count** $X$ and a **sample proportion** of successes $\hat{p} = X/n$ are

$$\mu_X = np \qquad \mu_{\hat{p}} = p$$

$$\sigma_X = \sqrt{np(1-p)} \qquad \sigma_{\hat{p}} = \sqrt{\frac{p(1-p)}{n}}$$

The sample proportion $\hat{p}$ is therefore an unbiased estimator of the population proportion $p$.

The **Normal approximation** to the binomial distribution says that if $X$ is a count having the $B(n, p)$ distribution, then when $n$ is large,

$$X \text{ is approximately } N(np, \sqrt{np(1-p)})$$

$$\hat{p} \text{ is approximately } N\left(p, \sqrt{\frac{p(1-p)}{n}}\right)$$

We will use these approximations when $np \geq 10$ and $n(1-p) \geq 10$. The **continuity correction** improves the accuracy of the Normal approximations.

The exact **binomial probability formula** is

$$P(X = k) = \binom{n}{k} p^k (1-p)^{n-k}$$

where the possible values of $X$ are $k = 0, 1, \ldots, n$. The binomial probability formula uses the **binomial coefficient**

$$\binom{n}{k} = \frac{n!}{k!\,(n-k)!}$$

Here the **factorial** $n!$ is

$$n! = n \times (n-1) \times (n-2) \times \cdots \times 3 \times 2 \times 1$$

for positive whole numbers $n$ and $0! = 1$. The binomial coefficient counts the number of ways of distributing $k$ successes among $n$ trials.

## SECTION 5.1 Exercises

*For Exercises 5.1 and 5.2, see page 314; for Exercises 5.3 and 5.4, see page 315; for Exercise 5.5, see page 319; for Exercise 5.6, see page 322; for Exercise 5.7, see page 326; and for Exercise 5.8, see page 330.*

*Most binomial probability calculations required in these exercises can be done by using Table C or the Normal approximation. Your instructor may request that you use the binomial probability formula or software. In exercises requiring the Normal approximation, you should use the continuity correction if you studied that topic.*

**5.9 What is wrong?** Explain what is wrong in each of the following scenarios.

(a) If you toss a fair coin three times and a head appears each time, then the next toss is more likely to be a tail than a head.

(b) If you toss a fair coin three times and a head appears each time, then the next toss is more likely to be a head than a tail.

(c) $\hat{p}$ is one of the parameters for a binomial distribution.

**5.10 What is wrong?** Explain what is wrong in each of the following scenarios.

(a) In the binomial setting $X$ is a proportion.

(b) The variance for a binomial count is $\sqrt{p(1-p)/n}$.

(c) The Normal approximation to the binomial distribution is always accurate when $n$ is greater than 1000.

**5.11 Should you use the binomial distribution?** In each situation below, is it reasonable to use a binomial distribution for the random variable $X$? Give reasons for your answer in each case. If a binomial distribution applies, give the values of $n$ and $p$.

(a) A poll of 200 college students asks whether or not you are usually irritable in the morning. $X$ is the number who reply that they are usually irritable in the morning.

(b) You toss a fair coin until a head appears. $X$ is the count of the number of tosses that you make.

(c) Most calls made at random by sample surveys don't succeed in talking with a live person. Of calls to New York City, only 1/12 succeed. A survey calls 500 randomly selected numbers in New York City. $X$ is the number that reach a live person.

**5.12 Should you use the binomial distribution?** In each situation below, is it reasonable to use a binomial distribution for the random variable $X$? Give reasons for your answer in each case.

(a) A random sample of students in a fitness study. $X$ is the mean systolic blood pressure of the sample.

(b) A manufacturer of running shoes picks a random sample of the production of shoes each day for a detailed inspection. Today's sample of 20 pairs of shoes includes 1 pair with a defect.

(c) A nutrition study chooses an SRS of college students. They are asked whether or not they usually eat at least five servings of fruits or vegetables per day. $X$ is the number who say that they do.

**5.13 Typographic errors.** Typographic errors in a text are either nonword errors (as when "the" is typed as "teh") or word errors that result in a real but incorrect word. Spell-checking software will catch nonword errors but not word errors. Human proofreaders catch 70% of word errors. You ask a fellow student to proofread an essay in which you have deliberately made 10 word errors.

(a) If the student matches the usual 70% rate, what is the distribution of the number of errors caught? What is the distribution of the number of errors missed?

(b) Missing 4 or more out of 10 errors seems a poor performance. What is the probability that a proofreader who catches 70% of word errors misses 4 or more out of 10?

**5.14 Visits to Web sites.** What kinds of Web sites do males aged 18 to 34 visit most often? Pornographic sites take first place, but about 50% of male Internet users in this age group visit an auction site such as eBay at least once a month.[2] Interview a random sample of 15 male Internet users aged 18 to 34.

(a) What is the distribution of the number who have visited an online auction site in the past month?

(b) What is the probability that at least 8 of the 15 have visited an auction site in the past month?

**5.15 Typographic errors.** Return to the proofreading setting of Exercise 5.13.

(a) What is the mean number of errors caught? What is the mean number of errors missed? You see that these two means must add to 10, the total number of errors.

(b) What is the standard deviation $\sigma$ of the number of errors caught?

(c) Suppose that a proofreader catches 90% of word errors, so that $p = 0.9$. What is $\sigma$ in this case? What is $\sigma$ if $p = 0.99$? What happens to the standard deviation of a binomial distribution as the probability of a success gets close to 1?

**5.16 Visits to Web sites.** Suppose that 50% of male Internet users aged 18 to 34 have visited an auction site at least once in the past month.

(a) If you interview 15 at random, what is the mean of the count $X$ who have visited an auction site? What is the mean of the proportion $\hat{p}$ in your sample who have visited an auction site?

(b) Repeat the calculations in (a) for samples of size 150 and 1500. What happens to the mean count of successes as the sample size increases? What happens to the mean proportion of successes?

**5.17 Typographic errors.** In the proofreading setting of Exercise 5.13, what is the smallest number of misses $m$ with $P(X \geq m)$ no larger than 0.05? You might consider $m$ or more misses as evidence that a proofreader actually catches fewer than 70% of word errors.

**5.18 Attitudes toward drinking and behavior studies.** Some of the methods in this section are approximations rather than exact probability results. We have given rules of thumb for safe use of these approximations.

(a) You are interested in attitudes toward drinking among the 75 members of a fraternity. You choose 30 members at random to interview. One question is "Have you had five or more drinks at one time during the last week?" Suppose that in fact 30% of the 75 members would say "Yes." Explain why you *cannot* safely use the $B(30, 0.3)$ distribution for the count $X$ in your sample who say "Yes."

(b) The National AIDS Behavioral Surveys found that 0.2% (that's 0.002 as a decimal fraction) of adult heterosexuals had both received a blood transfusion and had a sexual partner from a group at high risk of AIDS. Suppose that this national proportion holds for your region. Explain why you *cannot* safely use the Normal approximation for the sample proportion who fall in this group when you interview an SRS of 1000 adults.

**5.19 Random digits.** Each entry in a table of random digits like Table B has probability 0.1 of being a 0, and digits are independent of each other.

(a) What is the probability that a group of five digits from the table will contain at least one 5?

(b) What is the mean number of 5s in lines 40 digits long?

**5.20 Use the *Probability* applet.** The *Probability* applet simulates tosses of a coin. You can choose the number of tosses $n$ and the probability $p$ of a head. You can therefore use the applet to simulate binomial random variables.

The count of misclassified sales records in Example 5.8 (page 317) has the binomial distribution with $n = 15$ and $p = 0.08$. Set these values for the number of tosses and probability of heads in the applet. Table C shows that the probability of getting a sample with exactly 0 misclassified records is 0.2863. This is the long-run proportion of samples with exactly 1 bad record. Click "Toss" and "Reset" repeatedly to simulate 25 samples. Record the number of bad records (the count of heads) in each of the 25 samples. What proportion of the 25 samples had exactly 0 bad records? Remember that probability tells us only what happens in the long run.

**5.21 Inheritance of blood types.** Children inherit their blood type from their parents, with probabilities that reflect the parents' genetic makeup. Children of Juan and Maria each have probability 1/4 of having blood type A and inherit independently of each other. Juan and Maria plan to have 4 children; let $X$ be the number who have blood type A.

(a) What are $n$ and $p$ in the binomial distribution of $X$?

(b) Find the probability of each possible value of $X$, and draw a probability histogram for this distribution.

(c) Find the mean number of children with type A blood, and mark the location of the mean on your probability histogram.

**5.22 The ideal number of children.** "What do you think is the ideal number of children for a family to have?" A Gallup Poll asked this question of 1016 randomly chosen adults. Almost half (49%) thought two children was ideal.[3] Suppose that $p = 0.49$ is exactly true for the population of all adults. Gallup announced a margin of error of $\pm 3$ percentage points for this poll. What is the probability that the sample proportion $\hat{p}$ for an SRS of size $n = 1016$ falls between 0.46 and 0.52? You see that it is likely, but not certain, that polls like this give results that are correct within their margin of error. We will say more about margins of error in Chapter 6.

**5.23 Visiting a casino and betting on college sports.** A Gallup Poll finds that 30% of adults visited a casino in the past 12 months, and that 6% bet on college sports.[4] These results come from a random sample of 1011 adults. For an SRS of size $n = 1011$:

(a) What is the probability that the sample proportion $\hat{p}$ is between 0.28 and 0.32 if the population proportion is $p = 0.30$?

(b) What is the probability that the sample proportion $\hat{p}$ is between 0.04 and 0.08 if the population proportion is $p = 0.06$?

(c) How does the probability that $\hat{p}$ falls within $\pm 0.02$ of the true $p$ change as $p$ gets closer to 0?

**5.24 How do the results depend on the sample size?** Return to the Gallup Poll setting of Exercise 5.22. We are supposing that the proportion of all adults who think that two children is ideal is $p = 0.49$. What is the probability that a sample proportion $\hat{p}$ falls between 0.46 and 0.52 (that is, within $\pm 3$ percentage points of the true $p$) if the sample is an SRS of size $n = 300$? Of size $n = 5000$? Combine these results with your work in Exercise 5.22 to make a general statement about the effect of larger samples in a sample survey.

**5.25** CHALLENGE **A college alcohol study.** The Harvard College Alcohol Study finds that 67% of college students support efforts to "crack down on underage drinking." The study took a sample of almost 15,000 students, so the population proportion who support a crackdown is very close to $p = 0.67$.[5] The administration of your college surveys an SRS of 200 students and finds that 140 support a crackdown on underage drinking.

(a) What is the sample proportion who support a crackdown on underage drinking?

(b) If in fact the proportion of all students on your campus who support a crackdown is the same as the national 67%, what is the probability that the proportion in an SRS of 200 students is as large or larger than the result of the administration's sample?

(c) A writer in the student paper says that support for a crackdown is higher on your campus than nationally. Write a short letter to the editor explaining why the survey does not support this conclusion.

**5.26** CHALLENGE **How large a sample is needed?** The changing probabilities you found in Exercises 5.22 and 5.24 are due to the fact that the standard deviation of the sample proportion $\hat{p}$ gets smaller as the sample size $n$ increases. If the population proportion is $p = 0.49$, how large a sample is needed to reduce the standard deviation of $\hat{p}$ to $\sigma_{\hat{p}} = 0.004$? (The 68–95–99.7 rule then says that about 95% of all samples will have $\hat{p}$ within 0.01 of the true $p$.)

**5.27 A test for ESP.** In a test for ESP (extrasensory perception), the experimenter looks at cards that are hidden from the subject. Each card contains either a star, a circle, a wave, or a square. As the experimenter looks at each of 20 cards in turn, the subject names the shape on the card.

(a) If a subject simply guesses the shape on each card, what is the probability of a successful guess on a single card? Because the cards are independent, the count of successes in 20 cards has a binomial distribution.

(b) What is the probability that a subject correctly guesses at least 10 of the 20 shapes?

(c) In many repetitions of this experiment with a subject who is guessing, how many cards will the subject guess correctly on the average? What is the standard deviation of the number of correct guesses?

(d) A standard ESP deck actually contains 25 cards. There are five different shapes, each of which appears on 5 cards. The subject knows that the deck has this makeup. Is a binomial model still appropriate for the count of correct guesses in one pass through this deck? If so, what are $n$ and $p$? If not, why not?

**5.28 Admitting students to college.** A selective college would like to have an entering class of 950 students. Because not all students who are offered admission accept, the college admits more than 950 students. Past experience shows that about 75% of the students admitted will accept. The college decides to admit 1200 students. Assuming that students make their decisions independently, the number who accept has the $B(1200, 0.75)$ distribution. If this number is less than 950, the college will admit students from its waiting list.

(a) What are the mean and the standard deviation of the number $X$ of students who accept?

(b) The college does not want more than 950 students. What is the probability that more than 950 will accept?

(c) If the college decides to increase the number of admission offers to 1300, what is the probability that more than 950 will accept?

**5.29** CHALLENGE **Is the ESP result better than guessing?** When the ESP study of Exercise 5.27 discovers a subject whose performance appears to be better than guessing, the study continues at greater length. The experimenter looks at many cards bearing one of five shapes (star, square, circle, wave, and cross) in an order determined by random numbers. The subject cannot see the experimenter as he looks at each card in turn, in order to avoid any possible nonverbal clues. The answers of a subject who does not have ESP should be independent observations, each with probability 1/5 of success. We record 900 attempts.

(a) What are the mean and the standard deviation of the count of successes?

(b) What are the mean and standard deviation of the proportion of successes among the 900 attempts?

(c) What is the probability that a subject without ESP will be successful in at least 24% of 900 attempts?

(d) The researcher considers evidence of ESP to be a proportion of successes so large that there is only probability 0.01 that a subject could do this well or better by guessing. What proportion of successes must a subject have to meet this standard? (Example 1.32 shows how to do an inverse calculation for the Normal distribution that is similar to the type required here.)

**5.30** CHALLENGE **Scuba-diving trips.** The mailing list of an agency that markets scuba-diving trips to the Florida Keys contains 60% males and 40% females. The agency calls 30 people chosen at random from its list.

(a) What is the probability that 20 of the 30 are men? (Use the binomial probability formula.)

(b) What is the probability that the first woman is reached on the fourth call? (That is, the first 4 calls give MMMF.)

**5.31 Checking for problems with a sample survey.** One way of checking the effect of undercoverage, nonresponse, and other sources of error in a sample survey is to compare the sample with known demographic facts about the population. The 2000 census found that 23,772,494 of the 209,128,094 adults (aged 18 and over) in the United States called themselves "Black or African American."

(a) What is the population proportion $p$ of blacks among American adults?

(b) An opinion poll chooses 1200 adults at random. What is the mean number of blacks in such samples? (Explain the reasoning behind your calculation.)

(c) Use a Normal approximation to find the probability that such a sample will contain 100 or fewer blacks. Be sure to check that you can safely use the approximation.

**5.32** CHALLENGE **Show that these facts are true.** Use the definition of binomial coefficients to show that each of the following facts is true. Then restate each fact in words in terms of the number of ways that $k$ successes can be distributed among $n$ observations.

(a) $\binom{n}{n} = 1$ for any whole number $n \geq 1$.

(b) $\binom{n}{n-1} = n$ for any whole number $n \geq 1$.

(c) $\binom{n}{k} = \binom{n}{n-k}$ for any $n$ and $k$ with $k \leq n$.

**5.33 Multiple-choice tests.** Here is a simple probability model for multiple-choice tests. Suppose that each student has probability $p$ of correctly answering a question chosen at random from a universe of possible questions. (A strong student has a higher $p$ than a weak student.) The correctness of an answer to a question is independent of the correctness of answers to other questions. Jodi is a good student for whom $p = 0.85$.

(a) Use the Normal approximation to find the probability that Jodi scores 80% or lower on a 100-question test.

(b) If the test contains 250 questions, what is the probability that Jodi will score 80% or lower?

(c) How many questions must the test contain in order to reduce the standard deviation of Jodi's proportion of correct answers to half its value for a 100-item test?

(d) Laura is a weaker student for whom $p = 0.75$. Does the answer you gave in (c) for the standard deviation of Jodi's score apply to Laura's standard deviation also?

**5.34 Tossing a die.** You are tossing a balanced die that has probability 1/6 of coming up 1 on each toss. Tosses are independent. We are interested in how long we must wait to get the first 1.

(a) The probability of a 1 on the first toss is 1/6. What is the probability that the first toss is not a 1 and the second toss is a 1?

(b) What is the probability that the first two tosses are not 1s and the third toss is a 1? This is the probability that the first 1 occurs on the third toss.

(c) Now you see the pattern. What is the probability that the first 1 occurs on the fourth toss? On the fifth toss?

**5.35** CHALLENGE **The geometric distribution.** Generalize your work in Exercise 5.34. You have independent trials, each resulting in a success or a failure. The probability of a success is $p$ on each trial. The binomial distribution describes the count of successes in a fixed number of trials. Now the number of trials is not fixed; instead, continue until you get a success. The random variable $Y$ is the number of the trial on which the first success occurs. What are the possible values of $Y$? What is the probability $P(Y = k)$ for any of these values? (*Comment:* The distribution of the number of trials to the first success is called a **geometric distribution.**)

## 5.2 The Sampling Distribution of a Sample Mean

Counts and proportions are discrete random variables that describe categorical data. The statistics most often used to describe quantitative data, on the other hand, are continuous random variables. The sample mean, percentiles, and standard deviation are examples of statistics based on quantitative data. Statistical theory describes the sampling distributions of these statistics. In this section we will concentrate on the sample mean. Because sample means are just averages of observations, they are among the most common statistics.

EXAMPLE

**5.17 Sample means are approximately Normal.** Figure 5.8 illustrates two striking facts about the sampling distribution of a sample mean. Figure 5.8(a) displays the distribution of customer service call lengths for a bank service center for a month. There are more than 30,000 calls in this population.[6] (We omitted a few extreme outliers, calls that lasted more than 20 minutes.) The distribution is extremely skewed to the right. The population mean is $\mu = 173.95$ seconds.

Table 1.1 (page 8) contains the lengths of a sample of 80 calls from this population. The mean of these 80 calls is $\bar{x} = 196.6$ seconds. If we take more samples of size 80, we will get different values of $\bar{x}$. To find the sampling distribution of $\bar{x}$, take many random samples of size 80 and calculate $\bar{x}$

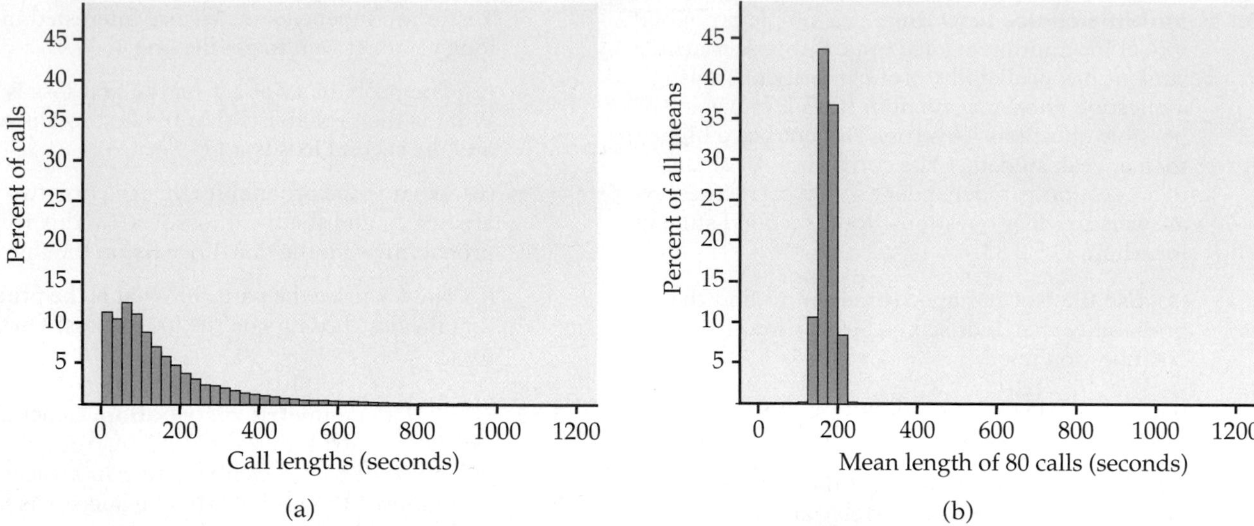


**FIGURE 5.8** (a) The distribution of lengths of all customer service calls received by a bank in a month. (b) The distribution of the sample means $\bar{x}$ for 500 random samples of size 80 from this population. The scales and histogram classes are exactly as in Figure 5.8(a).

for each sample. Figure 5.8(b) is the distribution of the values of $\bar{x}$ for 500 samples. The scales and choice of classes are exactly the same as in Figure 5.8(a), so that we can make a direct comparison.

The sample means are much less spread out than the individual call lengths. What is more, the distribution in Figure 5.8(b) is roughly symmetric rather than skewed. The Normal quantile plot in Figure 5.9 confirms that the distribution is close to Normal.

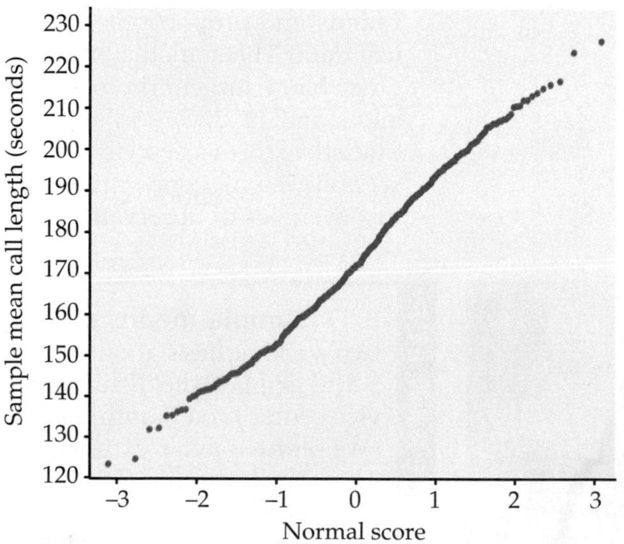


**FIGURE 5.9** Normal quantile plot of the 500 sample means in Figure 5.8(b). The distribution is close to Normal.

This example illustrates two important facts about sample means that we will discuss in this section.

FACTS ABOUT SAMPLE MEANS

1. Sample means are less variable than individual observations.
2. Sample means are more Normal than individual observations.

These two facts contribute to the popularity of sample means in statistical inference.

## The mean and standard deviation of $\bar{x}$

The sample mean $\bar{x}$ from a sample or an experiment is an estimate of the mean $\mu$ of the underlying population, just as a sample proportion $\hat{p}$ is an estimate of a population proportion $p$. The sampling distribution of $\bar{x}$ is determined by the design used to produce the data, the sample size $n$, and the population distribution.

Select an SRS of size $n$ from a population, and measure a variable $X$ on each individual in the sample. The $n$ measurements are values of $n$ random variables $X_1, X_2, \ldots, X_n$. A single $X_i$ is a measurement on one individual selected at random from the population and therefore has the distribution of the population. If the population is large relative to the sample, we can consider $X_1, X_2, \ldots, X_n$ to be independent random variables each having the same distribution. This is our probability model for measurements on each individual in an SRS.

The sample mean of an SRS of size $n$ is

$$\bar{x} = \frac{1}{n}(X_1 + X_2 + \cdots + X_n)$$

If the population has mean $\mu$, then $\mu$ is the mean of the distribution of each observation $X_i$. The addition rule for means of random variables,

LOOK BACK
**addition rule for means, page 278**

$$\mu_{\bar{x}} = \frac{1}{n}(\mu_{X_1} + \mu_{X_2} + \cdots + \mu_{X_n})$$
$$= \frac{1}{n}(\mu + \mu + \cdots + \mu) = \mu$$

That is, *the mean of $\bar{x}$ is the same as the mean of the population.* The sample mean $\bar{x}$ is therefore an unbiased estimator of the unknown population mean $\mu$.

The observations are independent, so the addition rule for variances also applies:

LOOK BACK
**addition rule for variances, page 282**

$$\sigma^2_{\bar{x}} = \left(\frac{1}{n}\right)^2 (\sigma^2_{X_1} + \sigma^2_{X_2} + \cdots + \sigma^2_{X_n})$$
$$= \left(\frac{1}{n}\right)^2 (\sigma^2 + \sigma^2 + \cdots + \sigma^2)$$
$$= \frac{\sigma^2}{n}$$

Just as in the case of a sample proportion $\hat{p}$, the variability of the sampling distribution of a sample mean decreases as the sample size grows. Because the standard deviation of $\bar{x}$ is $\sigma/\sqrt{n}$, it is again true that the standard deviation of the statistic decreases in proportion to the square root of the sample size. Here is a summary of these facts.

**MEAN AND STANDARD DEVIATION OF A SAMPLE MEAN**

Let $\bar{x}$ be the mean of an SRS of size $n$ from a population having mean $\mu$ and standard deviation $\sigma$. The mean and standard deviation of $\bar{x}$ are

$$\mu_{\bar{x}} = \mu$$

$$\sigma_{\bar{x}} = \frac{\sigma}{\sqrt{n}}$$

How accurately does a sample mean $\bar{x}$ estimate a population mean $\mu$? Because the values of $\bar{x}$ vary from sample to sample, we must give an answer in terms of the sampling distribution. We know that $\bar{x}$ is an unbiased estimator of $\mu$, so its values in repeated samples are not systematically too high or too low. Most samples will give an $\bar{x}$-value close to $\mu$ if the sampling distribution is concentrated close to its mean $\mu$. So the accuracy of estimation depends on the spread of the sampling distribution.

EXAMPLE

**5.18 Standard deviations for sample means of service call lengths.** The standard deviation of the population of service call lengths in Figure 5.8(a) is $\sigma = 184.81$ seconds. The length of a single call will often be far from the population mean. If we choose an SRS of 20 calls, the standard deviation of their mean length is

$$\sigma_{\bar{x}} = \frac{184.81}{\sqrt{20}} = 41.32 \text{ seconds}$$

Averaging over more calls reduces the variability and makes it more likely that $\bar{x}$ is close to $\mu$. Our sample size of 80 calls is 4 times 20, so the standard deviation will be half as large:

$$\sigma_{\bar{x}} = \frac{184.81}{\sqrt{80}} = 20.66 \text{ seconds}$$

**USE YOUR KNOWLEDGE**

**5.36 Find the mean and the standard deviation of the sampling distribution.** You take an SRS of size 25 from a population with mean 200 and standard deviation 10. Find the mean and standard deviation of the sampling distribution of your sample mean.

**5.37 The effect of increasing the sample size.** In the setting of the previous exercise, repeat the calculations for a sample size of 100. Explain the effect of the increase on the sample mean and standard deviation.

## The central limit theorem

We have described the center and spread of the probability distribution of a sample mean $\bar{x}$, but not its shape. The shape of the distribution of $\bar{x}$ depends on the shape of the population distribution. Here is one important case: if the population distribution is Normal, then so is the distribution of the sample mean.

> **SAMPLING DISTRIBUTION OF A SAMPLE MEAN**
>
> If a population has the $N(\mu, \sigma)$ distribution, then the sample mean $\bar{x}$ of $n$ independent observations has the $N(\mu, \sigma/\sqrt{n})$ distribution.

This is a somewhat special result. Many population distributions are not Normal. The service call lengths in Figure 5.8(a), for example, are strongly skewed. Yet Figures 5.8(b) and 5.9 show that means of samples of size 80 are close to Normal. One of the most famous facts of probability theory says that, for large sample sizes, the distribution of $\bar{x}$ is close to a Normal distribution. This is true no matter what shape the population distribution has, as long as the population has a finite standard deviation $\sigma$. This is the **central limit theorem.** It is much more useful than the fact that the distribution of $\bar{x}$ is exactly Normal if the population is exactly Normal.

central limit theorem

> **CENTRAL LIMIT THEOREM**
>
> Draw an SRS of size $n$ from any population with mean $\mu$ and finite standard deviation $\sigma$. When $n$ is large, the sampling distribution of the sample mean $\bar{x}$ is approximately Normal:
>
> $$\bar{x} \text{ is approximately } N\left(\mu, \frac{\sigma}{\sqrt{n}}\right)$$

EXAMPLE

**5.19 How close will the sample mean be to the population mean?** With the Normal distribution to work with, we can better describe how accurately a random sample of 80 calls estimates the mean length of all the calls in the population. The population standard deviation for the more than 30,000 calls in the population of Figure 5.8(a) is $\sigma = 184.81$ seconds. From Example 5.18 we know $\sigma_{\bar{x}} = 20.66$ seconds. By the 95 part of the 68–95–99.7 rule, 95% of all samples will have mean $\bar{x}$ within two standard deviations of $\mu$, that is, within $\pm 41.32$ seconds of $\mu$.

LOOK BACK
**68–95–99.7 rule, page 59**

### USE YOUR KNOWLEDGE

**5.38 Use the 68–95–99.7 rule.** You take an SRS of size 100 from a population with mean 200 and standard deviation 10. According to the central limit theorem, what is the approximate sampling distribution of the sample mean? Use the 95 part of the 68–95–99.7 rule to describe the variability of this sample mean.

For the sample size of $n = 80$ in Example 5.19, the sample mean is not very accurate. The population is very spread out, so the sampling distribution of $\bar{x}$ is still quite variable.

EXAMPLE

**5.20 How can we reduce the standard deviation?** In the setting of Example 5.19, if we want to reduce the standard deviation of $\bar{x}$ by a factor of 4, we must take a sample 16 times as large, $n = 16 \times 80$, or 1280. Then

$$\sigma_{\bar{x}} = \frac{184.81}{\sqrt{1280}} = 5.165 \text{ seconds}$$

For samples of size 1280, 95% of the sample means will be within twice 5.165, or 10.33 seconds, of the population mean $\mu$.

## USE YOUR KNOWLEDGE

**5.39 The effect of increasing the sample size.** In the setting of Exercise 5.38, suppose we increase to the sample size to 400. Use the 95 part of the 68–95–99.7 rule to describe the variability of this sample mean. Compare your results with those you found in Exercise 5.38.

Example 5.20 reminds us that if the population is very spread out, the $\sqrt{n}$ in the standard deviation of $\bar{x}$ implies that even large samples will not estimate the population mean accurately. But the big point of the example is that the central limit theorem allows us to use Normal probability calculations to answer questions about sample means even when the population distribution is not Normal. How large a sample size $n$ is needed for $\bar{x}$ to be close to Normal depends on the population distribution. More observations are required if the shape of the population distribution is far from Normal. Even for the very skewed call length population, however, samples of size 80 are large enough. Here is a more detailed example.

EXAMPLE

**5.21 The central limit theorem in action.** Figure 5.10 shows the central limit theorem in action for another very non-Normal population. Figure 5.10(a) displays the density curve of a single observation, that is, of the population. The distribution is strongly right-skewed, and the most probable outcomes are near 0. The mean $\mu$ of this distribution is 1, and its standard deviation $\sigma$ is also 1. This particular continuous distribution is called an **exponential distribution.** Exponential distributions are used as models for how long an electronic component will last and for the time required to serve a customer or repair a machine.

exponential distribution

Figures 5.10(b), (c), and (d) are the density curves of the sample means of 2, 10, and 25 observations from this population. As $n$ increases, the shape be-

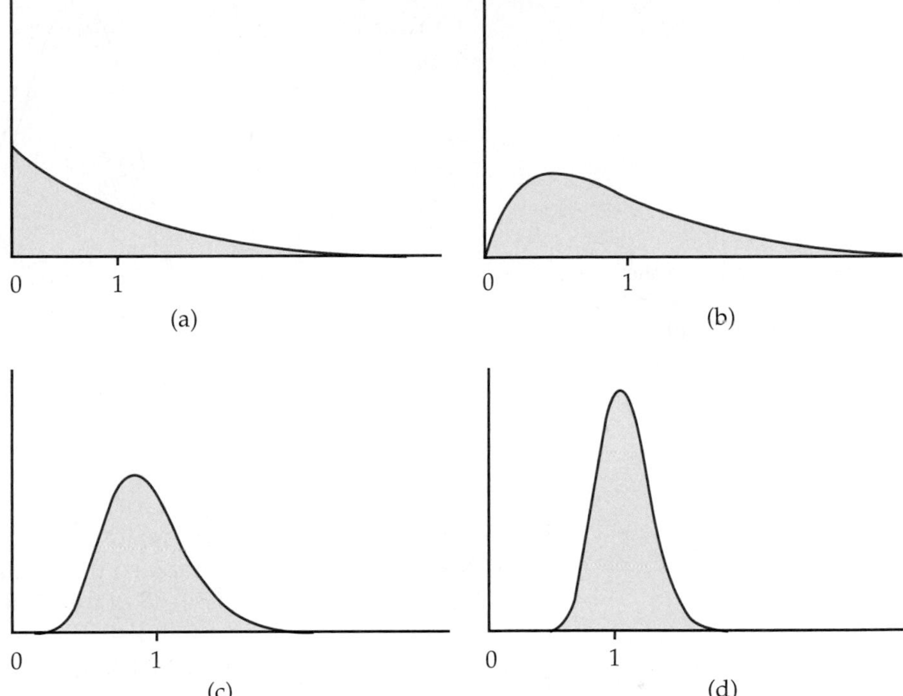


**FIGURE 5.10** The central limit theorem in action: the distribution of sample means from a strongly non-Normal population becomes more Normal as the sample size increases. (a) The distribution of 1 observation. (b) The distribution of $\bar{x}$ for 2 observations. (c) The distribution of $\bar{x}$ for 10 observations. (d) The distribution of $\bar{x}$ for 25 observations.

comes more Normal. The mean remains at $\mu = 1$, but the standard deviation decreases, taking the value $1/\sqrt{n}$. The density curve for 10 observations is still somewhat skewed to the right but already resembles a Normal curve having $\mu = 1$ and $\sigma = 1/\sqrt{10} = 0.32$. The density curve for $n = 25$ is yet more Normal. The contrast between the shapes of the population distribution and of the distribution of the mean of 10 or 25 observations is striking.

The *Central Limit Theorem* applet animates Figure 5.10. You can slide the sample size $n$ from 1 to 100 and watch both the exact density curve of $\bar{x}$ and the Normal approximation. As you increase $n$, the two curves move closer together.

EXAMPLE

**5.22 Preventive maintenance on an air-conditioning unit.** The time $X$ that a technician requires to perform preventive maintenance on an air-conditioning unit is governed by the exponential distribution whose density curve appears in Figure 5.10(a). The mean time is $\mu = 1$ hour and the standard deviation is $\sigma = 1$ hour. Your company operates 70 of these units. What is the probability that their average maintenance time exceeds 50 minutes?

The central limit theorem says that the sample mean time $\bar{x}$ (in hours) spent working on 70 units has approximately the Normal distribution with mean equal to the population mean $\mu = 1$ hour and standard deviation

$$\frac{\sigma}{\sqrt{70}} = \frac{1}{\sqrt{70}} = 0.12 \text{ hour}$$

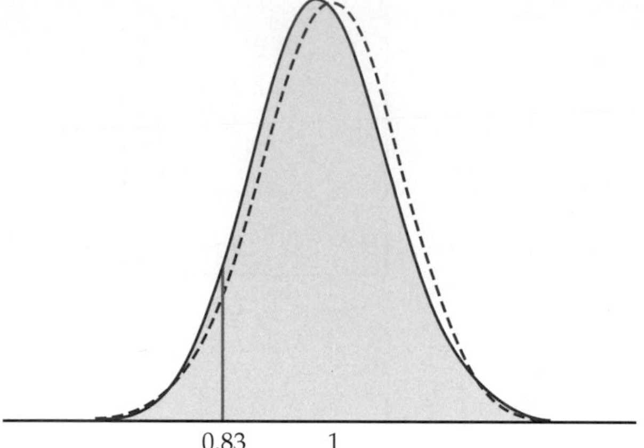


**FIGURE 5.11** The exact distribution (dashed) and the Normal approximation from the central limit theorem (solid) for the average time needed to maintain an air conditioner, for Example 5.22.

The distribution of $\bar{x}$ is therefore approximately $N(1, 0.12)$. Figure 5.11 shows this Normal curve (solid) and also the actual density curve of $\bar{x}$ (dashed).

Because 50 minutes is 50/60 of an hour, or 0.83 hour, the probability we want is $P(\bar{x} > 0.83)$. A Normal distribution calculation gives this probability as 0.9222. This is the area to the right of 0.83 under the solid Normal curve in Figure 5.11. The exactly correct probability is the area under the dashed density curve in the figure. It is 0.9294. The central limit theorem Normal approximation is off by only about 0.007.

## USE YOUR KNOWLEDGE

**5.40 Find a probability.** Refer to the example above. Find the probability that the mean time spent working on 70 units is less than 1.1 hours.

EXAMPLE

**5.23 Convert the results to the total maintenance time.** In Example 5.22 what can we say about the total maintenance time for 70 units? According to the central limit theorem

$$P(\bar{x} > 0.83) = 0.9222$$

We know that the sample mean is the total maintenance time divided by 70, so the event $\{\bar{x} > 0.83\}$ is the same as the event $\{70\bar{x} > 70(0.83)\}$. We can say that the probability is 0.9222 that the total maintenance time is $70(0.83) = 58.1$ hours or greater.

Figure 5.12 summarizes the facts about the sampling distribution of $\bar{x}$ in a way that emphasizes the big idea of a sampling distribution.

- Keep taking random samples of size $n$ from a population with mean $\mu$.
- Find the sample mean $\bar{x}$ for each sample.
- Collect all the $\bar{x}$'s and display their distribution.

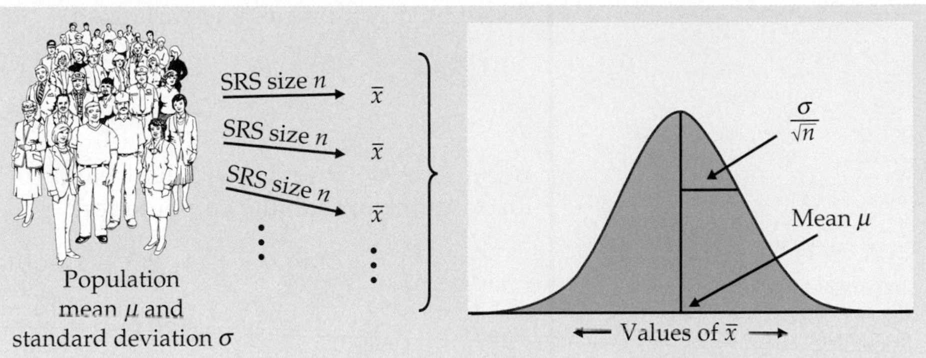


**FIGURE 5.12** The sampling distribution of a sample mean $\bar{x}$ has mean $\mu$ and standard deviation $\sigma/\sqrt{n}$. The distribution is Normal if the population distribution is Normal; it is approximately Normal for large samples in any case.

That's the sampling distribution of $\bar{x}$. Sampling distributions are the key to understanding statistical inference. Keep this figure in mind as you go forward.

## A few more facts

The central limit theorem is the big fact of this section. Here are three useful smaller facts related to our topic.

**The Normal approximation for sample proportions and counts is an example of the central limit theorem.** This is true because a sample proportion can be thought of as a sample mean. Recall the idea that we used to find the mean and variance of a binomial random variable $X$. We wrote the count $X$ as a sum

$$X = S_1 + S_2 + \cdots + S_n$$

of random variables $S_i$ that take the value 1 if a success occurs on the $i$th trial and the value 0 otherwise. The variables $S_i$ take only the values 0 and 1 and are far from Normal. The proportion $\hat{p} = X/n$ is the sample mean of the $S_i$ and, like all sample means, is approximately Normal when $n$ is large.

LOOK BACK
rules for means, page 278
rules for variances, page 282

The fact that the sample mean of an SRS from a Normal population has a Normal distribution is a special case of a more general fact: **any linear combination of independent Normal random variables is also Normally distributed.** That is, if $X$ and $Y$ are independent Normal random variables and $a$ and $b$ are any fixed numbers, $aX + bY$ is also Normally distributed, and so it is for any number of Normal variables. In particular, the sum or difference of independent Normal random variables has a Normal distribution. The mean and standard deviation of $aX + bY$ are found as usual from the addition rules for means and variances. These facts are often used in statistical calculations.

EXAMPLE

**5.24 Who will win?** Tom and George are playing in the club golf tournament. Their scores vary as they play the course repeatedly. Tom's score $X$ has the $N(110, 10)$ distribution, and George's score $Y$ varies from round to round according to the $N(100, 8)$ distribution. If they play independently, what is the probability that Tom will score lower than George and thus do better in the tournament? The difference $X - Y$ between their scores is Normally distributed, with mean and variance

$$\mu_{X-Y} = \mu_X - \mu_Y = 110 - 100 = 10$$
$$\sigma^2_{X-Y} = \sigma^2_X + \sigma^2_Y = 10^2 + 8^2 = 164$$

Because $\sqrt{164} = 12.8$, $X - Y$ has the $N(10, 12.8)$ distribution. Figure 5.13 illustrates the probability computation:

$$\begin{aligned} P(X < Y) &= P(X - Y < 0) \\ &= P\left(\frac{(X - Y) - 10}{12.8} < \frac{0 - 10}{12.8}\right) \\ &= P(Z < -0.78) = 0.2177 \end{aligned}$$

Although George's score is 10 strokes lower on the average, Tom will have the lower score in about one of every five matches.

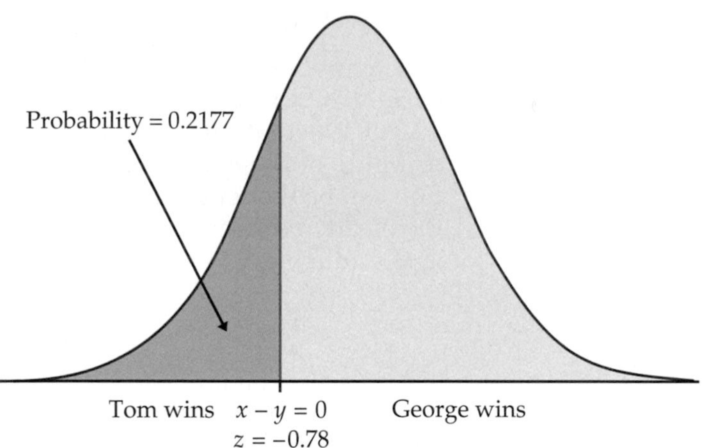


**FIGURE 5.13** The Normal probability calculation for Example 5.24.

**More general versions of the central limit theorem say that the distribution of a sum or average of many small random quantities is close to Normal.** This is true even if the quantities are not independent (as long as they are not too highly correlated) and even if they have different distributions (as long as no single random quantity is so large that it dominates the others). The central limit theorem suggests why the Normal distributions are common models for observed data. Any variable that is a sum of many small influences will have approximately a Normal distribution.

## BEYOND THE BASICS

### Weibull Distributions

Our discussion of sampling distributions has concentrated on the binomial model for count data and the Normal model for quantitative variables. These models are important in statistical practice, but simplicity also contributes to their popularity. The parameters $p$ in the binomial model and $\mu$ in the Normal model are easy to understand. To estimate them from data we use statistics $\hat{p}$ and $\bar{x}$ that are also easy to understand and that have simple sampling distributions.

Weibull distributions

There are many other probability distributions that are used to model data in various circumstances. The time that a product, such as a computer disk drive, lasts before failing rarely has a Normal distribution. Another class of continuous distributions, the **Weibull distributions,** is often used to model time to failure. For engineers studying the reliability of products, Weibull distributions are more common than Normal distributions.

EXAMPLE

**5.25 Weibull density curves.** Figure 5.14 shows the density curves of three members of the Weibull family. Each describes a different type of distribution for the time to failure of a product.

1. The top curve in Figure 5.14 is a model for *infant mortality.* Many of these products fail immediately. If they do not fail at once, then most last a long time. The manufacturer tests these products and ships only the ones that do not fail immediately.
2. The middle curve in Figure 5.14 is a model for *early failure.* These products do not fail immediately, but many fail early in their lives after they are in the hands of customers. This is disastrous—the product or the process that makes it must be changed at once.

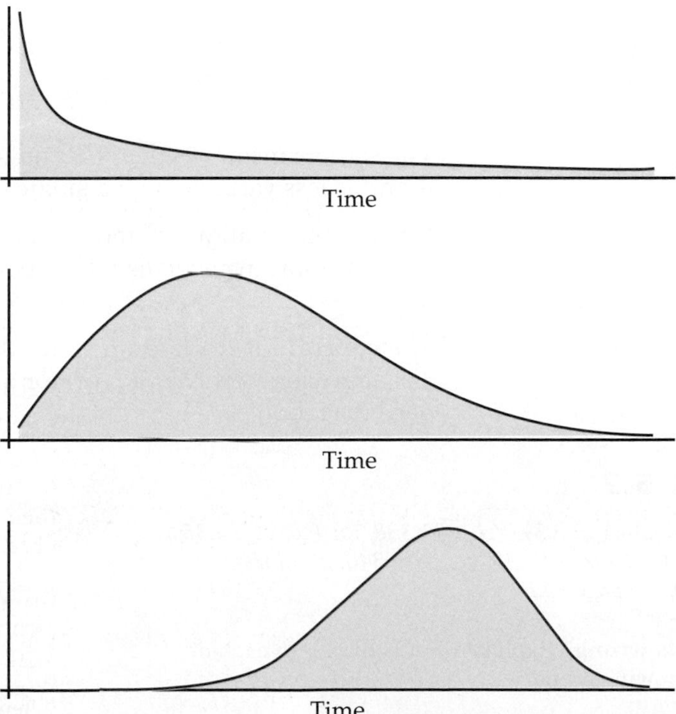


**FIGURE 5.14** Density curves for three members of the Weibull family of distributions, for Example 5.25.

3. The bottom curve in Figure 5.14 is a model for *old-age wearout.* Most of these products fail only when they begin to wear out, and then many fail at about the same age.

A manufacturer certainly wants to know to which of these classes a new product belongs. To find out, engineers operate a random sample of products until they fail. From the failure time data we can estimate the parameter (called the "shape parameter") that distinguishes among the three Weibull distributions in Figure 5.14. The shape parameter has no simple definition like that of a population proportion or mean, and it cannot be estimated by a simple statistic such as $\hat{p}$ or $\bar{x}$.

Two things save the situation. First, statistical theory provides general approaches for finding good estimates of any parameter. These general methods not only tell us how to use $\hat{p}$ and $\bar{x}$ in the binomial and Normal settings but also tell us how to estimate the Weibull shape parameter. Second, modern software can calculate the estimate from data even though there is no algebraic formula that we can write for the estimate. Statistical practice often relies on both mathematical theory and methods of computation more elaborate than the ones we will meet in this book. Fortunately, big ideas such as sampling distributions carry over to more complicated situations.[7]

## SECTION 5.2 Summary

The **sample mean** $\bar{x}$ of an SRS of size $n$ drawn from a large population with mean $\mu$ and standard deviation $\sigma$ has a sampling distribution with mean and standard deviation

$$\mu_{\bar{x}} = \mu$$

$$\sigma_{\bar{x}} = \frac{\sigma}{\sqrt{n}}$$

The sample mean $\bar{x}$ is therefore an unbiased estimator of the population mean $\mu$ and is less variable than a single observation.

Linear combinations of independent Normal random variables have Normal distributions. In particular, if the population has a Normal distribution, so does $\bar{x}$.

The **central limit theorem** states that for large $n$ the sampling distribution of $\bar{x}$ is approximately $N(\mu, \sigma/\sqrt{n})$ for any population with mean $\mu$ and finite standard deviation $\sigma$.

## SECTION 5.2 Exercises

*For Exercises 5.36 and 5.37, see page 338; for Exercise 5.38, see page 339; for Exercise 5.39, see page 340; and for Exercise 5.40, see page 342.*

**5.41 What is wrong?** Explain what is wrong in each of the following scenarios.

(a) If the standard deviation of a population is 10, then the variance of the mean for an SRS of 20 observations from this population will be $10/\sqrt{20}$.

(b) When taking SRS's from a population, larger sample sizes will have larger standard deviations.

(c) The mean of a sampling distribution changes when the sample size changes.

**5.42 Songs on an iPod.** An iPod has about 10,000 songs. The distribution of the play time for these songs is highly skewed. Assume that the standard deviation for the population is 300 seconds.

(a) What is the standard deviation of the average time when you take an SRS of 10 songs from this population?

(b) How many songs would you need to sample if you wanted the standard deviation of $\bar{x}$ to be 30 seconds?

**5.43 A grinding machine for auto axles.** An automatic grinding machine in an auto parts plant prepares axles with a target diameter $\mu = 40.135$ millimeters (mm). The machine has some variability, so the standard deviation of the diameters is $\sigma = 0.003$ mm. A sample of 4 axles is inspected each hour for process control purposes, and records are kept of the sample mean diameter. If the process mean is exactly equal to the target value, what will be the mean and standard deviation of the numbers recorded?

**5.44 Play times for songs on an iPod.** Averages of several measurements are less variable than individual measurements. Suppose the true mean duration of the play time for the songs in the iPod of Exercise 5.42 is 350 seconds.

(a) Sketch on the same graph the two Normal curves, for sampling a single song and for the mean of 10 songs.

(b) What is the probability that the sample mean differs from the population mean by more than 19 seconds when only 1 song is sampled?

(c) How does the probability that you calculated in part (b) change for the mean of an SRS of 10 songs?

**5.45 Axle diameters.** Averages are less variable than individual observations. Suppose that the axle diameters in Exercise 5.43 vary according to a Normal distribution. In that case, the mean $\bar{x}$ of an SRS of axles also has a Normal distribution.

(a) Make a sketch of the Normal curve for a single axle. Add the Normal curve for the mean of an SRS of 4 axles on the same sketch.

(b) What is the probability that the diameter of a single randomly chosen axle differs from the target value by 0.006 mm or more?

(c) What is the probability that the mean diameter of an SRS of 4 axles differs from the target value by 0.006 mm or more?

**5.46 Lightning strikes.** The number of lightning strikes on a square kilometer of open ground in a year has mean 6 and standard deviation 2.4. (These values are typical of much of the United States.) Counts of strikes on separate areas are independent. The National Lightning Detection Network uses automatic sensors to watch for lightning in an area of 10 square kilometers.

(a) What are the mean and standard deviation of the total number of lightning strikes observed?

(b) What are the mean and standard deviation of the mean number of strikes per square kilometer?

**5.47** CHALLENGE **Cholesterol levels of teenagers.** A study of the health of teenagers plans to measure the blood cholesterol level of an SRS of 13- to 16-year olds. The researchers will report the mean $\bar{x}$ from their sample as an estimate of the mean cholesterol level $\mu$ in this population.

(a) Explain to someone who knows no statistics what it means to say that $\bar{x}$ is an "unbiased" estimator of $\mu$.

(b) The sample result $\bar{x}$ is an unbiased estimator of the population truth $\mu$ no matter what size SRS the study chooses. Explain to someone who knows no statistics why a large sample gives more trustworthy results than a small sample.

**5.48 ACT scores of high school seniors.** The scores of high school seniors on the ACT college entrance examination in 2003 had mean $\mu = 20.8$ and standard deviation $\sigma = 4.8$. The distribution of scores is only roughly Normal.

(a) What is the approximate probability that a single student randomly chosen from all those taking the test scores 23 or higher?

(b) Now take an SRS of 25 students who took the test. What are the mean and standard deviation of the sample mean score $\bar{x}$ of these 25 students?

(c) What is the approximate probability that the mean score $\bar{x}$ of these students is 23 or higher?

(d) Which of your two Normal probability calculations in (a) and (c) is more accurate? Why?

**5.49 Gypsy moths threaten oak and aspen trees.** The gypsy moth is a serious threat to oak and aspen trees. A state agriculture department places traps throughout the state to detect the moths. When traps are checked periodically, the mean number of moths trapped is only 0.5, but some traps have several moths. The distribution of moth counts is discrete and strongly skewed, with standard deviation 0.7.

(a) What are the mean and standard deviation of the average number of moths $\bar{x}$ in 50 traps?

(b) Use the central limit theorem to find the probability that the average number of moths in 50 traps is greater than 0.6.

**5.50 Grades in an English course.** North Carolina State University posts the grade distributions for its courses online.[8] Students in one section of English 210 in the spring 2006 semester received 31% A's, 40% B's, 20% C's, 4% D's, and 5% F's.

(a) Using the common scale A = 4, B = 3, C = 2, D = 1, F = 0, take $X$ to be the grade of a randomly chosen English 210 student. Use the definitions of the mean (page 271) and standard deviation (page 280) for discrete random variables to find the mean $\mu$ and the standard deviation $\sigma$ of grades in this course.

(b) English 210 is a large course. We can take the grades of an SRS of 50 students to be independent of each other. If $\bar{x}$ is the average of these 50 grades, what are the mean and standard deviation of $\bar{x}$?

(c) What is the probability $P(X \geq 3)$ that a randomly chosen English 210 student gets a B or better? What is the approximate probability $P(\bar{x} \geq 3)$ that the grade for 50 randomly chosen English 210 students is B or better?

**5.51 Diabetes during pregnancy.** Sheila's doctor is concerned that she may suffer from gestational diabetes (high blood glucose levels during pregnancy). There is variation both in the actual glucose level and in the blood test that measures the level. A patient is classified as having gestational diabetes if the glucose level is above 140 milligrams per deciliter (mg/dl) one hour after a sugary drink is ingested. Sheila's measured glucose level one hour after ingesting the sugary drink varies according to the Normal distribution with $\mu$ = 125 mg/dl and $\sigma$ = 10 mg/dl.

(a) If a single glucose measurement is made, what is the probability that Sheila is diagnosed as having gestational diabetes?

(b) If measurements are made instead on 3 separate days and the mean result is compared with the criterion 140 mg/dl, what is the probability that Sheila is diagnosed as having gestational diabetes?

**5.52 A lottery payoff.** A \$1 bet in a state lottery's Pick 3 game pays \$500 if the three-digit number you choose exactly matches the winning number, which is drawn at random. Here is the distribution of the payoff $X$:

| Payoff $X$ | \$0 | \$500 |
|---|---|---|
| Probability | 0.999 | 0.001 |

Each day's drawing is independent of other drawings.

(a) What are the mean and standard deviation of $X$?

(b) Joe buys a Pick 3 ticket twice a week. What does the law of large numbers say about the average payoff Joe receives from his bets?

(c) What does the central limit theorem say about the distribution of Joe's average payoff after 104 bets in a year?

(d) Joe comes out ahead for the year if his average payoff is greater than \$1 (the amount he spent each day on a ticket). What is the probability that Joe ends the year ahead?

**5.53 Defining a high glucose reading.** In Exercise 5.51, Sheila's measured glucose level one hour after ingesting the sugary drink varies according to the Normal distribution with $\mu$ = 125 mg/dl and $\sigma$ = 10 mg/dl. What is the level $L$ such that there is probability only 0.05 that the mean glucose level of 3 test results falls above $L$ for Sheila's glucose level distribution?

**5.54 Flaws in carpets.** The number of flaws per square yard in a type of carpet material varies with mean 1.5 flaws per square yard and standard deviation 1.3 flaws per square yard. This population distribution cannot be Normal, because a count takes only whole-number values. An inspector studies 200 square yards of the material, records the number of flaws found in each square yard, and calculates $\bar{x}$, the mean number of flaws per square yard inspected. Use the central limit theorem to find the approximate probability that the mean number of flaws exceeds 2 per square yard.

**5.55 Weights of airline passengers.** In response to the increasing weight of airline passengers, the Federal Aviation Administration told airlines to assume that passengers average 190 pounds in the summer, including clothing and carry-on baggage. But passengers vary: the FAA gave a mean but not a standard deviation. A reasonable standard deviation is 35 pounds. Weights are not Normally distributed, especially when the population includes both men and women, but they are not very non-Normal. A commuter plane carries 25 passengers. What is the approximate probability that the total weight of the passengers exceeds 5200 pounds? (*Hint:* To apply the central limit theorem, restate the problem in terms of the mean weight.)

**5.56 Risks and insurance.** The idea of insurance is that we all face risks that are unlikely but carry high cost. Think of a fire destroying your home. So we form a group to share the risk: we all pay a small amount, and the insurance policy pays a large amount to those few of us whose homes burn down. An insurance company looks at the records for millions of homeowners and sees that the mean loss from fire in a year is $\mu = \$250$ per house and that the standard deviation of the loss is $\sigma = \$1000$. (The distribution of losses is extremely right-skewed: most people have \$0 loss, but a few have large losses.) The company plans to sell fire insurance for \$250 plus enough to cover its costs and profit.

(a) Explain clearly why it would be unwise to sell only 12 policies. Then explain why selling many thousands of such policies is a safe business.

(b) If the company sells 10,000 policies, what is the approximate probability that the average loss in a year will be greater than \$275?

**5.57 Returns on common stocks.** The distribution of annual returns on common stocks is roughly symmetric, but extreme observations are more frequent than in a Normal distribution. Because the distribution is not strongly non-Normal, the mean return over even a moderate number of years is close to Normal. Annual real returns on the Standard & Poor's 500 stock index over the period 1871 to 2004 have varied with mean 9.2% and standard deviation 20.6%. Andrew plans to retire in 45 years and is considering investing in stocks. What is the probability (assuming that the past pattern of variation continues) that the mean annual return on common stocks over the next 45 years will exceed 15%? What is the probability that the mean return will be less than 5%?

**5.58 Holes in engine blocks.** A hole in an engine block is 2.5 centimeters (cm) in diameter. Shafts manufactured to go through this hole must have 0.024 cm clearance for unforced fit. That is, shaft diameter cannot exceed 2.476 cm. The shafts vary in diameter according to the Normal distribution with mean 2.45 cm and standard deviation 0.01 cm.

(a) What percent of shafts will fit into the hole?

(b) Redo the problem assuming that the hole diameter also varies, independently of the shaft diameter, following the Normal distribution with mean 2.5 cm and standard deviation 0.01 cm. You must find the probability that the hole diameter exceeds the shaft diameter by at least 0.024 cm.

**5.59 Treatment of cotton fabrics.** "Durable press" cotton fabrics are treated to improve their recovery from wrinkles after washing. Unfortunately, the treatment also reduces the strength of the fabric. The breaking strength of untreated fabric is Normally distributed with mean 57 pounds and standard deviation 2.2 pounds. The same type of fabric after treatment has Normally distributed breaking strength with mean 30 pounds and standard deviation 1.6 pounds.[9] A clothing manufacturer tests 6 specimens of each fabric. All 12 strength measurements are independent.

(a) What is the probability that the mean breaking strength of the 6 untreated specimens exceeds 50 pounds?

(b) What is the probability that the mean breaking strength of the 6 untreated specimens is at least 25 pounds greater than the mean strength of the 6 treated specimens?

**5.60 Advertisements and brand image.** Many companies place advertisements to improve the image of their brand rather than to promote specific products. In a randomized comparative experiment, business students read ads that cited either the *Wall Street Journal* or the *National Enquirer* for important facts about a fictitious company. The students then rated the trustworthiness of the source on a 7-point scale. Suppose that in the population of all students scores for the *Journal* have mean 4.8 and standard deviation 1.5, while scores for the *Enquirer* have mean 2.4 and standard deviation 1.6.[10]

(a) There are 28 students in each group. Although individual scores are discrete, the mean score for a group of 28 will be close to Normal. Why?

(b) What are the means and standard deviations of the sample mean scores $\bar{y}$ for the *Journal* group and $\bar{x}$ for the *Enquirer* group?

(c) We can take all 56 scores to be independent because students are not told each other's scores. What is the distribution of the difference $\bar{y} - \bar{x}$ between the mean scores in the two groups?

(d) Find $P(\bar{y} - \bar{x} \geq 1)$.

**5.61** CHALLENGE **Treatment and control groups.** The two previous exercises illustrate a common setting for statistical inference. This exercise gives the general form of the sampling distribution needed in this setting. We have a sample of $n$ observations from a treatment group and an independent sample of $m$ observations from a control group. Suppose that the response to the treatment has the $N(\mu_X, \sigma_X)$ distribution and that the response of control subjects has the $N(\mu_Y, \sigma_Y)$ distribution. Inference about the difference $\mu_Y - \mu_X$ between the population means is based on the difference $\bar{y} - \bar{x}$ between the sample means in the two groups.

(a) Under the assumptions given, what is the distribution of $\bar{y}$? Of $\bar{x}$?

(b) What is the distribution of $\bar{y} - \bar{x}$?

**5.62** CHALLENGE **Investments in two funds.** Linda invests her money in a portfolio that consists of 70% Fidelity 500 Index Fund and 30% Fidelity Diversified International Fund. Suppose that in the long run the annual real return $X$ on the 500 Index Fund has mean 9% and standard deviation 19%, the annual real return $Y$ on the Diversified International Fund has mean 11% and standard deviation 17%, and the correlation between $X$ and $Y$ is 0.6.

(a) The return on Linda's portfolio is $R = 0.7X + 0.3Y$. What are the mean and standard deviation of $R$?

(b) The distribution of returns is typically roughly symmetric but with more extreme high and low observations than a Normal distribution. The average return over a number of years, however, is close to Normal. If Linda holds her portfolio for 20 years, what is the approximate probability that her average return is less than 5%?

(c) The calculation you just made is not overly helpful, because Linda isn't really concerned about the mean return $\bar{R}$. To see why, suppose that her portfolio returns 12% this year and 6% next year. The mean return for the two years is 9%. If Linda starts with \$1000, how much does she have at the end of the first year? At the end of the second year? How does this amount compare with what she would have if both years had the mean return, 9%? Over 20 years, there may be a large difference between the ordinary mean $\bar{R}$ and the *geometric mean*, which reflects the fact that returns in successive years multiply rather than add.

**5.63 Concrete blocks and mortar.** You are building a wall from precast concrete blocks. Standard "8 inch" blocks are $7\frac{5}{8}$ inches high to allow for a $\frac{3}{8}$ inch layer of mortar under each row of blocks. In practice, the height of a block-plus-mortar row varies according to a Normal distribution with mean 8 inches and standard deviation 0.1 inch. Heights of successive rows are independent. Your wall has four rows of blocks. What is the distribution of the height of the wall? What is the probability that the height differs from the design height of 32 inches by more than half an inch?

## CHAPTER 5 Exercises

**5.64** CHALLENGE **The effect of sample size on the standard deviation.** Assume that the standard deviation in a very large population is 100.

(a) Calculate the standard deviation for the sample mean for samples of size 1, 4, 25, 100, 250, 500, 1000, and 5000.

(b) Graph your results with the sample size on the $x$ axis and the standard deviation on the $y$ axis.

(c) Summarize the relationship between the sample size and the standard deviation that you showed in your graph.

**5.65 Auto accidents.** The probability that a randomly chosen driver will be involved in an accident in the next year is about 0.2. This is based on the proportion of millions of drivers who have accidents. "Accident" includes things like crumpling a fender in your own driveway, not just highway accidents. Carlos, David, Jermaine, Ramon, and Scott are college students who live together in an off-campus apartment. Last year, 3 of the 5 had accidents. What is the probability that 3 or more of 5 randomly chosen drivers have an accident in the same year? Why does your calculation not apply to drivers like the 5 students?

**5.66** CHALLENGE **SAT scores.** Example 4.37 (page 284) notes that the total SAT scores of high school seniors in a recent year had mean $\mu = 1026$ and standard deviation $\sigma = 209$. The distribution of SAT scores is roughly Normal.

(a) Julie scored 1110. If scores have a Normal distribution, what percentile of the distribution is this?

(b) Now consider the mean $\bar{x}$ of the scores of 80 randomly chosen students. If $\bar{x} = 1110$, what percentile of the sampling distribution of $\bar{x}$ is this?

(c) Which of your calculations, (a) or (b), is less accurate because SAT scores do not have an exactly Normal distribution? Explain your answer.

**5.67 Carpooling.** Although cities encourage carpooling to reduce traffic congestion, most vehicles carry only one person. For example, 70% of vehicles on the roads in the Minneapolis–St. Paul metropolitan area are occupied by just the driver.

(a) If you choose 12 vehicles at random, what is the probability that more than half (that is, 7 or more) carry just one person?

(b) If you choose 80 vehicles at random, what is the probability that more than half (that is, 41 or more) carry just one person?

**5.68 Common last names.** The Census Bureau says that the 10 most common names in the United States are (in order) Smith, Johnson, Williams, Jones, Brown, Davis, Miller, Wilson, Moore, and Taylor. These names account for 5.6% of all U.S. residents. Out of curiosity, you look at the authors of the textbooks for your current courses. There are 12 authors in all. Would you be surprised if none of the names of these authors were among the 10 most common? Give a probability to support your answer and explain the reasoning behind your calculation.

**5.69 Benford's law.** It is a striking fact that the first digits of numbers in legitimate records often follow a distribution known as Benford's law. Here it is:

| First digit | 1 | 2 | 3 | 4 | 5 | 6 | 7 | 8 | 9 |
|---|---|---|---|---|---|---|---|---|---|
| Proportion | 0.301 | 0.176 | 0.125 | 0.097 | 0.079 | 0.067 | 0.058 | 0.051 | 0.046 |

Fake records usually have fewer first digits 1, 2, and 3. What is the approximate probability, if Benford's law holds, that among 1000 randomly chosen invoices there are 560 or fewer in amounts with first digit 1, 2, or 3?

**5.70 Genetics of peas.** According to genetic theory, the blossom color in the second generation of a certain cross of sweet peas should be red or white in a 3:1 ratio. That is, each plant has probability 3/4 of having red blossoms, and the blossom colors of separate plants are independent.

(a) What is the probability that exactly 9 out of 12 of these plants have red blossoms?

(b) What is the mean number of red-blossomed plants when 120 plants of this type are grown from seeds?

(c) What is the probability of obtaining at least 80 red-blossomed plants when 120 plants are grown from seeds?

**5.71 The weight of a dozen eggs.** The weight of the eggs produced by a certain breed of hen is Normally distributed with mean 65 grams (g) and standard deviation 5 g. If cartons of such eggs can be considered to be SRSs of size 12 from the population of all eggs, what is the probability that the weight of a carton falls between 755 and 830 g?

**5.72 Losses of British aircraft in World War II.** Serving in a bomber crew in World War II was dangerous. The British estimated that the probability of an aircraft loss due to enemy action was 1/20 for each mission. A tour of duty for British airmen in Bomber Command was 30 missions. What is the probability that an airman would complete a tour of duty without being on an aircraft lost from enemy action?

**5.73 A survey of college women.** A sample survey interviews an SRS of 280 college women. Suppose (as is roughly true) that 70% of all college women have been on a diet within the last 12 months. What is the probability that 75% or more of the women in the sample have been on a diet?

**5.74 Plastic caps for motor oil containers.** A machine fastens plastic screw-on caps onto containers of motor oil. If the machine applies more torque than the cap can withstand, the cap will break. Both the torque applied and the strength of the caps vary. The capping-machine torque has the Normal distribution with mean 7.0 inch-pounds and standard deviation 0.9 inch-pounds. The cap strength (the torque that would break the cap) has the Normal distribution with mean 10.1 inch-pounds and standard deviation 1.2 inch-pounds.

(a) Explain why it is reasonable to assume that the cap strength and the torque applied by the machine are independent.

(b) What is the probability that a cap will break while being fastened by the capping machine?

**5.75 Colors of cashmere sweaters.** The unique colors of the cashmere sweaters your firm makes result from heating undyed yarn in a kettle with a dye liquor. The pH (acidity) of the liquor is critical for regulating dye uptake and hence the final color. There are 5 kettles, all of which receive dye liquor from a common source. Past data show that pH varies according to a Normal distribution with $\mu = 4.25$ and $\sigma = 0.135$. You use statistical process control to check the stability of the process. Twice each day, the pH of the liquor in each kettle is measured, each time giving a sample of size 5. The mean pH $\bar{x}$ is compared with "control limits" given by the 99.7 part of the 68–95–99.7 rule for Normal distributions, namely, $\mu_{\bar{x}} \pm 3\sigma_{\bar{x}}$. What are the numerical values of these control limits for $\bar{x}$?

5.76 CHALLENGE **Learning a foreign language.** Does delaying oral practice hinder learning a foreign language? Researchers randomly assigned 25 beginning students of Russian to begin speaking practice immediately and another 25 to delay speaking for 4 weeks. At the end of the semester both groups took a standard test of comprehension of spoken Russian. Suppose that in the population of all beginning students, the test scores for early speaking vary according to the $N(32, 6)$ distribution and scores for delayed speaking have the $N(29, 5)$ distribution.

(a) What is the sampling distribution of the mean score $\bar{x}$ in the early-speaking group in many repetitions of the experiment? What is the sampling distribution of the mean score $\bar{y}$ in the delayed-speaking group?

(b) If the experiment were repeated many times, what would be the sampling distribution of the difference $\bar{y} - \bar{x}$ between the mean scores in the two groups?

(c) What is the probability that the experiment will find (misleadingly) that the mean score for delayed speaking is at least as large as that for early speaking?

5.77 CHALLENGE **Summer employment of college students.** Suppose (as is roughly true) that 88% of college men and 82% of college women were employed last summer. A sample survey interviews SRSs of 400 college men and 400 college women. The two samples are of course independent.

(a) What is the approximate distribution of the proportion $\hat{p}_F$ of women who worked last summer? What is the approximate distribution of the proportion $\hat{p}_M$ of men who worked?

(b) The survey wants to compare men and women. What is the approximate distribution of the difference in the proportions who worked, $\hat{p}_M - \hat{p}_F$? Explain the reasoning behind your answer.

(c) What is the probability that in the sample a higher proportion of women than men worked last summer?

5.78 **Income of working couples.** A study of working couples measures the income $X$ of the husband and the income $Y$ of the wife in a large number of couples in which both partners are employed. Suppose that you knew the means $\mu_X$ and $\mu_Y$ and the variances $\sigma_X^2$ and $\sigma_Y^2$ of both variables in the population.

(a) Is it reasonable to take the mean of the total income $X + Y$ to be $\mu_X + \mu_Y$? Explain your answer.

(b) Is it reasonable to take the variance of the total income to be $\sigma_X^2 + \sigma_Y^2$? Explain your answer.

5.79 CHALLENGE **A random walk.** A particle moves along the line in a random walk. That is, the particle starts at the origin (position 0) and moves either right or left in independent steps of length 1. If the particle moves to the right with probability 0.6, its movement at the $i$th step is a random variable $X_i$ with distribution

$$P(X_i = 1) = 0.6$$
$$P(X_i = -1) = 0.4$$

The position of the particle after $k$ steps is the sum of these random movements,

$$Y = X_1 + X_2 + \cdots + X_k$$

Use the central limit theorem to find the approximate probability that the position of the particle after 500 steps is at least 200 to the right.

CHAPTER

6

# Introduction to Inference

Undergraduate student loan debt has been increasing steadily during the past decade. Is the debt becoming too much of a burden upon graduation? Example 6.4 discusses the average debt of undergraduate borrowers.



## Introduction

Statistical inference draws conclusions about a population or process based on sample data. It also provides a statement, expressed in terms of probability, of how much confidence we can place in our conclusions. Although there are many specific recipes for inference, there are only a few general types of statistical inference. This chapter introduces the two most common types: *confidence intervals* and *tests of significance*.

Because the underlying reasoning for these types of inference remains the same across different settings, this chapter considers a single simple setting: inference about the mean of a Normal population whose standard deviation is known. Later chapters will present the recipes for inference in other situations.

In this setting, we can address questions like:

- What is the average loan debt among undergraduate borrowers?
- What is the average miles per gallon (mpg) for a hybrid car?
- Is the average cholesterol level of undergraduate women at your university below the national average?

## Overview of Inference

The purpose of statistical inference is to draw conclusions from data. We have already examined data and arrived at conclusions many times. Formal inference emphasizes substantiating our conclusions by probability calculations. Probability allows us to take chance variation into account. Here is an example.

EXAMPLE

**6.1 Clustering of trees in a forest.** The Wade Tract in Thomas County, Georgia, is an old-growth forest of longleaf pine trees (*Pinus palustris*) that has survived in a relatively undisturbed state since before the settlement of the area by Europeans. Foresters who study these trees are interested in how the trees are distributed in the forest. Is there some sort of clustering, resulting in regions of the forest with more trees than others? Or are the tree locations random, resulting in no particular patterns?

Figure 6.1 gives a plot of the locations of all 584 longleaf pine trees in a 200-meter by 200-meter region in the Wade Tract.[1] Do the locations appear to be random, or do there appear to be clusters of trees? One approach to the analysis of these data indicates that a pattern as clustered as, or more clustered than, the one in Figure 6.1 would occur only 4% of the time if, in fact, the locations of longleaf pine trees in the Wade Tract are random. Because this chance is fairly small, we conclude that there is some clustering of these trees.

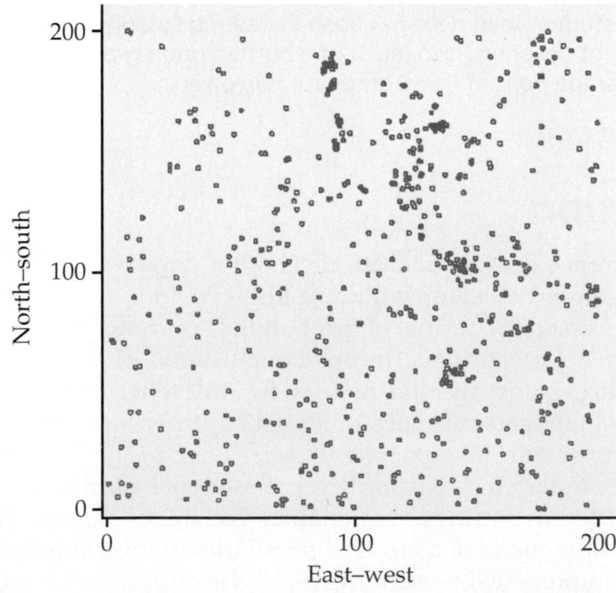


**FIGURE 6.1** The distribution of longleaf pine trees, for Example 6.1.

Our probability calculation helps us to distinguish between patterns that are consistent or inconsistent with the random location scenario. Here is an example comparing two drug treatments with a different conclusion.

EXAMPLE

**6.2 Effectiveness of a new drug.** Researchers want to know if a new drug is more effective than a placebo. Twenty patients receive the new drug, and 20 receive a placebo. Twelve (60%) of those taking the drug show improvement versus only 8 (40%) of the placebo patients.

Our unaided judgment would suggest that the new drug is better. However, probability calculations tell us that a difference this large or larger between the results in the two groups would occur about one time in five simply because of chance variation. Since this probability is not very small, it is better to conclude that the observed difference is due to chance rather than a real difference between the two treatments.

In this chapter we introduce the two most prominent types of formal statistical inference. Section 6.1 concerns *confidence intervals* for estimating the value of a population parameter. Section 6.2 presents *tests of significance,* which assess the evidence for a claim. Both types of inference are based on the sampling distributions of statistics. That is, both report probabilities that state *what would happen if we used the inference method many times.* This kind of probability statement is characteristic of standard statistical inference. Users of statistics must understand the nature of this reasoning and the meaning of the probability statements that appear, for example, in newspaper and journal articles as well as statistical software output.

**sampling distributions, page 215**

Because the methods of formal inference are based on sampling distributions, they require a probability model for the data. Trustworthy probability models can arise in many ways, but the model is most secure and inference is most reliable when the data are produced by a properly randomized design. *When you use statistical inference, you are acting as if the data come from a random sample or a randomized experiment.* If this is not true, your conclusions may be open to challenge. Do not be overly impressed by the complex details of formal inference. This elaborate machinery cannot remedy basic flaws in producing the data such as voluntary response samples and confounded experiments. Use the common sense developed in your study of the first three chapters of this book, and proceed to detailed formal inference only when you are satisfied that the data deserve such analysis.

The primary purpose of this chapter is to describe the reasoning used in statistical inference. We will discuss only a few specific inference techniques, and these require an unrealistic assumption: that we know the standard deviation $\sigma$. Later chapters will present inference methods for use in most of the settings we met in learning to explore data. There are libraries—both of books and of computer software—full of more elaborate statistical techniques. Informed use of any of these methods requires an understanding of the underlying reasoning. A computer will do the arithmetic, but you must still exercise judgment based on understanding.

## 6.1 Estimating with Confidence

The SAT test is a widely used measure of readiness for college study. Originally, there were two sections, one for verbal reasoning ability (SATV) and one for mathematical reasoning ability (SATM). In April 1995, section scores were *recentered* so that the mean is approximately 500 in a large "standardized group." This scale has been maintained since then so that scores have a constant interpretation.

LOOK BACK
**linear transformations, page 45**

In 2005, the College Board changed the test, renaming the verbal section "Critical Reading" and adding a third section on writing ability. These changes increased the total possible score to 2400, extended the exam an additional 35 minutes, and increased the cost to register for the exam by $12. The changes also raised concerns about the constant-interpretation assumption.

EXAMPLE

**6.3 Estimating the mean SATM score for seniors in California.** Suppose you want to estimate the mean SATM score for the more than 420,000 high school seniors in California. You know better than to trust data from the students who choose to take the SAT. Only about 45% of California students take the SAT. These self-selected students are planning to attend college and are not representative of all California seniors. At considerable effort and expense, you give the test to a simple random sample (SRS) of 500 California high school seniors. The mean score for your sample is $\bar{x} = 461$. What can you say about the mean score $\mu$ in the population of all 420,000 seniors?

The sample mean $\bar{x}$ is the natural estimator of the unknown population mean $\mu$. We know that $\bar{x}$ is an unbiased estimator of $\mu$. More important, the law of large numbers says that the sample mean must approach the population mean as the size of the sample grows. The value $\bar{x} = 461$ therefore appears to be a reasonable estimate of the mean score $\mu$ that all 420,000 students would achieve if they took the test. But how reliable is this estimate? A second sample would surely not give 461 again. Unbiasedness says only that there is no systematic tendency to underestimate or overestimate the truth. Could we plausibly get a sample mean of 410 or 510 on repeated samples? An estimate without an indication of its variability is of little value.

### Statistical confidence

Just as unbiasedness of an estimator concerns the center of its sampling distribution, questions about variation are answered by looking at the spread. We know that if the entire population of SAT scores has mean $\mu$ and standard deviation $\sigma$, then in repeated samples of size 500 the sample mean $\bar{x}$ follows the $N(\mu, \sigma/\sqrt{500})$ distribution. Let us suppose that we know that the standard deviation $\sigma$ of SATM scores in our California population is $\sigma = 100$. (This is not realistic. We will see in the next chapter how to proceed when $\sigma$ is not known. For now, we are more interested in statistical reasoning than in details of realistic methods.) In repeated sampling the sample mean $\bar{x}$ has a Normal distribution centered at the unknown population mean $\mu$ and having standard deviation

LOOK BACK
**distribution of the sample mean, page 339**

$$\sigma_{\bar{x}} = \frac{100}{\sqrt{500}} = 4.5$$

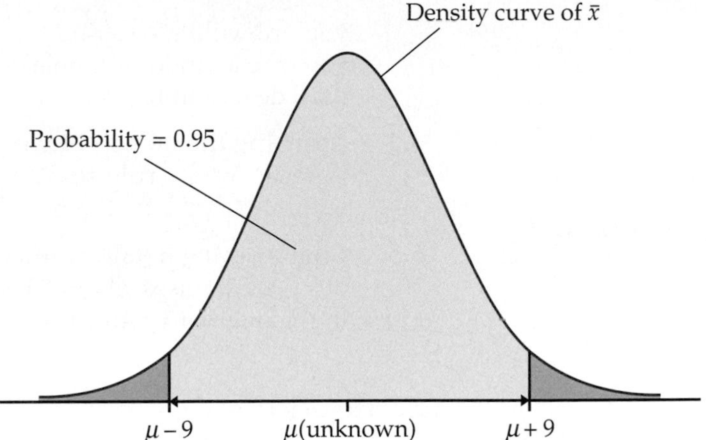


**FIGURE 6.2** $\bar{x}$ lies within $\pm 9$ of $\mu$ in 95% of all samples, so $\mu$ also lies within $\pm 9$ of $\bar{x}$ in those samples.

Now we are in business. Consider this line of thought, which is illustrated by Figure 6.2:

- The 68–95–99.7 rule says that the probability is about 0.95 that $\bar{x}$ will be within 9 points (two standard deviations of $\bar{x}$) of the population mean score $\mu$.
- To say that $\bar{x}$ lies within 9 points of $\mu$ is the same as saying that $\mu$ is within 9 points of $\bar{x}$.
- So 95% of all samples will capture the true $\mu$ in the interval from $\bar{x} - 9$ to $\bar{x} + 9$.

We have simply restated a fact about the sampling distribution of $\bar{x}$. *The language of statistical inference uses this fact about what would happen in the long run to express our confidence in the results of any one sample.* Our sample gave $\bar{x} = 461$. We say that we are *95% confident* that the unknown mean score for all California seniors lies between

$$\bar{x} - 9 = 461 - 9 = 452$$

and

$$\bar{x} + 9 = 461 + 9 = 470$$

Be sure you understand the grounds for our confidence. There are only two possibilities for our SRS:

1. The interval between 452 and 470 contains the true $\mu$.
2. The interval between 452 and 470 does not contain the true $\mu$.

We cannot know whether our sample is one of the 95% for which the interval $\bar{x} \pm 9$ catches $\mu$ or one of the unlucky 5% that does not catch $\mu$. The statement that we are 95% confident is shorthand for saying, "We arrived at these numbers by a method that gives correct results 95% of the time."

## USE YOUR KNOWLEDGE

**6.1** **How much do you spend on lunch?** The average amount you spend on a lunch during the week is not known. Based on past experience, you are willing to assume that the standard deviation is about \$2. If you take a random sample of 36 lunches, what is the value of the standard deviation for $\bar{x}$?

**6.2 Applying the 68–95–99.7 rule.** In the setting of the previous exercise, the 68–95–99.7 rule says that the probability is about 0.95 that $\bar{x}$ is within \$_______ of the population mean $\mu$. Fill in the blank.

**6.3 Constructing a 95% confidence interval.** In the setting of the previous two exercises, about 95% of all samples will capture the true mean in the interval $\bar{x}$ plus or minus \$_______. Fill in the blank.

## Confidence intervals

The interval of numbers between the values $\bar{x} \pm 9$ is called a *95% confidence interval* for $\mu$. Like most confidence intervals we will discuss, this one has the form

$$\text{estimate} \pm \text{margin of error}$$

margin of error

The estimate ($\bar{x}$ in this case) is our guess for the value of the unknown parameter. The **margin of error** (9 here) reflects how accurate we believe our guess is, based on the variability of the estimate, and how confident we are that the procedure will catch the true population mean $\mu$.

Figure 6.3 illustrates the behavior of 95% confidence intervals in repeated sampling. The center of each interval is at $\bar{x}$ and therefore varies from sample to sample. The sampling distribution of $\bar{x}$ appears at the top of the figure to show

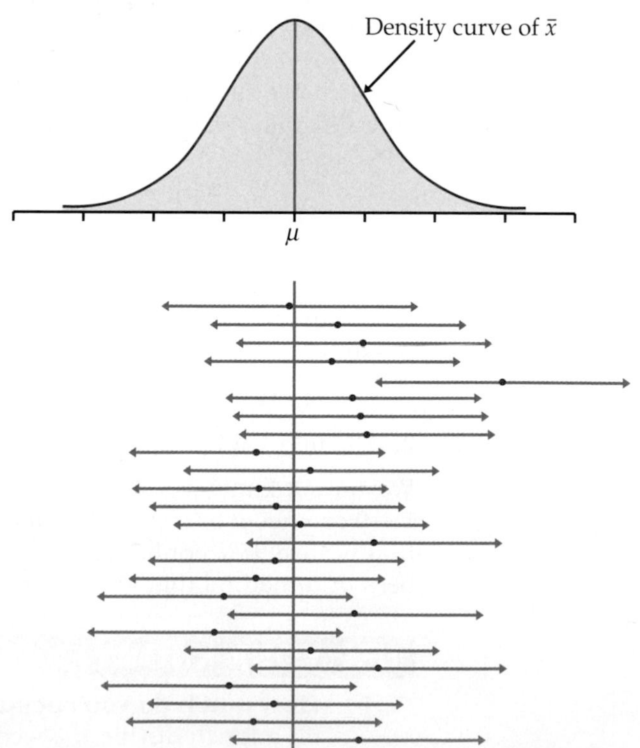


**FIGURE 6.3** Twenty-five samples from the same population gave these 95% confidence intervals. In the long run, 95% of all samples give an interval that covers $\mu$.

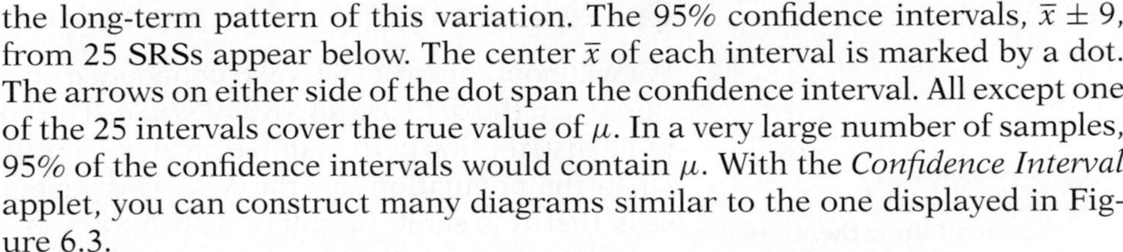

the long-term pattern of this variation. The 95% confidence intervals, $\bar{x} \pm 9$, from 25 SRSs appear below. The center $\bar{x}$ of each interval is marked by a dot. The arrows on either side of the dot span the confidence interval. All except one of the 25 intervals cover the true value of $\mu$. In a very large number of samples, 95% of the confidence intervals would contain $\mu$. With the *Confidence Interval* applet, you can construct many diagrams similar to the one displayed in Figure 6.3.

Statisticians have constructed confidence intervals for many different parameters based on a variety of designs for data collection. We will meet a number of these in later chapters. You need to know two important things about a confidence interval:

1. It is an interval of the form $(a, b)$, where $a$ and $b$ are numbers computed from the data.

2. It has a property called a confidence level that gives the probability of producing an interval that contains the unknown parameter.

Users can choose the confidence level, but 95% is the standard for most situations. Occasionally, 90% or 99% is used. We will use $C$ to stand for the confidence level in decimal form. For example, a 95% confidence level corresponds to $C = 0.95$.

### CONFIDENCE INTERVAL

A level $C$ **confidence interval** for a parameter is an interval computed from sample data by a method that has probability $C$ of producing an interval containing the true value of the parameter.

## USE YOUR KNOWLEDGE

**6.4** **80% confidence intervals.** The idea of an 80% confidence interval is that the interval captures the true parameter value in 80% of all samples. That's not high enough confidence for practical use, but 80% hits and 20% misses make it easy to see how a confidence interval behaves in repeated samples from the same population.

(a) Set the confidence level in the *Confidence Interval* applet to 80%. Click "Sample" to choose an SRS and calculate the confidence interval. Do this 10 times. How many of the 10 intervals captured the true mean $\mu$? How many missed?

(b) You see that we can't predict whether the next sample will capture $\mu$ or miss. The confidence level, however, tells us what percent will capture $\mu$ in the long run. Reset the applet and click "Sample 50" to get the confidence intervals from 50 SRSs. How many hit? Keep clicking "Sample 50" and record the percent of hits among 100, 200, 300, 400, 500, 600, 700, 800, and 1000 SRSs. As the number of samples increases, we expect the percent of captures to get closer to the confidence level, 80%.

## Confidence interval for a population mean

We will now construct a level $C$ confidence interval for the mean $\mu$ of a population when the data are an SRS of size $n$. The construction is based on the sampling distribution of the sample mean $\bar{x}$. This distribution is exactly $N(\mu, \sigma/\sqrt{n})$ when the population has the $N(\mu, \sigma)$ distribution. The central limit theorem says that this same sampling distribution is approximately correct for large samples whenever the population mean and standard deviation are $\mu$ and $\sigma$.

LOOK BACK
**central limit theorem, page 339**

Our construction of a 95% confidence interval for the mean SATM score began by noting that any Normal distribution has probability about 0.95 within ±2 standard deviations of its mean. To construct a level $C$ confidence interval we first catch the central $C$ area under a Normal curve. That is, we must find the number $z^*$ such that any Normal distribution has probability $C$ within $\pm z^*$ standard deviations of its mean. Because all Normal distributions have the same standardized form, we can obtain everything we need from the standard Normal curve. Figure 6.4 shows how $C$ and $z^*$ are related. Values of $z^*$ for many choices of $C$ appear in the row labeled $z^*$ at the bottom of Table D at the back of the book. Here are the most important entries from that row:

| $z^*$ | 1.645 | 1.960 | 2.576 |
|---|---|---|---|
| $C$ | 90% | 95% | 99% |

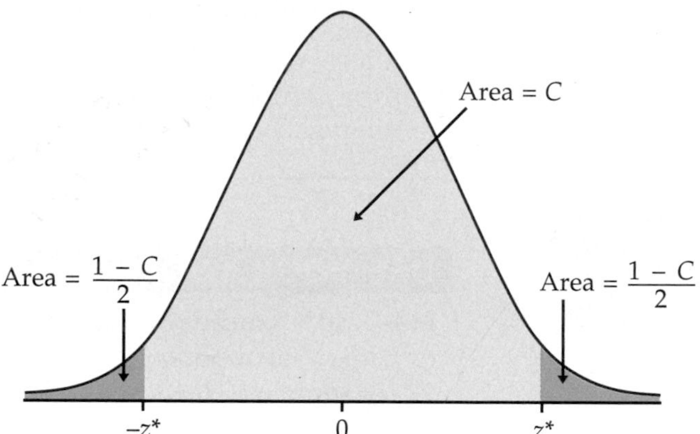


**FIGURE 6.4** The area between $-z^*$ and $z^*$ under the standard normal curve is $C$.

As Figure 6.4 reminds us, any Normal curve has probability $C$ between the point $z^*$ standard deviations below the mean and the point $z^*$ standard deviations above the mean. The sample mean $\bar{x}$ has the Normal distribution with mean $\mu$ and standard deviation $\sigma/\sqrt{n}$. So there is probability $C$ that $\bar{x}$ lies between

$$\mu - z^*\frac{\sigma}{\sqrt{n}} \quad \text{and} \quad \mu + z^*\frac{\sigma}{\sqrt{n}}$$

This is exactly the same as saying that the unknown population mean $\mu$ lies between

$$\bar{x} - z^*\frac{\sigma}{\sqrt{n}} \quad \text{and} \quad \bar{x} + z^*\frac{\sigma}{\sqrt{n}}$$

That is, there is probability $C$ that the interval $\bar{x} \pm z^*\sigma/\sqrt{n}$ contains $\mu$. That is our confidence interval. The estimate of the unknown $\mu$ is $\bar{x}$, and the margin of error is $z^*\sigma/\sqrt{n}$.

### CONFIDENCE INTERVAL FOR A POPULATION MEAN

Choose an SRS of size $n$ from a population having unknown mean $\mu$ and known standard deviation $\sigma$. The **margin of error** for a level $C$ confidence interval for $\mu$ is

$$m = z^* \frac{\sigma}{\sqrt{n}}$$

Here $z^*$ is the value on the standard Normal curve with area $C$ between the critical points $-z^*$ and $z^*$. The level $C$ **confidence interval** for $\mu$ is

$$\bar{x} \pm m$$

This interval is exact when the population distribution is Normal and is approximately correct when $n$ is large in other cases.

EXAMPLE

**6.4 Average debt of undergraduate borrowers.** The National Student Loan Survey collects data to examine questions related to the amount of money that borrowers owe. The survey selected a sample of 1280 borrowers who began repayment of their loans between four and six months prior to the study.[2] The mean of the debt for undergraduate study was \$18,900 and the standard deviation was about \$49,000. This distribution is clearly skewed but because our sample size is quite large, we can rely on the central limit theorem to assure us that the confidence interval based on the Normal distribution will be a good approximation. Let's compute a 95% confidence interval for the true mean debt for all borrowers. Although the standard deviation is estimated from the data collected, we will treat it as a known quantity for our calculations here.

For 95% confidence, we see from Table D that $z^* = 1.960$. The margin of error for the 95% confidence interval for $\mu$ is therefore

$$\begin{aligned} m &= z^* \frac{\sigma}{\sqrt{n}} \\ &= 1.960 \frac{49{,}000}{\sqrt{1280}} \\ &= 2684 \end{aligned}$$

We have computed the margin of error with more digits than we really need. Our mean is rounded to the nearest \$100, so we will do the same for the margin of error. Keeping additional digits would provide no additional useful information. Therefore, we will use $m = 2700$. The 95% confidence interval is

$$\begin{aligned} \bar{x} \pm m &= 18{,}900 \pm 2700 \\ &= (16{,}200,\ 21{,}600) \end{aligned}$$

We are 95% confident that the mean debt for all borrowers is between \$16,200 and \$21,600.

Suppose the researchers who designed the National Student Loan Survey had used a different sample size. How would this affect the confidence interval? We can answer this question by changing the sample size in our calculations and assuming that the mean and standard deviation are the same.

EXAMPLE

**6.5 How sample size affects the confidence interval.** Let's assume that the sample mean of the debt for undergraduate study is \$18,900 and the standard deviation is about \$49,000, as in Example 6.4. But suppose that the sample size is only 320. The margin of error for 95% confidence is

$$\begin{aligned} m &= z^* \frac{\sigma}{\sqrt{n}} \\ &= 1.960 \frac{49{,}000}{\sqrt{320}} \\ &= 5400 \end{aligned}$$

and the 95% confidence interval is

$$\begin{aligned} \bar{x} \pm m &= 18{,}900 \pm 5400 \\ &= (13{,}500,\ 24{,}300) \end{aligned}$$

Notice that the margin of error for this example is twice as large as the margin of error that we computed in Example 6.4. The only change that we made was to assume that the sample size is 320 rather than 1280. This sample size is exactly one-fourth of the original 1280. Thus, we double the margin of error when we reduce the sample size to one-fourth of the original value. Figure 6.5 illustrates the effect in terms of the intervals.

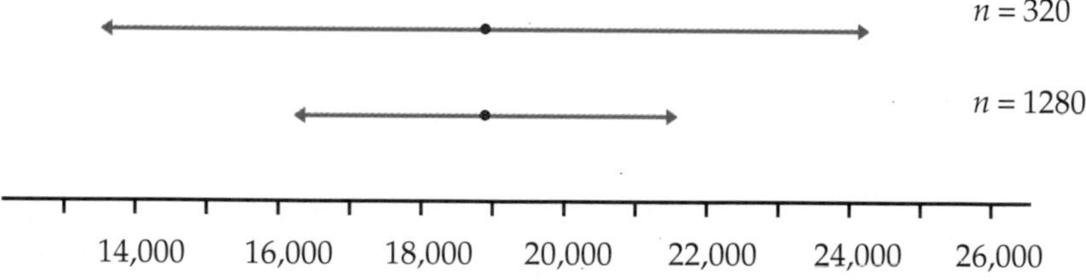


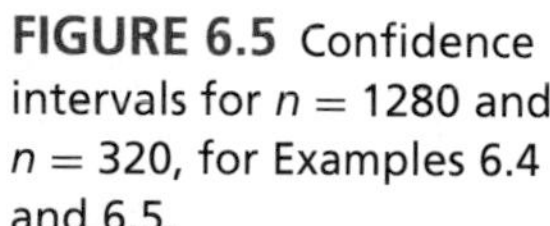

**FIGURE 6.5** Confidence intervals for $n = 1280$ and $n = 320$, for Examples 6.4 and 6.5.

The argument leading to the form of confidence intervals for the population mean $\mu$ rested on the fact that the statistic $\bar{x}$ used to estimate $\mu$ has a Normal distribution. Because many sample estimates have Normal distributions (at least approximately), it is useful to notice that the confidence interval has the form

$$\text{estimate} \pm z^* \sigma_{\text{estimate}}$$

The estimate based on the sample is the center of the confidence interval. The margin of error is $z^*\sigma_{\text{estimate}}$. The desired confidence level determines $z^*$ from Table D. The standard deviation of the estimate is found from a knowledge of the sampling distribution in a particular case. When the estimate is $\bar{x}$ from an SRS, the standard deviation of the estimate is $\sigma_{\text{estimate}} = \sigma/\sqrt{n}$.

## USE YOUR KNOWLEDGE

**6.5** **College freshmen anxiety level.** An SRS of 100 incoming freshmen was taken to look at their college anxiety level. The mean score of the sample was 83.5 (out of 100). Assuming a standard deviation of 4, give the 95% confidence interval for $\mu$, the average anxiety level among all freshmen.

**6.6** **Changing the confidence level.** In the setting of the previous exercise, would the margin of error for 99% confidence be larger or smaller? Verify your answer by performing the calculations.

## How confidence intervals behave

The margin of error $z^*\sigma/\sqrt{n}$ for the mean of a Normal population illustrates several important properties that are shared by all confidence intervals in common use. The user chooses the confidence level, and the margin of error follows from this choice. High confidence is desirable and so is a small margin of error. High confidence says that our method almost always gives correct answers. A small margin of error says that we have pinned down the parameter quite precisely.

Suppose that you calculate a margin of error and decide that it is too large. Here are your choices to reduce it:

- Use a lower level of confidence (smaller $C$).
- Increase the sample size (larger $n$).
- Reduce $\sigma$.

For most problems you would choose a confidence level of 90%, 95%, or 99%. So $z^*$ will be 1.645, 1.960, or 2.576. Figure 6.4 shows that $z^*$ will be smaller for lower confidence (smaller $C$). The bottom row of Table D also shows this. If $n$ and $\sigma$ are unchanged, a smaller $z^*$ leads to a smaller margin of error. Similarly, increasing the sample size $n$ reduces the margin of error for any fixed confidence level. The square root in the formula implies that we must multiply the number of observations by 4 in order to cut the margin of error in half. The standard deviation $\sigma$ measures the variation in the population. You can think of the variation among individuals in the population as noise that obscures the average value $\mu$. It is harder to pin down the mean $\mu$ of a highly variable population; that is why the margin of error of a confidence interval increases with $\sigma$. In practice, we can sometimes reduce $\sigma$ by carefully controlling the measurement process or by restricting our attention to only part of a large population.

EXAMPLE

**6.6 How confidence level affects the confidence interval.** Suppose that for the student loan data in Example 6.4, we wanted 99% confidence. Table D tells us that for 99% confidence, $z^* = 2.576$. The margin of error for 99% confidence based on 1280 observations is

$$\begin{aligned} m &= z^* \frac{\sigma}{\sqrt{n}} \\ &= 2.576 \frac{49{,}000}{\sqrt{1280}} \\ &= 3500 \end{aligned}$$

and the 99% confidence interval is

$$\begin{aligned} \bar{x} \pm m &= 18{,}900 \pm 3500 \\ &= (15{,}400,\ 22{,}400) \end{aligned}$$

Requiring 99%, rather than 95%, confidence has increased the margin of error from 2700 to 3500. Figure 6.6 compares the two intervals.

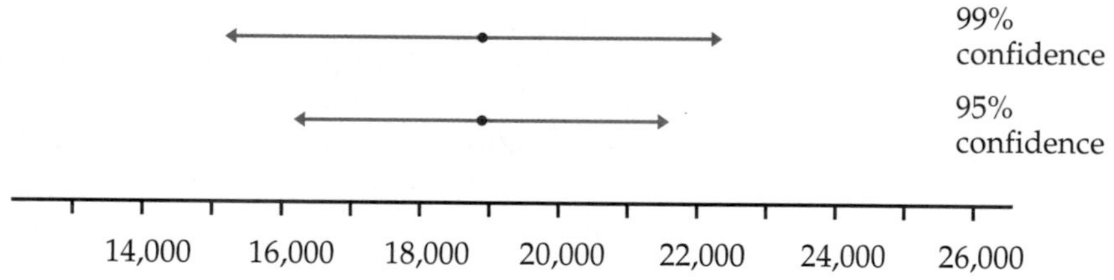


**FIGURE 6.6** Confidence intervals for Examples 6.4 and 6.6.

## Choosing the sample size

A wise user of statistics never plans data collection without at the same time planning the inference. You can arrange to have both high confidence and a small margin of error. The margin of error of the confidence interval for a population mean is

$$m = z^* \frac{\sigma}{\sqrt{n}}$$

To obtain a desired margin of error $m$, plug in the value of $\sigma$ and the value of $z^*$ for your desired confidence level, and solve for the sample size $n$. Here is the result.

### SAMPLE SIZE FOR DESIRED MARGIN OF ERROR

The confidence interval for a population mean will have a specified margin of error $m$ when the sample size is

$$n = \left( \frac{z^* \sigma}{m} \right)^2$$

This formula does not account for collection costs. In practice, taking observations costs time and money. The required sample size may be impossibly expensive. Do notice once again that it is the size of the *sample* that determines the margin of error. The size of the *population* (as long as the population is much larger than the sample) does not influence the sample size we need.

EXAMPLE

**6.7 How many students should we survey?** Suppose that we are planning a student loan survey similar to the one described in Example 6.4. If we want the margin of error to be \$2000 with 95% confidence, what sample size $n$ do we need? For 95% confidence, Table D gives $z^* = 1.960$. For $\sigma$ we will use the value from the previous study, \$49,000. If the margin of error is \$2000, we have

$$n = \left(\frac{z^*\sigma}{m}\right)^2 = \left(\frac{1.96 \times 49{,}000}{2000}\right)^2 = 2305.9$$

Because 2305 measurements will give a slightly wider interval than desired and 2306 measurements a slightly narrower interval, we could choose $n = 2306$. We need loan information from 2306 borrowers to determine an estimate of mean debt with the desired margin of error.

It is always safe to round *up* to the next higher whole number when finding $n$ because this will give us a smaller margin of error. The purpose of this calculation is to determine a sample size that is sufficient to provide useful results, but the determination of what is useful is a matter of judgment. Would we need a much larger sample size to obtain a margin of error of \$1500? Here is the calculation:

$$n = \left(\frac{z^*\sigma}{m}\right)^2 = \left(\frac{1.96 \times 49{,}000}{1500}\right)^2 = 4099.4$$

A sample of $n = 4100$ is much larger, and the costs of such a large sample may be prohibitive.

Unfortunately, the actual number of usable observations is often less than what we plan at the beginning of a study. This is particularly true of data collected in surveys but is an important consideration in most studies. Careful study designers often assume a nonresponse rate or dropout rate that specifies what proportion of the originally planned sample will fail to provide data. We use this information to calculate the sample size to be used at the start of the study. For example, if in the survey above, we expect only 25% of those contacted to respond, we would need to start with a sample size of $4 \times 2306 = 9224$ to obtain usable information from 2306 borrowers.

## USE YOUR KNOWLEDGE

**6.7** **Starting salaries.** You are planning a survey of starting salaries for recent marketing majors. In 2005, the average starting salary was reported to be \$37,832.[3] Assuming the standard deviation for this study is \$10,500, what sample size do you need to have a margin of error equal to \$900 with 95% confidence?

**6.8 Changes in sample size.** Suppose that in the setting of the previous exercise you have the resources to contact 1000 recent graduates. If all respond, will your margin of error be larger or smaller than \$900? What if only 45% respond? Verify your answers by performing the calculations.

## Some cautions

We have already seen that small margins of error and high confidence can require large numbers of observations. You should also be keenly aware that *any formula for inference is correct only in specific circumstances*. If the government required statistical procedures to carry warning labels like those on drugs, most inference methods would have long labels indeed. Our handy formula $\bar{x} \pm z^*\sigma/\sqrt{n}$ for estimating a population mean comes with the following list of warnings for the user:

- The data should be an SRS from the population. We are completely safe if we actually did a randomization and drew an SRS. We are not in great danger if the data can plausibly be thought of as independent observations from a population. That is the case in Examples 6.4 to 6.7, where we redefine our population to correspond to survey respondents.

- The formula is not correct for probability sampling designs more complex than an SRS. Correct methods for other designs are available. We will not discuss confidence intervals based on multistage or stratified samples. If you plan such samples, be sure that you (or your statistical consultant) know how to carry out the inference you desire.

- There is no correct method for inference from data haphazardly collected with bias of unknown size. Fancy formulas cannot rescue badly produced data.

- Because $\bar{x}$ is not resistant, outliers can have a large effect on the confidence interval. You should search for outliers and try to correct them or justify their removal before computing the interval. If the outliers cannot be removed, ask your statistical consultant about procedures that are not sensitive to outliers.

- If the sample size is small and the population is not Normal, the true confidence level will be different from the value $C$ used in computing the interval. Examine your data carefully for skewness and other signs of non-Normality. The interval relies only on the distribution of $\bar{x}$, which even for quite small sample sizes is much closer to Normal than that of the individual observations. When $n \geq 15$, the confidence level is not greatly disturbed by non-Normal populations unless extreme outliers or quite strong skewness are present. Our debt data in Example 6.4 are clearly skewed, but because of the large sample size, we are confident that the sample mean will be approximately Normal. We will discuss this issue in more detail in the next chapter.

- This interval $\bar{x} \pm z^*\sigma/\sqrt{n}$ assumes that the standard deviation $\sigma$ of the population is known. This unrealistic requirement renders the interval of little use

**standard deviation, page 40**

in statistical practice. We will learn in the next chapter what to do when $\sigma$ is unknown. If, however, the sample is large, the sample standard deviation $s$ will be close to the unknown $\sigma$. The interval $\bar{x} \pm z^* s/\sqrt{n}$ is then an approximate confidence interval for $\mu$.

The most important caution concerning confidence intervals is a consequence of the first of these warnings. *The margin of error in a confidence interval covers only random sampling errors.* The margin of error is obtained from the sampling distribution and indicates how much error can be expected because of chance variation in randomized data production. *Practical difficulties such as undercoverage and nonresponse in a sample survey cause additional errors. These errors can be larger than the random sampling error.* This often happens when the sample size is large (so that $\sigma/\sqrt{n}$ is small). Remember this unpleasant fact when reading the results of an opinion poll or other sample survey. The practical conduct of the survey influences the trustworthiness of its results in ways that are not included in the announced margin of error.

Every inference procedure that we will meet has its own list of warnings. Because many of the warnings are similar to those above, we will not print the full warning label each time. It is easy to state (from the mathematics of probability) conditions under which a method of inference is exactly correct. These conditions are *never* fully met in practice. For example, no population is exactly Normal. Deciding when a statistical procedure should be used in practice often requires judgment assisted by exploratory analysis of the data. Mathematical facts are therefore only a part of statistics. The difference between statistics and mathematics can be stated thus: mathematical theorems are true; statistical methods are often effective when used with skill.

Finally, you should understand what statistical confidence does not say. Based on our SRS in Example 6.3, we are 95% confident that the mean SATM score for the California students lies between 452 and 470. This says that this interval was calculated by a method that gives correct results in 95% of all possible samples. It does *not* say that the probability is 95% that the true mean falls between 452 and 470. No randomness remains after we draw a particular sample and compute the interval. The true mean either is or is not between 452 and 470. The probability calculations of standard statistical inference describe how often the *method*, not a particular sample, gives correct answers.

## USE YOUR KNOWLEDGE

**6.9** **Internet users.** A survey of users of the Internet found that males outnumbered females by nearly 2 to 1. This was a surprise, because earlier surveys had put the ratio of men to women closer to 9 to 1. Later in the article we find this information:

> *Detailed surveys were sent to more than 13,000 organizations on the Internet; 1,468 usable responses were received. According to Mr. Quarterman, the margin of error is 2.8 percent, with a confidence level of 95 percent.*[4]

Do you think that the small margin of error is a good measure of the accuracy of the survey's results? Explain your answer.

## BEYOND THE BASICS

### The Bootstrap

Confidence intervals are based on sampling distributions. In this section we have used the fact that the sampling distribution of $\bar{x}$ is $N(\mu, \sigma/\sqrt{n})$ when the data are an SRS from an $N(\mu, \sigma)$ population. If the data are not Normal, the central limit theorem tells us that this sampling distribution is still a reasonable approximation as long as the distribution of the data is not strongly skewed and there are no outliers. Even a fair amount of skewness can be tolerated when the sample size is large.

bootstrap

What if the population does not appear to be Normal and we have only a small sample? Then we do not know what the sampling distribution of $\bar{x}$ looks like. The **bootstrap** is a procedure for approximating sampling distributions when theory cannot tell us their shape.[5]

resample

The basic idea is to act as if our sample were the population. We take many samples from it. Each of these is called a **resample.** We calculate the mean $\bar{x}$ for each resample. We get different results from different resamples because we sample *with replacement.* An individual observation in the original sample can appear more than once in the resample.

For example, suppose we have four measurements of a student's daily time online last month (in minutes):

190.5 109.0 95.5 137.0

one resample could be

109.0 95.5 137.0 109.0

with $\bar{x} = 112.625$. Collect the $\bar{x}$'s from 1000 such resamples. Their distribution will be close to what we would get if we took 1000 samples from the entire population. We treat the distribution of $\bar{x}$'s from our 1000 resamples as if it were the sampling distribution. If we want a 95% confidence interval, for example, we can use the middle 95% of this distribution.

The bootstrap is practical only when you can use a computer to take 1000 or more samples quickly. It is an example of how the use of fast and easy computing is changing the way we do statistics. More details about the bootstrap can be found in Chapter 16.

## SECTION 6.1 Summary

The purpose of a **confidence interval** is to estimate an unknown parameter with an indication of how accurate the estimate is and of how confident we are that the result is correct.

Any confidence interval has two parts: an interval computed from the data and a confidence level. The interval often has the form

$$\text{estimate} \pm \text{margin of error}$$

The **confidence level** states the probability that the method will give a correct answer. That is, if you use 95% confidence intervals, in the long run 95% of your intervals will contain the true parameter value. When you apply the method once, you do not know if your interval gave a correct value (this happens 95% of the time) or not (this happens 5% of the time).

The **margin of error** for a level $C$ confidence interval for the mean $\mu$ of a Normal population with known standard deviation $\sigma$, based on an SRS of size $n$, is given by

$$m = z^* \frac{\sigma}{\sqrt{n}}$$

Here $z^*$ is obtained from the row labeled $z^*$ at the bottom of Table D. The probability is $C$ that a standard Normal random variable takes a value between $-z^*$ and $z^*$. The confidence interval is

$$\bar{x} \pm m$$

Other things being equal, the margin of error of a confidence interval decreases as

- the confidence level $C$ decreases,
- the sample size $n$ increases, and
- the population standard deviation $\sigma$ decreases.

The sample size $n$ required to obtain a confidence interval of specified margin of error $m$ for a Normal mean is

$$n = \left(\frac{z^*\sigma}{m}\right)^2$$

where $z^*$ is the critical point for the desired level of confidence.

A specific confidence interval recipe is correct only under specific conditions. The most important conditions concern the method used to produce the data. Other factors such as the form of the population distribution may also be important.

## SECTION 6.1 Exercises

*For Exercises 6.1 to 6.3, see pages 357 and 358; for Exercise 6.4, see page 359; for Exercises 6.5 and 6.6, see page 363; for Exercises 6.7 and 6.8, see pages 365 and 366; and for Exercise 6.9, see page 367.*

**6.10 Margin of error and the confidence interval.** A study based on a sample of size 25 reported a mean of 93 with a margin of error of 11 for 95% confidence.

(a) Give the 95% confidence interval.

(b) If you wanted 99% confidence for the same study, would your margin of error be greater than, equal to, or less than 11? Explain your answer.

**6.11 Changing the sample size.** Suppose that the sample mean is 50 and the standard deviation is assumed to be 5. Make a diagram similar to Figure 6.5 (page 362) that illustrates the effect of sample size on the width of a 95% interval. Use the following sample sizes: 10, 20, 40, and 80. Summarize what the diagram shows.

**6.12 Changing the confidence level.** A study with 36 observations had a mean of 70. Assume that the standard deviation is 12. Make a diagram similar to Figure 6.6 (page 364) that illustrates the effect of the confidence level on the width of the interval. Use 80%, 90%, 95%, and 99%. Summarize what the diagram shows.

**6.13 Importance of recreational sports.** The National Intramural-Recreational Sports Association (NIRSA) performed a study to look at the value of recreational sports on college campuses.[6] A total of 2673 students were asked to indicate how important (on a 10-point scale) each of 21 factors was in terms of their college satisfaction and success. The factor "recreational sports and activities" resulted in a mean score of 7.5. Assuming a standard deviation of 4.1, give the margin of error and find the 95% confidence interval for this sample.

**6.14 More on the importance of recreational sports.** Refer to Exercise 6.13. Repeat the calculations for a 99% confidence interval. How do the results compare with those in Exercise 6.13?

**6.15 Importance of quality professors.** Refer to Exercise 6.13. In this same sample, the factor "quality of professors and ability to interact with them" resulted in a mean score of 8.7. Assuming a standard deviation of 3.5, find the 95% confidence interval.

**6.16 Inference based on integer values.** Refer to Exercise 6.13. The data for this study are integer values between 1 and 10. Explain why the confidence interval based on the Normal distribution will be a good approximation.

**6.17 Mean serum TRAP in young women.** For many important processes that occur in the body, direct measurement of characteristics of the process is not possible. In many cases, however, we can measure a *biomarker*, a biochemical substance that is relatively easy to measure and is associated with the process of interest. Bone turnover is the net effect of two processes: the breaking down of old bone, called resorption, and the building of new bone, called formation. One biochemical measure of bone resorption is tartrate resistant acid phosphatase (TRAP), which can be measured in blood. In a study of bone turnover in young women, serum TRAP was measured in 31 subjects.[7] The units are units per liter (U/l). The mean was 13.2 U/l. Assume that the standard deviation is known to be 6.5 U/l. Give the margin of error and find a 95% confidence interval for the mean for young women represented by this sample.

**6.18 Mean OC in young women.** Refer to the previous exercise. A biomarker for bone formation measured in the same study was osteocalcin (OC), measured in the blood. The units are nanograms per milliliter (ng/ml). For the 31 subjects in the study the mean was 33.4 ng/ml. Assume that the standard deviation is known to be 19.6 ng/ml. Report the 95% confidence interval.

**6.19 Populations sampled and margins of error.** Consider the following two scenarios. (A) Take a simple random sample of 100 sophomore students at your college or university. (B) Take a simple random sample of 100 sophomore students in your major at your college or university. For each of these samples you will record the amount spent on textbooks used for classes during the fall semester. Which sample should have the smaller margin of error? Explain your answer.

**6.20 Apartment rental rates.** You want to rent an unfurnished one-bedroom apartment in Boston next year. The mean monthly rent for a random sample of 10 apartments advertised in the local newspaper is $1400. Assume that the standard deviation is $220. Find a 95% confidence interval for the mean monthly rent for unfurnished one-bedroom apartments available for rent in this community.

**6.21 More on apartment rental rates.** Refer to the previous exercise. Will the 95% confidence interval include approximately 95% of the rents of all unfurnished one-bedroom apartments in this area? Explain why or why not.

**6.22** CHALLENGE **Inference based on skewed data.** The mean OC for the 31 subjects in Exercise 6.18 was 33.4 ng/ml. In our calculations, we assumed that the standard deviation was known to be 19.6 ng/ml. Use the 68–95–99.7 rule from Chapter 1 (page 59) to find the approximate bounds on the values of OC that would include these percents of the population. If the assumed standard deviation is correct, it would appear that this distribution may be highly skewed. Why? (*Hint:* The measured values for a variable such as this are all positive.) Do you think that this skewness will invalidate the use of the Normal confidence interval in this case? Explain your answer.

**6.23 Average hours per week on the Internet.** The *Student Monitor* surveys 1200 undergraduates from 100 colleges semiannually to understand trends among college students.[8] Recently, the *Student Monitor* reported that the average amount of time spent per week on the Internet was 15.1 hours. Assume that the standard deviation is 5 hours.

(a) Give a 95% confidence interval for the mean time spent per week on the Internet.

(b) Is it true that 95% of the students surveyed reported weekly times that lie in the interval you found in part (a)? Explain your answer.

**6.24 Average minutes per week on the Internet.** Refer to the previous exercise.

(a) Give the mean and standard deviation in minutes.

(b) Calculate the 95% confidence interval in minutes from your answer to part (a).

(c) Explain how you could have directly calculated this interval from the 95% interval that you calculated in the previous exercise.

**6.25 Calories consumed by women in the U.S.** The mean number of calories consumed by women in the United States who are 19 to 30 years of age is $\mu = 1791$ calories per day. The standard deviation is 31 calories.[9] You will study a sample of 200 women in this age range, and one of the variables you will collect is calories consumed per day.

(a) What is the standard deviation of the sample mean $\bar{x}$?

(b) The 68–95–99.7 rule says that the probability is about 0.95 that $\bar{x}$ is within ______ calories of the population mean $\mu$. Fill in the blank.

(c) About 95% of all samples will capture the true mean of calories consumed per day in the interval $\bar{x}$ plus or minus ______ calories. Fill in the blank.

**6.26 Fuel efficiency.** Computers in some vehicles calculate various quantities related to performance. One of these is the fuel efficiency, or gas mileage, usually expressed as miles per gallon (mpg). For one vehicle equipped in this way, the mpg were recorded each time the gas tank was filled, and the computer was then reset.[10] Here are the mpg values for a random sample of 20 of these records:

| | | | | | | | | | |
|---|---|---|---|---|---|---|---|---|---|
| 41.5 | 50.7 | 36.6 | 37.3 | 34.2 | 45.0 | 48.0 | 43.2 | 47.7 | 42.2 |
| 43.2 | 44.6 | 48.4 | 46.4 | 46.8 | 39.2 | 37.3 | 43.5 | 44.3 | 43.3 |

Suppose that the standard deviation is known to be $\sigma = 3.5$ mpg.

(a) What is $\sigma_{\bar{x}}$, the standard deviation of $\bar{x}$?

(b) Give a 95% confidence interval for $\mu$, the mean mpg for this vehicle.

**6.27 Fuel efficiency in metric units.** In the previous exercise you found an estimate with a margin of error for the average miles per gallon. Convert your estimate and margin of error to the metric units kilometers per liter (kpl). To change mpg to kpl, use the facts that 1 mile = 1.609 kilometers and 1 gallon = 3.785 liters.

**6.28** APPLET **Percent coverage of 95% confidence interval.** The *Confidence Interval* applet lets you simulate large numbers of confidence intervals quickly. Select 95% confidence and then sample 50 intervals. Record the number of intervals that cover the true value (this appears in the "Hit" box in the applet). Press the reset button and repeat 30 times. Make a stemplot of the results and find the mean. Describe the results. If you repeated this experiment very many times, what would you expect the average number of hits to be?

**6.29 Required sample size for specifed margin of error.** A new bone study is being planned that will measure the biomarker TRAP described in Exercise 6.17. Using the value of $\sigma$ given there, 6.5 U/l, find the sample size required to provide an estimate of the mean TRAP with a margin of error of 2.0 U/l for 95% confidence.

**6.30** CHALLENGE **Adjusting required sample size for drop out.** Refer to the previous exercise. In similar previous studies, about 20% of the subjects drop out before the study is completed. Adjust your sample size requirement to have enough subjects at the end of the study to meet the margin of error criterion.

**6.31 Sample size needed for apartment rental rates.** How large a sample of one-bedroom apartments in Exercise 6.20 would be needed to estimate the mean $\mu$ within $\pm\$50$ with 90% confidence?

**6.32 Accuracy of a laboratory scale.** To assess the accuracy of a laboratory scale, a standard weight known to weigh 10 grams is weighed repeatedly. The scale readings are Normally distributed with unknown mean (this mean is 10 grams if the scale has no bias). The standard deviation of the scale readings is known to be 0.0002 gram.

(a) The weight is measured five times. The mean result is 10.0023 grams. Give a 98% confidence interval for the mean of repeated measurements of the weight.

(b) How many measurements must be averaged to get a margin of error of $\pm 0.0001$ with 98% confidence?

**6.33** CHALLENGE **More than one confidence interval.** As we prepare to take a sample and compute a 95% confidence interval, we know that the probability that the interval we compute will cover the parameter is 0.95. That's the meaning of 95% confidence. If we use several such intervals, however, our confidence that *all* of them give correct results is less than 95%. Suppose we take independent samples each month for five months and report a 95% confidence interval for each set of data.

(a) What is the probability that all five intervals cover the true means? This probability (expressed as a percent) is our overall confidence level for the five simultaneous statements.

(b) What is the probability that at least four of the five intervals cover the true means?

**6.34 Telemarketing wages.** An advertisement in the student newspaper asks you to consider working for a telemarketing company. The ad states, "Earn between $500 and $1000 per week." Do you think that the ad is describing a confidence interval? Explain your answer.

**6.35 Like your job?** A Gallup Poll asked working adults about their job satisfaction. One question was "All in all, which best describes how you feel about your job?" The possible answers were "love job," "like job," "dislike job," and "hate job." Fifty-nine percent of the sample responded that they liked their job. Material provided with the results of the poll noted:

*Results are based on telephone interviews with 1,001 national adults, aged 18 and older, conducted Aug. 8–11, 2005. For results based on the total sample of national adults, one can say with 95% confidence that the maximum margin of sampling error is ±3 percentage points.*[11]

The Gallup Poll uses a complex multistage sample design, but the sample percent has approximately a Normal sampling distribution.

(a) The announced poll result was 59% ± 3%. Can we be certain that the true population percent falls in this interval?

(b) Explain to someone who knows no statistics what the announced result 59% ± 3% means.

(c) This confidence interval has the same form we have met earlier:

$$\text{estimate} \pm z^* \sigma_{\text{estimate}}$$

What is the standard deviation $\sigma_{\text{estimate}}$ of the estimated percent?

(d) Does the announced margin of error include errors due to practical problems such as undercoverage and nonresponse?

## 6.2 Tests of Significance

The confidence interval is appropriate when our goal is to estimate population parameters. The second common type of inference is directed at a quite different goal: to assess the evidence provided by the data in favor of some claim about the population parameters.

### The reasoning of significance tests

A significance test is a formal procedure for comparing observed data with a hypothesis whose truth we want to assess. The hypothesis is a statement about the population parameters. The results of a test are expressed in terms of a probability that measures how well the data and the hypothesis agree. We use the following examples to illustrate these concepts.

EXAMPLE

**6.8 Debt levels of private and public college borrowers.** One purpose of the National Student Loan Survey described in Example 6.4 (page 361) is to compare the debt of different subgroups of students. For example, the 525 borrowers who last attended a private four-year college had a mean debt of $21,200, while those who last attended a public four-year college had a mean debt of $17,100. The difference of $4100 is fairly large, but we know that these numbers are estimates of the true means. If we took different samples, we would get different estimates. Can we conclude from these data that the average debt of borrowers who attended a private college is different than the average debt of borrowers who attended a public college?

One way to answer this question is to compute the probability of obtaining a difference as large or larger than the observed \$4100 assuming that, in fact, there is no difference in the true means. This probability is 0.17. Because this probability is not particularly small, we conclude that observing a difference of \$4100 is not very surprising when the true means are equal. The data do not provide evidence for us to conclude that the mean debts for private four-year borrowers and public four-year borrowers are different.

Here is an example with a different conclusion.

EXAMPLE

**6.9 Change in average debt levels between 1997 and 2002.** Another purpose of the National Student Loan Survey is to look for changes over time. For example, in 1997, the survey found that the mean debt for undergraduate study was \$11,400. How does this compare with the value of \$18,900 in the 2002 study? The difference is \$7500. As we learned in the previous example, an observed difference in means is not necessarily sufficient for us to conclude that the true means are different. Do the data provide evidence that there is an increase in borrowing? Again, we answer this question with a probability calculated under the assumption that there is *no difference in the true means*. The probability is 0.00004 of observing an increase in mean debt that is \$7500 or more when there really is no difference. Because this probability is so small, we have sufficient evidence in the data to conclude that there has been a change in borrowing between 1997 and 2002.

What are the key steps in these examples?

- We started each with a question about the difference between two mean debts. In Example 6.8, we compare private four-year borrowers with public four-year borrowers. In Example 6.9, we compare borrowers in 2002 with borrowers in 1997. In both cases, we ask whether or not the data are compatible with no difference, that is, a difference of \$0.
- Next we compared the data, \$4100 in the first case and \$7500 in the second, with the value that comes from the question, \$0.
- The results of the comparisons are probabilities, 0.17 in the first case and 0.00004 in the second.

The 0.17 probability is not particularly small, so we have no evidence to question the possibility that the true difference is zero. In the second case, however, the probability is quite small. Something that happens with probability 0.00004 occurs only about 4 times out of 100,000. In this case we have two possible explanations:

1. we have observed something that is very unusual, or
2. the assumption that underlies the calculation, no difference in mean debt, is not true.

Because this probability is so small, we prefer the second conclusion: there has been a change in the mean debt between 1997 and 2002.

The probabilities in Examples 6.8 and 6.9 are measures of the compatibility of the data (a difference in means of $4100 and $7500) with the *null hypothesis* that there is no difference in the true means. Figures 6.7 and 6.8 compare the two results graphically. For each a Normal curve centered at 0 is the sampling distribution. You can see that we are not particularly surprised to observe the difference $4100 in Figure 6.7, but the difference $7500 in Figure 6.8 is clearly an unusual observation. We will now consider some of the formal aspects of significance testing.

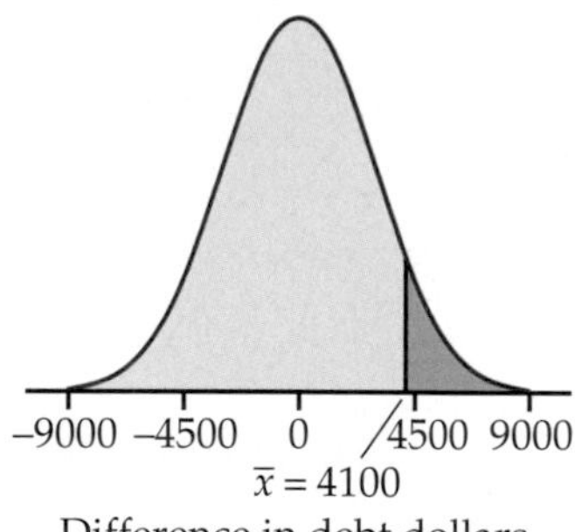


**FIGURE 6.7** Comparison of the sample mean in Example 6.8 relative to the null hypothesized value 0.

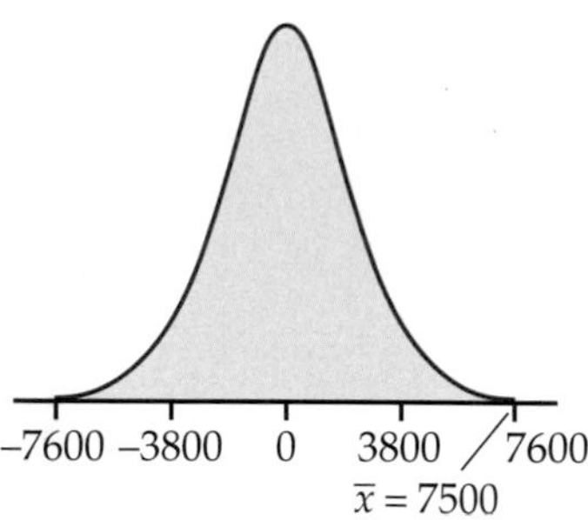


**FIGURE 6.8** Comparison of the sample mean in Example 6.9 relative to the null hypothesized value 0.

## Stating hypotheses

In Examples 6.8 and 6.9, we asked whether the difference in the observed means is reasonable if, in fact, there is no difference in the true means. To answer this, we begin by supposing that the statement following the "if" in the previous sentence is true. In other words, we suppose that the true difference is $0. We then ask whether the data provide evidence against the supposition we have made. If so, we have evidence in favor of an effect (the means are different) we are seeking. The first step in a test of significance is to state a claim that we will try to find evidence *against*.

> **NULL HYPOTHESIS**
>
> The statement being tested in a test of significance is called the **null hypothesis.** The test of significance is designed to assess the strength of the evidence against the null hypothesis. Usually the null hypothesis is a statement of "no effect" or "no difference."

We abbreviate "null hypothesis" as $H_0$. A null hypothesis is a statement about the population parameters. For example, our null hypothesis for Example 6.8 is

$$H_0\text{: there is no difference in the true means}$$

Note that the null hypothesis refers to the *true* means for all borrowers from either a four-year private or public college, including those for whom we do not have data.

alternative hypothesis

It is convenient also to give a name to the statement we hope or suspect is true instead of $H_0$. This is called the **alternative hypothesis** and is abbreviated as $H_a$. In Example 6.8, the alternative hypothesis states that the means are different. We write this as

$$H_a\text{: the true means are not the same}$$

*Hypotheses always refer to some populations or a model, not to a particular outcome. For this reason, we must state $H_0$ and $H_a$ in terms of population parameters.*

Because $H_a$ expresses the effect that we hope to find evidence *for,* we often begin with $H_a$ and then set up $H_0$ as the statement that the hoped-for effect is not present. Stating $H_a$ is often the more difficult task. It is not always clear, in particular, whether $H_a$ should be **one-sided** or **two-sided,** which refers to whether a parameter differs from its null hypothesis value in a specific direction or in either direction.

one-sided or two-sided alternatives

The alternative hypothesis should express the hopes or suspicions we bring to the data. *It is cheating to first look at the data and then frame $H_a$ to fit what the data show.* If you do not have a specific direction firmly in mind in advance, you must use a two-sided alternative. Moreover, some users of statistics argue that we should always use a two-sided alternative.

## USE YOUR KNOWLEDGE

**6.36 Food court survey.** The food court at your dormitory has been redesigned. A survey is planned to determine whether or not students think that the new design is an improvement. Sampled students will respond on a seven-point scale with scores less than 4 favoring the old design and scores greater than 4 favoring the new design (to varying degrees). State the null and alternative hypotheses that provide a framework for examining whether or not the new design is an improvement.

**6.37 DXA scanners.** A dual-energy X-ray absorptiometry (DXA) scanner is used to measure bone mineral density for people who may be at risk for osteoporosis. To ensure its accuracy, the company uses an object called a "phantom" that has known mineral density $\mu = 1.4$ grams per square centimeter. Once installed, the company scans the phantom 10 times and compares the sample mean reading $\bar{x}$ with the theoretical mean $\mu$ using a significance test. State the null and alternative hypotheses for this test.

## Test statistics

We will learn the form of significance tests in a number of common situations. Here are some principles that apply to most tests and that help in understanding these tests:

- The test is based on a statistic that estimates the parameter that appears in the hypotheses. Usually this is the same estimate we would use in a confidence interval for the parameter. When $H_0$ is true, we expect the estimate to take a value near the parameter value specified by $H_0$.
- Values of the estimate far from the parameter value specified by $H_0$ give evidence against $H_0$. The alternative hypothesis determines which directions count against $H_0$.
- To assess how far the estimate is from the parameter, standardize the estimate. In many common situations the test statistic has the form

$$z = \frac{\text{estimate} - \text{hypothesized value}}{\text{standard deviation of the estimate}}$$

Let's return to our student loan example.

EXAMPLE

**6.10 Debt levels of private and public college borrowers: the hypotheses.** In Example 6.8, the hypotheses are stated in terms of the difference in debt between borrowers who attended a private college and those who attended a public college:

$H_0$: there is no difference in the true means

$H_a$: there is a difference in the true means

Because $H_a$ is two-sided, large values of both positive and negative differences count as evidence against the null hypothesis.

test statistic

A **test statistic** measures compatibility between the null hypothesis and the data. We use it for the probability calculation that we need for our test of significance. It is a random variable with a distribution that we know.

EXAMPLE

**6.11 Debt levels of private and public college borrowers: the test statistic.** In Example 6.8, we can state the null hypothesis as $H_0$: the true mean difference is 0. The estimate of the difference is \$4100. Using methods that we will discuss in detail later, we can determine that the standard deviation of the estimate is \$3000. For this problem the test statistic is

$$z = \frac{\text{estimate} - \text{hypothesized value}}{\text{standard deviation of the estimate}}$$

For our data,

$$z = \frac{4100 - 0}{3000} = 1.37$$

We have observed a sample estimate that is about one and a third standard deviations away from the hypothesized value of the parameter. Because the sample sizes are sufficiently large for us to conclude that the distribution of the sample estimate is approximately Normal, the standardized test statistic $z$ will have approximately the $N(0, 1)$ distribution.

LOOK BACK
**Normal distribution, page 58**

We will use facts about the Normal distribution in what follows.

## *P*-values

If all test statistics were Normal, we could base our conclusions on the value of the $z$ test statistic. In fact, the Supreme Court of the United States has said that "two or three standard deviations" ($z = 2$ or 3) is its criterion for rejecting $H_0$ (see Exercise 6.42 on page 381), and this is the criterion used in most applications involving the law. Because not all test statistics are Normal, we translate the value of test statistics into a common language, the language of probability.

A test of significance finds the probability of getting an outcome *as extreme or more extreme than the actually observed outcome.* "Extreme" means "far from what we would expect if $H_0$ were true." The direction or directions that count as "far from what we would expect" are determined by $H_a$ and $H_0$.

In Example 6.8 we want to know if the debt of private college borrowers is different from the debt of public college borrowers. The difference we calculated based on our sample is \$4100, which corresponds to 1.37 standard deviations away from zero—that is, $z = 1.37$. Because we are using a two-sided alternative for this problem, the evidence against $H_0$ is measured by the probability that we observe a value of $Z$ as extreme or more extreme than 1.37. More formally, this probability is

$$P(Z \leq -1.37 \text{ or } Z \geq 1.37)$$

where $Z$ has the standard Normal distribution $N(0, 1)$.

> ### *P*-VALUE
>
> The probability, assuming $H_0$ is true, that the test statistic would take a value as extreme or more extreme than that actually observed is called the ***P*-value** of the test. The smaller the *P*-value, the stronger the evidence against $H_0$ provided by the data.

The key to calculating the *P*-value is the sampling distribution of the test statistic. For the problems we consider in this chapter, we need only the standard Normal distribution for the test statistic $z$.

EXAMPLE

**6.12 Debt levels of private and public college borrowers: the *P*-value.** In Example 6.11 we found that the test statistic for testing

$$H_0\text{: the true mean difference is 0}$$

versus

$$H_a\text{: there is a difference in the true means}$$

is

$$z = \frac{4100 - 0}{3000} = 1.37$$

If $H_0$ is true, then $z$ is a single observation from the standard Normal, $N(0, 1)$, distribution. Figure 6.9 illustrates this calculation. The *P*-value is the probability of observing a value of $Z$ at least as extreme as the one that we observed, $z = 1.37$. From Table A, our table of standard Normal probabilities, we find

$$P(Z \geq 1.37) = 1 - 0.9147 = 0.0853$$

The probability for being extreme in the negative direction is the same:

$$P(Z \leq -1.37) = 0.0853$$

So the *P*-value is

$$P = 2P(Z \geq 1.37) = 2(0.0853) = 0.1706$$

This is the value that was reported on page 373. There is a 17% chance of observing a difference as extreme as the $4100 in our sample if the true population difference is zero. The *P*-value tells us that our outcome is not particularly extreme, so we conclude that the data do not provide evidence that would cause us to doubt the validity of the null hypothesis.

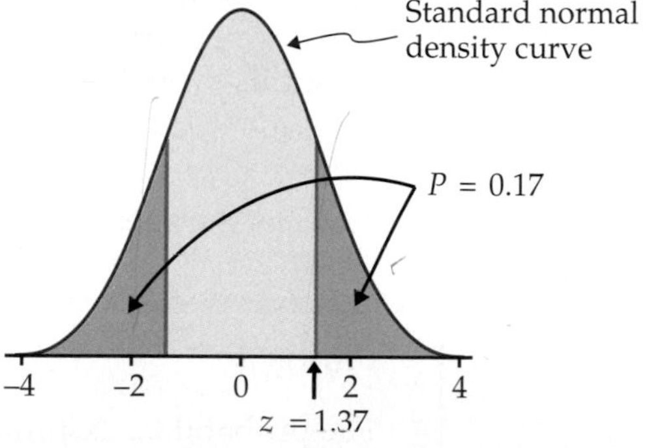


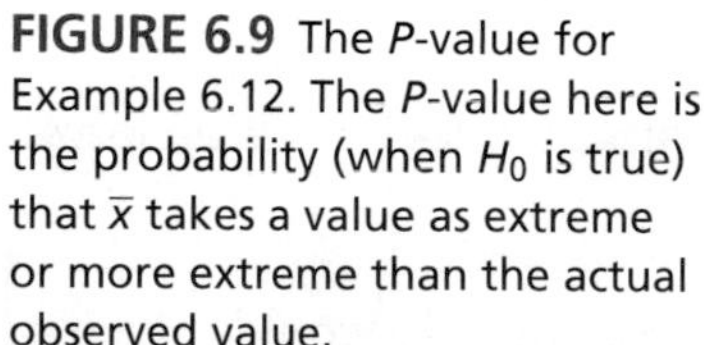
**FIGURE 6.9** The *P*-value for Example 6.12. The *P*-value here is the probability (when $H_0$ is true) that $\bar{x}$ takes a value as extreme or more extreme than the actual observed value.

## USE YOUR KNOWLEDGE

**6.38 Normal curve and the *P*-value.** A test statistic for a two-sided significance test for a population mean is $z = 2.7$. Sketch a standard Normal curve and mark this value of $z$ on it. Find the *P*-value and shade the appropriate areas under the curve to illustrate your calculations.

**6.39 More on the Normal curve and the *P*-value.** A test statistic for a two-sided significance test for a population mean is $z = -1.2$. Sketch a standard Normal curve and mark this value of $z$ on it. Find the *P*-value and shade the appropriate areas under the curve to illustrate your calculations.

## Statistical significance

We started our discussion of the reasoning of significance tests with the statement of null and alternative hypotheses. We then learned that a test statistic is the tool used to examine the compatibility of the observed data with the null hypothesis. Finally, we translated the test statistic into a *P*-value to quantify the evidence against $H_0$. One important final step is needed: to state our conclusion.

We can compare the *P*-value we calculated with a fixed value that we regard as decisive. This amounts to announcing in advance how much evidence against $H_0$ we will require to reject $H_0$. The decisive value of $P$ is called the **significance level.** It is commonly denoted by $\alpha$. If we choose $\alpha = 0.05$, we are requiring that the data give evidence against $H_0$ so strong that it would happen no more than 5% of the time (1 time in 20) when $H_0$ is true. If we choose $\alpha = 0.01$, we are insisting on stronger evidence against $H_0$, evidence so strong that it would appear only 1% of the time (1 time in 100) if $H_0$ is in fact true.

significance level

### STATISTICAL SIGNIFICANCE

If the *P*-value is as small or smaller than $\alpha$, we say that the data are **statistically significant at level $\alpha$.**

"Significant" in the statistical sense does not mean "important." The original meaning of the word is "signifying something." In statistics the term is used to indicate only that the evidence against the null hypothesis reached the standard set by $\alpha$. Significance at level 0.01 is often expressed by the statement "The results were significant ($P < 0.01$)." Here $P$ stands for the *P*-value. The *P*-value is more informative than a statement of significance because we can then assess significance at any level we choose. For example, a result with $P = 0.03$ is significant at the $\alpha = 0.05$ level but is not significant at the $\alpha = 0.01$ level.

A test of significance is a process for assessing the significance of the evidence provided by data against a null hypothesis. The four steps common to all tests of significance are as follows:

1. State the *null hypothesis* $H_0$ and the *alternative hypothesis* $H_a$. The test is designed to assess the strength of the evidence against $H_0$; $H_a$ is the statement that we will accept if the evidence enables us to reject $H_0$.
2. Calculate the value of the *test statistic* on which the test will be based. This statistic usually measures how far the data are from $H_0$.
3. Find the *P-value* for the observed data. This is the probability, calculated assuming that $H_0$ is true, that the test statistic will weigh against $H_0$ at least as strongly as it does for these data.

4. State a conclusion. One way to do this is to choose a *significance level* $\alpha$, how much evidence against $H_0$ you regard as decisive. If the *P*-value is less than or equal to $\alpha$, you conclude that the alternative hypothesis is true; if it is greater than $\alpha$, you conclude that the data do not provide sufficient evidence to reject the null hypothesis. Your conclusion is a sentence that summarizes what you have found by using a test of significance.

We will learn the details of many tests of significance in the following chapters. The proper test statistic is determined by the hypotheses and the data collection design. We use computer software or a calculator to find its numerical value and the *P*-value. The computer will not formulate your hypotheses for you, however. Nor will it decide if significance testing is appropriate or help you to interpret the *P*-value that it presents to you. The most difficult and important step is the last one: stating a conclusion.

EXAMPLE

**6.13 Debt levels of private and public college borrowers: significance.** In Example 6.12 we found that the *P*-value is 0.1706. There is a 17% chance of observing a difference as extreme as the \$4100 in our sample if the true population difference is zero. The *P*-value tells us that our outcome is not particularly extreme. We could report the result as "the data do not provide evidence that would cause us to conclude that there is a difference in student loan debt between private college borrowers and public college borrowers ($z = 1.37$, $P = 0.17$)."

If the *P*-value is small, we reject the null hypothesis. Here is an example.

EXAMPLE

**6.14 Change in mean debt levels: significance.** In Example 6.9 we found that the average debt has risen by \$7500 from 1997 to 2002. Since we would have a prior expectation that the debt would increase over this period because of rising costs of a college education, it is appropriate to use a one-sided alternative in this situation. So, our hypotheses are

$$H_0\text{: the true mean difference is 0}$$

versus

$$H_a\text{: the mean debt has increased between 1997 and 2002}$$

The standard deviation is \$1900 (again we defer details regarding this calculation), and the test statistic is

$$z = \frac{\text{estimate} - \text{hypothesized value}}{\text{standard deviation of the estimate}}$$

$$z = \frac{7500 - 0}{1900}$$

$$= 3.95$$

Because only increases in debt count against the null hypothesis, the one-sided alternative leads to the calculation of the *P*-value using the upper tail of the Normal distribution. The $P$-value is

$$P = P(Z \geq 3.95)$$
$$= 0.00004$$

The calculation is illustrated in Figure 6.10. There is about a 4 in 100,000 chance of observing a difference as large or larger than the \$7500 in our sample if the true population difference is zero. This $P$-value tells us that our outcome is extremely rare. We conclude that the null hypothesis must be false. Here is one way to report the result: "The data clearly show that the mean debt for college loans has increased between 1997 and 2002 ($z = 3.95$, $P < 0.001$)."

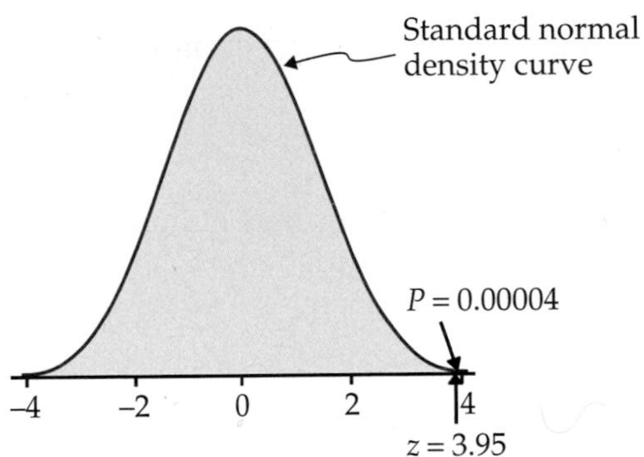


**FIGURE 6.10** The $P$-value for Example 6.14. The $P$-value here is the probability (when $H_0$ is true) that $\bar{x}$ takes a value as large or larger than the actual observed value.

Note that the calculated $P$-value for this example is 0.00004 but we reported the result as $P < 0.001$. The value 0.001, 1 in 1000, is sufficiently small to force a clear rejection of $H_0$. Standard practice is to report very small $P$-values as simply less than 0.001.

## USE YOUR KNOWLEDGE

**6.40 Finding significant $z$-scores.** Consider a significance test of the true mean based on an SRS of 30 observations from a Normal population. The alternative hypothesis is that the true mean is different from 1000. What values of the $z$ statistic are statistically significant at the $\alpha = 0.05$ level?

**6.41 More on finding significant $z$-scores.** Consider a significance test of the true mean based on an SRS of 30 observations from a Normal population. The alternative hypothesis is that the true mean is larger than 1000. What values of the $z$ statistic are statistically significant at the $\alpha = 0.05$ level?

**6.42 The Supreme Court speaks.** The Supreme Court has said that $z$-scores beyond $z^* = 2$ or 3 are generally convincing statistical evidence. For a two-sided test, what significance level corresponds to $z^* = 2$? To $z^* = 3$?

## Tests for a population mean

Our discussion has focused on the reasoning of statistical tests, and we have outlined the key ideas for one type of procedure. Here is a summary. We want to test the hypothesis that a parameter has a specified value. This is the null hypothesis. For a test of a population mean $\mu$, the null hypothesis is

$$H_0\text{: the true population mean is equal to } \mu_0$$

which often is expressed as

$$H_0: \mu = \mu_0$$

where $\mu_0$ is the specified value of $\mu$ that we would like to examine.

The test is based on data summarized as an estimate of the parameter. For a population mean this is the sample mean $\bar{x}$. Our test statistic measures the difference between the sample estimate and the hypothesized parameter in terms of standard deviations of the test statistic:

$$z = \frac{\text{estimate} - \text{hypothesized value}}{\text{standard deviation of the estimate}}$$

LOOK BACK
**distribution of sample mean, page 339**
**central limit theorem, page 339**

Recall from Chapter 5 that the standard deviation of $\bar{x}$ is $\sigma/\sqrt{n}$. Therefore, the test statistic is

$$z = \frac{\bar{x} - \mu_0}{\sigma/\sqrt{n}}$$

Again recall from Chapter 5 that, if the population is Normal, then $\bar{x}$ will be Normal and $z$ will have the standard Normal distribution when $H_0$ is true. By the central limit theorem both distributions will be approximately Normal when the sample size is large even if the population is not Normal.

Suppose we have calculated a test statistic $z = 1.7$. If the alternative is one-sided on the high side, then the $P$-value is the probability that a standard Normal random variable $Z$ takes a value as large or larger than the observed 1.7. That is,

$$\begin{aligned} P &= P(Z \geq 1.7) \\ &= 1 - P(Z < 1.7) \\ &= 1 - 0.9554 \\ &= 0.0446 \end{aligned}$$

Similar reasoning applies when the alternative hypothesis states that the true $\mu$ lies below the hypothesized $\mu_0$ (one-sided). When $H_a$ states that $\mu$ is simply unequal to $\mu_0$ (two-sided), values of $z$ away from zero in either direction count against the null hypothesis. The $P$-value is the probability that a standard Normal $Z$ is at least as far from zero as the observed $z$. Again, if the test statistic is $z = 1.7$, the two-sided $P$-value is the probability that $Z \leq -1.7$ or $Z \geq 1.7$. Because the standard Normal distribution is symmetric, we calculate this probability by finding $P(Z \geq 1.7)$ and *doubling* it:

$$\begin{aligned} P(Z \leq -1.7 \text{ or } Z \geq 1.7) &= 2P(Z \geq 1.7) \\ &= 2(1 - 0.9554) = 0.0892 \end{aligned}$$

We would make exactly the same calculation if we observed $z = -1.7$. It is the absolute value $|z|$ that matters, not whether $z$ is positive or negative. Here is a statement of the test in general terms.

### $z$ TEST FOR A POPULATION MEAN

To test the hypothesis $H_0: \mu = \mu_0$ based on an SRS of size $n$ from a population with unknown mean $\mu$ and known standard deviation $\sigma$, compute the test statistic

$$z = \frac{\bar{x} - \mu_0}{\sigma/\sqrt{n}}$$

In terms of a standard Normal random variable $Z$, the $P$-value for a test of $H_0$ against

$H_a: \mu > \mu_0$ is $P(Z \geq z)$

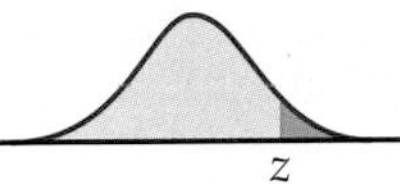


$H_a: \mu < \mu_0$ is $P(Z \leq z)$

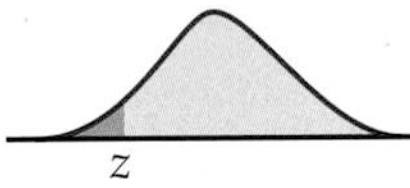


$H_a: \mu \neq \mu_0$ is $2P(Z \geq |z|)$

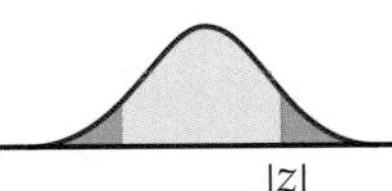


These $P$-values are exact if the population distribution is Normal and are approximately correct for large $n$ in other cases.

EXAMPLE

**6.15 Cholesterol level of sedentary female undergraduates.** In 1999, it was reported that the mean serum cholesterol level for female undergraduates was 168 mg/dl with a standard deviation of 27 mg/dl. A recent study at Baylor University investigated the lipid levels in a cohort of sedentary university students.[12] The mean total cholesterol level among $n = 71$ females was $\bar{x} = 173.7$. Is this evidence that cholesterol levels of sedentary students differ from the previously reported average?

The null hypothesis is "no difference" from the published mean $\mu_0 = 168$. The alternative is two-sided because the researcher did not have a particular direction in mind before examining the data. So the hypotheses about the unknown mean $\mu$ of the sedentary population are

$$H_0: \mu = 168$$
$$H_a: \mu \neq 168$$

As usual in this chapter, we make the unrealistic assumption that the population standard deviation is known, in this case that sedentary female students have the same $\sigma = 27$ as the general population of female undergraduates. The $z$ test requires that the 71 students in the sample are an SRS from the population of all sedentary female students. We check this assumption by asking how the data were produced. In this case, all participants were enrolled in a health class at Baylor, so there may be some concerns about whether the sample is an SRS. We will press on for now.

We compute the test statistic:

$$z = \frac{\bar{x} - \mu_0}{\sigma/\sqrt{n}} = \frac{173.7 - 168}{27/\sqrt{71}}$$
$$= 1.78$$

Figure 6.11 illustrates the $P$-value, which is the probability that a standard Normal variable $Z$ takes a value at least 1.78 away from zero. From Table A we find that this probability is

$$P = 2P(Z \geq 1.78) = 2(1 - 0.9625) = 0.075$$

That is, more than 7% of the time an SRS of size 71 from the general undergraduate female population would have a mean cholesterol level at least as far from 168 as that of the sedentary sample. The observed $\bar{x} = 173.7$ is therefore not strong evidence that the sedentary female undergraduate population differs from the general female undergraduate population.

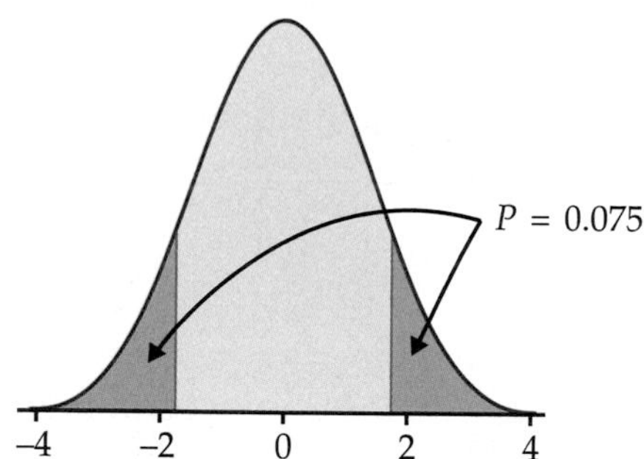


**FIGURE 6.11** The $P$-value for the two-sided test in Example 6.15.

The data in Example 6.15 do *not* establish that the mean cholesterol level $\mu$ for the sedentary population is 168. We sought evidence that $\mu$ differed from 168 and failed to find convincing evidence. That is all we can say. No doubt the mean cholesterol level of the entire sedentary population is not exactly equal to 168. A large enough sample would give evidence of the difference, even if it is very small. Tests of significance assess the evidence *against* $H_0$. If the evidence is strong, we can confidently reject $H_0$ in favor of the alternative. Failing to find evidence against $H_0$ means only that the data are consistent with $H_0$, not that we have clear evidence that $H_0$ is true.

EXAMPLE

**6.16 Significance test of the mean SATM score.** In a discussion of SAT Mathematics (SATM) scores, someone comments: "Because only a minority of California high school students take the test, the scores overestimate the ability of typical high school seniors. I think that if all seniors took the test, the mean score would be no more than 450." You decided to test this claim ($H_0$) and gave the SAT to an SRS of 500 seniors from California (Example 6.3). These students had a mean SATM score of $\bar{x} = 461$. Is this good evidence against this claim? Because the claim states the mean is "no more than 450," the alternative hypothesis is one-sided. The hypotheses are

$$H_0: \mu = 450$$

$$H_a: \mu > 450$$

As we did in the discussion following Example 6.3, we assume that $\sigma = 100$. The $z$ statistic is

$$z = \frac{\bar{x} - \mu_0}{\sigma/\sqrt{n}} = \frac{461 - 450}{100/\sqrt{500}}$$
$$= 2.46$$

Because $H_a$ is one-sided on the high side, large values of $z$ count against $H_0$. From Table A, we find that the $P$-value is

$$P = P(Z \geq 2.46) = 1 - 0.9931 = 0.0069$$

Figure 6.12 illustrates this $P$-value. A mean score as large as that observed would occur fewer than seven times in 1000 samples if the population mean were 450. This is convincing evidence that the mean SATM score for all California high school seniors is higher than 450.

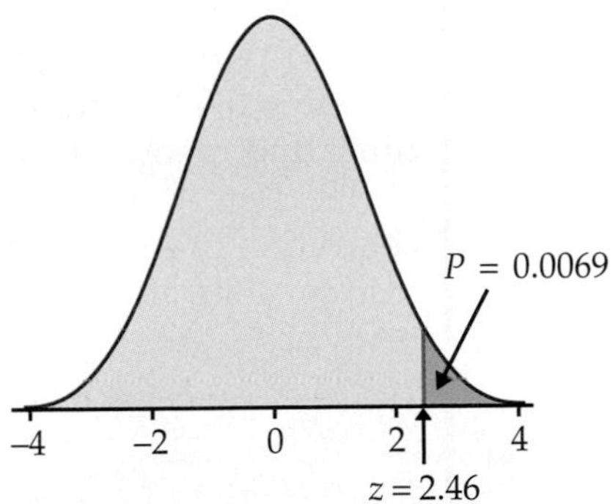


**FIGURE 6.12** The $P$-value for the one-sided test in Example 6.16.

## USE YOUR KNOWLEDGE

**6.43 Computing the test statistic and $P$-value.** You will perform a significance test of $H_0: \mu = 25$ based on an SRS of $n = 25$. Assume $\sigma = 5$.

(a) If $\bar{x} = 27$, what is the test statistic $z$?

(b) What is the $P$-value if $H_a$: $\mu > 25$?

(c) What is the $P$-value if $H_a$: $\mu \neq 25$?

**6.44 Testing a random number generator.** Statistical software has a "random number generator" that is supposed to produce numbers uniformly distributed between 0 to 1. If this is true, the numbers generated come from a population with $\mu = 0.5$. A command to generate 100 random numbers gives outcomes with mean $\bar{x} = 0.522$ and $s = 0.316$. Because the sample is reasonably large, take the population standard deviation also to be $\sigma = 0.316$. Do we have evidence that the mean of all numbers produced by this software is not 0.5?

## Two-sided significance tests and confidence intervals

Recall the basic idea of a confidence interval, discussed in the first section of this chapter. We constructed an interval that would include the true value of $\mu$ with a specified probability $C$. Suppose we use a 95% confidence interval ($C = 0.95$). Then the values of $\mu$ that are not in our interval would seem to be incompatible with the data. This sounds like a significance test with $\alpha = 0.05$ (or 5%) as our standard for drawing a conclusion. The following examples demonstrate that this is correct.

EXAMPLE

**6.17 Testing a pharmaceutical product.** The Deely Laboratory analyzes specimens of a pharmaceutical product to determine the concentration of the active ingredient. Such chemical analyses are not perfectly precise. Repeated measurements on the same specimen will give slightly different results. The results of repeated measurements follow a Normal distribution quite closely. The analysis procedure has no bias, so that the mean $\mu$ of the population of all measurements is the true concentration in the specimen. The standard deviation of this distribution is a property of the analytical procedure and is known to be $\sigma = 0.0068$ grams per liter. The laboratory analyzes each specimen three times and reports the mean result.

The Deely Laboratory has been asked to evaluate the claim that the concentration of the active ingredient in a specimen is 0.86 grams per liter. The true concentration is the mean $\mu$ of the population of repeated analyses. The hypotheses are

$$H_0: \mu = 0.86$$
$$H_a: \mu \neq 0.86$$

The lab chooses the 1% level of significance, $\alpha = 0.01$.

Three analyses of one specimen give concentrations

0.8403 0.8363 0.8447

The sample mean of these readings is

$$\bar{x} = \frac{0.8403 + 0.8363 + 0.8447}{3} = 0.8404$$

The test statistic is

$$z = \frac{\bar{x} - \mu_0}{\sigma/\sqrt{n}} = \frac{0.8404 - 0.86}{0.0068/\sqrt{3}} = -4.99 \text{ standard deviations}$$

Because the alternative is two-sided, the $P$-value is

$$P = 2P(Z \geq |-4.99|) = 2P(Z \geq 4.99)$$

We cannot find this probability in Table A. The largest value of $z$ in that table is 3.49. All that we can say from Table A is that $P$ is less than $2P(Z \geq 3.49) = 2(1 - 0.9998) = 0.0004$. If we use the bottom row of Table D, we find that the largest value of $z^*$ is 3.291, corresponding to a $P$-value of $1 - 0.999 = 0.001$. Software could be used to give an accurate value of the $P$-value. However, because the $P$-value is clearly less than the company's standard of 1%, we reject $H_0$.

Suppose we compute a 99% confidence interval for the same data.

EXAMPLE

**6.18 99% confidence interval for the mean concentration.** The 99% confidence interval for $\mu$ in Example 6.17 is

$$\bar{x} \pm z^* \frac{\sigma}{\sqrt{n}} = 0.8404 \pm 0.0101$$
$$= (0.8303, 0.8505)$$

The hypothesized value $\mu_0 = 0.86$ in Example 6.17 falls outside the confidence interval we computed in Example 6.18. We are therefore 99% confident that $\mu$ is *not* equal to 0.86, so we can reject

$$H_0\colon \mu = 0.86$$

at the 1% significance level. On the other hand, we cannot reject

$$H_0\colon \mu = 0.85$$

at the 1% level in favor of the two-sided alternative $H_a\colon \mu \neq 0.85$, because 0.85 lies inside the 99% confidence interval for $\mu$. Figure 6.13 illustrates both cases.

The calculation in Example 6.17 for a 1% significance test is very similar to the calculation for a 99% confidence interval. In fact, a two-sided test at significance level $\alpha$ can be carried out directly from a confidence interval with confidence level $C = 1 - \alpha$.

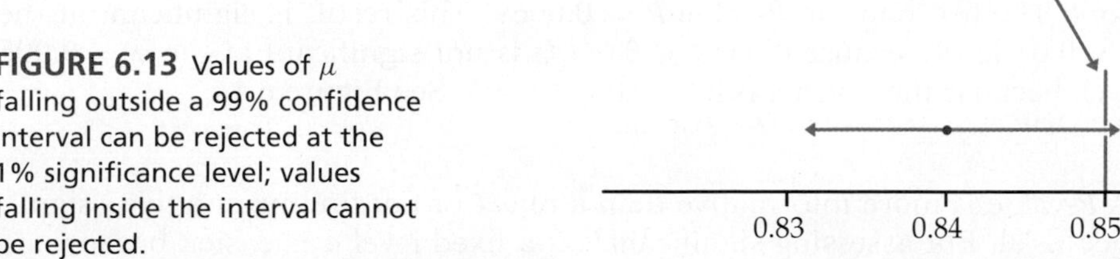

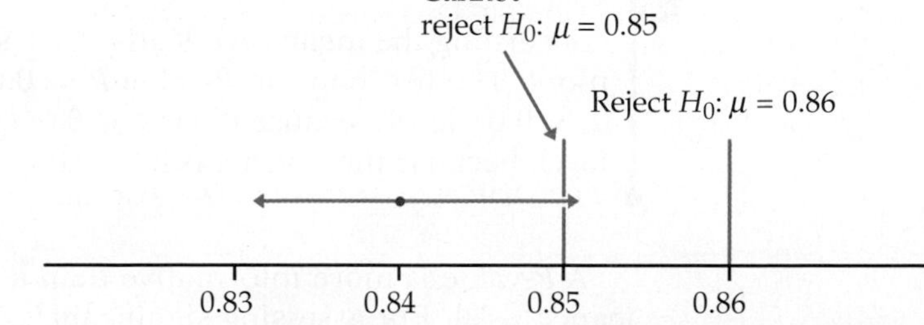


**FIGURE 6.13** Values of $\mu$ falling outside a 99% confidence interval can be rejected at the 1% significance level; values falling inside the interval cannot be rejected.

**TWO-SIDED SIGNIFICANCE TESTS AND CONFIDENCE INTERVALS**

A level $\alpha$ two-sided significance test rejects a hypothesis $H_0: \mu = \mu_0$ exactly when the value $\mu_0$ falls outside a level $1 - \alpha$ confidence interval for $\mu$.

**USE YOUR KNOWLEDGE**

**6.45 Two-sided significance tests and confidence intervals.** The *P*-value for a two-sided test of the null hypothesis $H_0: \mu = 30$ is 0.08.

(a) Does the 95% confidence interval include the value 30? Explain.

(b) Does the 90% confidence interval include the value 30? Explain.

**6.46 More on two-sided tests and confidence intervals.** A 95% confidence interval for a population mean is (57, 65).

(a) Can you reject the null hypothesis that $\mu = 68$ at the 5% significance level? Explain.

(b) Can you reject the null hypothesis that $\mu = 62$ at the 5% significance level? Explain.

## *P*-values versus fixed $\alpha$

The observed result in Example 6.17 was $z = -4.99$. The conclusion that this result is significant at the 1% level does not tell the whole story. The observed $z$ is far beyond the $z$ corresponding to 1%, and the evidence against $H_0$ is far stronger than 1% significance suggests. The *P*-value

$$2P(Z \geq 4.99) = 0.0000006$$

gives a better sense of how strong the evidence is. *The P-value is the smallest level $\alpha$ at which the data are significant.* Knowing the *P*-value allows us to assess significance at any level.

EXAMPLE

**6.19 Test of the mean SATM score: significance.** In Example 6.16, we tested the hypotheses

$$H_0: \mu = 450$$
$$H_a: \mu > 450$$

concerning the mean SAT Mathematics score $\mu$ of California high school seniors. The test had the *P*-value $P = 0.0069$. This result is significant at the $\alpha = 0.01$ level because $0.0069 \leq 0.01$. It is not significant at the $\alpha = 0.005$ level, because the *P*-value is larger than 0.005. See Figure 6.14.

A *P*-value is more informative than a reject-or-not finding at a fixed significance level. But assessing significance at a fixed level $\alpha$ is easier, because no

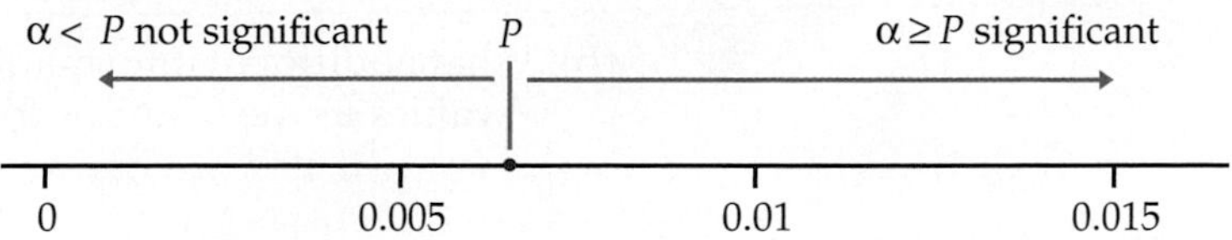


**FIGURE 6.14** An outcome with $P$-value $P$ is significant at all levels $\alpha$ at or above $P$ and is not significant at smaller levels $\alpha$.

probability calculation is required. You need only look up a number in a table. A value $z^*$ with a specified area to its right under the standard Normal curve is called a **critical value** of the standard Normal distribution. Because the practice of statistics almost always employs computer software that calculates $P$-values automatically, the use of tables of critical values is becoming outdated. We include the usual tables of critical values (such as Table D) at the end of the book for learning purposes and to rescue students without good computing facilities. The tables can be used directly to carry out fixed $\alpha$ tests. They also allow us to approximate $P$-values quickly without a probability calculation. The following example illustrates the use of Table D to find an approximate $P$-value.

critical value

EXAMPLE

**6.20 Debt levels of private and public college borrowers: assessing significance.** In Example 6.11 we found the test statistic $z = 1.37$ for testing the null hypothesis that there was no difference in the mean debt between borrowers who attended a private college and those who attended a public college. The alternative was two-sided. Under the null hypothesis, $z$ has a standard Normal distribution, and from the last row in Table D we can see that there is a 95% chance that $z$ is between $\pm 1.96$. Therefore, we reject $H_0$ in favor of $H_a$ whenever $z$ is outside this range. Since our calculated value is 1.37, we are within the range and we do not reject the null hypothesis at the 5% level of significance.

## USE YOUR KNOWLEDGE

**6.47 *P*-value and the significance level.** The $P$-value for a significance test is 0.026.

(a) Do you reject the null hypothesis at level $\alpha = 0.05$?

(b) Do you reject the null hypothesis at level $\alpha = 0.01$?

(c) Explain your answers.

**6.48 More on the *P*-value and the significance level.** The $P$-value for a significance test is 0.074.

(a) Do you reject the null hypothesis at level $\alpha = 0.05$?

(b) Do you reject the null hypothesis at level $\alpha = 0.01$?

(c) Explain your answers.

**6.49 One-sided and two-sided *P*-values.** The $P$-value for a two-sided significance test is 0.06.

(a) State the $P$-values for the one-sided tests.

(b) What additional information do you need to properly assign these $P$-values to the $>$ and $<$ (one-sided) alternatives?

## SECTION 6.2 Summary

A **test of significance** is intended to assess the evidence provided by data against a **null hypothesis $H_0$** in favor of an **alternative hypothesis $H_a$.**

The hypotheses are stated in terms of population parameters. Usually $H_0$ is a statement that no effect or no difference is present, and $H_a$ says that there is an effect or difference, in a specific direction (**one-sided alternative**) or in either direction (**two-sided alternative**).

The test is based on a **test statistic.** The ***P*-value** is the probability, computed assuming that $H_0$ is true, that the test statistic will take a value at least as extreme as that actually observed. Small $P$-values indicate strong evidence against $H_0$. Calculating $P$-values requires knowledge of the sampling distribution of the test statistic when $H_0$ is true.

If the $P$-value is as small or smaller than a specified value $\alpha$, the data are **statistically significant** at significance level $\alpha$.

Significance tests for the hypothesis $H_0: \mu = \mu_0$ concerning the unknown mean $\mu$ of a population are based on the **$z$ statistic:**

$$z = \frac{\bar{x} - \mu_0}{\sigma/\sqrt{n}}$$

The $z$ test assumes an SRS of size $n$, known population standard deviation $\sigma$, and either a Normal population or a large sample. $P$-values are computed from the Normal distribution (Table A). Fixed $\alpha$ tests use the table of **standard Normal critical values** (Table D).

## SECTION 6.2 Exercises

*For Exercises 6.36 and 6.37, see page 375; for Exercises 6.38 and 6.39, see pages 378 and 379; for Exercises 6.40 to 6.42, see page 381; for Exercises 6.43 and 6.44, see pages 385 and 386; for Exercises 6.45 and 6.46, see page 388; and for Exercises 6.47 to 6.49, see page 389.*

**6.50 What's wrong?** Here are several situations where there is an incorrect application of the ideas presented in this section. Write a short paragraph explaining what is wrong in each situation and why it is wrong.

(a) A random sample of size 20 is taken from a population that is assumed to have a standard deviation of 12. The standard deviation of the sample mean is 12/20.

(b) A researcher tests the following null hypothesis: $H_0: \bar{x} = 10$.

(c) A study with $\bar{x} = 48$ reports statistical significance for $H_a: \mu > 54$.

(d) A researcher tests the hypothesis $H_0: \mu = 50$ and concludes that the population mean is equal to 50.

**6.51 What's wrong?** Here are several situations where there is an incorrect application of the ideas presented in this section. Write a short paragraph explaining what is wrong in each situation and why it is wrong.

(a) A significance test rejected the null hypothesis that the sample mean is equal to 1500.

(b) A change is made that should improve student satisfaction with the way grades are processed. The null hypothesis, that there is an improvement, is tested versus the alternative, that there is no change.

(c) A study summary says that the results are statistically significant and the $P$-value is 0.99.

**6.52 Determining hypotheses.** State the appropriate null hypothesis $H_0$ and alternative hypothesis $H_a$ in each of the following cases.

(a) A 2002 study reported that 70% of students owned a cell phone. You plan to take an SRS of students to see if the percent has increased.

(b) The examinations in a large freshman chemistry class are scaled after grading so that the mean score is 72. The professor thinks that students who attend morning recitation sections will have a higher mean score than the class as a whole. Her students this semester can be considered a sample from the population of all students she might teach, so she compares their mean score with 72.

(c) The student newspaper at your college recently changed the format of their opinion page. You take a random sample of students and select those who regularly read the newspaper. They are asked to indicate their opinions on the changes using a five-point scale: −2 if the new format is much worse than the old, −1 if the new format is somewhat worse than the old, 0 if the new format is the same as the old, +1 if the new format is somewhat better than the old, and +2 if the new format is much better than the old.

**6.53 More on determining hypotheses.** State the null hypothesis $H_0$ and the alternative hypothesis $H_a$ in each case. Be sure to identify the parameters that you use to state the hypotheses.

(a) A university gives credit in French language courses to students who pass a placement test. The language department wants to know if students who get credit in this way differ in their understanding of spoken French from students who actually take the French courses. Experience has shown that the mean score of students in the courses on a standard listening test is 26. The language department gives the same listening test to a sample of 35 students who passed the credit examination to see if their performance is different.

(b) Experiments on learning in animals sometimes measure how long it takes a mouse to find its way through a maze. The mean time is 20 seconds for one particular maze. A researcher thinks that playing rap music will cause the mice to complete the maze faster. She measures how long each of 12 mice takes with the rap music as a stimulus.

(c) The average square footage of one-bedroom apartments in a new student-housing development is advertised to be 460 square feet. A student group thinks that the apartments are smaller than advertised. They hire an engineer to measure a sample of apartments to test their suspicion.

**6.54 Even more on determining hypotheses.** In each of the following situations, state an appropriate null hypothesis $H_0$ and alternative hypothesis $H_a$. Be sure to identify the parameters that you use to state the hypotheses. (We have not yet learned how to test these hypotheses.)

(a) A sociologist asks a large sample of high school students which academic subject they like best. She suspects that a higher percent of males than of females will name marketing as their favorite subject.

(b) An education researcher randomly divides sixth-grade students into two groups for physical education class. He teaches both groups basketball skills, using the same methods of instruction in both classes. He encourages Group A with compliments and other positive behavior but acts cool and neutral toward Group B. He hopes to show that positive teacher attitudes result in a higher mean score on a test of basketball skills than do neutral attitudes.

(c) An education researcher believes that among college students there is a negative correlation between credit card debt and self-esteem. To test this, she gathers credit card debt information and self-esteem data from a sample of students at your college.

**6.55 Translating research questions into hypotheses.** Translate each of the following research questions into appropriate $H_0$ and $H_a$.

(a) Census Bureau data show that the mean household income in the area served by a shopping mall is $62,500 per year. A market research firm questions shoppers at the mall to find out whether the mean household income of mall shoppers is higher than that of the general population.

(b) Last year, your company's service technicians took an average of 2.6 hours to respond to trouble calls from business customers who had purchased service contracts. Do this year's data show a different average response time?

**6.56 Computing the *P*-value.** A test of the null hypothesis $H_0: \mu = \mu_0$ gives test statistic $z = 1.34$.

(a) What is the *P*-value if the alternative is $H_a: \mu > \mu_0$?

(b) What is the *P*-value if the alternative is $H_a: \mu < \mu_0$?

(c) What is the $P$-value if the alternative is $H_a: \mu \neq \mu_0$?

**6.57 More on computing the *P*-value.** A test of the null hypothesis $H_0: \mu = \mu_0$ gives test statistic $z = -1.73$.

(a) What is the $P$-value if the alternative is $H_a: \mu > \mu_0$?

(b) What is the $P$-value if the alternative is $H_a: \mu < \mu_0$?

(c) What is the $P$-value if the alternative is $H_a: \mu \neq \mu_0$?

**6.58 A two-sided test and the confidence interval.** The $P$-value for a two-sided test of the null hypothesis $H_0: \mu = 30$ is 0.04.

(a) Does the 95% confidence interval include the value 30? Why?

(b) Does the 90% confidence interval include the value 30? Why?

**6.59 More on a two-sided test and the confidence interval.** A 90% confidence interval for a population mean is (23, 28).

(a) Can you reject the null hypothesis that $\mu = 24$ at the 10% significance level? Why?

(b) Can you reject the null hypothesis that $\mu = 30$ at the 10% significance level? Why?

**6.60 Use of bed nets.** A study found that the use of bed nets was associated with a lower prevalence of malarial infections in the Gambia.[13] A report of the study states that the significance is $P < 0.001$. Explain what this means in a way that could be understood by someone who has not studied statistics.

**6.61 Purity of a catalyst.** A new supplier offers a good price on a catalyst used in your production process. You compare the purity of this catalyst with that of the catalyst offered by your current supplier. The $P$-value for a test of "no difference" is 0.27. Can you be confident that the purity of the new product is the same as the purity of the product that you have been using? Discuss.

**6.62 Symbol of wealth in ancient China?** Every society has its own symbols of wealth and prestige. In ancient China, it appears that owning pigs was such a symbol. Evidence comes from examining burial sites. If the skulls of sacrificed pigs tend to appear along with expensive ornaments, that suggests that the pigs, like the ornaments, signal the wealth and prestige of the person buried. A study of burials from around 3500 B.C. concluded that, "there are striking differences in grave goods between burials with pig skulls and burials without them. . . . A test indicates that the two samples of total artifacts are significantly different at the 0.01 level."[14] Explain clearly why "significantly different at the 0.01 level" gives good reason to think that there really is a systematic difference between burials that contain pig skulls and those that lack them.

**6.63 Alcohol awareness among college students.** A study of alcohol awareness among college students reported a higher awareness for students enrolled in a health and safety class than for those enrolled in a statistics class.[15] The difference is described as being statistically significant. Explain what this means in simple terms and offer an explanation for why the health and safety students had a higher mean score.

**6.64 Change in California's eighth-grade average science score.** A report based on the 2005 National Assessment of Educational Progress (NAEP)[16] states that the average score on their science test for eighth-grade students in California is significantly higher than in 2000. A footnote states that comparisons (higher/lower/different) are determined by statistical tests with 0.05 as the level of significance. Explain what this means in language understandable to someone who knows no statistics. Do not use the word "significance" in your answer.

**6.65 More on the eighth-grade average science score.** The 2005 NAEP report described in the previous exercise states that the average score on their science test for eighth-grade students across the nation was not significantly different from the average score in 2000. A footnote states that comparisons (higher/lower/different) are determined by statistical tests with 0.05 as the level of significance. Explain what this means in language understandable to someone who knows no statistics. Do not use the word "significance" in your answer.

**6.66** CHALLENGE **Are the pine trees randomly distributed north to south?** In Example 6.1 we looked at the distribution of longleaf pine trees in the Wade Tract. One way to formulate hypotheses about whether or not the trees are randomly distributed in the tract is to examine the average location in the north-south direction. The values range from 0 to 200, so if the trees are uniformly distributed in this direction, any difference from the middle value (100) should be due to chance variation. The sample mean for the 584 trees in the tract is 99.74. A theoretical calculation based on the assumption that the trees are uniformly distributed gives a standard deviation of 58. Carefully state the null and alternative hypotheses in terms of this variable. Note that this requires that you translate the research question about the random distribution of the trees into specific statements about the mean of a probability distribution. Test your hypotheses, report your results, and write a short summary of what you have found.

6.67 CHALLENGE **Are the pine trees randomly distributed east to west?** Answer the questions in the previous exercise for the east-west direction, where the sample mean is 113.8.

6.68 **Who is the author?** Statistics can help decide the authorship of literary works. Sonnets by a certain Elizabethan poet are known to contain an average of $\mu = 8.9$ new words (words not used in the poet's other works). The standard deviation of the number of new words is $\sigma = 2.5$. Now a manuscript with 6 new sonnets has come to light, and scholars are debating whether it is the poet's work. The new sonnets contain an average of $\bar{x} = 10.2$ words not used in the poet's known works. We expect poems by another author to contain more new words, so to see if we have evidence that the new sonnets are not by our poet we test

$$H_0: \mu = 8.9$$

$$H_a: \mu > 8.9$$

Give the $z$ test statistic and its $P$-value. What do you conclude about the authorship of the new poems?

6.69 **Attitudes toward school.** The Survey of Study Habits and Attitudes (SSHA) is a psychological test that measures the motivation, attitude toward school, and study habits of students. Scores range from 0 to 200. The mean score for U.S. college students is about 115, and the standard deviation is about 30. A teacher who suspects that older students have better attitudes toward school gives the SSHA to 25 students who are at least 30 years of age. Their mean score is $\bar{x} = 132.2$.

(a) Assuming that $\sigma = 30$ for the population of older students, carry out a test of

$$H_0: \mu = 115$$

$$H_a: \mu > 115$$

Report the $P$-value of your test, and state your conclusion clearly.

(b) Your test in (a) required two important assumptions in addition to the assumption that the value of $\sigma$ is known. What are they? Which of these assumptions is most important to the validity of your conclusion in (a)?

6.70 **Calcium level in pregnant women in rural Guatemala.** The level of calcium in the blood in healthy young adults varies with mean about 9.5 milligrams per deciliter and standard deviation about $\sigma = 0.4$. A clinic in rural Guatemala measures the blood calcium level of 160 healthy pregnant women at their first visit for prenatal care. The mean is $\bar{x} = 9.57$. Is this an indication that the mean calcium level in the population from which these women come differs from 9.5?

(a) State $H_0$ and $H_a$.

(b) Carry out the test and give the $P$-value, assuming that $\sigma = 0.4$ in this population. Report your conclusion.

(c) Give a 95% confidence interval for the mean calcium level $\mu$ in this population. We are confident that $\mu$ lies quite close to 9.5. This illustrates the fact that a test based on a large sample ($n = 160$ here) will often declare even a small deviation from $H_0$ to be statistically significant.

6.71 **Are the mpg measurements similar?** Refer to Exercise 6.26 (page 371). In addition to the computer computing mpg, the driver also recorded the mpg by dividing the miles driven by the number of gallons at each fill-up. The following data are the differences between the computer's and the driver's calculations for that random sample of 20 records. The driver wants to determine if these calculations are different. Assume the standard deviation of a difference to be $\sigma = 3.0$.

| | | | | | | | | | |
|---|---|---|---|---|---|---|---|---|---|
| 5.0 | 6.5 | −0.6 | 1.7 | 3.7 | 4.5 | 8.0 | 2.2 | 4.9 | 3.0 |
| 4.4 | 0.1 | 3.0 | 1.1 | 1.1 | 5.0 | 2.1 | 3.7 | −0.6 | −4.2 |

(a) State the appropriate $H_0$ and $H_a$ to test this suspicion.

(b) Carry out the test. Give the $P$-value, and then interpret the result in plain language.

6.72 **Adjusting for changes in the value of the dollar.** In Example 6.14 (page 380), we found that the average student debt has risen between 1997 to 2002. In computing the difference, we did not adjust for differing values of the dollar. Using the fact that \$1 in 1997 was worth approximately \$1.12 in 2002, redo the test based on 2002 dollars. For simplicity, assume the standard deviation is unchanged.

**6.73 Level of nicotine in cigarettes.** According to data from the Tobacco Institute Testing Laboratory, Camel Lights King Size cigarettes contain an average of 1.4 milligrams of nicotine. An advocacy group commissions an independent test to see if the mean nicotine content is higher than the industry laboratory claims.

(a) What are $H_0$ and $H_a$?

(b) Suppose that the test statistic is $z = 2.36$. Is this result significant at the 5% level?

(c) Is the result significant at the 1% level?

**6.74** APPLET **Changes of $\bar{x}$ on significance.** The *Statistical Significance* applet illustrates statistical tests with a fixed level of significance for Normally distributed data with known standard deviation. Open the applet and keep the default settings for the null ($\mu = 0$) and the alternative ($\mu > 0$) hypotheses, the sample size ($n = 10$), the standard deviation ($\sigma = 1$), and the significance level ($\alpha = 0.05$). In the "I have data, and the observed $\bar{x}$ is $\bar{x} =$" box enter the value 1. Is the difference between $\bar{x}$ and $\mu_0$ significant at the 5% level? Repeat for $\bar{x}$ equal to 0.1, 0.2, 0.3, 0.4, 0.5, 0.6, 0.7, 0.8, 0.9. Make a table giving $\bar{x}$ and the results of the significance tests. What do you conclude?

**6.75** APPLET **Changes of $\alpha$ on significance.** Repeat the previous exercise with significance level $\alpha = 0.01$. How does the choice of $\alpha$ affect the values of $\bar{x}$ that are far enough away from $\mu_0$ to be statistically significant?

**6.76** APPLET **Changes of $\bar{x}$ on the *P*-value.** The *P-Value of a Test of Significance* applet illustrates *P*-values of significance tests for Normally distributed data with known standard deviation. Open the applet and keep the default settings for the null ($\mu = 0$) and the alternative ($\mu > 0$) hypotheses, the sample size ($n = 10$), the standard deviation ($\sigma = 1$), and the significance level ($\alpha = 0.05$). In the "I have data, and the observed $\bar{x}$ is $\bar{x} =$" box enter the value 1. What is the *P*-value? Repeat for $\bar{x}$ equal to 0.1, 0.2, 0.3, 0.4, 0.5, 0.6, 0.7, 0.8, 0.9. Make a table giving $\bar{x}$ and *P*-values. How does the *P*-value change as $\bar{x}$ moves farther away from $\mu_0$?

**6.77 Understanding levels of significance.** Explain in plain language why a significance test that is significant at the 1% level must always be significant at the 5% level.

**6.78 More on understanding levels of significance.** You are told that a significance test is significant at the 5% level. From this information can you determine whether or not it is significant at the 1% level? Explain your answer.

**6.79 Test statistic and levels of significance.** Consider a significance test for a null hypothesis versus a two-sided alternative with a $z$ test statistic. Give a value of $z$ that will give a result significant at the 0.5% level but not at the 0.1% level.

**6.80 Using Table D to find a *P*-value.** You have performed a two-sided test of significance and obtained a value of $z = 3.1$. Use Table D to find the approximate *P*-value for this test.

**6.81 More on using Table D to find a *P*-value.** You have performed a one-sided test of significance and obtained a value of $z = 0.35$. Use Table D to find the approximate *P*-value for this test.

**6.82 Using Table A and Table D to find a *P*-value.** Consider a significance test for a null hypothesis versus a two-sided alternative. Between what values from Table D does the *P*-value for an outcome $z = 1.37$ lie? Calculate the *P*-value using Table A, and verify that it lies between the values you found from Table D.

**6.83 More on using Table A and Table D to find a *P*-value.** Refer to the previous exercise. Find the *P*-value for $z = -1.37$.

## 6.3 Use and Abuse of Tests

Carrying out a test of significance is often quite simple, especially if the *P*-value is given effortlessly by a computer. Using tests wisely is not so simple. Each test is valid only in certain circumstances, with properly produced data being particularly important. The $z$ test, for example, should bear the same warning label that was attached in Section 6.1 to the corresponding confidence interval (page 366). Similar warnings accompany the other tests that we will learn. There are additional caveats that concern tests more than confidence intervals, enough to warrant this separate section. Some hesitation about the unthinking use of significance tests is a sign of statistical maturity.

The reasoning of significance tests has appealed to researchers in many fields, so that tests are widely used to report research results. In this setting $H_a$ is a "research hypothesis" asserting that some effect or difference is present. The null hypothesis $H_0$ says that there is no effect or no difference. A low $P$-value represents good evidence that the research hypothesis is true. Here are some comments on the use of significance tests, with emphasis on their use in reporting scientific research.

## Choosing a level of significance

The spirit of a test of significance is to give a clear statement of the degree of evidence provided by the sample against the null hypothesis. The $P$-value does this. It is common practice to report $P$-values and to describe results as statistically significant whenever $P \leq 0.05$. *However, there is no sharp border between "significant" and "not significant," only increasingly strong evidence as the P-value decreases.* Having both the $P$-value and the statement that we reject or fail to reject $H_0$ allows us to draw better conclusions from our data.

EXAMPLE

**6.21 Information provided by the *P*-value.** Suppose the test statistic for a two-sided significance test for a population mean is $z = 1.95$. From Table A we can calculate the $P$-value. It is

$$P = 2[1 - P(Z \leq 1.95)] = 2(1 - 0.9744) = 0.0512$$

We have failed to meet the standard of $\alpha = 0.05$. However, with the information provided by the $P$-value, we can see that the result just barely missed the standard. If the effect in question is interesting and potentially important, we might want to design another study with a larger sample to investigate it further.

Here is another example where the $P$-value provides useful information beyond that provided by the statement that we reject or fail to reject the null hypothesis.

EXAMPLE

**6.22 More on information provided by the *P*-value?** We have a test statistic of $z = -4.66$ for a two-sided significance test on a population mean. Software tells us that the $P$-value is 0.000003. This means that there are 3 chances in 1,000,000 of observing a sample mean this far or farther away from the null hypothesized value of $\mu$. This kind of event is virtually impossible if the null hypothesis is true. There is no ambiguity in the result; we can clearly reject the null hypothesis.

We frequently report small $P$-values such as that in the previous example as $P < 0.001$. This corresponds to a chance of 1 in 1000 and is sufficiently small to lead us to a clear rejection of the null hypothesis.

One reason for the common use of $\alpha = 0.05$ is the great influence of Sir R. A. Fisher, the inventor of formal statistical methods for analyzing experimental data. Here is his opinion on choosing a level of significance: "A scientific fact should be regarded as experimentally established only if a properly designed experiment *rarely fails* to give this level of significance."[17]

## What statistical significance does not mean

When a null hypothesis ("no effect" or "no difference") can be rejected at the usual level $\alpha = 0.05$, there is good evidence that an effect is present. But that effect can be extremely small. *When large samples are available, even tiny deviations from the null hypothesis will be significant.*

EXAMPLE

**6.23 It's significant. So what?** Suppose that we are testing the hypothesis of no correlation between two variables. With 400 observations, an observed correlation of only $r = 0.1$ is significant evidence at the $\alpha = 0.05$ level that the correlation in the population is not zero. The low significance level does not mean there is a strong association, only that there is strong evidence of some association. The proportion of the variability in one of the variables explained by the other is $r^2 = 0.01$, or 1%.

For practical purposes, we might well decide to ignore this association. *Statistical significance is not the same as practical significance.*

The remedy for attaching too much importance to statistical significance is to pay attention to the actual experimental results as well as to the *P*-value. Plot your data and examine them carefully. Beware of outliers. *The foolish user of statistics who feeds the data to a computer without exploratory analysis will often be embarrassed.* It is usually wise to give a confidence interval for the parameter in which you are interested. Confidence intervals are not used as often as they should be, while tests of significance are perhaps overused.

### USE YOUR KNOWLEDGE

**6.84 Is it significant?** More than 200,000 people worldwide take the GMAT examination each year as they apply for MBA programs. Their scores vary Normally with mean about $\mu = 525$ and standard deviation about $\sigma = 100$. One hundred students go through a rigorous training program designed to raise their GMAT scores. Test the following hypotheses about the training program

$$H_0: \mu = 525$$
$$H_a: \mu > 525$$

in each of the following situations:

(a) The students' average score is $\bar{x} = 541.4$. Is this result significant at the 5% level?

(b) The average score is $\bar{x} = 541.5$. Is this result significant at the 5% level?

(c) Explain how you would reconcile this difference in significance, especially if any increase greater than 15 points is considered a success.

## Don't ignore lack of significance

There is a tendency to conclude that there is no effect whenever a $P$-value fails to attain the usual 5% standard. A provocative editorial in the *British Medical Journal* entitled "Absence of Evidence Is Not Evidence of Absence" deals with this issue.[18] Here is one of the examples they cite.

EXAMPLE

**6.24 Interventions to reduce HIV-1 transmission.** A randomized trial of interventions for reducing transmission of HIV-1 reported an incident rate ratio of 1.00, meaning that the intervention group and the control group both had the same rate of HIV-1 infection. The 95% confidence interval was reported as 0.63 to 1.58.[19] The editorial notes that a summary of these results that says the intervention has no effect on HIV-1 infection is misleading. The confidence interval indicates that the intervention may be capable of achieving a 37% decrease in infection; it might also be harmful and produce a 58% increase in infection. Clearly, more data are needed to distinguish between these possibilities.

The situation can be worse. Research in some fields has rarely been published unless significance at the 0.05 level is attained.

EXAMPLE

**6.25 Journal survey of reported significance results.** A survey of four journals published by the American Psychological Association showed that of 294 articles using statistical tests, only 8 reported results that did not attain the 5% significance level.[20] It is very unlikely that these were the only 8 studies of scientific merit that did not attain significance at the 0.05 level. Manuscripts describing other studies were likely rejected because of a lack of statistical significance or were never submitted in the first place due to the expectation of rejection.

In some areas of research, small effects that are detectable only with large sample sizes can be of great practical significance. Data accumulated from a large number of patients taking a new drug may be needed before we can conclude that there are life-threatening consequences for a small number of people.

On the other hand, sometimes a meaningful result is not found significant.

EXAMPLE

**6.26 A meaningful but statistically insignificant result.** A sample of size 10 gave a correlation of $r = 0.5$ between two variables. The $P$-value is 0.102 for a two-sided significance test. In many situations, a correlation this large would be interesting and worthy of additional study. When it takes a lot of effort (say, in terms of time or money) to obtain samples, researchers often use small studies like these to gain interest from various funding sources. With financial support, a larger, more powerful study can then be run.

*Another important aspect of planning a study is to verify that the test you plan to use does have high probability of detecting an effect of the size you hope to find.* This probability is the *power* of the test. Power calculations are discussed in Section 6.4.

## Statistical inference is not valid for all sets of data

LOOK BACK
**design of experiments, page 181**

In Chapter 3, we learned that badly designed surveys or experiments often produce invalid results. *Formal statistical inference cannot correct basic flaws in the design.*

EXAMPLE

**6.27 English vocabulary and studying a foreign language.** There is no doubt that there is a significant difference in English vocabulary scores between high school seniors who have studied a foreign language and those who have not. But because the effect of actually studying a language is confounded with the differences between students who choose language study and those who do not, this statistical significance is hard to interpret. The most plausible explanation is that students who were already good at English chose to study another language. A randomized comparative experiment would isolate the actual effect of language study and so make significance meaningful. However, such an experiment probably could not be done.

Tests of significance and confidence intervals are based on the laws of probability. Randomization in sampling or experimentation ensures that these laws apply. *But we must often analyze data that do not arise from randomized samples or experiments. To apply statistical inference to such data, we must have confidence in a probability model for the data.* We can check a probability model by examining the data. If the Normal distribution model appears correct, we can apply the methods of this chapter to do inference about the mean $\mu$.

### USE YOUR KNOWLEDGE

**6.85 Home security systems.** A recent TV advertisement for home security systems said that homes without an alarm system are 3 times more likely to be broken into. Suppose this conclusion was obtained by examining an SRS of police records of break-ins and determining whether the percent of homes with alarm systems was significantly smaller than 50%. Explain why the significance of this study is suspect and propose an alternative study that would help clarify the importance of an alarm system.

## Beware of searching for significance

Statistical significance is an outcome much desired by researchers. It means (or ought to mean) that you have found an effect that you were looking for. *The reasoning behind statistical significance works well if you decide what effect you are seeking, design an experiment or sample to search for it, and use a test of significance to weigh the evidence you get.* But because a successful search for a new scientific phenomenon often ends with statistical significance, it is all too tempting to make significance itself the object of the search. There are several ways to do this, none of them acceptable in polite scientific society.

EXAMPLE

**6.28 Microarray studies.** In genomic experiments using microarrays, it is common to perform tens of thousands of significance tests. If each of these was examined separately and statistical significance declared for all that had *P*-values that pass the 0.05 standard, we would have quite a mess. In the absence of any real biological effects, we would expect that, by chance alone, approximately 5% of these tests will show statistical significance. Much research in genomics is directed toward appropriate ways to deal with this situation.[21]

We do not mean that searching data for suggestive patterns is not proper scientific work. It certainly is. Many important discoveries have been made by accident rather than by design. Exploratory analysis of data is an essential part of statistics. We do mean that the usual reasoning of statistical inference does not apply when the search for a pattern is successful. *You cannot legitimately test a hypothesis on the same data that first suggested that hypothesis.* The remedy is clear. Once you have a hypothesis, design a study to search specifically for the effect you now think is there. If the result of this study is statistically significant, you have real evidence.

## SECTION 6.3 Summary

*P*-values are more informative than the reject-or-not result of a fixed level $\alpha$ test. Beware of placing too much weight on traditional values of $\alpha$, such as $\alpha = 0.05$.

Very small effects can be highly significant (small *P*), especially when a test is based on a large sample. A statistically significant effect need not be practically important. Plot the data to display the effect you are seeking, and use confidence intervals to estimate the actual values of parameters.

On the other hand, lack of significance does not imply that $H_0$ is true, especially when the test has low power.

Significance tests are not always valid. Faulty data collection, outliers in the data, and testing a hypothesis on the same data that suggested the hypothesis can invalidate a test. Many tests run at once will probably produce some significant results by chance alone, even if all the null hypotheses are true.

## SECTION 6.3 Exercises

*For Exercise 6.84, see page 396; and for Exercise 6.85, see page 398.*

**6.86** **A role as a statistical consultant.** You are the statistical expert for a graduate student planning her PhD research. After you carefully present the mechanics of significance testing, she suggests using $\alpha = 0.25$ for the study because she would be more likely to obtain statistically significant results and she *really* needs significant results to graduate. Explain in simple terms why this would not be a good use of statistical methods.

**6.87** **What do you know?** A research report described two results that both achieved statistical significance at the 5% level. The *P*-value for the first is 0.049; for the second it is 0.00002. Do the *P*-values add any useful information beyond that conveyed by the statement that both results are statistically significant? Write a short paragraph explaining your views on this question.

6.88 **Interpreting the *P*-value.** A *P*-value of 0.90 is reported for a significance test for a population mean. Interpret this result.

6.89 **What a test of significance can answer.** Explain whether a test of significance can answer each of the following questions.

(a) Is the sample or experiment properly designed?

(b) Is the observed effect compatible with the null hypothesis?

(c) Is the observed effect important?

6.90 **Statistical versus practical significance.** A study with 7500 subjects reported a result that was statistically significant at the 5% level. Explain why this result might not be particularly large or important.

6.91 **More on statistical versus practical significance.** A study with 14 subjects reported a result that failed to achieve statistical significance at the 5% level. The *P*-value was 0.052. Write a short summary of how you would interpret these findings.

6.92 **Vitamin C and colds.** In a study to investigate whether vitamin C will prevent colds, 400 subjects are assigned at random to one of two groups. The experimental group takes a vitamin C tablet daily, while the control group takes a placebo. At the end of the experiment, the researchers calculate the difference between the percents of subjects in the two groups who were free of colds. This difference is statistically significant ($P = 0.03$) in favor of the vitamin C group. Can we conclude that vitamin C has a strong effect in preventing colds? Explain your answer.

6.93 **How far do rich parents take us?** How much education children get is strongly associated with the wealth and social status of their parents, termed "socioeconomic status," or SES. The SES of parents, however, has little influence on whether children who have graduated from college continue their education. One study looked at whether college graduates took the graduate admissions tests for business, law, and other graduate programs. The effects of the parents' SES on taking the LSAT test for law school were "both statistically insignificant and small."

(a) What does "statistically insignificant" mean?

(b) Why is it important that the effects were small in size as well as insignificant?

6.94 CHALLENGE **Find journal articles.** Find two journal articles that report results with statistical analyses. For each article, summarize how the results are reported and write a critique of the presentation. Be sure to include details regarding use of significance testing at a particular level of significance, *P*-values, and confidence intervals.

6.95 **Coaching for the SAT.** Every user of statistics should understand the distinction between statistical significance and practical importance. A sufficiently large sample will declare very small effects statistically significant. Let us suppose that SAT Mathematics (SATM) scores in the absence of coaching vary Normally with mean $\mu = 505$ and $\sigma = 100$. Suppose further that coaching may change $\mu$ but does not change $\sigma$. An increase in the SATM score from 505 to 508 is of no importance in seeking admission to college, but this unimportant change can be statistically very significant. To see this, calculate the *P*-value for the test of

$$H_0: \mu = 505$$

$$H_a: \mu > 505$$

in each of the following situations:

(a) A coaching service coaches 100 students; their SATM scores average $\bar{x} = 508$.

(b) By the next year, the service has coached 1000 students; their SATM scores average $\bar{x} = 508$.

(c) An advertising campaign brings the number of students coached to 10,000; their average score is still $\bar{x} = 508$.

6.96 **More on coaching for the SAT.** Give a 99% confidence interval for the mean SATM score $\mu$ after coaching in each part of the previous exercise. For large samples, the confidence interval says, "Yes, the mean score is higher after coaching, but only by a small amount."

6.97 **Property damage by tornadoes.** Table 1.5 (page 25) gives average property damage per year due to tornadoes for each of the states. Is it appropriate to use the statistical methods we discussed in this chapter for these data? Explain why or why not.

6.98 **When statistical inference is not valid.** Give an example of a set of data for which statistical inference is not valid. Explain your answer.

**6.99 When statistical inference is valid.** Give an example of an interesting set of data for which statistical inference is valid. Explain your answer.

**6.100** CHALLENGE **Predicting success of trainees.** What distinguishes managerial trainees who eventually become executives from those who, after expensive training, don't succeed and leave the company? We have abundant data on past trainees—data on their personalities and goals, their college preparation and performance, even their family backgrounds and their hobbies. Statistical software makes it easy to perform dozens of significance tests on these dozens of variables to see which ones best predict later success. We find that future executives are significantly more likely than washouts to have an urban or suburban upbringing and an undergraduate degree in a technical field.

Explain clearly why using these "significant" variables to select future trainees is not wise. Then suggest a follow-up study using this year's trainees as subjects that should clarify the importance of the variables identified by the first study.

**6.101 Searching for significance.** Give an example of a situation where searching for significance would lead to misleading conclusions.

**6.102 More on searching for significance.** You perform 1000 significance tests using $\alpha = 0.05$. Assuming that all null hypotheses are true, about how many of the test results would you expect to be statistically significant? Explain how you obtained your answer.

**6.103 Interpreting a very small *P*-value.** Assume that you are performing a large number of significance tests. Let $n$ be the number of these tests. How large would $n$ need to be for you to expect about one *P*-value to be 0.00001 or smaller? Use this information to write an explanation of how to interpret a result that has $P = 0.00001$ in this setting.

**6.104** CHALLENGE **An adjustment for multiple tests.** One way to deal with the problem of misleading *P*-values when performing more than one significance test is to adjust the criterion you use for statistical significance. The **Bonferroni procedure** does this in a simple way. If you perform 2 tests and want to use the $\alpha = 5\%$ significance level, you would require a *P*-value of $0.05/2 = 0.025$ to declare either one of the tests significant. In general, if you perform $k$ tests and want protection at level $\alpha$, use $\alpha/k$ as your cutoff for statistical significance. You perform 6 tests and obtain individual *P*-values 0.076, 0.042, 0.241, 0.008, 0.010, and <0.001. Which of these are statistically significant using the Bonferroni procedure with $\alpha = 0.05$?

**6.105** CHALLENGE **Significance using the Bonferroni procedure.** Refer to the previous problem. A researcher has performed 12 tests of significance and wants to apply the Bonferroni procedure with $\alpha = 0.05$. The calculated *P*-values are 0.041, 0.569, 0.050, 0.416, 0.001, 0.004, 0.256, 0.041, 0.888, 0.010, 0.002, and 0.433. Which of these tests reject their null hypotheses with this procedure?

# 6.4 Power and Inference as a Decision*

Although we prefer to use *P*-values rather than the reject-or-not view of the fixed $\alpha$ significance test, the latter view is very important for planning studies and for understanding statistical decision theory. We will discuss these two topics in this section.

## Power

Fixed level $\alpha$ significance tests are closely related to confidence intervals—in fact, we saw that a two-sided test can be carried out directly from a confidence interval. The significance level, like the confidence level, says how reliable the method is in repeated use. If we use 5% significance tests repeatedly when $H_0$

*Although the topics in this section are important in planning and interpreting significance tests, they can be omitted without loss of continuity.

is in fact true, we will be wrong (the test will reject $H_0$) 5% of the time and right (the test will fail to reject $H_0$) 95% of the time.

The ability of a test to detect that $H_0$ is false is measured by the probability that the test will reject $H_0$ when an alternative is true. The higher this probability is, the more sensitive the test is.

### POWER

The probability that a fixed level $\alpha$ significance test will reject $H_0$ when a particular alternative value of the parameter is true is called the **power** of the test to detect that alternative.

EXAMPLE

**6.29 Power of TBBMC significance test.** Can a 6-month exercise program increase the total body bone mineral content (TBBMC) of young women? A team of researchers is planning a study to examine this question. Based on the results of a previous study, they are willing to assume that $\sigma = 2$ for the percent change in TBBMC over the 6-month period. A change in TBBMC of 1% would be considered important, and the researchers would like to have a reasonable chance of detecting a change this large or larger. Is 25 subjects a large enough sample for this project?

We will answer this question by calculating the power of the significance test that will be used to evaluate the data to be collected. The calculation consists of three steps:

1. State $H_0$, $H_a$, the particular alternative we want to detect, and the significance level $\alpha$.
2. Find the values of $\bar{x}$ that will lead us to reject $H_0$.
3. Calculate the probability of observing these values of $\bar{x}$ when the alternative is true.

**Step 1** The null hypothesis is that the exercise program has no effect on TBBMC. In other words, the mean percent change is zero. The alternative is that exercise is beneficial; that is, the mean change is positive. Formally, we have

$$H_0: \mu = 0$$

$$H_a: \mu > 0$$

The alternative of interest is $\mu = 1\%$ increase in TBBMC. A 5% test of significance will be used.

**Step 2** The $z$ test rejects $H_0$ at the $\alpha = 0.05$ level whenever

$$z = \frac{\bar{x} - \mu_0}{\sigma/\sqrt{n}} = \frac{\bar{x} - 0}{2/\sqrt{25}} \geq 1.645$$

Be sure you understand why we use 1.645. Rewrite this in terms of $\bar{x}$:

$$\bar{x} \geq 1.645 \frac{2}{\sqrt{25}}$$

$$\bar{x} \geq 0.658$$

Because the significance level is $\alpha = 0.05$, this event has probability 0.05 of occurring *when the population mean $\mu$ is 0.*

**Step 3** The power against the alternative $\mu = 1\%$ is the probability that $H_0$ will be rejected *when in fact $\mu = 1\%$*. We calculate this probability by standardizing $\bar{x}$, using the value $\mu = 1$, the population standard deviation $\sigma = 2$, and the sample size $n = 25$. The power is

$$P(\bar{x} \geq 0.658 \text{ when } \mu = 1) = P\left(\frac{\bar{x} - \mu}{\sigma/\sqrt{n}} \geq \frac{0.658 - 1}{2/\sqrt{25}}\right)$$
$$= P(Z \geq -0.855) = 0.80$$

Figure 6.15 illustrates the power with the sampling distribution of $\bar{x}$ when $\mu = 1$. This significance test rejects the null hypothesis that exercise has no effect on TBBMC 80% of the time if the true effect of exercise is a 1% increase in TBBMC. If the true effect of exercise is a greater percent increase, the test will have greater power; it will reject with a higher probability.

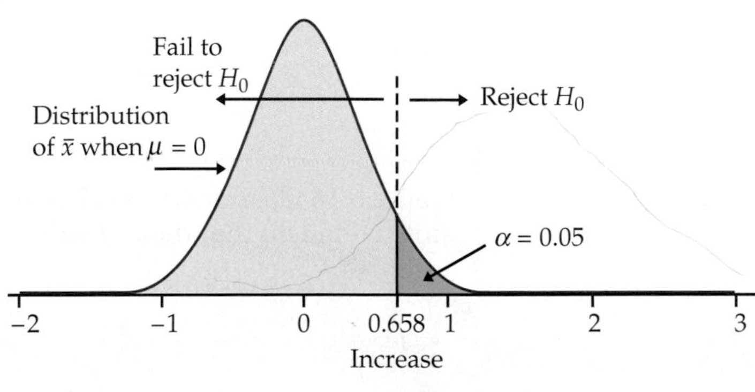


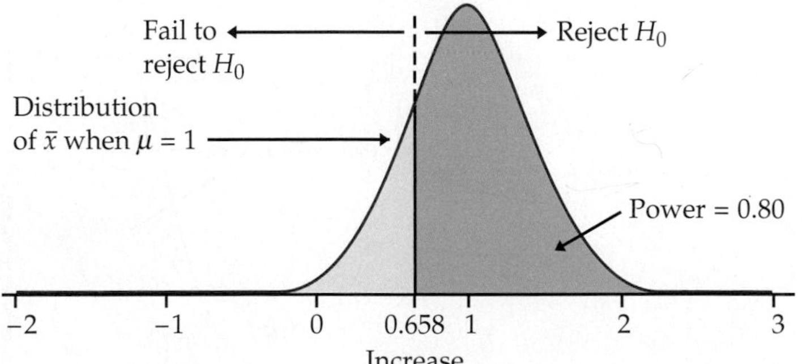


**FIGURE 6.15** The sampling distributions of $\bar{x}$ when $\mu = 0$ and when $\mu = 1$. The power is the probability that the test rejects $H_0$ when the alternative is true.

Here is another example of a power calculation, this time for a two-sided $z$ test.

EXAMPLE

**6.30 Power of the pharmaceutical product test.** Example 6.17 (page 386) presented a test of

$$H_0: \mu = 0.86$$
$$H_a: \mu \neq 0.86$$

at the 1% level of significance. What is the power of this test against the specific alternative $\mu = 0.845$?

The test rejects $H_0$ when $|z| \geq 2.576$. The test statistic is

$$z = \frac{\bar{x} - 0.86}{0.0068/\sqrt{3}}$$

Some arithmetic shows that the test rejects when either of the following is true:

$$z \geq 2.576 \qquad \text{(in other words, } \bar{x} \geq 0.870\text{)}$$
$$z \leq -2.576 \qquad \text{(in other words, } \bar{x} \leq 0.850\text{)}$$

These are disjoint events, so the power is the sum of their probabilities, *computed assuming that the alternative $\mu$ =0.845 is true.* We find that

$$P(\bar{x} \geq 0.87) = P\left(\frac{\bar{x} - \mu}{\sigma/\sqrt{n}} \geq \frac{0.87 - 0.845}{0.0068/\sqrt{3}}\right)$$
$$= P(Z \geq 6.37) \doteq 0$$
$$P(\bar{x} \leq 0.85) = P\left(\frac{\bar{x} - \mu}{\sigma/\sqrt{n}} \leq \frac{0.85 - 0.845}{0.0068/\sqrt{3}}\right)$$
$$= P(Z \leq 1.27) = 0.8980$$

Figure 6.16 illustrates this calculation. Because the power is about 0.9, we are quite confident that the test will reject $H_0$ when this alternative is true.

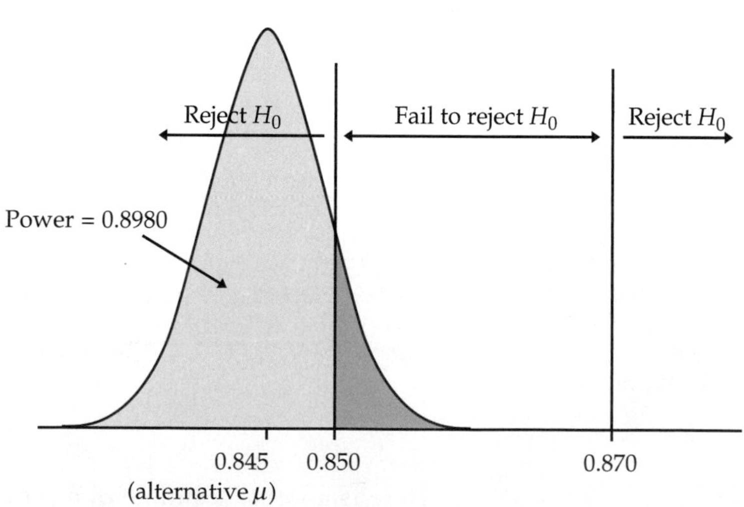


**FIGURE 6.16** The power for Example 6.30.

High power is desirable. Along with 95% confidence intervals and 5% significance tests, 80% power is becoming a standard. Many U.S. government agencies that provide research funds require that the sample size for the funded studies be sufficient to detect important results 80% of the time using a 5% test of significance.

## Increasing the power

Suppose you have performed a power calculation and found that the power is too small. What can you do to increase it? Here are four ways:

- Increase $\alpha$. A 5% test of significance will have a greater chance of rejecting the alternative than a 1% test because the strength of evidence required for rejection is less.
- Consider a particular alternative that is farther away from $\mu_0$. Values of $\mu$ that are in $H_a$ but lie close to the hypothesized value $\mu_0$ are harder to detect (lower power) than values of $\mu$ that are far from $\mu_0$.
- Increase the sample size. More data will provide more information about $\overline{x}$ so we have a better chance of distinguishing values of $\mu$.
- Decrease $\sigma$. This has the same effect as increasing the sample size: more information about $\mu$. Improving the measurement process and restricting attention to a subpopulation are two common ways to decrease $\sigma$.

Power calculations are important in planning studies. Using a significance test with low power makes it unlikely that you will find a significant effect even if the truth is far from the null hypothesis. A null hypothesis that is in fact false can become widely believed if repeated attempts to find evidence against it fail because of low power. The following example illustrates this point.

EXAMPLE

**6.31 Are stock markets efficient?** The "efficient market hypothesis" for the time series of stock prices says that future stock prices (when adjusted for inflation) show only random variation. No information available now will help us predict stock prices in the future, because the efficient working of the market has already incorporated all available information in the present price. Many studies have tested the claim that one or another kind of information is helpful. In these studies, the efficient market hypothesis is $H_0$, and the claim that prediction is possible is $H_a$. Almost all the studies have failed to find good evidence against $H_0$. As a result, the efficient market hypothesis is quite popular. But an examination of the significance tests employed finds that the power is generally low. Failure to reject $H_0$ when using tests of low power is not evidence that $H_0$ is true. As one expert says, "The widespread impression that there is strong evidence for market efficiency may be due just to a lack of appreciation of the low power of many statistical tests."[22]

## Inference as decision*

We have presented tests of significance as methods for assessing the strength of evidence against the null hypothesis. This assessment is made by the $P$-value, which is a probability computed under the assumption that $H_0$ is true. The alternative hypothesis (the statement we seek evidence for) enters the test only to help us see what outcomes count against the null hypothesis.

acceptance sampling

There is another way to think about these issues. Sometimes we are really concerned about making a decision or choosing an action based on our evaluation of the data. **Acceptance sampling** is one such circumstance. A producer of bearings and a skateboard manufacturer agree that each carload lot of bearings shall meet certain quality standards. When a carload arrives, the manufacturer chooses a sample of bearings to be inspected. On the basis of the sample outcome, the manufacturer will either accept or reject the carload. Let's examine how the idea of inference as a decision changes the reasoning used in tests of significance.

## Two types of error

Tests of significance concentrate on $H_0$, the null hypothesis. If a decision is called for, however, there is no reason to single out $H_0$. There are simply two hypotheses, and we must accept one and reject the other. It is convenient to call the two hypotheses $H_0$ and $H_a$, but $H_0$ no longer has the special status (the statement we try to find evidence against) that it had in tests of significance. In the acceptance sampling problem, we must decide between

$H_0$: the lot of bearings meets standards

$H_a$: the lot does not meet standards

on the basis of a sample of bearings.

We hope that our decision will be correct, but sometimes it will be wrong. There are two types of incorrect decisions. We can accept a bad lot of bearings, or we can reject a good lot. Accepting a bad lot injures the consumer, while rejecting a good lot hurts the producer. To help distinguish these two types of error, we give them specific names.

### TYPE I AND TYPE II ERRORS

If we reject $H_0$ (accept $H_a$) when in fact $H_0$ is true, this is a **Type I error.**
If we accept $H_0$ (reject $H_a$) when in fact $H_a$ is true, this is a **Type II error.**

The possibilities are summed up in Figure 6.17. If $H_0$ is true, our decision either is correct (if we accept $H_0$) or is a Type I error. If $H_a$ is true, our decision either is correct or is a Type II error. Only one error is possible at one time. Figure 6.18 applies these ideas to the acceptance sampling example.

*The purpose of this discussion is to clarify the reasoning of significance tests by contrast with a related type of reasoning. It can be omitted without loss of continuity.

| Decision based on sample | | Truth about the population: $H_0$ true | $H_a$ true |
|---|---|---|---|
| | Reject $H_0$ | Type I error | Correct decision |
| | Accept $H_0$ | Correct decision | Type II error |

**FIGURE 6.17** The two types of error in testing hypotheses.

| Decision based on sample | | Truth about the lot: Does meet standards | Does not meet standards |
|---|---|---|---|
| | Reject the lot | Type I error | Correct decision |
| | Accept the lot | Correct decision | Type II error |

**FIGURE 6.18** The two types of error in the acceptance sampling setting.

## Error probabilities

Any rule for making decisions is assessed in terms of the probabilities of the two types of error. This is in keeping with the idea that statistical inference is based on probability. We cannot (short of inspecting the whole lot) guarantee that good lots of bearings will never be rejected and bad lots never be accepted. But by random sampling and the laws of probability, we can say what the probabilities of both kinds of error are.

Significance tests with fixed level $\alpha$ give a rule for making decisions because the test either rejects $H_0$ or fails to reject it. If we adopt the decision-making way of thought, failing to reject $H_0$ means deciding that $H_0$ is true. We can then describe the performance of a test by the probabilities of Type I and Type II errors.

EXAMPLE

**6.32 Outer diameter of a skateboard bearing.** The mean outer diameter of a skateboard bearing is supposed to be 22.000 millimeters (mm). The outer diameters vary Normally with standard deviation $\sigma = 0.010$ mm. When a lot of the bearings arrives, the skateboard manufacturer takes an SRS of 5 bearings from the lot and measures their outer diameters. The manufacturer rejects the bearings if the sample mean diameter is significantly different from 22 at the 5% significance level.

This is a test of the hypotheses

$$H_0: \mu = 22$$

$$H_a: \mu \neq 22$$

To carry out the test, the manufacturer computes the $z$ statistic:

$$z = \frac{\bar{x} - 22}{0.01/\sqrt{5}}$$

and rejects $H_0$ if

$$z < -1.96 \quad \text{or} \quad z > 1.96$$

A Type I error is to reject $H_0$ when in fact $\mu = 22$.

What about Type II errors? Because there are many values of $\mu$ in $H_a$, we will concentrate on one value. The producer and the manufacturer agree that a lot of bearings with mean 0.015 cm away from the desired mean 22.000 should be rejected. So a particular Type II error is to accept $H_0$ when in fact $\mu = 22.015$.

Figure 6.19 shows how the two probabilities of error are obtained from the two sampling distributions of $\bar{x}$, for $\mu = 22$ and for $\mu = 22.015$. When $\mu = 22$, $H_0$ is true and to reject $H_0$ is a Type I error. When $\mu = 22.015$, accepting $H_0$ is a Type II error. We will now calculate these error probabilities.

The probability of a Type I error is the probability of rejecting $H_0$ when it is really true. In Example 6.32, this is the probability that $|z| \geq 1.96$ when $\mu = 22$. But this is exactly the significance level of the test. The critical value 1.96 was chosen to make this probability 0.05, so we do not have to compute it again. The definition of "significant at level 0.05" is that sample outcomes this extreme will occur with probability 0.05 when $H_0$ is true.

### SIGNIFICANCE AND TYPE I ERROR

The significance level $\alpha$ of any fixed level test is the probability of a Type I error. That is, $\alpha$ is the probability that the test will reject the null hypothesis $H_0$ when $H_0$ is in fact true.

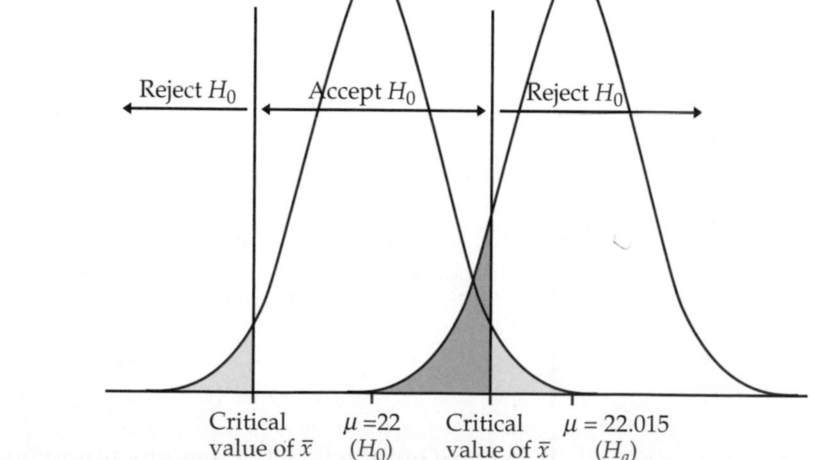


**FIGURE 6.19** The two error probabilities for Example 6.32. The probability of a Type I error (lighter yellow area) is the probability of rejecting $H_0$: $\mu = 22$ when in fact $\mu = 22$. The probability of a Type II error (blue area) is the probability of accepting $H_0$ when in fact $\mu = 22.015$.

The probability of a Type II error for the particular alternative $\mu = 22.015$ in Example 6.32 is the probability that the test will fail to reject $H_0$ when $\mu$ has this alternative value. The *power* of the test against the alternative $\mu = 22.015$ is just the probability that the test *does* reject $H_0$. By following the method of Example 6.30, we can calculate that the power is about 0.92. The probability of a Type II error is therefore $1 - 0.92$, or 0.08.

> **POWER AND TYPE II ERROR**
>
> The power of a fixed level test to detect a particular alternative is 1 minus the probability of a Type II error for that alternative.

The two types of error and their probabilities give another interpretation of the significance level and power of a test. The distinction between tests of significance and tests as rules for deciding between two hypotheses does not lie in the calculations but in the reasoning that motivates the calculations. In a test of significance we focus on a single hypothesis ($H_0$) and a single probability (the $P$-value). The goal is to measure the strength of the sample evidence against $H_0$. Calculations of power are done to check the sensitivity of the test. If we cannot reject $H_0$, we conclude only that there is not sufficient evidence against $H_0$, not that $H_0$ is actually true. If the same inference problem is thought of as a decision problem, we focus on two hypotheses and give a rule for deciding between them based on the sample evidence. We therefore must focus equally on two probabilities, the probabilities of the two types of error. We must choose one hypothesis and cannot abstain on grounds of insufficient evidence.

## The common practice of testing hypotheses

Such a clear distinction between the two ways of thinking is helpful for understanding. In practice, the two approaches often merge. We continued to call one of the hypotheses in a decision problem $H_0$. The common practice of *testing hypotheses* mixes the reasoning of significance tests and decision rules as follows:

1. State $H_0$ and $H_a$ just as in a test of significance.
2. Think of the problem as a decision problem, so that the probabilities of Type I and Type II errors are relevant.
3. Because of Step 1, Type I errors are more serious. So choose an $\alpha$ (significance level) and consider only tests with probability of Type I error no greater than $\alpha$.
4. Among these tests, select one that makes the probability of a Type II error as small as possible (that is, power as large as possible). If this probability is too large, you will have to take a larger sample to reduce the chance of an error.

Testing hypotheses may seem to be a hybrid approach. It was, historically, the effective beginning of decision-oriented ideas in statistics. An impressive mathematical theory of hypothesis testing was developed between 1928 and 1938 by Jerzy Neyman and Egon Pearson. The decision-making approach came later (1940s). Because decision theory in its pure form leaves you with two error probabilities and no simple rule on how to balance them, it has been used less often than either tests of significance or tests of hypotheses. Decision ideas have been applied in testing problems mainly by way of the Neyman-Pearson hypothesis-testing theory. That theory asks you first to choose $\alpha$, and the influence of Fisher often has led users of hypothesis testing comfortably back to $\alpha = 0.05$ or $\alpha = 0.01$. Fisher, who was exceedingly argumentative, violently attacked the Neyman-Pearson decision-oriented ideas, and the argument still continues.

## SECTION 6.4 Summary

The **power** of a significance test measures its ability to detect an alternative hypothesis. The power against a specific alternative is calculated as the probability that the test will reject $H_0$ when that alternative is true. This calculation requires knowledge of the sampling distribution of the test statistic under the alternative hypothesis. Increasing the size of the sample increases the power when the significance level remains fixed.

An alternative to significance testing regards $H_0$ and $H_a$ as two statements of equal status that we must decide between. This **decision theory** point of view regards statistical inference in general as giving rules for making decisions in the presence of uncertainty.

In the case of testing $H_0$ versus $H_a$, decision analysis chooses a decision rule on the basis of the probabilities of two types of error. A **Type I error** occurs if $H_0$ is rejected when it is in fact true. A **Type II error** occurs if $H_0$ is accepted when in fact $H_a$ is true.

In a fixed level $\alpha$ significance test, the significance level $\alpha$ is the probability of a Type I error, and the power against a specific alternative is 1 minus the probability of a Type II error for that alternative.

## SECTION 6.4 Exercises

**6.106 Make a recommendation.** Your manager has asked you to review a research proposal that includes a section on sample size justification. A careful reading of this section indicates that the power is 20% for detecting an effect that you would consider important. Write a short report for your manager explaining what this means and make a recommendation on whether or not this study should be run.

**6.107 Explain power and sample size.** Two studies are identical in all respects except for the sample sizes. Consider the power versus a particular sample size. Will the study with the larger sample size have more power or less power than the one with the smaller sample size? Explain your answer in terms that could be understood by someone with very little knowledge of statistics.

**6.108 Power for a different alternative.** The power for a two-sided test of the null hypothesis $\mu = 0$ versus the alternative $\mu = 5$ is 0.82. What is the power versus the alternative $\mu = -5$? Explain your answer.

**6.109 More on the power for a different alternative.** A one-sided test of the null hypothesis $\mu = 50$ versus the alternative $\mu = 60$ has power equal to 0.5. Will the power for the alternative $\mu = 70$ be higher or lower than 0.5? Draw a picture and use this to explain your answer.

**6.110** APPLET **Power of the random north-south distribution of trees test.** In Exercise 6.66 (page 392) you performed a two-sided significance test of the null hypothesis that the average north-south location of the longleaf pine trees sampled in the Wade Tract was $\mu = 100$. There were 584 trees in the sample and the standard deviation was assumed to be 58. The sample mean in that analysis was $\bar{x} = 99.74$. Use the *Power* applet to compute the power for the alternative $\mu = 99$ using a two-sided test at the 5% level of significance.

**6.111** APPLET **Power of the random east-west distribution of trees test.** Refer to the previous exercise. Note that in the east-west direction, the average location was 113.8. Use the *Power* applet to find the power for the alternative $\mu = 110$.

**6.112 Mail-order catalog sales.** You want to see if a redesign of the cover of a mail-order catalog will increase sales. A very large number of customers will receive the original catalog, and a random sample of customers will receive the one with the new cover. For planning purposes, you are willing to assume that the sales from the new catalog will be approximately Normal with $\sigma = 50$ dollars and that the mean for the original catalog will be $\mu = 25$ dollars. You decide to use a sample size of $n = 900$. You wish to test

$$H_0: \mu = 25$$

$$H_a: \mu > 25$$

You decide to reject $H_0$ if $\bar{x} > 26$.

(a) Find the probability of a Type I error, that is, the probability that your test rejects $H_0$ when in fact $\mu = 25$ dollars.

(b) Find the probability of a Type II error when $\mu = 28$ dollars. This is the probability that your test accepts $H_0$ when in fact $\mu = 28$.

(c) Find the probability of a Type II error when $\mu = 30$.

(d) The distribution of sales is not Normal, because many customers buy nothing. Why is it nonetheless reasonable in this circumstance to assume that the mean will be approximately Normal?

**6.113 Power of the mean SAT score test.** Example 6.16 (page 385) gives a test of a hypothesis about the SAT scores of California high school students based on an SRS of 500 students. The hypotheses are

$$H_0: \mu = 450$$

$$H_a: \mu > 450$$

Assume that the population standard deviation is $\sigma = 100$. The test rejects $H_0$ at the 1% level of significance when $z \geq 2.326$, where

$$z = \frac{\bar{x} - 450}{100/\sqrt{500}}$$

Is this test sufficiently sensitive to usually detect an increase of 10 points in the population mean SAT score? Answer this question by calculating the power of the test against the alternative $\mu = 460$.

**6.114** CHALLENGE **Choose the appropriate distribution.** You must decide which of two discrete distributions a random variable $X$ has. We will call the distributions $p_0$ and $p_1$. Here are the probabilities they assign to the values $x$ of $X$:

| $x$ | 0 | 1 | 2 | 3 | 4 | 5 | 6 |
|---|---|---|---|---|---|---|---|
| $p_0$ | 0.1 | 0.1 | 0.1 | 0.2 | 0.1 | 0.1 | 0.3 |
| $p_1$ | 0.3 | 0.1 | 0.1 | 0.2 | 0.1 | 0.1 | 0.1 |

You have a single observation on $X$ and wish to test

$$H_0: p_0 \text{ is correct}$$

$$H_a: p_1 \text{ is correct}$$

One possible decision procedure is to reject $H_0$ only if $X = 0$ or $X = 1$.

(a) Find the probability of a Type I error, that is, the probability that you reject $H_0$ when $p_0$ is the correct distribution.

(b) Find the probability of a Type II error.

**6.115 A Web-based business.** You are in charge of marketing for a Web site that offers automated medical diagnoses. The program will scan the results of routine medical tests (pulse rate, blood pressure, urinalysis, etc.) and either clear the patient or refer the case to a doctor. You are marketing the program for use as part of a preventive-medicine system to screen many thousands of persons who do not have specific medical complaints. The program makes a decision about each patient.

(a) What are the two hypotheses and the two types of error that the program can make? Describe the two types of error in terms of "false-positive" and "false-negative" test results.

(b) The program can be adjusted to decrease one error probability at the cost of an increase in the other error probability. Which error probability would you choose to make smaller, and why? (This is a matter of judgment. There is no single correct answer.)

## CHAPTER 6 Exercises

**6.116** CHALLENGE **Full-time employment and age.** A study of late adolescents and early adults reported average months of full-time employment for individuals aged 18 to 26.[23] Here are the means:

| Age | 18 | 19 | 20 | 21 | 22 | 23 | 24 | 25 | 26 |
|---|---|---|---|---|---|---|---|---|---|
| Months employed | 2.9 | 4.2 | 5.0 | 5.3 | 6.4 | 7.4 | 8.5 | 8.9 | 9.3 |

Assume that the standard deviation for each of these means is 4.5 months and that each sample size is 750.

(a) Calculate the 95% confidence interval for each mean.

(b) Plot the means versus age. Draw a vertical line through the first mean extending up to the upper confidence limit and down to the lower limit. At the ends of the line, draw a short dash. Do the same for each of the other means.

(c) Write a summary of what the data show. Note that in circumstances such as this, it is common practice not to make any adjustments for the fact that several confidence intervals are being reported. Be sure to include comments about this in your summary.

**6.117** CHALLENGE **Workers' perceptions about safety.** The Safety Climate Index (SCI) measures workers' perceptions about the safety of their work environment. A study of safe work practices of industrial workers reported mean SCI scores for workers classified by workplace size.[24] Here are the means:

| Workplace size | Fewer than 50 workers | 50 to 200 workers | More than 200 workers |
|---|---|---|---|
| Mean SCI | 67.23 | 70.37 | 74.83 |

Assume that the standard deviation is 19 and the sample sizes are all 180. (We will discuss ways to compare three means such as these in Chapter 12.)

(a) Calculate the 95% confidence interval for each mean.

(b) Plot the means versus workplace size. Draw a vertical line through the first mean extending up to the upper confidence limit and down to the lower limit. At the ends of the line, draw a short dash. Do the same for each of the other means.

(c) One way to adjust for the fact that we are reporting three confidence intervals is a procedure that uses a larger value of $z^*$ in the calculation of the margin of error. For this problem one recommendation would be to use $z^* = 2.40$. Repeat parts (a) and (b) making this adjustment.

(d) Summarize your results. Be sure to include comments on the effects of the adjustment on your results.

**6.118** APPLET CHALLENGE **Coverage percent of 95% confidence interval.** For this exercise you will use the *Confidence Interval* applet. Set the confidence level at 95% and click the "Sample" button 10 times to simulate 10 confidence intervals. Record the percent hit. Simulate another 10 intervals by clicking another 10 times (do not click the "Reset" button). Record the percent hit for your 20 intervals. Repeat the process of simulating 10 additional intervals and recording the results until you have a total of 200 intervals. Plot your results and write a summary of what you have found.

**6.119** APPLET CHALLENGE **Coverage percent of 90% confidence interval.** Refer to the previous exercise. Do the simulations and report the results for 90% confidence.

**6.120** CHALLENGE **Effect of sample size on significance.** You are testing the null hypothesis that $\mu = 0$ versus the alternative $\mu > 0$ using $\alpha = 0.05$. Assume $\sigma = 14$. Suppose $\bar{x} = 4$ and $n = 10$. Calculate the test statistic and its $P$-value. Repeat assuming the same value of $\bar{x}$ but with $n = 20$. Do the same for sample sizes of 30, 40, and 50. Plot the values of the test statistic versus the sample size. Do the same for the $P$-values. Summarize what this demonstration shows about the effect of the sample size on significance testing.

**6.121** CHALLENGE **Blood phosphorus level in dialysis patients.** Patients with chronic kidney failure may be treated by dialysis, using a machine that removes toxic wastes from the blood, a function normally performed by the kidneys. Kidney failure and dialysis can cause other changes, such as retention of phosphorus, that must be corrected by changes in diet. A study of the nutrition of dialysis patients measured the level of phosphorus in the blood of several patients on six occasions. Here are the data for one patient (in milligrams of phosphorus per deciliter of blood):[25]

5.4 5.2 4.5 4.9 5.7 6.3

The measurements are separated in time and can be considered an SRS of the patient's blood phosphorus level. Assume that this level varies Normally with $\sigma = 0.9$ mg/dl.

(a) Give a 95% confidence interval for the mean blood phosphorus level.

(b) The normal range of phosphorus in the blood is considered to be 2.6 to 4.8 mg/dl. Is there strong evidence that this patient has a mean phosphorus level that exceeds 4.8?

**6.122 Cellulose content in alfalfa hay.** An agronomist examines the cellulose content of a variety of alfalfa hay. Suppose that the cellulose content in the population has standard deviation $\sigma = 8$ milligrams per gram (mg/g). A sample of 15 cuttings has mean cellulose content $\bar{x} = 145$ mg/g.

(a) Give a 90% confidence interval for the mean cellulose content in the population.

(b) A previous study claimed that the mean cellulose content was $\mu = 140$ mg/g, but the agronomist believes that the mean is higher than that figure. State $H_0$ and $H_a$ and carry out a significance test to see if the new data support this belief.

(c) The statistical procedures used in (a) and (b) are valid when several assumptions are met. What are these assumptions?

**6.123 Odor threshold of future wine experts.** Many food products contain small quantities of substances that would give an undesirable taste or smell if they are present in large amounts. An example is the "off-odors" caused by sulfur compounds in wine. Oenologists (wine experts) have determined the odor threshold, the lowest concentration of a compound that the human nose can detect. For example, the odor threshold for dimethyl sulfide (DMS) is given in the oenology literature as 25 micrograms per liter of wine ($\mu$g/l). Untrained noses may be less sensitive, however. Here are the DMS odor thresholds for 10 beginning students of oenology:

31 31 43 36 23 34 32 30 20 24

Assume (this is not realistic) that the standard deviation of the odor threshold for untrained noses is known to be $\sigma = 7$ $\mu$g/l.

(a) Make a stemplot to verify that the distribution is roughly symmetric with no outliers. (A Normal quantile plot confirms that there are no systematic departures from Normality.)

(b) Give a 95% confidence interval for the mean DMS odor threshold among all beginning oenology students.

(c) Are you convinced that the mean odor threshold for beginning students is higher than the published threshold, 25 $\mu$g/l? Carry out a significance test to justify your answer.

**6.124** CHALLENGE **Where do you buy?** Consumers can purchase nonprescription medications at food stores, mass merchandise stores such as Kmart and Wal-Mart, or pharmacies. About 45% of consumers make such purchases at pharmacies. What accounts for the popularity of pharmacies, which often charge higher prices?

A study examined consumers' perceptions of overall performance of the three types of stores, using a long questionnaire that asked about such things as "neat and attractive store," "knowledgeable staff," and "assistance in choosing among various types of nonprescription medication." A performance score was based on 27 such questions. The subjects were 201 people chosen at random from the Indianapolis telephone directory. Here are the means and standard deviations of the performance scores for the sample:[26]

| Store type | $\bar{x}$ | $s$ |
|---|---|---|
| Food stores | 18.67 | 24.95 |
| Mass merchandisers | 32.38 | 33.37 |
| Pharmacies | 48.60 | 35.62 |

We do not know the population standard deviations, but a sample standard deviation $s$ from so large a sample is usually close to $\sigma$. Use $s$ in place of the unknown $\sigma$ in this exercise.

(a) What population do you think the authors of the study want to draw conclusions about? What population are you certain they can draw conclusions about?

(b) Give 95% confidence intervals for the mean performance for each type of store.

(c) Based on these confidence intervals, are you convinced that consumers think that pharmacies offer higher performance than the other types of stores? (In Chapter 12, we will study a statistical method for comparing means of several groups.)

**6.125 CEO pay.** A study of the pay of corporate chief executive officers (CEOs) examined the increase in cash compensation of the CEOs of 104 companies, adjusted for inflation, in a recent year. The mean increase in real compensation was $\bar{x} = 6.9\%$, and the standard deviation of the increases was $s = 55\%$. Is this good evidence that the mean real compensation $\mu$ of all CEOs increased that year? The hypotheses are

$$H_0: \mu = 0 \quad \text{(no increase)}$$

$$H_a: \mu > 0 \quad \text{(an increase)}$$

Because the sample size is large, the sample $s$ is close to the population $\sigma$, so take $\sigma = 55\%$.

(a) Sketch the Normal curve for the sampling distribution of $\bar{x}$ when $H_0$ is true. Shade the area that represents the $P$-value for the observed outcome $\bar{x} = 6.9\%$.

(b) Calculate the $P$-value.

(c) Is the result significant at the $\alpha = 0.05$ level? Do you think the study gives strong evidence that the mean compensation of all CEOs went up?

**6.126 Meaning of "statistically significant."** When asked to explain the meaning of "statistically significant at the $\alpha = 0.01$ level," a student says, "This means there is only probability 0.01 that the null hypothesis is true." Is this an essentially correct explanation of statistical significance? Explain your answer.

**6.127 More on the meaning of "statistically significant."** Another student, when asked why statistical significance appears so often in research reports, says, "Because saying that results are significant tells us that they cannot easily be explained by chance variation alone." Do you think that this statement is essentially correct? Explain your answer.

**6.128 Roulette.** A roulette wheel has 18 red slots among its 38 slots. You observe many spins and record the number of times that red occurs. Now you want to use these data to test whether the probability of a red has the value that is correct for a fair roulette wheel. State the hypotheses $H_0$ and $H_a$ that you will test. (We will describe the test for this situation in Chapter 8.)

**6.129** CHALLENGE **Simulation study of the confidence interval.** Use a computer to generate $n = 12$ observations from a Normal distribution with mean 25 and standard deviation 4: $N(25, 4)$. Find the 95% confidence interval for $\mu$. Repeat this process 100 times and then count the number of times that the confidence interval includes the value $\mu = 25$. Explain your results.

**6.130** CHALLENGE **Simulation study of a test of significance.** Use a computer to generate $n = 12$ observations from a Normal distribution with mean 25 and standard deviation 4: $N(25, 4)$. Test the null hypothesis that $\mu = 25$ using a two-sided significance test. Repeat this process 100 times and then count the number of times that you reject $H_0$. Explain your results.

**6.131** CHALLENGE **Another simulation study of a test of significance.** Use the same procedure for generating data as in the previous exercise. Now test the null hypothesis that $\mu = 23$. Explain your results.

**6.132** CHALLENGE **Older customer concerns in restaurants.** Persons aged 55 and over represented 21.3% of the U.S. population in the year 2000. This group is expected to increase to 30.5% by 2025. In terms of actual numbers of people, the increase is from 58.6 million to 101.4 million. Restauranteurs have found this market to be important and would like to make their businesses attractive to older customers. One study used a questionnaire to collect data from people aged 50 and over.[27] For one part of the analysis, individuals were classified into two age groups: 50 to 64 and 65 to 79. There were 267 people in the first group and 263 in the second. One set of items concerned ambiance, menu design, and service. A series of statements were rated on a 1 to 5 scale with 1 representing "strongly disagree" and 5 representing "strongly agree." In some cases the wording has been shortened in the table below. Here are the means:

| Statement | 50–64 | 65–79 |
|---|---|---|
| Ambiance: | | |
| Most restaurants are too dark | 2.75 | 2.93 |
| Most restaurants are too noisy | 3.33 | 3.43 |
| Background music is often too loud | 3.27 | 3.55 |
| Restaurants are too smoky | 3.17 | 3.12 |
| Tables are too small | 3.00 | 3.19 |
| Tables are too close together | 3.79 | 3.81 |
| Menu design: | | |
| Print size is not large enough | 3.68 | 3.77 |
| Glare makes menus difficult to read | 2.81 | 3.01 |
| Colors of menus make them difficult to read | 2.53 | 2.72 |

| Statement | 50–64 | 65–79 |
|---|---|---|
| Service: | | |
| It is difficult to hear the service staff | 2.65 | 3.00 |
| I would rather be served than serve myself | 4.23 | 4.14 |
| I would rather pay the server than a cashier | 3.88 | 3.48 |
| Service is too slow | 3.13 | 3.10 |

First examine the means of the people who are 50 to 64. Order the statements according to the means and describe the results. Then do the same for the older group. For each statement compute the $z$ statistic and the associated $P$-value for the comparison between the two groups. For these calculations you can assume that the standard deviation of the difference is 0.08, so $z$ is simply the difference in the means divided by 0.08. Note that you are performing 13 significance tests in this exercise. Keep this in mind when you interpret your results. Write a report summarizing your work.

6.133 CHALLENGE **Find published studies with confidence intervals.** Search the Internet or some journals that report research in your field and find two reports that provide an estimate with a margin of error or a confidence interval. For each report:

(a) Describe the method used to collect the data.

(b) Describe the variable being studied.

(c) Give the estimate and the confidence interval.

(d) Describe any practical difficulties that may have led to errors in addition to the sampling errors quantified by the margin of error.

CHAPTER

7

# Inference for Distributions

Some people feel that a full moon causes strange and aggressive behavior in people. Is there any scientific evidence to support this? Example 7.7 describes one such study.



## Introduction

We began our study of data analysis in Chapter 1 by learning graphical and numerical tools for describing the distribution of a single variable and for comparing several distributions. Our study of the practice of statistical inference begins in the same way, with inference about a single distribution and comparison of two distributions. Comparing more than two distributions requires more elaborate methods, which are presented in Chapters 12 and 13.

Two important aspects of any distribution are its center and spread. If the distribution is Normal, we describe its center by the mean $\mu$ and its spread by the standard deviation $\sigma$. In this chapter, we will meet confidence intervals and significance tests for inference about a population mean $\mu$ and for comparing the means or spreads of two populations. The previous chapter emphasized the reasoning of tests and confidence intervals; now we emphasize statistical practice, so we no longer assume that population standard deviations are known. The $t$ procedures for inference about means are among the most common statistical methods. Inference about the spreads, as we will see, poses some difficult practical problems.

The methods in this chapter will allow us to address questions like:

- Does cellular phone use, specifically the number of hours listening to music tracks, differ between cell phone users in the United States and the United Kingdom?
- Do male and female college students differ in terms of "social insight," the ability to appraise other people?
- Does the daily number of disruptive behaviors in dementia patients change when there is a full moon?

## 7.1 Inference for the Mean of a Population

LOOK BACK
**sampling distribution of $\bar{x}$, page 339**

Both confidence intervals and tests of significance for the mean $\mu$ of a Normal population are based on the sample mean $\bar{x}$, which estimates the unknown $\mu$. The sampling distribution of $\bar{x}$ depends on $\sigma$. This fact causes no difficulty when $\sigma$ is known. When $\sigma$ is unknown, however, we must estimate $\sigma$ even though we are primarily interested in $\mu$. The sample standard deviation $s$ is used to estimate the population standard deviation $\sigma$.

### The *t* distributions

Suppose that we have a simple random sample (SRS) of size $n$ from a Normally distributed population with mean $\mu$ and standard deviation $\sigma$. The sample mean $\bar{x}$ is then Normally distributed with mean $\mu$ and standard deviation $\sigma/\sqrt{n}$. When $\sigma$ is not known, we estimate it with the sample standard deviation $s$, and then we estimate the standard deviation of $\bar{x}$ by $s/\sqrt{n}$. This quantity is called the *standard error* of the sample mean $\bar{x}$ and we denote it by $\mathrm{SE}_{\bar{x}}$.

> STANDARD ERROR
>
> When the standard deviation of a statistic is estimated from the data, the result is called the **standard error** of the statistic. The standard error of the sample mean is
>
> $$\mathrm{SE}_{\bar{x}} = \frac{s}{\sqrt{n}}$$

The term "standard error" is sometimes used for the actual standard deviation of a statistic. The estimated value is then called the "estimated standard error." In this book we will use the term "standard error" only when the standard deviation of a statistic is estimated from the data. The term has this meaning in the output of many statistical computer packages and in research reports that apply statistical methods.

The standardized sample mean, or one-sample $z$ statistic,

$$z = \frac{\bar{x} - \mu}{\sigma/\sqrt{n}}$$

is the basis of the $z$ procedures for inference about $\mu$ when $\sigma$ is known. This statistic has the standard Normal distribution $N(0, 1)$. When we substitute the standard error $s/\sqrt{n}$ for the standard deviation $\sigma/\sqrt{n}$ of $\bar{x}$, the statistic does *not* have a Normal distribution. It has a distribution that is new to us, called a *t distribution.*

### THE $t$ DISTRIBUTIONS

Suppose that an SRS of size $n$ is drawn from an $N(\mu, \sigma)$ population. Then the **one-sample $t$ statistic**

$$t = \frac{\bar{x} - \mu}{s/\sqrt{n}}$$

has the **$t$ distribution** with $n - 1$ **degrees of freedom.**

A particular $t$ distribution is specified by giving the *degrees of freedom.* We use $t(k)$ to stand for the $t$ distribution with $k$ degrees of freedom. The degrees of freedom for this $t$ statistic come from the sample standard deviation $s$ in the denominator of $t$. We showed earlier that $s$ has $n - 1$ degrees of freedom. Thus, there is a different $t$ distribution for each sample size. There are also other $t$ statistics with different degrees of freedom, some of which we will meet later in this chapter.

LOOK BACK
**degrees of freedom, page 42**

The $t$ distributions were discovered in 1908 by William S. Gosset. Gosset was a statistician employed by the Guinness brewing company, which prohibited its employees from publishing their discoveries that were brewing related. In this case, the company let him publish under the pen name "Student" using an example that did not involve brewing. The $t$ distribution is often called "Student's $t$" in his honor.

The density curves of the $t(k)$ distributions are similar in shape to the standard Normal curve. That is, they are symmetric about 0 and are bell-shaped. Figure 7.1 compares the density curves of the standard Normal distribution and the $t$ distributions with 5 and 10 degrees of freedom. The similarity in shape is apparent, as is the fact that the $t$ distributions have more probability in the tails and less in the center. This greater spread is due to the extra variability caused by substituting the random variable $s$ for the fixed parameter $\sigma$. Figure 7.1 also shows that as the degrees of freedom $k$ increase, the $t(k)$ density curve gets closer to the $N(0, 1)$ curve. This reflects the fact that $s$ will likely be closer to $\sigma$ as the sample size increases.

Table D in the back of the book gives critical values $t^*$ for the $t$ distributions. For convenience, we have labeled the table entries both by the value of $p$ needed for significance tests and by the confidence level $C$ (in percent) required for confidence intervals. The standard Normal critical values are in the bottom row of entries and labeled $z^*$. As in the case of the Normal table (Table A), computer software often makes Table D unnecessary.

## USE YOUR KNOWLEDGE

**7.1** **Apartment rents.** You randomly choose 15 unfurnished one-bedroom apartments from a large number of advertisements in your local

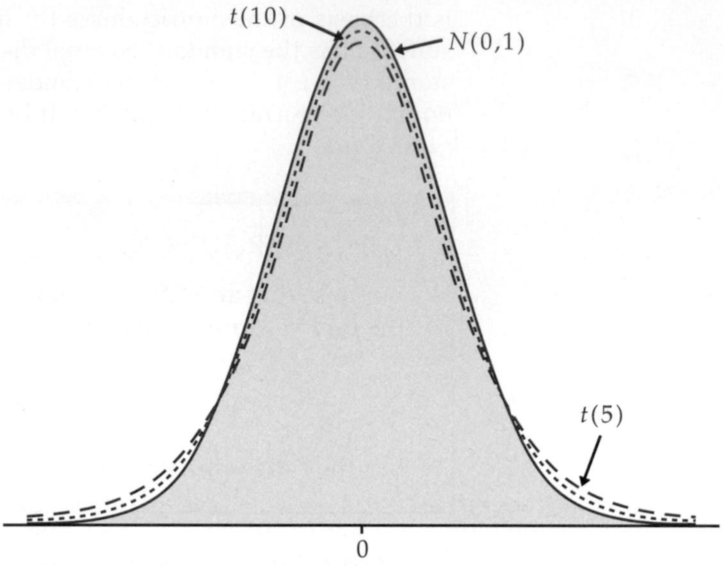

**FIGURE 7.1** Density curves for the standard Normal, $t(10)$, and $t(5)$ distributions. All are symmetric with center 0. The $t$ distributions have more probability in the tails than the standard Normal distribution.

newspaper. You calculate that their mean monthly rent is \$570 and their standard deviation is \$105.

(a) What is the standard error of the mean?

(b) What are the degrees of freedom for a one-sample $t$ statistic?

**7.2** **Finding critical $t^*$ values.** What critical value $t^*$ from Table D should be used to construct

(a) a 95% confidence interval when $n = 12$?

(b) a 99% confidence interval when $n = 24$?

(c) a 90% confidence interval when $n = 200$?

## The one-sample $t$ confidence interval

LOOK BACK
*z* **confidence interval, page 361**

With the $t$ distributions to help us, we can now analyze a sample from a Normal population with unknown $\sigma$. The one-sample $t$ confidence interval is similar in both reasoning and computational detail to the $z$ confidence interval of Chapter 6. There, the margin of error for the population mean was $z^*\sigma/\sqrt{n}$. Here, we replace $\sigma$ by its estimate $s$ and $z^*$ by $t^*$. This means that the margin of error for the population mean when we use the data to estimate $\sigma$ is $t^*s/\sqrt{n}$.

### THE ONE-SAMPLE $t$ CONFIDENCE INTERVAL

Suppose that an SRS of size $n$ is drawn from a population having unknown mean $\mu$. A level $C$ **confidence interval** for $\mu$ is

$$\bar{x} \pm t^* \frac{s}{\sqrt{n}}$$

where $t^*$ is the value for the $t(n-1)$ density curve with area $C$ between $-t^*$ and $t^*$. The quantity

$$t^* \frac{s}{\sqrt{n}}$$

is the **margin of error.** This interval is exact when the population distribution is Normal and is approximately correct for large $n$ in other cases.

EXAMPLE

**7.1 Listening to music on cell phones.** Founded in 1998, Telephia provides a wide variety of information on cellular phone use. In 2006, Telephia reported that, on average, United Kingdom (U.K.) subscribers with third-generation technology (3G) phones spent an average of 8.3 hours per month listening to full-track music on their cell phones.[1] Suppose we want to determine a 95% confidence interval for the U.S. average and draw the following random sample of size 8 from the U.S. population of 3G subscribers:

5 6 0 4 11 9 2 3

The sample mean is $\bar{x} = 5$ and the standard deviation is $s = 3.63$ with degrees of freedom $n - 1 = 7$. The standard error is

$$\text{SE}_{\bar{x}} = s/\sqrt{n} = 3.63/\sqrt{8} = 1.28$$

From Table D we find $t^* = 2.365$. The 95% confidence interval is

$$\begin{aligned} \bar{x} \pm t^* \frac{s}{\sqrt{n}} &= 5.0 \pm 2.365 \frac{3.63}{\sqrt{8}} \\ &= 5.0 \pm (2.365)(1.28) \\ &= 5.0 \pm 3.0 \\ &= (2.0, 8.0) \end{aligned}$$

We are 95% confident that the U.S. population's average time spent listening to full-track music on a cell phone is between 2.0 and 8.0 hours per month. Since this interval does not contain 8.3 hours, these data suggest that, on average, a U.S. subscriber listens to less full-track music.

In this example we have given the actual interval (2.0, 8.0) as our answer. Sometimes we prefer to report the mean and margin of error: the mean time is 5.0 hours per month with a margin of error of 3.0 hours.

The use of the $t$ confidence interval in Example 7.1 rests on assumptions that appear reasonable here. First, we assume our random sample is an SRS from the U.S. population of cell phone users. Second, we assume the distribution of listening times is Normal. With only 8 observations, this assumption cannot be effectively checked. In fact, because the listening time cannot be negative, we might expect this distribution to be skewed to the right. With these data, however, there are no extreme outliers to suggest a severe departure from Normality.

**USE YOUR KNOWLEDGE**

**7.3 More on apartment rents.** Recall Exercise 7.1 (page 419). Construct a 95% confidence interval for the mean monthly rent of all advertised one-bedroom apartments.

**7.4 90% versus 95% confidence interval.** If you were to use 90% confidence, rather than 95% confidence, would the margin of error be larger or smaller? Explain your answer.

## The one-sample $t$ test

LOOK BACK

**$z$ significance test, page 383**

Significance tests using the standard error are also very similar to the $z$ test that we studied in the last chapter.

### THE ONE-SAMPLE $t$ TEST

Suppose that an SRS of size $n$ is drawn from a population having unknown mean $\mu$. To test the hypothesis $H_0: \mu = \mu_0$ based on an SRS of size $n$, compute the one-sample $t$ statistic

$$t = \frac{\overline{x} - \mu_0}{s/\sqrt{n}}$$

In terms of a random variable $T$ having the $t(n-1)$ distribution, the $P$-value for a test of $H_0$ against

$H_a: \mu > \mu_0$ is $P(T \geq t)$

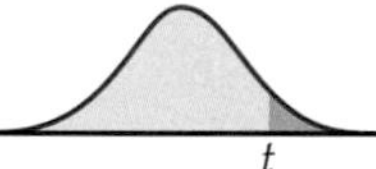


$H_a: \mu < \mu_0$ is $P(T \leq t)$

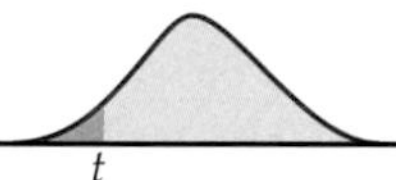


$H_a: \mu \neq \mu_0$ is $2P(T \geq |t|)$

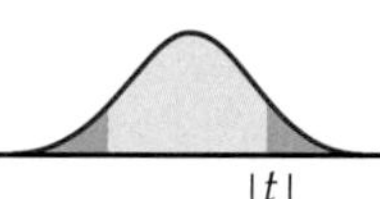


These $P$-values are exact if the population distribution is Normal and are approximately correct for large $n$ in other cases.

EXAMPLE

**7.2 Significance test for cell phone use.** Suppose that, for the U.S. data in Example 7.1, we want to test whether the U.S. average is different from the reported U.K. average. Specifically, we want to test

$$H_0: \mu = 8.3$$
$$H_a: \mu \neq 8.3$$

Recall that $n = 8$, $\bar{x} = 5.0$, and $s = 3.63$. The $t$ test statistic is

$$t = \frac{\bar{x} - \mu_0}{s/\sqrt{n}} = \frac{5.0 - 8.3}{3.63/\sqrt{8}}$$

$$= -2.57$$

df = 7

| $p$ | 0.02 | 0.01 |
|---|---|---|
| $t^*$ | 2.517 | 2.998 |

This means that the sample mean $\bar{x} = 5.0$ is slightly over 2.5 standard deviations away from the null hypothesized value $\mu = 8.3$. Because the degrees of freedom are $n - 1 = 7$, this $t$ statistic has the $t(7)$ distribution. Figure 7.2 shows that the $P$-value is $2P(T \geq 2.57)$, where $T$ has the $t(7)$ distribution. From Table D we see that $P(T \geq 2.517) = 0.02$ and $P(T \geq 2.998) = 0.01$. Therefore, we conclude that the $P$-value is between $2 \times 0.01 = 0.02$ and $2 \times 0.02 = 0.04$. Software gives the exact value as $P = 0.037$. These data are incompatible with a mean of 8.3 hours per month at the $\alpha = 0.05$ level.

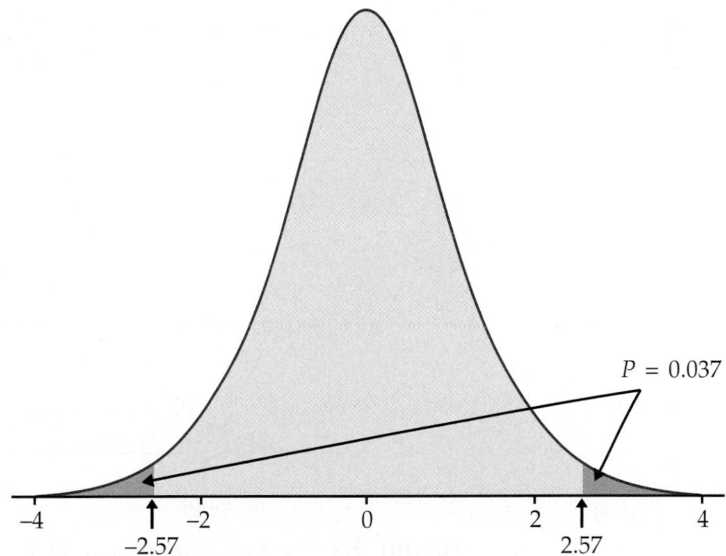


**FIGURE 7.2** The *P*-value for Example 7.2.

In this example we tested the null hypothesis $\mu = 8.3$ hours per month against the two-sided alternative $\mu \neq 8.3$ hours per month because we had no prior suspicion that the average in the United States would be larger or smaller. If we had suspected that the U.S. average would be smaller, we would have used a one-sided test. *It is wrong, however, to examine the data first and then decide to do a one-sided test in the direction indicated by the data.* If in doubt, use a two-sided test. In the present circumstance, however, we could use our results from Example 7.2 to justify a one-sided test for *another* sample from the same population.

EXAMPLE

**7.3 One-sided test for cell phone use.** For the cell phone problem described in the previous example, we want to test whether the U.S. average is smaller than the U.K. average. Here we test

$$H_0: \mu = 8.3$$

versus

$$H_a: \mu < 8.3$$

The $t$ test statistic does not change: $t = -2.57$. As Figure 7.3 illustrates, however, the $P$-value is now $P(T \leq -2.57)$, half of the value in the previous example. From Table D we can determine that $0.01 < P < 0.02$; software gives the exact value as $P = 0.0185$. At the $\alpha = 0.05$ level, we conclude that the U.S. average is smaller than the U.K. average.

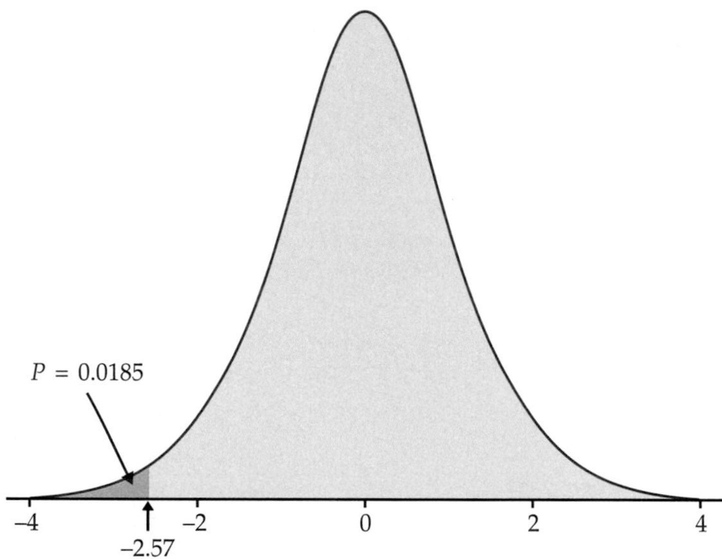


**FIGURE 7.3** The *P*-value for Example 7.3.

For small data sets, such as the one in Example 7.1, it is easy to perform the computations for confidence intervals and significance tests with an ordinary calculator. For larger data sets, however, we prefer to use software or a statistical calculator.

EXAMPLE

**7.4 Stock portfolio diversification?** An investor with a stock portfolio worth several hundred thousand dollars sued his broker and brokerage firm because lack of diversification in his portfolio led to poor performance. Table 7.1 gives the rates of return for the 39 months that the account was managed by the broker.[2] Figure 7.4 gives a histogram for these data and Figure 7.5 gives the Normal quantile plot. There are no outliers and the distribution shows no strong skewness. We are reasonably confident that the

**TABLE 7.1**

**Monthly rates of return on a portfolio (percent)**

| | | | | | | | |
|---|---|---|---|---|---|---|---|
| −8.36 | 1.63 | −2.27 | −2.93 | −2.70 | −2.93 | −9.14 | −2.64 |
| 6.82 | −2.35 | −3.58 | 6.13 | 7.00 | −15.25 | −8.66 | −1.03 |
| −9.16 | −1.25 | −1.22 | −10.27 | −5.11 | −0.80 | −1.44 | 1.28 |
| −0.65 | 4.34 | 12.22 | −7.21 | −0.09 | 7.34 | 5.04 | −7.24 |
| −2.14 | −1.01 | −1.41 | 12.03 | −2.56 | 4.33 | 2.35 | |

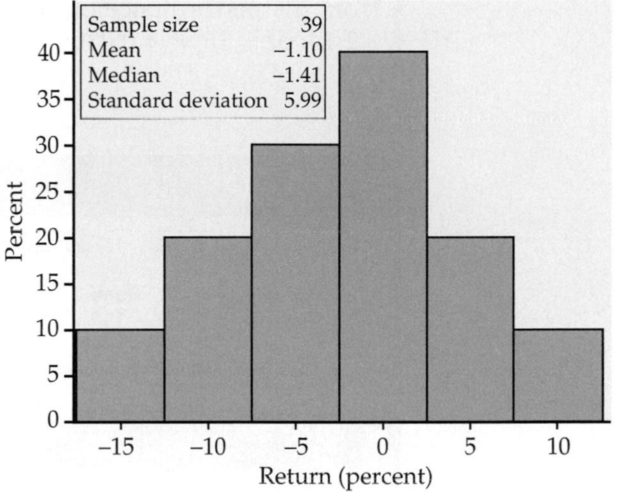


**FIGURE 7.4** Histogram for Example 7.4.

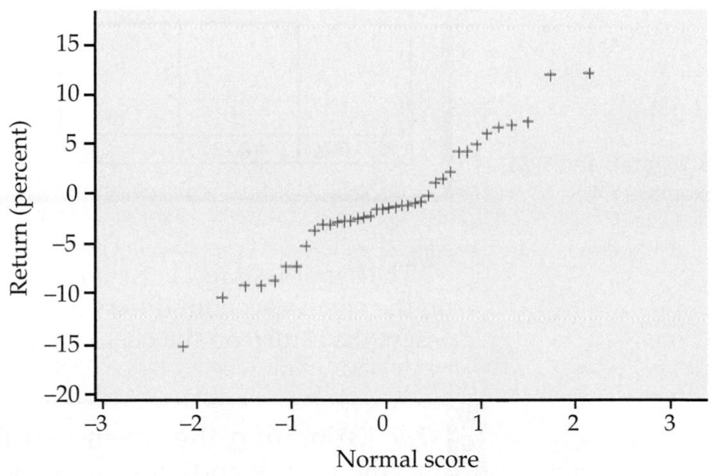


**FIGURE 7.5** Normal quantile plot for Example 7.4.

distribution of $\bar{x}$ is approximately Normal, and we proceed with our inference based on Normal theory.

The arbitration panel compared these returns with the average of the Standard and Poor's 500 stock index for the same period. Consider the 39 monthly returns as a random sample from the population of monthly returns the brokerage would generate if it managed the account forever. Are these returns compatible with a population mean of $\mu = 0.95\%$, the S&P 500 average? Our hypotheses are

$$H_0: \mu = 0.95$$
$$H_a: \mu \neq 0.95$$

Minitab and SPSS outputs appear in Figure 7.6. Output from other software will look similar.

Here is one way to report the conclusion: the mean monthly return on investment for this client's account was $\bar{x} = -1.1\%$. This differs significantly from the performance of the S&P 500 stock index for the same period ($t = -2.14$, df $= 38$, $P < 0.039$).

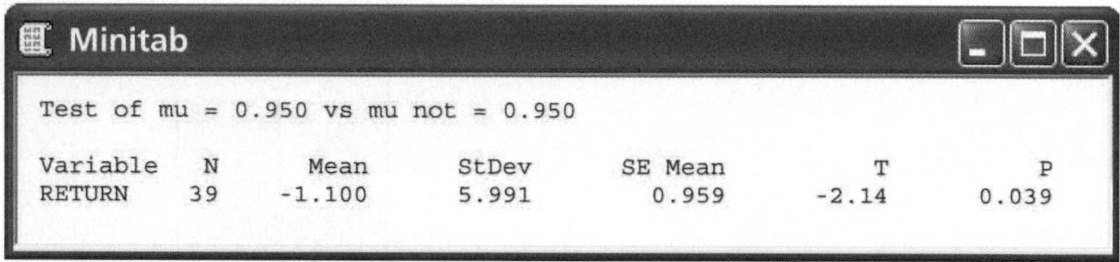

Minitab

Test of mu = 0.950 vs mu not = 0.950

| Variable | N | Mean | StDev | SE Mean | T | P |
|---|---|---|---|---|---|---|
| RETURN | 39 | -1.100 | 5.991 | 0.959 | -2.14 | 0.039 |

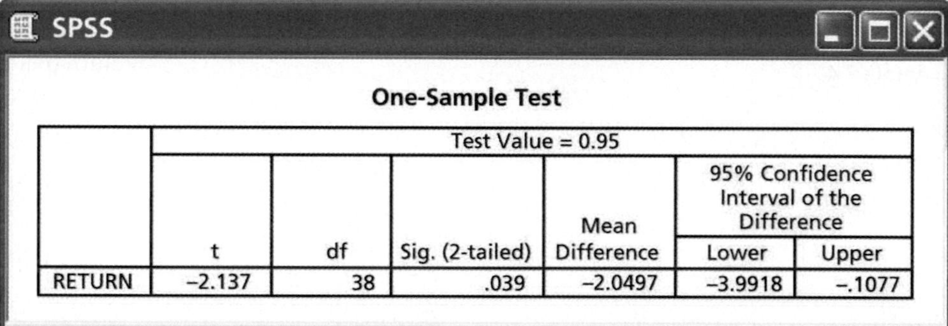

SPSS

**One-Sample Test**

| | Test Value = 0.95 | | | | | |
|---|---|---|---|---|---|---|
| | | | | | 95% Confidence Interval of the Difference | |
| | t | df | Sig. (2-tailed) | Mean Difference | Lower | Upper |
| RETURN | −2.137 | 38 | .039 | −2.0497 | −3.9918 | −.1077 |

**FIGURE 7.6** Minitab and SPSS output for Example 7.4.

The hypothesis test in Example 7.4 leads us to conclude that the mean return on the client's account differs from that of the S&P 500 stock index. Now let's assess the return on the client's account with a confidence interval.

EXAMPLE

**7.5 Estimating the mean monthly return.** The mean monthly return on the client's portfolio was $\bar{x} = -1.1\%$ and the standard deviation was $s = 5.99\%$. Figure 7.7 gives the Minitab, SPSS, and Excel outputs for a 95% confidence interval for the population mean $\mu$. Note that Excel gives the margin of error next to the label "Confidence Level (95.0%)" rather than the actual confidence interval. We see that the 95% confidence interval is $(-3.04, 0.84)$, or (from Excel) $-1.0997 \pm 1.9420$.

Because the S&P 500 return, 0.95%, falls outside this interval, we know that $\mu$ differs significantly from 0.95% at the $\alpha = 0.05$ level. Example 7.4 gave the actual $P$-value as $P = 0.039$.

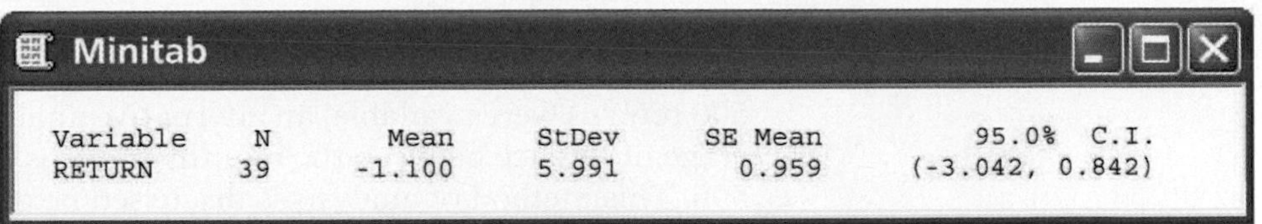

Minitab

| Variable | N | Mean | StDev | SE Mean | 95.0% C.I. |
|---|---|---|---|---|---|
| RETURN | 39 | -1.100 | 5.991 | 0.959 | (-3.042, 0.842) |

SPSS

Descriptives

| | | | Statistic | Std. Error |
|---|---|---|---|---|
| RETURN | MEAN | | –1.0997 | .9593 |
| | 95% Confidence Interval for Mean | Lower Bound | –3.0418 | |
| | | Upper Bound | .8423 | |

Excel

| | A | B |
|---|---|---|
| 1 | Mean | -1.09974359 |
| 2 | Standard Error | 0.95930991 |
| 3 | Standard Deviation | 5.990888471 |
| 4 | Count | 39 |
| 5 | Confidence Level(95.0%) | 1.942021452 |

**FIGURE 7.7** Minitab, SPSS, and Excel output for Example 7.5.

The confidence interval suggests that the broker's management of this account had a long-term mean somewhere between a loss of 3.04% and a gain of 0.84% per month. We are interested not in the actual mean but in the difference between the performance of the client's portfolio and that of the diversified S&P 500 stock index.

EXAMPLE

**7.6 Estimating the difference from a standard.** Following the analysis accepted by the arbitration panel, we are considering the S&P 500 monthly average return as a constant standard. (It is easy to envision scenarios where we would want to treat this type of quantity as random.) The difference between the mean of the investor's account and the S&P 500 is $\bar{x} - \mu = -1.10 - 0.95 = -2.05\%$. In Example 7.5 we found that the 95% confidence interval for the investor's account was $(-3.04,\ 0.84)$. To obtain the corresponding interval for the difference, subtract 0.95 from each of the endpoints. The resulting interval is $(-3.04 - 0.95, 0.84 - 0.95)$, or $(-3.99, -0.11)$. We conclude with 95% confidence that the underperformance was between $-3.99\%$ and $-0.11\%$. This interval is presented in the SPSS output of Figure 7.6. This estimate helps to set the compensation owed the investor.

The assumption that these 39 monthly returns represent an SRS from the population of monthly returns is certainly questionable. If the monthly S&P 500 returns were available, an alternative analysis would be to compare the average difference between the monthly returns for this account and for the S&P 500. This method of analysis is discussed next.

**USE YOUR KNOWLEDGE**

**7.5** **Significance test using the *t* distribution.** A test of a null hypothesis versus a two-sided alternative gives $t = 2.35$.

(a) The sample size is 15. Is the test result significant at the 5% level? Explain how you obtained your answer.

(b) The sample size is 6. Is the test result significant at the 5% level? Explain how you obtained your answer.

(c) Sketch the two *t* distributions to illustrate your answers.

**7.6** **Significance test for apartment rents.** Recall Exercise 7.1 (page 419). Does this SRS give good reason to believe that the mean rent of all advertised one-bedroom apartments is greater than $550? State the hypotheses, find the *t* statistic and its *P*-value, and state your conclusion.

**7.7** **Using software.** In Example 7.1 (page 421) we calculated the 95% confidence interval for the U.S. average of hours per month spent listening to full-track music on a cell phone. Use software to compute this interval and verify that you obtain the same interval.

## Matched pairs *t* procedures

The cell phone problem of Example 7.1 concerns only a single population. We know that comparative studies are usually preferred to single-sample investigations because of the protection they offer against confounding. For that reason, inference about a parameter of a single distribution is less common than comparative inference. One common comparative design, however, makes use of single-sample procedures. In a matched pairs study, subjects are matched in pairs and the outcomes are compared within each matched pair. The experimenter can toss a coin to assign two treatments to the two subjects in each pair. Matched pairs are also common when randomization is not possible. One situation calling for matched pairs is when observations are taken on the same subjects, under different conditions.

LOOK BACK
**matched pairs design, page 189**

EXAMPLE

**7.7 Does a full moon affect behavior?** Many people believe that the moon influences the actions of some individuals. A study of dementia patients in nursing homes recorded various types of disruptive behaviors every day for 12 weeks. Days were classified as moon days if they were in a three-day period centered at the day of the full moon. For each patient the average number of disruptive behaviors was computed for moon days and for all other days. The data for the 15 subjects whose behaviors were classified as aggressive are presented in Table 7.2.[3] The patients in this study are not a

**TABLE 7.2**

Aggressive behaviors of dementia patients

| Patient | Moon days | Other days | Difference | Patient | Moon days | Other days | Difference |
|---|---|---|---|---|---|---|---|
| 1 | 3.33 | 0.27 | 3.06 | 9 | 6.00 | 1.59 | 4.41 |
| 2 | 3.67 | 0.59 | 3.08 | 10 | 4.33 | 0.60 | 3.73 |
| 3 | 2.67 | 0.32 | 2.35 | 11 | 3.33 | 0.65 | 2.68 |
| 4 | 3.33 | 0.19 | 3.14 | 12 | 0.67 | 0.69 | −0.02 |
| 5 | 3.33 | 1.26 | 2.07 | 13 | 1.33 | 1.26 | 0.07 |
| 6 | 3.67 | 0.11 | 3.56 | 14 | 0.33 | 0.23 | 0.10 |
| 7 | 4.67 | 0.30 | 4.37 | 15 | 2.00 | 0.38 | 1.62 |
| 8 | 2.67 | 0.40 | 2.27 | | | | |

random sample of dementia patients. However, we examine their data in the hope that what we find is not unique to this particular group of individuals and applies to other patients who have similar characteristics.

To analyze these paired data, we first subtract the disruptive behaviors for moon days from the disruptive behaviors for other days. These 15 differences form a single sample. They appear in the "Difference" columns in Table 7.2. The first patient, for example, averaged 3.33 aggressive behaviors on moon days but only 0.27 aggressive behaviors on other days. The difference 3.33 − 0.27 = 3.06 is what we will use in our analysis.

Next, we examine the distribution of these differences. Figure 7.8 gives a stemplot of the differences. This plot indicates that there are three patients with very small differences but there are no indications of extreme outliers or strong skewness. We will proceed with our analysis using the Normality-based methods of this section.

```
4 | 4 4
3 | 1 1 1 6 7
2 | 1 3 4 7
1 |
0 | 0 1 1
```

**FIGURE 7.8** Stemplot of differences in aggressive behaviors for Examples 7.7 and 7.8.

To assess whether there is a difference in aggressive behaviors on moon days versus other days, we test

$$H_0\colon \mu = 0$$

$$H_a\colon \mu \neq 0$$

Here $\mu$ is the mean difference in aggressive behaviors, moon versus other days, for patients of this type. The null hypothesis says that aggressive behaviors occur at the same frequency for both types of days, and $H_a$ says that the behaviors on moon days are not the same as on other days.

The 15 differences have

$$\bar{x} = 2.433 \quad \text{and} \quad s = 1.460$$

The one-sample $t$ statistic is therefore

$$t = \frac{\bar{x} - 0}{s/\sqrt{n}} = \frac{2.433}{1.460/\sqrt{15}} = 6.45$$

**df = 14**

| $p$ | 0.001 | 0.0005 |
|---|---|---|
| $t^*$ | 3.787 | 4.140 |

The $P$-value is found from the $t(14)$ distribution (remember that the degrees of freedom are 1 less than the sample size). Table D shows that 6.45 lies beyond the upper 0.0005 critical value of the $t(14)$ distribution. Since we are using a two-sided alternative, we know that the $P$-value is less than two times this value, or 0.0010. Software gives a value that is much smaller, $P = 0.000015$. In practice, there is little difference between these two $P$-values; the data provide clear evidence in favor of the alternative hypothesis. A difference this large is very unlikely to occur by chance if there is, in fact, no effect of the moon on aggressive behaviors. In scholarly publications, the details of routine statistical procedures are omitted; our test would be reported in the form: "There was more aggressive behavior on moon days than on other days ($t = 6.45$, df $= 14$, $P < 0.001$)."

Note that we could have justified a one-sided alternative in this example. Based on previous research, we expect more aggressive behaviors on moon days, and the alternative $H_a: \mu > 0$ is reasonable in this setting. The choice of the alternative here, however, has no effect on the conclusion: from Table D we determine that $P$ is less than 0.0005; from software it is 0.000008. These are very small values and we would still report $P < 0.001$. *In most circumstances we cannot be absolutely certain about the direction and the safest strategy is to use the two-sided alternative.*

The results of the significance test allow us to conclude that dementia patients exhibit more aggressive behaviors in the days around a full moon. What are the implications of the study for the administrators who run the facilities where these patients live? For example, should they increase staff on these days? To make these kinds of decisions, an estimate of the magnitude of the problem, with a margin of error, would be helpful.

EXAMPLE

**7.8 95% confidence interval for the full-moon study.** A 95% confidence interval for the mean difference in aggressive behaviors per day requires the critical value $t^* = 2.145$ from Table D. The margin of error is

$$t^* \frac{s}{\sqrt{n}} = 2.145 \frac{1.460}{\sqrt{15}}$$
$$= 0.81$$

and the confidence interval is

$$\bar{x} \pm t^* \frac{s}{\sqrt{n}} = 2.43 \pm 0.81$$
$$= (1.62, 3.24)$$

The estimated average difference is 2.43 aggressive behaviors per day, with margin of error 0.81 for 95% confidence. The increase needs to be interpreted in terms of the baseline values. The average number of aggressive behaviors per day on other days is 0.59; on moon days it is 3.02. This is approximately a 400% increase. If aggressive behaviors require a substantial amount of attention by staff, then administrators should be aware of the increased level of these activities during the full-moon period. Additional staff may be needed.

The following are key points to remember concerning matched pairs:

1. A matched pairs analysis is called for when subjects are matched in pairs or there are two measurements or observations on each individual and we want to examine the difference.

2. For each pair or individual, use the difference between the two measurements as the data for your analysis.

3. Use the one-sample confidence interval and significance-testing procedures that we learned in this section.

Use of the $t$ procedures in Examples 7.7 and 7.8 faces several issues. First, no randomization is possible in a study like this. Our inference procedures assume that there is a process that generates these aggressive behaviors and that the process produces them at possibly different rates during the days near the full moon. Second, many of the patients in these nursing homes did not exhibit any disruptive behaviors. These were not included in our analysis. So our inference is restricted to patients who do exhibit disruptive behaviors.

A final difficulty is that the data show departures from Normality. In a matched pairs analysis, when the $t$ procedures are applied to the differences, we are assuming that the differences are Normally distributed. Figure 7.8 gives a stemplot of the differences. There are 3 patients with very small differences in aggressive behaviors while the other 12 have a large increase. We have a dilemma here similar to that in Example 7.1. *The data may not be Normal, and our sample size is very small.* We can try an alternative procedure that does not require the Normality assumption—but there is a price to pay. The alternative procedures have less power to detect differences. Despite these caveats, for Example 7.7 the $P$-value is so small that we are very confident that we have found an effect of the moon phase on behavior.

## USE YOUR KNOWLEDGE

**7.8 Comparison of two energy drinks.** Consider the following study to compare two popular energy drinks. Each drink was rated on a 0 to 100 scale, with 100 being the highest rating.

| Subject | 1 | 2 | 3 | 4 | 5 |
|---|---|---|---|---|---|
| Drink A | 43 | 79 | 66 | 88 | 78 |
| Drink B | 45 | 78 | 61 | 77 | 70 |

Is there a difference in preference? State appropriate hypotheses and carry out a matched pairs $t$ test for these data.

**7.9 95% confidence interval for the difference in energy drinks.** For the companies producing these drinks, the real question is how much difference there is between the two preferences. Use the data above to give a 95% confidence interval for the difference in preference between Drink A and Drink B.

## Robustness of the $t$ procedures

The results of one-sample $t$ procedures are exactly correct only when the population is Normal. Real populations are never exactly Normal. The usefulness of the $t$ procedures in practice therefore depends on how strongly they are affected by non-Normality. Procedures that are not strongly affected are called *robust*.

> **ROBUST PROCEDURES**
>
> A statistical inference procedure is called **robust** if the required probability calculations are insensitive to violations of the assumptions made.

The assumption that the population is Normal rules out outliers, so the presence of outliers shows that this assumption is not valid. The $t$ procedures are not robust against outliers, because $\bar{x}$ and $s$ are not resistant to outliers.

LOOK BACK
**resistant measure, page 32**

In Example 7.7, there are three patients with fairly low values of the difference. Whether or not these are outliers is a matter of judgment. If we rerun the analysis without these three patients, the $t$ statistic would increase to 11.89 and the $P$-value would be much lower. Careful inspection of the records may reveal some characteristic of these patients which distinguishes them from the others in the study. Without such information, it is difficult to justify excluding them from the analysis. *In general, we should be very cautious about discarding suspected outliers, particularly when they make up a substantial proportion of the data, as they do in this example.*

Fortunately, the $t$ procedures are quite robust against non-Normality of the population except in the case of outliers or strong skewness. Larger samples improve the accuracy of $P$-values and critical values from the $t$ distributions when the population is not Normal. This is true for two reasons:

LOOK BACK
**central limit theorem, page 339**
**law of large numbers, page 274**

1. The sampling distribution of the sample mean $\bar{x}$ from a large sample is close to Normal (that's the central limit theorem). Normality of the individual observations is of little concern when the sample is large.
2. As the sample size $n$ grows, the sample standard deviation $s$ will be an accurate estimate of $\sigma$ whether or not the population has a Normal distribution. This fact is closely related to the law of large numbers.

Constructing a Normal quantile plot, stemplot, or boxplot to check for skewness and outliers is an important preliminary to the use of $t$ procedures for small samples. For most purposes, the one-sample $t$ procedures can be safely used when $n \geq 15$ unless an outlier or clearly marked skewness is present. *Except in the case of small samples, the assumption that the data are an SRS from the population of interest is more crucial than the assumption that the population distribution is Normal.* Here are practical guidelines for inference on a single mean:[4]

- *Sample size less than 15:* Use $t$ procedures if the data are close to Normal. If the data are clearly non-Normal or if outliers are present, do not use $t$.
- *Sample size at least 15:* The $t$ procedures can be used except in the presence of outliers or strong skewness.

- *Large samples:* The $t$ procedures can be used even for clearly skewed distributions when the sample is large, roughly $n \geq 40$.

Consider, for example, some of the data we studied in Chapter 1. The breaking-strength data in Figure 1.34 (page 69) contain three outliers in a sample of size 23, which makes the use of $t$ procedures risky. The guinea pig survival times in Figure 1.35 (page 70) are strongly skewed to the right with no outliers. Since there are 72 observations, we could use the $t$ procedures here. On the other hand, many would prefer to use a transformation to make these data more nearly Normal. (See the material on inference for non-Normal populations on page 435 and in Chapter 16.) Figure 1.36 (page 71) gives the Normal quantile plot for 105 acidity measurements of rainwater. These data appear to be Normal and we would apply the $t$ procedures in this case.

## USE YOUR KNOWLEDGE

**7.10 Significance test for $CO_2$ emissions?** Consider the $CO_2$ emissions data presented in Figure 1.40 (page 76). Would you feel comfortable applying the $t$ procedures in this case? Explain your answer.

**7.11 Significance test for mounting holes data?** Consider data on the distance between mounting holes presented in Figure 1.41 (page 76). Would you feel comfortable applying the $t$ procedures in this case? Explain your answer.

## The power of the $t$ test*

The power of a statistical test measures its ability to detect deviations from the null hypothesis. In practice, we carry out the test in the hope of showing that the null hypothesis is false, so high power is important. The power of the one-sample $t$ test for a specific alternative value of the population mean $\mu$ is the probability that the test will reject the null hypothesis when the alternative value of the mean is true. To calculate the power, we assume a fixed level of significance, often $\alpha = 0.05$.

Calculation of the exact power of the $t$ test takes into account the estimation of $\sigma$ by $s$ and is a bit complex. But an approximate calculation that acts as if $\sigma$ were known is almost always adequate for planning a study. This calculation is very much like that for the $z$ test:

LOOK BACK

**power of the $z$ test, page 402**

1. Decide on a standard deviation, significance level, whether the test is one-sided or two-sided, and an alternative value of $\mu$ to detect.
2. Write the event that the test rejects $H_0$ in terms of $\bar{x}$.
3. Find the probability of this event when the population mean has this alternative value.

Consider Example 7.7, where we examined the effect of the moon on the aggressive behavior of dementia patients in nursing homes. Suppose that we wanted to perform a similar study in a different setting. How many patients should we include in our new study? To answer this question, we do a power calculation.

*This section can be omitted without loss of continuity.

In Example 7.7, we found $\bar{x} = 2.433$ and $s = 1.460$. Let's use $s = 1.5$ for our calculations. *It is always better to use a value of the standard deviation that is a little larger than what we expect than one that is smaller.* This may give a sample size that is a little larger than we need. We want to avoid a situation where we fail to find the effect that we are looking for because we did not have enough data. Let's use $\mu = 1.0$ as the alternative value to detect. We are very confident that the effect was larger than this in our previous study, and this amount of an increase in aggressive behavior would still be important to those who work in these facilities. Finally, based on the previous study, we can justify using a one-sided alternative; we expect the moon days to be associated with an increase in aggressive behavior.

EXAMPLE

**7.9 Computing the power of a *t* test.** Let's compute the power of the $t$ test for

$$H_0: \mu = 0$$
$$H_a: \mu > 0$$

when the alternative $\mu = 1.0$. We will use a 5% level of significance. The $t$ test with $n$ observations rejects $H_0$ at the 5% significance level if the $t$ statistic

$$t = \frac{\bar{x} - 0}{s/\sqrt{n}}$$

exceeds the upper 5% point of $t(n-1)$. Taking $n = 20$ and $s = 1.5$, the upper 5% point of $t(19)$ is 1.729. The event that the test rejects $H_0$ is therefore

$$t = \frac{\bar{x}}{1.5/\sqrt{20}} \geq 1.729$$
$$\bar{x} \geq 1.729 \frac{1.5}{\sqrt{20}}$$
$$\bar{x} \geq 0.580$$

The power is the probability that $\bar{x} \geq 0.580$ when $\mu = 1.0$. Taking $\sigma = 1.5$, this probability is found by standardizing $\bar{x}$:

$$\begin{aligned} P(\bar{x} \geq 0.580 \text{ when } \mu = 1.0) &= P\left(\frac{\bar{x} - 1.0}{1.5/\sqrt{20}} \geq \frac{0.580 - 1.0}{1.5/\sqrt{20}}\right) \\ &= P(Z \geq -1.25) \\ &= 1 - 0.1056 = 0.89 \end{aligned}$$

The power is 89% that we will detect an increase of 1.0 aggressive behaviors per day during moon days. This is sufficient power for most situations. For many studies, 80% is considered the standard value for desirable power. We could repeat the calculations for some smaller values of $n$ to determine the smallest value that would meet the 80% criterion.

Power calculations are used in planning studies to ensure that we have a reasonable chance of detecting effects of interest. They give us some guidance in selecting a sample size. In making these calculations, we need assumptions about the standard deviation and the alternative of interest. In our example we assumed that the standard deviation would be 1.5, but in practice we are hoping that the value will be somewhere around this value. Similarly, we have used a somewhat arbitrary alternative of 1.0. This is a guess based on the results of the previous study. *Beware of putting too much trust in fine details of the results of these calculations.* They serve as a guide, not a mandate.

## USE YOUR KNOWLEDGE

**7.12 Power and the alternative mean $\mu$.** If you were to repeat the power calculation in Example 7.9 for a value of $\mu$ that is smaller than 1, would you expect the power to be higher or lower than 89%? Why?

**7.13 More on power and the alternative mean $\mu$.** Verify your answer to the previous question by doing the calculation for the alternative $\mu = 0.75$.

## Inference for non-Normal populations*

We have not discussed how to do inference about the mean of a clearly non-Normal distribution based on a small sample. If you face this problem, you should consult an expert. Three general strategies are available:

1. In some cases a distribution other than a Normal distribution will describe the data well. There are many non-Normal models for data, and inference procedures for these models are available.
2. Because skewness is the chief barrier to the use of $t$ procedures on data without outliers, you can attempt to transform skewed data so that the distribution is symmetric and as close to Normal as possible. Confidence levels and $P$-values from the $t$ procedures applied to the transformed data will be quite accurate for even moderate sample sizes.
3. Use a **distribution-free** inference procedure. Such procedures do not assume that the population distribution has any specific form, such as Normal. Distribution-free procedures are often called **nonparametric procedures.** Chapter 15 discusses several of these procedures.

distribution-free procedures

nonparametric procedures

Each of these strategies can be effective, but each quickly carries us beyond the basic practice of statistics. We emphasize procedures based on Normal distributions because they are the most common in practice, because their robustness makes them widely useful, and (most important) because we are first of all concerned with understanding the principles of inference. We will therefore not discuss procedures for non-Normal continuous distributions. We will be content with illustrating by example the use of a transformation and of a simple distribution-free procedure.

**Transforming data** When the distribution of a variable is skewed, it often happens that a simple transformation results in a variable whose distribution is symmetric and even close to Normal. The most common transformation is the **logarithm,** or **log.** The logarithm tends to pull in the right tail of a distribution.

log transformation

*This section can be omitted without loss of continuity.

For example, the data 2, 3, 4, 20 show an outlier in the right tail. Their logarithms 0.30, 0.48, 0.60, 1.30 are much less skewed. Taking logarithms is a possible remedy for right-skewness. Instead of analyzing values of the original variable $X$, we first compute their logarithms and analyze the values of $\log X$. Here is an example of this approach.

EXAMPLE

**7.10 Length of audio files on an iPod.** Table 7.3 presents data on the length (in seconds) of audio files found on an iPod. There were a total of 10,003 audio files and 50 files were randomly selected using the "shuffle songs" command.[5] We would like to give a confidence interval for the average audio file length $\mu$ for this iPod.

A Normal quantile plot of the audio data from Table 7.3 (Figure 7.9) shows that the distribution is skewed to the right. Because there are no extreme outliers, the sample mean of the 50 observations will nonetheless have an approximately Normal sampling distribution. The $t$ procedures could be used for approximate inference. For more exact inference, we will seek to transform the data so that the distribution is more nearly Normal. Figure 7.10 is a Normal quantile plot of the logarithms of the time measurements. The transformed data are very close to Normal, so $t$ procedures will give quite exact results.

**TABLE 7.3**

Length (in seconds) of audio files sampled from an iPod

| | | | | | | |
|---|---|---|---|---|---|---|
| 240 | 316 | 259 | 46 | 871 | 411 | 1366 |
| 233 | 520 | 239 | 259 | 535 | 213 | 492 |
| 315 | 696 | 181 | 357 | 130 | 373 | 245 |
| 305 | 188 | 398 | 140 | 252 | 331 | 47 |
| 309 | 245 | 69 | 293 | 160 | 245 | 184 |
| 326 | 612 | 474 | 171 | 498 | 484 | 271 |
| 207 | 169 | 171 | 180 | 269 | 297 | 266 |
| 1847 | | | | | | |

The application of the $t$ procedures to the transformed data is straightforward. Call the original length values from Table 7.3 the variable $X$. The transformed data are values of $X_{new} = \log X$. In most software packages, it is an easy task to transform data in this way and then analyze the new variable.

EXAMPLE

**7.11 Software output of audio length data.** Analysis of the logs of the length values in Minitab produces the following output:

```
 N      MEAN      STDEV     SE MEAN     95.0 PERCENT C.I.
50    5.6315     0.6840      0.0967     ( 5.4371, 5.8259)
```

For comparison, the 95% $t$ confidence interval for the original mean $\mu$ is found from the original data as follows:

```
 N    MEAN   STDEV  SE MEAN   95.0 PERCENT C.I.
50   354.1   307.9     43.6    (266.6, 441.6)
```

The advantage of analyzing transformed data is that use of procedures based on the Normal distributions is better justified and the results are more exact.

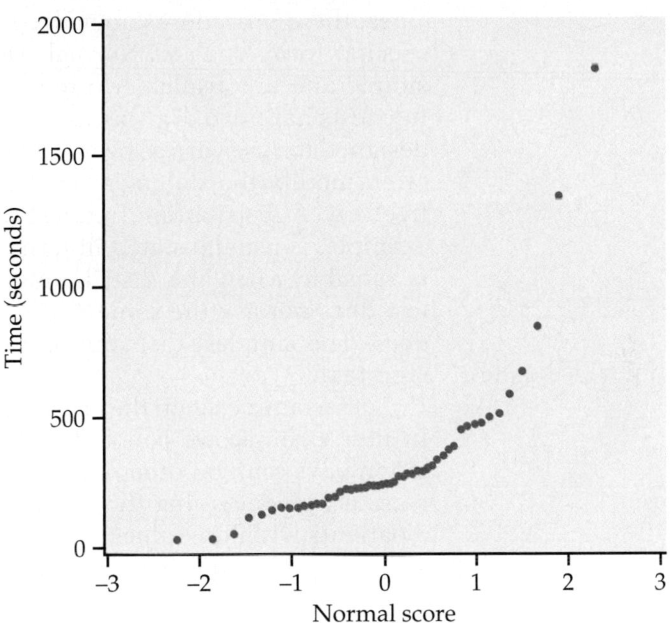


**FIGURE 7.9** Normal quantile plot of audio file length, for Example 7.10. The distribution is skewed to the right.

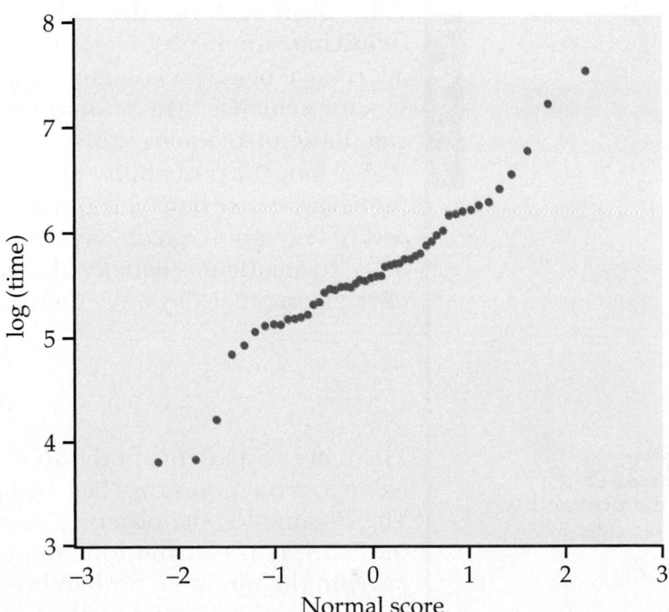


**FIGURE 7.10** Normal quantile plot of the logarithms of the audio file lengths, for Example 7.10. This distribution is close to Normal.

The disadvantage is that a confidence interval for the mean $\mu$ in the original scale (in our example, seconds) cannot be easily recovered from the confidence interval for the mean of the logs. One approach based on the lognormal distribution[6] results in an interval of (290.33, 428.30), which is much narrower than the $t$ interval.

**The sign test** Perhaps the most straightforward way to cope with non-Normal data is to use a *distribution-free,* or *nonparametric,* procedure. As the name indicates, these procedures do not require the population distribution to have any specific form, such as Normal. Distribution-free significance tests are quite simple and are available in most statistical software packages. Distribution-free tests have two drawbacks. First, they are generally less powerful than tests designed for use with a specific distribution, such as the $t$ test. Second, we must often modify the statement of the hypotheses in order to use a distribution-free test. A distribution-free test concerning the center of a distribution, for example, is usually stated in terms of the median rather than the mean. This is sensible when the distribution may be skewed. But the distribution-free test does not ask the same question (Has the mean changed?) that the $t$ test does. The simplest distribution-free test, and one of the most useful, is the **sign test.**

sign test

Let's examine again the aggressive-behavior data of Example 7.7 (page 428). In that example we concluded that there was more aggressive behavior on moon days than on other days. The stemplot given in Figure 7.8 was not very reassuring concerning the assumption that the data are Normal. There were 3 patients with low values that seemed to be somewhat different from the observations on the other 12 patients. How does the sign test deal with these data?

EXAMPLE

**7.12 Sign test for the full-moon effect.** The sign test is based on the following simple observation: of the 15 patients in our sample, 14 had more aggressive behaviors on moon days than on other days. This sounds like convincing evidence in favor of a moon effect on behavior, but we need to do some calculations to confirm this.

Let $p$ be the probability that a randomly chosen dementia patient will have more aggressive behaviors on moon days than on other days. The null hypothesis of "no moon effect" says that the moon days are no different from other days, so a patient is equally likely to have more aggressive behaviors on moon days as on other days. We therefore want to test

$$H_0: p = 1/2$$
$$H_a: p > 1/2$$

LOOK BACK
**binomial probability formula, page 329**

There are 15 patients in the study, so the number who have more aggressive behaviors on moon days has the binomial distribution $B(15, 1/2)$ if $H_0$ is true. The $P$-value for the observed count 14 is therefore $P(X \geq 14)$, where $X$ has the $B(15, 1/2)$ distribution. You can compute this probability with software or from the binomial probability formula:

$$\begin{aligned} P(X \geq 14) &= P(X = 14) + P(X = 15) \\ &= \binom{15}{14}\left(\frac{1}{2}\right)^{14}\left(\frac{1}{2}\right)^{1} + \binom{15}{15}\left(\frac{1}{2}\right)^{15}\left(\frac{1}{2}\right)^{0} \\ &= (15)\left(\frac{1}{2}\right)^{15} + \left(\frac{1}{2}\right)^{15} \\ &= 0.000488 \end{aligned}$$

Using Table C we would approximate this value as 0.0005. As in Example 7.7, there is very strong evidence in favor of an increase in aggressive behavior on moon days.

There are several varieties of sign test, all based on counts and the binomial distribution. The sign test for matched pairs (Example 7.12) is the most useful. The null hypothesis of "no effect" is then always $H_0{:}p = 1/2$. The alternative can be one-sided in either direction or two-sided, depending on the type of change we are looking for. The test gets its name from the fact that we look only at the signs of the differences, not their actual values.

### THE SIGN TEST FOR MATCHED PAIRS

Ignore pairs with difference 0; the number of trials $n$ is the count of the remaining pairs. The test statistic is the count $X$ of pairs with a positive difference. $P$-values for $X$ are based on the binomial $B(n, 1/2)$ distribution.

The matched pairs $t$ test in Example 7.7 tested the hypothesis that the mean of the distribution of differences (moon days minus other days) is 0. The sign test in Example 7.12 is in fact testing the hypothesis that the *median* of the differences is 0. If $p$ is the probability that a difference is positive, then $p = 1/2$ when the median is 0. This is true because the median of the distribution is the point with probability 1/2 lying to its right. As Figure 7.11 illustrates, $p > 1/2$ when the median is greater than 0, again because the probability to the right of the median is always 1/2. The sign test of $H_0{:}p = 1/2$ against $H_a{:}p > 1/2$ is a test of

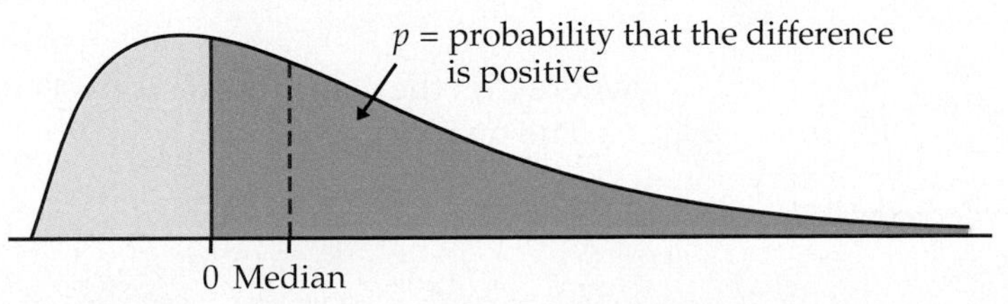


**FIGURE 7.11** Why the sign test tests the median difference: when the median is greater than 0, the probability $p$ of a positive difference is greater than 1/2, and vice versa.

$$H_0\text{: population median} = 0$$

$$H_a\text{: population median} > 0$$

The sign test in Example 7.12 makes no use of the actual differences—it just counts how many patients had more aggressive behaviors on moon days than on other days. Because the sign test uses so little of the available information, it is much less powerful than the $t$ test when the population is close to Normal. *It is better to use a test that is powerful when we believe our assumptions are approximately satisfied than a less powerful test with fewer assumptions.* There are other distribution-free tests that are more powerful than the sign test.[7]

## USE YOUR KNOWLEDGE

**7.14 Sign test for energy drink comparison.** Exercise 7.8 (page 431) gives data on the appeal of two popular energy drinks. Is there evidence that the medians are different? State the hypotheses, carry out the sign test, and report your conclusion.

## SECTION 7.1 Summary

Significance tests and confidence intervals for the mean $\mu$ of a Normal population are based on the sample mean $\bar{x}$ of an SRS. Because of the central limit theorem, the resulting procedures are approximately correct for other population distributions when the sample is large.

The standardized sample mean, or **one-sample $z$ statistic,**

$$z = \frac{\bar{x} - \mu}{\sigma/\sqrt{n}}$$

has the $N(0, 1)$ distribution. If the standard deviation $\sigma/\sqrt{n}$ of $\bar{x}$ is replaced by the **standard error** $s/\sqrt{n}$, the **one-sample $t$ statistic**

$$t = \frac{\bar{x} - \mu}{s/\sqrt{n}}$$

has the **$t$ distribution** with $n - 1$ degrees of freedom.

There is a $t$ distribution for every positive **degrees of freedom $k$.** All are symmetric distributions similar in shape to Normal distributions. The $t(k)$ distribution approaches the $N(0, 1)$ distribution as $k$ increases.

A level $C$ **confidence interval for the mean** $\mu$ of a Normal population is

$$\bar{x} \pm t^* \frac{s}{\sqrt{n}}$$

where $t^*$ is the value for the $t(n - 1)$ density curve with area $C$ between $-t^*$ and $t^*$. The quantity

$$t^* \frac{s}{\sqrt{n}}$$

is the **margin of error.**

Significance tests for $H_0: \mu = \mu_0$ are based on the $t$ statistic. $P$-values or fixed significance levels are computed from the $t(n-1)$ distribution.

These one-sample procedures are used to analyze **matched pairs** data by first taking the differences within the matched pairs to produce a single sample.

The $t$ procedures are relatively **robust** against non-Normal populations. The $t$ procedures are useful for non-Normal data when $15 \leq n < 40$ unless the data show outliers or strong skewness. When $n \geq 40$, the $t$ procedures can be used even for clearly skewed distributions.

The **power** of the $t$ test is calculated like that of the $z$ test, using an approximate value for both $\sigma$ and $s$.

Small samples from skewed populations can sometimes be analyzed by first applying a **transformation** (such as the logarithm) to obtain an approximately Normally distributed variable. The $t$ procedures then apply to the transformed data.

The **sign test** is a **distribution-free test** because it uses probability calculations that are correct for a wide range of population distributions.

The sign test for "no treatment effect" in matched pairs counts the number of positive differences. The $P$-value is computed from the $B(n, 1/2)$ distribution, where $n$ is the number of non-0 differences. The sign test is less powerful than the $t$ test in cases where use of the $t$ test is justified.

## SECTION 7.1 Exercises

*For Exercises 7.1 and 7.2, see pages 419 and 420; for Exercises 7.3 and 7.4, see page 422; for Exercises 7.5 to 7.7, see page 428; for Exercises 7.8 and 7.9, see page 431; for Exercises 7.10 and 7.11, see page 433; for Exercises 7.12 and 7.13, see page 435; and for Exercise 7.14, see page 440.*

**7.15 Finding the critical value $t^*$.** What critical value $t^*$ from Table D should be used to calculate the margin of error for a confidence interval for the mean of the population in each of the following situations?

(a) A 95% confidence interval based on $n = 15$ observations.

(b) A 95% confidence interval from an SRS of 25 observations.

(c) A 90% confidence interval from a sample of size 25.

(d) These cases illustrate how the size of the margin of error depends upon the confidence level and the sample size. Summarize these relationships.

**7.16 Distribution of the $t$ statistic.** Assume a sample size of $n = 20$. Draw a picture of the distribution of the $t$ statistic under the null hypothesis. Use Table D and your picture to illustrate the values of the test statistic that would lead to rejection of the null hypothesis at the 5% level for a two-sided alternative.

**7.17 More on the distribution of the $t$ statistic.** Repeat the previous exercise for the two situations where the alternative is one-sided.

**7.18 One-sided versus two-sided $P$-values.** Computer software reports $\bar{x} = 15.3$ and $P = 0.04$ for a $t$ test of $H_0: \mu = 0$ versus $H_a: \mu \neq 0$. Based on prior knowledge, you can justify testing the alternative $H_a: \mu > 0$. What is the $P$-value for your significance test?

**7.19 More on one-sided versus two-sided $P$-values.** Suppose that $\bar{x} = -15.3$ in the setting of the previous exercise. Would this change your $P$-value? Use a sketch of the distribution of the test statistic under the null hypothesis to illustrate and explain your answer.

**7.20 A one-sample $t$ test.** The one-sample $t$ statistic for testing

$$H_0: \mu = 10$$

$$H_a: \mu > 10$$

from a sample of $n = 20$ observations has the value $t = 2.10$.

(a) What are the degrees of freedom for this statistic?

(b) Give the two critical values $t^*$ from Table D that bracket $t$.

(c) Between what two values does the $P$-value of the test fall?

(d) Is the value $t = 2.10$ significant at the 5% level? Is it significant at the 1% level?

(e) If you have software available, find the exact $P$-value.

**7.21 Another one-sample $t$ test.** The one-sample $t$ statistic for testing

$$H_0: \mu = 60$$
$$H_a: \mu \neq 60$$

from a sample of $n = 24$ observations has the value $t = 2.40$.

(a) What are the degrees of freedom for $t$?

(b) Locate the two critical values $t^*$ from Table D that bracket $t$.

(c) Between what two values does the $P$-value of the test fall?

(d) Is the value $t = 2.40$ statistically significant at the 5% level? At the 1% level?

(e) If you have software available, find the exact $P$-value.

**7.22 A final one-sample $t$ test.** The one-sample $t$ statistic for testing

$$H_0: \mu = 20$$
$$H_a: \mu < 20$$

based on $n = 115$ observations has the value $t = -1.55$.

(a) What are the degrees of freedom for this statistic?

(b) Between what two values does the $P$-value of the test fall?

(c) If you have software available, find the exact $P$-value.

**7.23 Two-sided to one-sided $P$-value.** Most software gives $P$-values for two-sided alternatives. Explain why you cannot always divide these $P$-values by 2 to obtain $P$-values for one-sided alternatives.

**7.24** CHALLENGE **Fuel efficiency $t$ test.** Computers in some vehicles calculate various quantities related to performance. One of these is the fuel efficiency, or gas mileage, usually expressed as miles per gallon (mpg). For one vehicle equipped in this way, the mpg were recorded each time the gas tank was filled, and the computer was then reset.[8] Here are the mpg values for a random sample of 20 of these records:

| | | | | | | | | | |
|---|---|---|---|---|---|---|---|---|---|
| 41.5 | 50.7 | 36.6 | 37.3 | 34.2 | 45.0 | 48.0 | 43.2 | 47.7 | 42.2 |
| 43.2 | 44.6 | 48.4 | 46.4 | 46.8 | 39.2 | 37.3 | 43.5 | 44.3 | 43.3 |

(a) Describe the distribution using graphical methods. Is it appropriate to analyze these data using methods based on Normal distributions? Explain why or why not.

(b) Find the mean, standard deviation, standard error, and margin of error for 95% confidence.

(c) Report the 95% confidence interval for $\mu$, the mean mpg for this vehicle based on these data.

**7.25 Random distribution of trees $t$ test.** A study of 584 longleaf pine trees in the Wade Tract in Thomas County, Georgia, is described in Example 6.1 (page 354). For each tree in the tract, the researchers measured the diameter at breast height (DBH). This is the diameter of the tree at 4.5 feet and the units are centimeters (cm). Only trees with DBH greater than 1.5 cm were sampled. Here are the diameters of a random sample of 40 of these trees:

| | | | | | | | | | |
|---|---|---|---|---|---|---|---|---|---|
| 10.5 | 13.3 | 26.0 | 18.3 | 52.2 | 9.2 | 26.1 | 17.6 | 40.5 | 31.8 |
| 47.2 | 11.4 | 2.7 | 69.3 | 44.4 | 16.9 | 35.7 | 5.4 | 44.2 | 2.2 |
| 4.3 | 7.8 | 38.1 | 2.2 | 11.4 | 51.5 | 4.9 | 39.7 | 32.6 | 51.8 |
| 43.6 | 2.3 | 44.6 | 31.5 | 40.3 | 22.3 | 43.3 | 37.5 | 29.1 | 27.9 |

(a) Use a histogram or stemplot and a boxplot to examine the distribution of DBHs. Include a Normal quantile plot if you have the necessary software. Write a careful description of the distribution.

(b) Is it appropriate to use the methods of this section to find a 95% confidence interval for the mean DBH of all trees in the Wade Tract? Explain why or why not.

(c) Report the mean with the margin of error and the confidence interval. Write a short summary describing the meaning of the confidence interval.

(d) Do you think these results would apply to other similar trees in the same area? Give reasons for your answer.

**7.26 C-reactive protein in children.** C-reactive protein (CRP) is a substance that can be measured in the blood. Values increase substantially within 6 hours of an infection and reach a peak within 24 to 48 hours after. In adults, chronically high values have been linked to an increased risk of cardiovascular disease. In a study of apparently healthy children aged 6 to 60 months in Papua New Guinea, CRP was measured in 90 children.[9] The units are milligrams per liter (mg/l). Here are the data from a random sample of 40 of these children.

| | | | | | | | | | |
|---|---|---|---|---|---|---|---|---|---|
| 0.00 | 3.90 | 5.64 | 8.22 | 0.00 | 5.62 | 3.92 | 6.81 | 30.61 | 0.00 |
| 73.20 | 0.00 | 46.70 | 0.00 | 0.00 | 26.41 | 22.82 | 0.00 | 0.00 | 3.49 |
| 0.00 | 0.00 | 4.81 | 9.57 | 5.36 | 0.00 | 5.66 | 0.00 | 59.76 | 12.38 |
| 15.74 | 0.00 | 0.00 | 0.00 | 0.00 | 9.37 | 20.78 | 7.10 | 7.89 | 5.53 |

(a) Look carefully at the data above. Do you think that there are outliers or is this a skewed distribution? Now use a histogram or stemplot to examine the distribution. Write a short summary describing the distribution.

(b) Do you think that the mean is a good characterization of the center of this distribution? Explain why or why not.

(c) Find a 95% confidence interval for the mean CRP. Discuss the appropriateness of using this methodology for these data.

**7.27** CHALLENGE **More on C-reactive protein in children.** Refer to the previous exercise. With strongly skewed distributions such as this, we frequently reduce the skewness by taking a log transformation. We have a bit of a problem here, however, because some of the data are recorded as 0.00 and the logarithm of zero is not defined. For this variable, the value 0.00 is recorded whenever the amount of CRP in the blood is below the level that the measuring instrument is capable of detecting. The usual procedure in this circumstance is to add a small number to each observation before taking the logs. Transform these data by adding 1 to each observation and then taking the logarithm. Use the questions in the previous exercise as a guide to your analysis, and prepare a summary contrasting this analysis with the one that you performed in the previous exercise.

**7.28** CHALLENGE **Serum retinol in children.** In the Papua New Guinea study that provided the data for the previous two exercises, the researchers also measured serum retinol. A low value of this variable can be an indicator of vitamin A deficiency. Following are the data on the same sample of 40 children from this study. The units are micromoles per liter ($\mu$mol/l).

| | | | | | | | | | |
|---|---|---|---|---|---|---|---|---|---|
| 1.15 | 1.36 | 0.38 | 0.34 | 0.35 | 0.37 | 1.17 | 0.97 | 0.97 | 0.67 |
| 0.31 | 0.99 | 0.52 | 0.70 | 0.88 | 0.36 | 0.24 | 1.00 | 1.13 | 0.31 |
| 1.44 | 0.35 | 0.34 | 1.90 | 1.19 | 0.94 | 0.34 | 0.35 | 0.33 | 0.69 |
| 0.69 | 1.04 | 0.83 | 1.11 | 1.02 | 0.56 | 0.82 | 1.20 | 0.87 | 0.41 |

Analyze these data. Use the questions in the previous two exercises as a guide.

**7.29** **Do you feel lucky?** Children in a psychology study were asked to solve some puzzles and were then given feedback on their performance. Then they were asked to rate how luck played a role in determining their scores.[10] This variable was recorded on a 1 to 10 scale with 1 corresponding to very lucky and 10 corresponding to very unlucky. Here are the scores for 60 children:

| | | | | | | | | | | | | | | |
|---|---|---|---|---|---|---|---|---|---|---|---|---|---|---|
| 1 | 10 | 1 | 10 | 1 | 1 | 10 | 5 | 1 | 1 | 8 | 1 | 10 | 2 | 1 |
| 9 | 5 | 2 | 1 | 8 | 10 | 5 | 9 | 10 | 10 | 9 | 6 | 10 | 1 | 5 |
| 1 | 9 | 2 | 1 | 7 | 10 | 9 | 5 | 10 | 10 | 10 | 1 | 8 | 1 | 6 |
| 10 | 1 | 6 | 10 | 10 | 8 | 10 | 3 | 10 | 8 | 1 | 8 | 10 | 4 | 2 |

(a) Use graphical methods to display the distribution. Describe any unusual characteristics. Do you think that these would lead you to hesitate before using the Normality-based methods of this section?

(b) Give a 95% confidence interval for the mean luck score.

(c) The children in this study were volunteers whose parents agreed to have them participate in the study. To what extent do you think your results would apply to all similar children in this community?

**7.30** **Perceived organizational skills.** In a study of children with attention deficit hyperactivity disorder (ADHD), parents were asked to rate their child on a variety of items related to how well their child performs different tasks.[11] One item was "Has difficulty organizing work," rated on a five-point scale of 0 to 4 with 0 corresponding to "not at all" and 4 corresponding to "very much." The mean rating for 282 boys with ADHD was reported as 2.22 with a standard deviation of 1.03.

(a) Do you think that these data are Normally distributed? Explain why or why not.

(b) Is it appropriate to use the methods of this section to compute a 99% confidence interval? Explain why or why not.

(c) Find the 99% margin of error and the corresponding confidence interval. Write a sentence explaining the interval and the meaning of the 99% confidence level.

(d) The boys in this study were all evaluated at the Western Psychiatric Institute and Clinic at the University of Pittsburgh. To what extent do you think the results could be generalized to boys with ADHD in other locations?

**7.31 Confidence level and interval width.** Refer to the previous exercise. Compute the 90% and the 95% confidence intervals. Display the three intervals graphically and write a short explanation of the effect of the confidence level on the width of the interval using your display as an example.

**7.32** CHALLENGE **Food intake and weight gain.** If we increase our food intake, we generally gain weight. Nutrition scientists can calculate the amount of weight gain that would be associated with a given increase in calories. In one study, 16 nonobese adults, aged 25 to 36 years, were fed 1000 calories per day in excess of the calories needed to maintain a stable body weight. The subjects maintained this diet for 8 weeks, so they consumed a total of 56,000 extra calories.[12] According to theory, 3500 extra calories will translate into a weight gain of 1 pound. Therefore, we expect each of these subjects to gain 56,000/3500 = 16 pounds (lb). Here are the weights before and after the 8-week period expressed in kilograms (kg):

| Subject | 1 | 2 | 3 | 4 | 5 | 6 | 7 | 8 |
|---|---|---|---|---|---|---|---|---|
| Weight before | 55.7 | 54.9 | 59.6 | 62.3 | 74.2 | 75.6 | 70.7 | 53.3 |
| Weight after | 61.7 | 58.8 | 66.0 | 66.2 | 79.0 | 82.3 | 74.3 | 59.3 |
| **Subject** | **9** | **10** | **11** | **12** | **13** | **14** | **15** | **16** |
| Weight before | 73.3 | 63.4 | 68.1 | 73.7 | 91.7 | 55.9 | 61.7 | 57.8 |
| Weight after | 79.1 | 66.0 | 73.4 | 76.9 | 93.1 | 63.0 | 68.2 | 60.3 |

(a) For each subject, subtract the weight before from the weight after to determine the weight change.

(b) Find the mean and the standard deviation for the weight change.

(c) Calculate the standard error and the margin of error for 95% confidence. Report the 95% confidence interval in a sentence that explains the meaning of the 95%.

(d) Convert the mean weight gain in kilograms to mean weight gain in pounds. Because there are 2.2 kg per pound, multiply the value in kilograms by 2.2 to obtain pounds. Do the same for the standard deviation and the confidence interval.

(e) Test the null hypothesis that the mean weight gain is 16 lb. Be sure to specify the null and alternative hypotheses, the test statistic with degrees of freedom, and the $P$-value. What do you conclude?

(f) Write a short paragraph explaining your results.

**7.33 Food intake and NEAT.** Nonexercise activity thermogenesis (NEAT) provides a partial explanation for the results you found in the previous analysis. NEAT is energy burned by fidgeting, maintenance of posture, spontaneous muscle contraction, and other activities of daily living. In the study of the previous exercise, the 16 subjects increased their NEAT by 328 calories per day, on average, in response to the additional food intake. The standard deviation was 256.

(a) Test the null hypothesis that there was no change in NEAT versus the two-sided alternative. Summarize the results of the test and give your conclusion.

(b) Find a 95% confidence interval for the change in NEAT. Discuss the additional information provided by the confidence interval that is not evident from the results of the significance test.

**7.34 Potential insurance fraud?** Insurance adjusters are concerned about the high estimates they are receiving from Jocko's Garage. To see if the estimates are unreasonably high, each of 10 damaged cars was taken to Jocko's and to another garage and the estimates recorded. Here are the results:

| Car | 1 | 2 | 3 | 4 | 5 |
|---|---|---|---|---|---|
| Jocko's | 500 | 1550 | 1250 | 1300 | 750 |
| Other | 400 | 1500 | 1300 | 1300 | 800 |
| **Car** | **6** | **7** | **8** | **9** | **10** |
| Jocko's | 1000 | 1250 | 1300 | 800 | 2500 |
| Other | 800 | 1000 | 1100 | 650 | 2200 |

Test the null hypothesis that there is no difference between the two garages. Be sure to specify the null and alternative hypotheses, the test statistic with degrees of freedom, and the $P$-value. What do you conclude?

**7.35 Fuel efficiency comparison *t* test.** Refer to Exercise 7.24. In addition to the computer calculating mpg, the driver also recorded the mpg by dividing the miles driven by the amount of gallons at fill-up. The driver wants to determine if these calculations are different.

| Fill-up | 1 | 2 | 3 | 4 | 5 | 6 | 7 | 8 | 9 | 10 |
|---|---|---|---|---|---|---|---|---|---|---|
| Computer | 41.5 | 50.7 | 36.6 | 37.3 | 34.2 | 45.0 | 48.0 | 43.2 | 47.7 | 42.2 |
| Driver | 36.5 | 44.2 | 37.2 | 35.6 | 30.5 | 40.5 | 40.0 | 41.0 | 42.8 | 39.2 |
| **Fill-up** | 11 | 12 | 13 | 14 | 15 | 16 | 17 | 18 | 19 | 20 |
| Computer | 43.2 | 44.6 | 48.4 | 46.4 | 46.8 | 39.2 | 37.3 | 43.5 | 44.3 | 43.3 |
| Driver | 38.8 | 44.5 | 45.4 | 45.3 | 45.7 | 34.2 | 35.2 | 39.8 | 44.9 | 47.5 |

(a) State the appropriate $H_0$ and $H_a$.

(b) Carry out the test. Give the $P$-value, and then interpret the result.

**7.36 Level of phosphate in the blood.** The level of various substances in the blood of kidney dialysis patients is of concern because kidney failure and dialysis can lead to nutritional problems. A researcher performed blood tests on several dialysis patients on 6 consecutive clinic visits. One variable measured was the level of phosphate in the blood. Phosphate levels for an individual tend to vary Normally over time. The data on one patient, in milligrams of phosphate per deciliter (mg/dl) of blood, are given below:[13]

5.6 5.1 4.6 4.8 5.7 6.4

(a) Calculate the sample mean $\bar{x}$ and its standard error.

(b) Use the $t$ procedures to give a 90% confidence interval for this patient's mean phosphate level.

**7.37 More on the level of phosphate in the blood.** The normal range of values for blood phosphate levels is 2.6 to 4.8 mg/dl. The sample mean for the patient in the previous exercise falls above this range. Is this good evidence that the patient's mean level in fact falls above 4.8? State $H_0$ and $H_a$ and use the data in the previous exercise to carry out a $t$ test. Between which levels from Table D does the $P$-value lie? Are you convinced that the patient's phosphate level is higher than normal?

**7.38 A customer satisfaction survey.** Many organizations are doing surveys to determine the satisfaction of their customers. Attitudes toward various aspects of campus life were the subject of one such study conducted at Purdue University. Each item was rated on a 1 to 5 scale, with 5 being the highest rating. The average response of 1406 first-year students to "Feeling welcomed at Purdue" was 3.9 with a standard deviation of 0.98. Assuming that the respondents are an SRS, give a 90% confidence interval for the mean of all first-year students.

**7.39 Comparing operators of a DXA machine.** Dual-energy X-ray absorptiometry (DXA) is a technique for measuring bone health. One of the most common measures is total body bone mineral content (TBBMC). A highly skilled operator is required to take the measurements. Recently, a new DXA machine was purchased by a research lab and two operators were trained to take the measurements. TBBMC for eight subjects was measured by both operators.[14] The units are grams (g). A comparison of the means for the two operators provides a check on the training they received and allows us to determine if one of the operators is producing measurements that are consistently higher than the other. Here are the data:

| | Subject | | | | | | | |
|---|---|---|---|---|---|---|---|---|
| Operator | 1 | 2 | 3 | 4 | 5 | 6 | 7 | 8 |
| 1 | 1.328 | 1.342 | 1.075 | 1.228 | 0.939 | 1.004 | 1.178 | 1.286 |
| 2 | 1.323 | 1.322 | 1.073 | 1.233 | 0.934 | 1.019 | 1.184 | 1.304 |

(a) Take the difference between the TBBMC recorded for Operator 1 and the TBBMC for Operator 2. Describe the distribution of these differences.

(b) Use a significance test to examine the null hypothesis that the two operators have the same mean. Be sure to give the test statistic with its degrees of freedom, the $P$-value, and your conclusion.

(c) The sample here is rather small, so we may not have much power to detect differences of interest. Use a 95% confidence interval to provide a range of differences that are compatible with these data.

(d) The eight subjects used for this comparison were not a random sample. In fact, they were friends of the researchers whose ages and weights were similar to the types of people who would be measured with this DXA. Comment on the appropriateness of this procedure for selecting a sample, and discuss any consequences regarding the interpretation of the significance testing and confidence interval results.

**7.40 Another comparison of DXA machine operators.** Refer to the previous exercise. TBBMC measures the total amount of mineral in the bones. Another important variable is total body bone mineral density (TBBMD). This variable is calculated by dividing TBBMC by the area corresponding to bone in the DXA scan. The units are grams per squared centimeter (g/cm$^2$). Here are the TBBMD values for the same subjects:

| | Subject | | | | | | | |
|---|---|---|---|---|---|---|---|---|
| Operator | 1 | 2 | 3 | 4 | 5 | 6 | 7 | 8 |
| 1 | 4042 | 3703 | 2626 | 2673 | 1724 | 2136 | 2808 | 3322 |
| 2 | 4041 | 3697 | 2613 | 2628 | 1755 | 2140 | 2836 | 3287 |

Analyze these data using the questions in the previous exercise as a guide.

**7.41** CHALLENGE **Assessment of a foreign-language institute.** The National Endowment for the Humanities sponsors summer institutes to improve the skills of high school teachers of foreign languages. One such institute hosted 20 French teachers for 4 weeks. At the beginning of the period, the teachers were given the Modern Language Association's listening test of understanding of spoken French. After 4 weeks of immersion in French in and out of class, the listening test was given again. (The actual French spoken in the two tests was different, so that simply taking the first test should not improve the score on the second test.) The maximum possible score on the test is 36.[15] Here are the data:

| Teacher | Pretest | Posttest | Gain | Teacher | Pretest | Posttest | Gain |
|---|---|---|---|---|---|---|---|
| 1 | 32 | 34 | 2 | 11 | 30 | 36 | 6 |
| 2 | 31 | 31 | 0 | 12 | 20 | 26 | 6 |
| 3 | 29 | 35 | 6 | 13 | 24 | 27 | 3 |
| 4 | 10 | 16 | 6 | 14 | 24 | 24 | 0 |
| 5 | 30 | 33 | 3 | 15 | 31 | 32 | 1 |
| 6 | 33 | 36 | 3 | 16 | 30 | 31 | 1 |
| 7 | 22 | 24 | 2 | 17 | 15 | 15 | 0 |
| 8 | 25 | 28 | 3 | 18 | 32 | 34 | 2 |
| 9 | 32 | 26 | −6 | 19 | 23 | 26 | 3 |
| 10 | 20 | 26 | 6 | 20 | 23 | 26 | 3 |

To analyze these data, we first subtract the pretest score from the posttest score to obtain the improvement for each teacher. These 20 differences form a single sample. They appear in the "Gain" columns. The first teacher, for example, improved from 32 to 34, so the gain is $34 - 32 = 2$.

(a) State appropriate null and alternative hypotheses for examining the question of whether or not the course improves French spoken-language skills.

(b) Describe the gain data. Use numerical and graphical summaries.

(c) Perform the significance test. Give the test statistic, the degrees of freedom, and the *P*-value. Summarize your conclusion.

(d) Give a 95% confidence interval for the mean improvement.

**7.42 Length of calls to customer service center.** Refer to the lengths of calls to a customer service center in Table 1.1 (page 8). Give graphical and numerical summaries for these data. Compute a 95% confidence interval for the mean call length. Comment on the validity of your interval.

**7.43 IQ test scores.** Refer to the IQ test scores for fifth-grade students in Table 1.3 (page 13). Give numerical and graphical summaries of the data and compute a 95% confidence interval. Comment on the validity of the interval.

**7.44 Property damage due to tornadoes.** Table 1.5 (page 25) gives the average property damage per year due to tornadoes for each of the 50 states and Puerto Rico. It does not make sense to use the *t* procedures (or any other statistical procedures) to give a 95% confidence interval for the mean property damage per year due to tornadoes in the United States. Explain why not.

*The following exercises concern the optional material in the sections on the power of the t test and on non-Normal populations.*

**7.45 Sign test for potential insurance fraud.** The differences in the repair estimates in Exercise 7.34 can also be analyzed using a sign test. Set up the appropriate null and alternative hypotheses, carry out the test, and summarize the results. How do these results compare with those that you obtained in Exercise 7.34?

**7.46 Sign test for the comparison of operators.** The differences in the TBBMC measures in Exercise 7.39 can also be analyzed using a sign test. Set up the appropriate null and alternative hypotheses, carry out the test, and summarize the results. How do these results compare with those that you obtained in Exercise 7.39?

**7.47 Another sign test for the comparison of operators.** TBBMD values for the same subjects that you studied in the previous exercise are given in Exercise 7.40. Answer the questions given in the previous exercise for TBBMD.

**7.48 Sign test for assessment of foreign-language institute.** Use the sign test to assess whether the summer institute of Exercise 7.41 improves French listening skills. State the hypotheses, give the *P*-value using the binomial table (Table C), and report your conclusion.

**7.49 Sign test for fuel efficiency comparison.** Use the sign test to assess whether the computer calculates a higher mpg than the driver in Exercise 7.35. State the hypotheses, give the *P*-value using the binomial table (Table C), and report your conclusion.

**7.50 Insulation study.** A manufacturer of electric motors tests insulation at a high temperature (250°C) and records the number of hours until the insulation fails.[16] The data for 5 specimens are

446 326 372 377 310

The small sample size makes judgment from the data difficult, but engineering experience suggests that the logarithm of the failure time will have a Normal distribution. Take the logarithms of the 5 observations, and use $t$ procedures to give a 90% confidence interval for the mean of the log failure time for insulation of this type.

**7.51 Power of the comparison of DXA machine operators.** Suppose that the bone researchers in Exercise 7.39 wanted to be able to detect an alternative mean difference of 0.002. Find the power for this alternative for a sample size of 15. Use the standard deviation that you found in Exercise 7.39 for these calculations.

**7.52** CHALLENGE **Sample size calculations.** You are designing a study to test the null hypothesis that $\mu = 0$ versus the alternative that $\mu$ is positive. Assume that $\sigma$ is 10. Suppose that it would be important to be able to detect the alternative $\mu = 2$. Perform power calculations for a variety of sample sizes and determine how large a sample you would need to detect this alternative with power of at least 0.80.

**7.53** CHALLENGE **Determining the sample size.** Consider Example 7.9 (page 434). What is the minimum sample size needed for the power to be greater than 80% when $\mu = 1.0$?

## 7.2 Comparing Two Means

A nutritionist is interested in the effect of increased calcium on blood pressure. A psychologist wants to compare male and female college students' impressions of personality based on selected photographs. A bank wants to know which of two incentive plans will most increase the use of its credit cards. Two-sample problems such as these are among the most common situations encountered in statistical practice.

**TWO-SAMPLE PROBLEMS**

- The goal of inference is to compare the responses in two groups.
- Each group is considered to be a sample from a distinct population.
- The responses in each group are independent of those in the other group.

LOOK BACK
**randomized comparative experiment, page 183**

A two-sample problem can arise from a randomized comparative experiment that randomly divides the subjects into two groups and exposes each group to a different treatment. Comparing random samples separately selected from two populations is also a two-sample problem. Unlike the matched pairs designs studied earlier, there is no matching of the units in the two samples, and the two samples may be of different sizes. Inference procedures for two-sample data differ from those for matched pairs.

We can present two-sample data graphically by a back-to-back stemplot (for small samples) or by side-by-side boxplots (for larger samples). Now we will apply the ideas of formal inference in this setting. When both population distributions are symmetric, and especially when they are at least approximately Normal, a comparison of the mean responses in the two populations is most often the goal of inference.

We have two independent samples, from two distinct populations (such as subjects given a treatment and those given a placebo). The same variable is measured for both samples. We will call the variable $x_1$ in the first population and $x_2$ in the second because the variable may have different distributions in the two populations. Here is the notation that we will use to describe the two populations:

| Population | Variable | Mean | Standard deviation |
|---|---|---|---|
| 1 | $x_1$ | $\mu_1$ | $\sigma_1$ |
| 2 | $x_2$ | $\mu_2$ | $\sigma_2$ |

We want to compare the two population means, either by giving a confidence interval for $\mu_1 - \mu_2$ or by testing the hypothesis of no difference, $H_0: \mu_1 = \mu_2$.

Inference is based on two independent SRSs, one from each population. Here is the notation that describes the samples:

| Population | Sample size | Sample mean | Sample standard deviation |
|---|---|---|---|
| 1 | $n_1$ | $\bar{x}_1$ | $s_1$ |
| 2 | $n_2$ | $\bar{x}_2$ | $s_2$ |

Throughout this section, the subscripts 1 and 2 show the population to which a parameter or a sample statistic refers.

## The two-sample *z* statistic

The natural estimator of the difference $\mu_1 - \mu_2$ is the difference between the sample means, $\bar{x}_1 - \bar{x}_2$. If we are to base inference on this statistic, we must know its sampling distribution. First, the mean of the difference $\bar{x}_1 - \bar{x}_2$ is the difference of the means $\mu_1 - \mu_2$. This follows from the addition rule for means and the fact that the mean of any $\bar{x}$ is the same as the mean of the population. Because the samples are independent, their sample means $\bar{x}_1$ and $\bar{x}_2$ are independent random variables. The addition rule for variances says that the variance of the difference $\bar{x}_1 - \bar{x}_2$ is the sum of their variances, which is

**LOOK BACK**
**addition rule for means, page 278**
**addition rule for variances, page 282**

$$\frac{\sigma_1^2}{n_1} + \frac{\sigma_2^2}{n_2}$$

We now know the mean and variance of the distribution of $\bar{x}_1 - \bar{x}_2$ in terms of the parameters of the two populations. If the two population distributions are both Normal, then the distribution of $\bar{x}_1 - \bar{x}_2$ is also Normal. This is true because each sample mean alone is Normally distributed and because a difference of independent Normal random variables is also Normal.

EXAMPLE

**7.13 Heights of 10-year-old girls and boys.** A fourth-grade class has 12 girls and 8 boys. The children's heights are recorded on their 10th birthdays. What is the chance that the girls are taller than the boys? Of course, it is very unlikely that all of the girls are taller than all of the boys. We translate the question into the following: what is the probability that the mean height of the girls is greater than the mean height of the boys?

Based on information from the National Health and Nutrition Examination Survey,[17] we assume that the heights (in inches) of 10-year-old girls are $N(56.4, 2.7)$ and the heights of 10-year-old boys are $N(55.7, 3.8)$. The heights of the students in our class are assumed to be random samples from these populations. The two distributions are shown in Figure 7.12(a).

The difference $\bar{x}_1 - \bar{x}_2$ between the female and male mean heights varies in different random samples. The sampling distribution has mean

$$\mu_1 - \mu_2 = 56.4 - 55.7 = 0.7 \text{ inch}$$

and variance

$$\frac{\sigma_1^2}{n_1} + \frac{\sigma_2^2}{n_2} = \frac{2.7^2}{12} + \frac{3.8^2}{8} = 2.41$$

The standard deviation of the difference in sample means is therefore $\sqrt{2.41} = 1.55$ inches.

If the heights vary Normally, the difference in sample means is also Normally distributed. The distribution of the difference in heights is shown in Figure 7.12(b). We standardize $\bar{x}_1 - \bar{x}_2$ by subtracting its mean (0.7) and dividing by its standard deviation (1.55). Therefore, the probability that the

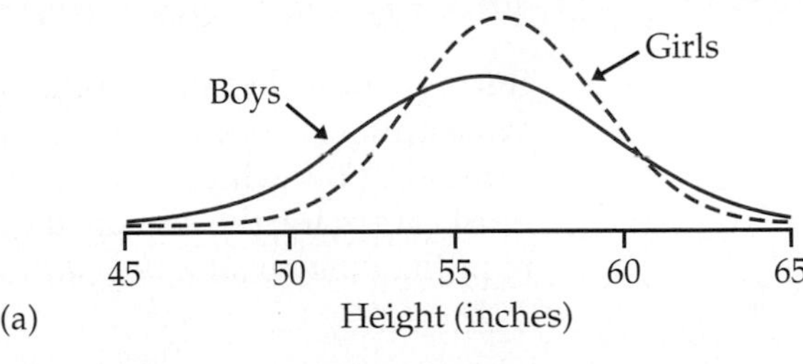


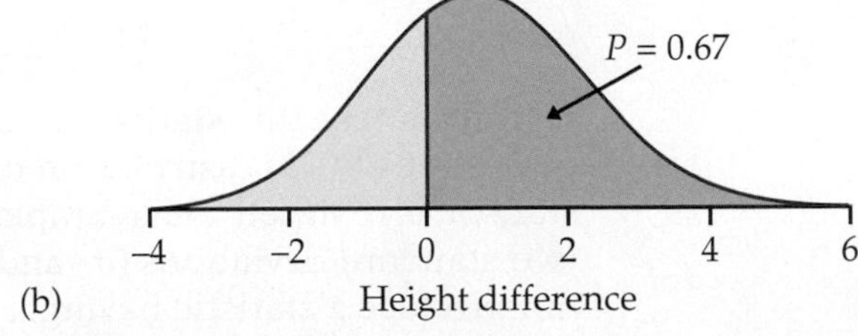


**FIGURE 7.12** Distributions for Example 7.13. (a) Distributions of heights of 10-year-old boys and girls. (b) Distribution of the difference between mean heights of 12 girls and 8 boys.

girls are taller than the boys is

$$P(\bar{x}_1 - \bar{x}_2 > 0) = P\left(\frac{(\bar{x}_1 - \bar{x}_2) - 0.7}{1.55} > \frac{0 - 0.7}{1.55}\right)$$
$$= P(Z > -0.45) = 0.67$$

Even though the population mean height of 10-year-old girls is greater than the population mean height of 10-year-old boys, the probability that the sample mean of the girls is greater than the sample mean of the boys in our class is only 67%. *Large samples are needed to see the effects of small differences.*

As Example 7.13 reminds us, any Normal random variable has the $N(0, 1)$ distribution when standardized. We have arrived at a new $z$ statistic.

### TWO-SAMPLE z STATISTIC

Suppose that $\bar{x}_1$ is the mean of an SRS of size $n_1$ drawn from an $N(\mu_1, \sigma_1)$ population and that $\bar{x}_2$ is the mean of an independent SRS of size $n_2$ drawn from an $N(\mu_2, \sigma_2)$ population. Then the **two-sample $z$ statistic**

$$z = \frac{(\bar{x}_1 - \bar{x}_2) - (\mu_1 - \mu_2)}{\sqrt{\dfrac{\sigma_1^2}{n_1} + \dfrac{\sigma_2^2}{n_2}}}$$

has the standard Normal $N(0, 1)$ sampling distribution.

In the unlikely event that both population standard deviations are known, the two-sample $z$ statistic is the basis for inference about $\mu_1 - \mu_2$. Exact $z$ procedures are seldom used, however, because $\sigma_1$ and $\sigma_2$ are rarely known. In Chapter 6, we discussed the one-sample $z$ procedures in order to introduce the ideas of inference. Here we move directly to the more useful $t$ procedures.

## The two-sample t procedures

Suppose now that the population standard deviations $\sigma_1$ and $\sigma_2$ are not known. We estimate them by the sample standard deviations $s_1$ and $s_2$ from our two samples. Following the pattern of the one-sample case, we substitute the standard errors for the standard deviations used in the two-sample $z$ statistic. The result is the *two-sample t statistic:*

$$t = \frac{(\bar{x}_1 - \bar{x}_2) - (\mu_1 - \mu_2)}{\sqrt{\dfrac{s_1^2}{n_1} + \dfrac{s_2^2}{n_2}}}$$

Unfortunately, this statistic does *not* have a $t$ distribution. A $t$ distribution replaces the $N(0, 1)$ distribution only when a single standard deviation ($\sigma$) in a $z$ statistic is replaced by its sample standard deviation ($s$). In this case, we replace two standard deviations ($\sigma_1$ and $\sigma_2$) by their estimates ($s_1$ and $s_2$), which does not produce a statistic having a $t$ distribution.

Nonetheless, we can approximate the distribution of the two-sample $t$ statistic by using the $t(k)$ distribution with an **approximation for the degrees of freedom $k$.** We use these approximations to find approximate values of $t^*$ for confidence intervals and to find approximate $P$-values for significance tests. Here are two approximations:

approximations for the degrees of freedom

1. Use a value of $k$ that is calculated from the data. In general, it will not be a whole number.
2. Use $k$ equal to the smaller of $n_1 - 1$ and $n_2 - 1$.

Most statistical software uses the first option to approximate the $t(k)$ distribution for two-sample problems unless the user requests another method. Use of this approximation without software is a bit complicated; we will give the details later in this section. If you are not using software, the second approximation is preferred. This approximation is appealing because it is conservative.[18] Margins of error for confidence intervals are a bit larger than they need to be, so the true confidence level is larger than $C$. For significance testing, the true $P$-values are a bit smaller than those we obtain from the approximation; for tests at a fixed significance level, we are a little less likely to reject $H_0$ when it is true. In practice, the choice of approximation rarely makes a difference in our conclusion.

## The two-sample *t* significance test

### THE TWO-SAMPLE *t* SIGNIFICANCE TEST

Suppose that an SRS of size $n_1$ is drawn from a Normal population with unknown mean $\mu_1$ and that an independent SRS of size $n_2$ is drawn from another Normal population with unknown mean $\mu_2$. To test the hypothesis $H_0$: $\mu_1 = \mu_2$, compute the **two-sample *t* statistic**

$$t = \frac{\bar{x}_1 - \bar{x}_2}{\sqrt{\dfrac{s_1^2}{n_1} + \dfrac{s_2^2}{n_2}}}$$

and use $P$-values or critical values for the $t(k)$ distribution, where the degrees of freedom $k$ are either approximated by software or are the smaller of $n_1 - 1$ and $n_2 - 1$.

EXAMPLE

**7.14 Directed reading activities assessment.** An educator believes that new directed reading activities in the classroom will help elementary school pupils improve some aspects of their reading ability. She arranges for a third-grade class of 21 students to take part in these activities for an eight-week period. A control classroom of 23 third-graders follows the same curriculum without the activities. At the end of the eight weeks, all students are given a Degree of Reading Power (DRP) test, which measures the aspects of reading ability that the treatment is designed to improve. The data appear in Table 7.4.[19]

**TABLE 7.4**

**DRP scores for third-graders**

| Treatment Group | | | | Control Group | | | |
|---|---|---|---|---|---|---|---|
| 24 | 61 | 59 | 46 | 42 | 33 | 46 | 37 |
| 43 | 44 | 52 | 43 | 43 | 41 | 10 | 42 |
| 58 | 67 | 62 | 57 | 55 | 19 | 17 | 55 |
| 71 | 49 | 54 | | 26 | 54 | 60 | 28 |
| 43 | 53 | 57 | | 62 | 20 | 53 | 48 |
| 49 | 56 | 33 | | 37 | 85 | 42 | |

First examine the data:

| Control | | Treatment |
|---|---|---|
| 970 | 1 | |
| 860 | 2 | 4 |
| 773 | 3 | 3 |
| 8632221 | 4 | 3334699 |
| 5543 | 5 | 23467789 |
| 20 | 6 | 127 |
| | 7 | 1 |
| 5 | 8 | |

A back-to-back stemplot suggests that there is a mild outlier in the control group but no deviation from Normality serious enough to forbid use of $t$ procedures. Separate Normal quantile plots for both groups (Figure 7.13) confirm that both are approximately Normal. The scores of the treatment group appear to be somewhat higher than those of the control group. The summary statistics are

| Group | $n$ | $\overline{x}$ | $s$ |
|---|---|---|---|
| Treatment | 21 | 51.48 | 11.01 |
| Control | 23 | 41.52 | 17.15 |

Because we hope to show that the treatment (Group 1) is better than the control (Group 2), the hypotheses are

$$H_0: \mu_1 = \mu_2$$

$$H_a: \mu_1 > \mu_2$$

The two-sample $t$ test statistic is

$$t = \frac{\overline{x}_1 - \overline{x}_2}{\sqrt{\dfrac{s_1^2}{n_1} + \dfrac{s_2^2}{n_2}}} = \frac{51.48 - 41.52}{\sqrt{\dfrac{11.01^2}{21} + \dfrac{17.15^2}{23}}}$$

$$= 2.31$$

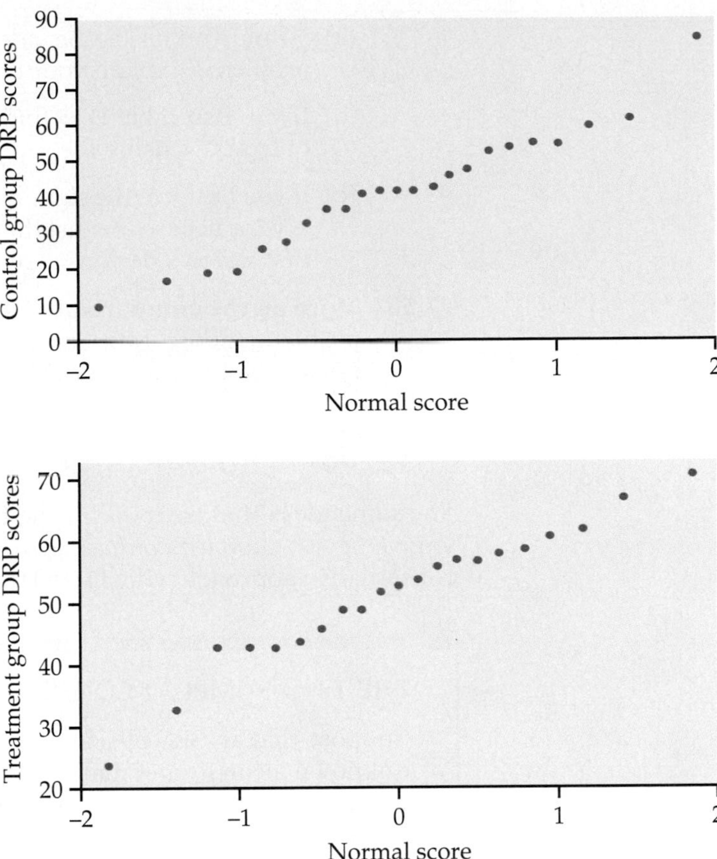


**FIGURE 7.13** Normal quantile plots of the DRP scores in Table 7.4.

The $P$-value for the one-sided test is $P(T \geq 2.31)$. Software gives the approximate $P$-value as 0.0132 and uses 37.9 as the degrees of freedom. For the second approximation, the degrees of freedom $k$ are equal to the smaller of

$$n_1 - 1 = 21 - 1 = 20 \quad \text{and} \quad n_2 - 1 = 23 - 1 = 22$$

Comparing 2.31 with the entries in Table D for 20 degrees of freedom, we see that $P$ lies between 0.02 and 0.01. The data strongly suggest that directed reading activity improves the DRP score ($t = 2.31$, df $= 20$, $P < 0.02$).

**df = 20**

| $p$ | 0.02 | 0.01 |
|---|---|---|
| $t^*$ | 2.197 | 2.528 |

Note that when we report a result such as this with $P < 0.02$, we imply that the result is *not* significant at the 0.01 level.

If your software gives $P$-values for only the two-sided alternative, $2P(T \geq |t|)$, you need to divide the reported value by 2 after checking that the means differ in the direction specified by the alternative hypothesis.

## USE YOUR KNOWLEDGE

**7.54 Comparison of two Web designs.** You want to compare the daily number of hits for two different Web designs that advertise your Internet business. You assign the next 50 days to either Design A or Design B, 25 days to each.

(a) Would you use a one-sided or two-sided significance test for this problem? Explain your choice.

(b) If you use Table D to find the critical value, what are the degrees of freedom using the second approximation?

(c) If you perform the significance test using $\alpha = 0.05$, how large (positive or negative) must the $t$ statistic be to reject the null hypothesis that the two designs result in the same average hits?

**7.55 More on the comparison of two Web designs.** Consider the previous problem. If the $t$ statistic for comparing the mean hits were 2.75, what $P$-value would you report? What would you conclude using $\alpha = 0.05$?

## The two-sample *t* confidence interval

The same ideas that we used for the two-sample $t$ significance tests also apply to give us *two-sample t confidence intervals.* We can use either software or the conservative approach with Table D to approximate the value of $t^*$.

### THE TWO-SAMPLE *t* CONFIDENCE INTERVAL

Suppose that an SRS of size $n_1$ is drawn from a Normal population with unknown mean $\mu_1$ and that an independent SRS of size $n_2$ is drawn from another Normal population with unknown mean $\mu_2$. The **confidence interval for $\mu_1 - \mu_2$** given by

$$(\bar{x}_1 - \bar{x}_2) \pm t^* \sqrt{\frac{s_1^2}{n_1} + \frac{s_2^2}{n_2}}$$

has confidence level at least $C$ no matter what the population standard deviations may be. Here, $t^*$ is the value for the $t(k)$ density curve with area $C$ between $-t^*$ and $t^*$. The value of the degrees of freedom $k$ is approximated by software or we use the smaller of $n_1 - 1$ and $n_2 - 1$.

To complete the analysis of the DRP scores we examined in Example 7.14, we need to describe the size of the treatment effect. We do this with a confidence interval for the difference between the treatment group and the control group means.

EXAMPLE

**7.15 How much improvement?** We will find a 95% confidence interval for the mean improvement in the entire population of third-graders. The interval is

$$(\bar{x}_1 - \bar{x}_2) \pm t^* \sqrt{\frac{s_1^2}{n_1} + \frac{s_2^2}{n_2}} = (51.48 - 41.52) \pm t^* \sqrt{\frac{11.01^2}{21} + \frac{17.15^2}{23}}$$

$$= 9.96 \pm 4.31t^*$$

Using software, the degrees of freedom are 37.9 and $t^* = 2.025$. This approximation gives

$$9.96 \pm (4.31 \times 2.025) = 9.96 \pm 8.72 = (1.2, 18.7)$$

The conservative approach uses the $t(20)$ distribution. Table D gives $t^* = 2.086$. With this approximation we have

$$9.96 \pm (4.31 \times 2.086) = 9.96 \pm 8.99 = (1.0, 18.9)$$

We can see that the conservative approach does, in fact, give a larger interval than the more accurate approximation used by software. However, the difference is pretty small.

We estimate the mean improvement to be about 10 points, but with a margin of error of almost 9 points with either method. Although we have good evidence of some improvement, the data do not allow a very precise estimate of the size of the average improvement.

The design of the study in Example 7.14 is not ideal. Random assignment of students was not possible in a school environment, so existing third-grade classes were used. The effect of the reading programs is therefore confounded with any other differences between the two classes. The classes were chosen to be as similar as possible—for example, in terms of the social and economic status of the students. Extensive pretesting showed that the two classes were on the average quite similar in reading ability at the beginning of the experiment. To avoid the effect of two different teachers, the researcher herself taught reading in both classes during the eight-week period of the experiment. We can therefore be somewhat confident that the two-sample test is detecting the effect of the treatment and not some other difference between the classes. This example is typical of many situations in which an experiment is carried out but randomization is not possible.

## USE YOUR KNOWLEDGE

**7.56 Two-sample $t$ confidence interval.** Assume $\bar{x}_1 = 100$, $\bar{x}_2 = 120$, $s_1 = 10$, $s_2 = 12$, $n_1 = 50$, and $n_2 = 50$. Find a 95% confidence interval for the difference in the corresponding values of $\mu$ using the second approximation for degrees of freedom. Does this interval include more or fewer values than a 99% confidence interval? Explain your answer.

**7.57 Another two-sample $t$ confidence interval.** Assume $\bar{x}_1 = 100$, $\bar{x}_2 = 120$, $s_1 = 10$, $s_2 = 12$, $n_1 = 10$, and $n_2 = 10$. Find a 95% confidence interval for the difference in the corresponding values of $\mu$ using the second approximation for degrees of freedom. Would you reject the null hypothesis that the population means are equal in favor of the two-sided alternative at significance level 0.05? Explain.

## Robustness of the two-sample procedures

The two-sample $t$ procedures are more robust than the one-sample $t$ methods. When the sizes of the two samples are equal and the distributions of the two populations being compared have similar shapes, probability values from the $t$ table are quite accurate for a broad range of distributions when the sample sizes are as small as $n_1 = n_2 = 5$.[20] When the two population distributions have different shapes, larger samples are needed. The guidelines for the use of one-sample $t$ procedures can be adapted to two-sample procedures by replacing "sample size" with the "sum of the sample sizes" $n_1 + n_2$. These guidelines are rather conservative, especially when the two samples are of equal size. *In planning a two-sample study, choose equal sample sizes if you can.* The two-sample $t$ procedures are most robust against non-Normality in this case, and the conservative probability values are most accurate.

Here is an example with moderately large sample sizes that are not equal. Even if the distributions are not Normal, we are confident that the sample means will be approximately Normal. The two-sample $t$ test is very robust in this case.

EXAMPLE

**7.16 Wheat prices.** The U.S. Department of Agriculture (USDA) uses sample surveys to produce important economic estimates. One pilot study estimated wheat prices in July and in September using independent samples of wheat producers in the two months. Here are the summary statistics, in dollars per bushel:[21]

| Month | $n$ | $\bar{x}$ | $s$ |
|---|---|---|---|
| September | 45 | \$3.61 | \$0.19 |
| July | 90 | \$2.95 | \$0.22 |

The September prices are higher on the average. But we have data from only a sample of producers each month. Can we conclude that national average prices in July and September are not the same? Or are these differences merely what we would expect to see due to random variation?

Because we did not specify a direction for the difference before looking at the data, we choose a two-sided alternative. The hypotheses are

$$H_0: \mu_1 = \mu_2$$

$$H_a: \mu_1 \neq \mu_2$$

Because the samples are moderately large, we can confidently use the $t$ procedures even though we lack the detailed data and so cannot verify the Normality condition.

The two-sample $t$ statistic is

$$t = \frac{\bar{x}_1 - \bar{x}_2}{\sqrt{\dfrac{s_1^2}{n_1} + \dfrac{s_2^2}{n_2}}} = \frac{3.61 - 2.95}{\sqrt{\dfrac{0.19^2}{45} + \dfrac{0.22^2}{90}}}$$

$$= 18.03$$

**df = 40**

| $p$ | 0.0005 |
|---|---|
| $t^*$ | 3.551 |

The conservative approach finds the $P$-value by comparing 18.03 to critical values for the $t(44)$ distribution because the smaller sample has 45 observations. We must double the table tail area $p$ because the alternative is two-sided. Table D does not have entries for 44 degrees of freedom. When this happens, we use the next smaller degrees of freedom. Our calculated value of $t$ is larger than the $p = 0.0005$ entry in the table. Doubling 0.0005, we conclude that the $P$-value is less than 0.001. The data give conclusive evidence that the mean wheat prices were higher in September than they were in July ($t = 18.03$, df $= 44$, $P < 0.001$).

In this example the exact $P$-value is very small because $t = 18$ says that the observed difference in means is 18 standard errors above the hypothesized difference of zero ($\mu_1 = \mu_2$). This is so unlikely that the probability is zero for all practical purposes. The difference in mean prices is not only highly significant but large enough (66 cents per bushel) to be important to producers.

In this and other examples, we can choose which population to label 1 and which to label 2. After inspecting the data, we chose September as Population 1 because this choice makes the $t$ statistic a positive number. This avoids any possible confusion from reporting a negative value for $t$. *Choosing the population labels is **not** the same as choosing a one-sided alternative after looking at the data.* Choosing hypotheses after seeing a result in the data is a violation of sound statistical practice.

## Inference for small samples

Small samples require special care. We do not have enough observations to examine the distribution shapes, and only extreme outliers stand out. The power of significance tests tends to be low, and the margins of error of confidence intervals tend to be large. Despite these difficulties, we can often draw important conclusions from studies with small sample sizes. If the size of an effect is as large as it was in the wheat price example, it should still be evident even if the $n$'s are small.

EXAMPLE

**7.17 More about wheat prices.** In the setting of Example 7.16, a quick survey collects prices from only 5 producers each month. The data are

| Month | Price of wheat ($/bushel) | | | | |
|---|---|---|---|---|---|
| September | $3.5900 | $3.6150 | $3.5950 | $3.5725 | $3.5825 |
| July | $2.9200 | $2.9675 | $2.9175 | $2.9250 | $2.9325 |

The prices are reported to the nearest quarter of a cent. First, examine the distributions with a back-to-back stemplot after rounding each price to the nearest cent.

| September | | July |
|---:|:---:|:---|
| | 2.9 | 22337 |
| | 3.0 | |
| | 3.1 | |
| | 3.2 | |
| | 3.3 | |
| | 3.4 | |
| 987 | 3.5 | |
| 20 | 3.6 | |

The pattern is clear. There is little variation among prices within each month, and the distributions for the two months are far apart relative to the within-month variation.

A significance test can confirm that the difference between months is too large to easily arise just by chance. We test

$$H_0: \mu_1 = \mu_2$$
$$H_a: \mu_1 \neq \mu_2$$

The price is higher in September ($t = 56.99$, df $= 7.55$, $P < 0.0001$). The difference in sample means is 65.9 cents.

Figure 7.14 gives outputs for this analysis from several software packages. Although the formats differ, the basic information is the same. All report the sample sizes, the sample means and standard deviations (or variances), the $t$ statistic, and its $P$-value. All agree that the $P$-value is very small, though some give more detail than others. Software often labels the groups in alphabetical order. In this example, July is then the first population and $t = -56.99$, the negative of our result. Always check the means first and report the statistic (you may need to change the sign) in an appropriate way. Be sure to also mention the size of the effect you observed, such as "The mean price for September was 65.9 cents higher than in July."

SAS

**The TTEST Procedure**

Statistics

| variable | month | N | Lower CL Mean | Mean | Upper CL Mean | Std Err |
|---|---|---|---|---|---|---|
| price | July | 5 | 2.9072 | 2.9325 | 2.9578 | 0.0091 |
| price | Sept | 5 | 3.5713 | 3.591 | 3.6107 | 0.0071 |
| price | Diff (1-2) | | -0.685 | -0.659 | -0.632 | 0.0116 |

T-Tests

| Variable | Method | Variances | DF | t Value | Pr > \|t\| |
|---|---|---|---|---|---|
| price | Pooled | Equal | 8 | -56.99 | <.0001 |
| price | Satterthwaite | Unequal | 7.55 | -56.99 | <.0001 |

**FIGURE 7.14** SAS, Excel, Minitab, and SPSS output for Example 7.17. *(continued)*

Excel

| | A | B | C |
|---|---|---|---|
| 1 | **t-Test: Two-Sample Assuming Unequal Variances** | | |
| 2 | | | |
| 3 | | *Variable 1* | *Variable 2* |
| 4 | Mean | 2.9325 | 3.591 |
| 5 | Variance | 0.000415625 | 0.000251875 |
| 6 | Observations | 5 | 5 |
| 7 | Hypothesized Mean Difference | 0 | |
| 8 | df | 8 | |
| 9 | t Stat | –56.9921639 | |
| 10 | P(T<=t) one-tail | 4.98679E–12 | |
| 11 | t Critical one-tail | 1.859548033 | |
| 12 | P(T<=t) two-tail | 9.97357E–12 | |
| 13 | t Critical two-tail | 2.306004133 | |

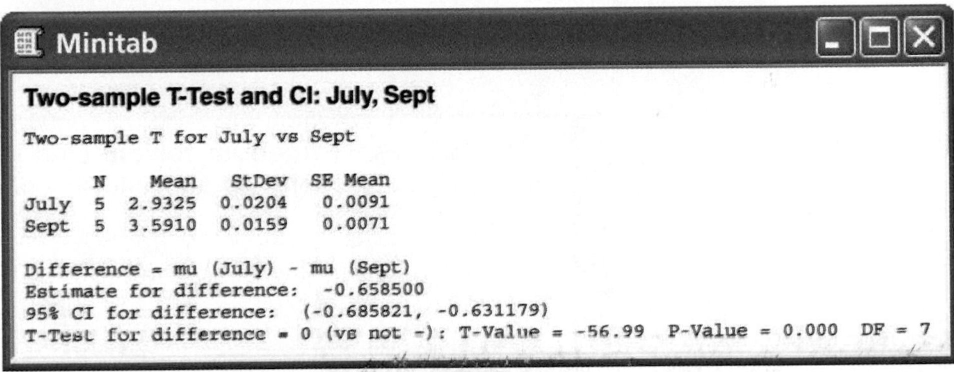

Minitab

**Two-sample T-Test and CI: July, Sept**

```
Two-sample T for July vs Sept

      N   Mean    StDev   SE Mean
July  5  2.9325  0.0204   0.0091
Sept  5  3.5910  0.0159   0.0071

Difference = mu (July) - mu (Sept)
Estimate for difference:  -0.658500
95% CI for difference:  (-0.685821, -0.631179)
T-Test for difference = 0 (vs not =): T-Value = -56.99  P-Value = 0.000  DF = 7
```

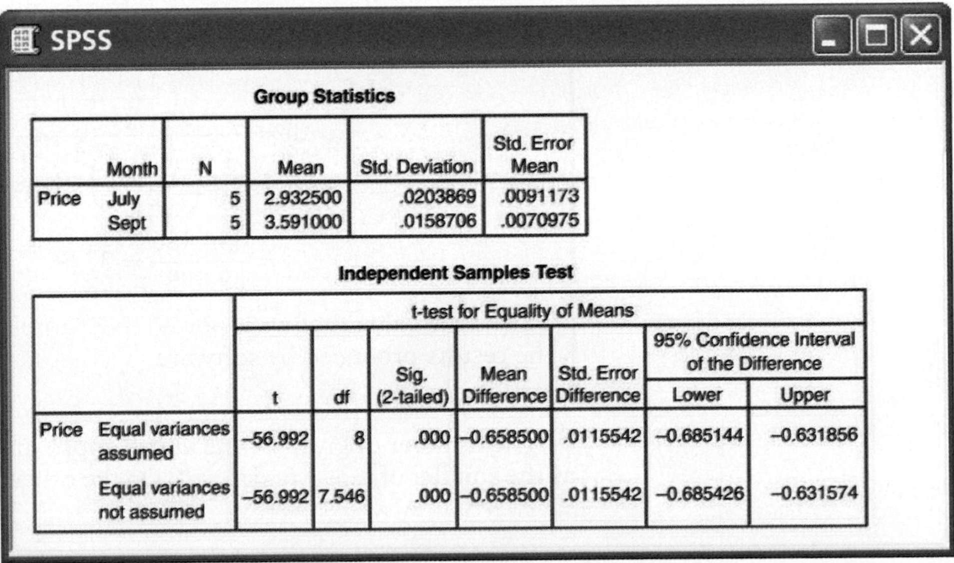

SPSS

**Group Statistics**

| | Month | N | Mean | Std. Deviation | Std. Error Mean |
|---|---|---|---|---|---|
| Price | July | 5 | 2.932500 | .0203869 | .0091173 |
| | Sept | 5 | 3.591000 | .0158706 | .0070975 |

**Independent Samples Test**

| | | t-test for Equality of Means | | | | | | |
|---|---|---|---|---|---|---|---|---|
| | | | | | | | 95% Confidence Interval of the Difference | |
| | | t | df | Sig. (2-tailed) | Mean Difference | Std. Error Difference | Lower | Upper |
| Price | Equal variances assumed | –56.992 | 8 | .000 | –0.658500 | .0115542 | –0.685144 | –0.631856 |
| | Equal variances not assumed | –56.992 | 7.546 | .000 | –0.658500 | .0115542 | –0.685426 | –0.631574 |

**FIGURE 7.14** *(Continued)* SAS, Excel, Minitab, and SPSS output for Example 7.17.

SPSS and SAS report the results of *two* $t$ procedures: a special procedure that assumes that the two population variances are equal and the general two-sample procedure that we have just studied. We don't recommend the "equal-variances" procedures, but we describe them later, in the section on pooled two-sample $t$ procedures.

## Software approximation for the degrees of freedom*

We noted earlier that the two-sample $t$ statistic does not have an exact $t$ distribution. Moreover, the exact distribution changes as the unknown population standard deviations $\sigma_1$ and $\sigma_2$ change. However, the distribution can be approximated by a $t$ distribution with degrees of freedom given by

$$\text{df} = \frac{\left(\dfrac{s_1^2}{n_1} + \dfrac{s_2^2}{n_2}\right)^2}{\dfrac{1}{n_1 - 1}\left(\dfrac{s_1^2}{n_1}\right)^2 + \dfrac{1}{n_2 - 1}\left(\dfrac{s_2^2}{n_2}\right)^2}$$

This is the approximation used by most statistical software. It is quite accurate when both sample sizes $n_1$ and $n_2$ are 5 or larger.

EXAMPLE

**7.18 Degrees of freedom for directed reading assessment.** For the DRP study of Example 7.14, the following table summarizes the data:

| Group | $n$ | $\bar{x}$ | $s$ |
|---|---|---|---|
| 1 | 21 | 51.48 | 11.01 |
| 2 | 23 | 41.52 | 17.15 |

For greatest accuracy, we will use critical points from the $t$ distribution with degrees of freedom given by the equation above:

$$\text{df} = \frac{\left(\dfrac{11.01^2}{21} + \dfrac{17.15^2}{23}\right)^2}{\dfrac{1}{20}\left(\dfrac{11.01^2}{21}\right)^2 + \dfrac{1}{22}\left(\dfrac{17.15^2}{23}\right)^2}$$

$$= \frac{344.486}{9.099} = 37.86$$

This is the value that we reported in Examples 7.14 and 7.15, where we gave the results produced by software.

The number df given by the above approximation is always at least as large as the smaller of $n_1 - 1$ and $n_2 - 1$. On the other hand, df is never larger than the

*This material can be omitted unless you are using statistical software and wish to understand what the software does.

sum $n_1 + n_2 - 2$ of the two individual degrees of freedom. The number of degrees of freedom is generally not a whole number. There is a $t$ distribution with any positive degrees of freedom, even though Table D contains entries only for whole-number degrees of freedom. When df is small and is not a whole number, interpolation between entries in Table D may be needed to obtain an accurate critical value or $P$-value. Because of this and the need to calculate df, we do not recommend regular use of this approximation if a computer is not doing the arithmetic. With a computer, however, the more accurate procedures are painless.

USE YOUR KNOWLEDGE

**7.58 Calculating the degrees of freedom.** Assume $s_1 = 10$, $s_2 = 12$, $n_1 = 20$, and $n_2 = 18$. Find the software approximate degrees of freedom.

## The pooled two-sample *t* procedures*

There is one situation in which a $t$ statistic for comparing two means has exactly a $t$ distribution. Suppose that the two Normal population distributions have the *same* standard deviation. In this case we need substitute only a single standard error in a $z$ statistic, and the resulting $t$ statistic has a $t$ distribution. We will develop the $z$ statistic first, as usual, and from it the $t$ statistic.

Call the common—and still unknown—standard deviation of both populations $\sigma$. Both sample variances $s_1^2$ and $s_2^2$ estimate $\sigma^2$. The best way to combine these two estimates is to average them with weights equal to their degrees of freedom. This gives more weight to the information from the larger sample, which is reasonable. The resulting estimator of $\sigma^2$ is

$$s_p^2 = \frac{(n_1 - 1)s_1^2 + (n_2 - 1)s_2^2}{n_1 + n_2 - 2}$$

pooled estimator of $\sigma^2$

This is called the **pooled estimator of $\sigma^2$** because it combines the information in both samples.

When both populations have variance $\sigma^2$, the addition rule for variances says that $\bar{x}_1 - \bar{x}_2$ has variance equal to the *sum* of the individual variances, which is

$$\frac{\sigma^2}{n_1} + \frac{\sigma^2}{n_2} = \sigma^2\left(\frac{1}{n_1} + \frac{1}{n_2}\right)$$

The standardized difference of means in this equal-variance case is therefore

$$z = \frac{(\bar{x}_1 - \bar{x}_2) - (\mu_1 - \mu_2)}{\sigma\sqrt{\dfrac{1}{n_1} + \dfrac{1}{n_2}}}$$

This is a special two-sample $z$ statistic for the case in which the populations have the same $\sigma$. Replacing the unknown $\sigma$ by the estimate $s_p$ gives a $t$ statistic.

*This section can be omitted if desired, but it should be read if you plan to read Chapters 12 and 13.

The degrees of freedom are $n_1 + n_2 - 2$, the sum of the degrees of freedom of the two sample variances. This statistic is the basis of the pooled two-sample $t$ inference procedures.

### THE POOLED TWO-SAMPLE $t$ PROCEDURES

Suppose that an SRS of size $n_1$ is drawn from a Normal population with unknown mean $\mu_1$ and that an independent SRS of size $n_2$ is drawn from another Normal population with unknown mean $\mu_2$. Suppose also that the two populations have the same standard deviation. A level $C$ confidence interval for $\mu_1 - \mu_2$ is

$$(\bar{x}_1 - \bar{x}_2) \pm t^* s_p \sqrt{\frac{1}{n_1} + \frac{1}{n_2}}$$

Here $t^*$ is the value for the $t(n_1 + n_2 - 2)$ density curve with area $C$ between $-t^*$ and $t^*$.

To test the hypothesis $H_0: \mu_1 = \mu_2$, compute the pooled two-sample $t$ statistic

$$t = \frac{\bar{x}_1 - \bar{x}_2}{s_p \sqrt{\frac{1}{n_1} + \frac{1}{n_2}}}$$

In terms of a random variable $T$ having the $t(n_1 + n_2 - 2)$ distribution, the $P$-value for a test of $H_0$ against

$$H_a: \mu_1 > \mu_2 \text{ is } P(T \geq t)$$

$$H_a: \mu_1 < \mu_2 \text{ is } P(T \leq t)$$

$$H_a: \mu_1 \neq \mu_2 \text{ is } 2P(T \geq |t|)$$

EXAMPLE

**7.19 Calcium and blood pressure.** Does increasing the amount of calcium in our diet reduce blood pressure? Examination of a large sample of people revealed a relationship between calcium intake and blood pressure, but such observational studies do not establish causation. Animal experiments, however, showed that calcium supplements do reduce blood pressure in rats, justifying an experiment with human subjects. A randomized comparative experiment gave one group of 10 black men a calcium supplement for 12 weeks. The control group of 11 black men received a placebo that appeared identical. (In fact, a block design with black and white men as the blocks was used. We will look only at the results for blacks, because the earlier survey suggested that calcium is more effective for blacks.) The experiment was double-blind. Table 7.5 gives the seated systolic (heart contracted) blood pressure for all subjects at the beginning and end of the 12-week period, in millimeters (mm) of mercury. Because the researchers were interested in de-

**TABLE 7.5**
Seated systolic blood pressure

| Calcium Group | | | Placebo Group | | |
|---|---|---|---|---|---|
| Begin | End | Decrease | Begin | End | Decrease |
| 107 | 100 | 7 | 123 | 124 | −1 |
| 110 | 114 | −4 | 109 | 97 | 12 |
| 123 | 105 | 18 | 112 | 113 | −1 |
| 129 | 112 | 17 | 102 | 105 | −3 |
| 112 | 115 | −3 | 98 | 95 | 3 |
| 111 | 116 | −5 | 114 | 119 | −5 |
| 107 | 106 | 1 | 119 | 114 | 5 |
| 112 | 102 | 10 | 114 | 112 | 2 |
| 136 | 125 | 11 | 110 | 121 | −11 |
| 102 | 104 | −2 | 117 | 118 | −1 |
| | | | 130 | 133 | −3 |

creasing blood pressure, Table 7.5 also shows the decrease for each subject. An increase appears as a negative entry.[22]

As usual, we first examine the data. To compare the effects of the two treatments, take the response variable to be the amount of the decrease in blood pressure. Inspection of the data reveals that there are no outliers. Normal quantile plots (Figure 7.15) give a more detailed picture. The calcium group has a somewhat short left tail, but there are no departures from Normality that will prevent use of $t$ procedures. To examine the question of the researchers who collected these data, we perform a significance test.

EXAMPLE

**7.20 Does increased calcium reduce blood pressure?** Take Group 1 to be the calcium group and Group 2 to be the placebo group. The evidence that calcium lowers blood pressure more than a placebo is assessed by testing

$$H_0: \mu_1 = \mu_2$$

$$H_a: \mu_1 > \mu_2$$

Here are the summary statistics for the decrease in blood pressure:

| Group | Treatment | $n$ | $\bar{x}$ | $s$ |
|---|---|---|---|---|
| 1 | Calcium | 10 | 5.000 | 8.743 |
| 2 | Placebo | 11 | −0.273 | 5.901 |

The calcium group shows a drop in blood pressure, and the placebo group has a small increase. The sample standard deviations do not rule out equal population standard deviations. A difference this large will often arise by chance in samples this small. We are willing to assume equal population standard

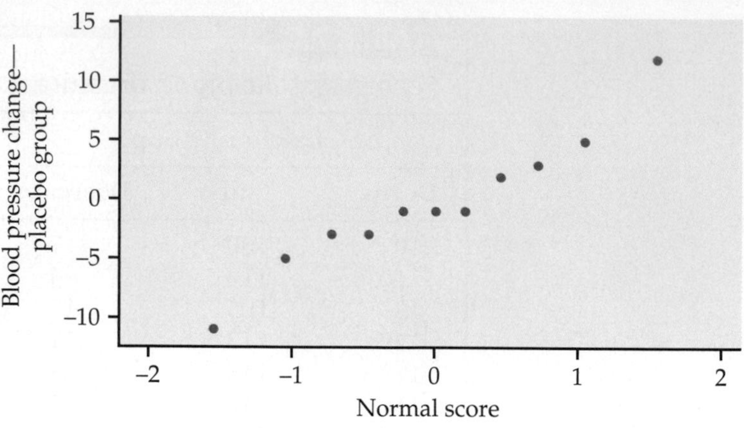


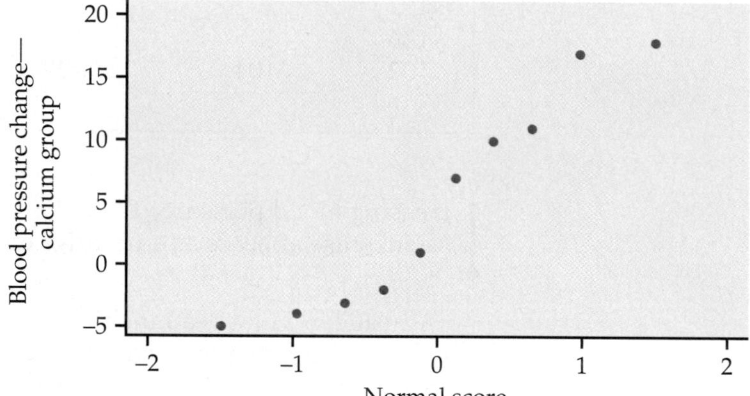


**FIGURE 7.15** Normal quantile plots of the change in blood pressure from Table 7.5.

deviations. The pooled sample variance is

$$s_p^2 = \frac{(n_1 - 1)s_1^2 + (n_2 - 1)s_2^2}{n_1 + n_2 - 2}$$

$$= \frac{(10 - 1)8.743^2 + (11 - 1)5.901^2}{10 + 11 - 2} = 54.536$$

so that

$$s_p = \sqrt{54.536} = 7.385$$

The pooled two-sample $t$ statistic is

$$t = \frac{\bar{x}_1 - \bar{x}_2}{s_p\sqrt{\frac{1}{n_1} + \frac{1}{n_2}}}$$

$$= \frac{5.000 - (-0.273)}{7.385\sqrt{\frac{1}{10} + \frac{1}{11}}}$$

$$= \frac{5.273}{3.227} = 1.634$$

**df = 19**

| $p$ | 0.10 | 0.05 |
|---|---|---|
| $t^*$ | 1.328 | 1.729 |

The $P$-value is $P(T \geq 1.634)$, where $T$ has the $t(19)$ distribution. From Table D we can see that $P$ falls between the $\alpha = 0.10$ and $\alpha = 0.05$ levels. Statistical software gives the exact value $P = 0.059$. The experiment found evidence that calcium reduces blood pressure, but the evidence falls a bit short of the traditional 5% and 1% levels.

Sample size strongly influences the $P$-value of a test. An effect that fails to be significant at a specified level $\alpha$ in a small sample can be significant in a larger sample. In the light of the rather small samples in Example 7.20, the evidence for some effect of calcium on blood pressure is rather good. The published account of the study combined these results for blacks with the results for whites and adjusted for pretest differences among the subjects. Using this more detailed analysis, the researchers were able to report the $P$-value $P = 0.008$.

Of course, a $P$-value is almost never the last part of a statistical analysis. To make a judgment regarding the size of the effect of calcium on blood pressure, we need a confidence interval.

EXAMPLE

**7.21 How different are the calcium and placebo groups?** We estimate that the effect of calcium supplementation is the difference between the sample means of the calcium and the placebo groups, $\bar{x}_1 - \bar{x}_2 = 5.273$ mm. A 90% confidence interval for $\mu_1 - \mu_2$ uses the critical value $t^* = 1.729$ from the $t(19)$ distribution. The interval is

$$(\bar{x}_1 - \bar{x}_2) \pm t^* s_p \sqrt{\frac{1}{n_1} + \frac{1}{n_2}} = [5.000 - (-0.273)] \pm (1.729)(7.385)\sqrt{\frac{1}{10} + \frac{1}{11}}$$

$$= 5.273 \pm 5.579$$

We are 90% confident that the difference in means is in the interval

$$(-0.306, 10.852)$$

The calcium treatment reduced blood pressure by about 5.3 mm more than a placebo on the average, but the margin of error for this estimate is 5.6 mm.

The pooled two-sample $t$ procedures are anchored in statistical theory and so have long been the standard version of the two-sample $t$ in textbooks. *But they require the assumption that the two unknown population standard deviations are equal.* As we shall see in Section 7.3, this assumption is hard to verify. The pooled $t$ procedures are therefore a bit risky. They are reasonably robust against both non-Normality and unequal standard deviations when the sample sizes are nearly the same. When the samples are quite different in size, the pooled $t$ procedures become sensitive to unequal standard deviations and should be used with caution unless the samples are large. Unequal standard deviations are quite common. In particular, it is not unusual for the spread of data to increase when the center gets larger. Statistical software often calculates both the pooled and the unpooled $t$ statistics, as in Figure 7.14.

**USE YOUR KNOWLEDGE**

**7.59 Wheat prices revisited.** Figure 7.14 (page 458) gives the outputs from four software packages for comparing prices received by wheat producers in July and September for small samples of 5 producers in each month. Some of the software reports both pooled and unpooled analyses. Which outputs give the pooled results? What are the pooled $t$ and its $P$-value?

**7.60 More on wheat prices.** The software outputs in Figure 7.14 give the *same value* for the pooled and unpooled $t$ statistics. Do some simple algebra to show that this is always true when the two sample sizes $n_1$ and $n_2$ are the same. In other cases, the two $t$ statistics usually differ.

## SECTION 7.2 Summary

Significance tests and confidence intervals for the difference of the means $\mu_1$ and $\mu_2$ of two Normal populations are based on the difference $\bar{x}_1 - \bar{x}_2$ of the sample means from two independent SRSs. Because of the central limit theorem, the resulting procedures are approximately correct for other population distributions when the sample sizes are large.

When independent SRSs of sizes $n_1$ and $n_2$ are drawn from two Normal populations with parameters $\mu_1$, $\sigma_1$ and $\mu_2$, $\sigma_2$ the **two-sample $z$ statistic**

$$z = \frac{(\bar{x}_1 - \bar{x}_2) - (\mu_1 - \mu_2)}{\sqrt{\dfrac{\sigma_1^2}{n_1} + \dfrac{\sigma_2^2}{n_2}}}$$

has the $N(0, 1)$ distribution.

The **two-sample $t$ statistic**

$$t = \frac{(\bar{x}_1 - \bar{x}_2) - (\mu_1 - \mu_2)}{\sqrt{\dfrac{s_1^2}{n_1} + \dfrac{s_2^2}{n_2}}}$$

does *not* have a $t$ distribution. However, good approximations are available.

**Conservative inference procedures** for comparing $\mu_1$ and $\mu_2$ are obtained from the two-sample $t$ statistic by using the $t(k)$ distribution with degrees of freedom $k$ equal to the smaller of $n_1 - 1$ and $n_2 - 1$.

**More accurate probability values** can be obtained by estimating the degrees of freedom from the data. This is the usual procedure for statistical software.

An approximate level $C$ **confidence interval** for $\mu_1 - \mu_2$ is given by

$$(\bar{x}_1 - \bar{x}_2) \pm t^* \sqrt{\frac{s_1^2}{n_1} + \frac{s_2^2}{n_2}}$$

Here, $t^*$ is the value for the $t(k)$ density curve with area $C$ between $-t^*$ and $t^*$, where $k$ is computed from the data by software or is the smaller of $n_1 - 1$ and

$n_2 - 1$. The quantity

$$t^* \sqrt{\frac{s_1^2}{n_1} + \frac{s_2^2}{n_2}}$$

is the **margin of error.**

Significance tests for $H_0: \mu_1 = \mu_2$ use the **two-sample *t* statistic**

$$t = \frac{\bar{x}_1 - \bar{x}_2}{\sqrt{\frac{s_1^2}{n_1} + \frac{s_2^2}{n_2}}}$$

The *P*-value is approximated using the $t(k)$ distribution where $k$ is estimated from the data using software or is the smaller of $n_1 - 1$ and $n_2 - 1$.

The guidelines for practical use of two-sample *t* procedures are similar to those for one-sample *t* procedures. Equal sample sizes are recommended.

If we can assume that the two populations have equal variances, **pooled two-sample *t* procedures** can be used. These are based on the **pooled estimator**

$$s_p^2 = \frac{(n_1 - 1)s_1^2 + (n_2 - 1)s_2^2}{n_1 + n_2 - 2}$$

of the unknown common variance and the $t(n_1 + n_2 - 2)$ distribution.

## SECTION 7.2 Exercises

*For Exercises 7.54 and 7.55, see pages 453 and 454; for Exercises 7.56 and 7.57, see page 455; for Exercise 7.58, see page 461; and for Exercises 7.59 and 7.60, see page 466.*

*In exercises that call for two-sample t procedures, you may use either of the two approximations for the degrees of freedom that we have discussed: the value given by your software or the smaller of $n_1 - 1$ and $n_2 - 1$. Be sure to state clearly which approximation you have used.*

**7.61 Comparison of blood lipid levels in males and females.** A recent study at Baylor University investigated the lipid levels in a cohort of sedentary university students.[23] A total of 108 students volunteered for the study and met the eligibility criteria. The following table summarizes the blood lipid levels, in milligrams per deciliter (mg/dl), of the participants broken down by gender:

| | Females ($n = 71$) | | Males ($n = 37$) | |
|---|---|---|---|---|
| | $\bar{x}$ | $s$ | $\bar{x}$ | $s$ |
| Total cholesterol | 173.70 | 34.79 | 171.81 | 33.24 |
| LDL | 96.38 | 29.78 | 109.44 | 31.05 |
| HDL | 61.62 | 13.75 | 46.47 | 7.94 |

(a) Is it appropriate to use the two-sample *t* procedures that we studied in this section to analyze these data for gender differences? Give reasons for your answer.

(b) Describe appropriate null and alternative hypotheses for comparing male and female total cholesterol levels.

(c) Carry out the significance test. Report the test statistic with the degrees of freedom and the *P*-value. Write a short summary of your conclusion.

(d) Find a 95% confidence interval for the difference between the two means. Compare the information given by the interval with the information given by the significance test.

(e) The participants in this study were all taking an introductory health class. To what extent do you think the results can be generalized to other populations?

**7.62 More on blood lipid levels.** Refer to the previous exercise. LDL is also known as "bad" cholesterol. Suppose the researchers wanted to test the hypothesis that LDL levels are higher in sedentary males than in sedentary females. Describe appropriate null and alternative hypotheses and carry out the significance test using $\alpha = 0.05$. Report the test statistic with the degrees of freedom and the *P*-value. Write a short summary of your conclusion.

**7.63** **Evaluating a multimedia program.** A multimedia program designed to improve dietary behavior among low-income women was evaluated by comparing women who were randomly assigned to intervention and control groups. The intervention was a 30-minute session in a computer kiosk in the Food Stamp office.[24] One of the outcomes was the score on a knowledge test taken about 2 months after the program. Here is a summary of the data:

| Group | $n$ | $\bar{x}$ | $s$ |
|---|---|---|---|
| Intervention | 165 | 5.08 | 1.15 |
| Control | 212 | 4.33 | 1.16 |

(a) The test had six multiple-choice items that were scored as correct or incorrect, so the total score was an integer between 0 and 6. Do you think that these data are Normally distributed? Explain why or why not.

(b) Is it appropriate to use the two-sample $t$ procedures that we studied in this section to analyze these data? Give reasons for your answer.

(c) Describe appropriate null and alternative hypotheses for evaluating the intervention. Some people would prefer a two-sided alternative in this situation while others would use a one-sided significance test. Give reasons for each point of view.

(d) Carry out the significance test. Report the test statistic with the degrees of freedom and the *P*-value. Write a short summary of your conclusion.

(e) Find a 95% confidence interval for the difference between the two means. Compare the information given by the interval with the information given by the significance test.

(f) The women in this study were all residents of Durham, North Carolina. To what extent do you think the results can be generalized to other populations?

**7.64** **Self-control and food.** Self-efficacy is a general concept that measures how well we think we can control different situations. In the study described in the previous exercise, the participants were asked, "How sure are you that you can eat foods low in fat over the next month?" The response was measured on a five-point scale with 1 corresponding to "not sure at all" and 5 corresponding to "very sure." Here is a summary of the self-efficacy scores obtained about 2 months after the intervention:

| Group | $n$ | $\bar{x}$ | $s$ |
|---|---|---|---|
| Intervention | 165 | 4.10 | 1.19 |
| Control | 212 | 3.67 | 1.12 |

Analyze the data using the questions in the previous exercise as a guide.

**7.65** **Dust exposure at work.** Exposure to dust at work can lead to lung disease later in life. One study measured the workplace exposure of tunnel construction workers.[25] Part of the study compared 115 drill and blast workers with 220 outdoor concrete workers. Total dust exposure was measured in milligram years per cubic meter (mg.y/m$^3$). The mean exposure for the drill and blast workers was 18.0 mg.y/m$^3$ with a standard deviation of 7.8 mg.y/m$^3$. For the outdoor concrete workers, the corresponding values were 6.5 mg.y/m$^3$ and 3.4 mg.y/m$^3$.

(a) The sample included all workers for a tunnel construction company who received medical examinations as part of routine health checkups. Discuss the extent to which you think these results apply to other similar types of workers.

(b) Use a 95% confidence interval to describe the difference in the exposures. Write a sentence that gives the interval and provides the meaning of 95% confidence.

(c) Test the null hypothesis that the exposures for these two types of workers are the same. Justify your choice of a one-sided or two-sided alternative. Report the test statistic, the degrees of freedom, and the *P*-value. Give a short summary of your conclusion.

(d) The authors of the article describing these results note that the distributions are somewhat skewed. Do you think that this fact makes your analysis invalid? Give reasons for your answer.

**7.66** **Not all dust is the same.** Not all dust particles that are in the air around us cause problems for our lungs. Some particles are too large and stick to other areas of our body before they can get to our lungs. Others are so small that we can breathe them in and out and they will not deposit on our lungs. The researchers in the study described in the previous exercise also measured respirable dust. This is dust that deposits in our lungs when we breathe it. For the drill and blast workers, the mean exposure to respirable dust was 6.3 mg.y/m$^3$ with a standard deviation of 2.8 mg.y/m$^3$. The corresponding values for the outdoor concrete workers were 1.4 mg.y/m$^3$ and 0.7 mg.y/m$^3$. Analyze these data using the questions in the previous exercise as a guide.

**7.67 Change in portion size.** A recent study of food portion sizes reported that over a 17-year period, the average size of a soft drink consumed by Americans aged 2 years and older increased from 13.1 ounces (oz) to 19.9 oz. The authors state that the difference is statistically significant with $P < 0.01$.[26] Explain what additional information you would need to compute a confidence interval for the increase, and outline the procedure that you would use for the computations. Do you think that a confidence interval would provide useful additional information? Explain why or why not.

**7.68 Beverage consumption.** The results in the previous exercise were based on two national surveys with a very large number of individuals. Here is a study that also looked at beverage consumption but the sample sizes are much smaller. One part of this study compared 20 children who were 7 to 10 years old with 5 who were 11 to 13.[27] The younger children consumed an average of 8.2 oz of sweetened drinks per day while the older ones averaged 14.5 oz. The standard deviations were 10.7 oz and 8.2 oz respectively.

(a) Do you think that it is reasonable to assume that these data are Normally distributed? Explain why or why not. (*Hint:* Think about the 68–95–99.7 rule.)

(b) Using the methods in this section, test the null hypothesis that the two groups of children consume equal amounts of sweetened drinks versus the two-sided alternative. Report all details of the significance-testing procedure with your conclusion.

(c) Give a 95% confidence interval for the difference in means.

(d) Do you think that the analyses performed in parts (b) and (c) are appropriate for these data? Explain why or why not.

(e) The children in this study were all participants in an intervention study at the Cornell Summer Day Camp at Cornell University. To what extent do you think that these results apply to other groups of children?

**7.69 What is wrong?** In each of the following situations explain what is wrong and why.

(a) A researcher wants to test $H_0: \bar{x}_1 = \bar{x}_2$ versus the two-sided alternative $H_a: \bar{x}_1 \neq \bar{x}_2$.

(b) A study recorded the IQ scores of 50 college freshmen. The scores of the 24 males in the study were compared with the scores of all 50 freshmen using the two-sample methods of this section.

(c) A two-sample $t$ statistic gave a $P$-value of 0.93. From this we can reject the null hypothesis with 90% confidence.

(d) A researcher is interested in testing the one-sided alternative $H_a: \mu_1 < \mu_2$. The significance test gave $t = 2.25$. Since the $P$-value for the two-sided alternative is 0.04, he concluded that his $P$-value was 0.02.

**7.70 Basic concepts.** For each of the following, answer the question and give a short explanation of your reasoning.

(a) A 95% confidence interval for the difference between two means is reported as $(-0.1, 1.5)$. What can you conclude about the results of a significance test of the null hypothesis that the population means are equal versus the two-sided alternative?

(b) Will larger samples generally give a larger or smaller margin of error for the difference between two sample means?

**7.71 More basic concepts.** For each of the following, answer the question and give a short explanation of your reasoning.

(a) A significance test for comparing two means gave $t = -3.69$ with 9 degrees of freedom. Can you reject the null hypothesis that the $\mu$'s are equal versus the two-sided alternative at the 5% significance level?

(b) Answer part (a) for the one-sided alternative that the difference in means is negative.

**7.72 Effect of the confidence level.** Assume $\bar{x}_1 = 100$, $\bar{x}_2 = 120$, $s_1 = 10$, $s_2 = 12$, $n_1 = 50$, and $n_2 = 40$. Find a 95% confidence interval for the difference in the corresponding values of $\mu$. Does this interval include more or fewer values than a 99% confidence interval? Explain your answer.

**7.73 Study design is important!** Recall Exercise 7.54 (page 453). You are concerned that day of the week may affect online sales. So to compare the two Web page designs, you choose two successive weeks in the middle of a month. You flip a coin to assign one Monday to the first design and the other Monday to the second. You repeat this for each of the seven days of the week. You now have 7 hit amounts for each design. It is *incorrect* to use the two-sample $t$

test to see if the mean hits differ for the two designs. Carefully explain why.

**7.74 New computer monitors?** The purchasing department has suggested that all new computer monitors for your company should be flat screens. You want data to assure you that employees will like the new screens. The next 20 employees needing a new computer are the subjects for an experiment.

(a) Label the employees 01 to 20. Randomly choose 10 to receive flat screens. The remaining 10 get standard monitors.

(b) After a month of use, employees express their satisfaction with their new monitors by responding to the statement "I like my new monitor" on a scale from 1 to 5, where 1 represents "strongly disagree," 2 is "disagree," 3 is "neutral," 4 is "agree," and 5 stands for "strongly agree." The employees with the flat screens have average satisfaction 4.8 with standard deviation 0.7. The employees with the standard monitors have average 3.0 with standard deviation 1.5. Give a 95% confidence interval for the difference in the mean satisfaction scores for all employees.

(c) Would you reject the null hypothesis that the mean satisfaction for the two types of monitors is the same versus the two-sided alternative at significance level 0.05? Use your confidence interval to answer this question. Explain why you do not need to calculate the test statistic.

**7.75 Why randomize?** Refer to the previous exercise. A coworker suggested that you give the flat screens to the next 10 employees who need new screens and the standard monitor to the following 10. Explain why your randomized design is better.

**7.76 Effect of storage time on vitamin C content.** Does bread lose its vitamins when stored? Small loaves of bread were prepared with flour that was fortified with a fixed amount of vitamins. After baking, the vitamin C content of two loaves was measured. Another two loaves were baked at the same time, stored for three days, and then the vitamin C content was measured. The units are milligrams per hundred grams of flour (mg/100 g).[28] Here are the data:

| | | |
|---|---|---|
| Immediately after baking: | 47.62 | 49.79 |
| Three days after baking: | 21.25 | 22.34 |

(a) When bread is stored, does it lose vitamin C? To answer this question, perform a two-sample $t$ test for these data. Be sure to state your hypotheses, the test statistic with degrees of freedom, and the $P$-value.

(b) Give a 90% confidence interval for the amount of vitamin C lost.

**7.77 Study design matters!** Suppose that the researchers in the previous exercise could have measured the same two loaves of bread immediately after baking and again after three days. Assume that the data given had come from this study design. (Assume that the values given in the previous exercise are for first loaf and second loaf from left to right.)

(a) Explain carefully why your analysis in the previous exercise is *not correct* now, even though the data are the same.

(b) Redo the analysis for the design based on measuring the same loaves twice.

**7.78 Another ingredient.** Refer to Exercise 7.76. The amount of vitamin E (in mg/100 g of flour) in the same loaves was also measured. Here are the data:

| | | |
|---|---|---|
| Immediately after baking: | 94.6 | 96.0 |
| Three days after baking: | 97.4 | 94.3 |

(a) When bread is stored, does it lose vitamin E? To answer this question, perform a two-sample $t$ test for these data. Be sure to state your hypotheses, the test statistic with degrees of freedom, and the $P$-value.

(b) Give a 90% confidence interval for the amount of vitamin E lost.

**7.79 Are the samples too small?** Refer to Exercises 7.76 and 7.78. Some people claim that significance tests with very small samples never lead to rejection of the null hypothesis. Discuss this claim using the results of these two exercises.

**7.80 Does ad placement matter?** Corporate advertising tries to enhance the image of the corporation. A study compared two ads from two sources, the *Wall Street Journal* and the *National Enquirer.* Subjects were asked to pretend that their company was considering a major investment in Performax, the fictitious sportswear firm in the ads. Each subject was asked to respond to the question "How trustworthy was the source in the sportswear company ad for Performax?" on a 7-point scale. Higher values indicated more trustworthiness.[29] Here is a summary of the results:

| Ad source | $n$ | $\bar{x}$ | $s$ |
|---|---|---|---|
| *Wall Street Journal* | 66 | 4.77 | 1.50 |
| *National Enquirer* | 61 | 2.43 | 1.64 |

(a) Compare the two sources of ads using a $t$ test. Be sure to state your null and alternative hypotheses, the test statistic with degrees of freedom, the $P$-value, and your conclusion.

(b) Give a 95% confidence interval for the difference.

(c) Write a short paragraph summarizing the results of your analyses.

**7.81** CHALLENGE **Size of trees in the northern and southern halves.** The study of 584 longleaf pine trees in the Wade Tract in Thomas County, Georgia, had several purposes. Are trees in one part of the tract more or less like trees in any other part of the tract or are there differences? In Example 6.1 (page 354) we examined how the trees were distributed in the tract and found that the pattern was not random. In this exercise we will examine the sizes of the trees. In Exercise 7.25 we analyzed the sizes, measured as diameter at breast height (DBH), for a random sample of 40 trees. Here we divide the tract into northern and southern halves and take random samples of 30 trees from each half. Here are the diameters in centimeters (cm) of the sampled trees:

| | | | | | | | | | | |
|---|---|---|---|---|---|---|---|---|---|---|
| | 27.8 | 14.5 | 39.1 | 3.2 | 58.8 | 55.5 | 25.0 | 5.4 | 19.0 | 30.6 |
| North | 15.1 | 3.6 | 28.4 | 15.0 | 2.2 | 14.2 | 44.2 | 25.7 | 11.2 | 46.8 |
| | 36.9 | 54.1 | 10.2 | 2.5 | 13.8 | 43.5 | 13.8 | 39.7 | 6.4 | 4.8 |
| | 44.4 | 26.1 | 50.4 | 23.3 | 39.5 | 51.0 | 48.1 | 47.2 | 40.3 | 37.4 |
| South | 36.8 | 21.7 | 35.7 | 32.0 | 40.4 | 12.8 | 5.6 | 44.3 | 52.9 | 38.0 |
| | 2.6 | 44.6 | 45.5 | 29.1 | 18.7 | 7.0 | 43.8 | 28.3 | 36.9 | 51.6 |

(a) Use a back-to-back stemplot and side-by-side boxplots to examine the data graphically. Describe the patterns in the data.

(b) Is it appropriate to use the methods of this section to compare the mean DBH of the trees in the north half of the tract with the mean DBH of trees in the south half? Give reasons for your answer.

(c) What are appropriate null and alternative hypotheses for comparing the two samples of tree DBHs? Give reasons for your choices.

(d) Perform the significance test. Report the test statistic, the degrees of freedom, and the $P$-value. Summarize your conclusion.

(e) Find a 95% confidence interval for the difference in mean DBHs. Explain how this interval provides additional information about this problem.

**7.82** CHALLENGE **Size of trees in the eastern and western halves.** The Wade Tract can also be divided into eastern and western halves. Here are the DBHs of 30 randomly selected longleaf pine trees from each half:

| | | | | | | | | | | |
|---|---|---|---|---|---|---|---|---|---|---|
| | 23.5 | 43.5 | 6.6 | 11.5 | 17.2 | 38.7 | 2.3 | 31.5 | 10.5 | 23.7 |
| East | 13.8 | 5.2 | 31.5 | 22.1 | 6.7 | 2.6 | 6.3 | 51.1 | 5.4 | 9.0 |
| | 43.0 | 8.7 | 22.8 | 2.9 | 22.3 | 43.8 | 48.1 | 46.5 | 39.8 | 10.9 |
| | 17.2 | 44.6 | 44.1 | 35.5 | 51.0 | 21.6 | 44.1 | 11.2 | 36.0 | 42.1 |
| West | 3.2 | 25.5 | 36.5 | 39.0 | 25.9 | 20.8 | 3.2 | 57.7 | 43.3 | 58.0 |
| | 21.7 | 35.6 | 30.9 | 40.6 | 30.7 | 35.6 | 18.2 | 2.9 | 20.4 | 11.4 |

Using the questions in the previous exercise, analyze these data.

**7.83 Sales of a small appliance across months.** A market research firm supplies manufacturers with estimates of the retail sales of their products from samples of retail stores. Marketing managers are prone to look at the estimate and ignore sampling error. Suppose that an SRS of 70 stores this month shows mean sales of 53 units of a small appliance, with standard deviation 15 units. During the same month last year, an SRS of 55 stores gave mean sales of 50 units, with standard deviation 18 units. An increase from 50 to 53 is a rise of 6%. The marketing manager is happy because sales are up 6%.

(a) Use the two-sample $t$ procedure to give a 95% confidence interval for the difference in mean number of units sold at all retail stores.

(b) Explain in language that the manager can understand why he cannot be certain that sales rose by 6%, and that in fact sales may even have dropped.

**7.84 An improper significance test.** A friend has performed a significance test of the null hypothesis that two means are equal. His report states that the null hypothesis is rejected in favor of the alternative that the first mean is larger than the second. In a presentation on his work, he notes that the first sample mean was larger than the second mean and this is why he chose this particular one-sided alternative.

(a) Explain what is wrong with your friend's procedure and why.

(b) Suppose he reported $t = 1.70$ with a $P$-value of 0.06. What is the correct $P$-value that he should report?

**7.85 Breast-feeding versus baby formula.** A study of iron deficiency among infants compared samples of infants following different feeding regimens. One group contained breast-fed infants, while the children in another group were fed a standard baby formula without any iron supplements. Here are summary results on blood hemoglobin levels at 12 months of age:[30]

| Group | $n$ | $\bar{x}$ | $s$ |
|---|---|---|---|
| Breast-fed | 23 | 13.3 | 1.7 |
| Formula | 19 | 12.4 | 1.8 |

(a) Is there significant evidence that the mean hemoglobin level is higher among breast-fed babies? State $H_0$ and $H_a$ and carry out a $t$ test. Give the $P$-value. What is your conclusion?

(b) Give a 95% confidence interval for the mean difference in hemoglobin level between the two populations of infants.

(c) State the assumptions that your procedures in (a) and (b) require in order to be valid.

*The following exercises concern optional material on the pooled two-sample t procedures and on the power of tests.*

**7.86 Revisiting the comparison of LDL levels for males and females.** In Exercise 7.62 (page 467), the LDL levels for males and females were compared using the two-sample $t$ procedures that do not assume equal standard deviations. Compare the means using a significance test and find the 95% confidence interval for the difference using the pooled methods. How do the results compare with those you obtained in Exercise 7.62?

**7.87 Revisiting the evaluation of a multimedia program.** In Exercise 7.63 (page 468), the knowledge test means for intervention and control groups were compared using the two-sample $t$ procedures that do not assume equal standard deviations. Examine the standard deviations for the two groups and verify that it is appropriate to use the pooled procedures for these data. Compare the means using a significance test and find the 95% confidence interval for the difference using the pooled methods. How do the results compare with those you obtained in Exercise 7.63?

**7.88 Revisiting self-control and food.** You used methods that do not require equal standard deviations when you analyzed the self-efficacy data in Exercise 7.64 (page 468). Can you justify using the pooled procedures for these data? Explain your answer. Analyze the data using these procedures and compare what you found in Exercise 7.64 with these results.

**7.89 Revisiting the size of trees.** Refer to the Wade Tract DBH data in Exercise 7.81 (page 471), where we compared a sample of trees from the northern half of the tract with a sample from the southern half. Because the standard deviations for the two samples are quite close, it is reasonable to analyze these data using the pooled procedures. Perform the significance test and find the 95% confidence interval for the difference in means using these methods. Summarize your results and compare them with what you found in Exercise 7.81.

**7.90 Revisiting the price of wheat.** Example 7.16 (page 456) gives summary statistics for prices received by wheat producers in September and July. The two sample standard deviations are very similar, so we may be willing to assume equal population standard deviations. Calculate the pooled $t$ test statistic and its degrees of freedom from the summary statistics. Use Table D to assess significance. How do your results compare with the unpooled analysis in the example?

**7.91 Computing the degrees of freedom.** Use the Wade Tract data in Exercise 7.81 to calculate the software approximation to the degrees of freedom using the formula on page 460. Verify your calculation with software.

**7.92 Again computing the degrees of freedom.** Use the Wade Tract data in Exercise 7.82 to calculate the software approximation to the degrees of freedom using the formula on page 460. Verify your calculation with software.

**7.93** CHALLENGE **Revisiting the dust exposure study.** The data on occupational exposure to dust that we analyzed in Exercise 7.65 (page 468) come from two groups of workers that are quite different in size. This complicates the issue regarding pooling because the sample that is larger will dominate the calculations.

(a) Calculate the degrees of freedom approximation using the formula for the degrees of freedom given on page 460. Then verify your calculations with software.

(b) Find the pooled estimate of the standard deviation. Write a short summary comparing it with the estimates of the standard deviations that come from each group.

(c) Find the standard error of the difference in sample means that you would use for the method that does not assume equal variances. Do the same for the pooled approach. Compare these two estimates with each other.

(d) Perform the significance test and find the 95% confidence interval using the pooled methods. How do these results compare with those you found in Exercise 7.65?

(e) Exercise 7.66 has data for the same workers but for respirable dust. Here the standard deviations differ more than those in Example 7.65 do. Answer parts (a) through (d) for these data. Write a summary of what you have found in this exercise.

**7.94** **Revisiting the effect of storage time on vitamin C.** The analysis of the loss of vitamin C when bread is stored in Exercise 7.76 (page 470) is a rather unusual case involving very small sample sizes. There are only two observations per condition (immediately after baking and three days later). When the samples are so small, we have very little information to make a judgment about whether the population standard deviations are equal. The potential gain from pooling is large when the sample sizes are very small. Assume that we will perform a two-sided test using the 5% significance level.

(a) Find the critical value for the unpooled $t$ test statistic that does not assume equal variances. Use the minimum of $n_1 - 1$ and $n_2 - 1$ for the degrees of freedom.

(b) Find the critical value for the pooled $t$ test statistic.

(c) How does comparing these critical values show an advantage of the pooled test?

# 7.3 Optional Topics in Comparing Distributions*

In this section we discuss three topics that are related to the material that we have already covered in this chapter. If we can do inference for means, it is natural to ask if we can do something similar for spread. The answer is yes, but there are many cautions. We also discuss robustness and show how to find the power for the two-sample $t$ test. If you plan to design studies, you should become familiar with this last topic.

## Inference for population spread

The two most basic descriptive features of a distribution are its center and spread. In a Normal population, these aspects are measured by the mean and the standard deviation. We have described procedures for inference about population means for Normal populations and found that these procedures are often useful for non-Normal populations as well. It is natural to turn next to inference about the standard deviations of Normal populations. Our recommendation here is short and clear: Don't do it without expert advice.

We will describe the $F$ test for comparing the spread of two Normal populations. *Unlike the t procedures for means, the F test and other procedures for standard deviations are extremely sensitive to non-Normal distributions.* This lack of robustness does not improve in large samples. It is difficult in practice to tell whether a significant $F$-value is evidence of unequal population spreads or simply evidence that the populations are not Normal. Consequently, we do not recommend use of inference about population standard deviations in basic statistical practice.[31]

*This section can be omitted without loss of continuity.

It was once common to test equality of standard deviations as a preliminary to performing the pooled two-sample $t$ test for equality of two population means. It is better practice to check the distributions graphically, with special attention to skewness and outliers, and to use the software-based two-sample $t$ that does not require equal standard deviations. In the words of one distinguished statistician, "To make a preliminary test on variances is rather like putting to sea in a rowing boat to find out whether conditions are sufficiently calm for an ocean liner to leave port!"[32]

## The *F* test for equality of spread

Because of the limited usefulness of procedures for inference about the standard deviations of Normal distributions, we will present only one such procedure. Suppose that we have independent SRSs from two Normal populations, a sample of size $n_1$ from $N(\mu_1, \sigma_1)$ and a sample of size $n_2$ from $N(\mu_2, \sigma_2)$. The population means and standard deviations are all unknown. The hypothesis of equal spread $H_0: \sigma_1 = \sigma_2$ is tested against $H_a: \sigma_1 \neq \sigma_2$ by a simple statistic, the ratio of the sample variances.

### THE *F* STATISTIC AND *F* DISTRIBUTIONS

When $s_1^2$ and $s_2^2$ are sample variances from independent SRSs of sizes $n_1$ and $n_2$ drawn from Normal populations, the $F$ statistic

$$F = \frac{s_1^2}{s_2^2}$$

has the $F$ distribution with $n_1 - 1$ and $n_2 - 1$ degrees of freedom when $H_0: \sigma_1 = \sigma_2$ is true.

*F* distributions

The ***F* distributions** are a family of distributions with two parameters: the degrees of freedom of the sample variances in the numerator and denominator of the $F$ statistic. The $F$ distributions are another of R. A. Fisher's contributions to statistics and are called $F$ in his honor. Fisher introduced $F$ statistics for comparing several means. We will meet these useful statistics in later chapters. The numerator degrees of freedom are always mentioned first. Interchanging the degrees of freedom changes the distribution, so the order is important. Our brief notation will be $F(j, k)$ for the $F$ distribution with $j$ degrees of freedom in the numerator and $k$ degrees of freedom in the denominator. The $F$ distributions are not symmetric but are right-skewed. The density curve in Figure 7.16 illustrates the shape. Because sample variances cannot be negative, the $F$ statistic takes only positive values and the $F$ distribution has no probability below 0. The peak of the $F$ density curve is near 1; values far from 1 in either direction provide evidence against the hypothesis of equal standard deviations.

Tables of $F$ critical values are awkward because a separate table is needed for every pair of degrees of freedom $j$ and $k$. Table E in the back of the book gives upper $p$ critical values of the $F$ distributions for $p = 0.10, 0.05, 0.025, 0.01$, and

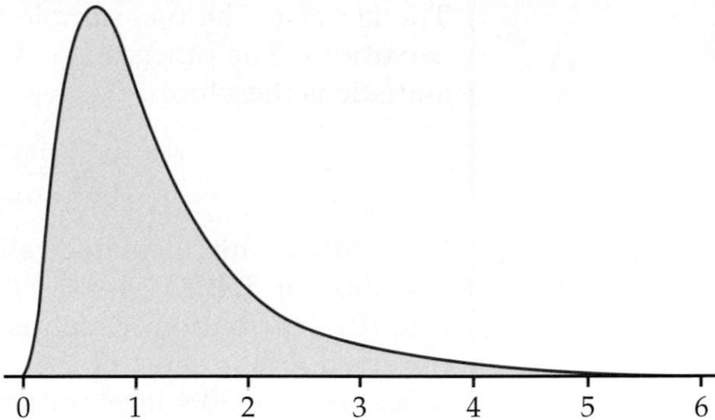


**FIGURE 7.16** The density curve for the $F(9, 10)$ distribution. The $F$ distributions are skewed to the right.

0.001. For example, these critical values for the $F(9, 10)$ distribution shown in Figure 7.16 are

| $p$ | 0.10 | 0.05 | 0.025 | 0.01 | 0.001 |
|---|---|---|---|---|---|
| $F^*$ | 2.35 | 3.02 | 3.78 | 4.94 | 8.96 |

The skewness of $F$ distributions causes additional complications. In the symmetric Normal and $t$ distributions, the point with probability 0.05 below it is just the negative of the point with probability 0.05 above it. This is not true for $F$ distributions. We therefore require either tables of both the upper and lower tails or means of eliminating the need for lower-tail critical values. Statistical software that eliminates the need for tables is plainly very convenient. If you do not use statistical software, arrange the $F$ test as follows:

1. Take the test statistic to be

$$F = \frac{\text{larger } s^2}{\text{smaller } s^2}$$

This amounts to naming the populations so that $s_1^2$ is the larger of the observed sample variances. The resulting $F$ is always 1 or greater.

2. Compare the value of $F$ with the critical values from Table E. Then *double* the probabilities obtained from the table to get the significance level for the two-sided $F$ test.

The idea is that we calculate the probability in the upper tail and double to obtain the probability of all ratios on either side of 1 that are at least as improbable as that observed. Remember that the order of the degrees of freedom is important in using Table E.

EXAMPLE

**7.22 Comparing calcium and placebo groups.** Example 7.19 (page 462) recounts a medical experiment comparing the effects of calcium and a placebo on the blood pressure of black men. The analysis (Example 7.20) employed the pooled two-sample $t$ procedures. Because these procedures require equal population standard deviations, it is tempting to first test

$$H_0: \sigma_1 = \sigma_2 \qquad H_a: \sigma_1 \neq \sigma_2$$

The larger of the two sample standard deviations is $s = 8.743$ from 10 observations. The other is $s = 5.901$ from 11 observations. The two-sided test statistic is therefore

$$F = \frac{\text{larger } s^2}{\text{smaller } s^2} = \frac{8.743^2}{5.901^2} = 2.20$$

We compare the calculated value $F = 2.20$ with critical points for the $F(9, 10)$ distribution. Table E shows that 2.20 is *less* than the 0.10 critical value of the $F(9, 10)$ distribution, which is $F^* = 2.35$. Doubling 0.10, we know that the observed $F$ falls short of the 0.20 significance level. The results are not significant at the 20% level (or any lower level). Statistical software shows that the exact upper-tail probability is 0.118, and hence $P = 0.236$. *If* the populations were Normal, the observed standard deviations would give little reason to suspect unequal population standard deviations. Because one of the populations shows some non-Normality, however, we cannot be fully confident of this conclusion.

## USE YOUR KNOWLEDGE

**7.95 The $F$ statistic.** The $F$ statistic $F = s_1^2/s_2^2$ is calculated from samples of size $n_1 = 16$ and $n_2 = 21$.

(a) What is the upper 5% critical value for this $F$?

(b) In a test of equality of standard deviations against the two-sided alternative, this statistic has the value $F = 2.45$. Is this value significant at the 10% level? Is it significant at the 5% level?

## Robustness of Normal inference procedures

We have claimed that

- The $t$ procedures for inference about means are quite robust against non-Normal population distributions. These procedures are particularly robust when the population distributions are symmetric and (for the two-sample case) when the two sample sizes are equal.
- The $F$ test and other procedures for inference about variances are so lacking in robustness as to be of little use in practice.

Simulations with a large variety of non-Normal distributions support these claims. One set of simulations was carried out with samples of size 25 and used significance tests with fixed level $\alpha = 0.05$. The three types of tests studied were the one-sample and pooled two-sample $t$ tests and the $F$ test for comparing two variances.

The robustness of the one-sample and two-sample $t$ procedures is remarkable. The true significance level remains between about 4% and 6% for a large range of populations. The $t$ test and the corresponding confidence intervals are among the most reliable tools that statisticians use. Remember, however, that outliers can greatly disturb the $t$ procedures. Also, two-sample procedures are less robust when the sample sizes are not similar.

*The lack of robustness of the tests for variances is equally remarkable.* The true significance levels depart rapidly from the target 5% as the population distribution departs from Normality. The two-sided $F$ test carried out with 5% critical values can have a true level of less than 1% or greater than 11% even in symmetric populations with no outliers. Results such as these are the basis for our recommendation that these procedures not be used.

## The power of the two-sample $t$ test

The two-sample $t$ test is one of the most used statistical procedures. Unfortunately, because of inadequate planning, users frequently fail to find evidence for the effects that they believe to be true. Power calculations should be part of the planning of any statistical study. Information from a pilot study or previous research is needed.

In Section 7.1 (optional material), we learned how to find an approximation for the power of the one-sample $t$ test. The basic concepts for the two-sample case are the same. Here, we give the exact method, which involves a new distribution, the **noncentral $t$ distribution.** To perform the calculations, we simply need software to calculate probabilities for this distribution.

noncentral $t$ distribution

We first present the method for the pooled two-sample $t$ test, where the parameters are $\mu_1$, $\mu_2$, and the common standard deviation $\sigma$. Modifications to get approximate results when we do not pool are then described.

To find the power for the pooled two-sample $t$ test, use the following steps. We consider only the case where the null hypothesis is $\mu_1 - \mu_2 = 0$.

1. Specify
   (a) an alternative value for $\mu_1 - \mu_2$ that you consider important to detect;
   (b) the sample sizes, $n_1$ and $n_2$;
   (c) the Type I error for a fixed significance level, $\alpha$;
   (d) a guess at the standard deviation, $\sigma$.
2. Find the degrees of freedom df $= n_1 + n_2 - 2$ and the value of $t^*$ that will lead to rejection of $H_0$.
3. Calculate the **noncentrality parameter**

noncentrality parameter

$$\delta = \frac{|\mu_1 - \mu_2|}{\sigma\sqrt{\dfrac{1}{n_1} + \dfrac{1}{n_2}}}$$

4. Find the power as the probability that a noncentral $t$ random variable with degrees of freedom df and noncentrality parameter $\delta$ will be greater than $t^*$. In SAS the command is 1-PROBT(tstar, df, delta). If you do not have software that can perform this calculation, you can approximate the power as the probability that a standard Normal random variable is greater than $t^* - \delta$, that is, $P(z > t^* - \delta)$, and use Table A.

Note that the denominator in the noncentrality parameter,

$$\sigma\sqrt{\frac{1}{n_1} + \frac{1}{n_2}}$$

is our guess at the standard error for the difference in the sample means. Therefore, if we wanted to assess a possible study in terms of the margin of error for the estimated difference, we would examine $t^*$ times this quantity.

If we do not assume that the standard deviations are equal, we need to guess both standard deviations and then combine these for our guess at the standard error:

$$\sqrt{\frac{\sigma_1^2}{n_1} + \frac{\sigma_2^2}{n_2}}$$

This guess is then used in the denominator of the noncentrality parameter. For the degrees of freedom, the conservative approximation is appropriate.

EXAMPLE

**7.23 Planning a new study of calcium versus placebo groups.** In Example 7.20 we examined the effect of calcium on blood pressure by comparing the means of a treatment group and a placebo group using a pooled two-sample $t$ test. The $P$-value was 0.059, failing to achieve the usual standard of 0.05 for statistical significance. Suppose that we wanted to plan a new study that would provide convincing evidence, say at the 0.01 level, with high probability. Let's examine a study design with 45 subjects in each group ($n_1 = n_2 = 45$). Based on our previous results we choose $\mu_1 - \mu_2 = 5$ as an alternative that we would like to be able to detect with $\alpha = 0.01$. For $\sigma$ we use 7.4, our pooled estimate from Example 7.20. The degrees of freedom are $n_1 + n_2 - 2 = 88$ and $t^* = 2.37$ for the significance test. The noncentrality parameter is

$$\delta = \frac{5}{7.4\sqrt{\frac{1}{45} + \frac{1}{45}}} = \frac{5}{1.56} = 3.21$$

Software gives the power as 0.7965, or 80%. The Normal approximation gives 0.7983, a very accurate result. With this choice of sample sizes we would expect the margin of error for a 95% confidence interval ($t^* = 1.99$) for the difference in means to be

$$t^* \times 7.4\sqrt{\frac{1}{45} + \frac{1}{45}} = 1.99 \times 1.56 = 3.1$$

With software it is very easy to examine the effects of variations on a study design. In the above example, we might want to examine the power for $\alpha = 0.05$ and the effects of reducing the sample sizes.

## USE YOUR KNOWLEDGE

**7.96 Power and $\mu_1 - \mu_2$.** If you repeat the calculation in Example 7.23 for other values of $\mu_1 - \mu_2$ that are larger than 5, would you expect the power to be higher or lower than 0.7965? Why?

**7.97 Power and the standard deviation.** If the true population standard deviation were 7.0 instead of the 7.4 hypothesized in Example 7.23, would the power for this new experiment be greater or smaller than 0.7965? Explain.

## SECTION 7.3 Summary

Inference procedures for comparing the standard deviations of two Normal populations are based on the **$F$ statistic,** which is the ratio of sample variances:

$$F = \frac{s_1^2}{s_2^2}$$

If an SRS of size $n_1$ is drawn from the $x_1$ population and an independent SRS of size $n_2$ is drawn from the $x_2$ population, the $F$ statistic has the **$F$ distribution** $F(n_1 - 1, n_2 - 1)$ if the two population standard deviations $\sigma_1$ and $\sigma_2$ are in fact equal.

The **$F$ test for equality of standard deviations** tests $H_0: \sigma_1 = \sigma_2$ versus $H_a: \sigma_1 \neq \sigma_2$ using the statistic

$$F = \frac{\text{larger } s^2}{\text{smaller } s^2}$$

and doubles the upper-tail probability to obtain the $P$-value.

The $t$ procedures are quite **robust** when the distributions are not Normal. The $F$ tests and other procedures for inference about the spread of one or more Normal distributions are so strongly affected by non-Normality that we do not recommend them for regular use.

The **power** of the pooled two-sample $t$ test is found by first computing the critical value for the significance test, the degrees of freedom, and the **noncentrality parameter** for the alternative of interest. These are used to find the power from the **$t$ distribution.** A Normal approximation works quite well. Calculating margins of error for various study designs and assumptions is an alternative procedure for evaluating designs.

## SECTION 7.3 Exercises

*For Exercise 7.95, see page 476; and for Exercises 7.96 and 7.97, see page 478.*

*In all exercises calling for use of the F test, assume that both population distributions are very close to Normal. The actual data are not always sufficiently Normal to justify use of the F test.*

**7.98 Comparison of standard deviations.** Here are some summary statistics from two independent samples from Normal distributions:

| Sample | $n$ | $s^2$ |
|---|---|---|
| 1 | 10 | 3.1 |
| 2 | 16 | 9.3 |

You want to test the null hypothesis that the two population standard deviations are equal versus the two-sided alternative at the 5% significance level.

(a) Calculate the test statistic.

(b) Find the appropriate value from Table E that you need to perform the significance test.

(c) What do you conclude?

**7.99 Revisiting the cholesterol comparison.** Compare the standard deviations of total cholesterol in Exercise 7.61 (page 467). Give the test statistic, the degrees of freedom, and the $P$-value. Write a short summary of your analysis, including comments on the assumptions for the test.

**7.100 An HDL comparison.** HDL is also known as "good" cholesterol. Compare the standard deviations of HDL in Exercise 7.61 (page 467). Give the test statistic, the degrees of freedom, and the *P*-value. Write a short summary of your analysis, including comments on the assumptions for the test.

**7.101** CHALLENGE **Revisiting the multimedia evaluation study.** Mean scores on a knowledge test are compared for two groups of women in Exercise 7.63 (page 468). Compare the standard deviations using an *F* test. What do you conclude? Comment on the Normal assumption for these data. These standard deviations are so close that we are not particularly surprised at the result of the significance test. Assume that the sample standard deviation in the intervention is the value given in Exercise 7.63 (1.15). How large would the standard deviation in the control group need to be to reject the null hypothesis of equal standard deviations at the 5% level?

**7.102 Revisiting the self-control and food study.** Compare the standard deviations of the self-efficacy scores in Exercise 7.64 (page 468). Give the test statistic, the degrees of freedom, and the *P*-value. Write a short summary of your analysis, including comments on the assumptions for the test.

**7.103 Revisiting the dust exposure study.** The two-sample problem in Exercise 7.65 (page 468) compares drill and blast workers with outdoor concrete workers with respect to the total dust that they are exposed to in the workplace. Here it may be useful to know whether or not the standard deviations differ in the two groups. Perform the *F* test and summarize the results. Are you concerned about the assumptions here? Explain why or why not.

**7.104 More on the dust exposure study.** Exercise 7.66 (page 468) is similar to Exercise 7.65, but the response variable here is exposure to dust particles that can enter and stay in the lungs. Compare the standard deviations with a significance test and summarize the results. Be sure to comment on the assumptions.

**7.105 Revisiting the size of trees in the north and south.** The diameters of trees in the Wade Tract for random samples selected from the north and south portions of the tract are compared in Exercise 7.81 (page 471). Are there statistically significant differences in the standard deviations for these two parts of the tract? Perform the significance test and summarize the results. Does the Normal assumption appear reasonable for these data?

**7.106 Revisiting the size of trees in the east and west.** Tree diameters for the east and west halves of the Wade Tract are compared in Exercise 7.82 (page 471). Using the questions in the previous exercise as a guide, analyze these data.

**7.107 Revisiting the storage time study.** We studied the loss of vitamin C when bread is stored in Exercise 7.76 (page 470). Recall that two loaves were measured immediately after baking and another two loaves were measured after three days of storage. These are very small sample sizes.

(a) Use Table E to find the value that the ratio of variances would have to exceed for us to reject the null hypothesis (at the 5% level) that the standard deviations are equal. What does this suggest about the power of the test?

(b) Perform the test and state your conclusion.

**7.108 Planning a study to compare tree size.** In Exercise 7.81 (page 471) DBH data for longleaf pine trees in two parts of the Wade Tract are compared. Suppose that you are planning a similar study where you will measure the diameters of longleaf pine trees. Based on Exercise 7.81, you are willing to assume that the standard deviation is 20 cm. Suppose that a difference in mean DBH of 10 cm or more would be important to detect. You will use a *t* statistic and a two-sided alternative for the comparison.

(a) Find the power if you randomly sample 20 trees from each area to be compared.

(b) Repeat the calculations for 60 trees in each sample.

(c) If you had to choose between the 20 and 60 trees per sample, which would be acceptable? Give reasons for your answer.

**7.109** CHALLENGE **More on planning a study to compare tree size.** Refer to the previous exercise. Find the two standard deviations from Exercise 7.81. Do the same for the data in Exercise 7.82, which is a similar setting. These are somewhat smaller than the assumed value that you used in the previous exercise. Explain why it is generally a better idea to assume a standard deviation that is larger than you expect than one that is smaller. Repeat the power calculations for some other reasonable values of $\sigma$ and comment on the impact of the size of $\sigma$ for planning the new study.

**7.110 Planning a study to compare ad placement.** Refer to Exercise 7.80 (page 470), where we compared trustworthy ratings for ads from two different publications. Suppose that you are planning a similar study using two different publications that are not expected to show the differences seen when comparing the *Wall Street Journal* with the *National Enquirer*. You would like to detect a difference of 1.5 points using a two-sided significance test with a 5% level of significance. Based on Exercise 7.80, it is reasonable to use 1.6 as the value of the standard deviation for planning purposes.

(a) What is the power if you use sample sizes similar to those used in the previous study, for example, 65 for each publication?

(b) Repeat the calculations for 100 in each group.

(c) What sample size would you recommend for the new study?

## CHAPTER 7 Exercises

**7.111 LSAT scores.** The scores of four senior roommates on the Law School Admission Test (LSAT) are

158, 168, 143, 155

Find the mean, the standard deviation, and the standard error of the mean. Is it appropriate to calculate a confidence interval based on these data? Explain why or why not.

**7.112 Converting a two-sided *P*-value.** You use statistical software to perform a significance test of the null hypothesis that two means are equal. The software reports *P*-values for the two-sided alternative. Your alternative is that the first mean is greater than the second mean.

(a) The software reports $t = 1.81$ with a *P*-value of 0.07. Would you reject $H_0$ with $\alpha = 0.05$? Explain your answer.

(b) The software reports $t = -1.81$ with a *P*-value of 0.07. Would you reject $H_0$ with $\alpha = 0.05$? Explain your answer.

**7.113 Degrees of freedom and confidence interval width.** As the degrees of freedom increase, the $t$ distributions get closer and closer to the $z$ ($N(0, 1)$) distribution. One way to see this is to look at how the value of $t^*$ for a 95% confidence interval changes with the degrees of freedom. Make a plot with degrees of freedom from 2 to 100 on the $x$ axis and $t^*$ on the $y$ axis. Draw a horizontal line on the plot corresponding to the value of $z^* = 1.96$. Summarize the main features of the plot.

**7.114 Degrees of freedom and $t^*$.** Refer to the previous exercise. Make a similar plot and summarize its features for the value of $t^*$ for a 90% confidence interval.

**7.115 Sample size and margin of error.** The margin of error for a confidence interval depends on the confidence level, the standard deviation, and the sample size. Fix the confidence level at 95% and the standard deviation at 1 to examine the effect of the sample size. Find the margin of error for sample sizes of 5 to 100 by 5s—that is, let $n = 5, 10, 15, \ldots, 100$. Plot the margins of error versus the sample size and summarize the relationship.

**7.116 More on sample size and margin of error.** Refer to the previous exercise. Make a similar plot and summarize its features for a 99% confidence interval.

**7.117** CHALLENGE **Alcohol consumption and body composition.** Individuals who consume large amounts of alcohol do not use the calories from this source as efficiently as calories from other sources. One study examined the effects of moderate alcohol consumption on body composition and the intake of other foods. Fourteen subjects participated in a crossover design where they either drank wine for the first 6 weeks and then abstained for the next 6 weeks or vice versa.[33] During the period when they drank wine, the subjects, on average, lost 0.4 kilograms (kg) of body weight; when they did not drink wine, they lost an average of 1.1 kg. The standard deviation of the difference between the weight lost under these two conditions is 8.6 kg. During the wine period, they consumed an average of 2589 calories; with no wine, the mean consumption was 2575. The standard deviation of the difference was 210.

(a) Compute the differences in means and the standard errors for comparing body weight and caloric intake under the two experimental conditions.

(b) A report of the study indicated that there were no significant differences in these two outcome measures. Verify this result for each measure, giving the test statistic, degrees of freedom, and the *P*-value.

(c) One concern with studies such as this, with a small number of subjects, is that there may not be sufficient power to detect differences that are potentially important. Address this question by computing 95% confidence intervals for the two measures and discuss the information provided by the intervals.

(d) Here are some other characteristics of the study. The study periods lasted for 6 weeks. All subjects were males between the ages of 21 and 50 years who weighed between 68 and 91 kilograms (kg). They were all from the same city. During the wine period, subjects were told to consume two 135 milliliter (ml) servings of red wine and no other alcohol. The entire 6-week supply was given to each subject at the beginning of the period. During the other period, subjects were instructed to refrain from any use of alcohol. All subjects reported that they complied with these instructions except for three subjects, who said that they drank no more than 3 to 4 12-ounce bottles of beer during the no-alcohol period. Discuss how these factors could influence the interpretation of the results.

**7.118 Healthy bones study.** Healthy bones are continually being renewed by two processes. Through bone formation, new bone is built; through bone resorption, old bone is removed. If one or both of these processes are disturbed, by disease, aging, or space travel, for example, bone loss can be the result. Osteocalcin (OC) is a biochemical marker for bone formation: higher levels of bone formation are associated with higher levels of OC. A blood sample is used to measure OC, and it is much less expensive to obtain than direct measures of bone formation. The units are milligrams of OC per milliliter of blood (mg/ml). One study examined various biomarkers of bone turnover.[34] Here are the OC measurements on 31 healthy females aged 11 to 32 years who participated in this study:

| | | | | | | | |
|---|---|---|---|---|---|---|---|
| 68.9 | 56.3 | 54.6 | 31.2 | 36.4 | 31.4 | 52.8 | 38.4 |
| 35.7 | 76.5 | 44.4 | 40.2 | 77.9 | 54.6 | 9.9 | 20.6 |
| 20.0 | 17.2 | 24.2 | 20.9 | 17.9 | 19.7 | 15.9 | 20.8 |
| 8.1 | 19.3 | 16.9 | 10.1 | 47.7 | 30.2 | 17.2 | |

(a) Display the data with a stemplot or histogram and a boxplot. Describe the distribution.

(b) Find a 95% confidence interval for the mean OC. Comment on the suitability of using this procedure for these data.

**7.119 More on the healthy bones study.** Refer to the previous exercise. Tartrate resistant acid phosphatase (TRAP) is a biochemical marker for bone resorption that is also measured in blood. Here are the TRAP measurements, in units per liter (U/l), for the same 31 females:

| | | | | | | | |
|---|---|---|---|---|---|---|---|
| 19.4 | 25.5 | 19.0 | 9.0 | 19.1 | 14.6 | 25.2 | 14.6 |
| 28.8 | 14.9 | 10.7 | 5.9 | 23.7 | 19.0 | 6.9 | 8.1 |
| 9.5 | 6.3 | 10.1 | 10.5 | 9.0 | 8.8 | 8.2 | 10.3 |
| 3.3 | 10.1 | 9.5 | 8.1 | 18.6 | 14.4 | 9.6 | |

(a) Display the data with a stemplot or histogram and a boxplot. Describe the distribution.

(b) Find a 95% confidence interval for the mean TRAP. Comment on the suitability of using this procedure for these data.

**7.120 Transforming the data.** Refer to Exercise 7.118 and the OC data for 31 females. Variables that measure concentrations such as this often have distributions that are skewed to the right. For this reason it is common to work with the logarithms of the measured values. Here are the OC values transformed with the (natural) log:

| | | | | | | | |
|---|---|---|---|---|---|---|---|
| 4.23 | 4.03 | 4.00 | 3.44 | 3.59 | 3.45 | 3.97 | 3.65 |
| 3.58 | 4.34 | 3.79 | 3.69 | 4.36 | 4.00 | 2.29 | 3.03 |
| 3.00 | 2.84 | 3.19 | 3.04 | 2.88 | 2.98 | 2.77 | 3.03 |
| 2.09 | 2.96 | 2.83 | 2.31 | 3.86 | 3.41 | 2.84 | |

(a) Display the data with a stemplot and a boxplot. Describe the distribution.

(b) Find a 95% confidence interval for the mean OC. Comment on the suitability of using this procedure for these data.

(c) Transform the mean and the endpoints of the confidence interval back to the original scale, mg/ml. Compare this interval with the one you computed in Exercise 7.118.

**7.121 More on transforming the data.** Refer to Exercise 7.119 and the TRAP data for 31 females. Variables that measure concentrations such as this often have distributions that are skewed to the right. For this reason it is common to work with the logarithms of the measured values. Here are the TRAP values transformed with the (natural) log:

| | | | | | | | |
|---|---|---|---|---|---|---|---|
| 2.97 | 3.24 | 2.94 | 2.20 | 2.95 | 2.68 | 3.23 | 2.68 |
| 3.36 | 2.70 | 2.37 | 1.77 | 3.17 | 2.94 | 1.93 | 2.09 |
| 2.25 | 1.84 | 2.31 | 2.35 | 2.20 | 2.17 | 2.10 | 2.33 |
| 1.19 | 2.31 | 2.25 | 2.09 | 2.92 | 2.67 | 2.26 | |

(a) Display the data with a stemplot and a boxplot. Describe the distribution.

(b) Find a 95% confidence interval for the mean TRAP. Comment on the suitability of using this procedure for these data.

(c) Transform the mean and the endpoints of the confidence interval back to the original scale, U/l. Compare this interval with the one you computed in Exercise 7.119.

**7.122** CHALLENGE **Analysis of tree size using the complete data set.** The data used in Exercises 7.25 (page 442), 7.81, and 7.82 (page 471) were obtained by taking simple random samples from the 584 longleaf pine trees that were measured in the Wade Tract. The entire data set is given in the LONGLEAF data set. More details about this data set can be found in the Data Appendix at the back of the book. Find the 95% confidence interval for the mean DBH using the entire data set, and compare this interval with the one that you calculated in Exercise 7.25. Write a report about these data. Include comments on the effect of the sample size on the margin of error, the distribution of the data, the appropriateness of the Normality-based methods for this problem, and the generalizability of the results to other similar stands of longleaf pine or other kinds of trees in this area of the United States and other areas.

**7.123** CHALLENGE **More on the complete tree size data set.** Use the LONGLEAF data set to repeat the calculations that you performed in Exercises 7.81 and 7.82. Discuss the effect of the sample size on the results.

**7.124** CHALLENGE **Even more on the complete tree size data set.** The DBH measures in the LONGLEAF data set do not appear to be Normally distributed. Make a histogram of the data and a Normal quantile plot if you have the software available. Mark the mean and the median on the histogram. Now, transform the data using a logarithm. Does this make the distribution appear to be Normal? Use the same graphical summaries with the mean and the median marked on the histogram. Write a summary of your conclusions, paying particular attention to the use of data such as these for inference using the methods based on Normal distributions.

**7.125** **Competitive prices?** A retailer entered into an exclusive agreement with a supplier who guaranteed to provide all products at competitive prices. The retailer eventually began to purchase supplies from other vendors who offered better prices. The original supplier filed a legal action claiming violation of the agreement. In defense, the retailer had an audit performed on a random sample of invoices. For each audited invoice, all purchases made from other suppliers were examined and the prices were compared with those offered by the original supplier. For each invoice, the percent of purchases for which the alternate supplier offered a lower price than the original supplier was recorded.[35] Here are the data:

| | | | | | | | | | | | |
|---|---|---|---|---|---|---|---|---|---|---|---|
| 0 | 100 | 0 | 100 | 33 | 34 | 100 | 48 | 78 | 100 | 77 | 100 | 38 |
| 68 | 100 | 79 | 100 | 100 | 100 | 100 | 100 | 100 | 89 | 100 | 100 | |

Report the average of the percents with a 95% margin of error. Do the sample invoices suggest that the original supplier's prices are not competitive on the average?

**7.126** **Weight-loss programs.** In a study of the effectiveness of weight-loss programs, 47 subjects who were at least 20% overweight took part in a group support program for 10 weeks. Private weighings determined each subject's weight at the beginning of the program and 6 months after the program's end. The matched pairs $t$ test was used to assess the significance of the average weight loss. The paper reporting the study said, "The subjects lost a significant amount of weight over time, $t(46) = 4.68$, $p < 0.01$." It is common to report the results of statistical tests in this abbreviated style.[36]

(a) Why was the matched pairs statistic appropriate?

(b) Explain to someone who knows no statistics but is interested in weight-loss programs what the practical conclusion is.

(c) The paper follows the tradition of reporting significance only at fixed levels such as $\alpha = 0.01$. In fact, the results are more significant than

"$p < 0.01$" suggests. What can you say about the *P*-value of the *t* test?

**7.127** CHALLENGE **Do women perform better in school?** Some research suggests that women perform better than men in school but men score higher on standardized tests. Table 1.9 (page 29) presents data on a measure of school performance, grade point average (GPA), and a standardized test, IQ, for 78 seventh-grade students. Do these data lend further support to the previously found gender differences? Give graphical displays of the data and describe the distributions. Use significance tests and confidence intervals to examine this question, and prepare a short report summarizing your findings.

**7.128** CHALLENGE **Self-concept and school performance.** Refer to the previous exercise. Although self-concept in this study was measured on a scale with values in the data set ranging from 20 to 80, many prefer to think of this kind of variable as having only two possible values: low self-concept or high self-concept. Find the median of the self-concept scores in Table 1.9 and define those students with scores at or below the median to be low-self-concept students and those with scores above the median to be high-self-concept students. Do high-self-concept students have grade point averages that are different from low-self-concept students? What about IQ? Prepare a report addressing these questions. Be sure to include graphical and numerical summaries and confidence intervals, and state clearly the details of significance tests.

**7.129 Behavior of pet owners.** On the morning of March 5, 1996, a train with 14 tankers of propane derailed near the center of the small Wisconsin town of Weyauwega. Six of the tankers were ruptured and burning when the 1700 residents were ordered to evacuate the town. Researchers study disasters like this so that effective relief efforts can be designed for future disasters. About half of the households with pets did not evacuate all of their pets. A study conducted after the derailment focused on problems associated with retrieval of the pets after the evacuation and characteristics of the pet owners. One of the scales measured "commitment to adult animals," and the people who evacuated all or some of their pets were compared with those who did not evacuate any of their pets. Higher scores indicate that the pet owner is more likely to take actions that benefit the pet.[37] Here are the data summaries:

| Group | $n$ | $\bar{x}$ | $s$ |
|---|---|---|---|
| Evacuated all or some pets | 116 | 7.95 | 3.62 |
| Did not evacuate any pets | 125 | 6.26 | 3.56 |

Analyze the data and prepare a short report describing the results.

**7.130 Occupation and diet.** Do various occupational groups differ in their diets? A British study of this question compared 98 drivers and 83 conductors of London double-decker buses.[38] The conductors' jobs require more physical activity. The article reporting the study gives the data as "Mean daily consumption (± se)." Some of the study results appear below:

| | Drivers | Conductors |
|---|---|---|
| Total calories | $2821 \pm 44$ | $2844 \pm 48$ |
| Alcohol (grams) | $0.24 \pm 0.06$ | $0.39 \pm 0.11$ |

(a) What does "se" stand for? Give $\bar{x}$ and $s$ for each of the four sets of measurements.

(b) Is there significant evidence at the 5% level that conductors consume more calories per day than do drivers? Use the two-sample *t* method to give a *P*-value, and then assess significance.

(c) How significant is the observed difference in mean alcohol consumption? Use two-sample *t* methods to obtain the *P*-value.

(d) Give a 95% confidence interval for the mean daily alcohol consumption of London double-decker bus conductors.

(e) Give a 99% confidence interval for the difference in mean daily alcohol consumption between drivers and conductors.

**7.131 Occupation and diet, continued (optional).** Use of the pooled two-sample *t* test is justified in part (b) of the previous exercise. Explain why. Find the *P*-value for the pooled *t* statistic, and compare with your result in the previous exercise.

**7.132 Conditions for inference.** The report cited in Exercise 7.130 says that the distribution of alcohol consumption among the individuals studied is "grossly skew."

(a) Do you think that this skewness prevents the use of the two-sample *t* test for equality of means? Explain your answer.

(b) (Optional) Do you think that the skewness of the distributions prevents the use of the $F$ test for equality of standard deviations? Explain your answer.

**7.133 More on conditions for inference.** Table 1.2 (page 10) gives literacy rates for men and women in 17 Islamic nations. Is it proper to apply the one-sample $t$ method to these data to give a 95% confidence interval for the mean literacy rate of Islamic men? Explain your answer.

**7.134** CHALLENGE **PCBs in fish.** Polychlorinated biphenyls (PCBs) are a collection of compounds that are no longer produced in the United States but are still found in the environment. Evidence suggests that they can cause harmful health effects when consumed. Because PCBs can accumulate in fish, efforts have been made to identify areas where fish contain excessive amounts so that recommendations concerning consumption limits can be made. There are over 200 types of PCBs. Data from the Environmental Protection Agency National Study of Residues in Lake Fish are given in the data set PCB. More details about this data set can be found in the Data Appendix. Various lakes in the United States were sampled and the amounts of PCBs in fish were measured. The variable PCB is the sum of the amounts of all PCBs found in the fish. The units are parts per billion (ppb).

(a) Use graphical and numerical summaries to describe the distribution of this variable. Include a histogram with the location of the mean and the median clearly marked.

(b) Do you think it is appropriate to use methods based on Normal distributions for these data? Explain why or why not.

(c) Find a 95% confidence interval for the mean. Will this interval contain approximately 95% of the observations in the data set? Explain your answer.

(d) Transform the PCB variable with a logarithm. Analyze the transformed data and summarize your results. Do you prefer to work with the raw data or with logs for this variable? Give reasons for your answer.

(e) Visit the Web site `http://epa.gov/waterscience/fishstudy/` to find details about how the data were collected. Write a summary describing these details and discuss how the results from this study can be generalized to other settings.

**7.135** CHALLENGE **PCBs in fish, continued.** Refer to the previous exercise. Not all types of PCBs are equally harmful. A scale has been developed to convert the raw amount of each type of PCB to a toxic equivalent (TEQ). The PCB data set contains a variable TEQPCB that is the total TEQ from all PCBs found in each sample. Using the questions in the previous exercise, analyze these data and summarize the results.

**7.136** CHALLENGE **Inference using the complete CRP data set.** In Exercise 7.26 (page 442) you analyzed the C-reactive protein (CRP) scores for a random sample of 40 children who participated in a study in Papua New Guinea. Serum retinol for the same children was analyzed in Exercise 7.28. Data for all 90 children who participated in the study are given in the data set PNG, described in the Data Appendix. Researchers who analyzed these data along with data from several other countries were interested in whether or not infections (as indicated by high CRP values) were associated with lower levels of serum retinol. A child with a value of CRP greater than 5.0 mg/l is classified as recently infected. Those whose CRP is less than or equal to 5.0 mg/l are not. Compare the serum retinol levels of the infected and noninfected children. Include graphical and numerical summaries, comments on all assumptions, and details of your analyses. Write a short report summarizing your results.

**7.137** CHALLENGE **More on using the complete CRP data set.** Refer to the previous exercise. The researchers in this study also measured $\alpha$1-acid glycoprotein (AGP). This protein is similar to CRP in that it is an indicator of infection. However, it rises more slowly than CRP and reaches a maximum 2 to 3 days after an infection. The units for AGP are grams per liter (g/l), and any value greater than 1.0 g/l is an indication of infection. Analyze the data on AGP in the data set PNG and write a report summarizing your results.

**7.138** CHALLENGE **Male and female CS students (optional).** Is there a difference between the average SAT scores of males and females? The CSDATA data set, described in the Data Appendix, gives the Math (SATM) and Verbal (SATV) scores for a group of 224 computer science majors. The variable SEX indicates whether each individual is male or female.

(a) Compare the two distributions graphically, and then use the two-sample $t$ test to compare the average SATM scores of males and females.

Is it appropriate to use the pooled *t* test for this comparison? Write a brief summary of your results and conclusions that refers to both versions of the *t* test and to the *F* test for equality of standard deviations. Also give a 95% confidence interval for the difference in the means.

(b) Answer part (a) for the SATV scores.

(c) The students in the CSDATA data set were all computer science majors who began college during a particular year. To what extent do you think that your results would generalize to (*i*) computer science students entering in different years, (*ii*) computer science majors at other colleges and universities, and (*iii*) college students in general?

**7.139 Different methods of teaching reading.** In the READING data set, described in the Data Appendix, the response variable Post3 is to be compared for three methods of teaching reading. The Basal method is the standard, or control, method, and the two new methods are DRTA and Strat. We can use the methods of this chapter to compare Basal with DRTA and Basal with Strat. Note that to make comparisons among three treatments it is more appropriate to use the procedures that we will learn in Chapter 12.

(a) Is the mean reading score with the DRTA method higher than that for the Basal method? Perform an analysis to answer this question, and summarize your results.

(b) Answer part (a) for the Strat method in place of DRTA.

**7.140 Sample size calculation (optional).** Example 7.13 (page 449) tells us that the mean height of 10-year-old girls is $N(56.4, 2.7)$ and for boys it is $N(55.7, 3.8)$. The null hypothesis that the mean heights of 10-year-old boys and girls are equal is clearly false. The difference in mean heights is $56.4 - 55.7 = 0.7$ inch. Small differences such as this can require large sample sizes to detect. To simplify our calculations, let's assume that the standard deviations are the same, say $\sigma = 3.2$, and that we will measure the heights of an equal number of girls and boys. How many would we need to measure to have a 90% chance of detecting the (true) alternative hypothesis?

**7.141** CHALLENGE **House prices.** How much more would you expect to pay for a home that has four bedrooms than for a home that has three? Here are some data for West Lafayette, Indiana.[39] These are the asking prices (in dollars) that the owners have set for their homes.

Four-bedroom homes:

| | | | | | |
|---|---|---|---|---|---|
| 121,900 | 139,900 | 157,000 | 159,900 | 176,900 | 224,900 |
| 235,000 | 245,000 | 294,000 | | | |

Three-bedroom homes:

| | | | | | |
|---|---|---|---|---|---|
| 65,500 | 79,900 | 79,900 | 79,900 | 82,900 | 87,900 |
| 94,000 | 97,500 | 105,000 | 111,900 | 116,900 | 117,900 |
| 119,900 | 122,900 | 124,000 | 125,000 | 126,900 | 127,900 |
| 127,900 | 127,900 | 132,900 | 145,000 | 145,500 | 157,500 |
| 194,000 | 205,900 | 259,900 | 265,000 | | |

(a) Plot the asking prices for the two sets of homes and describe the two distributions.

(b) Test the null hypothesis that the mean asking prices for the two sets of homes are equal versus the two-sided alternative. Give the test statistic with degrees of freedom, the *P*-value, and your conclusion.

(c) Would you consider using a one-sided alternative for this analysis? Explain why or why not.

(d) Give a 95% confidence interval for the difference in mean asking prices.

(e) These data are not SRSs from a population. Give a justification for use of the two-sample *t* procedures in this case.

**7.142** CHALLENGE **More on house prices.** Go to the Web site `www.realtor.com` and select two geographical areas of interest to you. You will compare the prices of similar types of homes in these two areas. State clearly how you define the areas and the type of homes. For example, you can use city names or zip codes to define the area and you can select single-family homes or condominiums. We view these homes as representative of the asking prices of homes for these areas at the time of your search. If the search gives a large number of homes, select a random sample. Be sure to explain exactly how you do this. Use the methods you have learned in this chapter to compare the asking prices. Be sure to include a graphical summary.

CHAPTER

8

# Inference for Proportions

Does a new medicine reduce the chance of getting a cold? A randomized comparative experiment is often used to answer this question. This chapter describes procedures for statistical inference when the response variable is Yes/No.



## Introduction

Many statistical studies produce counts rather than measurements. For example, the data from an opinion poll that asks a sample of adults whether they approve of the conduct of the president in office are the counts of "Yes," "No," and "Don't know." In an experiment that compares the effectiveness of four cold prevention treatments, the data are the number of subjects given each treatment and the number of subjects in each treatment group who catch a cold during the next month. Similarly, in a survey on driving behavior, the proportions of men and women who admit to shouting, cursing, or making gestures to other drivers in the last year are compared using count data. This chapter, and the next, present procedures for statistical inference in these settings.

The parameters we want to do inference about are population proportions. Just as in the case of inference about population means, we may be concerned with a single population or with comparing two populations. Inference about proportions in these one-sample and two-sample settings is very similar to inference about means, which we discussed in Chapter 7.

We begin in Section 8.1 with inference about a single population proportion. The statistical model for a count is then the binomial distribution, which we studied in Section 5.1. Section 8.2 concerns methods for comparing two proportions. Binomial distributions again play an important role.

## 8.1 Inference for a Single Proportion

We want to estimate the proportion $p$ of some characteristic, such as approval of the president's conduct in office, among the members of a large population. We select a simple random sample (SRS) of size $n$ from the population and record the count $X$ of "successes" (such as "Yes" answers to a question about the president). We will use "success" to represent the characteristic of interest. The sample proportion of successes $\hat{p} = X/n$ estimates the unknown population proportion $p$. If the population is much larger than the sample (say, at least 20 times as large), the count $X$ has approximately the binomial distribution $B(n, p)$.[1] In statistical terms, we are concerned with inference about the probability $p$ of a success in the binomial setting.

If the sample size $n$ is very small, we must base tests and confidence intervals for $p$ on the binomial distributions. These are awkward to work with because of the discreteness of the binomial distributions.[2] But we know that when the sample is large, both the count $X$ and the sample proportion $\hat{p}$ are approximately Normal. We will consider only inference procedures based on the Normal approximation. These procedures are similar to those for inference about the mean of a Normal distribution.

LOOK BACK
**Normal approximation for counts, page 323**

### Large-sample confidence interval for a single proportion

The unknown population proportion $p$ is estimated by the sample proportion $\hat{p} = X/n$. If the sample size $n$ is sufficiently large, $\hat{p}$ has approximately the Normal distribution, with mean $\mu_{\hat{p}} = p$ and standard deviation $\sigma_{\hat{p}} = \sqrt{p(1-p)/n}$. This means that approximately 95% of the time $\hat{p}$ will be within $2\sqrt{p(1-p)/n}$ of the unknown population proportion $p$.

LOOK BACK
**Normal approximation for proportions, page 323**
**standard error, page 418**

Note that the standard deviation $\sigma_{\hat{p}}$ depends upon the unknown parameter $p$. To estimate this standard deviation using the data, we replace $p$ in the formula by the sample proportion $\hat{p}$. As we did in Chapter 7, we use the term *standard error* for the standard deviation of a statistic that is estimated from data. Here is a summary of the procedure.

#### LARGE-SAMPLE CONFIDENCE INTERVAL FOR A POPULATION PROPORTION

Choose an SRS of size $n$ from a large population with unknown proportion $p$ of successes. The **sample proportion** is

$$\hat{p} = \frac{X}{n}$$

where $X$ is the number of successes. The **standard error of $\hat{p}$** is

$$SE_{\hat{p}} = \sqrt{\frac{\hat{p}(1-\hat{p})}{n}}$$

and the **margin of error** for confidence level $C$ is

$$m = z^*\text{SE}_{\hat{p}}$$

where $z^*$ is the value for the standard Normal density curve with area $C$ between $-z^*$ and $z^*$. An **approximate level $C$ confidence interval** for $p$ is

$$\hat{p} \pm m$$

Use this interval for 90%, 95%, or 99% confidence when the number of successes and the number of failures are both at least 15.

EXAMPLE

**8.1 Proportion of frequent binge drinkers.** Alcohol abuse has been described by college presidents as the number one problem on campus, and it is a major cause of death in young adults. How common is it? A survey of 13,819 students in U.S. four-year colleges collected information on drinking behavior and alcohol-related problems.[3]

The researchers defined "binge drinking" as having five or more drinks in a row for men and four or more drinks in a row for women. "Frequent binge drinking"was defined as binge drinking three or more times in the past two weeks. According to this definition, 3140 students were classified as frequent binge drinkers. The proportion of drinkers is

$$\hat{p} = \frac{3140}{13{,}819} = 0.227$$

To find a 95% confidence interval, first compute the standard error:

$$\begin{aligned}\text{SE}_{\hat{p}} &= \sqrt{\frac{\hat{p}(1-\hat{p})}{n}} \\ &= \sqrt{\frac{(0.227)(1-0.227)}{13{,}819}} \\ &= 0.00356\end{aligned}$$

Approximately 95% of the time, $\hat{p}$ will be within two standard errors ($2 \times 0.00356 = 0.00712$) of the true $p$. From Table A or D we find the value of $z^*$ to be 1.960. So the confidence interval is

$$\begin{aligned}\hat{p} \pm z^*\text{SE}_{\hat{p}} &= 0.227 \pm (1.960)(0.00356) \\ &= 0.227 \pm 0.007 \\ &= (0.220, 0.234)\end{aligned}$$

We estimate with 95% confidence that between 22.0% and 23.4% of college students are frequent binge drinkers. In other words, we estimate that 22.7% of college students are frequent binge drinkers, with a 95% confidence level margin of error of 0.7%.

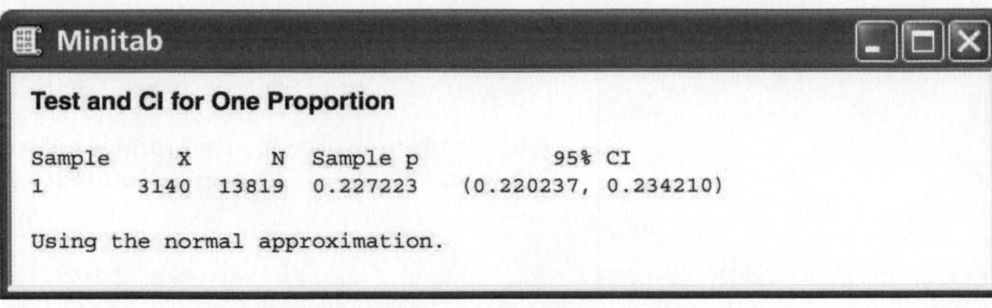

Minitab

**Test and CI for One Proportion**

```
Sample     X      N  Sample p           95% CI
1       3140  13819  0.227223   (0.220237, 0.234210)

Using the normal approximation.
```

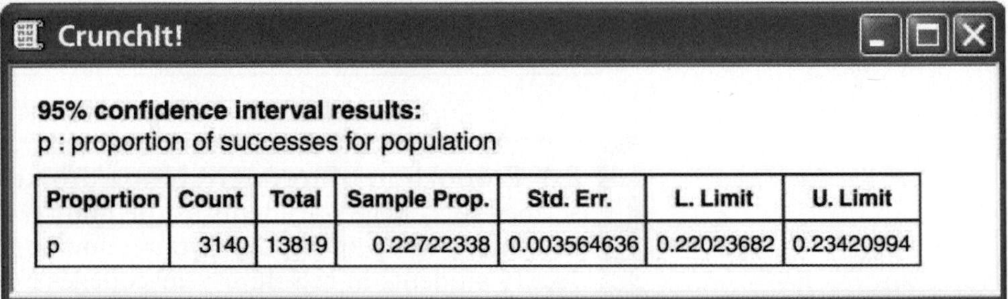

CrunchIt!

**95% confidence interval results:**
p : proportion of successes for population

| Proportion | Count | Total | Sample Prop. | Std. Err. | L. Limit | U. Limit |
|---|---|---|---|---|---|---|
| p | 3140 | 13819 | 0.22722338 | 0.003564636 | 0.22023682 | 0.23420994 |

**FIGURE 8.1** Minitab and CrunchIt! output for Example 8.1. By default, Minitab outputs an interval based on the binomial distribution. The large-sample confidence interval shown in the figure can be requested as an option.

Because the calculations for statistical inference for a single proportion are relatively straightforward, many software packages do not include them. Figure 8.1 gives output from Minitab and CrunchIt! for Example 8.1. As usual, the output reports more digits than are useful. *When you use software, be sure to think about how many digits are meaningful for your purposes.*

Remember that the margin of error in this confidence interval includes only random sampling error. There are other sources of error that are not taken into account. This survey used a design where the number of students sampled was proportional to the size of the college they attended; in this way we can treat the data as if we had an SRS. However, as is the case with many such surveys, we are forced to assume that the respondents provided accurate information. If the students did not answer the questions honestly, the results may be biased. Furthermore, we also have the typical problem of nonresponse. The response rate for this survey was 60%, a very good rate for surveys of this type. Do the students who did not respond have different drinking habits than those who did? If so, this is another source of bias.

We recommend the large-sample confidence interval for 90%, 95%, and 99% confidence whenever the number of successes and the number of failures are both at least 15. For smaller sample sizes, we recommend exact methods that use the binomial distribution. These are available as the default (for example, in Minitab and SAS) or as options in many statistical software packages and we do not cover them here. There is also an intermediate case between large samples and very small samples where a slight modification of the large-sample approach works quite well.[4] This method is called the "plus four" procedure and is described later.

## USE YOUR KNOWLEDGE

**8.1** **Owning a cell phone.** In a 2004 survey of 1200 undergraduate students throughout the United States, 89% of the respondents said they owned a cell phone.[5] For 90% confidence, what is the margin of error?

**8.2** **Importance of cell phone "features and functions."** In that same survey, one question asked what aspect was most important when buying a cell phone. "Features and functions" was the choice for 336 students. Give a 95% confidence interval for the proportion of U.S. students who find "features and functions" the most important aspect when buying a phone.

## BEYOND THE BASICS

### The Plus Four Confidence Interval for a Single Proportion

Computer studies reveal that confidence intervals based on the large-sample approach can be quite inaccurate when the number of successes and the number of failures are not at least 15. When this occurs, a simple adjustment to the confidence interval works very well in practice. The adjustment is based on assuming that the sample contains 4 additional observations, 2 of which are successes and 2 of which are failures. The estimator of the population proportion based on this *plus four* rule is

$$\tilde{p} = \frac{X+2}{n+4}$$

plus four estimate

This estimate was first suggested by Edwin Bidwell Wilson in 1927 and we call it the **plus four estimate.** The confidence interval is based on the $z$ statistic obtained by standardizing the plus four estimate $\tilde{p}$. Because $\tilde{p}$ is the sample proportion for our modified sample of size $n+4$, it isn't surprising that the distribution of $\tilde{p}$ is close to the Normal distribution with mean $p$ and standard deviation $\sqrt{p(1-p)/(n+4)}$. To get a confidence interval, we estimate $p$ by $\tilde{p}$ in this standard deviation to get the standard error of $\tilde{p}$. Here is the final result.

#### PLUS FOUR CONFIDENCE INTERVAL FOR A SINGLE PROPORTION

Choose an SRS of size $n$ from a large population with unknown proportion $p$ of successes. The **plus four estimate of the population proportion** is

$$\tilde{p} = \frac{X+2}{n+4}$$

where $X$ is the number of successes. The **standard error of $\tilde{p}$** is

$$SE_{\tilde{p}} = \sqrt{\frac{\tilde{p}(1-\tilde{p})}{n+4}}$$

and the **margin of error** for confidence level $C$ is

$$m = z^* SE_{\tilde{p}}$$

where $z^*$ is the value for the standard Normal density curve with area $C$ between $-z^*$ and $z^*$. An **approximate level $C$ confidence interval** for

$p$ is

$$\tilde{p} \pm m$$

Use this interval for 90%, 95%, or 99% confidence whenever the sample size is at least $n = 10$.

EXAMPLE

**8.2 Percent of equol producers.** Research has shown that there are many health benefits associated with a diet that contains soy foods. Substances in soy called isoflavones are known to be responsible for these benefits. When soy foods are consumed, some subjects produce a chemical called equol, and it is thought that production of equol is a key factor in the health benefits of a soy diet. Unfortunately, not all people are equol producers; there appear to be two distinct subpopulations: equol producers and equol nonproducers.[6]

A nutrition researcher planning some bone health experiments would like to include some equol producers and some nonproducers among her subjects. A preliminary sample of 12 female subjects were measured, and 4 were found to be equol producers. We would like to estimate the proportion of equol producers in the population from which this researcher will draw her subjects.

The plus four estimate of the proportion of equol producers is

$$\tilde{p} = \frac{4+2}{12+4} = \frac{6}{16} = 0.375$$

For a 95% confidence interval, we use Table D to find $z^* = 1.96$. We first compute the standard error

$$\begin{aligned} \text{SE}_{\tilde{p}} &= \sqrt{\frac{\tilde{p}(1-\tilde{p})}{n+4}} \\ &= \sqrt{\frac{(0.375)(1-0.375)}{16}} \\ &= 0.12103 \end{aligned}$$

and then the margin of error

$$\begin{aligned} m &= z^* \text{SE}_{\tilde{p}} \\ &= (1.96)(0.12103) \\ &= 0.237 \end{aligned}$$

So the confidence interval is

$$\begin{aligned} \tilde{p} \pm m &= 0.375 \pm 0.237 \\ &= (0.138, 0.612) \end{aligned}$$

We estimate with 95% confidence that between 14% and 61% of women from this population are equol producers.

If the true proportion of equol users is near 14%, the lower limit of this interval, there may not be a sufficient number of equol producers in the study if subjects are tested only after they are enrolled in the experiment. It may be necessary to determine whether or not a potential subject is an equol producer. The study could then be designed to have the same number of equol producers and nonproducers.

## Significance test for a single proportion

**LOOK BACK**
**Normal approximation for proportions, page 323**

Recall that the sample proportion $\hat{p} = X/n$ is approximately Normal, with mean $\mu_{\hat{p}} = p$ and standard deviation $\sigma_{\hat{p}} = \sqrt{p(1-p)/n}$. For confidence intervals, we substitute $\hat{p}$ for $p$ in the last expression to obtain the standard error. When performing a significance test, however, the null hypothesis specifies a value for $p$, and we assume that this is the true value when calculating the $P$-value. Therefore, when we test $H_0: p = p_0$, we substitute $p_0$ into the expression for $\sigma_{\hat{p}}$ and then standardize $\hat{p}$. Here are the details.

### LARGE-SAMPLE SIGNIFICANCE TEST FOR A POPULATION PROPORTION

Draw an SRS of size $n$ from a large population with unknown proportion $p$ of successes. To test the hypothesis $H_0: p = p_0$, compute the $z$ **statistic**

$$z = \frac{\hat{p} - p_0}{\sqrt{\dfrac{p_0(1-p_0)}{n}}}$$

In terms of a standard Normal random variable $Z$, the approximate $P$-value for a test of $H_0$ against

$H_a: p > p_0$ is $P(Z \geq z)$

$H_a: p < p_0$ is $P(Z \leq z)$

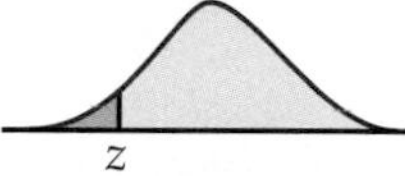


$H_a: p \neq p_0$ is $2P(Z \geq |z|)$

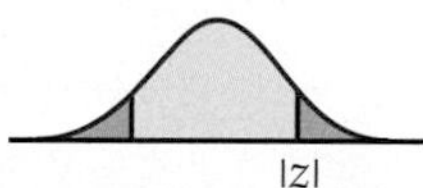


**LOOK BACK**
**sign test for matched pairs, page 439**

We recommend the large-sample $z$ significance test as long as the expected number of successes, $np_0$, and the expected number of failures, $n(1-p_0)$, are both at least 10. If this rule of thumb is not met, or if the population is less than 20 times as large as the sample, other procedures should be used. One such approach is to use the binomial distribution as we did with the sign test. Here is a large-sample example.

EXAMPLE

**8.3 Work stress.** According to the National Institute for Occupational Safety and Health,[7] job stress poses a major threat to the health of workers. A national survey of restaurant employees found that 75% said that work stress had a negative impact on their personal lives.[8] A sample of 100 employees of a restaurant chain finds that 68 answer "Yes" when asked, "Does work stress have a negative impact on your personal life?" Is this good reason to think that the proportion of all employees of this chain who would say "Yes" differs from the national proportion $p_0 = 0.75$?

To answer this question, we test

$$H_0: p = 0.75$$

$$H_a: p \neq 0.75$$

The expected numbers of "Yes" and "No" responses are $100 \times 0.75 = 75$ and $100 \times 0.25 = 25$. Both are greater than 10, so we can use the $z$ test. The test statistic is

$$z = \frac{\hat{p} - p_0}{\sqrt{\dfrac{p_0(1 - p_0)}{n}}} = \frac{0.68 - 0.75}{\sqrt{\dfrac{(0.75)(0.25)}{100}}} = -1.62$$

From Table A we find $P(Z \leq -1.62) = 0.0526$. The $P$-value is the area in both tails, $P = 2 \times 0.0526 = 0.1052$. Figure 8.2 displays the $P$-value as an area under the standard Normal curve. We conclude that the chain restaurant data are compatible with the survey results ($\hat{p} = 0.68$, $z = -1.62$, $P = 0.11$).

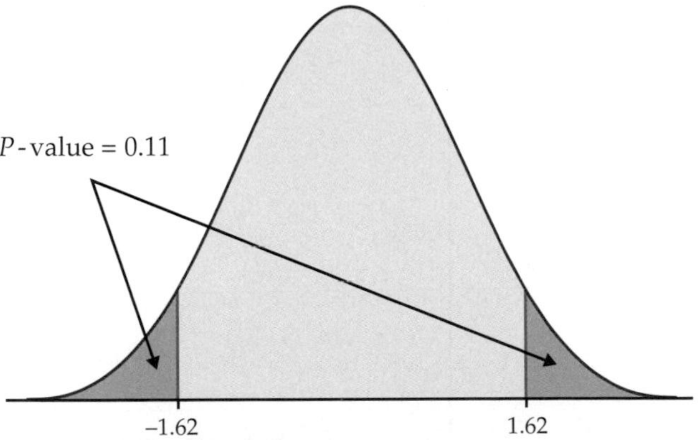


**FIGURE 8.2** The *P*-value for Example 8.3.

Figure 8.3 gives computer output from Minitab and CrunchIt! for this example. *Note that for some entries software gives many more digits than we need.* You should decide how many digits are important for your analysis. In general, we will round proportions to two digits, for example, 0.68, and nonsignificant $P$-values to two digits, for example, $P = 0.11$.

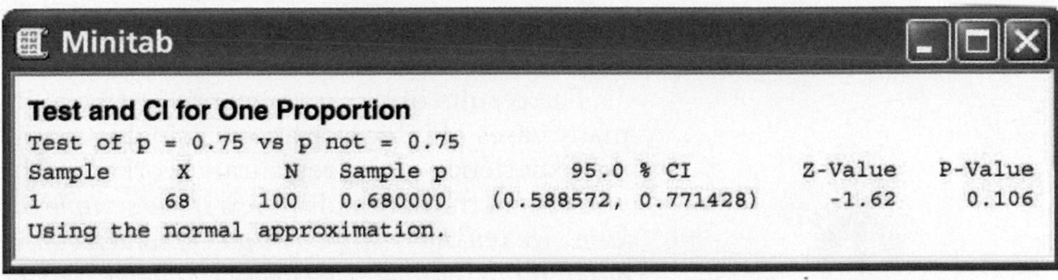


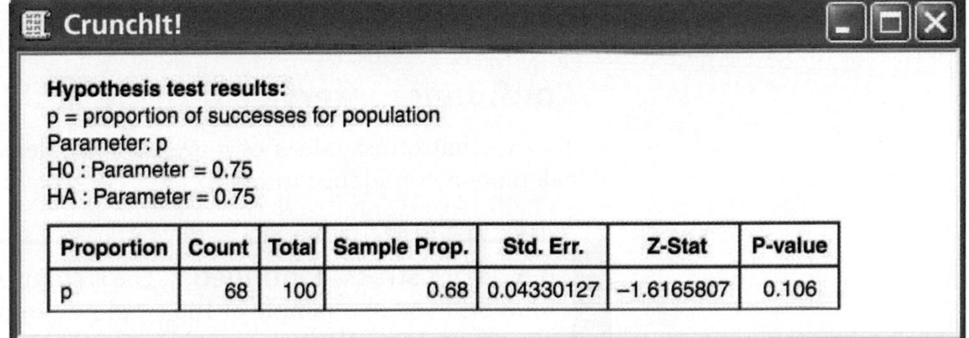


**FIGURE 8.3** Minitab and CrunchIt! output for Example 8.3. By default, Minitab performs a test using the binomial distribution. The large-sample significance test shown in the figure can be requested as an option.

In this example we have arbitrarily chosen to associate the response "Yes" that work stress has a negative impact on the respondent's personal life with success and "No" with failure. Suppose we reversed the choice. If we observed that 68 respondents said "Yes," then the other $100 - 68 = 32$ people said "No." Let's repeat the significance test with "No" as the success outcome. The national comparison value for the significance test is now 25%, the proportion in the national survey who responded "No."

EXAMPLE

**8.4 Work stress, revisited.** A sample of 100 restaurant workers were asked whether or not work stress had a negative impact on their personal lives and 32 of them responded "No." A large national survey reported that 25% of workers reported a negative impact. We test the null hypothesis

$$H_0: p = 0.25$$

against

$$H_a: p \neq 0.25$$

The test statistic is

$$z = \frac{\hat{p} - 0.25}{\sqrt{\frac{(0.25)(0.75)}{100}}} = \frac{0.32 - 0.25}{\sqrt{\frac{(0.25)(0.75)}{100}}} = 1.62$$

Using Table A, we find that $P = 0.11$.

When we interchanged "Yes" and "No" (or success and failure), we simply changed the sign of the test statistic $z$. The $P$-value remained the same. These facts are true in general. Our conclusion does not depend on an arbitrary choice of success and failure.

The results of our significance test have limited use in this example, as in many cases of inference about a single parameter. Of course, we do not expect the experience of the restaurant workers to be *exactly* the same as that of the workers in the national survey. *If the sample of restaurant workers is sufficiently large, we will have sufficient power to detect a very small difference. On the other hand, if our sample size is very small, we may be unable to detect differences that could be very important.* For these reasons we prefer to include a confidence interval as part of our analysis.

CAUTION

## Confidence intervals provide additional information

To see what other values of $p$ are compatible with the sample results, we will calculate a confidence interval.

EXAMPLE

**8.5 Work stress, continued.** The restaurant worker survey in Example 8.3 found that 68 of a sample of 100 employees agreed that work stress had a negative impact on their personal lives. That is, the sample size is $n = 100$ and the count of successes is $X = 68$. Because the number of successes and the number of failures are both at least 15, we will use the large-sample procedure to compute a 95% confidence interval. The sample proportion is

$$\hat{p} = \frac{X}{n} = \frac{68}{100} = 0.68$$

The standard error is

$$\text{SE}_{\hat{p}} = \sqrt{\frac{\hat{p}(1-\hat{p})}{n}} = \sqrt{\frac{(0.68)(1-0.68)}{100}} = 0.0466$$

The $z$ critical value for 95% confidence is $z^* = 1.96$, so the margin of error is

$$m = z^*\text{SE}_{\hat{p}} = (1.96)(0.0466) = 0.091$$

The confidence interval is

$$\hat{p} \pm m = 0.68 \pm (1.96)(0.0466) = 0.68 \pm 0.09 = (0.59, 0.77)$$

We are 95% confident that between 59% and 77% of the restaurant chain's employees feel that work stress is damaging their personal lives.

The confidence interval of Example 8.5 is much more informative than the significance test of Example 8.3. We have determined the values of $p$ that are consistent with the observed results. Note that the standard error used for the confidence interval is estimated from the data, whereas the denominator in the test statistic $z$ is based on the value assumed in the null hypothesis. A consequence of this fact is that the correspondence between the significance test result and the confidence interval is no longer exact. However, the correspondence is still very close. The confidence interval $(0.59, 0.77)$ gives an approximate range of $p_0$'s that would not be rejected by a test at the $\alpha = 0.05$ level of significance. We would not be surprised if the true proportion of restaurant workers who would say that work stress has a negative impact on their lives was as low as 60% or as high as 75%.

CAUTION

*We do not often use significance tests for a single proportion, because it is uncommon to have a situation where there is a precise* $p_0$ *that we want to test.* For physical experiments such as coin tossing or drawing cards from a well-shuffled deck, probability arguments lead to an ideal $p_0$. Even here, however, it can be argued, for example, that no real coin has a probability of heads *exactly* equal to 0.5. Data from past large samples can sometimes provide a $p_0$ for the null hypothesis of a significance test. In some types of epidemiology research, for example, "historical controls" from past studies serve as the benchmark for evaluating new treatments. Medical researchers argue about the validity of these approaches, because the past never quite resembles the present. In general, we prefer comparative studies whenever possible.

## USE YOUR KNOWLEDGE

**8.3** **Working while enrolled in school.** A 1993 nationwide survey by the National Center for Education Statistics reports that 72% of all undergraduates work while enrolled in school.[9] You decide to test whether this percent is different at your university. In your random sample of 100 students, 77 said they were currently working.

(a) Give the null and alternative hypotheses for this study.

(b) Carry out the significance test. Report the test statistic and $P$-value.

(c) Does it appear that the percent of students working at your university is different at the $\alpha = 0.05$ level?

**8.4** **Owning a cell phone, continued.** Refer to Exercise 8.1 (page 490). It was reported that cell phone ownership by undergraduate students in 2003 was 83%. Do the sample data in 2004 give good evidence that this percent has increased?

(a) Give the null and alternative hypotheses.

(b) Carry out the significance test. Report the test statistic and the $P$-value.

(c) State your conclusion using $\alpha = 0.05$.

## Choosing a sample size

**choosing the sample size, page 364**

In Chapter 6, we showed how to choose the sample size $n$ to obtain a confidence interval with specified margin of error $m$ for a Normal mean. Because we are using a Normal approximation for inference about a population proportion, sample size selection proceeds in much the same way.

Recall that the margin of error for the large-sample confidence interval for a population proportion is

$$m = z^*\mathrm{SE}_{\hat{p}} = z^*\sqrt{\frac{\hat{p}(1-\hat{p})}{n}}$$

Choosing a confidence level $C$ fixes the critical value $z^*$. The margin of error also depends on the value of $\hat{p}$ and the sample size $n$. Because we don't know the value of $\hat{p}$ until we gather the data, we must guess a value to use in the calculations. We will call the guessed value $p^*$. There are two common ways to get $p^*$:

1. Use the sample estimate from a pilot study or from similar studies done earlier.
2. Use $p^* = 0.5$. Because the margin of error is largest when $\hat{p} = 0.5$, this choice gives a sample size that is somewhat larger than we really need for the confidence level we choose. It is a safe choice no matter what the data later show.

Once we have chosen $p^*$ and the margin of error $m$ that we want, we can find the $n$ we need to achieve this margin of error. Here is the result.

### SAMPLE SIZE FOR DESIRED MARGIN OF ERROR

The level $C$ confidence interval for a proportion $p$ will have a margin of error approximately equal to a specified value $m$ when the sample size satisfies

$$n = \left(\frac{z^*}{m}\right)^2 p^*(1-p^*)$$

Here $z^*$ is the critical value for confidence $C$, and $p^*$ is a guessed value for the proportion of successes in the future sample.

The margin of error will be less than or equal to $m$ if $p^*$ is chosen to be 0.5. The sample size required when $p^* = 0.5$ is

$$n = \frac{1}{4}\left(\frac{z^*}{m}\right)^2$$

The value of $n$ obtained by this method is not particularly sensitive to the choice of $p^*$ when $p^*$ is fairly close to 0.5. However, if the value of $p$ is likely to be smaller than about 0.3 or larger than about 0.7, use of $p^* = 0.5$ may result in a sample size that is much larger than needed.

EXAMPLE

**8.6 Planning a survey of students.** A large university is interested in assessing student satisfaction with the overall campus environment. The plan is to distribute a questionnaire to an SRS of students, but before proceeding, the university wants to determine how many students to sample. The questionnaire asks about a student's degree of satisfaction with various student services, each measured on a five-point scale. The university is interested in the proportion $p$ of students who are satisfied (that is, who choose either "satisfied" or "very satisfied," the two highest levels on the five-point scale).

The university wants to estimate $p$ with 95% confidence and a margin of error less than or equal to 3%, or 0.03. For planning purposes, they are willing to use $p^* = 0.5$. To find the sample size required,

$$n = \frac{1}{4}\left(\frac{z^*}{m}\right)^2 = \frac{1}{4}\left[\frac{1.96}{0.03}\right]^2 = 1067.1$$

Round up to get $n = 1068$. (Always round up. Rounding down would give a margin of error slightly greater than 0.03.)

Similarly, for a 2.5% margin of error we have (after rounding up)

$$n = \frac{1}{4}\left[\frac{1.96}{0.025}\right]^2 = 1537$$

and for a 2% margin of error,

$$n = \frac{1}{4}\left[\frac{1.96}{0.02}\right]^2 = 2401$$

News reports frequently describe the results of surveys with sample sizes between 1000 and 1500 and a margin of error of about 3%. These surveys generally use sampling procedures more complicated than simple random sampling, so the calculation of confidence intervals is more involved than what we have studied in this section. The calculations in Example 8.6 nonetheless show in principle how such surveys are planned.

In practice, many factors influence the choice of a sample size. The following example illustrates one set of factors.

EXAMPLE

**8.7 Assessing interest in Pilates classes.** The Division of Recreational Sports (Rec Sports) at a major university is responsible for offering comprehensive recreational programs, services, and facilities to the students. Rec Sports is continually examining its programs to determine how well it is meeting the needs of the students. Rec Sports is considering adding some new programs and would like to know how much interest there is in a new exercise program based on the Pilates method.[10] They will take a survey of undergraduate students. In the past, they emailed short surveys to all undergraduate students. The response rate obtained in this way was about 5%. This time they will send emails to a simple random sample of the students and will follow up with additional emails and eventually a phone call to get a higher response rate. Because of limited staff and the work involved with the follow-up, they would like to use a sample size of about 200. One of the questions they will ask is "Have you ever heard about the Pilates method of exercise?"

The primary purpose of the survey is to estimate various sample proportions for undergraduate students. Will the proposed sample size of $n = 200$ be adequate to provide Rec Sports with the needed information? To address this question, we calculate the margins of error of 95% confidence intervals for various values of $\hat{p}$.

EXAMPLE

**8.8 Margins of error.** In the Rec Sports survey, the margin of error of a 95% confidence interval for any value of $\hat{p}$ and $n = 200$ is

$$\begin{aligned} m &= z^*\text{SE}_{\hat{p}} \\ &= 1.96\sqrt{\frac{\hat{p}(1-\hat{p})}{200}} \\ &= 0.139\sqrt{\hat{p}(1-\hat{p})} \end{aligned}$$

The results for various values of $\hat{p}$ are

| $\hat{p}$ | $m$ | $\hat{p}$ | $m$ |
|---|---|---|---|
| 0.05 | 0.030 | 0.60 | 0.068 |
| 0.10 | 0.042 | 0.70 | 0.064 |
| 0.20 | 0.056 | 0.80 | 0.056 |
| 0.30 | 0.064 | 0.90 | 0.042 |
| 0.40 | 0.068 | 0.95 | 0.030 |
| 0.50 | 0.070 | | |

Rec Sports judged these margins of error to be acceptable, and they used a sample size of 200 in their survey.

The table in Example 8.8 illustrates two points. First, the margins of error for $\hat{p} = 0.05$ and $\hat{p} = 0.95$ are the same. The margins of error will always be the same for $\hat{p}$ and $1 - \hat{p}$. This is a direct consequence of the form of the confidence interval. Second, the margin of error varies only between 0.064 and 0.070 as $\hat{p}$ varies from 0.3 to 0.7, and the margin of error is greatest when $\hat{p} = 0.5$, as we claimed earlier. It is true in general that the margin of error will vary relatively little for values of $\hat{p}$ between 0.3 and 0.7. Therefore, when planning a study, it is not necessary to have a very precise guess for $p$. If $p^* = 0.5$ is used and the observed $\hat{p}$ is between 0.3 and 0.7, the actual interval will be a little shorter than needed but the difference will be small.

*Again it is important to emphasize that these calculations consider only the effects of sampling variability that are quantified in the margin of error.* Other sources of error, such as nonresponse and possible misinterpretation of questions, are not included in the table of margins of error for Example 8.8. Rec Sports is trying to minimize these kinds of errors. They did a pilot study using a small group of current users of their facilities to check the wording of the questions, and they devised a careful plan to follow up with the students who did not respond to the initial email.

**USE YOUR KNOWLEDGE**

**8.5** **Confidence level and sample size.** Refer to Example 8.6 (page 499). Suppose the university was interested in a 90% confidence interval with margin of error 0.03. Would the required sample size be smaller or larger than 1068 students? Verify this by performing the calculation.

**8.6** **Calculating the sample size.** Refer to Exercise 8.3 (page 497). You plan to do a larger survey such that the 95% margin of error is no larger than 0.02. Using the results from the small survey of 100 students, what sample size would you use?

## SECTION 8.1 Summary

Inference about a population proportion $p$ from an SRS of size $n$ is based on the **sample proportion** $\hat{p} = X/n$. When $n$ is large, $\hat{p}$ has approximately the Normal distribution with mean $p$ and standard deviation $\sqrt{p(1-p)/n}$.

For large samples, the **margin of error for confidence level *C*** is

$$m = z^* \text{SE}_{\hat{p}}$$

where $z^*$ is the value for the standard Normal density curve with area $C$ between $-z^*$ and $z^*$, and the **standard error of $\hat{p}$** is

$$\text{SE}_{\hat{p}} = \sqrt{\frac{\hat{p}(1-\hat{p})}{n}}$$

The **level *C* large-sample confidence interval** is

$$\hat{p} \pm m$$

We recommend using this interval for 90%, 95% and 99% confidence whenever the number of successes and the number of failures are both at least 15. When sample sizes are smaller, alternative procedures such as the **plus four estimate of the population proportion** are recommended.

The **sample size** required to obtain a confidence interval of approximate margin of error $m$ for a proportion is found from

$$n = \left(\frac{z^*}{m}\right)^2 p^*(1-p^*)$$

where $p^*$ is a guessed value for the proportion, and $z^*$ is the standard Normal critical value for the desired level of confidence. To ensure that the margin of error of the interval is less than or equal to $m$ no matter what $\hat{p}$ may be, use

$$n = \frac{1}{4}\left(\frac{z^*}{m}\right)^2$$

Tests of $H_0: p = p_0$ are based on the **$z$ statistic**

$$z = \frac{\hat{p} - p_0}{\sqrt{\dfrac{p_0(1-p_0)}{n}}}$$

with $P$-values calculated from the $N(0, 1)$ distribution. Use this procedure when the expected number of successes, $np_0$, and the expected number of failures, $n(1 - p_0)$, are both at least 10.

## SECTION 8.1 Exercises

*For Exercises 8.1 and 8.2, see pages 490 and 491; for Exercises 8.3 and 8.4, see page 497; and for Exercises 8.5 and 8.6, see page 501.*

**8.7 Can we use the large-sample confidence interval?** In each of the following circumstances state whether you would use the large-sample confidence interval.

(a) $n = 50, X = 30$.

(b) $n = 90, X = 15$.

(c) $n = 10, X = 2$.

(d) $n = 60, X = 50$.

(e) $n = 25, X = 15$.

**8.8 More on whether to use the large-sample confidence interval.** In each of the following circumstances state whether you would use the large-sample confidence interval.

(a) $n = 8, X = 4$.

(b) $n = 500, X = 13$.

(c) $n = 40, X = 18$.

(d) $n = 15, X = 15$.

(e) $n = 50, X = 22$.

**8.9 What's wrong?** Explain what is wrong with each of the following:

(a) An approximate 99% confidence interval for an unknown proportion $p$ is $\hat{p}$ plus or minus its standard error.

(b) When performing a large-sample significance test for a population proportion, the $t$ distribution is used to compute the $P$-value.

(c) A significance test is used to evaluate $H_0: \hat{p} = 0.2$ versus the two-sided alternative.

**8.10 $\hat{p}$ and the Normal distribution.** Consider the binomial setting with $n = 60$ and $p = 0.4$.

(a) The sample proportion $\hat{p}$ will have a distribution that is approximately Normal. Give the mean and the standard deviation of this Normal distribution.

(b) Draw a sketch of this Normal distribution. Mark the location of the mean.

(c) Find a value $p^*$ for which the probability is 95% that $\hat{p}$ will be between $\pm p^*$. Mark these two values on your sketch.

**8.11 Gambling and college athletics.** Gambling is an issue of great concern to those involved in intercollegiate athletics. Because of this, the National Collegiate Athletic Association (NCAA) surveyed student-athletes concerning their gambling-related behaviors.[11] There were 5594 Division I male athletes in the survey. Of these, 3547 reported participation in some gambling behavior. This included playing cards, betting on games of skill, buying lottery tickets, and betting on sports.

(a) Find the sample proportion and the large-sample margin of error for 95% confidence. Explain in simple terms the meaning of the 95%.

(b) Because of the way that the study was designed to protect the anonymity of the student-athletes who responded, it was not possible to calculate the number of students who were asked to respond but did not. Does this fact affect the way that you interpret the results? Write a short paragraph explaining your answer.

**8.12 Gambling and female athletes.** In the study described in the previous exercise, 1447 out of a total of 3469 female student-athletes reported participation in some gambling activity.

(a) Use the large-sample methods to find an estimate of the true proportion with a 95% confidence interval.

(b) The margin of error for this sample is not the same as the margin of error calculated for the previous exercise. Explain why.

**8.13 Do students report Internet sources?** The National Survey of Student Engagement found that 87% of students report that their peers at least "sometimes" copy information from the Internet in their papers without reporting the source.[12] Assume that the sample size is 430,000.

(a) Find the margin of error for 99% confidence.

(b) Here are some items from the report that summarizes the survey. More than 430,000 students from 730 four-year colleges and universities participated. The average response rate was 43% and ranged from 15% to 89%. Institutions pay a participation fee of between $3000 and $7500 based on the size of their undergraduate enrollment. Discuss these as sources of error in this study. How do you think these errors would compare with the error that you calculated in part (a)?

**8.14 Do you enjoy driving your car?** The Pew Research Center recently polled $n = 1048$ U.S. drivers and found that 69% enjoyed driving their automobiles.[13]

(a) Construct a 95% confidence interval for the proportion of U.S. drivers who enjoy driving their automobiles.

(b) In 1991, a Gallup Poll reported this percent to be 79%. Using the data from this poll, test the claim that the percent of drivers who enjoy driving their cars has declined since 1991. Report the large-sample $z$ statistic and its $P$-value.

**8.15 Getting angry at other drivers.** Refer to Exercise 8.14. The same Pew Poll found that 38% of the respondents "shouted, cursed or made gestures to other drivers" in the last year.

(a) Construct a 95% confidence interval for the true proportion of U.S. drivers who did these actions in the last year.

(b) Does the fact that the respondent is self-reporting these actions affect the way that you interpret the results? Write a short paragraph explaining your answer.

**8.16 Cheating during a test.** A national survey of high school students conducted by the Josephson Institute of Ethics was sent to 37,328 students, and 24,142 were returned. One question asked students if they had cheated during a test in the last school year.[14] Of those who returned the survey, 9054 responded that they had cheated at least two times in the last year.

(a) What is the sample proportion of respondents who cheated at least twice?

(b) Compute the 95% confidence interval for the true proportion of students who have cheated on at least two tests in the last year.

(c) Compute the nonresponse rate for this study. Does this influence how you interpret these results? Write a short discussion of this issue.

**8.17** CHALLENGE **Long sermons.** The National Congregations Study collected data in a one-hour interview with a key informant—that is, a minister, priest, rabbi, or other staff person or leader.[15] One question asked concerned the length of the typical sermon. For this question 390 out of 1191 congregations reported that the typical sermon lasted more than 30 minutes.

(a) Use the large-sample inference procedures to estimate the true proportion for this question with a 95% confidence interval.

(b) The respondents to this question were not asked to use a stopwatch to record the lengths of a random sample of sermons at their congregations. They responded based on their impressions of the sermons. Do you think that ministers, priests, rabbis, or other staff persons or leaders might perceive sermon lengths differently from the people listening to the sermons? Discuss how your ideas would influence your interpretation of the results of this study.

**8.18 Confidence level and interval width.** Refer to Exercise 8.17. Would a 90% confidence interval be wider or narrower than the one that you found in that exercise? Verify your results by computing the interval.

**8.19 Student loans larger than $30,000.** A survey of 1280 student loan borrowers found that 192 had loans totaling more than $30,000 for their undergraduate education.[16] Give a 95% confidence interval for the proportion of all student loan borrowers who have loans of more than $30,000 for their undergraduate education.

**8.20 More on confidence level and interval width.** Refer to Exercise 8.19. Would a 99% confidence interval be wider or narrower than the one that you found in that exercise? Verify your results by computing the interval.

**8.21 Can we use the $z$ test?** In each of the following cases state whether or not the Normal approximation to the binomial should be used for a significance test on the population proportion $p$.

(a) $n = 30$ and $H_0: p = 0.2$.

(b) $n = 30$ and $H_0: p = 0.6$.

(c) $n = 100$ and $H_0: p = 0.5$.

(d) $n = 200$ and $H_0: p = 0.01$.

**8.22 Instant versus fresh-brewed coffee.** A matched pairs experiment compares the taste of instant versus fresh-brewed coffee. Each subject tastes two unmarked cups of coffee, one of each type, in random order and states which he or she prefers. Of the 40 subjects who participate in the study, 12 prefer the instant coffee. Let $p$ be the probability that a randomly chosen subject prefers fresh-brewed coffee to instant coffee. (In practical terms, $p$ is the proportion of the population who prefer fresh-brewed coffee.)

(a) Test the claim that a majority of people prefer the taste of fresh-brewed coffee. Report the large-sample $z$ statistic and its $P$-value.

(b) Draw a sketch of a standard Normal curve and mark the location of your $z$ statistic. Shade the appropriate area that corresponds to the $P$-value.

(c) Is your result significant at the 5% level? What is your practical conclusion?

**8.23 College students and dieting.** For a study of unhealthy eating behaviors, 267 college women aged 18 to 25 years were surveyed.[17] Of these, 69% reported that they had been on a diet sometime during the past year. Give a 95% confidence interval for the true proportion of college women aged 18 to 25 years in this population who dieted last year.

**8.24 High school students and dieting.** In the study described in the previous exercise, the researchers also surveyed 266 high school students who were 18 years old. In this sample 58.3% reported that they had dieted sometime in the past year. Give a 95% confidence interval for the true proportion of 18-year-old high school students in this population who were on a diet sometime during the past year.

**8.25 Pet ownership among older adults.** In a study of the relationship between pet ownership and physical activity in older adults,[18] 594 subjects reported that they owned a pet, while 1939 reported that they did not. Give a 95% confidence interval for the proportion of older adults in this population who are pet owners.

**8.26** CHALLENGE **Annual income of older adults.** In the study described in the previous exercise, 1434 subjects out of a total of 2533 reported that their annual income was $25,000 or more.

(a) Give a 95% confidence interval for the true proportion of subjects in this population with incomes of at least $25,000.

(b) Do you think that some respondents might not give truthful answers to a question about their income? Discuss the possible effects on your estimate and confidence interval.

(c) In the previous exercise, the question analyzed concerned pet ownership. Compare this question with the income question with respect to the possibility that the respondents were not truthful.

**8.27 Dogs sniffing out cancer.** A 2005 study by researchers set out to determine whether dogs could be trained to detect lung and breast cancer by sniffing exhaled breath samples.[19] For the breast cancer portion, breath samples from 6 cancer patients and 17 cancer-free volunteers were used. Each dog had to sniff five breath samples. 125 of the trials involved one cancer sample and four control samples. A correct response on these trials involved lying down next to the sample from the breast cancer patient. Collectively, the dogs correctly identified the cancer sample in 110 of these trials. Construct a 95% confidence interval for the true proportion of times these dogs will correctly identify a breast cancer sample.

**8.28** CHALLENGE **Bicycle accidents and alcohol.** In the United States approximately 900 people die in bicycle accidents each year. One study examined the records of 1711 bicyclists aged 15 or older who were fatally injured in bicycle accidents between 1987 and 1991 and were tested for alcohol. Of these, 542 tested positive for alcohol (blood alcohol concentration of 0.01% or higher).[20]

(a) Summarize the data with appropriate descriptive statistics.

(b) To do statistical inference for these data, we think in terms of a model where $p$ is a parameter that represents the probability that a tested bicycle rider is positive for alcohol. Find a 99% confidence interval for $p$.

(c) Can you conclude from your analysis of this study that alcohol causes fatal bicycle accidents? Explain.

(d) In this study 386 bicyclists had blood alcohol levels above 0.10%, a level defining legally drunk in many states at the time. Give a 99% confidence interval for the proportion who were legally drunk according to this criterion.

**8.29 Tossing a coin 10,000 times!** The South African mathematician John Kerrich, while a prisoner of war during World War II, tossed a coin 10,000 times and obtained 5067 heads.

(a) Is this significant evidence at the 5% level that the probability that Kerrich's coin comes up heads is not 0.5? Use a sketch of the standard Normal distribution to illustrate the $P$-value.

(b) Use a 95% confidence interval to find the range of probabilities of heads that would not be rejected at the 5% level.

**8.30** **Is there interest in a new product?** One of your employees has suggested that your company develop a new product. You decide to take a random sample of your customers and ask whether or not there is interest in the new product. The response is on a 1 to 5 scale with 1 indicating "definitely would not purchase"; 2, "probably would not purchase"; 3, "not sure"; 4, "probably would purchase"; and 5, "definitely would purchase." For an initial analysis, you will record the responses 1, 2, and 3 as "No" and 4 and 5 as "Yes." What sample size would you use if you wanted the 95% margin of error to be 0.15 or less?

**8.31** **More information is needed.** Refer to the previous exercise. Suppose that after reviewing the results of the previous survey, you proceeded with preliminary development of the product. Now you are at the stage where you need to decide whether or not to make a major investment to produce and market it. You will use another random sample of your customers but now you want the margin of error to be smaller. What sample size would you use if you wanted the 95% margin of error to be 0.075 or less?

**8.32** **Sample size needed for an evaluation.** You are planning an evaluation of a semester-long alcohol awareness campaign at your college. Previous evaluations indicate that about 25% of the students surveyed will respond "Yes" to the question "Did the campaign alter your behavior toward alcohol consumption?" How large a sample of students should you take if you want the margin of error for 95% confidence to be about 0.1?

**8.33** CHALLENGE **Sample size needed for an evaluation, continued.** The evaluation in the previous exercise will also have questions that have not been asked before, so you do not have previous information about the possible value of $p$. Repeat the calculation above for the following values of $p^*$: 0.1, 0.2, 0.3, 0.4, 0.5, 0.6, 0.7, 0.8, and 0.9. Summarize the results in a table and graphically. What sample size will you use?

**8.34** **Are the customers dissatisfied?** An automobile manufacturer would like to know what proportion of its customers are dissatisfied with the service received from their local dealer. The customer relations department will survey a random sample of customers and compute a 95% confidence interval for the proportion that are dissatisfied. From past studies, they believe that this proportion will be about 0.15. Find the sample size needed if the margin of error of the confidence interval is to be no more than 0.02.

## 8.2 Comparing Two Proportions

Because comparative studies are so common, we often want to compare the proportions of two groups (such as men and women) that have some characteristic. In the previous section we used data to estimate the proportion of college students who were frequent binge drinkers. Suppose we also wanted to compare the binge-drinking behaviors across years or of men and women college students. Our problem now concerns the comparison of two proportions.

We call the two groups being compared Population 1 and Population 2, and the two population proportions of "successes" $p_1$ and $p_2$. The data consist of two independent SRSs, of size $n_1$ from Population 1 and size $n_2$ from Population 2. The proportion of successes in each sample estimates the corresponding population proportion. Here is the notation we will use in this section:

| Population | Population proportion | Sample size | Count of successes | Sample proportion |
|---|---|---|---|---|
| 1 | $p_1$ | $n_1$ | $X_1$ | $\hat{p}_1 = X_1/n_1$ |
| 2 | $p_2$ | $n_2$ | $X_2$ | $\hat{p}_2 = X_2/n_2$ |

To compare the two populations, we use the difference between the two sample proportions:

$$D = \hat{p}_1 - \hat{p}_2$$

When both sample sizes are sufficiently large, the sampling distribution of the difference $D$ is approximately Normal.

Inference procedures for comparing proportions are $z$ procedures based on the Normal approximation and on standardizing the difference $D$. The first step is to obtain the mean and standard deviation of $D$. By the addition rule for means, the mean of $D$ is the difference of the means:

LOOK BACK
**addition rule for means, page 278**
**addition rule for variances, page 282**

$$\mu_D = \mu_{\hat{p}_1} - \mu_{\hat{p}_2} = p_1 - p_2$$

That is, the difference $D = \hat{p}_1 - \hat{p}_2$ between the sample proportions is an unbiased estimator of the population difference $p_1 - p_2$. Similarly, the addition rule for variances tells us that the variance of $D$ is the *sum* of the variances:

$$\begin{aligned} \sigma_D^2 &= \sigma_{\hat{p}_1}^2 + \sigma_{\hat{p}_2}^2 \\ &= \frac{p_1(1-p_1)}{n_1} + \frac{p_2(1-p_2)}{n_2} \end{aligned}$$

Therefore, when $n_1$ and $n_2$ are large, $D$ is approximately Normal with mean $\mu_D = p_1 - p_2$ and standard deviation

$$\sigma_D = \sqrt{\frac{p_1(1-p_1)}{n_1} + \frac{p_2(1-p_2)}{n_2}}$$

## Large-sample confidence interval for a difference in proportions

To obtain a confidence interval for $p_1 - p_2$, we once again replace the unknown parameters in the standard deviation by estimates to obtain an estimated standard deviation, or standard error. Here is the confidence interval we want.

### LARGE-SAMPLE CONFIDENCE INTERVAL FOR COMPARING TWO PROPORTIONS

Choose an SRS of size $n_1$ from a large population having proportion $p_1$ of successes and an independent SRS of size $n_2$ from another population having proportion $p_2$ of successes. The estimate of the difference in the population proportions is

$$D = \hat{p}_1 - \hat{p}_2$$

The **standard error of $D$** is

$$SE_D = \sqrt{\frac{\hat{p}_1(1-\hat{p}_1)}{n_1} + \frac{\hat{p}_2(1-\hat{p}_2)}{n_2}}$$

and the **margin of error** for confidence level $C$ is

$$m = z^* SE_D$$

where $z^*$ is the value for the standard Normal density curve with area $C$ between $-z^*$ and $z^*$. An **approximate level $C$ confidence interval** for $p_1 - p_2$ is

$$D \pm m$$

Use this method for 90%, 95%, or 99% confidence when the number of successes and the number of failures in each sample are at least 10.

In Example 8.1 (page 489) we estimated the proportion of college students who engage in frequent binge drinking. Are there characteristics of these students that relate to this behavior? For example, how similar is this behavior in men and women? This kind of follow-up question is natural in many studies like this one. We will first use a confidence interval to examine it. In the binge-drinking study, data were also summarized by gender:

| **Population** | $n$ | $X$ | $\hat{p} = X/n$ |
|---|---|---|---|
| 1 (men) | 5,348 | 1,392 | 0.260 |
| 2 (women) | 8,471 | 1,748 | 0.206 |
| Total | 13,819 | 3,140 | 0.227 |

In this table the $\hat{p}$ column gives the sample proportions of frequent binge drinkers. The last line gives the totals that we studied in Example 8.1.

EXAMPLE

**8.9 Gender and the proportion of frequent binge drinkers.** Let's find a 95% confidence interval for the difference between the proportions of men and of women who are frequent binge drinkers. Output from Minitab and CrunchIt! is given in Figure 8.4. To perform the computations using our formulas, we first find the difference in the proportions:

$$\begin{aligned} D &= \hat{p}_1 - \hat{p}_2 \\ &= 0.260 - 0.206 \\ &= 0.054 \end{aligned}$$

Then we calculate the standard error of $D$:

$$\begin{aligned} SE_D &= \sqrt{\frac{\hat{p}_1(1-\hat{p}_1)}{n_1} + \frac{\hat{p}_2(1-\hat{p}_2)}{n_2}} \\ &= \sqrt{\frac{(0.260)(0.740)}{5348} + \frac{(0.206)(0.794)}{8471}} \\ &= 0.00744 \end{aligned}$$

For 95% confidence, we have $z^* = 1.96$, so the margin of error is

$$\begin{aligned} m &= z^* SE_D = (1.96)(0.00744) \\ &= 0.015 \end{aligned}$$

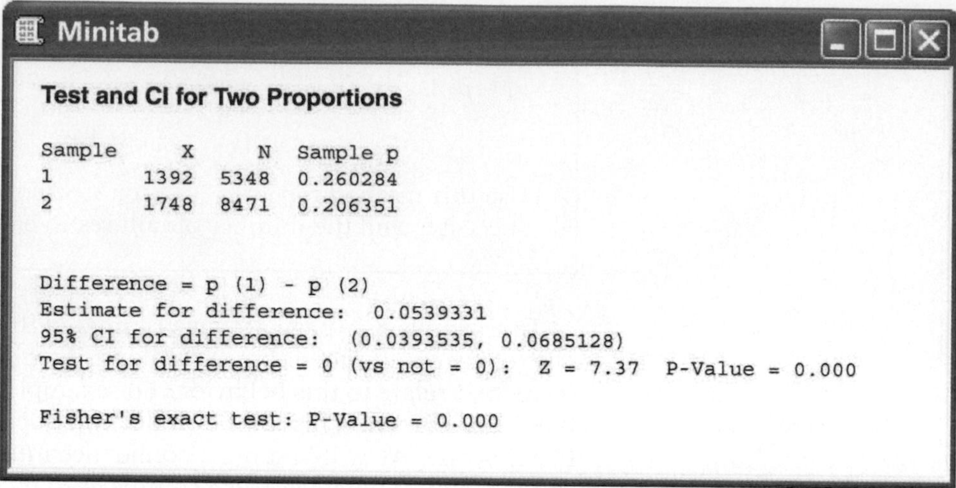

Minitab

**Test and CI for Two Proportions**

```
Sample      X     N  Sample p
1        1392  5348  0.260284
2        1748  8471  0.206351


Difference = p (1) - p (2)
Estimate for difference:  0.0539331
95% CI for difference:  (0.0393535, 0.0685128)
Test for difference = 0 (vs not = 0):  Z = 7.37  P-Value = 0.000

Fisher's exact test: P-Value = 0.000
```

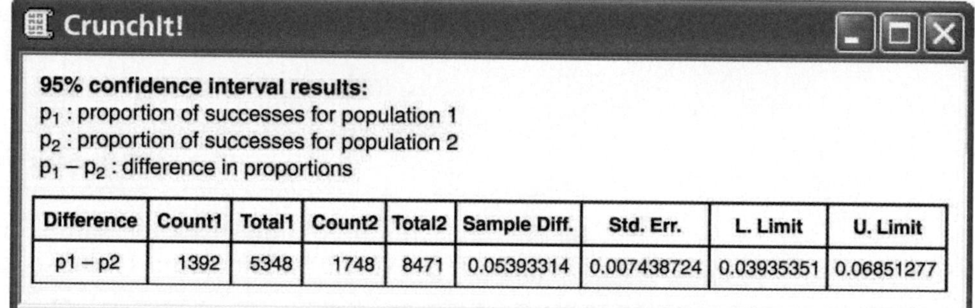

CrunchIt!

**95% confidence interval results:**
$p_1$ : proportion of successes for population 1
$p_2$ : proportion of successes for population 2
$p_1 - p_2$ : difference in proportions

| Difference | Count1 | Total1 | Count2 | Total2 | Sample Diff. | Std. Err. | L. Limit | U. Limit |
|---|---|---|---|---|---|---|---|---|
| p1 − p2 | 1392 | 5348 | 1748 | 8471 | 0.05393314 | 0.007438724 | 0.03935351 | 0.06851277 |

**FIGURE 8.4** Minitab and CrunchIt! output for Example 8.9.

The 95% confidence interval is

$$\begin{aligned} D \pm m &= 0.054 \pm 0.015 \\ &= (0.039, 0.069) \end{aligned}$$

With 95% confidence we can say that the difference in the proportions is between 0.039 and 0.069. Alternatively, we can report that the men are 5.4% more likely to be frequent binge drinkers, with a 95% margin of error of 1.5%.

In this example men and women were not sampled separately. The sample sizes are in fact random and reflect the gender distributions of the colleges that were randomly chosen. Two-sample significance tests and confidence intervals are still approximately correct in this situation. The authors of the report note that women are somewhat overrepresented partly because 6 of the 140 colleges in the study were all-women institutions.

In the example above we chose men to be the first population. Had we chosen women to be the first population, the estimate of the difference would be negative (−0.054). Because it is easier to discuss positive numbers, we generally choose the first population to be the one with the higher proportion.

## USE YOUR KNOWLEDGE

**8.35 Gender and commercial preference.** A study was designed to compare two energy drink commercials. Each participant was shown the commercials in random order and asked to select the better one. Commercial A was selected by 45 out of 100 women and 80 out of 140 men. Give an estimate of the difference in gender proportions that favored Commercial A. Also construct a large-sample 95% confidence interval for this difference.

**8.36 Gender and commercial preference, revisited.** Refer to Exercise 8.35. Construct a 95% confidence interval for the difference in proportions that favor Commercial B. Explain how you could have obtained these results from the calculations you did in Exercise 8.35.

## BEYOND THE BASICS

### Plus Four Confidence Interval for a Difference in Proportions

Just as in the case of estimating a single proportion, a small modification of the sample proportions can greatly improve the accuracy of confidence intervals.[21] As before, we add 2 successes and 2 failures to the actual data, but now we divide them equally between the two samples. That is, we *add 1 success and 1 failure to each sample*. We will again call the estimates produced by adding hypothetical observations plus four estimates. The plus four estimates of the two population proportions are

$$\tilde{p}_1 = \frac{X_1 + 1}{n_1 + 2} \quad \text{and} \quad \tilde{p}_2 = \frac{X_2 + 1}{n_2 + 2}$$

The estimated difference between the populations is

$$\tilde{D} = \tilde{p}_1 - \tilde{p}_2$$

and the standard deviation of $\tilde{D}$ is approximately

$$\sigma_{\tilde{D}} = \sqrt{\frac{p_1(1 - p_1)}{n_1 + 2} + \frac{p_2(1 - p_2)}{n_2 + 2}}$$

This is similar to the formula for $\sigma_D$, adjusted for the sizes of the modified samples.

To obtain a confidence interval for $p_1 - p_2$, we once again replace the unknown parameters in the standard deviation by estimates to obtain an estimated standard deviation, or standard error. Here is the confidence interval we want.

#### PLUS FOUR CONFIDENCE INTERVAL FOR COMPARING TWO PROPORTIONS

Choose an SRS of size $n_1$ from a large population having proportion $p_1$ of successes and an independent SRS of size $n_2$ from another population having proportion $p_2$ of successes. The **plus four estimate of the difference in proportions** is

$$\tilde{D} = \tilde{p}_1 - \tilde{p}_2$$

where

$$\tilde{p}_1 = \frac{X_1 + 1}{n_1 + 2} \qquad \tilde{p}_2 = \frac{X_2 + 1}{n_2 + 2}$$

The **standard error of $\tilde{D}$** is

$$SE_{\tilde{D}} = \sqrt{\frac{\tilde{p}_1(1-\tilde{p}_1)}{n_1+2} + \frac{\tilde{p}_2(1-\tilde{p}_2)}{n_2+2}}$$

and the **margin of error** for confidence level $C$ is

$$m = z^* SE_{\tilde{D}}$$

where $z^*$ is the value for the standard Normal density curve with area $C$ between $-z^*$ and $z^*$. An **approximate level $C$ confidence interval** for $p_1 - p_2$ is

$$\tilde{D} \pm m$$

Use this method for 90%, 95%, or 99% confidence when both sample sizes are at least 5.

EXAMPLE

**8.10 Gender and sexual maturity.** In studies that look for a difference between genders, a major concern is whether or not apparent differences are due to other variables that are associated with gender. Because boys mature more slowly than girls, a study of adolescents that compares boys and girls of the same age may confuse a gender effect with an effect of sexual maturity. The "Tanner score" is a commonly used measure of sexual maturity.[22] Subjects are asked to determine their score by placing a mark next to a rough drawing of an individual at their level of sexual maturity. There are five different drawings, so the score is an integer between 1 and 5.

A pilot study included 12 girls and 12 boys from a population that will be used for a large experiment. Four of the boys and three of the girls had Tanner scores of 4 or 5, a high level of sexual maturity. Let's find a 95% confidence interval for the difference between the proportions of boys and girls who have high (4 or 5) Tanner scores in this population. The numbers of successes and failures in both groups are not all at least 10, so the large-sample approach is not recommended. On the other hand, the sample sizes are both at least 5, so the plus four method is appropriate.

The plus four estimate of the population proportion for boys is

$$\tilde{p}_1 = \frac{X_1 + 1}{n_1 + 2} = \frac{4+1}{12+2} = 0.3571$$

For girls, the estimate is

$$\tilde{p}_2 = \frac{X_2 + 1}{n_2 + 2} = \frac{3 + 1}{12 + 2} = 0.2857$$

Therefore, the estimate of the difference is

$$\tilde{D} = \tilde{p}_1 - \tilde{p}_2 = 0.3571 - 0.2857 = 0.071$$

The standard error of $\tilde{D}$ is

$$\begin{aligned} SE_{\tilde{D}} &= \sqrt{\frac{\tilde{p}_1(1 - \tilde{p}_1)}{n_1 + 2} + \frac{\tilde{p}_2(1 - \tilde{p}_2)}{n_2 + 2}} \\ &= \sqrt{\frac{(0.3571)(1 - 0.3571)}{12 + 2} + \frac{(0.2857)(1 - 0.2857)}{12 + 2}} \\ &= 0.1760 \end{aligned}$$

For 95% confidence, $z^* = 1.96$ and the margin of error is

$$m = z^* SE_{\tilde{D}} = (1.96)(0.1760) = 0.345$$

The confidence interval is

$$\tilde{D} \pm m = 0.071 \pm 0.345 = (-0.274, 0.416)$$

With 95% confidence we can say that the difference in the proportions is between −0.274 and 0.416. Alternatively, we can report that the difference in the proportions of boys and girls with high Tanner scores in this population is 7.1% with a 95% margin of error of 34.5%.

The very large margin of error in this example indicates that either boys or girls could be more sexually mature in this population and that the difference could be quite large. *Although the interval includes the possibility that there is no difference, corresponding to $p_1 = p_2$ or $p_1 - p_2 = 0$, we must be very cautious about concluding that there is* ***no*** *difference in the proportions.* With small sample sizes such as these, the data do not provide us with a lot of information for our inference. This fact is expressed quantitatively through the very large margin of error.

## Significance test for a difference in proportions

Although we prefer to compare two proportions by giving a confidence interval for the difference between the two population proportions, it is sometimes useful to test the null hypothesis that the two population proportions are the same.

We standardize $D = \hat{p}_1 - \hat{p}_2$ by subtracting its mean $p_1 - p_2$ and then dividing by its standard deviation

$$\sigma_D = \sqrt{\frac{p_1(1 - p_1)}{n_1} + \frac{p_2(1 - p_2)}{n_2}}$$

If $n_1$ and $n_2$ are large, the standardized difference is approximately $N(0, 1)$. For the large-sample confidence interval we used sample estimates in place of the unknown population values in the expression for $\sigma_D$. Although this approach would lead to a valid significance test, we instead adopt the more common practice of replacing the unknown $\sigma_D$ with an estimate that takes into account our null hypothesis $H_0: p_1 = p_2$. If these two proportions are equal, then we can view all of the data as coming from a single population. Let $p$ denote the common value of $p_1$ and $p_2$; then the standard deviation of $D = \hat{p}_1 - \hat{p}_2$ is

$$\sigma_D = \sqrt{\frac{p(1-p)}{n_1} + \frac{p(1-p)}{n_2}}$$

$$= \sqrt{p(1-p)\left(\frac{1}{n_1} + \frac{1}{n_2}\right)}$$

We estimate the common value of $p$ by the overall proportion of successes in the two samples:

$$\hat{p} = \frac{\text{number of successes in both samples}}{\text{number of observations in both samples}} = \frac{X_1 + X_2}{n_1 + n_2}$$

pooled estimate of $p$

This estimate of $p$ is called the **pooled estimate** because it combines, or pools, the information from both samples.

To estimate $\sigma_D$ under the null hypothesis, we substitute $\hat{p}$ for $p$ in the expression for $\sigma_D$. The result is a standard error for $D$ that assumes $H_0: p_1 = p_2$:

$$\text{SE}_{Dp} = \sqrt{\hat{p}(1-\hat{p})\left(\frac{1}{n_1} + \frac{1}{n_2}\right)}$$

The subscript on $\text{SE}_{Dp}$ reminds us that we pooled data from the two samples to construct the estimate.

## SIGNIFICANCE TEST FOR COMPARING TWO PROPORTIONS

To test the hypothesis

$$H_0: p_1 = p_2$$

compute the **$z$ statistic**

$$z = \frac{\hat{p}_1 - \hat{p}_2}{\text{SE}_{Dp}}$$

where the **pooled standard error** is

$$\text{SE}_{Dp} = \sqrt{\hat{p}(1-\hat{p})\left(\frac{1}{n_1} + \frac{1}{n_2}\right)}$$

and where

$$\hat{p} = \frac{X_1 + X_2}{n_1 + n_2}$$

In terms of a standard Normal random variable $Z$, the $P$-value for a test of $H_0$ against

$H_a: p_1 > p_2$ is $P(Z \geq z)$

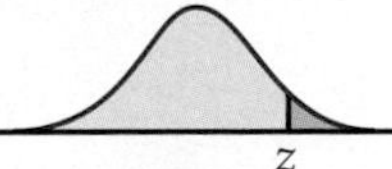


$H_a: p_1 < p_2$ is $P(Z \leq z)$

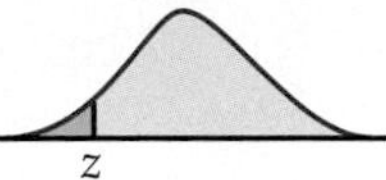


$H_a: p_1 \neq p_2$ is $2P(Z \geq |z|)$

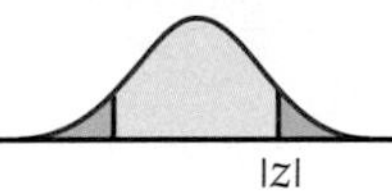


This $z$ test is based on the Normal approximation to the binomial distribution. As a general rule, we will use it when the number of successes and the number of failures in each of the samples are at least 5.

EXAMPLE

**8.11 Gender and the proportion of frequent binge drinkers: the $z$ test.** Are men and women college students equally likely to be frequent binge drinkers? We examine the survey data in Example 8.9 (page 507) to answer this question. Here is the data summary:

| Population | $n$ | $X$ | $\hat{p} = X/n$ |
|---|---|---|---|
| 1 (men) | 5,348 | 1,392 | 0.260 |
| 2 (women) | 8,471 | 1,748 | 0.206 |
| Total | 13,819 | 3,140 | 0.227 |

The sample proportions are certainly quite different, but we will perform a significance test to see if the difference is large enough to lead us to believe that the population proportions are not equal. Formally, we test the hypotheses

$$H_0: p_1 = p_2$$
$$H_a: p_1 \neq p_2$$

The pooled estimate of the common value of $p$ is

$$\hat{p} = \frac{1392 + 1748}{5348 + 8471} = \frac{3140}{13,819} = 0.227$$

Note that this is the estimate on the bottom line of the data summary above.

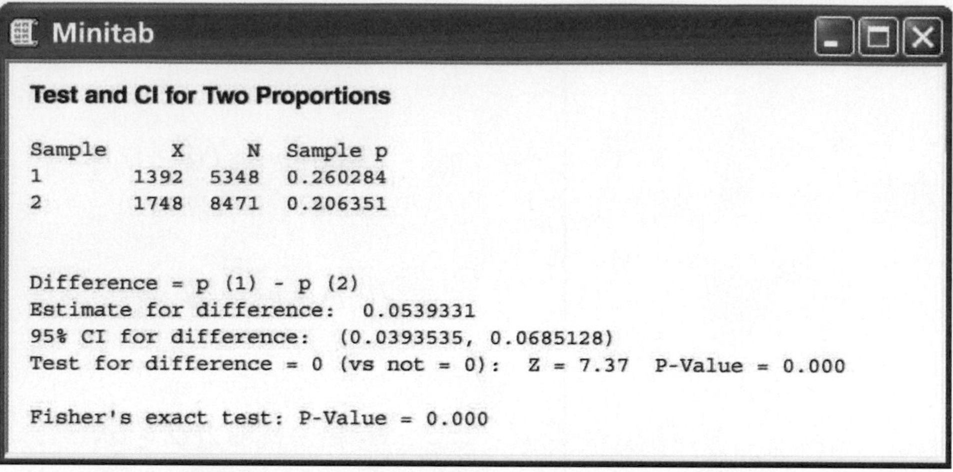

Minitab

**Test and CI for Two Proportions**

```
Sample      X     N  Sample p
1        1392  5348  0.260284
2        1748  8471  0.206351


Difference = p (1) - p (2)
Estimate for difference:  0.0539331
95% CI for difference:  (0.0393535, 0.0685128)
Test for difference = 0 (vs not = 0):  Z = 7.37  P-Value = 0.000

Fisher's exact test: P-Value = 0.000
```

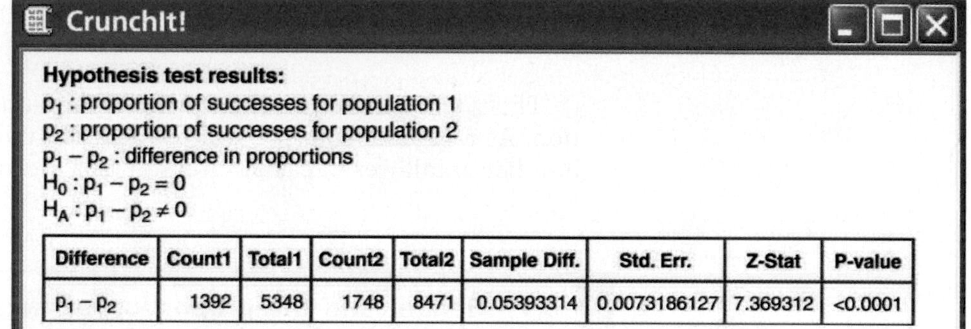

CrunchIt!

**Hypothesis test results:**
$p_1$ : proportion of successes for population 1
$p_2$ : proportion of successes for population 2
$p_1 - p_2$ : difference in proportions
$H_0 : p_1 - p_2 = 0$
$H_A : p_1 - p_2 \neq 0$

| Difference | Count1 | Total1 | Count2 | Total2 | Sample Diff. | Std. Err. | Z-Stat | P-value |
|---|---|---|---|---|---|---|---|---|
| $p_1 - p_2$ | 1392 | 5348 | 1748 | 8471 | 0.05393314 | 0.0073186127 | 7.369312 | <0.0001 |

**FIGURE 8.5** Minitab and CrunchIt! output for Example 8.11. With Minitab, the option "use pooled estimate of p for test" was selected.

The test statistic is calculated as follows:

$$\text{SE}_{Dp} = \sqrt{(0.227)(0.773)\left(\frac{1}{5348} + \frac{1}{8471}\right)} = 0.007316$$

$$z = \frac{\hat{p}_1 - \hat{p}_2}{\text{SE}_{Dp}} = \frac{0.260 - 0.206}{0.007316} = 7.37$$

The *P*-value is $2P(Z \geq 7.37)$. The largest value of $z$ in Table A is 3.49, so from this table we can conclude $P < 2 \times 0.0002 = 0.0004$. Most software reports this result as simply 0 or a very small number. Output from Minitab and CrunchIt! is given in Figure 8.5. Minitab reports the *P*-value as 0.000. This means that the calculated value is less than 0.0005; this is certainly a very small number. CrunchIt! gives $< 0.0001$. The exact value is not particularly important; it is clear that we should reject the null hypothesis. For most situations 0.001 (1 chance in 1000) is sufficiently small. We report: among college students in the study, 26.0% of the men and 20.6% of the women were frequent binge drinkers; the difference is statistically significant ($z = 7.37$, $P < 0.001$).

We could have argued that we expect the proportion to be higher for men than for women in this example. This would justify using the one-sided alternative $H_a: p_1 > p_2$. The *P*-value would be half of the value obtained for the two-sided test. Because the $z$ statistic is so large, this distinction is of no practical importance.

**USE YOUR KNOWLEDGE**

**8.37 Gender and commercial preference: the $z$ test.** Refer to Exercise 8.35 (page 509). Test that the proportions of women and men that liked Commercial A are the same versus the two-sided alternative at the 5% level.

**8.38 Changing the alternative hypothesis.** Refer to the previous exercise. Does your conclusion change if you test whether the proportion of men that favor Commercial A is larger than the proportion of females? Explain.

## BEYOND THE BASICS

### Relative Risk

We summarized the comparison of the frequent binge-drinking proportions for men and for women by reporting a confidence interval for the difference in Example 8.9. Another way to summarize the comparison is to view each sample proportion as the **risk** that a college student of that gender is a frequent binge drinker. We then compare these two risks with the ratio of the two proportions, which is called the **relative risk** (RR) in many applications. Note that a relative risk of 1 means that the two proportions, $\hat{p}_1$ and $\hat{p}_2$, are equal. The procedure for calculating confidence intervals for relative risk is based on the same kind of principles that we have studied, but the details are somewhat more complicated. Fortunately, we can leave the details to software and concentrate on interpretation and communication of the results.

risk

relative risk

EXAMPLE

**8.12 Gender and the proportion of frequent binge drinkers: the relative risk.** On page 507 we summarized the data on the proportions of men and women who are frequent binge drinkers with the following table:

| **Population** | $n$ | $X$ | $\hat{p} = X/n$ |
|---|---|---|---|
| 1 (men) | 5,348 | 1,392 | 0.260 |
| 2 (women) | 8,471 | 1,748 | 0.206 |
| Total | 13,819 | 3,140 | 0.227 |

The relative risk is

$$\text{RR} = \frac{\hat{p}_1}{\hat{p}_2} = \frac{0.260}{0.206} = 1.26$$

Software gives the 95% confidence interval as 1.19 to 1.34. Men are 1.26 times as likely as women to be frequent binge drinkers; the 95% confidence interval is (1.19, 1.34).

In this example the confidence interval appears to be symmetric about the estimate. If we reported the results with more accuracy (RR = 1.261, 95% confidence interval = 1.186 to 1.341), we would see that the interval is *not* symmetric, and this is true in general.

## SECTION 8.2 Summary

The **large-sample estimate of the difference in two population proportions** is

$$D = \hat{p}_1 - \hat{p}_2$$

where $\hat{p_1}$ and $\hat{p_2}$ are the sample proportions

$$\hat{p}_1 = \frac{X_1}{n_1} \quad \text{and} \quad \hat{p}_2 = \frac{X_2}{n_2}$$

The **standard error of the difference $D$** is

$$\text{SE}_D = \sqrt{\frac{\hat{p}_1(1-\hat{p}_1)}{n_1} + \frac{\hat{p}_2(1-\hat{p}_2)}{n_2}}$$

The **margin of error for confidence level $C$** is

$$m = z^*\text{SE}_D$$

where $z^*$ is the value for the standard Normal density curve with area $C$ between $-z^*$ and $z^*$. The **large-sample level $C$ confidence interval** is

$$D \pm m$$

We recommend using this interval for 90%, 95%, or 99% confidence when the number of successes and the number of failures in both samples are all at least 10. When sample sizes are smaller, alternative procedures such as the **plus four estimate of the difference in two population proportions** are recommended.

Significance tests of $H_0: p_1 = p_2$ use the **$z$ statistic**

$$z = \frac{\hat{p}_1 - \hat{p}_2}{\text{SE}_{Dp}}$$

with $P$-values from the $N(0, 1)$ distribution. In this statistic,

$$\text{SE}_{Dp} = \sqrt{\hat{p}(1-\hat{p})\left(\frac{1}{n_1} + \frac{1}{n_2}\right)}$$

and $\hat{p}$ is the **pooled estimate** of the common value of $p_1$ and $p_2$:

$$\hat{p} = \frac{X_1 + X_2}{n_1 + n_2}$$

**Relative risk** is the ratio of two sample proportions:

$$RR = \frac{\hat{p}_1}{\hat{p}_2}$$

Confidence intervals for relative risk are often used to summarize the comparison of two proportions.

## SECTION 8.2 Exercises

*For Exercises 8.35 and 8.36, see page 509; and for Exercises 8.37 and 8.38, see page 515.*

**8.39 Can we use the large-sample confidence interval?** In each of the following circumstances state whether you would use the large-sample confidence interval.

(a) $n_1 = 30$, $n_2 = 30$, $X_1 = 10$, and $X_2 = 15$.

(b) $n_1 = 15$, $n_2 = 10$, $X_1 = 10$, and $X_2 = 5$.

(c) $n_1 = 25$, $n_2 = 20$, $X_1 = 11$, and $X_2 = 8$.

(d) $n_1 = 40$, $n_2 = 40$, $X_1 = 20$, and $X_2 = 12$.

(e) $n_1 = 50$, $n_2 = 50$, $X_1 = 40$, and $X_2 = 45$.

**8.40 More on whether to use the large-sample confidence interval.** In each of the following circumstances state whether you would use the large-sample confidence interval.

(a) $n_1 = 25$, $n_2 = 25$, $X_1 = 12$, and $X_2 = 8$.

(b) $n_1 = 25$, $n_2 = 25$, $X_1 = 17$, and $X_2 = 12$.

(c) $n_1 = 60$, $n_2 = 30$, $X_1 = 30$, and $X_2 = 15$.

(d) $n_1 = 60$, $n_2 = 55$, $X_1 = 45$, and $X_2 = 37$.

(e) $n_1 = 200$, $n_2 = 100$, $X_1 = 128$, and $X_2 = 94$.

**8.41 Comparing cell phone ownership in 2003 and 2004.** In Exercise 8.4 (page 497), you were asked to compare the 2004 proportion of cell phone owners (89%) with the 2003 estimate (83%). It would be more appropriate to compare these two proportions using the methods of this section. Given that the sample size of each SRS is 1200 students, compare these two years with a significance test, and give an estimate of the difference in proportions of undergraduate cell phone owners with a 95% margin of error. Write a short summary of your results.

**8.42 $\hat{p}_1 - \hat{p}_2$ and the Normal distribution.** Suppose there are two binomial populations. For the first, the true proportion of successes is 0.3; for the second, it is 0.4. Consider taking independent samples from these populations, 50 from the first and 60 from the second.

(a) Find the mean and the standard deviation of the distribution of $\hat{p}_1 - \hat{p}_2$.

(b) This distribution is approximately Normal. Sketch this Normal distribution and mark the location of the mean.

(c) Find a value $d$ for which the probability is 0.95 that the difference in sample proportions is within $\pm d$. Mark these values on your sketch.

**8.43** CHALLENGE **Gender and gambling behaviors among student-athletes.** Gambling behaviors of Division I intercollegiate student-athletes were analyzed in Exercises 8.11 and 8.12 (page 502). Use the methods of this section to compare the males and females with a significance test, and give an estimate of the difference in proportions of student-athletes who participate in any gambling activity with a 95% margin of error. In Exercise 8.11 it is noted that we do not have any information available to assess nonresponse. Consider the possibility that the response rates differ by gender and by whether or not the person participates in any gambling activity. Write a short summary of how these differences might affect inference.

**8.44 Pet ownership and gender.** In the Health ABC Study, 595 subjects owned a pet and 1939 subjects did not.[23] Among the pet owners, there were 285 women; 1024 of the non–pet owners were women. Find the proportion of pet owners who were women. Do the same for the non–pet owners. Give a 95% confidence interval for the difference in the two proportions. (Be sure to let Population 1 correspond to the group with the higher proportion so that the difference will be positive.)

**8.45 Pet ownership and marital status.** Refer to the previous exercise. The 595 pet owners and 1939 non–pet owners were also classified according to whether or not they were married. For the pet owners, 53.3% were married, while for the non–pet owners, 57.7% were married. Find a 95% confidence interval for the difference. Write a short summary of your work.

**8.46 A comparison of the proportion of frequent binge drinkers.** In the published report on binge drinking that we used for Example 8.1, survey results from both 1993 and 1999 are presented. Using the table below, test whether the proportions of frequent binge drinkers are different at the 5% level. Also construct a 95% confidence interval for the difference. Write a short summary of your results.

| Year | $n$ | $X$ |
|---|---|---|
| 1993 | 14,995 | 2,973 |
| 1999 | 13,819 | 3,140 |

**8.47 A comparison of the proportion of frequent binge drinkers, revisited.** Refer to Exercise 8.46. Redo the exercise in terms of the proportion of nonfrequent binge drinkers in each classification. Explain how you could have obtained these results from the calculations you did in Exercise 8.46.

**8.48 Effects of reducing air pollution.** A study that evaluated the effects of a reduction in exposure to traffic-related air pollutants compared respiratory symptoms of 283 residents of an area with congested streets with 165 residents in a similar area where the congestion was removed because a bypass was constructed. The symptoms of the residents of both areas were evaluated at baseline and again a year after the bypass was completed.[24] For the residents of the congested streets, 17 reported that their symptoms of wheezing improved between baseline and one year later, while 35 of the residents of the bypass streets reported improvement.

(a) Find the two sample proportions.

(b) Report the difference in the proportions and the standard error of the difference.

(c) What are the appropriate null and alternative hypotheses for examining the question of interest? Be sure to explain your choice of the alternative hypothesis.

(d) Find the test statistic. Construct a sketch of the distribution of the test statistic under the assumption that the null hypothesis is true. Find the *P*-value and use your sketch to explain its meaning.

(e) Is no evidence of an effect the same as evidence that there is no effect? Use a 95% confidence interval to answer this question. Summarize your ideas in a way that could be understood by someone who has very little experience with statistics.

(f) The study was done in the United Kingdom. To what extent do you think that the results can be generalized to other circumstances?

**8.49 Downloading music from the Internet.** A 2005 survey of Internet users reported that 22% downloaded music onto their computers. The filing of lawsuits by the recording industry may be a reason why this percent has decreased from the estimate of 29% from a survey taken two years before.[25] Assume that the sample sizes are both 1421. Using a significance test, evaluate whether or not there has been a change in the percent of Internet users who download music. Provide all details for the test and summarize your conclusion. Also report a 95% confidence interval for the difference in proportions and explain what information is provided in the interval that is not in the significance test results.

**8.50 More on downloading music from the Internet.** Refer to the previous exercise. Suppose we are not exactly sure about the sizes of the samples. Redo the calculations for the significance test and the confidence interval under the following assumptions: (*i*) both sample sizes are 1000, (*ii*) both sample sizes are 1600, (*iii*) the first sample size is 1000 and the second is 1600. Summarize the effects of the sample sizes on the results.

**8.51 Who gets stock options?** Different kinds of companies compensate their key employees in different ways. Established companies may pay higher salaries, while new companies may offer stock options that will be valuable if the company succeeds. Do high-tech companies tend to offer stock options more often than other companies? One study looked at a random sample of 200 companies. Of these, 91 were listed in the *Directory of Public High Technology Corporations* and 109 were not listed. Treat these two groups as SRSs of high-tech and non-high-tech companies. Seventy-three of the high-tech companies and 75 of the non-high-tech companies offered incentive stock options to key employees.[26]

(a) Give a 95% confidence interval for the difference in the proportions of the two types of companies that offer stock options.

(b) Compare the two groups of companies with a significance test.

(c) Summarize your analysis and conclusions.

**8.52 Cheating during a test: 2002 versus 2004.** In Exercise 8.16, you examined the proportion of high school students who cheated on tests at least twice during the past year. Included in that study were the results for both 2002 and 2004. A reported 9054 out of 24,142 students said they cheated at least twice in 2004. A reported 5794 out of 12,121 students said they cheated at least twice in 2002. Give an estimate of the difference between these two proportions with a 90% confidence interval.

**8.53 Gender bias in textbooks.** To what extent do syntax textbooks, which analyze the structure of sentences, illustrate gender bias? A study of this question sampled sentences from 10 texts.[27] One part of the study examined the use of the words "girl," "boy," "man," and "woman." We will call the first two words *juvenile* and the last two *adult*. Is the proportion of female references that are juvenile (girl) equal to the proportion of male references that are juvenile (boy)? Here are data from one of the texts:

| Gender | $n$ | $X$(juvenile) |
|---|---|---|
| Female | 60 | 48 |
| Male | 132 | 52 |

(a) Find the proportion of juvenile references for females and its standard error. Do the same for the males.

(b) Give a 90% confidence interval for the difference and briefly summarize what the data show.

(c) Use a test of significance to examine whether the two proportions are equal.

**8.54** CHALLENGE **Bicycle accidents, alcohol, and gender.** In Exercise 8.28 (page 504) we examined the percent of fatally injured bicyclists tested for alcohol who tested positive. Here we examine the same data with respect to gender.

| Gender | $n$ | $X$(tested positive) |
|---|---|---|
| Female | 191 | 27 |
| Male | 1520 | 515 |

(a) Summarize the data by giving the estimates of the two population proportions and a 95% confidence interval for their difference.

(b) The standard error $SE_D$ contains a contribution from each sample, $\hat{p}_1(1-\hat{p}_1)/n_1$ and $\hat{p}_2(1-\hat{p}_2)/n_2$. Which of these contributes the larger amount to the standard error of the difference? Explain why.

(c) Use a test of significance to examine whether the two proportions are equal.

**8.55 Pet ownership and gender: the significance test.** In Exercise 8.44 (page 517) we compared the proportion of pet owners who were women with the proportion of non–pet owners who were women in the Health ABC Study. Use a significance test to make the comparison and summarize the results of your analysis.

**8.56 Pet ownership and marital status: the significance test.** In Exercise 8.45 (page 517) we compared the proportion of pet owners who were married with the proportion of non–pet owners who were married in the Health ABC Study. Use a significance test to make the comparison and summarize the results of your analysis.

## CHAPTER 8 Exercises

**8.57 What's wrong?** For each of the following, explain what is wrong and why.

(a) A 90% confidence interval for the difference in two proportions includes errors due to nonresponse.

(b) A $z$ statistic is used to test the null hypothesis that $H_0: \hat{p}_1 = \hat{p}_2$.

(c) If two sample proportions are equal, then the sample counts must be equal.

**8.58 Using a handheld phone while driving.** Refer to Exercise 8.14 (page 503). This same poll found that 58% of the respondents talked on a handheld phone while driving in the last year. Construct a 90% confidence interval for the proportion of U.S. drivers who talked on a handheld phone while driving in the last year.

**8.59 Gender and using a handheld phone while driving.** Refer to the previous exercise. In this same report, this percent was broken down into 59% for men and 56% for women. Assuming that, among the 1048 respondents, there were an equal number of men and women, construct a 95% confidence interval for the difference in these proportions.

**8.60** CHALLENGE **Even more on downloading music from the Internet.** The following quotation is from a recent survey of Internet users. The sample size for the survey was 1371. Since 18% of those surveyed said they download music, the sample size for this subsample is 247.

Among current music downloaders, 38% say they are downloading less because of the RIAA suits.... About a third of current music downloaders say they use peer-to-peer networks.... 24% of them say they swap files using email and instant messaging; 20% download files from music-related Web sites like those run by music magazines or musician homepages. And while online music services like iTunes are far from trumping the popularity of file-sharing networks, 17% of current music downloaders say they are using these paid services. Overall, 7% of Internet users say they have bought music at these new services at one time or another, including 3% who currently use paid services.[28]

(a) For each percent quoted, give the margin of error. You should express these in percents, as given in the quote.

(b) Rewrite the paragraph more concisely and include the margins of error.

(c) Pick either side A or side B below and give arguments in favor of the view that you select.

*(A) The margins of error should be included because they are necessary for the reader to properly interpret the results.*

*(B) The margins of error interfere with the flow of the important ideas. It would be better to just report one margin of error and say that all of the others are no greater than this number.*

If you choose view B, be sure to give the value of the margin of error that you report.

**8.61** CHALLENGE **Proportion of male heavy lottery players.** A study of state lotteries included a random digit dialing (RDD) survey conducted by the National Opinion Research Center (NORC). The survey asked 2406 adults about their lottery spending.[29] A total of 248 individuals were classified as "heavy" players. Of these, 152 were male. The study notes that 48.5% of U.S. adults are male. For this analysis, assume that the 248 heavy lottery players are a random sample of all heavy lottery players and that the margin of error for the 48.5% estimate of the percent of males in the U.S. adult population is so small that it can be neglected. Use a significance test to compare the proportion of males among heavy lottery players with the proportion of males in the U.S. adult population. Construct a 95% confidence interval for the proportion. Write a summary of what you have found. Be sure to comment on the possibility that some people may be reluctant to provide information about their lottery spending and how this might affect the results.

**8.62** CHALLENGE **Cell phone ownership: 2000 versus 2004.** Refer to Exercise 8.41 (page 517). The estimated proportion of undergraduates owning a phone in 2000 was 43%. We want to test whether the proportion of undergraduate cell phone owners has more than doubled in the last 4 years.

(a) Compute the quantity $\hat{p}_1 - 2\hat{p}_2$ where $\hat{p}_1$ is the 2004 estimate and $\hat{p}_2$ is the 2000 estimate.

(b) Using the rules for variances, compute the standard error of this estimate.

(c) Compute the $z$ statistic and $P$-value. What is your conclusion at the 5% level?

**8.63** CHALLENGE **More on the effects of reducing air pollution.** In Exercise 8.48 the effects of a reduction in air pollution on wheezing was examined by comparing the one-year change in symptoms in a group of residents who lived on congested streets with a group who lived in an area that had been congested but from which the congestion was removed when a bypass was built. The effect of the reduction in air pollution was assessed by comparing the proportions of residents in the two groups who reported that their wheezing symptoms improved. Here are some additional data from the same study:

| | Bypass | | Congested | |
|---|---|---|---|---|
| Symptom | $n$ | Improved | $n$ | Improved |
| Number of wheezing attacks | 282 | 45 | 163 | 21 |
| Wheezing disturbs sleep | 282 | 45 | 164 | 12 |
| Wheezing limits speech | 282 | 12 | 164 | 4 |
| Wheezing affects activities | 281 | 26 | 165 | 13 |
| Winter cough | 261 | 15 | 156 | 14 |
| Winter phlegm | 253 | 12 | 144 | 10 |
| Consulted doctor | 247 | 29 | 140 | 18 |

The table gives the number of subjects in each group and the number reporting improvement. So, for example, the proportion who reported improvement in the number of wheezing attacks was 21/163 in the congested group.

(a) The reported sample sizes vary from symptom to symptom. Give possible reasons for this and discuss the possible impact on the results.

(b) Calculate the difference in the proportions for each symptom. Make a table of symptoms ordered from highest to lowest based on these differences.

Include the estimates of the differences and the 95% confidence intervals in the table. Summarize your conclusions.

(c) Can you justify a one-sided alternative in this situation? Give reasons for your answer.

(d) Perform a significance test to compare the two groups for each of the symptoms. Summarize the results.

(e) Reanalyze the data using only the data from the bypass group. Give confidence intervals for the proportions that reported improved symptoms. Compare the conclusions that someone might make from these results with those you presented in part (b). Use your analyses of the data in this exercise to discuss the importance of a control group in studies such as this.

8.64 **"No Sweat" garment labels.** Following complaints about the working conditions in some apparel factories both in the United States and abroad, a joint government and industry commission recommended in 1998 that companies that monitor and enforce proper standards be allowed to display a "No Sweat" label on their products. Does the presence of these labels influence consumer behavior? A survey of U.S. residents aged 18 or older asked a series of questions about how likely they would be to purchase a garment under various conditions. For some conditions, it was stated that the garment had a "No Sweat" label; for others, there was no mention of such a label. On the basis of the responses, each person was classified as a "label user" or a "label nonuser."[30] There were 296 women surveyed. Of these, 63 were label users. On the other hand, 27 of 251 men were classified as users.

(a) Give a 95% confidence interval for the difference in the proportions.

(b) You would like to compare the women with the men. Set up appropriate hypotheses, and find the test statistic and the *P*-value. What do you conclude?

8.65 **Education of the customers.** To devise effective marketing strategies it is helpful to know the characteristics of your customers. A study compared demographic characteristics of people who use the Internet for travel arrangements and of people who do not.[31] Of 1132 Internet users, 643 had completed college. Among the 852 nonusers, 349 had completed college.

(a) Do users and nonusers differ significantly in the proportion of college graduates?

(b) Give a 95% confidence interval for the difference in the proportions.

8.66 **Income of the customers.** The study mentioned in the previous exercise also asked about income. Among Internet users, 493 reported income of less than \$50,000 and 378 reported income of \$50,000 or more. (Not everyone answered the income question.) The corresponding numbers for nonusers were 477 and 200. Perform a significance test to compare the incomes of users with nonusers and also give an estimate of the difference in proportions with a 95% margin of error.

8.67 **Nonresponse for the income question.** Refer to the previous two exercises. Give the total number of users and the total number of nonusers for the analysis of education. Do the same for the analysis of income. The difference is due to respondents who chose "Rather not say" for the income question. Give the proportions of "Rather not say" individuals for users and nonusers. Perform a significance test to compare these and give a 95% confidence interval for the difference. People are often reluctant to provide information about their income. Do you think that this amount of nonresponse for the income question is a serious limitation for this study?

8.68 **Improving the time to repair golf clubs.** The Ping Company makes custom-built golf clubs and competes in the \$4 billion golf equipment industry. To improve its business processes, Ping decided to seek ISO 9001 certification.[32] As part of this process, a study of the time it took to repair golf clubs that were sent to the company by mail determined that 16% of orders were sent back to the customers in 5 days or less. Ping examined the processing of repair orders and made changes. Following the changes, 90% of orders were completed within 5 days. Assume that each of the estimated percents is based on a random sample of 200 orders.

(a) How many orders were completed in 5 days or less before the changes? Give a 95% confidence interval for the proportion of orders completed in this time.

(b) Do the same for orders after the changes.

(c) Give a 95% confidence interval for the improvement. Express this both for a difference in proportions and for a difference in percents.

8.69 **Parental pressure to succeed in school.** A Pew Research Center Poll used telephone interviews to ask American adults if parents are pushing their kids too hard to succeed in school. Of those responding, 56% said parents are placing too little pressure on their children.[33] Assuming that this is an SRS of 1200 U.S. residents over the age of 18, give the 95% margin of error for this estimate.

**8.70 Brand loyalty and the Chicago Cubs.** According to literature on brand loyalty, consumers who are loyal to a brand are likely to consistently select the same product. This type of consistency could come from a positive childhood association. To examine brand loyalty among fans of the Chicago Cubs, 371 Cubs fans among patrons of a restaurant located in Wrigleyville were surveyed prior to a game at Wrigley Field, the Cubs' home field.[34] The respondents were classified as "die-hard fans" or "less loyal fans." Of the 134 die-hard fans, 90.3% reported that they had watched or listened to Cubs games when they were children. Among the 237 less loyal fans, 67.9% said that they had watched or listened as children.

(a) Find the numbers of die-hard Cubs fans who watched or listened to games when they were children. Do the same for the less loyal fans.

(b) Use a significance test to compare the die-hard fans with the less loyal fans with respect to their childhood experiences relative to the team.

(c) Express the results with a 95% confidence interval for the difference in proportions.

**8.71 Brand loyalty in action.** The study mentioned in the previous exercise found that two-thirds of the die-hard fans attended Cubs games at least once a month, but only 20% of the less loyal fans attended this often. Analyze these data using a significance test and a confidence interval. Write a short summary of your findings.

**8.72** CHALLENGE **More on gender bias in textbooks.** Refer to the study of gender bias and stereotyping described in Exercise 8.53 (page 519). Here are the counts of "girl," "woman," "boy," and "man" for all of the syntax texts studied. The one we analyzed in Exercise 8.53 was number 6.

| | Text number | | | | | | | | | |
|---|---|---|---|---|---|---|---|---|---|---|
| | 1 | 2 | 3 | 4 | 5 | 6 | 7 | 8 | 9 | 10 |
| Girl | 2 | 5 | 25 | 11 | 2 | 48 | 38 | 5 | 48 | 13 |
| Woman | 3 | 2 | 31 | 65 | 1 | 12 | 2 | 13 | 24 | 5 |
| Boy | 7 | 18 | 14 | 19 | 12 | 52 | 70 | 6 | 128 | 32 |
| Man | 27 | 45 | 51 | 138 | 31 | 80 | 2 | 27 | 48 | 95 |

For each text perform the significance test to compare the proportions of juvenile references for females and males. Summarize the results of the significance tests for the 10 texts studied. The researchers who conducted the study note that the authors of the last three texts are women, while the other seven texts were written by men. Do you see any pattern that suggests that the gender of the author is associated with the results?

**8.73** CHALLENGE **Even more on gender bias in textbooks.** Refer to the previous exercise. Let us now combine the categories "girl" with "woman" and "boy" with "man." For each text calculate the proportion of male references and test the hypothesis that male and female references are equally likely (that is, the proportion of male references is equal to 0.5). Summarize the results of your 10 tests. Is there a pattern that suggests a relation with the gender of the author?

**8.74** CHALLENGE **Changing majors during college.** In a random sample of 975 students from a large public university, it was found that 463 of the students changed majors during their college years.

(a) Give a 95% confidence interval for the proportion of students at this university who change majors.

(b) Express your results from (a) in terms of the *percent* of students who change majors.

(c) University officials concerned with counseling students are interested in the number of students who change majors rather than the proportion. The university has 37,500 undergraduate students. Convert the confidence interval you found in (a) to a confidence interval for the *number* of students who change majors during their college years.

**8.75 Gallup Poll study.** Go to the Gallup Poll Web site `http://www.galluppoll.com/` and find a poll that has several questions of interest to you. Summarize the results of the poll giving margins of error and comparisons of interest. (For this exercise, you may assume that the data come from an SRS.)

**8.76 Parental pressure and gender.** The Pew Research Center poll in Exercise 8.69 (page 521) also reported that 62% of the men and 51% of the women thought parents are placing too little pressure on their children to succeed in school. Assuming that the respondents were 52% women, compare the proportions with a significance test and give a 95% confidence interval for the difference. Write a summary of your results.

**8.77** CHALLENGE **Sample size and the *P*-value.** In this exercise we examine the effect of the sample size on the significance test for comparing two proportions. In each case suppose that $\hat{p}_1 = 0.5$ and $\hat{p}_2 = 0.4$, and take $n$ to be the common value of $n_1$ and $n_2$. Use the $z$ statistic to test $H_0: p_1 = p_2$ versus the alternative $H_a: p_1 \neq p_2$. Compute the statistic and the associated *P*-value for the following values of $n$: 40, 50, 80, 100, 400, 500, and 1000. Summarize the results in a table. Explain what you observe about the effect of the sample size on statistical significance when the sample proportions $\hat{p}_1$ and $\hat{p}_2$ are unchanged.

**8.78** CHALLENGE **Sample size and the margin of error.** In the first section of this chapter, we studied the effect of the sample size on the margin of error of the confidence interval for a single proportion. In this exercise we perform some calculations to observe this effect for the two-sample problem. Suppose that $\hat{p}_1 = 0.7$ and $\hat{p}_2 = 0.6$, and $n$ represents the common value of $n_1$ and $n_2$. Compute the 95% margins of error for the difference in the two proportions for $n = 40$, 50, 80, 100, 400, 500, and 1000. Present the results in a table and with a graph. Write a short summary of your findings.

**8.79** CHALLENGE **Calculating sample sizes for the two-sample problem.** For a single proportion the margin of error of a confidence interval is largest for any given sample size $n$ and confidence level $C$ when $\hat{p} = 0.5$. This led us to use $p^* = 0.5$ for planning purposes. The same kind of result is true for the two-sample problem. The margin of error of the confidence interval for the difference between two proportions is largest when $\hat{p}_1 = \hat{p}_2 = 0.5$. You are planning a survey and will calculate a 95% confidence interval for the difference in two proportions when the data are collected. You would like the margin of error of the interval to be less than or equal to 0.075. You will use the same sample size $n$ for both populations.

(a) How large a value of $n$ is needed?

(b) Give a general formula for $n$ in terms of the desired margin of error $m$ and the critical value $z^*$.

**8.80** **A corporate liability trial.** A major court case on the health effects of drinking contaminated water took place in the town of Woburn, Massachusetts. A town well in Woburn was contaminated by industrial chemicals. During the period that residents drank water from this well, there were 16 birth defects among 414 births. In years when the contaminated well was shut off and water was supplied from other wells, there were 3 birth defects among 228 births. The plaintiffs suing the firm responsible for the contamination claimed that these data show that the rate of birth defects was higher when the contaminated well was in use.[35] How statistically significant is the evidence? What assumptions does your analysis require? Do these assumptions seem reasonable in this case?

**8.81** CHALLENGE **Statistics and the law.** *Castaneda v. Partida* is an important court case in which statistical methods were used as part of a legal argument.[36] When reviewing this case, the Supreme Court used the phrase "two or three standard deviations" as a criterion for statistical significance. This Supreme Court review has served as the basis for many subsequent applications of statistical methods in legal settings. (The two or three standard deviations referred to by the Court are values of the $z$ statistic and correspond to *P*-values of approximately 0.05 and 0.0026.) In *Castaneda* the plaintiffs alleged that the method for selecting juries in a county in Texas was biased against Mexican Americans. For the period of time at issue, there were 181,535 persons eligible for jury duty, of whom 143,611 were Mexican Americans. Of the 870 people selected for jury duty, 339 were Mexican Americans.

(a) What proportion of eligible jurors were Mexican Americans? Let this value be $p_0$.

(b) Let $p$ be the probability that a randomly selected juror is a Mexican American. The null hypothesis to be tested is $H_0: p = p_0$. Find the value of $\hat{p}$ for this problem, compute the $z$ statistic, and find the *P*-value. What do you conclude? (A finding of statistical significance in this circumstance does not constitute proof of discrimination. It can be used, however, to establish a prima facie case. The burden of proof then shifts to the defense.)

(c) We can reformulate this exercise as a two-sample problem. Here we wish to compare the proportion of Mexican Americans among those selected as jurors with the proportion of Mexican Americans among those not selected as jurors. Let $p_1$ be the probability that a randomly selected juror is a Mexican American, and let $p_2$ be the probability that a randomly selected nonjuror is a Mexican American. Find the $z$ statistic and its *P*-value. How do your answers compare with your results in (b)?

**8.82** CHALLENGE **Home court advantage.** In many sports there is a home field or home court advantage. This means that the home team is more likely to win when playing at home than they are to win when playing at an opponent's field or court, all other things being equal. Go to the Web site of your favorite sports team and find the proportion of wins for home games and the proportion of wins for away games. Now consider these games to be a random sample of the process that generates wins and losses. A complete analysis of data like these requires methods that are beyond what we have studied, but the methods discussed in this chapter will give us a reasonable approximation. Examine the home court advantage for your team and write a summary of your results. Be sure to comment on the effect of the sample size.

**8.83** CHALLENGE **Attitudes toward student loan debt.** The National Student Loan Survey asked the student loan borrowers in their sample about attitudes toward debt.[37] Here are some of the questions they asked, with the percent who responded in a particular way:

(a) "To what extent do you feel burdened by your student loan payments?" 55.5% said they felt burdened.

(b) "If you could begin again, taking into account your current experience, what would you borrow?" 54.4% said they would borrow less.

(c) "Since leaving school, my education loans have not caused me more financial hardship than I had anticipated at the time I took out the loans." 34.3% disagreed.

(d) "Making loan payments is unpleasant but I know that the benefits of education loans are worth it." 58.9% agreed.

(e) "I am satisfied that the education I invested in with my student loan(s) was worth the investment for career opportunities." 58.9% agreed.

(f) "I am satisfied that the education I invested in with my student loan(s) was worth the investment for personal growth." 71.5% agreed.

Assume that the sample size is 1280 for all of these questions. Compute a 95% confidence interval for each of the questions, and write a short report about what student loan borrowers think about their debt.

CHAPTER

9

# Analysis of Two-Way Tables

There is growing evidence that early exposure to frightening movies is associated with lingering fright symptoms. Is this relationship different for boys and girls? Example 9.3 addresses this question.



## Introduction

We continue our study of methods for analyzing categorical data in this chapter. Inference about proportions in one-sample and two-sample settings was the focus of Chapter 8. We now study how to compare two or more populations when the response variable has two or more categories and how to test whether two categorical variables are independent. A single statistical test handles both of these cases.

The first section of this chapter gives the basics of statistical inference that are appropriate in this setting. An optional second section provides some technical details, and a goodness of fit test is presented in the last section. The methods in this chapter answer questions such as

- Are men and women equally likely to suffer lingering fear symptoms after watching scary movies like *Jaws* and *Poltergeist* at a young age?
- Does the style of a store's background music affect the purchase of French and Italian wine?
- Is vitamin A supplementation of young children in developing countries associated with a reduction in death rates?

## 9.1 Inference for Two-Way Tables

When we studied inference for two proportions in Chapter 8, we started summarizing the raw data by giving the number of observations in each population ($n$) and how many of these were classified as "successes" ($X$).

EXAMPLE

**9.1 Gender and the proportion of frequent binge drinkers.** In Example 8.9, we compared the proportions of male and female college students who engage in frequent binge drinking. The following table summarizes the data used in this comparison:

| Population | $n$ | $X$ | $\hat{p} = X/n$ |
|---|---|---|---|
| 1 (men) | 5,348 | 1,392 | 0.260 |
| 2 (women) | 8,471 | 1,748 | 0.206 |
| Total | 13,819 | 3,140 | 0.227 |

These data suggest that the men are 5.4% more likely to be frequent binge drinkers, with a 95% margin of error of 1.5%.

LOOK BACK
two-way table, page 142

In this chapter we consider a different summary of the data. Rather than recording just the count of binge drinkers, we record counts of all the outcomes in a two-way table.

EXAMPLE

**9.2 Two-way table of frequent binge drinking and gender.** Here is the two-way table classifying students by gender and whether or not they are frequent binge drinkers. The two categorical variables are "Frequent binge drinker," with values "Yes" and "No," and "Gender," with values "Men" and "Women." Since the objective is to compare the genders, we view "Gender" as an explanatory variable, and therefore, we make it the column variable. The row variable is a categorical response variable, "Frequent binge drinker."

| Two-way table for frequent binge drinking and gender | | | |
|---|---|---|---|
| | Gender | | |
| Frequent binge drinker | Men | Women | Total |
| Yes | 1,392 | 1,748 | 3,140 |
| No | 3,956 | 6,723 | 10,679 |
| Total | 5,348 | 8,471 | 13,819 |

The next example presents another two-way table.

EXAMPLE

**9.3 Lingering symptoms from frightening movies.** There is a growing body of literature demonstrating that early exposure to frightening movies is associated with lingering fright symptoms. As part of a class on media effects, college students were asked to write narrative accounts of their exposure to frightening movies before the age of 13. More than one-fourth of the respondents said that some of the fright symptoms were still present in waking life.[1] The following table breaks down these results by gender:

| Observed numbers of students | | | |
|---|---|---|---|
| | Gender | | |
| Ongoing fright symptoms | Men | Women | Total |
| Yes | 7 | 29 | 36 |
| No | 31 | 50 | 81 |
| Total | 38 | 79 | 117 |

The two categorical variables in this example are "Ongoing fright symptoms," with values "Yes" and "No," and "Gender," with values "Men" and "Women." Again we view "Gender" as an explanatory variable and "Ongoing fright symptoms" as a categorical response variable.

LOOK BACK
**conditional distributions, page 146**

In Chapter 2 we discussed two-way tables and the basics about joint, marginal, and conditional distributions. There we learned that the key to examining the relationship between two categorical variables is to look at conditional distributions. Figure 9.1 shows the output from CrunchIt! for the data of Example 9.3. Check this figure carefully. Be sure that you can identify the joint distribution, the marginal distributions, and the conditional distributions.

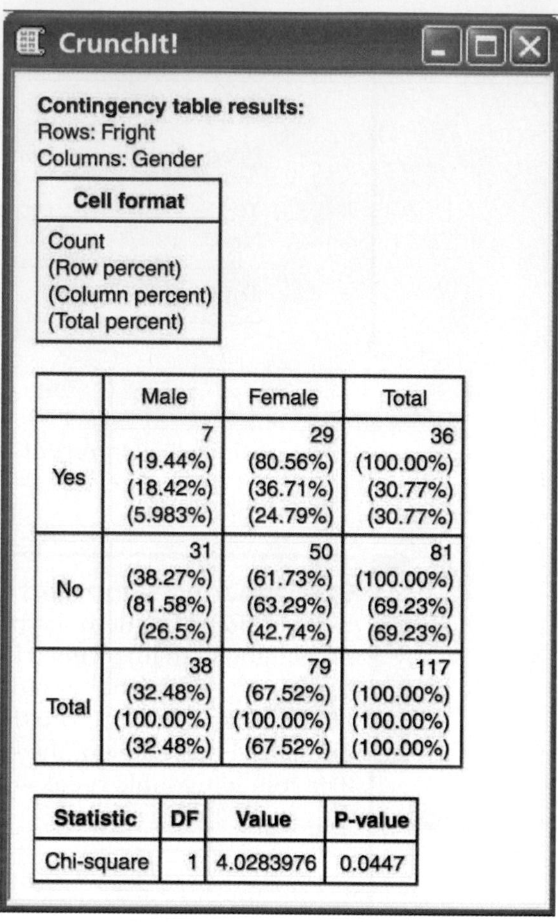

CrunchIt!

**Contingency table results:**
Rows: Fright
Columns: Gender

| Cell format |
|---|
| Count<br>(Row percent)<br>(Column percent)<br>(Total percent) |

| | Male | Female | Total |
|---|---|---|---|
| Yes | 7<br>(19.44%)<br>(18.42%)<br>(5.983%) | 29<br>(80.56%)<br>(36.71%)<br>(24.79%) | 36<br>(100.00%)<br>(30.77%)<br>(30.77%) |
| No | 31<br>(38.27%)<br>(81.58%)<br>(26.5%) | 50<br>(61.73%)<br>(63.29%)<br>(42.74%) | 81<br>(100.00%)<br>(69.23%)<br>(69.23%) |
| Total | 38<br>(32.48%)<br>(100.00%)<br>(32.48%) | 79<br>(67.52%)<br>(100.00%)<br>(67.52%) | 117<br>(100.00%)<br>(100.00%)<br>(100.00%) |

| Statistic | DF | Value | P-value |
|---|---|---|---|
| Chi-square | 1 | 4.0283976 | 0.0447 |

**FIGURE 9.1** CrunchIt! computer output for Example 9.3.

EXAMPLE

**9.4 Two-way table of ongoing fright symptoms and gender.** To compare the frequency of lingering fright symptoms across genders, we examine column percents. Here they are, rounded from the output for clarity:

| Column percents for gender | | |
|---|---|---|
| | Gender | |
| Ongoing fright symptoms | Male | Female |
| Yes | 18% | 37% |
| No | 82% | 63% |
| Total | 100% | 100% |

The "Total" row reminds us that 100% of the male and female students have been classified as having ongoing fright symptoms or not. (The sums sometimes differ slightly from 100% because of roundoff error.) The bar graph in Figure 9.2 compares the percents. The data reveal a clear relationship: 37% of the women have ongoing fright symptoms, as opposed to only 18% of the men.

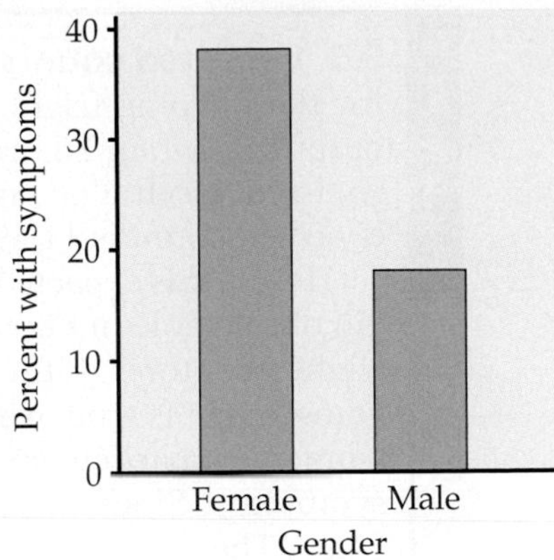


**FIGURE 9.2** Bar graph of the percents of male and female students with lingering fright symptoms, for Example 9.4.

The difference between the percents of students with lingering fears is reasonably large. A statistical test will tell us whether or not this difference can be plausibly attributed to chance. Specifically, if there is no association between gender and having ongoing fright symptoms, how likely is it that a sample would show a difference as large or larger than that displayed in Figure 9.2? In the remainder of this section we discuss the significance test to examine this question.

## The hypothesis: no association

The null hypothesis $H_0$ of interest in a two-way table is: There is *no association* between the row variable and the column variable. In Example 9.3, this null hypothesis says that gender and having ongoing fright symptoms are not related. The alternative hypothesis $H_a$ is that there is an association between these two variables. The alternative $H_a$ does not specify any particular direction for the association. For two-way tables in general, the alternative includes many different possibilities. Because it includes all sorts of possible associations, we cannot describe $H_a$ as either one-sided or two-sided.

In our example, the hypothesis $H_0$ that there is no association between gender and having ongoing fright symptoms is equivalent to the statement that the distributions of the ongoing fright symptoms variable are the same across the genders. For other two-way tables, where the columns correspond to independent samples from distinct populations, there are $c$ distributions for the row variable, one for each population. The null hypothesis then says that the $c$ distributions of the row variable are identical. The alternative hypothesis is that the distributions are not all the same.

## Expected cell counts

expected cell counts

To test the null hypothesis in $r \times c$ tables, we compare the observed cell counts with **expected cell counts** calculated under the assumption that the null hypothesis is true. A numerical summary of the comparison will be our test statistic.

EXAMPLE

**9.5 Expected counts from software.** The observed and expected counts for the ongoing fright symptoms example appear in the Minitab computer output shown in Figure 9.3. The expected counts are given as the second entry in each cell. For example, in the first cell the observed count is 7 and the expected count is 11.69.

How is this expected count obtained? Look at the percents in the right margin of the table in Figure 9.1. We see that 30.77% of all students had ongoing fright symptoms. If the null hypothesis of no relation between gender and ongoing fright is true, we expect this overall percent to apply to both men and women. In particular, we expect 30.77% of the men to have lingering fright symptoms. Since there are 38 men, the expected count is 30.77% of 38, or 11.69. The other expected counts are calculated in the same way.

Minitab

```
Rows: Symptom   Columns: Gender

        1_Male  2_Female     All

1_Yes        7        29      36
         11.69     24.31   36.00

2_No        31        50      81
         26.31     54.69   81.00

All         38        79     117
         38.00     79.00  117.00

Cell Contents:     Count
                   Expected count


Pearson Chi-Square = 4.028, DF = 1, P-Value = 0.045
```

**FIGURE 9.3** Minitab computer output for Example 9.5.

The reasoning of Example 9.5 leads to a simple formula for calculating expected cell counts. To compute the expected count of men with ongoing fright symptoms, we multiplied the proportion of students with fright symptoms (36/117) by the number of men (38). From Figures 9.1 and 9.3 we see that the numbers 36 and 38 are the row and column totals for the cell of interest and that 117 is $n$, the total number of observations for the table. The expected cell count is therefore the product of the row and column totals divided by the table total.

$$\text{expected cell count} = \frac{\text{row total} \times \text{column total}}{n}$$

## The chi-square test

To test the $H_0$ that there is no association between the row and column classifications, we use a statistic that compares the entire set of observed counts with the set of expected counts. To compute this statistic,

- First, take the difference between each observed count and its corresponding expected count, and square these values so that they are all 0 or positive.
- Since a large difference means less if it comes from a cell that is expected to have a large count, divide each squared difference by the expected count. This is a kind of standardization.
- Finally, sum over all cells.

LOOK BACK
**standardizing, page 61**

The result is called the *chi-square statistic* $X^2$. The chi-square statistic was invented by the English statistician Karl Pearson (1857–1936) in 1900, for purposes slightly different from ours. It is the oldest inference procedure still used in its original form. With the work of Pearson and his contemporaries at the beginning of the last century, statistics first emerged as a separate discipline.

### CHI-SQUARE STATISTIC

The **chi-square statistic** is a measure of how much the observed cell counts in a two-way table diverge from the expected cell counts. The formula for the statistic is

$$X^2 = \sum \frac{(\text{observed count} - \text{expected count})^2}{\text{expected count}}$$

where "observed" represents an observed cell count, "expected" represents the expected count for the same cell, and the sum is over all $r \times c$ cells in the table.

If the expected counts and the observed counts are very different, a large value of $X^2$ will result. Large values of $X^2$ provide evidence against the null hypothesis. To obtain a *P*-value for the test, we need the sampling distribution of $X^2$ under the assumption that $H_0$ (no association between the row and column variables) is true. We once again use an approximation, related to the Normal approximation for binomial distributions. The result is a new distribution, the **chi-square distribution,** which we denote by $\chi^2$ ($\chi$ is the lowercase Greek letter chi).

LOOK BACK
**Normal approximation for counts, page 323**

chi-square distribution $\chi^2$

Like the *t* distributions, the $\chi^2$ distributions form a family described by a single parameter, the degrees of freedom. We use $\chi^2$(df) to indicate a particular member of this family. Figure 9.4 displays the density curves of the $\chi^2(2)$ and $\chi^2(4)$ distributions. As the figure suggests, $\chi^2$ distributions take only positive

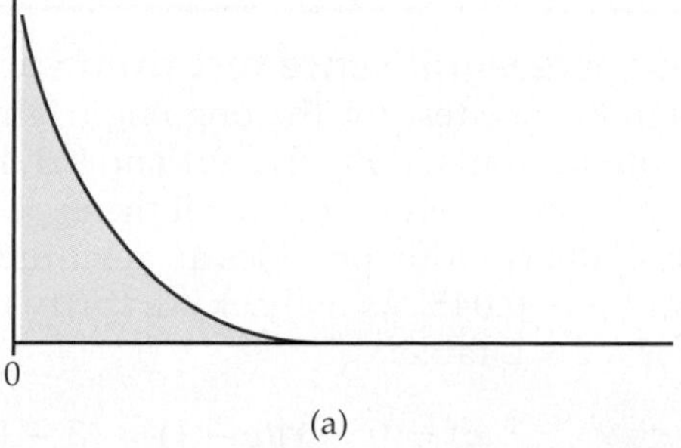


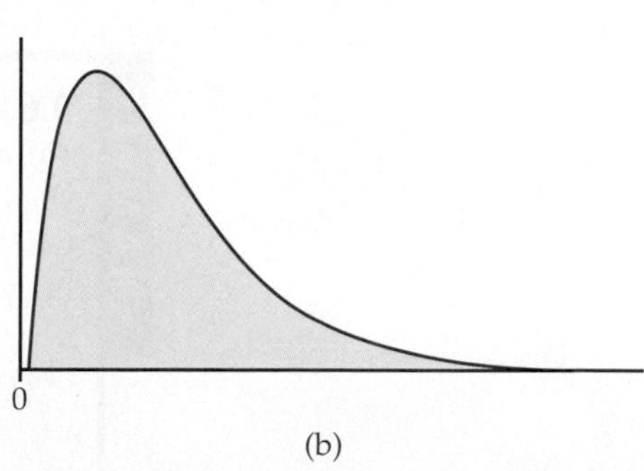


**FIGURE 9.4** (a) The $\chi^2(2)$ density curve. (b) The $\chi^2(4)$ density curve.

values and are skewed to the right. Table F in the back of the book gives upper critical values for the $\chi^2$ distributions.

### CHI-SQUARE TEST FOR TWO-WAY TABLES

The null hypothesis $H_0$ is that there is no association between the row and column variables in a two-way table. The alternative is that these variables are related.

If $H_0$ is true, the chi-square statistic $X^2$ has approximately a $\chi^2$ distribution with $(r-1)(c-1)$ degrees of freedom.

The $P$-value for the chi-square test is

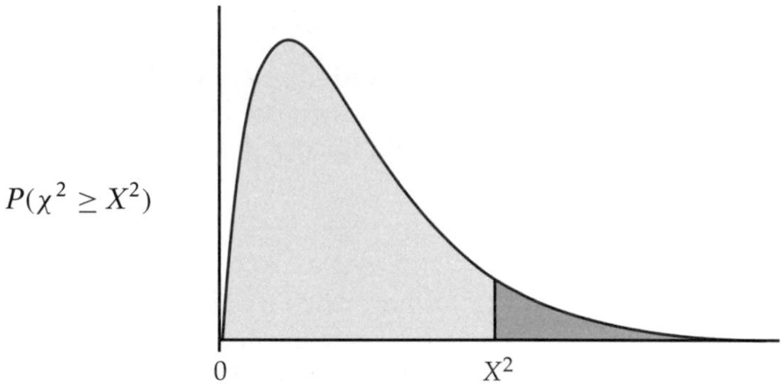


where $\chi^2$ is a random variable having the $\chi^2(\mathrm{df})$ distribution with $\mathrm{df} = (r-1)(c-1)$.

The chi-square test always uses the upper tail of the $\chi^2$ distribution, because any deviation from the null hypothesis makes the statistic larger. The approximation of the distribution of $X^2$ by $\chi^2$ becomes more accurate as the cell counts increase. Moreover, it is more accurate for tables larger than $2 \times 2$ tables. For tables larger than $2 \times 2$, we will use this approximation whenever the average of the expected counts is 5 or more and the smallest expected count is 1 or more. For $2 \times 2$ tables, we require that all four expected cell counts be 5 or more.[2]

EXAMPLE

**9.6 Chi-square significance test from software.** The results of the chi-square significance test for the ongoing fright symptoms example appear in the computer outputs in Figures 9.1 and 9.3, labeled Chi-square and Pearson Chi-Square respectively. Because all the expected cell counts are moderately large, the $\chi^2$ distribution provides an accurate $P$-value. We see that $X^2 = 4.03$, $\mathrm{df} = 1$, and $P = 0.045$. As a check we verify that the degrees of freedom are correct for a $2 \times 2$ table:

$$\mathrm{df} = (r-1)(c-1) = (2-1)(2-1) = 1$$

The chi-square test confirms that the data contain clear evidence against the null hypothesis that there is no relationship between gender and ongoing fright symptoms. Under $H_0$, the chance of obtaining a value of $X^2$ greater than or equal to the calculated value of 4.03 is small—fewer than 5 times in 100.

The test does not tell us what kind of relationship is present. It is up to us to see that the data show that women are more likely to have lingering fright symptoms. You should always accompany a chi-square test by percents such as those in Example 9.4 and Figure 9.2 and by a description of the nature of the relationship.

confounding, page 156

The observational study of Example 9.3 cannot tell us whether gender is a *cause* of lingering fright symptoms. The association may be explained by confounding with other variables. For example, other research has shown that there are gender differences in the social desirability of admitting fear.[3] *Our data don't allow us to investigate possible confounding variables.* Often a randomized comparative experiment can settle the issue of causation, but we cannot randomly assign gender to each student. The researcher who published the data of our example states merely that women are more likely to have lingering fright symptoms and that this conclusion is consistent with other studies.

## The chi-square test and the $z$ test

A comparison of the proportions of "successes" in two populations leads to a 2 × 2 table. We can compare two population proportions either by the chi-square test or by the two-sample $z$ test from Section 8.2. In fact, *these tests always give exactly the same result,* because the $X^2$ statistic is equal to the square of the $z$ statistic, and $\chi^2(1)$ critical values are equal to the squares of the corresponding $N(0, 1)$ critical values. The advantage of the $z$ test is that we can test either one-sided or two-sided alternatives. The chi-square test always tests the two-sided alternative. Of course, the chi-square test can compare more than two populations, whereas the $z$ test compares only two.

### USE YOUR KNOWLEDGE

**9.1 Comparison of conditional distributions.** Consider the following 2 × 2 table.

Observed counts

| | Explanatory variable | | |
|---|---|---|---|
| Response variable | 1 | 2 | |
| Yes | 75 | 95 | 170 |
| No | 125 | 105 | 230 |
| Total | 200 | 200 | 400 |

(a) Compute the conditional distribution of the response variable for each of the two explanatory-variable categories.

(b) Display the distributions graphically.

(c) Write a short paragraph describing the two distributions and how they differ.

**9.2 Expected cell counts and the chi-square test.** Refer to Exercise 9.1. You decide to use the chi-square test to compare these two conditional distributions.

(a) What is the expected count for the first cell (observed count is 75)?

(b) Computer software gives you $X^2 = 4.09$. What are the degrees of freedom for this statistic?

(c) Using Table F, give an appropriate bound on the *P*-value.

## BEYOND THE BASICS

### Meta-analysis

Policymakers wanting to make decisions based on research are sometimes faced with the problem of summarizing the results of many studies. These studies may show effects of different magnitudes, some highly significant and some not significant. What *overall conclusion* can we draw? **Meta-analysis** is a collection of statistical techniques designed to combine information from different but similar studies. Each individual study must be examined with care to ensure that its design and data quality are adequate. The basic idea is to compute a measure of the effect of interest for each study. These are then combined, usually by taking some sort of weighted average, to produce a summary measure for all of the studies. Of course, a confidence interval for the summary is included in the results. Here is an example.

meta-analysis

EXAMPLE

**9.7 Vitamin A saves lives of young children.** Vitamin A is often given to young children in developing countries to prevent night blindness. It was observed that children receiving vitamin A appear to have reduced death rates. To investigate the possible relationship between vitamin A supplementation and death, a large field trial with over 25,000 children was undertaken in Aceh Province of Indonesia. About half of the children were given large doses of vitamin A, and the other half were controls. In 1986, the researchers reported a 34% reduction in mortality (deaths) for the treated children who were 1 to 6 years old compared with the controls. Several additional studies were then undertaken. Most of the results confirmed the association: treatment of young children in developing countries with vitamin A reduces the death rate; but the size of the effect varied quite a bit.

How can we use the results of these studies to guide policy decisions? To address this question, a meta-analysis was performed on data available from eight studies.[4] Although the designs varied, each study provided a two-way table of counts. Here is the table for Aceh. A total of $n = 25{,}200$ children were enrolled in the study. Approximately half received vitamin A supplements.

One year after the start of the study, the number of children who died was determined.

| | Vitamin A | Control |
|---|---|---|
| Dead | 101 | 130 |
| Alive | 12,890 | 12,079 |
| Total | 12,991 | 12,209 |

LOOK BACK
**relative risk, page 515**

The summary measure chosen was the relative risk: the ratio formed by dividing the proportion of children who died in the vitamin A group by the proportion of children who died in the control group. For Aceh, the proportion who died in the vitamin A group was

$$\frac{101}{12{,}991} = 0.00777$$

or 7.7 per thousand; for the control group, the proportion who died was

$$\frac{130}{12{,}209} = 0.01065$$

or 10.6 per thousand. The relative risk is therefore

$$\frac{0.00777}{0.01065} = 0.73$$

Relative risk less than 1 means that the vitamin A group has the lower mortality rate.

The relative risks for the eight studies were

| | | | | | | | |
|---|---|---|---|---|---|---|---|
| 0.73 | 0.50 | 0.94 | 0.71 | 0.70 | 1.04 | 0.74 | 0.80 |

A meta-analysis combined these eight results to produce a relative risk estimate of 0.77 with a 95% confidence interval of (0.68, 0.88). That is, vitamin A supplementation reduced the mortality rate to 77% of its value in an untreated group. In other words, there is a 23% reduction in the mortality rate. The confidence interval does not include 1, so the null hypothesis of no effect (a relative risk of 1) can be clearly rejected. The researchers examined many variations of this meta-analysis, such as using different weights and leaving out one study at a time. These variations had little effect on the final estimate.

After these findings were published, large-scale programs to distribute high-potency vitamin A supplements were started. These programs have saved hundreds of thousands of lives since the meta-analysis was conducted and the arguments and uncertainties were resolved.

## SECTION 9.1 Summary

The **null hypothesis** for $r \times c$ tables of count data is that there is no relationship between the row variable and the column variable.

**Expected cell counts** under the null hypothesis are computed using the formula

$$\text{expected count} = \frac{\text{row total} \times \text{column total}}{n}$$

The null hypothesis is tested by the **chi-square statistic,** which compares the observed counts with the expected counts:

$$X^2 = \sum \frac{(\text{observed} - \text{expected})^2}{\text{expected}}$$

Under the null hypothesis, $X^2$ has approximately the $\chi^2$ distribution with $(r-1)(c-1)$ degrees of freedom. The $P$-value for the test is

$$P(\chi^2 \geq X^2)$$

where $\chi^2$ is a random variable having the $\chi^2(\text{df})$ distribution with df $=$ $(r-1)(c-1)$.

The chi-square approximation is adequate for practical use when the average expected cell count is 5 or greater and all individual expected counts are 1 or greater, except in the case of $2 \times 2$ tables. All four expected counts in a $2 \times 2$ table should be 5 or greater.

*The section we just completed assumed that you have access to software or a statistical calculator. If you do not, you now need to study the material on computations in the following optional section. All exercises appear at the end of the chapter.*

# 9.2 Formulas and Models for Two-Way Tables*

## Computations

The calculations required to analyze a two-way table are straightforward but tedious. In practice, we recommend using software, but it is possible to do the work with a calculator, and some insight can be gained by examining the details. Here is an outline of the steps required.

### COMPUTATIONS FOR TWO-WAY TABLES

**1.** Calculate descriptive statistics that convey the important information in the table. Usually these will be column or row percents.

*The analysis of two-way tables is based on computations that are a bit messy and on statistical models that require a fair amount of notation to describe. This section gives the details. By studying this material you will deepen your understanding of the methods described in this chapter, but this section is optional.

**2.** Find the expected counts and use these to compute the $X^2$ statistic.

**3.** Use chi-square critical values from Table F to find the approximate $P$-value.

**4.** Draw a conclusion about the association between the row and column variables.

The following example illustrates these steps.

EXAMPLE

**9.8 Background music and wine sales.** Market researchers know that background music can influence the mood and purchasing behavior of customers. One study in a supermarket in Northern Ireland compared three treatments: no music, French accordion music, and Italian string music. Under each condition, the researchers recorded the numbers of bottles of French, Italian, and other wine purchased.[5] Here is the two-way table that summarizes the data:

| | Music | | | |
|---|---|---|---|---|
| **Wine** | **None** | **French** | **Italian** | **Total** |
| French | 30 | 39 | 30 | 99 |
| Italian | 11 | 1 | 19 | 31 |
| Other | 43 | 35 | 35 | 113 |
| Total | 84 | 75 | 84 | 243 |

This is a $3 \times 3$ table, to which we have added the marginal totals obtained by summing across rows and columns. For example, the first-row total is $30 + 39 + 30 = 99$. The grand total, the number of bottles of wine in the study, can be computed by summing the row totals, $99 + 31 + 113 = 243$, or the column totals, $84 + 75 + 84 = 243$. *It is easy to make an error in these calculations, so it is a good idea to do both as a check on your arithmetic.*

## Computing conditional distributions

First, we summarize the observed relation between the music being played and the type of wine purchased. The researchers expected that music would influence sales, so music type is the explanatory variable and the type of wine purchased is the response variable. In general, the clearest way to describe this kind of relationship is to compare the conditional distributions of the response variable for each value of the explanatory variable. So we will compare the column percents that give the conditional distribution of purchases for each type of music played.

EXAMPLE

**9.9 Background music and wine sales: conditional distributions.** When no music was played, there were 84 bottles of wine sold. Of these, 30 were French wine. Therefore, the column proportion for this cell is

$$\frac{30}{84} = 0.357$$

That is, 35.7% of the wine sold was French when no music was played. Similarly, 11 bottles of Italian wine were sold under this condition, and this is 13.1% of the sales:

$$\frac{11}{84} = 0.131$$

In all, we calculate nine percents. Here are the results:

**Column percents for wine and music**

| | Music | | | |
|---|---|---|---|---|
| **Wine** | **None** | **French** | **Italian** | **Total** |
| French | 35.7 | 52.0 | 35.7 | 40.7 |
| Italian | 13.1 | 1.3 | 22.6 | 12.8 |
| Other | 51.2 | 46.7 | 41.7 | 46.5 |
| Total | 100.0 | 100.0 | 100.0 | 100.0 |

In addition to the conditional distributions of types of wine sold for each kind of music being played, the table also gives the marginal distribution of the types of wine sold. These percents appear in the rightmost column, labeled "Total."

The sum of the percents in each column should be 100, except for possible small roundoff errors. It is good practice to calculate each percent separately and then sum each column as a check. In this way we can find arithmetic errors that would not be uncovered if, for example, we calculated the column percent for the "Other" row by subtracting the sum of the percents for "French" and "Italian" from 100.

Figure 9.5 compares the distributions of types of wine sold for each of the three music conditions. There appears to be an association between the music played and the type of wine that customers buy. Sales of Italian wine are very low when French music is playing but are higher when Italian music or no music is playing. French wine is popular in this market, selling well under all music conditions but notably better when French music is playing.

Another way to look at these data is to examine the row percents. These fix a type of wine and compare its sales when different types of music are playing. Figure 9.6 displays these results. We see that more French wine is sold when French music is playing, and more Italian wine is sold when Italian music is playing. The negative effect of French music on sales of Italian wine is dramatic.

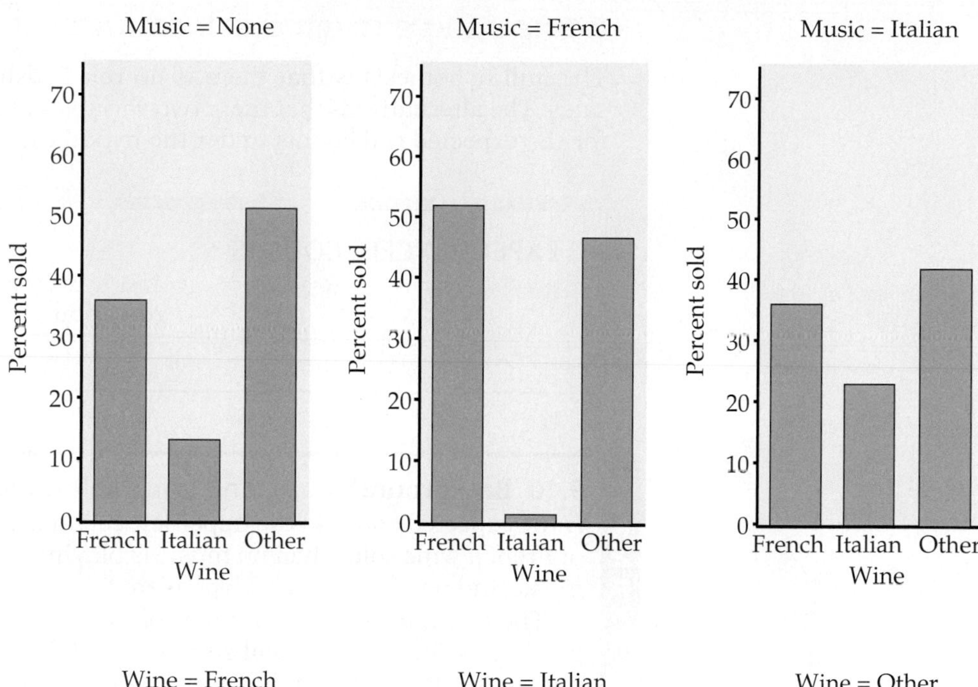


FIGURE 9.5 Comparison of the percents of different types of wine sold for different music conditions, for Example 9.8.

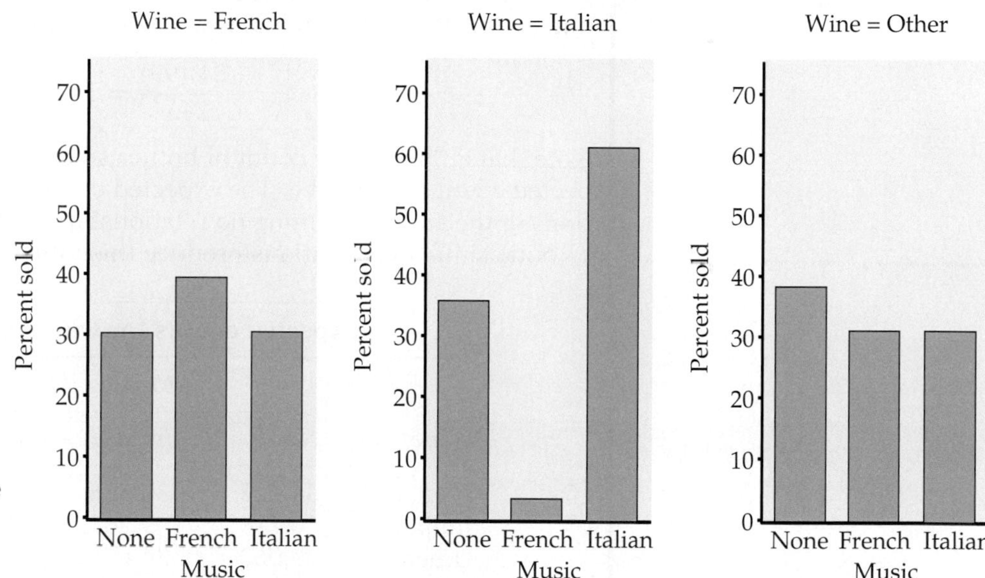


FIGURE 9.6 Comparison of the percents of different types of wine sold for different music conditions, for Example 9.8.

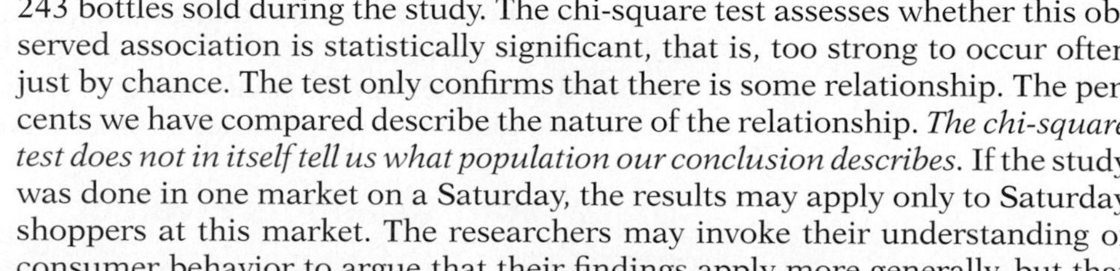

We observe a clear relationship between music type and wine sales for the 243 bottles sold during the study. The chi-square test assesses whether this observed association is statistically significant, that is, too strong to occur often just by chance. The test only confirms that there is some relationship. The percents we have compared describe the nature of the relationship. *The chi-square test does not in itself tell us what population our conclusion describes.* If the study was done in one market on a Saturday, the results may apply only to Saturday shoppers at this market. The researchers may invoke their understanding of consumer behavior to argue that their findings apply more generally, but that is beyond the scope of the statistical analysis.

## Computing expected cell counts

The null hypothesis is that there is no relationship between music and wine sales. The alternative is that these two variables are related. Here is the formula for the expected cell counts under the hypothesis of "no relationship."

**EXPECTED CELL COUNTS**

$$\text{expected count} = \frac{\text{row total} \times \text{column total}}{n}$$

EXAMPLE

**9.10 Background music and wine sales: expected cell counts.** What is the expected count in the upper-left cell in the table of Example 9.8, bottles of French wine sold when no music is playing, under the null hypothesis that music and wine sales are independent?

The column total, the number of bottles of wine sold when no music is playing, is 84. The row total shows that 99 bottles of French wine were sold during the study. The total sales were 243. The expected cell count is therefore

$$\frac{(84)(99)}{243} = 34.222$$

Note that although any count of bottles sold must be a whole number, an expected count need not be. The expected count is the mean over many repetitions of the study, assuming no relationship.

Nine similar calculations produce this table of expected counts:

**Expected counts for wine and music**

| | Music | | | |
|---|---|---|---|---|
| **Wine** | **None** | **French** | **Italian** | **Total** |
| French | 34.222 | 30.556 | 34.222 | 99.000 |
| Italian | 10.716 | 9.568 | 10.716 | 31.000 |
| Other | 39.062 | 34.877 | 39.062 | 113.001 |
| Total | 84.000 | 75.001 | 84.000 | 243.001 |

We can check our work by adding the expected counts to obtain the row and column totals, as in the table. These should be the same as those in the table of observed counts except for small roundoff errors, such as 113.001 rather than 113 for the total number of bottles of other wine sold.

## The $X^2$ statistic and its *P*-value

The expected counts are all large, so we proceed with the chi-square test. We compare the table of observed counts with the table of expected counts using the $X^2$ statistic.[6] We must calculate the term for each cell, then sum over all nine cells. For French wine with no music, the observed count is 30 bottles and the expected count is 34.222. The contribution to the $X^2$ statistic for this cell is therefore

$$\frac{(30-34.222)^2}{34.222} = 0.5209$$

The $X^2$ statistic is the sum of nine such terms:

$$\begin{aligned}
X^2 &= \sum \frac{(\text{observed} - \text{expected})^2}{\text{expected}} \\
&= \frac{(30-34.222)^2}{34.222} + \frac{(39-30.556)^2}{30.556} + \frac{(30-34.222)^2}{34.222} \\
&\quad + \frac{(11-10.716)^2}{10.716} + \frac{(1-9.568)^2}{9.568} + \frac{(19-10.716)^2}{10.716} \\
&\quad + \frac{(43-39.062)^2}{39.062} + \frac{(35-34.877)^2}{34.877} + \frac{(35-39.062)^2}{39.062} \\
&= 0.5209 + 2.3337 + 0.5209 + 0.0075 + 7.6724 + 6.4038 \\
&\quad + 0.3971 + 0.0004 + 0.4223 \\
&= 18.28
\end{aligned}$$

Because there are $r = 3$ types of wine and $c = 3$ music conditions, the degrees of freedom for this statistic are

$$\text{df} = (r-1)(c-1) = (3-1)(3-1) = 4$$

**df = 4**

| $p$ | 0.0025 | 0.001 |
|---|---|---|
| $\chi^2$ | 16.42 | 18.47 |

Under the null hypothesis that music and wine sales are independent, the test statistic $X^2$ has a $\chi^2(4)$ distribution. To obtain the $P$-value, look at the df = 4 row in Table F. The calculated value $X^2 = 18.28$ lies between the critical points for probabilities 0.0025 and 0.001. The $P$-value is therefore between 0.0025 and 0.001. Because the expected cell counts are all large, the $P$-value from Table F will be quite accurate. There is strong evidence ($X^2 = 18.28$, df = 4, $P < 0.0025$) that the type of music being played has an effect on wine sales.

The size and nature of the relationship between music and wine sales are described by row and column percents. These are displayed in Figures 9.5 and 9.6. Here is another way to look at the data: we see that just two of the nine terms that make up the chi-square sum contribute about 14 of the total $X^2 = 18.28$. Comparing the observed and expected counts in these two cells, we see that sales of Italian wine are much below expectation when French music is playing and much above expectation when Italian music is playing. We are led to a specific conclusion: sales of Italian wine are strongly affected by Italian and French music. Figure 9.6(b) displays this effect.

## Models for two-way tables

The chi-square test for the presence of a relationship between the two directions in a two-way table is valid for data produced from several different study designs. The precise statement of the null hypothesis "no relationship" in terms of population parameters is different for different designs. We now describe

two of these settings in detail. *An essential requirement is that each experimental unit or subject is counted only once in the data table.*

**Comparing several populations: the first model** Example 9.8 (wine sales in three environments) is an example of *separate and independent random samples* from each of $c$ populations. The $c$ columns of the two-way table represent the populations. There is a single categorical response variable, wine type. The $r$ rows of the table correspond to the values of the response variable.

We know that the $z$ test for comparing the two proportions of successes and the chi-square test for the $2 \times 2$ table are equivalent. The $r \times c$ table allows us to compare more than two populations or more than two categories of response, or both. In this setting, the null hypothesis "no relationship between column variable and row variable" becomes

$H_0$: The distribution of the response variable is the same in all $c$ populations.

Because the response variable is categorical, its distribution just consists of the probabilities of its $r$ possible values. The null hypothesis says that these probabilities (or population proportions) are the same in all $c$ populations.

EXAMPLE

**9.11 Background music and wine sales: comparing three populations.** In the market research study of Example 9.8, we compare three populations:

Population 1: bottles of wine sold when no music is playing

Population 2: bottles of wine sold when French music is playing

Population 3: bottles of wine sold when Italian music is playing

We have three samples, of sizes 84, 75, and 84, a separate sample from each population. The null hypothesis for the chi-square test is

$H_0$: The proportions of each wine type sold are the same in all three populations.

The parameters of the model are the proportions of the three types of wine that would be sold in each of the three environments. There are three proportions (for French wine, Italian wine, and other wine) for each environment.

More generally, if we take an independent SRS from each of $c$ populations and classify each outcome into one of $r$ categories, we have an $r \times c$ table of population proportions. There are $c$ different sets of proportions to be compared. There are $c$ groups of subjects, and a single categorical variable with $r$ possible values is measured for each individual.

### MODEL FOR COMPARING SEVERAL POPULATIONS USING TWO-WAY TABLES

Select independent SRSs from each of $c$ populations, of sizes $n_1, n_2, \ldots, n_c$. Classify each individual in a sample according to a categorical response variable with $r$ possible values. There are $c$ different probability distributions, one for each population.

The null hypothesis is that the distributions of the response variable are the same in all $c$ populations. The alternative hypothesis says that these $c$ distributions are not all the same.

**Testing independence: the second model** A second model for which our analysis of $r \times c$ tables is valid is illustrated by the ongoing fright symptoms study, Example 9.3. There, a *single* sample from a *single* population was classified according to two categorical variables.

EXAMPLE

**9.12 Ongoing fright symptoms and gender: testing independence.** The single population studied is college students. Each college student was classified according to the following categorical variables: "Ongoing fright symptoms," with possible responses "Yes" and "No," and "Gender," with possible responses "Men" and "Women." The null hypothesis for the chi-square test is

$$H_0\text{: "Ongoing fright symptoms" and "Gender" are independent.}$$

The parameters of the model are the probabilities for each of the four possible combinations of values of the row and column variables. If the null hypothesis is true, the multiplication rule for independent events says that these can be found as the products of outcome probabilities for each variable alone.

LOOK BACK
**multiplication rule, page 295**

More generally, take an SRS from a single population and record the values of two categorical variables, one with $r$ possible values and the other with $c$ possible values. The data are summarized by recording the numbers of individuals for each possible combination of outcomes for the two random variables. This gives an $r \times c$ table of counts. Each of these $r \times c$ possible outcomes has its own probability. The probabilities give the joint distribution of the two categorical variables.

LOOK BACK
**joint distribution, page 144**
**marginal distributions, page 145**

Each of the two categorical random variables has a distribution. These are the marginal distributions because they are the sums of the population proportions in the rows and columns.

The null hypothesis "no relationship" now states that the row and column variables are independent. The multiplication rule for independent events tells us that the joint probabilities are the products of the marginal probabilities.

EXAMPLE

**9.13 The joint distribution and the two marginal distributions.** The joint probability distribution gives a probability for each of the four cells in our $2 \times 2$ table of "Ongoing fright symptoms" and "Gender." The marginal distribution for "Ongoing fright symptoms" gives probabilities for each of the two possible categories; the marginal distribution for "Gender" gives probabilities for each of the two possible gender categories.

Independence between "Ongoing fright symptoms" and "Gender" implies that the joint distribution can be obtained by multiplying the appropriate terms from the two marginal distributions. For example, the probability that a randomly chosen college student has ongoing fright symptoms *and* is male is equal to the probability that the student has ongoing symptoms *times* the probability that the student is male. The hypothesis that "Ongoing fright symptoms" and "Gender" are independent says that the multiplication rule applies to *all* outcomes.

**MODEL FOR EXAMINING INDEPENDENCE IN TWO-WAY TABLES**

Select an SRS of size $n$ from a population. Measure two categorical variables for each individual.

The null hypothesis is that the row and column variables are independent. The alternative hypothesis is that the row and column variables are dependent.

## Concluding remarks

You can distinguish between the two models by examining the design of the study. In the independence model, there is a single sample. The column totals and row totals are random variables. The total sample size $n$ is set by the researcher; the column and row sums are known only after the data are collected. For the comparison-of-populations model, on the other hand, there is a sample from each of two or more populations. The column sums are the sample sizes selected at the design phase of the research. The null hypothesis in both models says that there is no relationship between the column variable and the row variable. The precise statement of the hypothesis differs, depending on the sampling design. Fortunately, *the test of the hypothesis of "no relationship" is the same for both models;* it is the chi-square test. There are yet other statistical models for two-way tables that justify the chi-square test of the null hypothesis "no relation," made precise in ways suitable for these models. Statistical methods related to the chi-square test also allow the analysis of three-way and higher-way tables of count data. You can find a discussion of these topics in advanced texts on categorical data.[7]

**USE YOUR KNOWLEDGE**

**9.3 Find the *P*-value.** For each of the following give the degrees of freedom and an appropriate bound on the $P$-value for the $X^2$ statistic.

(a) $X^2 = 2.5$ for a 2 by 2 table

(b) $X^2 = 6.5$ for a 2 by 2 table

(c) $X^2 = 16.3$ for a 3 by 5 table

(d) $X^2 = 16.3$ for a 5 by 3 table

**9.4 Frequent binge drinking and gender: the chi-square test.** Refer to Example 9.2 (page 526). Use the chi-square test to assess if frequent binge drinking is associated with gender. Also, compare the chi-square statistic with the $z$ statistic in Example 8.11 (page 513). State your conclusions.

## SECTION 9.2 Summary

For two-way tables we first compute percents or proportions that describe the relationship of interest. Then, we compute expected counts, the $X^2$ statistic, and the $P$-value.

Two different models for generating $r \times c$ tables lead to the chi-square test. In the first model, independent SRSs are drawn from each of $c$ populations, and each observation is classified according to a categorical variable with $r$ possible values. The null hypothesis is that the distributions of the row categorical variable are the same for all $c$ populations. In the second model, a single SRS is drawn from a population, and observations are classified according to two categorical variables having $r$ and $c$ possible values. In this model, $H_0$ states that the row and column variables are independent.

# 9.3 Goodness of Fit*

In the last two sections, we discussed the use of the chi-square test to compare categorical-variable distributions of $c$ populations. We now consider a slight variation on this scenario where we compare a sample from one population with a hypothesized distribution. Here is an example that illustrates the basic ideas.

EXAMPLE

**9.14 Vehicle collisions and cell phones.** Are you more likely to have a motor vehicle collision when using a cell phone? A study of 699 drivers who were using a cell phone when they were involved in a collision examined this question.[8] These drivers made 26,798 cell phone calls during a 14-month study period. Each of the 699 collisions was classified in various ways. Here are the numbers for each day of the week:

**Number of collisions by day of the week**

| Sun. | Mon. | Tue. | Wed. | Thu. | Fri. | Sat. | Total |
|---|---|---|---|---|---|---|---|
| 20 | 133 | 126 | 159 | 136 | 113 | 12 | 699 |

We have a total of 699 accidents involving drivers who were using a cell phone at the time of their accident. Let's explore the relationship between these accidents and the day of the week. Are the accidents equally likely to occur on any day of the week?

We can think of this table of counts as a one-way table with seven cells, each with a count of the number of accidents that occurred on the particular day

*This section is optional.

of the week. Our question is translated into a null hypothesis that motor vehicle accidents involving cell phone use are equally likely to occur on each of the seven days of the week. The alternative is that the probabilities vary from day to day. Our analysis of these data is very similar to the analyses of two-way tables that we studied in Section 9.1. We first compute expected counts. Since there are 699 accidents and 7 days, under the null hypothesis we expect one-seventh of the accidents to occur on each day. So the expected number of accidents for each day is $699/7 = 99.86$. Next, we construct a chi-square statistic that compares the actual numbers with the expected numbers:

$$X^2 = \sum \frac{(\text{observed count} - \text{expected count})^2}{\text{expected count}}$$

For Sunday, we have

$$\frac{(\text{observed count} - \text{expected count})^2}{\text{expected count}} = \frac{(20 - 99.86)^2}{99.86} = 63.86$$

and for Monday,

$$\frac{(\text{observed count} - \text{expected count})^2}{\text{expected count}} = \frac{(133 - 99.86)^2}{99.86} = 11.0$$

Performing the same calculations for the other days and summing the results over the seven cells gives

$$X^2 = 208.85$$

The degrees of freedom are 1 less than the number of cells, $\text{df} = 7 - 1 = 6$. We calculate the $P$-value using Table F or software. From Table F, we can determine $P < 0.0005$ and we conclude that these types of accidents are not equally likely to occur on each of the seven days of the week.

We have covered all of the basics for chi-square goodness of fit tests in our example. Now we will summarize the details for the general case.

## THE CHI-SQUARE GOODNESS OF FIT TEST

Data for $n$ observations of a categorical variable with $k$ possible outcomes are summarized as observed counts, $n_1, n_2, \ldots, n_k$ in $k$ cells. A null hypothesis specifies probabilities $p_1, p_2, \ldots, p_k$ for the possible outcomes.

For each cell, multiply the total number of observations $n$ by the specified probability to determine the expected counts:

$$\text{expected count} = np_i$$

The **chi-square statistic** measures how much the observed cell counts differ from the expected cell counts. The formula for the statistic is

$$X^2 = \sum \frac{(\text{observed count} - \text{expected count})^2}{\text{expected count}}$$

The degrees of freedom are $k - 1$, and $P$-values are computed from the chi-square distribution.

In our example about cell phones and motor vehicle accidents we examined the accident rates for all seven days of the week. A traffic expert would presumably tell us that accidents are less likely on weekends. Indeed, the two days contributing the largest amounts to the $X^2$ statistic are Saturday and Sunday. Let's examine the pattern of accidents for weekdays. *Note that this is not the same as making up hypotheses after looking at the data.* After some reflection about the data and the context of the problem, we have determined that it is reasonable to ask this question. Such decisions require careful judgment and the rationale must be clearly explained. More advanced methods of inference allow the examination of this kind of question within the framework of the overall analysis.[9]

EXAMPLE

**9.15 Vehicle collisions and cell phones: weekdays only.** Here are the data from Example 9.24 for weekdays only:

| Number of collisions by day of the week | | | | | |
|---|---|---|---|---|---|
| Mon. | Tue. | Wed. | Thu. | Fri. | Total |
| 133 | 126 | 159 | 136 | 113 | 667 |

We now have a total of 667 accidents because we are not considering the $32 = 20 + 12$ accidents that occurred on weekends. For five equally likely outcomes we have $p_i = 1/5$ for each. The expected count for each of the five days is $667(1/5) = 133.4$, and the contribution to the chi-square statistic for Monday is

$$\frac{(\text{observed count} - \text{expected count})^2}{\text{expected count}} = \frac{(133 - 133.4)^2}{133.4} = 0.0012$$

Summing the results for the five weekdays gives $X^2 = 8.49$ with $5 - 1 = 4$ degrees of freedom. The $P$-value is 0.08. For weekdays, the data do not provide evidence in support of differences due to the day of the week in the accident rates for motor vehicle collisions where the driver was using a cell phone.

Many software packages do not provide routines for computing the chi-square goodness of fit test. However, there is a very simple trick that can be used to produce the results from software that can analyze two-way tables. Make a two-way table where the first column contains $k$ cells with the observed counts. Make a second column with counts that correspond *exactly* to the probabilities specified by the null hypothesis, with a very large number of observations. For the problem in Example 9.15, we would have a second column with 10,000 accidents on each of the five days. When analyzed as a two-way table, software gives $X^2 = 8.37$ with $P = 0.08$, which agrees reasonably well with our exact calculations.

## USE YOUR KNOWLEDGE

**9.5 Distribution of M&M colors.** M&M Mars Company has varied the mix of colors for M&M's Milk Chocolate Candies over the years. These changes in color blends are the result of consumer preference tests. Most recently, the color distribution is reported to be 13% brown, 14% yellow, 13% red, 20% orange, 24% blue, and 16% green.[10] You open up a 14-ounce bag of M&M's and find 61 brown, 59 yellow, 49 red, 77 orange, 141 blue, and 88 green. Use a goodness of fit test to examine how well this bag fits the percents stated by the M&M Mars Company.

## SECTION 9.3 Summary

The **chi-square goodness of fit test** is used to compare the sample distribution of a categorical variable from a population with a hypothesized distribution. The data for $n$ observations with $k$ possible outcomes are summarized as observed counts $n_1, n_2, \dots, n_k$ in $k$ cells. The **null hypothesis** specifies probabilities $p_1, p_2, \dots, p_k$ for the possible outcomes.

The analysis of these data is similar to the analyses of two-way tables discussed in Section 9.1. For each cell, the **expected count** is determined by multiplying the total number of observations $n$ by the specified probability $p_i$. The null hypothesis is tested by the usual **chi-square statistic**, which compares the observed counts, $n_i$, with the expected counts. Under the null hypothesis, $X^2$ has approximately the $\chi^2$ distribution with df $= k - 1$.

## CHAPTER 9 Exercises

*For Exercises 9.1 and 9.2, see pages 533 and 534; for Exercises 9.3 and 9.4, see page 544; and for Exercise 9.5, see page 548.*

**9.6 Why not use a chi-square test?** As part of the study on ongoing fright symptoms due to exposure to horror movies at a young age, the following table was created based on the written responses from 119 students.[11] Explain why a chi-square test is not appropriate for this table.

**Percent of students who reported each problem**

| | Type of Problem | | | |
|---|---|---|---|---|
| | Bedtime | | Waking | |
| Movie or video | Short term | Enduring | Short term | Enduring |
| *Poltergeist* ($n = 29$) | 68 | 7 | 64 | 32 |
| *Jaws* ($n = 23$) | 39 | 4 | 83 | 43 |
| *Nightmare on Elm Street* ($n = 16$) | 69 | 13 | 37 | 31 |
| *Thriller* (music video) ($n = 16$) | 40 | 0 | 27 | 7 |
| *It* ($n = 24$) | 64 | 0 | 64 | 50 |
| *The Wizard of Oz* ($n = 12$) | 75 | 17 | 50 | 8 |
| *E.T.* ($n = 11$) | 55 | 0 | 64 | 27 |

**9.7 Age and time status of U.S. college students.** The Census Bureau provides estimates of numbers of people in the United States classified in various ways.[12] Let's look at college students. The following table gives us data to examine the relation between age and full-time or part-time status. The numbers in the table are expressed as thousands of U.S. college students.

**U.S. college students by age and status: October 2004**

| | Status | |
|---|---|---|
| Age | Full-time | Part-time |
| 15–19 | 3553 | 329 |
| 20–24 | 5710 | 1215 |
| 25–34 | 1825 | 1864 |
| 35 and over | 901 | 1983 |

(a) Give the joint distribution of age and status for this table.

(b) What is the marginal distribution of age? Display the results graphically.

(c) What is the marginal distribution of status? Display the results graphically.

(d) Compute the conditional distribution of age for each of the two status categories. Display the results graphically.

(e) Write a short paragraph describing the distributions and how they differ.

**9.8 Time status versus gender for the 20–24 age category.** Refer to Exercise 9.7. The table below breaks down the 20–24 age category by gender.

| | Gender | | |
|---|---|---|---|
| Status | Male | Female | Total |
| Full-time | 2719 | 2991 | 5710 |
| Part-time | 535 | 680 | 1215 |
| Total | 3254 | 3671 | 6925 |

(a) Compute the marginal distribution for gender. Display the results graphically.

(b) Compute the conditional distribution of status for males and for females. Display the results graphically and comment on how these distributions differ.

(c) If you wanted to test the null hypothesis that there is no difference between these two conditional distributions, what would the expected cell counts be for the full-time status row of the table?

(d) Computer software gives $X^2 = 5.17$. Using Table F, give an appropriate bound for the $P$-value and state your conclusions at the 5% level.

**9.9 Does using Rodham matter?** In April 2006, the Opinion Research Corporation conducted a telephone poll for CNN of 1012 adult Americans.[13] Half those polled were asked their opinion of Hillary Rodham Clinton. The other half were asked their opinion of Hillary Clinton. The table below summarizes the results. A chi-square test was used to determine if opinions differed based on the name.

| | Opinion | | | |
|---|---|---|---|---|
| Name | Favorable | Unfavorable | Never heard of | No opinion |
| Hillary Rodham Clinton | 50% | 42% | 2% | 6% |
| Hillary Clinton | 46% | 43% | 2% | 9% |

(a) Computer software gives $X^2 = 4.23$. Can we comfortably use the chi-square distribution to compute the $P$-value? Explain.

(b) What are the degrees of freedom for $X^2$?

(c) Give an appropriate bound for the $P$-value using Table F and state your conclusions.

**9.10 Waking versus bedtime symptoms.** As part of the study on ongoing fright symptoms due to exposure to horror movies at a young age, the following table was presented to describe the lasting impact these movies have had during bedtime and waking life:

| | Waking symptoms | |
|---|---|---|
| Bedtime symptoms | Yes | No |
| Yes | 36 | 33 |
| No | 33 | 17 |

(a) What percent of the students have lasting waking-life symptoms?

(b) What percent of the students have both waking-life and bedtime symptoms?

(c) Test whether there is an association between waking-life and bedtime symptoms. State the null and alternative hypotheses, the $X^2$ statistic, and the $P$-value.

**9.11 New treatment for cocaine addiction.** Cocaine addiction is difficult to overcome. Addicts have been reported to have a significant depletion of stimulating neurotransmitters and thus continue to take cocaine to avoid feelings of depression and anxiety. A 3-year study with 72 chronic cocaine users compared an antidepressant drug called desipramine with lithium and a placebo. (Lithium is a standard drug to treat cocaine addiction. A placebo is a substance containing no medication, used so that the effect of being in the study but not taking any drug can be seen.) One-third of the subjects, chosen at random, received each treatment.[14] Following are the results:

| Treatment | Cocaine relapse? Yes | No |
|---|---|---|
| Desipramine | 10 | 14 |
| Lithium | 18 | 6 |
| Placebo | 20 | 4 |

(a) Compare the effectiveness of the three treatments in preventing relapse using percents and a bar graph. Write a brief summary.

(b) Can we comfortably use the chi-square test to test the null hypothesis that there is no difference between treatments? Explain.

(c) Perform the significance test and summarize the results.

**9.12 Find the degrees of freedom and *P*-value.** For each of the following situations give the degrees of freedom and an appropriate bound on the *P*-value (give the exact value if you have software available) for the $X^2$ statistic for testing the null hypothesis of no association between the row and column variables.

(a) A 2 by 2 table with $X^2 = 1.25$.

(b) A 4 by 4 table with $X^2 = 18.34$.

(c) A 2 by 8 table with $X^2 = 24.21$.

(d) A 5 by 3 table with $X^2 = 12.17$.

**9.13 Can you construct the joint distribution from the marginal distributions?** Here are the row and column totals for a two-way table with two rows and two columns:

| | | |
|---|---|---|
| $a$ | $b$ | 50 |
| $c$ | $d$ | 150 |
| 100 | 100 | 200 |

Find *two different* sets of counts $a$, $b$, $c$, and $d$ for the body of the table. This demonstrates that the relationship between two variables cannot be obtained solely from the two marginal distributions of the variables.

**9.14 Construct a table with no association.** Construct a $3 \times 2$ table of counts where there is no apparent association between the row and column variables.

**9.15** CHALLENGE **Gender versus motivation for volunteer service.** A study examined patterns and characteristics of volunteer-service for young people from high school through early adulthood.[15] Here are some data that can be used to compare males and females on participation in unpaid volunteer service or community service and motivation for participation:

| Gender | Participants: Motivation: Strictly voluntary | Court-ordered | Other | Non-participants |
|---|---|---|---|---|
| Men | 31.9% | 2.1% | 6.3% | 59.7% |
| Women | 43.7% | 1.1% | 6.5% | 48.7% |

Note that the percents in each row sum to 100%.

(a) Graphically compare the volunteer-service profiles for men and women. Describe any differences that are striking.

(b) Find the proportion of men who volunteer. Do the same for women. Refer to the section on relative risk in Chapter 8 (page 515) and the discussion on page 535 of this chapter. Compute the relative risk of being a volunteer for females versus males. Write a clear sentence contrasting females and males using relative risk as your numerical summary.

**9.16** CHALLENGE **Gender versus motivation for volunteer service, continued.** Refer to the previous exercise. Recompute the table for volunteers only. To do this take the entries for each motivation and divide by the percent of volunteers. Do this separately for each gender. Verify that the percents sum to 100% for each gender. Give a graphical summary to compare the motivation of men and women who are volunteers. Compare this with your summary in part (a) of the previous exercise, and write a short paragraph describing similarities and differences in these two views of the data.

**9.17 Drinking status and class attendance.** As part of the 1999 College Alcohol Study, students who drank alcohol in the last year were asked if drinking ever resulted in missing a class.[16] The data are given in the following table:

| Missed a class | Drinking status: Nonbinger | Occasional binger | Frequent binger |
|---|---|---|---|
| No | 4617 | 2047 | 1176 |
| Yes | 446 | 915 | 1959 |

(a) Summarize the results of this table graphically and numerically.

(b) What is the marginal distribution of drinking status? Display the results graphically.

(c) Compute the relative risk of missing a class for occasional bingers versus nonbingers and for frequent bingers versus nonbingers. Summarize these results.

(d) Perform the chi-square test for this two-way table. Give the test statistic, degrees of freedom, the *P*-value, and your conclusion.

**9.18 Sexual imagery in magazine ads.** In what ways do advertisers in magazines use sexual imagery to appeal to youth? One study classified each of 1509 full-page or larger ads as "not sexual" or "sexual," according to the amount and style of the dress of the male or female model in the ad. The ads were also classified according to the target readership of the magazine.[17] Here is the two-way table of counts:

| | Magazine readership | | | |
|---|---|---|---|---|
| **Model dress** | **Women** | **Men** | **General interest** | **Total** |
| Not sexual | 351 | 514 | 248 | 1113 |
| Sexual | 225 | 105 | 66 | 396 |
| Total | 576 | 619 | 314 | 1509 |

(a) Summarize the data numerically and graphically.

(b) Perform the significance test that compares the model dress for the three categories of magazine readership. Summarize the results of your test and give your conclusion.

(c) All of the ads were taken from the March, July, and November issues of six magazines in one year. Discuss this fact from the viewpoint of the validity of the significance test and the interpretation of the results.

**9.19 Intended readership of ads with sexual imagery.** The ads in the study described in the previous exercise were also classified according to the age group of the intended readership. Here is a summary of the data:

| | Magazine readership age group | |
|---|---|---|
| **Model dress** | **Young adult** | **Mature adult** |
| Not sexual | 72.3% | 76.1% |
| Sexual | 27.2% | 23.9% |
| Number of ads | 1006 | 503 |

Using parts (a) and (b) in the previous exercise as a guide, analyze these data and write a report summarizing your work.

**9.20 Air pollution from a steel mill.** One possible effect of air pollution is genetic damage. A study designed to examine this problem exposed one group of mice to air near a steel mill and another group to air in a rural area and compared the numbers of mutations in each group.[18] Here are the data for a mutation at the *Hm-2* gene locus:

| | Location | |
|---|---|---|
| **Mutation** | **Steel mill air** | **Rural air** |
| Yes | 30 | 23 |
| No | | |
| Total | 96 | 150 |

(a) Fill in the missing entries in the table.

(b) Summarize the data numerically and graphically.

(c) Is there evidence to conclude that the location is related to the occurrence of mutations? Perform the significance test and summarize the results.

**9.21 Dieting trends among male and female undergraduates.** A recent study of undergraduates looked at gender differences in dieting trends.[19] There were 181 women and 105 men who participated in the survey. The table below summarizes whether a student tried a low-fat diet or not by gender:

| | Gender | |
|---|---|---|
| **Tried low-fat diet** | **Women** | **Men** |
| Yes | 35 | 8 |
| No | | |

(a) Fill in the missing cells of the table.

(b) Summarize the data numerically and graphically.

(c) Test that there is no association between gender and the likelihood of trying a low-fat diet. Summarize the results.

**9.22 Identity theft.** A study of identity theft looked at how well consumers protect themselves from this increasingly prevalent crime. The behaviors of 61 college students were compared with the behaviors of 59 nonstudents.[20] One of the questions was "When asked to create a password, I have used either my mother's maiden name, or my pet's name, or my birth date, or the last four digits of my social security number, or a series of consecutive numbers." For the students, 22 agreed with this statement while 30 of the nonstudents agreed.

(a) Display the data in a two-way table and perform the chi-square test. Summarize the results.

(b) Reanalyze the data using the methods for comparing two proportions that we studied in the previous chapter. Compare the results and verify that the chi-square statistic is the square of the $z$ statistic.

(c) The students in this study were junior and senior college students from two sections of a course in Internet marketing at a large northeastern university. The nonstudents were a group of individuals who were recruited to attend commercial focus groups on the West Coast conducted by a lifestyle marketing organization. Discuss how the method of selecting the subjects in this study relates to the conclusions that can be drawn from it.

**9.23** CHALLENGE **Student-athletes and gambling.** A survey of student-athletes that asked questions about gambling behavior classified students according to the National Collegiate Athletic Association (NCAA) division.[21] For male student-athletes, the percents who reported wagering on collegiate sports are given here along with the numbers of respondents in each division:

| Division | I | II | III |
|---|---|---|---|
| Percent | 17.2% | 21.0% | 24.4% |
| Number | 5619 | 2957 | 4089 |

(a) Use a significance test to compare the percents for the three NCAA divisions. Give details and a short summary of your conclusion.

(b) The percents in the table above are given in the NCAA report, but the numbers of male student-athletes in each division who responded to the survey question are estimated based on other information in the report. To what extent do you think this has an effect on the results? (*Hint:* Rerun your analysis a few times, with slightly different numbers of students but the same percents.)

(c) Some student-athletes may be reluctant to provide this kind of information, even in a survey where there is no possibility that they can be identified. Discuss how this fact may affect your conclusions.

(d) The chi-square test for this set of data assumes that the responses of the student-athletes are independent. However, some of the students are at the same school and even on the same team. Discuss how you think this might affect the results.

**9.24 Which model?** Refer to Exercises 9.17, 9.18, 9.20, and 9.23. For each, state whether you are comparing two or more populations (the first model for two-way tables) or testing independence between two categorical variables (the second model).

**9.25 Hummingbirds of Santa Lucia.** *E. jugularis* is a type of hummingbird that lives in the forest preserves of the Carribean island of Santa Lucia. The males and the females of this species have bills that are shaped somewhat differently. Researchers who study these birds thought that the bill shape might be related to the shape of the flowers that they visit for food. The researchers observed 49 females and 21 males. Of the females, 20 visited the flowers of *H. bihai*, while none of the males visited these flowers.[22] Display the data in a two-way table and perform the chi-square test. Summarize the results and give a brief statement of your conclusion. Your two-way table has a count of zero in one cell. Does this invalidate your significance test? Explain why or why not.

**9.26 Internet references in prominent journals.** The World Wide Web (WWW) has led to an enormous increase in the amount of information that is easily available to anyone with Internet access. References to Internet pages are becoming quite common in the scientific literature. One study examined Internet references in articles in three prominent journals: the *New England Journal of Medicine* (*NEJM*), the *Journal of the American Medical Association* (*JAMA*), and *Science*.[23] In one part of the study, Internet references were classified according to the top-level domain. Here are the data:

| Top-level domain | Journal *NEJM* | *JAMA* | *Science* |
|---|---|---|---|
| .gov | 41 | 103 | 111 |
| .org | 37 | 46 | 162 |
| .com | 6 | 17 | 14 |
| .edu | 4 | 8 | 47 |
| Other | 9 | 15 | 52 |

Analyze the data. Include numerical and graphical summaries as well as a significance test. Summarize your results and conclusions.

**9.27 Pet ownership and education level.** The Health, Aging, and Body Composition (Health ABC) study is a 10-year study of older adults. A research project based on this study examined the relationship between physical activity and pet ownership.[24] The data collected included information concerning pet owner characteristics and the type of pet owned. Here is a table of counts of subjects classified by pet ownership status and education level:

| Education level | Pet ownership status: Non–pet owners | Dog owners | Cat owners |
|---|---|---|---|
| Less than high school | 421 | 93 | 28 |
| High school graduate | 666 | 100 | 40 |
| Postsecondary | 845 | 135 | 99 |

Note that "Dog owners" and "Cat owners" designate individuals who own a dog only or a cat only, respectively. Individuals who own both a dog and a cat are not included in this table. Analyze the data. Include numerical and graphical summaries as well as a significance test. Summarize your results and conclusions.

**9.28 Pet ownership and gender.** Refer to the previous exercise. Here are similar data giving the relationship between pet ownership status and gender:

| Gender | Pet ownership status: Non–pet owners | Dog owners | Cat owners |
|---|---|---|---|
| Female | 1024 | 157 | 85 |
| Male | 915 | 171 | 82 |

Analyze the data. Include numerical and graphical summaries as well as a significance test. Summarize your results and conclusions.

**9.29 Changing majors.** A task force set up to examine retention of students in the majors that they chose when starting college examined data on transfers to other majors.[25] Here are some data giving counts of students classified by initial major and the area that they transferred to:

| Initial major | Area transferred to: Engineering | Management | Liberal arts | Other | Total |
|---|---|---|---|---|---|
| Biology | 13 | 25 | 158 | | 398 |
| Chemistry | 16 | 15 | 19 | | 114 |
| Mathematics | 3 | 11 | 20 | | 72 |
| Physics | 9 | 5 | 14 | | 61 |

Complete the table by computing the values for the "Other" column. Write a short paragraph explaining what conclusions you can draw about the relationship between initial major and area transferred to. Be sure to include numerical and graphical summaries as well as the details of your significance test.

**9.30 Secondhand stores.** Shopping at secondhand stores is becoming more popular and has even attracted the attention of business schools. A study of customers' attitudes toward secondhand stores interviewed samples of shoppers at two secondhand stores of the same chain in two cities. The breakdown of the respondents by gender is as follows:[26]

| Gender | City 1 | City 2 |
|---|---|---|
| Men | 38 | 68 |
| Women | 203 | 150 |
| Total | 241 | 218 |

Is there a significant difference between the proportions of women customers in the two cities?

(a) State the null hypothesis, find the sample proportions of women in both cities, do a two-sided $z$ test, and give a $P$-value using Table A.

(b) Calculate the $X^2$ statistic and show that it is the square of the $z$ statistic. Show that the $P$-value from Table F agrees (up to the accuracy of the table) with your result from (a).

(c) Give a 95% confidence interval for the difference between the proportions of women customers in the two cities.

**9.31 More on secondhand stores.** The study of shoppers in secondhand stores cited in the previous exercise also compared the income distributions of shoppers in the two stores. Here is the two-way table of counts:

| Income | City 1 | City 2 |
|---|---|---|
| Under $10,000 | 70 | 62 |
| $10,000 to $19,999 | 52 | 63 |
| $20,000 to $24,999 | 69 | 50 |
| $25,000 to $34,999 | 22 | 19 |
| $35,000 or more | 28 | 24 |

Verify that the $X^2$ statistic for this table is $X^2 = 3.955$. Give the degrees of freedom and the *P*-value. Is there good evidence that customers at the two stores have different income distributions?

**9.32** CHALLENGE **Cracks in veneer.** Many furniture pieces are built with veneer, a thin layer of fine wood that is fastened to less expensive wood products underneath. Face checks are cracks that sometimes develop in the veneer. When face checks appear, the furniture needs to be reconstructed. Because this is a fairly expensive process, researchers seek ways to minimize the occurrence of face checks by controlling the manufacturing process. In one study, the type of adhesive used was one of the factors examined.[27] Because of the way that the veneer is cut, it has two different sides, called loose and tight, either of which can face out. Here is a table giving the numbers of veneer panels with and without face checks for two different adhesives, PVA and UF. Separate columns are given for the loose side and the tight side.

| | Loose side | | Tight side | |
|---|---|---|---|---|
| | Face checks | | Face checks | |
| Adhesive | No | Yes | No | Yes |
| PVA | 10 | 54 | 44 | 20 |
| UF | 21 | 43 | 37 | 27 |

Analyze the data. Write a summary of your results concerning the relationship between the adhesive and the occurrence of face checks. Be sure to include numerical and graphical summaries as well as the details of your significance tests.

**9.33** CHALLENGE **Why are animals brought to animal shelters?** Euthanasia of healthy but unwanted pets by animal shelters is believed to be the leading cause of death for cats and dogs. A study designed to find factors associated with bringing a cat to an animal shelter compared data on cats that were brought to an animal shelter with data on cats from the same county that were not brought in.[28] One of the factors examined was the source of the cat: the categories were private owner or breeder, pet store, and other (includes born in home, stray, and obtained from a shelter). This kind of study is called a **case-control study** by epidemiologists. Here are the data:

| | Source | | |
|---|---|---|---|
| Group | Private | Pet store | Other |
| Cases | 124 | 16 | 76 |
| Controls | 219 | 24 | 203 |

The same researchers did a similar study for dogs.[29] The data are given in the following table:

| | Source | | |
|---|---|---|---|
| Group | Private | Pet store | Other |
| Cases | 188 | 7 | 90 |
| Controls | 518 | 68 | 142 |

(a) Analyze the data for the dogs and the cats separately. Be sure to include graphical and numerical summaries. Is there evidence to conclude that the source of the animal is related to whether or not the pet is brought to an animal shelter?

(b) Write a discussion comparing the results for the cats with those for the dogs.

(c) These data were collected using a telephone interview with pet owners in Mishawaka, Indiana. The animal shelter was run by the Humane Society of Saint Joseph County. The control group data were obtained by a random digit dialing telephone survey. Discuss how these facts relate to your interpretation of the results.

**9.34 Student loans.** A study of 865 college students found that 42.5% had student loans.[30] The students were randomly selected from the approximately 30,000 undergraduates enrolled in a large public university. The overall purpose of the study was to examine the effects of student loan burdens on the choice of a career. A student with a large debt may be more likely to choose a field where starting salaries are high so that the loan can more easily be repaid. The following table classifies the students by field of study and whether or not they have a loan:

| | Student loan | |
|---|---|---|
| **Field of study** | **Yes** | **No** |
| Agriculture | 32 | 35 |
| Child development and family studies | 37 | 50 |
| Engineering | 98 | 137 |
| Liberal arts and education | 89 | 124 |
| Management | 24 | 51 |
| Science | 31 | 29 |
| Technology | 57 | 71 |

Carry out a complete analysis of the association between having a loan and field of study, including a description of the association and an assessment of its statistical significance.

**9.35 Altruism and field of study.** In one part of the study described in the previous exercise, students were asked to respond to some questions regarding their interests and attitudes. Some of these questions form a scale called PEOPLE that measures altruism, or an interest in the welfare of others. Each student was classified as low, medium, or high on this scale. Is there an association between PEOPLE score and field of study? Here are the data:

| | PEOPLE score | | |
|---|---|---|---|
| **Field of study** | **Low** | **Medium** | **High** |
| Agriculture | 5 | 27 | 35 |
| Child development and family studies | 1 | 32 | 54 |
| Engineering | 12 | 129 | 94 |
| Liberal arts and education | 7 | 77 | 129 |
| Management | 3 | 44 | 28 |
| Science | 7 | 29 | 24 |
| Technology | 2 | 62 | 64 |

Analyze the data and summarize your results. Are there some fields of study that have very large or very small proportions of students in the high-PEOPLE category?

**9.36** CHALLENGE **"No Sweat" label.** Following complaints about the working conditions in some apparel factories both in the United States and abroad, a joint government and industry commission recommended in 1998 that companies that monitor and enforce proper standards be allowed to display a "No Sweat" label on their products. Does the presence of these labels influence consumer behavior? A survey of U.S. residents aged 18 or older asked a series of questions about how likely they would be to purchase a garment under various conditions. For some conditions, it was stated that the garment had a "No Sweat" label; for others, there was no mention of such a label. On the basis of the responses, each person was classified as a "label user" or a "label nonuser."[31] There were 296 women surveyed. Of these, 63 were label users. On the other hand, 27 of 251 men were classified as users.

(a) Construct the $2 \times 2$ table of counts for this problem. Include the marginal totals for your table.

(b) Use a $X^2$ statistic to examine the question of whether or not there is a relationship between gender and use of No Sweat labels. Give the test statistic, degrees of freedom, the $P$-value, and your conclusion.

(c) You examined this question using the methods of the previous chapter in Exercise 8.64 (page 521). Verify that if you square the $z$ statistic you calculated for that exercise, you obtain the $X^2$ statistic that you calculated for this exercise.

**9.37 Are Mexican Americans less likely to be selected as jurors?** Refer to Exercise 8.81 (page 523) concerning *Castaneda v. Partida,* the case where the Supreme Court review used the phrase "two or three standard deviations" as a criterion for statistical significance. Recall that there were 181,535 persons eligible for jury duty, of whom 143,611 were Mexican Americans. Of the 870 people selected for jury duty, 339 were Mexican Americans. We are interested in finding out if there is an association between being a Mexican American and being selected as a juror. Formulate this problem using a two-way table of counts. Construct the $2 \times 2$ table using the variables Mexican American or not and juror or not. Find the $X^2$ statistic and its $P$-value. Square the $z$ statistic that you obtained in Exercise 8.81 and verify that the result is equal to the $X^2$ statistic.

**9.38** CHALLENGE **More on why animals are brought to animal shelters.** Refer to Exercise 9.33 (page 554) concerning the case-control study of factors associated with bringing a cat to an animal shelter and the similar study for dogs. The last category for the source of the pet was given as "Other" and includes born in home, stray, and obtained from a shelter. The following two-way table lists these categories separately for cats:

| | Source | | | | |
|---|---|---|---|---|---|
| Group | Private | Pet store | Home | Stray | Shelter |
| Cases | 124 | 16 | 20 | 38 | 18 |
| Controls | 219 | 24 | 38 | 116 | 49 |

Here is the same breakdown for dogs:

| | Source | | | | |
|---|---|---|---|---|---|
| Group | Private | Pet store | Home | Stray | Shelter |
| Cases | 188 | 7 | 11 | 23 | 56 |
| Controls | 518 | 68 | 20 | 55 | 67 |

Analyze these 2 × 5 tables and compare the results with those that you obtained for the 2 × 3 tables in Exercise 9.33. With a large number of cells, the chi-square test sometimes does not have very much power.

**9.39** CHALLENGE **Evaluation of an herbal remedy.** A study designed to evaluate the effects of the herbal remedy *Echinacea purpurea* randomly assigned healthy children who were 2 to 11 years old to receive either echinacea or a placebo.[32] Each time a child had an upper respiratory infection (URI) treatment with echinacea or the placebo was given for the duration of the URI. The dose for the echinacea was based on the age of the child according to the recommendation of the manufacturer. The echinacea children had 329 URIs, while the placebo children had 367 URIs. For each URI many variables were measured. One of these was the parental assessment of the illness severity. Here are the data:

| | Group | |
|---|---|---|
| Parental assessment | Echinacea | Placebo |
| Mild | 153 | 170 |
| Moderate | 128 | 157 |
| Severe | 48 | 40 |

They also recorded the presence or absence of various types of *adverse events*. Here is a summary:

| | Group | |
|---|---|---|
| Adverse event | Echinacea | Placebo |
| Itchiness | 13 | 7 |
| Rash | 24 | 10 |
| "Hyper" behavior | 30 | 23 |
| Diarrhea | 38 | 34 |
| Vomiting | 22 | 21 |
| Headache | 33 | 24 |
| Stomachache | 52 | 41 |
| Drowsiness | 63 | 48 |
| Other | 63 | 48 |
| Any adverse event | 152 | 146 |

(a) Analyze the parental assessment data. Write a summary of your analysis and conclusion. Be sure to include graphical and numerical summaries.

(b) Analyze each adverse event. Display the results graphically in a single graph. Make a table of the relevant descriptive statistics.

(c) Use a statistical significance test to compare the echinacea URIs with those of the placebo URIs for each type of adverse event. Summarize the results in a table and write a short report giving your conclusions about the effect of echinacea on URIs in healthy children who are 2 to 11 years old. Explain why you need to analyze each type of adverse event separately rather than performing a chi-square test on the 10 × 2 table above.

(d) One concern about analyzing several outcome variables in situations like this is that we may be able to find statistical significance by chance if we look at a sufficiently large number of outcome variables. Explain why this is a concern in general but is not a concern that is important for the interpretation of the results of your analysis here.

(e) The authors of the paper describing these results note that the unit of analysis for their computations is the URI and not the child. They state that similar results were found using more sophisticated statistical methods. Based on the descriptive statistics that you have computed, are you inclined to agree or to disagree with this statement of the authors? Explain your answer.

(f) This study was published in the *Journal of the American Medical Association* (*JAMA*) and was criticized in an article that appeared in *Alternative & Complementary Therapies*.[33] Three herbalists gave responses to the original article. Among their criticisms were (*i*) the dose of echinacea was too low, (*ii*) the treatment should have been given before the URI, not at the onset of symptoms, (*iii*) we should be skeptical of any positive trials on pharmaceuticals or negative trials on natural remedies that are published in *JAMA*, and (*iv*) *Echinacea angustifolia* (not *E. purpurea*) should have been used in combination with other herbs. Discuss these criticisms and write a summary of your opinions regarding echinacea.

*The following exercises concern the optional material on goodness of fit discussed in Section 9.3.*

**9.40 Is there a random distribution of trees?** In Example 6.1 (page 354) we examined data concerning the longleaf pine trees in the Wade Tract and concluded that the distribution of trees in the tract was not random. Here is another way to examine the same question. First, we divide the tract into four equal parts, or quadrants, in the east-west direction. Call the four parts, $Q_1$ to $Q_4$. Then we take a random sample of 100 trees and count the number of trees in each quadrant. Here are the data:

| Quadrant | $Q_1$ | $Q_2$ | $Q_3$ | $Q_4$ |
|---|---|---|---|---|
| Count | 18 | 22 | 39 | 21 |

(a) If the trees are randomly distributed, we expect to find 25 trees in each quadrant. Why? Explain your answer.

(b) We do not really expect to get *exactly* 25 trees in each quadrant. Why? Explain your answer.

(c) Perform the goodness of fit test for these data to determine if these trees are randomly scattered. Write a short report giving the details of your analysis and your conclusion.

**9.41 Use of academic assistance services.** The 2005 National Survey of Student Engagement reported on the use of campus services during the first year of college. [34] In terms of academic assistance (for example tutoring, writing lab), 43% never used the services, 35% sometimes used the services, 15% often used the services, and 7% very often used the services. You decide to see if your large university has this same distribution. You survey first-year students and obtain the counts 79, 83, 36, and 12 respectively. Use a goodness of fit test to examine how well your university reflects the national average.

**9.42 Goodness of fit to a standard Normal distribution.** Computer software generated 500 random numbers that should look like they are from the standard Normal distribution. They are categorized into five groups: (1) less than or equal to −0.6, (2) greater than −0.6 and less than or equal to −0.1, (3) greater than −0.1 and less than or equal to 0.1, (4) greater than 0.1 and less than or equal to 0.6, and (5) greater than 0.6. The counts in the five groups are 139, 102, 41, 78, and 140, respectively. Find the probabilities for these five intervals using Table A. Then compute the expected number for each interval for a sample of 500. Finally, perform the goodness of fit test and summarize your results.

**9.43 More on the goodness of fit to a standard Normal distribution.** Refer to the previous exercise. Use software to generate your own sample of 500 standard Normal random variables, and perform the goodness of fit test. Choose a different set of intervals from the ones used in the previous exercise.

**9.44 Goodness of fit to the uniform distribution.** Computer software generated 500 random numbers that should look like they are from the uniform distribution on the interval 0 to 1 (see page 263). They are categorized into five groups: (1) less than or equal to 0.2, (2) greater than 0.2 and less than or equal to 0.4, (3) greater than 0.4 and less than or equal to 0.6, (4) greater than 0.6 and less than or equal to 0.8, and (5) greater than 0.8. The counts in the five groups are 114, 92, 108, 101, and 85, respectively. The probabilities for these five intervals are all the same. What is this probability? Compute the expected number for each interval for a sample of 500. Finally, perform the goodness of fit test and summarize your results.

**9.45 More on goodness of fit to the uniform distribution.** Refer to the previous exercise. Use software to generate your own sample of 500 uniform random variables on the interval from 0 to 1, and perform the goodness of fit test. Choose a different set of intervals from the ones used in the previous exercise.

**9.46** CHALLENGE **Suspicious results?** An instructor who assigned an exercise similar to the one described in the previous exercise received homework from a student who reported a *P*-value of 0.999. The instructor suspected that the student did not use the computer for the assignment but just made up some numbers for the homework. Why was the instructor suspicious? How would this scenario change if there were 1000 students in the class?

# Inference for Regression

Previously we looked at the average property damage per year due to tornadoes. What about the frequency of tornadoes? Has the annual number of reported tornadoes increased over time? See Exercises 10.23 and 10.24 for more details.



## Introduction

In this chapter we describe methods for inference when there is a single quantitative response variable and a single quantitative explanatory variable. The descriptive tools we learned in Chapter 2—scatterplots, least-squares regression, and correlation—are essential preliminaries to inference and also provide a foundation for confidence intervals and significance tests.

We first met the sample mean $\overline{x}$ in Chapter 1 as a measure of the center of a collection of observations. Later we learned that when the data are a random sample from a population, the sample mean is an estimate of the population mean $\mu$. In Chapters 6 and 7, we used $\overline{x}$ as the basis for confidence intervals and significance tests for inference about $\mu$.

Now we will follow the same approach for the problem of fitting straight lines to data. In Chapter 2 we met the least-squares regression line $\hat{y} = b_0 + b_1x$ as a description of a straight-line relationship between a response variable $y$ and an explanatory variable $x$. At that point we did not distinguish between sample and population. Now we will think of the least-squares line computed from a sample as an estimate of a *true* regression line for the population.

Following the common practice of using Greek letters for population parameters, we will write the population line as $\beta_0 + \beta_1x$. This notation reminds us that the intercept $b_0$ of the fitted line estimates the intercept $\beta_0$ of the population line, and the slope $b_1$ estimates the slope $\beta_1$.

The methods detailed in this chapter will help us answers questions such as:

- Is the trend in the annual number of tornadoes reported in the United States linear? If so, what is the average yearly increase in the number of tornadoes? How many are predicted for next year?
- What is the relationship between the selling price of a home and the number of bathrooms that it contains?
- Among North American universities, is there a strong correlation between the binge-drinking rate and the average price for a bottle of beer at establishments within a 2-mile radius of campus?

# 10.1 Simple Linear Regression

## Statistical model for linear regression

Simple linear regression studies the relationship between a response variable $y$ and a single explanatory variable $x$. We expect that different values of $x$ will produce different mean responses. We encountered a similar but simpler situation in Chapter 7 when we discussed methods for comparing two population means. Figure 10.1 illustrates the statistical model for a comparison of blood pressure change in two groups of experimental subjects, one group taking a calcium supplement and the other a placebo. We can think of the treatment (placebo or calcium) as the explanatory variable in this example. This model has two important parts:

- The mean change may be different in the two populations. These means are labeled $\mu_1$ and $\mu_2$ in Figure 10.1.
- Individual changes in blood pressure vary within each population according to a Normal distribution. The two Normal curves in Figure 10.1 describe the individual responses. These Normal distributions have the same spread, indicating that the population standard deviations are assumed to be equal.

In linear regression the explanatory variable $x$ is quantitative and can have many different values. Imagine, for example, giving different amounts $x$ of calcium to different groups of subjects. We can think of the values of $x$ as defining

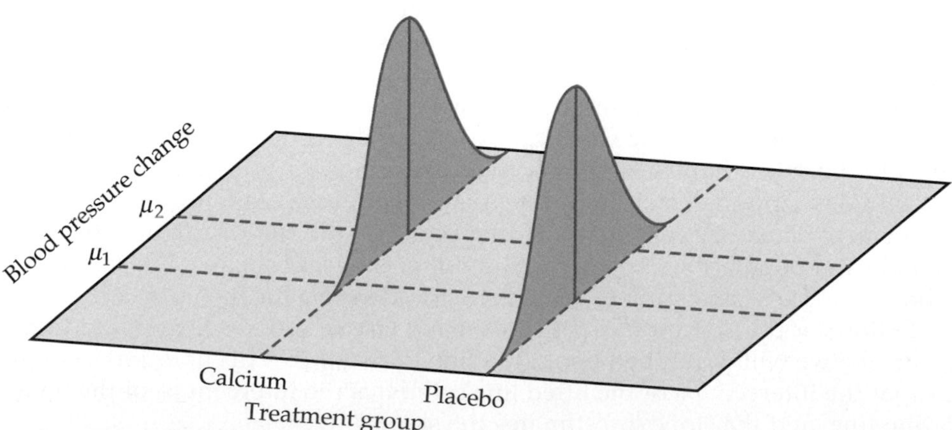


**FIGURE 10.1** The statistical model for comparing responses to two treatments; the mean response varies with the treatment.

subpopulations

different **subpopulations,** one for each possible value of $x$. Each subpopulation consists of all individuals in the population having the same value of $x$. If we conducted an experiment with five different amounts of calcium, we could view these values as defining five different subpopulations.

simple linear regression

The statistical model for simple linear regression also assumes that for each value of $x$ the observed values of the response variable $y$ are Normally distributed with a mean that depends on $x$. We use $\mu_y$ to represent these means. In general, the means $\mu_y$ can change as $x$ changes according to any sort of pattern. In **simple linear regression** we assume the means all lie on a line when plotted against $x$. To summarize, this model also has two important parts:

- The mean of the response variable $y$ changes as $x$ changes. The means all lie on a straight line. That is, $\mu_y = \beta_0 + \beta_1 x$.
- Individual responses of $y$ with the same $x$ vary according to a Normal distribution. These Normal distributions all have the same standard deviation.

This statistical model is pictured in Figure 10.2. Rather than just two means $\mu_1$ and $\mu_2$, we are interested in how the many means $\mu_y$ change as $x$ changes. The simple linear regression model assumes that they all lie on a line when plotted against $x$. The equation of the line is

$$\mu_y = \beta_0 + \beta_1 x$$

population regression line

with intercept $\beta_0$ and slope $\beta_1$. This is the **population regression line;** it describes how the mean response changes with $x$. The line in Figure 10.2 is the population regression line. Observed $y$'s will vary about these means. The three Normal curves show how the response $y$ will vary for three different values of the explanatory variable $x$. The model assumes that this variation, measured by the standard deviation $\sigma$, is the same for all values of $x$.

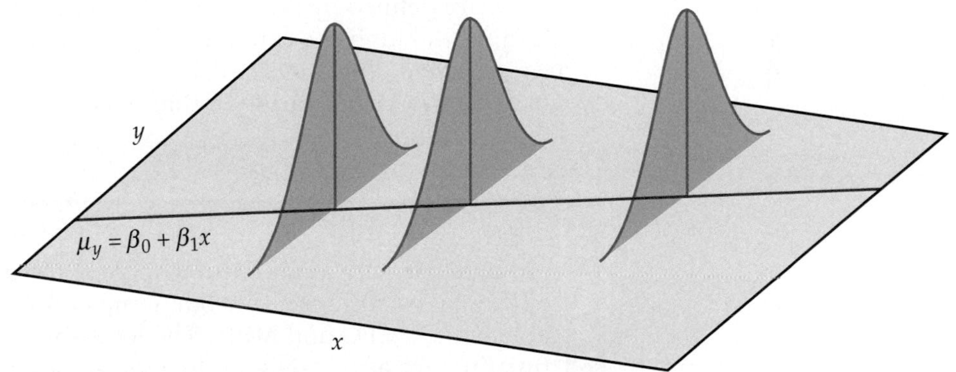


**FIGURE 10.2** The statistical model for linear regression; the mean response is a straight-line function of the explanatory variable.

## Data for simple linear regression

The data for a linear regression are observed values of $y$ and $x$. The model takes each $x$ to be a fixed known quantity. In practice, $x$ may not be exactly known. *If the error in measuring x is large, more advanced inference methods are needed.* The response $y$ to a given $x$ is a random variable. The linear regression model describes the mean and standard deviation of this random variable $y$. These unknown parameters must be estimated from the data.

LOOK BACK
**random variable, page 259**

We will use the following example to explain the fundamentals of simple linear regression. Because regression calculations in practice are always done by statistical software, we will rely on computer output for the arithmetic. In the next section, we give an example that illustrates how to do the work with a calculator if software is unavailable.

EXAMPLE

**10.1 Relationship between speed driven and fuel efficiency.** Computers in some vehicles calculate various quantities related to the vehicle's performance. One of these is the fuel efficiency, or gas mileage, expressed as miles per gallon (mpg). Another is the average speed in miles per hour (mph). For one vehicle equipped in this way, mpg and mph were recorded each time the gas tank was filled, and the computer was then reset.[1] How does the speed at which the vehicle is driven affect the fuel efficiency? There are 234 observations available. We will work with a simple random sample of size 60.

Before starting our analysis, it is appropriate to consider the extent to which our results can reasonably be generalized. Because we have a simple random sample from a population of size 234, we are on firm ground in making inferences about this particular vehicle. However, as a practical matter, no one really cares about this particular vehicle. Our results are interesting only if they can be applied to other similar vehicles that are driven under similar conditions. Our statistical modeling for this data set is concerned about the process by which speed affects the fuel efficiency. Although we would not expect the parameters that describe the relationship between speed and fuel efficiency to be *exactly* the same for similar vehicles, we would expect to find qualitatively similar results.

In the statistical model for predicting fuel efficiency from speed, subpopulations are defined by the explanatory variable, speed. For a particular value of speed, say 30 mph, we can think about operating this vehicle repeatedly at this average speed. Variation in driving conditions and the behavior of the driver would be sources of variation that would give different values of mpg for this subpopulation.

EXAMPLE

**10.2 Graphical display of the fuel efficiency relationship.** We start our analysis with a graphical display of the data. Figure 10.3 is a plot of fuel efficiency versus speed for our sample of 60 observations. We use the variable names MPG and MPH. The least-squares regression line and a smooth function are also shown in the plot. Although there is a positive association between MPG and MPH, the fit is not linear. The smooth function shows us that the relationship levels off somewhat with increasing speed.

*Always start with a graphical display of the data.* There is no point in trying to do statistical inference if our data do not, at least approximately, meet the assumptions that are the foundation for our inference. At this point we need to make a choice. One possibility would be to confine our interest to speeds that are 30 mph or less, a region where it appears that a line would be a good fit to the data. Another possibility is to make some sort of transformation that will

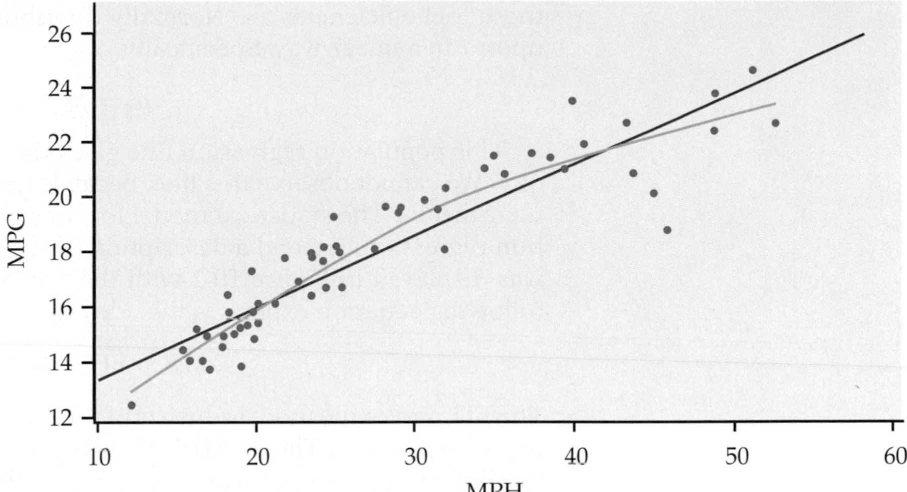


**FIGURE 10.3** Scatterplot of MPG versus MPH with a smooth function and the least-squares line, for Example 10.2.

make the relationship approximately linear for the entire set of data. We will choose the second option.

EXAMPLE

**10.3 Is this relationship linear?** One type of function that looks similar to the smooth-function fit in Figure 10.3 is a logarithm. Therefore, we will examine the effect of transforming speed by taking the natural logarithm. The result is shown in Figure 10.4. In this plot the smooth function and the line are quite close. We are satisfied that the relationship between the log of speed and fuel efficiency is approximately linear for this set of data.

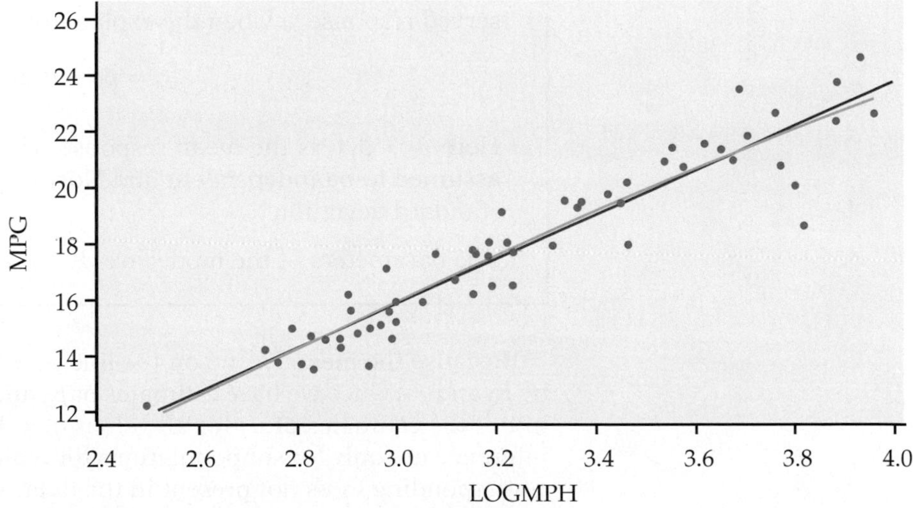


**FIGURE 10.4** Scatterplot of MPG versus logarithm of MPH with a smooth function and the least-squares line, for Example 10.3.

Now that we have an approximate linear relationship, we return to predicting fuel efficiency for different subpopulations, defined by the explanatory variable speed. Consider a particular value of speed, for example 30 mph, which in Figure 10.4 would be $x = \log(30) = 3.4$. Our statistical model assumes that these fuel efficiencies are Normally distributed with a mean $\mu_y$ that depends upon $x$ in a linear way. Specifically,

$$\mu_y = \beta_0 + \beta_1 x$$

This population regression line gives the mean fuel efficiency for all values of $x$. We cannot observe this line, because the observed responses $y$ vary about their means. The statistical model for linear regression consists of the population regression line and a description of the variation of $y$ about the line. This was displayed in Figure 10.2 with the line and the three Normal curves. The following equation expresses this idea in an equation:

$$\text{DATA} = \text{FIT} + \text{RESIDUAL}$$

The FIT part of the model consists of the subpopulation means, given by the expression $\beta_0 + \beta_1 x$. The RESIDUAL part represents deviations of the data from the line of population means. We assume that these deviations are Normally distributed with standard deviation $\sigma$. We use $\epsilon$ (the Greek letter epsilon) to stand for the RESIDUAL part of the statistical model. A response $y$ is the sum of its mean and a chance deviation $\epsilon$ from the mean. The deviations $\epsilon$ represent "noise," that is, variation in $y$ due to other causes that prevent the observed $(x, y)$-values from forming a perfectly straight line on the scatterplot.

### SIMPLE LINEAR REGRESSION MODEL

Given $n$ observations of the explanatory variable $x$ and the response variable $y$,

$$(x_1, y_1),\ (x_2, y_2), \dots,\ (x_n, y_n)$$

the **statistical model for simple linear regression** states that the observed response $y_i$ when the explanatory variable takes the value $x_i$ is

$$y_i = \beta_0 + \beta_1 x_i + \epsilon_i$$

Here $\beta_0 + \beta_1 x_i$ is the mean response when $x = x_i$. The deviations $\epsilon_i$ are assumed to be independent and Normally distributed with mean 0 and standard deviation $\sigma$.

The parameters of the model are $\beta_0$, $\beta_1$, and $\sigma$.

Because the means $\mu_y$ lie on the line $\mu_y = \beta_0 + \beta_1 x$, they are all determined by $\beta_0$ and $\beta_1$. Once we have estimates of $\beta_0$ and $\beta_1$, the linear relationship determines the estimates of $\mu_y$ for all values of $x$. Linear regression allows us to do inference not only for subpopulations for which we have data but also for those corresponding to $x$'s not present in the data. We will learn how to do inference about

- the slope $\beta_1$ and the intercept $\beta_0$ of the population regression line,
- the mean response $\mu_y$ for a given value of $x$, and
- an individual future response $y$ for a given value of $x$.

## Estimating the regression parameters

LOOK BACK
least-squares regression, page 112

The method of least squares presented in Chapter 2 fits a line to summarize a relationship between the observed values of an explanatory variable and a response variable. Now we want to use the least-squares line as a basis for inference about a population from which our observations are a sample. We can do this only when the statistical model just presented holds. In that setting, the slope $b_1$ and intercept $b_0$ of the least-squares line

$$\hat{y} = b_0 + b_1 x$$

estimate the slope $\beta_1$ and the intercept $\beta_0$ of the population regression line.

LOOK BACK
least-squares equations, page 114
correlation, page 102
unbiased estimator, page 217

Using the formulas from Chapter 2, the slope of the least-squares line is

$$b_1 = r\frac{s_y}{s_x}$$

and the intercept is

$$b_0 = \bar{y} - b_1\bar{x}$$

Here, $r$ is the correlation between $y$ and $x$, $s_y$ is the standard deviation of $y$, and $s_x$ is the standard deviation of $x$. Some algebra based on the rules for means of random variables (Section 4.4) shows that $b_0$ and $b_1$ are unbiased estimators of $\beta_0$ and $\beta_1$. Furthermore, $b_0$ and $b_1$ are Normally distributed with means $\beta_0$ and $\beta_1$ and standard deviations that can be estimated from the data. Normality of these sampling distributions is a consequence of the assumption that the $\epsilon_i$ are distributed Normally. A general form of the central limit theorem tells us that the distributions of $b_0$ and $b_1$ will still be approximately Normal even if the $\epsilon_i$ are not. On the other hand, outliers and influential observations can invalidate the results of inference for regression.

LOOK BACK
central limit theorem, page 339

The predicted value of $y$ for a given value $x^*$ of $x$ is the point on the least-squares line $\hat{y} = b_0 + b_1 x^*$. This is an unbiased estimator of the mean response $\mu_y$ when $x = x^*$. The **residual** is

residual

$$\begin{aligned} e_i &= \text{observed response} - \text{predicted response} \\ &= y_i - \hat{y}_i \\ &= y_i - b_0 - b_1 x_i \end{aligned}$$

The residuals $e_i$ correspond to the model deviations $\epsilon_i$. The $e_i$ sum to 0, and the $\epsilon_i$ come from a population with mean 0.

The remaining parameter to be estimated is $\sigma$, which measures the variation of $y$ about the population regression line. Because this parameter is the standard deviation of the model deviations, it should come as no surprise that we use the residuals to estimate it. As usual, we work first with the variance and take the square root to obtain the standard deviation. For simple linear regression, the estimate of $\sigma^2$ is the average squared residual

$$\begin{aligned} s^2 &= \frac{\sum e_i^2}{n-2} \\ &= \frac{\sum (y_i - \hat{y}_i)^2}{n-2} \end{aligned}$$

**sample variance, page 40**

degrees of freedom

We average by dividing the sum by $n - 2$ in order to make $s^2$ an unbiased estimate of $\sigma^2$. The sample variance of $n$ observations uses the divisor $n - 1$ for this same reason. The quantity $n - 2$ is called the **degrees of freedom** for $s^2$. The estimate of $\sigma$ is given by

$$s = \sqrt{s^2}$$

We will use statistical software to calculate the regression for predicting fuel efficiency with the log of speed for Example 10.3. In entering the data, we chose the names LOGMPH for the log of speed and MPG for fuel efficiency. *It is good practice to use names, rather than just x and y, to remind yourself which data the output describes.*

EXAMPLE

**10.4 Statistical software output for fuel efficiency.** Figure 10.5 gives the outputs for four commonly used statistical software packages and Excel. Other software will give similar information. The SPSS output reports estimates of our three parameters as $b_0 = -7.796$, $b_1 = 7.874$, and $s = 0.9995$. Be sure that you can find these entries in this output and the corresponding values in the other outputs.

The least-squares regression line is the straight line that is plotted in Figure 10.4. We would report it as

$$\widehat{\text{MPG}} = -7.80 + 7.87\text{LOGMPH}$$

with a model standard deviation of $s = 1.00$. Note that the number of digits provided varies with the software used and we have rounded off the values to three significant digits. *It is important to avoid cluttering up your report of the results of a statistical analysis with many digits that are not relevant.* Software often reports many more digits than are meaningful or useful.

The outputs contain other information that we will ignore for now. Computer outputs often give more information than we want or need. *The experienced user of statistical software learns to ignore the parts of the output that are not needed for the current problem.* This is done to reduce user frustration when a software package does not print out the particular statistics wanted for an analysis.

Now that we have fitted a line, we should examine the residuals for Normality and any remaining patterns in the data. We usually plot the residuals both against the case number (especially if this reflects the order in which the observations were collected) and against the explanatory variable. For this example, in place of case number, we prefer another variable that is similar but is recorded in a more useful scale. It is the total number of miles that the vehicle has been driven.

Figure 10.6 gives a plot of the residuals versus miles driven with a smooth-function fit. The smooth function suggests that the residuals increase slightly up to about 50,000 miles and then tend to decrease somewhat. With the data that we have for this example, it is difficult to decide if this effect is real or due to chance variation. It is not unreasonable to think that the vehicle performance decreases with age. Since the effect does not appear to be particularly large, we

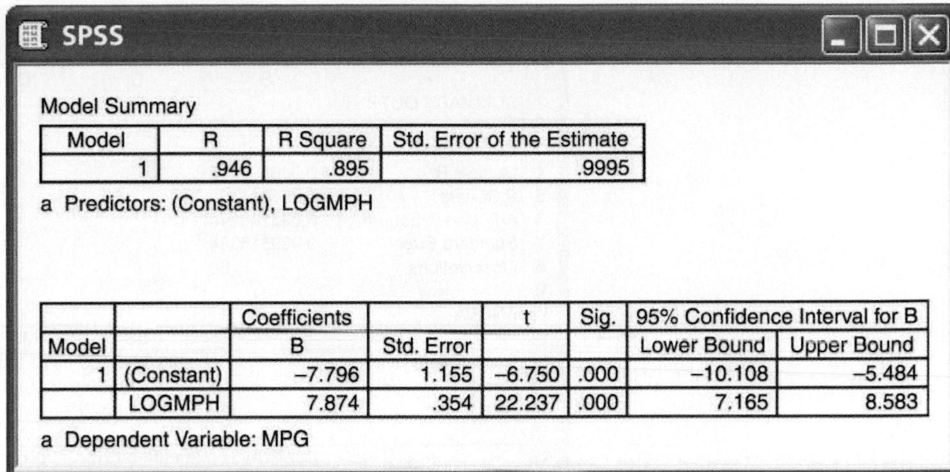
SPSS

Model Summary

| Model | R | R Square | Std. Error of the Estimate |
|---|---|---|---|
| 1 | .946 | .895 | .9995 |

a Predictors: (Constant), LOGMPH

| | | Coefficients | | t | Sig. | 95% Confidence Interval for B | |
|---|---|---|---|---|---|---|---|
| Model | | B | Std. Error | | | Lower Bound | Upper Bound |
| 1 | (Constant) | −7.796 | 1.155 | −6.750 | .000 | −10.108 | −5.484 |
| | LOGMPH | 7.874 | .354 | 22.237 | .000 | 7.165 | 8.583 |

a Dependent Variable: MPG

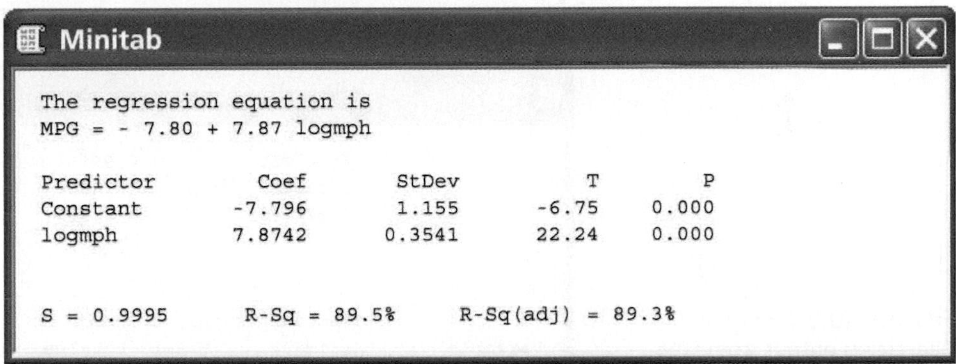
Minitab

```
The regression equation is
MPG = - 7.80 + 7.87 logmph

Predictor        Coef       StDev          T        P
Constant       -7.796       1.155      -6.75    0.000
logmph         7.8742      0.3541      22.24    0.000


S = 0.9995       R-Sq = 89.5%     R-Sq(adj) = 89.3%
```

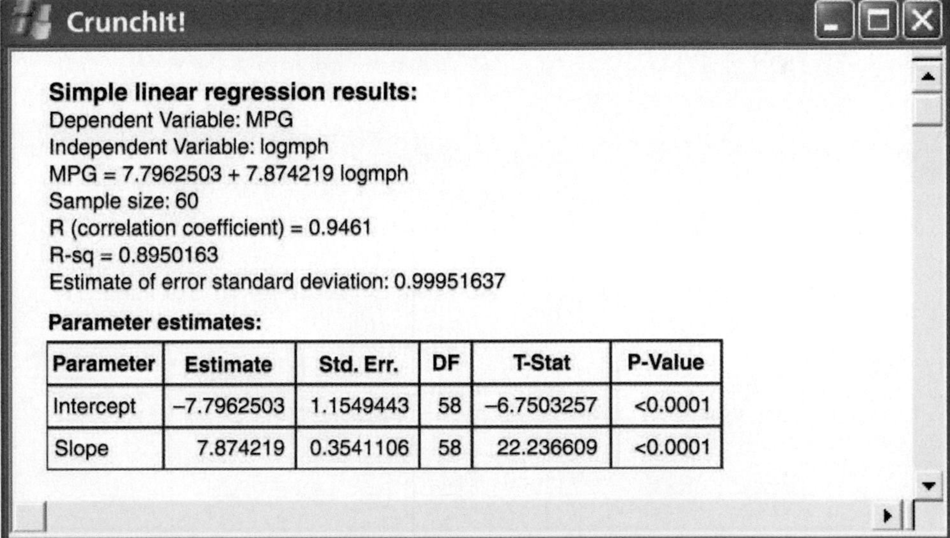
CrunchIt!

**Simple linear regression results:**
Dependent Variable: MPG
Independent Variable: logmph
MPG = 7.7962503 + 7.874219 logmph
Sample size: 60
R (correlation coefficient) = 0.9461
R-sq = 0.8950163
Estimate of error standard deviation: 0.99951637

**Parameter estimates:**

| Parameter | Estimate | Std. Err. | DF | T-Stat | P-Value |
|---|---|---|---|---|---|
| Intercept | −7.7962503 | 1.1549443 | 58 | −6.7503257 | <0.0001 |
| Slope | 7.874219 | 0.3541106 | 58 | 22.236609 | <0.0001 |

**FIGURE 10.5** Regression output from SPSS, Minitab, CrunchIt!, Excel, and SAS for the fuel efficiency example. *(continued)*

Excel

| | A | B | C | D | E | F | G |
|---|---|---|---|---|---|---|---|
| 1 | SUMMARY OUTPUT | | | | | | |
| 2 | | | | | | | |
| 3 | Regression Statistics | | | | | | |
| 4 | Multiple R | 0.946053015 | | | | | |
| 5 | R Square | 0.895016308 | | | | | |
| 6 | Adjusted R Square | 0.893206244 | | | | | |
| 7 | Standard Error | 0.999516364 | | | | | |
| 8 | Observations | 60 | | | | | |
| 9 | | | | | | | |
| 10 | ANOVA | | | | | | |
| 11 | | df | SS | MS | F | Significance F | |
| 12 | Regression | 1 | 493.9885883 | 493.9886 | 494.4668 | 4.50949E-30 | |
| 13 | Residual | 58 | 57.94391174 | 0.999033 | | | |
| 14 | Total | 59 | 551.9325 | | | | |
| 15 | | | | | | | |
| 16 | | Coefficients | Standard Error | tStet | P-value | Lower 95% | Upper 95% |
| 17 | intercept | –7.796250129 | 1.154944262 | –6.75033 | 7.69E-09 | –10.10812052 | –5.48437974 |
| 18 | logmph | 7.874219013 | 0.354110611 | 22.23661 | 4.51E-30 | 7.165390143 | 8583047883 |

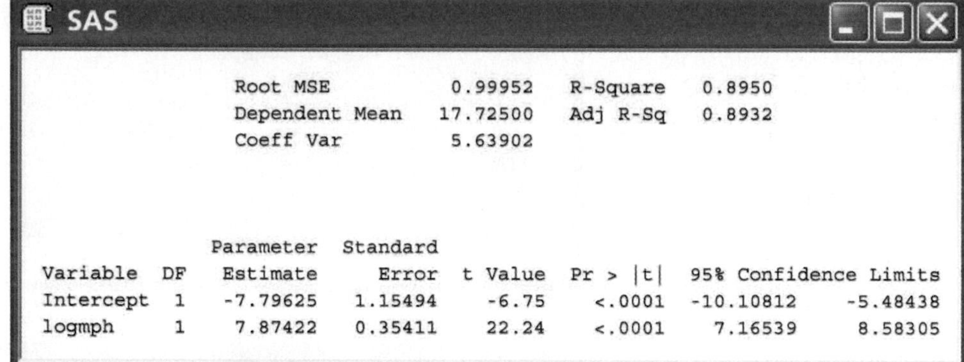

SAS

| | | | |
|---|---|---|---|
| Root MSE | 0.99952 | R-Square | 0.8950 |
| Dependent Mean | 17.72500 | Adj R-Sq | 0.8932 |
| Coeff Var | 5.63902 | | |

| Variable | DF | Parameter Estimate | Standard Error | t Value | Pr > \|t\| | 95% Confidence Limits | |
|---|---|---|---|---|---|---|---|
| Intercept | 1 | -7.79625 | 1.15494 | -6.75 | <.0001 | -10.10812 | -5.48438 |
| logmph | 1 | 7.87422 | 0.35411 | 22.24 | <.0001 | 7.16539 | 8.58305 |

**FIGURE 10.5** *(Continued)* Regression output from SPSS, Minitab, CrunchIt!, Excel, and SAS for the fuel efficiency example.

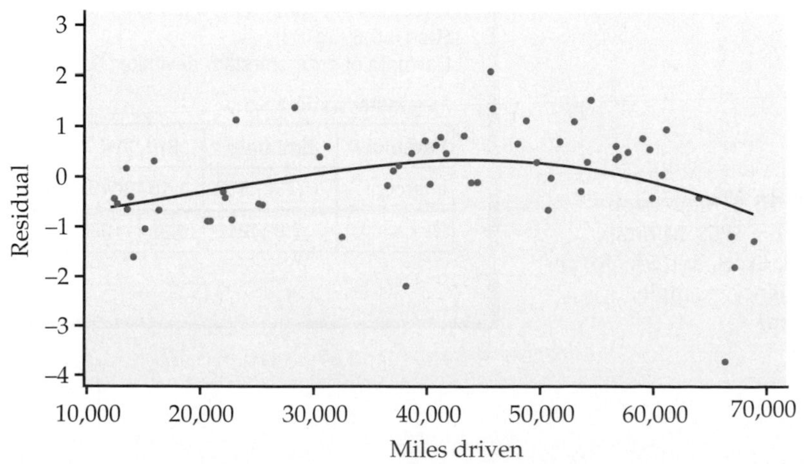


**FIGURE 10.6** Plot of residuals versus miles driven with a smooth function, for the fuel efficiency example.

will ignore it for the present analysis. With more data, however, it may be an interesting phenomenon to study.

The residuals are plotted versus the explanatory variable, log of mph (labeled LOGMPH), in Figure 10.7. No clear pattern is evident. There is one residual that is somewhat low, and we have seen it in all of our plots. Inspection of Figure 10.4 reveals that this observation does not appear to distort our least-squares regression line.

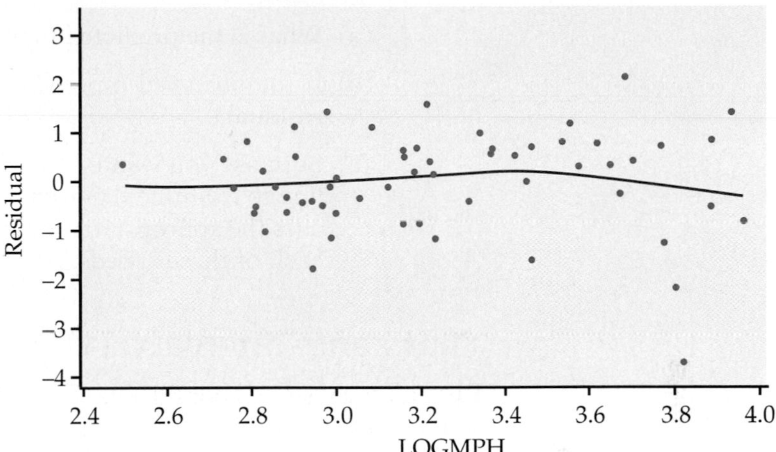


**FIGURE 10.7** Plot of residuals versus log of MPH with a smooth function, for the fuel efficiency example.

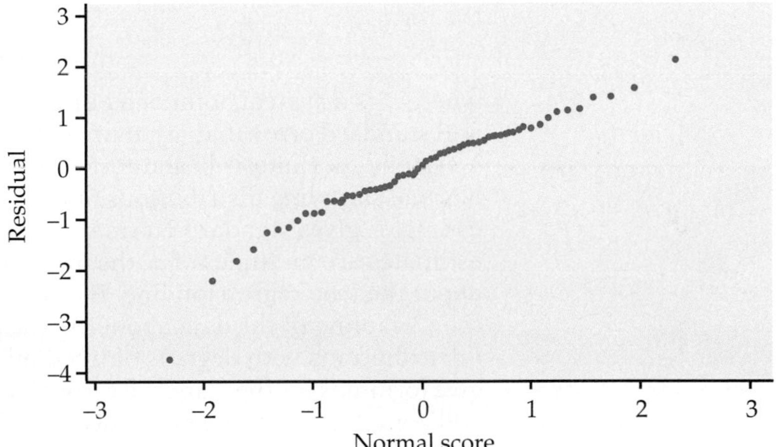


**FIGURE 10.8** Normal quantile plot of the residuals for the fuel efficiency example.

Finally, Figure 10.8 is a Normal quantile plot of the residuals. Because the plot looks fairly straight, we are confident that we do not have a serious violation of our assumption that the residuals are Normally distributed. Observe that the low outlier is also visible in this plot.

## USE YOUR KNOWLEDGE

**10.1 Understanding a linear regression model.** Consider a linear regression model with $\mu_y = 40.5 - 2.5x$ and standard deviation $\sigma = 2.0$.

(a) What is the slope of the population regression line?

(b) Explain clearly what this slope says about the change in the mean of $y$ for a change in $x$.

(c) What is the subpopulation mean when $x = 10$?

(d) Between what 2 values would approximately 95% of the observed responses, $y$, fall when $x = 10$?

**10.2** **More on speed's effect on fuel efficiency.** Refer to Example 10.4.

(a) What is the predicted mpg for the car when it averages 35 mph?

(b) If the observed mpg when $x = 35$ mph were 21.0, what is the residual?

(c) Suppose you wanted to use the estimated population regression line to examine the average mpg at 45, 55, 65, and 75 mph. Discuss the appropriateness of using the equation to predict mpg for each of these speeds.

## Confidence intervals and significance tests

Chapter 7 presented confidence intervals and significance tests for means and differences in means. In each case, inference rested on the standard errors of estimates and on $t$ distributions. Inference for the intercept and slope in a linear regression is similar in principle. For example, the confidence intervals have the form

$$\text{estimate} \pm t^* \text{SE}_{\text{estimate}}$$

where $t^*$ is a critical point of a $t$ distribution. It is the formulas for the estimate and standard error that are more complicated.

Confidence intervals and tests for the slope and intercept are based on the Normal sampling distributions of the estimates $b_1$ and $b_0$. Standardizing these estimates gives standard Normal $z$ statistics. The standard deviations of these estimates are multiples of $\sigma$, the model parameter that describes the variability about the true regression line. Because we do not know $\sigma$, we estimate it by $s$, the variability of the data about the least-squares line. When we do this, we get $t$ distributions with degrees of freedom $n - 2$, the degrees of freedom of $s$. We give formulas for the standard errors $\text{SE}_{b_1}$ and $\text{SE}_{b_0}$ in Section 10.2. For now we will concentrate on the basic ideas and let the computer do the computations.

### CONFIDENCE INTERVALS AND SIGNIFICANCE TESTS FOR REGRESSION SLOPE AND INTERCEPT

A **level $C$ confidence interval for the intercept** $\beta_0$ is

$$b_0 \pm t^* \text{SE}_{b_0}$$

A **level $C$ confidence interval for the slope** $\beta_1$ is

$$b_1 \pm t^* \text{SE}_{b_1}$$

In these expressions $t^*$ is the value for the $t(n - 2)$ density curve with area $C$ between $-t^*$ and $t^*$.

To test the hypothesis $H_0: \beta_1 = 0$, compute the **test statistic**

$$t = \frac{b_1}{\mathrm{SE}_{b_1}}$$

The **degrees of freedom** are $n - 2$. In terms of a random variable $T$ having the $t(n-2)$ distribution, the $P$-value for a test of $H_0$ against

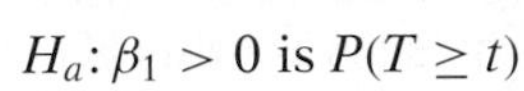

$H_a: \beta_1 > 0$ is $P(T \geq t)$

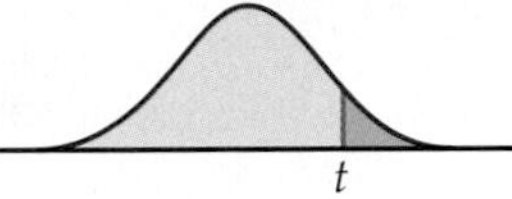


$H_a: \beta_1 < 0$ is $P(T \leq t)$

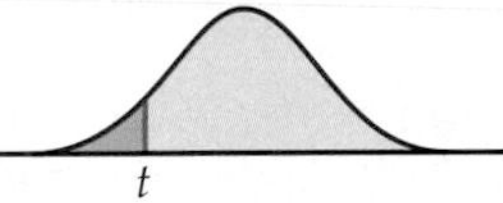


$H_a: \beta_1 \neq 0$ is $2P(T \geq |t|)$

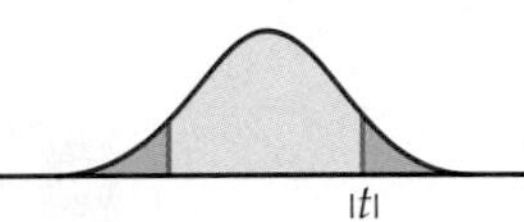


There is a similar significance test about the intercept $\beta_0$ that uses $\mathrm{SE}_{b_0}$ and the $t(n-2)$ distribution. Although computer outputs often include a test of $H_0: \beta_0 = 0$, this information usually has little practical value. From the equation for the population regression line, $\mu_y = \beta_0 + \beta_1 x$, we see that $\beta_0$ is the mean response corresponding to $x = 0$. In many practical situations, this subpopulation does not exist or is not interesting.

On the other hand, the test of $H_0: \beta_1 = 0$ is quite useful. When we substitute $\beta_1 = 0$ in the model, the $x$ term drops out and we are left with

$$\mu_y = \beta_0$$

This model says that the mean of $y$ does not vary with $x$. All of the $y$'s come from a single population with mean $\beta_0$, which we would estimate by $\bar{y}$. The hypothesis $H_0: \beta_1 = 0$ therefore says that there is no straight-line relationship between $y$ and $x$ and that linear regression of $y$ on $x$ is of no value for predicting $y$.

EXAMPLE

**10.5 Statistical software output, continued.** The computer outputs in Figure 10.5 for the fuel efficiency problem contain the information needed for inference about the regression slope and intercept. Let's look at the SPSS output. The column labeled Std. Error gives the standard errors of the estimates. The value of $\mathrm{SE}_{b_1}$ appears on the line labeled with the variable name for the explanatory variable, LOGMPH. It is given as 0.354. In a summary we would report that the regression coefficient for the log of speed is 7.87 with a standard error of 0.35.

The $t$ statistic and $P$-value for the test of $H_0: \beta_1 = 0$ against the two-sided alternative $H_a: \beta_1 \neq 0$ appear in the columns labeled t and Sig. We can verify the $t$ calculation from the formula for the standardized estimate:

$$t = \frac{b_1}{\mathrm{SE}_{b_1}} = \frac{7.874}{0.354} = 22.24$$

The $P$-value is given as 0.000. This is a rounded number and from that information we can conclude that $P < 0.0005$. The other outputs in Figure 10.5 also indicate that the $P$-value is very small. We will report the result as $P < 0.001$ because 1 chance in 1000 is sufficiently small for us to decisively reject the null hypothesis.

We have found a statistically significant linear relationship between fuel efficiency and log speed. The estimated slope is more than 22 standard deviations away from zero. Because this is extremely unlikely to happen if the true slope is zero, we have strong evidence for our claim. Note, however, that this is not the same as concluding that we have found a strong relationship between the response and explanatory variables in this example. *A very small P-value for the significance test for a zero slope does not necessarily imply that we have found a strong relationship.* A confidence interval will provide additional information about the relationship.

EXAMPLE

**10.6 Confidence interval for the slope.** A confidence interval for $\beta_1$ requires a critical value $t^*$ from the $t(n-2) = t(58)$ distribution. In Table D there are entries for 50 and 60 degrees of freedom. The values for these rows are very similar. To be conservative, we will use the larger critical value, for 50 degrees of freedom. Find the confidence level values at the bottom of the table. In the 95% confidence column the entry for 50 degrees of freedom is $t^* = 2.009$.

To compute the 95% confidence interval for $\beta_1$ we combine the estimate of the slope with the margin of error:

$$\begin{aligned} b_1 \pm t^* \mathrm{SE}_{b_1} &= 7.874 \pm (2.009)(0.354) \\ &= 7.874 \pm 0.711 \end{aligned}$$

The interval is (7.16, 8.58). This agrees with the value given by the software outputs that provide this information in Figure 10.5. We estimate that an increase of 1 in the logarithm of speed is associated with an increase of between 7.16 and 8.58 mpg.

To interpret the interval, it is useful to translate the statement back to the original mph scale. From Figure 10.4 we can see that the values for LOGMPH range from about 2.5 to 3.9. Let's translate the increase of 1 unit in LOGMPH to the mph scale by considering a change from 2.8 to 3.8. Since $\log(16.4) = 2.8$ and $\log(44.7) = 3.8$, the change corresponds to an increase in speed from 16.4 to 44.7 mph. An increase in average speed from 16.4 to 44.7 mph is associated with an increase of $7.8 \pm 0.7$ in mpg.

Note that the intercept in this example is not of practical interest. It estimates mpg when the logarithm of mph (that's $x$) is 0, a value that cannot occur. For this reason, we do not compute a confidence interval for $\beta_0$.

## Confidence intervals for mean response

For any specific value of $x$, say $x^*$, the mean of the response $y$ in this subpopulation is given by

$$\mu_y = \beta_0 + \beta_1 x^*$$

To estimate this mean from the sample, we substitute the estimates $b_0$ and $b_1$ for $\beta_0$ and $\beta_1$:

$$\hat{\mu}_y = b_0 + b_1 x^*$$

A confidence interval for $\mu_y$ adds to this estimate a margin of error based on the standard error $\text{SE}_{\hat{\mu}}$. (The formula for the standard error is given in Section 10.2.)

### CONFIDENCE INTERVAL FOR A MEAN RESPONSE

A **level *C* confidence interval for the mean response** $\mu_y$ when $x$ takes the value $x^*$ is

$$\hat{\mu}_y \pm t^* \text{SE}_{\hat{\mu}}$$

where $t^*$ is the value for the $t(n-2)$ density curve with area $C$ between $-t^*$ and $t^*$.

Many computer programs calculate confidence intervals for the mean response corresponding to each of the $x$-values in the data. Some can calculate an interval for any value $x^*$ of the explanatory variable. We will use a plot to illustrate these intervals.

EXAMPLE

**10.7 Confidence intervals for the mean response.** Figure 10.9 shows the upper and lower confidence limits on a graph with the data and the least-squares line. The 95% confidence limits appear as dashed curves. For any $x^*$, the confidence interval for the mean response extends from the lower dashed curve to the upper dashed curve. The intervals are narrowest for values of $x^*$ near the mean of the observed $x$'s and widen as $x^*$ moves away from $\bar{x}$.

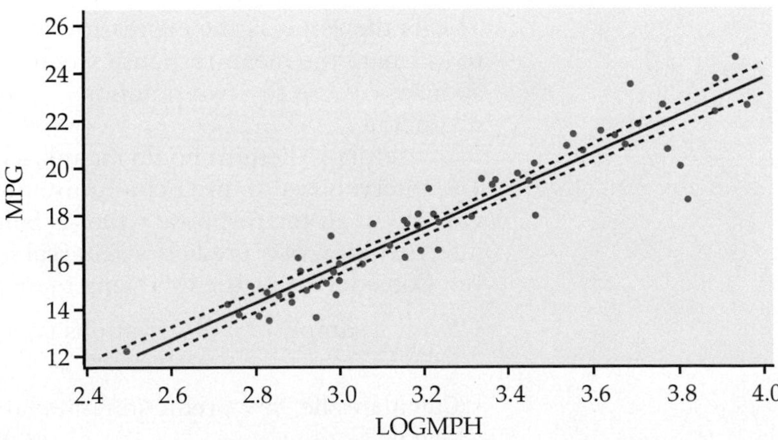


**FIGURE 10.9** The 95% confidence limits (dashed curves) for the mean response for the fuel efficiency example.

Some software will do these calculations directly if you input a value for the explanatory variable. Others will calculate the intervals for each value of $x$ in the data set. Creating a new data set with an additional observation with $x$ equal to the value of interest and $y$ missing will often work.

EXAMPLE

**10.8 Confidence interval for a speed of 30 mph.** Let's find the confidence interval for the mean response at 30 mph. We use $x = \log(30) = 3.4$ as the value for the explanatory variable. Our predicted fuel efficiency is

$$\begin{aligned}\widehat{\text{MPG}} &= -7.80 + 7.87\text{LOGMPH}\\ &= -7.80 + (7.87)(3.4)\\ &= 19.0\end{aligned}$$

Software tells us that the 95% confidence interval for the mean response is 18.7 to 19.3 mpg.

If we operated this vehicle many times under similar conditions at an average speed of 30 mph, we would expect the fuel efficiency to be between 18.7 and 19.3 mpg. Note that many of the observations in Figure 10.9 lie outside the confidence bands. *These confidence intervals do not tell us what mileage to expect for a single observation at a particular average speed such as 30 mph.* We need a different kind of interval for this purpose.

## Prediction intervals

In the last example, we predicted the mean fuel efficiency when the average speed is 30 mph. Suppose we now want to predict a future observation of fuel efficiency when the vehicle is driven at 30 mph under similar conditions. Our best guess at the fuel efficiency is what we obtained before using the regression equation, that is, 19.0 mpg. The margin of error, on the other hand, is larger because it is harder to predict an individual value than to predict the mean.

The predicted response $y$ for an individual case with a specific value $x^*$ of the explanatory variable $x$ is

$$\hat{y} = b_0 + b_1 x^*$$

This is the same as the expression for $\hat{\mu}_y$. That is, the fitted line is used both to estimate the mean response when $x = x^*$ and to predict a single future response. We use the two notations $\hat{\mu}_y$ and $\hat{y}$ to remind ourselves of these two distinct uses.

prediction interval

A useful prediction should include a margin of error to indicate its accuracy. The interval used to predict a future observation is called a **prediction interval.** Although the response $y$ that is being predicted is a random variable, the interpretation of a prediction interval is similar to that for a confidence interval. Consider doing the following many times:

- Draw a sample of $n$ observations $(x_i, y_i)$ and then one additional observation $(x^*, y)$.
- Calculate the 95% prediction interval for $y$ when $x = x^*$ using the sample of size $n$.

Then 95% of the prediction intervals will contain the value of $y$ for the additional observation. In other words, the probability that this method produces an interval that contains the value of a future observation is 0.95.

The form of the prediction interval is very similar to that of the confidence interval for the mean response. The difference is that the standard error $SE_{\hat{y}}$ used in the prediction interval includes both the variability due to the fact that the least-squares line is not exactly equal to the true regression line *and* the variability of the future response variable $y$ around the subpopulation mean. (The formula for $SE_{\hat{y}}$ appears in Section 10.2.)

## PREDICTION INTERVAL FOR A FUTURE OBSERVATION

A **level $C$ prediction interval for a future observation** on the response variable $y$ from the subpopulation corresponding to $x^*$ is

$$\hat{y} \pm t^* SE_{\hat{y}}$$

where $t^*$ is the value for the $t(n-2)$ density curve with area $C$ between $-t^*$ and $t^*$.

Again, we use a graph to illustrate the results.

EXAMPLE

**10.9 Prediction intervals for fuel efficiency.** Figure 10.10 shows the upper and lower prediction limits, along with the data and the least-squares line. The 95% prediction limits are indicated by the dashed curves. Compare this figure with Figure 10.9, which shows the 95% confidence limits drawn to the same scale. The upper and lower limits of the prediction intervals are farther from the least-squares line than are the confidence limits. This results in most, but not all, of the observations in Figure 10.10 lying within the prediction bands.

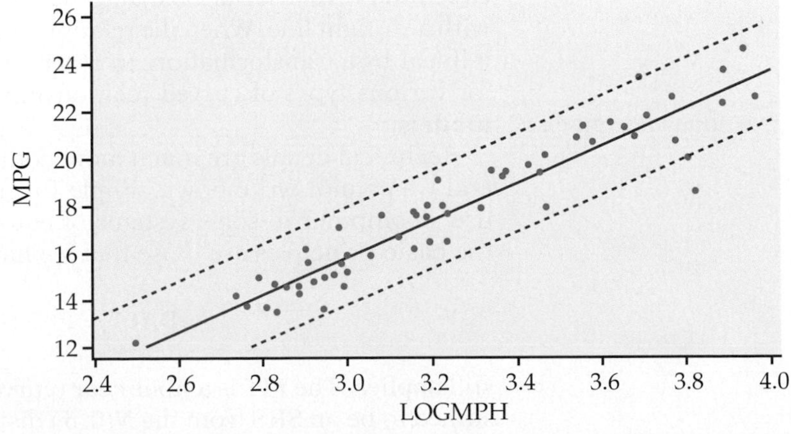


**FIGURE 10.10** The 95% prediction limits (dashed curves) for individual responses for the fuel efficiency example. Compare with Figure 10.9.

*The comparison of Figures 10.9 and 10.10 reminds us that the interval for a single future observation must be larger than an interval for the mean of its subpopulation.*

EXAMPLE

**10.10 Prediction interval for a speed of 30 mph.** Let's find the prediction interval for a future observation of fuel efficiency when the vehicle is driven at 30 mph. The predicted value is the same as the estimate of the mean response that we calculated in Example 10.8, that is, 19.0 mpg. Software tells us that the 95% prediction interval is 17.0 to 21.0 mpg. If we operated this vehicle a single time under similar conditions at an average speed of 30 mph, we would expect the fuel efficiency to be between 17.0 and 21.0 mpg.

## USE YOUR KNOWLEDGE

**10.3 Constructing confidence intervals for the mean response.** Refer to Example 10.6. For the following three changes in average speed, construct a 95% confidence interval for the change in mpg.

(a) LOGMPH increases from 2.9 to 3.9 (18.2 to 49.4 mph).

(b) LOGMPH decreases from 3.7 to 2.7 (40.4 to 14.9 mph).

(c) LOGMPH increases from 2.8 to 3.3 (16.4 to 27.1 mph).

**10.4 Standard error for the mean response.** Refer to Example 10.10. What is the standard error of $\hat{y}$ when $x = 30$ mph? Would you expect the standard error of $\hat{y}$ to be larger, smaller, or the same when $x = 40$ mph? Explain.

## BEYOND THE BASICS

### Nonlinear Regression

The regression model that we have studied assumes that the relationship between the response variable and the explanatory variable can be summarized with a straight line. When the relationship is not linear, we can sometimes make it linear by a transformation. In other circumstances, we use models that allow for various types of curved relationships. These models are called **nonlinear models.**

nonlinear models

Technical details are much more complicated for nonlinear models. In general we cannot write down simple formulas for the parameter estimates; we use a computer to solve systems of equations to find the estimates. However, the basic principles are those that we have already learned. For example,

$$\text{DATA} = \text{FIT} + \text{RESIDUAL}$$

still applies. The FIT is a nonlinear (curved) function, and the residuals are assumed to be an SRS from the $N(0, \sigma)$ distribution. The nonlinear function contains parameters that must be estimated from the data. Approximate standard errors for these estimates are part of the standard output provided by software. Here is an example.

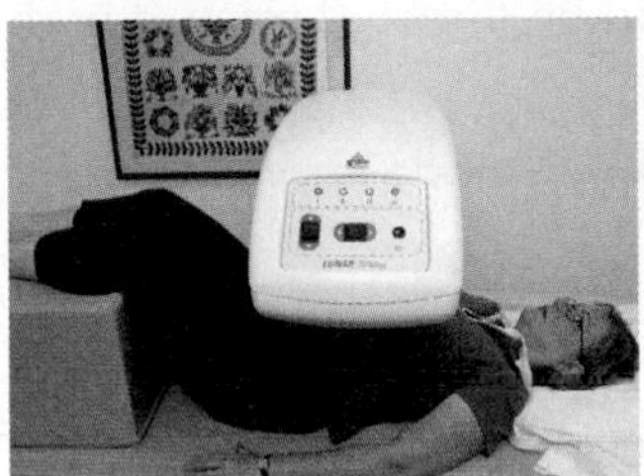

EXAMPLE

**10.11 Bone mass accumulation in young women.** As we age, our bones become weaker and are more likely to break. Osteoporosis (or weak bones) is the major cause of bone fractures in older women. Some researchers have studied this problem by looking at how and when bone mass is accumulated by young women. Understanding the relationship between age and bone mass is an important part of this approach to the problem.

Figure 10.11 displays data for a measure of bone strength, called "total body bone mineral density" (TBBMD), and age for a sample of 256 young women.[2] TBBMD is measured in grams per square centimeter (g/cm$^2$), and age is recorded in years. The solid curve is the nonlinear fit, and the dashed curves are 95% prediction limits. The fitted nonlinear equation is

$$\hat{y} = 1.162 \frac{e^{-1.162+0.28x}}{1 + e^{-1.162+0.28x}}$$

In this equation, $\hat{y}$ is the predicted value of TBBMD, the response variable; and $x$ is age, the explanatory variable. A straight line would not do a very good job of summarizing the relationship between TBBMD and age. At first, TBBMD increases with age, but then it levels off as age increases. The value of the function where it is level is called "peak bone mass"; it is a parameter in the nonlinear model. The estimate is 1.162 and the standard error is 0.008. Software gives the 95% confidence interval as (1.146, 1.178).

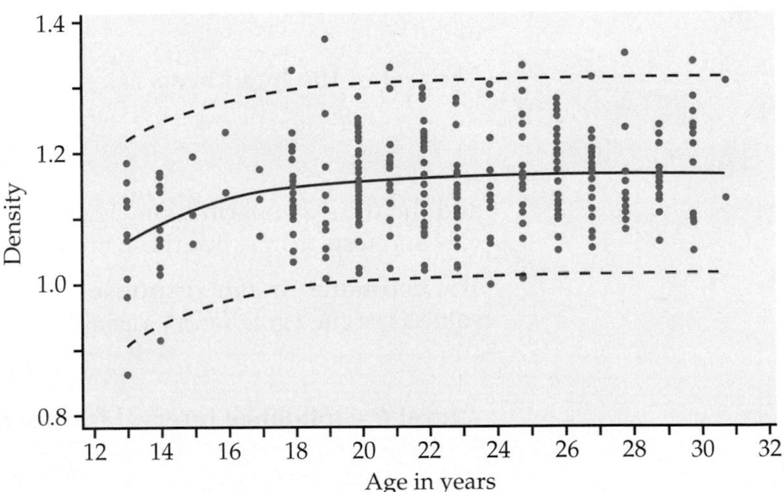


**FIGURE 10.11** Plot of total body bone mineral density versus age.

The long-range goals of the researchers who conducted this study include developing intervention programs (exercise and increasing calcium intake have been shown to be effective) for young women that will increase their TBBMD. What age groups should be the target of these interventions? The fitted nonlinear model can be used to obtain estimates of the age (with a standard error) at which any given percent of the peak bone mass is attained. We estimate that the age at which the population reaches 95% of peak bone mass is 16.2 years (SE = 1.1 years). For 99% of peak bone mass, the age is 22.1 years (SE = 2.5 years). Intervention programs should be directed toward high school- and college-aged women.

## SECTION 10.1 Summary

The statistical model for **simple linear regression** is

$$y_i = \beta_0 + \beta_1 x_i + \epsilon_i$$

where $i = 1, 2, \ldots, n$. The $\epsilon_i$ are assumed to be independent and Normally distributed with mean 0 and standard deviation $\sigma$. The **parameters** of the model are $\beta_0$, $\beta_1$, and $\sigma$.

The intercept and slope $\beta_0$ and $\beta_1$ are estimated by the intercept and slope of the **least-squares regression line,** $b_0$ and $b_1$. The parameter $\sigma$ is estimated by

$$s = \sqrt{\frac{\sum e_i^2}{n-2}}$$

where the $e_i$ are the **residuals**

$$e_i = y_i - \hat{y}_i$$

A **level $C$ confidence interval for $\beta_1$** is

$$b_1 \pm t^* \mathrm{SE}_{b_1}$$

where $t^*$ is the value for the $t(n-2)$ density curve with area $C$ between $-t^*$ and $t^*$.

The **test of the hypothesis** $H_0\colon \beta_1 = 0$ is based on the ***t* statistic**

$$t = \frac{b_1}{\mathrm{SE}_{b_1}}$$

and the $t(n-2)$ distribution. There are similar formulas for confidence intervals and tests for $\beta_0$, but these are meaningful only in special cases.

The **estimated mean response** for the subpopulation corresponding to the value $x^*$ of the explanatory variable is

$$\hat{\mu}_y = b_0 + b_1 x^*$$

A **level $C$ confidence interval for the mean response** is

$$\hat{\mu}_y \pm t^* \mathrm{SE}_{\hat{\mu}}$$

where $t^*$ is the value for the $t(n-2)$ density curve with area $C$ between $-t^*$ and $t^*$.

The **estimated value of the response variable** $y$ for a future observation from the subpopulation corresponding to the value $x^*$ of the explanatory variable is

$$\hat{y} = b_0 + b_1 x^*$$

A **level $C$ prediction interval** for the estimated response is

$$\hat{y} \pm t^* \mathrm{SE}_{\hat{y}}$$

where $t^*$ is the value for the $t(n-2)$ density curve with area $C$ between $-t^*$ and $t^*$.

*The following section contains material that you should study if you plan to read Chapter 11 on multiple regression. In addition, the section we just completed assumes that you have access to software or a statistical calculator. If you do not, you now need to study the material on computations in the following optional section. The exercises are given at the end of the chapter.*

# 10.2 More Detail about Simple Linear Regression*

In this section we study three optional topics. The first is analysis of variance for regression. If you plan to study the next chapter on multiple regression, you should study this material. The second topic concerns computations for regression inference. Here we present and illustrate the use of formulas for the inference procedures that we have just studied. Finally, we discuss inference for correlation.

## Analysis of variance for regression

analysis of variance

The usual computer output for regression includes additional calculations called **analysis of variance.** Analysis of variance, often abbreviated ANOVA, is essential for multiple regression (Chapter 11) and for comparing several means (Chapters 12 and 13). Analysis of variance summarizes information about the sources of variation in the data. It is based on the

$$\text{DATA} = \text{FIT} + \text{RESIDUAL}$$

framework.

The total variation in the response $y$ is expressed by the deviations $y_i - \bar{y}$. If these deviations were all 0, all observations would be equal and there would be no variation in the response. There are two reasons why the individual observations $y_i$ are not all equal to their mean $\bar{y}$.

1. The responses $y_i$ correspond to different values of the explanatory variable $x$ and will differ because of that. The fitted value $\hat{y}_i$ estimates the mean response for the specific $x_i$. The differences $\hat{y}_i - \bar{y}$ reflect the variation in mean response due to differences in the $x_i$. This variation is accounted for by the regression line, because the $\hat{y}$'s lie exactly on the line.
2. Individual observations will vary about their mean because of variation within the subpopulation of responses to a fixed $x_i$. This variation is represented by the residuals $y_i - \hat{y}_i$ that record the scatter of the actual observations about the fitted line.

*This material is optional.

The overall deviation of any $y$ observation from the mean of the $y$'s is the sum of these two deviations:

$$(y_i - \overline{y}) = (\hat{y}_i - \overline{y}) + (y_i - \hat{y}_i)$$

In terms of deviations, this equation expresses the idea that DATA = FIT + RESIDUAL.

Several times we have measured variation by an average of squared deviations. If we square each of the three deviations above and then sum over all $n$ observations, it is an algebraic fact that the sums of squares add:

$$\sum(y_i - \overline{y})^2 = \sum(\hat{y}_i - \overline{y})^2 + \sum(y_i - \hat{y}_i)^2$$

We rewrite this equation as

$$\text{SST} = \text{SSM} + \text{SSE}$$

where

$$\text{SST} = \sum(y_i - \overline{y})^2$$

$$\text{SSM} = \sum(\hat{y}_i - \overline{y})^2$$

$$\text{SSE} = \sum(y_i - \hat{y}_i)^2$$

sum of squares

The SS in each abbreviation stands for **sum of squares,** and the T, M, and E stand for total, model, and error, respectively. ("Error" here stands for deviations from the line, which might better be called "residual" or "unexplained variation.") The total variation, as expressed by SST, is the sum of the variation due to the straight-line model (SSM) and the variation due to deviations from this model (SSE). This partition of the variation in the data between two sources is the heart of analysis of variance.

If $H_0: \beta_1 = 0$ were true, there would be no subpopulations and all of the $y$'s should be viewed as coming from a single population with mean $\mu_y$. The variation of the $y$'s would then be described by the sample variance

$$s_y^2 = \frac{\sum(y_i - \overline{y})^2}{n - 1}$$

The numerator in this expression is SST. The denominator is the total degrees of freedom, or simply DFT.

Just as the total sum of squares SST is the sum of SSM and SSE, the total degrees of freedom DFT is the sum of DFM and DFE, the degrees of freedom for the model and for the error:

$$\text{DFT} = \text{DFM} + \text{DFE}$$

The model has one explanatory variable $x$, so the degrees of freedom for this source are DFM = 1. Because DFT = $n - 1$, this leaves DFE = $n - 2$ as the degrees of freedom for error. For each source, the ratio of the sum of squares to the degrees of freedom is called the **mean square,** or simply MS. The general formula for a mean square is

mean square

$$\text{MS} = \frac{\text{sum of squares}}{\text{degrees of freedom}}$$

Each mean square is an average squared deviation. MST is just $s_y^2$, the sample variance that we would calculate if all of the data came from a single population. MSE is also familiar to us:

$$\text{MSE} = s^2 = \frac{\sum(y_i - \hat{y}_i)^2}{n-2}$$

It is our estimate of $\sigma^2$, the variance about the population regression line.

### SUMS OF SQUARES, DEGREES OF FREEDOM, AND MEAN SQUARES

**Sums of squares** represent variation present in the responses. They are calculated by summing squared deviations. **Analysis of variance** partitions the total variation between two sources.

The sums of squares are related by the formula

$$\text{SST} = \text{SSM} + \text{SSE}$$

That is, the total variation is partitioned into two parts, one due to the model and one due to deviations from the model.

**Degrees of freedom** are associated with each sum of squares. They are related in the same way:

$$\text{DFT} = \text{DFM} + \text{DFE}$$

To calculate **mean squares,** use the formula

$$\text{MS} = \frac{\text{sum of squares}}{\text{degrees of freedom}}$$

interpretation of $r^2$

In Section 2.3 we noted that $r^2$ is the fraction of variation in the values of $y$ that is explained by the least-squares regression of $y$ on $x$. The sums of squares make this interpretation precise. Recall that SST = SSM + SSE. It is an algebraic fact that

$$r^2 = \frac{\text{SSM}}{\text{SST}} = \frac{\sum(\hat{y}_i - \bar{y})^2}{\sum(y_i - \bar{y})^2}$$

Because SST is the total variation in $y$ and SSM is the variation due to the regression of $y$ on $x$, this equation is the precise statement of the fact that $r^2$ is the fraction of variation in $y$ explained by $x$.

## The ANOVA $F$ test

$F$ statistic

The null hypothesis $H_0$: $\beta_1 = 0$ that $y$ is not linearly related to $x$ can be tested by comparing MSM with MSE. The ANOVA test statistic is an ***F* statistic,**

$$F = \frac{\text{MSM}}{\text{MSE}}$$

**LOOK BACK**
***F* distribution, page 474**
***F* test for equality of spread, page 474**

When $H_0$ is true, this statistic has an $F$ distribution with 1 degree of freedom in the numerator and $n - 2$ degrees of freedom in the denominator. These degrees of freedom are those of MSM and MSE. Just as there are many $t$ statistics, there are many $F$ statistics. The ANOVA $F$ statistic is not the same as the $F$ statistic of equality of spread.

When $\beta_1 \neq 0$, MSM tends to be large relative to MSE. So large values of $F$ are evidence against $H_0$ in favor of the two-sided alternative.

## ANALYSIS OF VARIANCE *F* TEST

In the simple linear regression model, the hypotheses

$$H_0: \beta_1 = 0$$

$$H_a: \beta_1 \neq 0$$

are tested by the ***F* statistic**

$$F = \frac{\text{MSM}}{\text{MSE}}$$

F

The $P$-value is the probability that a random variable having the $F(1, n-2)$ distribution is greater than or equal to the calculated value of the $F$ statistic.

The $F$ statistic tests the same null hypothesis as one of the $t$ statistics that we encountered earlier in this chapter, so it is not surprising that the two are related. It is an algebraic fact that $t^2 = F$ in this case. For linear regression with one explanatory variable, we prefer the $t$ form of the test because it more easily allows us to test one-sided alternatives and is closely related to the confidence interval for $\beta_1$.

ANOVA table

The ANOVA calculations are displayed in an *analysis of variance table*, often abbreviated **ANOVA table.** Here is the format of the table for simple linear regression:

| Source | Degrees of freedom | Sum of squares | Mean square | $F$ |
|---|---|---|---|---|
| Model | 1 | $\sum(\hat{y}_i - \bar{y})^2$ | SSM/DFM | MSM/MSE |
| Error | $n - 2$ | $\sum(y_i - \hat{y}_i)^2$ | SSE/DFE | |
| Total | $n - 1$ | $\sum(y_i - \bar{y})^2$ | SST/DFT | |

EXAMPLE

**10.12 Interpreting SPSS output for fuel efficiency.** The entire output generated by SPSS for the fuel efficiency data in Example 10.3 is given in Figure 10.12. Note that SPSS uses the labels Regression, Residual, and Total for the three sources of variation. We have called these Model, Error, and Total. Other statistical software packages may use slightly different labels. We round the calculated value of the $F$ statistic to 494.47; the $P$-value is given as 0.000. This is a rounded value and we can conclude that $P < 0.0005$. (The actual value is much less than this.) There is strong evidence against the null hypothesis that there is no relationship between fuel efficiency and the logarithm of speed. Now look at the output for the regression coefficients. The $t$ statistic for LOGMPH is given as 22.237, which we round to 22.24. If we square this number, we obtain the $F$ statistic (accurate up to roundoff error). The value of $r^2$ is also given in the output. Log of speed explains 89.5% of the variability in fuel efficiency.

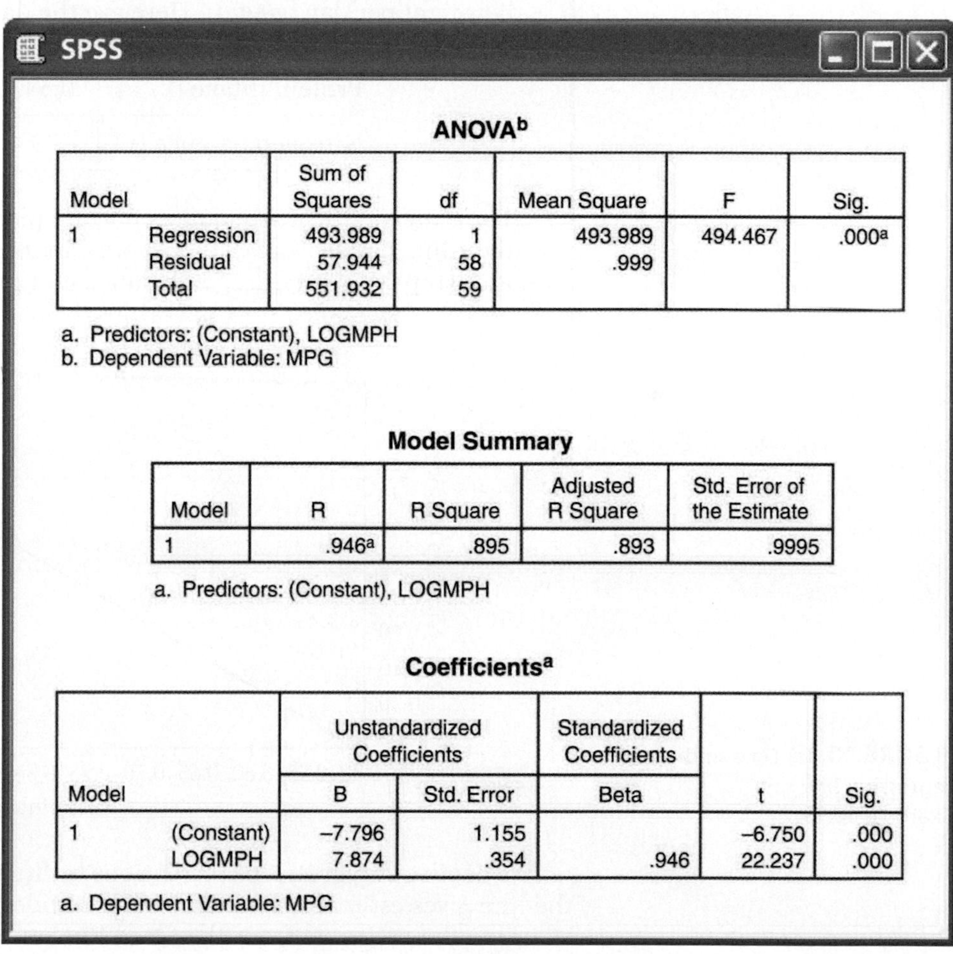

SPSS

**ANOVA[b]**

| Model | | Sum of Squares | df | Mean Square | F | Sig. |
|---|---|---|---|---|---|---|
| 1 | Regression | 493.989 | 1 | 493.989 | 494.467 | .000[a] |
| | Residual | 57.944 | 58 | .999 | | |
| | Total | 551.932 | 59 | | | |

a. Predictors: (Constant), LOGMPH
b. Dependent Variable: MPG

**Model Summary**

| Model | R | R Square | Adjusted R Square | Std. Error of the Estimate |
|---|---|---|---|---|
| 1 | .946[a] | .895 | .893 | .9995 |

a. Predictors: (Constant), LOGMPH

**Coefficients[a]**

| Model | | Unstandardized Coefficients | | Standardized Coefficients | t | Sig. |
|---|---|---|---|---|---|---|
| | | B | Std. Error | Beta | | |
| 1 | (Constant) | −7.796 | 1.155 | | −6.750 | .000 |
| | LOGMPH | 7.874 | .354 | .946 | 22.237 | .000 |

a. Dependent Variable: MPG

**FIGURE 10.12** Regression output with ANOVA table for Example 10.12.

## Calculations for regression inference

We recommend using statistical software for regression calculations. With time and care, however, the work is feasible with a calculator. We will use the following example to illustrate how to perform inference for regression analysis using a calculator.

EXAMPLE

**10.13 Protein requirements via nitrogen balance.** Nitrogen balance studies are used to determine protein requirements for people. Each subject is fed three different controlled diets during three separate experimental periods. The three diets are similar with regard to all nutrients except protein.

Nitrogen balance is the difference between the amount of nitrogen consumed and the amount lost in feces and urine and by other means. Since virtually all of the nitrogen in a diet comes from protein, nitrogen balance is an indicator of the amount of protein retained by the body. The protein requirement for an individual is the intake corresponding to a balance of zero.

Linear regression is used to model the relationship between nitrogen balance, measured in milligrams of nitrogen per kilogram of body weight per day (mg/kg/d), and protein intake, measured in grams of protein per kilogram of body weight per day (g/kg/d). Here are the data for one subject:[3]

| Protein intake ($x$) | 0.543 | 0.797 | 1.030 |
|---|---|---|---|
| Nitrogen balance ($y$) | −23.4 | 17.8 | 67.3 |

The data and the least-squares line are plotted in Figure 10.13. The strong straight-line pattern suggests that we can use linear regression to model the relationship between nitrogen balance and protein intake.

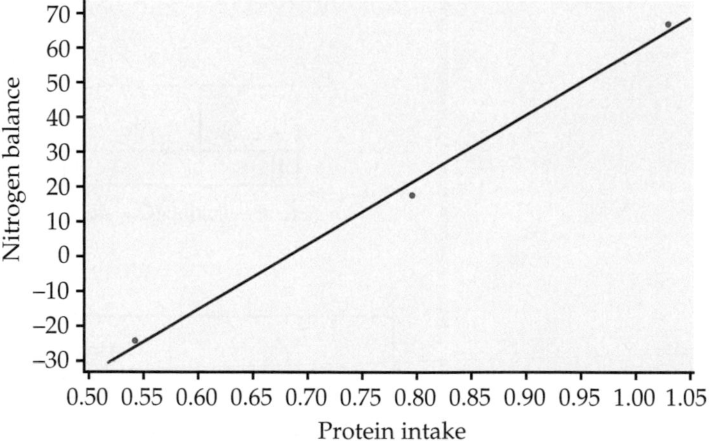


**FIGURE 10.13** Data and regression line for Example 10.13.

We begin our regression calculations by fitting the least-squares line. Fitting the line gives estimates $b_1$ and $b_0$ of the model parameters $\beta_1$ and $\beta_0$. Next we examine the residuals from the fitted line and obtain an estimate $s$ of the remaining parameter $\sigma$. These calculations are preliminary to inference. Finally, we use $s$ to obtain the standard errors needed for the various interval estimates and significance tests. *Roundoff errors that accumulate during these calculations can ruin the final results. Be sure to carry many significant digits and check your work carefully.*

**Preliminary calculations** After examining the scatterplot (Figure 10.13) to verify that the data show a straight-line pattern, we begin our calculations.

EXAMPLE

**10.14 Summary statistics for nitrogen balance study.** We start by making a table with the mean and standard deviation for each of the variables, the correlation, and the sample size. These calculations should be familiar from Chapters 1 and 2. Here is the summary:

| Variable | Mean | Standard deviation | Correlation | Sample size |
|---|---|---|---|---|
| Intake | $\bar{x} = 0.79000$ | $s_x = 0.24357545$ | $r = 0.99698478$ | $n = 3$ |
| N balance | $\bar{y} = 20.56667$ | $s_y = 45.4132506$ | | |

These quantities are the building blocks for our calculations.

We will need one additional quantity for the calculations to follow. It is the expression $\sum(x_i - \bar{x})^2$. We obtain this quantity as an intermediate step when we calculate $s_x$. You could also find it using the fact that $\sum(x_i - \bar{x})^2 = (n-1)s_x^2$. You should verify that the value for our example is

$$\sum(x_i - \bar{x})^2 = 0.118658$$

Our first task is to find the least-squares line. This is easy with the building blocks that we have assembled.

EXAMPLE

**10.15 Computing the least-squares regression line.** The slope of the least-squares line is

$$\begin{aligned} b_1 &= r\frac{s_y}{s_x} \\ &= 0.99698478\frac{45.4132506}{0.24357545} \\ &= 185.882 \end{aligned}$$

The intercept is

$$\begin{aligned} b_0 &= \bar{y} - b_1\bar{x} \\ &= 20.56667 - (185.882)(0.79000) \\ &= -126.280 \end{aligned}$$

The equation of the least-squares regression line is therefore

$$\hat{y} = -126 + 186x$$

This is the line shown in Figure 10.13.

We now have estimates of the first two parameters, $\beta_0$ and $\beta_1$, of our linear regression model. We now find the estimate of the third parameter, $\sigma$: the standard deviation $s$ about the fitted line. To do this we need to find the predicted values and then the residuals.

EXAMPLE

**10.16 Computing the predicted values and residuals.** The first observation is an intake of $x = 0.543$. The corresponding predicted value of nitrogen balance is

$$\begin{aligned}\hat{y}_1 &= b_0 + b_1 x_1 \\ &= -126.280 + (185.882)(0.543) \\ &= -25.346\end{aligned}$$

and the residual is

$$\begin{aligned}e_1 &= y_1 - \hat{y}_1 \\ &= -23.4 - (-25.346) \\ &= 1.946\end{aligned}$$

The residuals for the other intakes are calculated in the same way. You should verify that they are $-4.068$ and $2.122$.

Notice that the sum of these three residuals is zero. When doing these calculations by hand, it is always helpful to check that the sum of the residuals is zero.

EXAMPLE

**10.17 Computing $s^2$.** The estimate of $\sigma^2$ is $s^2$, the sum of the squares of the residuals divided by $n - 2$. The estimated standard deviation about the line is the square root of this quantity.

$$\begin{aligned}s^2 &= \frac{\sum e_i^2}{n-2} \\ &= \frac{(1.946)^2 + (-4.068)^2 + (2.122)^2}{1} \\ &= 24.84\end{aligned}$$

So the estimate of the standard deviation about the line is

$$s = \sqrt{24.84} = 4.984$$

Now that we have estimates of the three parameters of our model, we can proceed to the more detailed calculations needed for regression inference.

**Inference for slope and intercept** Confidence intervals and significance tests for the slope $\beta_1$ and intercept $\beta_0$ of the population regression line make use of the estimates $b_1$ and $b_0$ and their standard errors.

LOOK BACK
**rules for variances, page 282**

Some algebra using the rules for variances establishes that the standard deviation of $b_1$ is

$$\sigma_{b_1} = \frac{\sigma}{\sqrt{\sum(x_i - \bar{x})^2}}$$

Similarly, the standard deviation of $b_0$ is

$$\sigma_{b_0} = \sigma\sqrt{\frac{1}{n} + \frac{\bar{x}^2}{\sum(x_i - \bar{x})^2}}$$

To estimate these standard deviations, we need only replace $\sigma$ by its estimate $s$.

## STANDARD ERRORS FOR ESTIMATED REGRESSION COEFFICIENTS

The standard error of the slope $b_1$ of the least-squares regression line is

$$\text{SE}_{b_1} = \frac{s}{\sqrt{\sum(x_i - \bar{x})^2}}$$

The standard error of the intercept $b_0$ is

$$\text{SE}_{b_0} = s\sqrt{\frac{1}{n} + \frac{\bar{x}^2}{\sum(x_i - \bar{x})^2}}$$

The plot of the regression line with the data in Figure 10.13 *appears* to show a very strong relationship, but our sample size is very small. We assess the situation with a significance test for the slope.

EXAMPLE

**10.18 Testing the slope.** First we need the standard error of the estimated slope:

$$\begin{aligned} \text{SE}_{b_1} &= \frac{s}{\sqrt{\sum(x_i - \bar{x})^2}} \\ &= \frac{4.984}{\sqrt{0.118658}} \\ &= 14.47 \end{aligned}$$

To test

$$H_0: \beta_1 = 0$$
$$H_a: \beta_1 \neq 0$$

calculate the $t$ statistic:

$$\begin{aligned} t &= \frac{b_1}{\text{SE}_{b_1}} \\ &= \frac{185.882}{14.47} = 12.85 \end{aligned}$$

Using Table D with $n - 2 = 1$ degree of freedom, we conclude that $P < 0.05$. (The exact value obtained from software is 0.0494.) The data provide evidence in favor of a relationship between nitrogen balance and protein intake ($t = 12.85$, df $= 1$, $P < 0.05$).

Three things are important to note about this example. First, the sample size is very small. Even though the estimated slope is more than 12 standard deviations away from zero, we have only barely attained the 0.05 standard for statistical significance. *It is important to remember that we need to have a very large effect if we expect to detect it with a small sample size.* Second, we would, of course, prefer to have more than three observations for this analysis. However, for each diet, data are collected for about a month. Because the requirement is assumed to be a distribution rather than a single number for everyone, it is important to measure several subjects. Because of the enormous expense involved, researchers typically use only three levels of intake. Third, because we expect balance to increase with increasing intake, a one-sided significance test is justified in this setting.

The significance test tells us that the data provide sufficient information to conclude that intake and balance are related. We use the estimate $b_1$ and its confidence interval to further describe the relationship.

EXAMPLE

**10.19 Computing a 95% confidence interval for the slope.** For the protein requirement problem, let's find a 95% confidence interval for the slope $\beta_1$. The degrees of freedom are $n - 2 = 1$, so $t^*$ from Table D is 12.71. We compute

$$b_1 \pm t^*\mathrm{SE}_{b_1} = 185.882 \pm (12.71)(14.47)$$
$$= 186 \pm 184$$

The interval is (2, 370).

Note the effect of the small sample size on the critical value $t^*$. With one additional observation, it would decrease to 4.303.

In this example, the intercept $\beta_0$ does not have a meaningful interpretation. A protein intake of zero is theoretically possible, but we would not expect our linear model to be reasonable when extended to such an extreme value. For problems where inference for $\beta_0$ is appropriate, the calculations are performed in the same way as those for $\beta_1$. Note that there is a different formula for the standard error, however.

**Confidence intervals for the mean response and prediction intervals for a future observation** When we substitute a particular value $x^*$ of the explanatory variable into the regression equation and obtain a value of $\hat{y}$, we can view the result in two ways:

1. We have estimated the mean response $\mu_y$.
2. We have predicted a future value of the response $y$.

The margins of error for these two uses are often quite different. Prediction intervals for an individual response are wider than confidence intervals for estimating a mean response. We now proceed with the details of these calculations. Once again, standard errors are the essential quantities. And once again, these standard errors are multiples of $s$, our basic measure of the variability of the responses about the fitted line.

**STANDARD ERRORS FOR $\hat{\mu}$ AND $\hat{y}$**

The standard error of $\hat{\mu}$ is

$$SE_{\hat{\mu}} = s\sqrt{\frac{1}{n} + \frac{(x^* - \bar{x})^2}{\sum(x_i - \bar{x})^2}}$$

The standard error for predicting an individual response $\hat{y}$ is[4]

$$SE_{\hat{y}} = s\sqrt{1 + \frac{1}{n} + \frac{(x^* - \bar{x})^2}{\sum(x_i - \bar{x})^2}}$$

Note that the only difference between the formulas for these two standard errors is the extra 1 under the square root sign in the standard error for prediction. This standard error is larger due to the additional variation of individual responses about the mean response. It produces prediction intervals that are wider than the confidence intervals for the mean response.

For the nitrogen balance example, we can think about the mean balance that would result if a particular protein intake was consumed many times. The confidence interval for the mean response would provide an interval estimate of this population value. On the other hand, we might want to predict a future observation under conditions similar to those used in the study, that is, for a one-month period, at a particular intake level. A prediction interval attempts to capture this future observation.

EXAMPLE

**10.20 Confidence and prediction intervals.** Let's find a 95% confidence interval for the mean balance corresponding to an intake of 0.7 g/kg/d. The estimated mean balance is

$$\begin{aligned}\hat{\mu} &= b_0 + b_1x_1\\ &= -126.280 + (185.882)(0.7)\\ &= 3.837\end{aligned}$$

The standard error is

$$\begin{aligned}SE_{\hat{\mu}} &= s\sqrt{\frac{1}{n} + \frac{(x^* - \bar{x})^2}{\sum(x_i - \bar{x})^2}}\\ &= 4.984\sqrt{\frac{1}{3} + \frac{(0.70 - 0.79)^2}{0.118658}}\\ &= 3.158\end{aligned}$$

To find the 95% confidence interval we compute

$$\begin{aligned}\hat{\mu} \pm t^{*}\mathrm{SE}_{\hat{\mu}} &= 3.837 \pm (12.71)(3.158)\\ &= 3.837 \pm 40.138\\ &= 4 \pm 40\end{aligned}$$

The interval is −36 to 44 mg/kg/d of nitrogen.

Calculations for the prediction intervals are similar. The only difference is the use of the formula for $\mathrm{SE}_{\hat{y}}$ in place of $\mathrm{SE}_{\hat{\mu}}$.

Since the confidence interval for mean response includes the value 0, the corresponding intake 0.7 g/kg/d should be considered as a possible value for the intake requirement for this individual. Other intakes would also produce confidence intervals that would include the value of 0 for mean balance. Here is one method that is commonly used to determine a single value of the requirement for an individual.

EXAMPLE

**10.21 Estimating the protein requirement.** We define the estimated requirement for an individual to be the intake corresponding to zero balance using the fitted regression equation. To do this, we set the equation

$$\hat{\mu} = b_0 + b_1 x$$

equal to 0 and solve for the intake $x$. So,

$$\begin{aligned}x &= -b_0/b_1\\ &= -(-126.280)/185.882\\ &= 0.68\end{aligned}$$

The estimated protein requirement for this individual is 0.68 g/kg/d.

If we repeat these calculations using data collected on a large number of individuals, we can estimate the requirement distribution for a population. There are many interesting statistical issues related to this problem.[5]

## Inference for correlation

LOOK BACK
correlation, page 102

The correlation coefficient is a measure of the strength and direction of the linear association between two variables. Correlation does not require an explanatory-response relationship between the variables. We can consider the sample correlation $r$ as an estimate of the correlation in the population and base inference about the population correlation on $r$.

population correlation $\rho$

The correlation between the variables $x$ and $y$ when they are measured for every member of a population is the **population correlation.** As usual, we use Greek letters to represent population parameters. In this case $\rho$ (the Greek letter rho) is the population correlation. When $\rho = 0$, there is no linear association in the population. In the important case where the two variables $x$ and $y$ are both Normally distributed, the condition $\rho = 0$ is equivalent to the statement that $x$ and $y$ are independent. That is, there is no association of any kind between $x$ and $y$. (Technically, the condition required is that $x$ and $y$ be **jointly Normal.** This means that the distribution of $x$ is Normal and also that the conditional distribution of $y$, given any fixed value of $x$, is Normal.) We therefore may wish to test the null hypothesis that a population correlation is 0.

jointly Normal variables

## TEST FOR A ZERO POPULATION CORRELATION

To test the hypothesis $H_0$: $\rho = 0$, compute the $t$ statistic:

$$t = \frac{r\sqrt{n-2}}{\sqrt{1-r^2}}$$

where $n$ is the sample size and $r$ is the sample correlation.

In terms of a random variable $T$ having the $t(n-2)$ distribution, the $P$-value for a test of $H_0$ against

$H_a$: $\rho > 0$ is $P(T \geq t)$

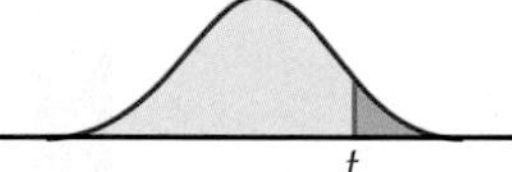


$H_a$: $\rho < 0$ is $P(T \leq t)$

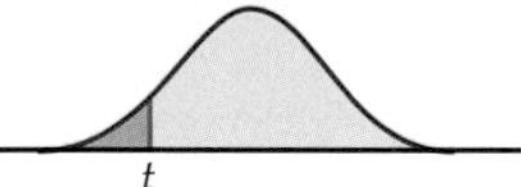


$H_a$: $\rho \neq 0$ is $2P(T \geq |t|)$

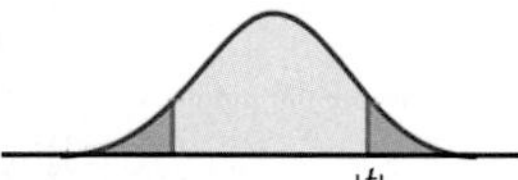


Most computer packages have routines for calculating correlations and some will provide the significance test for the null hypothesis that $\rho$ is zero.

EXAMPLE

**10.22 Correlation in the fuel efficiency study.** For the fuel efficiency example, the SPSS output appears in Figure 10.14. The sample correlation between fuel efficiency and the logarithm of speed is $r = 0.946$. SPSS calls this a Pearson correlation to distinguish it from other kinds of correlations that it can calculate. The $P$-value for a two-sided test of $H_0$: $\rho = 0$ is given as 0.000. This means that the actual $P$-value is less than 0.0005. We conclude that there is a nonzero correlation between MPG and LOGMPH.

If we wanted to test the one-sided alternative that the population correlation is negative, we divide the $P$-value in the output by 2, after checking that the sample coefficient is in fact negative.

If your software does not give the significance test, you can do the computations easily with a calculator.

SPSS

**Correlations**

| | | | |
|---|---|---|---|
| LOGMPH | Pearson Correlation | 1 | .946** |
| | Sig. (2-tailed) | . | .000 |
| | N | 60 | 60 |
| MPG | Pearson Correlation | .946** | 1 |
| | Sig. (2-tailed) | .000 | . |
| | N | 60 | 60 |

**. Correlation is significant at the 0.01 level (2-tailed).

**FIGURE 10.14** Correlation output for Example 10.22.

EXAMPLE

**10.23 Correlation test using a calculator.** The correlation between MPG and LOGMPH is $r = 0.946$. Recall that $n = 60$. The $t$ statistic for testing the null hypothesis that the population correlation is zero is

$$\begin{aligned} t &= \frac{r\sqrt{n-2}}{\sqrt{1-r^2}} \\ &= \frac{0.946\sqrt{60-2}}{\sqrt{1-(0.946)^2}} \\ &= 22.2 \end{aligned}$$

The degrees of freedom are $n - 2 = 58$. From Table D we conclude that $P < 0.0001$. This agrees with the SPSS output in Figure 10.14, where the $P$-value is given as 0.000. The data provide clear evidence that fuel efficiency and the log of speed are related.

There is a close connection between the significance test for a correlation and the test for the slope in a linear regression. Recall that

$$b_1 = r\frac{s_y}{s_x}$$

From this fact we see that if the slope is 0, so is the correlation, and vice versa. It should come as no surprise to learn that the procedures for testing $H_0\colon \beta_1 = 0$ and $H_0\colon \rho = 0$ are also closely related. In fact, the $t$ statistics for testing these hypotheses are numerically equal. That is,

$$\frac{b_1}{s_{b_1}} = \frac{r\sqrt{n-2}}{\sqrt{1-r^2}}$$

Check that this holds in both of our examples.

In our examples, the conclusion that there is a statistically significant correlation between the two variables would not come as a surprise to anyone familiar with the meaning of these variables. The significance test simply tells us whether or not there is evidence in the data to conclude that the population correlation is different from 0. The actual size of the correlation is of considerably more interest. We would therefore like to give a confidence interval for the population correlation. Unfortunately, most software packages do not perform this calculation. Because hand calculation of the confidence interval is very tedious, we do not give the method here.[6]

## USE YOUR KNOWLEDGE

**10.5** **Research and development spending.** The National Science Foundation collects data on the research and development spending by universities and colleges in the United States.[7] Here are the data for the years 1999 to 2001 (using 1996 dollars):

| Year | 1999 | 2000 | 2001 |
|---|---|---|---|
| Spending (billions of dollars) | 26.4 | 28.0 | 29.7 |

Do the following by hand or with a calculator and verify your results with a software package.

(a) Make a scatterplot that shows the increase in research and development spending over time. Does the pattern suggest that the spending is increasing linearly over time?

(b) Find the equation of the least-squares regression line for predicting spending from year. Add this line to your scatterplot.

(c) For each of the three years, find the residual. Use these residuals to calculate the standard error $s$.

(d) Write the regression model for this setting. What are your estimates of the unknown parameters in this model?

(e) Compute a 95% confidence interval for the slope and summarize what this interval tells you about the increase in spending over time.

## SECTION 10.2 Summary

The **ANOVA table** for a linear regression gives the degrees of freedom, sum of squares, and mean squares for the model, error, and total sources of variation. The **ANOVA *F* statistic** is the ratio MSM/MSE. Under $H_0: \beta_1 = 0$, this statistic has an $F(1, n - 2)$ distribution and is used to test $H_0$ versus the two-sided alternative.

The **square of the sample correlation** can be expressed as

$$r^2 = \frac{\text{SSM}}{\text{SST}}$$

and is interpreted as the proportion of the variability in the response variable $y$ that is explained by the explanatory variable $x$ in the linear regression.

The **standard errors** for $b_0$ and $b_1$ are

$$\mathrm{SE}_{b_0} = s\sqrt{\frac{1}{n} + \frac{\bar{x}^2}{\sum(x_i - \bar{x})^2}}$$

$$\mathrm{SE}_{b_1} = \frac{s}{\sqrt{\sum(x_i - \bar{x})^2}}$$

The **standard error** that we use for a confidence interval for the estimated mean response for the subpopulation corresponding to the value $x^*$ of the explanatory variable is

$$\mathrm{SE}_{\hat{\mu}} = s\sqrt{\frac{1}{n} + \frac{(x^* - \bar{x})^2}{\sum(x_i - \bar{x})^2}}$$

The **standard error** that we use for a prediction interval for a future observation from the subpopulation corresponding to the value $x^*$ of the explanatory variable is

$$\mathrm{SE}_{\hat{y}} = s\sqrt{1 + \frac{1}{n} + \frac{(x^* - \bar{x})^2}{\sum(x_i - \bar{x})^2}}$$

When the variables $y$ and $x$ are jointly Normal, the sample correlation is an estimate of the **population correlation $\rho$.** The test of $H_0$: $\rho = 0$ is based on the ***t* statistic**

$$t = \frac{r\sqrt{n-2}}{\sqrt{1-r^2}}$$

which has a $t(n-2)$ distribution under $H_0$. This test statistic is numerically identical to the $t$ statistic used to test $H_0$: $\beta_1 = 0$.

## CHAPTER 10 Exercises

*For Exercises 10.1 and 10.2, see pages 569 and 570; for Exercises 10.3 and 10.4, see page 576; and for Exercise 10.5, see page 593.*

**10.6** **What's wrong?** For each of the following, explain what is wrong and why.

(a) The slope describes the change in $x$ for a change in $y$.

(b) The population regression line is $y = b_0 + b_1x$.

(c) A 95% confidence interval for the mean response is the same width regardless of $x$.

**10.7** **What's wrong?** For each of the following, explain what is wrong and why.

(a) The parameters of the simple linear regression model are $b_0$, $b_1$, and $s$.

(b) To test $H_0$: $b_1 = 0$, use a $t$ test.

(c) For a particular value of the explanatory variable $x$, the confidence interval for the mean response will be wider than the prediction interval for a future observation.

**10.8** **95% confidence intervals for the slope.** Find a 95% confidence interval for the slope in each of the following settings:

(a) $n = 25$, $\hat{y} = 1.3 + 12.10x$, and $\mathrm{SE}_{b_1} = 6.31$

(b) $n = 25$, $\hat{y} = 13.0 + 6.10x$, and $\mathrm{SE}_{b_1} = 6.31$

(c) $n = 100$, $\hat{y} = 1.3 + 12.10x$, and $\mathrm{SE}_{b_1} = 6.31$

**10.9** **Significance test for the slope.** For each of the settings in the previous exercise, test the null hypothesis that the slope is zero versus the two-sided alternative.

**TABLE 10.1**

In-state tuition and fees (in dollars) for 32 public universities

| School | 2000 | 2005 | School | 2000 | 2005 | School | 2000 | 2005 |
|---|---|---|---|---|---|---|---|---|
| Penn State | 7,018 | 11,508 | Virginia | 4,335 | 7,370 | Iowa State | 3,132 | 5,634 |
| Pittsburgh | 7,002 | 11,436 | Indiana | 4,405 | 7,112 | Oregon | 3,819 | 5,613 |
| Michigan | 6,926 | 9,798 | Cal-Santa Barbara | 3,832 | 6,997 | Iowa | 3,204 | 5,612 |
| Rutgers | 6,333 | 9,221 | Texas | 3,575 | 6,972 | Washington | 3,761 | 5,610 |
| Illinois | 4,994 | 8,634 | Cal-Irvine | 3,970 | 6,770 | Nebraska | 3,450 | 5,540 |
| Minnesota | 4,877 | 8,622 | Cal-San Diego | 3,848 | 6,685 | Kansas | 2,725 | 5,413 |
| Michigan State | 5,432 | 8,108 | Cal-Berkeley | 4,047 | 6,512 | Colorado | 3,188 | 5,372 |
| Ohio State | 4,383 | 8,082 | UCLA | 3,698 | 6,504 | North Carolina | 2,768 | 4,613 |
| Maryland | 5,136 | 7,821 | Purdue | 3,872 | 6,458 | Arizona | 2,348 | 4,498 |
| Cal-Davis | 4,072 | 7,457 | Wisconsin | 3,791 | 6,284 | Florida | 2,256 | 3,094 |
| Missouri | 4,726 | 7,415 | Buffalo | 4,715 | 6,068 | | | |

**10.10 Public university tuition: 2000 versus 2005.** Table 10.1 shows the in-state undergraduate tuition and required fees for 34 public universities in 2000 and 2005.[8]

(a) Plot the data with the 2000 tuition on the $x$-axis and describe the relationship. Are there any outliers or unusual values? Does a linear relationship between the tuition in 2000 and 2005 seem reasonable?

(b) Run the simple linear regression and state the least-squares regression line.

(c) Obtain the residuals and plot them versus the 2000 tuition amount. Is there anything unusual in the plot?

(d) Do the residuals appear to be approximately Normal? Explain.

(e) Give the null and alternative hypotheses for examining the relationship between 2000 and 2005 tuition amounts.

(f) Write down the test statistic and $P$-value for the hypotheses stated in part (e). State your conclusions.

**10.11 More on public university tuition.** Refer to Exercise 10.10.

(a) Construct a 95% confidence interval for the slope. What does this interval tell you about the percent increase in tuition between 2000 and 2005?

(b) The tuition at Stat U was $5000 in 2000. What is the predicted tuition in 2005?

(c) Find a 95% prediction interval for the 2005 tuition at Stat U and summarize the results.

**10.12 Are the two fuel efficiency measurements similar?** Refer to Exercise 7.24. In addition to the computer calculating mpg, the driver also recorded the mpg by dividing the miles driven by the amount of gallons at fill-up. The driver wants to determine if these calculations are different.

| Fill-up | 1 | 2 | 3 | 4 | 5 | 6 | 7 | 8 | 9 | 10 |
|---|---|---|---|---|---|---|---|---|---|---|
| Computer | 41.5 | 50.7 | 36.6 | 37.3 | 34.2 | 45.0 | 48.0 | 43.2 | 47.7 | 42.2 |
| Driver | 36.5 | 44.2 | 37.2 | 35.6 | 30.5 | 40.5 | 40.0 | 41.0 | 42.8 | 39.2 |
| **Fill-up** | 11 | 12 | 13 | 14 | 15 | 16 | 17 | 18 | 19 | 20 |
| Computer | 43.2 | 44.6 | 48.4 | 46.4 | 46.8 | 39.2 | 37.3 | 43.5 | 44.3 | 43.3 |
| Driver | 38.8 | 44.5 | 45.4 | 45.3 | 45.7 | 34.2 | 35.2 | 39.8 | 44.9 | 47.5 |

(a) Consider the driver's mpg calculations as the explanatory variable. Plot the data and describe the relationship. Are there any outliers or unusual values? Does a linear relationship seem reasonable?

(b) Run the simple linear regression and state the least-squares regression line.

(c) Summarize the results. Does it appear that the computer and driver calculations are the same? Explain.

**10.13 Beer and blood alcohol.** How well does the number of beers a student drinks predict his or her blood alcohol content? Sixteen student volunteers at Ohio State University drank a randomly assigned number of 12-ounce cans of beer. Thirty minutes later, a police officer measured their blood alcohol content (BAC). Here are the data:[9]

| Student | 1 | 2 | 3 | 4 | 5 | 6 | 7 | 8 |
|---|---|---|---|---|---|---|---|---|
| Beers | 5 | 2 | 9 | 8 | 3 | 7 | 3 | 5 |
| BAC | 0.10 | 0.03 | 0.19 | 0.12 | 0.04 | 0.095 | 0.07 | 0.06 |
| **Student** | 9 | 10 | 11 | 12 | 13 | 14 | 15 | 16 |
| Beers | 3 | 5 | 4 | 6 | 5 | 7 | 1 | 4 |
| BAC | 0.02 | 0.05 | 0.07 | 0.10 | 0.085 | 0.09 | 0.01 | 0.05 |

The students were equally divided between men and women and differed in weight and usual drinking habits. Because of this variation, many students don't believe that number of drinks predicts blood alcohol well.

(a) Make a scatterplot of the data. Find the equation of the least-squares regression line for predicting blood alcohol from number of beers and add this line to your plot. What is $r^2$ for these data? Briefly summarize what your data analysis shows.

(b) Is there significant evidence that drinking more beers increases blood alcohol on the average in the population of all students? State hypotheses, give a test statistic and *P*-value, and state your conclusion.

(c) Steve thinks he can drive legally 30 minutes after he drinks 5 beers. The legal limit is BAC = 0.08. Give a 90% prediction interval. Can he be confident he won't be arrested if he drives and is stopped?

**10.14** CHALLENGE **Predicting water quality.** The index of biotic integrity (IBI) is a measure of the water quality in streams. IBI and land use measures for a collection of streams in the Ozark Highland ecoregion of Arkansas were collected as part of a study.[10] Table 10.2 gives the data for IBI and the area of the watershed in square kilometers for streams in the original sample with area less than or equal to 70 km$^2$.

(a) Use numerical and graphical methods to describe the variable IBI. Do the same for area. Summarize your results.

(b) Plot the data and describe the relationship. Are there any outliers or unusual patterns?

(c) Give the statistical model for simple linear regression for this problem.

(d) State the null and alternative hypotheses for examining the relationship between IBI and area.

(e) Run the simple linear regression and summarize the results.

(f) Obtain the residuals and plot them versus area. Is there anything unusual in the plot?

(g) Do the residuals appear to be approximately Normal? Give reasons for your answer.

(h) Do the assumptions for the analysis of these data using the model you gave in part (c) appear to be reasonable? Explain your answer.

**10.15** CHALLENGE **More on predicting water quality.** The researchers who conducted the study described in the previous exercise also recorded the percent of the watershed area that was forest for each of the streams. The data are given in Table 10.3. Analyze these data using the questions in the previous exercise as a guide.

**10.16 Comparing the analyses.** In Exercises 10.14 and 10.15, you used two different explanatory variables to predict IBI. Summarize the two analyses and compare the results. If you had to choose between the two explanatory variables for predicting IBI, which one would you prefer? Give reasons for your answer.

**10.17 How an outlier can affect statistical significance.** Consider the data in Table 10.3 and the relationship between IBI and the percent of watershed area that was forest. The relationship between these two variables is almost significant at the .05 level. In this exercise you will demonstrate the potential effect of an outlier on statistical significance. Investigate what happens when you decrease the IBI to 0.0 for (1) an observation with 0% forest and (2) an observation with 100% forest.

**TABLE 10.2**

Watershed area and index of biotic integrity

| Area | IBI | Area | IBI | Area | IBI | Area | IBI | Area | IBI |
|---|---|---|---|---|---|---|---|---|---|
| 21 | 47 | 29 | 61 | 31 | 39 | 32 | 59 | 34 | 72 |
| 34 | 76 | 49 | 85 | 52 | 89 | 2 | 74 | 70 | 89 |
| 6 | 33 | 28 | 46 | 21 | 32 | 59 | 80 | 69 | 80 |
| 47 | 78 | 8 | 53 | 8 | 43 | 58 | 88 | 54 | 84 |
| 10 | 62 | 57 | 55 | 18 | 29 | 19 | 29 | 39 | 54 |
| 49 | 78 | 9 | 71 | 5 | 55 | 14 | 58 | 9 | 71 |
| 23 | 33 | 31 | 59 | 18 | 81 | 16 | 71 | 21 | 75 |
| 32 | 64 | 10 | 41 | 26 | 82 | 9 | 60 | 54 | 84 |
| 12 | 83 | 21 | 82 | 27 | 82 | 23 | 86 | 26 | 79 |
| 16 | 67 | 26 | 56 | 26 | 85 | 28 | 91 | | |

**TABLE 10.3**
Percent forest and index of biotic integrity

| Forest | IBI | Forest | IBI | Forest | IBI | Forest | IBI | Forest | IBI |
|---|---|---|---|---|---|---|---|---|---|
| 0 | 47 | 0 | 61 | 0 | 39 | 0 | 59 | 0 | 72 |
| 0 | 76 | 3 | 85 | 3 | 89 | 7 | 74 | 8 | 89 |
| 9 | 33 | 10 | 46 | 10 | 32 | 11 | 80 | 14 | 80 |
| 17 | 78 | 17 | 53 | 18 | 43 | 21 | 88 | 22 | 84 |
| 25 | 62 | 31 | 55 | 32 | 29 | 33 | 29 | 33 | 54 |
| 33 | 78 | 39 | 71 | 41 | 55 | 43 | 58 | 43 | 71 |
| 47 | 33 | 49 | 59 | 49 | 81 | 52 | 71 | 52 | 75 |
| 59 | 64 | 63 | 41 | 68 | 82 | 75 | 60 | 79 | 84 |
| 79 | 83 | 80 | 82 | 86 | 82 | 89 | 86 | 90 | 79 |
| 95 | 67 | 95 | 56 | 100 | 85 | 100 | 91 | | |

Write a short summary of what you learn from this exercise.

**10.18 Predicting water quality for an area of 30 km$^2$.** Refer to Exercise 10.14.

(a) Find a 95% confidence interval for mean response corresponding to an area of 30 km$^2$.

(b) Find a 95% prediction interval for a future response.

(c) Write a short paragraph interpreting the meaning of the intervals in terms of Ozark Highland streams.

(d) Do you think that these results can be applied to other streams in Arkansas or in other states? Explain why or why not.

**10.19 Compare the predictions.** Case 21 in Table 10.2 and Table 10.3 corresponds to the same watershed area. For this case the area is 10 km$^2$ and the percent forest is 25%. A predicted index of biotic integrity based on area was computed in Exercise 10.14, while one based on percent forest was computed in Exercise 10.15. Compare these two estimates and explain why they differ. Use the idea of a prediction interval to interpret these results.

**10.20 U.S. versus overseas stock returns.** Returns on common stocks in the United States and overseas appear to be growing more closely correlated as economies become more interdependent. Suppose that the following population regression line connects the total annual returns (in percent) on two indexes of stock prices:

$$\text{MEAN OVERSEAS RETURN} = 4.6 + 0.67 \times \text{U.S. RETURN}$$

(a) What is $\beta_0$ in this line? What does this number say about overseas returns when the U.S. market is flat (0% return)?

(b) What is $\beta_1$ in this line? What does this number say about the relationship between U.S. and overseas returns?

(c) We know that overseas returns will vary in years having the same return on U.S. common stocks. Write the regression model based on the population regression line given above. What part of this model allows overseas returns to vary when U.S. returns remain the same?

**10.21** CHALLENGE **Breaking strength of wood.** Exercise 2.144 (page 163) gives the modulus of elasticity (MOE) and the modulus of rupture (MOR) for 32 plywood specimens. Because measuring MOR involves breaking the wood but measuring MOE does not, we would like to predict the destructive test result, MOR, using the nondestructive test result, MOE.

(a) Describe the distribution of MOR using graphical and numerical summaries. Do the same for MOE.

(b) Make a plot of the two variables. Which should be plotted on the $x$ axis? Give a reason for your answer.

(c) Give the statistical model for this analysis, run the analysis, summarize the results, and write a short summary of your conclusions.

(d) Examine the assumptions needed for the analysis. Are you satisfied that there are no serious violations that would cause you to question the validity of your conclusions?

**10.22 Breaking strength of wood, continued.** Refer to the previous exercise. Consider an MOE of 2,000,000.

(a) Interpret the confidence interval for mean response and the prediction interval for a future observation for this value of MOE.

(b) Which interval will include more values? Give a reason for your answer.

(c) (Optional) Calculate the two intervals.

**10.23 Are the number of tornadoes increasing?** *Storm Data* is a publication of the National Climatic Data Center that contains a listing of tornadoes, thunderstorms, floods, lightning, temperature extremes, and other weather phenomena. Table 10.4 summarizes the annual number of tornadoes in the United States between 1953 and 2005.[11]

(a) Make a plot of the total number of tornadoes by year. Does a linear trend over years appear reasonable?

(b) Are there any outliers or unusual patterns? Explain your answer.

(c) Run the simple linear regression and summarize the results, making sure to construct a 95% confidence interval for the average annual increase in the number of tornadoes.

(d) Obtain the residuals and plot them versus year. Is there anything unusual in the plot?

(e) Are the residuals Normal? Justify your answer.

**10.24 More on the number of tornadoes.** Refer to the previous exercise. The number of tornadoes in 2004 was much larger than expected under the linear model. Remove this observation and rerun the simple linear regression. Compare these results with the results of the previous exercise.

**10.25 CRP and serum retinol.** In Exercise 7.26 (page 442) we examined the distribution of C-reactive protein (CRP) in a sample of 40 children from Papua New Guinea. Serum retinol values for the same children were studied in Exercise 7.28. One important question that can be addressed with these data is whether or not infections, as indicated by CRP, cause a decrease in the measured values of retinol, low values of which indicate a vitamin A deficiency. The data are given in Table 10.5.

**TABLE 10.4**

Annual number of tornadoes in the United States between 1953 and 2005

| Year | Number of tornadoes | Year | Number of tornadoes | Year | Number of tornadoes | Year | Number of tornadoes |
|---|---|---|---|---|---|---|---|
| 1953 | 421 | 1967 | 926 | 1981 | 783 | 1995 | 1235 |
| 1954 | 550 | 1968 | 660 | 1982 | 1046 | 1996 | 1173 |
| 1955 | 593 | 1969 | 608 | 1983 | 931 | 1997 | 1148 |
| 1956 | 504 | 1970 | 653 | 1984 | 907 | 1998 | 1449 |
| 1957 | 856 | 1971 | 888 | 1985 | 684 | 1999 | 1340 |
| 1958 | 564 | 1972 | 741 | 1986 | 764 | 2000 | 1076 |
| 1959 | 604 | 1973 | 1102 | 1987 | 656 | 2001 | 1213 |
| 1960 | 616 | 1974 | 947 | 1988 | 702 | 2002 | 934 |
| 1961 | 697 | 1975 | 920 | 1989 | 856 | 2003 | 1372 |
| 1962 | 657 | 1976 | 835 | 1990 | 1133 | 2004 | 1819 |
| 1963 | 464 | 1977 | 852 | 1991 | 1132 | 2005 | 1194 |
| 1964 | 704 | 1978 | 788 | 1992 | 1298 | | |
| 1965 | 906 | 1979 | 852 | 1993 | 1176 | | |
| 1966 | 585 | 1980 | 866 | 1994 | 1082 | | |

**TABLE 10.5**

C-reactive protein and serum retinol

| CRP | Retinol | CRP | Retinol | CRP | Retinol | CRP | Retinol | CRP | Retinol |
|---|---|---|---|---|---|---|---|---|---|
| 0.00 | 1.15 | 30.61 | 0.97 | 22.82 | 0.24 | 5.36 | 1.19 | 0.00 | 0.83 |
| 3.90 | 1.36 | 0.00 | 0.67 | 0.00 | 1.00 | 0.00 | 0.94 | 0.00 | 1.11 |
| 5.64 | 0.38 | 73.20 | 0.31 | 0.00 | 1.13 | 5.66 | 0.34 | 0.00 | 1.02 |
| 8.22 | 0.34 | 0.00 | 0.99 | 3.49 | 0.31 | 0.00 | 0.35 | 9.37 | 0.56 |
| 0.00 | 0.35 | 46.70 | 0.52 | 0.00 | 1.44 | 59.76 | 0.33 | 20.78 | 0.82 |
| 5.62 | 0.37 | 0.00 | 0.70 | 0.00 | 0.35 | 12.38 | 0.69 | 7.10 | 1.20 |
| 3.92 | 1.17 | 0.00 | 0.88 | 4.81 | 0.34 | 15.74 | 0.69 | 7.89 | 0.87 |
| 6.81 | 0.97 | 26.41 | 0.36 | 9.57 | 1.90 | 0.00 | 1.04 | 5.53 | 0.41 |

(a) Examine the distributions of CRP and serum retinol. Use graphical and numerical methods.

(b) Forty percent of the CRP values are zero. Does this violate any assumption that we need to do a regression analysis using CRP to predict serum retinol? Explain your answer.

(c) Run the regression, summarize the results, and write a short paragraph explaining your conclusions.

(d) Explain the assumptions needed for your results to be valid. Examine the data with respect to these assumptions and report your results.

10.26 CHALLENGE **Osteocalcin and bone formation.** In Exercise 7.118 (page 482) we looked at the distribution of osteocalcin (OC), a biomarker for bone formation, in a sample of 31 healthy females aged 11 to 32 years. This biomarker is relatively inexpensive to measure, requiring only a single sample of blood. Measuring bone formation (VO+), on the other hand, is very expensive. Oral and intravenous administration of stable isotopes of calcium are needed, 25 blood samples over a period of two weeks are drawn, and the collection of all urine and fecal samples for two weeks is required. If a biomarker can reliably predict bone formation, then we could avoid the cost of the expensive VO+ measures. Studies designed to assess the effects of interventions intended to increase bone formation could include many more subjects if only the biomarker measurement is needed. The measured values of VO+ and OC for the 31 females in this study are given in Table 10.6.

(a) Use numerical and graphical summaries to describe the distributions of VO+ and OC.

(b) Plot the data. Give a reason for your choice of variables for the $x$ and $y$ axes. Describe the pattern and note any unusual observations. Do the assumptions needed for regression analysis appear to be approximately satisfied?

(c) Run the regression using OC to predict VO+. Summarize the results.

10.27 CHALLENGE **TRAP and bone resorption.** In Exercise 7.119 (page 482) we looked at the distribution of tartrate resistant acid phosphatase (TRAP), a biomarker for bone resorption. Table 10.7 gives values for this biomarker and a measure of bone resorption VO−. Analyze these data using the questions in the previous exercise as a guide.

10.28 **Transforming the data.** Refer to the OC and VO+ data in Exercise 10.26. For variables such as these, it is common to work with the logarithms of the

**TABLE 10.6**

VO+ and osteocalcin

| VO+ | OC | VO+ | OC | VO+ | OC | VO+ | OC |
|---|---|---|---|---|---|---|---|
| 476 | 8.1 | 1032 | 40.2 | 624 | 17.2 | 285 | 9.9 |
| 694 | 10.1 | 445 | 20.6 | 479 | 15.9 | 403 | 19.7 |
| 753 | 17.9 | 896 | 31.2 | 572 | 16.9 | 391 | 20.0 |
| 687 | 17.2 | 968 | 19.3 | 512 | 24.2 | 513 | 20.8 |
| 628 | 20.9 | 985 | 44.4 | 838 | 30.2 | 878 | 31.4 |
| 1100 | 38.4 | 1251 | 76.5 | 870 | 47.7 | 2221 | 54.6 |
| 1303 | 54.6 | 2545 | 36.4 | 1606 | 68.9 | 1126 | 77.9 |
| 1682 | 52.8 | 2240 | 56.3 | 1557 | 35.7 | | |

**TABLE 10.7**

VO− and TRAP

| VO− | TRAP | VO− | TRAP | VO− | TRAP | VO− | TRAP |
|---|---|---|---|---|---|---|---|
| 407 | 3.3 | 874 | 5.9 | 445 | 6.3 | 351 | 6.9 |
| 980 | 8.1 | 493 | 8.1 | 572 | 8.2 | 634 | 8.8 |
| 1028 | 9.0 | 1116 | 9.0 | 857 | 9.5 | 536 | 9.5 |
| 701 | 9.6 | 934 | 10.1 | 477 | 10.1 | 254 | 10.3 |
| 766 | 10.5 | 496 | 10.7 | 924 | 14.4 | 954 | 14.6 |
| 918 | 14.6 | 1065 | 14.9 | 722 | 18.6 | 1486 | 19.0 |
| 1018 | 19.0 | 2236 | 19.1 | 903 | 19.4 | 960 | 23.7 |
| 1251 | 25.2 | 1761 | 25.5 | 1446 | 28.8 | | |

measured values. Reanalyze these data using the logs of both OC and VO+. Summarize your results and compare them with those you obtained in Exercise 10.26.

**10.29 TRAP and bone resorption using logs.** Refer to the TRAP and VO– data in Exercise 10.27. Reanalyze these data using the logs of both TRAP and VO–. Summarize your results and compare them with those you obtained in Exercise 10.27.

**10.30 Reading test scores and IQ.** In Exercise 2.11 (page 95) you examined the relationship between a reading test score and an IQ score for a sample of 60 fifth-grade children.

(a) Run the regression and summarize the results of the significance tests.

(b) Rerun the analysis with the four possible outliers removed. Summarize your findings, paying particular attention to the effects of removing the outliers.

**10.31** CHALLENGE **Neuron responses.** Exercise 2.143 (page 163) gives data on neuron responses to pure tones and to monkey calls.

(a) Describe each variable graphically and numerically.

(b) Plot the data with the pure tone response on the $x$ axis and the monkey call response on the $y$ axis. Describe the relationship and mark the point with the largest residual and the point with the extreme value of the tone response.

(c) Analyze the entire set of 37 observations and summarize the results.

(d) Perform additional analyses to assess the effects of the two marked points on the results. Summarize your findings.

**10.32 School budget and number of students.** Suppose that there is a linear relationship between the number of students $x$ in an elementary school and the annual budget $y$. Write a population regression model to describe this relationship.

(a) Which parameter in your model is the fixed cost in the budget (for example, the salary of the principal and some administrative costs) that does not change as $x$ increases?

(b) Which parameter in your model shows how total cost changes when there are more students in the school? Do you expect this number to be greater than 0 or less than 0?

(c) Actual data from schools will not fit a straight line exactly. What term in your model allows variation among schools of the same size $x$?

**10.33 Stocks and bonds.** How is the flow of investors' money into stock mutual funds related to the flow of money into bond mutual funds? Here are data on the net new money flowing into stock and bond mutual funds in the years 1985 to 2000, in billions of dollars.[12] "Net" means that funds flowing out are subtracted from those flowing in. If more money leaves than arrives, the net flow will be negative. To eliminate the effect of inflation, all dollar amounts are in "real dollars" with constant buying power equal to that of a dollar in the year 2000.

| Year | 1985 | 1986 | 1987 | 1988 | 1989 | 1990 | 1991 | 1992 |
|---|---|---|---|---|---|---|---|---|
| Stocks | 12.8 | 34.6 | 28.8 | −23.3 | 8.3 | 17.1 | 50.6 | 97.0 |
| Bonds | 100.8 | 161.8 | 10.6 | −5.8 | −1.4 | 9.2 | 74.6 | 87.1 |
| **Year** | **1993** | **1994** | **1995** | **1996** | **1997** | **1998** | **1999** | **2000** |
| Stocks | 151.3 | 133.6 | 140.1 | 238.2 | 243.5 | 165.9 | 194.3 | 309.0 |
| Bonds | 84.6 | −72.0 | −6.8 | 3.3 | 30.0 | 79.2 | −6.2 | −48.0 |

(a) Make a scatterplot with cash flow into stock funds as the explanatory variable. Find the least-squares line for predicting net bond investments from net stock investments. What do the data suggest?

(b) Is there statistically significant evidence that there is some straight-line relationship between the flows of cash into bond funds and stock funds? (State hypotheses, give a test statistic and its $P$-value, and state your conclusion.)

(c) What fact about the scatterplot explains why the relationship described by the least-squares line is not significant?

**10.34 Math pretest predicts success?** Can a pretest on mathematics skills predict success in a statistics course? The 82 students in an introductory statistics class took a pretest at the beginning of the semester. The least-squares regression line for predicting the score $y$ on the final exam from the pretest score $x$ was $\hat{y} = 9.7 + 0.76x$. The standard error of $b_1$ was 0.44.

(a) Test the null hypothesis that there is no linear relationship between the pretest score and the score on the final exam against the two-sided alternative.

(b) Would you reject this null hypothesis versus the one-sided alternative that the slope is positive? Explain your answer.

**10.35 Severities of MA and HAV.** Metatarsus adductus (call it MA) is a turning in of the front part of the foot that is common in adolescents and usually corrects itself. Hallux abducto valgus (call it HAV) is a deformation of the big toe that is not common in youth and often requires surgery. Perhaps the severity of MA can help predict the severity of HAV. Table 2.2 (page 98) gives data on 38 consecutive patients who came to a medical center for HAV surgery.[13] Using X-rays, doctors measured the angle of deformity for both MA and HAV. They speculated that there is a positive association—more serious MA is associated with more serious HAV.

(a) Make a scatterplot of the data in Table 2.2. (Which is the explanatory variable?)

(b) Describe the form, direction, and strength of the relationship between MA angle and HAV angle. Are there any clear outliers in your graph?

(c) Give a statistical model that provides a framework for asking the question of interest for this problem.

(d) Translate the question of interest into null and alternative hypotheses.

(e) Test these hypotheses and write a short description of the results. Be sure to include the value of the test statistic, the degrees of freedom, the *P*-value, and a clear statement of what you conclude.

**10.36 More on MA and HAV.** Refer to the previous exercise. Give a 95% confidence interval for the slope. Explain how this interval can tell you what to conclude from a significance test for this parameter.

**10.37 Do wages rise with experience?** We assume that our wages will increase as we gain experience and become more valuable to our employers. Wages also increase because of inflation. By examining a sample of employees at a given point in time, we can look at part of the picture. How does length of service (LOS) relate to wages? Table 10.8 gives data on the LOS in months and wages for 60 women who work in Indiana banks. Wages are yearly total income divided by the number of weeks worked. We have multiplied wages by a constant for reasons of confidentiality.[14]

(a) Plot wages versus LOS. Describe the relationship. There is one woman with relatively high wages for her length of service. Circle this point and do not use it in the rest of this exercise.

(b) Find the least-squares line. Summarize the significance test for the slope. What do you conclude?

(c) State carefully what the slope tells you about the relationship between wages and length of service.

(d) Give a 95% confidence interval for the slope.

**TABLE 10.8**

Bank wages, length of service (LOS), and bank size

| Wages | LOS | Size | Wages | LOS | Size | Wages | LOS | Size |
|---|---|---|---|---|---|---|---|---|
| 48.3355 | 94 | Large | 64.1026 | 24 | Large | 41.2088 | 97 | Small |
| 49.0279 | 48 | Small | 54.9451 | 222 | Small | 67.9096 | 228 | Small |
| 40.8817 | 102 | Small | 43.8095 | 58 | Large | 43.0942 | 27 | Large |
| 36.5854 | 20 | Small | 43.3455 | 41 | Small | 40.7000 | 48 | Small |
| 46.7596 | 60 | Large | 61.9893 | 153 | Large | 40.5748 | 7 | Large |
| 59.5238 | 78 | Small | 40.0183 | 16 | Small | 39.6825 | 74 | Small |
| 39.1304 | 45 | Large | 50.7143 | 43 | Small | 50.1742 | 204 | Large |
| 39.2465 | 39 | Large | 48.8400 | 96 | Large | 54.9451 | 24 | Large |
| 40.2037 | 20 | Large | 34.3407 | 98 | Large | 32.3822 | 13 | Small |
| 38.1563 | 65 | Small | 80.5861 | 150 | Large | 51.7130 | 30 | Large |
| 50.0905 | 76 | Large | 33.7163 | 124 | Small | 55.8379 | 95 | Large |
| 46.9043 | 48 | Small | 60.3792 | 60 | Large | 54.9451 | 104 | Large |
| 43.1894 | 61 | Small | 48.8400 | 7 | Large | 70.2786 | 34 | Large |
| 60.5637 | 30 | Large | 38.5579 | 22 | Small | 57.2344 | 184 | Small |
| 97.6801 | 70 | Large | 39.2760 | 57 | Large | 54.1126 | 156 | Small |
| 48.5795 | 108 | Large | 47.6564 | 78 | Large | 39.8687 | 25 | Large |
| 67.1551 | 61 | Large | 44.6864 | 36 | Large | 27.4725 | 43 | Small |
| 38.7847 | 10 | Small | 45.7875 | 83 | Small | 67.9584 | 36 | Large |
| 51.8926 | 68 | Large | 65.6288 | 66 | Large | 44.9317 | 60 | Small |
| 51.8326 | 54 | Large | 33.5775 | 47 | Small | 51.5612 | 102 | Large |

**10.38 Do wages rise with experience?** Refer to the previous exercise. Analyze the data with the outlier included.

(a) How does this change the estimates of the parameters $\beta_0$, $\beta_1$, and $\sigma$?

(b) What effect does the outlier have on the results of the significance test for the slope?

(c) How has the width of the 95% confidence interval changed?

**10.39 Leaning Tower of Pisa.** The Leaning Tower of Pisa is an architectural wonder. Engineers concerned about the tower's stability have done extensive studies of its increasing tilt. Measurements of the lean of the tower over time provide much useful information. The following table gives measurements for the years 1975 to 1987. The variable "lean" represents the difference between where a point on the tower would be if the tower were straight and where it actually is. The data are coded as tenths of a millimeter in excess of 2.9 meters, so that the 1975 lean, which was 2.9642 meters, appears in the table as 642. Only the last two digits of the year were entered into the computer.[15]

| Year | 75 | 76 | 77 | 78 | 79 | 80 | 81 | 82 | 83 | 84 | 85 | 86 | 87 |
|---|---|---|---|---|---|---|---|---|---|---|---|---|---|
| Lean | 642 | 644 | 656 | 667 | 673 | 688 | 696 | 698 | 713 | 717 | 725 | 742 | 757 |

(a) Plot the data. Does the trend in lean over time appear to be linear?

(b) What is the equation of the least-squares line? What percent of the variation in lean is explained by this line?

(c) Give a 99% confidence interval for the average rate of change (tenths of a millimeter per year) of the lean.

**10.40 More on the Leaning Tower of Pisa.** Refer to the previous exercise.

(a) In 1918 the lean was 2.9071 meters. (The coded value is 71.) Using the least-squares equation for the years 1975 to 1987, calculate a predicted value for the lean in 1918. (Note that you must use the coded value 18 for year.)

(b) Although the least-squares line gives an excellent fit to the data for 1975 to 1987, this pattern did not extend back to 1918. Write a short statement explaining why this conclusion follows from the information available. Use numerical and graphical summaries to support your explanation.

**10.41 Predicting the lean in 2009.** Refer to the previous two exercises.

(a) How would you code the explanatory variable for the year 2009?

(b) The engineers working on the Leaning Tower of Pisa were most interested in how much the tower would lean if no corrective action was taken. Use the least-squares equation to predict the tower's lean in the year 2009.

(c) To give a margin of error for the lean in 2009, would you use a confidence interval for a mean response or a prediction interval? Explain your choice.

**10.42 Correlation between binge drinking and the average price of beer.** A recent study looked at 118 colleges to investigate the association between the binge-drinking rate and the average price for a bottle of beer at establishments within a 2-mile radius of campus.[16] A correlation of −0.36 was found. Explain this correlation.

**10.43 Is this relationship significant?** Refer to the previous exercise. Test the null hypothesis that the correlation between the binge-drinking rate and the average price for a bottle of beer within a 2-mile radius of campus is zero.

**10.44 Capacity of DRAM.** The capacity (bits) of the largest DRAM (dynamic random access memory) chips commonly available at retail has increased as follows:[17]

| Year | 1971 | 1980 | 1987 | 1993 | 1999 | 2000 |
|---|---|---|---|---|---|---|
| Bits | 1,024 | 64,000 | 1,024,000 | 16,384,000 | 256,000,000 | 512,000,000 |

(a) Make a scatterplot of the data. Growth is much faster than linear.

(b) Plot the logarithm of DRAM capacity against year. These points are close to a straight line.

(c) Regress the logarithm of DRAM capacity on year. Give a 95% confidence interval for the slope of the population regression line.

**10.45 Net flow in stock and bond funds.** Is there a nonzero correlation between net flow of money into stock mutual funds and into bond funds? Use the regression analysis you did in Exercise 10.33 (page 600) to answer this question with no additional calculations.

**10.46 Parental behavior and self-esteem.** Chinese students from public schools in Hong Kong were the subjects of a study designed to investigate the relationship between various measures of parental behavior and other variables. The sample size was 713. The data were obtained from questionnaires filled in by the students. One of the variables examined was parental control, an indication of the amount of control that the parents exercised over the behavior of the students. Another was the self-esteem of the students.[18]

(a) The correlation between parental control and self-esteem was $r = -0.19$. Calculate the $t$ statistic for testing the null hypothesis that the population correlation is 0.

(b) Find an approximate $P$-value for testing $H_0$ versus the two-sided alternative and report your conclusion.

**10.47 Completing an ANOVA table.** How are returns on common stocks in overseas markets related to returns in U.S. markets? Measure U.S. returns by the annual rate of return on the Standard & Poor's 500 stock index and overseas returns by the annual rate of return on the Morgan Stanley Europe, Australasia, Far East (EAFE) index. Both are recorded in percents. Regress the EAFE returns on the S&P 500 returns for the 30 years 1971 to 2000. Here is part of the Minitab output for this regression:

```
The regression equation is
EAFE = 4.76 + 0.663 S&P

Analysis of Variance

Source          DF       SS       MS      F       P
Regression       1   3445.9   3445.9   9.50   0.005
Residual Error
Total           29  13598.3
```

Complete the analysis of variance table by filling in the "Residual Error" row.

**10.48 Interpreting statistical software output.** Refer to the previous exercise. What are the values of the regression standard error $s$ and the squared correlation $r^2$?

**10.49 Standard error and confidence interval for the slope.** Refer to the previous two exercises. The standard deviation of the S&P 500 returns for these years is 16.45%. From this and your work in the previous exercise, find the standard error for the least-squares slope $b_1$. Give a 95% confidence interval for the slope $\beta_1$ of the population regression line.

**10.50 Quality of life in chronically ill patients.** Concern about the quality of life for chronically ill patients is becoming as important as treating their physical symptoms. The SF-36, a questionnaire for measuring the health quality of life, was given to 50 patients with chronic obstructive lung disease.[19] A correlation of 0.68 was reported between the component of the questionnaire called general health perceptions (GHP) and a measure of lung function called forced vital capacity (FVC), expressed as a percent of Normal. The mean and standard deviation of GHP are 43.5 and 20.3, and for FVC the values are 80.9 and 17.2.

(a) Find the equation of the least-squares line for predicting GHP from FVC.

(b) Give the results of the significance test for the null hypothesis that the slope is 0. (*Hint:* What is the relation between this test and the test for a zero correlation?)

**10.51 Significance test of the correlation.** A study reported a correlation $r = 0.5$ based on a sample size of $n = 20$; another reported the same correlation based on a sample size of $n = 10$. For each, perform the test of the null hypothesis that $\rho = 0$. Describe the results and explain why the conclusions are different.

**10.52 Verifying the effect of bank size.** Refer to the bank wages data given in Table 10.8 and described in Exercise 10.37 (page 601). The data also include a variable "Size," which classifies the bank as large or small. Obtain the residuals from the regression used to predict wages from LOS, and plot them versus LOS using different symbols for the large and small banks. Include on your plot a horizontal line at 0 (the mean of the residuals). Describe the important features of this plot. Explain what they indicate about wages in this set of data.

**10.53 SAT versus ACT.** The SAT and the ACT are the two major standardized tests that colleges use to evaluate candidates. Most students take just one of these tests. However, some students take both. Table 10.9 gives the scores of 60 students who did this. How can we relate the two tests?

(a) Plot the data with SAT on the $x$ axis and ACT on the $y$ axis. Describe the overall pattern and any unusual observations.

**TABLE 10.9**

SAT and ACT scores

| SAT | ACT | SAT | ACT | SAT | ACT | SAT | ACT |
|---|---|---|---|---|---|---|---|
| 1000 | 24 | 870 | 21 | 1090 | 25 | 800 | 21 |
| 1010 | 24 | 880 | 21 | 860 | 19 | 1040 | 24 |
| 920 | 17 | 850 | 22 | 740 | 16 | 840 | 17 |
| 840 | 19 | 780 | 22 | 500 | 10 | 1060 | 25 |
| 830 | 19 | 830 | 20 | 780 | 12 | 870 | 21 |
| 1440 | 32 | 1190 | 30 | 1120 | 27 | 1120 | 25 |
| 490 | 7 | 800 | 16 | 590 | 12 | 800 | 18 |
| 1050 | 23 | 830 | 16 | 990 | 24 | 960 | 27 |
| 870 | 18 | 890 | 23 | 700 | 16 | 880 | 21 |
| 970 | 21 | 880 | 24 | 930 | 22 | 1020 | 24 |
| 920 | 22 | 980 | 27 | 860 | 23 | 790 | 14 |
| 810 | 19 | 1030 | 23 | 420 | 21 | 620 | 18 |
| 1080 | 23 | 1220 | 30 | 800 | 20 | 1150 | 28 |
| 1000 | 19 | 1080 | 22 | 1140 | 24 | 970 | 20 |
| 1030 | 25 | 970 | 20 | 920 | 21 | 1060 | 24 |

(b) Find the least-squares regression line and draw it on your plot. Give the results of the significance test for the slope.

(c) What is the correlation between the two tests?

10.54 **SAT versus ACT, continued.** Refer to the previous exercise. Find the predicted value of ACT for each observation in the data set.

(a) What is the mean of these predicted values? Compare it with the mean of the ACT scores.

(b) Compare the standard deviation of the predicted values with the standard deviation of the actual ACT scores. If least-squares regression is used to predict ACT scores for a large number of students such as these, the average predicted value will be accurate but the variability of the predicted scores will be too small.

(c) Find the SAT score for a student who is one standard deviation above the mean ($z = (x - \bar{x})/s = 1$). Find the predicted ACT score and standardize this score. (Use the means and standard deviations from this set of data for these calculations.)

(d) Repeat part (c) for a student whose SAT score is one standard deviation below the mean ($z = -1$).

(e) What do you conclude from parts (c) and (d)? Perform additional calculations for different $z$'s if needed.

10.55 **Matching standardized scores.** Refer to the previous two exercises. An alternative to the least-squares method is based on matching standardized scores. Specifically, we set

$$\frac{(\hat{y} - \bar{y})}{s_y} = \frac{(x - \bar{x})}{s_x}$$

and solve for $y$. Let's use the notation $y = a_0 + a_1 x$ for this line. The slope is $a_1 = s_y/s_x$ and the intercept is $a_0 = \bar{y} - a_1\bar{x}$. Compare these expressions with the formulas for the least-squares slope and intercept (page 565).

(a) Using the data in Table 10.9, find the values of $a_0$ and $a_1$.

(b) Plot the data with the least-squares line and the new prediction line.

(c) Use the new line to find predicted ACT scores. Find the mean and the standard deviation of these scores. How do they compare with the mean and standard deviation of the ACT scores?

10.56 **Length, width, and weight of perch.** Here are data for 12 perch caught in a lake in Finland:[20]

| Weight (grams) | Length (cm) | Width (cm) | Weight (grams) | Length (cm) | Width (cm) |
|---|---|---|---|---|---|
| 5.9 | 8.8 | 1.4 | 300.0 | 28.7 | 5.1 |
| 100.0 | 19.2 | 3.3 | 300.0 | 30.1 | 4.6 |
| 110.0 | 22.5 | 3.6 | 685.0 | 39.0 | 6.9 |
| 120.0 | 23.5 | 3.5 | 650.0 | 41.4 | 6.0 |
| 150.0 | 24.0 | 3.6 | 820.0 | 42.5 | 6.6 |
| 145.0 | 25.5 | 3.8 | 1000.0 | 46.6 | 7.6 |

In this exercise we will examine different models for predicting weight.

(a) Run the regression using length to predict weight. Do the same using width as the explanatory variable. Summarize the results. Be sure to include the value of $r^2$.

(b) Plot weight versus length and weight versus width. Include the least-squares lines in these plots. Do these relationships appear to be linear? Explain your answer.

**10.57**  **Transforming the perch data.** Refer to the previous exercise.

(a) Try to find a better model using a transformation of length. One possibility is to use the square. Make a plot and perform the regression analysis. Summarize the results.

(b) Do the same for width.

**10.58**  **Creating a new explanatory variable.** Refer to the previous two exercises.

(a) Create a new variable that is the product of length and width. Make a plot and run the regression using this new variable. Summarize the results.

(b) Write a short report summarizing and comparing the different regression analyses that you performed in this exercise and the previous two exercises.

**10.59** **Index of biotic integrity.** Refer to the data on the index of biotic integrity and area in Exercise 10.14 (page 596) and the additional data on percent watershed area that was forest in Exercise 10.15. Find the correlations among these three variables, perform the test of statistical significance, and summarize the results. Which of these test results could have been obtained from the analyses that you performed in Exercises 10.14 and 10.15?

**10.60** **Food neophobia.** Food neophobia is a personality trait associated with avoiding unfamiliar foods. In one study of 564 children who were 2 to 6 years of age, food neophobia and the frequency of consumption of different types of food were measured.[21] Here is a summary of the correlations:

| Type of food | Correlation |
|---|---|
| Vegetables | −0.27 |
| Fruit | −0.16 |
| Meat | −0.15 |
| Eggs | −0.08 |
| Sweet/fatty snacks | 0.04 |
| Starchy staples | −0.02 |

Perform the significance test for each correlation and write a summary about food neophobia and the consumption of different types of food.

**10.61** **Personality traits and scores on the GRE.** A study reported correlations between several personality traits and scores on the Graduate Record Examination (GRE) for a sample of 342 test takers.[22] Here is a table of the correlations:

| | GRE score | | |
|---|---|---|---|
| Personality trait | Analytical | Quantitative | Verbal |
| Conscientiousness | −0.17 | −0.14 | −0.12 |
| Rationality | −0.06 | −0.03 | −0.08 |
| Ingenuity | −0.06 | −0.08 | −0.02 |
| Quickness | 0.21 | 0.15 | 0.26 |
| Creativity | 0.24 | 0.26 | 0.29 |
| Depth | 0.06 | 0.08 | 0.15 |

For each correlation, test the null hypothesis that the corresponding true correlation is zero. Reproduce the table and mark the correlations that have $P < 0.001$ with ***, those that have $P < 0.01$ with **, and those that have $P < 0.05$ with *. Some critics of standardized tests have suggested that the tests penalize students who are "deep thinkers" and those who are very creative. Others have suggested that students who work quickly do better on these tests. Write a summary of the results of your significance tests, taking into account these comments.

**10.62** **Resting metabolic rate and exercise.** Metabolic rate, the rate at which the body consumes energy, is important in studies of weight gain, dieting, and exercise. The table below gives data on the lean body mass and resting metabolic rate for 12 women and 7 men who are subjects in a study of dieting. Lean body mass, given in kilograms, is a person's weight leaving out all fat. Metabolic rate is measured in calories burned per 24 hours, the same calories used to describe the energy content of foods. The researchers believe that lean body mass is an important influence on metabolic rate.

| Subject | Sex | Mass | Rate | Subject | Sex | Mass | Rate |
|---|---|---|---|---|---|---|---|
| 1 | M | 62.0 | 1792 | 11 | F | 40.3 | 1189 |
| 2 | M | 62.9 | 1666 | 12 | F | 33.1 | 913 |
| 3 | F | 36.1 | 995 | 13 | M | 51.9 | 1460 |
| 4 | F | 54.6 | 1425 | 14 | F | 42.4 | 1124 |
| 5 | F | 48.5 | 1396 | 15 | F | 34.5 | 1052 |
| 6 | F | 42.0 | 1418 | 16 | F | 51.1 | 1347 |
| 7 | M | 47.4 | 1362 | 17 | F | 41.2 | 1204 |
| 8 | F | 50.6 | 1502 | 18 | M | 51.9 | 1867 |
| 9 | F | 42.0 | 1256 | 19 | M | 46.9 | 1439 |
| 10 | M | 48.7 | 1614 | | | | |

(a) Make a scatterplot of the data, using different symbols or colors for men and women. Summarize what you see in the plot.

(b) Run the regression to predict metabolic rate from lean body mass for the women in the sample and summarize the results. Do the same for the men.

10.63 CHALLENGE **Resting metabolic rate and exercise, continued.** Refer to the previous exercise. It is tempting to conclude that there is a strong linear relationship for the women but no relationship for the men. Let's look at this issue a little more carefully.

(a) Find the confidence interval for the slope in the regression equation that you ran for the females. Do the same for the males. What do these suggest about the possibility that these two slopes are the same? (The formal method for making this comparison is a bit complicated and is beyond the scope of this chapter.)

(b) Examine the formula for the standard error of the regression slope given on page 587. The term in the denominator is $\sqrt{\Sigma(x_i - \bar{x})^2}$. Find this quantity for the females; do the same for the males. How do these calculations help to explain the results of the significance tests?

(c) Suppose you were able to collect additional data for males. How would you use lean body mass in deciding which subjects to choose?

10.64 CHALLENGE **Inference over different ranges of *X*.** Think about what would happen if you analyzed a subset of a set of data by analyzing only data for a restricted range of values of the explanatory variable. What results would you expect to change? Examine your ideas by analyzing the fuel efficiency data studied in Example 10.4 (page 566) for only those cases with speed less than or equal to 30 mph. Note that this corresponds to 3.4 in the log scale. Now, let's do the same analysis with a restriction on the response variable. Run the analysis with only those cases with fuel efficiency less than or equal to 20 mpg. Write a summary comparing the effects of these two restrictions with each other and with the results in Example 10.4.

CHAPTER

11

# Multiple Regression

How well do high school achievement scores predict success in college? Example 11.1 addresses this question.



## Introduction

In Chapter 10 we presented methods for inference in the setting of a linear relationship between a response variable $y$ and a *single* explanatory variable $x$. In this chapter, we use *more than one* explanatory variable to explain or predict a single response variable. Many of the ideas that we encountered in our study of simple linear regression carry over to the multiple linear regression setting. For example, the descriptive tools we learned in Chapter 2—scatterplots, least-squares regression, and correlation—are still essential preliminaries to inference and also provide a foundation for confidence intervals and significance tests. However, the introduction of several explanatory variables leads to many additional considerations. In this short chapter we cannot explore all of these issues. Rather, we will outline some basic facts about inference in the multiple regression setting and then illustrate the analysis with a case study whose purpose was to predict success in college based on several high school achievement scores.

## 11.1 Inference for Multiple Regression

### Population multiple regression equation

The simple linear regression model assumes that the mean of the response variable $y$ depends on the explanatory variable $x$ according to a linear equation

$$\mu_y = \beta_0 + \beta_1 x$$

For any fixed value of $x$, the response $y$ varies Normally around this mean and has a standard deviation $\sigma$ that is the same for all values of $x$.

In the multiple regression setting, the response variable $y$ depends on $p$ explanatory variables, which we will denote by $x_1, x_2, \ldots, x_p$. The mean response depends on these explanatory variables according to a linear function

$$\mu_y = \beta_0 + \beta_1 x_1 + \beta_2 x_2 + \cdots + \beta_p x_p$$

Similar to simple linear regression, this expression is the population regression equation. We do not observe the mean response because the observed values of $y$ vary about their means. We can think of subpopulations of responses, each corresponding to a particular set of values for *all* of the explanatory variables $x_1, x_2, \ldots, x_p$. In each subpopulation, $y$ varies Normally with a mean given by the population regression equation. The regression model assumes that the standard deviation $\sigma$ of the responses is the same in all subpopulations.

EXAMPLE

**11.1 Predicting early success in college.** Our case study uses data collected at a large university on all first-year computer science majors in a particular year.[1] The purpose of the study was to attempt to predict success in the early university years. One measure of success was the cumulative grade point average (GPA) after three semesters. Among the explanatory variables recorded at the time the students enrolled in the university were average high school grades in mathematics (HSM), science (HSS), and English (HSE).

We will use high school grades to predict the response variable GPA. There are $p = 3$ explanatory variables: $x_1$ = HSM, $x_2$ = HSS, and $x_3$ = HSE. The high school grades are coded on a scale from 1 to 10, with 10 corresponding to A, 9 to A−, 8 to B+, and so on. These grades define the subpopulations. For example, the straight-C students are the subpopulation defined by HSM = 4, HSS = 4, and HSE = 4.

One possible multiple regression model for the subpopulation mean GPAs is

$$\mu_{\text{GPA}} = \beta_0 + \beta_1 \text{HSM} + \beta_2 \text{HSS} + \beta_3 \text{HSE}$$

For the straight-C subpopulation of students, the model gives the subpopulation mean as

$$\mu_{\text{GPA}} = \beta_0 + \beta_1 4 + \beta_2 4 + \beta_3 4$$

## Data for multiple regression

The data for a simple linear regression problem consist of observations $(x_i, y_i)$ of the two variables. Because there are several explanatory variables in multiple regression, the notation needed to describe the data is more elaborate. Each observation or case consists of a value for the response variable and for each of the explanatory variables. Call $x_{ij}$ the value of the $j$th explanatory variable for the $i$th case. The data are then

Case 1: $(x_{11}, x_{12}, \ldots, x_{1p}, y_1)$

Case 2: $(x_{21}, x_{22}, \ldots, x_{2p}, y_2)$

$\vdots$

Case $n$: $(x_{n1}, x_{n2}, \ldots, x_{np}, y_n)$

Here, $n$ is the number of cases and $p$ is the number of explanatory variables. Data are often entered into computer regression programs in this format. Each row is a case and each column corresponds to a different variable. The data for Example 11.1, with several additional explanatory variables, appear in this format in the CSDATA data set described in the Data Appendix.

## Multiple linear regression model

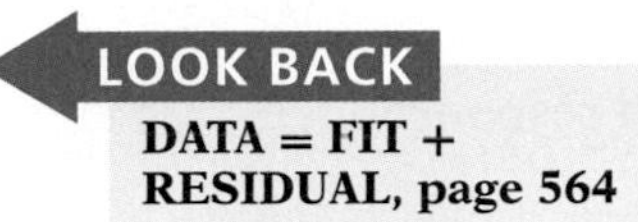


**DATA = FIT + RESIDUAL, page 564**

We combine the population regression equation and assumptions about variation to construct the multiple linear regression model. The subpopulation means describe the FIT part of our statistical model. The RESIDUAL part represents the variation of observations about the means. We will use the same notation for the residual that we used in the simple linear regression model. The symbol $\epsilon$ represents the deviation of an individual observation from its subpopulation mean. We assume that these deviations are Normally distributed with mean 0 and an unknown standard deviation $\sigma$ that does not depend on the values of the $x$ variables. *These are assumptions that we can check by examining the residuals in the same way that we did for simple linear regression.*

> MULTIPLE LINEAR REGRESSION MODEL
>
> The **statistical model for multiple linear regression** is
>
> $$y_i = \beta_0 + \beta_1 x_{i1} + \beta_2 x_{i2} + \cdots + \beta_p x_{ip} + \epsilon_i$$
>
> for $i = 1, 2, \ldots, n$.
>
> The **mean response** $\mu_y$ is a linear function of the explanatory variables:
>
> $$\mu_y = \beta_0 + \beta_1 x_1 + \beta_2 x_2 + \cdots + \beta_p x_p$$
>
> The **deviations** $\epsilon_i$ are independent and Normally distributed with mean 0 and standard deviation $\sigma$. In other words, they are an SRS from the $N(0, \sigma)$ distribution.
>
> The parameters of the model are $\beta_0$, $\beta_1$, $\beta_2, \ldots, \beta_p$, and $\sigma$.

The assumption that the subpopulation means are related to the regression coefficients $\beta$ by the equation

$$\mu_y = \beta_0 + \beta_1 x_1 + \beta_2 x_2 + \cdots + \beta_p x_p$$

implies that we can estimate all subpopulation means from estimates of the $\beta$'s. To the extent that this equation is accurate, we have a useful tool for describing how the mean of $y$ varies with the collection of $x$'s.

## Estimation of the multiple regression parameters

LOOK BACK
**least squares, page 112**

Similar to simple linear regression, we use the method of least squares to obtain estimators of the regression coefficients $\beta$. The details, however, are more complicated. Let

$$b_0, b_1, b_2, \ldots, b_p$$

denote the estimators of the parameters

$$\beta_0, \beta_1, \beta_2, \ldots, \beta_p$$

For the $i$th observation, the predicted response is

$$\hat{y}_i = b_0 + b_1 x_{i1} + b_2 x_{i2} + \cdots + b_p x_{ip}$$

LOOK BACK
**residual, page 565**

The $i$th residual, the difference between the observed and predicted response, is therefore

$$\begin{aligned} e_i &= \text{observed response} - \text{predicted response} \\ &= y_i - \hat{y}_i \\ &= y_i - b_0 - b_1 x_{i1} - b_2 x_{i2} - \cdots - b_p x_{ip} \end{aligned}$$

The method of least squares chooses the values of the $b$'s that make the sum of the squared residuals as small as possible. In other words, the parameter estimates $b_0, b_1, b_2, \ldots, b_p$ minimize the quantity

$$\sum (y_i - b_0 - b_1 x_{i1} - b_2 x_{i2} - \cdots - b_p x_{ip})^2$$

The formula for the least-squares estimates is complicated. We will be content to understand the principle on which it is based and to let software do the computations.

The parameter $\sigma^2$ measures the variability of the responses about the population regression equation. As in the case of simple linear regression, we estimate $\sigma^2$ by an average of the squared residuals. The estimator is

$$\begin{aligned} s^2 &= \frac{\sum e_i^2}{n - p - 1} \\ &= \frac{\sum (y_i - \hat{y}_i)^2}{n - p - 1} \end{aligned}$$

LOOK BACK
**degrees of freedom, page 566**

The quantity $n - p - 1$ is the degrees of freedom associated with $s^2$. The degrees of freedom equal the sample size, $n$, minus $p + 1$, the number of $\beta$'s we must estimate to fit the model. In the simple linear regression case there is just one explanatory variable, so $p = 1$ and the degrees of freedom are $n - 2$. To estimate $\sigma$ we use

$$s = \sqrt{s^2}$$

### USE YOUR KNOWLEDGE

**11.1** **Describing a multiple regression.** As part of a recent study titled "Predicting Success for Actuarial Students in Undergraduate Mathematics Courses," data from 106 Bryant College actuarial graduates were obtained.[2] The researchers were interested in describing how students' overall math grade point averages are explained by SAT Math and SAT Verbal scores, class rank, and Bryant College's mathematics placement score.

(a) What is the response variable?

(b) What is $n$, the number of cases?

(c) What is $p$, the number of explanatory variables?

(d) What are the explanatory variables?

**11.2** **Understanding the fitted regression line.** The fitted regression equation for a multiple regression is

$$\hat{y} = -1.4 + 2.6x_1 - 2.3x_2$$

(a) If $x_1 = 4$ and $x_2 = 2$, what is the predicted value of $y$?

(b) For the answer to part (a) to be valid, is it necessary that the values $x_1 = 4$ and $x_2 = 2$ correspond to a case in the data set? Explain why or why not.

(c) If you hold $x_2$ at a fixed value, what is the effect of an increase of two units in $x_1$ on the predicted value of $y$?

## Confidence intervals and significance tests for regression coefficients

We can obtain confidence intervals and perform significance tests for each of the regression coefficients $\beta_j$ as we did in simple linear regression. The standard errors of the $b$'s have more complicated formulas, but all are multiples of $s$. We again rely on statistical software to do the calculations.

### CONFIDENCE INTERVALS AND SIGNIFICANCE TESTS FOR $\beta_j$

A **level *C* confidence interval** for $\beta_j$ is

$$b_j \pm t^* \text{SE}_{b_j}$$

where $\text{SE}_{b_j}$ is the standard error of $b_j$ and $t^*$ is the value for the $t(n - p - 1)$ density curve with area $C$ between $-t^*$ and $t^*$.

To test the hypothesis $H_0$: $\beta_j = 0$, compute the ***t* statistic**

$$t = \frac{b_j}{\text{SE}_{b_j}}$$

In terms of a random variable $T$ having the $t(n - p - 1)$ distribution, the $P$-value for a test of $H_0$ against

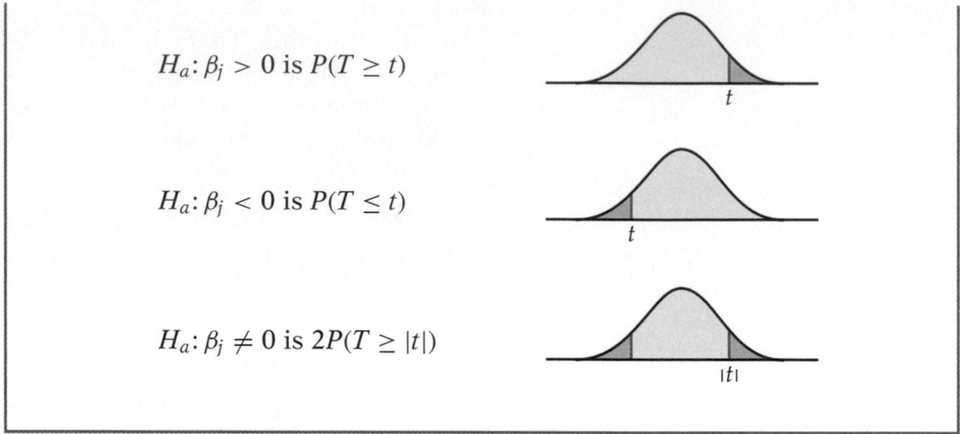


LOOK BACK
**confidence intervals for mean response, page 573**
**prediction intervals, page 575**

Because regression is often used for prediction, we may wish to use multiple regression models to construct confidence intervals for a mean response and prediction intervals for a future observation. The basic ideas are the same as in the simple linear regression case. In most software systems, the same commands that give confidence and prediction intervals for simple linear regression work for multiple regression. The only difference is that we specify a list of explanatory variables rather than a single variable. Modern software allows us to perform these rather-complex calculations without an intimate knowledge of all of the computational details. This frees us to concentrate on the meaning and appropriate use of the results.

## ANOVA table for multiple regression

LOOK BACK
**ANOVA *F* test, page 582**

In simple linear regression the $F$ test from the ANOVA table is equivalent to the two-sided $t$ test of the hypothesis that the slope of the regression line is 0. For multiple regression there is a corresponding ANOVA $F$ test, but it tests the hypothesis that *all* of the regression coefficients (with the exception of the intercept) are 0. Here is the general form of the ANOVA table for multiple regression:

| Source | Degrees of freedom | Sum of squares | Mean square | $F$ |
|---|---|---|---|---|
| Model | $p$ | $\sum(\hat{y}_i - \bar{y})^2$ | SSM/DFM | MSM/MSE |
| Error | $n - p - 1$ | $\sum(y_i - \hat{y}_i)^2$ | SSE/DFE | |
| Total | $n - 1$ | $\sum(y_i - \bar{y})^2$ | SST/DFT | |

LOOK BACK
**ANOVA table, page 582**

The ANOVA table is similar to that for simple linear regression. The degrees of freedom for the model increase from 1 to $p$ to reflect the fact that we now have $p$ explanatory variables rather than just one. As a consequence, the degrees of freedom for error decrease by the same amount. *It is always a good idea to calculate the degrees of freedom by hand and then check that your software agrees with your calculations. In this way you can verify that your software is using the number of cases and number of explanatory variables that you intended.*

The sums of squares represent sources of variation. Once again, both the sums of squares and their degrees of freedom add:

$$\text{SST} = \text{SSM} + \text{SSE}$$

$$\text{DFT} = \text{DFM} + \text{DFE}$$

The estimate of the variance $\sigma^2$ for our model is again given by the MSE in the ANOVA table. That is, $s^2$ = MSE.

**F statistic, page 582**

The ratio MSM/MSE is an $F$ statistic for testing the null hypothesis

$$H_0: \beta_1 = \beta_2 = \cdots = \beta_p = 0$$

against the alternative hypothesis

$$H_a: \text{at least one of the } \beta_j \text{ is not 0}$$

The null hypothesis says that none of the explanatory variables are predictors of the response variable when used in the form expressed by the multiple regression equation. The alternative states that at least one of them is a predictor of the response variable. As in simple linear regression, large values of $F$ give evidence against $H_0$. When $H_0$ is true, $F$ has the $F(p, n-p-1)$ distribution. The degrees of freedom for the $F$ distribution are those associated with the model and error in the ANOVA table. *A common error in the use of multiple regression is to assume that all of the regression coefficients are statistically different from zero whenever the F statistic has a small P-value. Be sure that you understand the difference between the F test and the t tests for individual coefficients.*

### ANALYSIS OF VARIANCE *F* TEST

In the multiple regression model, the hypothesis

$$H_0: \beta_1 = \beta_2 = \cdots = \beta_p = 0$$

is tested against the alternative hypothesis

$$H_a: \text{at least one of the } \beta_j \text{ is not 0}$$

by the analysis of variance $F$ statistic

$$F = \frac{\text{MSM}}{\text{MSE}}$$

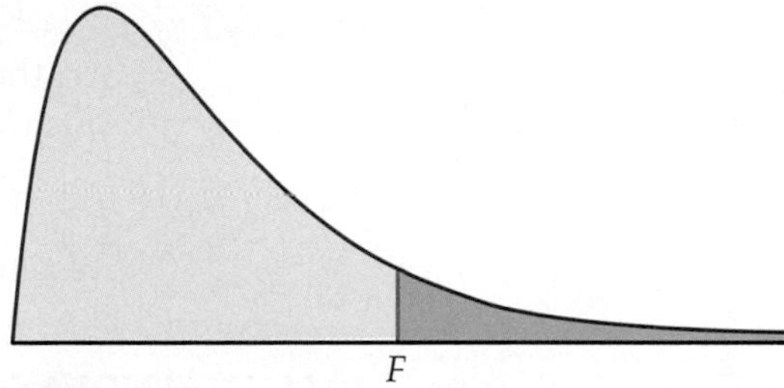


The $P$-value is the probability that a random variable having the $F(p, n-p-1)$ distribution is greater than or equal to the calculated value of the $F$ statistic.

## Squared multiple correlation $R^2$

For simple linear regression we noted that the square of the sample correlation could be written as the ratio of SSM to SST and could be interpreted as the proportion of variation in $y$ explained by $x$. A similar statistic is routinely calculated for multiple regression.

### THE SQUARED MULTIPLE CORRELATION

The statistic

$$R^2 = \frac{\text{SSM}}{\text{SST}} = \frac{\sum(\hat{y}_i - \bar{y})^2}{\sum(y_i - \bar{y})^2}$$

is the proportion of the variation of the response variable $y$ that is explained by the explanatory variables $x_1, x_2, \ldots, x_p$ in a multiple linear regression.

multiple correlation coefficient

Often, $R^2$ is multiplied by 100 and expressed as a percent. The square root of $R^2$, called the **multiple correlation coefficient,** is the correlation between the observations $y_i$ and the predicted values $\hat{y}_i$.

## USE YOUR KNOWLEDGE

**11.3 Significance tests for regression coefficients.** Recall Exercise 11.1 (page 610). Due to missing values for some students, only 86 students were used in the multiple regression analysis. The following table contains the estimated coefficients and standard errors:

| Variable | Estimate | SE |
|---|---|---|
| Intercept | −0.764 | 0.651 |
| SAT Math | 0.00156 | 0.00074 |
| SAT Verbal | 0.00164 | 0.00076 |
| High school rank | 1.470 | 0.430 |
| Bryant College placement | 0.889 | 0.402 |

(a) All the estimated coefficients for the explanatory variables are positive. Is this what you would expect? Explain.

(b) What are the degrees of freedom for the model and error?

(c) Test the significance of each coefficient and state your conclusions.

**11.4 ANOVA table for multiple regression.** Use the following information to perform the ANOVA $F$ test and compute $R^2$.

| Source | Degrees of freedom | Sum of squares |
|---|---|---|
| Model | | 175 |
| Error | 60 | |
| Total | 65 | 1015 |

# 11.2 A Case Study

## Preliminary analysis

In this section we illustrate multiple regression by analyzing the data from the study described in Example 11.1. The response variable is the cumulative GPA after three semesters for a group of computer science majors at a large university. The explanatory variables previously mentioned are average high school grades, represented by HSM, HSS, and HSE. We also examine the SAT Mathematics and SAT Verbal scores as explanatory variables. We have data for $n = 224$ students in the study. We use SAS to illustrate the outputs that are given by most software.

The first step in the analysis is to carefully examine each of the variables. Means, standard deviations, and minimum and maximum values appear in Figure 11.1. The minimum value for the SAT Mathematics (SATM) variable appears to be rather extreme; it is $(595 - 300)/86 = 3.43$ standard deviations below the mean. We do not discard this case at this time but will take care in our subsequent analyses to see if it has an excessive influence on our results. The mean for the SATM score is higher than the mean for the Verbal score (SATV), as we might expect for a group of computer science majors. The two standard deviations are about the same. The means of the three high school grade variables are similar, with the mathematics grades being a bit higher. The standard deviations for the high school grade variables are very close to each other. The mean GPA is 2.635 on a 4-point scale, with standard deviation 0.779.

SAS

| Variable | N | Mean | Std Dev | Minimum | Maximum |
|---|---|---|---|---|---|
| GPA | 224 | 2.6352232 | 0.7793949 | 0.1200000 | 4.0000000 |
| SATM | 224 | 595.2857143 | 86.4014437 | 300.0000000 | 800.0000000 |
| SATV | 224 | 504.5491071 | 92.6104591 | 285.0000000 | 760.0000000 |
| HSM | 224 | 8.3214286 | 1.6387367 | 2.0000000 | 10.0000000 |
| HSS | 224 | 8.0892857 | 1.6996627 | 3.0000000 | 10.0000000 |
| HSE | 224 | 8.0937500 | 1.5078736 | 3.0000000 | 10.0000000 |

**FIGURE 11.1** Descriptive statistics for the computer science student case study.

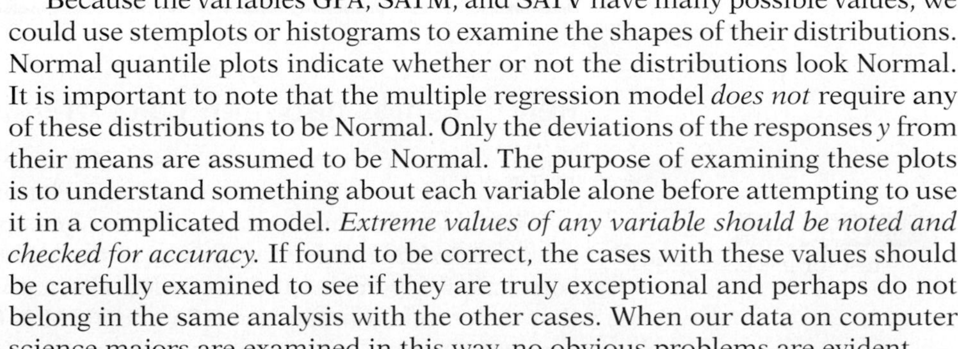

Because the variables GPA, SATM, and SATV have many possible values, we could use stemplots or histograms to examine the shapes of their distributions. Normal quantile plots indicate whether or not the distributions look Normal. It is important to note that the multiple regression model *does not* require any of these distributions to be Normal. Only the deviations of the responses $y$ from their means are assumed to be Normal. The purpose of examining these plots is to understand something about each variable alone before attempting to use it in a complicated model. *Extreme values of any variable should be noted and checked for accuracy.* If found to be correct, the cases with these values should be carefully examined to see if they are truly exceptional and perhaps do not belong in the same analysis with the other cases. When our data on computer science majors are examined in this way, no obvious problems are evident.

The high school grade variables HSM, HSS, and HSE have relatively few values and are best summarized by giving the relative frequencies for each possible value. The output in Figure 11.2 provides these summaries. The distributions are all skewed, with a large proportion of high grades (10 = A and 9 = A−). Again we emphasize that these distributions need not be Normal.

SAS

| hsm | Frequency | Percent | Cumulative Frequency | Cumulative Percent |
|---|---|---|---|---|
| 2 | 1 | 0.45 | 1 | 0.45 |
| 3 | 1 | 0.45 | 2 | 0.89 |
| 4 | 4 | 1.79 | 6 | 2.68 |
| 5 | 6 | 2.68 | 12 | 5.36 |
| 6 | 23 | 10.27 | 35 | 15.63 |
| 7 | 28 | 12.50 | 63 | 28.13 |
| 8 | 36 | 16.07 | 99 | 44.20 |
| 9 | 59 | 26.34 | 158 | 70.54 |
| 10 | 66 | 29.46 | 224 | 100.00 |

| hss | Frequency | Percent | Cumulative Frequency | Cumulative Percent |
|---|---|---|---|---|
| 3 | 1 | 0.45 | 1 | 0.45 |
| 4 | 7 | 3.13 | 8 | 3.57 |
| 5 | 9 | 4.02 | 17 | 7.59 |
| 6 | 24 | 10.71 | 41 | 18.30 |
| 7 | 42 | 18.75 | 83 | 37.05 |
| 8 | 31 | 13.84 | 114 | 50.89 |
| 9 | 50 | 22.32 | 164 | 73.21 |
| 10 | 60 | 26.79 | 224 | 100.00 |

| hse | Frequency | Percent | Cumulative Frequency | Cumulative Percent |
|---|---|---|---|---|
| 3 | 1 | 0.45 | 1 | 0.45 |
| 4 | 4 | 1.79 | 5 | 2.23 |
| 5 | 5 | 2.23 | 10 | 4.46 |
| 6 | 23 | 10.27 | 33 | 14.73 |
| 7 | 43 | 19.20 | 76 | 33.93 |
| 8 | 49 | 21.88 | 125 | 55.80 |
| 9 | 52 | 23.21 | 177 | 79.02 |
| 10 | 47 | 20.98 | 224 | 100.00 |

**FIGURE 11.2** The distributions of the high school grade variables.

## Relationships between pairs of variables

LOOK BACK
test for $\rho = 0$, page 591

The second step in our analysis is to examine the relationships between all pairs of variables. Scatterplots and correlations are our tools for studying two-variable relationships. The correlations appear in Figure 11.3. The output includes the *P*-value for the test of the null hypothesis that the population correlation is 0 versus the two-sided alternative for each pair. Thus, we see that the correlation between GPA and HSM is 0.44, with a *P*-value of 0.0001, whereas the correlation between GPA and SATV is 0.11, with a *P*-value of 0.087.

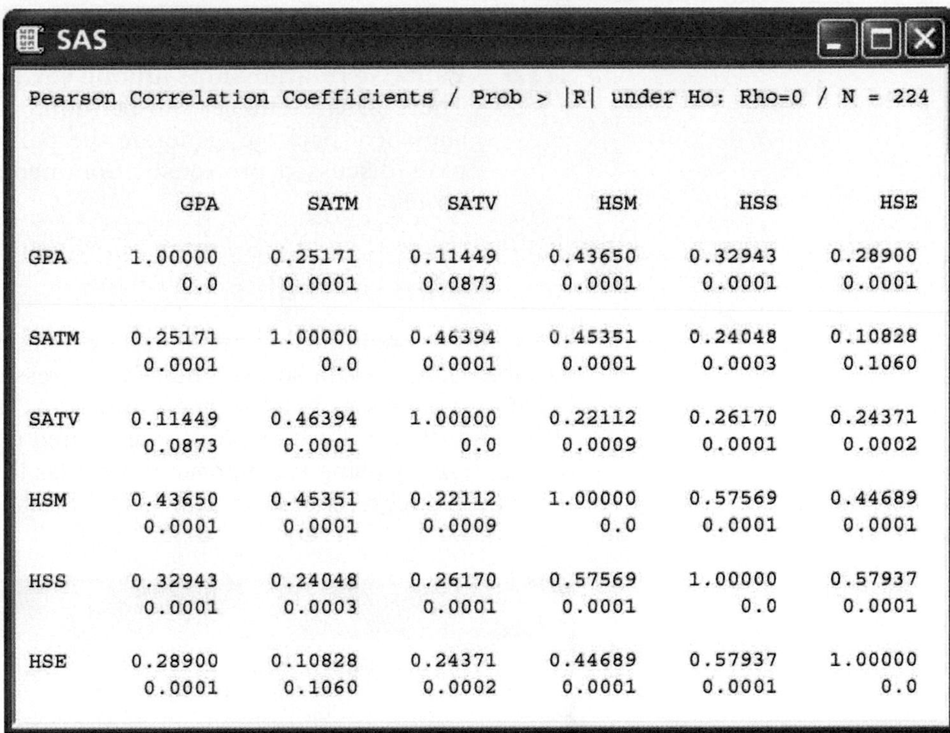

SAS

Pearson Correlation Coefficients / Prob > |R| under Ho: Rho=0 / N = 224

| | GPA | SATM | SATV | HSM | HSS | HSE |
|---|---|---|---|---|---|---|
| GPA | 1.00000 | 0.25171 | 0.11449 | 0.43650 | 0.32943 | 0.28900 |
| | 0.0 | 0.0001 | 0.0873 | 0.0001 | 0.0001 | 0.0001 |
| SATM | 0.25171 | 1.00000 | 0.46394 | 0.45351 | 0.24048 | 0.10828 |
| | 0.0001 | 0.0 | 0.0001 | 0.0001 | 0.0003 | 0.1060 |
| SATV | 0.11449 | 0.46394 | 1.00000 | 0.22112 | 0.26170 | 0.24371 |
| | 0.0873 | 0.0001 | 0.0 | 0.0009 | 0.0001 | 0.0002 |
| HSM | 0.43650 | 0.45351 | 0.22112 | 1.00000 | 0.57569 | 0.44689 |
| | 0.0001 | 0.0001 | 0.0009 | 0.0 | 0.0001 | 0.0001 |
| HSS | 0.32943 | 0.24048 | 0.26170 | 0.57569 | 1.00000 | 0.57937 |
| | 0.0001 | 0.0003 | 0.0001 | 0.0001 | 0.0 | 0.0001 |
| HSE | 0.28900 | 0.10828 | 0.24371 | 0.44689 | 0.57937 | 1.00000 |
| | 0.0001 | 0.1060 | 0.0002 | 0.0001 | 0.0001 | 0.0 |

**FIGURE 11.3** Correlations among the case study variables.

The first is statistically significant by any reasonable standard, and the second is marginally significant.

The high school grades all have higher correlations with GPA than do the SAT scores. As we might expect, math grades have the highest correlation ($r = 0.44$), followed by science grades (0.33) and then English grades (0.29). The two SAT scores have a rather high correlation with each other (0.46), and the high school grades also correlate well with each other (0.45 to 0.58). SATM correlates well with HSM (0.45), less well with HSS (0.24), and rather poorly with HSE (0.11). The correlations of SATV with the three high school grades are about equal, ranging from 0.22 to 0.26.

It is important to keep in mind that by examining pairs of variables we are seeking a better understanding of the data. *The fact that the correlation of a particular explanatory variable with the response variable does not achieve statistical significance does not necessarily imply that it will not be a useful (and statistically significant) predictor in a multiple regression.*

Numerical summaries such as correlations are useful, but plots are generally more informative when seeking to understand data. Plots tell us whether the numerical summary gives a fair representation of the data. For a multiple regression, each pair of variables should be plotted. For the six variables in our case study, this means that we should examine 15 plots. In general, there are $p + 1$ variables in a multiple regression analysis with $p$ explanatory variables, so that $p(p + 1)/2$ plots are required. Multiple regression is a complicated procedure. If we do not do the necessary preliminary work, we are in serious danger of producing useless or misleading results. We leave the task of making these plots as an exercise.

## USE YOUR KNOWLEDGE

**11.5 Pairwise relationships among variables in the CSDATA data set.** The CSDATA data set can be found in the Data Appendix. Using a statistical package, generate the pairwise correlations and scatterplots discussed previously. Comment on any unusual patterns or observations.

## Regression on high school grades

To explore the relationship between the explanatory variables and our response variable GPA, we run several multiple regressions. The explanatory variables fall into two classes. High school grades are represented by HSM, HSS, and HSE, and standardized tests are represented by the two SAT scores. We begin our analysis by using the high school grades to predict GPA. Figure 11.4 gives the multiple regression output.

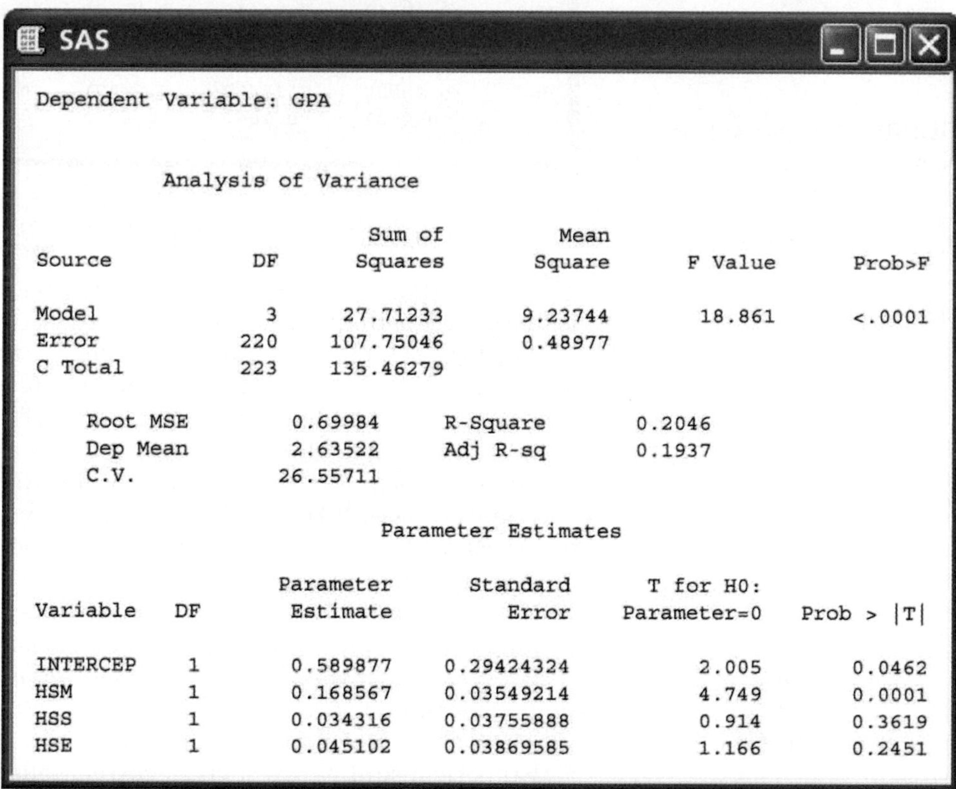

SAS

Dependent Variable: GPA

Analysis of Variance

| Source | DF | Sum of Squares | Mean Square | F Value | Prob>F |
|---|---|---|---|---|---|
| Model | 3 | 27.71233 | 9.23744 | 18.861 | <.0001 |
| Error | 220 | 107.75046 | 0.48977 | | |
| C Total | 223 | 135.46279 | | | |

| | | | |
|---|---|---|---|
| Root MSE | 0.69984 | R-Square | 0.2046 |
| Dep Mean | 2.63522 | Adj R-sq | 0.1937 |
| C.V. | 26.55711 | | |

Parameter Estimates

| Variable | DF | Parameter Estimate | Standard Error | T for H0: Parameter=0 | Prob > \|T\| |
|---|---|---|---|---|---|
| INTERCEP | 1 | 0.589877 | 0.29424324 | 2.005 | 0.0462 |
| HSM | 1 | 0.168567 | 0.03549214 | 4.749 | 0.0001 |
| HSS | 1 | 0.034316 | 0.03755888 | 0.914 | 0.3619 |
| HSE | 1 | 0.045102 | 0.03869585 | 1.166 | 0.2451 |

**FIGURE 11.4** Multiple regression output for regression using high school grades to predict GPA.

The output contains an ANOVA table, some additional descriptive statistics, and information about the parameter estimates. When examining any ANOVA table, it is a good idea to first verify the degrees of freedom. This ensures that we have not made some serious error in specifying the model for the software or in entering the data. Because there are $n = 224$ cases, we have $\text{DFT} = n - 1 = 223$. The three explanatory variables give $\text{DFM} = p = 3$ and $\text{DFE} = n - p - 1 = 223 - 3 = 220$.

The ANOVA $F$ statistic is 18.86, with a $P$-value of 0.0001. Under the null hypothesis

$$H_0: \beta_1 = \beta_2 = \beta_3 = 0$$

the $F$ statistic has an $F(3, 220)$ distribution. According to this distribution, the chance of obtaining an $F$ statistic of 18.86 or larger is 0.0001. We therefore conclude that at least one of the three regression coefficients for the high school grades is different from 0 in the population regression equation.

In the descriptive statistics that follow the ANOVA table we find that Root MSE is 0.6998. This value is the square root of the MSE given in the ANOVA table and is $s$, the estimate of the parameter $\sigma$ of our model. The value of $R^2$ is 0.20. That is, 20% of the observed variation in the GPA scores is explained by linear regression on high school grades. Although the $P$-value is very small, the model does not explain very much of the variation in GPA. Remember, a small $P$-value does not necessarily tell us that we have a large effect, particularly when the sample size is large.

From the Parameter Estimates section of the computer output we obtain the fitted regression equation

$$\widehat{\text{GPA}} = 0.590 + 0.169\text{HSM} + 0.034\text{HSS} + 0.045\text{HSE}$$

Let's find the predicted GPA for a student with an A− average in HSM, B+ in HSS, and B in HSE. The explanatory variables are HSM = 9, HSS = 8, and HSE = 7. The predicted GPA is

$$\begin{aligned}\widehat{\text{GPA}} &= 0.590 + 0.169(9) + 0.034(8) + 0.045(7)\\ &= 2.7\end{aligned}$$

Recall that the $t$ statistics for testing the regression coefficients are obtained by dividing the estimates by their standard errors. Thus, for the coefficient of HSM we obtain the $t$-value given in the output by calculating

$$t = \frac{b}{\text{SE}_b} = \frac{0.168567}{0.03549214} = 4.749$$

The $P$-values appear in the last column. Note that these $P$-values are for the two-sided alternatives. HSM has a $P$-value of 0.0001, and we conclude that the regression coefficient for this explanatory variable is significantly different from 0. The $P$-values for the other explanatory variables (0.36 for HSS and 0.25 for HSE) do not achieve statistical significance.

## Interpretation of results

The significance tests for the individual regression coefficients seem to contradict the impression obtained by examining the correlations in Figure 11.3. In that display we see that the correlation between GPA and HSS is 0.33 and the correlation between GPA and HSE is 0.29. The $P$-values for both of these correlations are 0.0001. In other words, if we used HSS alone in a regression to predict GPA, or if we used HSE alone, we would obtain statistically significant regression coefficients.

This phenomenon is not unusual in multiple regression analysis. Part of the explanation lies in the correlations between HSM and the other two explanatory variables. These are rather high (at least compared with the other correlations in Figure 11.3). The correlation between HSM and HSS is 0.58, and that between HSM and HSE is 0.45. Thus, when we have a regression model that contains all three high school grades as explanatory variables, there is considerable overlap of the predictive information contained in these variables. *The significance tests for individual regression coefficients assess the significance of each predictor variable assuming that all other predictors are included in the regression equation.* Given that we use a model with HSM and HSS as predictors, the coefficient of HSE is not statistically significant. Similarly, given that we have HSM and HSE in the model, HSS does not have a significant regression coefficient. HSM, however, adds significantly to our ability to predict GPA even after HSS and HSE are already in the model.

Unfortunately, we cannot conclude from this analysis that the *pair* of explanatory variables HSS and HSE contribute nothing significant to our model for predicting GPA once HSM is in the model. The impact of relations among the several explanatory variables on fitting models for the response is the most important new phenomenon encountered in moving from simple linear regression to multiple regression. We can only hint at the many complicated problems that arise.

## Residuals

As in simple linear regression, we should always examine the residuals as an aid to determining whether the multiple regression model is appropriate for the data. Because there are several explanatory variables, we must examine several residual plots. It is usual to plot the residuals versus the predicted values $\hat{y}$ and also versus each of the explanatory variables. Look for outliers, influential observations, evidence of a curved (rather than linear) relation, and anything else unusual. Again, we leave the task of making these plots as an exercise. The plots all appear to show more or less random noise around the center value of 0.

If the deviations $\epsilon$ in the model are Normally distributed, the residuals should be Normally distributed. Figure 11.5 presents a Normal quantile plot of the residuals. The distribution appears to be approximately Normal. There are many other specialized plots that help detect departures from the multiple regression model. Discussion of these, however, is more than we can undertake in this chapter.

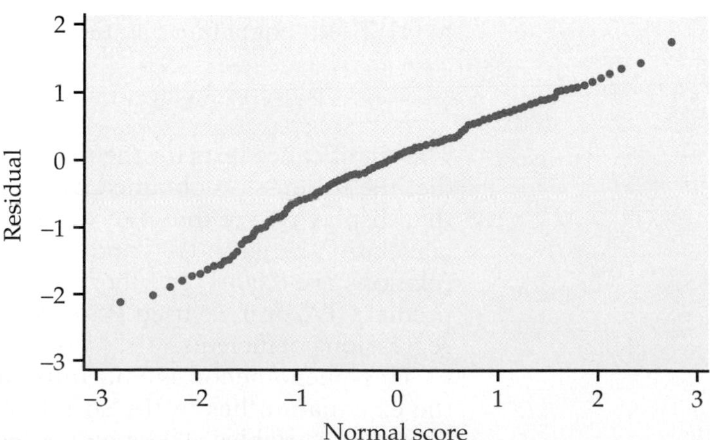


**FIGURE 11.5** Normal quantile plot of the residuals from the high school grades model. There are no important deviations from Normality.

## USE YOUR KNOWLEDGE

**11.6 Residual plots for the CSDATA analysis.** The CSDATA data set can be found in the Data Appendix. Using a statistical package, fit the linear model with HSM and HSE as predictors and obtain the residuals and predicted values. Plot the residuals versus the predicted values, HSM, and HSE. Are the residuals more or less randomly dispersed around zero? Comment on any unusual patterns.

## Refining the model

Because the variable HSS has the largest *P*-value of the three explanatory variables (see Figure 11.4) and therefore appears to contribute the least to our explanation of GPA, we rerun the regression using only HSM and HSE as explanatory variables. The SAS output appears in Figure 11.6. The *F* statistic indicates that we can reject the null hypothesis that the regression coefficients for the two explanatory variables are both 0. The *P*-value is still 0.0001. The value of $R^2$ has dropped very slightly compared with our previous run, from 0.2046 to 0.2016. Thus, dropping HSS from the model resulted in the loss of very little explanatory power. The measure *s* of variation about the fitted equation (Root MSE in the printout) is nearly identical for the two regressions, another indication that we lose very little dropping HSS. The *t* statistics for the individual regression coefficients indicate that HSM is still clearly significant ($P = 0.0001$), while the statistic for HSE is larger than before (1.747 versus 1.166) and approaches the traditional 0.05 level of significance ($P = 0.082$).

SAS

Dependent Variable: GPA

Analysis of Variance

| Source | DF | Sum of Squares | Mean Square | F Value | Prob>F |
|---|---|---|---|---|---|
| Model | 2 | 27.30349 | 13.65175 | 27.894 | <.0001 |
| Error | 221 | 108.15930 | 0.48941 | | |
| C Total | 223 | 135.46279 | | | |

| | | | |
|---|---|---|---|
| Root MSE | 0.69958 | R-Square | 0.2016 |
| Dep Mean | 2.63522 | Adj R-sq | 0.1943 |
| C.V. | 26.54718 | | |

Parameter Estimates

| Variable | DF | Parameter Estimate | Standard Error | T for H0: Parameter=0 | Prob > \|T\| |
|---|---|---|---|---|---|
| INTERCEP | 1 | 0.624228 | 0.29172204 | 2.140 | 0.0335 |
| HSM | 1 | 0.182654 | 0.03195581 | 5.716 | 0.0001 |
| HSE | 1 | 0.060670 | 0.03472914 | 1.747 | 0.0820 |

**FIGURE 11.6** Multiple regression output for regression using HSM and HSE to predict GPA.

Comparison of the fitted equations for the two multiple regression analyses tells us something more about the intricacies of this procedure. For the first run we have

$$\widehat{\text{GPA}} = 0.590 + 0.169\text{HSM} + 0.034\text{HSS} + 0.045\text{HSE}$$

whereas the second gives us

$$\widehat{\text{GPA}} = 0.624 + 0.183\text{HSM} + 0.061\text{HSE}$$

Eliminating HSS from the model changes the regression coefficients for all of the remaining variables and the intercept. This phenomenon occurs quite generally in multiple regression. *Individual regression coefficients, their standard errors, and significance tests are meaningful only when interpreted in the context of the other explanatory variables in the model.*

## Regression on SAT scores

We now turn to the problem of predicting GPA using the two SAT scores. Figure 11.7 gives the output. The fitted model is

$$\widehat{\text{GPA}} = 1.289 + 0.002283\text{SATM} - 0.000025\text{SATV}$$

The degrees of freedom are as expected: 2, 221, and 223. The $F$ statistic is 7.476, with a $P$-value of 0.0007. We conclude that the regression coefficients for SATM and SATV are not both 0. Recall that we obtained the $P$-value 0.0001 when we used high school grades to predict GPA. Both multiple regression equations are

SAS

Dependent Variable: GPA

Analysis of Variance

| Source | DF | Sum of Squares | Mean Square | F Value | Prob>F |
|---|---|---|---|---|---|
| Model | 2 | 8.58384 | 4.29192 | 7.476 | 0.0007 |
| Error | 221 | 126.87895 | 0.57411 | | |
| C Total | 223 | 135.46279 | | | |

| | | | |
|---|---|---|---|
| Root MSE | 0.75770 | R-Square | 0.0634 |
| Dep Mean | 2.63522 | Adj R-sq | 0.0549 |
| C.V. | 28.75287 | | |

Parameter Estimates

| Variable | DF | Parameter Estimate | Standard Error | T for H0: Parameter=0 | Prob > \|T\| |
|---|---|---|---|---|---|
| INTERCEP | 1 | 1.288677 | 0.37603684 | 3.427 | 0.0007 |
| SATM | 1 | 0.002283 | 0.00066291 | 3.444 | 0.0007 |
| SATV | 1 | -0.000024562 | 0.00061847 | -0.040 | 0.9684 |

**FIGURE 11.7** Multiple regression output for regression using SAT scores to predict GPA.

highly significant, but this obscures the fact that the two models have quite different explanatory power. For the SAT regression, $R^2 = 0.0634$, whereas for the high school grades model even with only HSM and HSE (Figure 11.6), we have $R^2 = 0.2016$, a value more than three times as large. *Stating that we have a statistically significant result is quite different from saying that an effect is large or important.*

Further examination of the output in Figure 11.7 reveals that the coefficient of SATM is significant ($t = 3.44$, $P = 0.0007$), and that for SATV is not ($t = -0.04$, $P = 0.9684$). For a complete analysis we should carefully examine the residuals. Also, we might want to run the analysis with SATM as the only explanatory variable.

## Regression using all variables

We have seen that either the high school grades or the SAT scores give a highly significant regression equation. The mathematics component of each of these groups of explanatory variables appears to be the key predictor. Comparing the values of $R^2$ for the two models indicates that high school grades are better predictors than SAT scores. Can we get a better prediction equation using all of the explanatory variables together in one multiple regression?

To address this question we run the regression with all five explanatory variables. The output appears in Figure 11.8. The $F$ statistic is 11.69, with a $P$-value of 0.0001, so at least one of our explanatory variables has a nonzero regression coefficient. This result is not surprising, given that we have already seen that HSM and SATM are strong predictors of GPA. The value of $R^2$ is 0.2115, not much higher than the value of 0.2046 that we found for the high school grades regression.

Examination of the $t$ statistics and the associated $P$-values for the individual regression coefficients reveals that HSM is the only one that is significant ($P = 0.0003$). That is, only HSM makes a significant contribution when it is added to a model that already has the other four explanatory variables. Once again it is important to understand that this result does not necessarily mean that the regression coefficients for the four other explanatory variables are *all* 0.

Figure 11.9 gives the Excel and Minitab multiple regression outputs for this problem. Although the format and organization of outputs differ among software packages, the basic results that we need are easy to find.

Many statistical software packages provide the capability for testing whether a collection of regression coefficients in a multiple regression model are *all* 0. We use this approach to address two interesting questions about this set of data. We did not discuss such tests in the outline that opened this section, but the basic idea is quite simple.

## Test for a collection of regression coefficients

In the context of the multiple regression model with all five predictors, we ask first whether or not the coefficients for the two SAT scores are both 0. In other words, do the SAT scores add any significant predictive information to that already contained in the high school grades? To be fair, we also ask the complementary question—do the high school grades add any significant predictive information to that already contained in the SAT scores?

```
Dependent Variable: GPA

              Analysis of Variance

                            Sum of          Mean
Source             DF      Squares        Square       F Value       Prob>F

Model               5     28.64364       5.72873        11.691        <.0001
Error             218    106.81914       0.49000
C Total           223    135.46279

    Root MSE          0.70000     R-Square         0.2115
    Dep Mean          2.63522     Adj R-sq         0.1934
    C.V.             26.56311

                           Parameter Estimates

                     Parameter        Standard        T for H0:
Variable    DF        Estimate           Error      Parameter=0     Prob > |T|

INTERCEP     1        0.326719      0.39999643            0.817         0.4149
SATM         1        0.000944      0.00068566            1.376         0.1702
SATV         1       -0.000408      0.00059189           -0.689         0.4915
HSM          1        0.145961      0.03926097            3.718         0.0003
HSS          1        0.035905      0.03779841            0.950         0.3432
HSE          1        0.055293      0.03956869            1.397         0.1637

Test: SAT      Numerator:        0.4657  DF:     2   F value:    0.9503
               Denominator:    0.489996  DF:   218   Prob>F:     0.3882

Test: HS       Numerator:        6.6866  DF:     3   F value:   13.6462
               Denominator:    0.489996  DF:   218   Prob>F:     0.0001
```

**FIGURE 11.8** Multiple regression output for regression using all variables to predict GPA.

The answers are given in the last two parts of the output in Figure 11.8. For the first test we see that $F = 0.9503$. Under the null hypothesis that the two SAT coefficients are 0, this statistic has an $F(2, 218)$ distribution and the *P*-value is 0.39. We conclude that the SAT scores are not significant predictors of GPA in a regression that already contains the high school scores as predictor variables. Recall that the model with just SAT scores has a highly significant *F* statistic. We now see that whatever predictive information is in the SAT scores can also be found in the high school grades. In this sense, the SAT scores are unnecessary.

The test statistic for the three high school grade variables is $F = 13.6462$. Under the null hypothesis that these three regression coefficients are 0, the statistic has an $F(3, 218)$ distribution and the *P*-value is 0.0001. We conclude that high school grades contain useful information for predicting GPA that is not contained in SAT scores.

Of course, our statistical analysis of these data does not imply that SAT scores are less useful than high school grades for predicting college grades for all groups of students. We have studied a select group of students—computer science majors—from a specific university. Generalizations to other situations are beyond the scope of inference based on these data alone.

Excel

| | A | B | C | D | E | F | G |
|---|---|---|---|---|---|---|---|
| 1 | SUMMARY OUTPUT | | | | | | |
| 2 | | | | | | | |
| 3 | *Regression Statistics* | | | | | | |
| 4 | Multiple R | 0.459837234 | | | | | |
| 5 | R Square | 0.211450282 | | | | | |
| 6 | Adjusted R Square | 0.193364279 | | | | | |
| 7 | Standard Error | 0.699997195 | | | | | |
| 8 | Observations | 224 | | | | | |
| 9 | | | | | | | |
| 10 | ANOVA | | | | | | |
| 11 | | *df* | *SS* | *MS* | *F* | *Significance F* | |
| 12 | Regression | 5 | 28.64364489 | 5.728729 | 11.69138 | 5.06E-10 | |
| 13 | Residual | 218 | 106.8191439 | 0.489996 | | | |
| 14 | Total | 223 | 135.4627888 | | | | |
| 15 | | | | | | | |
| 16 | | *Coefficients* | *Standard Error* | *tStat* | *P-value* | *Lower 95%* | *Upper 95%* |
| 17 | intercept | 0.326718739 | 0.399996431 | 0.816804 | 0.414932 | –0.461636967 | 1.115074446 |
| 18 | HSM | 0.14596108 | 0.039260974 | 3.717714 | 0.000256 | 0.068581358 | 0.223340801 |
| 19 | HSS | 0.03590532 | 0.037798412 | 0.949916 | 0.343207 | –0.03859183 | 0.11040247 |
| 20 | HSE | 0.055292581 | 0.039568691 | 1.397382 | 0.163719 | –0.022693622 | 0.133278785 |
| 21 | SATM | 0.000843593 | 0.000685657 | 1.376187 | 0.170176 | –0.000407774 | 0.002294959 |
| 22 | SATV | 0.00040785 | 0.000591893 | –0.68906 | 0.491518 | –0.001574415 | 0.00075816 |

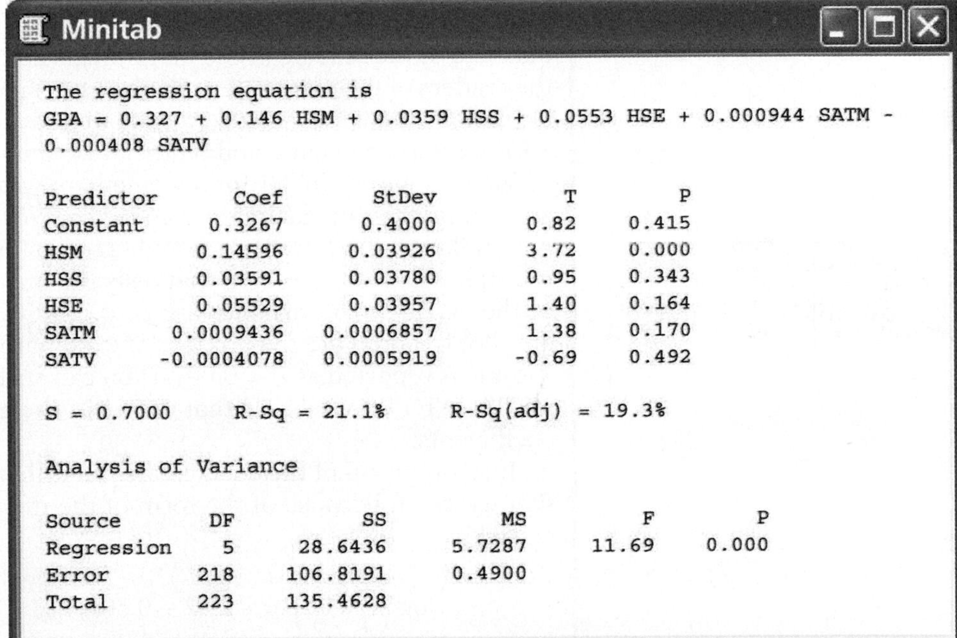
Minitab

```
The regression equation is
GPA = 0.327 + 0.146 HSM + 0.0359 HSS + 0.0553 HSE + 0.000944 SATM -
0.000408 SATV

Predictor        Coef        StDev         T       P
Constant       0.3267       0.4000      0.82   0.415
HSM           0.14596      0.03926      3.72   0.000
HSS           0.03591      0.03780      0.95   0.343
HSE           0.05529      0.03957      1.40   0.164
SATM        0.0009436    0.0006857      1.38   0.170
SATV       -0.0004078    0.0005919     -0.69   0.492

S = 0.7000      R-Sq = 21.1%      R-Sq(adj) = 19.3%

Analysis of Variance

Source       DF         SS        MS         F        P
Regression    5    28.6436    5.7287     11.69    0.000
Error       218   106.8191    0.4900
Total       223   135.4628
```

**FIGURE 11.9** Excel and Minitab multiple regression outputs for regression using all variables to predict GPA.

## BEYOND THE BASICS

### Multiple Logistic Regression

Many studies have yes/no or success/failure response variables. A surgery patient lives or dies; a consumer does or does not purchase a product after viewing an advertisement. Because the response variable in a multiple regression is assumed to have a Normal distribution, this methodology is not suitable for predicting such responses. However, there are models that apply the ideas of regression to response variables with only two possible outcomes.

logistic regression

One type of model that can be used is called **logistic regression.** We think in terms of a binomial model for the two possible values of the response variable and use one or more explanatory variables to explain the probability of success. Details are more complicated than those for multiple regression and are given in the supplemental Chapter 14 on this topic. However, the fundamental ideas are very much the same. Here is an example.

EXAMPLE

**11.2 Sexual imagery in advertisements.** Marketers sometimes use sexual imagery in advertisements targeted at teens and young adults. One study designed to examine this issue analyzed how models were dressed in 1509 ads in magazines read by young and mature adults.[3] The clothing of the models in the ads was classified as not sexual or sexual. Logistic regression was used to model the probability that the model's clothing was sexual as a function of four explanatory variables. Here, model clothing with values 1 for sexual and 0 for not sexual is the response variable.

The explanatory variables were $x_1$, a variable having the value 1 if the median age of the readers of the magazine is 20 to 29, and 0 if the median age of the readers of the magazine is 40 to 49; $x_2$, the gender of the model, coded as 1 for female and 0 for male; $x_3$, a code to indicate men's magazines with values 1 for a men's magazine and 0 otherwise; and $x_4$, a code to indicate women's magazines with values 1 for a women's magazine and 0 otherwise. Note that general-interest magazines are coded as 0 for both $x_3$ and $x_4$.

LOOK BACK
**chi-square distribution, page 531**

Similar to the $F$ test in multiple regression, there is a chi-square test for multiple logistic regression that tests the null hypothesis that *all* coefficients of the explanatory variables are zero. The value is $X^2 = 168.2$, and the degrees of freedom are the number of explanatory variables, 4 in this case. The $P$-value is reported as $P = 0.001$. (You can verify that it is less than 0.0005 using Table F.) We conclude that not all of the explanatory variables have zero coefficients.

Interpretation of the coefficients is a little more difficult in multiple logistic regression because of the form of the model. For our example, the fitted model is

$$\log\left(\frac{p}{1-p}\right) = -2.32 + 0.50x_1 + 1.31x_2 - 0.05x_3 + 0.45x_4$$

odds

The expression $p/(1-p)$ is the **odds** that the model is sexually dressed. Logistic regression models the "log odds" as a linear combination of the explanatory variables. Positive coefficients are associated with a higher probability that the model is dressed sexually. We see that ads in magazines with younger readers, female models, and women's magazines are more likely to show models dressed sexually.

In place of the $t$ tests for individual coefficients in multiple regression, chi-square tests, each with 1 degree of freedom, are used to test whether individual coefficients are zero. For reader age and model gender, $P < 0.01$, while for the indicator for women's magazines, $P < 0.05$. The indicator for men's magazines is not statistically significant.

## CHAPTER 11 Summary

The statistical model for **multiple linear regression** with response variable $y$ and $p$ explanatory variables $x_1, x_2, \ldots, x_p$ is

$$y_i = \beta_0 + \beta_1 x_{i1} + \beta_2 x_{i2} + \cdots + \beta_p x_{ip} + \epsilon_i$$

where $i = 1, 2, \ldots, n$. The $\epsilon_i$ are assumed to be independent and Normally distributed with mean 0 and standard deviation $\sigma$. The **parameters** of the model are $\beta_0, \beta_1, \beta_2, \ldots, \beta_p$, and $\sigma$.

The $\beta$'s are estimated by $b_0, b_1, b_2, \ldots, b_p$, which are obtained by the **method of least squares.** The parameter $\sigma$ is estimated by

$$s = \sqrt{\text{MSE}} = \sqrt{\frac{\sum e_i^2}{n - p - 1}}$$

where the $e_i$ are the **residuals,**

$$e_i = y_i - \hat{y}_i$$

A **level $C$ confidence interval for $\beta_j$** is

$$b_j \pm t^* \text{SE}_{b_j}$$

where $t^*$ is the value for the $t(n - p - 1)$ density curve with area $C$ between $-t^*$ and $t^*$.

The test of the hypothesis $H_0\colon \beta_j = 0$ is based on the **$t$ statistic**

$$t = \frac{b_j}{\text{SE}_{b_j}}$$

and the $t(n - p - 1)$ distribution.

The estimate $b_j$ of $\beta_j$ and the test and confidence interval for $\beta_j$ are all based on a specific multiple linear regression model. The results of all of these procedures change if other explanatory variables are added to or deleted from the model.

The **ANOVA table** for a multiple linear regression gives the degrees of freedom, sum of squares, and mean squares for the model, error, and total sources of variation. The **ANOVA $F$ statistic** is the ratio MSM/MSE and is used to test the null hypothesis

$$H_0\colon \beta_1 = \beta_2 = \cdots = \beta_p = 0$$

If $H_0$ is true, this statistic has an $F(p, n - p - 1)$ distribution.

The **squared multiple correlation** is given by the expression

$$R^2 = \frac{\text{SSM}}{\text{SST}}$$

and is interpreted as the proportion of the variability in the response variable $y$ that is explained by the explanatory variables $x_1, x_2, \ldots, x_p$ in the multiple linear regression.

## CHAPTER 11 Exercises

*For Exercises 11.1 and 11.2, see pages 610 and 611; for Exercises 11.3 and 11.4, see page 614; for Exercise 11.5, see page 618; and for Exercise 11.6, see page 621.*

**11.7 95% confidence intervals for regression coefficients.** In each of the following settings, give a 95% confidence interval for the coefficient of $x_1$.

(a) $n = 30, \hat{y} = 10.6 + 10.8x_1 + 7.9x_2$, $\text{SE}_{b_1} = 2.4$.

(b) $n = 53, \hat{y} = 10.6 + 10.8x_1 + 7.9x_2$, $\text{SE}_{b_1} = 2.4$.

(c) $n = 30, \hat{y} = 10.6 + 10.8x_1 + 7.9x_2 + 5.2x_3$, $\text{SE}_{b_1} = 2.4$.

(d) $n = 124, \hat{y} = 10.6 + 10.8x_1 + 7.9x_2 + 5.2x_3$, $\text{SE}_{b_1} = 2.4$.

**11.8 More on significance tests for regression coefficients.** For each of the settings in the previous exercise, test the null hypotheses that the coefficient of $x_1$ is zero versus the two-sided alternative.

**11.9 What's wrong?** In each of the following situations, explain what is wrong and why.

(a) In a multiple regression with a sample size of 40 and 4 explanatory variables, the test statistic for the null hypothesis $H_0$: $b_2 = 0$ is a $t$ statistic that follows the $t(35)$ distribution when the null hypothesis is true.

(b) The multiple correlation coefficient gives the proportion of the variation in the response variable that is explained by the explanatory variables.

(c) A small $P$-value for the ANOVA $F$ test implies all explanatory variables are statistically different from zero.

**11.10 What's wrong?** In each of the following situations, explain what is wrong and why.

(a) One of the assumptions for multiple regression is that the distribution of each explanatory variable is Normal.

(b) The smaller the $P$-value for the ANOVA $F$ test, the greater the explanatory power of the model.

(c) All explanatory variables that are significantly correlated with the response variable will have a statistically significant regression coefficient in the multiple regression model.

**11.11 Constructing the ANOVA table.** Seven explanatory variables are used to predict a response variable using a multiple regression. There are 140 observations.

(a) Write the statistical model that is the foundation for this analysis. Also include a description of all assumptions.

(b) Outline the analysis of variance table giving the sources of variation and numerical values for the degrees of freedom.

**11.12 More on constructing the ANOVA table.** A multiple regression analysis of 73 cases was performed with 5 explanatory variables. Suppose that SSM = 14.1 and SSE = 100.5.

(a) Find the value of the $F$ statistic for testing the null hypothesis that the coefficients of all of the explanatory variables are zero.

(b) What are the degrees of freedom for this statistic?

(c) Find bounds on the $P$-value using Table E. Show your work.

(d) What proportion of the variation in the response variable is explained by the explanatory variables?

**11.13 Childhood obsesity.** The prevalence of childhood obesity in industrialized nations is constantly rising. Since between 30% and 60% of obese children maintain their obesity into adulthood, there is great interest in better understanding the reasons for this rising trend. In one study, researchers looked at the relationship between a child's percent fat mass and several explanatory variables.[4] These were the percent of energy intake at dinner, each parent's body mass index (BMI), an index of energy intake validity (EI/BMR), and gender. The following table summarizes the results of the multiple regression analysis:

| | $b$ | $s(b)$ |
|---|---|---|
| Intercept | 5.13 | 3.03 |
| Gender (M = 0, F = 1) | 4.69 | 0.51 |
| Dinner (%) | 0.08 | 0.02 |
| EI/predicted BMR | −1.90 | 0.65 |
| Mother's BMI (kg/m²) | 0.23 | 0.07 |
| Father's BMI (kg/m²) | 0.27 | 0.09 |

In addition, it is reported that $R = 0.44$ and $F(5, 524) = 25.16$.

(a) How many children were used in this study?

(b) What percent of the variation in percent fat mass is explained by these explanatory variables?

(c) Interpret the sign of each of the regression coefficients given in the table. For EI/predicted BMR, data values ranged between 1.4 and 2.8 with a low value associated with underreporting of energy intake.

(d) Construct a 95% confidence interval for the difference in predicted percent fat mass when energy intake at dinner differs by 5% (assume all other variables are the same).

**11.14 Understanding the tests of significance.** Using a new software package, you ran a multiple regression. The output reported an $F$ statistic with $P < 0.05$, but none of the $t$ tests for the individual coefficients were significant ($P > 0.05$). Does this mean that there is something wrong with the software? Explain your answer.

**11.15 Predicting substance abuse.** What factors predict substance abuse among high school students? One study designed to answer this question collected data from 89 high school seniors in a suburban Florida high school.[5] One of the response variables was marijuana use, which was rated on a four-point scale. A multiple regression analysis used grade point average (GPA), popularity, and a depression score to predict marijuana use. The results were reported in a table similar to this:

| | $b$ | $t$ | $P$ |
|---|---|---|---|
| GPA | −0.597 | 4.55 | < 0.001 |
| Popularity | 0.340 | 2.69 | < 0.01 |
| Depression | 0.030 | 2.69 | < 0.01 |

A footnote to the table gives $R^2 = 0.34$, $F(3, 85) = 14.83$, and $P < 0.001$.

(a) State the null and alternative hypotheses that are tested by each of the $t$ statistics. Give the results of these significance tests.

(b) Interpret the sign of each of the regression coefficients given in the table.

(c) In the expression $F(3, 85)$, what do the numbers 3 and 85 represent?

(d) State the null and alternative hypotheses that are tested by the $F$ statistic. What is the conclusion?

(e) Each of the variables in this analysis was measured by having the students complete a questionnaire. Discuss how this might affect the results.

(f) How well do you think that these results can be applied to other populations of high school students?

**11.16** CHALLENGE **More on predicting substance abuse.** Refer to the previous exercise. The researchers also studied cigarette use, alcohol use, and cocaine use. Here is a summary of the results for the individual regression coefficients:

| | | $b$ | $t$ | $P$ |
|---|---|---|---|---|
| Cigarette | GPA | −0.340 | 2.16 | < 0.05 |
| | Popularity | 0.338 | 2.24 | < 0.05 |
| | Depression | 0.034 | 2.60 | < 0.05 |
| Alcohol | GPA | −0.321 | 3.83 | < 0.001 |
| | Popularity | 0.185 | 2.29 | < 0.05 |
| | Depression | 0.015 | 2.19 | < 0.05 |
| Cocaine | GPA | −0.583 | 5.99 | < 0.001 |
| | Popularity | −1.90 | 2.25 | < 0.05 |
| | Depression | 0.002 | 0.27 | Not sig. |

And here are other relevant results:

| Response variable | $R^2$ | $F$ | $P$ |
|---|---|---|---|
| Cigarette | 0.18 | 6.38 | < 0.001 |
| Alcohol | 0.27 | 10.37 | < 0.001 |
| Cocaine | 0.38 | 12.21 | < 0.001 |

Using the questions given in the previous exercise, summarize the results for each of these response variables. Then write a short essay comparing the results for the four different response variables.

**11.17 Demand for non-biotech cereals.** A study designed to determine how willing consumers are to pay a premium for non-biotech breakfast cereals (cereals that do not include gene-altered ingredients) included both U.S. and U.K. subjects.[6] The response variable was a measure of how much extra they would be willing to pay, and the explanatory variables included items related to perceived risks and benefits, demographic variables, and country. Country was coded as 0 for the U.K. subjects and 1 for the U.S. subjects. The parameter estimate for country was reported as −0.2304 with $t = -4.196$. The total number of subjects was 1810.

(a) Interpret the regression coefficient. Are subjects in the U.S. more willing to pay extra for non-biotech breakfast cereals than U.K. subjects, or are they less willing?

(b) Use the $t$ statistic to find a bound on the $P$-value. Explain the hypothesis tested by this statistic and summarize the result of the significance test.

(c) The U.S. data were collected using questionnaires that were sent to a nationally representative sample of 5200 households enrolled in the National Dairy Panel (NDP) Group. The response rate was 58%. The same questionnaire was used for an online survey of the 9000 U.K. customers enrolled in another NDP Group. The response rate was 28.5%. Several of the items used in the analysis included "Don't know" as a possible response. Respondents choosing this option were excluded from the analysis. Discuss the implications of these considerations on the results.

**11.18** CHALLENGE **Enjoyment of physical exercise.** Although the benefits of physical exercise are well known, most people do not exercise and many who start exercise programs drop out after a short time. A study designed to determine factors associated with exercise enjoyment collected data from 282 female volunteers who were participants in not-for-credit aerobic dance classes at two university centers.[7] Exercise enjoyment was the response variable, with a possible range of 18 to 136. Three explanatory variables were analyzed: satisfaction with the music used (range 4 to 28), satisfaction with the instructor (range 6 to 42), and identity, a variable that measured the extent to which the subject viewed herself as an exerciser. A table of correlations among the four variables was given, and the text noted that all were significant with $P < 0.01$. The coefficients for music (1.02), instructor (0.96), and identity (0.30) were given in another table, where it was noted that $R^2 = 0.33$.

(a) Can you give the fitted regression equation? If your answer is Yes, write the equation; if No, explain what additional information you would need.

(b) Does the fact that all of the correlations between the four variables are significant at $P < 0.01$ tell us that the regression coefficients for each of the three explanatory variables will be statistically significant? Explain your answer.

(c) The statistic for testing the null hypothesis that the population regression coefficients for the three explanatory variables are all zero is $F = 45.64$. Give the degrees of freedom for this statistic, and carry out the significance test. What do you conclude?

(d) What proportion of the variation in exercise enjoyment is explained by music, instructor, and identity?

(e) The authors of the study note that males were not included because there were too few of them in these classes. Do you think that these results would apply to males? Explain why or why not.

**11.19** **Nutrition labels for foods.** Labels providing nutrition facts give consumers information about the nutritional value of food products that they buy. A study of these labels collected data from 152 consumers who were sent information about a frozen chicken dinner. Each subject was asked to give an overall product nutrition score and also evaluated each of 10 nutrients on a 9-point scale, with higher values indicating that the product has a healthy value for the given nutrient. Composite scores for favorable nutrients (such as protein and fiber) and unfavorable nutrients (such as fat and sodium) were used in a multiple regression to predict the overall product nutrition score.[8] The following was reported in a table:

| Explanatory variables | $b$ | se | $t$ | Model $F$ | $R^2$ |
|---|---|---|---|---|---|
| | | | | 33.7** | 0.31 |
| Unfavorable nutrients | 0.82 | 0.12 | 6.8** | | |
| Favorable nutrients | 0.57 | 0.10 | 5.5** | | |
| Constant | 3.33 | 0.13 | 26.1** | | |

**$p < 0.01$

(a) What is the equation of the least-squares line?

(b) Give the null and alternative hypotheses associated with the entry labeled "Model $F$" and interpret this result.

(c) The column labeled "$t$" contains three entries. Explain what each of these means.

(d) What are the degrees of freedom associated with the $t$ statistics that you explained in part (c)?

**11.20 More on nutrition labels for foods.** The product used in the previous exercise was described by the researchers as a poor-nutrition product. The label information for this product had high values for unfavorable nutrients such as fat and low values for favorable nutrients such as fiber. The researchers who conducted this study collected a parallel set of data from subjects who were provided label information for a good-nutrition product. This label had low values for the unfavorable nutrients and high values for the favorable ones. The same type of multiple regression model was run for the 162 consumers who participated in this part of the study. Here are the regression results:

| Explanatory variables | $b$ | se | $t$ | Model $F$ | $R^2$ |
|---|---|---|---|---|---|
| | | | | 44.0** | 0.36 |
| Unfavorable nutrients | 0.86 | 0.12 | 6.9** | | |
| Favorable nutrients | 0.66 | 0.10 | 6.9** | | |
| Constant | 3.96 | 0.12 | 32.8** | | |

**$p < 0.01$

For this analysis, answer the questions in parts (a) to (d) of the previous exercise.

**11.21 Even more on nutrition labels for foods.** Refer to the previous two exercises. When the researchers planned these studies, they expected both unfavorable nutrients and favorable nutrients to be positively associated with the overall product nutrition score. They also expected the unfavorable nutrients to have a stronger effect. Examine the regression coefficients and the associated $t$ statistics for the two regression models. Then, use this information to discuss how well the researchers' expectations were fulfilled.

*The following five exercises use the data given in the next exercise.*

**11.22 Online stock trading.** Online stock trading has increased dramatically during the past several years. An article discussing this new method of investing provided data on the major Internet stock brokerages who provide this service.[9] Below are some data for the top 10 Internet brokerages. The variables are Mshare, the market share of the firm; Accts, the number of Internet accounts in thousands; and Assets, the total assets in billions of dollars. These firms are not a random sample from any population, but we will use multiple regression methods to develop statistical models that relate assets to the other two variables.

| ID | Broker | Mshare | Accts | Assets |
|---|---|---|---|---|
| 1 | Charles Schwab | 27.5 | 2500 | 219.0 |
| 2 | E*Trade | 12.9 | 909 | 21.1 |
| 3 | TD Waterhouse | 11.6 | 615 | 38.8 |
| 4 | Datek | 10.0 | 205 | 5.5 |
| 5 | Fidelity | 9.3 | 2300 | 160.0 |
| 6 | Ameritrade | 8.4 | 428 | 19.5 |
| 7 | DLJ Direct | 3.6 | 590 | 11.2 |
| 8 | Discover | 2.8 | 134 | 5.9 |
| 9 | Suretrade | 2.2 | 130 | 1.3 |
| 10 | National Discount Brokers | 1.3 | 125 | 6.8 |

(a) Plot assets versus accounts and describe the relationship.

(b) Perform a simple linear regression to predict assets from the number of accounts. Give the least-squares line and the results of the hypothesis test for the slope.

(c) Obtain the residuals from part (b) and plot them versus accounts. Describe the plot. What do you conclude?

(d) Construct a new variable that is the square of the number of accounts. Rerun the regression analysis with accounts and the square as explanatory variables. Summarize the results.

**11.23 Adjusting for correlated explanatory variables.** In the multiple regression you performed in the previous exercise, the $P$-value for the number of accounts was 0.8531, while the $P$-value for the square was 0.0070. Unless we have a strong theoretical reason for considering a model with a quadratic term and no linear term, we prefer not to do this. One problem with these two explanatory variables is that they are highly correlated. Here is a way to construct a version of the quadratic term that is less correlated with the linear term. We first find the mean for accounts, and then we subtract this value from accounts before squaring. The mean is 793.6, so the new quadratic explanatory variable will be $(\text{Accts} - 793.6)^2$. Run the multiple regression to predict assets using accounts and the new quadratic term. Compare these results with what you found in the previous exercise.

**11.24 Curvilinear relationship versus a couple of outliers.** To one person, the plot of assets versus the number of accounts indicates that the relationship is curved. Another person might see this as a linear relationship with two outliers. Identify the two outliers and rerun the linear regression and the multiple regression with the linear and quadratic terms. Summarize your results.

**11.25 Transforming the variables.** Sometimes we attempt to model curved relationships by transforming variables. Take the logarithm of assets and the logarithm of the number of accounts. Does the relationship between the logs appear to be approximately linear? Analyze the data and provide a summary of your results. Be sure to include plots along with the results of your statistical inference.

**11.26 Interpretation of coefficients in a multiple regression.** Recall that the relationship between an explanatory variable and a response variable can depend on what other explanatory variables are included in the model.

(a) Use a simple linear regression to predict assets using the number of accounts. Give the regression equation and the results of the significance test for the regression coefficient.

(b) Do the same using market share to predict assets.

(c) Run a multiple regression using both the number of accounts and market share to predict assets. Give the multiple regression equation and the results of the significance tests for the two regression coefficients.

(d) Compare the results in parts (a), (b), and (c). If you had to choose one of these three models, which one do you prefer? Give an explanation for your answer.

*The following three exercises use the RANKING data set described in the Data Appendix.*

**11.27 Annual ranking of world universities.** Let's consider developing a model to predict total score based on the peer review score (PEER), faculty-to-student ratio (FtoS), and citations-to-faculty ratio (CtoF).

(a) Using numerical and graphical summaries, describe the distribution of each explanatory variable.

(b) Using numerical and graphical summaries, describe the relationship between each pair of explanatory variables.

**11.28 Looking at the simple linear regressions.** Now let's look at the relationship between each explanatory variable and the total score.

(a) Generate scatterplots for each explanatory variable and the total score. Do these relationships all look linear?

(b) Compute the correlation between each explanatory variable and the total score. Are certain explanatory variables more strongly associated with the total score?

**11.29 Multiple linear regression model.** Now consider a regression model using all three explanatory variables.

(a) Write out the statistical model for this analysis, making sure to specify all assumptions.

(b) Run the multiple regression model and specify the fitted regression equation.

(c) Generate a 95% confidence interval for each coefficient. Should any of these intervals contain 0? Explain.

(d) What percent of the variation in total score is explained by this model? What is the estimate for $\sigma$?

**11.30 Predicting GPA of seventh-graders.** Refer to the educational data for 78 seventh-grade students given in Table 1.9 (page 29). We view GPA as the response variable. IQ, gender, and self-concept are the explanatory variables.

(a) Find the correlation between GPA and each of the explanatory variables. What percent of the total variation in student GPAs can be explained by the straight-line relationship for each of the explanatory variables?

(b) The importance of IQ in explaining GPA is not surprising. The purpose of the study is to assess the influence of self-concept on GPA. So we will include IQ in the regression model and ask, "How much does self-concept contribute to explaining GPA after the effect of IQ on GPA is taken into account?" Give a model that can be used to answer this question.

(c) Run the model and report the fitted regression equation. What percent of the variation in GPA is explained by the explanatory variables in your model?

(d) Translate the question of interest into appropriate null and alternative hypotheses about the model parameters. Give the value of the test statistic and its $P$-value. Write a short summary of your analysis with an emphasis on your conclusion.

*The following three exercises use the HAPPINESS data set described in the Data Appendix.*

**11.31 Predicting a nation's "average happiness" score.** Consider the following five variables for each nation: LSI, life-satisfaction score, an index of happiness; GINI, a measure of inequality in the distribution of income; CORRUPT, a measure of corruption in government; LIFE, the average life expectancy; and DEMOCRACY, a measure of civil and political liberties.

(a) Using numerical and graphical summaries, describe the distribution of each variable.

(b) Using numerical and graphical summaries, describe the relationship between each pair of variables.

**11.32 Building a multiple linear regression model.** Let's now build a model to predict the life-satisfaction score, LSI.

(a) Consider a simple linear regression using GINI as the explanatory variable. Run the regression and summarize the results. Be sure to check assumptions.

(b) Now consider a model using GINI and LIFE. Run the multiple regression and summarize the results. Again be sure to check assumptions.

(c) Now consider a model using GINI, LIFE, and DEMOCRACY. Run the multiple regression and summarize the results. Again be sure to check assumptions.

(d) Now consider a model using all four explanatory variables. Again summarize the results and check assumptions.

**11.33 Selecting from among several models.** Refer to the results from the previous exercise.

(a) Make a table giving the estimated regression coefficients, standard errors, $t$ statistics, and $P$-values.

(b) Describe how the coefficients and $P$-values change for the four models.

(c) Based on the table of coefficients, suggest another model. Run that model, summarize the results, and compare it with the other ones. Which model would you choose to explain LSI? Explain.

*The following six exercises use the BIOMARKERS data set described in the Data Appendix.*

**11.34 Bone formation and resorption.** Consider the following four variables: VO+, a measure of bone formation; VO−, a measure of bone resorption; OC, a biomarker of bone formation; and TRAP, a biomarker of bone resorption.

(a) Using numerical and graphical summaries, describe the distribution of each of these variables.

(b) Using numerical and graphical summaries, describe the relationship between each pair of variables in this set.

**11.35 Predicting bone formation.** Let's use regression methods to predict VO+, the measure of bone formation.

(a) Since OC is a biomarker of bone formation, we start with a simple linear regression using OC as the explanatory variable. Run the regression and summarize the results. Be sure to include an analysis of the residuals.

(b) Because the processes of bone formation and bone resorption are highly related, it is possible that there is some information in the bone resorption variables that can tell us something about bone formation. Use a model with both OC and TRAP, the biomarker of bone resorption, to predict VO+. Summarize the results. In the context of this model, it appears that TRAP is a better predictor of bone formation, VO+, than the biomarker of bone formation, OC. Is this view consistent with the pattern of relationships that you described in the previous exercise? One possible explanation is that, while all of these variables are highly related, TRAP is measured with more precision than OC.

**11.36 More on predicting bone formation.** Now consider a regression model for predicting VO+ using OC, TRAP, and VO−.

(a) Write out the statistical model for this analysis including all assumptions.

(b) Run the multiple regression to predict VO+ using OC, TRAP, and VO−. Summarize the results.

(c) Make a table giving the estimated regression coefficients, standard errors, and $t$ statistics with $P$-values for this analysis and the two that you ran in the previous exercise. Describe how the coefficients and the $P$-values differ for the three analyses.

(d) Give the percent of variation in VO+ explained by each of the three models and the estimate of $\sigma$. Give a short summary.

(e) The results you found in part (b) suggest another model. Run that model, summarize the results, and compare them with the results in (b).

**11.37** CHALLENGE **Predicting bone formation using transformed variables.** Because the distributions of VO+, VO−, OC, and TRAP tend to be skewed, it is common to work with logarithms rather than the measured values. Using the questions in the previous three exercises as a guide, analyze the log data.

**11.38** CHALLENGE **Predicting bone resorption.** Refer to Exercises 11.34 to 11.36. Answer these questions with the roles of VO+ and VO− reversed; that is, run models to predict VO−, with VO+ as an explanatory variable.

**11.39** CHALLENGE **Predicting bone resorption using transformed variables.** Refer to the previous exercise. Rerun using logs.

*The following eleven exercises use the PCB data set described in the Data Appendix.*

**11.40 Relationship among PCB congeners.** Production of polychlorinated biphenyls (PCBs) was banned in the United States in 1977, but because of their widespread use, these compounds are found in many species of fish. As a result, 38 states have issued advisories about limiting consumption of certain species caught in some areas. Specific advisories are targeted at pregnant and lactating women and all women who are of childbearing age. Unfortunately, there are 209 different varieties, called congeners, of PCB. Measurement of all of these congeners in a fish specimen is an expensive and time-consuming process. If the total amount of PCB in a specimen can be estimated with data collected on a few congeners, costs can be reduced and more specimens can be measured. Consider the following variables: PCB (the total amount of PCB) and four congeners, PCB52, PCB118, PCB138, and PCB180.

(a) Using numerical and graphical summaries, describe the distribution of each of these variables.

(b) Using numerical and graphical summaries, describe the relationship between each pair of variables in this set.

**11.41 Predicting the total amount of PCB.** Use the four congeners, PCB52, PCB118, PCB138, and PCB180, in a multiple regression to predict PCB.

(a) Write the statistical model for this analysis. Include all assumptions.

(b) Run the regression and summarize the results.

(c) Examine the residuals. Do they appear to be approximately Normal? When you plot them versus each of the explanatory variables, are any patterns evident?

**11.42 Adjusting analysis for potential outliers.** The examination of the residuals in part (c) of the previous exercise suggests that there may be two outliers, one with a high residual and one with a low residual.

(a) Because of safety issues, we are more concerned about underestimating PCB in a specimen than about overestimating. Give the specimen number for each of the two suspected outliers. Which one corresponds to an overestimate of PCB?

(b) Rerun the analysis with the two suspected outliers deleted, summarize these results, and compare them with those you obtained in the previous exercise.

**11.43 More on predicting the total amount of PCB.** Run a regression to predict PCB using the variables PCB52, PCB118, and PCB138. Note that this is similar to the analysis that you did in Exercise 11.41, with the change that PCB180 is not included as an explanatory variable.

(a) Summarize the results.

(b) In this analysis, the regression coefficient for PCB118 is not statistically significant. Give the estimate of the coefficient and the associated $P$-value.

(c) Find the estimate of the coefficient for PCB118 and the associated $P$-value for the model analyzed in Exercise 11.41.

(d) Using the results in parts (b) and (c), write a short paragraph explaining how the inclusion of other variables in a multiple regression can have an effect on the estimate of a particular coefficient and the results of the associated significance test.

**11.44 Multiple regression model for total TEQ.** Dioxins and furans are other classes of chemicals that can cause undesirable health effects similar to those caused by PCB. The three types of chemicals are combined using toxic equivalent scores (TEQs), which attempt to measure the health effects on a common scale. The PCB data set contains TEQs for PCB, dioxins, and furans. The variables are called TEQPCB, TEQDIOXIN, and TEQFURAN. The data set also includes the total TEQ, defined to be the sum of these three variables.

(a) Consider using a multiple regression to predict TEQ using the three components TEQPCB, TEQDIOXIN, and TEQFURAN as explanatory variables. Write the multiple regression model in the form

$$\text{TEQ} = \beta_0 + \beta_1\text{TEQPCB} + \beta_2\text{TEQDIOXIN}$$
$$+ \beta_3\text{TEQFURAN} + \epsilon$$

Give numerical values for the parameters $\beta_0$, $\beta_1$, $\beta_2$, and $\beta_3$.

(b) The multiple regression model assumes that the $\epsilon$'s are Normal with mean zero and standard deviation $\sigma$. What is the numerical value of $\sigma$?

(c) Use software to run this regression and summarize the results.

11.45 CHALLENGE **Multiple regression model for total TEQ, continued.** The information summarized in TEQ is used to assess and manage risks from these chemicals. For example, the World Health Organization (WHO) has established the tolerable daily intake (TDI) as 1 to 4 TEQs per kilogram of body weight per day. Therefore, it would be very useful to have a procedure for estimating TEQ using just a few variables that can be measured cheaply. Use the four PCB congeners, PCB52, PCB118, PCB138, and PCB180, in a multiple regression to predict TEQ. Give a description of the model and assumptions, summarize the results, examine the residuals, and write a summary of what you have found.

11.46 CHALLENGE **Predicting total amount of PCB using transformed variables.** Because distributions of variables such as PCB, the PCB congeners, and TEQ tend to be skewed, researchers frequently analyze the logarithms of the measured variables. Create a data set that has the logs of each of the variables in the PCB data set. Note that zero is a possible value for PCB126; most software packages will eliminate these cases when you request a log transformation.

(a) If you do not do anything about the 16 zero values of PCB126, what does your software do with these cases? Is there an error message of some kind?

(b) If you attempt to run a regression to predict the log of PCB using the log of PCB126 and the log of PCB52, are the cases with the zero values of PCB126 eliminated? Do you think that is a good way to handle this situation?

(c) The smallest nonzero value of PCB126 is 0.0052. One common practice when taking logarithms of measured values is to replace the zeros by one-half of the smallest observed value. Create a logarithm data set using this procedure; that is, replace the 16 zero values of PCB126 by 0.0026 before taking logarithms. Use numerical and graphical summaries to describe the distributions of the log variables.

11.47 CHALLENGE **Predicting total amount of PCB using transformed variables, continued.** Refer to the previous exercise.

(a) Use numerical and graphical summaries to describe the relationships between each pair of log variables.

(b) Compare these summaries with the summaries that you produced in Exercise 11.40 for the measured variables.

11.48 CHALLENGE **Even more on predicting total amount of PCB using transformed variables.** Use the log data set that you created in Exercise 11.46 to find a good multiple regression model for predicting the log of PCB. Use only log PCB variables for this analysis. Write a report summarizing your results.

11.49 CHALLENGE **Predicting total TEQ using transformed variables.** Use the log data set that you created in Exercise 11.46 to find a good multiple regression model for predicting the log of TEQ. Use only log PCB variables for this analysis. Write a report summarizing your results and comparing them with the results that you obtained in the previous exercise.

11.50 **Interpretation of coefficients in log PCB regressions.** Use the results of your analysis of the log PCB data in Exercise 11.48 to write an explanation of how regression coefficients, standard errors of regression coefficients, and tests of significance for explanatory variables can change depending on what other explanatory variables are included in the multiple regression analysis.

*The following nine exercises use the CHEESE data set described in the Data Appendix.*

**11.51 Describing the explanatory variables.** For each of the four variables in the CHEESE data set, find the mean, median, standard deviation, and interquartile range. Display each distribution by means of a stemplot and use a Normal quantile plot to assess Normality of the data. Summarize your findings. Note that when doing regressions with these data, we do not assume that these distributions are Normal. Only the residuals from our model need to be (approximately) Normal. The careful study of each variable to be analyzed is nonetheless an important first step in any statistical analysis.

**11.52 Pairwise scatterplots of the explanatory variables.** Make a scatterplot for each pair of variables in the CHEESE data set (you will have six plots). Describe the relationships. Calculate the correlation for each pair of variables and report the *P*-value for the test of zero population correlation in each case.

**11.53 Simple linear regression model of Taste.** Perform a simple linear regression analysis using Taste as the response variable and Acetic as the explanatory variable. Be sure to examine the residuals carefully. Summarize your results. Include a plot of the data with the least-squares regression line. Plot the residuals versus each of the other two chemicals. Are any patterns evident? (The concentrations of the other chemicals are lurking variables for the simple linear regression.)

**11.54 Another simple linear regression model of Taste.** Repeat the analysis of Exercise 11.53 using Taste as the response variable and H2S as the explanatory variable.

**11.55 The final simple linear regression model of Taste.** Repeat the analysis of Exercise 11.53 using Taste as the response variable and Lactic as the explanatory variable.

**11.56 Comparing the simple linear regression models.** Compare the results of the regressions performed in the three previous exercises. Construct a table with values of the $F$ statistic, its $P$-value, $R^2$, and the estimate $s$ of the standard deviation for each model. Report the three regression equations. Why are the intercepts in these three equations different?

**11.57 Multiple regression model of Taste.** Carry out a multiple regression using Acetic and H2S to predict Taste. Summarize the results of your analysis. Compare the statistical significance of Acetic in this model with its significance in the model with Acetic alone as a predictor (Exercise 11.53). Which model do you prefer? Give a simple explanation for the fact that Acetic alone appears to be a good predictor of Taste, but with H2S in the model, it is not.

**11.58 Another multiple regression model of Taste.** Carry out a multiple regression using H2S and Lactic to predict Taste. Comparing the results of this analysis with the simple linear regressions using each of these explanatory variables alone, it is evident that a better result is obtained by using both predictors in a model. Support this statement with explicit information obtained from your analysis.

**11.59 The final multiple regression model of Taste.** Use the three explanatory variables Acetic, H2S, and Lactic in a multiple regression to predict Taste. Write a short summary of your results, including an examination of the residuals. Based on all of the regression analyses you have carried out on these data, which model do you prefer and why?

**11.60 Finding a multiple regression model on the Internet.** Search the Internet to find an example of the use of multiple regression. Give the setting of the example, describe the data, give the model, and summarize the results. Explain why the use of multiple regression in this setting was appropriate or inappropriate.

CHAPTER 12

# One-Way Analysis of Variance

Which brand of tires lasts the longest under city driving conditions? The methods described in this chapter allow us to compare the average wear of each brand.



## Introduction

Many of the most effective statistical studies are comparative. For example, we may wish to compare customer satisfaction of men and women using an online fantasy football site or compare the responses to various treatments in a clinical trial. We display these comparisons with back-to-back stemplots or side-by-side boxplots, and we measure them with five-number summaries or with means and standard deviations.

When only two groups are compared, Chapter 7 provides the tools we need to answer the question "Is the difference between groups statistically significant?" Two-sample $t$ procedures compare the means of two Normal populations, and we saw that these procedures, unlike comparisons of spread, are sufficiently robust to be widely useful.

In this chapter, we will compare any number of means by techniques that generalize the two-sample $t$ and share its robustness and usefulness. These methods will allow us to address comparisons such as

- Which of 4 advertising offers mailed to sample households produces the highest dollar sales?
- Which of 10 brands of automobile tires wears longest?
- How long do cancer patients live under each of 3 therapies for their lung cancer?

## 12.1 Inference for One-Way Analysis of Variance

When comparing different populations or treatments, the data are subject to sampling variability. For example, we would not expect the same sales data if we mailed various advertising offers to a different sample of households. We therefore pose the question for inference in terms of the *mean* response. In Chapter 7 we met procedures for comparing the means of two populations. We are now ready to extend those methods to problems involving more than two populations. The statistical methodology for comparing several means is called **analysis of variance,** or simply **ANOVA.** In the sections that follow, we will examine the basic ideas and assumptions that are needed for ANOVA. Although the details differ, many of the concepts are similar to those discussed in the two-sample case.

LOOK BACK
**comparing two means, page 447**

ANOVA

We will consider two ANOVA techniques. When there is only one way to classify the populations of interest, we use **one-way ANOVA** to analyze the data. For example, to compare the survival times for three different lung cancer therapies we use one-way ANOVA. This chapter presents the details for one-way ANOVA.

one-way ANOVA

In many other comparison studies, there is more than one way to classify the populations. For the advertising study, the company may also consider mailing the offers using two different envelope styles. Will each offer draw more sales on the average when sent in an attention-grabbing envelope? Analyzing the effect of advertising offer and envelope layout together requires **two-way ANOVA.** This technique will be discussed in Chapter 13. While adding yet more factors necessitates even higher-way ANOVA techniques, most of the new ideas in ANOVA with more than one factor already appear in two-way ANOVA.

two-way ANOVA

### Data for one-way ANOVA

One-way analysis of variance is a statistical method for comparing several population means. We draw a simple random sample (SRS) from each population and use the data to test the null hypothesis that the population means are all equal. Consider the following two examples:

EXAMPLE

**12.1 Choosing the best magazine layout.** A magazine publisher wants to compare three different layouts for a magazine that will be offered for sale at supermarket checkout lines. She is interested in whether there is a layout that better catches shoppers' attention and results in more sales. To investigate, she randomly assigns each of 60 stores to one of the three layouts and records the number of magazines that are sold in a one-week period.

EXAMPLE

**12.2 Average age of bookstore customers.** How do five bookstores in the same city differ in the demographics of their customers? Are certain bookstores more popular among teenagers? Do upper-income shoppers tend to go to one store? A market researcher asks 50 customers of each store to respond to a questionnaire. Two variables of interest are the customer's age and income level.

These two examples are similar in that

- There is a single quantitative response variable measured on many units; the units are stores in the first example and customers in the second.
- The goal is to compare several populations: stores displaying three magazine layouts in the first example and customers of five bookstores in the second.

LOOK BACK
**observation versus experiment, page 175**

There is, however, an important difference. Example 12.1 describes an experiment in which stores are randomly assigned to layouts. Example 12.2 is an observational study in which customers are selected during a particular time period and not all agree to provide data. We will treat our samples of customers as random samples even though this is only approximately true.

In both examples, we will use ANOVA to compare the mean responses. The same ANOVA methods apply to data from random samples and to data from randomized experiments. *It is important to keep the data-production method in mind when interpreting the results. A strong case for causation is best made by a randomized experiment.*

## Comparing means

The question we ask in ANOVA is "Do all groups have the same population mean?" We will often use the term *groups* for the populations to be compared in a one-way ANOVA. To answer this question we compare the sample means. Figure 12.1 displays the sample means for Example 12.1. It appears that Layout 2 has the highest average sales. But is the observed difference in sample means just the result of chance variation? We should not expect sample means to be equal, even if the population means are all identical.

The purpose of ANOVA is to assess whether the observed differences among sample means are *statistically significant.* Could a variation among the three sample means this large be plausibly due to chance, or is it good evidence for a difference among the population means? This question can't be answered from the sample means alone. Because the standard deviation of a sample mean $\bar{x}$ is the population standard deviation $\sigma$ divided by $\sqrt{n}$, the answer also depends upon both the variation within the groups of observations and the sizes of the samples.

LOOK BACK
**standard deviation of $\bar{x}$, page 338**

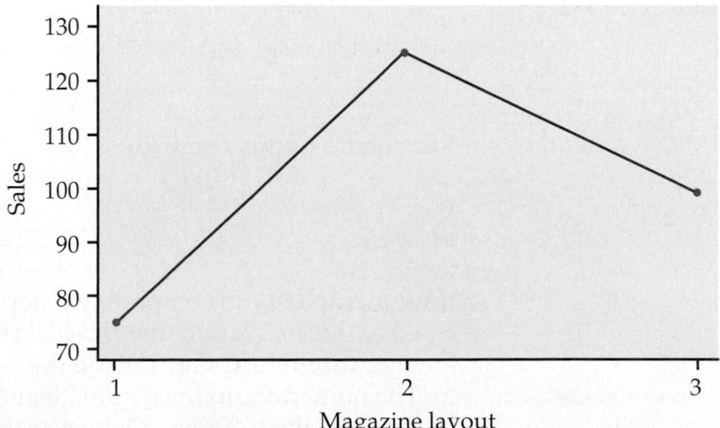


**FIGURE 12.1** Mean sales of magazines for three different magazine layouts.

Side-by-side boxplots help us see the within-group variation. Compare Figures 12.2(a) and 12.2(b). The sample medians are the same in both figures, but the large variation within the groups in Figure 12.2(a) suggests that the differences among the sample medians could be due simply to chance variation. The data in Figure 12.2(b) are much more convincing evidence that the populations differ. Even the boxplots omit essential information, however. To assess the observed differences, we must also know how large the samples are. Nonetheless, boxplots are a good preliminary display of the data. While ANOVA compares means and boxplots display medians, we expect the data to be approximately Normal and will consider a transformation if they are not. For distributions that are nearly symmetric, these two measures of center will be close together.

LOOK BACK
**transforming data, page 435**

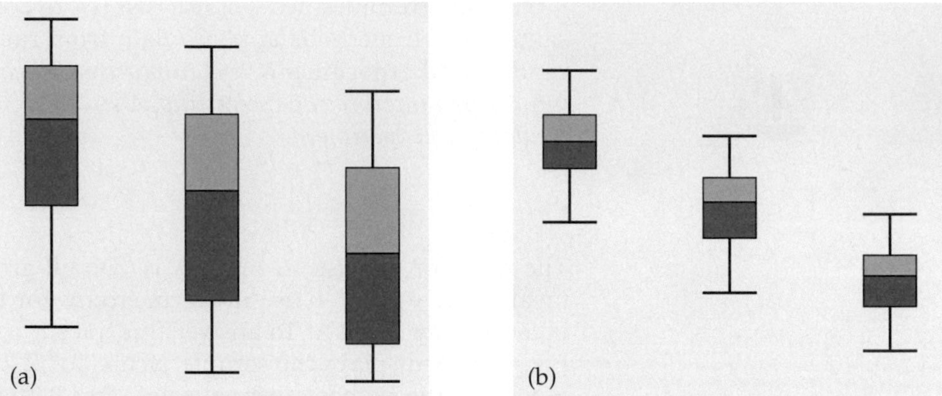


**FIGURE 12.2** (a) Side-by-side boxplots for three groups with large within-group variation. The differences among centers may be just chance variation. (b) Side-by-side boxplots for three groups with the same centers as in Figure 12.2(a) but with small within-group variation. The differences among centers are more likely to be significant.

## The two-sample *t* statistic

Two-sample *t* statistics compare the means of two populations. If the two populations are assumed to have equal but unknown standard deviations and the sample sizes are both equal to $n$, the $t$ statistic is

LOOK BACK
**pooled two-sample *t* statistic, page 462**

$$t = \frac{\bar{x} - \bar{y}}{s_p\sqrt{\frac{1}{n} + \frac{1}{n}}} = \frac{\sqrt{\frac{n}{2}}(\bar{x} - \bar{y})}{s_p}$$

The square of this $t$ statistic is

$$t^2 = \frac{\frac{n}{2}(\bar{x} - \bar{y})^2}{s_p^2}$$

If we use ANOVA to compare two populations, the ANOVA $F$ statistic is exactly equal to this $t^2$. We can therefore learn something about how ANOVA works by looking carefully at the statistic in this form.

between-group variation

The numerator in the $t^2$ statistic measures the variation **between** the groups in terms of the difference between their sample means $\bar{x}$ and $\bar{y}$. It includes a factor for the common sample size $n$. The numerator can be large because of a large difference between the sample means or because the sample sizes are large. The denominator measures the variation **within** groups by $s_p^2$, the pooled estimator of the common variance. If the within-group variation is small, the same variation between the groups produces a larger statistic and thus a more significant result.

within-group variation

Although the general form of the $F$ statistic is more complicated, the idea is the same. To assess whether several populations all have the same mean, we compare the variation *among* the means of several groups with the variation *within* groups. Because we are comparing variation, the method is called *analysis of variance.*

## An overview of ANOVA

ANOVA tests the null hypothesis that the population means are *all* equal. The alternative is that they are not all equal. This alternative could be true because all of the means are different or simply because one of them differs from the rest. This is a more complex situation than comparing just two populations. If we reject the null hypothesis, we need to perform some further analysis to draw conclusions about which population means differ from which others and by how much.

The computations needed for an ANOVA are more lengthy than those for the $t$ test. For this reason we generally use computer programs to perform the calculations. Automating the calculations frees us from the burden of arithmetic and allows us to concentrate on interpretation. *Complicated computations do not guarantee a valid statistical analysis. We should always start our ANOVA with a careful examination of the data using graphical and numerical summaries.*

EXAMPLE

**12.3 Workplace safety.** In a study of workplace safety, workers were asked to rate various elements of safety, and a composite score called the Safety Climate Index (SCI) was calculated.[1] The index is the sum of the responses to 10 different questions about safety. The response for each of these questions is an integer ranging from 0 to 10, so the SCI has values from 0 to 100. The workers were classified according to their job category as unskilled, skilled, and supervisor. Here is a summary of the data:

| Job category | $n$ | $\bar{x}$ | $s$ |
|---|---|---|---|
| Unskilled workers | 448 | 70.42 | 18.27 |
| Skilled workers | 91 | 71.21 | 18.83 |
| Supervisors | 51 | 80.51 | 14.58 |

Histograms and descriptive statistics for the three groups of workers are given in Figure 12.3. Note that the heights of the bars in the histograms are percents rather than counts. If we had used counts with the same scale on the $y$ axis, then the bars for the skilled workers and the supervisors would be very small because of the smaller sample sizes in these groups. Figure 12.4 gives side-by-side boxplots for these data. We see that the largest and the smallest possible values are present in the data. The distributions are somewhat skewed toward lower values. Our sample sizes, however, are sufficiently large that we are confident that the sample means are approximately Normal.

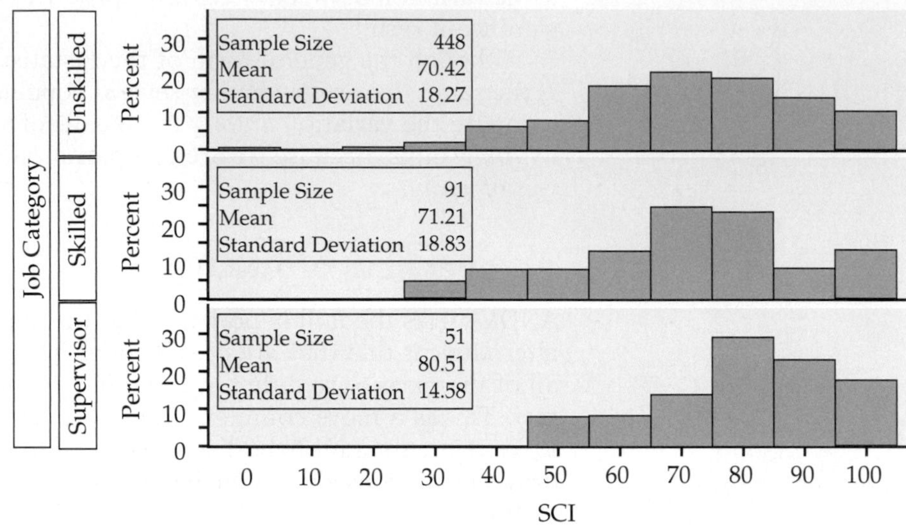


**FIGURE 12.3** Histograms and descriptive statistics for the worker safety example.

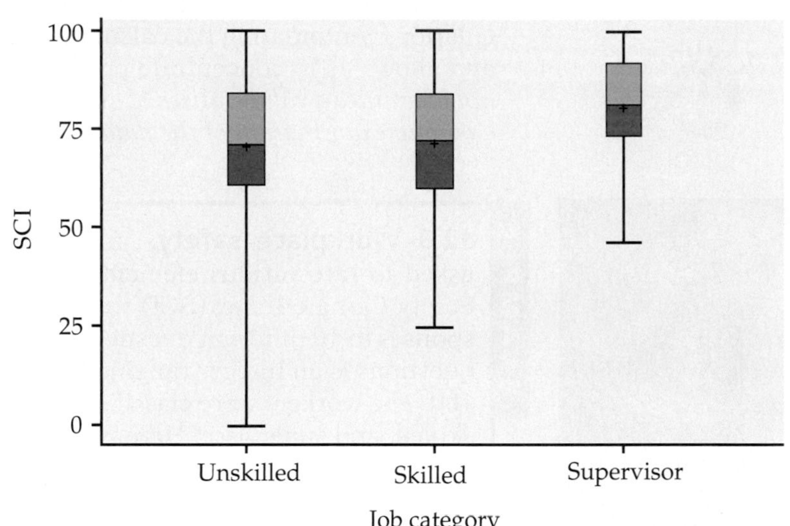


**FIGURE 12.4** Side-by-side boxplots for the worker safety example.

The three sample means are plotted in Figure 12.5. It appears that the means for the unskilled workers and the skilled workers are similar, while the supervisors have a higher mean. To apply ANOVA in this setting, we view the three samples that we have as three independent random samples from three distinct populations. Each of these populations has a mean and our inference asks questions about these means.

*Formulating a clear definition of the populations being compared with ANOVA can be difficult, as in our example. Often some expert judgment is required, and different consumers of the results may have differing opinions.* The workers in this study all worked in the same industry in a particular region. They certainly

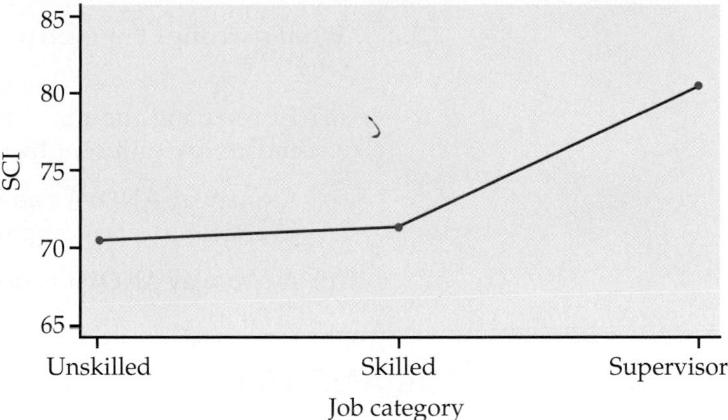


**FIGURE 12.5** SCI means for the worker safety example.

do represent some larger population of similar workers. We are more confident in generalizing our conclusions to similar populations when the results are clearly significant than when the level of significance just barely passes the standard of $P = 0.05$.

We first ask whether or not there is sufficient evidence in the data to conclude that the corresponding population means are not all equal. Our null hypothesis here states that the population mean SCI is the same for all three groups of workers. The alternative is that they are not all the same.

Our inspection of the data for our example suggests that the means for the skilled workers and the unskilled workers may be the same while the mean for the supervisors is higher. *Rejecting the null hypothesis that the means are all the same using ANOVA is not the same as concluding that all of the means are different from one another.* The ANOVA null hypothesis can be false in many different ways. Additional analysis is required to distinguish among these possibilities.

contrasts

When there are particular versions of the alternative hypothesis that are of interest, we use **contrasts** to examine them. In our example, we might want to compare the supervisors with all of the other workers. *Note that, to use contrasts, it is necessary that the questions of interest be formulated before examining the data. It is cheating to make up these questions after analyzing the data.*

multiple comparisons

If we have no specific relations among the means in mind before looking at the data, we instead use a **multiple-comparisons** procedure to determine which pairs of population means differ significantly. In later sections we will explore both contrasts and multiple comparisons in detail.

## USE YOUR KNOWLEDGE

**12.1** **What's wrong?** For each of the following, explain what is wrong and why.

(a) ANOVA tests the null hypothesis that the sample means are all equal.

(b) A strong case for causation is best made in an observational study.

(c) You use one-way ANOVA when the response variable has only two possible values.

**12.2** **What's wrong?** For each of the following, explain what is wrong and why.

(a) In rejecting the null hypothesis, one can conclude that all the means are different from each other.

(b) A one-way ANOVA can be used only when there are fewer than five means to be compared.

(c) A two-way ANOVA is used when comparing two populations.

## The ANOVA model

When analyzing data, the following equation reminds us that we look for an overall pattern and deviations from it:

$$\text{DATA} = \text{FIT} + \text{RESIDUAL}$$

LOOK BACK
**DATA = FIT + RESIDUAL, page 564**

In the regression model of Chapter 10, the FIT was the population regression line, and the RESIDUAL represented the deviations of the data from this line. We now apply this framework to describe the statistical models used in ANOVA. These models provide a convenient way to summarize the assumptions that are the foundation for our analysis. They also give us the necessary notation to describe the calculations needed.

LOOK BACK
**Normal distributions, page 58**

First, recall the statistical model for a random sample of observations from a single Normal population with mean $\mu$ and standard deviation $\sigma$. If the observations are

$$x_1, x_2, \ldots, x_n$$

we can describe this model by saying that the $x_j$ are an SRS from the $N(\mu, \sigma)$ distribution. Another way to describe the same model is to think of the $x$'s varying about their population mean. To do this, write each observation $x_j$ as

$$x_j = \mu + \epsilon_j$$

The $\epsilon_j$ are then an SRS from the $N(0, \sigma)$ distribution. Because $\mu$ is unknown, the $\epsilon$'s cannot actually be observed. This form more closely corresponds to our

$$\text{DATA} = \text{FIT} + \text{RESIDUAL}$$

way of thinking. The FIT part of the model is represented by $\mu$. It is the systematic part of the model, like the line in a regression. The RESIDUAL part is represented by $\epsilon_j$. It represents the deviations of the data from the fit and is due to random, or chance, variation.

There are two unknown parameters in this statistical model: $\mu$ and $\sigma$. We estimate $\mu$ by $\bar{x}$, the sample mean, and $\sigma$ by $s$, the sample standard deviation. The differences $e_j = x_j - \bar{x}$ are the sample residuals and correspond to the $\epsilon_j$ in the statistical model.

The model for one-way ANOVA is very similar. We take random samples from each of $I$ different populations. The sample size is $n_i$ for the $i$th population. Let $x_{ij}$ represent the $j$th observation from the $i$th population. The $I$ population means are the FIT part of the model and are represented by $\mu_i$. The random variation, or RESIDUAL, part of the model is represented by the deviations $\epsilon_{ij}$ of the observations from the means.

### THE ONE-WAY ANOVA MODEL

The **one-way ANOVA model** is

$$x_{ij} = \mu_i + \epsilon_{ij}$$

for $i = 1, \ldots, I$ and $j = 1, \ldots, n_i$. The $\epsilon_{ij}$ are assumed to be from an $N(0, \sigma)$ distribution. The **parameters of the model** are the population means $\mu_1, \mu_2, \ldots, \mu_I$ and the common standard deviation $\sigma$.

Note that the sample sizes $n_i$ may differ, but the standard deviation $\sigma$ is assumed to be the same in all of the populations. Figure 12.6 pictures this model for $I = 3$. The three population means $\mu_i$ are different, but the shapes of the three Normal distributions are the same, reflecting the assumption that all three populations have the same standard deviation.

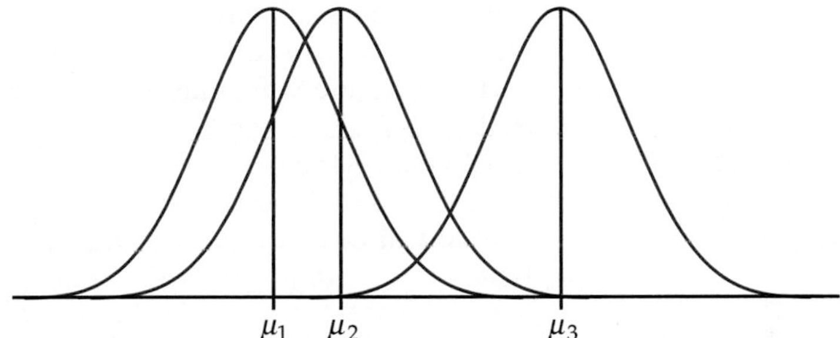


**FIGURE 12.6** Model for one-way ANOVA with three groups. The three populations have Normal distributions with the same standard deviation.

EXAMPLE

**12.4 ANOVA model for worker safety study.** In our worker safety example there are three groups of workers that we want to compare, so $I = 3$. The population means $\mu_1$, $\mu_2$, and $\mu_3$ are the mean SCI values for unskilled workers, for skilled workers, and for supervisors, respectively. The sample sizes $n_i$ are 448, 91, and 51.

The observation $x_{1,1}$ is the SCI score for the first unskilled worker. The data for the other unskilled workers are denoted by $x_{1,2}, x_{1,3}, \ldots, x_{1,448}$. Similarly, the data for the other two groups have a first subscript indicating the group and a second subscript indicating the worker in that group.

According to our model, the SCI for the first worker is $x_{1,1} = \mu_1 + \epsilon_{1,1}$, where $\mu_1$ is the average for *all* unskilled workers and $\epsilon_{1,1}$ is the chance variation due to this particular worker. The ANOVA model assumes that the $\epsilon_{ij}$ are independent and Normally distributed with mean 0 and standard deviation $\sigma$. We have clear evidence that the data are not Normal in our example. The values are numbers ranging from 0 to 100, and we saw some skewness for all three groups in Figures 12.3 and 12.4. However, because our inference is based on the sample means, which will be approximately Normal, we are not overly concerned about this violation of our assumptions.

It is common to use numerical subscripts to distinguish the different means, and some software requires that levels of factors in ANOVA be specified as numerical values. An alternative is to use subscripts that suggest the actual groups. In our example, we could replace $\mu_1$, $\mu_2$, and $\mu_3$ by $\mu_{UN}$, $\mu_{SK}$, and $\mu_{SU}$.

## Estimates of population parameters

The unknown parameters in the statistical model for ANOVA are the $I$ population means $\mu_i$ and the common population standard deviation $\sigma$. To estimate $\mu_i$ we use the sample mean for the $i$th group:

$$\bar{x}_i = \frac{1}{n_i} \sum_{j=1}^{n_i} x_{ij}$$

The residuals $e_{ij} = x_{ij} - \bar{x}_i$ reflect the variation about the sample means that we see in the data.

The ANOVA model assumes that the population standard deviations are all equal. If we have unequal standard deviations, we generally try to transform the data so that they are approximately equal. We might, for example, work with $\sqrt{x_{ij}}$ or $\log x_{ij}$. Fortunately, we can often find a transformation that *both* makes the group standard deviations more nearly equal and also makes the distributions of observations in each group more nearly Normal. If the standard deviations are markedly different and cannot be made similar by a transformation, inference requires different methods that are beyond the scope of this book.

LOOK BACK

***F* test for equality of spread, page 474**

Unfortunately, formal tests for the equality of standard deviations in several groups share the lack of robustness against non-Normality that we noted in Chapter 7 for the case of two groups. Because ANOVA procedures are not extremely sensitive to unequal standard deviations, we do *not* recommend a formal test of equality of standard deviations as a preliminary to the ANOVA. Instead, we will use the following rule as a guideline.

### RULE FOR EXAMINING STANDARD DEVIATIONS IN ANOVA

If the largest standard deviation is less than twice the smallest standard deviation, we can use methods based on the assumption of equal standard deviations, and our results will still be approximately correct.[2]

When we assume that the population standard deviations are equal, each sample standard deviation is an estimate of $\sigma$. To combine these into a single estimate, we use a generalization of the pooling method introduced in Chapter 7.

### POOLED ESTIMATOR OF $\sigma$

Suppose we have sample variances $s_1^2, s_2^2, \ldots, s_I^2$ from $I$ independent SRSs of sizes $n_1, n_2, \ldots, n_I$ from populations with common variance $\sigma^2$. The **pooled sample variance**

$$s_p^2 = \frac{(n_1 - 1)s_1^2 + (n_2 - 1)s_2^2 + \cdots + (n_I - 1)s_I^2}{(n_1 - 1) + (n_2 - 1) + \cdots + (n_I - 1)}$$

is an unbiased estimator of $\sigma^2$. The **pooled standard deviation**

$$s_p = \sqrt{s_p^2}$$

is the estimate of $\sigma$.

Pooling gives more weight to groups with larger sample sizes. If the sample sizes are equal, $s_p^2$ is just the average of the $I$ sample variances. *Note that $s_p$ is not the average of the I sample standard deviations.*

EXAMPLE

**12.5 Population estimates for worker safety study.** In the worker safety study there are $I = 3$ groups and the sample sizes are $n_1 = 448$, $n_2 = 91$, and $n_3 = 51$. The sample standard deviations are $s_1 = 18.27$, $s_2 = 18.83$, and $s_3 = 14.58$.

Because the largest standard deviation (18.83) is less than twice the smallest ($2 \times 14.58 = 29.16$), our rule indicates that we can use the assumption of equal population standard deviations.

The pooled variance estimate is

$$\begin{aligned} s_p^2 &= \frac{(n_1 - 1)s_1^2 + (n_2 - 1)s_2^2 + (n_3 - 1)s_3^2}{(n_1 - 1) + (n_2 - 1) + (n_3 - 1)} \\ &= \frac{(447)(18.27)^2 + (90)(18.83)^2 + (50)(14.58)^2}{447 + 90 + 50} \\ &= \frac{191{,}745}{587} = 326.7 \end{aligned}$$

The pooled standard deviation is

$$s_p = \sqrt{326.7} = 18.07$$

This is our estimate of the common standard deviation $\sigma$ of the SCI scores in the three populations of workers.

## USE YOUR KNOWLEDGE

**12.3 Computing the pooled standard deviation.** An experiment was run to compare three groups. The sample sizes were 25, 22, and 19, and the corresponding estimated standard deviations were 22, 20, and 18.

(a) Is it reasonable to use the assumption of equal standard deviations when we analyze these data? Give a reason for your answer.

(b) Give the values of the variances for the three groups.

(c) Find the pooled variance.

(d) What is the value of the pooled standard deviation?

**12.4 Visualizing the ANOVA model.** For each of the following situations, draw a picture of the ANOVA model similar to Figure 12.6 (page 645). Use numerical values for the $\mu_i$. To sketch the Normal curves, you may want to review the 68–95–99.7 rule on page 59.

(a) $\mu_1 = 15$, $\mu_2 = 16$, $\mu_3 = 21$, and $\sigma = 6$.

(b) $\mu_1 = 10$, $\mu_2 = 15$, $\mu_3 = 20$, $\mu_4 = 20.1$, and $\sigma = 2.5$.

(c) $\mu_1 = 15$, $\mu_2 = 16$, $\mu_3 = 21$, and $\sigma = 2$.

## Testing hypotheses in one-way ANOVA

LOOK BACK
**ANOVA table, page 582**

Comparison of several means is accomplished by using an $F$ statistic to compare the variation among groups with the variation within groups. We now show how the $F$ statistic expresses this comparison. Calculations are organized in an ANOVA table, which contains numerical measures of the variation among groups and within groups.

First we must specify our hypotheses for one-way ANOVA. As usual, $I$ represents the number of populations to be compared.

HYPOTHESES FOR ONE-WAY ANOVA

The **null and alternative hypotheses** for one-way ANOVA are

$$H_0: \mu_1 = \mu_2 = \cdots = \mu_I$$

$$H_a: \text{not all of the } \mu_i \text{ are equal}$$

We will now use our worker safety example to illustrate how to do a one-way ANOVA. Because the calculations are generally performed using statistical software, we focus on interpretation of the output.

EXAMPLE

**12.6 Reading software output.** Figure 12.7 gives descriptive statistics generated by SPSS for the ANOVA of the worker safety example. Summaries for each group are given on the first three lines. In addition to the sample size, the mean, and the standard deviation, this output also gives the minimum and maximum observed value, standard error of the mean and the 95% confidence interval for the mean of each group. The three sample means $\bar{x}_i$ given in the output are estimates of the three unknown population means $\mu_i$.

The output gives the estimates of the standard deviations for each group, the $s_i$, but does not provide $s_p$, the pooled estimate of the model standard deviation, $\sigma$. We could perform the calculation using a calculator, as we did in Example 12.5. We will see an easier way to obtain this quantity from the ANOVA table in Figure 12.8. Some software packages report $s_p$ as part of the standard ANOVA output. *Sometimes you are not sure whether or not a quantity given by*

SPSS

**Descriptives**

SCI

| | N | Mean | Std. Deviation | Std. Error | 95% Confidence Interval for Mean Lower Bound | 95% Confidence Interval for Mean Upper Bound | Minimum | Maximum |
|---|---|---|---|---|---|---|---|---|
| 1 | 448 | 70.42 | 18.269 | .863 | 68.73 | 72.12 | 0 | 100 |
| 2 | 91 | 71.21 | 18.830 | 1.974 | 67.29 | 75.13 | 25 | 100 |
| 3 | 51 | 80.51 | 14.577 | 2.041 | 76.41 | 84.61 | 46 | 100 |
| Total | 590 | 71.42 | 18.260 | .752 | 69.94 | 72.89 | 0 | 100 |

**FIGURE 12.7** Software output with descriptive statistics for the worker safety example.

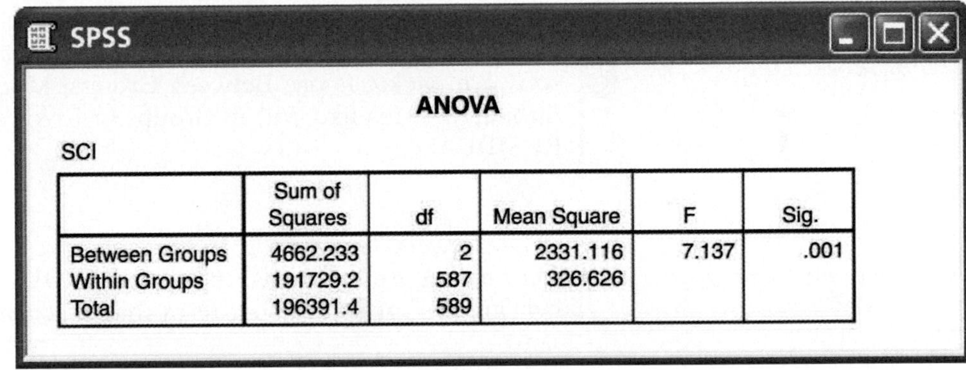

SPSS

**ANOVA**

SCI

| | Sum of Squares | df | Mean Square | F | Sig. |
|---|---|---|---|---|---|
| Between Groups | 4662.233 | 2 | 2331.116 | 7.137 | .001 |
| Within Groups | 191729.2 | 587 | 326.626 | | |
| Total | 196391.4 | 589 | | | |

**FIGURE 12.8** Software output giving the ANOVA table for the worker safety example.

*software is what you think it is. A good way to resolve this dilemma is to do a sample calculation with a simple example to check the numerical results.*

*Note that $s_p$ is not the standard deviation given in the Total row of Figure 12.7. This quantity is the standard deviation that we would obtain if we viewed the data as a single sample of 590 workers and ignored the possibility that the group means could be different.* As we have mentioned many times before, it is important to use care when reading and interpreting software output.

EXAMPLE

**12.7 Reading software output, continued.** Additional output generated by SPSS for the ANOVA of the worker safety example is given in Figure 12.8. We will discuss some details in the next section. For now, we observe that the results of our significance test are given in the last two columns of the output. The null hypothesis that the three population means are the same is tested by the statistic $F = 7.137$, and the associated $P$-value is reported as $P = 0.001$. The data provide clear evidence to support the claim that these three groups of workers have different mean SCI values.

## The ANOVA table

The information in an analysis of variance is organized in an ANOVA table. To understand the table, it is helpful to think in terms of our

$$\text{DATA} = \text{FIT} + \text{RESIDUAL}$$

view of statistical models. For one-way ANOVA, this corresponds to

$$x_{ij} = \mu_i + \epsilon_{ij}$$

We can think of these three terms as sources of variation. The ANOVA table separates the variation in the data into two parts: the part due to the fit and the remainder, which we call residual.

EXAMPLE

**12.8 ANOVA table for worker safety study.** The SPSS output in Figure 12.8 gives the sources of variation in the first column. Here, FIT is called Between Groups, RESIDUAL is called Within Groups, and DATA is the last entry, Total. Different software packages use different terms for these sources of variation but the basic concept is common to all. In place of FIT, some software packages use Between Groups, Model, or the name of the factor. Similarly, terms like Within Groups or Error are frequently used in place of RESIDUAL.

variation among groups

The Between Groups row in the table gives information related to the variation **among** group means. In writing ANOVA tables we will use the generic label "groups" or some other term that describes the factor being studied for this row.

variation within groups

The Within Groups row in the table gives information related to the variation **within** groups. We noted that the term "error" is frequently used for this source of variation, particularly for more general statistical models. This label is most appropriate for experiments in the physical sciences where the observations within a group differ because of measurement error. In business and the biological and social sciences, on the other hand, the within-group variation is often due to the fact that not all firms or plants or people are the same. This sort of variation is not due to errors and is better described as "residual" or "within-group" variation. Nevertheless, we will use the generic label "error" for this source of variation in writing ANOVA tables.

Finally, the Total row in the ANOVA table corresponds to the DATA term in our DATA = FIT + RESIDUAL framework. So, for analysis of variance,

$$\text{DATA} = \text{FIT} + \text{RESIDUAL}$$

translates into

$$\text{Total} = \text{Between Groups} + \text{Within Groups}$$

LOOK BACK
**sum of squares, page 580**

The second column in the software output given in Figure 12.8 is labeled Sum of Squares. As you might expect, each sum of squares is a sum of squared deviations. We use SSG, SSE, and SST for the entries in this column, corresponding to groups, error, and total. Each sum of squares measures a different type of variation. SST measures variation of the data around the overall mean, $x_{ij} - \overline{x}$. Variation of the group means around the overall mean $\overline{x}_i - \overline{x}$ is measured by SSG. Finally, SSE measures variation of each observation around its group mean, $x_{ij} - \bar{x}_i$.

EXAMPLE

**12.9 ANOVA table for worker safety study, continued.** The Sum of Squares column in Figure 12.8 gives the values for the three sums of squares.

$$\text{SST} = 196391.4$$
$$\text{SSG} = 4662.2$$
$$\text{SSE} = 191729.2$$

Verify that SST = SSG + SSE.

This fact is true in general. The total variation is always equal to the among-group variation plus the within-group variation. Note that software output frequently gives many more digits than we need, as in this case. In this example it appears that most of the variation is coming from within groups.

LOOK BACK
**degrees of freedom,**
**page 42**

Associated with each sum of squares is a quantity called the degrees of freedom. Because SST measures the variation of all $N$ observations around the overall mean, its degrees of freedom are DFT $= N - 1$. This is the same as the degrees of freedom for the ordinary sample variance with sample size $N$. Similarly, because SSG measures the variation of the $I$ sample means around the overall mean, its degrees of freedom are DFG $= I - 1$. Finally, SSE is the sum of squares of the deviations $x_{ij} - \bar{x}_i$. Here we have $N$ observations being compared with $I$ sample means, and DFE $= N - I$.

EXAMPLE

**12.10 Degrees of freedom for worker safety study.** In our worker safety example, we have $I = 3$ and $N = 590$. Therefore,

$$\text{DFT} = N - 1 = 590 - 1 = 589$$
$$\text{DFG} = I - 1 = 3 - 1 = 2$$
$$\text{DFE} = N - I = 590 - 3 = 587$$

These are the entries in the df column of Figure 12.8.

Note that the degrees of freedom add in the same way that the sums of squares add. That is, DFT = DFG + DFE.

LOOK BACK
**mean square,**
**page 581**

For each source of variation, the mean square is the sum of squares divided by the degrees of freedom. You can verify this by doing the divisions for the values given on the output in Figure 12.8.

SUMS OF SQUARES, DEGREES OF FREEDOM, AND MEAN SQUARES

**Sums of squares** represent variation present in the data. They are calculated by summing squared deviations. In one-way ANOVA there are three **sources of variation:** groups, error, and total. The sums of squares are related by the formula

$$\text{SST} = \text{SSG} + \text{SSE}$$

Thus, the total variation is composed of two parts, one due to groups and one due to error.

**Degrees of freedom** are related to the deviations that are used in the sums of squares. The degrees of freedom are related in the same way as the sums of squares are:

$$\text{DFT} = \text{DFG} + \text{DFE}$$

To calculate each **mean square,** divide the corresponding sum of squares by its degrees of freedom.

We can use the error mean square to find $s_p$, the pooled estimate of the parameter $\sigma$ of our model. It is true in general that

$$s_p^2 = \text{MSE} = \frac{\text{SSE}}{\text{DFE}}$$

In other words, the error mean square is an estimate of the within-group variance, $\sigma^2$. The estimate of $\sigma$ is therefore the square root of this quantity. So,

$$s_p = \sqrt{\text{MSE}}$$

EXAMPLE

**12.11 MSE for worker safety study.** From the SPSS output in Figure 12.8 we see that the MSE is reported as 326.626. The pooled estimate of $\sigma$ is therefore

$$\begin{aligned} s_p &= \sqrt{\text{MSE}} \\ &= \sqrt{326.626} = 18.07 \end{aligned}$$

## The *F* test

If $H_0$ is true, there are no differences among the group means. The ratio MSG/MSE is a statistic that is approximately 1 if $H_0$ is true and tends to be larger if $H_a$ is true. This is the ANOVA $F$ statistic. In our example, MSG = 2331.116 and MSE = 326.626, so the ANOVA $F$ statistic is

$$F = \frac{\text{MSG}}{\text{MSE}} = \frac{2331.116}{326.626} - 7.137$$

When $H_0$ is true, the $F$ statistic has an $F$ distribution that depends upon two numbers: the *degrees of freedom for the numerator* and the *degrees of freedom for the denominator.* These degrees of freedom are those associated with the mean squares in the numerator and denominator of the $F$ statistic. For one-way ANOVA, the degrees of freedom for the numerator are $\text{DFG} = I - 1$, and the degrees of freedom for the denominator are $\text{DFE} = N - I$. We use the notation $F(I - 1, N - I)$ for this distribution.

The *One-Way ANOVA* applet available on the Web site `www.whfreeman.com/ips` is an excellent way to see how the value of the $F$ statistic and the $P$-value depend upon the variability of the data within the groups and the differences between the means. See Exercises 12.18 and 12.19 for use of this applet.

### THE ANOVA $F$ TEST

To test the null hypothesis in a one-way ANOVA, calculate the ***F* statistic**

$$F = \frac{\text{MSG}}{\text{MSE}}$$

When $H_0$ is true, the $F$ statistic has the $F(I-1, N-I)$ distribution. When $H_a$ is true, the $F$ statistic tends to be large. We reject $H_0$ in favor of $H_a$ if the $F$ statistic is sufficiently large.

The ***P*-value** of the $F$ test is the probability that a random variable having the $F(I-1, N-I)$ distribution is greater than or equal to the calculated value of the $F$ statistic.

Tables of $F$ critical values are available for use when software does not give the $P$-value. Table E in the back of the book contains the $F$ critical values for probabilities $p = 0.100, 0.050, 0.025, 0.010$, and $0.001$. For one-way ANOVA we use critical values from the table corresponding to $I - 1$ degrees of freedom in the numerator and $N - I$ degrees of freedom in the denominator.

EXAMPLE

**12.12 The ANOVA $F$ test for the worker safety study.** In the study of worker safety, we found $F = 7.14$. (Note that it is standard practice to round $F$ statistics to two places after the decimal point.) There were three populations, so the degrees of freedom in the numerator are DFG $= I - 1 = 2$. For this example the degrees of freedom in the denominator are DFE $= N - I = 590 - 3 = 587$. In Table E we first find the column corresponding to 2 degrees of freedom in the numerator. For the degrees of freedom in the denominator, we see that there are entries for 200 and 1000. These entries are very close. To be conservative we use critical values corresponding to 200 degrees of freedom in the denominator since these are slightly larger.

| $p$ | Critical value |
|---|---|
| 0.100 | 2.33 |
| 0.050 | 3.04 |
| 0.025 | 3.76 |
| 0.010 | 4.71 |
| 0.001 | 7.15 |

We have $F = 7.14$. This is very close to the critical value for $P = 0.001$. Using the table, however, we can conclude only that $P < 0.010$ because our calculated $F$ does not exceed 7.15. (Note that the more accurate calculations performed by software indicated that, in fact, $P < 0.001$.) For this example, we reject $H_0$ and conclude that the population means are not all the same.

*When determining the P-value, remember that the F test is always one-sided because any differences among the group means tend to make F large.* The ANOVA $F$ test shares the robustness of the two-sample $t$ test. It is relatively insensitive to moderate non-Normality and unequal variances, especially when the sample sizes are similar.

The following display shows the general form of a one-way ANOVA table with the $F$ statistic. The formulas in the sum of squares column can be used for calculations in small problems. There are other formulas that are more efficient for hand or calculator use, but ANOVA calculations are usually done by computer software.

| Source | Degrees of freedom | Sum of squares | Mean square | $F$ |
|---|---|---|---|---|
| Groups | $I - 1$ | $\sum_{\text{groups}} n_i(\bar{x}_i - \bar{x})^2$ | SSG/DFG | MSG/MSE |
| Error | $N - I$ | $\sum_{\text{groups}} (n_i - 1)s_i^2$ | SSE/DFE | |
| Total | $N - 1$ | $\sum_{\text{obs}} (x_{ij} - \bar{x})^2$ | | |

coefficient of determination

One other item given by some software for ANOVA is worth noting. For an analysis of variance, we define the **coefficient of determination** as

$$R^2 = \frac{\text{SSG}}{\text{SST}}$$

LOOK BACK **multiple correlation squared, page 614**

The coefficient of determination plays the same role as the squared multiple correlation $R^2$ in a multiple regression. We can easily calculate the value from the ANOVA table entries.

EXAMPLE

**12.13 Coefficient of determination for the worker safety study.** The software-generated ANOVA table for the worker safety study is given in Figure 12.8. From that display, we see that SSG = 4662.233 and SST = 196,391.4. The coefficient of determination is

$$R^2 = \frac{\text{SSG}}{\text{SST}} = \frac{4662.233}{196{,}391.4} = 0.02$$

About 2% of the variation in SCI scores is explained by membership in the groups of workers: unskilled workers, skilled workers, and supervisors. The other 98% of the variation is due to worker-to-worker variation within each of the three groups. We can see this in the histograms of Figure 12.3. Each of the groups has a large amount of variation, and there is a substantial amount of overlap in the distributions. *The fact that we have strong evidence ($P < 0.001$)*

*against the null hypothesis that the three population means are not all the same does not tell us that the distributions of values are far apart.*

**USE YOUR KNOWLEDGE**

**12.5** **What's wrong?** For each of the following, explain what is wrong and why.

(a) Within-group variation is the variation in the data due to the differences in the sample means.

(b) The mean squares in an ANOVA table will add, that is, MST = MSG + MSE.

(c) The pooled estimate $s_p$ is a parameter of the ANOVA model.

**12.6** **Determining the critical value of *F*.** For each of the following situations, state how large the *F* statistic needs to be for rejection of the null hypothesis at the 0.05 level.

(a) Compare 5 groups with 3 observations per group.

(b) Compare 5 groups with 6 observations per group.

(c) Compare 5 groups with 9 observations per group.

(d) Summarize what you have learned about *F* distributions from this exercise.

## 12.2 Comparing the Means

### Contrasts

The ANOVA *F* test gives a general answer to a general question: are the differences among observed group means significant? Unfortunately, a small *P*-value simply tells us that the group means are not all the same. It does not tell us specifically which means differ from each other. Plotting and inspecting the means give us some indication of where the differences lie, but we would like to supplement inspection with formal inference.

In the ideal situation, specific questions regarding comparisons among the means are posed before the data are collected. We can answer specific questions of this kind and attach a level of confidence to the answers we give. We now explore these ideas through our worker safety example.

EXAMPLE

**12.14 Reporting the results.** In the worker safety study we compared the SCI scores for three groups of workers: unskilled workers, skilled workers, and supervisors. Let's use $\bar{x}_{UN}$, $\bar{x}_{SK}$, and $\bar{x}_{SU}$ to represent the three sample means and a similar notation for the population means. From Figure 12.7 we see that the three sample means are

$$\bar{x}_{UN} = 70.42, \ \bar{x}_{SK} = 71.21, \ \text{and} \ \bar{x}_{SU} = 80.51$$

The null hypothesis we tested was

$$H_0\colon \mu_{UN} = \mu_{SK} = \mu_{SU}$$

versus the alternative that the three population means are not all the same. We would report these results as $F(2,587) = 7.14$ with $P < 0.001$. (Note that we have given the degrees of freedom for the $F$ statistic in parentheses.) Because the $P$-value is very small, we conclude that the data provide clear evidence that the three population means are not all the same.

Having evidence that the three population means are not the same does not really tell us anything useful. We would really like our analysis to provide us with more specific information. The alternative hypothesis is true if

$$\mu_{\text{UN}} \neq \mu_{\text{SK}}$$

or if

$$\mu_{\text{UN}} \neq \mu_{\text{SU}}$$

or if

$$\mu_{\text{SK}} \neq \mu_{\text{SU}}$$

or if any combination of these statements is true. *When you reject the ANOVA null hypothesis, additional analyses are required to obtain useful results.*

Experts on safety in workplaces would suggest that supervisors face a very different safety environment than the other types of workers. Therefore, a reasonable question to ask is whether or not the mean of the supervisors is different from the mean of the others. We can take this question and translate it into a testable hypothesis.

EXAMPLE

**12.15 An additional comparison of interest.** To compare the supervisors with the other two groups of workers we construct the following null hypothesis:

$$H_{01}: \frac{1}{2}(\mu_{\text{UN}} + \mu_{\text{SK}}) = \mu_{\text{SU}}$$

We could use the two-sided alternative

$$H_{a1}: \frac{1}{2}(\mu_{\text{UN}} + \mu_{\text{SK}}) \neq \mu_{\text{SU}}$$

but we could also argue that the one-sided alternative

$$H_{a1}: \frac{1}{2}(\mu_{\text{UN}} + \mu_{\text{SK}}) < \mu_{\text{SU}}$$

is appropriate for this problem because we expect the unskilled workers and the skilled workers to have a work environment that is less safe than the supervisors' work environment.

In the example above we used $H_{01}$ and $H_{a1}$ to designate the null and alternative hypotheses. The reason for this is that there is a natural additional set of hypotheses that we should examine for this example. We use $H_{02}$ and $H_{a2}$ for these hypotheses.

EXAMPLE

**12.16 Another comparison of interest.** Do the data provide any evidence to support a conclusion that the unskilled workers and the skilled workers have different mean SCI scores? We translate this question into the following null and alternative hypotheses:

$$H_{02}: \mu_{\text{UN}} = \mu_{\text{SK}}$$
$$H_{a2}: \mu_{\text{UN}} \neq \mu_{\text{SK}}$$

Each of $H_{01}$ and $H_{02}$ says that a combination of population means is 0. These combinations of means are called contrasts because the coefficients sum to zero. We use $\psi$, the Greek letter psi, for contrasts among population means. For comparing the supervisors with the other two groups of workers, we have

$$\begin{aligned}\psi_1 &= -\frac{1}{2}(\mu_{\text{UN}} + \mu_{\text{SK}}) + \mu_{\text{SU}} \\ &= (-0.5)\mu_{\text{UN}} + (-0.5)\mu_{\text{SK}} + (1)\mu_{\text{SU}}\end{aligned}$$

and for comparing the unskilled workers with the skilled workers

$$\psi_2 = (1)\mu_{\text{UN}} + (-1)\mu_{\text{SK}}$$

In each case, the value of the contrast is 0 when $H_0$ is true. *Note that we have chosen to define the contrasts so that they will be positive when the alternative of interest (what we expect) is true. Whenever possible, this is a good idea because it makes some computations easier.*

sample contrast

A contrast expresses an effect in the population as a combination of population means. To estimate the contrast, form the corresponding **sample contrast** by using sample means in place of population means. Under the ANOVA assumptions, a sample contrast is a linear combination of independent Normal variables and therefore has a Normal distribution. We can obtain the standard error of a contrast by using the rules for variances. Inference is based on $t$ statistics. Here are the details.

LOOK BACK
**rules for variances, page 282**

CONTRASTS

A **contrast** is a combination of population means of the form

$$\psi = \sum a_i \mu_i$$

where the coefficients $a_i$ sum to 0. The corresponding **sample contrast** is

$$c = \sum a_i \bar{x}_i$$

The **standard error of *c*** is

$$\text{SE}_c = s_p \sqrt{\sum \frac{a_i^2}{n_i}}$$

To test the null hypothesis

$$H_0: \psi = 0$$

use the **$t$ statistic**

$$t = \frac{c}{\text{SE}_c}$$

with degrees of freedom DFE that are associated with $s_p$. The alternative hypothesis can be one-sided or two-sided.

A **level $C$ confidence interval for $\psi$** is

$$c \pm t^* \text{SE}_c$$

where $t^*$ is the value for the $t(\text{DFE})$ density curve with area $C$ between $-t^*$ and $t^*$.

LOOK BACK
**addition rule for means, page 278**

Because each $\bar{x}_i$ estimates the corresponding $\mu_i$, the addition rule for means tells us that the mean $\mu_c$ of the sample contrast $c$ is $\psi$. In other words, $c$ is an unbiased estimator of $\psi$. Testing the hypothesis that a contrast is 0 assesses the significance of the effect measured by the contrast. It is often more informative to estimate the size of the effect using a confidence interval for the population contrast.

EXAMPLE

**12.17 The contrast coefficients.** In our example the coefficients in the contrasts are

$$a_1 = -0.5, \; a_2 = -0.5, \; a_3 = 1 \text{ for } \psi_1$$
$$a_1 = 1, \; a_2 = -1, \; a_3 = 0 \text{ for } \psi_2$$

where the subscripts 1, 2, and 3 correspond to UN, SK, and SU. In each case the sum of the $a_i$ is 0. We look at inference for each of these contrasts in turn.

EXAMPLE

**12.18 Testing the first contrast of interest.** The sample contrast that estimates $\psi_1$ is

$$\begin{aligned} c_1 &= -\frac{1}{2}(\bar{x}_{\text{UN}} + \bar{x}_{\text{SK}}) + \bar{x}_{\text{SU}} \\ &= -(0.5)70.42 - (0.5)71.21 + 80.51 = 9.69 \end{aligned}$$

with standard error

$$\begin{aligned} \text{SE}_{c_1} &= 18.07\sqrt{\frac{(-0.5)^2}{448} + \frac{(-0.5)^2}{91} + \frac{(1)^2}{51}} \\ &= 2.74 \end{aligned}$$

The $t$ statistic for testing $H_{01}: \psi_1 = 0$ versus $H_{a1}: \psi_1 > 0$ is

$$t = \frac{c_1}{\text{SE}_{c_1}} = \frac{9.69}{2.74} = 3.54$$

Because $s_p$ has 587 degrees of freedom, software using the $t(587)$ distribution gives the one-sided $P$-value as $< 0.0001$. If we used Table D, we would conclude that $P < 0.0005$. The $P$-value is small, so there is strong evidence against $H_{01}$.

We have evidence to conclude that the mean SCI score for supervisors is higher than the average of the means for the unskilled workers and the skilled workers. The size of the difference can be described with a confidence interval.

EXAMPLE

**12.19 Confidence interval for the first contrast.** To find the 95% confidence interval for $\psi_1$, we combine the estimate with its margin of error:

$$
\begin{aligned}
c_1 \pm t^* \mathrm{SE}_{c_1} &= 9.69 \pm (1.984)(2.74) \\
&= 9.69 \pm 5.44
\end{aligned}
$$

The 1.984 is a conservative estimate of $t^*$ using 100 degrees of freedom. The interval is (4.25, 15.13). We are 95% confident that the difference is between 4.25 and 15.13 rating points.

We use the same method for the second contrast.

EXAMPLE

**12.20 Testing the second contrast of interest.** The second sample contrast, which compares the unskilled workers with the skilled workers, is

$$
\begin{aligned}
c_2 &= 70.42 - 71.21 \\
&= -0.79
\end{aligned}
$$

with standard error

$$
\begin{aligned}
\mathrm{SE}_{c_2} &= 18.07\sqrt{\frac{(1)^2}{448} + \frac{(-1)^2}{91}} \\
&= 2.08
\end{aligned}
$$

The $t$ statistic for assessing the significance of this contrast is

$$
t = \frac{-0.79}{2.08} = -0.38
$$

The $P$-value for the two-sided alternative is 0.706. The data do not provide us with evidence in favor of a difference in population mean SCI scores between unskilled workers and skilled workers.

Note that we have not concluded that there is *no* difference between the population means. A confidence interval will tell us what values of the population difference are compatible with the data.

EXAMPLE

**12.21 Confidence interval for the second contrast.** To find the 95% confidence interval for $\psi_2$, we combine the estimate with its margin of error:

$$c_2 \pm t^* \text{SE}_{c_2} = -0.79 \pm (1.984)(2.08)$$
$$= -0.79 \pm 4.13$$

The interval is $(-4.92, 3.34)$. With 95% confidence we state that the difference between the population means for these two groups of workers is between $-4.92$ and $3.34$.

SPSS output for contrasts is given in Figure 12.9. The results agree with the calculations that we performed in Examples 12.18 and 12.20 except for minor differences due to roundoff error in our calculations. Note that the output does not give the confidence intervals that we calculated in Examples 12.19 and 12.21. These are easily computed, however, from the contrast estimates and standard errors provided in the output.

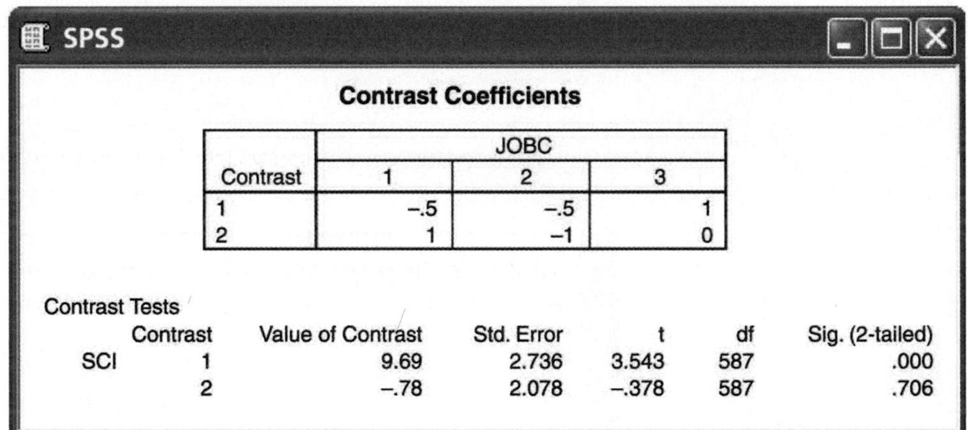
SPSS

Contrast Coefficients

| | JOBC | | |
|---|---|---|---|
| Contrast | 1 | 2 | 3 |
| 1 | -.5 | -.5 | 1 |
| 2 | 1 | -1 | 0 |

Contrast Tests

| | Contrast | Value of Contrast | Std. Error | t | df | Sig. (2-tailed) |
|---|---|---|---|---|---|---|
| SCI | 1 | 9.69 | 2.736 | 3.543 | 587 | .000 |
| | 2 | -.78 | 2.078 | -.378 | 587 | .706 |

**FIGURE 12.9** Software output giving the contrast analysis for the worker safety example.

Some statistical software packages report the test statistics associated with contrasts as $F$ statistics rather than $t$ statistics. These $F$ statistics are the squares of the $t$ statistics described above. As with much statistical software output, $P$-values for significance tests are reported for the two-sided alternative. *If the software you are using gives P-values for the two-sided alternative, and you are using the appropriate one-sided alternative, divide the reported P-value by 2.* In our example, we argued that a one-sided alternative was appropriate for the first contrast. The software reported the $P$-value as 0.000, so we can conclude $P < 0.0005$. Dividing this value by 2 has no effect on the conclusion.

Questions about population means are expressed as hypotheses about contrasts. A contrast should express a specific question that we have in mind when designing the study. When contrasts are formulated before seeing the data, *inference about contrasts is valid whether or not the ANOVA $H_0$ of equality of means is rejected.* Because the $F$ test answers a very general question, it is less powerful than tests for contrasts designed to answer specific questions. Specifying the important questions before the analysis is undertaken enables us to use this powerful statistical technique.

## Multiple comparisons

multiple-comparisons methods

In many studies, specific questions cannot be formulated in advance of the analysis. If $H_0$ is not rejected, we conclude that the population means are indistinguishable on the basis of the data given. On the other hand, if $H_0$ is rejected, we would like to know which pairs of means differ. **Multiple-comparisons methods** address this issue. It is important to keep in mind that multiple-comparisons methods are used only *after rejecting* the ANOVA $H_0$.

EXAMPLE

**12.22 Comparing each pair of groups.** Return once more to the worker safety data with three groups of workers. We can make three comparisons between pairs of means: unskilled workers versus skilled workers, unskilled workers versus supervisors, and skilled workers versus supervisors. We can write a $t$ statistic for each of these pairs. For example, the statistic

$$\begin{aligned} t_{12} &= \frac{\bar{x}_1 - \bar{x}_2}{s_p\sqrt{\dfrac{1}{n_1} + \dfrac{1}{n_2}}} \\ &= \frac{70.42 - 71.21}{18.07\sqrt{\dfrac{1}{448} + \dfrac{1}{91}}} \\ &= -0.38 \end{aligned}$$

compares populations 1 and 2. The subscripts on $t$ specify which groups are compared.

The $t$ statistics for the other two pairs are

$$\begin{aligned} t_{13} &= \frac{\bar{x}_1 - \bar{x}_3}{s_p\sqrt{\dfrac{1}{n_1} + \dfrac{1}{n_3}}} \\ &= \frac{70.42 - 80.51}{18.07\sqrt{\dfrac{1}{448} + \dfrac{1}{51}}} \\ &= -3.78 \end{aligned}$$

and

$$\begin{aligned} t_{23} &= \frac{\bar{x}_2 - \bar{x}_3}{s_p\sqrt{\dfrac{1}{n_2} + \dfrac{1}{n_3}}} \\ &= \frac{71.21 - 80.51}{18.07\sqrt{\dfrac{1}{91} + \dfrac{1}{51}}} \\ &= -2.94 \end{aligned}$$

We performed the first calculation when we analyzed the contrast $\psi_2 = \mu_1 - \mu_2$ in the previous section. These $t$ statistics are very similar to the pooled two-sample $t$ statistic for comparing two population means. The difference is that we now have more than two populations, so each statistic uses the pooled estimator $s_p$ from all groups rather than the pooled estimator from just the two groups being compared. This additional information about the common $\sigma$ increases the power of the tests. The degrees of freedom for all of these statistics are DFE = 587, those associated with $s_p$.

**pooled two-sample $t$ procedures, page 462**

Because we do not have any specific ordering of the means in mind as an alternative to equality, we must use a two-sided approach to the problem of deciding which pairs of means are significantly different.

### MULTIPLE COMPARISONS

To perform a **multiple-comparisons procedure,** compute ***t* statistics** for all pairs of means using the formula

$$t_{ij} = \frac{\bar{x}_i - \bar{x}_j}{s_p\sqrt{\dfrac{1}{n_i} + \dfrac{1}{n_j}}}$$

If

$$|t_{ij}| \geq t^{**}$$

we declare that the population means $\mu_i$ and $\mu_j$ are different. Otherwise, we conclude that the data do not distinguish between them. The value of $t^{**}$ depends upon which multiple-comparisons procedure we choose.

One obvious choice for $t^{**}$ is the upper $\alpha/2$ critical value for the $t$(DFE) distribution. This choice simply carries out as many separate significance tests of fixed level $\alpha$ as there are pairs of means to be compared. The procedure based on this choice is called the **least-significant differences method,** or simply LSD.

LSD method

*LSD has some undesirable properties, particularly if the number of means being compared is large.* Suppose, for example, that there are $I = 20$ groups and we use LSD with $\alpha = 0.05$. There are 190 different pairs of means. If we perform 190 $t$ tests, each with an error rate of 5%, our overall error rate will be unacceptably large. We expect about 5% of the 190 to be significant even if the corresponding population means are the same. Since 5% of 190 is 9.5, we expect 9 or 10 false rejections.

The LSD procedure fixes the probability of a false rejection for each single pair of means being compared. It does not control the overall probability of *some* false rejection among all pairs. Other choices of $t^{**}$ control possible errors in other ways. The choice of $t^{**}$ is therefore a complex problem, and a detailed discussion of it is beyond the scope of this text. Many choices for $t^{**}$ are used in practice. One major statistical package allows selection from a list of over a dozen choices.

Bonferroni method

We will discuss only one of these, called the **Bonferroni method.** Use of this procedure with $\alpha = 0.05$, for example, guarantees that the probability of *any* false rejection among all comparisons made is no greater than 0.05. This is much stronger protection than controlling the probability of a false rejection at 0.05 for *each separate* comparison.

EXAMPLE

**12.23 Applying the Bonferroni method.** We apply the Bonferroni multiple-comparisons procedure with $\alpha = 0.05$ to the data from the worker safety study. The value of $t^{**}$ for this procedure (from software or special tables) is 2.13. Of the statistics $t_{12} = -0.38$, $t_{13} = -3.78$, and $t_{23} = -2.94$ calculated in the beginning of this section, only $t_{13}$ and $t_{23}$ are significant. These two statistics compare supervisors with each of the other two groups.

Of course, we prefer to use software for the calculations.

EXAMPLE

**12.24 Interpreting software output.** The output generated by SPSS for Bonferroni comparisons appears in Figure 12.10. The software uses an asterisk to indicate that the difference in a pair of means is statistically significant. These results agree with the calculations that we performed in Examples 12.22 and 12.23. Note that each comparison is given twice in the output.

SPSS

**Multiple Comparisons**

Dependent Variable: SCI
Bonferroni

| (I) JOBC | (J) JOBC | Mean Difference (I – J) | Std. Error | Sig. | 95% Confidence Interval Lower Bound | 95% Confidence Interval Upper Bound |
|---|---|---|---|---|---|---|
| 1 | 2 | –.78 | 2.078 | 1.000 | –5.77 | 4.20 |
| | 3 | –10.09* | 2.671 | .001 | –16.50 | –3.67 |
| 2 | 1 | .78 | 2.078 | 1.000 | –4.20 | 5.77 |
| | 3 | –9.30* | 3.161 | .010 | –16.89 | –1.71 |
| 3 | 1 | 10.09* | 2.671 | .001 | 3.67 | 16.50 |
| | 2 | 9.30* | 3.161 | .010 | 1.71 | 16.89 |

*. The mean difference is significant at the .05 level.

**FIGURE 12.10** Software output giving the multiple-comparisons analysis for the worker safety example.

The data in the worker safety study provided a clear result: the supervisors have the highest mean SCI score, and we are unable to see a difference between the unskilled workers and the skilled workers. Unfortunately, this type of clarity does not always emerge from a multiple-comparisons analysis. For example, with three groups, we can (*a*) fail to detect a difference between Groups 1 and

2, (*b*) fail to detect a difference between Groups 2 and 3, and (*c*) conclude that Groups 1 and 3 are not the same. *This kind of apparent contradiction points out dramatically the nature of the conclusions of statistical tests of significance.* The conclusion appears to be illogical. If $\mu_1$ is the same as $\mu_2$ and $\mu_2$ is the same as $\mu_3$, doesn't it follow that $\mu_1$ is the same as $\mu_3$? Logically, the answer must be Yes.

Some of the difficulty can be resolved by noting the choice of words used. In describing the inferences, we talk about failing to detect a difference or concluding that two groups are different. In making logical statements, we say things like "is the same as." There is a big difference between the two modes of thought. Statistical tests ask, "Do we have adequate evidence to distinguish two means?" It is not illogical to conclude that we have sufficient evidence to distinguish $\mu_1$ from $\mu_3$, but not $\mu_1$ from $\mu_2$ or $\mu_2$ from $\mu_3$.

simultaneous confidence intervals

One way to deal with these difficulties of interpretation is to give confidence intervals for the differences. The intervals remind us that the differences are not known exactly. We want to give **simultaneous confidence intervals,** that is, intervals for *all* differences among the population means at once. Again, we must face the problem that there are many competing procedures—in this case, many methods of obtaining simultaneous intervals.

## SIMULTANEOUS CONFIDENCE INTERVALS FOR DIFFERENCES BETWEEN MEANS

**Simultaneous confidence intervals** for all differences $\mu_i - \mu_j$ between population means have the form

$$(\bar{x}_i - \bar{x}_j) \pm t^{**} s_p \sqrt{\frac{1}{n_i} + \frac{1}{n_j}}$$

The critical values $t^{**}$ are the same as those used for the multiple-comparisons procedure chosen.

The confidence intervals generated by a particular choice of $t^{**}$ are closely related to the multiple-comparisons results for that same method. If one of the confidence intervals includes the value 0, then that pair of means will not be declared significantly different, and vice versa.

EXAMPLE

**12.25 Interpreting software output, continued.** The SPSS output for the Bonferroni multiple-comparisons procedure given in Figure 12.10 includes the simultaneous 95% confidence intervals. We can see, for example, that the interval for $\mu_1 - \mu_2$ is $-5.77$ to 4.20. The fact that the interval includes 0 is consistent with the fact that we failed to detect a difference between these two means using this procedure. Note that the interval for $\mu_2 - \mu_1$ is also provided. This is not really a new piece of information, because it can be obtained from the other interval by reversing the signs and reversing the order, that is, $-4.20$ to 5.77. So, in fact, we really have only three intervals. Use of the Bonferroni procedure provides us with 95% confidence that *all three* intervals simultaneously contain the true values of the population mean differences.

## Software

We have used SPSS to illustrate the analysis of the worker safety data. Other statistical software gives similar output, and you should be able to read it without any difficulty.

EXAMPLE

**12.26 Customers' opinions of product quality.** Research suggests that customers think that a product is of high quality if it is heavily advertised. An experiment designed to explore this idea collected quality ratings (on a 1 to 7 scale) of a new line of take-home refrigerated entrées based on reading a magazine ad. Three groups were compared. The first group's ad included information that would undermine (U) the expected positive association between quality and advertising; the second group's ad contained information that would affirm (A) the association; and the third group was a control (C).[3] The data are given in Table 12.1. Outputs from SAS, Excel, Minitab, and the TI-83 are given in Figure 12.11.

TABLE 12.1

Quality ratings in three groups

| Group | Quality ratings | | | | | | | | | | | | | | | | | | | |
|---|---|---|---|---|---|---|---|---|---|---|---|---|---|---|---|---|---|---|---|---|
| Undermine | 6 | 5 | 5 | 5 | 4 | 5 | 4 | 6 | 5 | 5 | 5 | 5 | 3 | 3 | 5 | 4 | 5 | 5 | 5 | 4 |
| ($n = 55$) | 5 | 4 | 4 | 5 | 4 | 4 | 4 | 5 | 5 | 5 | 4 | 5 | 5 | 5 | 4 | 4 | 5 | 5 | 4 | 5 |
| | 5 | 4 | 4 | 5 | 4 | 3 | 4 | 5 | 5 | 5 | 3 | 4 | 4 | 4 | 4 | | | | | |
| Affirm | 4 | 6 | 4 | 6 | 5 | 5 | 5 | 6 | 4 | 5 | 5 | 5 | 4 | 6 | 6 | 5 | 5 | 7 | 4 | 6 |
| ($n = 36$) | 6 | 4 | 5 | 4 | 5 | 5 | 6 | 4 | 5 | 5 | 4 | 6 | 4 | 6 | 5 | 5 | | | | |
| Control | 5 | 4 | 5 | 6 | 5 | 7 | 5 | 6 | 7 | 5 | 7 | 5 | 4 | 5 | 4 | 4 | 6 | 6 | 5 | 6 |
| ($n = 36$) | 5 | 5 | 4 | 5 | 5 | 6 | 6 | 6 | 5 | 6 | 6 | 7 | 6 | 6 | 5 | 5 | | | | |

USE YOUR KNOWLEDGE

**12.7** **Why no multiple comparisons?** Any pooled two-sample $t$ problem can be run as a one-way ANOVA with $I = 2$. Explain why it is inappropriate to analyze the data using contrasts or multiple-comparisons procedures in this setting.

**12.8** **Growth of Douglas fir seedlings.** An experiment was conducted to compare the growth of Douglas fir seedlings under three different levels of vegetation control (0%, 50%, and 100%). Forty seedlings were randomized to each level of control. The resulting sample means for stem volume were 50, 75, and 120 cubic centimeters ($cm^3$) respectively with $s_p = 30$ $cm^3$. The researcher hypothesized that the average growth at 50% control would be less than the average of the 0% and 100% levels.

(a) What are the coefficients for testing this contrast?

(b) Perform the test. Do the data provide evidence to support this hypothesis?

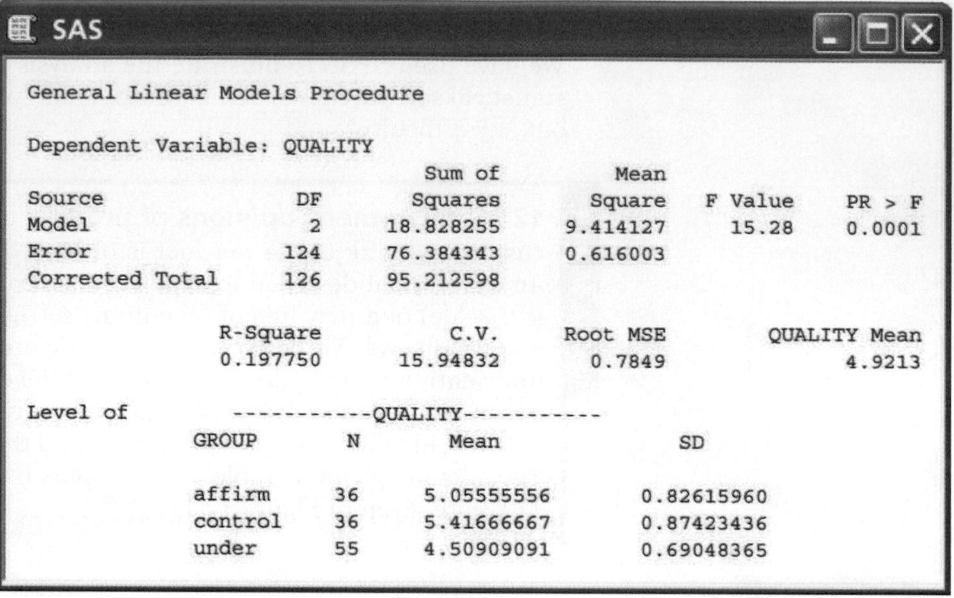

```
SAS

General Linear Models Procedure

Dependent Variable: QUALITY
                                Sum of          Mean
Source                 DF      Squares        Square    F Value    PR > F
Model                   2    18.828255      9.414127      15.28    0.0001
Error                 124    76.384343      0.616003
Corrected Total       126    95.212598

              R-Square          C.V.      Root MSE           QUALITY Mean
              0.197750      15.94832        0.7849                 4.9213

Level of          -----------QUALITY-----------
          GROUP        N         Mean              SD

          affirm      36    5.05555556      0.82615960
          control     36    5.41666667      0.87423436
          under       55    4.50909091      0.69048365
```

Excel

| | A | B | C | D | E | F | G |
|---|---|---|---|---|---|---|---|
| 1 | Anova Single Factor | | | | | | |
| 2 | | | | | | | |
| 3 | SUMMARY | | | | | | |
| 4 | *Groups* | *Count* | *Sum* | *Average* | *Variance* | | |
| 5 | Column 1 | 55 | 248 | 4.509091 | 0.476768 | | |
| 6 | Column 2 | 36 | 182 | 5.055556 | 0.68254 | | |
| 7 | Column 3 | 36 | 195 | 5.416667 | 0.764286 | | |
| 8 | | | | | | | |
| 9 | | | | | | | |
| 10 | ANOVA | | | | | | |
| 11 | *Source of Variation* | *SS* | *df* | *MS* | *F* | *P-value* | *F crit* |
| 12 | Between Groups | 18.82825 | 2 | 9.414127 | 15.28261 | 1.17E-06 | 3.069289 |
| 13 | Within Groups | 76.38434 | 124 | 0.616003 | | | |
| 14 | | | | | | | |
| 15 | Total | 95.2126 | 126 | | | | |

**FIGURE 12.11** SAS, Excel, Minitab, and TI-83 output for the advertising study in Example 12.26. *(continued)*

## Power*

Recall that the power of a test is the probability of rejecting $H_0$ when $H_a$ is in fact true. Power measures how likely a test is to detect a specific alternative. When planning a study in which ANOVA will be used for the analysis, it is important to perform power calculations to check that the sample sizes are adequate to detect differences among means that are judged to be important. Power calculations also help evaluate and interpret the results of studies in

*This section is optional.

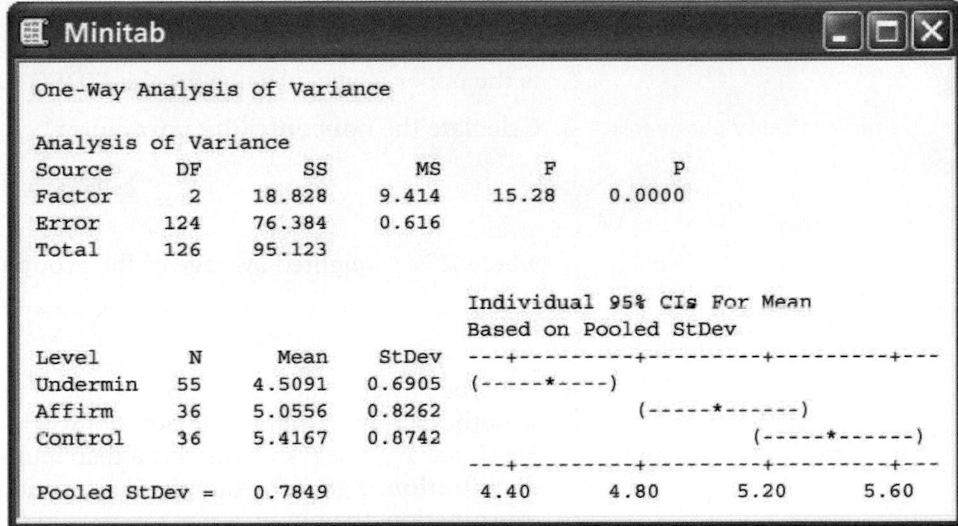

```
Minitab

One-Way Analysis of Variance

Analysis of Variance
Source     DF       SS       MS       F        P
Factor      2   18.828    9.414   15.28   0.0000
Error     124   76.384    0.616
Total     126   95.123
                                  Individual 95% CIs For Mean
                                  Based on Pooled StDev
Level      N     Mean   StDev   ---+---------+---------+---------+---
Undermin  55   4.5091  0.6905   (-----*----)
Affirm    36   5.0556  0.8262             (-----*-----)
Control   36   5.4167  0.8742                       (-----*------)
                                ---+---------+---------+---------+---
Pooled StDev =  0.7849            4.40      4.80      5.20      5.60
```

```
One-way ANOVA
 F=15.28260579
 p=1.1673896E-6
 Factor
  df=2
  SS=18.828255
  MS=9.4141275
 Error
  df=124
  SS=76.3843434
  MS=.61600277
 Sxp=.784858439
```

**FIGURE 12.11** *(Continued)* SAS, Excel, Minitab, and TI-83 output for the advertising study in Example 12.26.

which $H_0$ was not rejected. We sometimes find that the power of the test was so low against reasonable alternatives that there was little chance of obtaining a significant $F$.

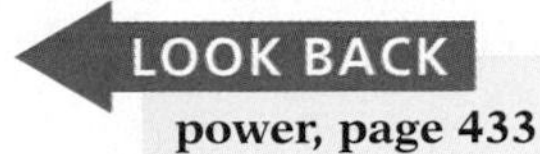


**power, page 433**

In Chapter 7 we found the power for the two-sample $t$ test. One-way ANOVA is a generalization of the two-sample $t$ test, so it is not surprising that the procedure for calculating power is quite similar. Here are the steps that are needed:

1. Specify
   (a) an alternative ($H_a$) that you consider important; that is, values for the true population means $\mu_1, \mu_2, \ldots, \mu_I$;
   (b) sample sizes $n_1, n_2, \ldots, n_I$; usually these will all be equal to the common value $n$;
   (c) a level of significance $\alpha$, usually equal to 0.05; and
   (d) a guess at the standard deviation $\sigma$.

2. Use the degrees of freedom DFG $= I - 1$ and DFE $= N - I$ to find the critical value that will lead to the rejection of $H_0$. This value, which we denote by $F^*$, is the upper $\alpha$ critical value for the $F$(DFG, DFE) distribution.

noncentrality parameter

3. Calculate the **noncentrality parameter**[4]

$$\lambda = \frac{\sum n_i(\mu_i - \overline{\mu})^2}{\sigma^2}$$

where $\overline{\mu}$ is a weighted average of the group means

$$\overline{\mu} = \sum \frac{n_i}{N}\mu_i$$

noncentral $F$ distribution

4. Find the power, which is the probability of rejecting $H_0$ when the alternative hypothesis is true, that is, the probability that the observed $F$ is greater than $F^*$. Under $H_a$, the $F$ statistic has a distribution known as the **noncentral $F$ distribution.** SAS, for example, has a function for this distribution. Using this function, the power is

$$\text{Power} = 1 - \text{PROBF}(F^*, \text{DFG}, \text{DFE}, \lambda)$$

Note that, if the $n_i$ are all equal to the common value $n$, then $\overline{\mu}$ is the ordinary average of the $\mu_i$ and

$$\lambda = \frac{n\sum(\mu_i - \overline{\mu})^2}{\sigma^2}$$

If the means are all equal (the ANOVA $H_0$), then $\lambda = 0$. The noncentrality parameter measures how unequal the given set of means is. Large $\lambda$ points to an alternative far from $H_0$, and we expect the ANOVA $F$ test to have high power. Software makes calculation of the power quite easy, but tables and charts are also available.

EXAMPLE

**12.27 Power of a reading comprehension study.** Suppose that a study on reading comprehension for three different teaching methods has 10 students in each group. How likely is this study to detect differences in the mean responses that would be viewed as important? A previous study performed in a different setting found sample means of 41, 47, and 44, and the pooled standard deviation was 7. Based on these results, we will use $\mu_1 = 41$, $\mu_2 = 47$, $\mu_3 = 44$, and $\sigma = 7$ in a calculation of power. The $n_i$ are equal, so $\overline{\mu}$ is simply the average of the $\mu_i$:

$$\overline{\mu} = \frac{41 + 47 + 44}{3} = 44$$

The noncentrality parameter is therefore

$$\begin{aligned}\lambda &= \frac{n\sum(\mu_i - \overline{\mu})^2}{\sigma^2} \\ &= \frac{(10)[(41-44)^2 + (47-44)^2 + (44-44)^2]}{49} \\ &= \frac{(10)(18)}{49} = 3.67\end{aligned}$$

Because there are three groups with 10 observations per group, DFG = 2 and DFE = 27. The critical value for $\alpha = 0.05$ is $F^* = 3.35$. The power is therefore

$$1 - \text{PROBF}(3.35, 2, 27, 3.67) = 0.3486$$

The chance that we reject the ANOVA $H_0$ at the 5% significance level is only about 35%.

If the assumed values of the $\mu_i$ in this example describe differences among the groups that the experimenter wants to detect, then we would want to use more than 10 subjects per group. Although $H_0$ is assumed to be false, the chance of rejecting it is only about 35%. This chance can be increased to acceptable levels by increasing the sample sizes.

EXAMPLE

**12.28 Changing the sample size.** To decide on an appropriate sample size for the experiment described in the previous example, we repeat the power calculation for different values of $n$, the number of subjects in each group. Here are the results:

| $n$ | DFG | DFE | $F^*$ | $\lambda$ | Power |
|---|---|---|---|---|---|
| 20 | 2 | 57 | 3.16 | 7.35 | 0.65 |
| 30 | 2 | 87 | 3.10 | 11.02 | 0.84 |
| 40 | 2 | 117 | 3.07 | 14.69 | 0.93 |
| 50 | 2 | 147 | 3.06 | 18.37 | 0.97 |
| 100 | 2 | 297 | 3.03 | 36.73 | $\approx 1$ |

With $n = 40$, the experimenters have a 93% chance of rejecting $H_0$ with $\alpha = 0.05$ and thereby demonstrating that the groups have different means. In the long run, 93 out of every 100 such experiments would reject $H_0$ at the $\alpha = 0.05$ level of significance. Using 50 subjects per group increases the chance of finding significance to 97%. With 100 subjects per group, the experimenters are virtually certain to reject $H_0$. The exact power for $n = 100$ is 0.99989. In most real-life situations the additional cost of increasing the sample size from 50 to 100 subjects per group would not be justified by the relatively small increase in the chance of obtaining statistically significant results.

## SECTION 12.2 Summary

**One-way analysis of variance (ANOVA)** is used to compare several population means based on independent SRSs from each population. The populations are assumed to be Normal with possibly different means and the same standard deviation.

To do an analysis of variance, first compute sample means and standard deviations for all groups. Side-by-side boxplots give an overview of the data. Examine Normal quantile plots (either for each group separately or for the residuals) to detect outliers or extreme deviations from Normality. Compute the ratio of the largest to the smallest sample standard deviation. If this ratio is less than 2 and the Normal quantile plots are satisfactory, ANOVA can be performed.

The **null hypothesis** is that the population means are *all* equal. The **alternative hypothesis** is true if there are *any* differences among the population means.

ANOVA is based on separating the total variation observed in the data into two parts: variation **among group means** and variation **within groups.** If the variation among groups is large relative to the variation within groups, we have evidence against the null hypothesis.

An **analysis of variance table** organizes the ANOVA calculations. **Degrees of freedom, sums of squares, and mean squares** appear in the table. The ***F* statistic** and its ***P*-value** are used to test the null hypothesis.

Specific questions formulated before examination of the data can be expressed as **contrasts.** Tests and confidence intervals for contrasts provide answers to these questions.

If no specific questions are formulated before examination of the data and the null hypothesis of equality of population means is rejected, **multiple-comparisons** methods are used to assess the statistical significance of the differences between pairs of means.

The **power** of the $F$ test depends upon the sample sizes, the variation among population means, and the within-group standard deviation.

## CHAPTER 12 Exercises

*For Exercises 12.1 and 12.2, see pages 643 and 644; for Exercises 12.3 and 12.4, see pages 647 and 648; for Exercises 12.5 and 12.6, see page 655; and for Exercises 12.7 and 12.8, see page 665.*

**12.9 Describing the ANOVA model.** For each of the following situations, identify the response variable and the populations to be compared, and give $I$, the $n_i$, and $N$.

(a) A poultry farmer is interested in reducing the cholesterol level in his marketable eggs. He wants to compare two different cholesterol-lowering drugs added to the hen's standard diet as well as an all-vegetarian diet. He assigns 25 of his hens to each of the three treatments.

(b) A researcher is interested in students' opinions regarding an additional annual fee to support non-income-producing varsity sports. Students were asked to rate their acceptance of this fee on a five-point scale. She received 94 responses, of which 31 were from students who attend varsity football or basketball games only, 18 were from students who also attend other varsity competitions, and 45 who did not attend any varsity games.

(c) A professor wants to evaluate the effectiveness of his teaching assistants. In one class period, the 42 students were randomly divided into three equal-sized groups, and each group was taught power calculations from one of the assistants. At the beginning of the next class, each student took a quiz on power calculations, and these scores were compared.

**12.10 Describing the ANOVA model, continued.** For each of the following situations, identify the response variable and the populations to be compared, and give $I$, the $n_i$, and $N$.

(a) A developer of a virtual-reality (VR) teaching tool for the deaf wants to compare the effectiveness of different navigation methods. A total of 40 children were available for the experiment, of which equal numbers were randomly assigned to use a joystick, wand, dancemat, or gesture-based pinch gloves. The time (in seconds) to complete a designed VR path is recorded for each child.

(b) To study the effects of pesticides on birds, an experimenter randomly (and equally) allocated 65 chicks to five diets (a control and four with a different pesticide included). After a month, the calcium content (milligrams) in a 1-centimeter length of bone from each chick was measured.

(c) A university sandwich shop wants to compare the effects of providing free food with a sandwich order on sales. The experiment will be conducted from 11:00 A.M. to 2:00 P.M. for the next 20 weekdays. On each day, customers will be offered one of the following: a free drink, free chips, a free cookie, or nothing. Each option will be offered 5 times.

**12.11 Determining the degrees of freedom.** Refer to Exercise 12.9. For each situation, give the following:

(a) Degrees of freedom for the model, for error, and for the total.

(b) Null and alternative hypotheses.

(c) Numerator and denominator degrees of freedom for the *F* statistic.

**12.12 Determining the degrees of freedom, continued.** Refer to Exercise 12.10. For each situation, give the following:

(a) Degrees of freedom for the model, for error, and for the total.

(b) Null and alternative hypotheses.

(c) Numerator and denominator degrees of freedom for the *F* statistic.

**12.13 Data collection and the interpretation of results.** Refer to Exercise 12.9. For each situation, discuss the method of obtaining the data and how this will affect the extent to which the results can be generalized.

**12.14 Data collection, continued.** Refer to Exercise 12.10. For each situation, discuss the method of obtaining the data and how this will affect the extent to which the results can be generalized.

**12.15 A one-way ANOVA example.** A study compared 4 groups with 8 observations per group. An *F* statistic of 3.33 was reported.

(a) Give the degrees of freedom for this statistic and the entries from Table E that correspond to this distribution.

(b) Sketch a picture of this *F* distribution with the information from the table included.

(c) Based on the table information, how would you report the *P*-value?

(d) Can you conclude that all pairs of means are different? Explain your answer.

**12.16 Calculating the ANOVA *F* test *P*-value.** For each of the following situations, find the degrees of freedom for the *F* statistic and then use Table E to approximate the *P*-value.

(a) Seven groups are being compared with 5 observations per group. The value of the *F* statistic is 2.31.

(b) Five groups are being compared with 11 observations per group. The value of the *F* statistic is 2.83.

(c) Six groups are being compared using 66 total observations. The value of the *F* statistic is 4.08.

**12.17 Calculating the ANOVA *F* test *P*-value, continued.** For each of the following situations, find the *F* statistic and the degrees of freedom. Then draw a sketch of the distribution under the null hypothesis and shade in the portion corresponding to the *P*-value. State how you would report the *P*-value.

(a) Compare 5 groups with 9 observations per group, MSE = 50, and MSG = 127.

(b) Compare 4 groups with 7 observations per group, SSG = 40, and SSE = 153.

**12.18** APPLET **The effect of increased variation within groups.** The *One-Way ANOVA* applet lets you see how the *F* statistic and the *P*-value depend on the variability of the data within groups and the differences among the means.

(a) The black dots are at the means of the three groups. Move these up and down until you get a configuration that gives a *P*-value of about 0.01. What is the value of the *F* statistic?

(b) Now increase the variation within the groups by dragging the mark on the pooled standard error scale to the right. Describe what happens to the *F* statistic and the *P*-value. Explain why this happens.

**12.19** APPLET **The effect of increased variation between groups.** Set the pooled standard error for the *One-Way ANOVA* applet at a middle value. Drag the black dots so that they are approximately equal.

(a) What is the *F* statistic? Give its *P*-value.

(b) Drag the mean of the second group up and the mean of the third group down. Describe the effect on the *F* statistic and its *P*-value. Explain why they change in this way.

**12.20 Calculating the pooled standard deviation.** An experiment was run to compare four groups. The sample sizes were 25, 28, 150, and 21, and the corresponding estimated standard deviations were 42, 38, 20, and 45.

(a) Is it reasonable to use the assumption of equal standard deviations when we analyze these data? Give a reason for your answer.

(b) Give the values of the variances for the four groups.

(c) Find the pooled variance.

(d) What is the value of the pooled standard deviation?

(e) Explain why your answer in part (d) is much closer to the standard deviation for the third group than to any of the other standard deviations.

**12.21 Sleep deprivation and reaction times.** Sleep deprivation experienced by physicians during residency training and the possible negative consequences are of concern to many in the health care community. One study of 33 resident anesthesiologists compared their changes from baseline in reaction times on four tasks.[5] Under baseline conditions, the physicians reported getting an average of 7.04 hours of sleep. While on duty, however, the average was 1.66 hours. For each of the tasks the researchers reported a statistically significant increase in the reaction time when the residents were working in a state of sleep deprivation.

(a) If each task is analyzed separately as the researchers did in their report, what is the appropriate statistical method to use? Explain your answer.

(b) Is it appropriate to use a one-way ANOVA with $I = 4$ to analyze these data? Explain why or why not.

**12.22** CHALLENGE **The two-sample $t$ test and one-way ANOVA.** Refer to the LDL level data in Exercise 7.61 (page 467). Find the two-sample pooled $t$ statistic for comparing men with women. Then formulate the problem as an ANOVA and report the results of this analysis. Verify that $F = t^2$.

**12.23 The importance of recreational sports to college satisfaction.** The National Intramural-Recreational Sports Association (NIRSA) performed a survey to look at the value of recreational sports on college campuses.[6] One of the questions asked each student to rate the importance of recreational sports to college satisfaction and success. Responses were on a 10-point scale with 1 indicating total lack of importance and 10 indicating very high importance. The following table summarizes these results:

| Class | $n$ | Mean score |
|---|---|---|
| Freshman | 724 | 7.6 |
| Sophomore | 536 | 7.6 |
| Junior | 593 | 7.5 |
| Senior | 437 | 7.3 |

(a) To compare the mean scores across classes, what are the degrees of freedom for the ANOVA $F$ statistic?

(b) The MSG = 11.806. If $s_p = 2.16$, what is the $F$ statistic?

(c) Give an approximate (from a table) or exact (from software) $P$-value. What do you conclude?

**12.24 Restaurant ambience and consumer behavior.** There have been numerous studies investigating the effects of restaurant ambience on consumer behavior. A recent study investigated the effects of musical genre on consumer spending.[7] At a single high-end restaurant in England over a 3-week period, there were a total of 141 participants; 49 of them were subjected to background pop music (for example, Britney Spears, Culture Club, and Ricky Martin) while dining, 44 to background classical music (for example, Vivaldi, Handel, and Strauss), and 48 to no background music. For each participant, the total food bill, adjusted for time spent dining, was recorded. The following table summarizes the means and standard deviations:

| Background music | Mean bill | $n$ | $s$ |
|---|---|---|---|
| Classical | 24.130 | 44 | 2.243 |
| Pop | 21.912 | 49 | 2.627 |
| None | 21.697 | 48 | 3.332 |
| Total | 22.531 | 141 | 2.969 |

(a) Plot the means versus the type of background music. Does there appear to be a difference in spending?

(b) Is it reasonable to assume that the variances are equal? Explain.

(c) The $F$ statistic is 10.62. Give the degrees of freedom and either an approximate (from a table) or an exact (from software) $P$-value. What do you conclude?

(d) Refer back to part (a). Without doing any formal analysis, describe the pattern in the means that is likely responsible for your conclusion in part (c).

(e) To what extent do you think the results of this study can be generalized to other settings? Give reasons for your answer.

**12.25 The effects of two stimulant drugs.** An experimenter was interested in investigating the effects of two stimulant drugs (labeled A and B). She divided 20 rats equally into 5 groups (placebo, Drug A low, Drug A high, Drug B low, and Drug B high) and, 20 minutes after injection of the drug, recorded each rat's activity level (higher score is more active). The following table summarizes the results:

| Treatment | $\bar{x}$ | $s^2$ |
|---|---|---|
| Placebo | 14.00 | 8.00 |
| Low A | 15.25 | 12.25 |
| High A | 18.25 | 12.25 |
| Low B | 16.75 | 6.25 |
| High B | 22.50 | 11.00 |

(a) Plot the means versus the type of treatment. Does there appear to be a difference in the activity level? Explain.

(b) Is it reasonable to assume that the variances are equal? Explain your answer, and if reasonable, compute $s_p$.

(c) Give the degrees of freedom for the $F$ statistic.

(d) The $F$ statistic is 4.35. Find the associated $P$-value and state your conclusions.

**12.26** CHALLENGE **Exam accommodations and end-of-term grades.** The Americans with Disabilities Act (ADA) requires that students with learning disabilities (LD) and/or attention deficit disorder (ADD) be given certain accommodations when taking examinations. One study designed to assess the effects of these accommodations examined the relationship between end-of-term grades and the number of accommodations given.[8] The researchers reported the mean grades with sample sizes and standard deviations versus the number of accommodations in a table similar to this:

| Accommodations | Mean grade | $n$ | $s$ |
|---|---|---|---|
| 0 | 2.7894 | 160 | 0.85035 |
| 1 | 2.8605 | 38 | 0.83068 |
| 2 | 2.5757 | 37 | 0.82745 |
| 3 | 2.6286 | 7 | 1.03072 |
| 4 | 2.4667 | 3 | 1.66233 |
| Total | 2.7596 | 245 | 0.85701 |

(a) Plot the means versus the number of accommodations. Is there a pattern evident?

(b) A large number of digits are reported for the means and the standard deviations. Do you think that all of these are necessary? Give reasons for your answer and describe how you would report these results.

(c) Should we pool to obtain an estimate of an assumed standard deviation for these data? Explain your answer and give the pooled estimate if your answer is Yes.

(d) The small numbers of observations with 3 or 4 accommodations lead to estimates that are highly variable in these groups compared with the other groups. Inclusion of groups with relatively few observations in an ANOVA can also lead to low power. We could eliminate these two levels from the analysis or we could combine them with the 37 observations in the group above to form a new group with 2 or more accommodations. Which of these options would you prefer? Give reasons for your answer.

(e) The 245 grades reported in the table were from a sample of 61 students who completed three, four, or five courses during a spring term at one college and who were qualified to receive accommodations. Students in the sample were self-identified, in the sense that they had to request qualification. Even when qualified, some students choose not to request accommodations for some or all of their courses. Based on these facts, would you advise that ANOVA methods be used for these data? Explain your answer. (The authors did not present the results of an ANOVA in their publication.)

(f) To what extent do you think the results of this study can be generalized to other settings? Give reasons for your answer.

(g) Most reasonable approaches to the analysis of these data would conclude that the data fail to provide evidence that the number of accommodations is related to the mean grades. Does this imply that the accommodations are not needed or does it suggest that they are effective? Discuss your answer.

**12.27 Do we experience emotions differently?** Do people from different cultures experience emotions differently? One study designed to examine this question collected data from 416 college students from five different cultures.[9] The participants were asked to record, on a 1 (never) to 7 (always) scale, how much of the time they typically felt eight specific emotions. These were averaged to produce the global emotion score for each participant. Here is a summary of this measure:

| Culture | $n$ | Mean ($s$) |
|---|---|---|
| European American | 46 | 4.39 (1.03) |
| Asian American | 33 | 4.35 (1.18) |
| Japanese | 91 | 4.72 (1.13) |
| Indian | 160 | 4.34 (1.26) |
| Hispanic American | 80 | 5.04 (1.16) |

Note that the convention of giving the standard deviations in parentheses after the means saves a great deal of space in a table such as this.

(a) From the information given, do you think that we need to be concerned that a possible lack of Normality in the data will invalidate the conclusions that we might draw using ANOVA to analyze the data? Give reasons for your answer.

(b) Is it reasonable to used a pooled standard deviation for these data? Why or why not?

(c) The ANOVA $F$ statistic was reported as 5.69. Give the degrees of freedom and either an approximate (from a table) or an exact (from software) $P$-value. Sketch a picture of the $F$ distribution that illustrates the $P$-value. What do you conclude?

(d) Without doing any additional formal analysis, describe the pattern in the means that appears to be responsible for your conclusion in part (c). Are there pairs of means that are quite similar?

**12.28** CHALLENGE **The emotion study, continued.** Refer to the previous exercise. The experimenters also measured emotions in some different ways. For a period of a week, each participant carried a device that sounded an alarm at random times during a 3-hour interval 5 times a day. When the alarm sounded, participants recorded several mood ratings indicating their emotions for the time immediately preceding the alarm. These responses were combined to form two variables: frequency, the number of emotions recorded, expressed as a percent; and intensity, an average of the intensity scores measured on a scale of 0 to 6. At the end of the 1-week experimental period, the subjects were asked to recall the percent of time that they experienced different emotions. This variable was called "recall." Here is a summary of the results:

| Culture | $n$ | Frequency mean ($s$) | Intensity mean ($s$) | Recall mean ($s$) |
|---|---|---|---|---|
| European American | 46 | 82.87 (18.26) | 2.79 (0.72) | 49.12 (22.33) |
| Asian American | 33 | 72.68 (25.15) | 2.37 (0.60) | 39.77 (23.24) |
| Japanese | 91 | 73.36 (22.78) | 2.53 (0.64) | 43.98 (22.02) |
| Indian | 160 | 82.71 (17.97) | 2.87 (0.74) | 49.86 (21.60) |
| Hispanic American | 80 | 92.25 (8.85) | 3.21 (0.64) | 59.99 (24.64) |
| $F$ statistic | | 11.89 | 13.10 | 7.06 |

(a) For each response variable state whether or not it is reasonable to use a pooled standard deviation to analyze these data. Give reasons for your answer.

(b) Give the degrees of freedom for the $F$ statistics and find the associated $P$-values. Summarize what you can conclude from these ANOVA analyses.

(c) Summarize the means, paying particular attention to similarities and differences across cultures and across variables. Include the means from the previous exercise in your summary.

(d) The European American and Asian American subjects were from the University of Illinois, the Japanese subjects were from two universities in Tokyo, the Indian subjects were from eight universities in or near Calcutta, and the Hispanic American subjects were from California State University at Fresno. Participants were paid $25 or an equivalent monetary incentive for the Japanese and Indians. Ads were posted on or near the campuses to recruit volunteers for the study. Discuss how these facts influence your conclusions and the extent to which you would generalize the results.

(e) The percents of female students in the samples were as follows: European American, 83%; Asian American, 67%; Japanese, 63%; Indian, 64%; and Hispanic American, 79%. Use a chi-square test to compare these proportions (see Section 9.2) and discuss how this information influences your interpretation of the results that you have found in this exercise.

**12.29 Storage time and the vitamin C content of bread.** Does bread lose its vitamins when stored? Small loaves of bread were prepared with flour that was fortified with a fixed amount of vitamins. After baking, the vitamin C content of two loaves was measured. Another two loaves were baked at the same time, stored for one day, and then the vitamin C content was measured. In a similar manner, two loaves were stored for three, five, and seven days before measurements were taken. The units are milligrams of vitamin C per hundred grams of flour (mg/100 g).[10] Here are the data:

| Condition | Vitamin C (mg/100 g) | |
|---|---|---|
| Immediately after baking | 47.62 | 49.79 |
| One day after baking | 40.45 | 43.46 |
| Three days after baking | 21.25 | 22.34 |
| Five days after baking | 13.18 | 11.65 |
| Seven days after baking | 8.51 | 8.13 |

(a) Give a table with sample size, mean, standard deviation, and standard error for each condition.

(b) Perform a one-way ANOVA for these data. Be sure to state your hypotheses, the test statistic with degrees of freedom, and the *P*-value.

(c) Summarize the data and the means with a plot. Use the plot and the ANOVA results to write a short summary of your conclusions.

**12.30 Storage time and vitamin C content, continued.** Refer to the previous exercise. Use the Bonferroni or another multiple-comparisons procedure to compare the group means. Summarize the results.

**12.31 Storage time and vitamin A and E content.** Refer to Exercise 12.29. Measurements of the amounts of vitamin A (beta-carotene) and vitamin E in each loaf are given below. Use the analysis of variance method to study the data for each of these vitamins.

| Condition | Vitamin A (mg/100 g) | | Vitamin E (mg/100 g) | |
|---|---|---|---|---|
| Immediately after baking | 3.36 | 3.34 | 94.6 | 96.0 |
| One day after baking | 3.28 | 3.20 | 95.7 | 93.2 |
| Three days after baking | 3.26 | 3.16 | 97.4 | 94.3 |
| Five days after baking | 3.25 | 3.36 | 95.0 | 97.7 |
| Seven days after baking | 3.01 | 2.92 | 92.3 | 95.1 |

**12.32 Storage time and vitamin A and E content, continued.** Refer to the previous exercise.

(a) Explain why it is inappropriate to perform a multiple-comparisons analysis for the vitamin E data.

(b) Perform the Bonferroni or another multiple-comparisons procedure for the vitamin A data and summarize the results.

**12.33** CHALLENGE **Summarizing the results of storage time on vitamin content.** Refer to Exercises 12.29 to 12.32. Write a report summarizing what happens to vitamins A, C, and E after bread is baked. Include appropriate statistical inference results and graphs.

**12.34 Air quality in poultry-processing plants.** The air in poultry-processing plants often contains fungus spores. If the ventilation is inadequate, this can affect the health of the workers. To measure the presence of spores, air samples are pumped to an agar plate, and "colony-forming units (CFUs)" are counted after an incubation period. Here are data from the "kill room" of a plant that slaughters 37,000 turkeys per day, taken at four seasons of the year. The units are CFUs per cubic meter of air.[11]

| Fall | Winter | Spring | Summer |
|---|---|---|---|
| 1231 | 384 | 2105 | 3175 |
| 1254 | 104 | 701 | 2526 |
| 1088 | 97 | 842 | 1090 |

(a) Examine the data using graphs and descriptive measures. How do airborne fungus spores vary with the seasons?

(b) Is the effect of season statistically significant?

**12.35** CHALLENGE **A comparison of tropical flower varieties.** Different varieties of the tropical flower *Heliconia* are fertilized by different species of hummingbirds. Over time, the lengths of the flowers and the form of the hummingbirds' beaks have evolved to match each other. Here are data on the lengths in millimeters of three varieties of these flowers on the island of Dominica:[12]

| *H. bihai* | | | | | | | |
|---|---|---|---|---|---|---|---|
| 47.12 | 46.75 | 46.81 | 47.12 | 46.67 | 47.43 | 46.44 | 46.64 |
| 48.07 | 48.34 | 48.15 | 50.26 | 50.12 | 46.34 | 46.94 | 48.36 |

| *H. caribaea* red | | | | | | | |
|---|---|---|---|---|---|---|---|
| 41.90 | 42.01 | 41.93 | 43.09 | 41.47 | 41.69 | 39.78 | 40.57 |
| 39.63 | 42.18 | 40.66 | 37.87 | 39.16 | 37.40 | 38.20 | 38.07 |
| 38.10 | 37.97 | 38.79 | 38.23 | 38.87 | 37.78 | 38.01 | |

| *H. caribaea* yellow | | | | | | | |
|---|---|---|---|---|---|---|---|
| 36.78 | 37.02 | 36.52 | 36.11 | 36.03 | 35.45 | 38.13 | 37.1 |
| 35.17 | 36.82 | 36.66 | 35.68 | 36.03 | 34.57 | 34.63 | |

Do a complete analysis that includes description of the data and a significance test to compare the mean lengths of the flowers for the three species.

12.36 CHALLENGE **Air quality in poultry-processing plants, continued.** Refer to Exercise 12.34. There is not sufficient information to examine the distributions in detail, but it is not unreasonable to expect count data such as these to be skewed. Reanalyze the data after taking logs of the CFU counts. Summarize your work and compare the results you have found here with what you obtained in Exercise 12.34.

12.37 CHALLENGE **Taking the log of the response variable.** The distributions of the flower lengths in Exercise 12.35 are somewhat skewed. Take logs of the lengths and reanalyze the data. Write a summary of your results and include a comparison with the results you found in Exercise 12.35.

12.38 CHALLENGE **Do poets die young?** According to William Butler Yeats, "She is the Gaelic muse, for she gives inspiration to those she persecutes. The Gaelic poets die young, for she is restless, and will not let them remain long on earth." One study designed to investigate this issue examined the age at death for writers from different cultures and genders.[13] Three categories of writers examined were novelists, poets, and nonfiction writers. The ages at death for female writers in these categories from North America are given in Table 12.2. Most of the writers are from the United States, but Canadian and Mexican writers are also included.

(a) Use graphical and numerical methods to describe the data.

(b) Examine the assumptions necessary for ANOVA. Summarize your findings.

(c) Run the ANOVA and report the results.

(d) Use a contrast to compare the poets with the two other types of writers. Do you think that the quotation from Yeats justifies the use of a one-sided alternative for examining this contrast? Explain your answer.

(e) Use another contrast to compare the novelists with the nonfiction writers. Explain your choice for an alternative hypothesis for this contrast.

(f) Use a multiple-comparisons procedure to compare the three means. How do the conclusions from this approach compare with those using the contrasts?

12.39 **Do isoflavones increase bone mineral density?** Kudzu is a plant that was imported to the United States from Japan and now covers over seven million acres in the South. The plant contains chemicals called isoflavones that have been shown to have beneficial effects on bones. One study used three groups of rats to compare a control group with rats that were fed either a low dose or a high dose of isoflavones from kudzu.[14] One of the outcomes examined was the bone mineral density

**TABLE 12.2**

Age at death for women writers

| Type | Age at death | | | | | | | | | | | | | | |
|---|---|---|---|---|---|---|---|---|---|---|---|---|---|---|---|
| Novels ($n = 67$) | 57 | 90 | 67 | 56 | 90 | 72 | 56 | 90 | 80 | 74 | 73 | 86 | 53 | 72 | 86 |
| | 82 | 74 | 60 | 79 | 80 | 79 | 77 | 64 | 72 | 88 | 75 | 79 | 74 | 85 | 71 |
| | 78 | 57 | 54 | 50 | 59 | 72 | 60 | 77 | 50 | 49 | 73 | 39 | 73 | 61 | 90 |
| | 77 | 57 | 72 | 82 | 54 | 62 | 74 | 65 | 83 | 86 | 73 | 79 | 63 | 72 | 85 |
| | 91 | 77 | 66 | 75 | 90 | 35 | 86 | | | | | | | | |
| Poems ($n = 32$) | 88 | 69 | 78 | 68 | 72 | 60 | 50 | 47 | 74 | 36 | 87 | 55 | 68 | 75 | 78 |
| | 85 | 69 | 38 | 58 | 51 | 72 | 58 | 84 | 30 | 79 | 90 | 66 | 45 | 70 | 48 |
| | 31 | 43 | | | | | | | | | | | | | |
| Nonfiction ($n = 24$) | 74 | 86 | 87 | 68 | 76 | 73 | 63 | 78 | 83 | 86 | 40 | 75 | 90 | 47 | 91 |
| | 94 | 61 | 83 | 75 | 89 | 77 | 86 | 66 | 97 | | | | | | |

in the femur (in grams per square centimeter). Here are the data:

| Treatment | Bone mineral density (g/cm²) | | | | | |
|---|---|---|---|---|---|---|
| Control | 0.228 | 0.207 | 0.234 | 0.220 | 0.217 | 0.228 |
| | 0.209 | 0.221 | 0.204 | 0.220 | 0.203 | 0.219 |
| | 0.218 | 0.245 | 0.210 | | | |
| Low dose | 0.211 | 0.220 | 0.211 | 0.233 | 0.219 | 0.233 |
| | 0.226 | 0.228 | 0.216 | 0.225 | 0.200 | 0.208 |
| | 0.198 | 0.208 | 0.203 | | | |
| High dose | 0.250 | 0.237 | 0.217 | 0.206 | 0.247 | 0.228 |
| | 0.245 | 0.232 | 0.267 | 0.261 | 0.221 | 0.219 |
| | 0.232 | 0.209 | 0.255 | | | |

(a) Use graphical and numerical methods to describe the data.

(b) Examine the assumptions necessary for ANOVA. Summarize your findings.

(c) Run the ANOVA and report the results.

(d) Use a multiple-comparisons method to compare the three groups.

(e) Write a short report explaining the effect of kudzu isoflavones on the femur of the rat.

**12.40 A consumer price promotion study.** If a supermarket product is offered at a reduced price frequently, do customers expect the price of the product to be lower in the future? This question was examined by researchers in a study conducted on students enrolled in an introductory management course at a large midwestern university. For 10 weeks 160 subjects received information about the products. The treatment conditions corresponded to the number of promotions (1, 3, 5, or 7) that were described during this 10-week period. Students were randomly assigned to four groups.[15] Table 12.3 gives the data.

(a) Make a Normal quantile plot for the data in each of the four treatment groups. Summarize the information in the plots and draw a conclusion regarding the Normality of these data.

(b) Summarize the data with a table containing the sample size, mean, standard deviation, and standard error for each group.

(c) Is the assumption of equal standard deviations reasonable here? Explain why or why not.

(d) Run the one-way ANOVA. Give the hypotheses tested, the test statistic with degrees of freedom, and the *P*-value. Summarize your conclusion.

**12.41 A consumer price promotion study, continued.** Refer to the previous exercise. Use the Bonferroni or another multiple-comparisons procedure to compare the group means. Summarize the results and support your conclusions with a graph of the means.

**12.42 Do piano lessons improve the spatial-temporal reasoning of preschool children?** The data in Table 12.4 contain the change in spatial-

TABLE 12.3

Price promotion data

| Number of promotions | Expected price (dollars) | | | | | | | | | |
|---|---|---|---|---|---|---|---|---|---|---|
| 1 | 3.78 | 3.82 | 4.18 | 4.46 | 4.31 | 4.56 | 4.36 | 4.54 | 3.89 | 4.13 |
| | 3.97 | 4.38 | 3.98 | 3.91 | 4.34 | 4.24 | 4.22 | 4.32 | 3.96 | 4.73 |
| | 3.62 | 4.27 | 4.79 | 4.58 | 4.46 | 4.18 | 4.40 | 4.36 | 4.37 | 4.23 |
| | 4.06 | 3.86 | 4.26 | 4.33 | 4.10 | 3.94 | 3.97 | 4.60 | 4.50 | 4.00 |
| 3 | 4.12 | 3.91 | 3.96 | 4.22 | 3.88 | 4.14 | 4.17 | 4.07 | 4.16 | 4.12 |
| | 3.84 | 4.01 | 4.42 | 4.01 | 3.84 | 3.95 | 4.26 | 3.95 | 4.30 | 4.33 |
| | 4.17 | 3.97 | 4.32 | 3.87 | 3.91 | 4.21 | 3.86 | 4.14 | 3.93 | 4.08 |
| | 4.07 | 4.08 | 3.95 | 3.92 | 4.36 | 4.05 | 3.96 | 4.29 | 3.60 | 4.11 |
| 5 | 3.32 | 3.86 | 4.15 | 3.65 | 3.71 | 3.78 | 3.93 | 3.73 | 3.71 | 4.10 |
| | 3.69 | 3.83 | 3.58 | 4.08 | 3.99 | 3.72 | 4.41 | 4.12 | 3.73 | 3.56 |
| | 3.25 | 3.76 | 3.56 | 3.48 | 3.47 | 3.58 | 3.76 | 3.57 | 3.87 | 3.92 |
| | 3.39 | 3.54 | 3.86 | 3.77 | 4.37 | 3.77 | 3.81 | 3.71 | 3.58 | 3.69 |
| 7 | 3.45 | 3.64 | 3.37 | 3.27 | 3.58 | 4.01 | 3.67 | 3.74 | 3.50 | 3.60 |
| | 3.97 | 3.57 | 3.50 | 3.81 | 3.55 | 3.08 | 3.78 | 3.86 | 3.29 | 3.77 |
| | 3.25 | 3.07 | 3.21 | 3.55 | 3.23 | 2.97 | 3.86 | 3.14 | 3.43 | 3.84 |
| | 3.65 | 3.45 | 3.73 | 3.12 | 3.82 | 3.70 | 3.46 | 3.73 | 3.79 | 3.94 |

**TABLE 12.4**

**Piano lesson data**

| Lessons | Scores | | | | | | | | | |
|---|---|---|---|---|---|---|---|---|---|---|
| Piano | 2 | 5 | 7 | −2 | 2 | 7 | 4 | 1 | 0 | 7 |
| | 3 | 4 | 3 | 4 | 9 | 4 | 5 | 2 | 9 | 6 |
| | 0 | 3 | 6 | −1 | 3 | 4 | 6 | 7 | −2 | 7 |
| | −3 | 3 | 4 | 4 | | | | | | |
| Singing | 1 | −1 | 0 | 1 | −4 | 0 | 0 | 1 | 0 | −1 |
| Computer | 0 | 1 | 1 | −3 | −2 | 4 | −1 | 2 | 4 | 2 |
| | 2 | 2 | −3 | −3 | 0 | 2 | 0 | −1 | 3 | −1 |
| None | 5 | −1 | 7 | 0 | 4 | 0 | 2 | 1 | −6 | 0 |
| | 2 | −1 | 0 | −2 | | | | | | |

temporal reasoning (after treatment minus before treatment) of 34 children who took piano lessons, 10 who took singing lessons, 20 who had some computer instruction, and 14 who received no extra lessons.

(a) Make a table giving the sample size, the mean, the standard deviation, and the standard error for each group.

(b) Analyze the data using one-way analysis of variance. State the null and alternative hypotheses, the test statistic with degrees of freedom, the *P*-value, and your conclusion.

**12.43 The piano lessons study, continued.** Refer to the previous exercise. Use the Bonferroni or another multiple-comparisons procedure to compare the group means. Summarize the results and support your conclusions with a graph of the means.

**12.44 More on the piano lessons study.** The researchers in Exercise 12.42 based their research on a biological argument for a causal link between music and spatial-temporal reasoning. Therefore, it is natural to test the contrast that compares the mean of the piano lesson group with the average of the three other means. Construct this contrast, perform the significance test, and summarize the results.

**12.45 How long should an infant be breast-fed?** Recommendations regarding how long infants in developing countries should be breast-fed are controversial. If the nutritional quality of the breast milk is inadequate because the mothers are malnourished, then there is risk of inadequate nutrition for the infant. On the other hand, the introduction of other foods carries the risk of infection from contamination. Further complicating the situation is the fact that companies that produce infant formulas and other foods benefit when these foods are consumed by large numbers of customers. One question related to this controversy concerns the amount of energy intake for infants who have other foods introduced into the diet at different ages. Part of one study compared the energy intakes, measured in kilocalories per day (kcal/d), for infants who were breast-fed exclusively for 4, 5, or 6 months.[16] Here are the data:

| Breast-fed for: | Energy intake (kcal/d) | | | | | | |
|---|---|---|---|---|---|---|---|
| 4 months | 499 | 620 | 469 | 485 | 660 | 588 | 675 |
| | 517 | 649 | 209 | 404 | 738 | 628 | 609 |
| | 617 | 704 | 558 | 653 | 548 | | |
| 5 months | 490 | 395 | 402 | 177 | 475 | 617 | 616 |
| | 587 | 528 | 518 | 370 | 431 | 518 | 639 |
| | 368 | 538 | 519 | 506 | | | |
| 6 months | 585 | 647 | 477 | 445 | 485 | 703 | 528 |
| | 465 | | | | | | |

(a) Make a table giving the sample size, mean, and standard deviation for each group of infants. Is it reasonable to pool the variances?

(b) Run the analysis of variance. Report the *F* statistic with its degrees of freedom and *P*-value. What do you conclude?

**12.46 Breast-feeding study, continued.** Refer to the previous exercise.

(a) Examine the residuals. Is the Normality assumption reasonable for these data?

(b) Explain why you do not need to use a multiple-comparisons procedure for these data.

**12.47 Exercise and healthy bones.** Many studies have suggested that there is a link between exercise and healthy bones. Exercise stresses the bones and this causes them to get stronger. One study examined the effect of jumping on the bone density of growing rats.[17] There were three treatments: a control with no jumping, a low-jump condition (the jump height was 30 centimeters), and a high-jump condition (60 centimeters). After 8 weeks of 10 jumps per day, 5 days per week, the bone density of the rats (expressed in mg/cm$^3$) was measured. Here are the data:

| Group | Bone density (mg/cm$^3$) | | | | | | | | | |
|---|---|---|---|---|---|---|---|---|---|---|
| Control | 611 | 621 | 614 | 593 | 593 | 653 | 600 | 554 | 603 | 569 |
| Low jump | 635 | 605 | 638 | 594 | 599 | 632 | 631 | 588 | 607 | 596 |
| High jump | 650 | 622 | 626 | 626 | 631 | 622 | 643 | 674 | 643 | 650 |

(a) Make a table giving the sample size, mean, and standard deviation for each group of rats. Is it reasonable to pool the variances?

(b) Run the analysis of variance. Report the $F$ statistic with its degrees of freedom and $P$-value. What do you conclude?

**12.48 Exercise and healthy bones, continued.** Refer to the previous exercise.

(a) Examine the residuals. Is the Normality assumption reasonable for these data?

(b) Use the Bonferroni or another multiple-comparisons procedure to determine which pairs of means differ significantly. Summarize your results in a short report. Be sure to include a graph.

**12.49 Does the type of cooking pot affect iron content?** Iron-deficiency anemia is the most common form of malnutrition in developing countries, affecting about 50% of children and women and 25% of men. Iron pots for cooking foods had traditionally been used in many of these countries, but they have been largely replaced by aluminum pots, which are cheaper and lighter. Some research has suggested that food cooked in iron pots will contain more iron than food cooked in other types of pots. One study designed to investigate this issue compared the iron content of some Ethiopian foods cooked in aluminum, clay, and iron pots.[18] One of the foods was *yesiga wet'*, beef cut into small pieces and prepared with several Ethiopian spices. The iron content of four samples of *yesiga wet'* cooked in each of the three types of pots is given below. The units are milligrams of iron per 100 grams of cooked food.

| Type of pot | Iron (mg/100 g food) | | | |
|---|---|---|---|---|
| Aluminum | 1.77 | 2.36 | 1.96 | 2.14 |
| Clay | 2.27 | 1.28 | 2.48 | 2.68 |
| Iron | 5.27 | 5.17 | 4.06 | 4.22 |

(a) Make a table giving the sample size, mean, and standard deviation for each type of pot. Is it reasonable to pool the variances? Note that with the small sample sizes in this experiment, we expect a large amount of variability in the sample standard deviations.

(b) Run the analysis of variance. Report the $F$ statistic with its degrees of freedom and $P$-value. What do you conclude?

**12.50 The cooking pot study, continued.** Refer to the previous exercise.

(a) Examine the residuals. Is the Normality assumption reasonable for these data?

(b) Use the Bonferroni or another multiple-comparisons procedure to determine which pairs of means differ significantly. Summarize your results in a short report. Be sure to include a graph.

**12.51 A comparison of different types of scaffold material.** One way to repair serious wounds is to insert some material as a scaffold for the body's repair cells to use as a template for new tissue. Scaffolds made from extracellular material (ECM) are particularly promising for this purpose. Because they are made from biological material, they serve as an effective scaffold and are then resorbed. Unlike biological material that includes cells, however, they do not trigger tissue rejection reactions in the body. One study compared 6 types of scaffold material.[19] Three of these were ECMs and the other three were made of inert materials. There were three mice used per scaffold type. The response measure was the percent of glucose phosphated isomerase (Gpi) cells in the region of the wound. A large value is good, indicating that there are many bone marrow cells sent by the body to repair the tissue.

| Material | Gpi (%) | | |
|---|---|---|---|
| ECM1 | 55 | 70 | 70 |
| ECM2 | 60 | 65 | 65 |
| ECM3 | 75 | 70 | 75 |
| MAT1 | 20 | 25 | 25 |
| MAT2 | 5 | 10 | 5 |
| MAT3 | 10 | 15 | 10 |

(a) Make a table giving the sample size, mean, and standard deviation for each of the six types of material. Is it reasonable to pool the variances? Note that the sample sizes are small and the data are rounded.

(b) Run the analysis of variance. Report the $F$ statistic with its degrees of freedom and $P$-value. What do you conclude?

**12.52 A comparison of different types of scaffold material, continued.** Refer to the previous exercise.

(a) Examine the residuals. Is the Normality assumption reasonable for these data?

(b) Use the Bonferroni or another multiple-comparisons procedure to determine which pairs of means differ significantly. Summarize your results in a short report. Be sure to include a graph.

(c) Use a contrast to compare the three ECM materials with the three other materials. Summarize your conclusions. How do these results compare with those that you obtained from the multiple-comparisons procedure in part (b)?

**12.53 Two contrasts of interest for the stimulant study.** Refer to Exercise 12.25 (page 673). There are two comparisons of interest to the experimenter. They are (1) Placebo versus the average of the 2 low-dose treatments; and (2) the difference between High A and Low A versus the difference between High B and Low B.

(a) Express each contrast in terms of the means ($\mu$'s) of the treatments.

(b) Give estimates with standard errors for each of the contrasts.

(c) Perform the significance tests for the contrasts. Summarize the results of your tests and your conclusions.

**12.54 A dandruff study.** Analysis of variance methods are often used in clinical trials where the goal is to assess the effectiveness of one or more treatments for a particular medical condition. One such study compared three treatments for dandruff and a placebo. The treatments were 1% pyrithione zinc shampoo (PyrI), the same shampoo but with instructions to shampoo two times (PyrII), 2% ketoconazole shampoo (Keto), and a placebo shampoo (Placebo). After six weeks of treatment, eight sections of the scalp were examined and given a score that measured the amount of scalp flaking on a 0 to 10 scale. The response variable was the sum of these eight scores. An analysis of the baseline flaking measure indicated that randomization of patients to treatments was successful in that no differences were found between the groups. At baseline there were 112 subjects in each of the three treatment groups and 28 subjects in the Placebo group. During the clinical trial, 3 dropped out from the PyrII group and 6 from the Keto group. No patients dropped out of the other two groups. The data are given in the DANDRUFF data set described in the Data Appendix.

(a) Find the mean, standard deviation, and standard error for the subjects in each group. Summarize these, along with the sample sizes, in a table and make a graph of the means.

(b) Run the analysis of variance on these data. Write a short summary of the results and your conclusion. Be sure to include the hypotheses tested, the test statistic with degrees of freedom, and the $P$-value.

**12.55 The dandruff study, continued.** Refer to the previous exercise.

(a) Plot the residuals versus case number (the first variable in the data set). Describe the plot. Is there any pattern that would cause you to question the assumption that the data are independent?

(b) Examine the standard deviations for the four treatment groups. Is there a problem with the assumption of equal standard deviations for ANOVA in this data set? Explain your answer.

(c) Prepare Normal quantile plots for each treatment group. What do you conclude from these plots?

(d) Obtain the residuals from the analysis of variance and prepare a Normal quantile plot of these. What do you conclude?

**12.56 Comparing each pair of dandruff treatments.** Refer to Exercise 12.54. Use the Bonferroni or another multiple-comparisons procedure that your software provides to compare the individual group means in the dandruff study. Write a short summary of your conclusions.

12.57 **Testing several contrasts from the dandruff study.** Refer to Exercise 12.54. There are several natural contrasts in this experiment that describe comparisons of interest to the experimenters. They are (1) Placebo versus the average of the three treatments; (2) Keto versus the average of the two Pyr treatments; and (3) PyrI versus PyrII.

(a) Express each of these three contrasts in terms of the means ($\mu$'s) of the treatments.

(b) Give estimates with standard errors for each of the contrasts.

(c) Perform the significance tests for the contrasts. Summarize the results of your tests and your conclusions.

12.58 CHALLENGE **Changing the response variable of the storage time study.** Refer to Exercise 12.29 (page 675), where we studied the effects of storage on the vitamin C content of bread. In this experiment 64 mg of vitamin C per 100 g of flour was added to the flour that was used to make each loaf.

(a) Convert the vitamin C amounts (mg/100 g) to percents of the amounts originally in the loaves by dividing the amounts in Exercise 12.29 by 64 and multiplying by 100. Calculate the transformed means, standard deviations, and standard errors and summarize them with the sample sizes in a table.

(b) Explain how you could have calculated the table entries directly from the table you gave in part (a) of Exercise 12.29.

(c) Analyze the percents using analysis of variance. Compare the test statistic, degrees of freedom, $P$-value, and conclusion you obtain here with the corresponding values that you found in Exercise 12.29.

12.59 **More on changing the response variable of the storage time study.** Refer to the previous exercise and Exercise 12.31 (page 675). The flour used to make the loaves contained 5 mg of vitamin A per 100 g of flour and 100 mg of vitamin E per 100 g of flour. Summarize the effects of transforming the data to percents for all three vitamins.

12.60 CHALLENGE **Linear transformation of the response variable.** Refer to the previous exercise. Can you suggest a general conclusion regarding what happens to the test statistic, degrees of freedom, $P$-value, and conclusion when you perform analysis of variance on data that have been transformed by multiplying the raw data by a constant and then adding another constant? (That is, if $y$ is the original data, we analyze $y^*$, where $y^* = a + by$ and $a$ and $b \neq 0$ are constants.)

12.61 CHALLENGE **Comparing three levels of reading comprehension instruction.** A study of reading comprehension in children compared three methods of instruction.[20] The three methods of instruction are called Basal, DRTA, and Strategies. As is common in such studies, several pretest variables were measured before any instruction was given. One purpose of the pretest was to see if the three groups of children were similar in their comprehension skills. The READING data set described in the Data Appendix gives two pretest measures that were used in this study. Use one-way ANOVA to analyze these data and write a summary of your results.

12.62 CHALLENGE **More on the reading comprehension study.** In the study described in the previous exercise, Basal is the traditional method of teaching, while DRTA and Strategies are two innovative methods based on similar theoretical considerations. The READING data set includes three response variables that the new methods were designed to improve. Analyze these variables using ANOVA methods. Be sure to include multiple comparisons or contrasts as needed. Write a report summarizing your findings.

12.63 CHALLENGE **More on the price promotion study.** Refer to the price promotion study that we examined in Exercise 12.40 (page 677). The explanatory variable in this study is the number of price promotions in a 10-week period, with possible values of 1, 3, 5, and 7. When using analysis of variance, we treat the explanatory variable as categorical. An alternative analysis is to use simple linear regression. Perform this analysis and summarize the results. Plot the residuals from the regression model versus the number of promotions. What do you conclude?

12.64 **Overall standard deviation versus the pooled standard deviation.** The last line of the summary table given in Exercise 12.26 (page 673) gives the mean and the standard deviation for all of the data combined. Compare this standard deviation with the pooled standard deviation that you would use as an estimate of the model standard deviation. Explain why you would expect this standard deviation to be larger than the pooled standard deviation.

**12.65** **Search the Internet.** Search the Internet or your library to find a study that is interesting to you and that used one-way ANOVA to analyze the data. First describe the question or questions of interest and then give the details of how ANOVA was used to provide answers. Be sure to include how the study authors examined the assumptions for the analysis. Evaluate how well the authors used ANOVA in this study. If your evaluation finds the analysis deficient, make suggestions for how it could be improved.

**12.66** **A power calculation exercise (optional).** In Example 12.27 (page 668) the power calculation indicated that there was a fairly small chance of detecting the alternative given. Redo the calculations for the alternative $\mu_1 = 40$, $\mu_2 = 47$, and $\mu_3 = 43$. Do you think that the choice of 10 students per treatment is adequate for this alternative?

**12.67** **Planning another emotions study.** Scores on an emotional scale were compared for five different cultures in Exercise 12.27 (page 674). Suppose that you are planning a new study using the same outcome variable. Your study will use European American, Asian American, and Hispanic American students from a large university.

(a) Explain how you would select the students to participate in your study.

(b) (Optional) Use the data from Exercise 12.27 to perform power calculations to determine sample sizes for your study.

(c) Write a report that could be understood by someone with limited background in statistics and that describes your proposed study and why you think it is likely that you will obtain interesting results.

**12.68** **Planning another isoflavone study.** Exercise 12.39 (page 676) gave data for a bone health study that examined the effect of isoflavones on rat bone mineral density. In this study there were three groups. Controls received a placebo, and the other two groups received either a low or a high dose of isoflavones from kudzu. You are planning a similar study of a new kind of isoflavone. Use the results of the study described in Exercise 12.39 to plan your study. Write a proposal explaining why your study should be funded.

**12.69** **Planning another restaurant ambience study.** Exercise 12.24 (page 672) gave data for a study that examined the effect of background music on total food spending at a high-end restaurant. You are planning a similar study but intend to look at total food spending at a more casual restaurant. Use the results of the study described in Exercise 12.24 to plan your study.

CHAPTER

# Two-Way Analysis of Variance

13

Can the consumption of red palm oil decrease the occurrence and severity of malaria in children? See Example 13.3 for more details.



## Introduction

The $t$ procedures of Chapter 7 compare the means of two populations. We generalized these procedures in Chapter 12 so that we could compare the means of several populations. In this chapter, we move from one-way ANOVA to two-way ANOVA. Two-way ANOVA compares the means of populations that can be classified in two ways or the mean responses in two-factor experiments.

Many of the key concepts are similar to those of one-way ANOVA, but the presence of more than one classification factor also introduces some new ideas. We once more assume that the data are approximately Normal and that groups may have different means but the same standard deviation; we again pool to estimate the variance; and we again use $F$ statistics for significance tests. The major difference between one-way and two-way ANOVA is in the FIT part of the model. We will carefully study this term, and we will find much that is both new and useful. This will allow us to address comparisons such as the following:

- Can an increase in the consumption of red palm oil reduce the occurrence and severity of malaria in both male and female children living in Nigeria?
- What effects do the floral morphologies of male and female jack-in-the-pulpit plants have on herbivory?
- Do calcium supplements prevent bone loss in elderly people with and without adequate vitamin D?

## 13.1 The Two-Way ANOVA Model

We begin with a discussion of the advantages of the two-way ANOVA design and illustrate these with some examples. Then we discuss the model and the assumptions.

### Advantages of two-way ANOVA

In one-way ANOVA, we classify populations according to one categorical variable, or factor. In the two-way ANOVA model, there are two factors, each with its own number of levels. When we are interested in the effects of two factors, a two-way design offers great advantages over several single-factor studies. Several examples will illustrate these advantages.

EXAMPLE

**13.1 Design 1: Choosing the best magazine layout and cover.** In Example 12.1, a magazine publisher wants to compare three different magazine layouts. To do this, she plans to randomly assign the three design layouts equally among 60 supermarkets. The number of magazines sold during a one-week period is the outcome variable.

Now suppose a second experiment is planned for the following week to compare four different covers for the magazine. A similar experimental design will be used, with the four covers randomly assigned among the same 60 supermarkets.

Here is a picture of the design of the first experiment with the sample sizes:

| Layout | $n$ |
|---|---|
| 1 | 20 |
| 2 | 20 |
| 3 | 20 |
| Total | 60 |

And this represents the second experiment:

| Cover | $n$ |
|---|---|
| 1 | 15 |
| 2 | 15 |
| 3 | 15 |
| 4 | 15 |
| Total | 60 |

In the first experiment 20 stores were assigned to each level of the factor for a total of 60 stores. In the second experiment 15 stores were assigned to each level of the factor for a total of 60 stores. The total amount of time for the two experiments is two weeks. Each experiment will be analyzed using one-way ANOVA. The factor in the first experiment is magazine layout with three levels, and the factor in the second experiment is magazine cover with four levels. Let's now consider combining the two experiments into one.

EXAMPLE

**13.2 Design 2: Choosing the best magazine layout and cover.** Suppose we use a two-way approach for the magazine design problem. There are two factors, layout and cover. Since layout has three levels and cover has four levels, this is a $3 \times 4$ design. This gives a total of 12 possible combinations of layout and cover. With a total of 60 stores, we could assign each combination of layout and cover to 5 stores. The number of magazines sold during a one-week period is the outcome variable.

Here is a picture of the two-way design with the sample sizes:

| | Cover | | | | |
|---|---|---|---|---|---|
| Layout | 1 | 2 | 3 | 4 | Total |
| 1 | 5 | 5 | 5 | 5 | 20 |
| 2 | 5 | 5 | 5 | 5 | 20 |
| 3 | 5 | 5 | 5 | 5 | 20 |
| Total | 15 | 15 | 15 | 15 | 60 |

cell

Each combination of the factors in a two-way design corresponds to a **cell.** The $3 \times 4$ ANOVA for the magazine experiment has twelve cells, each corresponding to a particular combination of layout and cover.

With the two-way design for layout, notice that we have 20 stores assigned to each level, the same as what we had for the one-way experiment for layout alone. Similarly, there are 15 stores assigned to each level of cover. Thus, the two-way design gives us the same amount of information for estimating the sales for each level of each factor as we had with the two one-way designs. The difference is that we can collect all of the information in only one week. By combining the two factors into one experiment, we have increased our efficiency by reducing the amount of data to be collected by half.

EXAMPLE

**13.3 Can increased palm oil consumption reduce malaria?** Malaria is a serious health problem causing an estimated 2.7 million deaths per year, mostly in Africa.[1] Some research suggests that vitamin A can reduce episodes of malaria in young children. Red palm oil is a good source of vitamin A and is readily available in Nigeria, a country where malaria accounts for about 30% of the deaths of young children. Can an increase in the consumption of red palm oil reduce the occurrence and severity of malaria in this region?[2]

To design a study to answer this question we first need to determine an appropriate target group. Since malaria is a serious problem for young children, we will concentrate on children who are 2 to 5 years of age. A supplement will be prepared that contains either a placebo, a low dose of red palm oil, or a high dose of red palm oil. Because boys and girls may differ in exposure to malaria and the response to the red palm oil supplement, our design should also take gender into account. Let's consider a two-way ANOVA for this study.

EXAMPLE

**13.4 Implementing the two-way ANOVA design.** The factors for our two-way ANOVA are red palm oil with three levels and gender with two levels. There are $3 \times 2 = 6$ cells in our study. Suppose we recruit 75 boys and 75 girls. We will then randomly assign 25 of each gender to each of the red palm oil levels. The outcome variable will be the amount of an acute-phase protein in the blood that measures the severity of infection.

Here is a table that summarizes the design:

| | Gender | | |
|---|---|---|---|
| **Red palm oil** | **Girls** | **Boys** | **Total** |
| Placebo | 25 | 25 | 50 |
| Low dose | 25 | 25 | 50 |
| High dose | 25 | 25 | 50 |
| Total | 75 | 75 | 150 |

This example illustrates another advantage of two-way designs. Although we are primarily interested in the possible benefit of red palm oil, we included gender in the design because we thought that there might be differences between the boys and the girls. Consider an alternative one-way design where we assign 150 children to the three levels of red palm oil and ignore gender. With this design we will have the same number of children at each of the red palm oil levels, so in this way it is similar to our two-way design. However, suppose that there are, in fact, differences between boys and girls. In this case, the one-way ANOVA would assign this variation to the RESIDUAL (within groups) part of the model. In the two-way ANOVA, gender is included as a factor, and therefore this variation is included in the FIT part of the model. Whenever we can move variation from RESIDUAL to FIT, we reduce the $\sigma$ of our model and increase the power of our tests.

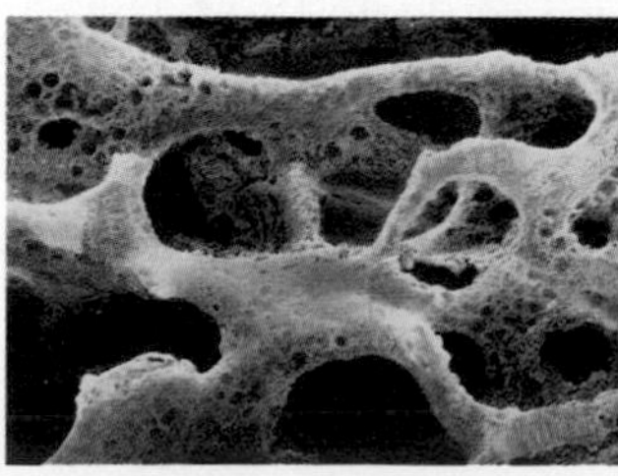

EXAMPLE

**13.5 Vitamin D and osteoporosis.** Osteoporosis is a disease primarily of the elderly. People with osteoporosis have low bone mass and an increased risk of bone fractures. Over 10 million people in the United States, 1.4 million Canadians, and many millions throughout the world have this disease. Adequate calcium in the diet is necessary for strong bones, but vitamin D is also needed for the body to efficiently use calcium. High doses of calcium in the diet will not prevent osteoporosis unless there is adequate vitamin D. Exposure of the skin to the ultraviolet rays in sunlight enables our bodies to make vitamin D. However, elderly people often avoid sunlight, and in northern areas such as Canada, there is not sufficient ultraviolet light to make vitamin D, particularly in the winter months.

Suppose we wanted to see if calcium supplements will increase bone mass (or prevent a decrease in bone mass) in an elderly Canadian population. Because of the vitamin D complication we will make this a factor in our design.

EXAMPLE

**13.6 Designing the osteoporosis study.** We will use a $2 \times 2$ design for our osteoporosis study. The two factors are calcium and vitamin D. The levels of each factor will be zero (placebo) and an amount that is expected to be adequate, 800 mg/day for calcium and 300 international units per day (IU/day) for vitamin D. Women between the ages of 70 and 80 will be recruited as subjects. Bone mineral density (BMD) will be measured at the beginning of the study, and supplements will be taken for one year. The change in BMD over the one-year period is the outcome variable. We expect a dropout rate of 20% and we would like to have about 20 subjects providing data in each group at the end of the study. We will therefore recruit 100 subjects and randomly assign 25 to each treatment combination.

Here is a table that summarizes the design with the sample sizes at baseline:

| | Vitamin D | | |
|---|---|---|---|
| Calcium | Placebo | 300 IU/day | Total |
| Placebo | 25 | 25 | 50 |
| 800 mg/day | 25 | 25 | 50 |
| Total | 50 | 50 | 100 |

This example illustrates a third reason for using two-way designs. The effectiveness of the calcium supplement on BMD depends on having adequate vitamin D. We call this an **interaction.** In contrast, the average values for the calcium effect and the vitamin D effect are represented as **main effects.** The two-way model represents FIT as the sum of a main effect for each of the two factors *and* an interaction. One-way designs that vary a single factor and hold other factors fixed cannot discover interactions. We will discuss interactions more fully in a later section.

interaction

main effects

These examples illustrate several reasons why two-way designs are preferable to one-way designs.

ADVANTAGES OF TWO-WAY ANOVA

**1.** It is more efficient to study two factors simultaneously rather than separately.

> **2.** We can reduce the residual variation in a model by including a second factor thought to influence the response.
>
> **3.** We can investigate interactions between factors.

These considerations also apply to study designs with more than two factors. We will be content to explore only the two-way case. The choice of sampling or experimental design is fundamental to any statistical study. *Factors and levels must be carefully selected by an individual or team who understands both the statistical models and the issues that the study will address.*

## The two-way ANOVA model

When discussing two-way models in general, we will use the labels A and B for the two factors. For particular examples and when using statistical software, it is better to use meaningful names for these categorical variables. Thus, in Example 13.2 we would say that the factors are layout and cover, and in Example 13.4 we would say the factors are dosage and gender.

The numbers of levels of the factors are often used to describe the model. Again using our earlier examples, we would say Example 13.2 represents a $3 \times 4$ ANOVA and Example 13.4 illustrates a $3 \times 2$ ANOVA. In general, Factor A will have $I$ levels and Factor B will have $J$ levels. Therefore, we call the general two-way problem an $I \times J$ ANOVA.

In a two-way design every level of A appears in combination with every level of B, so that $I \times J$ groups are compared. The sample size for level $i$ of Factor A and level $j$ of Factor B is $n_{ij}$.[3] The total number of observations is

$$N = \sum n_{ij}$$

> ### ASSUMPTIONS FOR TWO-WAY ANOVA
>
> We have independent SRSs of size $n_{ij}$ from each of $I \times J$ Normal populations. The population means $\mu_{ij}$ may differ, but all populations have the same standard deviation $\sigma$. The $\mu_{ij}$ and $\sigma$ are unknown parameters.
>
> Let $x_{ijk}$ represent the $k$th observation from the population having Factor A at level $i$ and Factor B at level $j$. The statistical model is
>
> $$x_{ijk} = \mu_{ij} + \epsilon_{ijk}$$
>
> for $i = 1, \ldots, I$ and $j = 1, \ldots, J$ and $k = 1, \ldots, n_{ij}$, where the deviations $\epsilon_{ijk}$ are from an $N(0, \sigma)$ distribution.

LOOK BACK
**one-way model, page 645**

Similar to the one-way model, the FIT part is the group means $\mu_{ij}$, and the RESIDUAL part is the deviations $\epsilon_{ijk}$ of the individual observations from their group means. To estimate a group mean $\mu_{ij}$ we use the sample mean of the observations in the samples from this group:

$$\bar{x}_{ij} = \frac{1}{n_{ij}} \sum_k x_{ijk}$$

The $k$ below the $\sum$ means that we sum the $n_{ij}$ observations that belong to the $(i, j)$th sample.

The RESIDUAL part of the model contains the unknown $\sigma$. We calculate the sample variances for each SRS and pool these to estimate $\sigma^2$:

$$s_p^2 = \frac{\sum(n_{ij} - 1)s_{ij}^2}{\sum(n_{ij} - 1)}$$

Just as in one-way ANOVA, the numerator in this fraction is SSE and the denominator is DFE. Also, DFE is the total number of observations minus the number of groups. That is, DFE = $N - IJ$. The estimator of $\sigma$ is $s_p$.

## Main effects and interactions

In this section we will further explore the FIT part of the two-way ANOVA, which is represented in the model by the population means $\mu_{ij}$. The two-way design gives some structure to the set of means $\mu_{ij}$.

So far, because we have independent samples from each of $I \times J$ groups, we have presented the problem as a one-way ANOVA with $IJ$ groups. Each population mean $\mu_{ij}$ is estimated by the corresponding sample mean $\overline{x}_{ij}$, and we can calculate sums of squares and degrees of freedom as in one-way ANOVA. In accordance with the conventions used by many computer software packages, we use the term *model* when discussing the sums of squares and degrees of freedom calculated as in one-way ANOVA with $IJ$ groups. Thus, SSM is a model sum of squares constructed from deviations of the form $\overline{x}_{ij} - \overline{x}$, where $\overline{x}$ is the average of all of the observations and $\overline{x}_{ij}$ is the mean of the $(i, j)$th group. Similarly, DFM is simply $IJ - 1$.

In two-way ANOVA, the terms SSM and DFM can be further broken down into terms corresponding to a main effect for A, a main effect for B, and an AB interaction. Each of SSM and DFM is then a sum of terms:

$$\text{SSM} = \text{SSA} + \text{SSB} + \text{SSAB}$$

and

$$\text{DFM} = \text{DFA} + \text{DFB} + \text{DFAB}$$

The term SSA represents variation among the means for the different levels of Factor A. Because there are $I$ such means, DFA $= I - 1$ degrees of freedom. Similarly, SSB represents variation among the means for the different levels of Factor B, with DFB $= J - 1$.

Interactions are a bit more involved. We can see that SSAB, which is SSM − SSA − SSB, represents the variation in the model that is not accounted for by the main effects. By subtraction we see that its degrees of freedom are

$$\begin{aligned} \text{DFAB} &= (IJ - 1) - (I - 1) - (J - 1) \\ &= (I - 1)(J - 1) \end{aligned}$$

There are many kinds of interactions. The easiest way to study them is through examples.

EXAMPLE

**13.7 Investigating differences in soft drink consumption.** There is a general consensus that food portions have been increasing, but there is little scientific evidence that documents this change. One study used data from three nationally representative surveys to examine this issue. More than 63,380 individuals provided data for these three surveys. Three time points were examined: 1978, 1991, and 1996.[4] Here are the means for the number of calories per portion in soft drinks consumed at home and in sit-down restaurants:

| | Year | | | |
|---|---|---|---|---|
| Location | 1978 | 1991 | 1996 | Mean |
| Home | 130 | 133 | 158 | 140 |
| Restaurant | 125 | 126 | 155 | 135 |
| Mean | 127 | 129 | 156 | 137 |

The table includes averages of the means in the rows and columns. For example, in 1978 the mean of calories in soft drinks consumed at home and in restaurants is

$$\frac{130 + 125}{2} = 127.5$$

which is rounded to 127 in the table. Similarly, the corresponding value for 1996 is

$$\frac{158 + 155}{2} = 156.5$$

marginal means

which is rounded to 156 in the table. These averages are called **marginal means** (because of their location at the *margins* of such tabulations). The grand mean can be obtained by averaging either set of marginal means. *It is always a good idea to do both as a check on your arithmetic.*

Figure 13.1 is a plot of the group means. From the plot we see that the soft drinks consumed at home have about 3 to 7 more calories per serving than

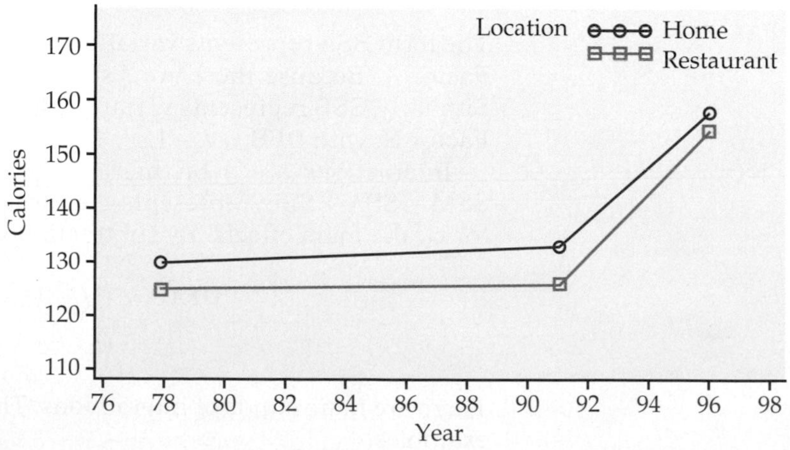

**FIGURE 13.1** Plot of the mean calories of soft drinks consumed at home and in restaurants in 1978, 1991, and 1996, for Example 13.7.

soft drinks consumed in restaurants for the three years. In statistical language, there is a main effect for location. We also see that the means for 1978 and 1991 are similar but there is a large increase by 1996. This is the main effect of year. These main effects can be described by differences between the marginal means. For example, the mean for 1978 is 127 calories, it increases by 2 calories to 129 calories in 1991, and then it jumps by 27 calories to 156 calories in 1996.

*To examine two-way ANOVA data for a possible interaction, always construct a plot similar to Figure 13.1.* In this plot, we see that the patterns of means over the years are similar for home and restaurant, with the restaurant means being about 5 calories lower than the home means. The two profiles are roughly parallel. This is another way of saying that there is no clear interaction between location and year evident in these data.

When no interaction is present, the marginal means provide a reasonable description of the two-way table of means. *On the other hand, if there is an interaction, then the marginal means do not tell the whole story.* Here is an example that illustrates this point.

EXAMPLE

**13.8 Soft drink consumption, continued.** The surveys described in the previous example also obtained data on soft drinks consumed in fast-food restaurants. Here are the data from the previous example with the fast-food means added:

| | Year | | | |
|---|---|---|---|---|
| **Location** | **1978** | **1991** | **1996** | **Mean** |
| Home | 130 | 133 | 158 | 140 |
| Restaurant | 125 | 126 | 155 | 135 |
| Fast-food | 131 | 143 | 191 | 155 |
| Mean | 129 | 134 | 168 | 143 |

Including the fast-food restaurants changes the marginal means for years and the overall mean. Figure 13.2 is a plot of the group means.

In this figure we see that home and the fast-food restaurants were quite similar in 1978. The increase from 1978 to 1991 was somewhat similar, with fast food increasing a little more than home. The most noticeable feature of the plot is the very large jump for the fast-food restaurants by 1996. The mean calories in soft drinks increased from 131 in 1978 to 191 in 1996, an increase of about 46%. This change is thought to be closely related to the trend toward "super-sizing" food portions at fast-food restaurants.

The three patterns of means in Figure 13.2 are clearly not all parallel. The change over time differs for the fast-food restaurants. We have an interaction between location and year. *However, the presence of an interaction does not necessarily mean that the main effects are uninformative.* The calories in soft drinks have increased from 1978 to 1991 and from 1991 to 1996. Similarly, there is a tendency for the means to be lowest in restaurants, a little higher at home, and highest at fast-food restaurants.

**FIGURE 13.2** Plot of the mean calories of soft drinks consumed at home, in restaurants, and in fast-food restaurants, in 1978, 1991, and 1996, for Example 13.8.

Interactions come in many shapes and forms. *When we find an interaction, a careful examination of the means is needed to properly interpret the data.* Simply stating that interactions are significant tells us little. Plots of the group means are essential.

EXAMPLE

**13.9 Eating in groups.** Some research has shown that people eat more when they eat in groups. One possible mechanism for this phenomenon is that they may spend more time eating when in a larger group. A study designed to examine this idea measured the length of time spent (in minutes) eating lunch in different settings.[5] Here are some data from this study:

| | Number of people eating | | | | | |
|---|---|---|---|---|---|---|
| **Lunch setting** | **1** | **2** | **3** | **4** | **5 or more** | **Mean** |
| Workplace | 12.6 | 23.0 | 33.0 | 41.1 | 44.0 | 30.7 |
| Fast-food restaurant | 10.7 | 18.2 | 18.4 | 19.7 | 21.9 | 17.8 |
| Mean | 11.6 | 20.6 | 25.7 | 30.4 | 32.9 | 24.2 |

Figure 13.3 gives the plot of the means for this example. The patterns are not parallel, so it appears that we have an interaction. Meals take longer when there are more people present, but this phenomenon is much greater for the meals consumed at work. For fast-food eating, the meal durations are fairly similar when there is more than one person present.

A different kind of interaction is present in the next example. Here, we must be very cautious in our interpretation of the main effects since one of them can lead to a distorted conclusion.

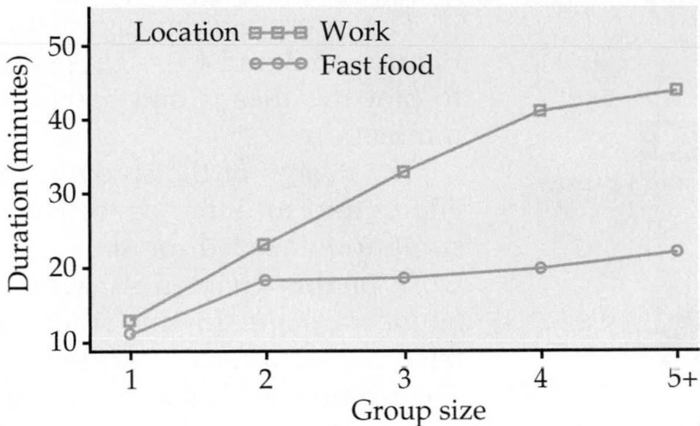


**FIGURE 13.3** Plot of mean meal duration versus lunch setting and group size, for Example 13.9.

EXAMPLE

**13.10 We got the beat?** When we hear music that is familiar to us, we can quickly pick up the beat and our mind synchronizes with the music. However, if the music is unfamiliar, it takes us longer to synchronize. In a study that investigated the theoretical framework for this phenomenon, French and Tunisian nationals listened to French and Tunisian music.[6] Each subject was asked to tap in time with the music being played. A synchronization score, recorded in milliseconds, measured how well the subjects synchronized with the music. A higher score indicates better synchronization. Six songs of each music type were used. Here are the means:

| | Music | | |
|---|---|---|---|
| **Nationality** | **French** | **Tunisian** | **Mean** |
| French | 950 | 750 | 850 |
| Tunisian | 760 | 1090 | 925 |
| Mean | 855 | 920 | 887 |

The means are plotted in Figure 13.4. In the study the researchers were not interested in main effects. Their theory predicted the interaction that we see in the figure. Subjects synchronize better with music from their own culture.

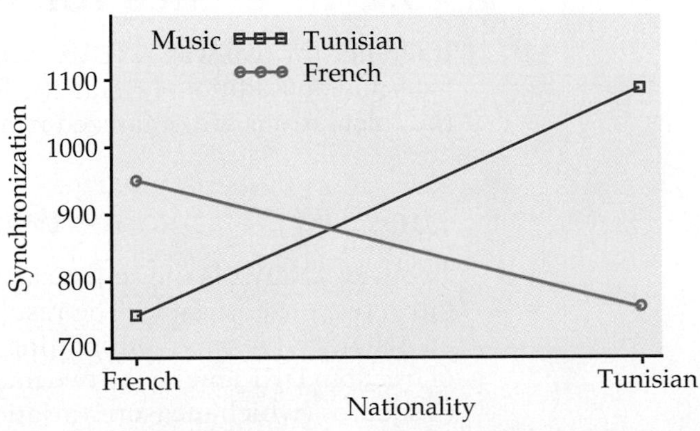


**FIGURE 13.4** Plot of mean synchronization score versus type of music for French and Tunisian nationals, for Example 13.10.

The interaction in Figure 13.4 is very different from those that we saw in Figures 13.2 and 13.3. These examples illustrate the point that it is necessary to plot the means and carefully describe the patterns when interpreting an interaction.

**matched pairs $t$ test, page 428**

The design of the study in Example 13.10 allows us to examine two main effects and an interaction. However, this setting does not meet all of the assumptions needed for statistical inference using the two-way ANOVA framework of this chapter. *As with one-way ANOVA, we require that observations be independent.* In this study, we have a design that has each subject contributing data for two types of music, so these two scores will be dependent. The framework is similar to the matched pairs setting. The design is called a **repeated-measures design.** More advanced texts on statistical methods cover this important design.

repeated-measures design

## USE YOUR KNOWLEDGE

**13.1 What's wrong?** For each of the following, explain what is wrong and why.

(a) A two-way ANOVA is used when there are two outcome variables.

(b) In a $3 \times 3$ ANOVA each level of Factor A appears with only two levels of Factor B.

(c) The FIT part of the model in a two-way ANOVA represents the variation that is sometimes called error or residual.

**13.2 What's wrong?** For each of the following, explain what is wrong and why.

(a) You can perform a two-way ANOVA only when the sample sizes are the same in all cells.

(b) The estimate $s_p^2$ is obtained by pooling the marginal sample variances.

(c) When interaction is present, the main effects are always uninformative.

# 13.2 Inference for Two-Way ANOVA

Inference for two-way ANOVA involves $F$ statistics for each of the two main effects and an additional $F$ statistic for the interaction. As with one-way ANOVA, the calculations are organized in an ANOVA table.

## The ANOVA table for two-way ANOVA

Two-way ANOVA is the statistical analysis for a two-way design with a quantitative response variable. The results of a two-way ANOVA are summarized in an ANOVA table based on splitting the total variation SST and the total degrees of freedom DFT among the two main effects and the interaction. Both the sums of squares (which measure variation) and the degrees of freedom add:

$$\text{SST} = \text{SSA} + \text{SSB} + \text{SSAB} + \text{SSE}$$

$$\text{DFT} = \text{DFA} + \text{DFB} + \text{DFAB} + \text{DFE}$$

The sums of squares are always calculated in practice by statistical software. *When the $n_{ij}$ are not all equal, some methods of analysis can give sums of squares that do not add.* From each sum of squares and its degrees of freedom we find the mean square in the usual way:

$$\text{mean square} = \frac{\text{sum of squares}}{\text{degrees of freedom}}$$

The significance of each of the main effects and the interaction is assessed by an $F$ statistic that compares the variation due to the effect of interest with the within-group variation. Each $F$ statistic is the mean square for the source of interest divided by MSE. Here is the general form of the two-way ANOVA table:

| Source | Degrees of freedom | Sum of squares | Mean square | $F$ |
|---|---|---|---|---|
| A | $I - 1$ | SSA | SSA/DFA | MSA/MSE |
| B | $J - 1$ | SSB | SSB/DFB | MSB/MSE |
| AB | $(I - 1)(J - 1)$ | SSAB | SSAB/DFAB | MSAB/MSE |
| Error | $N - IJ$ | SSE | SSE/DFE | |
| Total | $N - 1$ | SST | | |

There are three null hypotheses in two-way ANOVA, with an $F$ test for each. We can test for significance of the main effect of A, the main effect of B, and the AB interaction. *It is generally good practice to examine the test for interaction first, since the presence of a strong interaction may influence the interpretation of the main effects.* Be sure to plot the means as an aid to interpreting the results of the significance tests.

## SIGNIFICANCE TESTS IN TWO-WAY ANOVA

To test the main effect of A, use the $F$ statistic

$$F_{\text{A}} = \frac{\text{MSA}}{\text{MSE}}$$

To test the main effect of B, use the $F$ statistic

$$F_{\text{B}} = \frac{\text{MSB}}{\text{MSE}}$$

To test the interaction of A and B, use the $F$ statistic

$$F_{\text{AB}} = \frac{\text{MSAB}}{\text{MSE}}$$

If the effect being tested is zero, the calculated $F$ statistic has an $F$ distribution with numerator degrees of freedom corresponding to the effect and denominator degrees of freedom equal to DFE. Large values of the $F$ statistic lead to rejection of the null hypothesis. The $P$-value is the probability that a random variable having the corresponding $F$ distribution is greater than or equal to the calculated value.

The following example illustrates how to do a two-way ANOVA. As with the one-way ANOVA, we focus our attention on interpretation of the computer output.

EXAMPLE

**13.11 A study of cardiovascular risk factors.** A study of cardiovascular risk factors compared runners who averaged at least 15 miles per week with a control group described as "generally sedentary." Both men and women were included in the study.[7] The design is a 2 × 2 ANOVA with the factors group and gender. There were 200 subjects in each of the four combinations. One of the variables measured was the heart rate after 6 minutes of exercise on a treadmill. SAS computer analysis produced the outputs in Figure 13.5 and Figure 13.6.

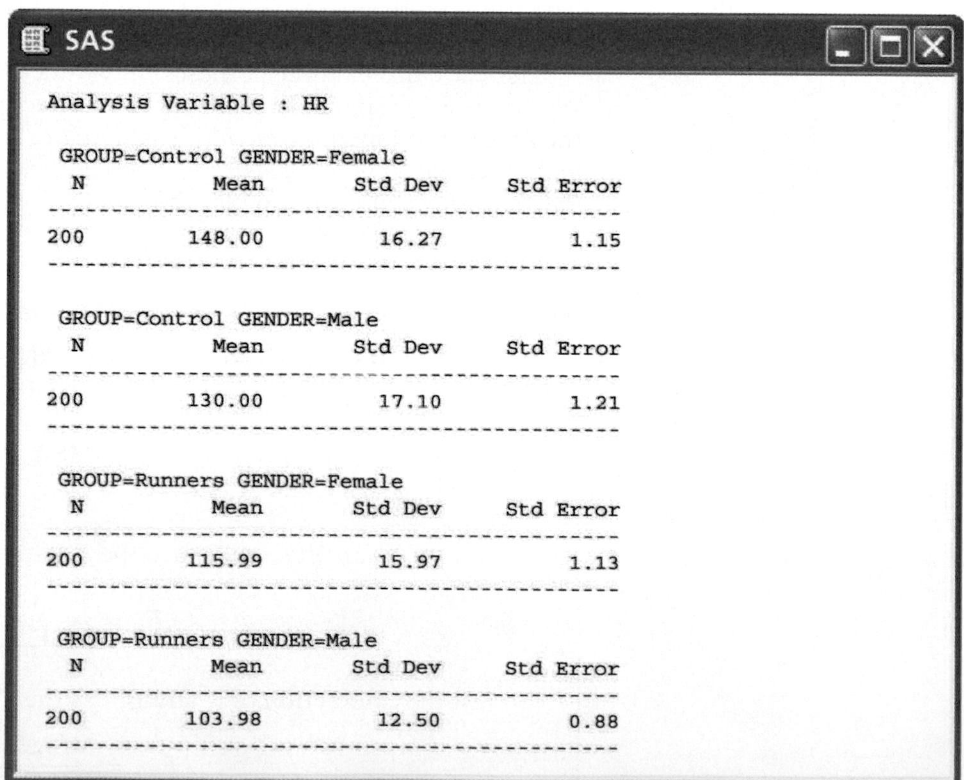

```
SAS

Analysis Variable : HR

 GROUP=Control GENDER=Female
  N           Mean         Std Dev      Std Error
 ------------------------------------------------
 200        148.00          16.27          1.15
 ------------------------------------------------

 GROUP=Control GENDER=Male
  N           Mean         Std Dev      Std Error
 ------------------------------------------------
 200        130.00          17.10          1.21
 ------------------------------------------------

 GROUP=Runners GENDER=Female
  N           Mean         Std Dev      Std Error
 ------------------------------------------------
 200        115.99          15.97          1.13
 ------------------------------------------------

 GROUP=Runners GENDER=Male
  N           Mean         Std Dev      Std Error
 ------------------------------------------------
 200        103.98          12.50          0.88
 ------------------------------------------------
```

**FIGURE 13.5** Summary statistics for heart rates in the four groups of a 2 × 2 ANOVA, for Example 13.11.

SAS

General Linear Models Procedure
Dependent Variable: HR

| Source | DF | Sum of Squares | Mean Square | F Value | Pr > F |
|---|---|---|---|---|---|
| Model | 3 | 215256.09 | 71752.03 | 296.35 | 0.0001 |
| Error | 796 | 192729.83 | 242.12 | | |
| Corrected Total | 799 | 407985.92 | | | |

| R-square | C.V. | Root MSE | HR Mean |
|---|---|---|---|
| 0.527607 | 12.49924 | 15.560 | 124.49 |

| Source | DF | Type I SS | Mean Square | F Value | Pr > F |
|---|---|---|---|---|---|
| GROUP | 1 | 168432.08 | 168432.08 | 695.65 | 0.0001 |
| GENDER | 1 | 45030.00 | 45030.00 | 185.98 | 0.0001 |
| GROUP*GENDER | 1 | 1794.01 | 1794.01 | 7.41 | 0.0066 |

**FIGURE 13.6** Two-way ANOVA output for heart rates, for Example 13.11.

We begin with the usual preliminary examination. From Figure 13.5 we see that the ratio of the largest to the smallest standard deviation is less than 2. Therefore, we are not concerned about violating the assumption of equal population standard deviations. Normal quantile plots (not shown) do not reveal any outliers, and the data appear to be reasonably Normal.

The ANOVA table at the top of the output in Figure 13.6 is in effect a one-way ANOVA with four groups: female control, female runner, male control, and male runner. In this analysis Model has 3 degrees of freedom, and Error has 796 degrees of freedom. The $F$ test and its associated $P$-value for this analysis refer to the hypothesis that all four groups have the same population mean. We are interested in the main effects and interaction, so we ignore this test.

The sums of squares for the group and gender main effects and the group-by-gender interaction appear at the bottom of Figure 13.6 under the heading Type I SS. These sum to the sum of squares for Model. Similarly, the degrees of freedom for these sums of squares sum to the degrees of freedom for Model. Two-way ANOVA splits the variation among the means (expressed by the Model sum of squares) into three parts that reflect the two-way layout.

Because the degrees of freedom are all 1 for the main effects and the interaction, the mean squares are the same as the sums of squares. The $F$ statistics for the three effects appear in the column labeled F Value, and the $P$-values are under the heading Pr > F. For the group main effect, we verify the calculation of $F$ as follows:

$$F = \frac{\text{MSG}}{\text{MSE}} = \frac{168{,}432}{242.12} = 695.65$$

All three effects are statistically significant. The group effect has the largest $F$, followed by the gender effect and then the group-by-gender interaction. To interpret these results, we examine the plot of means with bars indicating one standard error in Figure 13.7. Note that the standard errors are quite small due to the large sample sizes. The significance of the main effect for group is due to the fact that the controls have higher average heart rates than the runners for both genders. This is the largest effect evident in the plot.

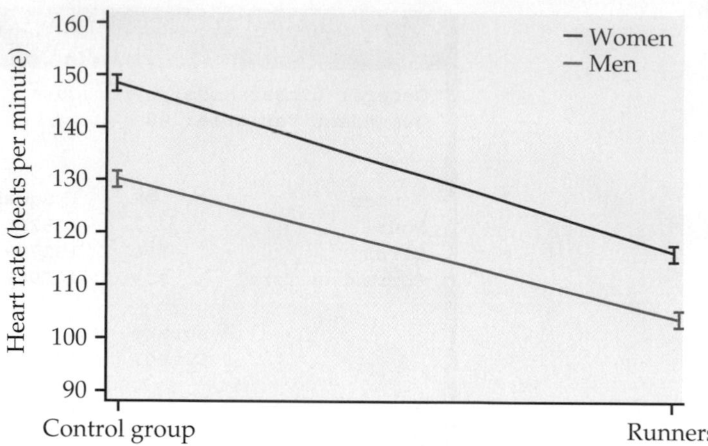


**FIGURE 13.7** Plot of the group means with standard errors for heart rates in the 2 × 2 ANOVA, for Example 13.11.

The significance of the main effect for gender is due to the fact that the females have higher heart rates than the men in both groups. The differences are not as large as those for the group effect, and this is reflected in the smaller value of the *F* statistic.

The analysis indicates that a complete description of the average heart rates requires consideration of the interaction in addition to the main effects. The two lines in the plot are not parallel. This interaction can be described in two ways. The female-male difference in average heart rates is greater for the controls than for the runners. Alternatively, the difference in average heart rates between controls and runners is greater for women than for men. As the plot suggests, the interaction is not large. It is statistically significant because there were 800 subjects in the study.

Two-way ANOVA output for other software is similar to that given by SAS. Figure 13.8 gives the analysis of the heart rate data using Excel and Minitab.

## SECTION 13.2 Summary

**Two-way analysis of variance** is used to compare population means when populations are classified according to two factors.

ANOVA assumes that the populations are Normal with possibly different means and the same standard deviation and that independent SRSs are drawn from each population.

As with one-way ANOVA, preliminary analysis includes examination of means, standard deviations, and Normal quantile plots. **Marginal means** are calculated by taking averages of the cell means across rows and columns. Pooling is used to estimate the within-group variance.

ANOVA separates the total variation into parts for the **model** and **error.** The model variation is separated into parts for each of the **main effects** and the **interaction.**

The calculations are organized into an **ANOVA table.** *F* statistics and *P*-values are used to test hypotheses about the main effects and the interaction.

Careful inspection of the means is necessary to interpret significant main effects and interactions. Plots are a useful aid.

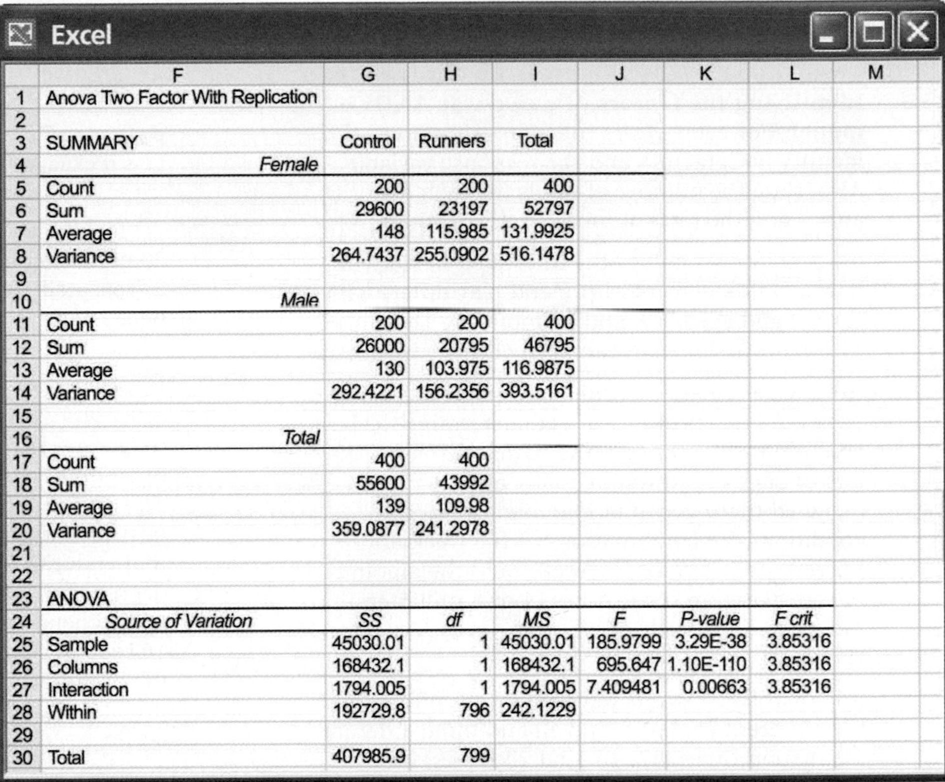

Excel

| | F | G | H | I | J | K | L | M |
|---|---|---|---|---|---|---|---|---|
| 1 | Anova Two Factor With Replication | | | | | | | |
| 2 | | | | | | | | |
| 3 | SUMMARY | Control | Runners | Total | | | | |
| 4 | *Female* | | | | | | | |
| 5 | Count | 200 | 200 | 400 | | | | |
| 6 | Sum | 29600 | 23197 | 52797 | | | | |
| 7 | Average | 148 | 115.985 | 131.9925 | | | | |
| 8 | Variance | 264.7437 | 255.0902 | 516.1478 | | | | |
| 9 | | | | | | | | |
| 10 | *Male* | | | | | | | |
| 11 | Count | 200 | 200 | 400 | | | | |
| 12 | Sum | 26000 | 20795 | 46795 | | | | |
| 13 | Average | 130 | 103.975 | 116.9875 | | | | |
| 14 | Variance | 292.4221 | 156.2356 | 393.5161 | | | | |
| 15 | | | | | | | | |
| 16 | *Total* | | | | | | | |
| 17 | Count | 400 | 400 | | | | | |
| 18 | Sum | 55600 | 43992 | | | | | |
| 19 | Average | 139 | 109.98 | | | | | |
| 20 | Variance | 359.0877 | 241.2978 | | | | | |
| 21 | | | | | | | | |
| 22 | | | | | | | | |
| 23 | ANOVA | | | | | | | |
| 24 | *Source of Variation* | *SS* | *df* | *MS* | *F* | *P-value* | *F crit* | |
| 25 | Sample | 45030.01 | 1 | 45030.01 | 185.9799 | 3.29E-38 | 3.85316 | |
| 26 | Columns | 168432.1 | 1 | 168432.1 | 695.647 | 1.10E-110 | 3.85316 | |
| 27 | Interaction | 1794.005 | 1 | 1794.005 | 7.409481 | 0.00663 | 3.85316 | |
| 28 | Within | 192729.8 | 796 | 242.1229 | | | | |
| 29 | | | | | | | | |
| 30 | Total | 407985.9 | 799 | | | | | |

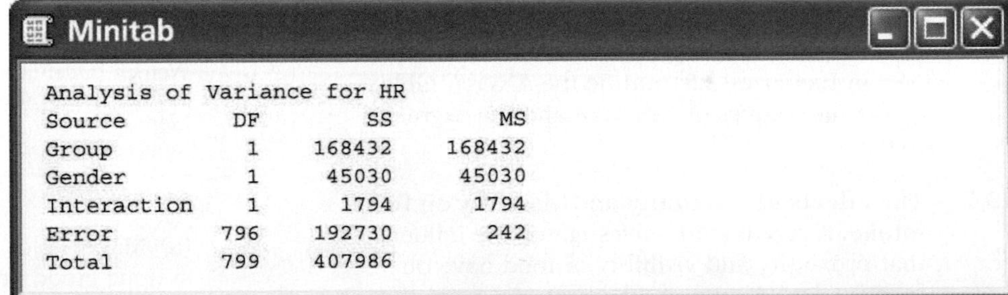

Minitab

Analysis of Variance for HR

| Source | DF | SS | MS |
|---|---|---|---|
| Group | 1 | 168432 | 168432 |
| Gender | 1 | 45030 | 45030 |
| Interaction | 1 | 1794 | 1794 |
| Error | 796 | 192730 | 242 |
| Total | 799 | 407986 | |

**FIGURE 13.8** Excel and Minitab two-way ANOVA output for the heart rate study, for Example 13.11.

## CHAPTER 13 Exercises

*For Exercises 13.1 and 13.2, see page 694.*

**13.3 Describing a two-way ANOVA model.** A 2 × 3 ANOVA was run with 6 observations per cell.

(a) Give the degrees of freedom for the *F* statistic that is used to test for interaction in this analysis and the entries from Table E that correspond to this distribution.

(b) Sketch a picture of this distribution with the information from the table included.

(c) The calculated value of the *F* statistic is 2.73. How would you report the *P*-value?

(d) Would you expect a plot of the means to look parallel? Explain your answer.

**13.4 Determining the critical value of *F*.** For each of the following situations, state how large the *F* statistic needs to be for rejection of the null hypothesis at the 5% level. Sketch each distribution and indicate the region where you would reject.

(a) The main effect for the first factor in a 3 × 5 ANOVA with 3 observations per cell.

(b) The interaction in a 3 × 3 ANOVA with 3 observations per cell.

(c) The interaction in a 2 × 2 ANOVA with 51 observations per cell.

**13.5 Identifying the factors of a two-way ANOVA model.** For each of the following situations, identify both factors and the response variable. Also, state the number of levels for each factor ($I$ and $J$) and the total number of observations ($N$).

(a) A child psychologist is interested in studying how a child's percent of pretend play differs with gender and age (4, 8, and 12 months). There are 11 infants assigned to each cell of the experiment.

(b) Brewers malt is produced from germinating barley. A homebrewer wants to determine the best conditions to germinate the barley. A total of 30 lots of barley seed were equally and randomly assigned to 10 germination conditions. The conditions are combinations of the week after harvest (1, 3, 6, 9, or 12 weeks) and the amount of water used in the process (4 or 8 milliliters). The percent of seeds germinating is the outcome variable.

(c) A virologist wants to compare the effects of two different media (A and B) and three different incubation times (12, 18, and 24 hours) on the growth of the Ebola virus. She plans on doing four replicates of each combination.

**13.6 Determining the degrees of freedom.** For each part in Exercise 13.5, outline the ANOVA table, giving the sources of variation and the degrees of freedom.

**13.7 The effects of proximity and visibility on food intake.** A recent study investigated the influence that proximity and visibility of food have on food intake.[8] A total of 40 secretaries from the University of Illinois participated in the study. A candy dish full of individually wrapped chocolates was placed either at the desk of the participant or at a location 2 meters from the participant. The candy dish was either a clear (candy visible) or opaque (candy not visible) covered bowl. After a week, the researchers noted not only the number of candies consumed per day but also the self-reported number of candies consumed by each participant. The table at the top of the next column summarizes the mean difference between these two values (reported minus actual).

(a) Make a plot of the means and describe the patterns that you see. Does the plot suggest an interaction between visibility and proximity?

(b) This study actually took 4 weeks, with each participant being observed at each treatment combination in a random order. Explain why a "repeated-measures" design like this may be beneficial.

| | Visibility | |
|---|---|---|
| Proximity | Clear | Opaque |
| Proximate | −1.2 | −0.8 |
| Less proximate | 0.5 | 0.4 |

**13.8 Hypotension and endurance exercise.** In sedentary individuals, low blood pressure (hypotension) often occurs after a single bout of aerobic exercise and lasts nearly two hours. This can cause dizziness, light-headedness, and possibly fainting upon standing. It is thought that endurance exercise training can reduce the degree of postexercise hypotension. To test this, researchers studied 16 endurance-trained and 16 sedentary men and women.[9] The following table summarizes the postexercise systolic arterial pressure (mmHg) after 60 minutes of upright cycling:

| Group | $n$ | $\bar{x}$ | Std. error |
|---|---|---|---|
| Women, sedentary | 8 | 100.7 | 3.4 |
| Women, endurance | 8 | 105.3 | 3.6 |
| Men, sedentary | 8 | 114.2 | 3.8 |
| Men, endurance | 8 | 110.2 | 2.3 |

(a) Make a plot similar to Figure 13.4 with the systolic blood pressure on the $y$ axis and training level on the $x$ axis. Describe the pattern you see.

(b) From the table, one can show that SSA = 677.12, SSB = 0.72, SSAB = 147.92, and SSE = 2478 where A is the gender effect and B is the training level. Construct the ANOVA table with $F$ statistics and degrees of freedom, and state your conclusions regarding main effects and interaction.

(c) The researchers also measured the before-exercise systolic blood pressure of the participants and looked at a model that incorporated both the pre- and postexercise values. Explain why it is likely beneficial to incorporate both measurements in the study.

**13.9 Evaluation of an intervention program.** The National Crime Victimization Survey estimates

**TABLE 13.1**

Safety behaviors of abused women

| Behavior | Intervention group (%) | | | Control group (%) | | |
|---|---|---|---|---|---|---|
| | Baseline | 3 months | 6 months | Baseline | 3 months | 6 months |
| Hide money | 68.0 | 60.0 | 62.7 | 60.0 | 37.8 | 35.1 |
| Hide extra keys | 52.7 | 76.0 | 68.9 | 53.3 | 33.8 | 39.2 |
| Abuse code to alert family | 30.7 | 74.7 | 60.0 | 22.7 | 27.0 | 43.2 |
| Hide extra clothing | 37.3 | 73.6 | 52.7 | 42.7 | 32.9 | 27.0 |
| Asked neighbors to call police | 49.3 | 73.0 | 66.2 | 32.0 | 45.9 | 40.5 |
| Know Social Security number | 93.2 | 93.2 | 100.0 | 89.3 | 93.2 | 98.6 |
| Keep rent, utility receipts | 75.3 | 95.5 | 89.4 | 70.3 | 84.7 | 80.9 |
| Keep birth certificates | 84.0 | 90.7 | 93.3 | 77.3 | 90.4 | 93.2 |
| Keep driver's license | 93.3 | 93.3 | 97.3 | 94.7 | 95.9 | 98.6 |
| Keep telephone numbers | 96.0 | 98.7 | 100.0 | 90.7 | 97.3 | 100.0 |
| Removed weapons | 50.0 | 70.6 | 38.5 | 40.7 | 23.8 | 5.9 |
| Keep bank account numbers | 81.0 | 94.3 | 96.2 | 76.2 | 85.5 | 94.4 |
| Keep insurance policy number | 70.9 | 90.4 | 89.7 | 68.3 | 84.2 | 94.8 |
| Keep marriage license | 71.1 | 92.3 | 84.6 | 63.3 | 73.2 | 80.0 |
| Hide valuable jewelry | 78.7 | 84.5 | 83.9 | 74.0 | 75.0 | 80.3 |

that there were over 400,000 violent crimes committed against women by their intimate partner that resulted in physical injury. An intervention study designed to increase safety behaviors of abused women compared the effectiveness of six telephone intervention sessions with a control group of abused women who received standard care. Fifteen different safety behaviors were examined.[10] One of the variables analyzed was the total number of behaviors (out of 15) that each woman performed. Here is a summary of the means of this variable at baseline (just before the first telephone call) and at follow-up 3 and 6 months later:

| Group | Time | | |
|---|---|---|---|
| | Baseline | 3 months | 6 months |
| Intervention | 10.4 | 12.5 | 11.9 |
| Control | 9.6 | 9.9 | 10.4 |

(a) Find the marginal means. Are they useful for understanding the results of this study?

(b) Plot the means. Do you think there is an interaction? Describe the meaning of an interaction for this study.

(*Note:* This exercise is from a repeated-measures design, and the data are not particularly Normal because they are counts with values from 1 to 15. Although we cannot use the methods in this chapter for statistical inference in this setting, the example does illustrate ideas about interactions.)

**13.10** CHALLENGE **More on the assessment of an intervention program.** Refer to the previous exercise. Table 13.1 gives the percents of women who responded that they performed each of the 15 safety behaviors studied.

(a) Summarize these data graphically. Do you think that your graphical display is more effective than Table 13.1 for describing the results of this study? Explain why or why not.

(b) Note any particular patterns in the data that would be important to someone who would use these results to design future intervention programs for abused women.

(c) The study was conducted "at a family violence unit of a large urban District Attorney's Office that serves an ethnically diverse population of three million citizens." To what extent do you think that this fact limits the conclusions that can be drawn?

**13.11 The acceptability of lying.** Lying is a common component of all human relationships. To investigate the acceptability of lying under various scenarios, researchers questioned 229 high school students from a West Coast public high school and 261 college students from a state university in the Midwest.[11] As part of the questioning, participants were asked to read a vignette in which the protagonist lies to his or her parents and to evaluate the acceptability of lying on a 4-point scale (1 = totally unacceptable,

4 = totally acceptable). Each participant was randomly assigned to read the vignette with either a female or male protagonist. The following table summarizes the mean response across age and protagonist.

| | Age | |
|---|---|---|
| Protagonist | H.S. | Col. |
| Male | 2.25 | 2.18 |
| Female | 2.35 | 1.82 |

(a) Plot the means and describe the pattern that you see.

(b) Suppose the $F$ statistic for the interaction was 3.26. What are the degrees of freedom for this statistic and the approximate $P$-value? Is there a significant interaction?

(c) This study involved participants from one high school and one college. To what extent do you think this limits the generalizability of the conclusions? Explain.

13.12 CHALLENGE **The effects of peer pressure on mathematics achievement.** Researchers were interested in comparing the relationship between high achievement in mathematics and peer pressure across several countries.[12] They hypothesized that in countries where high achievement is not valued highly, considerable peer pressure may exist. A questionnaire was distributed to 14-year-olds from three countries (Germany, Canada, and Israel). One of the questions asked students to rate how often they fear being called a nerd or teacher's pet on a 4-point scale (1 = never, 4 = frequently). The following table summarizes the response:

| Country | Gender | $n$ | $\bar{x}$ |
|---|---|---|---|
| Germany | Female | 336 | 1.62 |
| Germany | Male | 305 | 1.39 |
| Israel | Female | 205 | 1.87 |
| Israel | Male | 214 | 1.63 |
| Canada | Female | 301 | 1.91 |
| Canada | Male | 304 | 1.88 |

(a) The $P$-values for the interaction and the main effects of country and gender are 0.016, 0.068, and 0.108, respectively. Using the table and $P$-values, summarize the results both graphically and numerically.

(b) The researchers contend that Germany does not value achievement as highly as Canada and Israel. Do the results from (a) allow you to address their primary hypothesis? Explain.

(c) The students were also asked to indicate their current grade in mathematics on a 6-point scale (1 = excellent, 6 = insufficient). How might both responses be used to address the researchers' primary hypothesis?

13.13 **What can you conclude?** Analysis of data for a 3 × 2 ANOVA with 5 observations per cell gave the $F$ statistics in the following table:

| Effect | $F$ |
|---|---|
| A | 1.53 |
| B | 3.87 |
| AB | 2.94 |

What can you conclude from the information given?

13.14 **What can you conclude?** A study reported the following results for data analyzed using the methods that we studied in this chapter:

| Effect | $F$ | $P$-value |
|---|---|---|
| A | 7.50 | 0.006 |
| B | 18.66 | 0.001 |
| AB | 6.14 | 0.011 |

(a) What can you conclude from the information given?

(b) What additional information would you need to write a summary of the results for this study?

13.15 **Brand familiarity and repetitive advertising.** Does repetition of an advertising message increase its effectiveness? One theory suggests that there are two phases in the process. In the first phase, called "wearin," negative or unfamiliar views are transformed into positive views. In the second phase, called "wearout," the effectiveness of the ad is decreased because of boredom or other factors. One study designed to investigate this theory examined two factors. The first was familiarity of the ad, with two levels, familiar and unfamiliar; the second was repetition, with three levels, 1, 2, and 3.[13] One of the response variables collected was attitude toward the ad. This variable was the average of four items, each measured on a seven-point scale, anchored by bad–good, low quality–high quality, unappealing–appealing, and unpleasant–pleasant. Here are the means for attitude:

| | Repetition | | |
|---|---|---|---|
| **Familiarity** | **1** | **2** | **3** |
| Familiar | 4.56 | 4.73 | 5.24 |
| Unfamiliar | 4.14 | 5.26 | 4.41 |

(a) Make a plot of the means and describe the patterns that you see.

(b) Does the plot suggest that there is an interaction between familiarity and repetition? If your answer is Yes, describe the interaction.

**13.16** CHALLENGE **More on brand familiarity and repetitive advertising.** Refer to the previous exercise. In settings such as this, researchers collect data for several response variables. For this study, they also constructed variables that were called attitude toward the brand, total thoughts, support arguments, and counterarguments. Here are the means:

| | Attitude to brand | | | Total | | |
|---|---|---|---|---|---|---|
| | Repetition | | | Repetition | | |
| **Familiarity** | **1** | **2** | **3** | **1** | **2** | **3** |
| Familiar | 4.67 | 4.65 | 5.06 | 1.33 | 1.93 | 2.55 |
| Unfamiliar | 3.94 | 4.79 | 4.26 | 1.52 | 3.06 | 3.17 |

| | Support | | | Counter | | |
|---|---|---|---|---|---|---|
| | Repetition | | | Repetition | | |
| **Familiarity** | **1** | **2** | **3** | **1** | **2** | **3** |
| Familiar | 0.63 | 0.67 | 0.98 | 0.54 | 0.70 | 0.49 |
| Unfamiliar | 0.76 | 1.40 | 0.64 | 0.52 | 0.75 | 1.14 |

For each of the four response variables, give a graphical summary of the means. Use this summary to discuss any interactions that are evident. Write a short report summarizing the effect of repetition on the response variables measured, using the data in this exercise and the previous one.

**13.17 Estimating the within-group variance.** Refer to the previous exercise. Here are the standard deviations for attitude toward brand:

| | Repetition | | |
|---|---|---|---|
| **Familiarity** | **1** | **2** | **3** |
| Familiar | 1.16 | 1.46 | 1.16 |
| Unfamiliar | 1.39 | 1.22 | 1.42 |

Find the pooled estimate of the standard deviation for these data. Use the rule for examining standard deviations in ANOVA from Chapter 12 (page 646) to determine if it is reasonable to use a pooled standard deviation for the analysis of these data.

**13.18 More on estimating the within-group variance.** Refer to Exercise 13.16. Here are the standard deviations for total thoughts:

| | Repetition | | |
|---|---|---|---|
| **Familiarity** | **1** | **2** | **3** |
| Familiar | 1.63 | 1.42 | 1.52 |
| Unfamiliar | 1.64 | 2.16 | 1.59 |

Find the pooled estimate of the standard deviation for these data. Use the rule for examining standard deviations in ANOVA from Chapter 12 (page 646) to determine if it is reasonable to use a pooled standard deviation for the analysis of these data.

**13.19 Interpreting the results.** Refer to Exercises 13.15 and 13.16. The subjects were 94 adult staff members at a West Coast university. They watched a half-hour local news show from a different state that included the ads. The selected ads were judged to be "good" by some experts and had been shown in regions other than where the study was conducted. The real names of the products were replaced by either familiar or unfamiliar brand names by a professional video editor. The ads were pretested and no one in the pretest sample suggested that the ads were not real. Discuss each of these facts in terms of how you would interpret the results of this study.

**13.20 Assessing the Normality assumption.** Refer to Exercises 13.15 and 13.16. The ratings for this study were each measured on a seven-point scale, anchored by bad–good, low quality–high quality, unappealing–appealing, and unpleasant–pleasant. The results presented were averaged over three ads for different products: a bank, women's clothing, and a health care plan. Write a short report summarizing the Normality assumption for two-way ANOVA and the extent to which it is reasonable for the analysis of these data.

**13.21 The effect of chromium on insulin metabolism.** The amount of chromium in the diet has an effect on the way the body processes insulin. In an experiment designed to study this phenomenon, four diets were fed to male rats. There were two factors. Chromium had two levels: low (L) and normal (N). The rats were allowed to eat as much as they wanted (M) or the total amount that they could eat was restricted (R). We call the second factor Eat. One of the variables measured was the amount of an enzyme called GITH.[14] The means for this response variable appear in the following table:

| | Eat | |
|---|---|---|
| **Chromium** | **M** | **R** |
| L | 4.545 | 5.175 |
| N | 4.425 | 5.317 |

(a) Make a plot of the mean GITH for these diets, with the factor Chromium on the $x$ axis and GITH on the $y$ axis. For each Eat group, connect the points for the two Chromium means.

(b) Describe the patterns you see. Does the amount of chromium in the diet appear to affect the GITH mean? Does restricting the diet rather than letting the rats eat as much as they want appear to have an effect? Is there an interaction?

(c) Compute the marginal means. Compute the differences between the M and R diets for each level of Chromium. Use this information to summarize numerically the patterns in the plot.

**13.22 Changing your major.** A study of undergraduate computer science students examined changes in major after the first year.[15] The study examined the fates of 256 students who enrolled as first-year computer science students in the same fall semester. The students were classified according to gender and their declared major at the beginning of the second year. For convenience we use the labels CS for computer science majors, EO for engineering and other science majors, and O for other majors. The explanatory variables included several high school grade summaries coded as 10 = A, 9 = A−, etc. Here are the mean high school mathematics grades for these students:

| | Major | | |
|---|---|---|---|
| **Gender** | **CS** | **EO** | **O** |
| Males | 8.68 | 8.35 | 7.65 |
| Females | 9.11 | 9.36 | 8.04 |

Describe the main effects and interaction using appropriate graphs and calculations.

**13.23 More on changing your major.** The mean SAT Mathematics scores for the students in the previous exercise are summarized in the following table:

| | Major | | |
|---|---|---|---|
| **Gender** | **CS** | **EO** | **O** |
| Males | 628 | 618 | 589 |
| Females | 582 | 631 | 543 |

Summarize the results of this study using appropriate plots and calculations to describe the main effects and interaction.

**13.24 Designing a study.** The students studied in the previous two exercises were enrolled at a large midwestern university more than two decades ago. Discuss how you would conduct a similar study at a college or university of your choice today. Include a description of all variables that you would collect for your study.

**13.25 A comparison of different types of scaffold material.** One way to repair serious wounds is to insert some material as a scaffold for the body's repair cells to use as a template for new tissue. Scaffolds made from extracellular material (ECM) are particularly promising for this purpose. Because they are made from biological material, they serve as an effective scaffold and are then resorbed. Unlike biological material that includes cells, however, they do not trigger tissue rejection reactions in the body. One study compared 6 types of scaffold material.[16] Three of these were ECMs and the other three were made of inert materials. There were three mice used per scaffold type. The response measure was the percent of glucose phosphated isomerase (Gpi) cells in the region of the wound. A large value is good, indicating that there are many bone marrow cells sent by the body to repair the tissue. In Exercise 12.51 (page 679) we analyzed the data for rats whose tissues were measured 4 weeks after the repair. The experiment included additional groups of rats who received the same types of scaffold but were measured at different times. Here are the data for 4 weeks and 8 weeks after the repair:

| | Gpi (%) | | | | | |
|---|---|---|---|---|---|---|
| **Material** | **4 weeks** | | | **8 weeks** | | |
| ECM1 | 55 | 70 | 70 | 60 | 65 | 65 |
| ECM2 | 60 | 65 | 65 | 60 | 70 | 60 |
| ECM3 | 75 | 70 | 75 | 70 | 80 | 70 |
| MAT1 | 20 | 25 | 25 | 15 | 25 | 25 |
| MAT2 | 5 | 10 | 5 | 10 | 5 | 5 |
| MAT3 | 10 | 15 | 10 | 5 | 15 | 10 |

(a) Make a table giving the sample size, mean, and standard deviation for each of the material-by-time combinations. Is it reasonable to pool the variances? Because the sample sizes in this experiment are very small, we expect a large amount of variability in the sample standard deviations. Although they vary more than we would prefer, we will proceed with the ANOVA.

(b) Make a plot of the means. Describe the main features of the plot.

(c) Run the analysis of variance. Report the $F$ statistics with degrees of freedom and $P$-values for each of the main effects and the interaction. What do you conclude? Write a short paragraph summarizing the results of your analysis.

**13.26** CHALLENGE **A comparison of different types of scaffold material, continued.** Refer to the previous exercise. Here are the data that were collected at 2 weeks, 4 weeks, and 8 weeks:

| | Gpi (%) | | | | | | | | |
|---|---|---|---|---|---|---|---|---|---|
| **Material** | **2 weeks** | | | **4 weeks** | | | **8 weeks** | | |
| ECM1 | 70 | 75 | 65 | 55 | 70 | 70 | 60 | 65 | 65 |
| ECM2 | 60 | 65 | 70 | 60 | 65 | 65 | 60 | 70 | 60 |
| ECM3 | 80 | 60 | 75 | 75 | 70 | 75 | 70 | 80 | 70 |
| MAT1 | 50 | 45 | 50 | 20 | 25 | 25 | 15 | 25 | 25 |
| MAT2 | 5 | 10 | 15 | 5 | 10 | 5 | 10 | 5 | 5 |
| MAT3 | 30 | 25 | 25 | 10 | 15 | 10 | 5 | 15 | 10 |

Rerun the analyses that you performed in the previous exercise. How does the addition of the data for 2 weeks change the conclusions? Write a summary comparing these results with those in the previous exercise.

**13.27 Analysis using multiple one-way ANOVAs.** Refer to the previous exercise. Analyze the data for each time period separately using a one-way ANOVA. Use a multiple-comparisons procedure where needed. Summarize the results.

**13.28** CHALLENGE **Does the type of cooking pot affect iron content?** Iron-deficiency anemia is the most common form of malnutrition in developing countries, affecting about 50% of children and women and 25% of men. Iron pots for cooking foods had traditionally been used in many of these countries, but they have been largely replaced by aluminum pots, which are cheaper and lighter. Some research has suggested that food cooked in iron pots will contain more iron than food cooked in other types of pots. One study designed to investigate this issue compared the iron content of some Ethiopian foods cooked in aluminum, clay, and iron pots.[17] In Exercise 12.49 (page 679), we analyzed the iron content of *yesiga wet'*, beef cut into small pieces and prepared with several Ethiopian spices. The researchers who conducted this study also examined the iron content of *shiro wet'*, a legume-based mixture of chickpea flour and Ethiopian spiced pepper, and *ye-atkilt allych'a*, a lightly spiced vegetable casserole. In the table below, these three foods are labeled meat, legumes, and vegetables. Four samples of each food were cooked in each type of pot. The iron in the food is measured in milligrams of iron per 100 grams of cooked food. Here are the data:

| | Iron content | | | |
|---|---|---|---|---|
| **Type of pot** | **Meat** | | | |
| Aluminum | 1.77 | 2.36 | 1.96 | 2.14 |
| Clay | 2.27 | 1.28 | 2.48 | 2.68 |
| Iron | 5.27 | 5.17 | 4.06 | 4.22 |
| **Type of pot** | **Legumes** | | | |
| Aluminum | 2.40 | 2.17 | 2.41 | 2.34 |
| Clay | 2.41 | 2.43 | 2.57 | 2.48 |
| Iron | 3.69 | 3.43 | 3.84 | 3.72 |
| **Type of pot** | **Vegetables** | | | |
| Aluminum | 1.03 | 1.53 | 1.07 | 1.30 |
| Clay | 1.55 | 0.79 | 1.68 | 1.82 |
| Iron | 2.45 | 2.99 | 2.80 | 2.92 |

(a) Make a table giving the sample size, mean, and standard deviation for each type of pot. Is it reasonable to pool the variances? Although the standard deviations vary more than we would like, this is partially due to the small sample sizes and we will proceed with the analysis of variance.

(b) Plot the means. Give a short summary of how the iron content of foods depends upon the cooking pot.

(c) Run the analysis of variance. Give the ANOVA table, the $F$ statistics with degrees of freedom and $P$-values, and your conclusions regarding the hypotheses about main effects and interactions.

**13.29 Interpreting the results.** Refer to the previous exercise. Although there is a statistically significant interaction, do you think that these data support the conclusion that foods cooked in iron pots contain more iron than foods cooked in aluminum or clay pots? Discuss.

**13.30 Analysis using a one-way ANOVA.** Refer to Exercise 13.28. Rerun the analysis as a one-way ANOVA with 9 groups and 4 observations per group. Report the results of the $F$ test. Examine differences in means using a multiple-comparisons procedure. Summarize your results and compare them with those you obtained in Exercise 13.28.

**13.31 Examination of a drilling process.** One step in the manufacture of large engines requires that holes of very precise dimensions be drilled. The tools that do the drilling are regularly examined and are adjusted to ensure that the holes meet the required specifications. Part of the examination involves measurement of the diameter of the drilling tool. A team studying the variation in the sizes of the drilled holes selected this measurement procedure as a possible cause of variation in the drilled holes. They decided to use a designed experiment as one part of this examination. Some of the data are given in Table 13.2. The diameters in millimeters (mm) of five tools were measured by the same operator at three times (8:00 A.M., 11:00 A.M., and 3:00 P.M.). Three measurements were taken on each tool at each time. The person taking the measurements could not tell which tool was being measured, and the measurements were taken in random order.[18]

**TABLE 13.2**

Tool diameter data

| Tool | Time | Diameter (mm) | | |
|---|---|---|---|---|
| 1 | 1 | 25.030 | 25.030 | 25.032 |
| 1 | 2 | 25.028 | 25.028 | 25.028 |
| 1 | 3 | 25.026 | 25.026 | 25.026 |
| 2 | 1 | 25.016 | 25.018 | 25.016 |
| 2 | 2 | 25.022 | 25.020 | 25.018 |
| 2 | 3 | 25.016 | 25.016 | 25.016 |
| 3 | 1 | 25.005 | 25.008 | 25.006 |
| 3 | 2 | 25.012 | 25.012 | 25.014 |
| 3 | 3 | 25.010 | 25.010 | 25.008 |
| 4 | 1 | 25.012 | 25.012 | 25.012 |
| 4 | 2 | 25.018 | 25.020 | 25.020 |
| 4 | 3 | 25.010 | 25.014 | 25.018 |
| 5 | 1 | 24.996 | 24.998 | 24.998 |
| 5 | 2 | 25.006 | 25.006 | 25.006 |
| 5 | 3 | 25.000 | 25.002 | 24.999 |

(a) Make a table of means and standard deviations for each of the 5 × 3 combinations of the two factors.

(b) Plot the means and describe how the means vary with tool and time. Note that we expect the tools to have slightly different diameters. These will be adjusted as needed. It is the process of measuring the diameters that is important.

(c) Use a two-way ANOVA to analyze these data. Report the test statistics, degrees of freedom, and $P$-values for the significance tests.

(d) Write a short report summarizing your results.

**13.32 Examination of a drilling process, continued.** Refer to the previous exercise. Multiply each measurement by 0.04 to convert from millimeters to inches. Redo the plots and rerun the ANOVA using the transformed measurements. Summarize what parts of the analysis have changed and what parts have remained the same.

**13.33 A price promotion study.** How does the frequency that a supermarket product is promoted at a discount affect the price that customers expect to pay for the product? Does the percent reduction also affect this expectation? These questions were examined by researchers in a study conducted on students enrolled in an introductory management course at a large midwestern university. For 10 weeks 160 subjects received information about the products. The treatment conditions corresponded to the number of promotions (1, 3, 5, or 7) that were described during this 10-week period and the percent that the product was discounted (10%, 20%, 30%, and 40%). Ten students were randomly

**TABLE 13.3**
Expected price data

| Number of promotions | Percent discount | Expected price ($) | | | | | | | | | |
|---|---|---|---|---|---|---|---|---|---|---|---|
| 1 | 40 | 4.10 | 4.50 | 4.47 | 4.42 | 4.56 | 4.69 | 4.42 | 4.17 | 4.31 | 4.59 |
| 1 | 30 | 3.57 | 3.77 | 3.90 | 4.49 | 4.00 | 4.66 | 4.48 | 4.64 | 4.31 | 4.43 |
| 1 | 20 | 4.94 | 4.59 | 4.58 | 4.48 | 4.55 | 4.53 | 4.59 | 4.66 | 4.73 | 5.24 |
| 1 | 10 | 5.19 | 4.88 | 4.78 | 4.89 | 4.69 | 4.96 | 5.00 | 4.93 | 5.10 | 4.78 |
| 3 | 40 | 4.07 | 4.13 | 4.25 | 4.23 | 4.57 | 4.33 | 4.17 | 4.47 | 4.60 | 4.02 |
| 3 | 30 | 4.20 | 3.94 | 4.20 | 3.88 | 4.35 | 3.99 | 4.01 | 4.22 | 3.70 | 4.48 |
| 3 | 20 | 4.88 | 4.80 | 4.46 | 4.73 | 3.96 | 4.42 | 4.30 | 4.68 | 4.45 | 4.56 |
| 3 | 10 | 4.90 | 5.15 | 4.68 | 4.98 | 4.66 | 4.46 | 4.70 | 4.37 | 4.69 | 4.97 |
| 5 | 40 | 3.89 | 4.18 | 3.82 | 4.09 | 3.94 | 4.41 | 4.14 | 4.15 | 4.06 | 3.90 |
| 5 | 30 | 3.90 | 3.77 | 3.86 | 4.10 | 4.10 | 3.81 | 3.97 | 3.67 | 4.05 | 3.67 |
| 5 | 20 | 4.11 | 4.35 | 4.17 | 4.11 | 4.02 | 4.41 | 4.48 | 3.76 | 4.66 | 4.44 |
| 5 | 10 | 4.31 | 4.36 | 4.75 | 4.62 | 3.74 | 4.34 | 4.52 | 4.37 | 4.40 | 4.52 |
| 7 | 40 | 3.56 | 3.91 | 4.05 | 3.91 | 4.11 | 3.61 | 3.72 | 3.69 | 3.79 | 3.45 |
| 7 | 30 | 3.45 | 4.06 | 3.35 | 3.67 | 3.74 | 3.80 | 3.90 | 4.08 | 3.52 | 4.03 |
| 7 | 20 | 3.89 | 4.45 | 3.80 | 4.15 | 4.41 | 3.75 | 3.98 | 4.07 | 4.21 | 4.23 |
| 7 | 10 | 4.04 | 4.22 | 4.39 | 3.89 | 4.26 | 4.41 | 4.39 | 4.52 | 3.87 | 4.70 |

assigned to each of the $4 \times 4 = 16$ treatments.[19] Table 13.3 gives the data.

(a) Summarize the means and standard deviations in a table and plot the means. Summarize the main features of the plot.

(b) Analyze the data with a two-way ANOVA. Report the results of this analysis.

(c) Using your plot and the ANOVA results, prepare a short report explaining how the expected price depends on the number of promotions and the percent of the discount.

**13.34 Analysis using a one-way ANOVA.** Refer to the previous exercise. Rerun the analysis as a one-way ANOVA with $4 \times 4 = 16$ treatments. Summarize the results of this analysis. Use a multiple-comparisons procedure to describe combinations of number of promotions and percent discounts that are similar or different.

**13.35 Do left-handed people live shorter lives than right-handed people?** A study of this question examined a sample of 949 death records and contacted next of kin to determine handedness.[20] Note that there are many possible definitions of "left-handed." The researchers examined the effects of different definitions on the results of their analysis and found that their conclusions were not sensitive to the exact definition used. For the results presented here, people were defined to be right-handed if they wrote, drew, and threw a ball with the right hand. All others were defined to be left-handed. People were classified by gender (female or male) and handedness (left or right), and a $2 \times 2$ ANOVA was run with the age at death as the response variable. The $F$ statistics were 22.36 (handedness), 37.44 (gender), and 2.10 (interaction). The following marginal mean ages at death (in years) were reported: 77.39 (females), 71.32 (males), 75.00 (right-handed), and 66.03 (left-handed).

(a) For each of the $F$ statistics given above find the degrees of freedom and an approximate $P$-value. Summarize the results of these tests.

(b) Using the information given, write a short summary of the results of the study.

**13.36 A radon exposure study.** Scientists believe that exposure to the radioactive gas radon is associated with some types of cancers in the respiratory system. Radon from natural sources is present in many homes in the United States. A group of researchers decided to study the problem in dogs because dogs get similar types of cancers and are exposed to environments similar to those of their owners. Radon detectors are available for home monitoring but the researchers wanted to obtain actual measures of the exposure of a sample of dogs. To do this they placed the detectors in holders and attached them to the collars of the dogs. One problem was that the holders might in some way affect the radon readings. The researchers therefore devised a laboratory experiment to study the effects of the holders. Detectors from four series of production were available, so they used a two-way ANOVA design (series with 4 levels and holder with 2, representing the presence or absence of a holder). All detectors were exposed to the same radon source and the radon measure in picocuries per liter was recorded.[21] The $F$ statistic for the effect of series is 7.02, for holder it is 1.96, and for the interaction it is 1.24. The sample size is 69.

(a) Using Table E or statistical software find approximate $P$-values for the three test statistics. Summarize the results of these three significance tests.

(b) The mean radon readings for the four series were 330, 303, 302, and 295. The results of the significance test for series were of great concern to the researchers. Explain why.

**13.37 A comparison of plant species under low water conditions.** The PLANTS1 data set in the Data Appendix gives the percent of nitrogen in four different species of plants grown in a laboratory. The species are *Leucaena leucocephala*, *Acacia saligna*, *Prosopis juliflora*, and *Eucalyptus citriodora*. The researchers who collected these data were interested in commercially growing these plants in parts of the country of Jordan where there is very little rainfall. To examine the effect of water, they varied the amount per day from 50 millimeters (mm) to 650 mm in 100 mm increments. There were nine plants per species-by-water combination. Because the plants are to be used primarily for animal food, with some parts that can be consumed by people, a high nitrogen content is very desirable.

(a) Find the means for each species-by-water combination. Plot these means versus water for the four species, connecting the means for each species by lines. Describe the overall pattern.

(b) Find the standard deviations for each species-by-water combination. Is it reasonable to pool the standard deviations for this problem? Note that with sample sizes of size 9, we expect these standard deviations to be quite variable.

(c) Run the two-way analysis of variance. Give the results of the hypothesis tests for the main effects and the interaction.

**13.38 Examination of the residuals.** Refer to the previous exercise. Examine the residuals. Are there any unusual patterns or outliers? If you think that there are one or more points that are somewhat extreme, rerun the two-way analysis without these observations. Does this change the results in any substantial way?

**13.39 Analysis using multiple one-way ANOVAs.** Refer to Exercise 13.37. Run a separate one-way analysis of variance for each water level. If there is evidence that the species are not all the same, use a multiple-comparisons procedure to determine which pairs of species are significantly different. In what way, if any, do the differences appear to vary by water level? Write a short summary of your conclusions.

**13.40 More on the analysis using multiple one-way ANOVAs.** Refer to Exercise 13.37. Run a separate one-way analysis of variance for each species and summarize the results. Since the amount of water is a quantitative factor, we can also analyze these data using regression. Run simple linear regressions separately for each species to predict nitrogen percent from water. Use plots to determine whether or not a line is a good way to approximate this relationship. Summarize the regression results and compare them with the one-way ANOVA results.

**13.41 Another comparison of plant species under low water conditions.** Refer to Exercise 13.37. Additional data collected by the same researchers according to a similar design are given in the PLANTS2 data set in the Data Appendix. Here, there are two response variables. They are fresh biomass and dry biomass. High values for both of these variables are desirable. The same four species and seven levels of water are used for this experiment. Here, however, there are four plants per species-by-water combination. Analyze each of the response variables in the PLANTS2 data set using the outline from Exercise 13.37.

**13.42 Examination of the residuals.** Perform the tasks described in Exercise 13.38 for the two response variables in the PLANTS2 data set.

**13.43 Analysis using multiple one-way ANOVAs.** Perform the tasks described in Exercise 13.39 for the two response variables in the PLANTS2 data set.

**13.44 More on the analysis using multiple one-way ANOVAs.** Perform the tasks described in Exercise 13.40 for the two response variables in the PLANTS2 data set.

**13.45 Are insects more attracted to male plants?** Some scientists wanted to determine if there are gender-related differences in the level of herbivory in the jack-in-the-pulpit, a spring-blooming perennial plant common in deciduous forests. A study was conducted in southern Maryland at forests associated with the Smithsonian Environmental Research Center (SERC).[22] To determine the effects of flowering and floral characteristics on herbivory, the researchers altered the floral morphology of male and female plants. The three levels of floral characteristics were (1) the spathes were completely removed; (2) in females, a gap was created in the base of the spathe, and in males, the gap was closed; (3) plants were not altered (control). The percent of leaf area damaged by thrips (an order of insects) between early May and mid-June was recorded for each of 30 plants per combination of sex and floral characteristic. A table of means and standard deviations (in parentheses) is shown below:

| | Floral characteristic level | | |
|---|---|---|---|
| Gender | 1 | 2 | 3 |
| Males | 0.11 (0.081) | 1.28 (0.088) | 1.63 (0.382) |
| Females | 0.02 (0.002) | 0.58 (0.321) | 0.20 (0.035) |

(a) Give the degrees of freedom for the $F$ statistics that are used to test for gender, floral characteristic, and the interaction.

(b) Describe the main effects and interaction using appropriate graphs.

(c) The researchers used the natural logarithm of percent area as the response in their analysis. Using the relationship between the means and standard deviations, explain why this was done.

**13.46 Change-of-majors study: HSS.** Refer to the data given for the change-of-majors study in the data set MAJORS described in the Data Appendix. Analyze the data for HSS, the high school science grades. Your analysis should include a table of sample sizes, means, and standard deviations; Normal quantile plots; a plot of the means; and a two-way ANOVA using sex and major as the factors. Write a short summary of your conclusions.

**13.47 Change-of-majors study: HSE.** Refer to the data given for the change-of-majors study in the data set MAJORS described in the Data Appendix. Analyze the data for HSE, the high school English grades. Your analysis should include a table of sample sizes, means, and standard deviations; Normal quantile plots; a plot of the means; and a two-way ANOVA using sex and major as the factors. Write a short summary of your conclusions.

**13.48 Change-of-majors study: GPA.** Refer to the data given for the change-of-majors study in the data set MAJORS described in the Data Appendix. Analyze the data for GPA, the college grade point average. Your analysis should include a table of sample sizes, means, and standard deviations; Normal quantile plots; a plot of the means; and a two-way ANOVA using sex and major as the factors. Write a short summary of your conclusions.

**13.49 Change-of-majors study: SATV.** Refer to the data given for the change-of-majors study in the data set MAJORS described in the Data Appendix. Analyze the data for SATV, the SAT Verbal score. Your analysis should include a table of sample sizes, means, and standard deviations; Normal quantile plots; a plot of the means; and a two-way ANOVA using sex and major as the factors. Write a short summary of your conclusions.

**13.50 Search the Internet.** Search the Internet or your library to find a study that is interesting to you and uses a two-way ANOVA to analyze the data. First describe the question or questions of interest and then give the details of how ANOVA was used to provide answers. Be sure to include how the study authors examined the assumptions for the analysis. Evaluate how well the authors used ANOVA in this study. If your evaluation finds the analysis deficient, make suggestions for how it could be improved.

CHAPTER

14

# Logistic Regression

Will a patient live or die after being admitted to a hospital? Logistic regression can be used to model categorical outcomes such as this.



## Introduction

The simple and multiple linear regression methods we studied in Chapters 10 and 11 are used to model the relationship between a quantitative response variable and one or more explanatory variables. A key assumption for these models is that the deviations from the model fit are Normally distributed. In this chapter we describe similar methods that are used when the response variable has only two possible values.

Our response variable has only two values: success or failure, live or die, acceptable or not. If we let the two values be 1 and 0, the mean is the proportion of ones, $p = P(\text{success})$. With $n$ independent observations, we have the *binomial setting*. What is *new* here is that we have data on an *explanatory variable* $x$. We study how $p$ depends on $x$. For example, suppose we are studying whether a patient lives, ($y = 1$) or dies ($y = 0$) after being admitted to a hospital. Here, $p$ is the probability that a patient lives, and possible explanatory variables include (a) whether the patient is in good condition or in poor condition, (b) the type of medical problem that the patient has, and (c) the age of the patient. Note that the explanatory variables can be either categorical or quantitative. Logistic regression is a statistical method for describing these kinds of relationships.[1]


LOOK BACK

**binomial setting, page 314**

# 14.1 The Logistic Regression Model

## Binomial distributions and odds

In Chapter 5 we studied binomial distributions and in Chapter 8 we learned how to do statistical inference for the proportion $p$ of successes in the binomial setting. We start with a brief review of some of these ideas that we will need in this chapter.

EXAMPLE

**14.1 College students and binge drinking.** Example 8.1 (page 489) describes a survey of 13,819 four-year college students. The researchers were interested in estimating the proportion of students who are frequent binge drinkers. A male student who reports drinking five or more drinks in a row, or a female student who reports drinking four or more drinks in a row, three or more times in the past two weeks is called a frequent binge drinker. In the notation of Chapter 5, $p$ is the proportion of frequent binge drinkers in the entire population of college students in four-year colleges. The number of frequent binge drinkers in an SRS of size $n$ has the binomial distribution with parameters $n$ and $p$. The sample size is $n = 13{,}819$ and the number of frequent binge drinkers in the sample is 3140. The sample proportion is

$$\hat{p} = \frac{3140}{13{,}819} = 0.2272$$

odds

Logistic regressions work with **odds** rather than proportions. The odds are simply the ratio of the proportions for the two possible outcomes. If $\hat{p}$ is the proportion for one outcome, then $1 - \hat{p}$ is the proportion for the second outcome:

$$\text{odds} = \frac{\hat{p}}{1 - \hat{p}}$$

A similar formula for the population odds is obtained by substituting $p$ for $\hat{p}$ in this expression.

EXAMPLE

**14.2 Odds of being a binge drinker.** For the binge-drinking data the proportion of frequent binge drinkers in the sample is $\hat{p} = 0.2272$, so the proportion of students who are not frequent binge drinkers is

$$1 - \hat{p} = 1 - 0.2272 = 0.7728$$

Therefore, the odds of a student being a frequent binge drinker are

$$\begin{aligned} \text{odds} &= \frac{\hat{p}}{1 - \hat{p}} \\ &= \frac{0.2272}{0.7728} \\ &= 0.29 \end{aligned}$$

When people speak about odds, they often round to integers or fractions. Since 0.29 is approximately 1/3, we could say that the odds that a college student is a frequent binge drinker are 1 to 3. In a similar way, we could describe the odds that a college student is *not* a frequent binge drinker as 3 to 1.

**USE YOUR KNOWLEDGE**

**14.1 Odds of drawing a heart.** If you deal one card from a standard deck, the probability that the card is a heart is 0.25. Find the odds of drawing a heart.

**14.2 Given the odds, find the probability.** If you know the odds, you can find the probability by solving the equation for odds given above for the probability. So, $\hat{p} = \text{odds}/(\text{odds} + 1)$. If the odds of an outcome are 2 (or 2 to 1), what is the probability of the outcome?

## Odds for two samples

In Example 8.9 (page 507) we compared the proportions of frequent binge drinkers among men and women college students using a confidence interval. The proportion for men is 0.260 (26.0%), and the proportion for women is 0.206 (20.6%). The difference is 0.054, and the 95% confidence interval is (0.039, 0.069). We can summarize this result by saying, "The proportion of frequent binge drinkers is 5.4% higher among men than among women."

Another way to analyze these data is to use logistic regression. The explanatory variable is gender, a categorical variable. To use this in a regression (logistic or otherwise), we need to use a numeric code. The usual way to do this is with an **indicator variable.** For our problem we will use an indicator of whether or not the student is a man:

indicator variable

$$x = \begin{cases} 1 & \text{if the student is a man} \\ 0 & \text{if the student is a woman} \end{cases}$$

The response variable is the proportion of frequent binge drinkers. For use in a logistic regression, we perform two transformations on this variable. First, we convert to odds. For men,

$$\begin{aligned} \text{odds} &= \frac{\hat{p}}{1 - \hat{p}} \\ &= \frac{0.260}{1 - 0.260} \\ &= 0.351 \end{aligned}$$

Similarly, for women we have

$$\begin{aligned} \text{odds} &= \frac{\hat{p}}{1 - \hat{p}} \\ &= \frac{0.206}{1 - 0.206} \\ &= 0.259 \end{aligned}$$

**USE YOUR KNOWLEDGE**

**14.3 Energy drink commercials.** A study was designed to compare two energy drink commercials. Each participant was shown the commercials, A and B, in random order and asked to select the better one. There were 100 women and 140 men who participated in the study. Commercial A was selected by 45 women and by 80 men. Find the odds of selecting Commercial A for the men. Do the same for the women.

**14.4 Find the odds.** Refer to the previous exercise. Find the odds of selecting Commercial B for the men. Do the same for the women.

## Model for logistic regression

In simple linear regression we modeled the mean $\mu$ of the response variable $y$ as a linear function of the explanatory variable: $\mu = \beta_0 + \beta_1 x$. With logistic regression we are interested in modeling the mean of the response variable $p$ in terms of an explanatory variable $x$. We could try to relate $p$ and $x$ through the equation $p = \beta_0 + \beta_1 x$. Unfortunately, this is not a good model. As long as $\beta_1 \neq 0$, extreme values of $x$ will give values of $\beta_0 + \beta_1 x$ that are inconsistent with the fact that $0 \leq p \leq 1$.

log odds

The logistic regression solution to this difficulty is to transform the odds $(p/(1-p))$ using the natural logarithm. We use the term **log odds** for this transformation. We model the log odds as a linear function of the explanatory variable:

$$\log\left(\frac{p}{1-p}\right) = \beta_0 + \beta_1 x$$

Figure 14.1 graphs the relationship between $p$ and $x$ for some different values of $\beta_0$ and $\beta_1$. For logistic regression we use *natural* logarithms. There are tables of natural logarithms, and many calculators have a built-in function for this transformation. As we did with linear regression, we use $y$ for the response variable.

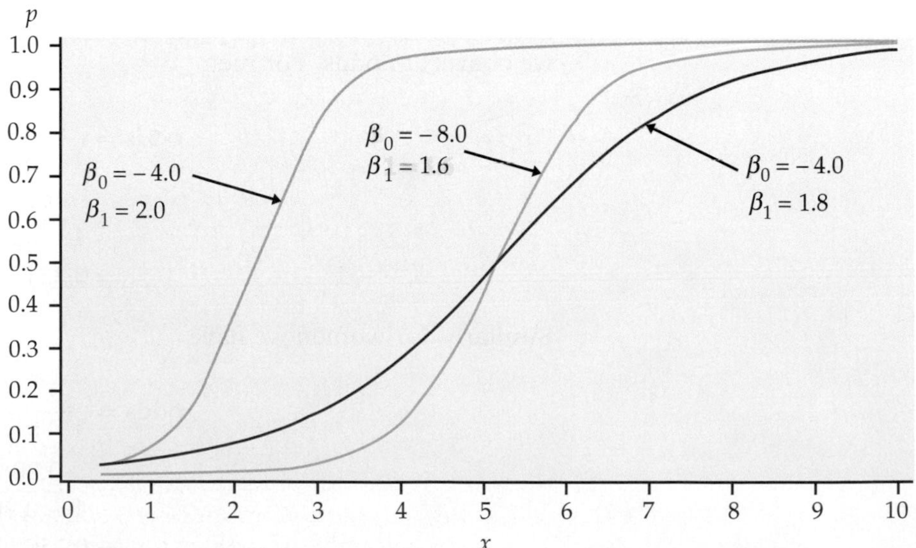


**FIGURE 14.1** Plot of $p$ versus $x$ for different logistic regression models.

So for men,

$$y = \log(\text{odds}) = \log(0.351) = -1.05$$

and for women,

$$y = \log(\text{odds}) = \log(0.259) = -1.35$$

## USE YOUR KNOWLEDGE

**14.5** **Find the odds.** Refer to Exercise 14.3. Find the log odds for the men and the log odds for the women.

**14.6** **Find the odds.** Refer to Exercise 14.4. Find the log odds for the men and the log odds for the women.

In these expressions for the log odds we use $y$ as the observed value of the response variable, the log odds of being a frequent binge drinker. We are now ready to build the logistic regression model.

### LOGISTIC REGRESSION MODEL

The **statistical model for logistic regression** is

$$\log\left(\frac{p}{1-p}\right) = \beta_0 + \beta_1 x$$

where $p$ is a binomial proportion and $x$ is the explanatory variable. The parameters of the logistic model are $\beta_0$ and $\beta_1$.

EXAMPLE

**14.3 Model for binge drinking.** For our binge-drinking example, there are $n = 13{,}819$ students in the sample. The explanatory variable is gender, which we have coded using an indicator variable with values $x = 1$ for men and $x = 0$ for women. The response variable is also an indicator variable. Thus, the student is either a frequent binge drinker or not a frequent binge drinker. Think of the process of randomly selecting a student and recording the value of $x$ and whether or not the student is a frequent binge drinker. The model says that the probability ($p$) that this student is a frequent binge drinker depends upon the student's gender ($x = 1$ or $x = 0$). So there are two possible values for $p$, say $p_{\text{men}}$ and $p_{\text{women}}$.

Logistic regression with an indicator explanatory variable is a very special case. It is important because many multiple logistic regression analyses focus on one or more such variables as the primary explanatory variables of interest. For now, we use this special case to understand a little more about the model.

The logistic regression model specifies the relationship between $p$ and $x$. Since there are only two values for $x$, we write both equations. For men,

$$\log\left(\frac{p_{\text{men}}}{1-p_{\text{men}}}\right) = \beta_0 + \beta_1$$

and for women,

$$\log\left(\frac{p_{\text{women}}}{1-p_{\text{women}}}\right) = \beta_0$$

Note that there is a $\beta_1$ term in the equation for men because $x = 1$, but it is missing in the equation for women because $x = 0$.

## Fitting and interpreting the logistic regression model

In general, the calculations needed to find estimates $b_0$ and $b_1$ for the parameters $\beta_0$ and $\beta_1$ are complex and require the use of software. When the explanatory variable has only two possible values, however, we can easily find the estimates. This simple framework also provides a setting where we can learn what the logistic regression parameters mean.

EXAMPLE

**14.4 Log odds for binge drinking.** In the binge-drinking example, we found the log odds for men,

$$y = \log\left(\frac{\hat{p}_{\text{men}}}{1-\hat{p}_{\text{men}}}\right) = -1.05$$

and for women,

$$y = \log\left(\frac{\hat{p}_{\text{women}}}{1-\hat{p}_{\text{women}}}\right) = -1.35$$

The logistic regression model for men is

$$\log\left(\frac{p_{\text{men}}}{1-p_{\text{men}}}\right) = \beta_0 + \beta_1$$

and for women it is

$$\log\left(\frac{p_{\text{women}}}{1-p_{\text{women}}}\right) = \beta_0$$

To find the estimates of $b_0$ and $b_1$, we match the male and female model equations with the corresponding data equations. Thus, we see that the estimate of the intercept $b_0$ is simply the log(odds) for the women:

$$b_0 = -1.35$$

and the slope is the difference between the log(odds) for the men and the log(odds) for the women:

$$b_1 = -1.05 - (-1.35) = 0.30$$

The fitted logistic regression model is

$$\log(\text{odds}) = -1.35 + 0.30x$$

The slope in this logistic regression model is the difference between the log(odds) for men and the log(odds) for women. Most people are not comfortable thinking in the log(odds) scale, so interpretation of the results in terms of the regression slope is difficult. Usually, we apply a transformation to help us. With a little algebra, it can be shown that

$$\frac{\text{odds}_{\text{men}}}{\text{odds}_{\text{women}}} = e^{0.30} = 1.34$$

odds ratio

The transformation $e^{0.30}$ undoes the logarithm and transforms the logistic regression slope into an **odds ratio,** in this case, the ratio of the odds that a man is a frequent binge drinker to the odds that a woman is a frequent binge drinker. In other words, we can multiply the odds for women by the odds ratio to obtain the odds for men:

$$\text{odds}_{\text{men}} = 1.34 \times \text{odds}_{\text{women}}$$

In this case, the odds for men are 1.34 times the odds for women.

Notice that we have chosen the coding for the indicator variable so that the regression slope is positive. This will give an odds ratio that is greater than 1. Had we coded women as 1 and men as 0, the signs of the parameters would be reversed, the fitted equation would be log(odds) $= 1.35 - 0.30x$, and the odds ratio would be $e^{-0.30} = 0.74$. The odds for women are 74% of the odds for men.

## USE YOUR KNOWLEDGE

**14.7 Find the logistic regression equation and the odds ratio.** Refer to Exercises 14.3 and 14.5. Find the logistic regression equation and the odds ratio.

**14.8 Find the logistic regression equation and the odds ratio.** Refer to Exercises 14.4 and 14.6. Find the logistic regression equation and the odds ratio.

Logistic regression with an explanatory variable having two values is a very important special case. Here is an example where the explanatory variable is quantitative.

EXAMPLE

**14.5 Predict whether or not the taste of the cheese is acceptable.** The CHEESE data set described in the Data Appendix includes a response variable called "Taste" that is a measure of the quality of the cheese in the opinions of several tasters. For this example, we will classify the cheese as acceptable (tasteok $= 1$) if Taste $\geq 37$ and unacceptable (tasteok $= 0$) if Taste $< 37$. This is our response variable. The data set contains three explanatory variables: "Acetic," "H2S," and "Lactic." Let's use Acetic as the explanatory variable. The model is

$$\log\left(\frac{p}{1-p}\right) = \beta_0 + \beta_1 x$$

where $p$ is the probability that the cheese is acceptable and $x$ is the value of Acetic. The model for estimated log odds fitted by software is

$$\log(\text{odds}) = b_0 + b_1 x = -13.71 + 2.25x$$

The odds ratio is $e^{b_1} = 9.48$. This means that if we increase the acetic acid content $x$ by one unit, we increase the odds that the cheese will be acceptable by about 9.5 times.

## 14.2 Inference for Logistic Regression

Statistical inference for logistic regression is very similar to statistical inference for simple linear regression. We calculate estimates of the model parameters and standard errors for these estimates. Confidence intervals are formed in the usual way, but we use standard Normal $z^*$-values rather than critical values from the $t$ distributions. The ratio of the estimate to the standard error is the basis for hypothesis tests. Often the test statistics are given as the squares of these ratios, and in this case the $P$-values are obtained from the chi-square distributions with 1 degree of freedom.

### Confidence Intervals and Significance Tests

#### CONFIDENCE INTERVALS AND SIGNIFICANCE TESTS FOR LOGISTIC REGRESSION PARAMETERS

A **level *C* confidence interval for the slope** $\beta_1$ is

$$b_1 \pm z^* \text{SE}_{b_1}$$

The ratio of the odds for a value of the explanatory variable equal to $x + 1$ to the odds for a value of the explanatory variable equal to $x$ is the **odds ratio.**

A **level *C* confidence interval for the odds ratio** $e^{\beta_1}$ is obtained by transforming the confidence interval for the slope

$$(e^{b_1 - z^* \text{SE}_{b_1}}, \; e^{b_1 + z^* \text{SE}_{b_1}})$$

In these expressions $z^*$ is the value for the standard Normal density curve with area $C$ between $-z^*$ and $z^*$.

To test the hypothesis $H_0\colon \beta_1 = 0$, compute the **test statistic**

$$z = \frac{b_1}{\text{SE}_{b_1}}$$

The $P$-value for the significance test of $H_0$ against $H_a\colon \beta_1 \neq 0$ is computed using the fact that, when the null hypothesis is true, $z$ has approximately a standard Normal distribution.

Wald statistic

The statistic $z$ is sometimes called a **Wald statistic.** Output from some statistical software reports the significance test result in terms of the square of the $z$ statistic.

$$X^2 = z^2$$

chi-square statistic

This statistic is called a **chi-square statistic.** When the null hypothesis is true, it has a distribution that is approximately a $\chi^2$ distribution with 1 degree of freedom, and the $P$-value is calculated as $P(\chi^2 \geq X^2)$. Because the square of a standard Normal random variable has a $\chi^2$ distribution with 1 degree of freedom, the $z$ statistic and the chi-square statistic give the same results for statistical inference.

We have expressed the hypothesis-testing framework in terms of the slope $\beta_1$ because this form closely resembles what we studied in simple linear regression. In many applications, however, the results are expressed in terms of the odds ratio. A slope of 0 is the same as an odds ratio of 1, so we often express the null hypothesis of interest as "the odds ratio is 1." This means that the two odds are equal and the explanatory variable is not useful for predicting the odds.

EXAMPLE

**14.6 Software output.** Figure 14.2 gives the output from SPSS and SAS for a different binge-drinking example that is similar to the one in Example 14.4. The parameter estimates are given as $b_0 = -1.5869$ and $b_1 = 0.3616$. The standard errors are 0.0267 and 0.0388. A 95% confidence interval for the slope is

$$\begin{aligned} b_1 \pm z^* \text{SE}_{b_1} &= 0.3616 \pm (1.96)(0.0388) \\ &= 0.3616 \pm 0.0760 \end{aligned}$$

We are 95% confident that the slope is between 0.2856 and 0.4376. The output provides the odds ratio 1.436 but does not give the confidence interval. This is easy to compute from the interval for the slope:

$$\begin{aligned} (e^{b_1 - z^* \text{SE}_{b_1}}, e^{b_1 + z^* \text{SE}_{b_1}}) &= (e^{0.2855}, e^{0.4376}) \\ &= (1.33, 1.55) \end{aligned}$$

For this problem we would report, "College men are more likely to be frequent binge drinkers than college women (odds ratio = 1.44, 95% CI = 1.33 to 1.55)."

In applications such as these, it is standard to use 95% for the confidence coefficient. With this convention, the confidence interval gives us the result of testing the null hypothesis that the odds ratio is 1 for a significance level of 0.05. If the confidence interval does not include 1, we reject $H_0$ and conclude that the odds for the two groups are different; if the interval does include 1, the data do not provide enough evidence to distinguish the groups in this way.

The following example is typical of many applications of logistic regression. Here there is a designed experiment with five different values for the explanatory variable.

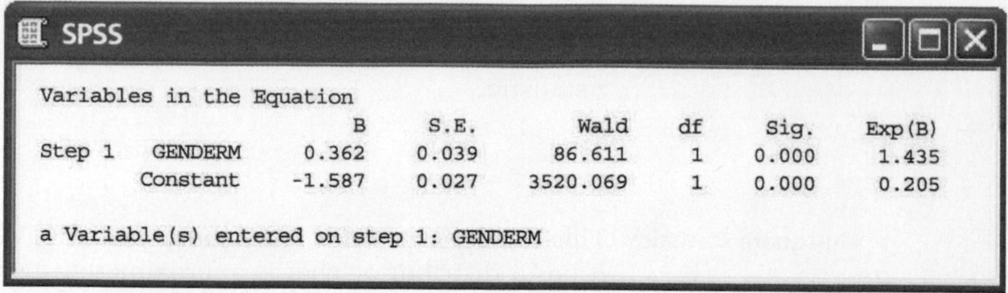

SPSS

Variables in the Equation

| | | B | S.E. | Wald | df | Sig. | Exp(B) |
|---|---|---|---|---|---|---|---|
| Step 1 | GENDERM | 0.362 | 0.039 | 86.611 | 1 | 0.000 | 1.435 |
| | Constant | -1.587 | 0.027 | 3520.069 | 1 | 0.000 | 0.205 |

a Variable(s) entered on step 1: GENDERM

SAS

The LOGISTIC Procedure

Analysis of Maximum Likelihood Estimates

| Parameter | DF | Estimate | Standard Error | Wald Chi-Square | Pr>ChiSq |
|---|---|---|---|---|---|
| Intercept | 1 | -1.5868 | 0.0267 | 3520.3120 | <0.001 |
| genderm | 1 | 0.3617 | 0.0388 | 86.6811 | <0.001 |

Odds Ratio Estimates

| Effect | Point Estimate | 95% Wald Confidence Limits | |
|---|---|---|---|
| genderm | 1.436 | 1.330 | 1.549 |

**FIGURE 14.2** Logistic regression output from SPSS and SAS for binge-drinking data, for Example 14.6.

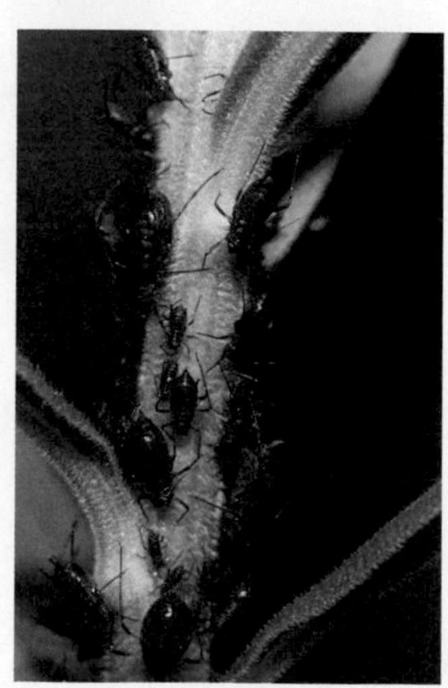

EXAMPLE

**14.7 An insecticide for aphids.** An experiment was designed to examine how well the insecticide rotenone kills an aphid, called *Macrosiphoniella sanborni*, that feeds on the chrysanthemum plant.[2] The explanatory variable is the concentration (in log of milligrams per liter) of the insecticide. At each concentration, approximately 50 insects were exposed. Each insect was either killed or not killed. We summarize the data using the number killed. The response variable for logistic regression is the log odds of the proportion killed. Here are the data:

| Concentration (log) | Number of insects | Number killed |
|---|---|---|
| 0.96 | 50 | 6 |
| 1.33 | 48 | 16 |
| 1.63 | 46 | 24 |
| 2.04 | 49 | 42 |
| 2.32 | 50 | 44 |

If we transform the response variable (by taking log odds) and use least squares, we get the fit illustrated in Figure 14.3. The logistic regression fit is given in Figure 14.4. It is a transformed version of Figure 14.3 with the fit calculated using the logistic model.

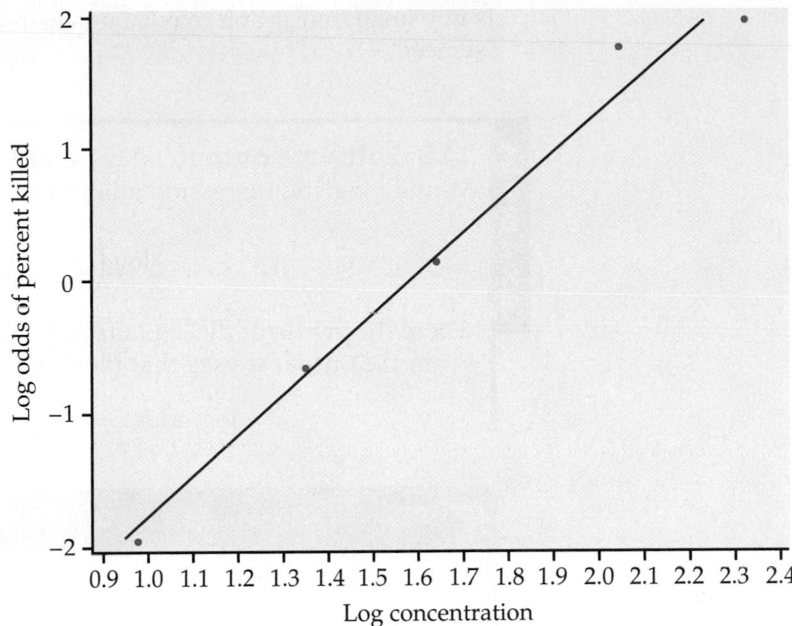


**FIGURE 14.3** Plot of log odds of percent killed versus log concentration for the insecticide data, for Example 14.7.

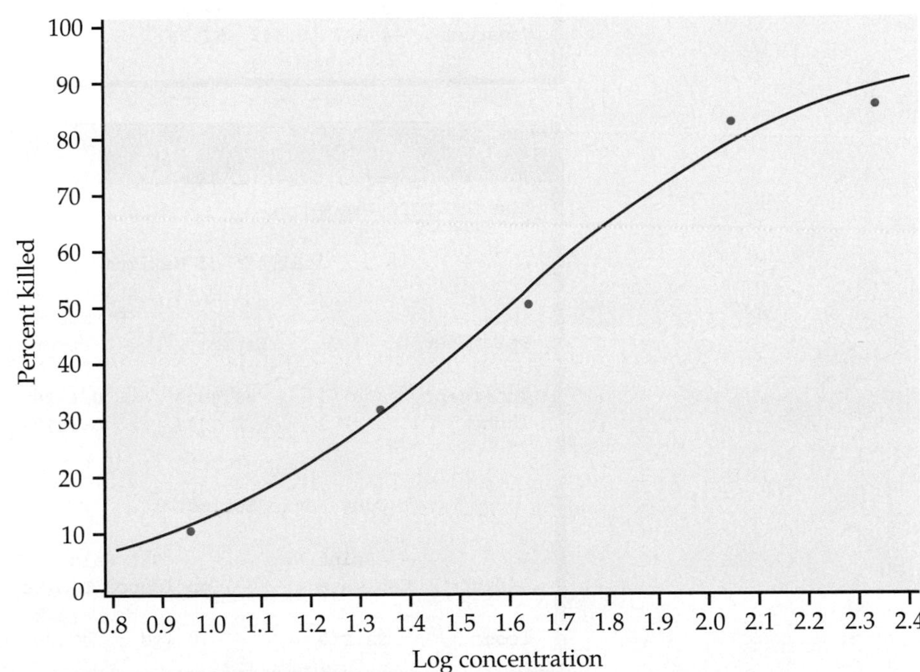


**FIGURE 14.4** Plot of the percent killed versus log concentration with the logistic regression fit for the insecticide data, for Example 14.7.

One of the major themes of this text is that we should present the results of a statistical analysis with a graph. For the insecticide example we have done this with Figure 14.4 and the results appear to be convincing. But suppose that rotenone has no ability to kill *Macrosiphoniella sanborni*. What is the chance that we would observe experimental results at least as convincing as what we observed if this supposition were true? The answer is the *P*-value for the test of the null hypothesis that the logistic regression slope is zero. If this *P*-value is not small, our graph may be misleading. Statistical inference provides what we need.

EXAMPLE

**14.8 Software output.** Figure 14.5 gives the output from SPSS, SAS, and Minitab logistic regression analysis of the insecticide data. The model is

$$\log\left(\frac{p}{1-p}\right) = \beta_0 + \beta_1 x$$

where the values of the explanatory variable $x$ are 0.96, 1.33, 1.63, 2.04, 2.32. From the output we see that the fitted model is

$$\log(\text{odds}) = b_0 + b_1 x = -4.89 + 3.10x$$

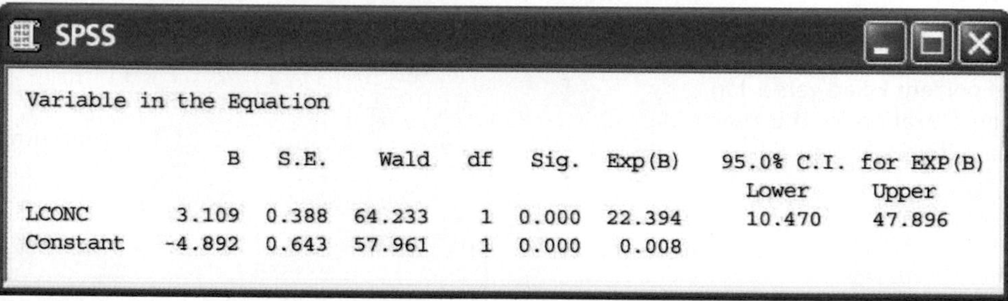

SPSS

Variable in the Equation

| | B | S.E. | Wald | df | Sig. | Exp(B) | 95.0% C.I. for EXP(B) Lower | Upper |
|---|---|---|---|---|---|---|---|---|
| LCONC | 3.109 | 0.388 | 64.233 | 1 | 0.000 | 22.394 | 10.470 | 47.896 |
| Constant | -4.892 | 0.643 | 57.961 | 1 | 0.000 | 0.008 | | |

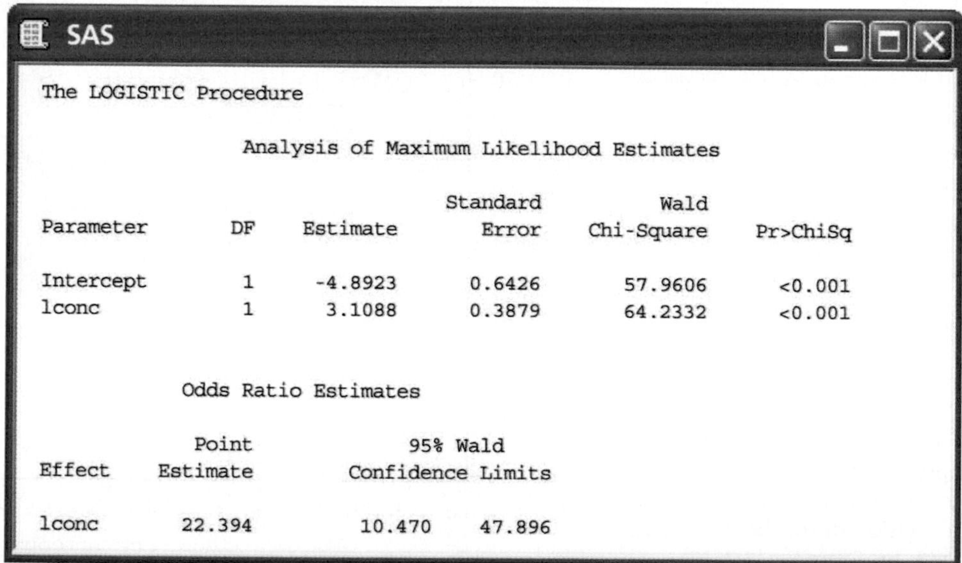

SAS

The LOGISTIC Procedure

Analysis of Maximum Likelihood Estimates

| Parameter | DF | Estimate | Standard Error | Wald Chi-Square | Pr>ChiSq |
|---|---|---|---|---|---|
| Intercept | 1 | -4.8923 | 0.6426 | 57.9606 | <0.001 |
| lconc | 1 | 3.1088 | 0.3879 | 64.2332 | <0.001 |

Odds Ratio Estimates

| Effect | Point Estimate | 95% Wald Confidence Limits | |
|---|---|---|---|
| lconc | 22.394 | 10.470 | 47.896 |

Minitab

Logistic Regression Table

| Predictor | Coef | StDev | Z | P | Odds Ratio | 95% CI Lower | Upper |
|---|---|---|---|---|---|---|---|
| Constant | -4.8923 | 0.6426 | -7.61 | 0.000 | | | |
| lconc | 3.1088 | 0.3879 | 8.01 | 0.000 | 22.39 | 10.47 | 47.90 |

**FIGURE 14.5** Logistic regression output from SPSS, SAS, and Minitab for the insecticide data, for Example 14.8.

This is the fit that we plotted in Figure 14.4. The null hypothesis that $\beta_1 = 0$ is clearly rejected ($X^2 = 64.23$, $P < 0.001$). We calculate a 95% confidence interval for $\beta_1$ using the estimate $b_1 = 3.1035$ and its standard error $\text{SE}_{b_1} = 0.3877$ given in the output:

$$b_1 \pm z^*\text{SE}_{b_1} = 3.1088 \pm (1.96)(0.3879)$$
$$= 3.1088 \pm 0.7603$$

We are 95% confident that the true value of the slope is between 2.34 and 3.86.

The odds ratio is given on the Minitab output as 22.39. An increase of one unit in the log concentration of insecticide ($x$) is associated with a 22-fold increase in the odds that an insect will be killed. The confidence interval for the odds is obtained from the interval for the slope:

$$(e^{b_1+z^*\text{SE}_{b_1}}, e^{b_1-z^*\text{SE}_{b_1}}) = (e^{2.3485}, e^{3.8691})$$
$$= (10.47, 47.90)$$

Note again that the test of the null hypothesis that the slope is 0 is the same as the test of the null hypothesis that the odds are 1. If we were reporting the results in terms of the odds, we could say, "The odds of killing an insect increase by a factor of 22.4 for each unit increase in the log concentration of insecticide ($X^2 = 64.23$, $P < 0.001$; 95% CI = 10.5 to 47.9)."

In Example 14.5 we studied the problem of predicting whether or not the taste of cheese was acceptable using Acetic as the explanatory variable. We now revisit this example and show how statistical inference is an important part of the conclusion.

EXAMPLE

**14.9 Software output.** Figure 14.6 gives the output from Minitab for a logistic regression analysis using Acetic as the explanatory variable. The fitted model is

$$\log(\text{odds}) = b_0 + b_1 x = -13.71 + 2.25x$$

This agrees up to rounding with the result reported in Example 14.5.

From the output we see that because $P = 0.029$, we can reject the null hypothesis that $\beta_1 = 0$. The value of the test statistic is $X^2 = 4.79$ with 1 degree of freedom. We use the estimate $b_1 = 2.249$ and its standard error $\text{SE}_{b_1} = 1.027$ to compute the 95% confidence interval for $\beta_1$:

Minitab

Logistic Regression Table

| Predictor | Coef | StDev | Z | P | Odds Ratio | 95% CI Lower | 95% CI Upper |
|---|---|---|---|---|---|---|---|
| Constant | -13.705 | 5.932 | -2.31 | 0.21 | | | |
| acetic | 2.249 | 1.027 | 2.19 | 0.029 | 9.48 | 1.27 | 70.96 |

**FIGURE 14.6** Logistic regression output from Minitab for the cheese data with Acetic as the explanatory variable, for Example 14.9.

$$b_1 \pm z^* \text{SE}_{b_1} = 2.249 \pm (1.96)(1.027)$$
$$= 2.249 \pm 2.0131$$

Our estimate of the slope is 2.25 and we are 95% confident that the true value is between 0.24 and 4.26. For the odds ratio, the estimate on the output is 9.48. The 95% confidence interval is

$$(e^{b_1 + z^* \text{SE}_{b_1}}, e^{b_1 - z^* \text{SE}_{b_1}}) = (e^{0.23588}, e^{4.26212})$$
$$= (1.27, 70.96)$$

We estimate that increasing the acetic acid content of the cheese by one unit will increase the odds that the cheese will be acceptable by about 9 times. The data, however, do not give us a very accurate estimate. The odds ratio could be as small as a little more than 1 or as large as 71 with 95% confidence. We have evidence to conclude that cheeses with higher concentrations of acetic acid are more likely to be acceptable, but establishing the true relationship accurately would require more data.

## Multiple logistic regression

multiple logistic regression

The cheese example that we just considered naturally leads us to the next topic. The data set includes three variables: Acetic, H2S, and Lactic. We examined the model where Acetic was used to predict the odds that the cheese was acceptable. Do the other explanatory variables contain additional information that will give us a better prediction? We use **multiple logistic regression** to answer this question. Generating the computer output is easy, just as it was when we generalized simple linear regression with one explanatory variable to multiple linear regression with more than one explanatory variable in Chapter 11. The statistical concepts are similar, although the computations are more complex. Here is the example.

EXAMPLE

**14.10 Software output.** As in Example 14.9, we predict the odds that the cheese is acceptable. The explanatory variables are Acetic, H2S, and Lactic. Figure 14.7 gives the outputs from SPSS, SAS, and Minitab for this analysis. The fitted model is

$$\log(\text{odds}) = b_0 + b_1 \,\text{Acetic} + b_2 \,\text{H2S} + b_3 \,\text{Lactic}$$
$$= -14.26 + 0.58 \,\text{Acetic} + 0.68 \,\text{H2S} + 3.47 \,\text{Lactic}$$

When analyzing data using multiple regression, we first examine the hypothesis that all of the regression coefficients for the explanatory variables are zero. We do the same for logistic regression. The hypothesis

$$H_0\colon \beta_1 = \beta_2 = \beta_3 = 0$$

is tested by a chi-square statistic with 3 degrees of freedom. For Minitab, this is given in the last line of the output and the statistic is called "G." The value is $G = 16.33$ and the $P$-value is 0.001. We reject $H_0$ and conclude that one or more of the explanatory variables can be used to predict the odds that the

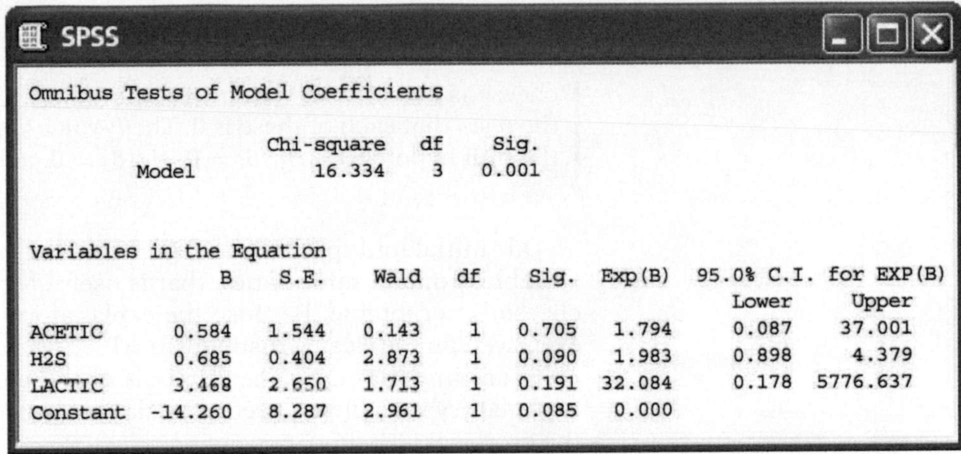

**SPSS**

Omnibus Tests of Model Coefficients

| | Chi-square | df | Sig. |
|---|---|---|---|
| Model | 16.334 | 3 | 0.001 |

Variables in the Equation

| | B | S.E. | Wald | df | Sig. | Exp(B) | 95.0% C.I. for EXP(B) Lower | Upper |
|---|---|---|---|---|---|---|---|---|
| ACETIC | 0.584 | 1.544 | 0.143 | 1 | 0.705 | 1.794 | 0.087 | 37.001 |
| H2S | 0.685 | 0.404 | 2.873 | 1 | 0.090 | 1.983 | 0.898 | 4.379 |
| LACTIC | 3.468 | 2.650 | 1.713 | 1 | 0.191 | 32.084 | 0.178 | 5776.637 |
| Constant | -14.260 | 8.287 | 2.961 | 1 | 0.085 | 0.000 | | |

**SAS**

Testing Global Null Hypothesis: BETA = 0

| Test | Chi-Square | DF | Pr>ChiSq |
|---|---|---|---|
| Likelihood Ratio | 16.3344 | 3 | 0.0010 |

Analysis of Maximum Likelihood Estimates

| Parameter | DF | Estimate | Standard Error | Wald Chi-Square | Pr>ChiSq |
|---|---|---|---|---|---|
| Intercept | 1 | -14.2604 | 8.2869 | 2.9613 | 0.0853 |
| acetic | 1 | 0.5845 | 1.5442 | 0.1433 | 0.7051 |
| h2s | 1 | 0.6848 | 0.4040 | 2.8730 | 0.0901 |
| lactic | 1 | 3.4684 | 2.6497 | 1.7135 | 0.1905 |

Odds Ratio Estimates

| Effect | Point Estimate | 95% Wald Confidence Limits | |
|---|---|---|---|
| acetic | 1.794 | 10.087 | 37.004 |
| h2s | 1.983 | 0.898 | 4.379 |
| lactic | 32.086 | 0.178 | >999.999 |

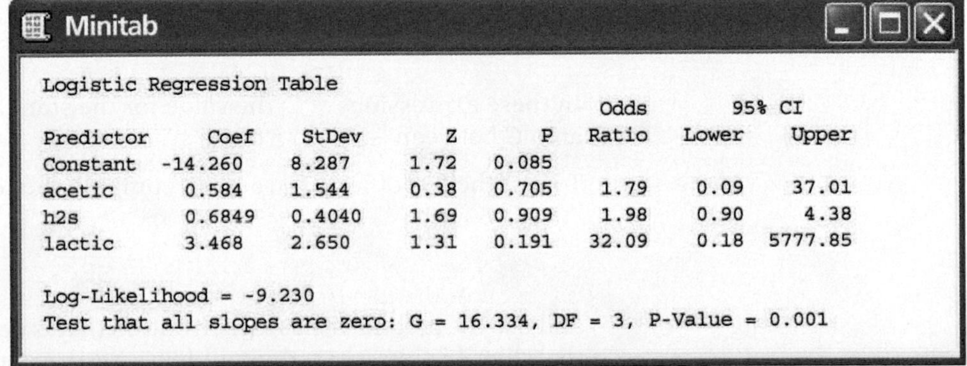

**Minitab**

Logistic Regression Table

| Predictor | Coef | StDev | Z | P | Odds Ratio | 95% CI Lower | Upper |
|---|---|---|---|---|---|---|---|
| Constant | -14.260 | 8.287 | 1.72 | 0.085 | | | |
| acetic | 0.584 | 1.544 | 0.38 | 0.705 | 1.79 | 0.09 | 37.01 |
| h2s | 0.6849 | 0.4040 | 1.69 | 0.909 | 1.98 | 0.90 | 4.38 |
| lactic | 3.468 | 2.650 | 1.31 | 0.191 | 32.09 | 0.18 | 5777.85 |

Log-Likelihood = -9.230

Test that all slopes are zero: G = 16.334, DF = 3, P-Value = 0.001

**FIGURE 14.7** Logistic regression output from SPSS, SAS, and Minitab for the cheese data with Acetic, H2S, and Lactic as the explanatory variables, for Example 14.10.

cheese is acceptable. We now examine the coefficients for each variable and the tests that each of these is 0. The $P$-values are 0.71, 0.09, and 0.19. None of the null hypotheses, $H_0\colon \beta_1 = 0$, $H_0\colon \beta_2 = 0$, and $H_0\colon \beta_3 = 0$, can be rejected.

Our initial multiple logistic regression analysis told us that the explanatory variables contain information that is useful for predicting whether or not the cheese is acceptable. Because the explanatory variables are correlated, however, we cannot clearly distinguish which variables or combinations of variables are important. Further analysis of these data using subsets of the three explanatory variables is needed to clarify the situation. We leave this work for the exercises.

## SECTION 14.2 Summary

If $\hat{p}$ is the sample proportion, then the **odds** are $\hat{p}/(1-\hat{p})$, the ratio of the proportion of times the event happens to the proportion of times the event does not happen.

The **logistic regression model** relates the log of the odds to the explanatory variable:

$$\log\left(\frac{p_i}{1-p_i}\right) = \beta_0 + \beta_1 x_i$$

where the response variables for $i = 1, 2, \ldots, n$ are independent binomial random variables with parameters 1 and $p_i$; that is, they are independent with distributions $B(1, p_i)$. The explanatory variable is $x$.

The **parameters** of the logistic model are $\beta_0$ and $\beta_1$.

The **odds ratio** is $e^{\beta_1}$, where $\beta_1$ is the slope in the logistic regression model.

A **level *C* confidence interval for the intercept** $\beta_0$ is

$$b_0 \pm z^* \text{SE}_{b_0}$$

A **level *C* confidence interval for the slope** $\beta_1$ is

$$b_1 \pm z^* \text{SE}_{b_1}$$

A **level *C* confidence interval for the odds ratio** $e^{\beta_1}$ is obtained by transforming the confidence interval for the slope

$$(e^{b_1 - z^* \text{SE}_{b_1}}, e^{b_1 + z^* \text{SE}_{b_1}})$$

In these expressions $z^*$ is the value for the standard Normal density curve with area $C$ between $-z^*$ and $z^*$.

To test the hypothesis $H_0\colon \beta_1 = 0$, compute the **test statistic**

$$z = \frac{b_1}{\text{SE}_{b_1}}$$

and use the fact that $z$ has a distribution that is approximately the standard Normal distribution when the null hypothesis is true. This statistic is sometimes called the **Wald statistic.** An alternative equivalent procedure is to report the square of $z$,

$$X^2 = z^2$$

This statistic has a distribution that is approximately a $\chi^2$ distribution with 1 degree of freedom, and the $P$-value is calculated as $P(\chi^2 \geq X^2)$. This is the same as testing the null hypothesis that the odds ratio is 1.

In **multiple logistic regression** the response variable has two possible values, as in logistic regression, but there can be several explanatory variables.

## CHAPTER 14 Exercises

*For Exercises 14.1 and 14.2, see page 713; for Exercises 14.3 and 14.4, see page 714; for Exercises 14.5 and 14.6, see page 715; and for Exercises 14.7 and 14.8, see page 717.*

**14.9 What's wrong?** For each of the following, explain what is wrong and why.

(a) For a multiple logistic regression with 6 explanatory variables, the null hypothesis that the regression coefficients of all of the explanatory variables are zero is tested with an $F$ test.

(b) In logistic regression with one explanatory variable we can use a chi-square statistic to test the null hypothesis $H_0 : b_1 = 0$ versus a two-sided alternative.

(c) For a logistic regression we assume that the error term in our model has a Normal distribution.

**14.10 Find the logistic regression equation and the odds ratio.** A study of 170 franchise firms classified each firm as to whether it was successful or not and whether or not it had an exclusive territory.[3] Here are the data:

Observed numbers of firms

| | Exclusive territory | | |
|---|---|---|---|
| Success | Yes | No | Total |
| Yes | 108 | 15 | 123 |
| No | 34 | 13 | 47 |
| Total | 142 | 28 | 170 |

(a) What proportion of the exclusive-territory firms are successful?

(b) Find the proportion for the firms that do not have exclusive territories.

(c) Convert the proportion you found in part (a) to odds. Do the same for the proportion you found in part (b).

(d) Find the log of each of the odds that you found in part (c).

**14.11 "No Sweat" labels on clothing.** Following complaints about the working conditions in some apparel factories both in the United States and abroad, a joint government and industry commission recommended in 1998 that companies that monitor and enforce proper standards be allowed to display a "No Sweat" label on their products. Does the presence of these labels influence consumer behavior?

A survey of U.S. residents aged 18 or older asked a series of questions about how likely they would be to purchase a garment under various conditions. For some conditions, it was stated that the garment had a "No Sweat" label; for others, there was no mention of such a label. On the basis of the responses, each person was classified as a "label user" or a "label nonuser."[4] Suppose we want to examine the data for a possible gender effect. Here are the data for comparing women and men:

| Gender | $n$ | Number of label users |
|---|---|---|
| Women | 296 | 63 |
| Men | 251 | 27 |

(a) For each gender find the proportion of label users.

(b) Convert each of the proportions that you found in part (a) to odds.

(c) Find the log of each of the odds that you found in part (b).

**14.12 Exclusive territories for franchises.** Refer to Exercise 14.10. Use $x = 1$ for the exclusive territories and $x = 0$ for the other territories.

(a) Find the estimates $b_0$ and $b_1$.

(b) Give the fitted logistic regression model.

(c) What is the odds ratio for exclusive territory versus no exclusive territory?

**14.13 "No Sweat" labels on clothing.** Refer to Exercise 14.11. Use $x = 1$ for women and $x = 0$ for men.

(a) Find the estimates $b_0$ and $b_1$.

(b) Give the fitted logistic regression model.

(c) What is the odds ratio for women versus men?

**14.14** CHALLENGE **Interpret the fitted model.** If we apply the exponential function to the fitted model in Example 14.9, we get

$$\text{odds} = e^{-13.71+2.25x} = e^{-13.71} \times e^{2.25x}$$

Show that, for any value of the quantitative explanatory variable $x$, the odds ratio for increasing $x$ by 1,

$$\frac{\text{odds}_{x+1}}{\text{odds}_x}$$

is $e^{2.25} = 9.49$. This justifies the interpretation given after Example 14.9.

**14.15 Give a 99% confidence interval for $\beta_1$.** Refer to Example 14.8. Suppose that you wanted to report a 99% confidence interval for $\beta_1$. Show how you would use the information provided in the outputs shown in Figure 14.5 to compute this interval.

**14.16 Give a 99% confidence interval for the odds ratio.** Refer to Example 14.8 and the outputs in Figure 14.5. Using the estimate $b_1$ and its standard error, find the 95% confidence interval for the odds ratio and verify that this agrees with the interval given by the software.

**14.17** CHALLENGE **$z$ and the $X^2$ statistic.** The Minitab output in Figure 14.5 does not give the value of $X^2$. The column labeled "Z" provides similar information.

(a) Find the value under the heading "Z" for the predictor lconc. Verify that Z is simply the estimated coefficient divided by its standard error. This is a $z$ statistic that has approximately the standard Normal distribution if the null hypothesis (slope 0) is true.

(b) Show that the square of $z$ is $X^2$. The two-sided $P$-value for $z$ is the same as $P$ for $X^2$.

(c) Draw sketches of the standard Normal and the chi-square distribution with 1 degree of freedom. (*Hint:* You can use the information in Table F to sketch the chi-square distribution.) Indicate the value of the $z$ and the $X^2$ statistics on these sketches and use shading to illustrate the $P$-value.

**14.18 Sexual imagery in magazine ads.** Exercise 9.18 (page 551) presents some results of a study about how advertisers use sexual imagery to appeal to young people. The clothing worn by the model in each of 1509 ads was classified as "not sexual" or "sexual" based on a standardized criterion. A logistic regression was used to describe the probability that the clothing in the ad was "not sexual" as a function of several explanatory variables. Here are some of the reported results:

| Explanatory variable | $b$ | Wald ($z$) statistic |
|---|---|---|
| Reader age | 0.50 | 13.64 |
| Model gender | 1.31 | 72.15 |
| Men's magazines | −0.05 | 0.06 |
| Women's magazines | 0.45 | 6.44 |
| Constant | −2.32 | 135.92 |

Reader age is coded as 0 for young adult and 1 for mature adult. Therefore, the coefficient of 0.50 for this explanatory variable suggests that the probability that the model clothing is *not* sexual is higher when the target reader age is mature adult. In other words, the model clothing is more likely to be sexual when the target reader age is young adult. Model gender is coded as 0 for female and 1 for male. The explanatory variable men's magazines is 1 if the intended readership is men and is 0 for women's magazines and magazines intended for both men and women. Women's magazines is coded similarly.

(a) State the null and alternative hypotheses for each of the explanatory variables.

(b) Perform the significance tests associated with the Wald statistics.

(c) Interpret the sign of each of the statistically significant coefficients in terms of the probability that the model clothing is sexual.

(d) Write an equation for the fitted logistic regression model.

**14.19 Interpret the odds ratios.** Refer to the previous exercise. The researchers also reported odds ratios with 95% confidence intervals for this logistic regression model. Here is a summary:

| Explanatory variable | Odds ratio | 95% Confidence Limits Lower | Upper |
|---|---|---|---|
| Reader age | 1.65 | 1.27 | 2.16 |
| Model gender | 3.70 | 2.74 | 5.01 |
| Men's magazines | 0.96 | 0.67 | 1.37 |
| Women's magazines | 1.57 | 1.11 | 2.23 |

(a) Explain the relationship between the confidence intervals reported here and the results of the Wald $z$ significance tests that you found in the previous exercise.

(b) Interpret the results in terms of the odds ratios.

(c) Write a short summary explaining the results. Include comments regarding the usefulness of the fitted coefficients versus the odds ratios.

**14.20 What purchases will be made?** A poll of 811 adults aged 18 or older asked about purchases that they intended to make for the upcoming holiday season.[5] One of the questions asked what kind of gift they intended to buy for the person on whom they intended to spend the most. Clothing was the first choice of 487 people.

(a) What proportion of adults said that clothing was their first choice?

(b) What are the odds that an adult will say that clothing is his or her first choice?

(c) What proportion of adults said that something other than clothing was their first choice?

(d) What are the odds that an adult will say that something other than clothing is his or her first choice?

(e) How are your answers to parts (a) through (d) related?

**14.21 High-tech companies and stock options.** Different kinds of companies compensate their key employees in different ways. Established companies may pay higher salaries, while new companies may offer stock options that will be valuable if the company succeeds. Do high-tech companies tend to offer stock options more often than other companies? One study looked at a random sample of 200 companies. Of these, 91 were listed in the *Directory of Public High Technology Corporations*, and 109 were not listed. Treat these two groups as SRSs of high-tech and non-high-tech companies. Seventy-three of the high-tech companies and 75 of the non-high-tech companies offered incentive stock options to key employees.[6]

(a) What proportion of the high-tech companies offer stock options to their key employees? What are the odds?

(b) What proportion of the non-high-tech companies offer stock options to their key employees? What are the odds?

(c) Find the odds ratio using the odds for the high-tech companies in the numerator. Describe the result in a few sentences.

**14.22 High-tech companies and stock options.** Refer to the previous exercise.

(a) Find the log odds for the high-tech firms. Do the same for the non-high-tech firms.

(b) Define an explanatory variable $x$ to have the value 1 for high-tech firms and 0 for non-high-tech firms. For the logistic model, we set the log odds equal to $\beta_0 + \beta_1 x$. Find the estimates $b_0$ and $b_1$ for the parameters $\beta_0$ and $\beta_1$.

(c) Show that the odds ratio is equal to $e^{b_1}$.

**14.23 High-tech companies and stock options.** Refer to Exercises 14.21 and 14.23. Software gives 0.3347 for the standard error of $b_1$.

(a) Find the 95% confidence interval for $\beta_1$.

(b) Transform your interval in (a) to a 95% confidence interval for the odds ratio.

(c) What do you conclude?

**14.24 High-tech companies and stock options.** Refer to Exercises 14.21 to 14.23. Repeat the calculations assuming that you have twice as many observations with the same proportions. In other words, assume that there are 182 high-tech firms and 218 non-high-tech firms. The numbers of firms offering stock options are 146 for the high-tech group and 150 for the non-high-tech group. The standard error of $b_1$ for this scenario is 0.2366. Summarize your results, paying particular attention to what remains the same and what is different from what you found in Exercises 14.21 to 14.23.

**14.25 High blood pressure and cardiovascular disease.** There is much evidence that high blood pressure is associated with increased risk of death from cardiovascular disease. A major study of this association examined 3338 men with high blood pressure and 2676 men with low blood pressure. During the period of the study, 21 men in the low-blood-pressure group and 55 in the high-blood-pressure group died from cardiovascular disease.

(a) Find the proportion of men who died from cardiovascular disease in the high-blood-pressure group. Then calculate the odds.

(b) Do the same for the low-blood-pressure group.

(c) Now calculate the odds ratio with the odds for the high-blood-pressure group in the numerator. Describe the result in words.

**14.26 Gender bias in syntax textbooks.** To what extent do syntax textbooks, which analyze the structure of sentences, illustrate gender bias? A study of this question sampled sentences from 10 texts.[7] One part of the study examined the use of the words "girl," "boy," "man," and "woman." We will call the first two words juvenile and the last two adult. Here are data from one of the texts:

| Gender | $n$ | $X$(juvenile) |
|---|---|---|
| Female | 60 | 48 |
| Male | 132 | 52 |

(a) Find the proportion of the female references that are juvenile. Then transform this proportion to odds.

(b) Do the same for the male references.

(c) What is the odds ratio for comparing the female references to the male references? (Put the female odds in the numerator.)

**14.27 High blood pressure and cardiovascular disease.** Refer to the study of cardiovascular disease and blood pressure in Exercise 14.25. Computer output for a logistic regression analysis of these data gives the estimated slope $b_1 = 0.7505$ with standard error $\text{SE}_{b_1} = 0.2578$.

(a) Give a 95% confidence interval for the slope.

(b) Calculate the $X^2$ statistic for testing the null hypothesis that the slope is zero and use Table F to find an approximate $P$-value.

(c) Write a short summary of the results and conclusions.

**14.28 Gender bias in syntax textbooks.** The data from the study of gender bias in syntax textbooks given in Exercise 14.26 are analyzed using logistic regression. The estimated slope is $b_1 = 1.8171$ and its standard error is $\text{SE}_{b_1} = 0.3686$.

(a) Give a 95% confidence interval for the slope.

(b) Calculate the $X^2$ statistic for testing the null hypothesis that the slope is zero and use Table F to find an approximate $P$-value.

(c) Write a short summary of the results and conclusions.

**14.29 High blood pressure and cardiovascular disease.** The results describing the relationship between blood pressure and cardiovascular disease are given in terms of the change in log odds in Exercise 14.27.

(a) Transform the slope to the odds and the 95% confidence interval for the slope to a 95% confidence interval for the odds.

(b) Write a conclusion using the odds to describe the results.

**14.30 Gender bias in syntax textbooks.** The gender bias in syntax textbooks is described in the log odds scale in Exercise 14.28.

(a) Transform the slope to the odds and the 95% confidence interval for the slope to a 95% confidence interval for the odds.

(b) Write a conclusion using the odds to describe the results.

**14.31 Reducing the number of workers.** To be competitive in global markets, many corporations are undertaking major reorganizations. Often these involve "downsizing" or a "reduction in force" (RIF), where substantial numbers of employees are terminated. Federal and various state laws require that employees be treated equally regardless of their age. In particular, employees over the age of 40 years are in a "protected" class, and many allegations of discrimination focus on comparing employees over 40 with their younger coworkers. Here are the data for a recent RIF:

| | Over 40 | |
|---|---|---|
| **Terminated** | **No** | **Yes** |
| Yes | 7 | 41 |
| No | 504 | 765 |

(a) Write the logistic regression model for this problem using the log odds of a RIF as the response variable and an indicator for over and under 40 years of age as the explanatory variable.

(b) Explain the assumption concerning binomial distributions in terms of the variables in this exercise. To what extent do you think that these assumptions are reasonable?

(c) Software gives the estimated slope $b_1 = 1.3504$ and its standard error $SE_{b_1} = 0.4130$. Transform the results to the odds scale. Summarize the results and write a short conclusion.

(d) If additional explanatory variables were available, for example, a performance evaluation, how would you use this information to study the RIF?

**14.32 Repair times for golf clubs.** The Ping Company makes custom-built golf clubs and competes in the $4 billion golf equipment industry. To improve its business processes, Ping decided to seek ISO 9001 certification.[8] As part of this process, a study of the time it took to repair golf clubs sent to the company by mail determined that 16% of orders were sent back to the customers in 5 days or less. Ping examined the processing of repair orders and made changes. Following the changes, 90% of orders were completed within 5 days. Assume that each of the estimated percents is based on a random sample of 200 orders. Use logistic regression to examine how the odds that an order will be filled in 5 days or less has improved. Write a short report summarizing your results.

**14.33 Education level of customers.** To devise effective marketing strategies it is helpful to know the characteristics of your customers. A study compared demographic characteristics of people who use the Internet for travel arrangements and of people who do not.[9] Of 1132 Internet users, 643 had completed college. Among the 852 nonusers, 349 had completed college. Model the log odds of using the Internet to make travel arrangements with an indicator variable for having completed college as the explanatory variable. Summarize your findings.

**14.34 Income level of customers.** The study mentioned in the previous exercise also asked about income. Among Internet users, 493 reported income of less than $50,000 and 378 reported income of $50,000 or more. (Not everyone answered the income question.) The corresponding numbers for nonusers were 477 and 200. Repeat the analysis using an indicator variable for income of $50,000 or more as the explanatory variable. What do you conclude?

**14.35 Alcohol use and bicycle accidents.** A study of alcohol use and deaths due to bicycle accidents collected data on a large number of fatal accidents.[10] For each of these, the individual who died was classified according to whether or not there was a positive test for alcohol and by gender. Here are the data:

| Gender | $n$ | $X$(tested positive) |
|---|---|---|
| Female | 191 | 27 |
| Male | 1520 | 515 |

Use logistic regression to study the question of whether or not gender is related to alcohol use in people who are fatally injured in bicycle accidents.

**14.36 The amount of acetic acid predicts the taste of cheese.** In Examples 14.5 and 14.9, we analyzed data from the CHEESE data set described in the Data Appendix. In those examples, we used Acetic as the explanatory variable. Run the same analysis using H2S as the explanatory variable.

**14.37 What about lactic acid?** Refer to the previous exercise. Run the same analysis using Lactic as the explanatory variable.

**14.38** CHALLENGE **Compare the analyses.** For the cheese data analyzed in Examples 14.9, 14.10, and the two exercises above, there are three explanatory variables. There are three different logistic regressions that include two explanatory variables. Run these. Summarize the results of these analyses, the ones using each explanatory variable alone, and the one using all three explanatory variables together. What do you conclude?

*The following four exercises use the CSDATA data set described in the Data Appendix. We examine models for relating success as measured by the GPA to several explanatory variables. In Chapter 11 we used multiple regression methods for our analysis. Here, we define an indicator variable, say HIGPA, to be 1 if the GPA is 3.0 or better and 0 otherwise.*

**14.39** CHALLENGE **Use high school grades to predict high grade point averages.** Use a logistic regression to predict HIGPA using the three high school grade summaries as explanatory variables.

(a) Summarize the results of the hypothesis test that the coefficients for all three explanatory variables are zero.

(b) Give the coefficient for high school math grades with a 95% confidence interval. Do the same for the two other predictors in this model.

(c) Summarize your conclusions based on parts (a) and (b).

**14.40** CHALLENGE **Use SAT scores to predict high grade point averages.** Use a logistic regression to predict HIGPA using the two SAT scores as explanatory variables.

(a) Summarize the results of the hypothesis test that the coefficients for both explanatory variables are zero.

(b) Give the coefficient for the SAT Math score with a 95% confidence interval. Do the same for the SAT Verbal score.

(c) Summarize your conclusions based on parts (a) and (b).

**14.41** CHALLENGE **Use high school grades and SAT scores to predict high grade point averages.** Run a logistic regression to predict HIGPA using the three high school grade summaries and the two SAT scores as explanatory variables. We want to produce an analysis that is similar to that done for the case study in Chapter 11.

(a) Test the null hypothesis that the coefficients of the three high school grade summaries are zero; that is, test $H_0\colon \beta_{\text{HSM}} = \beta_{\text{HSS}} = \beta_{\text{HSE}} = 0$.

(b) Test the null hypothesis that the coefficients of the two SAT scores are zero; that is, test $H_0\colon \beta_{\text{SATM}} = \beta_{\text{SATV}} = 0$.

(c) What do you conclude from the tests in (a) and (b)?

**14.42** CHALLENGE **Is there an effect of gender?** In this exercise we investigate the effect of gender on the odds of getting a high GPA.

(a) Use gender to predict HIGPA using a logistic regression. Summarize the results.

(b) Perform a logistic regression using gender and the two SAT scores to predict HIGPA. Summarize the results.

(c) Compare the results of parts (a) and (b) with respect to how gender relates to HIGPA. Summarize your conclusions.

**14.43** CHALLENGE **An example of Simpson's paradox.** Here is an example of Simpson's paradox, *the reversal of the direction of a comparison or an association when data from several groups are combined to form a single group*. The data concern two hospitals, A and B, and whether or not patients undergoing surgery died or survived. Here are the data for all patients:

| | Hospital A | Hospital B |
|---|---|---|
| Died | 63 | 16 |
| Survived | 2037 | 784 |
| Total | 2100 | 800 |

And here are the more detailed data where the patients are categorized as being in good condition or poor condition:

| Good condition | | |
|---|---|---|
| | **Hospital A** | **Hospital B** |
| Died | 6 | 8 |
| Survived | 594 | 592 |
| Total | 600 | 600 |
| **Poor condition** | | |
| | **Hospital A** | **Hospital B** |
| Died | 57 | 8 |
| Survived | 1443 | 192 |
| Total | 1500 | 200 |

(a) Use a logistic regression to model the odds of death with hospital as the explanatory variable. Summarize the results of your analysis and give a 95% confidence interval for the odds ratio of Hospital A relative to Hospital B.

(b) Rerun your analysis in (a) using hospital and the condition of the patient as explanatory variables. Summarize the results of your analysis and give a 95% confidence interval for the odds ratio of Hospital A relative to Hospital B.

(c) Explain Simpson's paradox in terms of your results in parts (a) and (b).

CHAPTER

15

# Nonparametric Tests

Is corn yield reduced by the presence of the common weed called lamb's-quarter? See Example 15.1 for more details.



## Introduction

The most commonly used methods for inference about the means of quantitative response variables assume that the variables in question have Normal distributions in the population or populations from which we draw our data. In practice, of course, no distribution is exactly Normal. Fortunately, our usual methods for inference about population means (the one-sample and two-sample $t$ procedures and analysis of variance) are quite **robust.** That is, the results of inference are not very sensitive to moderate lack of Normality, especially when the samples are reasonably large. Some practical guidelines for taking advantage of the robustness of these methods appear in Chapter 7.

robustness

What can we do if plots suggest that the data are clearly not Normal, especially when we have only a few observations? This is not a simple question. Here are the basic options:

1. If lack of Normality is due to **outliers,** it may be legitimate to remove the outliers. An outlier is an observation that may not come from the same population as the others. Equipment failure that produced a bad measurement, for example, entitles you to remove the outlier and analyze the remaining data. If the outlier appears to be "real data," you can base inference on statistics that are more resistant than $\bar{x}$ and $s$. Options 4 and 5 below allow this.

outlier

transforming data

2. Sometimes we can **transform** our data so that their distribution is more nearly Normal. Transformations such as the logarithm that pull in the long tail of right-skewed distributions are particularly helpful. Example 7.10 (page 436) illustrates use of the logarithm. A detailed discussion of transformations appears in the extra material entitled *Transforming Relationships* available on the text CD and Web site.

other standard distributions

3. In some settings, **other standard distributions** replace the Normal distributions as models for the overall pattern in the population. We mentioned in Chapter 5 (page 344) that the Weibull distributions are common models for the lifetimes in service of equipment in statistical studies of reliability. There are inference procedures for the parameters of these distributions that replace the $t$ procedures when we use specific non-Normal models.

bootstrap methods
permutation tests

4. Modern **bootstrap methods** and **permutation tests** do not require Normality or any other specific form of sampling distribution. Moreover, you can base inference on resistant statistics such as the trimmed mean. We recommend these methods unless the sample is so small that it may not represent the population well. Chapter 16 gives a full discussion.

nonparametric methods

rank tests

5. Finally, there are other **nonparametric methods** that do not require any specific form for the distribution of the population. Unlike bootstrap and permutation methods, common nonparametric methods do not make use of the actual values of the observations. The *sign test* (page 438) works with *counts* of observations. This chapter presents **rank tests** based on the *rank* (place in order) of each observation in the set of all the data.

continuous distribution

This chapter concerns rank tests that are designed to replace the $t$ tests and one-way analysis of variance when the Normality conditions for those tests are not met. Figure 15.1 presents an outline of the standard tests (based on Normal distributions) and the rank tests that compete with them. All of these tests require that the population or populations have **continuous distributions.** That is, each distribution must be described by a density curve that allows observations to take any value in some interval of outcomes. The Normal curves are one shape of density curve. Rank tests allow curves of any shape.

The rank tests we will study concern the *center* of a population or populations. When a population has at least roughly a Normal distribution, we describe its center by the mean. The "Normal tests" in Figure 15.1 all test hy-

| Setting | Normal test | Rank test |
|---|---|---|
| One sample | One-sample $t$ test<br>Section 7.1 | Wilcoxon signed rank test<br>Section 15.2 |
| Matched pairs | Apply one-sample test to differences within pairs | |
| Two independent samples | Two-sample $t$ test<br>Section 7.2 | Wilcoxon rank sum test<br>Section 15.1 |
| Several independent samples | One-way ANOVA $F$ test<br>Chapter 12 | Kruskal-Wallis test<br>Section 15.3 |

**FIGURE 15.1** Comparison of tests based on Normal distributions with nonparametric tests for similar settings.

potheses about population means. When distributions are strongly skewed, we often prefer the median to the mean as a measure of center. In simplest form, the hypotheses for rank tests just replace mean by median.

We devote a section of this chapter to each of the rank procedures. Section 15.1, which discusses the most common of these tests, also contains general information about rank tests. The kind of assumptions required, the nature of the hypotheses tested, the big idea of using ranks, and the contrast between exact distributions for use with small samples and approximations for use with larger samples are common to all rank tests. Sections 15.2 and 15.3 more briefly describe other rank tests.

## 15.1 The Wilcoxon Rank Sum Test

Two-sample problems (see Section 7.2) are among the most common in statistics. The most useful nonparametric significance test compares two distributions. Here is an example of this setting.

EXAMPLE

**15.1 Weeds and corn yield.** Does the presence of small numbers of weeds reduce the yield of corn? Lamb's-quarter is a common weed in corn fields. A researcher planted corn at the same rate in 8 small plots of ground, then weeded the corn rows by hand to allow no weeds in 4 randomly selected plots and exactly 3 lamb's-quarter plants per meter of row in the other 4 plots. Here are the yields of corn (bushels per acre) in each of the plots:[1]

| Weeds per meter | Yield (bu/acre) | | | |
|---|---|---|---|---|
| 0 | 166.7 | 172.2 | 165.0 | 176.9 |
| 3 | 158.6 | 176.4 | 153.1 | 156.0 |

Normal quantile plots (Figure 15.2) suggest that the data may be right-skewed. The samples are too small to assess Normality adequately or to rely

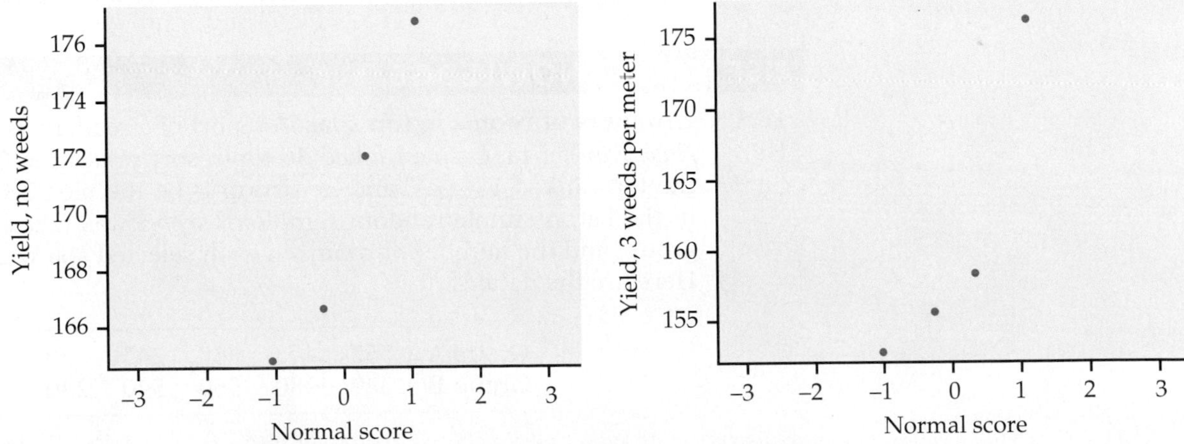


**FIGURE 15.2** Normal quantile plots of corn yields from plots with no weeds (left) and with 3 weeds per meter of row (right), for Example 15.1.

on the robustness of the two-sample $t$ test. We prefer to use a test that does not require Normality.

## The rank transformation

We first rank all 8 observations together. To do this, arrange them in order from smallest to largest:

153.1 156.0 158.6 **165.0** **166.7** **172.2** 176.4 **176.9**

The boldface entries in the list are the yields with no weeds present. We see that four of the five highest yields come from that group, suggesting that yields are higher with no weeds. The idea of rank tests is to look just at position in this ordered list. To do this, replace each observation by its order, from 1 (smallest) to 8 (largest). These numbers are the *ranks:*

| Yield | 153.1 | 156.0 | 158.6 | **165.0** | **166.7** | **172.2** | 176.4 | **176.9** |
|---|---|---|---|---|---|---|---|---|
| Rank | 1 | 2 | 3 | **4** | **5** | **6** | 7 | **8** |

### RANKS

To rank observations, first arrange them in order from smallest to largest. The **rank** of each observation is its position in this ordered list, starting with rank 1 for the smallest observation.

Moving from the original observations to their ranks is a transformation of the data, like moving from the observations to their logarithms. The rank transformation retains only the ordering of the observations and makes no other use of their numerical values. Working with ranks allows us to dispense with specific assumptions about the shape of the distribution, such as Normality.

### USE YOUR KNOWLEDGE

**15.1** **Numbers of rooms in top spas.** A report of a readers' poll in *Condé Nast Traveler* magazine ranked 36 top resort spas.[2] Let Group A be the top-ranked 18 spas, and let Group B be the next 18 rated spas in the list. A simple random sample of size 5 was taken from each group, and the number of rooms in each selected spa was recorded. Here are the data:

| Group A | 552 | 448 | 68 | 243 | 30 |
|---|---|---|---|---|---|
| Group B | 329 | 780 | 560 | 540 | 240 |

Rank all of the observations together and make a list of the ranks for Group A and Group B.

**15.2 The effect of Spa Bellagio on the result.** Refer to the previous exercise. Spa Bellagio in Las Vegas is one of the spas in Group B. Suppose this spa had been the second spa selected in the random sample for Group B. Replace the observation 780 in Group B by 4003, the number of rooms in Spa Bellagio. Use the modified data to make a list of the ranks for Groups A and B combined. What changes?

## The Wilcoxon rank sum test

If the presence of weeds reduces corn yields, we expect the ranks of the yields from plots with weeds to be smaller as a group than the ranks from plots without weeds. We might compare the *sums* of the ranks from the two treatments:

| Treatment | Sum of ranks |
|---|---|
| No weeds | 23 |
| Weeds | 13 |

These sums measure how much the ranks of the weed-free plots as a group exceed those of the weedy plots. In fact, the sum of the ranks from 1 to 8 is always equal to 36, so it is enough to report the sum for one of the two groups. If the sum of the ranks for the weed-free group is 23, the ranks for the other group must add to 13 because $23 + 13 = 36$. If the weeds have no effect, we would expect the sum of the ranks in each group to be 18 (half of 36). Here are the facts we need in a more general form that takes account of the fact that our two samples need not be the same size.

### THE WILCOXON RANK SUM TEST

Draw an SRS of size $n_1$ from one population and draw an independent SRS of size $n_2$ from a second population. There are $N$ observations in all, where $N = n_1 + n_2$. Rank all $N$ observations. The sum $W$ of the ranks for the first sample is the **Wilcoxon rank sum statistic.** If the two populations have the same continuous distribution, then $W$ has mean

$$\mu_W = \frac{n_1(N+1)}{2}$$

and standard deviation

$$\sigma_W = \sqrt{\frac{n_1 n_2 (N+1)}{12}}$$

The **Wilcoxon rank sum test** rejects the hypothesis that the two populations have identical distributions when the rank sum $W$ is far from its mean.*

*This test was invented by Frank Wilcoxon (1892–1965) in 1945. Wilcoxon was a chemist who encountered statistical problems in his work at the research laboratories of American Cyanimid Company.

In the corn yield study of Example 15.1, we want to test

$$H_0\text{: no difference in distribution of yields}$$

against the one-sided alternative

$$H_a\text{: yields are systematically higher in weed-free plots}$$

Our test statistic is the rank sum $W = 23$ for the weed-free plots.

## USE YOUR KNOWLEDGE

**15.3 Hypotheses and test statistic for top spas.** Refer to Exercise 15.1. State appropriate null and alternative hypotheses for this setting and calculate the value of $W$, the test statistic.

**15.4 Effect of Spa Bellagio on the test statistic.** Refer to Exercise 15.2. Using the altered data, state appropriate null and alternative hypotheses and calculate the value of $W$, the test statistic.

EXAMPLE

**15.2 Perform the significance test.** In Example 15.1, $n_1 = 4$, $n_2 = 4$, and there are $N = 8$ observations in all. The sum of ranks for the weed-free plots has mean

$$\mu_W = \frac{n_1(N+1)}{2} = \frac{(4)(9)}{2} = 18$$

and standard deviation

$$\sigma_W = \sqrt{\frac{n_1 n_2(N+1)}{12}} = \sqrt{\frac{(4)(4)(9)}{12}} = \sqrt{12} = 3.464$$

Although the observed rank sum $W = 23$ is higher than the mean, it is only about 1.4 standard deviations higher. We now suspect that the data do not give strong evidence that yields are higher in the population of weed-free corn.

The $P$-value for our one-sided alternative is $P(W \geq 23)$, the probability that $W$ is at least as large as the value for our data when $H_0$ is true.

To calculate the $P$-value $P(W \geq 23)$, we need to know the sampling distribution of the rank sum $W$ when the null hypothesis is true. This distribution depends on the two sample sizes $n_1$ and $n_2$. Tables are therefore a bit unwieldy, though you can find them in handbooks of statistical tables. Most statistical software will give you $P$-values, as well as carry out the ranking and calculate $W$. However, some software gives only approximate $P$-values. You must learn what your software offers.

EXAMPLE

**15.3 Software output.** Figure 15.3 shows the output from software that calculates the exact sampling distribution of $W$. We see that the sum of the ranks in the weed-free group is $W = 23$, with $P$-value $P = 0.100$ against the one-sided alternative that weed-free plots have higher yields. There is some evidence that weeds reduce yield, considering that we have data from only four plots for each treatment. The evidence does not, however, reach the levels usually considered convincing.

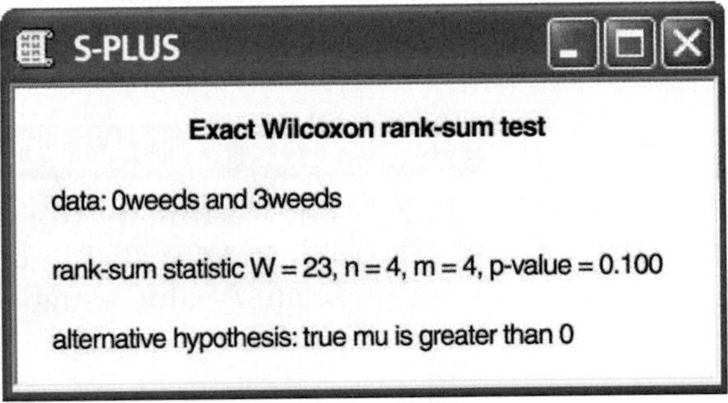


**FIGURE 15.3** Output from the S-PLUS statistical software for the data in Example 15.1. The program uses the exact distribution for W when the samples are small and there are no tied observations.

It is worth noting that the two-sample $t$ test gives essentially the same result as the Wilcoxon test in Example 15.3 ($t = 1.554$, $P = 0.0937$). A permutation test (Chapter 16) for the sample means gives $P = 0.084$. It is in fact somewhat unusual to find a strong disagreement among the conclusions reached by these tests.

## The Normal approximation

The rank sum statistic $W$ becomes approximately Normal as the two sample sizes increase. We can then form yet another $z$ statistic by standardizing $W$:

$$z = \frac{W - \mu_W}{\sigma_W} = \frac{W - n_1(N+1)/2}{\sqrt{n_1 n_2 (N+1)/12}}$$

LOOK BACK
**continuity correction, page 326**

Use standard Normal probability calculations to find $P$-values for this statistic. Because $W$ takes only whole-number values, the *continuity correction* improves the accuracy of the approximation.

EXAMPLE

**15.4 The continuity correction.** The standardized rank sum statistic $W$ in our corn yield example is

$$z = \frac{W - \mu_W}{\sigma_W} = \frac{23 - 18}{3.464} = 1.44$$

We expect $W$ to be larger when the alternative hypothesis is true, so the approximate $P$-value is

$$P(Z \geq 1.44) = 0.0749$$

The continuity correction acts as if the whole number 23 occupies the entire interval from 22.5 to 23.5. We calculate the *P*-value $P(W \geq 23)$ as $P(W \geq 22.5)$ because the value 23 is included in the range whose probability we want. Here is the calculation:

$$\begin{aligned} P(W \geq 22.5) &= P\left(\frac{W - \mu_W}{\sigma_W} \geq \frac{22.5 - 18}{3.464}\right) \\ &= P(Z \geq 1.30) \\ &= 0.0968 \end{aligned}$$

The continuity correction gives a result closer to the exact value $P = 0.100$.

## USE YOUR KNOWLEDGE

**15.5 The *P*-value for top spas.** Refer to Exercises 15.1 and 15.3. Find $\mu_W$, $\sigma_W$, and the standardized rank sum statistic. Then give an approximate *P*-value using the Normal approximation. What do you conclude?

**15.6 The effect of Spa Bellagio on the *P*-value.** Refer to Exercises 15.2 and 15.4. Answer the questions for Exercise 15.5 using the altered data.

We recommend always using either the exact distribution (from software or tables) or the continuity correction for the rank sum statistic *W*. The exact distribution is safer for small samples. As Example 15.4 illustrates, however, the Normal approximation with the continuity correction is often adequate.

EXAMPLE

Mann-Whitney test

**15.5 Software output.** Figure 15.4 shows the output for our data from two more statistical programs. Minitab offers only the Normal approximation, and it refers to the **Mann-Whitney test.** This is an alternative form of the Wilcoxon rank sum test. SAS carries out both the exact and the approximate tests. SAS calls the rank sum *S* rather than *W* and gives the mean 18 and standard deviation 3.464 as well as the *z* statistic 1.299 (using the continuity correction). SAS gives the approximate two-sided *P*-value as 0.1939, so the one-sided result is half this, $P = 0.0970$. This agrees with Minitab and (up to a small roundoff error) with our result in Example 15.4. This approximate *P*-value is close to the exact result $P = 0.100$, given by SAS and in Figure 15.3.

## What hypotheses does Wilcoxon test?

Our null hypothesis is that weeds do not affect yield. Our alternative hypothesis is that yields are lower when weeds are present. If we are willing to assume that yields are Normally distributed, or if we have reasonably large samples, we use the two-sample *t* test for means. Our hypotheses then become

$$H_0\colon \mu_1 = \mu_2$$
$$H_a\colon \mu_1 > \mu_2$$

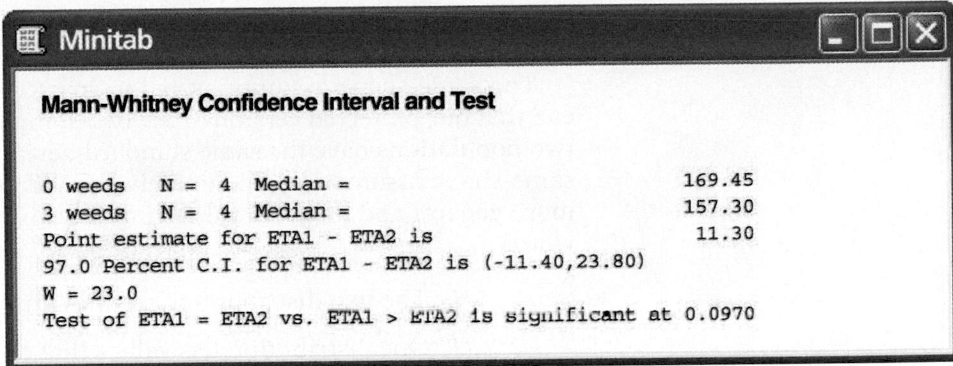

Minitab

**Mann-Whitney Confidence Interval and Test**

```
0 weeds    N =  4  Median =                         169.45
3 weeds    N =  4  Median =                         157.30
Point estimate for ETA1 - ETA2 is                    11.30
97.0 Percent C.I. for ETA1 - ETA2 is (-11.40,23.80)
W = 23.0
Test of ETA1 = ETA2 vs. ETA1 > ETA2 is significant at 0.0970
```

(a)

SAS

```
              Wilcoxon Scores (Rank Sums) for Variable YIELD
                     Classified by Variable WEEDS

                   Sum of    Expected      Std Dev         Mean
WEEDS      N       Scores    Under H0     Under H0        Score

0          4         23.0        18.0   3.46410162   5.75000000
3          4         13.0        18.0   3.46410162   3.25000000


                     Wilcoxon Two-Sample Test

        Statistic (S)                            23.000
        Exact P-Values
        One-Sided Pr >= S                        0.1000
        Two-Sided Pr >= |S - Mean|               0.2000

        Normal Approximation (with Continuity Correction of .5)
        Z =                                     1.29904
        Prob > |Z|                               0.1939
```

(b)

**FIGURE 15.4** Output from the Minitab and SAS statistical software for the data in Example 15.1. (a) Minitab uses the Normal approximation for the distribution of $W$. (b) SAS gives both the exact and approximate values.

When the distributions may not be Normal, we might restate the hypotheses in terms of population medians rather than means:

$$H_0\text{: median}_1 = \text{median}_2$$

$$H_a\text{: median}_1 > \text{median}_2$$

*The Wilcoxon rank sum test does test hypotheses about population medians, but only if an additional assumption is met: both populations must have distributions of the same shape.* That is, the density curve for corn yields with 3 weeds per meter looks exactly like that for no weeds except that it may slide to a different location on the scale of yields. The Minitab output in Figure 15.4(a) states the hypotheses in terms of population medians (which it calls "ETA") and also gives a confidence interval for the difference between the two population medians.

The same-shape assumption is too strict to be reasonable in practice. Recall that our preferred version of the two-sample $t$ test does not require that the two populations have the same standard deviation—that is, it does not make a same-shape assumption. Fortunately, the Wilcoxon test also applies in a much more general and more useful setting. It tests hypotheses that we can state in words as

$H_0$: The two distributions are the same.

$H_a$: One distribution has values that are systematically larger.

Here is a more exact statement of the "systematically larger" alternative hypothesis. Take $X_1$ to be corn yield with no weeds and $X_2$ to be corn yield with 3 weeds per meter. These yields are random variables. That is, every time we plant a plot with no weeds, the yield is a value of the variable $X_1$. The probability that the yield is more than 160 bushels per acre when no weeds are present is $P(X_1 > 160)$. If weed-free yields are "systematically larger" than those with weeds, yields higher than 160 should be more likely with no weeds. That is, we should have

$$P(X_1 > 160) > P(X_2 > 160)$$

The alternative hypothesis says that this inequality holds not just for 160 but for *any* yield we care to specify. No weeds always puts more probability "to the right" of whatever yield we are interested in.[3]

This exact statement of the hypotheses we are testing is a bit awkward. The hypotheses really are "nonparametric" because they do not involve any specific parameter such as the mean or median. If the two distributions do have the same shape, the general hypotheses reduce to comparing medians. Many texts and computer outputs state the hypotheses in terms of medians, sometimes ignoring the same-shape requirement. We recommend that you express the hypotheses in words rather than symbols. "Yields are systematically higher in weed-free plots" is easy to understand and is a good statement of the effect that the Wilcoxon test looks for.

## Ties

The exact distribution for the Wilcoxon rank sum is obtained assuming that all observations in both samples take different values. This allows us to rank them all. In practice, however, we often find observations tied at the same value. What shall we do? The usual practice is to *assign all tied values the* **average** *of the ranks they occupy.* Here is an example with 6 observations:

average ranks

| Observation | 153 | 155 | 158 | 158 | 161 | 164 |
|---|---|---|---|---|---|---|
| Rank | 1 | 2 | 3.5 | 3.5 | 5 | 6 |

The tied observations occupy the third and fourth places in the ordered list, so they share rank 3.5.

The exact distribution for the Wilcoxon rank sum $W$ changes if the data contain ties. Moreover, the standard deviation $\sigma_W$ must be adjusted if ties are present. The Normal approximation can be used after the standard deviation is adjusted. Statistical software will detect ties, make the necessary adjustment, and switch to the Normal approximation. In practice, software is required if you want to use rank tests when the data contain tied values.

It is sometimes useful to use rank tests on data that have very many ties because the scale of measurement has only a few values. Here is an example.

EXAMPLE

**15.6 Food safety.** Food sold at outdoor fairs and festivals may be less safe than food sold in restaurants because it is prepared in temporary locations and often by volunteer help. What do people who attend fairs think about the safety of the food served? One study asked this question of people at a number of fairs in the Midwest:

> How often do you think people become sick because of food they consume prepared at outdoor fairs and festivals?

The possible responses were:

1 = very rarely
2 = once in a while
3 = often
4 = more often than not
5 = always

In all, 303 people answered the question. Of these, 196 were women and 107 were men. Is there good evidence that men and women differ in their perceptions about food safety at fairs?[4]

We should first ask if the subjects in Example 15.6 are a random sample of people who attend fairs, at least in the Midwest. The researcher visited 11 different fairs. She stood near the entrance and stopped every 25th adult who passed. Because no personal choice was involved in choosing the subjects, we can reasonably treat the data as coming from a random sample. (As usual, there was some nonresponse, which could create bias.)

Here are the data, presented as a two-way table of counts:

| | Response | | | | | |
|---|---|---|---|---|---|---|
| | 1 | 2 | 3 | 4 | 5 | Total |
| Female | 13 | 108 | 50 | 23 | 2 | 196 |
| Male | 22 | 57 | 22 | 5 | 1 | 107 |
| Total | 35 | 165 | 72 | 28 | 3 | 303 |

Comparing row percents shows that the women in the sample are more concerned about food safety than the men:

| | Response | | | | | |
|---|---|---|---|---|---|---|
| | 1 | 2 | 3 | 4 | 5 | Total |
| Female | 6.6% | 55.1% | 25.5% | 11.7% | 1.0% | 100% |
| Male | 20.6% | 53.3% | 20.6% | 4.7% | 1.0% | 100% |

Is the difference between the genders statistically significant?

We might apply the chi-square test (Chapter 9). It is highly significant ($X^2 = 16.120$, df = 4, $P = 0.0029$). Although the chi-square test answers our general question, it ignores the ordering of the responses and so does not use all of the available information. We would really like to know whether men or women are more concerned about the safety of the food served. This question depends on the ordering of responses from least concerned to most concerned. We can use the Wilcoxon test for the hypotheses:

$H_0$: Men and women do not differ in their responses.

$H_a$: One of the two genders gives systematically larger responses than the other.

The alternative hypothesis is two-sided. Because the responses can take only five values, there are very many ties. All 35 people who chose "very rarely" are tied at 1, and all 165 who chose "once in a while" are tied at 2.

EXAMPLE

**15.7 Software output.** Figure 15.5 gives software output for the Wilcoxon test. The rank sum for men (using average ranks for ties) is $W = 14{,}059.5$. The standardized value is $z = -3.33$, with two-sided $P$-value $P = 0.0009$. There is very strong evidence of a difference. Women are more concerned than men about the safety of food served at fairs.

With more than 100 observations in each group and no outliers, we might use the two-sample $t$ even though responses take only five values. In fact, the results for Example 15.6 are $t = 3.3655$ with $P = 0.0009$. The $P$-value for the two-sample $t$ is the same as that for the Wilcoxon test. There is, however, another reason to prefer the rank test in this example. The $t$ statistic treats the response values 1 through 5 as meaningful numbers. In particular, the possible responses are treated as though they are equally spaced. The difference between "very rarely" and "once in a while" is the same as the difference between "once in a while" and "often." This may not make sense. The rank test, on the other hand, uses only the order of the responses, not their actual values. The responses are arranged in order from least to most concerned about safety, so the rank test makes sense. *Some statisticians avoid using t procedures when there is not a fully meaningful scale of measurement.*

## Rank, *t*, and permutation tests

The two-sample $t$ procedures are the most common method for comparing the centers of two populations based on random samples from each. The Wilcoxon

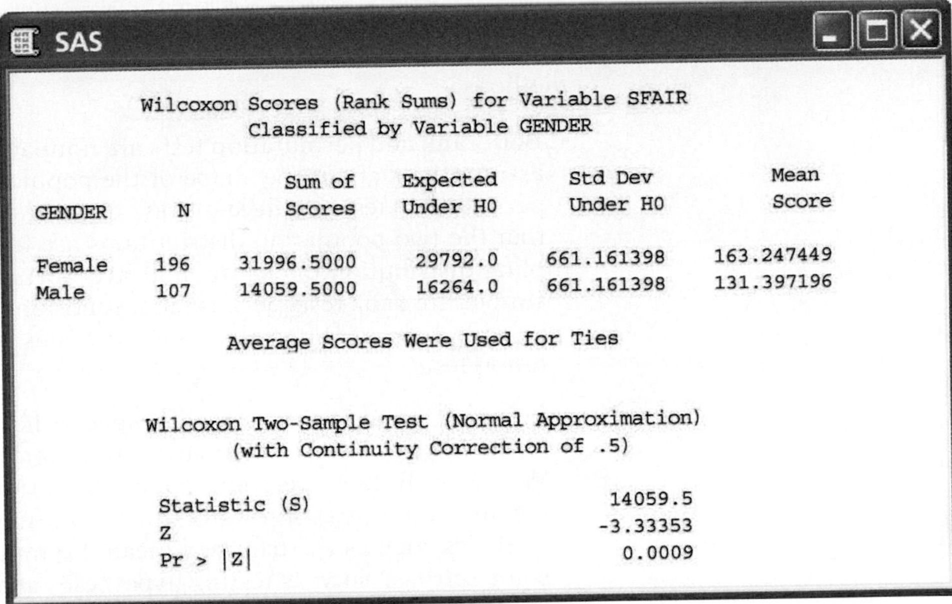

Wilcoxon Scores (Rank Sums) for Variable SFAIR
Classified by Variable GENDER

| GENDER | N | Sum of Scores | Expected Under H0 | Std Dev Under H0 | Mean Score |
|---|---|---|---|---|---|
| Female | 196 | 31996.5000 | 29792.0 | 661.161398 | 163.247449 |
| Male | 107 | 14059.5000 | 16264.0 | 661.161398 | 131.397196 |

Average Scores Were Used for Ties

Wilcoxon Two-Sample Test (Normal Approximation)
(with Continuity Correction of .5)

| | |
|---|---|
| Statistic (S) | 14059.5 |
| Z | -3.33353 |
| Pr > \|Z\| | 0.0009 |

**FIGURE 15.5** Output from SAS for the food safety study of Example 15.6. The approximate two-sided *P*-value is 0.0009.

rank sum test is a competing procedure that does not start from the condition that the populations have Normal distributions. Permutation tests (Chapter 16) also avoid the need for Normality. Tests based on Normality, rank tests, and permutation tests apply in many other settings as well. How do these three approaches compare in general?

First consider rank tests versus traditional tests based on Normal distributions. Both are available in almost all statistical software.

- Moving from the actual data values to their ranks allows us to find an exact sampling distribution for rank statistics such as the Wilcoxon rank sum $W$ when the null hypothesis is true. (Most software will do this only if there are no ties and if the samples are quite small.) When our samples are small, are truly random samples from the populations, and show non-Normal distributions of the same shape, the Wilcoxon test is more reliable than the two-sample $t$ test. In practice, the robustness of $t$ procedures implies that we rarely encounter data that require nonparametric procedures to obtain reasonably accurate *P*-values. The $t$ and $W$ tests give very similar results in our examples. Nonetheless, many statisticians would not use a $t$ test in Example 15.6 because the response variable gives only the order of the responses.

- Normal tests compare means and are accompanied by simple confidence intervals for means or differences between means. When we use rank tests to compare medians, we can also give confidence intervals for medians. However, the usefulness of rank tests is clearest in settings when they do not simply compare medians—see the discussion "What hypotheses does Wilcoxon test?" Rank methods emphasize tests, not confidence intervals.

- Inference based on ranks is largely restricted to simple settings. Normal inference extends to methods for use with complex experimental designs and multiple regression, but nonparametric tests do not. We stress Normal inference in part because it leads to more advanced statistics.

If you have read Chapter 16 and use software that makes permutation tests available to you, you will also want to compare rank tests with resampling methods.

- Both rank and permutation tests are nonparametric. That is, they require no assumptions about the shape of the population distribution. A two-sample permutation test has the same null hypothesis as the Wilcoxon rank sum test: that the two population distributions are identical. Calculation of the sampling distribution under the null hypothesis is similar for both tests but is simpler for rank tests because it depends only on the sizes of the samples. As a result, software often gives exact *P*-values for rank tests but not for permutation tests.
- Permutation tests have the advantage of flexibility. They allow wide choice of the statistic used to compare two samples, an advantage over both $t$ and Wilcoxon. In fact, we could apply the permutation test method to sample means (imitating $t$) or to rank sums (imitating Wilcoxon), as well as to other statistics such as the trimmed mean. Permutation tests are not available in some settings, such as testing hypotheses about a single population, though bootstrap confidence intervals do allow resampling tests in these settings. Permutation tests are available for multiple regression and some other quite elaborate settings.
- An important advantage of resampling methods over both Normal and rank procedures is that we can get bootstrap confidence intervals for the parameter corresponding to whatever statistic we choose for the permutation test. If the samples are very small, however, bootstrap confidence intervals may be unreliable because the samples don't represent the population well enough to provide a good basis for bootstrapping.

In general, both Normal distribution methods and resampling methods are more useful than rank tests. *If you are familiar with resampling, we recommend rank tests only for very small samples, and even then only if your software gives exact P-values for rank tests but not for permutation tests.*

## SECTION 15.1 Summary

**Nonparametric tests** do not require any specific form for the distribution of the population from which our samples come.

**Rank tests** are nonparametric tests based on the **ranks** of observations, their positions in the list ordered from smallest (rank 1) to largest. Tied observations receive the average of their ranks.

The **Wilcoxon rank sum test** compares two distributions to assess whether one has systematically larger values than the other. The Wilcoxon test is based on the **Wilcoxon rank sum statistic** $W$, which is the sum of the ranks of one of the samples. The Wilcoxon test can replace the **two-sample** $t$ **test.**

***P*-values** for the Wilcoxon test are based on the sampling distribution of the rank sum statistic $W$ when the null hypothesis (no difference in distributions) is true. You can find *P*-values from special tables, software, or a Normal approximation (with continuity correction).

## SECTION 15.1 Exercises

*For Exercises 15.1 and 15.2, see pages 736 and 737; for Exercises 15.3 and 15.4, see page 738; and for Exercises 15.5 and 15.6, see page 740.*

*Statistical software is very helpful in doing these exercises. If you do not have access to software, base your work on the Normal approximation with continuity correction.*

**15.7 Storytelling and the use of language.** A study of early childhood education asked kindergarten students to retell two fairy tales that had been read to them earlier in the week. The 10 children in the study included 5 high-progress readers and 5 low-progress readers. Each child told two stories. Story 1 had been read to them; Story 2 had been read and also illustrated with pictures. An expert listened to a recording of the children and assigned a score for certain uses of language. Here are the data:[5]

| Child | Progress | Story 1 score | Story 2 score |
|---|---|---|---|
| 1 | high | 0.55 | 0.80 |
| 2 | high | 0.57 | 0.82 |
| 3 | high | 0.72 | 0.54 |
| 4 | high | 0.70 | 0.79 |
| 5 | high | 0.84 | 0.89 |
| 6 | low | 0.40 | 0.77 |
| 7 | low | 0.72 | 0.49 |
| 8 | low | 0.00 | 0.66 |
| 9 | low | 0.36 | 0.28 |
| 10 | low | 0.55 | 0.38 |

Is there evidence that the scores of high-progress readers are higher than those of low-progress readers when they retell a story they have heard without pictures (Story 1)?

(a) Make Normal quantile plots for the 5 responses in each group. Are any major deviations from Normality apparent?

(b) Carry out a two-sample $t$ test. State hypotheses and give the two sample means, the $t$ statistic and its $P$-value, and your conclusion.

(c) Carry out the Wilcoxon rank sum test. State hypotheses and give the rank sum $W$ for high-progress readers, its $P$-value, and your conclusion. Do the $t$ and Wilcoxon tests lead you to different conclusions?

**15.8 Repeat the analysis for Story 2.** Repeat the analysis of Exercise 15.7 for the scores when children retell a story they have heard and seen illustrated with pictures (Story 2).

**15.9 Do the calculations by hand.** Use the data in Exercise 15.7 for children telling Story 2 to carry out by hand the steps in the Wilcoxon rank sum test.

(a) Arrange the 10 observations in order and assign ranks. There are no ties.

(b) Find the rank sum $W$ for the five high-progress readers. What are the mean and standard deviation of $W$ under the null hypothesis that low-progress and high-progress readers do not differ?

(c) Standardize $W$ to obtain a $z$ statistic. Do a Normal probability calculation with the continuity correction to obtain a one-sided $P$-value.

(d) The data for Story 1 contain tied observations. What ranks would you assign to the 10 scores for Story 1?

**15.10 Weeds and corn yield.** The corn yield study of Example 15.1 also examined yields in four plots having 9 lamb's-quarter plants per meter of row. The yields (bushels per acre) in these plots were

162.8 142.4 162.7 162.4

There is a clear outlier, but rechecking the results found that this is the correct yield for this plot. The outlier makes us hesitant to use $t$ procedures because $\bar{x}$ and $s$ are not resistant.

(a) Is there evidence that 9 weeds per meter reduces corn yields when compared with weed-free corn? Use the Wilcoxon rank sum test with the data above and some of the data from Example 15.1 to answer this question.

(b) Compare the results from (a) with those from the two-sample $t$ test for these data.

(c) Now remove the low outlier 142.4 from the data for 9 weeds per meter. Repeat both the Wilcoxon and $t$ analyses. By how much did the outlier reduce the mean yield in its group? By how much did it increase the standard deviation? Did it have a practically important impact on your conclusions?

**15.11 Decay of polyester fabrics in landfills.** How quickly do synthetic fabrics such as polyester decay in landfills? A researcher buried polyester strips in the soil for different lengths of time, then dug up the strips and measured the force required to break them. Breaking strength is easy to measure and is a good indicator of decay. Lower strength means the fabric has decayed. Part of the study involved burying 10 polyester strips in well-drained soil in the summer. Five of the strips, chosen at random, were dug up after 2 weeks; the other 5 were dug up after 16 weeks. Here are the breaking strengths in pounds:[6]

| | | | | | |
|---|---|---|---|---|---|
| 2 weeks | 118 | 126 | 126 | 120 | 129 |
| 16 weeks | 124 | 98 | 110 | 140 | 110 |

(a) Make a back-to-back stemplot. Does it appear reasonable to assume that the two distributions have the same shape?

(b) Is there evidence that breaking strengths are lower for strips buried longer?

**15.12 Learning math through subliminal messages.** A "subliminal" message is below our threshold of awareness but may nonetheless influence us. Can subliminal messages help students learn math? A group of students who had failed the mathematics part of the City University of New York Skills Assessment Test agreed to participate in a study to find out. All received a daily subliminal message, flashed on a screen too rapidly to be consciously read. The treatment group of 10 students was exposed to "Each day I am getting better in math." The control group of 8 students was exposed to a neutral message, "People are walking on the street." All students participated in a summer program designed to raise their math skills, and all took the assessment test again at the end of the program. Here are data on the subjects' scores before and after the program:[7]

| Treatment Group | | Control Group | |
|---|---|---|---|
| Pretest | Posttest | Pretest | Posttest |
| 18 | 24 | 18 | 29 |
| 18 | 25 | 24 | 29 |
| 21 | 33 | 20 | 24 |
| 18 | 29 | 18 | 26 |
| 18 | 33 | 24 | 38 |
| 20 | 36 | 22 | 27 |
| 23 | 34 | 15 | 22 |
| 23 | 36 | 19 | 31 |
| 21 | 34 | | |
| 17 | 27 | | |

(a) The study design was a randomized comparative experiment. Outline this design.

(b) Compare the gain in scores in the two groups, using a graph and numerical descriptions. Does it appear that the treatment group's scores rose more than the scores for the control group?

(c) Apply the Wilcoxon rank sum test to the posttest versus pretest differences. Note that there are some ties. What do you conclude?

**15.13 Effects of logging in Borneo.** "Conservationists have despaired over destruction of tropical rainforest by logging, clearing, and burning." These words begin a report on a statistical study of the effects of logging in Borneo.[8] Here are data on the number of tree species in 12 unlogged forest plots and 9 similar plots logged 8 years earlier:

| Unlogged | | Logged | |
|---|---|---|---|
| 22 | 18 | 17 | 4 |
| 22 | 20 | 18 | 14 |
| 15 | 21 | 18 | 15 |
| 13 | 13 | 15 | 10 |
| 19 | 13 | 12 | |
| 19 | 15 | | |

(a) Make a back-to-back stemplot of the data. Does there appear to be a difference in species counts for logged and unlogged plots?

(b) Does logging significantly reduce the number of species in a plot after 8 years? State hypotheses, do a Wilcoxon test, and state your conclusion.

**15.14 Improved methods for teaching reading.** Do new "directed reading activities" improve the reading ability of elementary school students, as measured by their Degree of Reading Power (DRP) score? A study assigns students at random to either the new method (treatment group, 21 students) or traditional teaching methods (control group, 23 students). Here are the DRP scores at the end of the study:[9]

| Treatment group | | | | Control group | | | |
|---|---|---|---|---|---|---|---|
| 24 | 61 | 59 | 46 | 42 | 33 | 46 | 37 |
| 43 | 44 | 52 | 43 | 43 | 41 | 10 | 42 |
| 58 | 67 | 62 | 57 | 55 | 19 | 17 | 55 |
| 71 | 49 | 54 | | 26 | 54 | 60 | 28 |
| 43 | 53 | 57 | | 62 | 20 | 53 | 48 |
| 49 | 56 | 33 | | 37 | 85 | 42 | |

For these data the two-sample $t$ test (Example 7.14) gives $P = 0.013$ and a permutation test based on the difference of means (Example 16.12) gives $P = 0.015$. Both of these tests are based on the difference of sample means. Does the Wilcoxon test, based on rank sums rather than means, give a similar $P$-value?

**15.15 Food safety.** Example 15.12 describes a study of the attitudes of people attending outdoor fairs about the safety of the food served at such locations. The full data set with the responses of 300 people to several questions is in the file *eg15_012*. The variables in this data set are (in order)

subject hfair sfair sfast srest gender

The variable "sfair" contains the responses described in the example concerning safety of food served at outdoor fairs and festivals. The variable "srest" contains responses to the same question asked about food served in restaurants. The variable "gender" contains 1 if the respondent is a woman, 2 if he is a man. We saw that women are more concerned than men about the safety of food served at fairs. Is this also true for restaurants?

**15.16 Compare fairs with restaurants.** The data file used in Example 15.6 and Exercise 15.15 contains 303 rows, one for each of the 303 respondents. Each row contains the responses of one person to several questions. We wonder if people are more concerned about the safety of food served at fairs than they are about the safety of food served at restaurants. Explain carefully why we *cannot* answer this question by applying the Wilcoxon rank sum test to the variables "sfair" and "srest."

**15.17 Attitudes toward secondhand stores.** To study customers' attitudes toward secondhand stores, researchers interviewed samples of shoppers at two secondhand stores of the same chain in two cities. Here are data on the incomes of shoppers at the two stores, presented as a two-way table of counts:[10]

| Income | City 1 | City 2 |
|---|---|---|
| Under $10,000 | 70 | 62 |
| $10,000 to $19,999 | 52 | 63 |
| $20,000 to $24,999 | 69 | 50 |
| $25,000 to $34,999 | 22 | 19 |
| $35,000 or more | 28 | 24 |

(a) Is there a relationship between city and income? Use the chi-square test to answer this question.

(b) The chi-square test ignores the ordering of the income categories. The data file *ex15_11* contains data on the 459 shoppers in this study. The first variable is the city (City1 or City2) and the second is the income as it is coded in the table above. Is there good evidence that shoppers in one city have systematically higher incomes than in the other?

## 15.2 The Wilcoxon Signed Rank Test

We use the one-sample $t$ procedures for inference about the mean of one population or for inference about the mean difference in a matched pairs setting. The matched pairs setting is more important because good studies are generally comparative. We will now meet a rank test for this setting.

EXAMPLE

**15.8 Storytelling and reading.** A study of early childhood education asked kindergarten students to retell two fairy tales that had been read to them earlier in the week. Each child told two stories. The first had been read to them, and the second had been read but also illustrated with pictures. An expert listened to a recording of the children and assigned a score for certain uses of language. Here are the data for five "low-progress" readers in a pilot study:[11]

| Child | 1 | 2 | 3 | 4 | 5 |
|---|---|---|---|---|---|
| Story 2 | 0.77 | 0.49 | 0.66 | 0.28 | 0.38 |
| Story 1 | 0.40 | 0.72 | 0.00 | 0.36 | 0.55 |
| Difference | 0.37 | −0.23 | 0.66 | −0.08 | −0.17 |

We wonder if illustrations improve how the children retell a story. We would like to test the hypotheses

$H_0$: Scores have the same distribution for both stories.

$H_a$: Scores are systematically higher for Story 2.

Because this is a matched pairs design, we base our inference on the differences. The matched pairs $t$ test gives $t = 0.635$ with one-sided $P$-value $P = 0.280$. Displays of the data (Figure 15.6) suggest some lack of Normality. We would therefore like to use a rank test.

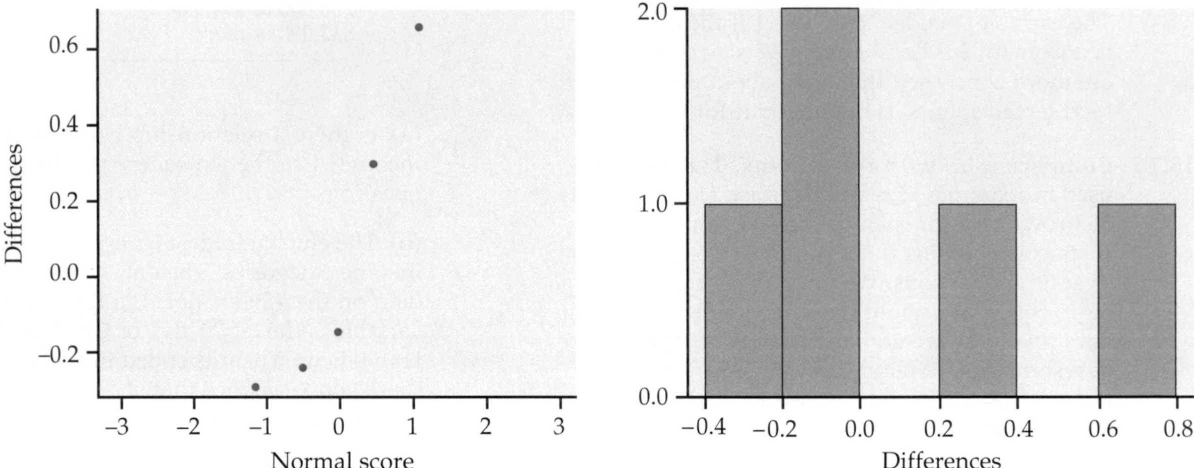


**FIGURE 15.6** Normal quantile plot and histogram for the five differences in Example 15.8.

Positive differences in Example 15.8 indicate that the child performed better telling Story 2. If scores are generally higher with illustrations, the positive differences should be farther from zero in the positive direction than the negative differences are in the negative direction. We therefore compare the **absolute values** of the differences, that is, their magnitudes without a sign. Here they are, with boldface indicating the positive values:

absolute value

**0.37** 0.23 **0.66** 0.08 0.17

Arrange these in increasing order and assign ranks, keeping track of which values were originally positive. Tied values receive the average of their ranks. If there are differences of zero, discard them before ranking.

| Absolute value | 0.08 | 0.17 | 0.23 | **0.37** | **0.66** |
|---|---|---|---|---|---|
| Rank | 1 | 2 | 3 | **4** | **5** |

The test statistic is the sum of the ranks of the positive differences. (We could equally well use the sum of the ranks of the negative differences.) This is the *Wilcoxon signed rank statistic*. Its value here is $W^+ = 9$.

### THE WILCOXON SIGNED RANK TEST FOR MATCHED PAIRS

Draw an SRS of size $n$ from a population for a matched pairs study and take the differences in responses within pairs. Rank the absolute values of these differences. The sum $W^+$ of the ranks for the positive differences is the **Wilcoxon signed rank statistic.** If the distribution of the responses is not affected by the different treatments within pairs, then $W^+$ has mean

$$\mu_{W^+} = \frac{n(n+1)}{4}$$

and standard deviation

$$\sigma_{W^+} = \sqrt{\frac{n(n+1)(2n+1)}{24}}$$

The **Wilcoxon signed rank test** rejects the hypothesis that there are no systematic differences within pairs when the rank sum $W^+$ is far from its mean.

## USE YOUR KNOWLEDGE

**15.18 Services provided by top spas.** The readers' poll in *Condé Nast Traveler* magazine that ranked 36 top resort spas and that was described in Exercise 15.1 also reported scores on Diet/Cuisine and on Program/Facilities. Here are the scores for a random sample of 7 spas that ranked in the top 18:

| Spa | 1 | 2 | 3 | 4 | 5 | 6 | 7 |
|---|---|---|---|---|---|---|---|
| Diet/Cuisine | 90.9 | 92.3 | 88.6 | 81.8 | 85.7 | 88.9 | 81.0 |
| Program/Facilities | 93.8 | 92.3 | 91.4 | 95.0 | 89.2 | 88.2 | 81.8 |

Is food, expressed by the Diet/Cuisine score, more important than activities, expressed as the Program/Facilities score, for a top ranking? Formulate this question in terms of null and alternative hypotheses. Then compute the differences and find the value of the Wilcoxon signed rank statistic, $W^+$.

**15.19 Scores for lower-ranked spas.** Refer to the previous exercise. Here are the scores for a random sample of 7 spas that ranked between 19 and 36:

| Spa | 1 | 2 | 3 | 4 | 5 | 6 | 7 |
|---|---|---|---|---|---|---|---|
| Diet/Cuisine | 77.3 | 85.7 | 84.2 | 85.3 | 83.7 | 84.6 | 78.5 |
| Program/Facilities | 95.7 | 78.0 | 87.2 | 85.3 | 93.6 | 76.0 | 86.3 |

Answer the questions from the previous exercise for this setting.

EXAMPLE

**15.9 Software output.** In the storytelling study of Example 15.8, $n = 5$. If the null hypothesis (no systematic effect of illustrations) is true, the mean of the signed rank statistic is

$$\mu_{W^+} = \frac{n(n+1)}{4} = \frac{(5)(6)}{4} = 7.5$$

Our observed value $W^+ = 9$ is only slightly larger than this mean. The one-sided $P$-value is $P(W^+ \geq 9)$.

Figure 15.7 displays the output of two statistical programs. We see from Figure 15.7(a) that the one-sided $P$-value for the Wilcoxon signed rank test with $n = 5$ observations and $W^+ = 9$ is $P = 0.4062$. This result differs from the $t$ test result $P = 0.280$, but both tell us that this very small sample gives no evidence that seeing illustrations improves the storytelling of low-progress readers.

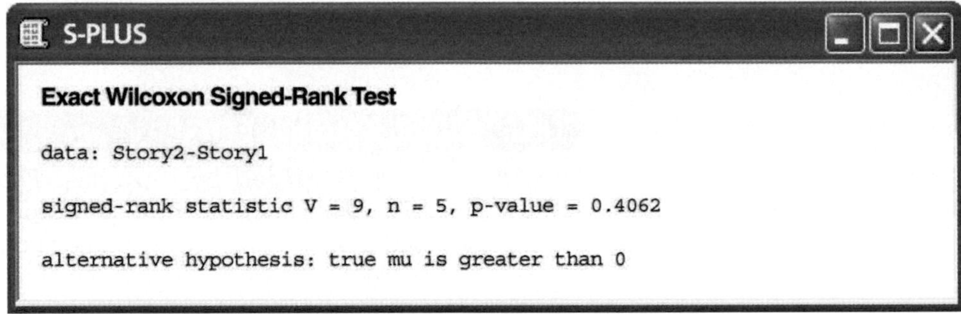

S-PLUS

**Exact Wilcoxon Signed-Rank Test**

```
data: Story2-Story1

signed-rank statistic V = 9, n = 5, p-value = 0.4062

alternative hypothesis: true mu is greater than 0
```

(a)

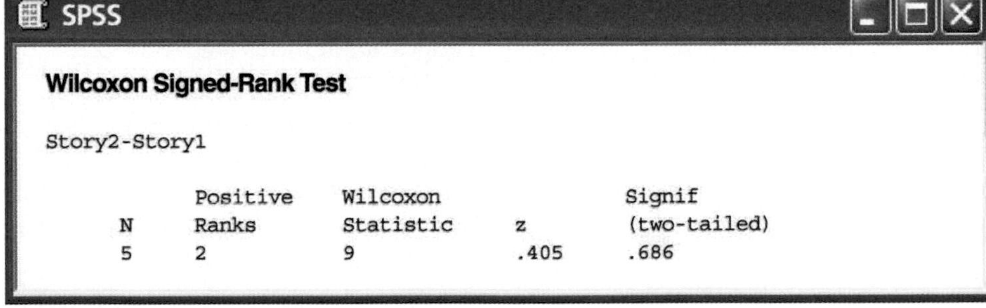

SPSS

**Wilcoxon Signed-Rank Test**

Story2-Story1

| N | Positive Ranks | Wilcoxon Statistic | z | Signif (two-tailed) |
|---|---|---|---|---|
| 5 | 2 | 9 | .405 | .686 |

(b)

**FIGURE 15.7** Output from (a) S-PLUS and (b) SPSS for the storytelling study of Example 15.8. S-PLUS reports the exact $P$-value, $P = 0.4062$. SPSS uses the Normal approximation without the continuity correction and so gives a less accurate $P$-value, $P = 0.343$ (one-sided).

## The Normal approximation

The distribution of the signed rank statistic when the null hypothesis (no difference) is true becomes approximately Normal as the sample size becomes large. We can then use Normal probability calculations (with the continuity correction) to obtain approximate $P$-values for $W^+$. Let's see how this works in the storytelling example, even though $n = 5$ is certainly not a large sample.

EXAMPLE

**15.10 The Normal approximation.** For $n = 5$ observations, we saw in Example 15.9 that $\mu_{W^+} = 7.5$. The standard deviation of $W^+$ under the null hypothesis is

$$\sigma_{W^+} = \sqrt{\frac{n(n+1)(2n+1)}{24}} = \sqrt{\frac{(5)(6)(11)}{24}} = \sqrt{13.75} = 3.708$$

The continuity correction calculates the $P$-value $P(W^+ \geq 9)$ as $P(W^+ \geq 8.5)$, treating the value $W^+ = 9$ as occupying the interval from 8.5 to 9.5. We find the Normal approximation for the $P$-value by standardizing and using the standard Normal table:

$$P(W^+ \geq 8.5) = P\left(\frac{W^+ - 7.5}{3.708} \geq \frac{8.5 - 7.5}{3.708}\right) = P(Z \geq 0.27) = 0.394$$

Despite the small sample size, the Normal approximation gives a result quite close to the exact value $P = 0.4062$. Figure 15.7(b) shows that the approximation is much less accurate without the continuity correction. *This output reminds us not to trust software unless we know exactly what it does.*

## USE YOUR KNOWLEDGE

**15.20 Significance test for top-ranked spas.** Refer to Exercise 15.18. Find $\mu_{W^+}$, $\sigma_{W^+}$, and the Normal approximation for the $P$-value for the Wilcoxon signed rank test.

**15.21 Significance test for lower-ranked spas.** Refer to Exercise 15.19. Find $\mu_{W^+}$, $\sigma_{W^+}$, and the Normal approximation for the $P$-value for the Wilcoxon signed rank test.

## Ties

Ties among the absolute differences are handled by assigning average ranks. A tie *within* a pair creates a difference of zero. Because these are neither positive nor negative, the usual procedure simply drops such pairs from the sample. *This amounts to dropping observations that favor the null hypothesis (no difference). If there are many ties, the test may be biased in favor of the alternative hypothesis.* As in the case of the Wilcoxon rank sum, ties complicate finding a $P$-value. Most software no longer provides an exact distribution for the signed rank statistic $W^+$, and the standard deviation $\sigma_{W^+}$ must be adjusted for the ties before we can use the Normal approximation. Software will do this. Here is an example.

EXAMPLE

**15.11 Golf scores of a women's golf team.** Here are the golf scores of 12 members of a college women's golf team in two rounds of tournament play. (A golf score is the number of strokes required to complete the course, so that low scores are better.)

| Player | 1 | 2 | 3 | 4 | 5 | 6 | 7 | 8 | 9 | 10 | 11 | 12 |
|---|---|---|---|---|---|---|---|---|---|---|---|---|
| Round 2 | 94 | 85 | 89 | 89 | 81 | 76 | 107 | 89 | 87 | 91 | 88 | 80 |
| Round 1 | 89 | 90 | 87 | 95 | 86 | 81 | 102 | 105 | 83 | 88 | 91 | 79 |
| Difference | 5 | −5 | 2 | −6 | −5 | −5 | 5 | −16 | 4 | 3 | −3 | 1 |

Negative differences indicate better (lower) scores on the second round. We see that 6 of the 12 golfers improved their scores. We would like to test the hypotheses that in a large population of collegiate women golfers

$H_0$: Scores have the same distribution in Rounds 1 and 2.

$H_a$: Scores are systematically lower or higher in Round 2.

A Normal quantile plot of the differences (Figure 15.8) shows some irregularity and a low outlier. We will use the Wilcoxon signed rank test.

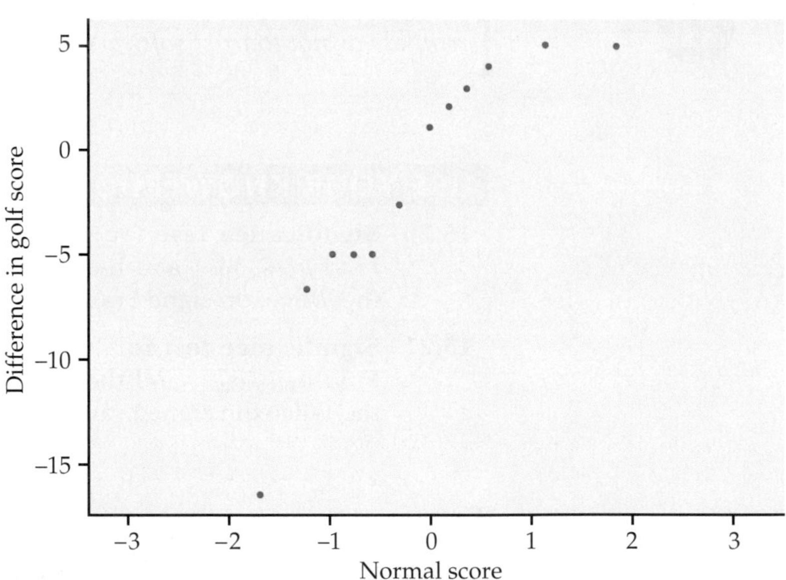


**FIGURE 15.8** Normal quantile plot of the differences in scores for two rounds of a golf tournament, for Example 15.11.

The absolute values of the differences, with boldface indicating those that were negative, are

5 **5** 2 **6** **5** **5** 5 **16** 4 3 **3** 1

Arrange these in increasing order and assign ranks, keeping track of which values were originally negative. Tied values receive the average of their ranks.

| Absolute value | 1 | 2 | **3** | 3 | 4 | **5** | 5 | **5** | 5 | **5** | **6** | **16** |
|---|---|---|---|---|---|---|---|---|---|---|---|---|
| Rank | 1 | 2 | **3.5** | 3.5 | 5 | **8** | 8 | **8** | 8 | **8** | **11** | **12** |

The Wilcoxon signed rank statistic is the sum of the ranks of the negative differences. (We could equally well use the sum of the ranks of the positive differences.) Its value is $W^+ = 50.5$.

EXAMPLE

**15.12 Software output.** Here are the two-sided $P$-values for the Wilcoxon signed rank test for the golf score data from several statistical programs:

| Program | $P$-value |
|---|---|
| Minitab | $P = 0.388$ |
| SAS | $P = 0.388$ |
| S-PLUS | $P = 0.384$ |
| SPSS | $P = 0.363$ |

All lead to the same practical conclusion: these data give no evidence for a systematic change in scores between rounds. However, the $P$-values reported differ a bit from program to program. The reason for the variations is that the programs use slightly different versions of the approximate calculations needed when ties are present. The exact result depends on which of these variations the programmer chooses to use.

For these data, the matched pairs $t$ test gives $t = 0.9314$ with $P = 0.3716$. Once again, $t$ and $W^+$ lead to the same conclusion.

## SECTION 15.2 Summary

The **Wilcoxon signed rank test** applies to matched pairs studies. It tests the null hypothesis that there is no systematic difference within pairs against alternatives that assert a systematic difference (either one-sided or two-sided).

The test is based on the **Wilcoxon signed rank statistic $W^+$**, which is the sum of the ranks of the positive (or negative) differences when we rank the absolute values of the differences. The **matched pairs $t$ test** and the **sign test** are alternative tests in this setting.

***P*-values** for the signed rank test are based on the sampling distribution of $W^+$ when the null hypothesis is true. You can find $P$-values from special tables, software, or a Normal approximation (with continuity correction).

## SECTION 15.2 Exercises

*For Exercises 15.18 and 15.19, see page 751; and for Exercises 15.20 and 15.21, see page 753.*

*Statistical software is very helpful in doing these exercises. If you do not have access to software, base your work on the Normal approximation with continuity correction.*

**15.22 Carbon dioxide and plant growth.** The concentration of carbon dioxide ($CO_2$) in the atmosphere is increasing rapidly due to our use of fossil fuels. Because plants use $CO_2$ to fuel photosynthesis, more $CO_2$ may cause trees and other plants to grow faster. An elaborate apparatus allows researchers to pipe extra $CO_2$ to a 30-meter circle of forest. They set up three pairs of circles in different parts of a forest in North Carolina. One of each pair received extra $CO_2$ for an entire growing season, and the other received ambient air. The response variable is the average growth in base area for trees in a circle, as a fraction of the starting area. Here are the data for one growing season:[12]

| Pair | Control | Treatment |
|---|---|---|
| 1 | 0.06528 | 0.08150 |
| 2 | 0.05232 | 0.06334 |
| 3 | 0.04329 | 0.05936 |

(a) Summarize the data. Does it appear that growth was faster in the treated plots?

(b) The researchers used a matched pairs $t$ test to see if the data give good evidence of faster growth in the treated plots. State hypotheses, carry out the test, and state your conclusion.

(c) The sample is so small that we cannot assess Normality. To be safe, we might use the Wilcoxon signed rank test. Carry out this test and report your result.

(d) The tests lead to very different conclusions. The primary reason is the lack of power of rank tests for very small samples. Explain to someone who knows no statistics what this means.

**15.23 Heart rate and exercise.** A student project asked subjects to step up and down for three minutes and measured their heart rates before and after the exercise. Here are data for five subjects and two treatments: stepping at a low rate (14 steps per minute) and at a medium rate (21 steps per minute). For each subject, we give the resting heart rate (beats per minute) and the heart rate at the end of the exercise.[13]

| | Low Rate | | Medium Rate | |
|---|---|---|---|---|
| Subject | Resting | Final | Resting | Final |
| 1 | 60 | 75 | 63 | 84 |
| 2 | 90 | 99 | 69 | 93 |
| 3 | 87 | 93 | 81 | 96 |
| 4 | 78 | 87 | 75 | 90 |
| 5 | 84 | 84 | 90 | 108 |

Does exercise at the low rate raise heart rate significantly? State hypotheses in terms of the median increase in heart rate and apply the Wilcoxon signed rank test. What do you conclude?

**15.24 Compare exercise at a medium rate with exercise at a low rate.** Do the data from the previous exercise give good reason to think that stepping at the medium rate increases heart rates more than stepping at the low rate?

(a) State hypotheses in terms of comparing the median increases for the two treatments. What is the proper rank test for these hypotheses?

(b) Carry out your test and state a conclusion.

**15.25 The full moon and behavior.** Can the full moon influence behavior? A study observed 15 nursing-home patients with dementia. The number of incidents of aggressive behavior was recorded each day for 12 weeks. Call a day a "moon day" if it is the day of a full moon or the day before or after a full moon. Here are the average numbers of aggressive incidents for moon days and other days for each subject:[14]

| Patient | Moon days | Other days |
|---|---|---|
| 1 | 3.33 | 0.27 |
| 2 | 3.67 | 0.59 |
| 3 | 2.67 | 0.32 |
| 4 | 3.33 | 0.19 |
| 5 | 3.33 | 1.26 |
| 6 | 3.67 | 0.11 |
| 7 | 4.67 | 0.30 |
| 8 | 2.67 | 0.40 |
| 9 | 6.00 | 1.59 |
| 10 | 4.33 | 0.60 |
| 11 | 3.33 | 0.65 |
| 12 | 0.67 | 0.69 |
| 13 | 1.33 | 1.26 |
| 14 | 0.33 | 0.23 |
| 15 | 2.00 | 0.38 |

The matched pairs $t$ test (Example 7.7) gives $P < 0.000015$ and a permutation test (Example 16.14) gives $P = 0.0001$. Does the Wilcoxon signed rank test, based on ranks rather than means, agree that there is strong evidence that there are more aggressive incidents on moon days?

**15.26 A summer language institute for teachers.** A matched pairs study of the effect of a summer language institute on the ability of teachers to comprehend spoken French had these improvements in scores between the pretest and the posttest for 20 teachers:

| | | | | | | | | | |
|---|---|---|---|---|---|---|---|---|---|
| 2 | 0 | 6 | 6 | 3 | 3 | 2 | 3 | −6 | 6 |
| 6 | 6 | 3 | 0 | 1 | 1 | 0 | 2 | 3 | 3 |

(Exercise 7.41 applies the $t$ test to these data; Exercise 16.59 applies a permutation test based on the means.) Show the assignment of ranks and the calculation of the signed rank statistic $W^+$ for these data. Remember that zeros are dropped from the data before ranking, so that $n$ is the number of nonzero differences within pairs.

**15.27 Food safety.** Example 15.6 describes a study of the attitudes of people attending outdoor fairs about the safety of the food served at such locations. The full data set is available on the text CD and Web site as the file *eg15_006*. It contains the responses of 303 people to several questions. The variables in this data set are (in order)

subject hfair sfair sfast srest gender

The variable "sfair" contains responses to the safety question described in Example 15.6. The variable "srest" contains responses to the same question asked about food served in restaurants. We suspect that restaurant food will appear safer than food served outdoors at a fair. Do the data give good evidence for this suspicion? (Give descriptive measures, a test statistic and its $P$-value, and your conclusion.) Why might we hesitate to accept a small $P$-value as good evidence against $H_0$ for these data?

**15.28 Use of latex gloves by nurses.** How often do nurses use latex gloves during procedures for which glove use is recommended? A matched pairs study observed nurses (without their knowledge) before and after a presentation on the importance of glove use. Here are the proportions of procedures for which each nurse wore gloves:[15]

| Nurse | Before | After |
|---|---|---|
| 1 | 0.500 | 0.857 |
| 2 | 0.500 | 0.833 |
| 3 | 1.000 | 1.000 |
| 4 | 0.000 | 1.000 |
| 5 | 0.000 | 1.000 |
| 6 | 0.000 | 1.000 |
| 7 | 1.000 | 1.000 |
| 8 | 0.000 | 1.000 |
| 9 | 0.000 | 0.667 |
| 10 | 0.167 | 1.000 |
| 11 | 0.000 | 0.750 |
| 12 | 0.000 | 1.000 |
| 13 | 0.000 | 1.000 |
| 14 | 1.000 | 1.000 |

Is there good evidence that glove use increased after the presentation?

**15.29 Radon detectors.** How accurate are radon detectors of a type sold to homeowners? To answer this question, university researchers placed 12 detectors in a chamber that exposed them to 105 picocuries per liter (pCi/l) of radon.[16] The detector readings are as follows:

| | | | | | |
|---|---|---|---|---|---|
| 91.9 | 97.8 | 111.4 | 122.3 | 105.4 | 95.0 |
| 103.8 | 99.6 | 96.6 | 119.3 | 104.8 | 101.7 |

We wonder if the median reading differs significantly from the true value 105.

(a) Graph the data, and comment on skewness and outliers. A rank test is appropriate.

(b) We would like to test hypotheses about the median reading from home radon detectors:

$$H_0\colon \text{median} = 105$$

$$H_a\colon \text{median} \neq 105$$

To do this, apply the Wilcoxon signed rank statistic to the differences between the observations and 105. (This is the one-sample version of the test.) What do you conclude?

**15.30 Vitamin C in wheat-soy blend.** The U.S. Agency for International Development provides large quantities of wheat-soy blend (WSB) for development programs and emergency relief in countries throughout the world. One study collected data on the vitamin C content of 27 bags of WSB at the factory and five months later in Haiti.[17] Here are the data:

| Sample | 1 | 2 | 3 | 4 | 5 |
|---|---|---|---|---|---|
| Before | 73 | 79 | 86 | 88 | 78 |
| After | 20 | 27 | 29 | 36 | 17 |

We want to know if vitamin C has been lost during transportation and storage. Describe what the data show about this question. Then use a rank test to see whether there has been a significant loss.

**15.31 Weight gains with an extra 1000 calories per day.** Exercise 7.32 (page 444) presents these data on the weight gains (in kilograms) of adults who were fed an extra 1000 calories per day for 8 weeks:[18]

| Subject | Weight Before | Weight After |
|---|---|---|
| 1 | 55.7 | 61.7 |
| 2 | 54.9 | 58.8 |
| 3 | 59.6 | 66.0 |
| 4 | 62.3 | 66.2 |
| 5 | 74.2 | 79.0 |
| 6 | 75.6 | 82.3 |
| 7 | 70.7 | 74.3 |
| 8 | 53.3 | 59.3 |
| 9 | 73.3 | 79.1 |
| 10 | 63.4 | 66.0 |
| 11 | 68.1 | 73.4 |
| 12 | 73.7 | 76.9 |
| 13 | 91.7 | 93.1 |
| 14 | 55.9 | 63.0 |
| 15 | 61.7 | 68.2 |
| 16 | 57.8 | 60.3 |

(a) Use a rank test to test the null hypothesis that the median weight gain is 16 pounds, as theory suggests. What do you conclude?

(b) If your software allows, give a 95% confidence interval for the median weight gain in the population.

## 15.3 The Kruskal-Wallis Test*

We have now considered alternatives to the matched pairs and two-sample $t$ tests for comparing the magnitude of responses to two treatments. To compare more than two treatments, we use one-way analysis of variance (ANOVA) if the distributions of the responses to each treatment are at least roughly Normal and have similar spreads. What can we do when these distribution requirements are violated?

EXAMPLE

**15.13 Weeds and corn yield.** Lamb's-quarter is a common weed that interferes with the growth of corn. A researcher planted corn at the same rate in 16 small plots of ground, then randomly assigned the plots to four groups. He weeded the plots by hand to allow a fixed number of lamb's-quarter plants to grow in each meter of corn row. These numbers were 0, 1, 3, and 9 in the four groups of plots. No other weeds were allowed to grow, and all plots received identical treatment except for the weeds. Here are the yields of corn (bushels per acre) in each of the plots:[19]

| Weeds per meter | Corn yield | Weeds per meter | Corn yield | Weeds per meter | Corn yield | Weeds per meter | Corn yield |
|---|---|---|---|---|---|---|---|
| 0 | 166.7 | 1 | 166.2 | 3 | 158.6 | 9 | 162.8 |
| 0 | 172.2 | 1 | 157.3 | 3 | 176.4 | 9 | 142.4 |
| 0 | 165.0 | 1 | 166.7 | 3 | 153.1 | 9 | 162.7 |
| 0 | 176.9 | 1 | 161.1 | 3 | 156.0 | 9 | 162.4 |

*Because this test is an alternative to the one-way analysis of variance $F$ test, you should first read Chapter 12.

The summary statistics are

| Weeds | $n$ | Mean | Std. dev. |
|---|---|---|---|
| 0 | 4 | 170.200 | 5.422 |
| 1 | 4 | 162.825 | 4.469 |
| 3 | 4 | 161.025 | 10.493 |
| 9 | 4 | 157.575 | 10.118 |

The sample standard deviations do not satisfy our rule of thumb that for safe use of ANOVA the largest should not exceed twice the smallest. Normal quantile plots (Figure 15.9) show that outliers are present in the yields for 3 and 9 weeds per meter. These are the correct yields for their plots, so we have no justification for removing them. We may want to use a rank test.

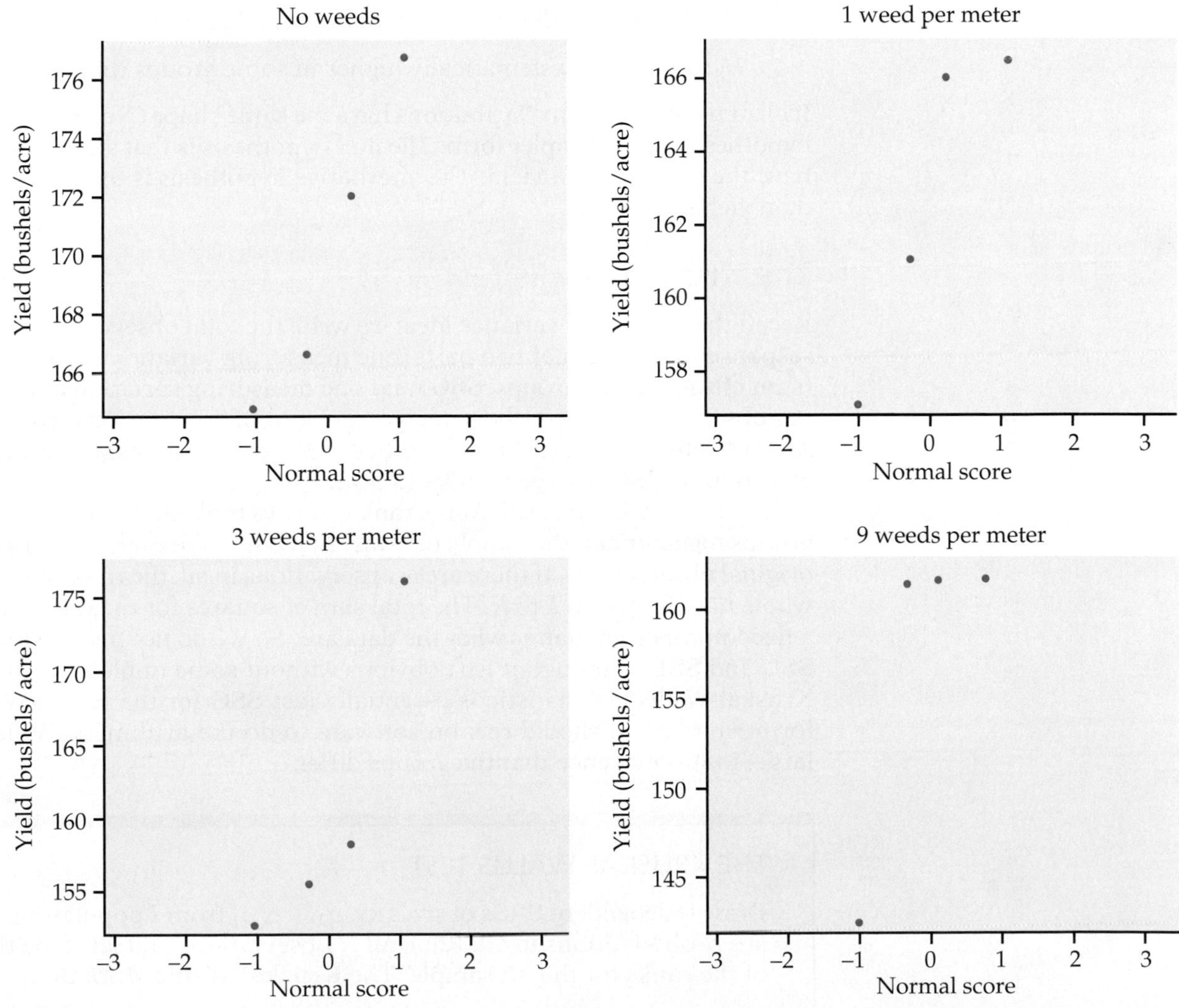


**FIGURE 15.9** Normal quantile plots for the corn yields in the four treatment groups in Example 15.13.

## Hypotheses and assumptions

The ANOVA $F$ test concerns the means of the several populations represented by our samples. For Example 15.13, the ANOVA hypotheses are

$$H_0: \mu_0 = \mu_1 = \mu_3 = \mu_9$$

$$H_a: \text{not all four means are equal}$$

For example, $\mu_0$ is the mean yield in the population of all corn planted under the conditions of the experiment with no weeds present. The data should consist of four independent random samples from the four populations, all Normally distributed with the same standard deviation.

The *Kruskal-Wallis test* is a rank test that can replace the ANOVA $F$ test. The assumption about data production (independent random samples from each population) remains important, but we can relax the Normality assumption. We assume only that the response has a continuous distribution in each population. The hypotheses tested in our example are

$H_0$: Yields have the same distribution in all groups.

$H_a$: Yields are systematically higher in some groups than in others.

If all of the population distributions have the same shape (Normal or not), these hypotheses take a simpler form. The null hypothesis is that all four populations have the same *median* yield. The alternative hypothesis is that not all four median yields are equal.

## The Kruskal-Wallis test

Recall the analysis of variance idea: we write the total observed variation in the responses as the sum of two parts, one measuring variation among the groups (sum of squares for groups, SSG) and one measuring variation among individual observations within the same group (sum of squares for error, SSE). The ANOVA $F$ test rejects the null hypothesis that the mean responses are equal in all groups if SSG is large relative to SSE.

The idea of the Kruskal-Wallis rank test is to rank all the responses from all groups together and then apply one-way ANOVA to the ranks rather than to the original observations. If there are $N$ observations in all, the ranks are always the whole numbers from 1 to $N$. The total sum of squares for the ranks is therefore a fixed number no matter what the data are. So we do not need to look at both SSG and SSE. Although it isn't obvious without some unpleasant algebra, the Kruskal-Wallis test statistic is essentially just SSG for the ranks. We give the formula, but you should rely on software to do the arithmetic. When SSG is large, that is evidence that the groups differ.

### THE KRUSKAL-WALLIS TEST

Draw independent SRSs of sizes $n_1, n_2, \ldots, n_I$ from $I$ populations. There are $N$ observations in all. Rank all $N$ observations and let $R_i$ be the sum of the ranks for the $i$th sample. The **Kruskal-Wallis statistic** is

$$H = \frac{12}{N(N+1)} \sum \frac{R_i^2}{n_i} - 3(N+1)$$

When the sample sizes $n_i$ are large and all $I$ populations have the same continuous distribution, $H$ has approximately the chi-square distribution with $I - 1$ degrees of freedom.

The **Kruskal-Wallis test** rejects the null hypothesis that all populations have the same distribution when $H$ is large.

We now see that, like the Wilcoxon rank sum statistic, the Kruskal-Wallis statistic is based on the sums of the ranks for the groups we are comparing. The more different these sums are, the stronger is the evidence that responses are systematically larger in some groups than in others.

The exact distribution of the Kruskal-Wallis statistic $H$ under the null hypothesis depends on all the sample sizes $n_1$ to $n_I$, so tables are awkward. The calculation of the exact distribution is so time-consuming for all but the smallest problems that even most statistical software uses the chi-square approximation to obtain $P$-values. As usual, there is no usable exact distribution when there are ties among the responses. We again assign average ranks to tied observations.

EXAMPLE

**15.14 Perform the significance test.** In Example 15.13, there are $I = 4$ populations and $N = 16$ observations. The sample sizes are equal, $n_i = 4$. The 16 observations arranged in increasing order, with their ranks, are

| Yield | 142.4 | 153.1 | 156.0 | 157.3 | 158.6 | 161.1 | 162.4 | 162.7 |
|---|---|---|---|---|---|---|---|---|
| Rank | 1 | 2 | 3 | 4 | 5 | 6 | 7 | 8 |
| Yield | 162.8 | 165.0 | 166.2 | 166.7 | 166.7 | 172.2 | 176.4 | 176.9 |
| Rank | 9 | 10 | 11 | 12.5 | 12.5 | 14 | 15 | 16 |

There is one pair of tied observations. The ranks for each of the four treatments are

| Weeds | Ranks | | | | Rank sums |
|---|---|---|---|---|---|
| 0 | 10 | 12.5 | 14 | 16 | 52.5 |
| 1 | 4 | 6 | 11 | 12.5 | 33.5 |
| 3 | 2 | 3 | 5 | 15 | 25.0 |
| 9 | 1 | 7 | 8 | 9 | 25.0 |

The Kruskal-Wallis statistic is therefore

$$\begin{aligned} H &= \frac{12}{N(N+1)} \sum \frac{R_i^2}{n_i} - 3(N+1) \\ &= \frac{12}{(16)(17)} \left( \frac{52.5^2}{4} + \frac{33.5^2}{4} + \frac{25^2}{4} + \frac{25^2}{4} \right) - (3)(17) \\ &= \frac{12}{272}(1282.125) - 51 \\ &= 5.56 \end{aligned}$$

Referring to the table of chi-square critical points (Table F) with df = 3, we find that the $P$-value lies in the interval $0.10 < P < 0.15$. This small experiment suggests that more weeds decrease yield but does not provide convincing evidence that weeds have an effect.

Figure 15.10 displays the output from the SAS statistical software, which gives the results $H = 5.5725$ and $P = 0.1344$. The software makes a small adjustment for the presence of ties that accounts for the slightly larger value of $H$. The adjustment makes the chi-square approximation more accurate. It would be important if there were many ties.

As an option, SAS will calculate the exact $P$-value for the Kruskal-Wallis test. The result for Example 15.14 is $P = 0.1299$. This result required more than an hour of computing time. Fortunately, the chi-square approximation is quite accurate. The ordinary ANOVA $F$ test gives $F = 1.73$ with $P = 0.2130$. Although the practical conclusion is the same, ANOVA and Kruskal-Wallis do not agree closely in this example. The rank test is more reliable for these small samples with outliers.

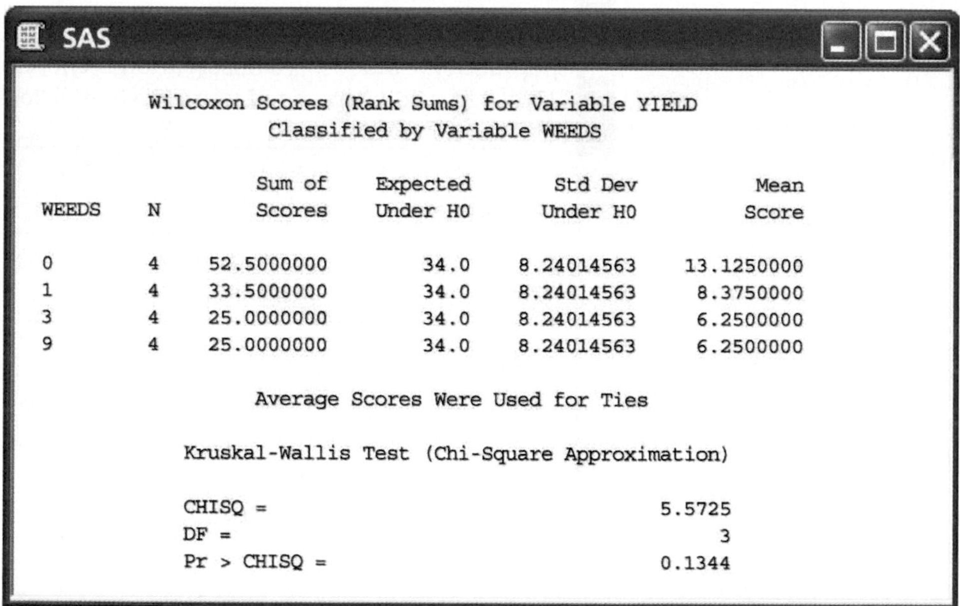
SAS

Wilcoxon Scores (Rank Sums) for Variable YIELD
Classified by Variable WEEDS

| WEEDS | N | Sum of Scores | Expected Under H0 | Std Dev Under H0 | Mean Score |
|---|---|---|---|---|---|
| 0 | 4 | 52.5000000 | 34.0 | 8.24014563 | 13.1250000 |
| 1 | 4 | 33.5000000 | 34.0 | 8.24014563 | 8.3750000 |
| 3 | 4 | 25.0000000 | 34.0 | 8.24014563 | 6.2500000 |
| 9 | 4 | 25.0000000 | 34.0 | 8.24014563 | 6.2500000 |

Average Scores Were Used for Ties

Kruskal-Wallis Test (Chi-Square Approximation)

| | |
|---|---|
| CHISQ = | 5.5725 |
| DF = | 3 |
| Pr > CHISQ = | 0.1344 |

**FIGURE 15.10** Output from SAS for the Kruskal-Wallis test applied to the data in Example 15.14. SAS uses the chi-square approximation to obtain a $P$-value.

## SECTION 15.3 Summary

The **Kruskal-Wallis test** compares several populations on the basis of independent random samples from each population. This is the **one-way analysis of variance** setting.

The null hypothesis for the Kruskal-Wallis test is that the distribution of the response variable is the same in all the populations. The alternative hypothesis is that responses are systematically larger in some populations than in others.

The **Kruskal-Wallis statistic *H*** can be viewed in two ways. It is essentially the result of applying one-way ANOVA to the ranks of the observations. It is also a comparison of the sums of the ranks for the several samples.

When the sample sizes are not too small and the null hypothesis is true, $H$ for comparing $I$ populations has approximately the chi-square distribution with $I - 1$ degrees of freedom. We use this approximate distribution to obtain $P$-values.

## SECTION 15.3 Exercises

*Statistical software is needed to do these exercises without unpleasant hand calculations. If you do not have access to software, find the Kruskal-Wallis statistic H by hand and use the chi-square table to get approximate P-values.*

**15.32 Vitamins in bread.** Does bread lose its vitamins when stored? Here are data on the vitamin C content (milligrams per 100 grams of flour) in bread baked from the same recipe and stored for 1, 3, 5, or 7 days.[20] The 10 observations are from 10 different loaves of bread.

| Condition | Vitamin C (mg/100 g) | |
|---|---|---|
| Immediately after baking | 47.62 | 49.79 |
| One day after baking | 40.45 | 43.46 |
| Three days after baking | 21.25 | 22.34 |
| Five days after baking | 13.18 | 11.65 |
| Seven days after baking | 8.51 | 8.13 |

The loss of vitamin C over time is clear, but with only 2 loaves of bread for each storage time we wonder if the differences among the groups are significant.

(a) Use the Kruskal-Wallis test to assess significance, then write a brief summary of what the data show.

(b) Because there are only 2 observations per group, we suspect that the common chi-square approximation to the distribution of the Kruskal-Wallis statistic may not be accurate. The exact $P$-value (from the SAS software) is $P = 0.0011$. Compare this with your $P$-value from (a). Is the difference large enough to affect your conclusion?

**15.33 Jumping and strong bones.** Many studies suggest that exercise causes bones to get stronger. One study examined the effect of jumping on the bone density of growing rats. Ten rats were assigned to each of three treatments: a 60-centimeter "high jump," a 30-centimeter "low jump," and a control group with no jumping. Here are the bone densities (in milligrams per cubic centimeter) after 8 weeks of 10 jumps per day:[21]

| Group | Bone density ($mg/cm^3$) | | | | |
|---|---|---|---|---|---|
| Control | 611 | 621 | 614 | 593 | 593 |
| | 653 | 600 | 554 | 603 | 569 |
| Low jump | 635 | 605 | 638 | 594 | 599 |
| | 632 | 631 | 588 | 607 | 596 |
| High jump | 650 | 622 | 626 | 626 | 631 |
| | 622 | 643 | 674 | 643 | 650 |

(a) The study was a randomized comparative experiment. Outline the design of this experiment.

(b) Make side-by-side stemplots for the three groups, with the stems lined up for easy comparison. The distributions are a bit irregular but not strongly non-Normal. We would usually use analysis of variance to assess the significance of the difference in group means.

(c) Do the Kruskal-Wallis test. Explain the distinction between the hypotheses tested by Kruskal-Wallis and ANOVA.

(d) Write a brief statement of your findings. Include a numerical comparison of the groups as well as your test result.

**15.34 Detecting insects in farm fields.** To detect the presence of harmful insects in farm fields, we can put up boards covered with a sticky material and examine the insects trapped on the boards. Which colors attract insects best? Experimenters placed six boards of each of four colors at random locations in a field of oats and measured the number of cereal leaf beetles trapped. Here are the data:[22]

| Color | Insects trapped | | | | | |
|---|---|---|---|---|---|---|
| Lemon yellow | 45 | 59 | 48 | 46 | 38 | 47 |
| White | 21 | 12 | 14 | 17 | 13 | 17 |
| Green | 37 | 32 | 15 | 25 | 39 | 41 |
| Blue | 16 | 11 | 20 | 21 | 14 | 7 |

Because the samples are small, we will apply a nonparametric test.

(a) What hypotheses does ANOVA test? What hypotheses does Kruskal-Wallis test?

(b) Find the median number of beetles trapped by boards of each color. Which colors appear more effective? Use the Kruskal-Wallis test to see if there are significant differences among the colors. What do you conclude?

**15.35 Do the calculations by hand.** Exercise 15.34 gives data on the counts of insects attracted by boards of four different colors. Carry out the Kruskal-Wallis test by hand, following these steps.

(a) What are $I$, the $n_i$, and $N$?

(b) Arrange the counts in order and assign ranks. Be careful about ties. Find the sum of the ranks $R_i$ for each color.

(c) Calculate the Kruskal-Wallis statistic $H$. How many degrees of freedom should you use for the chi-square approximation to its null distribution? Use the chi-square table to give an approximate $P$-value.

**15.36 Decay of polyester fabric in landfills.** Here are the breaking strengths (in pounds) of strips of polyester fabric buried in the ground for several lengths of time:[23]

| Time | Breaking strength | | | | |
|---|---|---|---|---|---|
| 2 weeks | 118 | 126 | 126 | 120 | 129 |
| 4 weeks | 130 | 120 | 114 | 126 | 128 |
| 8 weeks | 122 | 136 | 128 | 146 | 140 |
| 16 weeks | 124 | 98 | 110 | 140 | 110 |

Breaking strength is a good measure of the extent to which the fabric has decayed.

(a) Find the standard deviations of the 4 samples. They do not meet our rule of thumb for applying ANOVA. In addition, the sample buried for 16 weeks contains an outlier. We will use the Kruskal-Wallis test.

(b) Find the medians of the four samples. What are the hypotheses for the Kruskal-Wallis test, expressed in terms of medians?

(c) Carry out the test and report your conclusion.

**15.37 Food safety.** Example 15.6 describes a study of the attitudes of people attending outdoor fairs about the safety of the food served at such locations. The full data set is available on the text CD and Web site as the file *eg15_006*. It contains the responses of 303 people to several questions. The variables in this data set are (in order)

subject hfair sfair sfast srest gender

The variable "sfair" contains responses to the safety question described in Example 15.6. The variables "srest" and "sfast" contain responses to the same question asked about food served in restaurants and in fast-food chains. Explain carefully why we *cannot* use the Kruskal-Wallis test to see if there are systematic differences in perceptions of food safety in these three locations.

**15.38 Logging in Borneo.** In Exercise 15.13 you compared the number of tree species in plots of land in a tropical rainforest that had never been logged with similar plots nearby that had been logged 8 years earlier. The researchers also counted species in plots that had been logged just 1 year earlier. Here are the counts of species:[24]

| Plot type | Species count | | | | | |
|---|---|---|---|---|---|---|
| Unlogged | 22 | 18 | 22 | 20 | 15 | 21 |
| | 13 | 13 | 19 | 13 | 19 | 15 |
| Logged 1 year ago | 11 | 11 | 14 | 7 | 18 | 15 |
| | 15 | 12 | 13 | 2 | 15 | 8 |
| Logged 8 years ago | 17 | 4 | 18 | 14 | 18 | 15 |
| | 15 | 10 | 12 | | | |

(a) Use side-by-side stemplots to compare the distributions of number of species per plot for the three groups of plots. Are there features that might prevent use of ANOVA? Also give the median number of species per plot in the three groups.

(b) Use the Kruskal-Wallis test to compare the distributions of species counts. State hypotheses, the test statistic and its $P$-value, and your conclusions.

**15.39 Heart disease and smoking.** In a study of heart disease in male federal employees, researchers classified 356 volunteer subjects according to their socioeconomic status (SES) and their smoking habits. There were three categories of SES: high, middle, and low. Individuals were asked whether they were current smokers, former smokers, or had never smoked. Here are the data, as a two-way table of counts:[25]

| SES | Never (1) | Former (2) | Current (3) |
|---|---|---|---|
| High | 68 | 92 | 51 |
| Middle | 9 | 21 | 22 |
| Low | 22 | 28 | 43 |

The data for all 356 subjects are stored in the file *ex15_29* on the text CD and Web site. Smoking behavior is stored numerically as 1, 2, or 3 using the codes given in the column headings above.

(a) Higher-SES people in the United States smoke less as a group than lower-SES people. Do these data show a relationship of this kind? Give percents that back your statements.

(b) Apply the chi-square test to see if there is a significant relationship between SES and smoking behavior.

(c) The chi-square test ignores the ordering of the responses. Use the Kruskal-Wallis test (with many ties) to test the hypothesis that some SES classes smoke systematically more than others.

## CHAPTER 15 Exercises

**15.40 Response times for telephone repair calls.** Exercise 16.55 (page 16-53) presents data on the time required for the telephone company Verizon to respond to repair calls from its own customers and from customers of a CLEC, another phone company that pays Verizon to use its local lines. Here are the data, which are rounded to the nearest hour:

| Verizon | | | | | | | | | | | |
|---|---|---|---|---|---|---|---|---|---|---|---|
| 1 | 1 | 1 | 1 | 2 | 2 | 1 | 1 | 1 | 1 | 2 | 2 |
| 1 | 1 | 1 | 1 | 2 | 2 | 1 | 1 | 1 | 1 | 2 | 3 |
| 1 | 1 | 1 | 1 | 2 | 3 | 1 | 1 | 1 | 1 | 2 | 3 |
| 1 | 1 | 1 | 1 | 2 | 3 | 1 | 1 | 1 | 1 | 2 | 3 |
| 1 | 1 | 1 | 1 | 2 | 3 | 1 | 1 | 1 | 1 | 2 | 4 |
| 1 | 1 | 1 | 1 | 2 | 5 | 1 | 1 | 1 | 1 | 2 | 5 |
| 1 | 1 | 1 | 1 | 2 | 6 | 1 | 1 | 1 | 1 | 2 | 8 |
| 1 | 1 | 1 | 1 | 2 | 15 | 1 | 1 | 1 | 2 | 2 | |

| CLEC | | | | | | | | | |
|---|---|---|---|---|---|---|---|---|---|
| 1 | 1 | 5 | 5 | 5 | 1 | 5 | 5 | 5 | 5 |

(a) Does Verizon appear to give CLEC customers the same level of service as its own customers? Compare the data using graphs and descriptive measures and express your opinion.

(b) We would like to see if times are significantly longer for CLEC customers than for Verizon customers. Why would you hesitate to use a $t$ test for this purpose? Carry out a rank test. What can you conclude?

**15.41 Selling prices of three- and four-bedroom homes.** Exercise 7.141 (page 486) reports data on the selling prices of 9 four-bedroom houses and 28 three-bedroom houses in West Lafayette, Indiana. We wonder if there is a difference between the average prices of three- and four-bedroom houses in this community.

(a) Make a Normal quantile plot of the prices of three-bedroom houses. What kind of deviation from Normality do you see?

(b) The $t$ tests are quite robust. State the hypotheses for the proper $t$ test, carry out the test, and present your results, including appropriate data summaries.

(c) Carry out a nonparametric test. Once more state the hypotheses tested and present your results for both the test and the data summaries that should go with it.

**15.42 Air in a turkey-processing plant.** The air in poultry-processing plants often contains fungus spores. If the ventilation is inadequate, this can affect the health of the workers. To measure the presence of spores, air samples are pumped to an agar plate and "colony-forming units (CFUs)" are counted after an incubation period. Here are data from the "kill room" of a plant that slaughters 37,000 turkeys per day, taken at four seasons of the year. The units are CFUs per cubic meter of air.[26]

| Fall | Winter | Spring | Summer |
|---|---|---|---|
| 1231 | 384 | 2105 | 3175 |
| 1254 | 104 | 701 | 2526 |
| 752 | 251 | 2947 | 1763 |
| 1088 | 97 | 842 | 1090 |

(a) Examine the data using graphs and descriptive measures. How do airborne fungus spores vary with the seasons?

(b) Is the effect of season statistically significant?

**15.43** CHALLENGE **Plants and hummingbirds.** Different varieties of the tropical flower *Heliconia* are fertilized by different species of hummingbirds. Over time, the lengths of the flowers and the form of the hummingbirds' beaks have evolved to match each other. Here are data on the lengths in millimeters of three varieties of these flowers on the island of Dominica:[27]

| *H. bihai* | | | | | |
|---|---|---|---|---|---|
| 47.12 | 46.75 | 46.81 | 47.12 | 46.67 | 47.43 |
| 46.44 | 46.64 | 48.07 | 48.34 | 48.15 | 50.26 |
| 50.12 | 46.34 | 46.94 | 48.36 | | |
| ***H. caribaea* red** | | | | | |
| 41.90 | 42.01 | 41.93 | 43.09 | 41.47 | 41.69 |
| 39.78 | 40.57 | 39.63 | 42.18 | 40.66 | 37.87 |
| 39.16 | 37.40 | 38.20 | 38.07 | 38.10 | 37.97 |
| 38.79 | 38.23 | 38.87 | 37.78 | 38.01 | |
| ***H. caribaea* yellow** | | | | | |
| 36.78 | 37.02 | 36.52 | 36.11 | 36.03 | 35.45 |
| 38.13 | 37.10 | 35.17 | 36.82 | 36.66 | 35.68 |
| 36.03 | 34.57 | 34.63 | | | |

Do a complete analysis that includes description of the data and a rank test for the significance of the differences in lengths among the three species.

*Iron-deficiency anemia is the most common form of malnutrition in developing countries. Does the type of cooking pot affect the iron content of food? We have data from a study in Ethiopia that measured the iron content (milligrams per 100 grams of food) for three types of food cooked in each of three types of pots:*[28]

| Type of pot | Iron Content | | | |
|---|---|---|---|---|
| | | Meat | | |
| Aluminum | 1.77 | 2.36 | 1.96 | 2.14 |
| Clay | 2.27 | 1.28 | 2.48 | 2.68 |
| Iron | 5.27 | 5.17 | 4.06 | 4.22 |
| | | Legumes | | |
| Aluminum | 2.40 | 2.17 | 2.41 | 2.34 |
| Clay | 2.41 | 2.43 | 2.57 | 2.48 |
| Iron | 3.69 | 3.43 | 3.84 | 3.72 |
| | | Vegetables | | |
| Aluminum | 1.03 | 1.53 | 1.07 | 1.30 |
| Clay | 1.55 | 0.79 | 1.68 | 1.82 |
| Iron | 2.45 | 2.99 | 2.80 | 2.92 |

*Exercises 15.44 to 15.46 use these data.*

**15.44 Cooking vegetables in different pots.** Does the vegetable dish vary in iron content when cooked in aluminum, clay, and iron pots?

(a) What do the data appear to show? Check the conditions for one-way ANOVA. Which requirements are a bit dubious in this setting?

(b) Instead of ANOVA, do a rank test. Summarize your conclusions about the effect of pot material on the iron content of the vegetable dish.

**15.45 Cooking meat and legumes in aluminum and clay pots.** There appears to be little difference between the iron content of food cooked in aluminum pots and food cooked in clay pots. Is there a significant difference between the iron content of meat cooked in aluminum and clay? Is the difference between aluminum and clay significant for legumes? Use rank tests.

**15.46 Iron in food cooked in iron pots.** The data show that food cooked in iron pots has the highest iron content. They also suggest that the three types of food differ in iron content. Is there significant evidence that the three types of food differ in iron content when all are cooked in iron pots?

**15.47** CHALLENGE **Multiple comparisons for plants and hummingbirds.** As in ANOVA, we often want to carry out a **multiple-comparisons** procedure following a Kruskal-Wallis test to tell us *which* groups differ significantly.[29] Here is a simple method: If we carry out $k$ tests at fixed significance level $0.05/k$, the probability of *any* false rejection among the $k$ tests is always no greater than 0.05. That is, to get overall significance level 0.05 for all of $k$ comparisons, do each individual comparison at the $0.05/k$ level. In Exercise 15.43 you found a significant difference among the lengths of three varieties of the flower *Heliconia.* Now we will explore multiple comparisons.

(a) Write down all of the pairwise comparisons we can make, for example, *bihai* versus *caribaea* red. There are three possible pairwise comparisons.

(b) Carry out three Wilcoxon rank sum tests, one for each of the three pairs of flower varieties. What are the three two-sided *P*-values?

(c) For purposes of multiple comparisons, any of these three tests is significant if its *P*-value is no greater than $0.05/3 = 0.0167$. Which pairs differ significantly at the overall 0.05 level?

**15.48** CHALLENGE **Multiple comparisons for the turkey-processing plant.** Exercise 15.47 outlines how to use the Wilcoxon rank sum test several times for multiple comparisons with overall significance level 0.05 for all comparisons together. In Exercise 15.42 you found that the airborne fungus spore counts in a turkey-processing plant differ significantly among the seasons of the year. At the overall 0.05 level, which pairs of seasons differ significantly? (*Hint:* There are 6 possible pairwise comparisons among 4 seasons.)

# DATA APPENDIX

Some of the computer exercises in the text refer to 16 relatively large data sets that are on the CD that accompanies this text. The CD also contains data for many other exercises and examples.

Background information for each of the 16 data sets is presented below. For most, the first five cases are given here.

## 1 BIOMARKERS

Text Reference: Exercises 7.118–7.121, 10.26–10.29, and 11.34–11.39

Healthy bones are continually being renewed by two processes. Through bone formation, new bone is built; through bone resorption, old bone is removed. If one or both of these processes is disturbed, by disease, aging, or space travel, for example, bone loss can be the result. The variables VO+ and VO− measure bone formation and bone resorption, respectively. Osteocalcin (OC) is a biochemical marker for bone formation: higher levels of bone formation are associated with higher levels of OC. A blood sample is used to measure OC, and it is much less expensive to obtain than direct measures of bone formation. The units are milligrams of OC per milliliter of blood (mg/ml). Similarly, tartrate resistant acid phosphatase (TRAP) is a biochemical marker for bone resorption that is also measured in blood. It is measured in units per liter (U/l). These variables were measured in a study of 31 healthy women aged 11 to 32 years. The results were published in C. M. Weaver et al., "Quantification of biochemical markers of bone turnover by kinetic measures of bone formation and resorption in young healthy females," *Journal of Bone and Mineral Research,* 12 (1997), pp. 1714–1720. Variables with the first letter "L" are the logarithms of the measured variables. The data were provided by Linda McCabe. The first five cases are given in the table below.

| VO+ | VO− | OC | LOC | TRAP | LTRAP | LVO+ | LVO− |
|---|---|---|---|---|---|---|---|
| 1606 | 903 | 68.9 | 4.233 | 19.4 | 2.965 | 7.382 | 6.806 |
| 2240 | 1761 | 56.3 | 4.031 | 25.5 | 3.239 | 7.714 | 7.474 |
| 2221 | 1486 | 54.6 | 4.000 | 19.0 | 2.944 | 7.706 | 7.304 |
| 896 | 1116 | 31.2 | 3.440 | 9.0 | 2.197 | 6.798 | 7.018 |
| 2545 | 2236 | 36.4 | 3.595 | 19.1 | 2.950 | 7.842 | 7.712 |

## 2 BRFSS

Text Reference: Exercises 2.136, 2.137, and 2.166

With support from the Centers for Disease Control and Prevention (CDC), the Behavioral Risk Factor Surveillance System (BRFSS) conducts the world's largest, ongoing telephone survey of health conditions and risk behaviors in the United States. The prevalence of various health risk factors by state is summarized on the CDC Web site, `www.cdc.gov/brfss`. The data set BRFSS contains data on 29 demographic characteristics and risk factors for each state. The demographic characteristics are age (percents aged 18 to 24, 25 to 34, 35 to 44, 45 to 54, 55 to 64, and 65 or over), education (less than high school, high school or GED, some post–high school, college), income (less than \$15,000, \$15,000 to \$25,000, \$25,000 to \$35,000, \$35,000 to \$50,000, \$50,000 or more), and percent female. Risk factors are body mass index (BMI is weight in kilograms divided by the square of height in meters; the classifications are less than 25, 25 to 30, and

30 or more; 18.5 to 24.9 is considered normal, 25 to 29.9 is overweight, and 30 or over is obese), alcohol consumption (at least one drink within the last 30 days, heavy is more than two drinks per day for men and more than one drink per day for women, binge is five or more drinks [four for women] on one occasion during the past 30 days), physical exercise (at least 10 minutes at a time during a usual week), fruits and vegetables (eat at least five servings per day), physical activity (30 or more minutes of moderate physical activity five or more days per week or vigorous physical activity for 20 or more minutes three or more days per week), and smoking (every day, some days, former smoker, never smoked).

## 3 CHEESE

Text Reference: Examples 14.5 and 14.9; Exercises 11.51–11.59 and 14.36–14.38

As cheddar cheese matures, many chemical processes take place. The taste of matured cheese is related to the concentration of several chemicals in the final product. In a study of cheddar cheese from the LaTrobe Valley of Victoria, Australia, samples of cheese were analyzed for their chemical composition and were subjected to taste tests.

Data for one type of cheese-manufacturing process appear below. The variable "Case" is used to number the observations from 1 to 30. "Taste" is the response variable of interest. The taste scores were obtained by combining the scores from several tasters.

Three of the chemicals whose concentrations were measured were acetic acid, hydrogen sulfide, and lactic acid. For acetic acid and hydrogen sulfide (natural) log transformations were taken. Thus, the explanatory variables are the transformed concentrations of acetic acid ("Acetic") and hydrogen sulfide ("H2S") and the untransformed concentration of lactic acid ("Lactic"). These data are based on experiments performed by G. T. Lloyd and E. H. Ramshaw of the CSIRO Division of Food Research, Victoria, Australia. Some results of the statistical analyses of these data are given in G. P. McCabe, L. McCabe, and A. Miller, "Analysis of taste and chemical composition of cheddar cheese, 1982–83 experiments," CSIRO Division of Mathematics and Statistics Consulting Report VT85/6; and in I. Barlow et al., "Correlations and changes in flavour and chemical parameters of cheddar cheeses during maturation," *Australian Journal of Dairy Technology,* 44 (1989), pp. 7–18. The table below gives the data for the first five cases.

| Case | Taste | Acetic | H2S | Lactic |
|---|---|---|---|---|
| 01 | 12.3 | 4.543 | 3.135 | 0.86 |
| 02 | 20.9 | 5.159 | 5.043 | 1.53 |
| 03 | 39.0 | 5.366 | 5.438 | 1.57 |
| 04 | 47.9 | 5.759 | 7.496 | 1.81 |
| 05 | 5.6 | 4.663 | 3.807 | 0.99 |

## 4 CSDATA

Text Reference: Example 11.1; Exercises 1.173, 2.155, 3.93, 3.94, 7.138, 11.5, 11.6, and 14.39–14.42

The computer science department of a large university was interested in understanding why a large proportion of their first-year students failed to graduate as computer science majors. An examination of records from the registrar indicated that most of the attrition occurred during the first three semesters. Therefore, they decided to study all first-year students entering their program in a particular year and to follow their progress for the first three semesters.

The variables studied included the grade point average after three semesters and a collection of variables that would be available as students entered their program. These included scores on standardized tests such as the SATs and high school grades in various subjects. The individuals who conducted the study were also interested in examining differences between men and women in this program. Therefore, sex was included as a variable.

Data on 224 students who began study as computer science majors in a particular year were analyzed. A few exceptional cases were excluded, such as students who did not have complete data available on the variables of interest (a few students were admitted who did not take the SATs). Data for the first five students appear below. There are eight variables for each student. OBS is a variable used to identify the student. The data files kept by the registrar identified students by social security number, but for this study they were simply given a number from 1 to 224. The grade point average after three semesters is the variable GPA. This university uses a four-point scale, with A corresponding to 4, B to 3, C to 2, etc. A straight-A student has a 4.00 GPA.

The high school grades included in the data set are the variables HSM, HSS, and HSE. These correspond to average high school grades in math, science, and English. High schools use different grading systems (some high schools have a grade higher than A for honors courses), so the university's task in constructing these variables is not easy. The researchers were willing to accept the university's judgment and used its values. High school grades were recorded on a scale from 1 to 10, with 10 corresponding to A, 9 to A−, 8 to B+, etc.

The SAT scores are SATM and SATV, corresponding to the Mathematics and Verbal parts of the SAT. Gender was recorded as 1 for men and 2 for women. This is an arbitrary code. For software packages that can use alphanumeric variables (that is, values do not have to be numbers), it is more convenient to use M and F or Men and Women as values for the sex variable. With this kind of user-friendly capability, you do not have to remember who are the 1s and who are the 2s.

Results of the study are reported in P. F. Campbell and G. P. McCabe, "Predicting the success of freshmen in a computer science major," *Communications of the ACM,* 27 (1984), pp. 1108–1113. The table below gives data for the first five students.

| OBS | GPA | HSM | HSS | HSE | SATM | SATV | SEX |
|---|---|---|---|---|---|---|---|
| 001 | 3.32 | 10 | 10 | 10 | 670 | 600 | 1 |
| 002 | 2.26 | 6 | 8 | 5 | 700 | 640 | 1 |
| 003 | 2.35 | 8 | 6 | 8 | 640 | 530 | 1 |
| 004 | 2.08 | 9 | 10 | 7 | 670 | 600 | 1 |
| 005 | 3.38 | 8 | 9 | 8 | 540 | 580 | 1 |

## 5 DANDRUFF

Text Reference: Exercises 12.54–12.57

The DANDRUFF data set is based on W. L. Billhimer et al., "Results of a clinical trial comparing 1% pyrithione zinc and 2% ketoconazole shampoos," *Cosmetic Dermatology,* 9 (1996), pp. 34–39. The study reported in this paper is a clinical trial that compared three treatments for dandruff and a placebo. The treatments were 1% pyrithione zinc shampoo (PyrI), the same shampoo but with instructions to shampoo two times (PyrII), 2% ketoconazole shampoo (Keto), and a placebo shampoo (Placebo). After six weeks of treatment, eight sections of the scalp were examined and given a score that measured the amount of scalp flaking on a 0 to 10 scale. The response variable was the sum of these eight scores. An analysis of the baseline flaking measure indicated that randomization of patients to treatments was successful in that no differences were found between the groups. At baseline there were 112 subjects in each of the three treatment groups and 28 subjects in the Placebo group. During the clinical trial 3 dropped out from the PyrII group and 6 from the Keto group. No patients dropped out of the other two groups.

Summary statistics given in the paper were used to generate random data that give the same conclusions. Here are the first five cases:

| OBS | Treatment | Flaking |
|---|---|---|
| 001 | PyrI | 17 |
| 002 | PyrI | 16 |
| 003 | PyrI | 18 |
| 004 | PyrI | 17 |
| 005 | PyrI | 18 |

## 6 HAPPINESS

Text Reference: Exercises 11.31–11.33

Is a person living in Nation X more likely to be "happier" than a person living in Nation Y? If the answer is Yes, what country-level factors are associated with this happiness? The data set HAPPINESS is a fusion of two online resources that contain information to address these and other similar questions.

The *World Database of Happiness* is an online registry of scientific research on the subjective appreciation of life. It is available at `worlddatabaseofhappiness.eur.nl` and is directed by Dr. Ruut Veenhoven, Erasmus University Rotterdam. One inventory presents the "average happiness" score for various nations between 1995 and 2005. This average is based on individual responses from numerous general population surveys to a general life satisfaction (well-being) question. Scores ranged between 0 (dissatisfied) to 10 (satisfied). These responses are coded LSI in the HAPPINESS data set.

The NationMaster Web site, `www.nationmaster.com`, contains a collection of statistics associated with various nations. For the HAPPINESS data set, the factors considered are the GINI Index, which measures the degree of inequality in the distribution of income (higher score = greater inequality); the degree of corruption in government (higher score = less corruption); the degree of democracy (higher score = more political liberties); and average life expectancy.

Here are the first five cases:

| Country | LSI | GINI | CORRUPT | DEMOCRACY | LIFE |
|---|---|---|---|---|---|
| Albania | 4.4 | 28.2 | 2.5 | 2.5 | 77.43 |
| Algeria | 5.2 | 35.3 | 2.8 | 1.5 | 73.26 |
| Argentina | 6.8 | 52.2 | 2.8 | 5.5 | 76.12 |
| Armenia | 3.7 | 41.3 | 2.9 | 3.0 | 71.84 |
| Australia | 7.7 | 35.2 | 8.8 | 6.0 | 80.50 |

## 7 LONGLEAF

Text Reference: Example 6.1; Exercises 6.66, 6.67, 6.110, 6.111, 7.25, 7.81, 7.82, 7.89, 7.105, 7.106, 7.108, 7.109, 7.122–7.124, 9.40, 16.20, and 16.63

The Wade Tract in Thomas County, Georgia, is an old-growth forest of longleaf pine trees (*Pinus palustris*) that has survived in a relatively undisturbed state since before the settlement of the area by Europeans. This data set includes observations on 584 longleaf pine trees in a 200-meter by 200-meter region in the Wade Tract and is described in Noel Cressie, *Statistics for Spatial Data,* Wiley, 1993. The variable NS is the location within the region in the north-south direction, EW is the location within the region in the east-west direction, and DBH is the diameter of the tree at breast height, measured in centimeters. Here are the first five cases:

| NS | EW | DBH |
|---|---|---|
| 200.0 | 8.8 | 32.9 |
| 199.3 | 10.0 | 53.5 |
| 193.6 | 22.4 | 68.0 |
| 167.7 | 35.6 | 17.7 |
| 183.9 | 45.4 | 36.9 |

## 8 MAJORS

Text Reference: Exercises 13.22–13.24 and 13.46–13.49

See the description of the CSDATA data set for background information on the study behind this data set. In this data set, the variables described for CSDATA are given with an additional variable "Maj" that specifies the student's major field of study at the end of three semesters. The codes 1, 2, and 3 correspond to Computer Science, Engineering and Other Sciences, and Other. All available data were used in the analyses performed, which resulted in sample sizes that were unequal in the six sex-by-major groups.

For a one-way ANOVA this causes no particular problems. However, for a two-way ANOVA several complications arise when the sample sizes are unequal. A detailed discussion of these complications is beyond the scope of this text. To avoid these difficulties and still use these interesting data, simulated data based on the results of this study are given on the data disk. ANOVA based on these simulated data gives the same qualitative conclusions as those obtained with the original data. Here are the first five cases:

| OBS | SEX | Maj | SATM | SATV | HSM | HSS | HSE | GPA |
|---|---|---|---|---|---|---|---|---|
| 001 | 1 | 1 | 640 | 530 | 8 | 6 | 8 | 2.35 |
| 002 | 1 | 1 | 670 | 600 | 9 | 10 | 7 | 2.08 |
| 003 | 1 | 1 | 600 | 400 | 8 | 8 | 7 | 3.21 |
| 004 | 1 | 1 | 570 | 480 | 7 | 7 | 6 | 2.34 |
| 005 | 1 | 1 | 510 | 530 | 6 | 8 | 8 | 1.40 |

## 9 PCB

Text Reference: Exercises 7.134, 7.135, and 11.40–11.50

Polychlorinated biphenyls (PCBs) are a collection of compounds that are no longer produced in the United States but are still found in the environment. Evidence suggests that they can cause harmful health effects when consumed. Because PCBs can accumulate in fish, efforts have been made to identify areas where fish contain excessive amounts so that recommendations concerning consumption limits can be made. There are over 200 types of PCBs. Data from the Environmental Protection Agency National Study of Residues in Lake Fish are given in the data set PCB. Various lakes in the United States were sampled and the amounts of PCBs in fish were measured. The variable PCB is the sum of the amounts of all PCBs found in the fish, while the other variables with the prefix PCB are particular types of PCBs. The units are parts per billion (ppb). Not all types of PCBs are equally harmful. A scale has been developed to convert the raw amount of each type of PCB to a toxic equivalent (TEQ). TEQPCB is the total TEQ from all PCBs found in each sample. Dioxins and furans are other contaminants that were measured in these samples. The variables TEQDIOXIN and TEQFURAN give the TEQ of these, while TEQ is the total of TEQPCB, TEQDIOXIN, and TEQFURAN. More information is available online at `epa.gov/waterscience/fishstudy/`. This data set was provided by

Joanne Lasrado, Purdue University Department of Foods and Nutrition. Here are the first five cases:

| PCB138 | PCB153 | PCB180 | PCB28 | PCB52 | PCB126 | PCB118 |
|---|---|---|---|---|---|---|
| 1.46 | 1.59 | 0.738 | 0.421 | 0.532 | 0.0000 | 0.720 |
| 0.64 | 0.92 | 0.664 | 0.025 | 0.030 | 0.0000 | 0.236 |
| 3.29 | 3.90 | 1.150 | 0.076 | 0.134 | 0.0000 | 1.540 |
| 3.94 | 5.44 | 1.330 | 0.152 | 0.466 | 0.0055 | 1.940 |
| 3.18 | 3.65 | 2.140 | 0.116 | 0.243 | 0.0059 | 1.470 |

| PCB | TEQ | TEQPCB | TEQDIOXIN | TEQFURAN |
|---|---|---|---|---|
| 19.9959 | 0.93840 | 0.18892 | 0.60948 | 0.1400 |
| 6.0996 | 0.96881 | 0.06837 | 0.90044 | 0.0000 |
| 24.9655 | 0.97992 | 0.32992 | 0.62000 | 0.0300 |
| 37.4436 | 0.99850 | 0.92350 | 0.00500 | 0.0700 |
| 30.1830 | 1.01654 | 0.92654 | 0.00000 | 0.0900 |

## 10 PLANTS1

Text Reference: Exercises 13.37–13.40

These data were collected by Maher Tadros, Purdue University Department of Forestry and Natural Resources, under the direction of Professor Andrew Gillespie. Maher is from Jordan, a Middle Eastern country where there is very little rainfall in many areas. His research concerns four species of plants that may be suitable for commercial development in his country. Products produced by these species can be used as feed for animals and in some cases for humans. The four species of plants are *Leucaena leucocephala*, *Acacia saligna*, *Prosopis juliflora*, and *Eucalyptus citriodora*. A major research question concerns how well these species can tolerate drought stress.

PLANTS1 gives data for a laboratory experiment performed by Maher at Purdue University. Seven different amounts of water were given daily to plants of each species. For each of the species-by-water combinations, there were nine plants. The response variable is the percent of the plant that consists of nitrogen. A high nitrogen content is desirable for plant products that are used for food.

The actual experiment performed to collect these data had an additional factor that is not given in the data set. The $4 \times 7$, species-by-water combinations were actually run with three plants per combination. This design was then repeated three times. The repeat factor is often called a replicate, or rep, and is a standard part of most well-designed experiments of this type. For our purposes we ignore this additional factor and analyze the design as a $4 \times 7$ two-way ANOVA with 9 observations per treatment combination. The first five cases are listed below. The four species, *Leucaena leucocephala*, *Acacia saligna*, *Prosopis juliflora*, and *Eucalyptus citriodora,* are coded 1 to 4. The water levels, 50, 150, 250, 350, 450, 550, and 650 millimeters, are coded 1 to 7.

| OBS | Species | Water | pctnit |
|---|---|---|---|
| 001 | 1 | 1 | 3.644 |
| 002 | 1 | 1 | 3.500 |
| 003 | 1 | 1 | 3.509 |
| 004 | 1 | 1 | 3.137 |
| 005 | 1 | 1 | 3.100 |

## 11 PLANTS2

Text Reference:
Exercises 13.41–13.44

PLANTS2 gives data for a second experiment conducted by Maher Tadros in a lab at Purdue University. As in PLANTS1, there are the same four species of plants and the same seven levels of water. Here, however, there are four plants per species-by-water combination. The two response variables in the data set are fresh biomass and dry biomass. High values for these response variables indicate that the plants of the given species are resistant to drought at the given water level. Here are the first five cases:

| OBS | Species | Water | fbiomass | dbiomass |
|---|---|---|---|---|
| 001 | 1 | 1 | 105.13 | 37.65 |
| 002 | 1 | 1 | 138.95 | 48.85 |
| 003 | 1 | 1 | 90.05 | 38.85 |
| 004 | 1 | 1 | 102.25 | 36.91 |
| 005 | 1 | 1 | 207.90 | 74.35 |

## 12 PNG

Text Reference:
Exercises 7.26–7.28, 7.136, 7.137, 10.25, and 16.7

A randomized double-blind placebo-controlled trial to assess the benefits of giving high-dose vitamin A supplements to young children was performed in the North Wosera District of East Sepik Province in Papua New Guinea. The results of the trial are reported in F. Rosales et al., "Relation of serum retinol to acute phase proteins and malarial morbidity in Papua New Guinea children," *American Journal of Clinical Nutrition,* 71 (2000), pp. 1580–1588. The data here were collected at the start of the study and were provided by Francisco Rosales, Department of Nutritional Sciences, Pennsylvania State University. AGEY is age in years, RETINOL is serum retinol, CRP is C-reactive protein, AGP is $\alpha$1-acid glycoprotein, and ACT is $\alpha$1-antichymotrypsin. Serum retinol is a measure of vitamin A status, and CRP, AGP, and ACT are acute-phase proteins that have high values when there is an infection. Here are the first five cases:

| ID | AGEY | RETINOL | CRP | AGP | ACT |
|---|---|---|---|---|---|
| 1 | 4.33 | 0.31 | 3.49 | 0.61 | 0.28 |
| 2 | 3.33 | 0.26 | 8.64 | 0.82 | 0.34 |
| 3 | 2.50 | 0.24 | 33.59 | 0.73 | 0.36 |
| 4 | 3.33 | 0.67 | 3.47 | 1.05 | 0.34 |
| 5 | 4.83 | 0.59 | 0.00 | 0.55 | 0.08 |

## 13 RANKING

Text Reference:
Exercises 11.27–11.29

Since 2004, the *Times* Higher Education Supplement has provided an annual ranking of the world universities. A total score for each university is calculated based on the following scores: Peer Review (40%), Faculty-to-Student Ratio (20%), Citations-to-Faculty Ratio (20%), Recruiter Review (10%), Percent International Faculty (5%), Percent International Student (5%). The percents represent the contributions of each score to the total.

For our purposes here, we will assume these weights are unknown and will focus on the development of a model for the total score based on the first three explanatory variables. The report includes a table for the top 200 universities. The RANKING data set is a random sample of 75 of these universities. This is not a random sample of all universities but for our purposes here we will consider it to be.

## 14 READING

Text Reference: Exercises 7.139, 12.61, and 12.62

Jim Baumann and Leah Jones, Purdue University College of Education, conducted a study to compare three methods of teaching reading comprehension. The 66 students who participated in the study were randomly assigned to the methods (22 to each). The standard practice of comparing new methods with a traditional one was used in this study. The traditional method is called Basal and the two innovative methods are called DRTA and Strat.

In the data set the variable Subject is used to identify the individual students. The values are 1 to 66. The method of instruction is indicated by the variable Group, with values B, D, and S, corresponding to Basal, DRTA, and Strat. Two pretests and three posttests were given to all students. These are the variables Pre1, Pre2, Post1, Post2, and Post3. Data for the first five subjects are given below.

| Subject | Group | Pre1 | Pre2 | Post1 | Post2 | Post3 |
|---|---|---|---|---|---|---|
| 01 | B | 4 | 3 | 5 | 4 | 41 |
| 02 | B | 6 | 5 | 9 | 5 | 41 |
| 03 | B | 9 | 4 | 5 | 3 | 43 |
| 04 | B | 12 | 6 | 8 | 5 | 46 |
| 05 | B | 16 | 5 | 10 | 9 | 46 |

## 15 RUNNERS

Text Reference: Example 13.11; Exercises 1.81 and 1.143

A study of cardiovascular risk factors compared runners who averaged at least 15 miles per week with a control group described as "generally sedentary." Both men and women were included in the study. The data set was constructed based on information provided in P. D. Wood et al., "Plasma lipoprotein distributions in male and female runners," in P. Milvey (ed.), *The Marathon: Physiological, Medical, Epidemiological, and Psychological Studies,* New York Academy of Sciences, 1977.

The study design is a 2 × 2 ANOVA with the factors group and gender. There were 200 subjects in each of the four combinations. The variables are Id, a numeric subject identifier; Group, with values "Control" and "Runners"; Gender, with values Female and Male; and HeartRate, heart rate (beats per minute) after the subject ran for 6 minutes on a treadmill. Here are the data for the first five subjects:

| Id | Group | Gender | Beats |
|---|---|---|---|
| 1 | Control | Female | 159 |
| 2 | Control | Female | 183 |
| 3 | Control | Female | 140 |
| 4 | Control | Female | 140 |
| 5 | Control | Female | 125 |

## 16 WORKERS

Text Reference: Exercises 1.82–1.84, 1.172

Each March, the Bureau of Labor Statistics carries out an Annual Demographic Supplement to its monthly Current Population Survey. The data set WORKERS contains data about 71,076 people from one of these surveys. We included all people between the ages of 25 and 64 who have worked but whose main work experience is not in agriculture. Moreover, we combined the 16 levels of education in the BLS survey to form 6 levels.

There are five variables in the data set. Age is age in years. Education is the highest level of education a person has reached, with the following values: 1 = did not reach high school; 2 = some high school but no high school diploma; 3 = high school diploma; 4 = some college but no bachelor's degree (this includes people with an associate degree); 5 = bachelor's degree; 6 = postgraduate degree (master's, professional, or doctorate). Sex is coded as 1 = male and 2 = female. Total income is income from all sources. Note that income can be less than zero in some cases. Job class is a categorization of the person's main work experience, with 5 = private sector (outside households); 6 = government; 7 = self-employed. Here are the first five cases:

| Age | Education | Sex | Total income | Job class |
|---|---|---|---|---|
| 25 | 2 | 2 | 7,234 | 5 |
| 25 | 5 | 1 | 37,413 | 5 |
| 25 | 4 | 2 | 29,500 | 5 |
| 25 | 3 | 2 | 13,500 | 5 |
| 25 | 4 | 1 | 17,660 | 6 |

The first individual is a 25-year-old female who did not graduate from high school, works in the private sector, and had $7234 of income.

An Excel copy of the WORKERS data set is posted on the companion Web site, www.whfreeman.com/ips6e.

# TABLES

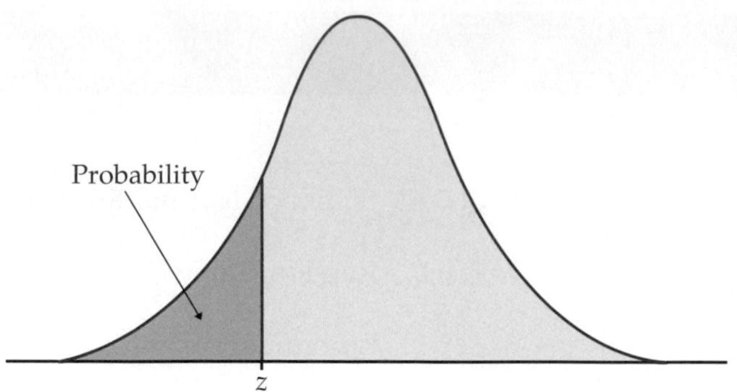


Table entry for $z$ is the area under the standard normal curve to the left of $z$.

**TABLE A**

**Standard normal probabilities**

| $z$ | .00 | .01 | .02 | .03 | .04 | .05 | .06 | .07 | .08 | .09 |
|---|---|---|---|---|---|---|---|---|---|---|
| −3.4 | .0003 | .0003 | .0003 | .0003 | .0003 | .0003 | .0003 | .0003 | .0003 | .0002 |
| −3.3 | .0005 | .0005 | .0005 | .0004 | .0004 | .0004 | .0004 | .0004 | .0004 | .0003 |
| −3.2 | .0007 | .0007 | .0006 | .0006 | .0006 | .0006 | .0006 | .0005 | .0005 | .0005 |
| −3.1 | .0010 | .0009 | .0009 | .0009 | .0008 | .0008 | .0008 | .0008 | .0007 | .0007 |
| −3.0 | .0013 | .0013 | .0013 | .0012 | .0012 | .0011 | .0011 | .0011 | .0010 | .0010 |
| −2.9 | .0019 | .0018 | .0018 | .0017 | .0016 | .0016 | .0015 | .0015 | .0014 | .0014 |
| −2.8 | .0026 | .0025 | .0024 | .0023 | .0023 | .0022 | .0021 | .0021 | .0020 | .0019 |
| −2.7 | .0035 | .0034 | .0033 | .0032 | .0031 | .0030 | .0029 | .0028 | .0027 | .0026 |
| −2.6 | .0047 | .0045 | .0044 | .0043 | .0041 | .0040 | .0039 | .0038 | .0037 | .0036 |
| −2.5 | .0062 | .0060 | .0059 | .0057 | .0055 | .0054 | .0052 | .0051 | .0049 | .0048 |
| −2.4 | .0082 | .0080 | .0078 | .0075 | .0073 | .0071 | .0069 | .0068 | .0066 | .0064 |
| −2.3 | .0107 | .0104 | .0102 | .0099 | .0096 | .0094 | .0091 | .0089 | .0087 | .0084 |
| −2.2 | .0139 | .0136 | .0132 | .0129 | .0125 | .0122 | .0119 | .0116 | .0113 | .0110 |
| −2.1 | .0179 | .0174 | .0170 | .0166 | .0162 | .0158 | .0154 | .0150 | .0146 | .0143 |
| −2.0 | .0228 | .0222 | .0217 | .0212 | .0207 | .0202 | .0197 | .0192 | .0188 | .0183 |
| −1.9 | .0287 | .0281 | .0274 | .0268 | .0262 | .0256 | .0250 | .0244 | .0239 | .0233 |
| −1.8 | .0359 | .0351 | .0344 | .0336 | .0329 | .0322 | .0314 | .0307 | .0301 | .0294 |
| −1.7 | .0446 | .0436 | .0427 | .0418 | .0409 | .0401 | .0392 | .0384 | .0375 | .0367 |
| −1.6 | .0548 | .0537 | .0526 | .0516 | .0505 | .0495 | .0485 | .0475 | .0465 | .0455 |
| −1.5 | .0668 | .0655 | .0643 | .0630 | .0618 | .0606 | .0594 | .0582 | .0571 | .0559 |
| −1.4 | .0808 | .0793 | .0778 | .0764 | .0749 | .0735 | .0721 | .0708 | .0694 | .0681 |
| −1.3 | .0968 | .0951 | .0934 | .0918 | .0901 | .0885 | .0869 | .0853 | .0838 | .0823 |
| −1.2 | .1151 | .1131 | .1112 | .1093 | .1075 | .1056 | .1038 | .1020 | .1003 | .0985 |
| −1.1 | .1357 | .1335 | .1314 | .1292 | .1271 | .1251 | .1230 | .1210 | .1190 | .1170 |
| −1.0 | .1587 | .1562 | .1539 | .1515 | .1492 | .1469 | .1446 | .1423 | .1401 | .1379 |
| −0.9 | .1841 | .1814 | .1788 | .1762 | .1736 | .1711 | .1685 | .1660 | .1635 | .1611 |
| −0.8 | .2119 | .2090 | .2061 | .2033 | .2005 | .1977 | .1949 | .1922 | .1894 | .1867 |
| −0.7 | .2420 | .2389 | .2358 | .2327 | .2296 | .2266 | .2236 | .2206 | .2177 | .2148 |
| −0.6 | .2743 | .2709 | .2676 | .2643 | .2611 | .2578 | .2546 | .2514 | .2483 | .2451 |
| −0.5 | .3085 | .3050 | .3015 | .2981 | .2946 | .2912 | .2877 | .2843 | .2810 | .2776 |
| −0.4 | .3446 | .3409 | .3372 | .3336 | .3300 | .3264 | .3228 | .3192 | .3156 | .3121 |
| −0.3 | .3821 | .3783 | .3745 | .3707 | .3669 | .3632 | .3594 | .3557 | .3520 | .3483 |
| −0.2 | .4207 | .4168 | .4129 | .4090 | .4052 | .4013 | .3974 | .3936 | .3897 | .3859 |
| −0.1 | .4602 | .4562 | .4522 | .4483 | .4443 | .4404 | .4364 | .4325 | .4286 | .4247 |
| 0.0 | .5000 | .4960 | .4920 | .4880 | .4840 | .4801 | .4761 | .4721 | .4681 | .4641 |

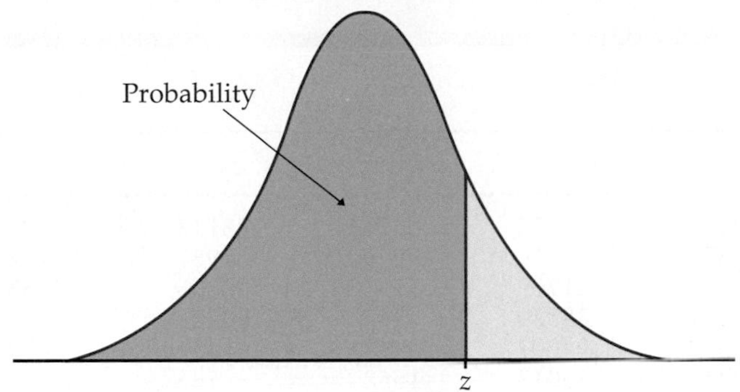


Table entry for $z$ is the area under the standard normal curve to the left of $z$.

**TABLE A**

## Standard normal probabilities (continued)

| $z$ | .00 | .01 | .02 | .03 | .04 | .05 | .06 | .07 | .08 | .09 |
|---|---|---|---|---|---|---|---|---|---|---|
| 0.0 | .5000 | .5040 | .5080 | .5120 | .5160 | .5199 | .5239 | .5279 | .5319 | .5359 |
| 0.1 | .5398 | .5438 | .5478 | .5517 | .5557 | .5596 | .5636 | .5675 | .5714 | .5753 |
| 0.2 | .5793 | .5832 | .5871 | .5910 | .5948 | .5987 | .6026 | .6064 | .6103 | .6141 |
| 0.3 | .6179 | .6217 | .6255 | .6293 | .6331 | .6368 | .6406 | .6443 | .6480 | .6517 |
| 0.4 | .6554 | .6591 | .6628 | .6664 | .6700 | .6736 | .6772 | .6808 | .6844 | .6879 |
| 0.5 | .6915 | .6950 | .6985 | .7019 | .7054 | .7088 | .7123 | .7157 | .7190 | .7224 |
| 0.6 | .7257 | .7291 | .7324 | .7357 | .7389 | .7422 | .7454 | .7486 | .7517 | .7549 |
| 0.7 | .7580 | .7611 | .7642 | .7673 | .7704 | .7734 | .7764 | .7794 | .7823 | .7852 |
| 0.8 | .7881 | .7910 | .7939 | .7967 | .7995 | .8023 | .8051 | .8078 | .8106 | .8133 |
| 0.9 | .8159 | .8186 | .8212 | .8238 | .8264 | .8289 | .8315 | .8340 | .8365 | .8389 |
| 1.0 | .8413 | .8438 | .8461 | .8485 | .8508 | .8531 | .8554 | .8577 | .8599 | .8621 |
| 1.1 | .8643 | .8665 | .8686 | .8708 | .8729 | .8749 | .8770 | .8790 | .8810 | .8830 |
| 1.2 | .8849 | .8869 | .8888 | .8907 | .8925 | .8944 | .8962 | .8980 | .8997 | .9015 |
| 1.3 | .9032 | .9049 | .9066 | .9082 | .9099 | .9115 | .9131 | .9147 | .9162 | .9177 |
| 1.4 | .9192 | .9207 | .9222 | .9236 | .9251 | .9265 | .9279 | .9292 | .9306 | .9319 |
| 1.5 | .9332 | .9345 | .9357 | .9370 | .9382 | .9394 | .9406 | .9418 | .9429 | .9441 |
| 1.6 | .9452 | .9463 | .9474 | .9484 | .9495 | .9505 | .9515 | .9525 | .9535 | .9545 |
| 1.7 | .9554 | .9564 | .9573 | .9582 | .9591 | .9599 | .9608 | .9616 | .9625 | .9633 |
| 1.8 | .9641 | .9649 | .9656 | .9664 | .9671 | .9678 | .9686 | .9693 | .9699 | .9706 |
| 1.9 | .9713 | .9719 | .9726 | .9732 | .9738 | .9744 | .9750 | .9756 | .9761 | .9767 |
| 2.0 | .9772 | .9778 | .9783 | .9788 | .9793 | .9798 | .9803 | .9808 | .9812 | .9817 |
| 2.1 | .9821 | .9826 | .9830 | .9834 | .9838 | .9842 | .9846 | .9850 | .9854 | .9857 |
| 2.2 | .9861 | .9864 | .9868 | .9871 | .9875 | .9878 | .9881 | .9884 | .9887 | .9890 |
| 2.3 | .9893 | .9896 | .9898 | .9901 | .9904 | .9906 | .9909 | .9911 | .9913 | .9916 |
| 2.4 | .9918 | .9920 | .9922 | .9925 | .9927 | .9929 | .9931 | .9932 | .9934 | .9936 |
| 2.5 | .9938 | .9940 | .9941 | .9943 | .9945 | .9946 | .9948 | .9949 | .9951 | .9952 |
| 2.6 | .9953 | .9955 | .9956 | .9957 | .9959 | .9960 | .9961 | .9962 | .9963 | .9964 |
| 2.7 | .9965 | .9966 | .9967 | .9968 | .9969 | .9970 | .9971 | .9972 | .9973 | .9974 |
| 2.8 | .9974 | .9975 | .9976 | .9977 | .9977 | .9978 | .9979 | .9979 | .9980 | .9981 |
| 2.9 | .9981 | .9982 | .9982 | .9983 | .9984 | .9984 | .9985 | .9985 | .9986 | .9986 |
| 3.0 | .9987 | .9987 | .9987 | .9988 | .9988 | .9989 | .9989 | .9989 | .9990 | .9990 |
| 3.1 | .9990 | .9991 | .9991 | .9991 | .9992 | .9992 | .9992 | .9992 | .9993 | .9993 |
| 3.2 | .9993 | .9993 | .9994 | .9994 | .9994 | .9994 | .9994 | .9995 | .9995 | .9995 |
| 3.3 | .9995 | .9995 | .9995 | .9996 | .9996 | .9996 | .9996 | .9996 | .9996 | .9997 |
| 3.4 | .9997 | .9997 | .9997 | .9997 | .9997 | .9997 | .9997 | .9997 | .9997 | .9998 |

**TABLE B**

## Random digits

| Line | | | | | | | | |
|---|---|---|---|---|---|---|---|---|
| 101 | 19223 | 95034 | 05756 | 28713 | 96409 | 12531 | 42544 | 82853 |
| 102 | 73676 | 47150 | 99400 | 01927 | 27754 | 42648 | 82425 | 36290 |
| 103 | 45467 | 71709 | 77558 | 00095 | 32863 | 29485 | 82226 | 90056 |
| 104 | 52711 | 38889 | 93074 | 60227 | 40011 | 85848 | 48767 | 52573 |
| 105 | 95592 | 94007 | 69971 | 91481 | 60779 | 53791 | 17297 | 59335 |
| 106 | 68417 | 35013 | 15529 | 72765 | 85089 | 57067 | 50211 | 47487 |
| 107 | 82739 | 57890 | 20807 | 47511 | 81676 | 55300 | 94383 | 14893 |
| 108 | 60940 | 72024 | 17868 | 24943 | 61790 | 90656 | 87964 | 18883 |
| 109 | 36009 | 19365 | 15412 | 39638 | 85453 | 46816 | 83485 | 41979 |
| 110 | 38448 | 48789 | 18338 | 24697 | 39364 | 42006 | 76688 | 08708 |
| 111 | 81486 | 69487 | 60513 | 09297 | 00412 | 71238 | 27649 | 39950 |
| 112 | 59636 | 88804 | 04634 | 71197 | 19352 | 73089 | 84898 | 45785 |
| 113 | 62568 | 70206 | 40325 | 03699 | 71080 | 22553 | 11486 | 11776 |
| 114 | 45149 | 32992 | 75730 | 66280 | 03819 | 56202 | 02938 | 70915 |
| 115 | 61041 | 77684 | 94322 | 24709 | 73698 | 14526 | 31893 | 32592 |
| 116 | 14459 | 26056 | 31424 | 80371 | 65103 | 62253 | 50490 | 61181 |
| 117 | 38167 | 98532 | 62183 | 70632 | 23417 | 26185 | 41448 | 75532 |
| 118 | 73190 | 32533 | 04470 | 29669 | 84407 | 90785 | 65956 | 86382 |
| 119 | 95857 | 07118 | 87664 | 92099 | 58806 | 66979 | 98624 | 84826 |
| 120 | 35476 | 55972 | 39421 | 65850 | 04266 | 35435 | 43742 | 11937 |
| 121 | 71487 | 09984 | 29077 | 14863 | 61683 | 47052 | 62224 | 51025 |
| 122 | 13873 | 81598 | 95052 | 90908 | 73592 | 75186 | 87136 | 95761 |
| 123 | 54580 | 81507 | 27102 | 56027 | 55892 | 33063 | 41842 | 81868 |
| 124 | 71035 | 09001 | 43367 | 49497 | 72719 | 96758 | 27611 | 91596 |
| 125 | 96746 | 12149 | 37823 | 71868 | 18442 | 35119 | 62103 | 39244 |
| 126 | 96927 | 19931 | 36089 | 74192 | 77567 | 88741 | 48409 | 41903 |
| 127 | 43909 | 99477 | 25330 | 64359 | 40085 | 16925 | 85117 | 36071 |
| 128 | 15689 | 14227 | 06565 | 14374 | 13352 | 49367 | 81982 | 87209 |
| 129 | 36759 | 58984 | 68288 | 22913 | 18638 | 54303 | 00795 | 08727 |
| 130 | 69051 | 64817 | 87174 | 09517 | 84534 | 06489 | 87201 | 97245 |
| 131 | 05007 | 16632 | 81194 | 14873 | 04197 | 85576 | 45195 | 96565 |
| 132 | 68732 | 55259 | 84292 | 08796 | 43165 | 93739 | 31685 | 97150 |
| 133 | 45740 | 41807 | 65561 | 33302 | 07051 | 93623 | 18132 | 09547 |
| 134 | 27816 | 78416 | 18329 | 21337 | 35213 | 37741 | 04312 | 68508 |
| 135 | 66925 | 55658 | 39100 | 78458 | 11206 | 19876 | 87151 | 31260 |
| 136 | 08421 | 44753 | 77377 | 28744 | 75592 | 08563 | 79140 | 92454 |
| 137 | 53645 | 66812 | 61421 | 47836 | 12609 | 15373 | 98481 | 14592 |
| 138 | 66831 | 68908 | 40772 | 21558 | 47781 | 33586 | 79177 | 06928 |
| 139 | 55588 | 99404 | 70708 | 41098 | 43563 | 56934 | 48394 | 51719 |
| 140 | 12975 | 13258 | 13048 | 45144 | 72321 | 81940 | 00360 | 02428 |
| 141 | 96767 | 35964 | 23822 | 96012 | 94591 | 65194 | 50842 | 53372 |
| 142 | 72829 | 50232 | 97892 | 63408 | 77919 | 44575 | 24870 | 04178 |
| 143 | 88565 | 42628 | 17797 | 49376 | 61762 | 16953 | 88604 | 12724 |
| 144 | 62964 | 88145 | 83083 | 69453 | 46109 | 59505 | 69680 | 00900 |
| 145 | 19687 | 12633 | 57857 | 95806 | 09931 | 02150 | 43163 | 58636 |
| 146 | 37609 | 59057 | 66967 | 83401 | 60705 | 02384 | 90597 | 93600 |
| 147 | 54973 | 86278 | 88737 | 74351 | 47500 | 84552 | 19909 | 67181 |
| 148 | 00694 | 05977 | 19664 | 65441 | 20903 | 62371 | 22725 | 53340 |
| 149 | 71546 | 05233 | 53946 | 68743 | 72460 | 27601 | 45403 | 88692 |
| 150 | 07511 | 88915 | 41267 | 16853 | 84569 | 79367 | 32337 | 03316 |

**TABLE B**

## Random digits (continued)

| Line | | | | | | | | |
|---|---|---|---|---|---|---|---|---|
| 151 | 03802 | 29341 | 29264 | 80198 | 12371 | 13121 | 54969 | 43912 |
| 152 | 77320 | 35030 | 77519 | 41109 | 98296 | 18984 | 60869 | 12349 |
| 153 | 07886 | 56866 | 39648 | 69290 | 03600 | 05376 | 58958 | 22720 |
| 154 | 87065 | 74133 | 21117 | 70595 | 22791 | 67306 | 28420 | 52067 |
| 155 | 42090 | 09628 | 54035 | 93879 | 98441 | 04606 | 27381 | 82637 |
| 156 | 55494 | 67690 | 88131 | 81800 | 11188 | 28552 | 25752 | 21953 |
| 157 | 16698 | 30406 | 96587 | 65985 | 07165 | 50148 | 16201 | 86792 |
| 158 | 16297 | 07626 | 68683 | 45335 | 34377 | 72941 | 41764 | 77038 |
| 159 | 22897 | 17467 | 17638 | 70043 | 36243 | 13008 | 83993 | 22869 |
| 160 | 98163 | 45944 | 34210 | 64158 | 76971 | 27689 | 82926 | 75957 |
| 161 | 43400 | 25831 | 06283 | 22138 | 16043 | 15706 | 73345 | 26238 |
| 162 | 97341 | 46254 | 88153 | 62336 | 21112 | 35574 | 99271 | 45297 |
| 163 | 64578 | 67197 | 28310 | 90341 | 37531 | 63890 | 52630 | 76315 |
| 164 | 11022 | 79124 | 49525 | 63078 | 17229 | 32165 | 01343 | 21394 |
| 165 | 81232 | 43939 | 23840 | 05995 | 84589 | 06788 | 76358 | 26622 |
| 166 | 36843 | 84798 | 51167 | 44728 | 20554 | 55538 | 27647 | 32708 |
| 167 | 84329 | 80081 | 69516 | 78934 | 14293 | 92478 | 16479 | 26974 |
| 168 | 27788 | 85789 | 41592 | 74472 | 96773 | 27090 | 24954 | 41474 |
| 169 | 99224 | 00850 | 43737 | 75202 | 44753 | 63236 | 14260 | 73686 |
| 170 | 38075 | 73239 | 52555 | 46342 | 13365 | 02182 | 30443 | 53229 |
| 171 | 87368 | 49451 | 55771 | 48343 | 51236 | 18522 | 73670 | 23212 |
| 172 | 40512 | 00681 | 44282 | 47178 | 08139 | 78693 | 34715 | 75606 |
| 173 | 81636 | 57578 | 54286 | 27216 | 58758 | 80358 | 84115 | 84568 |
| 174 | 26411 | 94292 | 06340 | 97762 | 37033 | 85968 | 94165 | 46514 |
| 175 | 80011 | 09937 | 57195 | 33906 | 94831 | 10056 | 42211 | 65491 |
| 176 | 92813 | 87503 | 63494 | 71379 | 76550 | 45984 | 05481 | 50830 |
| 177 | 70348 | 72871 | 63419 | 57363 | 29685 | 43090 | 18763 | 31714 |
| 178 | 24005 | 52114 | 26224 | 39078 | 80798 | 15220 | 43186 | 00976 |
| 179 | 85063 | 55810 | 10470 | 08029 | 30025 | 29734 | 61181 | 72090 |
| 180 | 11532 | 73186 | 92541 | 06915 | 72954 | 10167 | 12142 | 26492 |
| 181 | 59618 | 03914 | 05208 | 84088 | 20426 | 39004 | 84582 | 87317 |
| 182 | 92965 | 50837 | 39921 | 84661 | 82514 | 81899 | 24565 | 60874 |
| 183 | 85116 | 27684 | 14597 | 85747 | 01596 | 25889 | 41998 | 15635 |
| 184 | 15106 | 10411 | 90221 | 49377 | 44369 | 28185 | 80959 | 76355 |
| 185 | 03638 | 31589 | 07871 | 25792 | 85823 | 55400 | 56026 | 12193 |
| 186 | 97971 | 48932 | 45792 | 63993 | 95635 | 28753 | 46069 | 84635 |
| 187 | 49345 | 18305 | 76213 | 82390 | 77412 | 97401 | 50650 | 71755 |
| 188 | 87370 | 88099 | 89695 | 87633 | 76987 | 85503 | 26257 | 51736 |
| 189 | 88296 | 95670 | 74932 | 65317 | 93848 | 43988 | 47597 | 83044 |
| 190 | 79485 | 92200 | 99401 | 54473 | 34336 | 82786 | 05457 | 60343 |
| 191 | 40830 | 24979 | 23333 | 37619 | 56227 | 95941 | 59494 | 86539 |
| 192 | 32006 | 76302 | 81221 | 00693 | 95197 | 75044 | 46596 | 11628 |
| 193 | 37569 | 85187 | 44692 | 50706 | 53161 | 69027 | 88389 | 60313 |
| 194 | 56680 | 79003 | 23361 | 67094 | 15019 | 63261 | 24543 | 52884 |
| 195 | 05172 | 08100 | 22316 | 54495 | 60005 | 29532 | 18433 | 18057 |
| 196 | 74782 | 27005 | 03894 | 98038 | 20627 | 40307 | 47317 | 92759 |
| 197 | 85288 | 93264 | 61409 | 03404 | 09649 | 55937 | 60843 | 66167 |
| 198 | 68309 | 12060 | 14762 | 58002 | 03716 | 81968 | 57934 | 32624 |
| 199 | 26461 | 88346 | 52430 | 60906 | 74216 | 96263 | 69296 | 90107 |
| 200 | 42672 | 67680 | 42376 | 95023 | 82744 | 03971 | 96560 | 55148 |

## TABLE C

### Binomial probabilities

Entry is $P(X = k) = \binom{n}{k} p^k (1-p)^{n-k}$

| | | *p* | | | | | | | | |
|---|---|---|---|---|---|---|---|---|---|---|
| *n* | *k* | .01 | .02 | .03 | .04 | .05 | .06 | .07 | .08 | .09 |
| 2 | 0 | .9801 | .9604 | .9409 | .9216 | .9025 | .8836 | .8649 | .8464 | .8281 |
| | 1 | .0198 | .0392 | .0582 | .0768 | .0950 | .1128 | .1302 | .1472 | .1638 |
| | 2 | .0001 | .0004 | .0009 | .0016 | .0025 | .0036 | .0049 | .0064 | .0081 |
| 3 | 0 | .9703 | .9412 | .9127 | .8847 | .8574 | .8306 | .8044 | .7787 | .7536 |
| | 1 | .0294 | .0576 | .0847 | .1106 | .1354 | .1590 | .1816 | .2031 | .2236 |
| | 2 | .0003 | .0012 | .0026 | .0046 | .0071 | .0102 | .0137 | .0177 | .0221 |
| | 3 | | | | .0001 | .0001 | .0002 | .0003 | .0005 | .0007 |
| 4 | 0 | .9606 | .9224 | .8853 | .8493 | .8145 | .7807 | .7481 | .7164 | .6857 |
| | 1 | .0388 | .0753 | .1095 | .1416 | .1715 | .1993 | .2252 | .2492 | .2713 |
| | 2 | .0006 | .0023 | .0051 | .0088 | .0135 | .0191 | .0254 | .0325 | .0402 |
| | 3 | | | .0001 | .0002 | .0005 | .0008 | .0013 | .0019 | .0027 |
| | 4 | | | | | | | | | .0001 |
| 5 | 0 | .9510 | .9039 | .8587 | .8154 | .7738 | .7339 | .6957 | .6591 | .6240 |
| | 1 | .0480 | .0922 | .1328 | .1699 | .2036 | .2342 | .2618 | .2866 | .3086 |
| | 2 | .0010 | .0038 | .0082 | .0142 | .0214 | .0299 | .0394 | .0498 | .0610 |
| | 3 | | .0001 | .0003 | .0006 | .0011 | .0019 | .0030 | .0043 | .0060 |
| | 4 | | | | | | .0001 | .0001 | .0002 | .0003 |
| | 5 | | | | | | | | | |
| 6 | 0 | .9415 | .8858 | .8330 | .7828 | .7351 | .6899 | .6470 | .6064 | .5679 |
| | 1 | .0571 | .1085 | .1546 | .1957 | .2321 | .2642 | .2922 | .3164 | .3370 |
| | 2 | .0014 | .0055 | .0120 | .0204 | .0305 | .0422 | .0550 | .0688 | .0833 |
| | 3 | | .0002 | .0005 | .0011 | .0021 | .0036 | .0055 | .0080 | .0110 |
| | 4 | | | | | .0001 | .0002 | .0003 | .0005 | .0008 |
| | 5 | | | | | | | | | |
| | 6 | | | | | | | | | |
| 7 | 0 | .9321 | .8681 | .8080 | .7514 | .6983 | .6485 | .6017 | .5578 | .5168 |
| | 1 | .0659 | .1240 | .1749 | .2192 | .2573 | .2897 | .3170 | .3396 | .3578 |
| | 2 | .0020 | .0076 | .0162 | .0274 | .0406 | .0555 | .0716 | .0886 | .1061 |
| | 3 | | .0003 | .0008 | .0019 | .0036 | .0059 | .0090 | .0128 | .0175 |
| | 4 | | | | .0001 | .0002 | .0004 | .0007 | .0011 | .0017 |
| | 5 | | | | | | | | .0001 | .0001 |
| | 6 | | | | | | | | | |
| | 7 | | | | | | | | | |
| 8 | 0 | .9227 | .8508 | .7837 | .7214 | .6634 | .6096 | .5596 | .5132 | .4703 |
| | 1 | .0746 | .1389 | .1939 | .2405 | .2793 | .3113 | .3370 | .3570 | .3721 |
| | 2 | .0026 | .0099 | .0210 | .0351 | .0515 | .0695 | .0888 | .1087 | .1288 |
| | 3 | .0001 | .0004 | .0013 | .0029 | .0054 | .0089 | .0134 | .0189 | .0255 |
| | 4 | | | .0001 | .0002 | .0004 | .0007 | .0013 | .0021 | .0031 |
| | 5 | | | | | | | .0001 | .0001 | .0002 |
| | 6 | | | | | | | | | |
| | 7 | | | | | | | | | |
| | 8 | | | | | | | | | |

## TABLE C

**Binomial probabilities** (continued)

Entry is $P(X = k) = \binom{n}{k} p^k (1 - p)^{n-k}$

| | | *p* | | | | | | | | |
|---|---|---|---|---|---|---|---|---|---|---|
| *n* | *k* | .10 | .15 | .20 | .25 | .30 | .35 | .40 | .45 | .50 |
| 2 | 0 | .8100 | .7225 | .6400 | .5625 | .4900 | .4225 | .3600 | .3025 | .2500 |
| | 1 | .1800 | .2550 | .3200 | .3750 | .4200 | .4550 | .4800 | .4950 | .5000 |
| | 2 | .0100 | .0225 | .0400 | .0625 | .0900 | .1225 | .1600 | .2025 | .2500 |
| 3 | 0 | .7290 | .6141 | .5120 | .4219 | .3430 | .2746 | .2160 | .1664 | .1250 |
| | 1 | .2430 | .3251 | .3840 | .4219 | .4410 | .4436 | .4320 | .4084 | .3750 |
| | 2 | .0270 | .0574 | .0960 | .1406 | .1890 | .2389 | .2880 | .3341 | .3750 |
| | 3 | .0010 | .0034 | .0080 | .0156 | .0270 | .0429 | .0640 | .0911 | .1250 |
| 4 | 0 | .6561 | .5220 | .4096 | .3164 | .2401 | .1785 | .1296 | .0915 | .0625 |
| | 1 | .2916 | .3685 | .4096 | .4219 | .4116 | .3845 | .3456 | .2995 | .2500 |
| | 2 | .0486 | .0975 | .1536 | .2109 | .2646 | .3105 | .3456 | .3675 | .3750 |
| | 3 | .0036 | .0115 | .0256 | .0469 | .0756 | .1115 | .1536 | .2005 | .2500 |
| | 4 | .0001 | .0005 | .0016 | .0039 | .0081 | .0150 | .0256 | .0410 | .0625 |
| 5 | 0 | .5905 | .4437 | .3277 | .2373 | .1681 | .1160 | .0778 | .0503 | .0313 |
| | 1 | .3280 | .3915 | .4096 | .3955 | .3602 | .3124 | .2592 | .2059 | .1563 |
| | 2 | .0729 | .1382 | .2048 | .2637 | .3087 | .3364 | .3456 | .3369 | .3125 |
| | 3 | .0081 | .0244 | .0512 | .0879 | .1323 | .1811 | .2304 | .2757 | .3125 |
| | 4 | .0004 | .0022 | .0064 | .0146 | .0284 | .0488 | .0768 | .1128 | .1562 |
| | 5 | | .0001 | .0003 | .0010 | .0024 | .0053 | .0102 | .0185 | .0312 |
| 6 | 0 | .5314 | .3771 | .2621 | .1780 | .1176 | .0754 | .0467 | .0277 | .0156 |
| | 1 | .3543 | .3993 | .3932 | .3560 | .3025 | .2437 | .1866 | .1359 | .0938 |
| | 2 | .0984 | .1762 | .2458 | .2966 | .3241 | .3280 | .3110 | .2780 | .2344 |
| | 3 | .0146 | .0415 | .0819 | .1318 | .1852 | .2355 | .2765 | .3032 | .3125 |
| | 4 | .0012 | .0055 | .0154 | .0330 | .0595 | .0951 | .1382 | .1861 | .2344 |
| | 5 | .0001 | .0004 | .0015 | .0044 | .0102 | .0205 | .0369 | .0609 | .0937 |
| | 6 | | | .0001 | .0002 | .0007 | .0018 | .0041 | .0083 | .0156 |
| 7 | 0 | .4783 | .3206 | .2097 | .1335 | .0824 | .0490 | .0280 | .0152 | .0078 |
| | 1 | .3720 | .3960 | .3670 | .3115 | .2471 | .1848 | .1306 | .0872 | .0547 |
| | 2 | .1240 | .2097 | .2753 | .3115 | .3177 | .2985 | .2613 | .2140 | .1641 |
| | 3 | .0230 | .0617 | .1147 | .1730 | .2269 | .2679 | .2903 | .2918 | .2734 |
| | 4 | .0026 | .0109 | .0287 | .0577 | .0972 | .1442 | .1935 | .2388 | .2734 |
| | 5 | .0002 | .0012 | .0043 | .0115 | .0250 | .0466 | .0774 | .1172 | .1641 |
| | 6 | | .0001 | .0004 | .0013 | .0036 | .0084 | .0172 | .0320 | .0547 |
| | 7 | | | | .0001 | .0002 | .0006 | .0016 | .0037 | .0078 |
| 8 | 0 | .4305 | .2725 | .1678 | .1001 | .0576 | .0319 | .0168 | .0084 | .0039 |
| | 1 | .3826 | .3847 | .3355 | .2670 | .1977 | .1373 | .0896 | .0548 | .0313 |
| | 2 | .1488 | .2376 | .2936 | .3115 | .2965 | .2587 | .2090 | .1569 | .1094 |
| | 3 | .0331 | .0839 | .1468 | .2076 | .2541 | .2786 | .2787 | .2568 | .2188 |
| | 4 | .0046 | .0185 | .0459 | .0865 | .1361 | .1875 | .2322 | .2627 | .2734 |
| | 5 | .0004 | .0026 | .0092 | .0231 | .0467 | .0808 | .1239 | .1719 | .2188 |
| | 6 | | .0002 | .0011 | .0038 | .0100 | .0217 | .0413 | .0703 | .1094 |
| | 7 | | | .0001 | .0004 | .0012 | .0033 | .0079 | .0164 | .0312 |
| | 8 | | | | | .0001 | .0002 | .0007 | .0017 | .0039 |

*(Continued)*

## TABLE C

### Binomial probabilities (continued)

Entry is $P(X = k) = \binom{n}{k} p^k (1 - p)^{n-k}$

| | | *p* | | | | | | | | |
|---|---|---|---|---|---|---|---|---|---|---|
| *n* | *k* | .01 | .02 | .03 | .04 | .05 | .06 | .07 | .08 | .09 |
| 9 | 0 | .9135 | .8337 | .7602 | .6925 | .6302 | .5730 | .5204 | .4722 | .4279 |
| | 1 | .0830 | .1531 | .2116 | .2597 | .2985 | .3292 | .3525 | .3695 | .3809 |
| | 2 | .0034 | .0125 | .0262 | .0433 | .0629 | .0840 | .1061 | .1285 | .1507 |
| | 3 | .0001 | .0006 | .0019 | .0042 | .0077 | .0125 | .0186 | .0261 | .0348 |
| | 4 | | | .0001 | .0003 | .0006 | .0012 | .0021 | .0034 | .0052 |
| | 5 | | | | | | .0001 | .0002 | .0003 | .0005 |
| | 6 | | | | | | | | | |
| | 7 | | | | | | | | | |
| | 8 | | | | | | | | | |
| | 9 | | | | | | | | | |
| 10 | 0 | .9044 | .8171 | .7374 | .6648 | .5987 | .5386 | .4840 | .4344 | .3894 |
| | 1 | .0914 | .1667 | .2281 | .2770 | .3151 | .3438 | .3643 | .3777 | .3851 |
| | 2 | .0042 | .0153 | .0317 | .0519 | .0746 | .0988 | .1234 | .1478 | .1714 |
| | 3 | .0001 | .0008 | .0026 | .0058 | .0105 | .0168 | .0248 | .0343 | .0452 |
| | 4 | | | .0001 | .0004 | .0010 | .0019 | .0033 | .0052 | .0078 |
| | 5 | | | | | .0001 | .0001 | .0003 | .0005 | .0009 |
| | 6 | | | | | | | | | .0001 |
| | 7 | | | | | | | | | |
| | 8 | | | | | | | | | |
| | 9 | | | | | | | | | |
| | 10 | | | | | | | | | |
| 12 | 0 | .8864 | .7847 | .6938 | .6127 | .5404 | .4759 | .4186 | .3677 | .3225 |
| | 1 | .1074 | .1922 | .2575 | .3064 | .3413 | .3645 | .3781 | .3837 | .3827 |
| | 2 | .0060 | .0216 | .0438 | .0702 | .0988 | .1280 | .1565 | .1835 | .2082 |
| | 3 | .0002 | .0015 | .0045 | .0098 | .0173 | .0272 | .0393 | .0532 | .0686 |
| | 4 | | .0001 | .0003 | .0009 | .0021 | .0039 | .0067 | .0104 | .0153 |
| | 5 | | | | .0001 | .0002 | .0004 | .0008 | .0014 | .0024 |
| | 6 | | | | | | | .0001 | .0001 | .0003 |
| | 7 | | | | | | | | | |
| | 8 | | | | | | | | | |
| | 9 | | | | | | | | | |
| | 10 | | | | | | | | | |
| | 11 | | | | | | | | | |
| | 12 | | | | | | | | | |
| 15 | 0 | .8601 | .7386 | .6333 | .5421 | .4633 | .3953 | .3367 | .2863 | .2430 |
| | 1 | .1303 | .2261 | .2938 | .3388 | .3658 | .3785 | .3801 | .3734 | .3605 |
| | 2 | .0092 | .0323 | .0636 | .0988 | .1348 | .1691 | .2003 | .2273 | .2496 |
| | 3 | .0004 | .0029 | .0085 | .0178 | .0307 | .0468 | .0653 | .0857 | .1070 |
| | 4 | | .0002 | .0008 | .0022 | .0049 | .0090 | .0148 | .0223 | .0317 |
| | 5 | | | .0001 | .0002 | .0006 | .0013 | .0024 | .0043 | .0069 |
| | 6 | | | | | | .0001 | .0003 | .0006 | .0011 |
| | 7 | | | | | | | | .0001 | .0001 |
| | 8 | | | | | | | | | |
| | 9 | | | | | | | | | |
| | 10 | | | | | | | | | |
| | 11 | | | | | | | | | |
| | 12 | | | | | | | | | |
| | 13 | | | | | | | | | |
| | 14 | | | | | | | | | |
| | 15 | | | | | | | | | |

## TABLE C

### Binomial probabilities (continued)

Entry is $P(X = k) = \binom{n}{k} p^k (1 - p)^{n-k}$

| | | *p* | | | | | | | | |
|---|---|---|---|---|---|---|---|---|---|---|
| *n* | *k* | .10 | .15 | .20 | .25 | .30 | .35 | .40 | .45 | .50 |
| 9 | 0 | .3874 | .2316 | .1342 | .0751 | .0404 | .0207 | .0101 | .0046 | .0020 |
| | 1 | .3874 | .3679 | .3020 | .2253 | .1556 | .1004 | .0605 | .0339 | .0176 |
| | 2 | .1722 | .2597 | .3020 | .3003 | .2668 | .2162 | .1612 | .1110 | .0703 |
| | 3 | .0446 | .1069 | .1762 | .2336 | .2668 | .2716 | .2508 | .2119 | .1641 |
| | 4 | .0074 | .0283 | .0661 | .1168 | .1715 | .2194 | .2508 | .2600 | .2461 |
| | 5 | .0008 | .0050 | .0165 | .0389 | .0735 | .1181 | .1672 | .2128 | .2461 |
| | 6 | .0001 | .0006 | .0028 | .0087 | .0210 | .0424 | .0743 | .1160 | .1641 |
| | 7 | | | .0003 | .0012 | .0039 | .0098 | .0212 | .0407 | .0703 |
| | 8 | | | | .0001 | .0004 | .0013 | .0035 | .0083 | .0176 |
| | 9 | | | | | | .0001 | .0003 | .0008 | .0020 |
| 10 | 0 | .3487 | .1969 | .1074 | .0563 | .0282 | .0135 | .0060 | .0025 | .0010 |
| | 1 | .3874 | .3474 | .2684 | .1877 | .1211 | .0725 | .0403 | .0207 | .0098 |
| | 2 | .1937 | .2759 | .3020 | .2816 | .2335 | .1757 | .1209 | .0763 | .0439 |
| | 3 | .0574 | .1298 | .2013 | .2503 | .2668 | .2522 | .2150 | .1665 | .1172 |
| | 4 | .0112 | .0401 | .0881 | .1460 | .2001 | .2377 | .2508 | .2384 | .2051 |
| | 5 | .0015 | .0085 | .0264 | .0584 | .1029 | .1536 | .2007 | .2340 | .2461 |
| | 6 | .0001 | .0012 | .0055 | .0162 | .0368 | .0689 | .1115 | .1596 | .2051 |
| | 7 | | .0001 | .0008 | .0031 | .0090 | .0212 | .0425 | .0746 | .1172 |
| | 8 | | | .0001 | .0004 | .0014 | .0043 | .0106 | .0229 | .0439 |
| | 9 | | | | | .0001 | .0005 | .0016 | .0042 | .0098 |
| | 10 | | | | | | | .0001 | .0003 | .0010 |
| 12 | 0 | .2824 | .1422 | .0687 | .0317 | .0138 | .0057 | .0022 | .0008 | .0002 |
| | 1 | .3766 | .3012 | .2062 | .1267 | .0712 | .0368 | .0174 | .0075 | .0029 |
| | 2 | .2301 | .2924 | .2835 | .2323 | .1678 | .1088 | .0639 | .0339 | .0161 |
| | 3 | .0852 | .1720 | .2362 | .2581 | .2397 | .1954 | .1419 | .0923 | .0537 |
| | 4 | .0213 | .0683 | .1329 | .1936 | .2311 | .2367 | .2128 | .1700 | .1208 |
| | 5 | .0038 | .0193 | .0532 | .1032 | .1585 | .2039 | .2270 | .2225 | .1934 |
| | 6 | .0005 | .0040 | .0155 | .0401 | .0792 | .1281 | .1766 | .2124 | .2256 |
| | 7 | | .0006 | .0033 | .0115 | .0291 | .0591 | .1009 | .1489 | .1934 |
| | 8 | | .0001 | .0005 | .0024 | .0078 | .0199 | .0420 | .0762 | .1208 |
| | 9 | | | .0001 | .0004 | .0015 | .0048 | .0125 | .0277 | .0537 |
| | 10 | | | | | .0002 | .0008 | .0025 | .0068 | .0161 |
| | 11 | | | | | | .0001 | .0003 | .0010 | .0029 |
| | 12 | | | | | | | | .0001 | .0002 |
| 15 | 0 | .2059 | .0874 | .0352 | .0134 | .0047 | .0016 | .0005 | .0001 | .0000 |
| | 1 | .3432 | .2312 | .1319 | .0668 | .0305 | .0126 | .0047 | .0016 | .0005 |
| | 2 | .2669 | .2856 | .2309 | .1559 | .0916 | .0476 | .0219 | .0090 | .0032 |
| | 3 | .1285 | .2184 | .2501 | .2252 | .1700 | .1110 | .0634 | .0318 | .0139 |
| | 4 | .0428 | .1156 | .1876 | .2252 | .2186 | .1792 | .1268 | .0780 | .0417 |
| | 5 | .0105 | .0449 | .1032 | .1651 | .2061 | .2123 | .1859 | .1404 | .0916 |
| | 6 | .0019 | .0132 | .0430 | .0917 | .1472 | .1906 | .2066 | .1914 | .1527 |
| | 7 | .0003 | .0030 | .0138 | .0393 | .0811 | .1319 | .1771 | .2013 | .1964 |
| | 8 | | .0005 | .0035 | .0131 | .0348 | .0710 | .1181 | .1647 | .1964 |
| | 9 | | .0001 | .0007 | .0034 | .0116 | .0298 | .0612 | .1048 | .1527 |
| | 10 | | | .0001 | .0007 | .0030 | .0096 | .0245 | .0515 | .0916 |
| | 11 | | | | .0001 | .0006 | .0024 | .0074 | .0191 | .0417 |
| | 12 | | | | | .0001 | .0004 | .0016 | .0052 | .0139 |
| | 13 | | | | | | .0001 | .0003 | .0010 | .0032 |
| | 14 | | | | | | | | .0001 | .0005 |
| | 15 | | | | | | | | | |

*(Continued)*

## TABLE C

Binomial probabilities (continued)

| $n$ | $k$ | .01 | .02 | .03 | .04 | .05 | .06 | .07 | .08 | .09 |
|---|---|---|---|---|---|---|---|---|---|---|
| 20 | 0 | .8179 | .6676 | .5438 | .4420 | .3585 | .2901 | .2342 | .1887 | .1516 |
| | 1 | .1652 | .2725 | .3364 | .3683 | .3774 | .3703 | .3526 | .3282 | .3000 |
| | 2 | .0159 | .0528 | .0988 | .1458 | .1887 | .2246 | .2521 | .2711 | .2818 |
| | 3 | .0010 | .0065 | .0183 | .0364 | .0596 | .0860 | .1139 | .1414 | .1672 |
| | 4 | | .0006 | .0024 | .0065 | .0133 | .0233 | .0364 | .0523 | .0703 |
| | 5 | | | .0002 | .0009 | .0022 | .0048 | .0088 | .0145 | .0222 |
| | 6 | | | | .0001 | .0003 | .0008 | .0017 | .0032 | .0055 |
| | 7 | | | | | | .0001 | .0002 | .0005 | .0011 |
| | 8 | | | | | | | | .0001 | .0002 |
| | 9 | | | | | | | | | |
| | 10 | | | | | | | | | |
| | 11 | | | | | | | | | |
| | 12 | | | | | | | | | |
| | 13 | | | | | | | | | |
| | 14 | | | | | | | | | |
| | 15 | | | | | | | | | |
| | 16 | | | | | | | | | |
| | 17 | | | | | | | | | |
| | 18 | | | | | | | | | |
| | 19 | | | | | | | | | |
| | 20 | | | | | | | | | |

| $n$ | $k$ | .10 | .15 | .20 | .25 | .30 | .35 | .40 | .45 | .50 |
|---|---|---|---|---|---|---|---|---|---|---|
| 20 | 0 | .1216 | .0388 | .0115 | .0032 | .0008 | .0002 | .0000 | .0000 | .0000 |
| | 1 | .2702 | .1368 | .0576 | .0211 | .0068 | .0020 | .0005 | .0001 | .0000 |
| | 2 | .2852 | .2293 | .1369 | .0669 | .0278 | .0100 | .0031 | .0008 | .0002 |
| | 3 | .1901 | .2428 | .2054 | .1339 | .0716 | .0323 | .0123 | .0040 | .0011 |
| | 4 | .0898 | .1821 | .2182 | .1897 | .1304 | .0738 | .0350 | .0139 | .0046 |
| | 5 | .0319 | .1028 | .1746 | .2023 | .1789 | .1272 | .0746 | .0365 | .0148 |
| | 6 | .0089 | .0454 | .1091 | .1686 | .1916 | .1712 | .1244 | .0746 | .0370 |
| | 7 | .0020 | .0160 | .0545 | .1124 | .1643 | .1844 | .1659 | .1221 | .0739 |
| | 8 | .0004 | .0046 | .0222 | .0609 | .1144 | .1614 | .1797 | .1623 | .1201 |
| | 9 | .0001 | .0011 | .0074 | .0271 | .0654 | .1158 | .1597 | .1771 | .1602 |
| | 10 | | .0002 | .0020 | .0099 | .0308 | .0686 | .1171 | .1593 | .1762 |
| | 11 | | | .0005 | .0030 | .0120 | .0336 | .0710 | .1185 | .1602 |
| | 12 | | | .0001 | .0008 | .0039 | .0136 | .0355 | .0727 | .1201 |
| | 13 | | | | .0002 | .0010 | .0045 | .0146 | .0366 | .0739 |
| | 14 | | | | | .0002 | .0012 | .0049 | .0150 | .0370 |
| | 15 | | | | | | .0003 | .0013 | .0049 | .0148 |
| | 16 | | | | | | | .0003 | .0013 | .0046 |
| | 17 | | | | | | | | .0002 | .0011 |
| | 18 | | | | | | | | | .0002 |
| | 19 | | | | | | | | | |
| | 20 | | | | | | | | | |

Table entry for $p$ and $C$ is the critical value $t^*$ with probability $p$ lying to its right and probability $C$ lying between $-t^*$ and $t^*$.

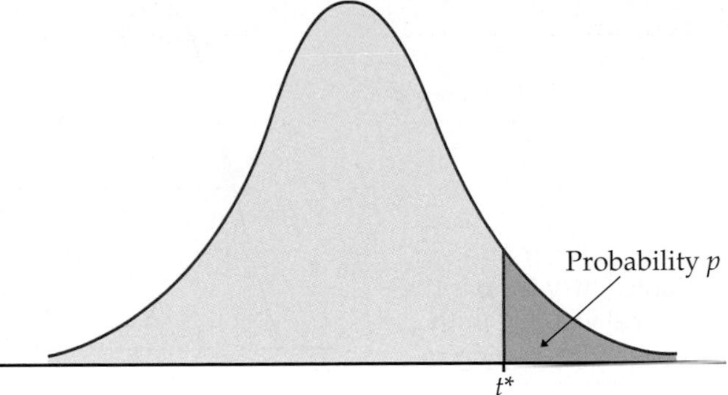


## TABLE D

### $t$ distribution critical values

| | Upper-tail probability $p$ | | | | | | | | | | | |
|---|---|---|---|---|---|---|---|---|---|---|---|---|
| df | .25 | .20 | .15 | .10 | .05 | .025 | .02 | .01 | .005 | .0025 | .001 | .0005 |
| 1 | 1.000 | 1.376 | 1.963 | 3.078 | 6.314 | 12.71 | 15.89 | 31.82 | 63.66 | 127.3 | 318.3 | 636.6 |
| 2 | 0.816 | 1.061 | 1.386 | 1.886 | 2.920 | 4.303 | 4.849 | 6.965 | 9.925 | 14.09 | 22.33 | 31.60 |
| 3 | 0.765 | 0.978 | 1.250 | 1.638 | 2.353 | 3.182 | 3.482 | 4.541 | 5.841 | 7.453 | 10.21 | 12.92 |
| 4 | 0.741 | 0.941 | 1.190 | 1.533 | 2.132 | 2.776 | 2.999 | 3.747 | 4.604 | 5.598 | 7.173 | 8.610 |
| 5 | 0.727 | 0.920 | 1.156 | 1.476 | 2.015 | 2.571 | 2.757 | 3.365 | 4.032 | 4.773 | 5.893 | 6.869 |
| 6 | 0.718 | 0.906 | 1.134 | 1.440 | 1.943 | 2.447 | 2.612 | 3.143 | 3.707 | 4.317 | 5.208 | 5.959 |
| 7 | 0.711 | 0.896 | 1.119 | 1.415 | 1.895 | 2.365 | 2.517 | 2.998 | 3.499 | 4.029 | 4.785 | 5.408 |
| 8 | 0.706 | 0.889 | 1.108 | 1.397 | 1.860 | 2.306 | 2.449 | 2.896 | 3.355 | 3.833 | 4.501 | 5.041 |
| 9 | 0.703 | 0.883 | 1.100 | 1.383 | 1.833 | 2.262 | 2.398 | 2.821 | 3.250 | 3.690 | 4.297 | 4.781 |
| 10 | 0.700 | 0.879 | 1.093 | 1.372 | 1.812 | 2.228 | 2.359 | 2.764 | 3.169 | 3.581 | 4.144 | 4.587 |
| 11 | 0.697 | 0.876 | 1.088 | 1.363 | 1.796 | 2.201 | 2.328 | 2.718 | 3.106 | 3.497 | 4.025 | 4.437 |
| 12 | 0.695 | 0.873 | 1.083 | 1.356 | 1.782 | 2.179 | 2.303 | 2.681 | 3.055 | 3.428 | 3.930 | 4.318 |
| 13 | 0.694 | 0.870 | 1.079 | 1.350 | 1.771 | 2.160 | 2.282 | 2.650 | 3.012 | 3.372 | 3.852 | 4.221 |
| 14 | 0.692 | 0.868 | 1.076 | 1.345 | 1.761 | 2.145 | 2.264 | 2.624 | 2.977 | 3.326 | 3.787 | 4.140 |
| 15 | 0.691 | 0.866 | 1.074 | 1.341 | 1.753 | 2.131 | 2.249 | 2.602 | 2.947 | 3.286 | 3.733 | 4.073 |
| 16 | 0.690 | 0.865 | 1.071 | 1.337 | 1.746 | 2.120 | 2.235 | 2.583 | 2.921 | 3.252 | 3.686 | 4.015 |
| 17 | 0.689 | 0.863 | 1.069 | 1.333 | 1.740 | 2.110 | 2.224 | 2.567 | 2.898 | 3.222 | 3.646 | 3.965 |
| 18 | 0.688 | 0.862 | 1.067 | 1.330 | 1.734 | 2.101 | 2.214 | 2.552 | 2.878 | 3.197 | 3.611 | 3.922 |
| 19 | 0.688 | 0.861 | 1.066 | 1.328 | 1.729 | 2.093 | 2.205 | 2.539 | 2.861 | 3.174 | 3.579 | 3.883 |
| 20 | 0.687 | 0.860 | 1.064 | 1.325 | 1.725 | 2.086 | 2.197 | 2.528 | 2.845 | 3.153 | 3.552 | 3.850 |
| 21 | 0.686 | 0.859 | 1.063 | 1.323 | 1.721 | 2.080 | 2.189 | 2.518 | 2.831 | 3.135 | 3.527 | 3.819 |
| 22 | 0.686 | 0.858 | 1.061 | 1.321 | 1.717 | 2.074 | 2.183 | 2.508 | 2.819 | 3.119 | 3.505 | 3.792 |
| 23 | 0.685 | 0.858 | 1.060 | 1.319 | 1.714 | 2.069 | 2.177 | 2.500 | 2.807 | 3.104 | 3.485 | 3.768 |
| 24 | 0.685 | 0.857 | 1.059 | 1.318 | 1.711 | 2.064 | 2.172 | 2.492 | 2.797 | 3.091 | 3.467 | 3.745 |
| 25 | 0.684 | 0.856 | 1.058 | 1.316 | 1.708 | 2.060 | 2.167 | 2.485 | 2.787 | 3.078 | 3.450 | 3.725 |
| 26 | 0.684 | 0.856 | 1.058 | 1.315 | 1.706 | 2.056 | 2.162 | 2.479 | 2.779 | 3.067 | 3.435 | 3.707 |
| 27 | 0.684 | 0.855 | 1.057 | 1.314 | 1.703 | 2.052 | 2.158 | 2.473 | 2.771 | 3.057 | 3.421 | 3.690 |
| 28 | 0.683 | 0.855 | 1.056 | 1.313 | 1.701 | 2.048 | 2.154 | 2.467 | 2.763 | 3.047 | 3.408 | 3.674 |
| 29 | 0.683 | 0.854 | 1.055 | 1.311 | 1.699 | 2.045 | 2.150 | 2.462 | 2.756 | 3.038 | 3.396 | 3.659 |
| 30 | 0.683 | 0.854 | 1.055 | 1.310 | 1.697 | 2.042 | 2.147 | 2.457 | 2.750 | 3.030 | 3.385 | 3.646 |
| 40 | 0.681 | 0.851 | 1.050 | 1.303 | 1.684 | 2.021 | 2.123 | 2.423 | 2.704 | 2.971 | 3.307 | 3.551 |
| 50 | 0.679 | 0.849 | 1.047 | 1.299 | 1.676 | 2.009 | 2.109 | 2.403 | 2.678 | 2.937 | 3.261 | 3.496 |
| 60 | 0.679 | 0.848 | 1.045 | 1.296 | 1.671 | 2.000 | 2.099 | 2.390 | 2.660 | 2.915 | 3.232 | 3.460 |
| 80 | 0.678 | 0.846 | 1.043 | 1.292 | 1.664 | 1.990 | 2.088 | 2.374 | 2.639 | 2.887 | 3.195 | 3.416 |
| 100 | 0.677 | 0.845 | 1.042 | 1.290 | 1.660 | 1.984 | 2.081 | 2.364 | 2.626 | 2.871 | 3.174 | 3.390 |
| 1000 | 0.675 | 0.842 | 1.037 | 1.282 | 1.646 | 1.962 | 2.056 | 2.330 | 2.581 | 2.813 | 3.098 | 3.300 |
| $z^*$ | 0.674 | 0.841 | 1.036 | 1.282 | 1.645 | 1.960 | 2.054 | 2.326 | 2.576 | 2.807 | 3.091 | 3.291 |
| | 50% | 60% | 70% | 80% | 90% | 95% | 96% | 98% | 99% | 99.5% | 99.8% | 99.9% |
| | Confidence level $C$ | | | | | | | | | | | |

Table entry for $p$ is the critical value $F^*$ with probability $p$ lying to its right.

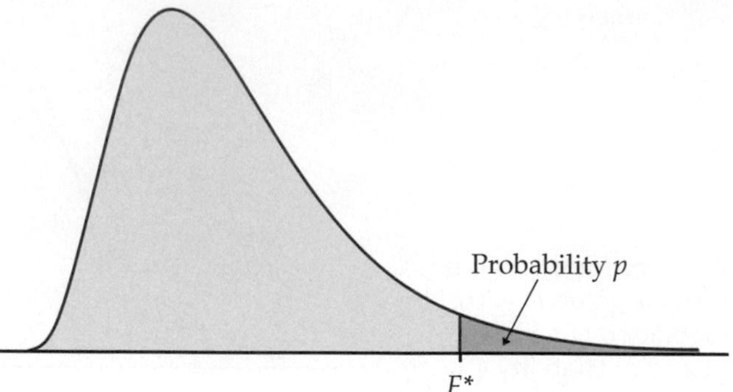


**TABLE E**

## *F* critical values

| Degrees of freedom in the denominator | $p$ | Degrees of freedom in the numerator 1 | 2 | 3 | 4 | 5 | 6 | 7 | 8 | 9 |
|---|---|---|---|---|---|---|---|---|---|---|
| 1 | .100 | 39.86 | 49.50 | 53.59 | 55.83 | 57.24 | 58.20 | 58.91 | 59.44 | 59.86 |
| | .050 | 161.45 | 199.50 | 215.71 | 224.58 | 230.16 | 233.99 | 236.77 | 238.88 | 240.54 |
| | .025 | 647.79 | 799.50 | 864.16 | 899.58 | 921.85 | 937.11 | 948.22 | 956.66 | 963.28 |
| | .010 | 4052.2 | 4999.5 | 5403.4 | 5624.6 | 5763.6 | 5859.0 | 5928.4 | 5981.1 | 6022.5 |
| | .001 | 405284 | 500000 | 540379 | 562500 | 576405 | 585937 | 592873 | 598144 | 602284 |
| 2 | .100 | 8.53 | 9.00 | 9.16 | 9.24 | 9.29 | 9.33 | 9.35 | 9.37 | 9.38 |
| | .050 | 18.51 | 19.00 | 19.16 | 19.25 | 19.30 | 19.33 | 19.35 | 19.37 | 19.38 |
| | .025 | 38.51 | 39.00 | 39.17 | 39.25 | 39.30 | 39.33 | 39.36 | 39.37 | 39.39 |
| | .010 | 98.50 | 99.00 | 99.17 | 99.25 | 99.30 | 99.33 | 99.36 | 99.37 | 99.39 |
| | .001 | 998.50 | 999.00 | 999.17 | 999.25 | 999.30 | 999.33 | 999.36 | 999.37 | 999.39 |
| 3 | .100 | 5.54 | 5.46 | 5.39 | 5.34 | 5.31 | 5.28 | 5.27 | 5.25 | 5.24 |
| | .050 | 10.13 | 9.55 | 9.28 | 9.12 | 9.01 | 8.94 | 8.89 | 8.85 | 8.81 |
| | .025 | 17.44 | 16.04 | 15.44 | 15.10 | 14.88 | 14.73 | 14.62 | 14.54 | 14.47 |
| | .010 | 34.12 | 30.82 | 29.46 | 28.71 | 28.24 | 27.91 | 27.67 | 27.49 | 27.35 |
| | .001 | 167.03 | 148.50 | 141.11 | 137.10 | 134.58 | 132.85 | 131.58 | 130.62 | 129.86 |
| 4 | .100 | 4.54 | 4.32 | 4.19 | 4.11 | 4.05 | 4.01 | 3.98 | 3.95 | 3.94 |
| | .050 | 7.71 | 6.94 | 6.59 | 6.39 | 6.26 | 6.16 | 6.09 | 6.04 | 6.00 |
| | .025 | 12.22 | 10.65 | 9.98 | 9.60 | 9.36 | 9.20 | 9.07 | 8.98 | 8.90 |
| | .010 | 21.20 | 18.00 | 16.69 | 15.98 | 15.52 | 15.21 | 14.98 | 14.80 | 14.66 |
| | .001 | 74.14 | 61.25 | 56.18 | 53.44 | 51.71 | 50.53 | 49.66 | 49.00 | 48.47 |
| 5 | .100 | 4.06 | 3.78 | 3.62 | 3.52 | 3.45 | 3.40 | 3.37 | 3.34 | 3.32 |
| | .050 | 6.61 | 5.79 | 5.41 | 5.19 | 5.05 | 4.95 | 4.88 | 4.82 | 4.77 |
| | .025 | 10.01 | 8.43 | 7.76 | 7.39 | 7.15 | 6.98 | 6.85 | 6.76 | 6.68 |
| | .010 | 16.26 | 13.27 | 12.06 | 11.39 | 10.97 | 10.67 | 10.46 | 10.29 | 10.16 |
| | .001 | 47.18 | 37.12 | 33.20 | 31.09 | 29.75 | 28.83 | 28.16 | 27.65 | 27.24 |
| 6 | .100 | 3.78 | 3.46 | 3.29 | 3.18 | 3.11 | 3.05 | 3.01 | 2.98 | 2.96 |
| | .050 | 5.99 | 5.14 | 4.76 | 4.53 | 4.39 | 4.28 | 4.21 | 4.15 | 4.10 |
| | .025 | 8.81 | 7.26 | 6.60 | 6.23 | 5.99 | 5.82 | 5.70 | 5.60 | 5.52 |
| | .010 | 13.75 | 10.92 | 9.78 | 9.15 | 8.75 | 8.47 | 8.26 | 8.10 | 7.98 |
| | .001 | 35.51 | 27.00 | 23.70 | 21.92 | 20.80 | 20.03 | 19.46 | 19.03 | 18.69 |
| 7 | .100 | 3.59 | 3.26 | 3.07 | 2.96 | 2.88 | 2.83 | 2.78 | 2.75 | 2.72 |
| | .050 | 5.59 | 4.74 | 4.35 | 4.12 | 3.97 | 3.87 | 3.79 | 3.73 | 3.68 |
| | .025 | 8.07 | 6.54 | 5.89 | 5.52 | 5.29 | 5.12 | 4.99 | 4.90 | 4.82 |
| | .010 | 12.25 | 9.55 | 8.45 | 7.85 | 7.46 | 7.19 | 6.99 | 6.84 | 6.72 |
| | .001 | 29.25 | 21.69 | 18.77 | 17.20 | 16.21 | 15.52 | 15.02 | 14.63 | 14.33 |

Table entry for $p$ is the critical value $F^*$ with probability $p$ lying to its right.

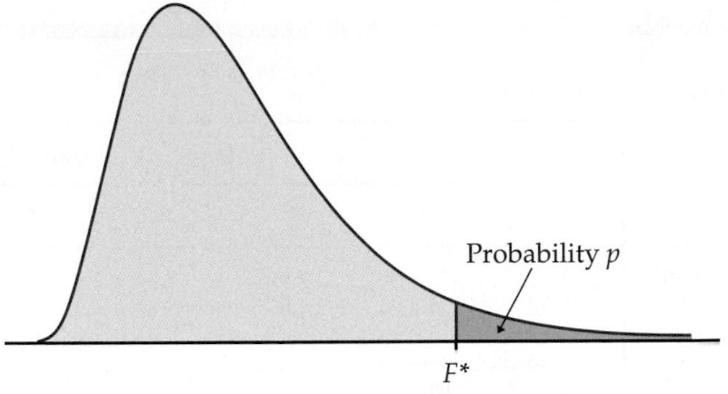


**TABLE E**

## $F$ critical values (continued)

| Degrees of freedom in the numerator | | | | | | | | | | |
|---|---|---|---|---|---|---|---|---|---|---|
| 10 | 12 | 15 | 20 | 25 | 30 | 40 | 50 | 60 | 120 | 1000 |
| 60.19 | 60.71 | 61.22 | 61.74 | 62.05 | 62.26 | 62.53 | 62.69 | 62.79 | 63.06 | 63.30 |
| 241.88 | 243.91 | 245.95 | 248.01 | 249.26 | 250.10 | 251.14 | 251.77 | 252.20 | 253.25 | 254.19 |
| 968.63 | 976.71 | 984.87 | 993.10 | 998.08 | 1001.4 | 1005.6 | 1008.1 | 1009.8 | 1014.0 | 1017.7 |
| 6055.8 | 6106.3 | 6157.3 | 6208.7 | 6239.8 | 6260.6 | 6286.8 | 6302.5 | 6313.0 | 6339.4 | 6362.7 |
| 605621 | 610668 | 615764 | 620908 | 624017 | 626099 | 628712 | 630285 | 631337 | 633972 | 636301 |
| 9.39 | 9.41 | 9.42 | 9.44 | 9.45 | 9.46 | 9.47 | 9.47 | 9.47 | 9.48 | 9.49 |
| 19.40 | 19.41 | 19.43 | 19.45 | 19.46 | 19.46 | 19.47 | 19.48 | 19.48 | 19.49 | 19.49 |
| 39.40 | 39.41 | 39.43 | 39.45 | 39.46 | 39.46 | 39.47 | 39.48 | 39.48 | 39.49 | 39.50 |
| 99.40 | 99.42 | 99.43 | 99.45 | 99.46 | 99.47 | 99.47 | 99.48 | 99.48 | 99.49 | 99.50 |
| 999.40 | 999.42 | 999.43 | 999.45 | 999.46 | 999.47 | 999.47 | 999.48 | 999.48 | 999.49 | 999.50 |
| 5.23 | 5.22 | 5.20 | 5.18 | 5.17 | 5.17 | 5.16 | 5.15 | 5.15 | 5.14 | 5.13 |
| 8.79 | 8.74 | 8.70 | 8.66 | 8.63 | 8.62 | 8.59 | 8.58 | 8.57 | 8.55 | 8.53 |
| 14.42 | 14.34 | 14.25 | 14.17 | 14.12 | 14.08 | 14.04 | 14.01 | 13.99 | 13.95 | 13.91 |
| 27.23 | 27.05 | 26.87 | 26.69 | 26.58 | 26.50 | 26.41 | 26.35 | 26.32 | 26.22 | 26.14 |
| 129.25 | 128.32 | 127.37 | 126.42 | 125.84 | 125.45 | 124.96 | 124.66 | 124.47 | 123.97 | 123.53 |
| 3.92 | 3.90 | 3.87 | 3.84 | 3.83 | 3.82 | 3.80 | 3.80 | 3.79 | 3.78 | 3.76 |
| 5.96 | 5.91 | 5.86 | 5.80 | 5.77 | 5.75 | 5.72 | 5.70 | 5.69 | 5.66 | 5.63 |
| 8.84 | 8.75 | 8.66 | 8.56 | 8.50 | 8.46 | 8.41 | 8.38 | 8.36 | 8.31 | 8.26 |
| 14.55 | 14.37 | 14.20 | 14.02 | 13.91 | 13.84 | 13.75 | 13.69 | 13.65 | 13.56 | 13.47 |
| 48.05 | 47.41 | 46.76 | 46.10 | 45.70 | 45.43 | 45.09 | 44.88 | 44.75 | 44.40 | 44.09 |
| 3.30 | 3.27 | 3.24 | 3.21 | 3.19 | 3.17 | 3.16 | 3.15 | 3.14 | 3.12 | 3.11 |
| 4.74 | 4.68 | 4.62 | 4.56 | 4.52 | 4.50 | 4.46 | 4.44 | 4.43 | 4.40 | 4.37 |
| 6.62 | 6.52 | 6.43 | 6.33 | 6.27 | 6.23 | 6.18 | 6.14 | 6.12 | 6.07 | 6.02 |
| 10.05 | 9.89 | 9.72 | 9.55 | 9.45 | 9.38 | 9.29 | 9.24 | 9.20 | 9.11 | 9.03 |
| 26.92 | 26.42 | 25.91 | 25.39 | 25.08 | 24.87 | 24.60 | 24.44 | 24.33 | 24.06 | 23.82 |
| 2.94 | 2.90 | 2.87 | 2.84 | 2.81 | 2.80 | 2.78 | 2.77 | 2.76 | 2.74 | 2.72 |
| 4.06 | 4.00 | 3.94 | 3.87 | 3.83 | 3.81 | 3.77 | 3.75 | 3.74 | 3.70 | 3.67 |
| 5.46 | 5.37 | 5.27 | 5.17 | 5.11 | 5.07 | 5.01 | 4.98 | 4.96 | 4.90 | 4.86 |
| 7.87 | 7.72 | 7.56 | 7.40 | 7.30 | 7.23 | 7.14 | 7.09 | 7.06 | 6.97 | 6.89 |
| 18.41 | 17.99 | 17.56 | 17.12 | 16.85 | 16.67 | 16.44 | 16.31 | 16.21 | 15.98 | 15.77 |
| 2.70 | 2.67 | 2.63 | 2.59 | 2.57 | 2.56 | 2.54 | 2.52 | 2.51 | 2.49 | 2.47 |
| 3.64 | 3.57 | 3.51 | 3.44 | 3.40 | 3.38 | 3.34 | 3.32 | 3.30 | 3.27 | 3.23 |
| 4.76 | 4.67 | 4.57 | 4.47 | 4.40 | 4.36 | 4.31 | 4.28 | 4.25 | 4.20 | 4.15 |
| 6.62 | 6.47 | 6.31 | 6.16 | 6.06 | 5.99 | 5.91 | 5.86 | 5.82 | 5.74 | 5.66 |
| 14.08 | 13.71 | 13.32 | 12.93 | 12.69 | 12.53 | 12.33 | 12.20 | 12.12 | 11.91 | 11.72 |

*(Continued)*

## TABLE E

### *F* critical values (continued)

| Degrees of freedom in the denominator | *p* | Degrees of freedom in the numerator 1 | 2 | 3 | 4 | 5 | 6 | 7 | 8 | 9 |
|---|---|---|---|---|---|---|---|---|---|---|
| 8 | .100 | 3.46 | 3.11 | 2.92 | 2.81 | 2.73 | 2.67 | 2.62 | 2.59 | 2.56 |
| | .050 | 5.32 | 4.46 | 4.07 | 3.84 | 3.69 | 3.58 | 3.50 | 3.44 | 3.39 |
| | .025 | 7.57 | 6.06 | 5.42 | 5.05 | 4.82 | 4.65 | 4.53 | 4.43 | 4.36 |
| | .010 | 11.26 | 8.65 | 7.59 | 7.01 | 6.63 | 6.37 | 6.18 | 6.03 | 5.91 |
| | .001 | 25.41 | 18.49 | 15.83 | 14.39 | 13.48 | 12.86 | 12.40 | 12.05 | 11.77 |
| 9 | .100 | 3.36 | 3.01 | 2.81 | 2.69 | 2.61 | 2.55 | 2.51 | 2.47 | 2.44 |
| | .050 | 5.12 | 4.26 | 3.86 | 3.63 | 3.48 | 3.37 | 3.29 | 3.23 | 3.18 |
| | .025 | 7.21 | 5.71 | 5.08 | 4.72 | 4.48 | 4.32 | 4.20 | 4.10 | 4.03 |
| | .010 | 10.56 | 8.02 | 6.99 | 6.42 | 6.06 | 5.80 | 5.61 | 5.47 | 5.35 |
| | .001 | 22.86 | 16.39 | 13.90 | 12.56 | 11.71 | 11.13 | 10.70 | 10.37 | 10.11 |
| 10 | .100 | 3.29 | 2.92 | 2.73 | 2.61 | 2.52 | 2.46 | 2.41 | 2.38 | 2.35 |
| | .050 | 4.96 | 4.10 | 3.71 | 3.48 | 3.33 | 3.22 | 3.14 | 3.07 | 3.02 |
| | .025 | 6.94 | 5.46 | 4.83 | 4.47 | 4.24 | 4.07 | 3.95 | 3.85 | 3.78 |
| | .010 | 10.04 | 7.56 | 6.55 | 5.99 | 5.64 | 5.39 | 5.20 | 5.06 | 4.94 |
| | .001 | 21.04 | 14.91 | 12.55 | 11.28 | 10.48 | 9.93 | 9.52 | 9.20 | 8.96 |
| 11 | .100 | 3.23 | 2.86 | 2.66 | 2.54 | 2.45 | 2.39 | 2.34 | 2.30 | 2.27 |
| | .050 | 4.84 | 3.98 | 3.59 | 3.36 | 3.20 | 3.09 | 3.01 | 2.95 | 2.90 |
| | .025 | 6.72 | 5.26 | 4.63 | 4.28 | 4.04 | 3.88 | 3.76 | 3.66 | 3.59 |
| | .010 | 9.65 | 7.21 | 6.22 | 5.67 | 5.32 | 5.07 | 4.89 | 4.74 | 4.63 |
| | .001 | 19.69 | 13.81 | 11.56 | 10.35 | 9.58 | 9.05 | 8.66 | 8.35 | 8.12 |
| 12 | .100 | 3.18 | 2.81 | 2.61 | 2.48 | 2.39 | 2.33 | 2.28 | 2.24 | 2.21 |
| | .050 | 4.75 | 3.89 | 3.49 | 3.26 | 3.11 | 3.00 | 2.91 | 2.85 | 2.80 |
| | .025 | 6.55 | 5.10 | 4.47 | 4.12 | 3.89 | 3.73 | 3.61 | 3.51 | 3.44 |
| | .010 | 9.33 | 6.93 | 5.95 | 5.41 | 5.06 | 4.82 | 4.64 | 4.50 | 4.39 |
| | .001 | 18.64 | 12.97 | 10.80 | 9.63 | 8.89 | 8.38 | 8.00 | 7.71 | 7.48 |
| 13 | .100 | 3.14 | 2.76 | 2.56 | 2.43 | 2.35 | 2.28 | 2.23 | 2.20 | 2.16 |
| | .050 | 4.67 | 3.81 | 3.41 | 3.18 | 3.03 | 2.92 | 2.83 | 2.77 | 2.71 |
| | .025 | 6.41 | 4.97 | 4.35 | 4.00 | 3.77 | 3.60 | 3.48 | 3.39 | 3.31 |
| | .010 | 9.07 | 6.70 | 5.74 | 5.21 | 4.86 | 4.62 | 4.44 | 4.30 | 4.19 |
| | .001 | 17.82 | 12.31 | 10.21 | 9.07 | 8.35 | 7.86 | 7.49 | 7.21 | 6.98 |
| 14 | .100 | 3.10 | 2.73 | 2.52 | 2.39 | 2.31 | 2.24 | 2.19 | 2.15 | 2.12 |
| | .050 | 4.60 | 3.74 | 3.34 | 3.11 | 2.96 | 2.85 | 2.76 | 2.70 | 2.65 |
| | .025 | 6.30 | 4.86 | 4.24 | 3.89 | 3.66 | 3.50 | 3.38 | 3.29 | 3.21 |
| | .010 | 8.86 | 6.51 | 5.56 | 5.04 | 4.69 | 4.46 | 4.28 | 4.14 | 4.03 |
| | .001 | 17.14 | 11.78 | 9.73 | 8.62 | 7.92 | 7.44 | 7.08 | 6.80 | 6.58 |
| 15 | .100 | 3.07 | 2.70 | 2.49 | 2.36 | 2.27 | 2.21 | 2.16 | 2.12 | 2.09 |
| | .050 | 4.54 | 3.68 | 3.29 | 3.06 | 2.90 | 2.79 | 2.71 | 2.64 | 2.59 |
| | .025 | 6.20 | 4.77 | 4.15 | 3.80 | 3.58 | 3.41 | 3.29 | 3.20 | 3.12 |
| | .010 | 8.68 | 6.36 | 5.42 | 4.89 | 4.56 | 4.32 | 4.14 | 4.00 | 3.89 |
| | .001 | 16.59 | 11.34 | 9.34 | 8.25 | 7.57 | 7.09 | 6.74 | 6.47 | 6.26 |
| 16 | .100 | 3.05 | 2.67 | 2.46 | 2.33 | 2.24 | 2.18 | 2.13 | 2.09 | 2.06 |
| | .050 | 4.49 | 3.63 | 3.24 | 3.01 | 2.85 | 2.74 | 2.66 | 2.59 | 2.54 |
| | .025 | 6.12 | 4.69 | 4.08 | 3.73 | 3.50 | 3.34 | 3.22 | 3.12 | 3.05 |
| | .010 | 8.53 | 6.23 | 5.29 | 4.77 | 4.44 | 4.20 | 4.03 | 3.89 | 3.78 |
| | .001 | 16.12 | 10.97 | 9.01 | 7.94 | 7.27 | 6.80 | 6.46 | 6.19 | 5.98 |
| 17 | .100 | 3.03 | 2.64 | 2.44 | 2.31 | 2.22 | 2.15 | 2.10 | 2.06 | 2.03 |
| | .050 | 4.45 | 3.59 | 3.20 | 2.96 | 2.81 | 2.70 | 2.61 | 2.55 | 2.49 |
| | .025 | 6.04 | 4.62 | 4.01 | 3.66 | 3.44 | 3.28 | 3.16 | 3.06 | 2.98 |
| | .010 | 8.40 | 6.11 | 5.19 | 4.67 | 4.34 | 4.10 | 3.93 | 3.79 | 3.68 |
| | .001 | 15.72 | 10.66 | 8.73 | 7.68 | 7.02 | 6.56 | 6.22 | 5.96 | 5.75 |

## TABLE E

### *F* critical values (continued)

| Degrees of freedom in the numerator | | | | | | | | | | |
|---|---|---|---|---|---|---|---|---|---|---|
| 10 | 12 | 15 | 20 | 25 | 30 | 40 | 50 | 60 | 120 | 1000 |
| 2.54 | 2.50 | 2.46 | 2.42 | 2.40 | 2.38 | 2.36 | 2.35 | 2.34 | 2.32 | 2.30 |
| 3.35 | 3.28 | 3.22 | 3.15 | 3.11 | 3.08 | 3.04 | 3.02 | 3.01 | 2.97 | 2.93 |
| 4.30 | 4.20 | 4.10 | 4.00 | 3.94 | 3.89 | 3.84 | 3.81 | 3.78 | 3.73 | 3.68 |
| 5.81 | 5.67 | 5.52 | 5.36 | 5.26 | 5.20 | 5.12 | 5.07 | 5.03 | 4.95 | 4.87 |
| 11.54 | 11.19 | 10.84 | 10.48 | 10.26 | 10.11 | 9.92 | 9.80 | 9.73 | 9.53 | 9.36 |
| 2.42 | 2.38 | 2.34 | 2.30 | 2.27 | 2.25 | 2.23 | 2.22 | 2.21 | 2.18 | 2.16 |
| 3.14 | 3.07 | 3.01 | 2.94 | 2.89 | 2.86 | 2.83 | 2.80 | 2.79 | 2.75 | 2.71 |
| 3.96 | 3.87 | 3.77 | 3.67 | 3.60 | 3.56 | 3.51 | 3.47 | 3.45 | 3.39 | 3.34 |
| 5.26 | 5.11 | 4.96 | 4.81 | 4.71 | 4.65 | 4.57 | 4.52 | 4.48 | 4.40 | 4.32 |
| 9.89 | 9.57 | 9.24 | 8.90 | 8.69 | 8.55 | 8.37 | 8.26 | 8.19 | 8.00 | 7.84 |
| 2.32 | 2.28 | 2.24 | 2.20 | 2.17 | 2.16 | 2.13 | 2.12 | 2.11 | 2.08 | 2.06 |
| 2.98 | 2.91 | 2.85 | 2.77 | 2.73 | 2.70 | 2.66 | 2.64 | 2.62 | 2.58 | 2.54 |
| 3.72 | 3.62 | 3.52 | 3.42 | 3.35 | 3.31 | 3.26 | 3.22 | 3.20 | 3.14 | 3.09 |
| 4.85 | 4.71 | 4.56 | 4.41 | 4.31 | 4.25 | 4.17 | 4.12 | 4.08 | 4.00 | 3.92 |
| 8.75 | 8.45 | 8.13 | 7.80 | 7.60 | 7.47 | 7.30 | 7.19 | 7.12 | 6.94 | 6.78 |
| 2.25 | 2.21 | 2.17 | 2.12 | 2.10 | 2.08 | 2.05 | 2.04 | 2.03 | 2.00 | 1.98 |
| 2.85 | 2.79 | 2.72 | 2.65 | 2.60 | 2.57 | 2.53 | 2.51 | 2.49 | 2.45 | 2.41 |
| 3.53 | 3.43 | 3.33 | 3.23 | 3.16 | 3.12 | 3.06 | 3.03 | 3.00 | 2.94 | 2.89 |
| 4.54 | 4.40 | 4.25 | 4.10 | 4.01 | 3.94 | 3.86 | 3.81 | 3.78 | 3.69 | 3.61 |
| 7.92 | 7.63 | 7.32 | 7.01 | 6.81 | 6.68 | 6.52 | 6.42 | 6.35 | 6.18 | 6.02 |
| 2.19 | 2.15 | 2.10 | 2.06 | 2.03 | 2.01 | 1.99 | 1.97 | 1.96 | 1.93 | 1.91 |
| 2.75 | 2.69 | 2.62 | 2.54 | 2.50 | 2.47 | 2.43 | 2.40 | 2.38 | 2.34 | 2.30 |
| 3.37 | 3.28 | 3.18 | 3.07 | 3.01 | 2.96 | 2.91 | 2.87 | 2.85 | 2.79 | 2.73 |
| 4.30 | 4.16 | 4.01 | 3.86 | 3.76 | 3.70 | 3.62 | 3.57 | 3.54 | 3.45 | 3.37 |
| 7.29 | 7.00 | 6.71 | 6.40 | 6.22 | 6.09 | 5.93 | 5.83 | 5.76 | 5.59 | 5.44 |
| 2.14 | 2.10 | 2.05 | 2.01 | 1.98 | 1.96 | 1.93 | 1.92 | 1.90 | 1.88 | 1.85 |
| 2.67 | 2.60 | 2.53 | 2.46 | 2.41 | 2.38 | 2.34 | 2.31 | 2.30 | 2.25 | 2.21 |
| 3.25 | 3.15 | 3.05 | 2.95 | 2.88 | 2.84 | 2.78 | 2.74 | 2.72 | 2.66 | 2.60 |
| 4.10 | 3.96 | 3.82 | 3.66 | 3.57 | 3.51 | 3.43 | 3.38 | 3.34 | 3.25 | 3.18 |
| 6.80 | 6.52 | 6.23 | 5.93 | 5.75 | 5.63 | 5.47 | 5.37 | 5.30 | 5.14 | 4.99 |
| 2.10 | 2.05 | 2.01 | 1.96 | 1.93 | 1.91 | 1.89 | 1.87 | 1.86 | 1.83 | 1.80 |
| 2.60 | 2.53 | 2.46 | 2.39 | 2.34 | 2.31 | 2.27 | 2.24 | 2.22 | 2.18 | 2.14 |
| 3.15 | 3.05 | 2.95 | 2.84 | 2.78 | 2.73 | 2.67 | 2.64 | 2.61 | 2.55 | 2.50 |
| 3.94 | 3.80 | 3.66 | 3.51 | 3.41 | 3.35 | 3.27 | 3.22 | 3.18 | 3.09 | 3.02 |
| 6.40 | 6.13 | 5.85 | 5.56 | 5.38 | 5.25 | 5.10 | 5.00 | 4.94 | 4.77 | 4.62 |
| 2.06 | 2.02 | 1.97 | 1.92 | 1.89 | 1.87 | 1.85 | 1.83 | 1.82 | 1.79 | 1.76 |
| 2.54 | 2.48 | 2.40 | 2.33 | 2.28 | 2.25 | 2.20 | 2.18 | 2.16 | 2.11 | 2.07 |
| 3.06 | 2.96 | 2.86 | 2.76 | 2.69 | 2.64 | 2.59 | 2.55 | 2.52 | 2.46 | 2.40 |
| 3.80 | 3.67 | 3.52 | 3.37 | 3.28 | 3.21 | 3.13 | 3.08 | 3.05 | 2.96 | 2.88 |
| 6.08 | 5.81 | 5.54 | 5.25 | 5.07 | 4.95 | 4.80 | 4.70 | 4.64 | 4.47 | 4.33 |
| 2.03 | 1.99 | 1.94 | 1.89 | 1.86 | 1.84 | 1.81 | 1.79 | 1.78 | 1.75 | 1.72 |
| 2.49 | 2.42 | 2.35 | 2.28 | 2.23 | 2.19 | 2.15 | 2.12 | 2.11 | 2.06 | 2.02 |
| 2.99 | 2.89 | 2.79 | 2.68 | 2.61 | 2.57 | 2.51 | 2.47 | 2.45 | 2.38 | 2.32 |
| 3.69 | 3.55 | 3.41 | 3.26 | 3.16 | 3.10 | 3.02 | 2.97 | 2.93 | 2.84 | 2.76 |
| 5.81 | 5.55 | 5.27 | 4.99 | 4.82 | 4.70 | 4.54 | 4.45 | 4.39 | 4.23 | 4.08 |
| 2.00 | 1.96 | 1.91 | 1.86 | 1.83 | 1.81 | 1.78 | 1.76 | 1.75 | 1.72 | 1.69 |
| 2.45 | 2.38 | 2.31 | 2.23 | 2.18 | 2.15 | 2.10 | 2.08 | 2.06 | 2.01 | 1.97 |
| 2.92 | 2.82 | 2.72 | 2.62 | 2.55 | 2.50 | 2.44 | 2.41 | 2.38 | 2.32 | 2.26 |
| 3.59 | 3.46 | 3.31 | 3.16 | 3.07 | 3.00 | 2.92 | 2.87 | 2.83 | 2.75 | 2.66 |
| 5.58 | 5.32 | 5.05 | 4.78 | 4.60 | 4.48 | 4.33 | 4.24 | 4.18 | 4.02 | 3.87 |

*(Continued)*

## TABLE E

### *F* critical values (continued)

| Degrees of freedom in the denominator | *p* | 1 | 2 | 3 | 4 | 5 | 6 | 7 | 8 | 9 |
|---|---|---|---|---|---|---|---|---|---|---|
| | | Degrees of freedom in the numerator | | | | | | | | |
| 18 | .100 | 3.01 | 2.62 | 2.42 | 2.29 | 2.20 | 2.13 | 2.08 | 2.04 | 2.00 |
| | .050 | 4.41 | 3.55 | 3.16 | 2.93 | 2.77 | 2.66 | 2.58 | 2.51 | 2.46 |
| | .025 | 5.98 | 4.56 | 3.95 | 3.61 | 3.38 | 3.22 | 3.10 | 3.01 | 2.93 |
| | .010 | 8.29 | 6.01 | 5.09 | 4.58 | 4.25 | 4.01 | 3.84 | 3.71 | 3.60 |
| | .001 | 15.38 | 10.39 | 8.49 | 7.46 | 6.81 | 6.35 | 6.02 | 5.76 | 5.56 |
| 19 | .100 | 2.99 | 2.61 | 2.40 | 2.27 | 2.18 | 2.11 | 2.06 | 2.02 | 1.98 |
| | .050 | 4.38 | 3.52 | 3.13 | 2.90 | 2.74 | 2.63 | 2.54 | 2.48 | 2.42 |
| | .025 | 5.92 | 4.51 | 3.90 | 3.56 | 3.33 | 3.17 | 3.05 | 2.96 | 2.88 |
| | .010 | 8.18 | 5.93 | 5.01 | 4.50 | 4.17 | 3.94 | 3.77 | 3.63 | 3.52 |
| | .001 | 15.08 | 10.16 | 8.28 | 7.27 | 6.62 | 6.18 | 5.85 | 5.59 | 5.39 |
| 20 | .100 | 2.97 | 2.59 | 2.38 | 2.25 | 2.16 | 2.09 | 2.04 | 2.00 | 1.96 |
| | .050 | 4.35 | 3.49 | 3.10 | 2.87 | 2.71 | 2.60 | 2.51 | 2.45 | 2.39 |
| | .025 | 5.87 | 4.46 | 3.86 | 3.51 | 3.29 | 3.13 | 3.01 | 2.91 | 2.84 |
| | .010 | 8.10 | 5.85 | 4.94 | 4.43 | 4.10 | 3.87 | 3.70 | 3.56 | 3.46 |
| | .001 | 14.82 | 9.95 | 8.10 | 7.10 | 6.46 | 6.02 | 5.69 | 5.44 | 5.24 |
| 21 | .100 | 2.96 | 2.57 | 2.36 | 2.23 | 2.14 | 2.08 | 2.02 | 1.98 | 1.95 |
| | .050 | 4.32 | 3.47 | 3.07 | 2.84 | 2.68 | 2.57 | 2.49 | 2.42 | 2.37 |
| | .025 | 5.83 | 4.42 | 3.82 | 3.48 | 3.25 | 3.09 | 2.97 | 2.87 | 2.80 |
| | .010 | 8.02 | 5.78 | 4.87 | 4.37 | 4.04 | 3.81 | 3.64 | 3.51 | 3.40 |
| | .001 | 14.59 | 9.77 | 7.94 | 6.95 | 6.32 | 5.88 | 5.56 | 5.31 | 5.11 |
| 22 | .100 | 2.95 | 2.56 | 2.35 | 2.22 | 2.13 | 2.06 | 2.01 | 1.97 | 1.93 |
| | .050 | 4.30 | 3.44 | 3.05 | 2.82 | 2.66 | 2.55 | 2.46 | 2.40 | 2.34 |
| | .025 | 5.79 | 4.38 | 3.78 | 3.44 | 3.22 | 3.05 | 2.93 | 2.84 | 2.76 |
| | .010 | 7.95 | 5.72 | 4.82 | 4.31 | 3.99 | 3.76 | 3.59 | 3.45 | 3.35 |
| | .001 | 14.38 | 9.61 | 7.80 | 6.81 | 6.19 | 5.76 | 5.44 | 5.19 | 4.99 |
| 23 | .100 | 2.94 | 2.55 | 2.34 | 2.21 | 2.11 | 2.05 | 1.99 | 1.95 | 1.92 |
| | .050 | 4.28 | 3.42 | 3.03 | 2.80 | 2.64 | 2.53 | 2.44 | 2.37 | 2.32 |
| | .025 | 5.75 | 4.35 | 3.75 | 3.41 | 3.18 | 3.02 | 2.90 | 2.81 | 2.73 |
| | .010 | 7.88 | 5.66 | 4.76 | 4.26 | 3.94 | 3.71 | 3.54 | 3.41 | 3.30 |
| | .001 | 14.20 | 9.47 | 7.67 | 6.70 | 6.08 | 5.65 | 5.33 | 5.09 | 4.89 |
| 24 | .100 | 2.93 | 2.54 | 2.33 | 2.19 | 2.10 | 2.04 | 1.98 | 1.94 | 1.91 |
| | .050 | 4.26 | 3.40 | 3.01 | 2.78 | 2.62 | 2.51 | 2.42 | 2.36 | 2.30 |
| | .025 | 5.72 | 4.32 | 3.72 | 3.38 | 3.15 | 2.99 | 2.87 | 2.78 | 2.70 |
| | .010 | 7.82 | 5.61 | 4.72 | 4.22 | 3.90 | 3.67 | 3.50 | 3.36 | 3.26 |
| | .001 | 14.03 | 9.34 | 7.55 | 6.59 | 5.98 | 5.55 | 5.23 | 4.99 | 4.80 |
| 25 | .100 | 2.92 | 2.53 | 2.32 | 2.18 | 2.09 | 2.02 | 1.97 | 1.93 | 1.89 |
| | .050 | 4.24 | 3.39 | 2.99 | 2.76 | 2.60 | 2.49 | 2.40 | 2.34 | 2.28 |
| | .025 | 5.69 | 4.29 | 3.69 | 3.35 | 3.13 | 2.97 | 2.85 | 2.75 | 2.68 |
| | .010 | 7.77 | 5.57 | 4.68 | 4.18 | 3.85 | 3.63 | 3.46 | 3.32 | 3.22 |
| | .001 | 13.88 | 9.22 | 7.45 | 6.49 | 5.89 | 5.46 | 5.15 | 4.91 | 4.71 |
| 26 | .100 | 2.91 | 2.52 | 2.31 | 2.17 | 2.08 | 2.01 | 1.96 | 1.92 | 1.88 |
| | .050 | 4.23 | 3.37 | 2.98 | 2.74 | 2.59 | 2.47 | 2.39 | 2.32 | 2.27 |
| | .025 | 5.66 | 4.27 | 3.67 | 3.33 | 3.10 | 2.94 | 2.82 | 2.73 | 2.65 |
| | .010 | 7.72 | 5.53 | 4.64 | 4.14 | 3.82 | 3.59 | 3.42 | 3.29 | 3.18 |
| | .001 | 13.74 | 9.12 | 7.36 | 6.41 | 5.80 | 5.38 | 5.07 | 4.83 | 4.64 |
| 27 | .100 | 2.90 | 2.51 | 2.30 | 2.17 | 2.07 | 2.00 | 1.95 | 1.91 | 1.87 |
| | .050 | 4.21 | 3.35 | 2.96 | 2.73 | 2.57 | 2.46 | 2.37 | 2.31 | 2.25 |
| | .025 | 5.63 | 4.24 | 3.65 | 3.31 | 3.08 | 2.92 | 2.80 | 2.71 | 2.63 |
| | .010 | 7.68 | 5.49 | 4.60 | 4.11 | 3.78 | 3.56 | 3.39 | 3.26 | 3.15 |
| | .001 | 13.61 | 9.02 | 7.27 | 6.33 | 5.73 | 5.31 | 5.00 | 4.76 | 4.57 |

## TABLE E

*F* critical values (continued)

| Degrees of freedom in the numerator | | | | | | | | | | |
|---|---|---|---|---|---|---|---|---|---|---|
| 10 | 12 | 15 | 20 | 25 | 30 | 40 | 50 | 60 | 120 | 1000 |
| 1.98 | 1.93 | 1.89 | 1.84 | 1.80 | 1.78 | 1.75 | 1.74 | 1.72 | 1.69 | 1.66 |
| 2.41 | 2.34 | 2.27 | 2.19 | 2.14 | 2.11 | 2.06 | 2.04 | 2.02 | 1.97 | 1.92 |
| 2.87 | 2.77 | 2.67 | 2.56 | 2.49 | 2.44 | 2.38 | 2.35 | 2.32 | 2.26 | 2.20 |
| 3.51 | 3.37 | 3.23 | 3.08 | 2.98 | 2.92 | 2.84 | 2.78 | 2.75 | 2.66 | 2.58 |
| 5.39 | 5.13 | 4.87 | 4.59 | 4.42 | 4.30 | 4.15 | 4.06 | 4.00 | 3.84 | 3.69 |
| 1.96 | 1.91 | 1.86 | 1.81 | 1.78 | 1.76 | 1.73 | 1.71 | 1.70 | 1.67 | 1.64 |
| 2.38 | 2.31 | 2.23 | 2.16 | 2.11 | 2.07 | 2.03 | 2.00 | 1.98 | 1.93 | 1.88 |
| 2.82 | 2.72 | 2.62 | 2.51 | 2.44 | 2.39 | 2.33 | 2.30 | 2.27 | 2.20 | 2.14 |
| 3.43 | 3.30 | 3.15 | 3.00 | 2.91 | 2.84 | 2.76 | 2.71 | 2.67 | 2.58 | 2.50 |
| 5.22 | 4.97 | 4.70 | 4.43 | 4.26 | 4.14 | 3.99 | 3.90 | 3.84 | 3.68 | 3.53 |
| 1.94 | 1.89 | 1.84 | 1.79 | 1.76 | 1.74 | 1.71 | 1.69 | 1.68 | 1.64 | 1.61 |
| 2.35 | 2.28 | 2.20 | 2.12 | 2.07 | 2.04 | 1.99 | 1.97 | 1.95 | 1.90 | 1.85 |
| 2.77 | 2.68 | 2.57 | 2.46 | 2.40 | 2.35 | 2.29 | 2.25 | 2.22 | 2.16 | 2.09 |
| 3.37 | 3.23 | 3.09 | 2.94 | 2.84 | 2.78 | 2.69 | 2.64 | 2.61 | 2.52 | 2.43 |
| 5.08 | 4.82 | 4.56 | 4.29 | 4.12 | 4.00 | 3.86 | 3.77 | 3.70 | 3.54 | 3.40 |
| 1.92 | 1.87 | 1.83 | 1.78 | 1.74 | 1.72 | 1.69 | 1.67 | 1.66 | 1.62 | 1.59 |
| 2.32 | 2.25 | 2.18 | 2.10 | 2.05 | 2.01 | 1.96 | 1.94 | 1.92 | 1.87 | 1.82 |
| 2.73 | 2.64 | 2.53 | 2.42 | 2.36 | 2.31 | 2.25 | 2.21 | 2.18 | 2.11 | 2.05 |
| 3.31 | 3.17 | 3.03 | 2.88 | 2.79 | 2.72 | 2.64 | 2.58 | 2.55 | 2.46 | 2.37 |
| 4.95 | 4.70 | 4.44 | 4.17 | 4.00 | 3.88 | 3.74 | 3.64 | 3.58 | 3.42 | 3.28 |
| 1.90 | 1.86 | 1.81 | 1.76 | 1.73 | 1.70 | 1.67 | 1.65 | 1.64 | 1.60 | 1.57 |
| 2.30 | 2.23 | 2.15 | 2.07 | 2.02 | 1.98 | 1.94 | 1.91 | 1.89 | 1.84 | 1.79 |
| 2.70 | 2.60 | 2.50 | 2.39 | 2.32 | 2.27 | 2.21 | 2.17 | 2.14 | 2.08 | 2.01 |
| 3.26 | 3.12 | 2.98 | 2.83 | 2.73 | 2.67 | 2.58 | 2.53 | 2.50 | 2.40 | 2.32 |
| 4.83 | 4.58 | 4.33 | 4.06 | 3.89 | 3.78 | 3.63 | 3.54 | 3.48 | 3.32 | 3.17 |
| 1.89 | 1.84 | 1.80 | 1.74 | 1.71 | 1.69 | 1.66 | 1.64 | 1.62 | 1.59 | 1.55 |
| 2.27 | 2.20 | 2.13 | 2.05 | 2.00 | 1.96 | 1.91 | 1.88 | 1.86 | 1.81 | 1.76 |
| 2.67 | 2.57 | 2.47 | 2.36 | 2.29 | 2.24 | 2.18 | 2.14 | 2.11 | 2.04 | 1.98 |
| 3.21 | 3.07 | 2.93 | 2.78 | 2.69 | 2.62 | 2.54 | 2.48 | 2.45 | 2.35 | 2.27 |
| 4.73 | 4.48 | 4.23 | 3.96 | 3.79 | 3.68 | 3.53 | 3.44 | 3.38 | 3.22 | 3.08 |
| 1.88 | 1.83 | 1.78 | 1.73 | 1.70 | 1.67 | 1.64 | 1.62 | 1.61 | 1.57 | 1.54 |
| 2.25 | 2.18 | 2.11 | 2.03 | 1.97 | 1.94 | 1.89 | 1.86 | 1.84 | 1.79 | 1.74 |
| 2.64 | 2.54 | 2.44 | 2.33 | 2.26 | 2.21 | 2.15 | 2.11 | 2.08 | 2.01 | 1.94 |
| 3.17 | 3.03 | 2.89 | 2.74 | 2.64 | 2.58 | 2.49 | 2.44 | 2.40 | 2.31 | 2.22 |
| 4.64 | 4.39 | 4.14 | 3.87 | 3.71 | 3.59 | 3.45 | 3.36 | 3.29 | 3.14 | 2.99 |
| 1.87 | 1.82 | 1.77 | 1.72 | 1.68 | 1.66 | 1.63 | 1.61 | 1.59 | 1.56 | 1.52 |
| 2.24 | 2.16 | 2.09 | 2.01 | 1.96 | 1.92 | 1.87 | 1.84 | 1.82 | 1.77 | 1.72 |
| 2.61 | 2.51 | 2.41 | 2.30 | 2.23 | 2.18 | 2.12 | 2.08 | 2.05 | 1.98 | 1.91 |
| 3.13 | 2.99 | 2.85 | 2.70 | 2.60 | 2.54 | 2.45 | 2.40 | 2.36 | 2.27 | 2.18 |
| 4.56 | 4.31 | 4.06 | 3.79 | 3.63 | 3.52 | 3.37 | 3.28 | 3.22 | 3.06 | 2.91 |
| 1.86 | 1.81 | 1.76 | 1.71 | 1.67 | 1.65 | 1.61 | 1.59 | 1.58 | 1.54 | 1.51 |
| 2.22 | 2.15 | 2.07 | 1.99 | 1.94 | 1.90 | 1.85 | 1.82 | 1.80 | 1.75 | 1.70 |
| 2.59 | 2.49 | 2.39 | 2.28 | 2.21 | 2.16 | 2.09 | 2.05 | 2.03 | 1.95 | 1.89 |
| 3.09 | 2.96 | 2.81 | 2.66 | 2.57 | 2.50 | 2.42 | 2.36 | 2.33 | 2.23 | 2.14 |
| 4.48 | 4.24 | 3.99 | 3.72 | 3.56 | 3.44 | 3.30 | 3.21 | 3.15 | 2.99 | 2.84 |
| 1.85 | 1.80 | 1.75 | 1.70 | 1.66 | 1.64 | 1.60 | 1.58 | 1.57 | 1.53 | 1.50 |
| 2.20 | 2.13 | 2.06 | 1.97 | 1.92 | 1.88 | 1.84 | 1.81 | 1.79 | 1.73 | 1.68 |
| 2.57 | 2.47 | 2.36 | 2.25 | 2.18 | 2.13 | 2.07 | 2.03 | 2.00 | 1.93 | 1.86 |
| 3.06 | 2.93 | 2.78 | 2.63 | 2.54 | 2.47 | 2.38 | 2.33 | 2.29 | 2.20 | 2.11 |
| 4.41 | 4.17 | 3.92 | 3.66 | 3.49 | 3.38 | 3.23 | 3.14 | 3.08 | 2.92 | 2.78 |

*(Continued)*

## TABLE E

### *F* critical values (continued)

| | | Degrees of freedom in the numerator | | | | | | | | |
|---|---|---|---|---|---|---|---|---|---|---|
| Degrees of freedom in the denominator | $p$ | 1 | 2 | 3 | 4 | 5 | 6 | 7 | 8 | 9 |
| 28 | .100 | 2.89 | 2.50 | 2.29 | 2.16 | 2.06 | 2.00 | 1.94 | 1.90 | 1.87 |
| | .050 | 4.20 | 3.34 | 2.95 | 2.71 | 2.56 | 2.45 | 2.36 | 2.29 | 2.24 |
| | .025 | 5.61 | 4.22 | 3.63 | 3.29 | 3.06 | 2.90 | 2.78 | 2.69 | 2.61 |
| | .010 | 7.64 | 5.45 | 4.57 | 4.07 | 3.75 | 3.53 | 3.36 | 3.23 | 3.12 |
| | .001 | 13.50 | 8.93 | 7.19 | 6.25 | 5.66 | 5.24 | 4.93 | 4.69 | 4.50 |
| 29 | .100 | 2.89 | 2.50 | 2.28 | 2.15 | 2.06 | 1.99 | 1.93 | 1.89 | 1.86 |
| | .050 | 4.18 | 3.33 | 2.93 | 2.70 | 2.55 | 2.43 | 2.35 | 2.28 | 2.22 |
| | .025 | 5.59 | 4.20 | 3.61 | 3.27 | 3.04 | 2.88 | 2.76 | 2.67 | 2.59 |
| | .010 | 7.60 | 5.42 | 4.54 | 4.04 | 3.73 | 3.50 | 3.33 | 3.20 | 3.09 |
| | .001 | 13.39 | 8.85 | 7.12 | 6.19 | 5.59 | 5.18 | 4.87 | 4.64 | 4.45 |
| 30 | .100 | 2.88 | 2.49 | 2.28 | 2.14 | 2.05 | 1.98 | 1.93 | 1.88 | 1.85 |
| | .050 | 4.17 | 3.32 | 2.92 | 2.69 | 2.53 | 2.42 | 2.33 | 2.27 | 2.21 |
| | .025 | 5.57 | 4.18 | 3.59 | 3.25 | 3.03 | 2.87 | 2.75 | 2.65 | 2.57 |
| | .010 | 7.56 | 5.39 | 4.51 | 4.02 | 3.70 | 3.47 | 3.30 | 3.17 | 3.07 |
| | .001 | 13.29 | 8.77 | 7.05 | 6.12 | 5.53 | 5.12 | 4.82 | 4.58 | 4.39 |
| 40 | .100 | 2.84 | 2.44 | 2.23 | 2.09 | 2.00 | 1.93 | 1.87 | 1.83 | 1.79 |
| | .050 | 4.08 | 3.23 | 2.84 | 2.61 | 2.45 | 2.34 | 2.25 | 2.18 | 2.12 |
| | .025 | 5.42 | 4.05 | 3.46 | 3.13 | 2.90 | 2.74 | 2.62 | 2.53 | 2.45 |
| | .010 | 7.31 | 5.18 | 4.31 | 3.83 | 3.51 | 3.29 | 3.12 | 2.99 | 2.89 |
| | .001 | 12.61 | 8.25 | 6.59 | 5.70 | 5.13 | 4.73 | 4.44 | 4.21 | 4.02 |
| 50 | .100 | 2.81 | 2.41 | 2.20 | 2.06 | 1.97 | 1.90 | 1.84 | 1.80 | 1.76 |
| | .050 | 4.03 | 3.18 | 2.79 | 2.56 | 2.40 | 2.29 | 2.20 | 2.13 | 2.07 |
| | .025 | 5.34 | 3.97 | 3.39 | 3.05 | 2.83 | 2.67 | 2.55 | 2.46 | 2.38 |
| | .010 | 7.17 | 5.06 | 4.20 | 3.72 | 3.41 | 3.19 | 3.02 | 2.89 | 2.78 |
| | .001 | 12.22 | 7.96 | 6.34 | 5.46 | 4.90 | 4.51 | 4.22 | 4.00 | 3.82 |
| 60 | .100 | 2.79 | 2.39 | 2.18 | 2.04 | 1.95 | 1.87 | 1.82 | 1.77 | 1.74 |
| | .050 | 4.00 | 3.15 | 2.76 | 2.53 | 2.37 | 2.25 | 2.17 | 2.10 | 2.04 |
| | .025 | 5.29 | 3.93 | 3.34 | 3.01 | 2.79 | 2.63 | 2.51 | 2.41 | 2.33 |
| | .010 | 7.08 | 4.98 | 4.13 | 3.65 | 3.34 | 3.12 | 2.95 | 2.82 | 2.72 |
| | .001 | 11.97 | 7.77 | 6.17 | 5.31 | 4.76 | 4.37 | 4.09 | 3.86 | 3.69 |
| 100 | .100 | 2.76 | 2.36 | 2.14 | 2.00 | 1.91 | 1.83 | 1.78 | 1.73 | 1.69 |
| | .050 | 3.94 | 3.09 | 2.70 | 2.46 | 2.31 | 2.19 | 2.10 | 2.03 | 1.97 |
| | .025 | 5.18 | 3.83 | 3.25 | 2.92 | 2.70 | 2.54 | 2.42 | 2.32 | 2.24 |
| | .010 | 6.90 | 4.82 | 3.98 | 3.51 | 3.21 | 2.99 | 2.82 | 2.69 | 2.59 |
| | .001 | 11.50 | 7.41 | 5.86 | 5.02 | 4.48 | 4.11 | 3.83 | 3.61 | 3.44 |
| 200 | .100 | 2.73 | 2.33 | 2.11 | 1.97 | 1.88 | 1.80 | 1.75 | 1.70 | 1.66 |
| | .050 | 3.89 | 3.04 | 2.65 | 2.42 | 2.26 | 2.14 | 2.06 | 1.98 | 1.93 |
| | .025 | 5.10 | 3.76 | 3.18 | 2.85 | 2.63 | 2.47 | 2.35 | 2.26 | 2.18 |
| | .010 | 6.76 | 4.71 | 3.88 | 3.41 | 3.11 | 2.89 | 2.73 | 2.60 | 2.50 |
| | .001 | 11.15 | 7.15 | 5.63 | 4.81 | 4.29 | 3.92 | 3.65 | 3.43 | 3.26 |
| 1000 | .100 | 2.71 | 2.31 | 2.09 | 1.95 | 1.85 | 1.78 | 1.72 | 1.68 | 1.64 |
| | .050 | 3.85 | 3.00 | 2.61 | 2.38 | 2.22 | 2.11 | 2.02 | 1.95 | 1.89 |
| | .025 | 5.04 | 3.70 | 3.13 | 2.80 | 2.58 | 2.42 | 2.30 | 2.20 | 2.13 |
| | .010 | 6.66 | 4.63 | 3.80 | 3.34 | 3.04 | 2.82 | 2.66 | 2.53 | 2.43 |
| | .001 | 10.89 | 6.96 | 5.46 | 4.65 | 4.14 | 3.78 | 3.51 | 3.30 | 3.13 |

## TABLE E

### *F* critical values (continued)

Degrees of freedom in the numerator

| 10 | 12 | 15 | 20 | 25 | 30 | 40 | 50 | 60 | 120 | 1000 |
|---|---|---|---|---|---|---|---|---|---|---|
| 1.84 | 1.79 | 1.74 | 1.69 | 1.65 | 1.63 | 1.59 | 1.57 | 1.56 | 1.52 | 1.48 |
| 2.19 | 2.12 | 2.04 | 1.96 | 1.91 | 1.87 | 1.82 | 1.79 | 1.77 | 1.71 | 1.66 |
| 2.55 | 2.45 | 2.34 | 2.23 | 2.16 | 2.11 | 2.05 | 2.01 | 1.98 | 1.91 | 1.84 |
| 3.03 | 2.90 | 2.75 | 2.60 | 2.51 | 2.44 | 2.35 | 2.30 | 2.26 | 2.17 | 2.08 |
| 4.35 | 4.11 | 3.86 | 3.60 | 3.43 | 3.32 | 3.18 | 3.09 | 3.02 | 2.86 | 2.72 |
| 1.83 | 1.78 | 1.73 | 1.68 | 1.64 | 1.62 | 1.58 | 1.56 | 1.55 | 1.51 | 1.47 |
| 2.18 | 2.10 | 2.03 | 1.94 | 1.89 | 1.85 | 1.81 | 1.77 | 1.75 | 1.70 | 1.65 |
| 2.53 | 2.43 | 2.32 | 2.21 | 2.14 | 2.09 | 2.03 | 1.99 | 1.96 | 1.89 | 1.82 |
| 3.00 | 2.87 | 2.73 | 2.57 | 2.48 | 2.41 | 2.33 | 2.27 | 2.23 | 2.14 | 2.05 |
| 4.29 | 4.05 | 3.80 | 3.54 | 3.38 | 3.27 | 3.12 | 3.03 | 2.97 | 2.81 | 2.66 |
| 1.82 | 1.77 | 1.72 | 1.67 | 1.63 | 1.61 | 1.57 | 1.55 | 1.54 | 1.50 | 1.46 |
| 2.16 | 2.09 | 2.01 | 1.93 | 1.88 | 1.84 | 1.79 | 1.76 | 1.74 | 1.68 | 1.63 |
| 2.51 | 2.41 | 2.31 | 2.20 | 2.12 | 2.07 | 2.01 | 1.97 | 1.94 | 1.87 | 1.80 |
| 2.98 | 2.84 | 2.70 | 2.55 | 2.45 | 2.39 | 2.30 | 2.25 | 2.21 | 2.11 | 2.02 |
| 4.24 | 4.00 | 3.75 | 3.49 | 3.33 | 3.22 | 3.07 | 2.98 | 2.92 | 2.76 | 2.61 |
| 1.76 | 1.71 | 1.66 | 1.61 | 1.57 | 1.54 | 1.51 | 1.48 | 1.47 | 1.42 | 1.38 |
| 2.08 | 2.00 | 1.92 | 1.84 | 1.78 | 1.74 | 1.69 | 1.66 | 1.64 | 1.58 | 1.52 |
| 2.39 | 2.29 | 2.18 | 2.07 | 1.99 | 1.94 | 1.88 | 1.83 | 1.80 | 1.72 | 1.65 |
| 2.80 | 2.66 | 2.52 | 2.37 | 2.27 | 2.20 | 2.11 | 2.06 | 2.02 | 1.92 | 1.82 |
| 3.87 | 3.64 | 3.40 | 3.14 | 2.98 | 2.87 | 2.73 | 2.64 | 2.57 | 2.41 | 2.25 |
| 1.73 | 1.68 | 1.63 | 1.57 | 1.53 | 1.50 | 1.46 | 1.44 | 1.42 | 1.38 | 1.33 |
| 2.03 | 1.95 | 1.87 | 1.78 | 1.73 | 1.69 | 1.63 | 1.60 | 1.58 | 1.51 | 1.45 |
| 2.32 | 2.22 | 2.11 | 1.99 | 1.92 | 1.87 | 1.80 | 1.75 | 1.72 | 1.64 | 1.56 |
| 2.70 | 2.56 | 2.42 | 2.27 | 2.17 | 2.10 | 2.01 | 1.95 | 1.91 | 1.80 | 1.70 |
| 3.67 | 3.44 | 3.20 | 2.95 | 2.79 | 2.68 | 2.53 | 2.44 | 2.38 | 2.21 | 2.05 |
| 1.71 | 1.66 | 1.60 | 1.54 | 1.50 | 1.48 | 1.44 | 1.41 | 1.40 | 1.35 | 1.30 |
| 1.99 | 1.92 | 1.84 | 1.75 | 1.69 | 1.65 | 1.59 | 1.56 | 1.53 | 1.47 | 1.40 |
| 2.27 | 2.17 | 2.06 | 1.94 | 1.87 | 1.82 | 1.74 | 1.70 | 1.67 | 1.58 | 1.49 |
| 2.63 | 2.50 | 2.35 | 2.20 | 2.10 | 2.03 | 1.94 | 1.88 | 1.84 | 1.73 | 1.62 |
| 3.54 | 3.32 | 3.08 | 2.83 | 2.67 | 2.55 | 2.41 | 2.32 | 2.25 | 2.08 | 1.92 |
| 1.66 | 1.61 | 1.56 | 1.49 | 1.45 | 1.42 | 1.38 | 1.35 | 1.34 | 1.28 | 1.22 |
| 1.93 | 1.85 | 1.77 | 1.68 | 1.62 | 1.57 | 1.52 | 1.48 | 1.45 | 1.38 | 1.30 |
| 2.18 | 2.08 | 1.97 | 1.85 | 1.77 | 1.71 | 1.64 | 1.59 | 1.56 | 1.46 | 1.36 |
| 2.50 | 2.37 | 2.22 | 2.07 | 1.97 | 1.89 | 1.80 | 1.74 | 1.69 | 1.57 | 1.45 |
| 3.30 | 3.07 | 2.84 | 2.59 | 2.43 | 2.32 | 2.17 | 2.08 | 2.01 | 1.83 | 1.64 |
| 1.63 | 1.58 | 1.52 | 1.46 | 1.41 | 1.38 | 1.34 | 1.31 | 1.29 | 1.23 | 1.16 |
| 1.88 | 1.80 | 1.72 | 1.62 | 1.56 | 1.52 | 1.46 | 1.41 | 1.39 | 1.30 | 1.21 |
| 2.11 | 2.01 | 1.90 | 1.78 | 1.70 | 1.64 | 1.56 | 1.51 | 1.47 | 1.37 | 1.25 |
| 2.41 | 2.27 | 2.13 | 1.97 | 1.87 | 1.79 | 1.69 | 1.63 | 1.58 | 1.45 | 1.30 |
| 3.12 | 2.90 | 2.67 | 2.42 | 2.26 | 2.15 | 2.00 | 1.90 | 1.83 | 1.64 | 1.43 |
| 1.61 | 1.55 | 1.49 | 1.43 | 1.38 | 1.35 | 1.30 | 1.27 | 1.25 | 1.18 | 1.08 |
| 1.84 | 1.76 | 1.68 | 1.58 | 1.52 | 1.47 | 1.41 | 1.36 | 1.33 | 1.24 | 1.11 |
| 2.06 | 1.96 | 1.85 | 1.72 | 1.64 | 1.58 | 1.50 | 1.45 | 1.41 | 1.29 | 1.13 |
| 2.34 | 2.20 | 2.06 | 1.90 | 1.79 | 1.72 | 1.61 | 1.54 | 1.50 | 1.35 | 1.16 |
| 2.99 | 2.77 | 2.54 | 2.30 | 2.14 | 2.02 | 1.87 | 1.77 | 1.69 | 1.49 | 1.22 |

Table entry for $p$ is the critical value $(\chi^2)^*$ with probability $p$ lying to its right.

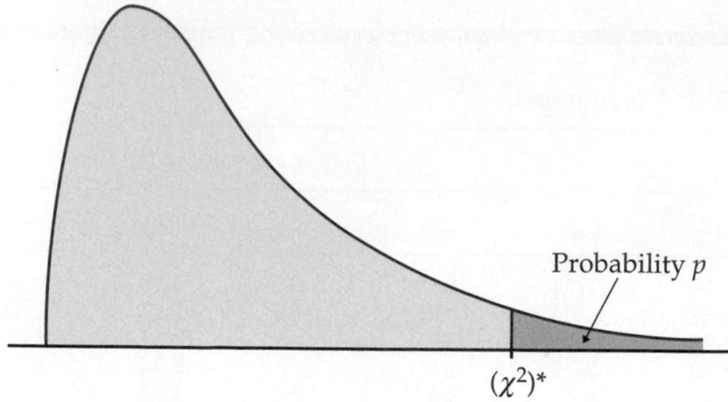


## TABLE F

### $\chi^2$ distribution critical values

| | Tail probability $p$ | | | | | | | | | | | |
|---|---|---|---|---|---|---|---|---|---|---|---|---|
| df | .25 | .20 | .15 | .10 | .05 | .025 | .02 | .01 | .005 | .0025 | .001 | .0005 |
| 1 | 1.32 | 1.64 | 2.07 | 2.71 | 3.84 | 5.02 | 5.41 | 6.63 | 7.88 | 9.14 | 10.83 | 12.12 |
| 2 | 2.77 | 3.22 | 3.79 | 4.61 | 5.99 | 7.38 | 7.82 | 9.21 | 10.60 | 11.98 | 13.82 | 15.20 |
| 3 | 4.11 | 4.64 | 5.32 | 6.25 | 7.81 | 9.35 | 9.84 | 11.34 | 12.84 | 14.32 | 16.27 | 17.73 |
| 4 | 5.39 | 5.99 | 6.74 | 7.78 | 9.49 | 11.14 | 11.67 | 13.28 | 14.86 | 16.42 | 18.47 | 20.00 |
| 5 | 6.63 | 7.29 | 8.12 | 9.24 | 11.07 | 12.83 | 13.39 | 15.09 | 16.75 | 18.39 | 20.51 | 22.11 |
| 6 | 7.84 | 8.56 | 9.45 | 10.64 | 12.59 | 14.45 | 15.03 | 16.81 | 18.55 | 20.25 | 22.46 | 24.10 |
| 7 | 9.04 | 9.80 | 10.75 | 12.02 | 14.07 | 16.01 | 16.62 | 18.48 | 20.28 | 22.04 | 24.32 | 26.02 |
| 8 | 10.22 | 11.03 | 12.03 | 13.36 | 15.51 | 17.53 | 18.17 | 20.09 | 21.95 | 23.77 | 26.12 | 27.87 |
| 9 | 11.39 | 12.24 | 13.29 | 14.68 | 16.92 | 19.02 | 19.68 | 21.67 | 23.59 | 25.46 | 27.88 | 29.67 |
| 10 | 12.55 | 13.44 | 14.53 | 15.99 | 18.31 | 20.48 | 21.16 | 23.21 | 25.19 | 27.11 | 29.59 | 31.42 |
| 11 | 13.70 | 14.63 | 15.77 | 17.28 | 19.68 | 21.92 | 22.62 | 24.72 | 26.76 | 28.73 | 31.26 | 33.14 |
| 12 | 14.85 | 15.81 | 16.99 | 18.55 | 21.03 | 23.34 | 24.05 | 26.22 | 28.30 | 30.32 | 32.91 | 34.82 |
| 13 | 15.98 | 16.98 | 18.20 | 19.81 | 22.36 | 24.74 | 25.47 | 27.69 | 29.82 | 31.88 | 34.53 | 36.48 |
| 14 | 17.12 | 18.15 | 19.41 | 21.06 | 23.68 | 26.12 | 26.87 | 29.14 | 31.32 | 33.43 | 36.12 | 38.11 |
| 15 | 18.25 | 19.31 | 20.60 | 22.31 | 25.00 | 27.49 | 28.26 | 30.58 | 32.80 | 34.95 | 37.70 | 39.72 |
| 16 | 19.37 | 20.47 | 21.79 | 23.54 | 26.30 | 28.85 | 29.63 | 32.00 | 34.27 | 36.46 | 39.25 | 41.31 |
| 17 | 20.49 | 21.61 | 22.98 | 24.77 | 27.59 | 30.19 | 31.00 | 33.41 | 35.72 | 37.95 | 40.79 | 42.88 |
| 18 | 21.60 | 22.76 | 24.16 | 25.99 | 28.87 | 31.53 | 32.35 | 34.81 | 37.16 | 39.42 | 42.31 | 44.43 |
| 19 | 22.72 | 23.90 | 25.33 | 27.20 | 30.14 | 32.85 | 33.69 | 36.19 | 38.58 | 40.88 | 43.82 | 45.97 |
| 20 | 23.83 | 25.04 | 26.50 | 28.41 | 31.41 | 34.17 | 35.02 | 37.57 | 40.00 | 42.34 | 45.31 | 47.50 |
| 21 | 24.93 | 26.17 | 27.66 | 29.62 | 32.67 | 35.48 | 36.34 | 38.93 | 41.40 | 43.78 | 46.80 | 49.01 |
| 22 | 26.04 | 27.30 | 28.82 | 30.81 | 33.92 | 36.78 | 37.66 | 40.29 | 42.80 | 45.20 | 48.27 | 50.51 |
| 23 | 27.14 | 28.43 | 29.98 | 32.01 | 35.17 | 38.08 | 38.97 | 41.64 | 44.18 | 46.62 | 49.73 | 52.00 |
| 24 | 28.24 | 29.55 | 31.13 | 33.20 | 36.42 | 39.36 | 40.27 | 42.98 | 45.56 | 48.03 | 51.18 | 53.48 |
| 25 | 29.34 | 30.68 | 32.28 | 34.38 | 37.65 | 40.65 | 41.57 | 44.31 | 46.93 | 49.44 | 52.62 | 54.95 |
| 26 | 30.43 | 31.79 | 33.43 | 35.56 | 38.89 | 41.92 | 42.86 | 45.64 | 48.29 | 50.83 | 54.05 | 56.41 |
| 27 | 31.53 | 32.91 | 34.57 | 36.74 | 40.11 | 43.19 | 44.14 | 46.96 | 49.64 | 52.22 | 55.48 | 57.86 |
| 28 | 32.62 | 34.03 | 35.71 | 37.92 | 41.34 | 44.46 | 45.42 | 48.28 | 50.99 | 53.59 | 56.89 | 59.30 |
| 29 | 33.71 | 35.14 | 36.85 | 39.09 | 42.56 | 45.72 | 46.69 | 49.59 | 52.34 | 54.97 | 58.30 | 60.73 |
| 30 | 34.80 | 36.25 | 37.99 | 40.26 | 43.77 | 46.98 | 47.96 | 50.89 | 53.67 | 56.33 | 59.70 | 62.16 |
| 40 | 45.62 | 47.27 | 49.24 | 51.81 | 55.76 | 59.34 | 60.44 | 63.69 | 66.77 | 69.70 | 73.40 | 76.09 |
| 50 | 56.33 | 58.16 | 60.35 | 63.17 | 67.50 | 71.42 | 72.61 | 76.15 | 79.49 | 82.66 | 86.66 | 89.56 |
| 60 | 66.98 | 68.97 | 71.34 | 74.40 | 79.08 | 83.30 | 84.58 | 88.38 | 91.95 | 95.34 | 99.61 | 102.7 |
| 80 | 88.13 | 90.41 | 93.11 | 96.58 | 101.9 | 106.6 | 108.1 | 112.3 | 116.3 | 120.1 | 124.8 | 128.3 |
| 100 | 109.1 | 111.7 | 114.7 | 118.5 | 124.3 | 129.6 | 131.1 | 135.8 | 140.2 | 144.3 | 149.4 | 153.2 |

# Answers to Odd-Numbered Exercises

## CHAPTER 1

**1.1** Exam1 = 95, Exam2 = 98, Final = 96.

**1.3** Cases: apartments. Five variables: rent (quantitative), cable (categorical), pets (categorical), bedrooms (quantitative), distance to campus (quantitative).

**1.5** Scores are slightly left-skewed; most range from 70 to the low 90s.

**1.7** The larger classes hide a lot of detail.

**1.9** A stemplot or histogram can be used; the distribution is left-skewed, centered near 80, and spread from 55 to 98.

**1.13** For example, heart rate before and after exercise, number of sit-ups, time to run 100 m.

**1.15** For example, blue is by far the most popular choice; 70% of respondents chose 3 of the 10 options.

**1.17** **(a)** 232 total respondents; 4.31%, 41.81%, 30.17%, 15.52%, 6.03%, 2.16%. **(c)** For example, 87.5% of the group were between 19 and 50. **(d)** The age-group classes do not have equal width.

**1.21** **(a)** Alaska, 5.7%; Florida, 17.6%. **(b)** Symmetric, centered near 13%, spread from 8.5% to 15.6%.

**1.23** 359 mg/dl is an outlier; only four are in the desired range.

**1.25** Roughly symmetric, centered near 7, spread from 2 to 13.

**1.27** There are three peaks; presumably, the lowest group includes public institutions; the highest group, exclusive private schools like Harvard; and the middle group, other private schools.

**1.29** **(a)** The default histogram has 25 intervals (but the applet may have been revised after these answers were written). The nine-class histogram might not be an exact match. **(b)** Click-and-drag the mouse to the far right to see the maximum number of classes. **(c)** Opinions of which histogram is "best" will vary.

**1.31** Both plots show random fluctuation. Pasadena temperatures show an upward trend. Redding temperatures are initially similar to Pasadena's but dropped in the mid-1980s.

**1.35** Right-skewed, centered near 5 or 6, spread from 0 to 18, no outliers.

**1.37** Top-left histogram, 4; top right, 2; bottom left, 1; bottom right, 3.

**1.39** Use a stemplot or a histogram. Right-skewed, center near 30 or 40 thousand barrels, with two or three high outliers; apart from these, the numbers are spread from 2 to 118.2 thousand barrels.

**1.41** **(a)** Most people will "round" their answers when asked to give an estimate like this, and some may exaggerate. **(b)** The stemplots and midpoints (175 for women, 120 for men) suggest that women (claim to) study more than men.

**1.43** **(a)** Four variables: GPA, IQ, and self-concept are quantitative; gender is categorical. **(c)** Skewed to the left, centered near 7.8, spread from 0.5 to 10.8. **(d)** There is more variability among the boys; in fact, there seem to be two groups of boys: those with GPAs below 5 and those with GPAs above 5.

**1.45** Skewed to the left, centered near 59.5; most scores are between 35 and 73, with a few below that and one high score of 80 (probably not quite an outlier).

**1.47** $\bar{x} = 82.1$.

**1.49** $Q_1 = 75$, $Q_3 = 92$.

**1.51** Use the five-number summary from Exercise 1.50 (55, 75, 82.5, 92, 98).

**1.53** $s^2 \doteq 157.43$ and $s \doteq 12.55$.

**1.55** 950/4 = 237.5 points.

**1.57** **(a)** Min = $Q_1$ = 0, $M$ = 5.085, $Q_3$ = 9.47, Max = 73.2. **(d)** The distribution is sharply right-skewed. The histogram seems to convey the distribution better.

**1.59** Min = 0.24, $Q_1$ = 0.355, $M$ = 0.76, $Q_3$ = 1.03, Max = 1.9. The distribution is right-skewed. A histogram or stemplot reveals an important feature not evident from a boxplot: this distribution has two peaks.

**1.61** Five-number summary: $4123, $15,717, $20,072, $27,957.5, $29,875. This and the boxplot do not reveal the three groups of schools visible in the histogram.

**1.63** **(a)** Five-number summary (1999 dollars): 0, 2.14, 10.64, 40.96, 88.6. The large gaps between the higher numbers show the skew. **(b)** The *IQR* is 38.82; no observations fall below −56.09 or above 99.19. **(c)** The mean is 21.95 (1999 dollars); the right-skew makes it much larger than the median.

**1.65** This distribution would almost surely be strongly skewed to the right.

**1.67** $\bar{x}$ = $62,500; seven of the eight employees earned less than the mean. $M$ = $25,000.

**1.69** The mean rises to $87,500, while the median is unchanged.

**1.71** Means are not the appropriate measure of center for skewed distributions.

**1.73** **(a)** pH: $\bar{x} \doteq 5.4256$ and $s \doteq 0.5379$. Density: $\bar{x} \doteq 5.4479$ and $s \doteq 0.2209$. **(b)** Medians: 5.44 and 5.46.

**1.75** The mean and median always agree for two observations.

**1.77 (a)** Place the new point at the current median.

**1.79 (a)** *Bihai:* $\bar{x} \doteq 47.5975$, $s \doteq 1.2129$. Red: $\bar{x} \doteq 39.7113$, $s \doteq 1.7988$. Yellow: $\bar{x} \doteq 36.1800$, $s \doteq 0.9753$ (all in mm). **(b)** *Bihai* and red appear to be right-skewed (although it is difficult to tell with such small samples).

**1.81** All four distributions are symmetric with no outliers. The comparison should include at least the four means: 147, 130, 116, and 103 bpm. The average heart rate for runners is about 30 bpm less than the average sedentary rate.

**1.83** Positions 748 and 14,211; about \$13,000 and \$137,000.

**1.85** The five-number summary is 1, 3, 4, 5, and 12 letters.

**1.87** Take six or more numbers, with the largest number much larger than $Q_3$.

**1.89 (a)** Any set of four identical numbers works. **(b)** 0, 0, 20, 20 is the only possible answer.

**1.91** Five-number summary: 43, 82.5, 102.5, 151.5, and 598 days. Typical guinea pig lifetimes were 82.5 to 151.5 days, but some live quite a bit longer.

**1.93** Multiply each value by 0.03937.

**1.95** 80, 99, 114, and 178 days.

**1.97** Full data set: $\bar{x} = 141.8$ and $M = 102.5$ days. 10% trimmed mean: $\bar{x}^* = 118.16$ days. 20% trimmed mean: $\bar{x}^{**} = 111.68$ days.

**1.99** 470 to 674.

**1.101** $z = 0.55$.

**1.103** Using Table A, the proportion below 600 is 0.7088, and the proportion at or above is 0.2912.

**1.105** About 655.9.

**1.109 (a)** 1/4. **(b)** 0.25. **(c)** 0.5.

**1.111 (a)** Mean C, median B. **(b)** Mean A, median A. **(c)** Mean A, median B.

**1.113 (a)** 0.6826; the 68–95–99.7 rule gives 0.68. **(b)** 0.9544 (compared to 0.95); 0.9974 (compared to 0.997).

**1.115 (a)** 327 to 345 days. **(b)** 16%.

**1.117** Women, $z = 2.96$; men, $z = 0.96$.

**1.119** $\bar{x} = 5.4256$ and $s = 0.5379$; 67.62% within $\bar{x} - s$ and $\bar{x} + s$, 95.24% within $\bar{x} \pm 2s$, and all within $\bar{x} \pm 3s$.

**1.121 (a)** 0.0287. **(b)** 0.9713. **(c)** 0.0606. **(d)** 0.9107.

**1.123 (a)** 1.0364. **(b)** 0.2533.

**1.125** About 2.5%.

**1.127** Jacob's score ($z = -0.7917$) is higher than Emily's ($z = -1.6555$).

**1.129** About 1383.

**1.131** 21st percentile.

**1.133** 850 and below.

**1.135** 850, 973, 1079, and 1202.

**1.137** Above 240 mg/dl: about 31%. Between 200 and 240 mg/dl: about 41%.

**1.139 (a)** About 5.2%. **(b)** About 54.7%. **(c)** More than 279 days.

**1.141 (a)** About 1.35. **(b)** 1.35.

**1.143** The plot is nearly linear. Because heart rate is measured in whole numbers, there is a slight "step" appearance to the graph.

**1.145** The plot is reasonably close to a line, apart from the stair-step appearance, presumably due to limited accuracy of the measuring instrument.

**1.147** The plot suggests no major deviations from Normality, although the three lowest measurements don't quite fall in line with the other points.

**1.149** The first plot is nearly linear; the other two each show a low value.

**1.151** Histograms will suggest (but not exactly match) Figure 1.35. The uniform distribution does not extend as low or as high as a Normal distribution.

**1.153** The given description is true on the average, but the curves (and a few calculations) give a more complete picture. For example, a score of about 675 is about the 97.5th percentile for both genders, so the top boys and girls have very similar scores.

**1.155** Slightly right-skewed, with one (or more) high outliers. Five-number summary: 22, 23.735, 24.31, 24.845, and 28.55 hours.

**1.157** Many (but fewer than half) of the students were 19.

**1.159** Women's weights are right-skewed.

**1.161 (a)** About 20% of low-income and 33% of high-income households. **(b)** The majority of low-income households, but only about 7% of high-income households, consist of one person. One-person households often have less income because the people are young and have no job or have only recently started working.

**1.163** No to both questions; no summary can exactly describe a distribution that can include any number of values.

**1.165 (a)** The number of home runs by the major league leader each year declines sharply and steadily. **(b)** Ruth led for a longer time and appears to have set a new standard for other players.

**1.167 (a)** $x_{\text{new}} = -50 + 2x$. **(b)** $x_{\text{new}} = -49.0909 + 1.8182x$. **(c)** David (106) is higher than Nancy (92.7). **(d)** About 62% of third-graders and 36% of sixth-graders.

**1.169 (a)** The plot would be linear in the middle. The outliers would show up as a point in the lower left below the line and a point in the upper right above the line.

**1.173** Men seem to have higher SATM scores than women; women generally have higher GPAs. All four distributions are close to Normal; female SATM and both GPA sets have low outliers.

## CHAPTER 2

**2.1** Students.

**2.3** Cases: cups of Mocha Frappuccino. Variables: size and price (both quantitative).

**2.5** **(a)** "Month" (the passage of time) explains changes in temperature (not vice versa). **(b)** Temperature increases linearly with time (about 10 degrees per month); the relationship is strong.

**2.7** **(a)** The second test happens before the final exam. **(b)** The plot shows a weak positive association. **(c)** Students' study habits are more established by the middle of the term.

**2.9** **(a)** Explanatory: age. Response: weight. **(b)** Explore the relationship. **(c)** Explanatory: number of bedrooms. Response: price. **(d)** Explanatory: amount of sugar. Response: sweetness. **(e)** Explore the relationship.

**2.11** **(a)** In general, we expect more intelligent children to be better readers, and less intelligent children to be weaker readers. The plot does show this positive association. **(b)** These four have moderate IQs but poor reading scores. **(c)** Roughly linear but weak (much scatter).

**2.13** **(a)** The response variable (estimated level) can take only the values 1, 2, 3, 4, 5. **(b)** The association is (weakly) positive. **(c)** The estimate is 4, which is an overestimate; that child had the lowest score on the test.

**2.15** **(a)** Areas with many breeding pairs would correspondingly have more males that might potentially return. **(c)** The theory suggests a negative association; the scatterplot shows this.

**2.17** A fairly strong, positive, linear association; social exclusion does appear to trigger a pain response.

**2.19** **(a)** The Lakers and the Knicks are high in both variables (but fit the pattern). The Grizzlies, Cavaliers, and Rockets have slightly higher value than their revenues would suggest. The association is positive and linear. **(b)** The Lakers and Knicks still stand out, as do the Bulls and Trailblazers, but the association is quite weak. Revenue is a better predictor of value.

**2.21** **(b)** The association is linear and positive, and is stronger for women. Males typically have larger values for both variables.

**2.23** The plot shows a fairly steady rate of improvement until the mid-1980s, with much slower progress after that.

**2.25** **(a)** Both show fairly steady improvement. Women have made more rapid progress, but their progress seems to have slowed, while men's records may be dropping more rapidly in recent years. **(b)** The data support the first claim but do not seem to support the second.

**2.27** **(b)** Technology. **(c)** No; positive/negative only make sense when both variables are quantitative.

**2.29** **(a)** Price is explanatory; the plot shows a positive linear association. **(b)** $r = 0.955$.

**2.31** **(a)** $r = 0.5194$. **(b)** This correlation is much larger (farther from 0) than the first.

**2.33** The correlation increases.

**2.35** Closest to 0.6.

**2.37** No; units do not affect correlation.

**2.39** **(a)** Value and revenue: $r_1 = 0.9265$. Value and income: $r_2 = 0.2107$. This agrees with conclusions from the scatterplots. **(b)** Without Portland, $r_2 = 0.3469$. The removal of this point makes the scatterplot appear (slightly) more linear.

**2.41** **(a)** Positive, but not close to 1. **(b)** $r = 0.5653$. **(c)** $r$ would not change; it does not tell us that the men were generally taller. **(d)** $r$ would not change. **(e)** 1.

**2.43** **(a)** $r = \pm 1$ for a line. **(c)** Leave some space above your vertical stack. **(d)** The curve must be higher at the right than at the left.

**2.45** **(a)** The Insight seems to fit the line. **(b)** Without: 0.9757. With: 0.9934. The Insight increases the strength of the association.

**2.47** $r = 1$ for a positively sloped line.

**2.49** There is little linear association between research and teaching—for example, knowing a professor is a good researcher gives little information about whether she is a good or bad teacher.

**2.51** Both relationships are somewhat linear; GPA/IQ ($r = 0.6337$) is stronger than GPA/self-concept ($r = 0.5418$). The two students with the lowest GPAs stand out in both plots; a few others stand out in at least one plot. Generally speaking, removing these points raises $r$, except for the lower-left point in the self-concept plot.

**2.53** This line lies almost entirely above the points in the scatterplot.

**2.55** The first and last predictions would not be trustworthy.

**2.57** **(a)** $y = 35$. **(b)** $y$ increases by 5. **(c)** 10.

**2.59** **(b)** Final $= 60.5 + 0.614\times$ Second.

**2.61** The regression line should be similar.

**2.63** **(a)** 4.2255 km$^3$ per year. **(b)** $-271$ km$^3$. **(c)** 617 km$^3$; the prediction error is about 63 km$^3$. **(d)** There are high spikes in the time plot in these two years.

**2.65** **(b)** $\bar{x} = 1997.6667$, $\bar{y} = 272.1667$, $s_x = 6.0222$, $s_y = 6.0470$, $r = 0.9739$, and $\hat{y} = -1681 + 0.9779x$. This line explains about 95% of the variation.

**2.67** **(a)** $\hat{y} = 0.06078x - 0.1261$. **(b)** $\hat{y} \doteq 0$ (the formula gives $-0.0045$). **(c)** 77%.

**2.69** For the slower flow rate (8903), icicles grow at 0.158 cm per minute; for 8905, the growth rate is 0.0911 cm/min.

**2.71** No; the scatterplot shows little or no association, and regression explains only 1.3% of the variation in stock return.

**2.73** For metabolic rate on body mass, the slope is 26.9 cal/day per kg. For body mass on metabolic rate, the slope is 0.0278 kg per cal/day.

**2.75** $\hat{y} = 33.67 + 0.54x$. The predicted height is 69.85 inches.

**2.77** **(a)** $\bar{x} = 95$ and $s_x = 53.3854$ minutes; $\bar{y} = 12.6611$ and $s_y = 8.4967$ cm; $r = 0.9958$ (no units). **(b)** $\bar{y} = 32.1591$ and $s_y = 21.5816$ inches; $r$ is unchanged. **(c)** 0.4025 inches/minute.

**2.79** $r = 0.40$.

**2.81** **(b)** The slope is 0.000051; the plot suggests a horizontal line (slope 0). **(c)** Storing the oil doesn't help, as the total toxin level does not change over time.

**2.83** The sum is 0.01.

**2.85** **(a)** The plot is curved (low at the beginning and end of the year, high in the middle). **(b)** $\hat{y} = 39.392 + 1.4832x$; it does not fit well. **(c)** Residuals are negative for January through March and October through December, and positive from April to September. **(d)** A similar pattern would be expected in any city that is subject to seasonal temperature variation. **(e)** Seasons in the Southern Hemisphere are reversed.

**2.87** **(b)** No, because the pattern is not linear. **(c)** The sum is 0.01. The first two and last four residuals are negative, and those in the middle are positive.

**2.89** The variation of individual data increases the scatter, thus decreasing the strength of the relationship.

**2.91** For example, a student who in the past might have received a grade of B (and a lower SAT score) now receives an A (but has a lower SAT score than an A student in the past).

**2.93** $r = 0.08795$ and $b = 0.000811$ kg/cal.

**2.95** **(a)** Player 7's point is influential. **(b)** The first line omits Player 7.

**2.97** **(b)** $\hat{y} = 6.47 + 1.01x$. **(c)** The largest residuals are the Porsche Boxster (2.365) and Lamborghini Murcielago (−2.545). **(d)** The Insight pulls the line toward its point.

**2.99** Without the Insight, $\hat{y} = 4.87 + 1.11x$. For city mileages between 10 and 30 MPG, the difference in predicted highway mileage (with or without the Insight) is no more than 1.4 MPG, so the Insight is not very influential; it falls near the line suggested by the other points.

**2.101** **(a)** Drawing the "best line" by eye is a very inaccurate process.

**2.103** The plot should show a positive association when either group of points is viewed separately and should show a large number of bachelor's degree economists in business and graduate degree economists in academia.

**2.105** 1684 are binge drinkers; 8232 are not.

**2.107** $8232/17{,}096 \doteq 0.482$.

**2.109** $1630/7180 \doteq 0.227$.

**2.111** **(a)** About 3,388,000. **(b)** 0.207, 0.024; 0.320, 0.071; 0.104, 0.104; 0.046, 0.125. **(c)** 0.230, 0.391, 0.208, 0.171. **(d)** 0.677, 0.323.

**2.113** Full-time: 0.305, 0.472, 0.154, 0.069. Part-time: 0.073, 0.220, 0.321, 0.386.

**2.115** Two-year FT: 0.479, 0.521. Two-year PT: 0.458, 0.542. Four-year FT: 0.466, 0.534. Four-year PT: 0.394, 0.606. Graduate school: 0.455, 0.545. Vocational school: 0.539, 0.461.

**2.117** Start by setting $a$ equal to any number from 0 to 200.

**2.119** **(a)** 51.1%. **(b)** Small, 41.7%; medium, 51.7%; large, 60.0%. **(d)** Small, 39.8%; medium, 33.0%; large, 27.3%.

**2.121** Success (nonrelapse) rates were 58.3% (desipramine), 25.0% (lithium), and 16.7% (placebo).

**2.123** Age is one lurking variable: married men are generally older.

**2.125** No; self-confidence and improving fitness could be a common response to some other personality trait, or high self-confidence could make a person more likely to join the exercise program.

**2.127** Students with music experience may have other advantages (wealthier parents, better school systems, etc.).

**2.129** The diagram should show that either chemical exposure or time standing up or both or neither affect miscarriages.

**2.131** Spending more time watching TV means that less time is spent on other activities; this may suggest lurking variables.

**2.133** **(a)** Given two groups of the same age, where one group walks and the other does not, the walkers are half as likely to die in (say) the next year. **(b)** Men who choose to walk might also choose (or have chosen, earlier in life) other habits and behaviors that reduce mortality.

**2.135** A school that accepts weaker students but graduates a higher-than-expected number of them would have a positive residual, while a school with a stronger incoming class but a lower-than-expected graduation rate would have a negative residual. It seems reasonable to measure school quality by how much benefit students receive from attending the school.

**2.137** **(a)** The scatterplot shows a moderate positive association. **(b)** The regression line ($y = 1.1353x + 4.5503$) fits the overall trend. **(c)** For example, a state whose point falls above the line has a higher percent of college graduates than we would expect based on the percent who eat 5 servings of fruits and vegetables. **(d)** No; association is not evidence of causation.

**2.141** These results support the idea (the slope is negative), but the relationship is only moderately strong ($r^2 = 0.34$).

**2.143 (a)** One possible measure is mean response: 106.2 spikes/second for pure tones, 176.6 spikes/second for monkey calls. **(b)** $\hat{y} = 93.9 + 0.778x$. The third point has the largest residual. The first point is an outlier in the $x$ direction. **(c)** The correlation drops only slightly (from 0.6386 to 0.6101) when the third point is removed; it drops more drastically (to 0.4793) without the first point. **(d)** Without the first point, the line is $\hat{y} = 101 + 0.693x$; without the third point, it is $\hat{y} = 98.4 + 0.679x$.

**2.145** Based on the quantile plot, the distribution is close to Normal.

**2.147 (a)** Lines appear to fit the data well; there do not appear to be any outliers or influential points. **(b)** 18.9 $ft^3$ before, 15.7 $ft^3$ after. **(c)** 770.4 $ft^3$ before, 634.8 $ft^3$ after. **(d)** About $50.44.

**2.149 (a)** $\hat{y} = 259.58 - 19.464x$; the relationship appears to be curved. **(b)** Either $\hat{y} = 5.9732 - 0.2184x$ or $\hat{y} = 2.5941 - 0.09486x$; the relationship appears to be linear.

**2.151** It is now more common for these stocks to rise and fall together.

**2.153** Number of firefighters and amount of damage are common responses to the seriousness of the fire.

**2.155** $\hat{y} = 1.28 + 0.00227x$, $r = 0.252$, and $r^2 = 6.3\%$. By itself, SATM does not give reliable predictions of GPA.

**2.157** A notably higher percent of women are "strictly voluntary" participants.

**2.159 (a)** Males: 490 admitted, 310 not. Females: 400 admitted, 300 not. **(b)** Males: 61.25% admitted. Females: 57.14% admitted. **(c)** Business school: 66.67% of males, 66.67% of females. Law school: 45% of males, 50% of females. **(d)** Most male applicants apply to the business school, where admission is easier. More women apply to the law school, which is more selective.

**2.161** First- and second-year: A has 8.3% small classes; B has 17.1% small. Upper-level: A has 77.5% small; B has 83.3% small.

**2.163** There is some suggestion that sexual ads are more common in magazines intended for young-adult readers, but the difference in percents is fairly small.

**2.165 (a)** Wagering on collegiate sports appears to be more common in Division II, and even more in Division III. **(b)** Even with smaller sample sizes (1000 or more), the estimates should be fairly accurate (barring dishonest responses). **(c)** Our conclusion might not hold for the true percents.

## CHAPTER 3

**3.1** Any group of friends is unlikely to include a representative cross section of all students.

**3.3** A computer programmer (and his friends) are not representative of all young people.

**3.7** This is an observational study. Explanatory variable: cell phone usage. Response variable: presence/absence of brain cancer.

**3.9** An experiment: each subject is (presumably randomly) assigned to a treatment group. Explanatory variable: teaching method. Response variable: change in each student's test score.

**3.11** Experimental units: food samples. Treatments: radiation exposure. Response variable: lipid oxidation. Factor: radiation exposure. Levels: nine different levels of radiation. It is likely that different lipids react to radiation differently.

**3.13** Those who volunteer to use the software may be better students (or worse).

**3.17 (a)** Students in the front rows have a different classroom experience from those in the back. (And if they chose their own seats, those who choose seats in the front may be different from those who choose back seats.) **(b)** There is no control group. **(c)** It is hard to compare different classes (zoology and botany) in different semesters.

**3.19** Those evaluating the exams should not know which teaching approach was used, and the students should not be told that they are being taught using the new (or old) method.

**3.21** Possible response variables include increase in weight or height, number of leaves, etc.

**3.23** Experimental units: pine tree seedlings. Factor: amount of light. Treatments: full light, or shaded to 5% of normal. Response variable: dry weight at end of study.

**3.25** Subjects: adults from selected households. Factors: level of identification and offer of survey results. Six treatments: interviewer's name/university name/both names, with or without results. Response variable: whether or not the interview is completed.

**3.27** Assign 9 subjects to each treatment. The first three groups are 03, 22, 29, 26, 01, 12, 11, 31, 21; 32, 30, 09, 23, 07, 27, 20, 06, 33; 05, 16, 28, 10, 18, 13, 25, 19, 04.

**3.29 (a)** Randomly assign 7 rats to each group. **(b)** Group 1 includes rats 16, 04, 21, 19, 07, 10, and 13. Group 2 is 15, 05, 09, 08, 18, 03, and 01.

**3.31** Assign 6 schools to each treatment group. Choose 16, 21, 06, 12, 02, 04 for Group 1; 14, 15, 23, 11, 09, 03 for Group 2; 07, 24, 17, 22, 01, 13 for Group 3; and the rest for Group 4.

**3.33 (a)** There are three factors (roller type, dyeing cycle time, and temperature), yielding eight treatments and requiring 24 fabric specimens.

**3.35 (a)** Population = 1 to 150, sample size 25, then click "Reset" and "Sample." **(b)** Without resetting, click "Sample" again.

**3.37** The first design is an experiment, while the second is an observational study; with the first, any difference in colon health between the two groups could be attributed to the treatment (bee pollen or not).

**3.39 (a)** See the definitions in this chapter.

**3.41** There are nine treatments. Choose the number of letters in each group, and send them out at random times over several weeks.

**3.43** Each player runs through the sequence (100 yards, four times) once with oxygen and once without (on different days to allow full recovery); observe the difference in times on the final run. We choose 12, 13, 04, 18, 19, 16, 02, 08, 17, 10 for the oxygen-first group.

**3.45 (a)** Randomly assign half the girls to get high-calcium punch; the other half will get low-calcium punch. Observe how each group processes the calcium. **(b)** Half receive high-calcium punch first; the rest get low-calcium punch first. For each subject, compute the difference in the response variable for each level. Matched pairs designs give more precise results. **(c)** The first five subjects are 38, 44, 18, 33, and 46.

**3.47 (a)** A block design. **(c)** Such results would rarely have occurred by chance if vitamin C were ineffective.

**3.49** Population: area forest owners. Sample: the 772 forest owners contacted. Response rate: 348/772 = 45%. Additionally, we would like to know the sample design (among other things).

**3.53 (a)** The content of a single chapter is not random; choose random words from random pages. **(b)** Students who are registered for a 7:30 class might have different characteristics from those who avoid such classes. **(c)** Alphabetic order is not random; for example, some last names occur more often in some ethnic groups.

**3.55 (a)** Population: U.S. adults. Sample size: 1001. **(b)** Note that polls like this sometimes report results only for "those expressing an opinion." One can argue for either approach.

**3.57** 12, 14, 11, 16, and 08.

**3.59** Population = 1 to 200, sample size 25, then click "Reset" and "Sample."

**3.61** 39 (block 3020), 10 (block 2003), 07 (block 2000), 11 (block 2004), and 20 (block 3001).

**3.63** The simplest method is to assign the labels 0 to 5 to blocks in Group 1, then choose one of those blocks; use the last two digits of the blocks in Group 2, and choose two of those, etc.

**3.65** Each student has chance 1/45 of being selected, but it is not an SRS, because the only possible samples have exactly one name from the first 45, one name from the second 45, and so on.

**3.67** Assign labels 01 to 36 for the Climax 1 group, 01 to 72 for the Climax 2 group, and so on, then choose (from Table B) 12, 32, 13, 04; 51, 44, 72, 32, 18, 19, 40; 24, 28, 23; and 29, 12, 16, 25.

**3.69** Each student has a 10% chance, but the only possible samples are those with 3 older and 2 younger students.

**3.71** The higher no-answer was probably the second period—more families are likely to be gone for vacations, etc.

**3.73 (a)** 1260 responses. **(b)** 50.08%, 44.76%, and 5.16%. **(c)** We have the opinions only of those who visit this site and feel strongly enough to respond.

**3.75** The first wording brought the higher numbers in favor of a tax cut.

**3.79** Population: undergraduate college students. Sample: 2036 students.

**3.81** The larger sample would have less sampling variability.

**3.83 (a)** Population: college students. Sample: 17,096 students. **(b)** Population: restaurant workers. Sample: 100 workers. **(c)** Population: longleaf pine trees. Sample: 584 trees.

**3.85 (a)** Smaller sample sizes give less information about the population. **(b)** The margin of error was so large that the results could not be viewed as an accurate reflection of the population of Cubans.

**3.87 (a)** The variability will be the same for all states. **(b)** There would be less variability for states with larger samples.

**3.89** The histograms should be centered at about 0.6 and 0.2.

**3.93 (a)** Mean GPA 2.6352, standard deviation 0.7794.

**3.95 (a)** If, for example, eight heads are observed, then $\hat{p} = \frac{8}{20} = 0.4 = 40\%$.

**3.97 (a)** A nonscientist might raise different viewpoints and concerns from those considered by scientists.

**3.105** Interviews conducted in person cannot be anonymous.

**3.113 (a)** 00001 through 14959. **(b)** 03638, 07871, 12193.

**3.115 (a)** A matched pairs design. **(b)** A stratified sample survey. **(c)** A block design.

**3.117** This is an experiment, because treatments are assigned. Explanatory variable: price history (steady or fluctuating). Response variable: price the subject expects to pay.

**3.121** Randomly choose the order in which the treatments (gear and steepness combination) are tried.

**3.123 (a)** One possibility: full-time undergraduate students in the fall term on a list provided by the registrar. **(b)** One possibility: a stratified sample with 125 students from each year. **(c)** Nonresponse might be higher with mailed (or emailed) questionnaires; telephone interviews exclude some students and may require repeated calling for those who are not home; face-to-face interviews might be too costly. The topic might also be subject to response bias.

**3.125 (a)** Factors: storage method (fresh, room temperature for one month, refrigerated for one month) and preparation method (cooked immediately or after one hour). This makes six treatments (storage/preparation combinations). Response variables: tasters' color and flavor ratings. **(b)** Randomly allocate $n$ potatoes to each of the 6 groups, then compare ratings. **(c)** For each taster, randomly choose the order in which the fries are tasted.

**3.127** Parents who fail to return the consent form may be more likely to place less priority on education.

## CHAPTER 4

**4.1** The proportion of heads is 0.55.

**4.3** If you hear music (or talking) one time, you will almost certainly hear the same thing for several more checks after that.

**4.11** One possibility: from 36 to 90 inches (largest and smallest numbers could vary but should include all possible heights).

**4.13** 0.77.

**4.15** 0.523.

**4.17** 1/4.

**4.19** There are 6 possible outcomes: {link1, link2, link3, link4, link5, leave}.

**4.21 (a)** 0.45, so the sum equals 1. **(b)** 0.56.

**4.23 (a)** The probabilities sum to 2. **(b)** Legitimate (for a nonstandard deck). **(c)** Legitimate (for a nonstandard die).

**4.25 (a)** 0.28. **(b)** 0.88.

**4.27** 0.1333 for 1, 1/6 for 2 through 5, and 0.2 for 6.

**4.29** Take each blood type probability and multiply by 0.84 and by 0.16. For example, the probability for O-positive blood is $(0.45)(0.84) = 0.378$.

**4.31 (a)** 1/38 (all are equally likely). **(b)** 18/38. **(c)** 12/38.

**4.33 (a)** 10,000. **(b)** 0.3439.

**4.35** About 0.38 (0.3773).

**4.37** The two events (being 75 or older and being a woman) are probably not independent.

**4.39 (a)** 0.2746. **(b)** 0.35. **(c)** 0.545.

**4.41** Observe that $P(A \text{ and } B^c) = P(A) - P(A \text{ and } B) = P(A) - P(A)P(B)$.

**4.43 (a)** Either B or O. **(b)** $P(\text{B}) = 0.75$, $P(\text{O}) = 0.25$.

**4.45 (a)** 0.25. **(b)** 0.015625; 0.140625.

**4.47** Possible values: 0, 1, 2. Probabilities: 1/4, 1/2, 1/4.

**4.49 (a)** Continuous. **(b)** Discrete. **(c)** Discrete.

**4.51 (b)** $P(X \geq 1) = 0.9$. **(c)** "No more than two nonword errors." $P(X \leq 2) = 0.7$; $P(X < 2) = 0.4$.

**4.53** Rented housing typically has fewer rooms and has a sharply skewed distribution.

**4.55 (a)** $P(X \geq 6) = 0.658$. **(b)** "The unit has more than 6 rooms." $P(X > 6) = 0.434$. **(c)** Pay attention to whether you have "greater than" or "greater than or equal to."

**4.57** $Y$ can be 1, 2, 3, . . . , 12, each with probability 1/12.

**4.59 (a)** 0, 1, 2, or 3. **(b)** DDD (probability 0.3890); DDF, DFD, and FDD (probability 0.1439); FFD, FDF, and DFF (probability 0.0532); FFF (probability 0.0197). **(c)** The probabilities for $W$ are 0.3890, 0.4316, 0.1597, and 0.0197.

**4.61 (a)** 0.4. **(b)** 0.4. **(c)** "Equal to" has no effect on the probability.

**4.63 (a)** The height should be 1/2. **(b)** 0.75. **(c)** 0.55. **(d)** 0.55.

**4.65** Very close to 1.

**4.67** Possible values: \$0 and \$1. Probabilities: 0.5 and 0.5. Mean: \$0.50.

**4.69** $\mu_Y = 95$.

**4.71** $\sigma_X^2 = 1$ and $\sigma_X = 1$.

**4.73** 2.88.

**4.75** For owner-occupied units, the mean is 6.284 rooms; for rented units, it is 4.187 rooms.

**4.77** The owned-unit distribution is more spread out; $\sigma_o \doteq 1.6399$ and $\sigma_r \doteq 1.3077$ rooms.

**4.79 (a)** $\mu_1 = \sigma_1 = 0.5$. **(b)** $\mu_4 = 2$ and $\sigma_4 = 1$.

**4.81** Mean 14 cm and standard deviation 0.0042 cm.

**4.83 (a)** Not independent. **(b)** Independent.

**4.85** Show that $\sigma_{X+Y}^2 = (\sigma_X + \sigma_Y)^2$.

**4.87** If one of the 10 homes were lost, it would cost more than the collected premiums. For many policies, the average claim should be close to \$300.

**4.89 (a)** 0.99749. **(b)** \$623.22.

**4.91** $\mu_R = 11.116\%$ and $\sigma_R \doteq 15.9291\%$.

**4.93** $\mu_R = 10.184\%$ and $\sigma_R \doteq 12.3442\%$.

**4.95** $4/6 = 2/3$.

**4.97 (a)** 0.21. **(b)** 0.3392. **(c)** 0.42. **(d)** A's are more common in HHS than overall.

**4.99** 0.27.

**4.101 (a)** The four probabilities sum to 1. **(b)** 0.77. **(c)** 0.7442. **(d)** The events are not independent.

**4.103 (a)** The four entries are 0.2684, 0.3416, 0.1599, 0.2301. **(b)** 0.5975.

**4.105** For example, the probability of selecting a female student is 0.5717; the probability that she comes from a 4-year institution is 0.5975.

**4.107** $P(A \mid B) \doteq 0.3142$. If $A$ and $B$ were independent, then $P(A \mid B)$ would equal $P(A)$.

**4.109 (a)** $P(A^c) = 0.69$. **(b)** $P(A \text{ and } B) = 0.08$.

**4.111 (a)** 0.6364. **(b)** Not independent.

**4.113** 1.

**4.115** $P(B \mid C) = 1/3$. $P(C \mid B) = 0.2$.

**4.117** **(a)** $P(M) \doteq 0.4124$. **(b)** $P(B \mid M) \doteq 0.6670$. **(c)** $P(M \text{ and } B) \doteq 0.2751$.

**4.119** **(a)** $P(L) \doteq 0.6722$. **(b)** $P(L \mid C) \doteq 0.7837$. **(c)** Not independent.

**4.121** Retired persons are more likely than other adults to have not completed high school.

**4.123** 75%.

**4.125** **(a)** Her brother has allele type *aa*, and he got one allele from each parent. **(b)** $P(aa) = 0.25$, $P(Aa) = 0.5$, $P(AA) = 0.25$. **(c)** $P(AA \mid \text{not } aa) = 1/3$, $P(Aa \mid \text{not } aa) = 2/3$.

**4.127** 1.

**4.129** 0.9333.

**4.131** **(a)** $P(A) = 1/6$ and $P(B) = 5/12$. **(b)** $P(A) = 1/6$ and $P(B) = 5/12$. **(c)** $P(A) = 5/12$ and $P(B) = 5/18$. **(d)** $P(A) = 5/12$ and $P(B) = 5/18$.

**4.133** Each mean payoff is $10; for a $10 bet, the net gain is $0.

**4.135** **(a)** All probabilities are greater than or equal to 0, and their sum is 1. **(b)** 0.61. **(c)** Both probabilities are 0.39.

**4.137** **(a)** 0.9389. **(b)** $1.85. **(c)** $16.1513.

**4.139** 0.005352.

**4.141** For example: $P(\text{2-year} \mid \text{Public}) = 0.3759$, while $P(\text{2-year} \mid \text{Private}) = 0.7528$.

**4.145** About 22.8%.

**4.147** $P(Y < 1/2 \mid Y > X) = 1/4$.

## CHAPTER 5

**5.1** **(a)** $n = 1500$. **(b)** The "Yes" count seems most reasonable, but either count is defensible. **(c)** $X = 525$ (or $X = 975$). **(d)** $\hat{p} = 0.35$ (or $\hat{p} = 0.65$).

**5.3** $B(20, 0.5)$.

**5.5** **(a)** $P(X = 0) = 0.2401$ and $P(X \geq 3) = 0.0837$. **(b)** $P(X = 4) = 0.2401$ and $P(X \leq 1) = 0.0837$. **(c)** The number of "failures" in a $B(4, 0.3)$ distribution has a $B(4, 0.7)$ distribution.

**5.7** **(a)** About 0.95. **(b)** About 0.68.

**5.9** **(a)** Separate flips are independent (coins have no "memory"). **(b)** Separate flips are independent (coins have no "memory"). **(c)** $\hat{p}$ can vary from one set of observed data to another; it is not a parameter.

**5.11** **(a)** A $B(200, p)$ distribution seems reasonable ($p$ not known). **(b)** Not binomial (no fixed $n$). **(c)** A $B(500, 1/12)$ distribution.

**5.13** **(a)** Caught: $B(10, 0.7)$. Missed: $B(10, 0.3)$. **(b)** 0.3503.

**5.15** **(a)** 7 errors caught, 3 missed. **(b)** 1.4491 errors. **(c)** With $p = 0.9$, 0.9487 errors; with $p = 0.99$, 0.3146 errors. $\sigma$ decreases toward 0 as $p$ approaches 1.

**5.17** $m = 6$.

**5.19** **(a)** 0.4095. **(b)** 4.

**5.21** **(a)** $n = 4$ and $p = 0.25$. **(b)** The probabilities are 0.3164, 0.4219, 0.2109, 0.0469, 0.0039. **(c)** $\mu = 1$ child.

**5.23** **(a)** 0.8354. **(b)** 0.9926. **(c)** It increases.

**5.25** **(a)** 0.7. **(b)** 0.1841.

**5.27** **(a)** $p = 0.25$. **(b)** 0.0139. **(c)** $\mu = 5$ and $\sigma = 1.9365$ successes. **(d)** No; the trials would not be independent.

**5.29** **(a)** $\mu = 180$ and $\sigma = 12$ successes. **(b)** For $\hat{p}$, $\mu = 0.2$ and $\sigma = 0.01333$. **(c)** About 0.0013. **(d)** 0.2310 or higher.

**5.31** **(a)** 0.1137. **(b)** 136.4 blacks. **(c)** About 0.0005.

**5.33** **(a)** 0.0808. **(b)** 0.0136. **(c)** 400. **(d)** Yes.

**5.35** Possible values: 1, 2, 3, .... $P(Y = k) = pq^{k-1}$.

**5.37** $\mu_{\bar{x}} = 200$, $\sigma_{\bar{x}} = 1$.

**5.39** About 95% of the time, $\bar{x}$ is between 199 and 201.

**5.41** **(a)** "Variance" should be "standard deviation." **(b)** Standard deviation decreases with increasing sample size. **(c)** $\mu_{\bar{x}}$ always equals $\mu$.

**5.43** 40.135 mm and 0.0015 mm.

**5.45** **(b)** About 0.05. **(c)** Nearly 0.

**5.47** **(a)** $\bar{x}$ is not systematically higher than or lower than $\mu$. **(b)** With large samples, $\bar{x}$ is more likely to be close to $\mu$.

**5.49** **(a)** 0.5 and 0.099 moths. **(b)** About 0.16.

**5.51** **(a)** 0.0668. **(b)** 0.0047.

**5.53** About 133.2 mg/dl.

**5.55** About 0.0052.

**5.57** 0.0294, 0.0853.

**5.59** **(a)** Nearly 1. **(b)** About 0.9641.

**5.61** **(a)** $\bar{y}$ is $N(\mu_Y, \sigma_Y/\sqrt{m})$ and $\bar{x}$ is $N(\mu_X, \sigma_X/\sqrt{n})$. **(b)** $N\left(\mu_Y - \mu_X, \sqrt{\sigma_Y^2/m + \sigma_X^2/n}\right)$.

**5.63** $N(32, 0.2)$; 0.0124.

**5.65** 0.0579; these five students were not randomly chosen from the population of all drivers.

**5.67** **(a)** 0.8822. **(b)** 0.9999.

**5.69** 0.0034.

**5.71** 0.9231.

**5.73** 0.0336.

**5.75** 4.0689 to 4.4311.

**5.77** **(a)** $\hat{p}_F$: $N(0.82, 0.01921)$. $\hat{p}_M$: $N(0.88, 0.01625)$. **(b)** $N(0.06, 0.02516)$. **(c)** About 0.0087.

**5.79** $P(Z \geq 4.56) \doteq 0$.

## CHAPTER 6

**6.1** $\sigma_{\bar{x}} \doteq \$0.33$.

**6.3** $0.67.

**6.5** 82.716 to 84.284.

**6.7** $n = 523$.

**6.9** The organizations that did not respond are (obviously) not represented in the results.

**6.11** Margins of error: 3.0990, 2.1913, 1.5495, 1.0957; interval width decreases with increasing sample size.

**6.13** Margin of error, 0.1554; interval, 7.3446 to 7.6554.

**6.15** 8.5673 to 8.8327.

**6.17** Margin of error, 2.29 U/l; interval, 10.91 to 15.49 U/l.

**6.19** Scenario B has a smaller margin of error; less variability in a single major.

**6.21** No, this is a range of values for the mean rent, not for individual rents.

**6.23** **(a)** 14.8171 to 15.3829 hours. **(b)** No, this is a range of values for the mean time spent, not for individual times.

**6.25** **(a)** 2.1920 cal/day. **(b)** 4.3841 cal/day. **(c)** 4.3841 (or 4.2694) cal/day.

**6.27** $\bar{x} = 18.3515$ kpl; the margin of error is 0.6521 kpl.

**6.29** $n = 41$.

**6.31** $n = 53$.

**6.33** **(a)** 0.7738. **(b)** 0.9774.

**6.35** **(a)** Not certain (only 95% confident). **(b)** We obtained the interval 56% to 62% by a method that gives a correct result 95% of the time. **(c)** About 1.5%. **(d)** No (only random sampling error).

**6.37** $H_0$: $\mu = 1.4$ g/cm$^2$; $H_a$: $\mu \neq 1.4$ g/cm$^2$.

**6.39** $P = 0.2302$.

**6.41** $z > 1.645$.

**6.43** **(a)** $z = 2$. **(b)** $P = 0.0228$. **(c)** $P = 0.0456$.

**6.45** **(a)** Yes. **(b)** No.

**6.47** **(a)** Yes. **(b)** No. **(c)** To reject, we need $P < \alpha$.

**6.49** **(a)** $P = 0.03$ and $P = 0.97$. **(b)** We need to know whether the observed data (for example, $\bar{x}$) are consistent with $H_a$. (If they are, use the smaller $P$-value.)

**6.51** **(a)** Population mean, not sample mean. **(b)** $H_0$ should be that there is no change. **(c)** A small $P$-value is needed for significance.

**6.53** **(a)** $H_0$: $\mu = 26$; $H_a$: $\mu \neq 26$. **(b)** $H_0$: $\mu = 20$ seconds; $H_a$: $\mu < 20$ seconds. **(c)** $H_0$: $\mu = 460$ ft$^2$; $H_a$: $\mu < 460$ ft$^2$.

**6.55** **(a)** $H_0$: $\mu = \$62,500$; $H_a$: $\mu > \$62,500$. **(b)** $H_0$: $\mu = 2.6$ hr; $H_a$: $\mu \neq 2.6$ hr.

**6.57** **(a)** $P = 0.9582$. **(b)** $P = 0.0418$. **(c)** $P = 0.0836$.

**6.59** **(a)** No, because 24 is in the interval. **(b)** Yes, because 30 is not in the interval.

**6.61** You have little reason to doubt that the purity is the same (although you cannot be completely sure).

**6.63** The difference was large enough that it would rarely arise by chance. Health issues related to alcohol use are probably discussed in the health and safety class.

**6.65** The difference in average scores from 2000 to 2005 was so small that it could have occurred by chance even if population mean scores had not changed in that time.

**6.67** $H_0$: $\mu = 100$; $H_a$: $\mu \neq 100$; $z = 5.75$; significant ($P < 0.0001$).

**6.69** **(a)** $z = 2.87$, $P = 0.0021$. **(b)** The important assumption is that this is an SRS. We also assume a Normal distribution, but this is not crucial provided there are no outliers and little skewness.

**6.71** **(a)** $H_0$: $\mu = 0$ mpg; $H_a$: $\mu \neq 0$ mpg, where $\mu$ is the mean difference. **(b)** $z \doteq 4.07$, which gives a very small $P$-value.

**6.73** **(a)** $H_0$: $\mu = 1.4$ mg; $H_a$: $\mu > 1.4$ mg. **(b)** Yes. **(c)** Yes.

**6.75** $\bar{x} = 0.8$ is significant, but 0.7 is not. Smaller $\alpha$ means that $\bar{x}$ must be farther away.

**6.77** Something that occurs "less than once in 100 repetitions" also occurs "less than 5 times in 100 repetitions."

**6.79** Any $z$ with $2.807 < |z| < 3.291$.

**6.81** $P > 0.25$.

**6.83** $0.1 < P < 0.2$; $P = 0.1706$.

**6.85** In order to determine the effectiveness of alarm systems, we need to know the percent of all homes with alarm systems and the percent of burglarized homes with alarm systems.

**6.87** The first test was barely significant at $\alpha = 0.05$, while the second was significant at any reasonable $\alpha$.

**6.89** A significance test answers only Question b.

**6.91** With a larger sample, we might have significant results.

**6.93** **(a)** The differences observed might occur by chance even if SES had no effect. **(b)** This tells us that the statistically insignificant test results did not occur merely because of a small sample size.

**6.95** **(a)** $P = 0.3821$. **(b)** $P = 0.1711$. **(c)** $P = 0.0013$.

**6.97** No, we have information about the entire population in question.

**6.103** $n$ should be about 100,000.

**6.105** Reject the fifth ($P = 0.001$), sixth ($P = 0.004$), and eleventh ($P = 0.002$).

**6.107** Larger samples give more power.

**6.109** Higher; larger differences are easier to detect.

**6.111** Power: about 0.99.

**6.113** Power: 0.4681.

**6.115** **(a)** $H_0$: Patient is healthy; $H_a$: Patient is ill. Type I error: sending a healthy patient to the doctor. Type II error: clearing a patient who is ill.

**6.117** **(a)** 64.45 to 70.01; 67.59 to 73.15; 72.05 to 77.61. **(c)** 63.83 to 70.63; 66.97 to 73.77; 71.43 to 78.23. **(d)** With the larger margin of error, the intervals overlap more.

**6.121** **(a)** 4.61 to 6.05 mg/dl. **(b)** $z = 1.45$, $P = 0.0735$; not significant.

**6.123** **(b)** 26.0614 to 34.7386 $\mu$g/l. **(c)** $z = 2.44$, $P = 0.0073$.

**6.125** **(a)** Under $H_0$, $\bar{x}$ has an $N(0\%, 5.3932\%)$ distribution. **(b)** $z = 1.28$, $P = 0.1003$. **(c)** Not significant.

**6.127** It is essentially correct.

**6.129** Find $\bar{x}$, then take $\bar{x} \pm 1.96(4/\sqrt{12}) = \bar{x} \pm 2.2632$.

**6.131** Find $\bar{x}$, then compute $z = (\bar{x} - 23)/(4/\sqrt{12})$. Reject $H_0$ based on your chosen significance level.

## CHAPTER 7

**7.1** **(a)** \$27.1109. **(b)** 14.

**7.3** \$511.85 to \$628.15.

**7.5** **(a)** Yes. **(b)** No.

**7.9** −1.8866 to 11.0866.

**7.11** The sample size should be sufficient to overcome any non-Normality. One might question the independence of 90 consecutive measurements.

**7.13** The power is about 0.6950.

**7.15** **(a)** $t^* = 2.145$. **(b)** $t^* = 2.064$. **(c)** $t^* = 1.711$. **(d)** $t^*$ decreases with increasing sample size, and increases with increasing confidence.

**7.17** $t^* = 1.729$ (or $-1.729$).

**7.19** For the alternative $\mu < 0$, the answer would be the same ($P = 0.02$). For the alternative $\mu > 0$, the answer would be $P = 0.98$.

**7.21** **(a)** df = 23. **(b)** $2.177 < t < 2.500$. **(c)** $0.02 < P < 0.04$. **(d)** Significant at 5% but not at 1%. **(e)** $P = 0.0248$.

**7.23** It depends on whether $\bar{x}$ is on the appropriate side of $\mu_0$.

**7.25** **(a)** Distribution is not Normal; it has two peaks and one large value. **(b)** Maybe; we have a large sample but a small population. **(c)** $27.29 \pm 5.717$, or 21.57 to 33.01 cm. **(d)** One could argue for either answer.

**7.27** **(a)** Right-skewed, but less than the original data, and no clear outliers. **(b)** No; for skewed distributions, the median is more appropriate. **(c)** $1.495 \pm 0.449$, or 1.05 to 1.94. After undoing the transformation, this is about 1.85 to 5.99 mg/l.

**7.29** **(a)** Not Normal (lots of 1s and 10s), but no outliers. **(b)** 4.92 to 6.88. **(c)** Because this is not a random sample, it may not represent other children well.

**7.31** 90%: 2.119 to 2.321. 95%: 2.099 to 2.341. Width increases with confidence level.

**7.33** **(a)** $t = 5.13$, df = 15, $P < 0.001$. **(b)** With 95% confidence, the mean NEAT increase is between 192 and 464 calories.

**7.35** **(a)** $H_0$: $\mu_c = \mu_d$; $H_a$: $\mu_c \neq \mu_d$. **(b)** $t \doteq 4.358$, $P \doteq 0.0003$; we reject $H_0$.

**7.37** $H_0$: $\mu = 4.8$; $H_a$: $\mu > 4.8$ mg/dl; $t = 2.086$, df = 5, $P = 0.046$.

**7.39** **(a)** The differences are spread from −0.018 to 0.020 g. **(b)** $t = -0.347$, df = 7, $P = 0.7388$. **(c)** −0.0117 to 0.0087 g. **(d)** They may be representative of future subjects, but the results are suspect because this is not a random sample.

**7.41** **(a)** $H_0$: $\mu = 0$; $H_a$: $\mu > 0$. **(b)** Slightly left-skewed; $\bar{x} = 2.5$ and $s = 2.893$. **(c)** $t = 3.865$, df = 19, $P = 0.00052$. **(d)** 1.146 to 3.854.

**7.43** $\bar{x} = 114.98$, $s = 14.80$; 111.14 to 118.82. This might adequately describe the mean IQ at this school, but the sample could not be considered representative of all fifth-graders.

**7.45** For the sign test, $P = 0.0898$; not quite significant, unlike Exercise 7.34.

**7.47** $H_0$: median = 0; $H_a$: median ≠ 0; $P = 0.7266$. This is similar to the $t$ test $P$-value.

**7.49** $H_0$: median = 0; $H_a$: median > 0; $P = 0.0013$.

**7.51** Reject $H_0$ if $|\bar{x}| \geq 0.00677$. The power is about 7%.

**7.53** $n \geq 16$. (The power is about 0.794 when $n = 15$.)

**7.55** $0.01 < P < 0.02$.

**7.57** −31.1735 to −8.8265; reject $H_0$.

**7.59** SPSS and SAS give both results; the pooled $t$ is −56.99, which has a tiny $P$-value.

**7.61** **(a)** Assuming we have SRSs from each population, this seems reasonable. **(b)** $H_0$: $\mu_f = \mu_m$; $H_a$: $\mu_f \neq \mu_m$. **(c)** $t \doteq 0.276$, $P \doteq 0.78$. **(d)** −11.75 to 15.53 (df $\doteq$ 76.1) or −12.1 to 15.9 (df = 30) mg/dl. **(e)** It might not be appropriate to treat these students as SRSs from larger populations.

**7.63** **(a)** Not Normal, because all numbers are integers. **(b)** Yes; we have two large samples, with no outliers. **(c)** $H_0$: $\mu_1 = \mu_2$; $H_a$: $\mu_1 >$ (or $\neq$) $\mu_2$. **(d)** $t = 6.258$, df = 354 or 164, $P < 0.0001$. **(e)** 0.51 to 0.99 (regardless of df). **(f)** This may not generalize well to other areas of the country.

**7.65** **(a)** This may be near enough to an SRS if this company's working conditions were similar to those of other workers. **(b)** 9.99 to 13.01 mg.y/m$^3$. **(c)** $t = 15.08$, $P < 0.0001$ with either df = 137 or 114. **(d)** The sample sizes are large enough that skewness should not matter.

**7.67** You need either sample sizes and standard deviations or degrees of freedom and a more accurate value for the $P$-value. The confidence interval will give us useful information about the magnitude of the difference.

**7.69** **(a)** Hypotheses should involve $\mu_1$ and $\mu_2$. **(b)** The samples are not independent. **(c)** We need $P$ to be small (for example, less than 0.10) to reject $H_0$. **(d)** $t$ should be negative.

**7.71** **(a)** Yes (in fact, $P \doteq 0.005$). **(b)** Yes ($P \doteq 0.0025$).

**7.73** This is a matched pairs design.

**7.75** The next 10 employees who need screens might not be an independent group—perhaps they all come from the same department, for example.

**7.77** **(a)** This is now a matched pairs design. **(b)** $t = -49.83$, df $= 1$, $P = 0.0064$.

**7.79** Small samples may lead to rejection of $H_0$ if the evidence is very strong.

**7.81** **(a)** The north distribution (five-number summary 2.2, 10.2, 17.05, 39.1, 58.8 cm) is right-skewed, while the south distribution (2.6, 26.1, 37.70, 44.6, 52.9) is left-skewed. **(b)** The methods of this section seem to be appropriate. **(c)** $H_0$: $\mu_n = \mu_s$; $H_a$: $\mu_n \neq \mu_s$. **(d)** $t = -2.63$ with df $= 55.7$ ($P = 0.011$) or df $= 29$ ($P = 0.014$). **(e)** Either $-19.09$ to $-2.57$ or $-19.26$ to $-2.40$ cm.

**7.83** **(a)** Either $-3.06$ to 9.06 units (df $= 50$) or $-2.98$ to 8.98 units (df $= 104.6$). **(b)** Random fluctuation may account for the difference in the two averages.

**7.85** **(a)** $H_0$: $\mu_B = \mu_F$; $H_a$: $\mu_B > \mu_F$; $t = 1.654$, $P = 0.053$ (df $= 37.6$) or $P = 0.058$ (df $= 18$). **(b)** $-0.2$ to 2.0. **(c)** We need two independent SRSs from Normal populations.

**7.87** $s_p = 1.1556$; $t = 6.251$ (df $= 375$); $P < 0.0001$; 0.51 to 0.99. All results are nearly the same as in Exercise 7.63.

**7.89** $s_p = 15.96$; $t = -2.629$ (df $= 58$); $P = 0.0110$; $-19.08$ to $-2.58$ cm. All results are nearly the same as in Exercise 7.81.

**7.91** df $= 55.725$.

**7.93** **(a)** df $= 137.066$. **(b)** $s_p = 5.332$ (slightly closer to $s_2$, from the larger sample). **(c)** With no assumption, $SE_1 = 0.7626$; with the pooled method, $SE_2 = 0.6136$. **(d)** $t = 18.74$, df $= 333$, $P < 0.0001$. $t$ and df are larger, so the evidence is stronger (although it was quite strong before). **(e)** df $= 121.503$; $s_p = 1.734$; $SE_1 = 0.2653$ and $SE_2 = 0.1995$. $t = 24.56$, df $= 333$, $P < 0.0001$.

**7.95** **(a)** $F^* = 2.20$. **(b)** Significant at the 10% level but not at the 5% level.

**7.97** A smaller $\sigma$ would yield more power.

**7.99** $F = 1.095$ with df $= 70$ and 36; $P = 0.7794$. We do not know if the distributions are Normal, so this test may not be reliable.

**7.101** $F = 1.017$ with df $= 211$ and 164; $P = 0.9114$. The distributions are not Normal, so this test may not be reliable (although the conclusion is reasonable). To reject at the 5% level, $s_2$ would need to be at least 1.39 (based on Table E) or 1.33 (software).

**7.103** $F = 5.263$ with df $= 114$ and 219; $P < 0.0001$. The authors described the distributions as somewhat skewed, so the Normality assumption may be violated.

**7.105** $F = 1.506$ with df $= 29$ and 29; $P = 0.2757$. The stemplots in Exercise 7.81 did not appear to be Normal.

**7.107** **(a)** $F^* = 647.79$; this is a low-power test. **(b)** $F \doteq 3.96$; do not reject $H_0$.

**7.109** Using a larger $\sigma$ for planning the study is advisable because it provides a conservative (safe) estimate of the power.

**7.111** $\bar{x} = 156$, $s \doteq 10.30$, $s_{\bar{x}} \doteq 5.15$. We cannot consider these four scores to be an SRS.

**7.113** As df increases, $t^*$ approaches 1.96.

**7.115** Margins of error decrease with increasing sample size.

**7.117** **(a)** Body weight: mean $-0.7$ kg, SE 2.298 kg. Caloric intake: mean 14 cal, SE 56.125 cal. **(b)** $t_1 = -0.305$ (body weight) and $t_2 = 0.249$ (caloric intake), both with df $= 13$; both $P$-values are about 0.8. **(c)** $-5.66$ to 4.26 kg and $-107.23$ to 135.23 cal.

**7.119** **(a)** Somewhat right-skewed with no extreme outliers. **(b)** 10.85 to 15.64 U/l.

**7.121** **(a)** Slightly left-skewed. **(b)** 2.285 to 2.650. **(c)** 9.82 to 14.15 U/l.

**7.123** For north/south differences: $t = 7.15$, df $= 575.4$ or 283, $P < 0.0001$; the confidence interval is 7.52 to 13.22 cm. For east/west differences: $t = 3.69$, df $= 472.7$ or 230, $P < 0.0005$; the confidence interval is 2.68 to 8.78 cm. With larger samples, $t$ increases, $P$ decreases, and the intervals shrink.

**7.125** 78.3% ± 13.8%, or 64.5% to 92.1%.

**7.127** GPA: $t = -0.91$, df $= 74.9$ ($P = 0.1839$) or 30 ($0.15 < P < 0.20$). Confidence interval: $-1.33$ to 0.5. IQ: $t = 1.64$, df $= 56.9$ ($P = 0.0503$) or 30 ($0.05 < P < 0.10$). Confidence interval: $-1.12$ to 11.36.

**7.129** $t = 3.65$, df $= 237.0$ or 115, $P < 0.0005$. 95% confidence interval for the difference: 0.78 to 2.60.

**7.131** $t = -0.3533$, df $= 179$, $P = 0.3621$.

**7.133** No; what we have is nothing like an SRS.

**7.135** **(a)** $\bar{x} = 1.7182$ and $s = 1.3428$. Distribution is right-skewed from 0.068 to 5.417. **(b)** The sample size should be large enough to overcome the skewness. **(c)** 1.396 to 2.041. No; this is an interval for the mean, not for individual observations. **(d)** The transformed distribution is left-skewed, with mean 0.1542 and standard deviation 1.0176. A 95% confidence interval for the mean is $-0.090$ to 0.399 (0.914 to 1.490, after undoing the logarithms).

**7.137** $\bar{x} = 0.8043$ and $s = 0.2765$ g/l; distribution is right-skewed. 95% confidence interval: 0.746 to 0.862 g/l.

**7.139** Basal: $\bar{x} = 41.0455$, $s = 5.6356$. DRTA: $\bar{x} = 46.7273$, $s = 7.3884$. Strat: $\bar{x} = 44.2727$, $s = 5.7668$. **(a)** $t \doteq 2.87$, $P < 0.005$. Confidence interval for difference: 1.7 to 9.7 points. **(b)** $t \doteq 1.88$, $P < 0.05$. Confidence interval for difference: $-0.24$ to 6.7 points.

**7.141** **(a)** Both distributions are right-skewed; four-bedroom homes are generally more expensive. The top three prices from the three-bedroom distribution qualify as outliers. **(b)** $t \doteq -3.08$ with either df $= 12.1$ ($P = 0.0095$) or df $= 8$ ($P = 0.0151$); we reject $H_0$. **(c)** It would be reasonable to guess that $\mu_3 < \mu_4$. **(d)** \$19,182 to \$111,614 (df = 12.1) or \$16,452 to \$114,344 (df = 8). **(e)** It seems that these houses should be a fair representation of three- and four-bedroom houses in West Lafayette.

## CHAPTER 8

**8.1** 0.01486.

**8.3** **(a)** $H_0: p = 0.72$; $H_a: p \neq 0.72$. **(b)** $z \doteq 1.11$, $P = 0.2670$. **(c)** No.

**8.5** A smaller sample is needed for 90% confidence; $n = 752$.

**8.7** **(a)** Yes. **(b)** Yes. **(c)** No. **(d)** No. **(e)** No.

**8.9** **(a)** Margin of error equals $z^*$ times standard error. **(b)** Use Normal distributions for proportions. **(c)** $H_0$ should refer to $p$, not $\hat{p}$.

**8.11** **(a)** $\hat{p} = 0.6341$, 0.6214 to 0.6467. This interval was found using a procedure that includes the correct proportion 95% of the time. **(b)** We do not know if those who did respond can reliably represent those who did not.

**8.13** **(a)** $\pm 0.001321$. **(b)** Other sources of error are much more significant than sampling error.

**8.15** **(a)** 0.3506 to 0.4094. **(b)** Yes; some respondents might not admit to such behavior.

**8.17** **(a)** $\hat{p} = 0.3275$; 0.3008 to 0.3541. **(b)** Speakers and listeners probably perceive sermon length differently.

**8.19** 0.1304 to 0.1696.

**8.21** **(a)** No. **(b)** Yes. **(c)** Yes. **(d)** No.

**8.23** 0.6345 to 0.7455.

**8.25** 0.2180 to 0.2510.

**8.27** 0.8230 to 0.9370.

**8.29** **(a)** $z = 1.34$, $P = 0.1802$. **(b)** 0.4969 to 0.5165.

**8.31** $n = 171$ or 172.

**8.33** The sample sizes are 35, 62, 81, 93, 97, 93, 81, 62, and 35; take $n = 97$.

**8.35** $\hat{p}_m - \hat{p}_w = 0.1214$; the interval is −0.0060 to 0.2488.

**8.37** $z \doteq 1.86$, $P = 0.0629$.

**8.39** **(a)** Yes. **(b)** No. **(c)** No. **(d)** Yes. **(e)** No.

**8.41** $z \doteq 4.24$, $P < 0.0001$. Confidence interval: 0.0323 to 0.0877.

**8.43** $z = 20.18$, so $P$ is tiny. Confidence interval: 0.1962 to 0.2377. Nonresponse error could render this interval and test result meaningless.

**8.45** −0.0017 to 0.0897.

**8.47** $z \doteq 6.01$, $P < 0.0001$; confidence interval is 0.0195 to 0.0384 (all the same as in Exercise 8.46).

**8.49** $z = 4.28$ and $P < 0.0001$. Confidence interval: 0.0381 to 0.1019.

**8.51** **(a)** −0.0053 to 0.2335. **(b)** $z \doteq 1.83$, $P = 0.0336$. **(c)** We have fairly strong evidence that high-tech companies are more likely to offer stock options, but the difference in proportions could be very small or as large as 23%.

**8.53** **(a)** $\hat{p}_f = 0.8$, SE $\doteq 0.05164$; $\hat{p}_m = 0.3939$, SE $\doteq 0.04253$. **(b)** 0.2960 to 0.5161.

**8.55** $z = 2.10$, $P = 0.0360$.

**8.57** **(a)** Confidence intervals account for only sampling error. **(b)** $H_0$ should refer to $p_1$ and $p_2$. **(c)** Only if $n_1 = n_2$.

**8.59** −0.0298 to 0.0898.

**8.61** $\hat{p} = 0.6129$, $z = 4.03$, $P < 0.0001$; confidence interval is 0.5523 to 0.6735.

**8.63** **(a)** People have different symptoms; for example, not all who wheeze consult a doctor. **(b)** Sleep: 0.0864, 0.0280 to 0.1448. Number: 0.0307, −0.0361 to 0.0976. Speech: 0.0182, −0.0152 to 0.0515. Activities: 0.0137, −0.0395 to 0.0670. Doctor: −0.0112, −0.0796 to 0.0573. Phlegm: −0.0220, −0.0711 to 0.0271. Cough: −0.0323, −0.0853 to 0.0207. **(c)** It is reasonable to expect that the bypass proportions would be higher. **(d)** In the same order: $z = 2.64$, $P = 0.0042$; $z = 0.88$, $P = 0.1897$; $z = 0.99$, $P = 0.1600$; $z = 0.50$, $P = 0.3100$; $z = -0.32$, $P = 0.6267$; $z = -0.92$, $P = 0.8217$; $z = -1.25$, $P = 0.8950$. **(e)** 95% confidence interval for sleep improvement: 0.1168 to 0.2023. Part (b) showed improvement relative to control group, which is a better measure of the effect of the bypass.

**8.65** **(a)** $z = 6.98$, $P < 0.0001$. **(b)** 0.1145 to 0.2022.

**8.67** Education: 1132 users, 852 nonusers. Income: 871 users, 677 nonusers. For users, $\hat{p}_1 = 0.2306$; for nonusers, $\hat{p}_2 = 0.2054$. $z = 1.34$, $P = 0.1802$; −0.0114 to 0.0617. The lack of response about income makes the conclusions for Exercise 8.66 suspect.

**8.69** The margin of error is ±2.8%.

**8.71** $z = 8.95$, $P < 0.0001$; 0.3720 to 0.5613.

**8.73** All $\hat{p}$-values are greater than 0.5. Texts 3, 7, and 8 have (respectively) $z = 0.82$, $P = 0.4122$; $z = 3.02$, $P = 0.0025$; and $z = 2.10$, $P = 0.0357$. For the other texts, $z \geq 4.64$ and $P < 0.00005$.

**8.77** $z$: 0.90, 1.01, 1.27, 1.42, 2.84, 3.18, 4.49. $P$: 0.3681, 0.3125, 0.2041, 0.1556, 0.0045, 0.0015, 0.0000.

**8.79** **(a)** $n = 342$. **(b)** $n = (z^*/m)^2/2$.

**8.81** **(a)** $p_0 = 0.7911$. **(b)** $\hat{p} = 0.3897$, $z = -29.1$, $P$ is tiny. **(c)** $\hat{p}_1 = 0.3897$, $\hat{p}_2 = 0.7930$, $z = -29.2$, $P$ is tiny.

**8.83** **(a)** 0.5278 to 0.5822. **(b)** 0.5167 to 0.5713. **(c)** 0.3170 to 0.3690. **(d)** 0.5620 to 0.6160. **(e)** 0.5620 to 0.6160. **(f)** 0.6903 to 0.7397.

## CHAPTER 9

**9.1** **(a)** Given Explanatory = 1: 37.5% Yes, 62.5% No. Given Explanatory = 2: 47.5% Yes, 52.5% No. **(c)** When Explanatory = 2, "Yes" and "No" are nearly evenly split.

**9.3 (a)** $0.10 < P < 0.15$. **(b)** $0.01 < P < 0.02$. **(c)** $0.025 < P < 0.05$. **(d)** $0.025 < P < 0.05$.

**9.5** $X^2 \doteq 15.2$, df = 5, $0.005 < P < 0.01$.

**9.7 (a)** 0.2044, 0.0189; 0.3285, 0.0699; 0.1050, 0.1072; 0.0518, 0.1141. **(b)** 0.2234, 0.3984, 0.2123, 0.1659. **(c)** 0.6898, 0.3102. **(d)** Full-time students: 0.2964, 0.4763, 0.1522, 0.0752. Part-time students: 0.0610, 0.2254, 0.3458, 0.3678.

**9.9 (a)** Yes, this seems to satisfy the assumptions. **(b)** df = 3. **(c)** $0.20 < P < 0.25$.

**9.11 (a)** Success (nonrelapse) rates were 58.3% (desipramine), 25.0% (lithium), and 16.7% (placebo). **(b)** Yes; this seems to satisfy the assumptions. **(c)** $X^2 = 10.5$, df = 2, $P = 0.005$.

**9.13** Start by setting $a$ equal to any number from 0 to 50.

**9.15 (a)** A notably higher percent of women are "strictly voluntary" participants. **(b)** 40.3% of men and 51.3% of women are participants; the relative risk is 1.27.

**9.17 (a)** For example, among nonbingers, only 8.8% have missed class, while 30.9% of occasional and 62.5% of frequent bingers have missed class. **(b)** 45.37% of subjects were nonbingers, 26.54% were occasional bingers, and 28.09% were frequent bingers. **(c)** Occasional versus nonbingers: 3.5068. Frequent versus nonbingers: 7.0937. **(d)** $X^2 \doteq 2672$, df = 2, $P$ is tiny.

**9.19 (b)** $X^2 = 2.591$, df = 1, $P = 0.108$.

**9.21 (a)** 146 women/No, 97 men/No. **(b)** For example, 19.34% of women, versus 7.62% of men, have tried low-fat diets. **(c)** $X^2 = 7.143$, df = 1, $P = 0.008$.

**9.23 (a)** $X^2 = 76.7$, df = 2, $P < 0.0001$. **(b)** Even with much smaller numbers of students, $P$ is still very small. **(c)** Our conclusion might not hold for the true percents. **(d)** Lack of independence could cause the estimated percents to be too large or too small.

**9.25** $X^2 = 12.0$, df = 1, $P = 0.001$. The smallest expected count is 6, so the test is valid.

**9.27** $X^2 = 23.1$, df = 4, $P < 0.0005$. Dog owners have less education, and cat owners more, than we would expect if there were no relationship between pet ownership and educational level.

**9.29** The missing entries are 202, 64, 38, 33. $X^2 = 50.5$, df = 9, $P < 0.0005$. The largest contributions to $X^2$ come from chemistry/engineering, physics/engineering, and biology/liberal arts (more than expected), and biology/engineering and chemistry/liberal arts (less than expected).

**9.31** $X^2 = 3.955$, df = 4, $P = 0.413$.

**9.33 (a)** Cats: $X^2 = 6.611$, df = 2, $P = 0.037$. Dogs: $X^2 = 26.939$, df = 2, $P < 0.0005$. **(b)** Dogs from pet stores are less likely to go to a shelter, while "other source" dogs are more likely to go. **(c)** The control group data should be reasonably like an SRS.

**9.35** $X^2 = 43.487$, df = 12, $P < 0.0005$. Science has a large proportion of low-scoring students, while liberal arts/education has a large proportion of high-scoring students.

**9.37** $X^2 = 852.433$, df = 1, $P < 0.0005$.

**9.39 (a)** $X^2 = 2.506$, df = 2, $P = 0.286$. **(b)** Divide each echinacea count by 337 and each placebo count by 370. **(c)** The only significant results are for rash ($z = 2.74$, $P = 0.0061$), drowsiness ($z = 2.09$, $P = 0.0366$), and other ($z = 2.09$, $P = 0.0366$). A $10 \times 2$ table would not be appropriate, because each URI could have multiple adverse events. **(d)** All results are unfavorable to echinacea, so we are not concerned with having detected "false-positives." **(e)** We do not have independent observations, but we would expect the dependence to have the same effect on both groups, so our conclusions should be fairly reliable.

**9.41** $X^2 = 3.781$, df = 3, $P = 0.2861$.

## CHAPTER 10

**10.1 (a)** −2.5. **(b)** When $x$ increases by 1, $\mu_y$ decreases by 2.5. **(c)** 15.5. **(d)** 11.5 and 19.5.

**10.3 (a)** An increase of 7.16 to 8.58 mpg. **(b)** A decrease of 7.16 to 8.58 mpg. **(c)** An increase of 3.58 to 4.29 mpg.

**10.5 (a)** The plot suggests a linear increase. **(b)** $\hat{y} = -3271.9667 + 1.65x$. **(c)** Residuals: 0.01667, −0.03333, 0.01667. $s \doteq 0.04082$. **(d)** Given $x$ (the year), spending comes from an $N(\mu_y, \sigma)$ distribution, where $\mu_y = \beta_0 + \beta_1 x$. Estimates: $b_0 \doteq -3271.9667$, $b_1 \doteq 1.65$, $s \doteq 0.04082$. **(e)** With 95% confidence, R&D spending increases from 1.283 to 2.017 billion dollars per year.

**10.7 (a)** $\beta_0$, $\beta_1$, and $\sigma$. **(b)** $H_0$ should refer to $\beta_1$. **(c)** The confidence interval will be narrower than the prediction interval.

**10.9 (a)** $t = 1.92$, $P = 0.0677$. **(b)** $t = 0.97$, $P = 0.3437$. **(c)** $t = 1.92$, $P = 0.0581$.

**10.11 (a)** 1.2095 to 1.5765; a \$1 difference in tuition in 2000 changes 2005 tuition by between \$1.21 and \$1.58, so we estimate that tuition increased by 21% to 58%. **(b)** \$8024. **(c)** \$6717 to \$9331.

**10.13 (a)** $\hat{y} = -0.0127 + 0.0180x$, $r^2 \doteq 80.0\%$. **(b)** $H_0$: $\beta_1 = 0$; $H_a$: $\beta_1 \neq 0$; $t = 7.48$, $P < 0.0001$. **(c)** The predicted mean is 0.07712; the interval is 0.06808 to 0.08616.

**10.15 (a)** $x$ (percent forested) is right-skewed; $\bar{x} = 39.3878\%$, $s_x = 32.2043\%$. $y$ (IBI) is left-skewed; $\bar{y} = 65.9388$, $s_y = 18.2796$. **(b)** A weak positive association, with more scatter in $y$ for small $x$. **(c)** $y_i = \beta_0 + \beta_1 x_i + \epsilon_i$, $i = 1, 2, \ldots, 49$; $\epsilon_i$ are independent $N(0, \sigma)$ variables. **(d)** $H_0$: $\beta_1 = 0$; $H_a$: $\beta_1 \neq 0$. **(e)** $\widehat{\text{IBI}} = 59.9 + 0.153\,\text{Area}$; $s = 17.79$. For testing the hypotheses in (d), $t = 1.92$ and $P = 0.061$. **(f)** Residual plot shows a slight curve. **(g)** Residuals are left-skewed.

**10.17** The first change decreases $P$ (that is, the relationship is more significant) because it accentuates the positive association. The second change weakens the association, so $P$ increases (the relationship is less significant).

**10.19** Using area = 10, $\hat{y} \doteq 57.52$; using forest = 25, $\hat{y} \doteq 63.74$. Both predictions have a lot of uncertainty (the prediction intervals are about 70 units wide).

**10.21** **(a)** Both distributions are fairly symmetric. For $x$ (MOE), $\bar{x} = 1,799,180$ and $s_x = 329,253$; for $y$ (MOR), $\bar{y} = 11,185$ and $s_y = 1980$. **(b)** Put MOE on the $x$ axis. **(c)** $y_i = \beta_0 + \beta_1 x_i + \epsilon_i$, $i = 1, 2, \ldots, 32$; $\epsilon_i$ are independent $N(0, \sigma)$ variables. $\widehat{\text{MOR}} = 2653 + 0.00474\,\text{MOE}$, $s = 1238$, $t = 7.02$, $P < 0.0001$. **(d)** Assumptions appear to be met.

**10.23** **(a)** The plot is roughly linear and increasing. **(b)** The number of tornadoes in 2004 is noticeably high (1819). **(c)** $\hat{y} \doteq -28,516 + 14.86x$; the confidence interval for the slope is 11.79 to 17.93 tornadoes per year. **(d)** Apart from the large residual for 2004, there are no striking features in the plot. **(e)** The 2004 residual is an outlier; the other residuals appear to be roughly Normal.

**10.25** **(a)** $x$ (CRP) is sharply right-skewed with high outliers; $\bar{x} = 10.0322$ and $s_x = 16.5632$. $y$ (retinol) is slightly right-skewed; $\bar{y} = 0.7648$ and $s_y = 0.3949$. **(b)** No; no assumption is made about $x$-values. **(c)** $\widehat{\text{Retinol}} = 0.843 - 0.00780\,\text{CRP}$, $s = 0.3781$, $t = -2.13$, $P = 0.039$. **(d)** The high outliers in CRP are influential; residuals are right-skewed rather than Normal.

**10.27** **(a)** Both distributions are slightly right-skewed. TRAP has $\bar{x} = 13.2484$ and $s_x = 6.5282$; VO– has $\bar{y} = 889.1935$ and $s_y = 427.6161$. **(b)** Put TRAP on the $x$ axis. A moderate positive association. **(c)** $\hat{y} = 301 + 44.4x$, $s = 319.7$, $t = 4.97$, $P < 0.0001$.

**10.29** Both distributions are slightly more symmetric than before. LOGTRAP has $\bar{x} = 2.4674$ and $s_x = 0.4979$; LOGVO– has $\bar{y} = 6.6815$ and $s_y = 0.4832$. A scatterplot shows a moderate positive association, with one low point that might be influential. $\hat{y} = 5.091 + 0.6446x$, $s = 0.3674$, $t = 4.78$, $P < 0.0001$.

**10.31** **(a)** Both variables are right-skewed. Pure tones: $\bar{x} = 106.2$ and $s = 91.76$ spikes/second. Monkey calls: $\bar{y} = 176.6$ and $s_y = 111.85$ spikes/second. **(b)** A moderate positive association; the third point has the largest residual; the first point is an outlier for tone response. **(c)** $\widehat{\text{CALL}} = 93.9 + 0.778\,\text{TONE}$, $s = 87.30$, $t = 4.91$, $P < 0.0001$. **(d)** Without the first point, $\hat{y} = 101 + 0.693x$, $s = 88.14$, $t = 3.18$. Without the third point, $\hat{y} = 98.4 + 0.679x$, $s = 80.69$, $t = 4.49$. With neither, $\hat{y} = 116 + 0.466x$, $s = 79.46$, $t = 2.21$.

**10.33** **(a)** Scatterplot shows a weak negative association. $\widehat{\text{Bonds}} = 53.4 - 0.196\,\text{Stocks}$, $s = 59.88$. **(b)** $H_0$: $\beta_1 = 0$; $H_a$: $\beta_1 \neq 0$; $t = -1.27$, $P = 0.226$. **(c)** The scatterplot shows a lot of variation, so $s$ is large and $t$ is small.

**10.35** **(a)** MA angle is explanatory. **(b)** A moderate-to-weak positive linear association, with one clear outlier. **(c)** $y_i = \beta_0 + \beta_1 x_i + \epsilon_i$, $i = 1, 2, \ldots, 38$; $\epsilon_i$ are independent $N(0, \sigma)$ variables. **(d)** $H_0$: $\beta_1 = 0$; $H_a$: $\beta_1 > 0$. **(e)** $t = 1.90$, df = 36, $P = 0.033$.

**10.37** **(a)** Aside from the one high point, there is a moderate positive association. **(b)** $\widehat{\text{Wages}} = 43.4 + 0.0733\,\text{LOS}$, $t = 2.85$, $P = 0.006$. **(c)** Wages rise an average of 0.0733 wage units per week of service. **(d)** 0.0218 to 0.1247.

**10.39** **(a)** It appears to be quite linear. **(b)** $\widehat{\text{Lean}} = -61.12 + 9.3187\,\text{Year}$; $r^2 = 98.8\%$. **(c)** 8.36 to 10.28 tenths of a millimeter/year.

**10.41** **(a)** $x = 109$. **(b)** 2.9955 m. **(c)** Use a prediction interval.

**10.43** $t \doteq -4.16$, df = 116, $P < 0.0001$.

**10.45** We cannot reject $H_0$: $\rho = 0$, because $t = -1.27$, $P = 0.226$.

**10.47** df = 28, SSE = 10,152.4, MSE = 362.6.

**10.49** The standard error is 0.2150; the confidence interval is 0.223 to 1.103.

**10.51** $n = 20$: $t = 2.45$, df = 18, $P = 0.0248$. $n = 10$: $t = 1.63$, df = 8, $P = 0.1411$.

**10.53** **(a)** Strong positive linear association with one outlier (SAT 420, ACT 21). **(b)** $\widehat{\text{ACT}} = 1.63 + 0.0214\,\text{SAT}$, $t = 10.78$, $P < 0.0005$. **(c)** $r = 0.8167$.

**10.55** **(a)** $a_1 = 0.02617$, $a_0 = -2.7522$. **(c)** Mean 21.1333 and standard deviation 4.7137—the same as for the ACT scores.

**10.57** **(a)** For squared length: $\widehat{\text{Weight}} = -118 + 0.497\,\text{SQLEN}$, $r^2 = 0.977$. **(b)** For squared width: $\widehat{\text{Weight}} = -99.0 + 18.7\,\text{SQWID}$, $r^2 = 0.965$.

**10.59** IBI and area: $r = 0.4459$, $t = 3.42$, $P = 0.001$ (from Exercise 10.14). IBI and percent forested: $r = 0.2698$, $t = 1.92$, $P = 0.061$ (Exercise 10.15). Area and percent forested: $r = -0.2571$, $t = -1.82$, $P = 0.074$.

**10.61** $P < 0.001$: All creativity, quickness/analytical, and quickness/verbal. $P < 0.01$: Conscientiousness/analytical, conscientiousness/quantitative, quickness/quantitative, and depth/verbal. $P < 0.05$: Conscientiousness/verbal.

**10.63** **(a)** 95% confidence interval for women: 14.73 to 33.33. For men: −9.47 to 42.97. These intervals overlap quite a bit. **(b)** For women: 22.78. For men: 16.38. The women's slope standard error is smaller in part because it is divided by a large number. **(c)** Choose men with a wider variety of lean body masses.

## CHAPTER 11

**11.1** **(a)** Math GPA. **(b)** $n = 106$. **(c)** $p = 4$. **(d)** SAT Math, SAT Verbal, class rank, and mathematics placement score.

**11.3** **(a)** Math GPA should increase when any explanatory variable increases. **(b)** DFM = 4, DFE = 81. **(c)** All four coefficients are significantly different from 0 (although the intercept is not).

**11.5** The correlations are found in Figure 11.3. The scatterplots for the pairs with the largest correlations are easy to pick out. The whole-number scale for high school grades causes point clusters in those scatterplots.

**11.7** **(a)** 5.8752 to 15.7248. **(b)** 5.9784 to 15.6216. **(c)** 5.8656 to 15.7344. **(d)** 6.0480 to 15.5520.

**11.9** **(a)** $H_0$ should refer to $\beta_2$. **(b)** Squared multiple correlation. **(c)** Small $P$ implies that at least one coefficient is different from 0.

**11.11** **(a)** $y_i = \beta_0 + \beta_1 x_{i1} + \beta_2 x_{i2} + \cdots + \beta_7 x_{i7} + \epsilon_i$, where $i = 1, 2, \ldots, 140$; $\epsilon_i$ are independent $N(0, \sigma)$ random variables. **(b)** Model (df = 7), error (df = 132), and total (df = 139).

**11.13** **(a)** 530 children. **(b)** 19.36%. **(c)** Predicted fat mass is higher for females, those who take in higher percents of energy at dinner, children of parents with higher BMIs, and those with underreported intake (low values of EI/predicted BMR). **(d)** Construct a 95% confidence interval for that coefficient, then multiply by 5: percent fat mass differs by between 0.20 and 0.60.

**11.15** **(a)** $H_0$: $\beta_1 = 0$; $H_a$: $\beta_1 \neq 0$ ($t = 4.55$, $P < 0.001$). $H_0$: $\beta_2 = 0$; $H_a$: $\beta_2 \neq 0$ ($t = 2.69$, $P < 0.01$). $H_0$: $\beta_3 = 0$; $H_a$: $\beta_3 \neq 0$ ($t = 2.69$, $P < 0.01$). **(b)** Marijuana use decreases with increasing GPA and increases with popularity and depression. **(c)** The degrees of freedom of the $F$ statistic. **(d)** $H_0$ ($\beta_1 = \beta_2 = \beta_3 = 0$) is rejected in favor of $H_a$: at least one $\beta_i$ is nonzero. **(e)** Students may have lied (or erred) in their responses. **(f)** We cannot assume that students are the same everywhere.

**11.17** **(a)** U.S. subjects are less willing to pay more. **(b)** $P < 0.001$ for testing $H_0$: $\beta_1 = 0$; $H_a$: $\beta_1 \neq 0$. **(c)** The U.K. response rate is much lower than the U.S. rate, and "don't know" responses are not much better than no response at all.

**11.19** **(a)** $\widehat{\text{Score}} = 3.33 + 0.82\,\text{Unfav} + 0.57\,\text{Fav}$. **(b)** We reject $H_0$: $\beta_1 = \beta_2 = 0$ in favor of $H_a$: at least one $\beta_i$ is nonzero. **(c)** The estimates of $\beta_0$, $\beta_1$, and $\beta_2$ are all significantly different from 0. **(d)** df = 149.

**11.21** All coefficients are positive, so the associations are positive, as expected. The unfavorable coefficients are larger, so they have a stronger effect.

**11.23** The coefficient of the new quadratic term is the same as for the old quadratic term, but the constant and coefficient of accounts have changed (and are now significantly different from 0).

**11.25** The log plot appears to be reasonably linear. $\hat{y} = -5.06 + 1.29x$, $t = 6.96$, $P < 0.0005$.

**11.27** **(a)** For example, all three distributions are right-skewed. CtoF has a high outlier of 100 (the next largest value is 55). **(b)** PEER and CtoF are positively correlated ($r = 0.382$); the other two correlations are very small.

**11.29** **(a)** $y_i = \beta_0 + \beta_1 x_{i1} + \beta_2 x_{i2} + \beta_3 x_{i3} + \epsilon_i$, where $\epsilon_i$ are independent $N(0, \sigma)$ random variables. **(b)** $\widehat{\text{Score}} = 3.62 + 0.690\,\text{PEER} + 0.254\,\text{FtoS} + 0.259\,\text{CtoF}$. **(c)** PEER: 0.6418 to 0.7383. FtoS: 0.2160 to 0.2914. CtoF: 0.1929 to 0.3255. None contain 0, because all coefficients are significantly different from 0. **(d)** $R^2 \doteq 88.1\%$, $s \doteq 5.108$.

**11.31** **(a)** For example, all distributions are skewed to varying degrees—GINI and CORRUPT to the right, the other three to the left. CORRUPT and DEMOCRACY have the most skewness. **(b)** LSI and GINI have a very small correlation (0.028); all other pairs have moderate to large correlations. GINI is negatively correlated to the other three variables, while all other correlations are positive.

**11.33** **(a)** Refer to your regression output. **(b)** For example, the $t$ statistic for the GINI coefficient grows from $t = 0.25$ ($P = 0.805$) to $t = 4.13$ ($P < 0.0005$). The DEMOCRACY $t$ is 3.65 in the third model ($P < 0.0005$) but drops to 0.69 ($P = 0.491$) in the fourth model. **(c)** A good choice is to use GINI, LIFE, and CORRUPT: all three coefficients are significant, and $R^2 = 53.5\%$ is nearly the same as the fourth model from Exercise 11.32.

**11.35** **(a)** Plot suggests greater variation in VO+ for large OC. $\widehat{\text{VO+}} = 334 + 19.5\,\text{OC}$, $t = 4.73$, $P < 0.0005$. Plot of residuals against OC is slightly curved. **(b)** $\widehat{\text{VO+}} = 58 + 6.41\,\text{OC} + 53.9\,\text{TRAP}$. Coefficient of OC is not significantly different from 0 ($t = 1.25$, $P = 0.221$), but coefficient of TRAP is ($t = 3.50$, $P = 0.002$). This is consistent with the correlations found in Exercise 11.34.

**11.37** The correlations are 0.840 (LVO+ and LVO–), 0.774 (LVO+ and LOC), and 0.755 (LVO+, LTRAP). Regression equations, $t$ statistics, $R^2$, and $s$ for each model: $\widehat{\text{LVO+}} = 4.38 + 0.706\,\text{LOC}$; $t = 6.58$, $P < 0.0005$; $R^2 = 0.599$, $s = 0.3580$. $\widehat{\text{LVO+}} = 4.26 + 0.430\,\text{LOC} + 0.424\,\text{LTRAP}$; $t = 2.56$, $P = 0.016$; $t = 2.06$, $P = 0.048$; $R^2 = 0.652$, $s = 0.3394$. $\widehat{\text{LVO+}} = 0.872 + 0.392\,\text{LOC} + 0.028\,\text{LTRAP} + 0.672\,\text{LVO−}$; $t = 3.40$, $P = 0.002$; $t = 0.18$, $P = 0.862$; $t = 5.71$, $P < 0.0005$; $R^2 = 0.842$, $s = 0.2326$. As before, this suggests a model without LTRAP: $\widehat{\text{LVO+}} = 0.832 + 0.406\,\text{LOC} + 0.682\,\text{LVO−}$; $t = 4.93$, $P < 0.0005$; $t = 6.57$, $P < 0.0005$; $R^2 = 0.842$, $s = 0.2286$.

**11.39** Regression equations, $t$ statistics, $R^2$, and $s$ for each model: $\widehat{\text{LVO−}} = 5.21 + 0.441\,\text{LOC}$; $t = 3.59$, $P = 0.001$; $R^2 = 0.308$, $s = 0.4089$. $\widehat{\text{LVO−}} = 5.04 + 0.057\,\text{LOC} + 0.590\,\text{LTRAP}$; $t = 0.31$, $P = 0.761$; $t = 2.61$, $P = 0.014$; $R^2 = 0.443$, $s = 0.3732$. $\widehat{\text{LVO−}} = 1.57 - 0.293\,\text{LOC} + 0.245\,\text{LTRAP} + 0.813\,\text{LVO+}$; $t = -2.08$, $P = 0.047$; $t = 1.47$, $P = 0.152$; $t = 5.71$, $P < 0.0005$; $R^2 = 0.748$, $s = 0.2558$. $\widehat{\text{LVO−}} = 1.31 - 0.188\,\text{LOC} + 0.890\,\text{LVO+}$; $t = -1.52$, $P = 0.140$; $t = 6.57$, $P < 0.0005$; $R^2 = 0.728$, $s = 0.2611$.

**11.41** **(a)** $y_i = \beta_0 + \beta_1 x_{i1} + \beta_2 x_{i2} + \beta_3 x_{i3} + \beta_4 x_{i4} + \epsilon_i$, where $i = 1, 2, \ldots, 69$; $\epsilon_i$ are independent $N(0, \sigma)$ random variables. **(b)** $\widehat{\text{PCB}} = 0.94 + 11.9x_1 + 3.76x_2 + 3.88x_3 + 4.18x_4$. All coefficients are significantly different from 0, although the constant 0.937 is not ($t = 0.76$, $P = 0.449$).

$R^2 = 0.989$, $s = 6.382$. **(c)** The residuals appear to be roughly Normal, but with two outliers. There are no clear patterns when plotted against the explanatory variables.

**11.43 (a)** $\widehat{\text{PCB}} = -1.02 + 12.6x_1 + 0.313x_2 + 8.25x_3$, $R^2 = 0.973$, $s = 9.945$. **(b)** $b_2 = 0.313$, $P = 0.708$. **(c)** In Exercise 11.41, $b_2 = 3.76$, $P < 0.0005$.

**11.45** The model is $y_i = \beta_0 + \beta_1 x_{i1} + \beta_2 x_{i2} + \beta_3 x_{i3} + \beta_4 x_{i4} + \epsilon_i$, where $i = 1, 2, \ldots, 69$; $\epsilon_i$ are independent $N(0, \sigma)$ random variables. Regression gives $\widehat{\text{TEQ}} = 1.06 - 0.097x_1 + 0.306x_2 + 0.106x_3 - 0.0039x_4$, with $R^2 = 0.677$. Only the constant (1.06) and the PCB118 coefficient (0.306) are significantly different from 0. Residuals are slightly right-skewed and show no clear patterns when plotted with the explanatory variables.

**11.47 (a)** The correlations are all positive, ranging from 0.227 (LPCB28 and LPCB180) to 0.956 (LPCB and LPCB138). LPCB28 has one outlier (specimen 39) when plotted with the other variables; except for that point, all scatterplots appear fairly linear. **(b)** All correlations are higher with the transformed data.

**11.49** It appears that a good model is LPCB126 and LPCB28 ($R^2 = 0.768$).

**11.51** $\bar{x}$, $M$, $s$, and $IQR$ for each variable: Taste: 24.53, 20.95, 16.26, 23.9. Acetic: 5.498, 5.425, 0.571, 0.656. H2S: 5.942, 5.329, 2.127, 3.689. Lactic: 1.442, 1.450, 0.3035, 0.430. None of the variables show striking deviations from Normality. Taste and H2S are slightly right-skewed, and Acetic has two peaks. There are no outliers.

**11.53** $\widehat{\text{Taste}} = -61.5 + 15.6\,\text{Acetic}$; $t = 3.48$, $P = 0.002$. The residuals seem to have a Normal distribution but are positively associated with both H2S and Lactic.

**11.55** $\widehat{\text{Taste}} = -29.9 + 37.7\,\text{Lactic}$; $t = 5.25$, $P < 0.0005$. The residuals seem to have a Normal distribution; there are no striking patterns for residuals against the other variables.

**11.57** $\widehat{\text{Taste}} = -26.9 + 3.80\,\text{Acetic} + 5.15\,\text{H2S}$. For the coefficient of Acetic, $t = 0.84$ and $P = 0.406$. This model is not much better than the model with H2S alone; Acetic and H2S are correlated ($r = 0.618$), so Acetic does not add significant information if H2S is included.

**11.59** $\widehat{\text{Taste}} = -28.9 + 0.33\,\text{Acetic} + 3.91\,\text{H2S} + 19.7\,\text{Lactic}$. The coefficient of Acetic is not significantly different from 0 ($P = 0.942$). Residuals of this regression appear to be Normally distributed and show no patterns in scatterplots with the explanatory variables. It appears that the H2S/Lactic model is best.

## CHAPTER 12

**12.1 (a)** $H_0$ says the population means are all equal. **(b)** Experiments are best for establishing causation. **(c)** ANOVA is used when the explanatory variable has two (or more) values.

**12.3 (a)** Yes: $22/18 \doteq 1.22 < 2$. **(b)** 484, 400, and 324. **(c)** 410.2857. **(d)** 20.2555.

**12.5 (a)** This is the description of *between*-group variation. **(b)** The *sum of* squares will add. **(c)** $\sigma$ is a parameter.

**12.7** Assuming the $t$ (ANOVA) test establishes that the means are different, contrasts and multiple comparisons provide no further useful information.

**12.9 (a)** Response: egg cholesterol level. Populations: chickens with different diets or drugs. $I = 3$, $n_1 = n_2 = n_3 = 25$, $N = 75$. **(b)** Response: rating on five-point scale. Populations: the three groups of students. $I = 3$, $n_1 = 31$, $n_2 = 18$, $n_3 = 45$, $N = 94$. **(c)** Response: quiz score. Populations: students in each TA group. $I = 3$, $n_1 = n_2 = n_3 = 14$, $N = 42$.

**12.11** For all three situations, we test $H_0$: $\mu_1 = \mu_2 = \mu_3$; $H_a$: at least one mean is different. **(a)** DFM 2, DFE 72, DFT 74. $F(2, 72)$. **(b)** DFM 2, DFE 91, DFT 93. $F(2, 91)$. **(c)** DFM 2, DFE 39, DFT 41. $F(2, 39)$.

**12.13 (a)** This sounds like a fairly well designed experiment, so the results should at least apply to this farmer's breed of chicken. **(b)** It would be good to know what proportion of the total student body falls in each of these groups—that is, is anyone overrepresented in this sample? **(c)** Effectiveness teaching one topic (power calculations) might not reflect overall effectiveness.

**12.15 (a)** df 3 and 28; $2.95 < F < 3.63$. **(c)** $0.025 < P < 0.050$. **(d)** We can conclude that at least one mean is different.

**12.17 (a)** df 4 and 40; $F = 2.54$, $0.050 < P < 0.100$. **(b)** df 3 and 24; $F \doteq 2.09$, $P > 0.100$.

**12.19 (a)** $F$ can be made very small (close to 0), and $P$ close to 1. **(b)** $F$ increases, and $P$ decreases.

**12.21 (a)** Matched pairs $t$ methods; we examine the change in reaction time for each subject. **(b)** No; we do not have 4 independent samples.

**12.23 (a)** df 3 and 2286. **(b)** $F \doteq 2.5304$. **(c)** $P = 0.0555$.

**12.25 (a)** Activity seems to increase with both drugs, and Drug B appears to have a greater effect. **(b)** Yes; the standard deviation ratio is 1.4. $s_p \doteq 3.154$. **(c)** df = 4 and 15. **(d)** $0.01 < P < 0.025$ ($P = 0.0156$).

**12.27 (a)** The variation in sample size is some cause for concern, but there can be no extreme outliers in a 1-to-7 scale, so ANOVA is probably reliable. **(b)** Yes: $1.26/1.03 = 1.22 < 2$. **(c)** $F(4, 405)$, $P = 0.0002$. **(d)** Hispanic Americans are highest, Japanese are in the middle, the other three are lowest.

**12.29 (a)** Immediate: $n = 2$, $\bar{x} = 48.705$, $s = 1.534$, $\text{SE}_{\bar{x}} = 1.085$ mg/100 g. One day: $n = 2$, $\bar{x} = 41.955$, $s = 2.128$, $\text{SE}_{\bar{x}} = 1.505$ mg/100 g. Three days: $n = 2$, $\bar{x} = 21.795$, $s = 0.771$, $\text{SE}_{\bar{x}} = 0.545$ mg/100 g. Five days: $n = 2$, $\bar{x} = 12.415$, $s = 1.082$, $\text{SE}_{\bar{x}} = 0.765$ mg/100 g. Seven days: $n = 2$, $\bar{x} = 8.320$, $s = 0.269$, $\text{SE}_{\bar{x}} = 0.190$ mg/100 g.

**(b)** $H_0$: $\mu_1 = \mu_2 = \cdots = \mu_5$; $H_a$: not all $\mu_i$ are equal. $F = 367.74$ with df 4 and 5; $P < 0.0005$, so we reject $H_0$. **(c)** Vitamin C content decreases over time.

**12.31** Vitamin A—Immediate: $n = 2$, $\bar{x} = 3.350$, $s = 0.01414$, $SE_{\bar{x}} = 0.010$ mg/100 g. One day: $n = 2$, $\bar{x} = 3.240$, $s = 0.05657$, $SE_{\bar{x}} = 0.040$ mg/100 g. Three days: $n = 2$, $\bar{x} = 3.210$, $s = 0.07071$, $SE_{\bar{x}} = 0.050$ mg/100 g. Five days: $n = 2$, $\bar{x} = 3.305$, $s = 0.07778$, $SE_{\bar{x}} = 0.055$ mg/100 g. Seven days: $n = 2$, $\bar{x} = 2.965$, $s = 0.06364$, $SE_{\bar{x}} = 0.045$ mg/100 g. Vitamin E—Immediate: $n = 2$, $\bar{x} = 95.30$, $s = 0.9900$, $SE_{\bar{x}} = 0.700$ mg/100 g. One day: $n = 2$, $\bar{x} = 94.45$, $s = 1.7678$, $SE_{\bar{x}} = 1.250$ mg/100 g. Three days: $n = 2$, $\bar{x} = 95.85$, $s = 2.1920$, $SE_{\bar{x}} = 1.550$ mg/100 g. Five days: $n = 2$, $\bar{x} = 96.35$, $s = 1.9092$, $SE_{\bar{x}} = 1.350$ mg/100 g. Seven days: $n = 2$, $\bar{x} = 93.70$, $s = 1.9799$, $SE_{\bar{x}} = 1.400$ mg/100 g. For vitamin A, $F = 12.09$, $P = 0.009$; we conclude that vitamin A content changes over time. For vitamin E, $F = 0.69$, $P = 0.630$; we cannot reject $H_0$.

**12.35** *H. bihai* and *H. caribaea* red distributions are slightly skewed. *H. bihai*: $n = 16$, $\bar{x} = 47.597$, $s = 1.213$ mm. *H. caribaea* red: $n = 23$, $\bar{x} = 39.711$, $s = 1.799$ mm. *H. caribaea* yellow: $n = 15$, $\bar{x} = 36.180$, $s = 0.975$ mm. This just meets our rule for standard deviations. ANOVA gives $F = 259.12$, df 2 and 51, $P < 0.0005$, so we conclude the means are different.

**12.37** *H. bihai*: $n = 16$, $\bar{x} = 3.8625$, $s = 0.0251$. *H. caribaea* red: $n = 23$, $\bar{x} = 3.6807$, $s = 0.0450$. *H. caribaea* yellow: $n = 15$, $\bar{x} = 3.5882$, $s = 0.0270$. ANOVA gives $F = 244.27$, df 2 and 51, $P < 0.0005$, so we conclude the means are different.

**12.39** **(a)** All three distributions show no particular skewness. Control: $n = 15$, $\bar{x} = 0.21887$, $s = 0.01159$ g/cm$^2$. Low dose: $n = 15$, $\bar{x} = 0.21593$, $s = 0.01151$ g/cm$^2$. High dose: $n = 15$, $\bar{x} = 0.23507$, $s = 0.01877$ g/cm$^2$. **(b)** All three distributions appear to be nearly Normal. **(c)** $F = 7.72$, df 2 and 42, $P = 0.001$. **(d)** For Bonferroni, $t^{**} = 2.49$ and MSD $= 0.0131$. The high-dose mean is significantly different from the other two. **(e)** High doses increase BMD.

**12.41** For Bonferroni, $t^{**} = 2.67$ and MSD $= 0.1476$. All six differences are significant.

**12.43** For Bonferroni, $t^{**} = 2.71$ and $s_p \doteq 2.7348$. The Piano mean is significantly higher than the other three, but the other three means are not significantly different.

**12.45** **(a)** Four months: $n = 19$, $\bar{x} = 570.0$, $s = 123.0$ kcal/d. Five months: $n = 18$, $\bar{x} = 483.0$, $s = 112.9$ kcal/d. Six months: $n = 8$, $\bar{x} = 541.9$, $s = 94.0$ kcal/d. Pooling is reasonable. **(b)** $F = 2.72$, df 2 and 42, $P = 0.078$. We cannot reject $H_0$.

**12.47** **(a)** Control: $n = 10$, $\bar{x} = 601.10$, $s = 27.36$ mg/cm$^3$. Low jump: $n = 10$, $\bar{x} = 612.50$, $s = 19.33$ mg/cm$^3$. High jump: $n = 10$, $\bar{x} = 638.70$, $s = 16.59$ mg/cm$^3$. Pooling is reasonable. **(b)** $F = 7.98$, df 2 and 27, $P = 0.002$. We conclude that not all means are equal.

**12.49** **(a)** Aluminum: $n = 4$, $\bar{x} = 2.0575$, $s = 0.2520$ mg/100 g. Clay: $n = 4$, $\bar{x} = 2.1775$, $s = 0.6213$ mg/100 g. Iron: $n = 4$, $\bar{x} = 4.6800$, $s = 0.6283$ mg/100 g. Pooling is risky because $0.6283/0.2520 = 2.49 > 2$. **(b)** $F = 31.16$, df 2 and 9, $P < 0.0005$. We cautiously conclude that the means are not the same.

**12.51** **(a)** ECM1: $n = 3$, $\bar{x} = 65.0\%$, $s = 8.66\%$. ECM2: $n = 3$, $\bar{x} = 63.33\%$, $s = 2.89\%$. ECM3: $n = 3$, $\bar{x} = 73.33\%$, $s = 2.89\%$. MAT1: $n = 3$, $\bar{x} = 23.33\%$, $s = 2.89\%$. MAT2: $n = 3$, $\bar{x} = 6.67\%$, $s = 2.89\%$. MAT3: $n = 3$, $\bar{x} = 11.67\%$, $s = 2.89\%$. Pooling is risky because $8.66/2.89 > 2$. **(b)** $F = 137.94$, df 5 and 12, $P < 0.0005$. We conclude that the means are not the same.

**12.53** **(a)** $\psi_1 = \mu_1 - 0.5\mu_2 - 0.5\mu_4$ and $\psi_2 = \mu_2 - \mu_3 - (\mu_4 - \mu_5)$. **(b)** $SE_{c_1} \doteq 1.9316$ and $SE_{c_2} \doteq 3.1544$. **(c)** Neither contrast is significant ($t_1 \doteq -1.035$ and $t_2 \doteq 0.872$).

**12.55** **(a)** The plot shows granularity (which varies between groups), but that should not make us question independence; it is due to the fact that the scores are all integers. **(b)** The ratio of the largest to the smallest standard deviations is less than 2. **(c)** Apart from the granularity, the quantile plots are reasonably straight. **(d)** Again, apart from the granularity, the quantile plots look pretty good.

**12.57** **(a)** $\psi_1 = (\mu_1 + \mu_2 + \mu_3)/3 - \mu_4$, $\psi_2 = (\mu_1 + \mu_2)/2 - \mu_3$, $\psi_3 = \mu_1 - \mu_2$. **(b)** The pooled standard deviation is $s_p = 1.1958$. $SE_{c_1} \doteq 0.2355$, $SE_{c_2} \doteq 0.1413$, $SE_{c_3} \doteq 0.1609$. **(c)** Testing $H_0$: $\psi_i = 0$; $H_a$: $\psi_i \neq 0$ for each contrast, we find $c_1 = -12.51$, $t_1 = -53.17$, $P_1 < 0.0005$; $c_2 = 1.269$, $t_2 = 8.98$, $P_2 < 0.0005$; $c_3 = 0.191$, $t_3 = 1.19$, $P_3 \doteq 0.2359$. The Placebo mean is significantly higher than the average of the other three, while the Keto mean is significantly lower than the average of the two Pyr means. The difference between the Pyr means is not significant (meaning the second application of the shampoo is of little benefit).

**12.59** **(a)** Vitamin A—Immediate: $n = 2$, $\bar{x} = 67.0\%$, $s = 0.2828\%$, $SE_{\bar{x}} = 0.2\%$. One day: $n = 2$, $\bar{x} = 64.8\%$, $s = 1.1314\%$, $SE_{\bar{x}} = 0.8\%$. Three days: $n = 2$, $\bar{x} = 64.2\%$, $s = 1.4142\%$, $SE_{\bar{x}} = 1.0\%$. Five days: $n = 2$, $\bar{x} = 66.1\%$, $s = 1.5556\%$, $SE_{\bar{x}} = 1.1\%$. Seven days: $n = 2$, $\bar{x} = 59.3\%$, $s = 1.2728\%$, $SE_{\bar{x}} = 0.9\%$. The transformation has no effect on vitamin E, since the number of milligrams remaining is also the percent of the original 100 mg. Although the SS and MS entries for vitamin A are different from those of Exercise 12.31, everything else is the same: $F = 12.09$, df 4 and 5, $P = 0.009$. So we (again) reject $H_0$ and conclude that vitamin A content decreases over time. Since the vitamin E numbers are unchanged, we again fail to reject $H_0$ ($F = 0.69$, df 4 and 5, $P = 0.630$).

**12.61** All distributions are reasonably Normal, and standard deviations are close enough to justify pooling. For PRE1, $F = 1.13$, df 2 and 63, $P = 0.329$. For PRE2, $F = 0.11$, df 2 and 63, $P = 0.895$. Neither set of pretest scores suggests a difference in means.

**12.63** $\hat{y} = 4.36 - 0.116x$. The regression is significant (that is, the slope is significantly different from 0): $t = -13.31$, df = 158, $P < 0.0005$. Regression explains $r^2 = 52.9\%$ of the variation in expected price. (This is similar to the ANOVA value: $R^2 = 53.5\%$.) The granularity of the "Number of promotions" observations makes interpreting the plot a bit tricky. For five promotions, the residuals seem more likely to be negative, while for three promotions, the residuals are weighted toward the positive side. This suggests that a linear model may not be appropriate.

**12.67 (b)** Answers will vary with choice of $H_a$ and desired power. For example, with $\mu_1 = \mu_2 = 4.4$, $\mu_3 = 5$, $\sigma = 1.2$, three samples of size 75 will produce power 0.89.

**12.69** The design can be similar, although the types of music might be different. Bear in mind that spending at a casual restaurant will likely be less than at the restaurants examined in Exercise 12.24; this might also mean that the standard deviations could be smaller. Decide how big a difference in mean spending you would want to detect, then do some power computations.

## CHAPTER 13

**13.1 (a)** Two-way ANOVA is used when there are two explanatory variables. **(b)** Each level of A should occur with all three levels of B. **(c)** The RESIDUAL part of the model represents the error.

**13.3 (a)** df 3 and 30. **(c)** $0.05 < P < 0.10$. **(d)** The plot would look somewhat parallel, because the interaction term is not significantly different from 0.

**13.5 (a)** Factors: gender ($I = 2$) and age ($J = 3$). Response: percent of pretend play. $N = 66$. **(b)** Factors: time after harvest ($I = 5$) and amount of water ($J = 2$). Response: percent of seeds germinating. $N = 30$. **(c)** Factors: media type ($I = 2$) and incubation time ($J = 3$). Response: growth of virus. $N = 24$.

**13.7 (a)** The plot suggests a possible interaction. **(b)** By subjecting the same individual to all four treatments, rather than four individuals to one treatment each, we reduce the variability.

**13.9 (a)** Intervention, 11.6; control, 9.967. Baseline, 10.0; 3 months, 11.2; 6 months, 11.15. Overall, 10.783. The row means suggest that the intervention group showed more improvement than the control group. **(b)** Interaction means that the mean number of actions changes differently over time for the two groups.

**13.11 (a)** High school and college results are very similar for male protagonists, but college students appear to be less accepting of lying female protagonists. **(b)** df = 1 and 486; $0.05 < P < 0.10$ (software: 0.0716). **(c)** The results might generalize to high schools and universities that are similar geographically and demographically, but it would be risky to assume they apply to, for example, East Coast high schools or universities.

**13.13** There were no significant effects (although B and AB are close): $F_A$ has df = 2 and 24 and $P = 0.2369$. $F_B$ has df = 1 and 24 and $P = 0.0608$. $F_{AB}$ has df = 2 and 24 and $P = 0.0721$.

**13.15 (a)** Familiar increases with repetition and has the higher rating for 1 and 3 repetitions, while Unfamiliar is higher for 2 repetitions. **(b)** The interaction is that Unfamiliar does better with 2 repetitions and worse for 1 and 3.

**13.17** $s_p = 1.308$. Pooling is reasonable, as $1.46/1.16 = 1.26 < 2$.

**13.19** For example, are opinions of university employees (and/or West Coast residents) similar to those of other groups? What do experts consider to be "good" ads? Did the content of the news show affect responses to the ads?

**13.21 (b)** There seems to be a fairly large difference between the means based on how much the rats were allowed to eat but not very much difference based on the chromium level. There may be an interaction: the NM mean is lower than the LM mean, while the NR mean is higher than the LR mean. **(c)** L mean: 4.86. N mean: 4.871. M mean: 4.485. R mean: 5.246. LR minus LM: 0.63. NR minus NM: 0.892. Mean GITH levels are lower for M than for R; there is not much difference between L and N. The difference between M and R is greater among rats who had normal chromium levels in their diets (N).

**13.23** The "Other" category had the lowest mean SATM score for both genders; this is apparent from a graph of the means as well as from the marginal means (CS, 605; EO, 624.5; O, 566). Males had higher mean scores in CS and O, while females were slightly higher in EO; this seems to be an interaction. Overall, the marginal means are 611.7 (males) and 585.3 (females).

**13.25 (a)** $n = 3$ for all combinations. The means and standard deviations are $\bar{x}_{E1,4} = 65.00$, $s_{E1,4} = 8.66$; $\bar{x}_{E1,8} = 63.33$, $s_{E1,8} = 2.89$; $\bar{x}_{E2,4} = 63.33$, $s_{E2,4} = 2.89$; $\bar{x}_{E2,8} = 63.33$, $s_{E2,8} = 5.77$; $\bar{x}_{E3,4} = 73.33$, $s_{E3,4} = 2.89$; $\bar{x}_{E3,8} = 73.33$, $s_{E3,8} = 5.77$; $\bar{x}_{M1,4} = 23.33$, $s_{M1,4} = 2.89$; $\bar{x}_{M1,8} = 21.67$, $s_{M1,8} = 5.77$; $\bar{x}_{M2,4} = 6.67$, $s_{M2,4} = 2.89$; $\bar{x}_{M2,8} = 6.67$, $s_{M2,8} = 2.89$; $\bar{x}_{M3,4} = 11.67$, $s_{M3,4} = 2.89$; $\bar{x}_{M3,8} = 10.00$, $s_{M3,8} = 5.00$. Apart from the first standard deviation, the ratio is 2 for the rest. **(b)** ECM means are all higher than MAT means. Time and interaction effects are not clearly suggested. **(c)** Only Material ($F = 251.26$, df 5 and 24, $P < 0.0005$) is significant. For Time, $F = 0.29$, df 1 and 24, $P = 0.595$. For interaction, $F = 0.06$, df 5 and 24, $P = 0.998$.

**13.27** For each time period, there is a significant difference among materials. The pooled standard deviations are 6.236, 4.410, and 4.859. For Bonferroni and $\alpha = 0.05$, $t^{**} = 3.65$, so the MSDs are 18.6, 13.1, and 14.5. The only ECM/MAT difference that is not significant is ECM2 and MAT1 at 2 weeks.

**13.29** Yes, the iron-pot means are the highest, and $F$ for testing the effect of the pot type is very large.

**13.31** **(a)** In the order listed in the table: $\bar{x}_{11} = 25.0307$, $s_{11} = 0.0011541$; $\bar{x}_{12} = 25.0280$, $s_{12} = 0$; $\bar{x}_{13} = 25.0260$, $s_{13} = 0$; $\bar{x}_{21} = 25.0167$, $s_{21} = 0.0011541$; $\bar{x}_{22} = 25.0200$, $s_{22} = 0.0019999$; $\bar{x}_{23} = 25.0160$, $s_{23} = 0$; $\bar{x}_{31} = 25.0063$, $s_{31} = 0.0015275$; $\bar{x}_{32} = 25.0127$, $s_{32} = 0.0011552$; $\bar{x}_{33} = 25.0093$, $s_{33} = 0.0011552$; $\bar{x}_{41} = 25.0120$, $s_{41} = 0$; $\bar{x}_{42} = 25.0193$, $s_{42} = 0.0011552$; $\bar{x}_{43} = 25.0140$, $s_{43} = 0.0039997$; $\bar{x}_{51} = 24.9973$, $s_{51} = 0.0011541$; $\bar{x}_{52} = 25.0060$, $s_{52} = 0$; $\bar{x}_{53} = 25.0003$, $s_{53} = 0.0015277$. **(b)** Except for Tool 1, mean diameter is highest at Time 2. Tool 1 had the highest mean diameters, followed by Tool 2, Tool 4, Tool 3, and Tool 5. **(c)** $F_A = 412.98$, df 4 and 30, $P < 0.0005$. $F_B = 43.61$, df 2 and 30, $P < 0.0005$. $F_{AB} = 7.65$, df 8 and 30, $P < 0.0005$. **(d)** There is strong evidence of a difference in mean diameter among the tools (A) and among the times (B). There is also an interaction (specifically, Tool 1's mean diameters changed differently over time compared to the other tools).

**13.33** **(a)** In the order listed in the table: $\bar{x}_{1,40} = 4.423$, $s_{1,40} = 0.1848$; $\bar{x}_{1,30} = 4.225$, $s_{1,30} = 0.3856$; $\bar{x}_{1,20} = 4.689$, $s_{1,20} = 0.2331$; $\bar{x}_{1,10} = 4.920$, $s_{1,10} = 0.1520$; $\bar{x}_{3,40} = 4.284$, $s_{3,40} = 0.2040$; $\bar{x}_{3,30} = 4.097$, $s_{3,30} = 0.2346$; $\bar{x}_{3,20} = 4.524$, $s_{3,20} = 0.2707$; $\bar{x}_{3,10} = 4.756$, $s_{3,10} = 0.2429$; $\bar{x}_{5,40} = 4.058$, $s_{5,40} = 0.1760$; $\bar{x}_{5,30} = 3.890$, $s_{5,30} = 0.1629$; $\bar{x}_{5,20} = 4.251$, $s_{5,20} = 0.2648$; $\bar{x}_{5,10} = 4.393$, $s_{5,10} = 0.2685$; $\bar{x}_{7,40} = 3.780$, $s_{7,40} = 0.2144$; $\bar{x}_{7,30} = 3.760$, $s_{7,30} = 0.2618$; $\bar{x}_{7,20} = 4.094$, $s_{7,20} = 0.2407$; $\bar{x}_{7,10} = 4.269$, $s_{7,10} = 0.2699$. The mean expected price decreases as percent discount increases and also as the number of promotions increases. **(b)** With A = number of promotions and B = percent discount, $F_A = 47.73$, df 3 and 144, $P < 0.0005$. $F_B = 47.42$, df 3 and 144, $P < 0.0005$. $F_{AB} = 0.44$, df 9 and 144, $P = 0.912$. **(c)** Both main effects are significant, but the interaction is not.

**13.35** **(a)** All three $F$-values have df 1 and 945, the $P$-values are $< 0.001$, $< 0.001$, and 0.1477. Gender and handedness both have significant effects on mean lifetime, but there is no interaction. **(b)** Women live about 6 years longer than men (on the average), while right-handed people average 9 more years of life than left-handed people. Handedness affects both genders in the same way, and vice versa.

**13.37** **(a)** and **(b)** The first three means and standard deviations are $\bar{x}_{1,1} = 3.2543$, $s_{1,1} = 0.2287$; $\bar{x}_{1,2} = 2.7636$, $s_{1,2} = 0.0666$; $\bar{x}_{1,3} = 2.8429$, $s_{1,3} = 0.2333$. The standard deviations range from 0.0666 to 0.3437, for a ratio of 5.16—larger than we like. **(c)** For Plant, $F = 1301.32$, df 3 and 224, $P < 0.0005$. For Water, $F = 9.76$, df 6 and 224, $P < 0.0005$. For interaction, $F = 5.97$, df 18 and 224, $P < 0.0005$.

**13.39** The seven $F$ statistics are 184.05, 115.93, 208.87, 218.37, 220.01, 174.14, and 230.17, all with df 3 and 32 and $P < 0.0005$.

**13.41** Fresh: Plant $F = 81.45$, df 3 and 84, $P < 0.0005$; Water $F = 43.71$, df 6 and 84, $P < 0.0005$; interaction $F = 1.79$, df 18 and 84, $P = 0.040$. Dry: Plant $F = 79.93$, df 3 and 84, $P < 0.0005$; Water $F = 44.79$, df 6 and 84, $P < 0.0005$; interaction $F = 2.22$, df 18 and 84, $P = 0.008$.

**13.43** Fresh: The seven $F$ statistics are 15.88, 11.81, 62.08, 10.83, 22.62, 8.20, and 10.81, all with df 3 and 12 and $P \le 0.003$. Fresh: The seven $F$ statistics are 8.14, 26.26, 22.58, 11.86, 21.38, 14.77, and 8.66, all with df 3 and 12 and $P \le 0.003$.

**13.45** **(a)** Gender: df 1 and 174. Floral characteristic: df 2 and 174. Interaction: df 2 and 174. **(b)** Damage to males was higher for all characteristics. For males, damage was higher under characteristic level 3, while for females, the highest damage occurred at level 2. **(c)** Three of the standard deviations are at least half as large as the means. Because the response variable (leaf damage) had to be nonnegative, this suggests that these distributions are right-skewed.

**13.47** Men in CS: $n = 39$, $\bar{x} = 7.79487$, $s = 1.50752$. Men in EOS: $n = 39$, $\bar{x} = 7.48718$, $s = 2.15054$. Men in Other: $n = 39$, $\bar{x} = 7.41026$, $s = 1.56807$. Women in CS: $n = 39$, $\bar{x} = 8.84615$, $s = 1.13644$. Women in EOS: $n = 39$, $\bar{x} = 9.25641$, $s = 0.75107$. Women in Other: $n = 39$, $\bar{x} = 8.61539$, $s = 1.16111$. The means suggest that females have higher HSE grades than males. For a given gender, there is not too much difference among majors. Normal quantile plots show no great deviations from Normality, apart from the granularity of the grades (most evident among women in EO). In the ANOVA, only the effect of gender is significant ($F = 50.32$, df 1 and 228, $P < 0.0005$).

**13.49** Men in CS: $n = 39$, $\bar{x} = 526.949$, $s = 100.937$. Men in EOS: $n = 39$, $\bar{x} = 507.846$, $s = 57.213$. Men in Other: $n = 39$, $\bar{x} = 487.564$, $s = 108.779$. Women in CS: $n = 39$, $\bar{x} = 543.385$, $s = 77.654$. Women in EOS: $n = 39$, $\bar{x} = 538.205$, $s = 102.209$. Women in Other: $n = 39$, $\bar{x} = 465.026$, $s = 82.184$. The means suggest that students who stay in the sciences have higher mean SATV scores than those who end up in the "Other" group. Female CS and EO students have higher scores than males in those majors, but males have the higher mean in the Other group. Normal quantile plots suggests some right-skewness in the "Women in CS" group and also some non-Normality in the tails of the "Women in EO" group. Other groups look reasonably Normal. In the ANOVA, only the effect of major is significant ($F = 9.32$, df 2 and 228, $P < 0.0005$).

## CHAPTER 14

**14.1** 1 to 3.

**14.3** For men: 4 to 3 (or 1.3333). For women: 9 to 11 (or 0.8182).

**14.5** For men: 0.2877. For women: −0.2007.

**14.7** If $x = 1$ for men and 0 for women, log(odds) $= -0.2007 + 0.4884x$. (If vice versa, log(odds) $= 0.2877 - 0.4884x$.) The odds ratio is 1.6296 (or 0.6136).

**14.9 (a)** Use a chi-square test with df = 6. **(b)** $H_0$ should refer to $\beta_1$. **(c)** The logistic regression model has no error term.

**14.11 (a)** $\hat{p}_w \doteq 0.2128$ and $\hat{p}_m \doteq 0.1076$. **(b)** $\text{Odds}_w \doteq 0.2704$ and $\text{odds}_m \doteq 0.1205$. **(c)** $\text{Log(odds}_w) \doteq -1.3079$ and $\text{log(odds}_m) \doteq -2.1158$.

**14.13 (a)** $b_0 \doteq -2.1158$ and $b_1 \doteq 0.8079$. **(b)** Log(odds) $= -2.1158 + 0.8079x$. **(c)** $e^{b_1} \doteq 2.2432$.

**14.15** 2.1096 to 4.1080.

**14.17 (a)** $z \doteq 8.01$. **(b)** $z^2 \doteq 64.23$.

**14.19 (a)** The men's magazine confidence interval includes 1, consistent with failing to reject $H_0$. For other explanatory variables, the interval does not include 1. **(b)** The odds that the model's clothing is not sexual are 1.27 to 2.16 times higher for magazines targeted at mature adults, 2.74 to 5.01 times higher when the model is male, and 1.11 to 2.23 times higher for magazines aimed at women. **(c)** For example, it is easier to interpret the odds ratio than the regression coefficients because it is hard to think in terms of a log-odds scale.

**14.21 (a)** $\hat{p}_{hi} \doteq 0.8022$ and $\text{odds}_{hi} = 4.0\overline{5}$. **(b)** $\hat{p}_{non} \doteq 0.6881$ and $\text{odds}_{non} \doteq 2.2059$. **(c)** The odds ratio is 1.8385.

**14.23 (a)** −0.0470 to 1.2650. **(b)** 0.9540 to 3.5430. **(c)** The interval for $\beta_1$ contains 0 (or the interval for the odds ratio contains 1), so we cannot reject $\beta_1 = 0$ at the 5% level.

**14.25 (a)** $\hat{p}_{hi} \doteq 0.01648$ and $\text{odds}_{hi} \doteq 0.01675$, or about 1 to 60. **(b)** $\hat{p}_{lo} \doteq 0.00785$ and $\text{odds}_{lo} \doteq 0.00791$, or about 1 to 126. **(c)** The odds ratio is 2.1181.

**14.27 (a)** 0.2452 to 1.2558. **(b)** $X^2 \doteq 8.47$, $0.0025 < P < 0.005$. **(c)** We have strong evidence of a difference in risk between the two groups.

**14.29 (a)** The estimated odds ratio is 2.1181; the odds-ratio interval is 1.28 to 3.51. **(b)** We are 95% confident that the odds of death from cardiovascular disease are about 1.3 to 3.5 times greater in the high-blood-pressure group.

**14.31 (a)** Log(odds) $= \beta_0 + \beta_1 x$, where $x = 1$ if the person is over 40, and 0 if the person is under 40. **(b)** $p_i$ is the probability that the $i$th person is terminated; this model assumes that the probability of termination depends on age (over/under 40). **(c)** The estimated odds ratio is 3.859. A 95% confidence interval for $b_1$ is 0.5409 to 2.1599. The odds of being terminated are 1.7 to 8.7 times greater for those over 40. **(d)** Use a multiple logistic regression model.

**14.33** $\hat{p}_c \doteq 0.6482$ and $\text{odds}_c \doteq 1.8424$; $\hat{p}_n \doteq 0.4929$ and $\text{odds}_n \doteq 0.9722$. With the model log(odds) $= \beta_0 + \beta_1 x$, where $x = 1$ for college graduates, we estimate $b_0 \doteq -0.0282$ and $b_1 \doteq 0.6393$. The odds of a college graduate using the Internet for travel arrangements are about 1.90 times higher than those for a noncollege graduate.

**14.35** For women: $\hat{p}_f \doteq 0.1414$, $\text{odds}_f \doteq 0.1646$, and $\text{log(odds}_f) \doteq -1.8040$. For men: $\hat{p}_m \doteq 0.3388$, $\text{odds}_m \doteq 0.5124$, and $\text{log(odds}_m) \doteq -0.6686$. If $x = 1$ for males, log(odds) $= -1.8040 + 1.1355x$, and the odds ratio is 3.1126.

**14.37** The fitted model is log(odds) $= -10.7799 + 6.3319x$ ($p$ is the probability that the cheese is acceptable, and $x$ is the value of Lactic). We estimate that the odds ratio increases by a factor of 562.22 for every unit increase in Lactic. For testing $\beta_1 = 0$, $X^2 = 6.66$ ($P = 0.0098$). We are 95% confident that $\beta_1$ is in the interval 1.5236 to 11.1402 and that the odds ratio increases by a factor between 4.5889 and about 68.884 for each unit increase in Lactic.

**14.39 (a)** $X^2 \doteq 33.65$ (df = 3), $P = 0.0001$. **(b)** Log(odds) $= -6.053 + 0.3710\,\text{HSM} + 0.2489\,\text{HSS} + 0.03605\,\text{HSE}$. 95% confidence intervals: 0.1158 to 0.6262, −0.0010 to 0.4988, and −0.2095 to 0.2816. **(c)** Only the coefficient of HSM is significantly different from 0, though HSS may also be useful.

**14.41 (a)** $X^2 \doteq 19.2256$, df = 3, $P = 0.0002$. **(b)** $X^2 \doteq 3.4635$, df = 2, $P = 0.1770$. **(c)** High school grades (especially HSM and, to a lesser extent, HSS) are useful, while SAT scores are not.

**14.43 (a)** Log(odds) $= -3.892 + 0.4157\,\text{Hospital}$, using 1 for Hospital A and 0 for Hospital B. The slope is not significantly different from 0 ($z = -1.47$ or $X^2 = 2.16$, $P = 0.1420$). A 95% confidence interval for $\beta_1$ is −0.1392 to 0.9706. The odds ratio is 1.515, with confidence interval 0.87 to 2.64. **(b)** Log(odds) $= -3.109 - 0.1320\,\text{Hospital} - 1.266\,\text{Condition}$, using 1 for Hospital A and 0 for Hospital B, and 1 for good condition and 0 for poor. The odds ratio is 0.8764, with confidence interval 0.48 to 1.60. **(c)** In the model with Hospital alone, the slope was positive and the odds ratio was greater than 1. When Condition is added to the model, the Hospital coefficient is negative and the odds ratio is less than 1.

## CHAPTER 15

**15.1** Group A ranks are 1, 2, 4, 6, and 8.

**15.3** To test whether the two groups of spas have the same distribution, we find $W = 21$.

**15.5** $\mu_W = 27.5$ and $\sigma_W \doteq 4.7871$. $z \doteq -1.36$, which gives $P \doteq 0.1738$; with the continuity correction, $z \doteq -1.25$, for which $P \doteq 0.2112$.

**15.7 (a)** Child 8's score may be a low outlier. **(b)** For testing $H_0$: $\mu_1 = \mu_2$; $H_a$: $\mu_1 > \mu_2$, $\bar{x}_1 = 0.676$, $\bar{x}_2 = 0.406$, and $t = 2.062$, which gives $P = 0.0447$. **(c)** We test $H_0$: scores for both groups are identically distributed; $H_a$: high-progress children systematically score higher. $W = 36$ and $P \doteq 0.0473$ or 0.0463, similar to the conclusion reached in (b).

**15.9 (a)** The 5 high-progress readers have ranks 8, 9, 4, 7, and 10. **(b)** $W = 38$; under $H_0$, $\mu_W = 27.5$ and $\sigma_W \doteq 4.7871$. **(c)** $z \doteq 2.09$, $P = 0.0183$. **(d)** The tied observations have ranks 4.5 and 8.5.

**15.11 (a)** The 16-week distribution is much more spread out. **(b)** $W = 33$ and $P \doteq 0.1481$.

**15.13 (a)** Unlogged plots appear to have higher species counts. **(b)** $W = 159$ and $P \doteq 0.0298$.

**15.15** Counts and percents suggest that women give higher ratings. $W = 32{,}267.5$ and $P = 0.0003$.

**15.17 (a)** $X^2 = 3.955$, df $= 4$, $P = 0.413$. **(b)** $W = 56{,}370$ and $P \doteq 0.5$.

**15.19** We test $H_0$: food scores and activities scores have the same distribution; $H_a$: food scores are higher. Ranking food minus activities differences gives $W^+ = 6$ and $P = 0.853$.

**15.21** One difference was 0, so use $n = 6$ differences: $\mu_{W^+} \doteq 10.5$, $\sigma_{W^+} \doteq 4.7697$, and $P = P(W^+ \geq 5.5) \doteq 0.8531$.

**15.23** We examine the heart rate increase (final minus resting) from low-rate exercise; our hypotheses are $H_0$: median $= 0$; $H_a$: median $> 0$. $W^+ = 10$ and $P \doteq 0.0505$.

**15.25** $W^+ = 119$ and $P < 0.0005$.

**15.27** The mean and median of sfair − srest are 0.5149 and 0.5. For the one-sided alternative (food at fairs is systematically rated higher [less safe] than restaurant food), $W^+ = 10{,}850.5$ and $P < 0.0005$.

**15.29 (a)** The distribution is right-skewed but has no outliers. **(b)** $W^+ = 31$ and $P = 0.556$.

**15.31 (a)** $W^+ = 0$ and $P = 0$. **(b)** 3.75 to 5.90 kg (different software might produce different results).

**15.33 (a)** The diagram should show 10 rats assigned to each group; apply the treatments, then observe bone density. **(b)** Stemplots suggest greater density for high-jump rats, and a greater spread for the control group. **(c)** $H = 10.66$ and $P = 0.005$. ANOVA assumes Normal distributions with the same standard deviation and tests whether the means are all equal. Kruskal-Wallis tests whether the distributions are the same (but not necessarily Normal). **(d)** There is strong evidence that the high-jump group has the highest average rank (and the highest density), the low-jump group is in the middle, and the control group is lowest.

**15.35 (a)** $I = 4$, $n_i = 6$, $N = 24$. **(b)** Yellow: $17 + 20 + 21 + 22 + 23 + 24 = 127$. White: $3 + 4 + 5.5 + 9.5 + 9.5 + 12.5 = 44$. Green: $7 + 14 + 15 + 16 + 18 + 19 = 89$. Blue: $1 + 2 + 5.5 + 8 + 11 + 12.5 = 40$. **(c)** $H = 16.95\bar{3}$; df $= 3$; $0.0005 < P < 0.001$.

**15.37** The Kruskal-Wallis test needs two or more independent samples.

**15.39 (a)** The data support this: 32.2% of high-SES subjects have never smoked, compared to 17.3% and 23.7% of middle- and low-SES subjects. Also, only 24.2% of high-SES subjects are current smokers, versus 42.3% and 46.2% of those in the other groups. **(b)** $X^2 = 18.510$, df $= 4$, $P = 0.001$. **(c)** $H = 12.72$, df $= 2$, $P = 0.002$ (adjusted for ties: $H = 14.43$, $P = 0.001$).

**15.41 (a)** Clearly right-skewed, with high outliers. **(b)** $H_0$: $\mu_3 = \mu_4$; $H_a$: $\mu_3 \neq \mu_4$. $t \doteq -3.08$ with either df $= 12.1$ ($P = 0.0095$) or df $= 8$ ($P = 0.0151$). **(c)** $H_0$: medians equal; $H_a$: medians different. $W = 447$ and $P \doteq 0.0028$.

**15.43** $H = 45.35$, df $= 2$, $P < 0.0005$.

**15.45** For meat, $W = 15$ and $P = 0.4705$, and for legumes, $W = 10.5$ and $P = 0.0433$ (or 0.0421, adjusted for ties).

**15.47 (a)** *Bihai*-red, *bihai*-yellow, and red-yellow. **(b)** $W_1 = 504$, $W_2 = 376$, $W_3 = 614$. All $P$-values are reported as 0 to four decimal places. **(c)** All three comparisons are significant at the overall 0.05 level (and would even be significant at the overall 0.01 level).

# NOTES AND DATA SOURCES

## TO TEACHERS: About This Book

1. The committee's report is George Cobb, "Teaching statistics," in L. A. Steen (ed.), *Heeding the Call for Change: Suggestions for Curricular Action,* Mathematical Association of America, 1990, pp. 3–43. A summary has been endorsed by the ASA Board of Directors.

2. See, for example, the evaluation of the current state of the first course in statistics, with discussion by leading statisticians from industry as well as academia, in David S. Moore and discussants, "New pedagogy and new content: the case of statistics," *International Statistical Review,* 65 (1997), pp. 123–165.

3. A. Agresti and B. A. Coull, "Approximate is better than 'exact' for interval estimation of binomial proportions," *The American Statistician,* 52 (1998), pp. 119–126. Alan Agresti and Brian Caffo, "Simple and effective confidence intervals for proportions and differences of proportions result from adding two successes and two failures," *The American Statistician,* 45 (2000), pp. 280–288.

4. Lawrence D. Brown, Tony Cai, and Anirban DasGupta, "Confidence intervals for a binomial proportion and asymptotic expansions," *Annals of Statistics,* 30 (2002), pp. 160–201. For an overview and compelling evidence of the weakness of the traditional intervals see by the same authors "Interval estimation for a binomial proportion," *Statistical Science,* 16 (2001), pp. 101–133.

5. A detailed discussion appears in George Cobb and David S. Moore, "Mathematics, statistics, and teaching," *American Mathematical Monthly,* 104 (1997), pp. 801–823. Readers interested in relations between mathematics and statistics as fields might also look at David S. Moore and George Cobb, "Statistics and mathematics: tension and cooperation," *American Mathematical Monthly,* 107 (2000), pp. 615–630.

## TO STUDENTS: What Is Statistics?

1. Federal Interagency Forum on Child and Family Studies, *America's Children in Brief: Key National Indicators of Well-Being,* 2004, `childstats.gov/americaschildren`.

2. The rise of statistics from the physical, life, and behavioral sciences is described in detail by S. M. Stigler, *The History of Statistics: The Measurement of Uncertainty before 1900,* Harvard-Belknap, 1986.

## CHAPTER 1

1. See the National Service-Learning Clearinghouse Web site at `www.servicelearning.org/index.php`.

2. From the Fatal Accident Reporting System Web site, `www-fars.nhtsa.dot.gov`.

3. Data for 2000, collected by the Current Population Survey and reported in the 2002 *Statistical Abstract of the United States.*

4. Haipeng Shen, "Nonparametric regression for problems involving lognormal distributions," PhD thesis, University of Pennsylvania, 2003. Thanks to Haipeng Shen and Larry Brown for sharing the data.

5. United Nations data found at `earthtrends.wri.org`.

6. James T. Fleming, "The measurement of children's perception of difficulty in reading materials," *Research in the Teaching of English,* 1 (1967), pp. 136–156.

7. Wayne Nelson, "Theory and applications of hazard plotting for censored failure data," *Technometrics,* 14 (1972), pp. 945–966.

8. Read from a graph in Peter A. Raymond and Jonathan J. Cole, "Increase in the export of alkalinity from North America's largest river," *Science,* 301 (2003), pp. 88–91.

9. Monthly gasoline price index from the Consumer Price Index, from the Bureau of Labor Statistics, `www.bls.gov`, converted into dollars.

10. From the Color Assignment Web site of Joe Hallock, `www.joehallock.com/edu/COM498/index.html`.

11. U.S. Environmental Protection Agency, *Municipal Solid Waste in the United States: 2000 Facts and Figures,* document EPA530-R-02-001, 2002.

12. Robyn Greenspan, "The deadly duo: spam and viruses, October 2003," found at `cyberatlas.internet.com`.

13. National Center for Education Statistics, NEDRC Table Library, at `nces.ed.gov/surveys/npsas/table_library`.

14. Debora L. Arsenau, "Comparison of diet management instruction for patients with non-insulin dependent diabetes mellitus: learning activity package vs. group instruction," MS thesis, Purdue University, 1993.

**15.** Data from Gary Community School Corporation, courtesy of Celeste Foster, Department of Education, Purdue University.

**16.** National Climatic Data Center, storm events data base. See `sciencepolicy.colorado.edu/sourcebook/tornadoes.html`.

**17.** Found online at `earthtrends.wri.org`.

**18.** Data from the U.S. Historical Climatology Network, archived at `www.co2science.org`. (Despite claims made on this site, temperatures at most U.S. locations show a gradual increase over the past century.)

**19.** National Oceanic and Atmospheric Administration, `www.noaa.gov`.

**20.** We thank Heeseung Roh Ryu for supplying the data, from Heeseung Roh Ryu, Roseann M. Lyle, and George P. McCabe, "Factors associated with weight concerns and unhealthy eating patterns among young Korean females," *Eating Disorders,* 11 (2003), pp. 129–141.

**21.** This exercise was provided by Nicolas Fisher.

**22.** J. Marcus Jobe and Hutch Jobe, "A statistical approach for additional infill development," *Energy Exploration and Exploitation,* 18 (2000), pp. 89–103.

**23.** S. M. Stigler, "Do robust estimators work with real data?" *Annals of Statistics,* 5 (1977), pp. 1055–1078.

**24.** T. Bjerkedal, "Acquisition of resistance in guinea pigs infected with different doses of virulent tubercle bacilli," *American Journal of Hygiene,* 72 (1960), pp. 130–148.

**25.** Data provided by Darlene Gordon, Purdue University.

**26.** U.S. Environmental Protection Agency, *Model Year 2004 Fuel Economy Guide,* found online at `www.fueleconomy.gov`.

**27.** Noel Cressie, *Statistics for Spatial Data,* Wiley, 1993.

**28.** Data provided by Francisco Rosales of the Department of Nutritional Sciences, Pennsylvania State University. See Rosales et al., "Relation of serum retinol to acute phase proteins and malarial morbidity in Papua New Guinea children," *American Journal of Clinical Nutrition,* 71 (2000), pp. 1580–1588.

**29.** Data provided by Betsy Hoza, Department of Psychological Sciences, University of Vermont.

**30.** Douglas Fore, "Do we have a retirement crisis in America?" TIAA-CREF Institute, *Research Dialogue*, No. 77 (2003). The data are for the year 2001.

**31.** *Extreme Weather Sourcebook 2001,* found online at `sciencepolicy.colorado.edu/sourcebook`.

**32.** We thank Ethan J. Temeles of Amherst College for providing the data. His work is described in Ethan J. Temeles and W. John Kress, "Adaptation in a plant-hummingbird association," *Science,* 300 (2003), pp. 630–633.

**33.** We thank Charles Cannon of Duke University for providing the data. The study report is C. H. Cannon, D. R. Peart, and M. Leighton, "Tree species diversity in commercially logged Bornean rainforest," *Science,* 281 (1998), pp. 1366–1367.

**34.** These graphs, like most others in this book, were produced by S-Plus, a professional statistical software package. Both graphs in Figure 1.24 used the same algorithm. The ability to fit normal curves is widespread in statistical software, and general "density estimators" are present in most professional software.

**35.** Information about the Indiana Statewide Testing for Educational Progress program can be found at `www.doe.state.in.us/istep/`.

**36.** Results for 1988 to 1991 from a large sample survey, reported in National Center for Health Statistics, *Health, United States, 1995,* 1996.

**37.** Data provided by Charles Hicks, Purdue University.

**38.** Data are from the Open Accessible Space Information System for New York City. See `www.oasisnyc.net`.

**39.** We thank C. Robertson McClung of Dartmouth College for supplying the data. The study is reported in Todd P. Michael et al., "Enhanced fitness conferred by naturally occurring variation in the circadian clock," *Science,* 302 (2003), pp. 1049–1053.

**40.** Julie Reinhart and Paul Schneider, "Student satisfaction, self-efficacy, and the perception of the two-way audio/visual distance learning environment," *Quarterly Review of Distance Education,* 2 (2001), pp. 357–365.

**41.** From `www.isp-planet.com`.

**42.** From the Current Population Survey, 2003 annual demographic supplement, found online at `www.census.gov`.

**43.** James W. Grier, "Ban of DDT and subsequent recovery of reproduction of bald eagles," *Science,* 218 (1982), pp. 1232–1235.

## CHAPTER 2

**1.** G. J. Patronek, D. L. Waters, and L. T. Glickman, "Comparative longevity of pet dogs and humans: implications for gerontology research," *Journal of Gerontology: Biological Sciences*, 52A (1997), pp. B171–B178.

**2.** Pernille Monberg, "The state of the Irish Wolfhound," at the Web site `www.wolfhouse.dk`.

**3.** Figure 2.1 displays data for 2003, from the College Board Web site, `www.collegeboard.com`.

**4.** A sophisticated treatment of improvements and additions to scatterplots is W. S. Cleveland and R. McGill, "The many faces of a scatterplot," *Journal of the American Statistical Association,* 79 (1984), pp. 807–822.

**5.** Data from National Institute of Standards and Technology, *Engineering Statistics Handbook,* `www.itl.nist.gov/div898/handbook`. The analysis there does not comment on the bias of field measurements.

**6.** Todd W. Anderson, "Predator responses, prey refuges, and density-dependent mortality of a marine fish," *Ecology,* 81 (2001), pp. 245–257.

**7.** The motorcycle crash test data are often used to test scatterplot smoothers. They appear in, for example, W. Hardle, *Applied Nonparametric Regression,* Cambridge University Press, 1990. The smoother used to produce Figure 2.5 is called "lowess." A full account appears in section 4.6 of J. M. Chambers, W. S. Cleveland, B. Kleiner, and P. A. Tukey, *Graphical Methods for Data Analysis,* Wadsworth, 1983.

**8.** James T. Fleming, "The measurement of children's perception of difficulty in reading materials," *Research in the Teaching of English,* 1 (1967), pp. 136–156.

**9.** Christer G. Wiklund, "Food as a mechanism of density-dependent regulation of breeding numbers in the merlin *Falco columbarius,*" *Ecology,* 82 (2001), pp. 860–867.

**10.** Data from a plot in Naomi I. Eisenberger, Matthew D. Lieberman, and Kipling D. Williams, "Does rejection hurt? An fMRI study of social exclusion," *Science,* 302 (2003), pp. 290–292.

**11.** We thank C. Robertson McClung of Dartmouth College for supplying the data. The study is reported in Todd P. Michael et al., "Enhanced fitness conferred by naturally occurring variation in the circadian clock," *Science,* 302 (2003), pp. 1049–1053.

**12.** *Forbes,* February 16, 2004.

**13.** Alan S. Banks et al., "Juvenile hallux abducto valgus association with metatarsus adductus," *Journal of the American Podiatric Medical Association,* 84 (1994), pp. 219–224.

**14.** Based on T. N. Lam, "Estimating fuel consumption from engine size," *Journal of Transportation Engineering,* 111 (1985), pp. 339–357. The data for 10 to 50 km/h are measured; those for 60 and higher are calculated from a model given in the paper and are therefore smoothed.

**15.** From the Web site of the International Olympic Committee, `www.olympic.org`.

**16.** N. Maeno et al., "Growth rates of icicles," *Journal of Glaciology,* 40 (1994), pp. 319–326.

**17.** Data provided by Matthew Moore.

**18.** Compiled from Fidelity data in the *Fidelity Insight* newsletter, 20, No. 1 (2004).

**19.** A careful study of this phenomenon is W. S. Cleveland, P. Diaconis, and R. McGill, "Variables on scatterplots look more highly correlated when the scales are increased," *Science,* 216 (1982), pp. 1138–1141.

**20.** Data from a plot in Timothy G. O'Brien and Margaret F. Kinnaird, "Caffeine and conservation," *Science,* 300 (2003), p. 587.

**21.** See Note 18.

**22.** Data from a plot in James A. Levine, Norman L. Eberhardt, and Michael D. Jensen, "Role of nonexercise activity thermogenesis in resistance to fat gain in humans," *Science,* 283 (1999), pp. 212–214.

**23.** E. P. Hubble, "A relation between distance and radial velocity among extra-galactic nebulae," *Proceedings of the National Academy of Sciences,* 15 (1929), pp. 168–173.

**24.** Data from G. A. Sacher and E. F. Staffelt, "Relation of gestation time to brain weight for placental mammals: implications for the theory of vertebrate growth," *American Naturalist,* 108 (1974), pp. 593–613. We found these data in F. L. Ramsey and D. W. Schafer, *The Statistical Sleuth: A Course in Methods of Data Analysis,* Duxbury, 1997.

**25.** From the Web site of the National Center for Education Statistics, `www.nces.ed.gov`.

**26.** Data from plots in Chaido Lentza-Rizos, Elizabeth J. Avramides, and Rosemary A. Roberts, "Persistence of fenthion residues in olive oil," *Pesticide Science,* 40 (1994), pp. 63–69.

**27.** Debora L. Arsenau, "Comparison of diet management instruction for patients with non–insulin dependent diabetes mellitus: learning activity package vs. group instruction," MS thesis, Purdue University, 1993.

**28.** Gannett News Service article appearing in the *Lafayette (Ind.) Journal and Courier,* April 23, 1994.

**29.** This example is drawn from M. Goldstein, "Preliminary inspection of multivariate data," *The American Statistician,* 36 (1982), pp. 358–362.

**30.** The facts in Example 2.26 come from Nancy W. Burton and Leonard Ramist, *Predicting Success in College: Classes Graduating since 1980,* Research Report No. 2001-2, The College Board, 2001.

**31.** Zeinab E. M. Afifi,"Principal components analysis of growth of Nahya infants: size, velocity and two physique factors," *Human Biology,* 57 (1985), pp. 659–669.

**32.** D. A. Kurtz (ed.), *Trace Residue Analysis,* American Chemical Society Symposium Series No. 284, 1985, Appendix.

**33.** Data from a plot in Feng Sheng Hu et al., "Cyclic variation and solar forcing of Holocene climate in the Alaskan subarctic," *Science,* 301 (2003), pp. 1890–1893.

**34.** R. C. Nelson, C. M. Brooks, and N. L. Pike, "Biomechanical comparison of male and female distance runners," in P. Milvy (ed.), *The Marathon: Physiological, Medical, Epidemiological, and Psychological Studies,* New York Academy of Sciences, 1977, pp. 793–807.

**35.** Frank J. Anscombe, "Graphs in statistical analysis," *The American Statistician,* 27 (1973), pp. 17–21.

**36.** Results of this survey are reported in Henry Wechsler et al., "Health and behavioral consequences of binge drinking in college," *Journal of the American Medical Association,* 272 (1994), pp. 1672–1677.

**37.** You can find a clear and comprehensive discussion of numerical measures of association for categorical data in Chapter 3 of A. M. Liebetrau, *Measures of Association,* Sage Publications, 1983.

**38.** These data, from reports submitted by airlines to the Department of Transportation, appear in A. Barnett, "How numbers can trick you," *Technology Review,* October 1994, pp. 38–45.

**39.** See the U.S. Bureau of the Census Web site at `www.census.gov/population/socdemo/school/` for these and similar data.

**40.** See Note 39.

**41.** F. D. Blau and M. A. Ferber, "Career plans and expectations of young women and men," *Journal of Human Resources,* 26 (1991), pp. 581–607.

**42.** D. M. Barnes, "Breaking the cycle of addiction," *Science,* 241 (1988), pp. 1029–1030.

**43.** Laura L. Calderon et al., "Risk factors for obesity in Mexican-American girls: dietary factors, anthropometric factors, physical activity, and hours of television viewing," *Journal of the American Dietetic Association,* 96 (1996), pp. 1177–1179.

**44.** Saccharin was on the National Institutes of Health's list of suspected carcinogens from 1977 to 2000 but remained in wide use with a warning label. It was officially cleared in 2000 and the label was removed.

**45.** A detailed study of this correlation appears in E. M. Remolona, P. Kleinman, and D. Gruenstein, "Market returns and mutual fund flows," *Federal Reserve Bank of New York Economic Policy Review,* 3, No. 2 (1997), pp. 33–52.

**46.** M. S. Linet et al., "Residential exposure to magnetic fields and acute lymphoblastic leukemia in children," *New England Journal of Medicine,* 337 (1997), pp. 1–7.

**47.** *The Health Consequences of Smoking: 1983,* U.S. Public Health Service, 1983.

**48.** Kevin Phillips, *Wealth and Democracy,* Broadway Books, 2002, p. 151.

**49.** David E. Bloom and David Canning, "The health and wealth of nations," *Science,* 287 (2000), pp. 1207–1208.

**50.** Contributed by Marigene Arnold, Kalamazoo College.

**51.** D. E. Powers and D. A. Rock, *Effects of Coaching on SAT I: Reasoning Test Scores,* Educational Testing Service Research Report 98-6, College Entrance Examination Board, 1998.

**52.** Information about this procedure was provided by Samuel Flanigan of *U.S. News & World Report.* See `www.usnews.com/usnews/rankguide/rghome.htm` for a description of the variables used to construct the ranks and for the most recent ranks.

**53.** See `www.cdc.gov/brfss/`. The data set BRFSS described in the Data Appendix contains several variables from this source.

**54.** Oskar Kindvall, "Habitat heterogeneity and survival in a bush cricket metapopulation," *Ecology,* 77 (1996), pp. 207–214.

**55.** H. Bradley Perry, "Analyzing growth stocks: what's a good growth rate?" *AAII Journal,* 13 (1991), pp. 7–10.

**56.** Data from a plot in Josef P. Rauschecker, Biao Tian, and Marc Hauser, "Processing of complex sounds in the macaque nonprimary auditory cortex," *Science,* 268 (1995), pp. 111–114. The paper states that there are $n = 41$ observations, but only $n = 37$ can be read accurately from the plot.

**57.** We thank Zhiyong Cai of Texas A&M University for providing the data. The full study is reported in Zhiyong Cai et al., "Predicting strength of matched sets of test specimens," *Wood and Fiber Science,* 30 (1998), pp. 396–404.

**58.** Data provided by Robert Dale, Purdue University.

**59.** See Note 34.

**60.** S. Chatterjee and B. Price, *Regression Analysis by Example,* Wiley, 1977.

**61.** Data provided by Jim Bateman and Michael Hunt, Purdue University.

**62.** "Dancing in step," *Economist*, March 22, 2001.

**63.** Gary Smith, "Do statistics test scores regress toward the mean?" *Chance*, 10, No. 4 (1997), pp. 42–45.

**64.** Data from National Science Foundation Science Resources Studies Division, *Data Brief*, 12 (1996), p. 1.

**65.** Based on data in Mike Planty et al., "Volunteer service by young people from high school through early adulthood," National Center for Educational Statistics Report NCES 2004–365.

**66.** Although these data are fictitious, similar, though less simple, situations occur. See P. J. Bickel and J. W. O'Connell, "Is there a sex bias in graduate admissions?" *Science*, 187 (1975), pp. 398–404.

**67.** Tom Reichert, "The prevalence of sexual imagery in ads targeted to young adults," *Journal of Consumer Affairs*, 37 (2003), pp. 403–412.

**68.** George R. Milne, "How well do consumers protect themselves from identity theft?" *Journal of Consumer Affairs*, 37 (2003), pp. 388–402.

**69.** Based on information in "NCAA 2003 national study of collegiate sports wagering and associated health risks," which can be found at the NCAA Web site, `www.ncaa.org`.

## CHAPTER 3

**1.** Caffeine equivalents are taken from `www.geocities.com/SiliconValley/9692/joltcola.html`.

**2.** See the NORC Web pages at `www.norc.uchicago.edu`.

**3.** J. E. Muscat et al., "Handheld cellular telephone use and risk of brain cancer," *Journal of the American Medical Association*, 284 (2000), pp. 3001–3007.

**4.** See Jeffrey G. Johnson et al., "Television viewing and aggressive behavior during adolescence and adulthood," *Science*, 295 (2002), pp. 2468–2471. The authors use statistical adjustments to control for the effects of a number of lurking variables. The association between TV viewing and aggression remains significant.

**5.** National Institute of Child Health and Human Development, Study of Early Child Care. The article appears in the July 2003 issue of *Child Development*.

**6.** The quotation is from the summary on the NICHD Web site, `www.nichd.nih.gov`.

**7.** M. K. Campbell et al., "A tailored multimedia nutrition education pilot program for low-income women receiving food assistance," *Health Education Research*, 14 (1999), pp. 257–267.

**8.** For a full description of the STAR program and its follow-up studies, go to `www.heros-inc.org/star.htm`.

**9.** Simplified from Arno J. Rethans, John L. Swasy, and Lawrence J. Marks, "Effects of television commercial repetition, receiver knowledge, and commercial length: a test of the two-factor model," *Journal of Marketing Research*, 23 (February 1986), pp. 50–61.

**10.** Based on an experiment performed by Jake Gandolph under the direction of Professor Lisa Mauer in the Purdue University Department of Food Science.

**11.** Based on an experiment performed by Evan Whalen under the direction of Professor Patrick Connolly in the Purdue University Department of Computer Graphics Technology.

**12.** L. L. Miao, "Gastric freezing: an example of the evaluation of medical therapy by randomized clinical trials," in J. P. Bunker, B. A. Barnes, and F. Mosteller (eds.), *Costs, Risks, and Benefits of Surgery*, Oxford University Press, 1977, pp. 198–211.

**13.** Simplified from David L. Strayer, Frank A. Drews, and William A. Johnston, "Cell phone–induced failures of visual attention during simulated driving," *Journal of Experimental Psychology: Applied*, 9 (2003), pp. 23–32. See Example 3.17 for more detail.

**14.** John H. Kagel, Raymond C. Battalio, and C. G. Miles, "Marijuana and work performance: results from an experiment," *Journal of Human Resources*, 15 (1980), pp. 373–395. A general discussion of failures of blinding is Dean Ferguson et al., "Turning a blind eye: the success of blinding reported in a random sample of randomised, placebo controlled trials," *British Medical Journal*, 328 (2004), p. 432.

**15.** Martin Enserink, "Fickle mice highlight test problems," *Science*, 284 (1999), pp. 1599–1600. There is a full report of the study in the same issue.

**16.** Based on a study conducted by Brent Ladd, a water quality specialist with the Purdue University Department of Agricultural and Biological Engineering.

**17.** Based on a study conducted by Sandra Simonis under the direction of Professor Jon Harbor from the Purdue University Earth and Atmospheric Sciences Department.

**18.** See the first citation in Note 14.

**19.** Geetha Thiagarajan et al., "Antioxidant properties of green and black tea, and their potential ability to retard the progression of eye lens cataract," *Experimental Eye Research*, 73 (2001), pp. 393–401.

**20.** Joel Brockner et al., "Layoffs, equity theory, and work performance: further evidence of the impact of survivor guilt," *Academy of Management Journal,* 29 (1986), pp. 373–384.

**21.** Simplified from Sanjay K. Dhar, Claudia González-Vallejo, and Dilip Soman, "Modeling the effects of advertised price claims: tensile versus precise pricing," *Marketing Science,* 18 (1999), pp. 154–177.

**22.** Data from the source in the preceding note.

**23.** Carol A. Warfield, "Controlled-release morphine tablets in patients with chronic cancer pain," *Cancer,* 82 (1998), pp. 2299–2306.

**24.** R. C. Shelton et al., "Effectiveness of St. John's wort in major depression," *Journal of the American Medical Association,* 285 (2001), pp. 1978–1986.

**25.** K. Wang, Y. Li, and J. Erickson, "A new look at the Monday effect," *Journal of Finance,* 52 (1997), pp. 2171–2186.

**26.** Based on Evan H. DeLucia et al., "Net primary production of a forest ecosystem with experimental $CO_2$ enhancement," *Science,* 284 (1999), pp. 1177–1179. The investigators used the block design.

**27.** E. M. Peters et al., "Vitamin C supplementation reduces the incidence of postrace symptoms of upper-respiratory tract infection in ultramarathon runners," *American Journal of Clinical Nutrition,* 57 (1993), pp. 170–174.

**28.** Based on a study conducted by Tammy Younts directed by Professor Deb Bennett of the Purdue University Department of Educational Studies. For more information about Reading Recovery, see `www.readingrecovery.org/`.

**29.** Based on a study conducted by Rajendra Chaini under the direction of Professor Bill Hoover of the Purdue University Department of Forestry and Natural Resources.

**30.** From the Hot Ringtones list at `www.billboard.com/` on August 4, 2006.

**31.** From the Rock Songs list at `www.billboard.com/` on August 4, 2006.

**32.** From the online version of the Bureau of Labor Statistics *Handbook of Methods,* modified April 17, 2003, at `www.bls.gov`. The details of the design are more complicated than our text describes.

**33.** For more detail on the material of this section and complete references, see P. E. Converse and M. W. Traugott, "Assessing the accuracy of polls and surveys," *Science,* 234 (1986), pp. 1094–1098.

**34.** The nonresponse rate for the CPS comes from the source cited in Note 32. The GSS reports its response rate on its Web site, `www.norc.org/projects/gensoc.asp`. The Pew study is described in Gregory Flemming and Kimberly Parker, "Race and reluctant respondents: possible consequences of non-response for pre-election surveys," Pew Research Center for the People and the Press, 1997, found at `www.people-press.org`.

**35.** George Bishop, "Americans' belief in God," *Public Opinion Quarterly,* 63 (1999), pp. 421–434.

**36.** For more detail on the limits of memory in surveys, see N. M. Bradburn, L. J. Rips, and S. K. Shevell, "Answering autobiographical questions: the impact of memory and inference on surveys," *Science,* 236 (1987), pp. 157–161.

**37.** Robert F. Belli et al., "Reducing vote overreporting in surveys: social desirability, memory failure, and source monitoring," *Public Opinion Quarterly,* 63 (1999), pp. 90–108.

**38.** Sex: Tom W. Smith, "The *JAMA* controversy and the meaning of sex," *Public Opinion Quarterly,* 63 (1999), pp. 385–400. Welfare: from a *New York Times*/CBS News poll reported in the *New York Times,* July 5, 1992. Scotland: "All set for independence?" *Economist,* September 12, 1998. Many other examples appear in T. W. Smith, "That which we call welfare by any other name would smell sweeter," *Public Opinion Quarterly,* 51 (1987), pp. 75–83.

**39.** Data from `poll.gallup.com` on August 8, 2006.

**40.** Giuliana Coccia, "An overview of non-response in Italian telephone surveys," *Proceedings of the 99th Session of the International Statistical Institute, 1993,* Book 3, pp. 271–272.

**41.** From *CIS Boletin 9, Spaniards' Economic Awareness,* found online at `www.cis.es/ingles/opinion/economia.htm`.

**42.** The survey question and result are reported in Trish Hall, "Shop? Many say 'Only if I must,' " *New York Times,* November 28, 1990.

**43.** From *Drawing the Line: Sexual Harassment on Campus,* a report from the American Association of University Women (AAUW) Educational Foundation published in 2006. See `www.aauw.org`.

**44.** Adam Nagourney and Janet Elder, "New York Times CBS Poll: What Hispanics Believe," found online at `www.Hispanic.cc`.

**45.** Warren McIsaac and Vivek Goel, "Is access to physician services in Ontario equitable?" Institute for Clinical Evaluative Sciences in Ontario, October 18, 1993.

**46.** John C. Bailar III, "The real threats to the integrity of science," *Chronicle of Higher Education,* April 21, 1995, pp. B1–B2.

**47.** The difficulties of interpreting guidelines for informed consent and for the work of institutional review boards in medical research are a main theme of Beverly Woodward, "Challenges to human subject protections in U.S. medical research," *Journal of the American Medical Association,* 282 (1999), pp. 1947–1952. The references in this paper point to other discussions.

**48.** Quotation from the *Report of the Tuskegee Syphilis Study Legacy Committee,* May 20, 1996. A detailed history is James H. Jones, *Bad Blood: The Tuskegee Syphilis Experiment,* Free Press, 1993.

**49.** Dr. Hennekens's words are from an interview in the Annenberg/Corporation for Public Broadcasting video series *Against All Odds: Inside Statistics.*

**50.** R. D. Middlemist, E. S. Knowles, and C. F. Matter, "Personal space invasions in the lavatory: suggestive evidence for arousal," *Journal of Personality and Social Psychology,* 33 (1976), pp. 541–546.

**51.** Based on Stephen A. Woodbury and Robert G. Spiegelman, "Bonuses to workers and employers to reduce unemployment: randomized trials in Illinois," *American Economic Review,* 77 (1987), pp. 513–530.

**52.** Based on Christopher Anderson, "Measuring what works in health care," *Science,* 263 (1994), pp. 1080–1082.

## CHAPTER 4

**1.** An informative and entertaining account of the origins of probability theory is Florence N. David, *Games, Gods and Gambling,* Charles Griffin, 1962.

**2.** D. A. Redelmeier and R. J. Tibshirani, "Association between cellular-telephone calls and motor vehicle collisions," *New England Journal of Medicine,* 336 (1997), pp. 453–458.

**3.** You can find a mathematical explanation of Benford's law in Ted Hill, "The first-digit phenomenon," *American Scientist,* 86 (1996), pp. 358–363; and Ted Hill, "The difficulty of faking data," *Chance,* 12, No. 3 (1999), pp. 27–31. Applications in fraud detection are discussed in the second paper by Hill and in Mark A. Nigrini, "I've got your number," *Journal of Accountancy,* May 1999, available online at `www.aicpa.org/pubs/jofa/joaiss.htm`.

**4.** Royal Statistical Society news release, "Royal Statistical Society concerned by issues raised in Sally Clark case," October 23, 2001, at `www.rss.org.uk`. For background, see an editorial and article in the *Economist,* January 22, 2004. The editorial is titled "The probability of injustice."

**5.** This is one of several tests discussed in Bernard M. Branson, "Rapid HIV testing: 2003 update," a presentation by the Centers for Disease Control and Prevention, found online at `www.cdc.gov`.

**6.** From Statistics Canada, `www.statcan.ca`.

**7.** Robyn Greenspan, "The deadly duo: spam and viruses, October 2003," found online at `cyberatlas.internet.com`.

**8.** Robert P. Dellavalle et al., "Going, going, gone: lost Internet references," *Science,* 302 (2003), pp. 787–788.

**9.** Corey Kilgannon, "When New York is on the end of the line," *New York Times,* November 7, 1999.

**10.** A. Tversky and D. Kahneman, "Extensional versus intuitive reasoning: the conjunction fallacy in probability judgment," *Psychological Review,* 90 (1983), pp. 293–315.

**11.** We use $\bar{x}$ both for the random variable, which takes different values in repeated sampling, and for the numerical value of the random variable in a particular sample. Similarly, $s$ and $\hat{p}$ stand both for random variables and for specific values. This notation is mathematically imprecise but statistically convenient.

**12.** We will consider only the case in which $X$ takes a finite number of possible values. The same ideas, implemented with more advanced mathematics, apply to random variables with an infinite but still countable collection of values.

**13.** From the Web site of the North Carolina State University Department of Registration and Records, `www.ncsu.edu/class/grades`.

**14.** The probability 0.12 is based on a small study at one university: Michele L. Head, "Examining college students' ethical values," Consumer Science and Retailing honors project, Purdue University, 2003.

**15.** From the Census Bureau's 1998 American Housing Survey.

**16.** The mean of a continuous random variable $X$ with density function $f(x)$ can be found by integration:

$$\mu_X = \int x f(x)dx$$

This integral is a kind of weighted average, analogous to the discrete-case mean

$$\mu_X = \sum xP(X = x)$$

The variance of a continuous random variable $X$ is the average squared deviation of the values of $X$ from their mean, found by the integral

$$\sigma_X^2 = \int (x - \mu)^2 f(x)dx$$

**17.** See A. Tversky and D. Kahneman, "Belief in the law of small numbers," *Psychological Bulletin,* 76 (1971), pp. 105–110, and other writings of these authors for a full account of our misperception of randomness.

**18.** Probabilities involving runs can be quite difficult to compute. That the probability of a run of three or more heads in 10 independent tosses of a fair coin is $(1/2) + (1/128) = 0.508$ can be found by clever counting. A general treatment using advanced methods appears in Section XIII.7 of William Feller, *An Introduction to Probability Theory and Its Applications,* vol. 1, 3rd ed., Wiley, 1968.

**19.** R. Vallone and A. Tversky, "The hot hand in basketball: on the misperception of random sequences," *Cognitive Psychology,* 17 (1985), pp. 295–314. A later series of articles that debate the independence question is A. Tversky and T. Gilovich, "The cold facts about the 'hot hand' in basketball," *Chance,* 2, No. 1 (1989), pp. 16–21; P. D. Larkey, R. A. Smith, and J. B. Kadane, "It's OK to believe in the 'hot hand,'" *Chance,* 2, No. 4 (1989), pp. 22–30; and A. Tversky and T. Gilovich, "The 'hot hand': statistical reality or cognitive illusion?" *Chance,* 2, No. 4 (1989), pp. 31–34.

**20.** Information found online at www.collegeboard.com.

**21.** Commissioners' Standard Ordinary Task Force final report, 2002, American Academy of Actuaries, found online at www.actuary.org/life/cso_0702.htm.

**22.** Means and standard deviations from Fidelity Investments, www.fidelity.com. Correlations from the *Fidelity Insight* newsletter, www.fidelityinsight.com. The correlations concern an unspecified period, and other online sources give different correlations, so these should be regarded as approximate at best.

**23.** Table 4.1 closely follows the grade distributions for these three schools at the University of New Hampshire for the fall of 2000, found in a self-study document at www.unh.edu/neasc. The counts of grades mirror the proportion of UNH students in these schools. The table is simplified to refer to a university with only these three schools.

**24.** Information about Internet users comes from sample surveys carried out by the Pew Internet and American Life Project, found online at www.pewinternet.org. The music-downloading data were collected in 2003.

**25.** These probabilities come from studies by the sociologist Harry Edwards, reported in the *New York Times,* February 25, 1986.

**26.** Online chat rooms

**27.** From www.irs.gov/taxstats.

**28.** From the 2006 *Statistical Abstract of the United States,* Table 491.

**29.** From the Bureau of Labor Statistics, found online at www.bls.gov/data.

**30.** See nces.ed.gov/programs/digest.

**31.** See Note 30.

**32.** See Note 24.

## CHAPTER 5

**1.** S. A. Rahimtoola, "Outcomes 15 years after valve replacement with a mechanical vs. a prosthetic valve, final report of the Veterans Administration randomized trial," American College of Cardiology, found online at www.acc.org/education/online/trials/acc2000/15yr.htm.

**2.** John Schwartz, "Leisure pursuits of today's young men," *New York Times,* March 29, 2004. The source cited is comScore Media Matrix.

**3.** Lydia Saad, "Small families are most Americans' ideal," Gallup poll press release, March 29, 2004, found online at www.gallup.com.

**4.** Jeffrey M. Jones, "Gambling a common activity for Americans," Gallup Poll press release, March 24, 2004, found online at www.gallup.com.

**5.** Henry Wechsler et al., *Binge Drinking on America's Colleges Campuses,* Harvard School of Public Health, 2001.

**6.** Haipeng Shen, "Nonparametric regression for problems involving lognormal distributions," PhD thesis, University of Pennsylvania, 2003. Thanks to Haipeng Shen and Larry Brown for sharing the data.

**7.** Statistical methods for dealing with time-to-failure data, including the Weibull model, are presented in Wayne Nelson, *Applied Life Data Analysis,* Wiley, 1982.

**8.** From the Web site of the North Carolina State University Department of Registration and Records, www.ncsu.edu/class/grades.

**9.** Values based on Sherri A. Buzinski, "The effect of position of methylation on the performance properties of durable press treated fabrics," CSR490 honors paper, Purdue University, 1985.

**10.** Values based on David Hon-Kuen Chu, "A test of corporate advertising using the elaboration likelihood model," MS thesis, Purdue University, 1993.

## CHAPTER 6


**1.** See Noel Cressie, *Statistics for Spatial Data,* Wiley, 1993. The significance test result that we report is one of several that could be used to address this question. See pp. 607–609 of the Cressie book for more details.

**2.** Based on information reported in Sandy Baum and Marie O'Malley, "College on credit: how borrowers perceive their education debt," Nellie Mae Corporation (2003), found online at `www.nelliemae.com/library/research_10.html`.

**3.** Average starting salary taken from a 2004–2005 spring salary survey from the National Association of Colleges and Employers (NACE).

**4.** P. H. Lewis, "Technology" column, *New York Times,* May 29, 1995.

**5.** The standard reference here is Bradley Efron and Robert J. Tibshirani, *An Introduction to the Bootstrap,* Chapman Hall, 1993. A less technical overview is in Bradley Efron and Robert J. Tibshirani, "Statistical data analysis in the computer age," *Science,* 253 (1991), pp. 390–395.

**6.** "Value of Recreational Sports on College Campuses," National Intramural-Recreational Sports Association (2002).

**7.** C. M. Weaver et al., "Quantification of biochemical markers of bone turnover by kinetic measures of bone formation and resorption in young healthy females," *Journal of Bone and Mineral Research,* 12 (1997), pp. 1714–1720.

**8.** 2006 press release from *The Student Monitor* located at `www.studentmonitor.com/press/10.pdf`.

**9.** *Dietary Reference Intakes for Energy, Carbohydrate, Fiber, Fat, Fatty Acids, Cholesterol, Protein, and Amino Acids (Macronutrients),* Food and Nutrition Board, Institute of Medicine, 2002.

**10.** The vehicle is a 2002 Toyota Prius owned by the third author.

**11.** Chris Chambers, "Americans skeptical about mergers of big companies," Gallup News Service, November 1, 2000. Found at `www.gallup.com`.

**12.** Rodney G. Bowden et al., "Lipid levels in a cohort of sedentary university students", *Internet Journal of Cardiovascular Research,* 2, No. 2 (2005).

**13.** Some analyses of the data collected by the Gambian National Impregnated Bednet Program are reported in M. C. Thompson et al., "A spatial model of malaria risk in the Gambia: predicting the impact of insecticide treated and untreated bednets on malaria infection," in *Proceedings of the Spatial Information Research Centre's 10th Colloquium,* University of Otago, New Zealand, 1998, pp. 321–328.

**14.** Seung-Ok Kim, "Burials, pigs, and political prestige in neolithic China," *Current Anthropology,* 35 (1994), pp. 119–141.

**15.** These data were collected in connection with the Purdue Police Alcohol Student Awareness Program run by Police Officer D. A. Larson.

**16.** National Assessment of Educational Progress, *The Nation's Report Card,* 2005.

**17.** R. A. Fisher, "The arrangement of field experiments," *Journal of the Ministry of Agriculture of Great Britain,* 33 (1926), p. 504, quoted in Leonard J. Savage, "On rereading R. A. Fisher," *Annals of Statistics,* 4 (1976), p. 471. Fisher's work is described in a biography by his daughter: Joan Fisher Box, *R. A. Fisher: The Life of a Scientist,* Wiley, 1978.

**18.** The editorial was written by Phil Anderson. See *British Medical Journal,* 328 (2004), pp. 476–477. A letter to the editor on this topic by Doug Altman and J. Martin Bland appeared shortly after. See "Confidence intervals illuminate absence of evidence," *British Medical Journal,* 328 (2004), pp. 1016–1017.

**19.** A. Kamali et al., "Syndromic management of sexually-transmitted infections and behavior change interventions on transmission of HIV-1 in rural Uganda: a community randomised trial," *Lancet,* 361 (2003), pp. 645–652.

**20.** T. D. Sterling, "Publication decisions and their possible effects on inferences drawn from tests of significance—or vice versa," *Journal of the American Statistical Association,* 54 (1959), pp. 30–34. Related comments appear in J. K. Skipper, A. L. Guenther, and G. Nass, "The sacredness of 0.05: a note concerning the uses of statistical levels of significance in social science," *American Sociologist,* 1 (1967), pp. 16–18.

**21.** For a good overview of these issues, see the Editor's Invited Article, "Gene expression data: the technology and statistical analysis," by Bruce A. Craig, Michael A. Black, and Rebecca W. Doerge, in *Journal of Agricultural, Biological, and Environmental Statistics,* 8 (2003), pp. 1–28.

**22.** Robert J. Schiller, "The volatility of stock market prices," *Science,* 235 (1987), pp. 33–36.

**23.** Sabrina Oesterle et al., "Volunteerism during the transition to adulthood: a life course perspective," *Social Forces,* 83 (2004), pp. 1123–1149.

**24.** Based on A. M. Garcia et al., "Why do workers behave unsafely at work? Determinants of safe work practices in industrial workers," *Occupational and Environmental Medicine,* 61 (2004), pp. 239–246.

25. Joan M. Susic, "Dietary phosphorus intakes, urinary and peritoneal phosphate excretion and clearance in continuous ambulatory peritoneal dialysis patients," MS thesis, Purdue University, 1985.

26. From a study by Mugdha Gore and Joseph Thomas, Purdue University School of Pharmacy.

27. Barbara A. Almanza, Richard Ghiselli, and William Jaffe, "Foodservice design and aging baby boomers: importance and perception of physical amenities in restaurants," *Foodservice Research International,* 12 (2000), pp. 25–40.

## CHAPTER 7

1. Average hours per month listening to full-track music obtained from a June 28, 2006, press release of Telephia's Q2 2006 3G U.K. Report, at `www.telephia.com`.

2. C. Don Wiggins, "The legal perils of 'underdiversification'—a case study," *Personal Financial Planning,* 1, No. 6 (1999), pp. 16–18.

3. These data were collected as part of a larger study of dementia patients conducted by Nancy Edwards, School of Nursing, and Alan Beck, School of Veterinary Medicine, Purdue University.

4. These recommendations are based on extensive computer work. See, for example, Harry O. Posten, "The robustness of the one-sample $t$-test over the Pearson system," *Journal of Statistical Computation and Simulation,* 9 (1979), pp. 133–149; and E. S. Pearson and N. W. Please, "Relation between the shape of population distribution and the robustness of four simple test statistics," *Biometrika,* 62 (1975), pp. 223–241.

5. The data were obtained on August 24, 2006, from an iPod owned by George McCabe, Jr.

6. Method described in Naihua Duan, "Smearing estimate: a nonparametric retransformation method," *Journal of the American Statistical Association,* 78 (1983), pp. 605–610.

7. You can find a practical discussion of distribution-free inference in Myles Hollander and Douglas A. Wolfe, *Nonparametric Statistical Methods,* 2nd ed., Wiley, 1999.

8. The vehicle is a 2002 Toyota Prius owned by the third author.

9. Data provided by Francisco Rosales, Pennsylvania State University, Department of Nutritional Sciences. See Rosales et al., "Relation of serum retinol to acute phase proteins and malarial morbidity in Papua New Guinea children," *American Journal of Clinical Nutrition,* 71 (2000), pp. 1580–1588.

10. Data provided by Betsy Hoza, Purdue University, Department of Psychological Sciences.

11. David R. Pillow et al., "Confirmatory factor analysis examining attention deficit hyperactivity disorder symptoms and other childhood disruptive behaviors," *Journal of Abnormal Child Psychology,* 26 (1998), pp. 293–309.

12. James A. Levine, Norman L. Eberhardt, and Michael D. Jensen, "Role of nonexercise activity thermogenesis in resistance to fat gain in humans," *Science,* 283 (1999), pp. 212–214. Data for this study are available from the *Science* Web site, `www.sciencemag.org`.

13. Joan M. Susic, "Dietary phosphorus intakes, urinary and peritoneal phosphate excretion and clearance in continuous ambulatory peritoneal dialysis patients," MS thesis, Purdue University, 1985.

14. These data were collected in connection with a bone health study at Purdue University and were provided by Linda McCabe.

15. Data provided by Joseph A. Wipf, Purdue University, Department of Foreign Languages and Literatures.

16. Wayne Nelson, *Applied Life Data Analysis,* Wiley, 1982, p. 471.

17. Summary information can be found at the National Center for Health Statistics Web site, `www.cdc.gov/nchs/nhanes.htm`.

18. Detailed information about the conservative $t$ procedures can be found in Paul Leaverton and John J. Birch, "Small sample power curves for the two sample location problem," *Technometrics,* 11 (1969), pp. 299–307; Henry Scheffé, "Practical solutions of the Behrens-Fisher problem," *Journal of the American Statistical Association,* 65 (1970), pp. 1501–1508; and D. J. Best and J. C. W. Rayner, "Welch's approximate solution for the Behrens-Fisher problem," *Technometrics,* 29 (1987), pp. 205–210.

19. This example is adapted from Maribeth C. Schmitt, "The effects of an elaborated directed reading activity on the metacomprehension skills of third graders," PhD thesis, Purdue University, 1987.

20. See the extensive simulation studies in Harry O. Posten, "The robustness of the two-sample $t$ test over the Pearson system," *Journal of Statistical Computation and Simulation,* 6 (1978), pp. 295–311.

21. Based on R. G. Hood, "Results of the prices received by farmers for grain—quality assurance project," U.S. Department of Agriculture Report SRB–95–07, 1995.

**22.** This study is reported in Roseann M. Lyle et al., "Blood pressure and metabolic effects of calcium supplementation in normotensive white and black men," *Journal of the American Medical Association,* 257 (1987), pp. 1772–1776. The individual measurements in Table 7.5 were provided by Dr. Lyle.

**23.** Rodney G. Bowden et al., "Lipid levels in a cohort of sedentary university students," *Internet Journal of Cardiovascular Research,* 2, No. 2 (2005).

**24.** M. K. Campbell et al., "A tailored multimedia nutrition education pilot program for low-income women receiving food assistance," *Health Education Research,* 14 (1999), pp. 257–267.

**25.** B. Bakke et al., "Cumulative exposure to dust and gases as determinants of lung function decline in tunnel construction workers," *Occupational Environmental Medicine,* 61 (2004), pp. 262–269.

**26.** Samara Joy Nielsen and Barry M. Popkin, "Patterns and trends in food portion sizes, 1977–1998," *Journal of the American Medical Association,* 289 (2003), pp. 450–453.

**27.** Gordana Mrdjenovic and David A. Levitsky, "Nutritional and energetic consequences of sweetened drink consumption in 6- to 13-year-old children," *Journal of Pediatrics,* 142 (2003), pp. 604–610.

**28.** The data were provided by Helen Park. See Helen Park et al., "Fortifying bread with each of three antioxidants," *Cereal Chemistry,* 74 (1997), pp. 202–206.

**29.** David Han-Kuen Chu, "A test of corporate advertising using the elaboration likelihood model," MS thesis, Purdue University, 1993.

**30.** M. F. Picciano and R. H. Deering, "The influence of feeding regimens on iron status during infancy," *American Journal of Clinical Nutrition,* 33 (1980), pp. 746–753.

**31.** The problem of comparing spreads is difficult even with advanced methods. Common distribution-free procedures do not offer a satisfactory alternative to the $F$ test, because they are sensitive to unequal shapes when comparing two distributions. A good introduction to the available methods is W. J. Conover, M. E. Johnson, and M. M. Johnson, "A comparative study of tests for homogeneity of variances, with applications to outer continental shelf bidding data," *Technometrics,* 23 (1981), pp. 351–361. Modern resampling procedures often work well. See Dennis D. Boos and Colin Brownie, "Bootstrap methods for testing homogeneity of variances," *Technometrics,* 31 (1989), pp. 69–82.

**32.** G. E. P. Box, "Non-normality and tests on variances," *Biometrika,* 40 (1953), pp. 318–335. The quotation appears on page 333.

**33.** Based on Loren Cordain et al., "Influence of moderate daily wine consumption on body weight regulation and metabolism in healthy free-living males," *Journal of the American College of Nutrition,* 16 (1997), pp. 134–139.

**34.** C. M. Weaver et al., "Quantification of biochemical markers of bone turnover by kinetic measures of bone formation and resorption in young healthy females," *Journal of Bone and Mineral Research,* 12 (1997), pp. 1714–1720. Data provided by Linda McCabe.

**35.** This exercise is based on events that are real. The data and details have been altered to protect the privacy of the individuals involved.

**36.** Based loosely on D. R. Black et al., "Minimal interventions for weight control: a cost-effective alternative," *Addictive Behaviors,* 9 (1984), pp. 279–285.

**37.** These data were provided by Professor Sebastian Heath, Purdue University, School of Veterinary Medicine.

**38.** J. W. Marr and J. A. Heady, "Within- and between-person variation in dietary surveys: number of days needed to classify individuals," *Human Nutrition: Applied Nutrition,* 40A (1986), pp. 347–364.

**39.** Data taken from the Web site `www.assist-2-sell.com` on July 9, 2001.

## CHAPTER 8

**1.** The actual distribution of $X$ based on an SRS from a finite population is the *hypergeometric distribution*. Details regarding this distribution can be found in William G. Cochran, *Sampling Techniques*, 3rd ed., Wiley, 1977.

**2.** Details of exact binomial procedures can be found in Myles Hollander and Douglas Wolfe, *Nonparametric Statistical Methods*, 2nd ed., Wiley, 1999.

**3.** Results of this survey are reported in Henry Wechsler et al., "College binge drinking in the 1990s: a continuing problem," *Journal of American College Health*, 48 (2000), pp. 199–210.

**4.** A. Agresti and B. A. Coull, "Approximate is better than 'exact' for interval estimation of binomial proportions," *The American Statistician*, 52 (1998), pp. 119–126. A detailed theoretical study is Lawrence D. Brown, Tony Cai, and Anirban DasGupta, "Confidence intervals for a binomial proportion and asymptotic expansions," *Annals of Statistics*, 30 (2002), pp. 160–201.

**5.** 2005 press release from *The Student Monitor,* available online at `www.studentmonitor.com/press/02.pdf`.

**6.** See K. D. R. Setchell et al., "The clinical importance of the metabolite equol—a clue to the effectiveness of soy and its isoflavones," *Journal of Nutrition*, 132 (2002), pp. 3577–3584.

**7.** National Institute for Occupational Safety and Health, *Stress at Work*, 2000, available online at `www.cdc.gov/niosh/stresswk.html`.

**8.** Results of this survey were reported in *Restaurant Business*, September 15, 1999, pp. 45–49.

**9.** Stephanie Cuccaro-Alamin and Susan P. Choy, *Postsecondary Financing Strategies: How Undergraduates Combine Work, Borrowing, and Attendance*, NCES 98088, U.S. Department of Education, National Center for Education Statistics, 1998.

**10.** See, for example, `www.pilatesmethodalliance.org/`.

**11.** Based on information in "NCAA 2003 national study of collegiate sports wagering and associated health risks," which can be found at the NCAA Web site, `www.ncaa.org`.

**12.** From the "National Survey of Student Engagement, the college student report," 2003, available online at `www.indiana.edu~nsse/`.

**13.** Results from the 2006 Pew Research Center report "Americans and their cars: is the romance on the skids?"

**14.** From "Report Card 2004: The Ethics of American Youth," 2004, available online at `www.josephsoninstitute.org/reportcard/`.

**15.** Information about the survey can be found online at `saint-denis.library.arizona.edu/natcong/`.

**16.** Based on information reported in Sandy Baum and Marie O'Malley, "College on credit: how borrowers perceive their education debt," Nellie Mae Corporation, 2003, found online at `www.nelliemae.com/library/research_10.html`.

**17.** Heeseung Roh Ryu, Roseann M. Lyle, and George P. McCabe, "Factors associated with weight concerns and unhealthy eating patterns among young Korean females," *Eating Disorders*, 11 (2003), pp. 129–141.

**18.** Data from Roland J. Thorpe, Jr., and based on his analysis of "Health ABC," a 10-year longitudinal study of older adults supported by the Laboratory of Epidemiology, Demography, and Biometry of the National Institute on Aging. Additional analyses are given in his PhD thesis, "Relationship between pet ownership, physical activity, and human health in an elderly population," Purdue University, 2004.

**19.** Michael McCulloch et al., "Diagnostic accuracy of canine scent detection in early- and late-stage lung and breast cancers," *Integrative Cancer Therapies*, 5, No. 1 (2006), pp. 30–39.

**20.** Guohua Li and Susan P. Baker, "Alcohol in fatally injured bicyclists," *Accident Analysis and Prevention*, 26 (1994), pp. 543–548.

**21.** See Alan Agresti and Brian Caffo, "Simple and effective confidence intervals for proportions and differences of proportions result from adding two successes and two failures," *The American Statistician*, 45 (2000), pp. 280–288. The plus four interval is a bit conservative (true coverage probability is higher than the confidence level) when $p_1$ and $p_2$ are equal and close to 0 or 1, but the traditional interval is much less accurate and has the fatal flaw that the true coverage probability is *less* than the confidence level.

**22.** J. M Tanner, "Physical growth and development," in J. O. Forfar and G. C Arneil, *Textbook of Paediatrics*, 3rd ed., Churchill Livingston, 1984, pp. 1–292.

**23.** See Note 18.

**24.** M. L. Burr et al., "Effects on respiratory health of a reduction in air pollution from vehicle exhaust emissions," *Occupational and Environmental Medicine*, 61 (2004), pp. 212–218.

**25.** Mary Madden and Lee Rainie, "Music and video downloading moves beyond P2P," PEW Internet Project Data Memo, 2005, available online at `www.pewinternet.org/`.

**26.** Based on Greg Clinch, "Employee compensation and firms' research and development activity," *Journal of Accounting Research*, 29 (1991), pp. 59–78.

**27.** Monica Macaulay and Colleen Brice, "Don't touch my projectile: gender bias and stereotyping in syntactic examples," *Language*, 73, No. 4 (1997), pp. 798–825. The first part of the title is a direct quote from one of the texts.

**28.** Lee Rainie et al., "The state of music downloading and file-sharing online," PEW Internet Project and Comscore Media Metrix Data Memo, 2004, available online at `www.pewinternet.org/`.

**29.** Charles T. Clotfelter et al., "State lotteries at the turn of the century: report to the National Gambling Impact Study Commission," 1999.

**30.** Marsha A. Dickson, "Utility of no sweat labels for apparel customers: profiling label users and predicting their purchases," *Journal of Consumer Affairs*, 35 (2001), pp. 96–119.

**31.** Karin Weber and Weley S. Roehl, "Profiling people searching for and purchasing travel products on the World Wide Web," *Journal of Travel Research*, 37 (1999), pp. 291–298.

**32.** Based on Robert T. Driescher, "A quality swing with Ping," *Quality Progress*, August 2001, pp. 37–41.

**33.** Based on a Pew Research poll conducted in March–May 2006, available online at `pewresearch.org/`.

**34.** Dennis N. Bristow and Richard J. Sebastian, "Holy cow! Wait till next year! A closer look at the brand loyalty of Chicago Cubs baseball fans," *Journal of Consumer Marketing,* 18 (2001), pp. 256–275.

**35.** S. W. Lagakos, B. J. Wessen, and M. Zelen, "An analysis of contaminated well water and health effects in Woburn, Massachusetts," *Journal of the American Statistical Association*, 81 (1986), pp. 583–596. This case is the basis for the movie *A Civil Action*.

**36.** This case is discussed in D. H. Kaye and M. Aickin (eds.), *Statistical Methods in Discrimination Litigation*, Marcel Dekker, 1986; and D. C. Baldus and J. W. L. Cole, *Statistical Proof of Discrimination*, McGraw-Hill, 1980.

**37.** See Note 16.

## CHAPTER 9

**1.** J. Cantor, "Long-term memories of frightening media often include lingering trauma symptoms," poster paper presented at the Association for Psychological Science Convention, New York, May 26, 2006.

**2.** When the expected cell counts are small, it is best to use a test based on the exact distribution rather than the chi-square approximation, particularly for $2 \times 2$ tables. Many statistical software systems offer an "exact" test as well as the chi-square test for $2 \times 2$ tables.

**3.** E. Y. Peck, "Gender differences in film-induced fear as a function of type of emotion measure and stimulus content: a meta-analysis and laboratory study," PhD thesis, University of Wisconsin—Madison, 1999.

**4.** The full report of the study appeared in George H. Beaton et al., "Effectiveness of vitamin A supplementation in the control of young child morbidity and mortality in developing countries," United Nations ACC/SCN State-of-the-Art Series, Nutrition Policy Discussion Paper No. 13, 1993.

**5.** C. M. Ryan, C. A. Northrup-Clewes, B. Knox, and D. I. Thurnham, "The effect of in-store music on consumer choice of wine," *Proceedings of the Nutrition Society,* 57 (1998), p. 1069A.

**6.** An alternative formula that can be used for hand or calculator computations is

$$X^2 = \sum \frac{(\text{observed})^2}{\text{expected}} - n$$

**7.** See, for example, Alan Agresti, *Categorical Data Analysis,* Wiley, 1990.

**8.** D. A. Redelmeier and R. J. Tibshirani, "Association between cellular-telephone calls and motor vehicle collisions," *New England Journal of Medicine,* 336 (1997), pp. 453–458.

**9.** See Note 7.

**10.** See the M&M Mars Company Web site at `us.mms.com/us/about/products` for this and other information.

**11.** See Note 1.

**12.** See the U.S. Bureau of the Census Web site at `www.census.gov/population/socdemo/school` for these and similar data.

**13.** See `i.a.cnn.net/cnn/2006/images/04/27/rel11e.pdf` for a summary of the poll.

**14.** Data from D. M. Barnes, "Breaking the cycle of addiction," *Science,* 241 (1988), pp. 1029–1030.

**15.** Based on data in Mike Planty et al., "Volunteer service by young people from high school through early adulthood," National Center for Educational Statistics Report NCES 2004–365.

**16.** Results of this survey are reported in Henry Wechsler et al., "College binge drinking in the 1990s: a continuing problem," *Journal of American College Health,* 48 (2000), pp. 199–210.

**17.** Tom Reichert, "The prevalence of sexual imagery in ads targeted to young adults," *Journal of Consumer Affairs,* 37 (2003), pp. 403–412.

**18.** Christopher Somers et al., "Air pollution induces heritable DNA mutations," *Proceedings of the National Academy of Sciences,* 99 (2002), pp. 15904–15907.

**19.** S. R. Davy, B. A. Benes , and J. A. Driskell, "Sex differences in dieting trends, eating habits, and nutrition beliefs of a group of midwestern college students," *Journal of the American Dietetic Association,* 10 (2006), pp. 1673–1677.

**20.** George R. Milne, "How well do consumers protect themselves from identity theft?" *Journal of Consumer Affairs,* 37 (2003), pp. 388–402.

**21.** Based on information in "NCAA 2003 national study of collegiate sports wagering and associated health risks," which can be found at the NCAA Web site, `www.ncaa.org`.

**22.** Ethan J. Temeles and W. John Kress, "Adaption in a plant-hummingbird association," *Science,* 300 (2003), pp. 630–633.

**23.** Dellavalle et al., "Going, going, gone: lost Internet references," *Science,* 302 (2003), pp. 787–788.

**24.** Data from Roland J. Thorpe, Jr., and based on his analysis of "Health ABC," a 10-year longitudinal study of older adults supported by the Laboratory of Epidemiology, Demography, and Biometry of the National Institute on Aging. Additional analyses are given in his PhD thesis, "Relationship between pet ownership, physical activity, and human health in an elderly population," Purdue University, 2004.

**25.** These data are from an Undergraduate Task Force study of student retention in the Purdue University College of Science.

**26.** William D. Darley, "Store-choice behavior for pre-owned merchandise," *Journal of Business Research,* 27 (1993), pp. 17–31.

**27.** Daniel Cassens et al., "Face check development in veneered furniture panels," *Forest Products Journal,* 53 (2003), pp. 79–86.

**28.** Gary J. Patronek et al., "Risk factors for relinquishment of cats to an animal shelter," *Journal of the American Veterinary Medical Association,* 209 (1996), pp. 582–588.

**29.** Gary J. Patronek et al., "Risk factors for relinquishment of dogs to an animal shelter," *Journal of the American Veterinary Medical Association,* 209 (1996), pp. 572–581.

**30.** Data provided by Susan Prohofsky, from her PhD thesis, "Selection of undergraduate major: the influence of expected costs and expected benefits," Purdue University, 1991.

**31.** Marsha A. Dickson, "Utility of no sweat labels for apparel customers: profiling label users and predicting their purchases," *Journal of Consumer Affairs,* 35 (2001), pp. 96–119.

**32.** James A. Taylor et al., "Efficacy and safety of echinacea in treating upper respiratory tract infections in children," *Journal of the American Medical Association,* 290 (2003), pp. 2824–2830.

**33.** Karen L. Dean, "Herbalists respond to *JAMA* echinacea study," *Alternative & Complementary Therapies,* 10 (2004), pp. 11–12.

**34.** The report can be found at `nsse.iub.edu/pdf/NSSE2005_annual_report.pdf`.

## CHAPTER 10

**1.** The vehicle is a 1997 Pontiac Transport Van owned by the second author.

**2.** The data were provided by Linda McCabe and were collected as part of a large study of women's bone health and another study of calcium kinetics, both directed by Professor Connie Weaver of Purdue University, Department of Foods and Nutrition. Results related to this example are reported in Dorothy Teegarden et al., "Peak bone mass in young women," *Journal of Bone and Mineral Research,* 10 (1995), pp. 711–715.

**3.** These data were provided by Professor Wayne Campbell of the Purdue University Department of Foods and Nutrition.

**4.** This quantity is the estimated standard deviation of $\hat{y} - y$, not the estimated standard deviation of $\hat{y}$ alone.

**5.** For more information about nutrient requirements, see the Institute of Medicine publications on Dietary Reference Intakes available at `www.nap.edu`.

**6.** The method is described in Chapter 2 of M. Kutner et al., *Applied Linear Statistical Models,* 5th ed., Irwin, 2005.

**7.** National Science Foundation, Division of Science Resources Statistics, Survey of Research and Development Expenditures and Universities and Colleges, Fiscal Year 2001.

**8.** 2005–2006 Tuition and Required Fees Report, University of Missouri; and the 2000–2001 Tuition and Required Fees Report, University of Missouri.

**9.** These are some of the data from the EESEE story "Blood Alcohol Content," found on the text Web site.

**10.** Based on Dan Dauwalter's master's thesis in the Department of Forestry and Natural Resources at Purdue University. More information is available in Daniel C. Dauwalter et al., "An index of biotic integrity for fish assemblages in Ozark Highland streams of Arkansas," *Southeastern Naturalist,* 2 (2003), pp. 447–468. These data were provided by Emmanuel Frimpong.

**11.** Data available at `www.ncdc.noaa.gov/oa/climate/sd`.

**12.** Net cash flow data from Sean Collins, *Mutual Fund Assets and Flows in 2000,* Investment Company Institute, 2001. Found online at `www.ici.org`. The raw data were converted to real dollars using annual average values of the Consumer Price Index.

**13.** Alan S. Banks et al., "Juvenile hallux abducto valgus association with metatarsus adductus," *Journal of the American Podiatric Medical Association,* 84 (1994), pp. 219–224.

**14.** The data were provided by Professor Shelly MacDermid, Department of Child Development and Family Studies, Purdue University, from a study reported in S. M. MacDermid et al., "Is small beautiful? Work-family tension, work conditions, and organizational size," *Family Relations,* 44 (1994), pp. 159–167.

**15.** G. Geri and B. Palla, "Considerazioni sulle più recenti osservazioni ottiche alla Torre Pendente di Pisa," *Estratto dal Bollettino della Società Italiana di Topografia e Fotogrammetria,* 2 (1988), pp. 121–135. Professor Julia Mortera of the University of Rome provided valuable assistance with the translation.

**16.** M. Kuo et al. "The marketing of alcohol to college students: the role of low prices and special promotions," *American Journal of Preventive Medicine,* 25, No. 3 (2003), pp. 204–211.

**17.** Data for the first four years from Manuel Castells, *The Rise of the Network Society,* 2nd ed., Blackwell, 2000, p. 41.

**18.** This study is reported in S. Lau and P. C. Cheung, "Relations between Chinese adolescents' perception of parental control and organization and their perception of parental warmth," *Developmental Psychology,* 23 (1987), pp. 726–729.

**19.** Donald A. Mahler and John I. Mackowiak, "Evaluation of the short-form 36-item questionnaire to measure health-related quality of life in patients with COLD," *Chest,* 107 (1995), pp. 1585–1589.

**20.** Data on a sample of 12 of 56 perch in a data set contributed to the *Journal of Statistics Education* data archive `www.amstat.org/publications/jse/` by Juha Puranen, University of Helsinki.

**21.** L. Cooke et al., "Relationship between parental report of food neophobia and everyday food consumption in 2–6-year-old children," *Appetite,* 41 (2003), pp. 205–206.

**22.** D. E. Powers and J. K. Kaufman, "Do standardized multiple choice tests penalize deep-thinking or creative students?" Educational Testing Service Research Report RR–02–15 (2002).

## CHAPTER 11

**1.** Results of the study are reported in P. F. Campbell and G. P. McCabe, "Predicting the success of freshmen in a computer science major," *Communications of the ACM,* 27 (1984), pp. 1108–1113.

**2.** R. M. Smith and P. A. Schumacher, "Predicting success for actuarial students in undergraduate mathematics courses," *College Student Journal,* 39, No. 1 (2005), pp. 165–177.

**3.** Tom Reichert, "The prevalence of sexual imagery in ads targeted to young adults," *Journal of Consumer Affairs,* 37 (2003), pp. 403–412.

**4.** C. Maffeis et al., "Distribution of food intake as a risk factor for childhood obesity," *International Journal of Obesity,* 24 (2000), pp. 75–80.

**5.** Miguel A. Diego et al., "Academic performance, popularity, and depression predict adolescent substance abuse," *Adolescence,* 38 (2003), pp. 35–42.

**6.** Wanki Moon and Siva K. Balasubramanian, "Willingness to pay for non-biotech foods in the U.S. and the U.K.," *Journal of Consumer Affairs,* 37 (2003), pp. 317–339. The authors used a maximum likelihood method to estimate the parameters of their model, but the methods described in this chapter are appropriate for the interpretation of the results given in this exercise.

**7.** Steven R. Wininger and David Pargman, "Assessment of factors associated with exercise enjoyment," *Journal of Music Therapy,* 40 (2003), pp. 57–73.

**8.** This study is reported in Scot Burton et al., "Implications of accurate usage of nutrition facts panel information for food product evaluations and purchase intentions," *Journal of the Academy of Marketing Science,* 27 (1999), pp. 470–480.

**9.** Alan Levinsohn, "Online brokerage, the new core account?" *ABA Banking Journal,* September 1999, pp. 34–42.

## CHAPTER 12

**1.** Based on A. M. Garcia et al., "Why do workers behave unsafely at work? Determinants of safe work practices in industrial workers," *Occupational and Environmental Medicine,* 61 (2004), pp. 239–246.

**2.** This rule is intended to provide a general guideline for deciding when serious errors may result by applying ANOVA procedures. When the sample sizes in each group are very small, the sample variances will tend to vary much more than when the sample sizes are large. In this case, the rule may be a little too conservative. For unequal sample sizes, particular difficulties can arise when a relatively small sample size is associated with a population having a relatively large standard deviation. Careful judgment is needed in all cases. By considering $P$-values rather than fixed level $\alpha$ testing, judgments in ambiguous cases can more easily be made; for example, if the $P$-value is very small, say 0.001, then it is probably safe to reject $H_0$ even if there is a fair amount of variation in the sample standard deviations.

**3.** This example is based on Amna Kirmani and Peter Wright, "Money talks: perceived advertising expense and expected product quality," *Journal of Consumer Research,* 16 (1989), pp. 344–353.

**4.** Several different definitions for the noncentrality parameter of the noncentral $F$ distribution are in use. When

$I = 2$, the $\lambda$ defined here is equal to the square of the noncentrality parameter $\delta$ that we used for the two-sample $t$ test in Chapter 7. Many authors prefer $\phi = \sqrt{\lambda/I}$. We have chosen to use $\lambda$ because it is the form needed for the SAS function PROBF.

**5.** P. Bartel et al., "Attention and working memory in resident anaesthetists after night duty: group and individual effects," *Occupational and Environmental Medicine,* 61 (2004), pp. 167–170.

**6.** "Value of Recreational Sports on College Campuses," National Intramural-Recreational Sports Association (2002).

**7.** Adrian C. North et al., "The effect of musical style on restaurant consumers' spending," *Environment and Behavior,* 35 (2003), pp. 712–718.

**8.** Based on Jack K. Trammel, "The impact of academic accommodations on final grades in a post secondary setting," *Journal of College Reading and Learning,* 34 (2003), pp. 76–90.

**9.** Christie N. Scollon et al., "Emotions across cultures and methods," *Journal of Cross-cultural Psychology,* 35 (2004), pp. 304–326.

**10.** These data were provided by Helen Park; see H. Park et al., "Fortifying bread with each of three antioxidants," *Cereal Chemistry,* 74 (1997), pp. 202–206.

**11.** Michael Wayne Peugh, "Field investigation of ventilation and air quality in duck and turkey slaughter plants," MS thesis, Purdue University, 1996.

**12.** We thank Ethan J. Temeles of Amherst College for providing the data. His work is described in Ethan J. Temeles and W. John Kress, "Adaptation in a plant-hummingbird association," *Science,* 300 (2003), pp. 630–633.

**13.** The data were provided by James Kaufman. The study is described in James C. Kaufman, "The cost of the muse: poets die young," *Death Studies,* 27 (2003), pp. 813–821. The quotation from Yeats appears in this article.

**14.** The experiment was performed in Connie Weaver's lab in the Purdue University Department of Foods and Nutrition. The data were provided by Berdine Martin and Yong Jiang.

**15.** Based on M. U. Kalwani and C. K. Yim, "Consumer price and promotion expectations: an experimental study," *Journal of Marketing Research,* 29 (1992), pp. 90–100.

**16.** Based on a study by J. E. Stuff and B. L. Nichols reported in Chelsea Lutter, "Recommended length of exclusive breast-feeding, age of introduction of complementary foods and the weaning dilemma," World Health Organization, 1992.

**17.** Data provided by Jo Welch of the Purdue University Department of Foods and Nutrition.

**18.** Based on A. A. Adish et al., "Effect of consumption of food cooked in iron pots on iron status and growth of young children: a randomised trial," *The Lancet,* 353 (1999), pp. 712–716.

**19.** Steve Badylak et al., "Marrow-derived cells populate scaffolds composed of xenogeneic extracellular matrix," *Experimental Hematology,* 29 (2001), pp. 1310–1318.

**20.** This exercise is based on data provided from a study conducted by Jim Baumann and Leah Jones of the Purdue University School of Education.

## CHAPTER 13

**1.** See `www.malaria.org/` for more information about malaria.

**2.** This example is based on the work of David I. Thurnham, of the Northern Ireland Center for Diet and Health, and Delana A. Adelekan of Obafemi Awolowo University, Ile-Ife, Nigeria.

**3.** We present the two-way ANOVA model and analysis for the general case in which the sample sizes may be unequal. If the sample sizes vary a great deal, serious complications can arise. There is no longer a single standard ANOVA analysis. Most computer packages offer several options for the computation of the ANOVA table when cell counts are unequal. When the counts are approximately equal, all methods give essentially the same results.

**4.** Samara Joy Nielsen and Barry M. Popkin, "Patterns and trends in food portion sizes, 1977–1998," *Journal of the American Medical Association,* 289 (2003), pp. 450–453.

**5.** The summaries in Example 13.9 are from Rick Bell and Patricia L. Pliner, "Time to eat: the relationship between the number of people eating and meal duration in three lunch settings," *Appetite,* 41 (2003), pp. 215–218.

**6.** Karolyn Drake and Jamel Ben El Hine, "Synchronizing with music: intercultural differences," *Annals of the New York Academy of Sciences,* 99 (2003), pp. 429–437.

**7.** Example 13.11 is based on a study described in P. D. Wood et al., "Plasma lipoprotein distributions in male and female runners," in P. Milvey (ed.), *The Marathon: Physiological, Medical, Epidemiological, and Psychological Studies,* New York Academy of Sciences, 1977.

**8.** Brian Wansink et al., "The office candy dish: proximity's influence on estimated and actual consumption," *International Journal of Obesity,* 30 (2006), pp. 871–875.

**9.** Annette N. Senitko et al., "Influence of endurance exercise training status and gender on postexercise hypotension," *Journal of Applied Physiology,* 92, pp. 2368–2374.

**10.** Judith McFarlane et al., "An intervention to increase safety behaviors of abused women," *Nursing Research,* 51 (2002), pp. 347–354.

**11.** Lene Arnett Jensen et al., "The right to do wrong: Lying to parents among adolescents and emerging adults," *Journal of Youth and Adolescence,* 33 (2004), pp. 101–112.

**12.** Klaus Boehnke et al., "On the interrelation of peer climate and school performance in mathematics: a German-Canadian-Israeli comparison of 14-year-old school students," in B. N. Setiadi, A. Supratiknya, W. J. Lonner, and Y. H. Poortinga (eds.), *Ongoing Themes in Psychology and Culture (Online Ed.),* Melbourne, FL: International Association for Cross-Cultural Psychology.

**13.** Margaret C. Campbell and Kevin Lane Keller, "Brand familiarity and advertising repetition effects," *Journal of Consumer Research,* 30 (2003), pp. 292–304.

**14.** Data provided by Julie Hendricks and V. J. K. Liu, Purdue University, Department of Foods and Nutrition.

**15.** Results of the study are reported in P. F. Campbell and G. P. McCabe, "Predicting the success of freshmen in a computer science major," *Communications of the ACM,* 27 (1984), pp. 1108–1113.

**16.** S. Badylak et al., "Marrow-derived cells populate scaffolds composed of xenogeneic extracellular matrix," *Experimental Hematology,* 29 (2001), pp. 1310–1318.

**17.** Based on A. A. Adish et al., "Effect of consumption of food cooked in iron pots on iron status and growth of young children: a randomised trial," *Lancet,* 353 (1999), pp. 712–716.

**18.** Based on a problem from Renée A. Jones and Regina P. Becker, Purdue University, Department of Statistics.

**19.** Based on M. U. Kalwani and C. K. Yim, "Consumer price and promotion expectations: an experimental study," *Journal of Marketing Research,* 29 (1992), pp. 90–100.

**20.** For a summary of this study and other research in this area, see Stanley Coren and Diane F. Halpern, "Left-handedness: a marker for decreased survival fitness," *Psychological Bulletin,* 109 (1991), pp. 90–106.

**21.** Data provided by Neil Zimmerman, Purdue University School of Health Sciences.

**22.** I. C. Feller et al., "Sex-biased herbivory in jack-in-the-pulpit (*Arisaema triphyllum*) by a specialist thrips (*Heterothrips arisaemae*)," *Proceedings of the 7th International Thysanoptera Conference*, Reggio Callabrio, Italy, pp. 163–172.

## CHAPTER 14

**1.** Logistic regression models for the general case where there are more than two possible values for the response variable have been developed. These are considerably more complicated and are beyond the scope of our present study. For more information on logistic regression, see A. Agresti, *An Introduction to Categorical Data Analysis,* 2nd ed., Wiley, 2002; and D. W. Hosmer and S. Lemeshow, *Applied Logistic Regression,* 2nd ed., Wiley, 2000.

**2.** This example is taken from a classic text written by a contemporary of R. A. Fisher, the person who developed many of the fundamental ideas of statistical inference that we use today. The reference is D. J. Finney, *Probit Analysis,* Cambridge University Press, 1947. Although not included in the analysis, it is important to note that the experiment included a control group that received no insecticide. No aphids died in this group. We have chosen to call the response "dead." In Finney's book the category is described as "apparently dead, moribund, or so badly affected as to be unable to walk more than a few steps." This is an early example of the need to make careful judgments when defining variables to be used in a statistical analysis. An insect that is "unable to walk more than a few steps" is unlikely to eat very much of a chrysanthemum plant!

**3.** From P. Azoulay and S. Shane, "Entrepreneurs, contracts, and the failure of young firms," *Management Science,* 47 (2001), pp. 337–358.

**4.** Marsha A. Dickson, "Utility of no sweat labels for apparel customers: profiling label users and predicting their purchases," *Journal of Consumer Affairs,* 35 (2001), pp. 96–119.

**5.** The poll is part of the American Express Retail Index Project and is reported in *Stores*, December 2000, pp. 38–40.

**6.** Based on Greg Clinch, "Employee compensation and firms' research and development activity," *Journal of Accounting Research,* 29 (1991), pp. 59–78.

**7.** Monica Macaulay and Colleen Brice, "Don't touch my projectile: gender bias and stereotyping in syntactic examples," *Language,* 73, no. 4 (1997), pp. 798–825.

**8.** Based on Robert T. Driescher, "A quality swing with Ping," *Quality Progress,* August 2001, pp. 37–41.

**9.** Karin Weber and Weley S. Roehl, "Profiling people searching for and purchasing travel products on the World Wide Web," *Journal of Travel Research,* 37 (1999), pp. 291–298.

**10.** Guohua Li and Susan P. Baker, "Alcohol in fatally injured bicyclists," *Accident Analysis and Prevention,* 26 (1994), pp. 543–548.

## CHAPTER 15

**1.** Data provided by Sam Phillips, Purdue University.

**2.** From the April 2007 issue of *Condé Nast Traveler* magazine.

**3.** For purists, here is the precise definition: $X_1$ is *stochastically larger* than $X_2$ if

$$P(X_1 > a) \geq P(X_2 > a)$$

for all $a$, with strict inequality for at least one $a$. The Wilcoxon rank sum test is effective against this alternative in the sense that the power of the test approaches 1 (that is, the test becomes more certain to reject the null hypothesis) as the number of observations increases.

**4.** Huey Chern Boo, "Consumers' perceptions and concerns about safety and healthfulness of food served at fairs and festivals," MS thesis, Purdue University, 1997.

**5.** Data provided by Susan Stadler, Purdue University.

**6.** Sapna Aneja, "Biodeterioration of textile fibers in soil," MS thesis, Purdue University, 1994.

**7.** Data provided by Warren Page, New York City Technical College, from a study done by John Hudesman.

**8.** Data provided by Charles Cannon, Duke University. The study report is C. H. Cannon, D. R. Peart, and M. Leighton, "Tree species diversity in commercially logged Bornean rainforest," *Science*, 281 (1998), pp. 1366–1367.

**9.** This example is adapted from Maribeth C. Schmitt, "The effects of an elaborated directed reading activity on the metacomprehension skills of third graders," PhD dissertation, Purdue University, 1987.

**10.** William D. Darley, "Store-choice behavior for preowned merchandise," *Journal of Business Research*, 27 (1993), pp. 17–31.

**11.** See Note 5.

**12.** Data for 1998 provided by Jason Hamilton, University of Illinois. The study report is Evan H. DeLucia et al., "Net primary production of a forest ecosystem with experimental $CO_2$ enhancement," *Science*, 284 (1999), pp. 1177–1179.

**13.** Simplified from the EESEE story "Stepping Up Your Heart Rate," on the CD.

**14.** These data were collected as part of a larger study of dementia patients conducted by Nancy Edwards, School of Nursing, and Alan Beck, School of Veterinary Medicine, Purdue University.

**15.** L. Friedland et al., "Effect of educational program on compliance with glove use in a pediatric emergency department," *American Journal of Diseases of Childhood*, 146 (1992), pp. 1355–1358.

**16.** Data provided by Diana Schellenberg, Purdue University School of Health Sciences.

**17.** These data are from "Results report on the vitamin C pilot program," prepared by SUSTAIN (Sharing United States Technology to Aid in the Improvement of Nutrition) for the U.S. Agency for International Development. The report was used by the Committee on International Nutrition of the National Academy of Sciences/Institute of Medicine to make recommendations on whether or not the vitamin C content of food commodities used in U.S. food aid programs should be increased. The program was directed by Peter Ranum and Françoise Chomé.

**18.** James A. Levine et al., "Role of nonexercise activity thermogenesis in resistance to fat gain in humans," *Science*, 283 (1999), pp. 212–214. Data for this study are available from the *Science* Web site, `www.sciencemag.org`.

**19.** See Note 1.

**20.** Data provided by Helen Park. See H. Park et al., "Fortifying bread with each of three antioxidants," *Cereal Chemistry*, 74 (1997), pp. 202–206.

**21.** Data provided by Jo Welch, Purdue University Department of Foods and Nutrition.

**22.** Modified from M. C. Wilson and R. E. Shade, "Relative attractiveness of various luminescent colors to the cereal leaf beetle and the meadow spittlebug," *Journal of Economic Entomology*, 60 (1967), pp. 578–580.

**23.** See Note 6.

**24.** See Note 8.

**25.** Ray H. Rosenman et al., "A 4-year prospective study of the relationship of different habitual vocational physical activity to risk and incidence of ischemic heart disease in volunteer male federal employees," in P. Milvey (ed.), *The Marathon: Physiological, Medical, Epidemiological and Psychological Studies*, New York Academy of Sciences, 301 (1977), pp. 627–641.

**26.** Michael Wayne Peugh, "Field investigation of ventilation and air quality in duck and turkey slaughter plants," MS thesis, Purdue University, 1996.

**27.** We thank Ethan J. Temeles of Amherst College for providing the data. His work is described in Ethan J. Temeles and W. John Kress, "Adaptation in a plant-hummingbird association," *Science*, 300 (2003), pp. 630–633.

**28.** Based on A. A. Adish et al., "Effect of consumption of food cooked in iron pots on iron status and growth of young children: a randomised trial," *The Lancet,* 353 (1999), pp. 712–716.

**29.** For more details on multiple comparisons, see M. Hollander and D. A. Wolfe, *Nonparametric Statistical Methods,* 2nd ed., Wiley, 1999. This book is a useful reference on applied aspects of nonparametric inference in general.

# PHOTO CREDITS

## CHAPTER 1

**p. 1** Holly Harris/Getty
**p. 2** Jan Erik Posth/Picturequest
**p. 17** © Brian Tolbert/CORBIS
**p. 18** USGS
**p. 33** Getty Images/Collection Mix: Subjects RM
**p. 43** White Packert/The Image Bank/Getty

## CHAPTER 2

**p. 83** AP Photo/Rick Rycroft
**p. 90** © Robert Harding Picture Library Ltd/Alamy
**p. 91** © Phillip Colla/SeaPics.com
**p. 91** © Richard Herrmann/SeaPics.com
**p. 105** © Stephane Reix/For Picture/CORBIS
**p. 116** Jeffrey Newman (UC Berkeley), NASA
**p. 158** Christian Michaels/Getty

## CHAPTER 3

**p. 171** © Brand X Pictures/Alamy
**p. 178** © Michael Newman/PhotoEdit
**p. 178** Elizabeth Crews/The Image Works
**p. 182** Creatas Images/JupiterImages
**p. 190** Dr. Palle Pedersen
**p. 198** Ariel Skelley/Blend/JupiterImages
**p. 201** © Bob Daemmrich/The Image Works
**p. 213** Stephen Chernin/Getty Images
**p. 220** © Brian E. Small/VIREO

## CHAPTER 4

**p. 237** Flipchip/LasVegasVegas.com
**p. 239** 2006 Getty Images
**p. 252** © Cut and Deal Ltd/Alamy
**p. 254** James Gathany/CDC
**p. 278** Kim Taylor/Nature Pl.com
**p. 283** © The Photo Works
**p. 293** Flipchip/LasVegasVegas.com

## CHAPTER 5

**p. 311** Getty Images
**p. 313** BananaStock/Picturequest
**p. 315** Antonia Reeve/Photo Researchers
**p. 321** © Michael Newman/PhotoEdit
**p. 335** Howard Grey/Getty
**p. 341** Robin Nelson/PhotoEdit

## CHAPTER 6

**p. 353** Image by © Images.com/CORBIS
**p. 354** USDA Forest Service Archives
**p. 356** © Gary Conner/PhotoEdit
**p. 361** Mike Lane/Cagle Cartoons, Inc.
**p. 407** Photo by The Photo Works.

## CHAPTER 7

**p. 417** Steve Satushek/Getty
**p. 418** Steve Satushek/Getty
**p. 436** Courtesy powerbooktrance
**p. 449** Robert Warren/Getty Images
**p. 451** Getty Images/Digital Vision

## CHAPTER 8

**p. 487** © Enzinger
**p. 492** Courtesy Minnesota Soybean Organization
**p. 494** David Woolley/Getty Images
**p. 499** Yellow Dog Productions/Getty

## CHAPTER 9

**p. 525** © Mel Yates/AgeFotostock
**p. 537** Jack Hobhouse/Alamy
**p. 545** Mark Scott/Getty
**p. 548** © Richard Levine/Alamy

## CHAPTER 10

**p. 559** © Chuck Doswell/Visuals Unlimited
**p. 562** © General Motors Corp. Used with permission. GM Media Archives
**p. 577** Custom Medical Stock Photo

## CHAPTER 11

**p. 607** Bill Losh/Getty
**p. 626** © The Photo Works

## CHAPTER 12

**p. 637** Rich Chenet/Car & Driver Magazine
**p. 638** maveric2003/Flickr
**p. 641** Steve Dunwell/Index Stock
**p. 665** © Gary Braasch/CORBIS

## CHAPTER 13

**p. 683** © John Van Hasselt/CORBIS
**p. 685** © John Van Hasselt/CORBIS
**p. 686** Professor Pietro M. Motta/Photo Researchers
**p. 691** Joel Rafkin/PhotoEdit
**p. 692** © Banana Stock

## CHAPTER 14

**p. 711** [p. 14-1] Index Stock/JupiterImages
**p. 717** [p. 14-7] BrandX Pictures/photolibrary
**p. 720** [p. 14-10] Rod Planck/Photo Researchers

## CHAPTER 15

**p. 733** [p. 15-1] Jim Wark/Airphoto
**p. 735** [p. 15-3] Nigel Cattlin/Photo Researchers
**p. 743** [p. 15-11] © Jeff Greenberg/PhotoEdit
**p. 749** [p. 15-17] © lizabeth Crews/The Image Works
**p. 754** [p. 15-22] iconsportsmedia.com

## CHAPTER 16

**p. 16-1** Greg Epperson/Index Stock Imagery/Jupiter Images
**p. 16-13** © Brad Perks Lightscapes/Alamy
**p. 16-20** AP Photo/Mark Humphrey
**p. 16-49** click/MorgueFile

## CHAPTER 17

**p. 17-1** © A. T. Willett/Alamy
**p. 17-4** Michael Rosenfeld/Getty
**p. 17-7** Darryl Leniuk/Masterfile
**p. 17-2** © Jeff Greenberg/The Image Works

# INDEX

Table entry for *z* is the area under the standard normal curve to the left of *z*.

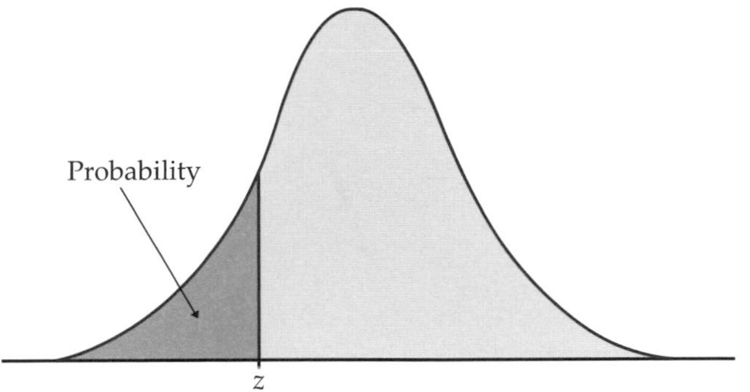


## TABLE A

### Standard normal probabilities

| z | .00 | .01 | .02 | .03 | .04 | .05 | .06 | .07 | .08 | .09 |
|---|---|---|---|---|---|---|---|---|---|---|
| −3.4 | .0003 | .0003 | .0003 | .0003 | .0003 | .0003 | .0003 | .0003 | .0003 | .0002 |
| −3.3 | .0005 | .0005 | .0005 | .0004 | .0004 | .0004 | .0004 | .0004 | .0004 | .0003 |
| −3.2 | .0007 | .0007 | .0006 | .0006 | .0006 | .0006 | .0006 | .0005 | .0005 | .0005 |
| −3.1 | .0010 | .0009 | .0009 | .0009 | .0008 | .0008 | .0008 | .0008 | .0007 | .0007 |
| −3.0 | .0013 | .0013 | .0013 | .0012 | .0012 | .0011 | .0011 | .0011 | .0010 | .0010 |
| −2.9 | .0019 | .0018 | .0018 | .0017 | .0016 | .0016 | .0015 | .0015 | .0014 | .0014 |
| −2.8 | .0026 | .0025 | .0024 | .0023 | .0023 | .0022 | .0021 | .0021 | .0020 | .0019 |
| −2.7 | .0035 | .0034 | .0033 | .0032 | .0031 | .0030 | .0029 | .0028 | .0027 | .0026 |
| −2.6 | .0047 | .0045 | .0044 | .0043 | .0041 | .0040 | .0039 | .0038 | .0037 | .0036 |
| −2.5 | .0062 | .0060 | .0059 | .0057 | .0055 | .0054 | .0052 | .0051 | .0049 | .0048 |
| −2.4 | .0082 | .0080 | .0078 | .0075 | .0073 | .0071 | .0069 | .0068 | .0066 | .0064 |
| −2.3 | .0107 | .0104 | .0102 | .0099 | .0096 | .0094 | .0091 | .0089 | .0087 | .0084 |
| −2.2 | .0139 | .0136 | .0132 | .0129 | .0125 | .0122 | .0119 | .0116 | .0113 | .0110 |
| −2.1 | .0179 | .0174 | .0170 | .0166 | .0162 | .0158 | .0154 | .0150 | .0146 | .0143 |
| −2.0 | .0228 | .0222 | .0217 | .0212 | .0207 | .0202 | .0197 | .0192 | .0188 | .0183 |
| −1.9 | .0287 | .0281 | .0274 | .0268 | .0262 | .0256 | .0250 | .0244 | .0239 | .0233 |
| −1.8 | .0359 | .0351 | .0344 | .0336 | .0329 | .0322 | .0314 | .0307 | .0301 | .0294 |
| −1.7 | .0446 | .0436 | .0427 | .0418 | .0409 | .0401 | .0392 | .0384 | .0375 | .0367 |
| −1.6 | .0548 | .0537 | .0526 | .0516 | .0505 | .0495 | .0485 | .0475 | .0465 | .0455 |
| −1.5 | .0668 | .0655 | .0643 | .0630 | .0618 | .0606 | .0594 | .0582 | .0571 | .0559 |
| −1.4 | .0808 | .0793 | .0778 | .0764 | .0749 | .0735 | .0721 | .0708 | .0694 | .0681 |
| −1.3 | .0968 | .0951 | .0934 | .0918 | .0901 | .0885 | .0869 | .0853 | .0838 | .0823 |
| −1.2 | .1151 | .1131 | .1112 | .1093 | .1075 | .1056 | .1038 | .1020 | .1003 | .0985 |
| −1.1 | .1357 | .1335 | .1314 | .1292 | .1271 | .1251 | .1230 | .1210 | .1190 | .1170 |
| −1.0 | .1587 | .1562 | .1539 | .1515 | .1492 | .1469 | .1446 | .1423 | .1401 | .1379 |
| −0.9 | .1841 | .1814 | .1788 | .1762 | .1736 | .1711 | .1685 | .1660 | .1635 | .1611 |
| −0.8 | .2119 | .2090 | .2061 | .2033 | .2005 | .1977 | .1949 | .1922 | .1894 | .1867 |
| −0.7 | .2420 | .2389 | .2358 | .2327 | .2296 | .2266 | .2236 | .2206 | .2177 | .2148 |
| −0.6 | .2743 | .2709 | .2676 | .2643 | .2611 | .2578 | .2546 | .2514 | .2483 | .2451 |
| −0.5 | .3085 | .3050 | .3015 | .2981 | .2946 | .2912 | .2877 | .2843 | .2810 | .2776 |
| −0.4 | .3446 | .3409 | .3372 | .3336 | .3300 | .3264 | .3228 | .3192 | .3156 | .3121 |
| −0.3 | .3821 | .3783 | .3745 | .3707 | .3669 | .3632 | .3594 | .3557 | .3520 | .3483 |
| −0.2 | .4207 | .4168 | .4129 | .4090 | .4052 | .4013 | .3974 | .3936 | .3897 | .3859 |
| −0.1 | .4602 | .4562 | .4522 | .4483 | .4443 | .4404 | .4364 | .4325 | .4286 | .4247 |
| 0.0 | .5000 | .4960 | .4920 | .4880 | .4840 | .4801 | .4761 | .4721 | .4681 | .4641 |

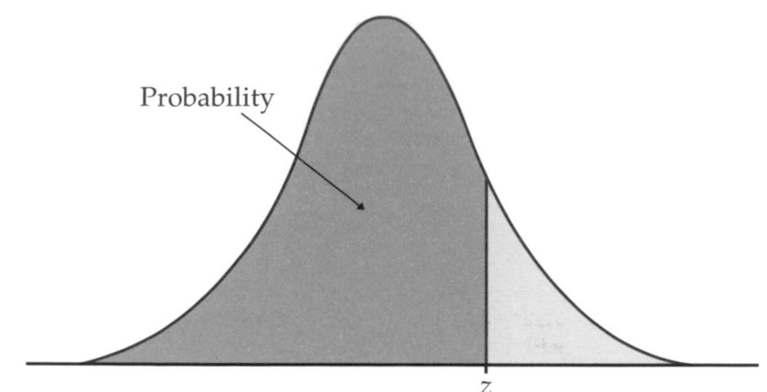


Table entry for $z$ is the area under the standard normal curve to the left of $z$.

**TABLE A**

## Standard normal probabilities (continued)

| $z$ | .00 | .01 | .02 | .03 | .04 | .05 | .06 | .07 | .08 | .09 |
|---|---|---|---|---|---|---|---|---|---|---|
| 0.0 | .5000 | .5040 | .5080 | .5120 | .5160 | .5199 | .5239 | .5279 | .5319 | .5359 |
| 0.1 | .5398 | .5438 | .5478 | .5517 | .5557 | .5596 | .5636 | .5675 | .5714 | .5753 |
| 0.2 | .5793 | .5832 | .5871 | .5910 | .5948 | .5987 | .6026 | .6064 | .6103 | .6141 |
| 0.3 | .6179 | .6217 | .6255 | .6293 | .6331 | .6368 | .6406 | .6443 | .6480 | .6517 |
| 0.4 | .6554 | .6591 | .6628 | .6664 | .6700 | .6736 | .6772 | .6808 | .6844 | .6879 |
| 0.5 | .6915 | .6950 | .6985 | .7019 | .7054 | .7088 | .7123 | .7157 | .7190 | .7224 |
| 0.6 | .7257 | .7291 | .7324 | .7357 | .7389 | .7422 | .7454 | .7486 | .7517 | .7549 |
| 0.7 | .7580 | .7611 | .7642 | .7673 | .7704 | .7734 | .7764 | .7794 | .7823 | .7852 |
| 0.8 | .7881 | .7910 | .7939 | .7967 | .7995 | .8023 | .8051 | .8078 | .8106 | .8133 |
| 0.9 | .8159 | .8186 | .8212 | .8238 | .8264 | .8289 | .8315 | .8340 | .8365 | .8389 |
| 1.0 | .8413 | .8438 | .8461 | .8485 | .8508 | .8531 | .8554 | .8577 | .8599 | .8621 |
| 1.1 | .8643 | .8665 | .8686 | .8708 | .8729 | .8749 | .8770 | .8790 | .8810 | .8830 |
| 1.2 | .8849 | .8869 | .8888 | .8907 | .8925 | .8944 | .8962 | .8980 | .8997 | .9015 |
| 1.3 | .9032 | .9049 | .9066 | .9082 | .9099 | .9115 | .9131 | .9147 | .9162 | .9177 |
| 1.4 | .9192 | .9207 | .9222 | .9236 | .9251 | .9265 | .9279 | .9292 | .9306 | .9319 |
| 1.5 | .9332 | .9345 | .9357 | .9370 | .9382 | .9394 | .9406 | .9418 | .9429 | .9441 |
| 1.6 | .9452 | .9463 | .9474 | .9484 | .9495 | .9505 | .9515 | .9525 | .9535 | .9545 |
| 1.7 | .9554 | .9564 | .9573 | .9582 | .9591 | .9599 | .9608 | .9616 | .9625 | .9633 |
| 1.8 | .9641 | .9649 | .9656 | .9664 | .9671 | .9678 | .9686 | .9693 | .9699 | .9706 |
| 1.9 | .9713 | .9719 | .9726 | .9732 | .9738 | .9744 | .9750 | .9756 | .9761 | .9767 |
| 2.0 | .9772 | .9778 | .9783 | .9788 | .9793 | .9798 | .9803 | .9808 | .9812 | .9817 |
| 2.1 | .9821 | .9826 | .9830 | .9834 | .9838 | .9842 | .9846 | .9850 | .9854 | .9857 |
| 2.2 | .9861 | .9864 | .9868 | .9871 | .9875 | .9878 | .9881 | .9884 | .9887 | .9890 |
| 2.3 | .9893 | .9896 | .9898 | .9901 | .9904 | .9906 | .9909 | .9911 | .9913 | .9916 |
| 2.4 | .9918 | .9920 | .9922 | .9925 | .9927 | .9929 | .9931 | .9932 | .9934 | .9936 |
| 2.5 | .9938 | .9940 | .9941 | .9943 | .9945 | .9946 | .9948 | .9949 | .9951 | .9952 |
| 2.6 | .9953 | .9955 | .9956 | .9957 | .9959 | .9960 | .9961 | .9962 | .9963 | .9964 |
| 2.7 | .9965 | .9966 | .9967 | .9968 | .9969 | .9970 | .9971 | .9972 | .9973 | .9974 |
| 2.8 | .9974 | .9975 | .9976 | .9977 | .9977 | .9978 | .9979 | .9979 | .9980 | .9981 |
| 2.9 | .9981 | .9982 | .9982 | .9983 | .9984 | .9984 | .9985 | .9985 | .9986 | .9986 |
| 3.0 | .9987 | .9987 | .9987 | .9988 | .9988 | .9989 | .9989 | .9989 | .9990 | .9990 |
| 3.1 | .9990 | .9991 | .9991 | .9991 | .9992 | .9992 | .9992 | .9992 | .9993 | .9993 |
| 3.2 | .9993 | .9993 | .9994 | .9994 | .9994 | .9994 | .9994 | .9995 | .9995 | .9995 |
| 3.3 | .9995 | .9995 | .9995 | .9996 | .9996 | .9996 | .9996 | .9996 | .9996 | .9997 |
| 3.4 | .9997 | .9997 | .9997 | .9997 | .9997 | .9997 | .9997 | .9997 | .9997 | .9998 |